International Russian Conference on Materials Science and Metallurgical Technology (RusMetalCon 2020)

IOP Conference Series: Materials Science and Engineering
Volume 969

Chelyabinsk, Russia
22 - 24 September 2020

ISBN: 978-1-7138-4789-2
ISSN: 1757-8981

Printed from e-media with permission by:

Curran Associates, Inc.
57 Morehouse Lane
Red Hook, NY 12571

Some format issues inherent in the e-media version may also appear in this print version.

Institute of Physics
Dirac House, Temple Back
Bristol BS1 6BE UK

Phone: 44 1 17 929 7481
Fax: 44 1 17 920 0979

techtracking@iop.org

Additional copies of this publication are available from:

Curran Associates, Inc.
57 Morehouse Lane
Red Hook, NY 12571 USA
Phone: 845-758-0400
Fax: 845-758-2633
Email: curran@proceedings.com
Web: www.proceedings.com

TABLE OF CONTENTS

PREFACE

International Russian Conference on Materials Science and Metallurgical Technology has taken place on September 22-24, 2020.

The conference was carried out under financial support of the South Ural State University (national research university). The conference was organized to maintain discus and the presentation of research papers, summarizing research in technical disciplines in universities, research institutes, large industrial enterprises, scientific and industrial associations of the Russian Federation and the World, as well as foreign authors, and research results obtained on the personal initiative of the authors. The aim of the conference is the exchange of information between scientists and practitioners of the most promising areas of research and achievements in the field of material science and metallurgical technologies.

The conference was organized in 7 sections, including:

1. Materials: design, synthesis and characterization;
2. Technologies of pyrometallurgical processes;
3. Technologies of hydrometallurgical processes;
4. Modern problems of foundry production;
5. Metal forming: rolling, stamping, forging, pressing;
6. Technologies of welding production;
7. Additive manufacturing.

The spread of the coronavirus COVID-19 made adjustments to our live in this year, including the organization of our conference. Participants could take part in the conference as in a traditional face-to-face format and as a format of video conference or of poster presentation remotely. Everyone had an opportunity to watch the poster presentation and ask questions in chat, and speakers were able to give their answers and comments.

Organizing committee would like to express our sincere appreciation to everybody who has contributed to the conference. Heartfelt thanks are due to authors, reviewers, participants and to all the team of organizers for their support and enthusiasm which granted success to the conference.

Conference Chair, prof. Andrey A. Radionov.

IOP Conf. Series: Materials Science and Engineering **969** (2020) 011002 doi:10.1088/1757-899X/969/1/011002

Peer review declaration of RusMetalCon 2020

All papers published in this volume of IOP Conference Series: Materials Science and Engineering have been peer reviewed through processes administered by the Editors. Reviews were conducted by expert referees to the professional and scientific standards expected of a proceedings journal published by IOP Publishing.

- **Type of peer review:** Double-blind

- **Conference submission management system:** Open Conference Systems http://conferences.i-publ.ru/index.php/RusMetalCon/

- **Number of submissions received:** 192

- **Number of submissions sent for review:** 167

- **Number of submissions accepted:** 108

- **Acceptance Rate (Number of Submissions Accepted / Number of Submissions Received X 100):** 56

- **Average number of reviews per paper: 2**

- **Total number of reviewers involved: 32**

- **Any additional info on review process:** The first assessment of an article will be by Editor. If the article has not sufficient qualities, it means: «Reject» or «Changes requested». If the article has sufficient qualities, it means: «Sent to Review Process». The second assessment of the article will be by conference reviewers. Double Blind Review system has been applied. Participants in the review process shall treat the contents of conference submissions under review as confidential information not to be disclosed to others before publication. No one with access to a submission shall make any inappropriate use of the special knowledge that access providers. The reviewers are chosen for their high qualifications and objectivity regarding a particular article. Reviewers make recommendations: accepted / revise is required / not accepted. On the basis of review assessments, Editor will make the final decision and inform authors.

- **Contact person for queries:**
  Conference Chair prof. Radionov A.A.

Austenite to Ferrite Transformation in Fe–9 %Cr Alloys. II. Determination of Growth Rates

D A Mirzaev, K Yu Okishev and I V Lapina

Department of Materials Science, South Ural State University, 76, pr. Lenina, Chelyabinsk, 454080, Russia

E-mail: okishevki@susu.ru

Abstract. The paper continues theoretical analysis of austenite to ferrite transformation in an Fe–9 %Cr alloy demonstrating a typical site-saturation behaviour under the condition of diffusionless growth of α phase. In the previous paper the *K* coefficient in the Avrami equations being the asymptotes of Cahn's solution of the site-saturation transformation kinetics was used that contained a constant activation energy indicating the small value of the work of critical nucleus formation. This approach permitted to determine activation energy of transformation that was found to be equal to about 300 kJ/mole. In the present paper a new method of finding Cahn's equation coefficients from bending point at isothermal curves plotted on double logarithmic scale is proposed. Actually this bending point is the intercept of the two above-mentioned asymptotes of the Cahn's curve, and equating the formulae describing theses asymptotes allows to find the coefficients in question. Given the average austenite grain size, it permits to find numerical values of growth rates. The activation energy thus acquires the exact physical meaning of the energy barrier at the interphase interface; its value was found to be 286 kJ/mole, which is close to previous results. The growth rates were thus determined as a function of temperature and the TTT diagram for the alloy is calculated that shows close correspondence with experiment.

1. Introduction

In the first part of this paper [1] we presented an analysis of the experimental data for the austenite→ferrite transformation in a Fe–9.24 %Cr with 0.04 %C using the conventional approach based on Avrami equation

$$f(t)=1-\exp\left(-Kt^{n}\right) \quad (1)$$

where *f* is the transformed volume fraction, *t* is time, and *K* and *n* are coefficients. It is to be reminded that this alloy exhibits diffusionless mode of transformation due to boundary lines of the $\alpha+\gamma$ region having a common horizontal tangent near this concentration [2], and was repeatedly used as a model material to study the diffusionless transformation kinetics [3–9].

Coefficient *n* was taken both as independent on time (and temperature) of transformation and as a variable parameter according to Cahn's theory of nucleation site saturation on grain boundaries [10, 11], that replaces (1) with a more complex equation with two coefficients, K_4 and K_1:

$$f(t)=1-\exp\left(-K_1 t\cdot\int_0^1\left(1-\exp\left(-\frac{2K_4}{K_1}\left(2x^3-3x^2+1\right)t^3\right)\right)dx\right) \quad (2)$$

that asymptotically tends to

$$f(t)=1-\exp\left(-K_4t^4\right) \text{ and } f(t)=1-\exp\left(-K_1t\right) \tag{3}$$

at early and late stages of transformation, respectively. Hence, not only *K* but also *n* does change in the course of transformation from 4 to 1. This change was clearly seen on experimental plots.

If all *K*'s are attached the form

$$K_i^{1/n}=k_i\left(\frac{T_s-T}{T}\right)^{m_i}\exp\left(-\frac{U_i}{RT}\right) \tag{4}$$

substantiated in our previous works [12, 13], where T_s is the upper temperature limit of transformation, then all parameters can be found from relevant plots. The key feature of this equation is that the activation energies U_i are assumed independent on temperature due to small value of critical nucleus formation work [14–17].

For both considered models, the activation energy *U* was similar: 280–315 kJ/mole. Exponent *m* varied in the range 1.03–1.15 which is very close to the value $m = 1$ following from the theory of diffusionless growth [11, 18].

However, the finally calculated TTT diagrams of ferrite formation had a noticeable discrepancy with experimental ones, especially at temperatures remote from the "nose".

In this paper the analysis will be continued.

2. Another Method of Finding Cahn's Coefficients from Experimental Curves

In [1], the coefficients K_4 and K_1 were taken from linear interpolation of initial and late parts of experimental plots on the scale ln(–ln(1–*f*)) — ln *t*. The values of *U* and *m* were then taken from the asymptotes of plots of ln *t* versus 1/*T* at *T*→0, that are equal to *U*/*R*. The shortcoming of this method is that the asymptote is determined not always precisely, which introduces some arbitrariness to the obtained values. So another method to find *U* and *m* can be proposed.

As it can be seen from figure 1 of [1], experimental plots on double logarithmic scale usually have a break that corresponds to saturation of nucleation sites and transfer from the first to the second line obeying equation (3). So for the break point (t^*, f^*) the following equation should fulfill:

$$4\ln t^*+\ln K_4=\ln t^*+\ln K_1\text{, or } \ln t^*=\frac{1}{3}\ln K_1-\frac{1}{3}\ln K_4. \tag{5}$$

Substituting this to equation (3), one can find that

$$\ln\left(-\ln(1-f^*)\right)=\frac{2}{3}\ln K_1-\frac{1}{3}\ln K_4, \tag{6}$$

$$\ln K_4=\ln\left(-\ln(1-f^*)\right)-4\ln t^* \text{ and } \ln K_1=\ln\left(-\ln(1-f^*)\right)-\ln t^*. \tag{7}$$

In terms of coefficients $\alpha_S=\left(\frac{\pi}{3}I_S\omega^2\right)^{1/3}$, $\beta_S=2\frac{S}{V}\omega\cdot\frac{1}{\alpha_S}$, where I_S is nucleation rate per unit grain face area, ω is growth rate and S/V is grain face area per unit volume [19], that were used in [1],

$$K_4=\frac{1}{2}\alpha_S^4\beta_S,\ K_1=\alpha_S\beta_S, \tag{8}$$

and thus

$$\ln\alpha_S=-\ln t^*+\frac{1}{3}\ln 2,\ \ln\beta_S=\ln\left(-\ln(1-f^*)\right)-\frac{1}{3}\ln 2. \tag{9}$$

If we use again equation (4) for temperature dependence of K_4 and K_1 then the plots of $\ln K_i^{1/n} - m_i \ln \frac{T_s - T}{T}$ against (1/*T*) are expected to be linear with the slope of (-*U*/*R*):

$$\ln K_i^{1/n} - m_i \ln \frac{T_s - T}{T} = \ln k_i - \frac{U_i}{RT}. \tag{10}$$

These plots for the Fe–9 %Cr alloy in question are presented in figure 1. As in [1], T_s was accepted 815 °C, exponent $m = 1$, and Avrami exponents n were taken from figure 1 of the paper [1] for corresponding part of the isothermal curve. One can see that experimental points lie really on straight lines. The slopes of these lines yield $U = 230$ and 286 kJ/mole for K_4 and K_1, respectively. The latter value is closer to those obtained by other methods [1].

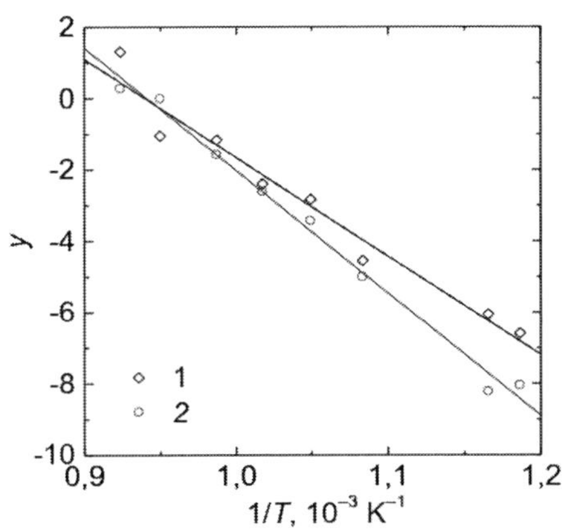

Figure 1. Plots of $y = \ln K_i^{1/n} - \ln \frac{T_s - T}{T}$ vs. $\frac{1}{T}$. *1* — for K_4, *2* — for K_1 (calculated using equations (7)).

3. Finding Growth Rates

It is worth recalling that from the original Cahn's theory [10] the product $K_1 = \alpha_s \beta_s$ is equal to

$$K_1 = \alpha_S \beta_S = 2\frac{S}{V}\omega. \tag{11}$$

It can easily be shown that the ratio (2*SD*/*V*) for cubic octahedra filling the volume without gaps and intersections equals $\eta = 6.696$ (*D* is the "diameter" of a cubic octahedron taken as a distance between its opposite square faces). If we assume *D* as the average grain diameter on the microsection, then equation (11) may be re-written in the form

$$\omega = \frac{D}{\eta} K_1 \text{ or } \omega = \frac{D}{\eta} \alpha_S \beta_S. \tag{12}$$

Analysis of micrographs of the alloy after partial transformation (some of them were presented in [8]) gives *D* for the alloy considered to be approximately 75 μm. This permits to calculate quantitative values of the growth rate ω for each transformation temperature.

Since ω for the diffusionless transformation is described by an equation similar to equation (4) [11]:

$$\omega = k_\omega \frac{T_s - T}{T} \exp\left(-\frac{U}{RT}\right), \tag{13}$$

then the plots of $\ln \omega - \ln((T_s - T)/T)$ against (1/*T*) are to be linear with the slope of (-*U*/*R*), where *U* is no more some "effective" activation energy of transformation, but has an exact physical meaning of the activation energy for an atom to cross the γ/α interphase interface.

Calculated values of ω (both by the first of equations(12) using the values obtained from the break on isothermal curves and by the second of equations (12) using the values of $\alpha_S\beta_S$ from the previous analysis done in [1]) and $\ln\omega - \ln((T_s - T)/T)$ as a function of reverse temperature are shown in figures 2, *a* and *b*. One can see from figure 2, *a* that ω demonstrates a characteristic behaviour with a "nose" at approximately 780 °C, and calculated curves coincide well with the experimental ones. Figure 2, *b* shows that the plots are actually linear, and the activation energy of growth is equal to 286 and 272 kJ/mole for the values of ω obtained by the first and the second of equations (12) respectively. This is very close to the activation energy of self-diffusion in γ iron of 284 kJ/mole [20].

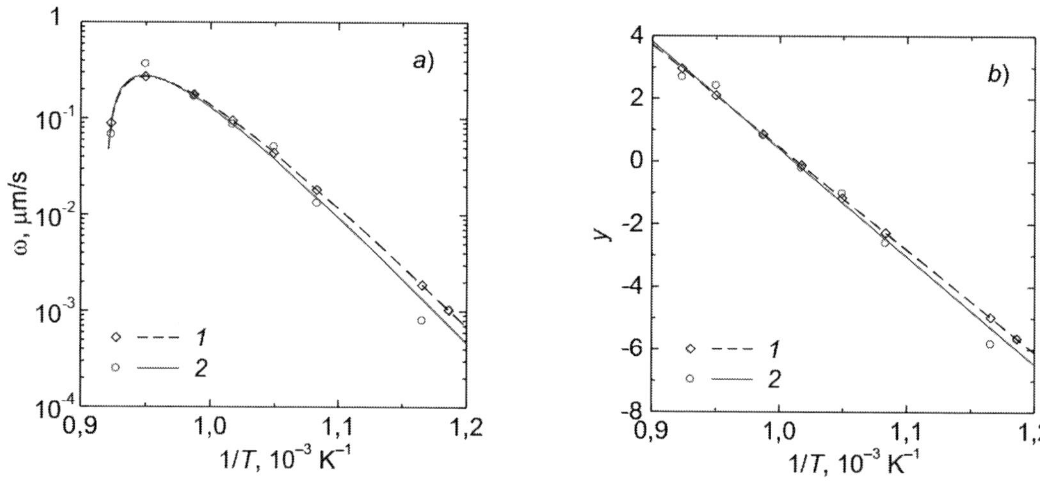

Figure 2. Growth rate ω (*a*) and the value of $y = \ln\omega - \ln((T_s - T)/T)$ (*b*) as a function of reverse temperature. *1* – values obtained in previous analysis [1] and calculated using the second of equations (12); *2* – values obtained in this work and calculated using the first of equations (12). Points are derived from experimental data, lines are calculated (see text).

4. Calculation of TTT Diagram

At last, let us try and use the new method of finding the coefficients of Cahn's equation to construct a complete TTT diagram. For simplicity, we will not use the complete form of this equation (2), but limit ourselves to two asymptotes (3). The time at which the transfer from $n = 4$ to $n = 1$ is to be done can be calculated using equations (5) and (4):

$$\ln t^* = \frac{1}{3}\ln\frac{K_1}{K_4} = \frac{1}{3}\ln\frac{k_1\left(\frac{T_s - T}{T}\right)\exp\left(-\frac{U_1}{RT}\right)}{\left[k_4\left(\frac{T_s - T}{T}\right)\exp\left(-\frac{U_4}{RT}\right)\right]^4} = \ln\left[\left(\frac{k_1}{k_4^4}\right)^{1/3}\left(\frac{T_s - T}{T}\right)^{-1}\exp\left(-\frac{\frac{1}{3}U_1 - \frac{4}{3}U_4}{RT}\right)\right], \quad (14)$$

The equation to calculate a TTT diagram will then look as follows:

$$f(t) = \begin{cases} 1 - \exp\left(-K_4 t^4\right) \text{ if } t \le t^*; \\ 1 - \exp\left(-K_1 t\right) \text{ if } t > t^*. \end{cases} \quad (15)$$

Figure 3 presents the results of calculation. Figure 3, *a* reproduces results of calculation from previous work (Figure 3, *b* in [1]), when α_S and β_S coefficients were determined with their separate activation energies from the $\ln t$ vs. T^{-1} plots. Figure 3, *b* is based on the results of the present work where activation energies for K_4 and K_1 coefficients were determined from the intersection of straight lines on double logarithmic scale ($\ln k_4 = 25.97$, $\ln k_1 = 32.33$; $U_4 = 230$, $U_1 = 286$ kJ/mole).

One can see that the TTT diagram calculated with the values obtained by the method described in this work is in much better correspondence with experimental one, even though the exact Cahn's solution was replaced by simple equation (15).

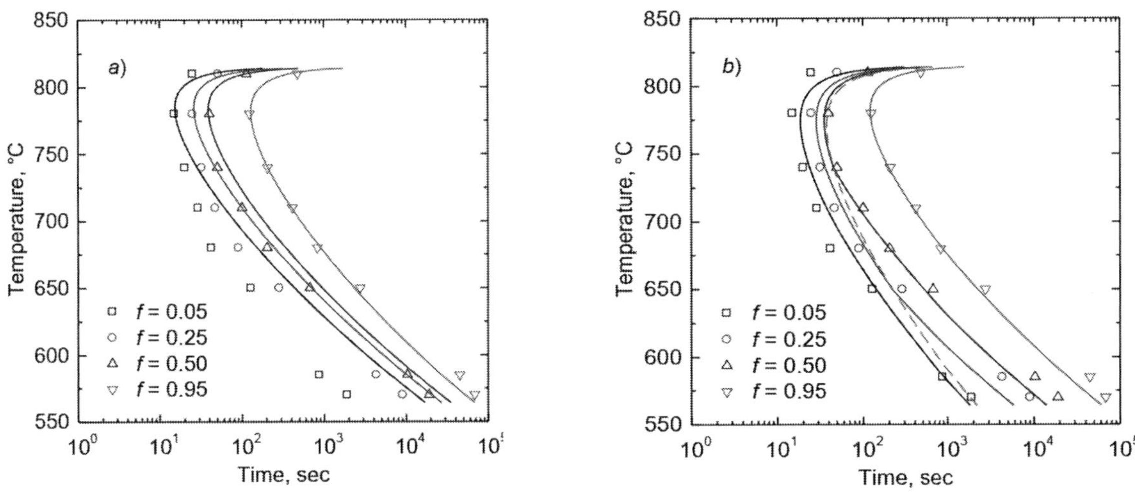

Figure 3. TTT diagrams for the Fe–9.24 %Cr alloy. Points are experimental data [7], lines are calculated: *a* — by exact Cahn's solution with the parameter values obtained in [1], *b* — by approximate equation (15) with the parameter values obtained in this work (see text). Dashed line on plot *b* indicates the time t^* at which the n exponent changes from 4 to 1.

5. Summary

1. A new method of finding Cahn's equation coefficients by bending point at isothermal curves plotted on double logarithmic scale is proposed.
2. Numerical values of growth rates in the Fe–9.24 %Cr are determined.
3. The TTT diagram for the alloy is calculated that shows close correspondence with experiment.

6. References

[1] Mirzaev D A, Okishev K Yu, Mirzoev A A, Makovetskii A N 2019 *Materials Science Forum* **946** 210 DOI: 10.4028/www.scientific.net/MSF.946.210

[2] Hansen M and Anderko K 1958 *Constitution of Binary Alloys* (New York: McGraw-Hill Book Co.)

[3] Kogan L I and Entin R I 1951 *Problems of Metallography and Metal Physics* (Moscow: Metallurgizdat Publ.) p 204

[4] Blanter M E and Mashkov A K 1959 *Metal Science and Heat Treatment of Metals* **1** 7 DOI: 10.1007/BF00807920

[5] Entin R I 1962 *Decomposition of Austenite by Diffusional Processes* (New York: Interscience) p 295

[6] Bee J V and Honeycombe R W K 1978 *Metallurgical Transactions* A **9** 587 DOI: 10.1007/BF02646416

[7] Mirzaev D A, Okishev K Yu, Schastlivtsev V M, Mirzoev A A, Yakovleva I L and Karzunov S E 1998 *The Physics of Metals and Metallography* **86** 590

[8] Yakovleva I L, Mirzaev D A, Schastlivtsev V M, Okishev K Yu and Umova V M 2000 *Metal Science and Heat Treatment* **42** 337 DOI: 10.1007/BF02725311

[9] Okishev K Yu 2012 *Bull. South Ural State Univ.* **15(18)** 116

[10] Cahn J W 1956 *Acta Metallurgica* **4** 449 DOI: 10.1016/0001-6160(56)90041-4

[11] Christian J W 2002 *The theory of transformations in metals and alloys (Part I)* (Elsevier Science Ltd.)

[12] Mirzayev D A, Okishev K Yu and Mirzayeva K D 2002 *Proc. of the Chelyabinsk Scientific Center* **4** 28 Online. Available: http://csc.ac.ru/news/2002_4/2002_4_3_4.pdf
[13] Mirzaev D A, Okishev K Yu and Mirzaeva K D 2013 *Materials Performance and Characterization* **2** 134 DOI: 10.1520/MPC20120023
[14] Johnson W C, White C L, Marth P E, Ruf P K, Tuominen S M, Wade K D, Russell K C and Aaronson H I 1975 *Metallurgical Transactions* A **6** 911–919 DOI: 10.1007/BF02672315
[15] Lange III W F, Enomoto M and Aaronson H I 1988 *Metallurgical Transactions* A **19** 427 DOI: 10.1007/BF02649256
[16] Aaronson H I 1999 *Lectures on the Theory of Phase Transformations* (TMS Publ.) chapter 4 pp 165–230.
[17] Offerman S E, van Dijk N H, Sietsma J, Grigull S, Lauridsen E M, Margulies L, and Poulsen H F 2002 *Science* **298** 1003 DOI: 10.1126/science.1076681
[18] Lyubov B Ya 1969 *Kinetic Theory of Phase Transformations* (Moscow: Metallurgiya Publ.)
[19] Saltykov S A 1974 *Stereometrische Metallographie* (Leipzig: VEB Deutscher Verlag für Grundstoffindustrie)
[20] Smithells C I 1976 *Metals reference book* (London and Boston: Butterworth and Co. Ltd)

IOP Conf. Series: Materials Science and Engineering **969** (2020) 012002 doi:10.1088/1757-899X/969/1/012002

Special Case of Destruction of the End Mill Made of High-Speed Steel Powder

E P Nikolaeva

Irkutsk National Research Technical University, 83, Lermontov str., Irkutsk, 664074, Russia

E-mail: nicolaeva-ep@yandex.ru

Abstract. Comparing with cast steels high-speed steel powder S390 Böhler MICROCLEAN® has better properties, which is determined by the technology of its manufacture: this steel is high-alloyed, contains an increased amount of carbon and carbide-forming metals. In contrast to cast steels, its microstructure has no carbide inhomogeneity. Despite many advantages and long life of steel S390, explained by its balanced microstructure and chemical content, sometimes early tool failures occur. The article provides the grounds for tool failure connected with the peculiarities of heat treatment conduct. The authors establish the influence of a heat treatment mode on the number and content of carbide phases. The authors give recommendations for increasing the tool service life. This article gives the results of the steel macro- and microstructure research. In the course of research hardness was measured and the content of chemical elements was determined. The phase composition was defined with the help of the X-ray diffractometer Shimadzu XRD-6000. The amount of retained austenite was defined by means of the X-ray method of "four peaks" at the diffractometer Xstress 3000. The distribution of chemical elements was studied with the help of a microprobe X-ray analyzer Camebax SX50 while the surface topography was studied with the help of a scanning electronic microscope with the 100-3000-fold magnification.

1. Introduction

High-speed steels (HSS) are frequently used for manufacturing metal treatment facilities. For example, HSS are used to make end mills for aviation material machining [1, 2]. The mill resistance is influenced by physical&chemical processes occurring in the cutting area, high contact voltages, intense heat action due to the emission of heat during cutting [3–5]. When a tool is heated, its microstructure is subject to undesirable changes because of the abrasive action of the treated material. The mill material is influenced by high cutting forces, thus, significant temperature causes softening and increased cutting edge wear.

The metal cutting tool material should meet a number of requirements: high hardness and red-hardness. Also, it should have sufficient flexure strength, compressive strength and shearing strength. But as a rule, one and the same HSS grade cannot have all these properties at the same time. The cause is a non-homogeneous microstructure of cast HSS [6–8].

The methods of hardening treatment of a cutting tool can be divided into 5 groups: superficial plastic deformation (SPD), heat treatment, superficial alloying, coating, combined treatment. For a metal cutting tool superficial plastic hardening is not applied frequently as it causes irregular stress field and roughness increase. Hardening methods, based upon deformative action, are most frequently applied for engineering materials [9–13]. Heat treatment hardening remains the main part of a technical process during the tool manufacture from HSS [6, 8].

At heat treatment hardness is the parameter to be measured most frequently. Hardness is the main steel characteristic used for evaluating the steel quality. Heat treatment, conducted to obtain the maximum hardness, causes significant decrease in crack resistance. When there are structural stress concentrators inside a tool (grooves, cavities, holes for delivering lubricating cooling liquid) unstable failure occurs more frequently.

Some technological properties of HSS alongside with the mechanical ones are predetermined by the method of steel manufacture. Powder technology for manufacturing high-speed steels provided the solution for many problems, caused by structural imperfections. Powder HSS of the new generation are highly-alloyed and contain the increased amount of carbon, their microstructure has no carbide inhomogeneity.

The design of end mills, intended for high-efficiency treatment of high-duty aviation materials, takes into account the end mill geometry as well as the need of internal cooling during cutting, vibration influence on the precision and quality of the treated part surface [14–18]. Tool manufacture is followed by quality control [6, 19, 20]. The combination of chisel steel powder with optimal geometry of a cutting tool should facilitate the efficiency increase. However, sometimes tool failure occurs. The purpose of this research is to find out the causes of destruction of the end mill, made of steel S390.

2. Materials and methods

This paper studies the end mills with a wave-shaped cutting edge, used for rough machining of the parts made of titanium alloys and high-resistance steels [1, 2, 8]. Mills were withdrawn from a technical process after 10 and 20 minutes.

The mills were made of steel powder S390 Böhler MICROCLEAN®. According to the data of optical emission&spectral analysis, the mill parts contain the elements (% wt): 1.6 C; 10.6 W; 7.7 Co; 5 V; 4.7 Cr; 1.9 Mo; 0.5 Si; by 0.3 Mn and Ni; 0.02 P; less than 0.01 S.

During the mill manufacture its material was hardened by heat treatment. At first, workpieces (bars) were annealing to remove residual stress at the temperature 600°C for 2 h. The hardening used after produce end mill which shape is needed. Due to low heat conductivity before quenching HSS was subject to two pre-heatings for temperature equalization along a whole mill section and conducting phase changes for a whole volume. Heating temperature: 1st heating – 560°C, for 2 h; 2d heating – 860°C, for 12 min. Final heating was conducted in a salt bath. The hardening temperature is 1,185°C, hold time – 4 min. 10 sec, cooling - in the oil, heated up to 60°C. Next was to three-stage tempering at the temperature 560°C, each tempering lasted 2 hours.

To study the microstructure, the authors selected the fractures of end and inside mill parts in the region of a cooling channel. The sites of sample cutting for microstructure research are shown in figure 1. At the end part one can observe the tracks of plastic deformation and the cutting edge fracture occurred because of high contact stresses (figure 1, a).

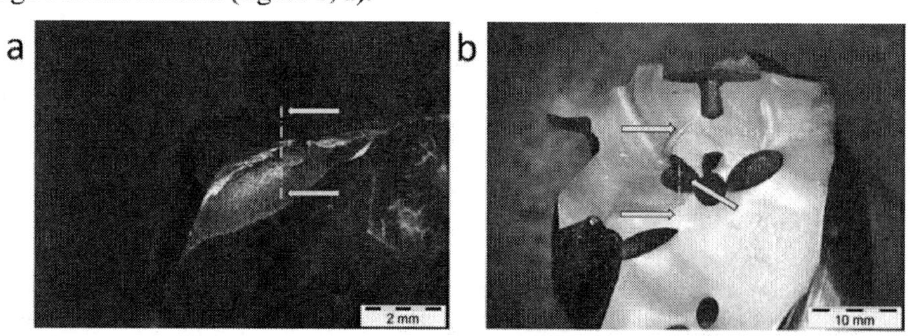

Figure 1. Stereoscopic Picture: (a) cutting edge of the mill end part; (b) fracture surface.

On the fracture of the mill internal part one can observe beach marks and cracks, propagating from the surface of cooling channels (figure 1, b).

The samples were prepared for their microstructure research with the help of a kit for sample preparation, which contains a cutoff machine Discotom-10, abrasive and polishing system Tegramin-25, STRUERS. Micro-structure research was conducted with the help of the microscope OLYMPUS GX51 with the 500-fold magnification.

The distribution of chemical elements was studied on a microprobe X-ray analyzer Camebax SX50. The examination of the sample surface was conducted with the help of a scanning electronic microscope at 100–3.000-fold magnification. The study of sample topography was conducted with the application of secondary electrons (SEI), chemical composition – in back-scattered electrons (BSE) as well as in characteristic X-ray beams of elements. The analysis was conducted with the help of wave X-ray spectrometers and energy-dispersive spectrometer KEVEX.

Finding out the phase composition of high-speed steels was conducted by means of the X-ray diffractometer Shimadzu XRD-6000 in Cu-K-alpha radiation.

Retained austenite was defined by the diffractometer Xstress 3000 with the "method of four peaks". The authors compared integrated intensities of the diffraction peaks of austenite and martensite in the range of angles 2 theta 79.1...156.4 degrees [21]. The measurements and calculations were conducted automatically by the software of the tool XTronik. The peculiarities of measurements: instead of the shooting mode "modified" [5, 6, 11, 12, 22, 23], used at the measurement of residual stresses in a superficial layer, the authors used the mode "omega". The authors applied a spheric detector arc (figure 2).

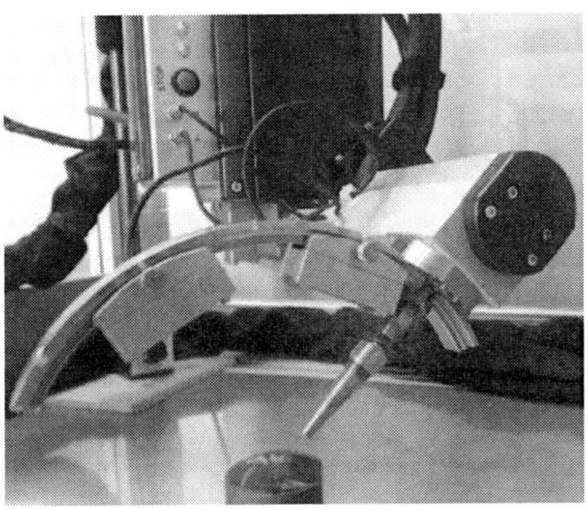

Figure 2. Diffractometer Xstress 3000 G3/G3R, omega mode.

At the treatment of diffraction peaks Lorentz correction was made. On the basis of the measured intensity the authors calculate the content of retained austenite according to the formula 1:

$$V_\gamma = \frac{(1-V_C)\cdot((q)^{-1}\cdot\sum_{j=1}^{q}((I_{\gamma j})\cdot(R_{\gamma j})^{-1})}{(p)^{-1}\cdot\sum_{i-1}^{p}((I_{\alpha i})\cdot(R_{\alpha i})^{-1})+(q)^{-1}\cdot\sum_{j=1}^{q}((I_{\gamma j})\cdot(R_{\gamma j})^{-1})} \quad (1)$$

where V_γ – volume fraction of austenite phase; V_c – volume fraction of carbide; q – number of austenite peaks (hkl); $I_{\gamma j}$ – integrated intensity of specific (hkl) austenite diffraction peak; $R_{\gamma j}$ – parameter of the theoretical austenite integrated intensity; p – number of martensite peaks (hkl); $I_{\alpha i}$ – integrated intensity of specific (hkl) diffraction martensite peak; $R_{\alpha i}$ – parameter of theoretical integrated intensity of martensite.

The parameter of theoretical integrated intensity (R) depends on the interplanar spacing (hkl), the Bragg angle, crystal structure and composition of the phase being measured. It is calculated by the main principles (see the standard ASTM E975-13).

IOP Conf. Series: Materials Science and Engineering **969** (2020) 012002 doi:10.1088/1757-899X/969/1/012002

3. Results and Discussion

The steel hardness, measured after tempering (66.5…67.5 HRC) falls in the range suitable for end mills, intended for highly efficient metal machining. At the microslice of the end surface of the cutting edge one can observe the sections of partially strained metal. The formed build-ups of the machined material have the thickness from 6 um to 36 um (figure 3a).

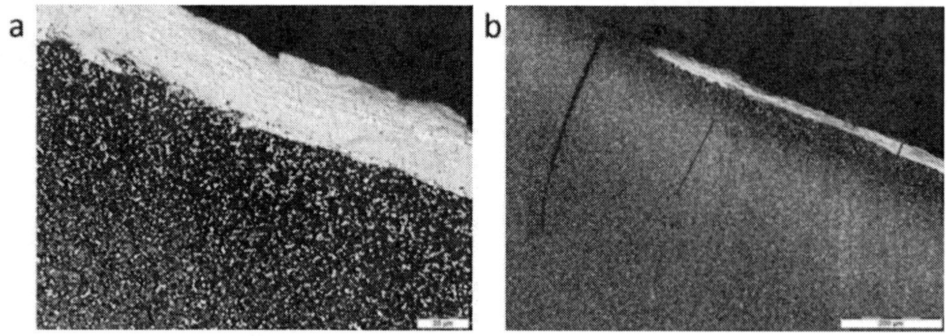

Figure 3. Mill cutting edge: (a) treated material adhesion; (b) region of heat influence and crack in the superficial layer.

The region of heat influence and microcracks, not characteristic for the material (figure 3, б), are clearly observable. The initiation of micro-cracks, developed in the direction from the surface towards the core, is obviously caused by excessive stresses during cutting.

The steel microstructure consists of fine-needled martensite quenching and multiple disperse carbides characteristic for steel powder. The steel structure is homogeneous, finely dispersed with the uniform distribution of carbides. Carbides have a round shape, equiaxial, small - their sizes are approximately 1.5 um. The sample microstructure is normal for the HSS heat treatment mode “quenching and three-stage tempering”. Cooling channels are covered with scale and remnants from heat treatment process (figure 4, b).

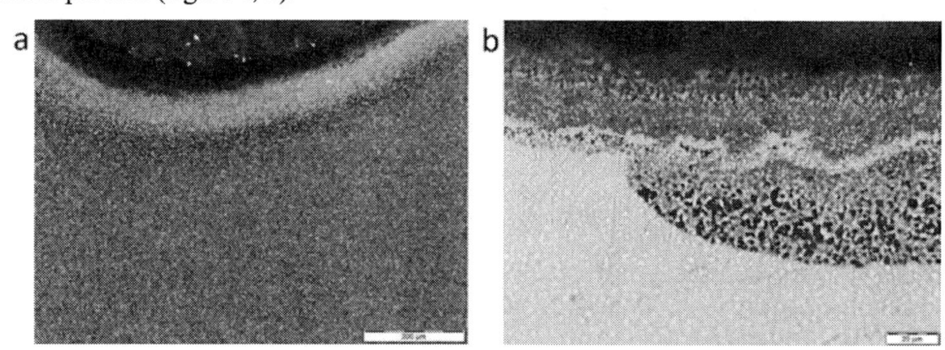

Figure 4. Cooling channels surface: a) scale and oxidation products are evident; (b) polished specimen, oxidation products and corrosion are evident in cooling channels.

The content of retained austenite does not exceed a permitted norm and falls within the range 1.4…1.8%. A larger part of analyzed chemical elements (Cr, Ni, Co) is contained in martensite. The distribution of carbide-forming alloying elements (W, Mo. V) is uniform.

The phase composition of the steel S390 is defined after various quenching temperatures. It is defined that martensite saturation with the alloyed elements increases at the growth of quenching temperatures from 1172 to 1193°C. Figure 5 shows the comparison of the diffractograms of steel S390 quenched at various temperatures and tempered at 560°C.

Irrespective of the quenching temperature in steel S390 the X-ray phase analysis detects ordered vanadium carbide – V_8C_7 [24], complex tungsten carbide – Fe_3W_3C и Co_3W_3C [25]. Cobalt is detected in the composition of complex tungsten carbides – Co_3W_3C and molybdene – $Co_6Mo_6C_2$

[26]. At the quenching temperature 1193°C steel S390 demonstrates the increase in the share of a stronger carbide-forming element – vanadium (V_8C_7), increase in the carbide share $Co_6Mo_6C_2$ and decrease in the amount of tungsten carbides.

Irrespective of the quenching temperature in steel S390 the X-ray phase analysis detects ordered vanadium carbide – V_8C_7, complex tungsten carbide – Fe_3W_3C и Co_3W_3C. Cobalt is detected in the composition of complex tungsten carbides – Co_3W_3C and molybdene – $Co_6Mo_6C_2$ [26]. At the quenching temperature 1193°C steel S390 demonstrates the increase in the share of a stronger carbide-forming element – vanadium (V_8C_7), increase in the carbide share $Co_6Mo_6C_2$ and decrease in the amount of tungsten carbides.

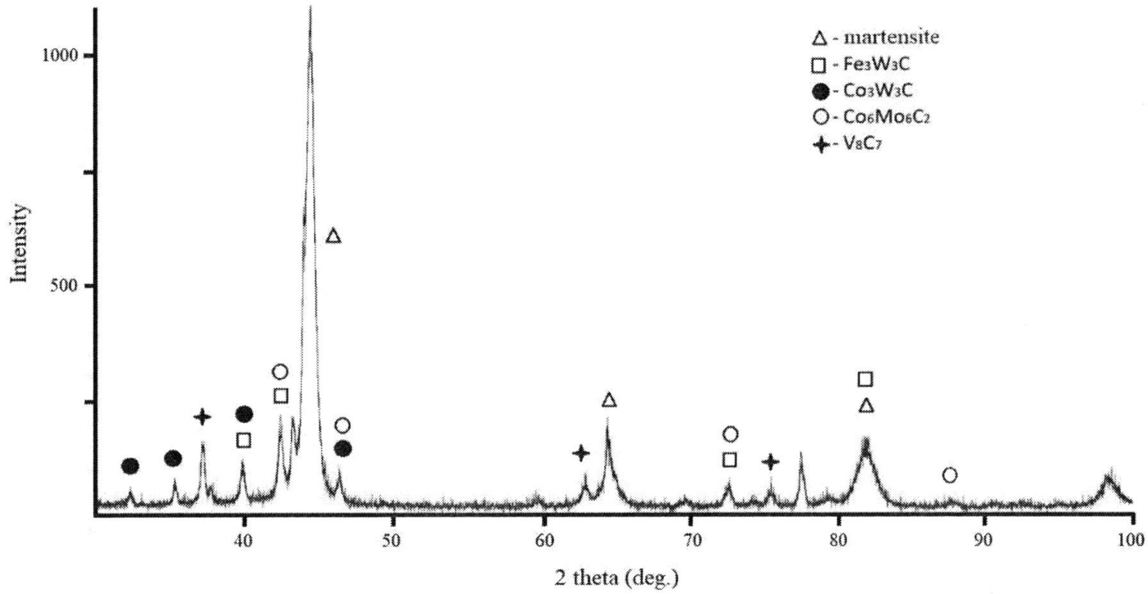

Figure 5. Diffractograms of steel S390 after heat treatment: red line – quenching temperature 1182°C; blue line – quenching temperature 1193°C.

In this steel no carbides of the type $Me_{23}C_6$, Me_7C_3, Me_2C are detected (these carbides are most frequently found in cast high-speed steels) [6, 25]. It is likely that some deviations from the recommended heat treatment modes [20] facilitate the occurrence of dispersed carbides of the specific composition in the microstructure of steel S390.

4. Conclusion

The conducted research make the authors draw a conclusion that one of the causes of early failure of the end mill is the state of material of cooling channel superficial layers. At all studied fragments along the inside surfaces of cooling channels one observes a fragile zone with the increased carbon content. Carburization occurs at the stage of hardening heat steel treatment. Carburization occurred because of oil use during hardening. In addition to carburization in these regions, one can also observe massive layers of oxides (scale) there. Hard and fragile carburization surface in combination with oxides (scale) depositions - these inside superficial defects are the cause of micro-cracks formation and, subsequently, the cause of tool destruction during high-speed treatment. Potential optimization of the service life of end mills, made of high-speed steel grades, consist in conducting heat treatment eliminating the formation of any deposits or carburization regions along the inside surfaces of cooling channels. This can be implemented at the application of modern heat equipment with cooling in vacuum or inert gas. Mechanical damage of the cutting edge during tool operation under high external load occurs because of high temperature action. Such temperatures are caused by friction and adhesion of the treated material. In this case one should select more favourable conditions of treatment under which it is possible to provide sufficient tool cooling for preventing defect occurrence.

References

[1] Savilov A V, Nikulin D S, Nikolaeva E P and Rodygina A E 2013 Current state of manufacturing high performance cutting tools from powdered metal high-speed steels and hard alloys *Proc. of ISTU* 26–33

[2] Bogdanov K V, Nikulin D S, Savilov A V, Nikolaeva E P and Rodyigina A E 2013 Production of high-performance cutting tools in IAP conditions. *Sci., and Tech.* 91–5

[3] Girshov V L and Topolyansky P A 2009 *Metalworking* **1(49)** 43–48

[4] Girshov V L, Korotkikh M T, Krupnova I V and Korobeinikov N V 2011 *Metalworking* **4(64)** 44–48

[5] Serebrennikova A G, Nikolaeva E P, Savilov A V, Timofeev S A and Pyatykh A S 2018 *IOP Conf. Series: Journal of Physics* **944**

[6] Nikolaeva E P and Vlasov D B 2017 Effect of heat treatment conditions on structure and properties of high-speed steel *IOP Conf. Series: Mat. Sci. and Eng.* **177**

[7] Tsemenko V N, Girshov V L and Mazurov S A 2011 *Metalworking* **5(65)** 38–40

[8] Nikolaeva E P and Nikulin D S 2016 The application of innovative means for quality control of the high-speed steel tools *Mod. Tech. Syst. An. Mod.* 73–80

[9] Vulykh N V 2019 Microprofile Model Form Changing Research at Axisymmetric Deformation with Account of Scale Factor *Lect. N. in Mech. Eng.* 1161–1168.

[10] Vulykh N V 2020 Centrifugal rolling of flexible shafts for achieving best possible roughness of the surface *Lect. N. in Mech. Eng.* 1079–1088.

[11] Nikolaeva E, Gridasova E and Gerasimov V 2015 The application of X-ray diffraction and Barkhausen noise for studying of shot peened constructional steel 30HGS *Izv. SSC RAS* **17** 125–132

[12] Gridasova E V and Nikolayeva E P 2015 Research of the improved steel 30HGSA. processed shot. X-ray diffraction analysis and Barkhausen noise measurements *Proc. of ISTU* 76–83

[13] Nikolaeva E P 2013 Application the Barkhausen noises method for control the hardening of details by superficial plastic deformation *Izv. SSC RAS* **15** 428–431

[14] Nikolaev A Yu 2017 Simulation of the plain milling process *IOP Conf. Series: Mat. Sci. and Eng.* **177**

[15] Savilov A V, Nikolaev D Yu and Nikolaev A Yu 2015 Study Of Instrumental Adjustment Imbalance Effect On Milling Output Parameters *Proc. of ISTU* **7(102)** 81–91

[16] Pyatykh A S, Savilov A V and Nikolaev A Yu 2016 Investigation of chucks for drilling precise holes *Sys. Met. Tech.* **4(32)** 70–74

[17] Nikolaev A Yu 2017 Simulation of the process of planar milling *Proc. of ISTU* 217–220

[18] Nikolaev A Yu 2019 Effect of tool setup imbalance on machined surface quality at end milling *Proc. of Ir. STU V* **23** **4(147)** 699–709

[19] Nikolaeva E P and Nikolaev A Y 2012 Application of magnetic methods for quality control of products from tool steels *Proc. of Ir. STU* 338–344

[20] Luchkina Ya V and Nikolayeva E P 2015 Determination of retained austenite by X-ray structural analysis in high-speed steel R6M5K5MP *Proc. of ISTU* 76–83

[21] Nikolaeva E P and Nikolaev A Y 2019 Measurement of residual stresses in thin-sheet welded constructions of low-alloyed steel *Lect. N. in Mech. Eng.* 2419–2426

[22] Nikolaeva E P and Mashukov A N 2017 Evaluation of Residual Stresses in High-Pressure Valve Seat Surfacing *Chem. and Pet. Eng.* **53(7–8)** 459–463

[23] Nikolaeva E P 2016 Structure Investigation of the Constructional Steel St3ps After Argon-Arc Plasma Treatment *Mat. Sci. Forum* **870** 500–506

IOP Conf. Series: Materials Science and Engineering **969** (2020) 012002 doi:10.1088/1757-899X/969/1/012002

[24] Gusev A I, Kurlov A S, Bobrikov I A and Balagurov A M 2015 V8C7–δ Superstructure In Nonstoichiometric Vanadium Carbide Powders *JETP Letters* **102** **3** 154–160
[25] Skakov M K, Rakhadilov B K, Karipbaeva G S and Manapbaeva A B 2014 Structural-phase state of R6M5 high-speed steel after thermal processing *Bulletin of KazNU The series is physical* **1(48)** 53–59
[26] Stulov Y V and Kuznetsov S A 2011 Standard Rate Constants of the Charge Transfer in the Cr(III)/Cr(II) Redox Couple in NaCl-KCl-CrCl3 and NaCl-KCl-K3CrF6 *Salt Melts Rus. Met.* **2** 158–163

Heat Resistance of Nickel-Manganese-Molybdenum Steel with Boron-Nitride Hardening

E N Eremin, A S Losev and S A Borodikhin

Machine-Building Institute, Omsk State Technical University, 11, Mira avenue, Omsk, 644050, Russia

E-mail: weld_techn@mail.ru

Abstract. The metal of a coating made of steel N8G6M3FTB, deposited by a flux-cored wire containing boron nitride, titanium diboride and zirconium diboride, was studied. The metal was subjected to surface oxidation at a temperature of 900 °C. It was found that the average weight gain of the metal scale of such a coating at 900 °C is 0.0087 kg/(m^2·h), which is 2.24 times less than that of steel 30Kh2V8F. It was revealed that the basis of metal scale, along with hematite and magnetite, are compounds with good protective properties. The results obtained show that the coating of steel N8G6M3FTB alloyed by boride compounds can be used for applying on the surface of parts operating at high temperatures.

1. Introduction

In the conditions of fierce competition in the market for rolled products, much attention is paid to its quality and production efficiency, which largely depend on the state of the metallurgical tool. Knives are one of the most worn parts of metallurgical equipment, which involves the processes of hot deformation of alloys subject to abrasive wear at elevated temperatures. The change of such a tool, mainly due to wear of the cutting edges, is a frequent and very time-consuming operation that restrains the operation of the entire mill.

Improving the tool life can be achieved by surfacing the work surfaces with wear-resistant materials [1–4]. At the same time, the resistance of such a tool at high temperatures depends not only on the metal resistance to abrasive wear, but also on the composition and structure of oxide films formed on the surface [5]. Due to its high hardness, which remains at hot working temperatures, the scale has significant abrasive properties, which enhances tool wear. The main element that increases the scale resistance of steels is chromium [6–8]. With an increase in the chromium content in iron, the onset of intense oxidation of steel shifts toward higher temperatures. High heat resistance (scaling resistance) of chromium-containing steels is due to the formation of very strong refractory chromium oxides or spinel oxides of the $FeCr_2O_4$ type on their surface [9]. Therefore, at metallurgical enterprises, knives are widely used, the cutting edge of which is deposited by steel of type 3Kh2B8. At the same time, continuous improvement of rolling production, an increase in the rolling speed and productivity of mills, and an increase in the share of rolled products from alloyed and special steels require an improvement in the properties of the tool. This is directly related to the introduction of new materials and surfacing technologies, providing an increase in the resource of a metallurgical tool subject to thermal power.

IOP Conf. Series: Materials Science and Engineering **969** (2020) 012003 doi:10.1088/1757-899X/969/1/012003

It is known that to harden a tool operating at high temperatures, it is advantageous to use composite surfacing alloys in the structure of which refractory particles of carbides, borides, and nitrides serve as hardening phases [10, 11]. From this point of view, the processes of applying wear-resistant coatings from economically alloyed martensitic-steel steels deposited by flux-cored wire containing boride compounds are finding wider application [12, 13].

Previous studies of metal deposited by flux-cored wire N8G6M3FTB found that such metal as a result of aging has a significant hardening effect and can be used for wear-resistant surfacing [14]. However, the features of oxidation of such a metal at high temperatures, containing a complex of compounds of boron nitride, titanium diborides, and zirconium, have not been studied.

In this regard, the aim of the work is to study the scale formation and changes in the structure and phase composition of steel N8G6M3FTB with boride-nitride hardening as a result of exposure to high temperatures in the air.

2. Objects and methods of research

The object of the research was cast steel of the coating deposited by flux-cored wire based on steel N8G6M3FTB alloyed with a complex of boron compounds (1.0% BN+2.5% TiB_2+1.0% ZrB_2). The calculated composition of the experimental cored wire is shown in table 1.

Table 1. The calculated composition of the experimental cored wire, %.

Ni	Mn	Mo	FeV	FeTi	FeNb	BN	TiB_2	ZrB_2	Na_2SiF_6	Fe
8.5	6	3.5	3.0	5.2	3.3	1.0	2.5	1.0	0.3	9.0

For comparison, the metal obtained by surfacing of 30Kh2V8F chrome-tungsten steel was widely used for the manufacture of a hot metal deformation technological tool.

Tests of the experimental composition of the coating metal for scale resistance (heat resistance) were carried out in a calm air atmosphere according to GOST 6130-71 [15]. An increase in the mass of samples at a temperature of 900 °C during a test period of 25 hours was chosen as a characteristic of scale resistance.

Metallographic studies of oxidized samples were carried out on a JEOL JCM – 5700 scanning electron microscope with a JED-2300 energy dispersive spectrometer.

X-ray phase analysis of oxidation products was carried out on a Shimadzu XRD-7000 multifunctional X-ray diffractometer. To process and analyze diffraction spectra, the Match! Software package version 3.8.1.151 was used.

3. Results of the experiments and discussion

The change in the mass gain of the metal deposited by N8G6M3FTB flux-cored wire alloyed with boride compounds and the metal deposited by 30Kh2V8F wire depending on the exposure time in an oxidizing atmosphere is shown in figure 1.

The graph shows a jump in the increase in the mass of scale on the surface of the metal N8G6M3FTB allowed with boride compounds occurs during the first hour. Over the next four hours, the increase in the mass of scale is negligible. After five hours of testing, the growth rate of scale increases, but to a much lesser extent than in the first hour of testing, and in the future this dependence is almost straightforward. So, if in 1 hour of testing the increase in the mass of scale was about 0.0389 kg/m^2, then in 5 hours it was only 0.0636 kg/m^2, and after 25 hours only 0.2166 kg/m^2. The thickness of the scale after 1 hour of testing at 900 °C is only 32.36...36.71 microns, and after 5 hours –55.69...65.50 microns. When a sample of metal N8G6M3FTB alloyed with boride compounds is cooled after the last 25 hours of testing, partial detachment of the scale from the surface occurs due to internal stresses. The thickness of the remaining part of the scale is 55.32...73.34 microns.

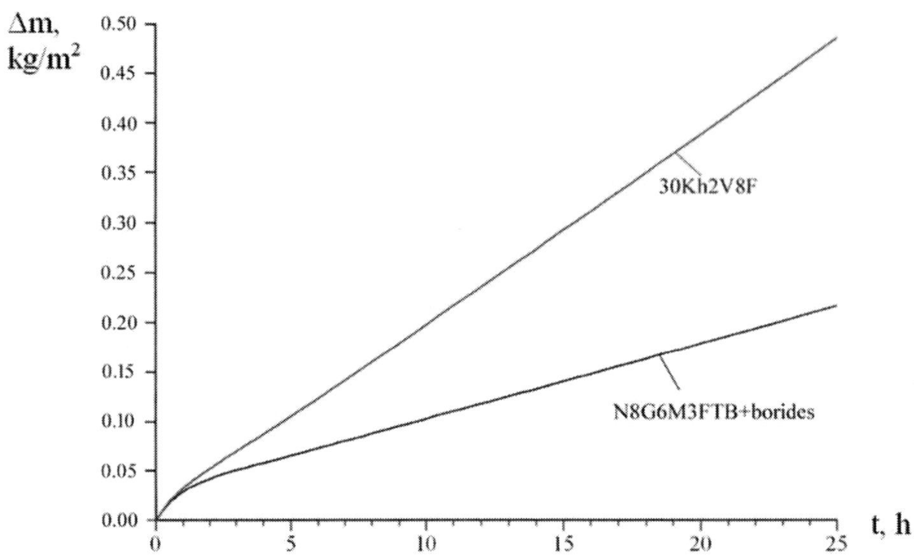

Figure 1. Change in the mass gain Δm of the metal N8G6M3FTB with borides and 30Kh2V8F from the exposure time at 900 °C.

The scale formation indices of steel 30Kh2V8F are significantly worse than the metal N8G6M3FTB alloyed with boride compounds. So if in 1 hour of testing the increase in the mass of scale was only about 0.0397 kg/m^2, in 5 hours it already amounted to 0.0986 kg/m^2, and after 25 hours it was 0.4858 kg/m^2. The thickness of the scale after 1 hour of testing at 900 °C is 44.17...52.29 microns, and after 5 hours - 70.07...81.17 microns. On the surface of a sample of 30Kh2V8F steel of the past 25 hours, the thickness of the oxide layer has increased significantly and is already 491.67...519.37 microns.

To identify the causes of increased scale resistance of the metal deposited by flux-cored wire N8G6M3FTB alloyed with boride compounds, the structure and phase composition of the deposited metal and scale were studied.

The microstructure of the thin section of the lateral surface of the N8G6M3FTB metal alloyed by boride compounds with a scale obtained after holding for 25 hours with the location of the scan points is shown in figure 2.

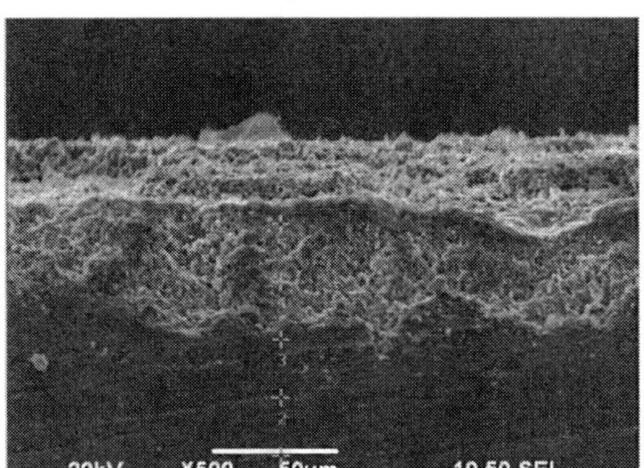

Figure 2. The lateral surface of a thin section of a metal of composition N8G6M3FTB alloyed by boride compounds with an oxide layer after a scale test at 900 °C - 25 hours with the location of the scan areas at points.

The figure shows that the structure of the scale consists of separate layers: the outer light and the inner darker. Almost uniform oxidation of the metal surface occurs, since a line of contact between the metal and the scale is traced. The interface is not perfectly flat. Roughnesses and gradations are observed due to different rates of phase formation reactions. Since the rates of chemical reactions at the phase boundary are quite high, their growth is completely determined by the supply of atoms to the phase boundary due to diffusion. Therefore, the dimensions of each phase depend on the diffusion coefficients, the relationship between them, and also on the concentrations of elements at its boundaries [16–19].

The chemical composition of the scanned areas is presented in table. 2.

Table 2. The chemical composition of the scanned metal regions of the composition N8G6M3FTB alloyed by boride compounds after a scale test at 900 °C - 25 hours.

Point №*	N	O	V	Ti	Mn	Fe	Ni	Zr	Nb	Mo
1	0	0	4.17	4.33	5.97	68.99	10.99	0.25	1.19	4.11
2	0	0	7.75	2.79	3.96	57.78	7.37	2.75	9.55	8.05
3	0.37	1.57	5.96	4.29	3.45	64.00	10.95	0.36	5.19	3.86
4	9.75	11.76	5.07	3.92	2.58	34.12	12.17	1.15	2.8	16.68
5	4.52	6.36	3.85	2.19	1.24	16.02	8.54	2.93	3.54	50.81
6	14.23	17.18	2.12	1.9	3.44	51.22	2.98	1.53	4.18	1.22

*Scan points 1, 2 correspond to the base metal; point 3 of the transition zone and points 4, 5, 6 of the scale.

The results of chemical analysis show that the concentration of the main alloying elements Nb, Mo, V, Zr in the metal increases as it approaches the transition layer (points 1, 2), and the concentration of Ni, Ti, and Mn decreases slightly. In the transition layer (point 3) there is a high concentration of Ni, Nb, Mo, V, Ti, and oxygen and nitrogen appear. Approaching the scale surface (points 4, 5), the concentration of Mo and Zr increases, and the concentrations of Mn, Ti, V, Nb, Ni decrease. Directly at the surface (point 6), all the main alloying elements are observed, and the concentrations of oxygen and nitrogen are maximum.

The results obtained indicate the formation of a weak chemical microinhomogeneity in the deposited coating. The high concentration of the main alloying elements is due to their low activity in diffusion processes when exposed to high temperatures [20].

The typical structure of the outer surface portion of the scale with the location of the scan points is shown in figure 3. The chemical composition of the scanned objects is shown in table 3.

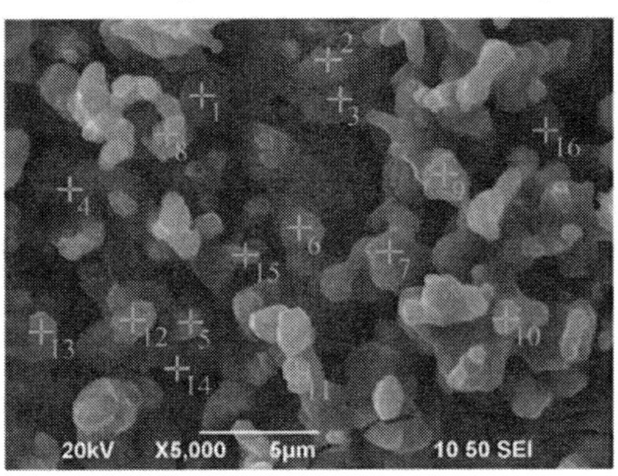

Figure 3. The surface of the oxide layer of a metal of composition N8G6M3FTB alloyed by boride compounds after a scale test at 900 °C - 25 hours with the location of the scan areas at points

It is seen that the surface layer of the scale is a chaotic mixture of rather densely packed particles of rounded shape with sizes up to 5 microns. All particles contain oxygen and nitrogen. Most particles contain all major alloying elements. However, there are particles that do not contain either Zr, or Mo, or Nb, or all of these elements (points 1, 4, 7, 10, 11, 14, 15, 16).

An X-ray analysis of a metal of the composition N8G6M3FTB alloyed with boride compounds after holding it for 25 hours at a temperature of 900 °C showed that the scale consists not only of the oxide phase, but also of the remnants of non-oxidized grains of solid solution and nitrides, borides and intermetallides. The scale base along with hematite Fe2O3, magnetite Fe3O4 and MnO are various compounds of the main alloying elements such as NbB2, Fe3C0,279N1,116, Ni0,6V0,4, Ni0,906Ti0,094, BN, V4O7, ZrN, TiC, ZrO2, NbBO4, Fe3C.

Table 3. The chemical composition of the scanned areas of the surface of the oxide layer of the metal with the composition N8G6M3FTB alloyed by boride compounds after a scale test at 900 °C - 25 hours.

Point №	N	O	V	Ti	Mn	Fe	Ni	Zr	Nb	Mo
1	17.23	21.95	3.31	1.21	9.51	43.66	0.02	0.98	2.13	0
2	11.65	11.88	4.05	0.91	9.06	58.71	0.84	0.85	1.54	0.51
3	0.7	1.12	2.26	0.96	11.42	80.71	2.74	0	0	0.09
4	7.5	9.19	3.25	1.38	10.45	63.33	0.41	1.57	2.92	0
5	12.55	16.4	3.28	1.33	9.36	51.03	0.84	1.97	1.96	1.28
6	10.75	14.05	3.44	1.54	9.13	55.81	0.09	0.18	2.62	2.39
7	7.74	9.59	5.17	2.17	9.25	64.54	0.46	0.53	0.55	0
8	13.51	17.81	3.11	1.04	8.09	47.23	2.39	1.43	3.52	1.87
9	16.67	21.63	2.68	0.84	8.72	44.36	0.21	1.96	2.49	0.44
10	15.52	19.03	2.22	0.91	13.77	47.25	0.79	0	0.51	0
11	8.71	11.41	2.99	1.31	12.97	61.4	1.21	0	0	0
12	11.72	14.44	2.74	1.14	14.43	48.85	0.6	2.02	2.89	1.17
13	6.38	8.1	3.79	1.63	10.11	66.49	0.48	0.56	1.47	0.99
14	1.59	1.57	3.95	1.43	10.98	79.82	0.66	0	0	0
15	3.15	3.4	4.54	2.04	11.03	73.87	0.77	0	1.2	0
16	0.51	0.66	5.97	2.09	11.39	78.52	0.86	0	0	0

Thus, the presence of compounds with high protective properties in the scale composition determines the increased heat resistance of the deposited metal. The increase in the mass of the scale of such a metal is 2.24 times less than that of the metal deposited by 30Kh2B8F wire. The results show that metal deposited by a flux-cored wire PPN8G6M3FTB alloyed with a complex of boron compounds (1.0% BN+2.5% TiB2+1.0% ZrB2) can be used to apply wear-resistant coatings to tools operating at high temperatures.

4. Conclusion

The coating of steel N8G6M3FTB with boride-nitride hardening is subject to slight surface oxidation. The average weight gain of the metal scale of such a coating at 900 °C is 0.0087 kg/(m^2·h). Such a coating can be used for applying on the surface of parts operating under conditions of abrasive wear at high temperatures up to 900 °C.

References

[1] Sokolov G N and Lysak V I 2005 *Deposition of wear-resistant alloys on pressing dies and tools for hot deformation of steels* (Volgograd: VolgSTU) p 248

[2] Shekhter S Ya and Shvartser A Ya 1981 *Surfacing of parts of metallurgical equipment. Directory* (Moscow: Metallurgy) p 160

[3] Mikhaylitsyn S V, Sheksheev M A, and Yaroslavtsev A V 2017 *Restoration and hardening ofmachine parts* (Magnitogorsk: NMSTU) p 179

[4] Balanovsky A E, Grechneva M V, and Chupin Yu 2019 *Strengthening and restoration technologies of welding and surfacing: a training manual* (Irkutsk: INRTU) p 142

[5] Tolstov I A, Pryakhin A V, and Nikolaev A M 1990 *Improving the performance of the hot deformation tool* (Moscow: Metallurgy) p 143
[6] Oryshchenko A S 2011 *Heat Resistant, heat resistant alloys* (St. Petersburg: Science) p 191
[7] Wang H L, Yu W S, and Shen S P 2019 *Science China – Technological Sciences* **62** 8 1246–54
[8] Singh R. 2019 *Metals* **9** 9 DOI: 10.3390/met9090942
[9] Young D J 2016 *High Temperature oxidation and corrosion of metals* (Kidlington: Elsevier Science LTD) p 733
[10] Sheenko I N, Oreshkin V D, and Repkin Yu D 1970 *Modern surfacing materials* (Kiev: Science thought) p 238
[11] Zhong L, Xiang C, Yan-xiang L, and Kai-hua H 2009 *J. of Iron and Steel Research* **16** 3 37–42
[12] Artem'ev A A, Sokolov G N, Dubtsov Yu N, and Lysak V I 2011 *News of higher educational institutions. Powder metallurgy and functional coatings* **2** 44–48
[13] Eremin E N 2013 *Welding International* **27** 2 144–146
[14] Eremin E N, Losev A S, Ponomarev I A, Borodikhin S A, and Matalasova A E 2019 *AIP Conference Proceedings: Oil and Gas Engineering* **2141** 040004–1–6
[15] GOST 6130-71 1990 *Metals. Methods of determining heat resistance* (Moscow, Standards publishing) p 53
[16] Wagner C 1969 *Acta metallurgica* **17** 2 99–107
[17] Clarke D R 2003 *Acta metallurgica* **51** 5 1393–1407
[18] Sherman Q C, Voorhees P W, Marks L D 2019 *Acta metallurgica* **181** 584–594
[19] Kirkaldy J S, Zia Ul Hag, Brown L S 1963 *Trans ASM* **56** 834–863
[20] Anastasiadi G P 1991 *Formation of chemical microhomogeneity in foundry alloys* (St. Petersburg: Polytechnic) p 148

Acknowledgments
This research was made possible through the Russian Science Foundation grant (project №17-19-01224).

IOP Conf. Series: Materials Science and Engineering **969** (2020) 012004 doi:10.1088/1757-899X/969/1/012004

Investigation of the Effect of Sulfide Inclusions on the Hydrogen Separation of Metal

O A Nasibullina and R G Rizvanov

Department of Petroleum Engineering Technology, Ufa State Petroleum Technological University, 1, Kosmonavtov str., Ufa, 450062, Russia

E-mail: ksu33@bk.ru

Abstract. In the presence of an increased concentration of hydrogen, premature destruction of the metal occurs. Hydrogen brittleness is caused by imperfection of the crystal lattice of metals. Distinguish between hydrogen corrosion caused by molecular hydrogen and hydrogen corrosion caused by atomic hydrogen. The presence of hydrogen in a metal increases the fragility of all metals, without exception. Hydrogen accumulates in traps (vacancies, dislocations). The segregations of carbon and other impurity atoms at grain boundaries enhance the ability of iron to capture hydrogen. Atomic hydrogen resulting from electrochemical reactions penetrates steel at normal temperature, causing it to stratify. Molecular hydrogen generated during the processing of hydrocarbons, electrolysis of water, violation of welding technology, penetrates into steel only at temperatures above 200 °C. The aim of the work was to determine non-metallic inclusions using the metallographic method. The authors also determined the presence of sulfide accumulation in the studied metal of the pipe. It is noted that the amount of sulfide inclusions is approximately the same, both near the metal stratification and on the base metal of the sample.

1. Introduction

Various kinds of defects may appear on the pipeline during operation, such as cracks [1–2], ulcers [3–8], scratches, etc. It is known that in the presence of an increased concentration of hydrogen, premature destruction of the metal occurs, regardless of temperature, but fracture is distinguished at high and low temperatures and pressures. At high temperatures, hydrogen corrosion occurs, which takes carbon atoms from the metal, thereby changing the mechanical properties. At low temperatures, only hydrogen embrittlement occurs [2, 3], in which hydrogen penetrates into the metal in a confined space, in a tank, in a vessel or in a cylinder, etc. Hydrogen penetrates into the metal by the diffusion mechanism, and the more intensively, the stronger the stretched zone, then it accumulates in the metal and when a certain concentration is reached, the mechanical properties of the metal change.

Hydrogen brittleness is caused by imperfection of the crystal lattice of metals [8–20]. The presence of hydrogen in a metal increases the fragility of all metals, without exception.

2. Experimental procedures

The paper presents the results of a study on samples of steel grade 09G2S, selected from the focal zone of destruction of the gas pipeline. There were bundles on the pipe. The defect is believed to have arisen due to hydrogen separation within the metal. To identify the causes of stratification of the metal, the metal was cut with defects, size 40x13x5 mm. The steel 09G2S under study is low-alloyed silica-manganese structural. The mechanical properties of steel, according to GOST 19282-73, are presented in table 1.

Table 1. Mechanical properties of steel 09G2S at room temperature.

Impact strength KC_U, J/cm^2	Strength limit $\sigma_в$ MPa	Elongation at break δ, %	Yield strength, $\sigma_т$ MPa
64	490	17	365

In the presence of defects, hydrogen is retained in the metal with the formation of a brittle solid solution (strength increases, fracture toughness decreases), metal stratification along the segregation floor, blistering (hydrogen or etching bubbles). Proton capture sites are called hydrogen traps (concentrators). They can be point (vacancies) or linear (dislocations). In the latter case, during plastic deformation, the movement of dislocations will slow down, which leads to an increase in strength and embrittlement. Hydrogen traps are also non-metallic and intermetallic inclusions, grain boundaries. Hydrogen is located in the tetra- and octahedral position in figure 1.

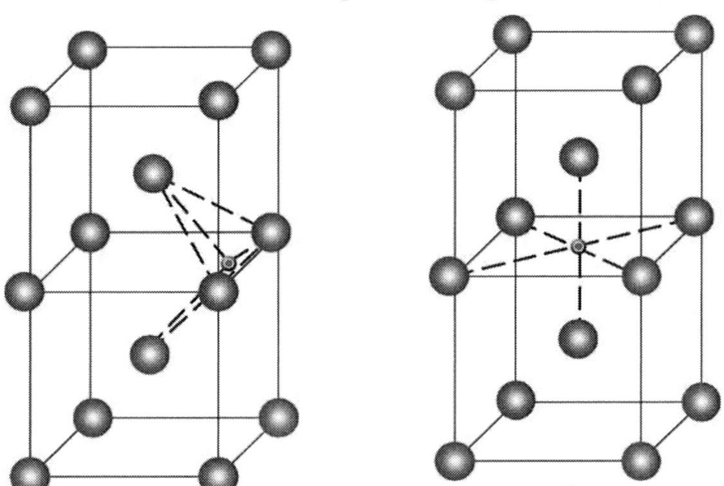

a b

Figure 1. The location of hydrogen in the metal; a – tetrahedron; b – octahedron; ● – metal atoms, ◉ – hydrogen atoms.

The diagram of the formation of the cavity inside the steel is shown in figure 1. Protonated hydrogen encounters in its path an inclusion that does not have a metal bond, i.e. generalized electrons are missing. He takes his electron from the electron gas and enters the inclusion crystal lattice, where molecular hydrogen is formed. The hydrogen pressure in the metal reaches tens, and according to other sources, hundreds of MPa. The resulting mechanical stresses lead to the formation of cracks.

The results of measuring the parameters of atomic hydrogen penetration into iron and steel by the Devanathan-Stachursky method (visible diffusion coefficients D_0 [m^2/s], activation energy Q [kJ/mol]) are given in Table 2.

Table 2. Diffusion characteristics of hydrogen penetration into iron and steel.

Material	Temperature	D_0	Q
Pure iron	350 K	$1 - 2.5 \cdot 10^{-7}$	7,6
	Ambient temperature	$0.5 - 1.2 \cdot 10^{-7}$	4 – 7
	298 K	$7 \cdot 10^{-9}$	
Real steel		$\sim 10^{-11}$	

As follows from the table, the apparent diffusion coefficients D_0 of pure iron with a low density of dislocations and real steels differ by several orders of magnitude due to the presence of more traps.

Some metals in the presence of hydrogen corrode by the mechanism of chemical corrosion. In this case, weight loss due to the formation of hydrides is observed. Moreover, hydrides are formed at elevated temperatures.

A number of authors hypothesize that the hydrogen separation of the metal is confined to the sites of segregation of sulfide inclusions. To clarify the validity of this hypothesis, the distribution of sulfide inclusions was determined. They were detected by the Baumann method. The sulfide inclusions were detected by the Bumann method. An etchant is required for this method [5 ml H2S04; 100 ml of H20]. This method was proposed by Baumann in 1906. It is the simplest imprinting method for detecting sulfur distribution and is therefore widely used.

The method is based on the following reactions: sulfuric acid reacts with iron and manganese sulfides in steel to form hydrogen sulfide by reaction:

$$(Fe, Mn)S + H_2SO_4 \rightarrow (Fe, Mn)SO_4 + H_2S\uparrow \quad (1)$$

the hydrogen sulfide formed interacts with silver halides AgHal (Hal - Br, I), which is in the emulsion of photo paper with the formation of brown silver sulfide by the reaction:

$$H_2S + 2AgHal \rightarrow Ag_2S + 2HHal. \quad (2)$$

The method is implemented as follows: Conventional non-fixed photo paper in daylight is impregnated with dilute sulfuric acid and then excess acid is removed. Then the paper is pressed with an emulsion layer for approximately 2 minutes to a previously sanded and degreased sample, and a thin press is used to avoid slipping when using glossy paper. The paper is thoroughly washed, fixed, washed again with water and dried. From the freshly prepared surface of the sample, you can get 2 to 3 prints, which even with a longer one will still be less intense than the first.

Photo paper should be as thin as possible. This rule applies to all fingerprinting methods. Photo paper, especially bromine silver with a clear gradation of shades, is best suited. Immersion in dilute sulfuric acid lasts about 2 minutes; if the duration of the paper is too short, it absorbs an insufficient amount of liquid; if it is too long, gelatin swells very much, which leads to an unsharp picture due to increased diffusion of the reagents.

The clarity of the print depends on the degree of roughness of the surface of the sample and the print media [14]. Even if you manage to get good prints using the Baumann method of rather rough surfaces (filed or pre-sanded), you should avoid too much surface deformation during sample preparation. On prints obtained from surfaces treated by cutting, on dark areas, traces of processing can be detected.

Defects of the investigated surface, such as cracks, pores, and others, due to their increased corrosion by acid due to the large surface, often lead to an incorrect interpretation of the print. It is necessary to maintain the optimal sulfuric acid content in the solution - about 5%.

3. Results and discussion

To determine the amount of sulfide inclusions in the test sample, "Photocont" photo paper was impregnated with a solution of 5 ml of H_2SO_4; 100 ml of H_2O. After impregnation, excess acid was removed with a cotton swab. The impregnated photo paper was rolled onto the sample using a photographic roller, kept on the surface of the sample for 10 minutes, washed under a stream of cold water and dried. The obtained sulfur imprint was examined using a stereoscopic microscope MBS 10 at a magnification of 100x. The number of inclusions was calculated using an ocular mesh. Photographs of prints are shown in figures 2–3.

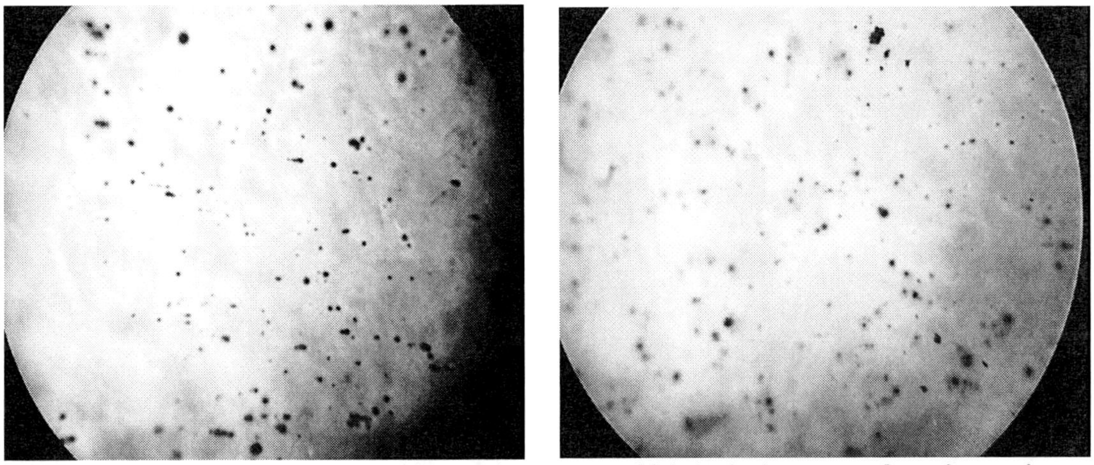

Sulfide inclusions at the top of one of the cracks, ×100

Sulfide inclusions away from the crack, ×100

Figure 2. Distribution of sulfide inclusions.

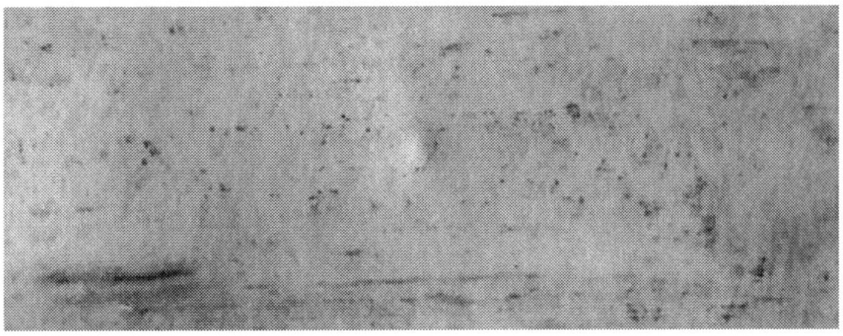

Figure 3. Macromodel of the fingerprint by the Bauman method.

As can be seen from the figures 2–3, the number of sulfide inclusions does not exceed the first or second point according to GOST 1778-70 (ISO 4967-79) "Steel. Metallographic methods for the determination of non-metallic inclusions". That is, there is no confinement of the foci of hydrogen separation to areas of metal contaminated with sulfide inclusions. It should be noted that the number of sulfide inclusions is approximately the same, both near the metal stratification and on the base metal of the sample. This phenomenon suggests that sulfide inclusions do not affect the process of metal separation.

4. Conclusion

Studies of the distribution of sulfide inclusions were carried out according to the Bauman method, which is the simplest method for obtaining fingerprints to detect the distribution of sulfur and is therefore widely used. It was established that the number of sulfide inclusions does not exceed the first point according to GOST 1778-70 (ISO 4967-79). It was noted that the number of sulfide inclusions is approximately the same, both near the metal stratification and on the base metal of the sample, this fact proves that sulfide inclusions do not affect the metal stratification process.

References

[1] Nasibullina O A, Gareev A G and Rizvanov R G 2018 *Solid State Phenomena* **284** 1302–1306
[2] Gaysin E Sh, Frolov Y A and Nasibullina O A 2020 *IOP Conf. Ser.: Earth and Environ. Sci.* **459** 032055
[3] Nasibullina O A and Tyusenkov A S 2019 *IOP Conf. Ser.: Mater. Sci. Eng.* **537** 022018
[4] Abdullin I G and Gareev A G 1993 *Physico-chemical mechanics of materials* **29(5)** 97–98

[5] Yakhin A V, Karetnikov D V, Rizvanov R G, Abakacheva E M and Gareev A G 2019 *Chemical and Petroleum Engineering* **54(11–12)** 801–805
[6] Zaripov M Z, Fairushin A M and Karetnikov D V 2019 *Materials Science Forum* **946** 883–888
[7] Faritov A T, Rozhdestvenskii Yu G, Yamshchikova S A, Minnikhanova E R and Tyusenkov A S 2016 *Russian Metallurgy (Metally)* **11** 1035–1041
[8] Kuzeev I R, Ibragimov I G, Bayazitov M I, Davydov S N and Khairudinov I R 1985 *Chem. and Technology of Fuels and Oils* **22(3)** 111–113
[9] Tyusenkov A S, Rubtsov A V and Tlyasheva R R 2017 *Solid State Phenomena* **265** 868–872
[10] Vakulenko M V and Zharinova N V 2020 *IOP Conf. Ser.: Mater. Sci. Eng.* **862** 062022
[11] Nasibullina O A and Gareev A G 2019 *Materials Science Forum* **946** 20–24
[12] Abdullin I G and Gareev A G 1994 *Materials Science* **29(5)** 539–541
[13] Zharinova N V and Vakulenko M V 2020 *IOP Conf. Ser.: Mater. Sci. Eng.* **862** 032014
[14] Abdulganiev M R, Khasanov R N and Gafarova V A 2019 *Materials Science Forum* **945** 569–573
[15] Krioni N K, Mingazheva A A, Kononova A Yu, Mingazhev A D and Gafarova V A 2017 *Solid State Phenomena* **265** 215–220
[16] Mingazhev A D, Kuzeev I R and Gafarova V A 2020 *J. Phys.: Conf. Ser.* **1515** 042074
[17] Tyusenkov A S and Nasibullina O A 2019 *IOP Conf. Ser.: Mater. Sci. Eng.* **687** 066016
[18] Latypov O R, Bugai D E and Boev E V 2015 *Chemical and Petrol. Eng.***51** 283–285
[19] Nasibullina O A and Tyusenkov A S 2019 *IOP Conf. Ser.: Mater. Sci. Eng.* **537** 022023

Acknowledgments
The authors are grateful to Alexei Gabdullovich Gareev, Professor of the Ufa State Petroleum Technological University (Russian Federation) for his advice and assistance in conducting experiments.

Microstructure and Mechanical Properties of the Wire Arc Additively Manufactured Sv-08G2S Steel

A A Kulikov, A E Balanovskiy and M V Grechneva

Irkutsk National Research Technical University, 83, Lermontov street, 664074, Irkutsk, Russian Federation

Email: the.tosha2013@gmail.com

Abstract. Wire arc additive manufacturing enables using almost any grade of metal for the additive production of parts with complex shapes and large sizes. As an increasingly growing number of materials are being used for WAAM, it is necessary to conduct systematic research on the properties of various metals and alloys. This article presents a study of the microstructure and mechanical properties of the wire arc additively manufactured Sv-08G2S steel. Once the wall-shaped sample was deposited, it was cooled in the air at room temperature. The sample for microstructure studies was cut across the central line of the deposited layers. In the sample, three sectors were conditionally distinguished: the lower sector, deposited on a cold substrate, has a ferrite structure with inclusions of lamellar perlite; the middle sector has a purely ferritic structure; the upper sector has a bainite structure. The microstructure study was followed by hardness measurements performed in all sectors of the sample. The hardness values confirmed the presence of different structures in all sectors, the hardest of which was the upper sector with a bainite structure. The last stage of the study was the tensile testing of the sample performed on a test breaking machine. The obtained values of the ultimate tensile strength and yield strength of additively deposited steel are within the range and even exceed the characteristics of wrought low-carbon steels. Therefore, the conducted research proved that the additive manufacturing of parts made of Sv-08G2S steel enables achieving the necessary quality and provides the mechanical properties required for structural steels.

1. Introduction

The popularity of additive manufacturing (AM) and its rapid development are primarily related to the necessity to develop modern cost-effective manufacturing technologies. AM enables high-performance manufacturing of parts with complex geometric shapes. Another significant advantage of AM is a high material utilization rate. One of the latest technologies for 3D printing of metal products is wire arc additive manufacturing (WAAM). This technology combines industrial robots with off-the-shelf welding equipment. Thanks to the use of welding wires, the WAAM technology has a huge potential for printing almost any metal alloys [1–8]. Compared with the traditional subtractive manufacturing (machining), WAAM has a higher performance (40-60% higher), which helps to reduce the time of the finishing machining by 15-20%, depending on the size of the product [9].

The heat source is an electric arc generated between the substrate and the metal wire [10–13]. Welding wires are used as a feedstock material, which is much cheaper than metal powders used in other additive technologies [14-17]. Along with printing new parts, WAAM can also be used for repair and restoration operations [18].

The production of structurally strong, high-quality, reliable parts requires an understanding of the existing technological processes, underlying physical processes, raw materials, process control methods, and the causes of various defects and ways to eliminate them. In order to produce parts with pre-set properties, it is necessary to understand the principles of the microstructure evolution and mechanical properties of metals during the thermal impact.

As steels are still the most common industrial materials, they are of great interest to the WAAM community. On the one hand, the allotropy of iron-based alloys combined with high-temperature gradients makes it possible to produce products with a unique microstructure. On the other hand, alloys that may have different phase compositions depending on the cooling rate, such as martensite and retained austenite in steels with dispersion hardening will be sensitive to different thermal welding cycles. Therefore, steels are needed to be carefully studied for application in WAAM.

This article presents the following studies of the wire arc additively manufactured Sv-08G2S steel: macrographic and micrographic analysis of the microstructure of steel using a scanning electron microscope, Brinell hardness test, tensile testing of samples to determine the yield strength and ultimate tensile strength.

2. Materials and Methods

A Lorch SpeedPulse S3 mobil welding machine was used for the deposition of the sample. The industrial robot KUKA KR 210 R2700 prime was used as a manipulator of the welding torch. The following modes were used for deposition: welding current 100 A, voltage 26 V, wire feed rate 4.5 m/min, printing speed 3 mm/sec, stickout distance 14 mm. To protect the molten metal from air contaminants a shielding CO2 gas with a flow rate of 10 l/min was used.

The wall was used as a sample (see figure 1) deposited on a Fe37-3FN structural carbon steel plate of 16 mm thickness. The wall has the following dimensions: length 100 mm, height 27 mm, the number of layers 13.

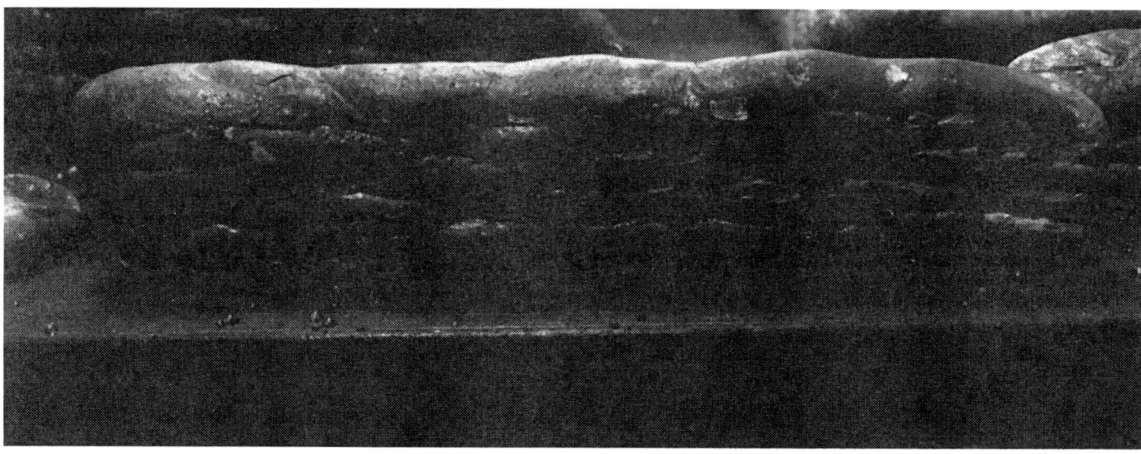

Figure 1. The deposited wall.

The Sv-08G2S steel welding wire was used as a feedstock material. The chemical composition of this wire is presented in table 1. The sample was cooled in the air at room temperature.

Table 1. Chemical composition of SV-08G2S wire.

C	Si	Mn	Ni	S	P	Cr	N
0.06	0.8	1.9	0.2	0.02	0.02	0.1	0.01

The sample for microstructure studies was cut across the central line of the deposited layers. Optical microscopy of the mechanically polished sample was performed in a cross-section. The sample was etched with Keller's reagent (2 ml HF, 3 ml HCl, and 5 ml HNO3 in 190 ml of water) to reveal the structure of the metal and grains. Electron microscopic studies were carried out on the basis of the Irkutsk National Research Technical University using a scanning electron microscope JIB-4501 JEOL multi-beam system equipped with an electronic and ion gun JIB-4501, completed with a nitrogen-free system of energy-dispersive microanalysis. The results were analysed using the Channel 5 software package developed by Oxford Instruments. For these purposes, the test sample has additionally passed the electro-polishing stage. The size of the scanning area for a sample was 620 microns. The grain was taken as an area surrounded by large-angle borders, i.e. the value of the limit angle was set to 15°.

The macro- and microanalysis were followed by the Brinell hardness testing along the length of the sample. The measurements present results of evaluating the hardness of sectors of the sample from the first layer to the last. The HBRV – 187.5 hardness testing machine was used for measurements. The sample was used under a load of 7kH for 10 seconds. The HBRV – 187.5 hardness testing machine uses a rotational type of load change mechanism, as well as an optical measurement display system and a measuring microscope mounted on the device body. Polished (Ra < 0.04 microns) balls made of a SHX15 steel with a nominal diameter of 5 mm were used as penetrators. The CDM-20 tensile testing machine was used for tensile testing of the sample.

3. Microstructural analysis

Preparation of the sample (see figure 2) for microstructural analysis includes following stages:

- rough grinding;
- final grinding;
- polishing;
- etching.

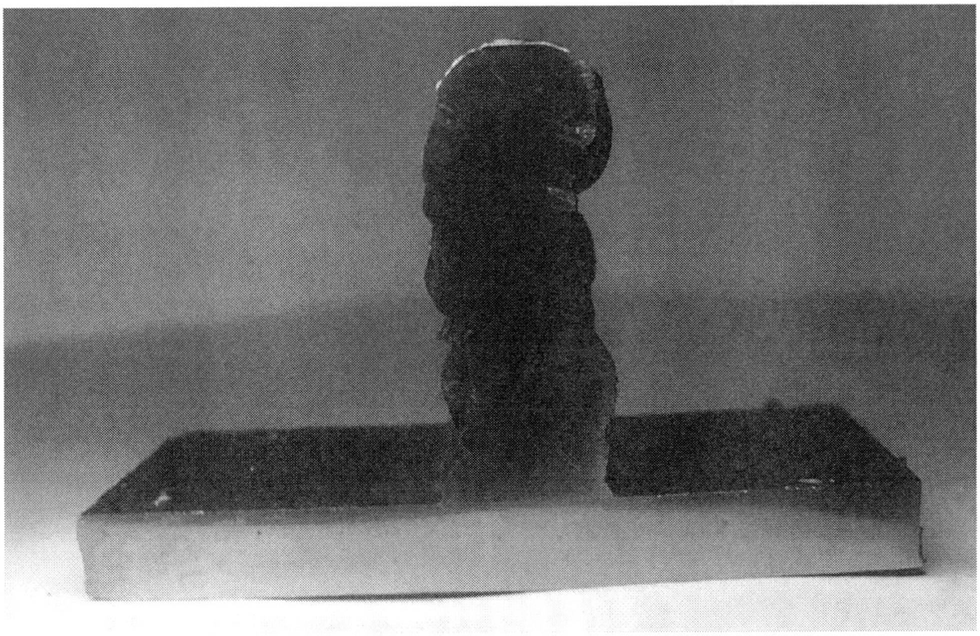

Figure 2. Sample for microstructural analysis.

Rough grinding is performed for the leveling of a surface of the sample using an engineer's file. Then a final grinding follows. Final grinding is performed to minimize surface imperfections using abrasive paper of decreasing grain size. Corundum, carborundum, and other solids are used as abrasives deposited on abrasive papers.

The next stage of preparation is polishing, which is necessary for the final alignment of the surface of the sample. Polishing is carried out using a polishing machine, the working body of which is a rotating disk covered with a fabric.

Etching of the polished sample is performed to reveal a complete picture of the microstructure (grain shape and size, presence of phases, structural components). Before etching, the surface of the sample is degreased with alcohol. The sample is etched with Keller's reagent (2 ml HF, 3 ml HCl, and 5 ml HNO_3 in 190 ml of water) for 10-20 seconds. After etching, the sample is washed with water and dried with filter paper.

The sample for microstructural analysis can be divided into 3 sectors: lower, middle and upper (see figure 3). The lower sector contacts the cold surface of the substrate metal and experiences a certain heat drop. The metal of the middle sector is characterized by a lower temperature difference since the weld beads of this sector are deposited on the preheated metal of the previous beads.

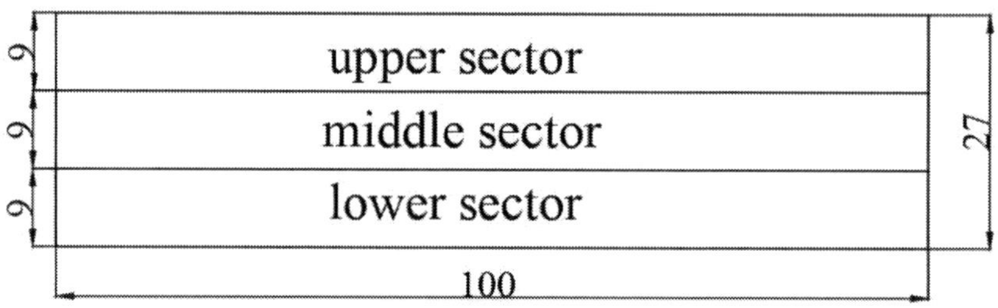

Figure 3. Sectors of the sample.

The metal of the upper sector is characterized by a sharper temperature difference since it is in a contact with room temperature air. Figure 4 shows the macrostructure of the Sv-08G2S steel in three different sectors.

Figure 4. Macrostructure of the Sv-08G2S steel in various sectors.

The macrostructure of the upper sector is significantly different from the macrostructure of the other two sectors. The metal of the lower and middle sectors is characterized by an almost equiaxed structure, while the metal of the upper sector has a lamellar structure. Figure 5 presents the transition zone between the lower and middle sectors. The difference in microstructure and grain size is clearly visible on the transition line.

Figure 5. Transition zone between the lower and middle sectors.

The microstructure of this transition zone is characterized by a mixture of coarse grains of the lower sector with fine grains of the middle sector. Figure 6 shows the microstructure of the lower sector. Equiaxed grains with inclusions of thin lamellas are observed in this sector [19, 20].

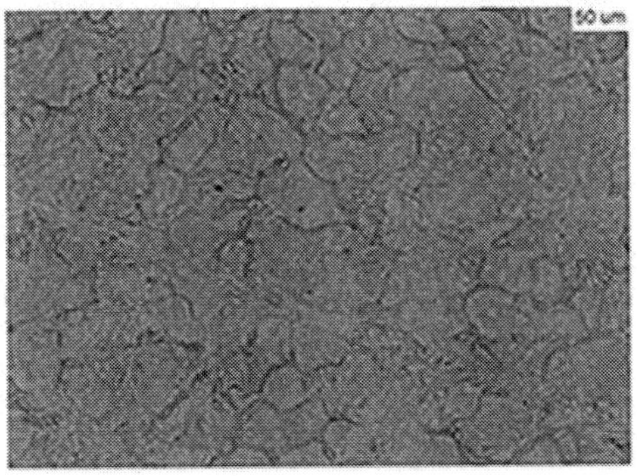

Figure 6. The microstructure of the lower sector (ferrite+perlite).

Since the metal is deposited on a cold substrate, the first beads experience a significant temperature difference. This contributes to the formation of a microstructure with ferrite grains and inclusions of lamellar perlite. The formation of ferrite with an equiaxed grain shape with inclusions of lamellar perlite is typical for low-carbon steel, which is the welding wire Sv-08G2S.

Figure 7 shows the microstructure of the middle sector. The middle sector is characterized by equiaxed ferrite grains without any inclusions. The microstructure shows that the ferrite of the middle sector has a more coarse-grained structure than in the lower sector. This is due to a higher temperature difference in the lower sector compared with the middle one.

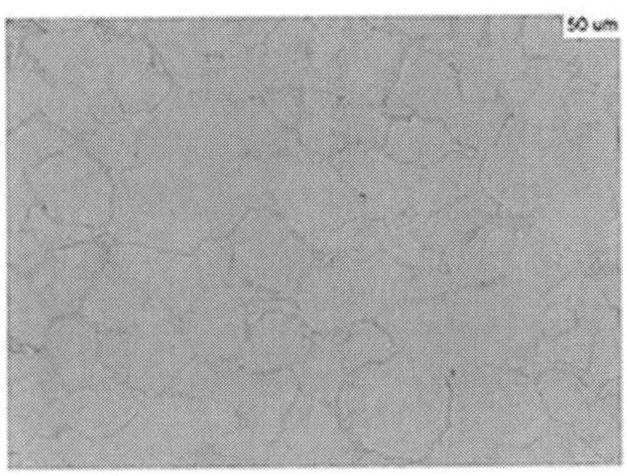

Figure 7. The microstructure of the middle sector (ferrite).

The metal of the middle sector cools slower than that of the upper sector, so its thermal gradient is lower than that in the lower sector. The metal of the lower sector also cools slower than the metal of the upper sector, which results in the formation of ferrite. The presence of lamellar perlite is explained by a less sharp temperature drop due to the presence of a cold substrate compared with air cooling in the upper sector. Figure 8 shows the microstructure of the upper sector.

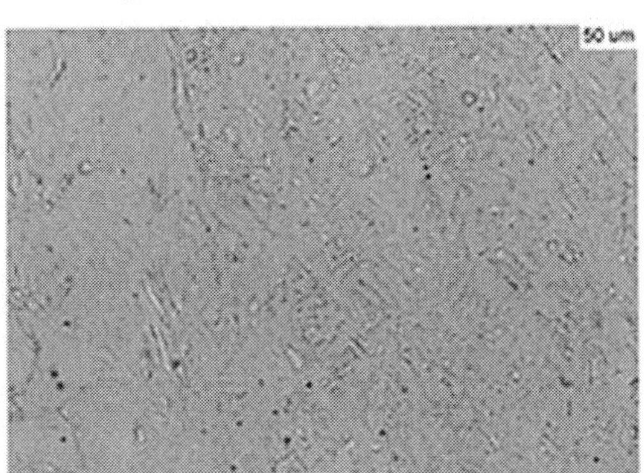

Figure 8. The microstructure of the upper sector (bainite).

As mentioned earlier, it is in this area that a higher temperature drop is observed, so the metal microstructure in this sector is significantly different from others and has a predominantly lamellar structure. In particular, it should be noted that this microstructure is bainite, which is a mixture of ferrite and cementite. Since cooling begins at a temperature about 70 °C above the critical temperature for the steel used, the growth of ferrite particles devoid of carbon occurs. Since the carbon in ferrite is contained in smaller amounts, it passes to the upper sections and the lamellas are filled with carbon.

It is also worth noting that due to the effect of reheating during the deposition of layers, it is possible to achieve the formation of different microstructures along the height of the sample. This enables adjusting the microstructure between ferrite and bainite to obtain the necessary mechanical properties of the product by alternating cooling cycles after the deposition of one or more layers.

4. Hardness testing

Figure 9 shows a diagram with Brinell hardness values. The results obtained show the hardness values in the direction from the lower to the upper sector. The highest values of hardness are observed in the sector with the bainite structure. It is worth noting the almost complete absence of pores, which is a significant advantage in comparison with other additive technologies used for the production of metal products.

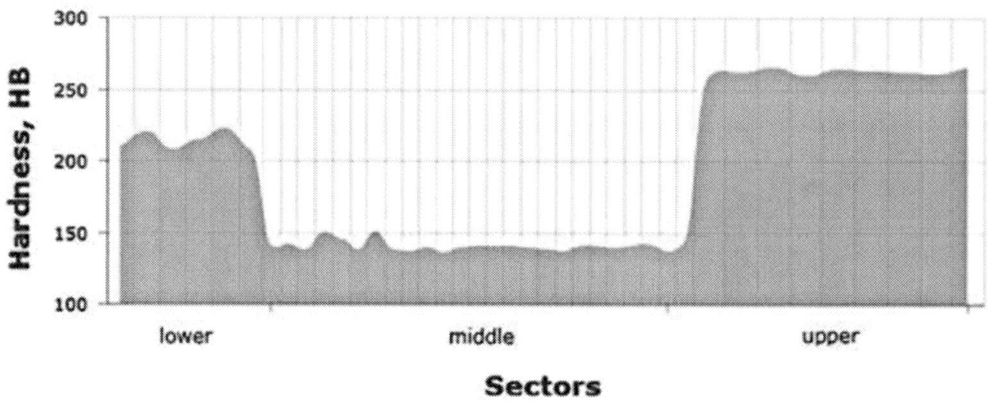

Figure 9. Hardness diagram.

The results of hardness measurements confirm the formation of various structures in the sample sectors. Thus, the lower sector, where the structure of ferrite with inclusions of lamellar perlite was formed, has a hardness of 200-220 HB. Higher hardness values of the lower sector compared to the middle sector occur due to inclusions of lamellar perlite.

The middle sector has a purely ferritic structure, which is proved by the lowest hardness values compared with all other sectors. Ferrite is soft and plastic. Ferrite hardness values are in the range of 130-150 HB.

The highest hardness values (260 HB) were found in the upper sector. The bainite structure provides high hardness and strength with high plasticity.

5. Tensile testing

The tensile diagram (see figure 10) shows the dependence of the elongation of the sample on the longitudinal tensile force.

The diagram has four distinctive areas:

OA – proportionality area (0-350 MPa);
AB – elasticity area (350-360 MPa);
BC – yield point area (360-405 MPa);
CD – hardening area (405-577 MPa);
DE – failure area (577-432 MPa).

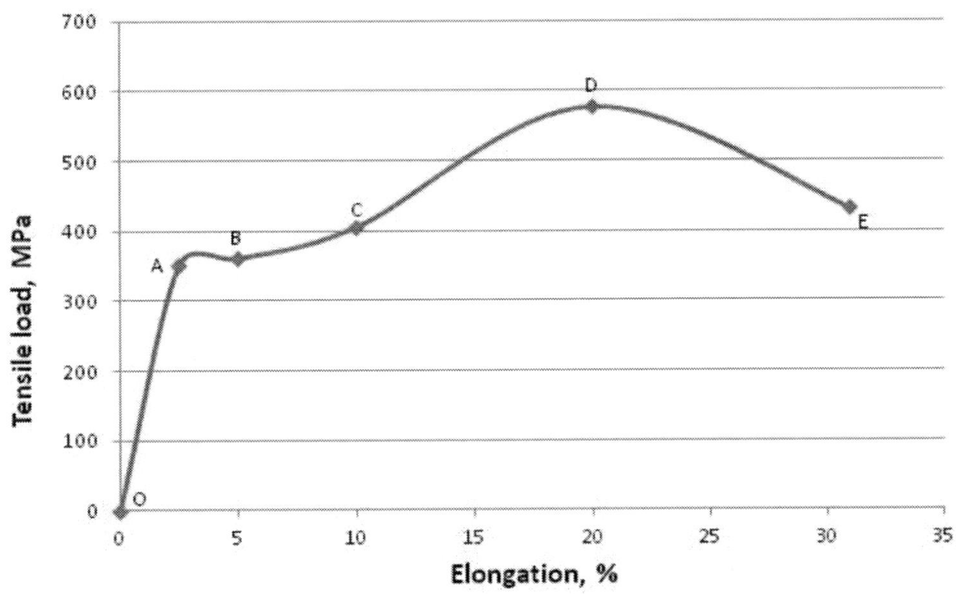

Figure 10. Tensile diagram.

At the very beginning of the tensile test, the tensile force, and therefore the deformation of the sample, is zero, so the diagram begins at the intersection of the respective axes (point O).

In the OA area, the diagram is drawn as a straight line. This indicates that in this segment of the diagram, the deformations grow in proportion to the increasing load.

Once point A passed, the diagram abruptly changes its direction and in the AB area the line runs almost parallel to the stretching axis for some time, that is, the deformations increase at almost the same load value.

At this point, irreversible changes begin to occur in the metal of the sample. The crystal lattice of the metal is rebuilt. At the same time, the effect of self-strengthening is observed.

After increasing the strength of the sample material, the diagram goes up again (BC area) and at point D the tensile force reaches the maximum value. At this point, a local thinning appears in the working part of the test sample, caused by violations of the material structure (the formation of voids, microcracks, etc.).

Due to thinning, and therefore reducing the cross-sectional area of the sample, the tensile force required to stretch it is reduced, and the curve of the diagram goes down.

The sample breaks at point E which is the thinning zone.

Mechanical properties of the WAAM Sv-08G2S steel and wrought low carbon steels with comparable chemical compositions are shown in table 2.

Table 2. Mechanical properties of steels.

Steel grade	Test temperature, °C	YS, MPA	UTS, MPa	Elongation, %	Contraction, %
WAAM Sv-08G2S	20	405	577	31	59
09G2S	20				
09G2T	20	265-385	430-490	17-31	56-63
09G2DT	20				
10G2S1	20	335	485	35	75

As can be seen from the table, the mechanical properties of the wire arc additively manufactured Sv-08G2S steel meet properties of conventional wrought low-carbon steels and even exceed them.

6. Conclusion

Based on the results of the study of the macro and microstructure and mechanical properties of the wire arc additively manufactured Sv-08G2S steel, the following conclusions can be drawn:

- The structure of the Sv-08G2S metal can be divided into three sectors depending on the formed microstructure. The lower sector is characterized by a ferritic structure with inclusions of lamellar perlite. The middle sector is characterized by a purely ferritic structure with equiaxed grains. The upper sector is characterized by a lamellar structure of bainite.
- Differences in the microstructure of the sectors are due to various temperature differences and heat input experienced by different deposited beads. In particular, the metal of the upper sector experiences a stronger temperature drop due to the difference in air temperature and the temperature of the lower layers.
- The lower sector is characterized by the presence of smaller particles, so the metal in this sector experiences stronger temperature differences compared to the metal in the middle sector, which is characterized by a larger grain.
- The results of the Brinell hardness study confirm the presence of various microstructures. The hardest was the upper sector, which has a bainite structure.
- The results of tensile testing of samples made of Sv-08G2S steel showed mechanical properties that are comparable and even superior to those of conventional wrought low-carbon steels.
- Thus, this steel can be used for wire arc additive manufacturing of steel parts, since the resulting mechanical properties meet the requirements for structural steels.

Also, based on the results of the research, it is possible to develop a technology to obtain the necessary structure of ferrite/bainite in accordance with the requirements for the manufacturing part. Obtaining the necessary microstructure can be achieved by alternating cooling cycles and deposition cycles.

References

[1] Wu B, Pan Z, Ding D, Cuiuri D 2018 A review of the wire arc additive manufacturing of metals: properties, defects and quality improvement. *Journal of Manufacturing Processes* **35** 127–139 DOI: 10.1016/j.jmapro.2018.08.001

[2] Shi X, Ma S, Liu C, Wu Q, Lu J 2017 Selective laser melting-wire arc additive manufacturing hybrid fabrication of Ti-6Al-4V alloy: Microstructure and mechanical properties *Materials Science and Engineering: A* **684** 196–204 Online. Available: https://doi.org/10.1016/j.msea.2016.12.065

[3] Horgar A, Fostervoll H, Nyhus B, Ren X 2018 Additive manufacturing using WAAM with AA5183 wire *Journal of Materials Processing Technology* **259** 68–74 Online. Available: https://doi.org/10.1016/j.jmatprotec.2018.04.014

[4] Zhang C, Li Y, Gao M, Zeng X 2018 Wire arc additive manufacturing of Al-6Mg alloy using variable polarity cold metal transfer arc as power source *Materials Science and Engineering: A* **711** 415–423 Online. Available: https://doi.org/10.1016/j.msea.2017.11.084

[5] Wang J, Pan Z, Yang G, Han J 2019 Location dependence of microstructure, phase transformation temperature and mechanical properties on Ni-rich NiTi alloy fabricated by wire arc additive manufacturing *Materials Science and Engineering: A* **749** 218–222 Online. Available: https://doi.org/10.1016/j.msea.2019.02.029

[6] Ding D, Pan Z, van Duin S, Li H 2016 Fabricating Superior NiAl Bronze Components through Wire Arc Additive Manufacturing *Materials* **9(8)** 652 Online. Available: https://doi.org/10.3390/ma9080652

[7] Wu B, Qiu Z, Pan Z, Carpenter K, Wang T, Ding D 2020 Enhanced interface strength in steel-nickel bimetallic component fabricated using wire arc additive manufacturing with

interweaving deposition strategy *Journal of Materials Science & Technology* **52** 226–234 Online. Available: https://doi.org/10.1016/j.jmst.2020.04.019

[8] Ahsan M R U, Tanvir A N M, Bates B, 2020 Heat-treatment effects on a bimetallic additively-manufactured structure (BAMS) of the low-carbon steel and austenitic-stainless steel *Additive Manufacturing* **32** 101036 Online. Available: https://doi.org/10.1016/j.addma.2020.101036

[9] Williams S W, Martina F, Addison A C, Ding J 2016 Wire Arc Additive Manufacturing *Materials Science and Technology* **32(7)** 641–647 DOI: 10.1179/1743284715y.0000000073

[10] Dickens P, Pridham M, Cobb R, Gibson I, Dixon G 1992 *Rapid prototyping using 3-D welding* (DTIC Document)

[11] Spencer J, Dickens P, Wykes C 1998 Rapid prototyping of metal parts by three-di-mensional welding *Proc Inst Mech Eng Part B J Eng Manuf* **212** 175–182

[12] Kovacevic R and Beardsley H 1998 Process control of 3D welding as a droplet-based Rapid prototyping technique *Proc. of the SFF Symposium Univ. of Texas at Austin* 57–64

[13] Dwivedi R and Kovacevic R 2004 Automated torch path planning using polygon subdivisionfor solid freeform fabrication based on welding *J Manuf Syst* **23** 278–291

[14] Song Y A, Park S, Choi D, Jee H 2005 3D welding and milling: part I–a direct approach for freeform fabrication of metallic prototypes *Int J Mach Tools Manuf* **45** 1057–1062

[15] Song Y A, Park S, Chae S-W 2005 3D welding and milling: part II—optimization of the 3D welding process using an experimental design approach *Int J Mach Tools Manuf* **45** 1063–1069

[16] Zhang Y, Chen Y, Li P, Male A T 2003 Weld deposition-based rapid prototyping: a pre-liminary study *J Mater Process Technol* **135** 347–357

[17] Zhang Y M, Li P, Chen Y, Male A T 2002 Automated system for welding-based rapid prototyping *Mechatronics* **12** 37–53

[18] Marenych O, Kostryzhev A, Shen C, Pan Z 2019 Precipitation Strengthening in Ni–Cu Alloys Fabricated Using Wire Arc Additive Manufacturing Technology *Metals* **9(1)** 105 DOI: 10.3390/met9010105

[19] Donghong Ding, Zengxi Pan, Dominic Cuiuri, Huijun Li 2015 A multi-bead overlapping model for robotic wire and arc additive manufacturing *Robot Cim-Int Manuf* **31** 101–110

[20] Filippo Montevecchi, Niccolò Grossi, Hisataka Takagi, Antonio Scippa 2016 Cutting Forces Analysis in Additive Manufactured AISI H13 Alloy *Procedia CIRP* **46** 476–479 Online. Available: http://dx.doi.org/10.1016/j.procir.2016.04.034.

Acknowledgments

Anton A. Kulikov prepared the steel sample, conducted the microstructural analysis, mechanical properties testing, and wrote the paper. Maria V. Grechneva contributed to the microstructure and mechanical properties analysis. Andrei E. Balanovskiy carried out the overall project management and participated in the discussion of the results.

IOP Conf. Series: Materials Science and Engineering **969** (2020) 012006 doi:10.1088/1757-899X/969/1/012006

X-ray Diffraction Analysis by Williamson-Hall Method of Strontium Hexaferrite Lattice Features after Mechanical Milling

I N Egorov[1], S I Egorova[2] and N Ya Egorov[1]

[1]Southern federal university, 105/42, Bolshaya Sadovaya Str., Rostov-on-Don, 344006, Russia
[2]Don State Technical University, Gagarin square 1, Rostov-on-Don, 344000, Russia

E-mail: iegorov@sfedu.ru

Abstract. The paper presents the result of experimental studies of strontium hexaferrite milling in beater mill. SEM studies showed that milling of strontium hexaferrite in beater mill with formation of magneto fluidized bed leads to significant intensification of the process. Williamson-Hall method was used to determine structural characteristics changes in the strontium hexaferrite powder from milling time in magneto fluidized bed. Carried out studies showed that milling intensification is significant only to a certain dispersity of processed powder. However, during milling with formation of magneto fluidized bed, decrease of particle sizes and structural characteristics changes speed happens at finer disperse compound.

1. Introduction

Application of ferromagnetic powders in powder metallurgy makes high demands not only to powder disperse compound but also to its structural characteristics. Mechanical milling is often applied to produce powders of defined dispersity. In order to reduce sizes of initial material particles one use various kinds of ball, beater, planetary, jet mills [1-3]. In most cases the problem is not only to obtain fine powder, but to do it with minimal time and energy consumption. This leads to necessity of development of rational milling methods [4–5].

Milling process of ferromagnetic materials can be intensified by fluidized bed [6], created by rotatory magnetic field [7] or by perpendicular alternating gradient and constant magnetic fields [8]. Milling of coarse ferromagnetic materials in beater mill with application of alternating gradient and constant magnetic fields with mutually perpendicular induction lines provides more intense milling of powder than just milling in beater mill. Due to significant reduce of milling time with magneto fluidized bed, obtaining powder of defined dispersity becomes more energy effective [9]. Impact actions that effect powder particles in beater mill change sizes of structural elements and increase lattice micro strain level. However, during milling in beater mill with magneto fluidized bed, the speed of structural characteristics changes increases due to self-milling caused by impacts between particles [10]. For this reason, studies of state of structural powder characteristics are an important task when optimizing powder processing conditions. Presently the methods of X-ray diffraction (XRD) analysis are widely and effectively used in studies of structural characteristics of particulate materials [11].

The goal of current work was to establish patterns of changes of strontium hexaferrite structural characteristics ($SrFe_{12}O_{19}$) during milling in beater mill with formation of magneto fluidized bed by means of X-ray diffraction and Williamson-Hall method.

2. Experiment

Coarse strontium ferrite with average particle size 1558.5 μm and 1476.9 particle median size was milled in beater mill by rotating with 15 krpm beaters. Constant and alternating magnetic fields with mutually perpendicular induction lines effected on milled material to form magneto fluidized bed. Induction of constant one was 15.3 mT and alternating one had 90 mT/m gradient and 50 Hz frequency. Analysis of particle size distribution was carried out by SEM Ziess Supra 25. Structural characteristics of particulate material were studied by X-ray diffraction method on Shimadzu XRD – 7000 diffractometer with CuKα emitting anode with 1.5406 Å wavelength with Bregg-Brentano focusing (θ–2θ). Diffraction peaks were analyzed using PowderCell 2.3 [12] based on full profile Rietveld refinement method [13]. Diffraction studies were carried out at room temperature in step scan mode (0,03° step) covering 2θ interval from 5° to 90°.

One important task of powder materials structural characteristics studies is obtaining results that describe structural state in the best way. In order to get coherent scattering regions (CSR) sizes and values of micro strain, one widely uses various XRD methods [14-16].

3. Results and discussion

During milling in magneto fluidized bed ferromagnetic particle is effected by a pair of forces trying to turn its magnetic moment in the field's direction and a moving force with value proportional to magnetic field induction gradient. Particles do translational motion with frequent abrupt change of speed by direction. Such motion leads to holding of powder in the area of rotating beaters and increase number of interparticle collisions intensifying milling process and affecting structural characteristics of particles. Milling intensification is proved by SEM studies of particle size distributions dependencies from milling time. Dependencies of strontium hexaferrite average particle sizes and median values from milling time without electromagnetic effect and with magneto fluidized bed are presented on Figure 1. Milling just by rotating beaters for 30 minutes provides decrease of average particle size to 13.9 μm and median to 9.6 μm (Figure 1 a, c). After 120 minutes milling average particle size and median became 9.2 μm and 6.4 μm correspondingly. Milling with formation of magneto fluidized bed with 90 mT/m induction gradient of the alternating magnetic field during 30 minutes gave average particle size 2.3 μm and median 1.6 μm (Figure 1 b, d). After 120 minutes average particle size became 0.57 μm and median 0.46 μm.

So, after 30 minutes milling without electromagnetic effect average particle size decreased in 112 times and with magneto fluidized bed – in 677 times. Figure 1 shows that further milling in both regimes goes with decreased intensity.

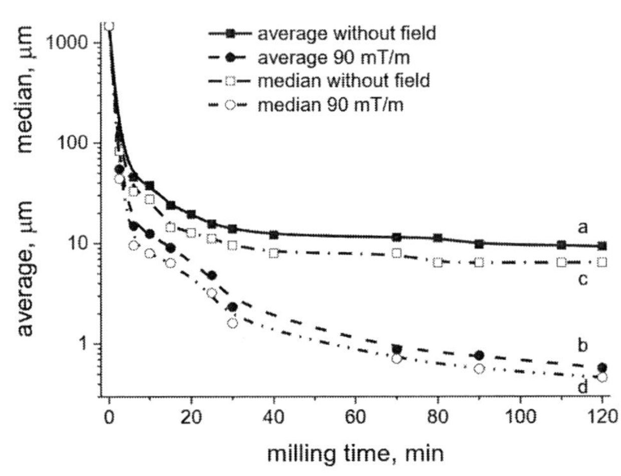

Figure 1. Dependency of strontium hexaferrite powder average particle size and median value from milling time without electromagnetic field (a, c) and in magneto fluidized bed formed by alternating magnetic field with 90 mT/m induction gradient (b, d).

Magneto fluidization of particulate system during milling process significantly effects on powder structural characteristics. Analysis of diffraction peaks widening allows determination of CSR sizes and lattice micro strain $<\varepsilon>=\Delta d/d_{hkl}$, where Δd is average change of interplanar distance d_{hkl} caused by lattice defects. Diffractogram fragments of initial strontium hexaferrite and after milling during 10, 30 and 120 minutes are presented on Figure 2. These XRD peaks widens as milling time increases. Such widening can be caused by both lattice microstrain and changes of CSR sizes. Williamson-Hall method allows to separate CSR sizes and micro strain contributions to peak widening [17–21]. 30 peaks with relative intensity more than 5% were taken for analysis.

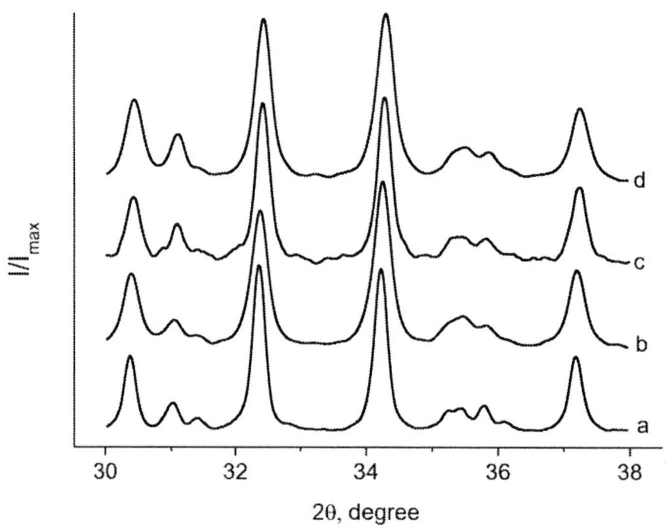

Figure 2. Fragment of relative intensity diffractograms of particulate strontium hexaferrite: initial (a), milled in magneto fluidized bed with 90 mT/m induction gradient for 10 minutes (b), 30 minutes (c) and 120 minutes (d).

The biggest and the smallest values of CSR sizes (D) and lattice micro strain (ε) can be obtained when approximating peaks with Lorentz and Gauss functions. In case of Lorentz approximation, full peak widening consists of size (CSR) and strain (ε) widening factors $\beta=\beta_{size}+\beta_{strain}$. In case of Gauss approximation - $\beta=\beta^2_{size}+\beta^2_{strain}$.

The equations have the following form:
for the case of Lorentz approximation

$$\beta=\left(\frac{\lambda}{D\cos(\theta)}\right)+\left(4\varepsilon\cdot tg(\theta)\right) \tag{1}$$

for Gauss approximation

$$\beta^2=\left(\frac{\lambda}{D\cos(\theta)}\right)^2+\left(4\varepsilon\cdot tg(\theta)\right)^2 \tag{2}$$

If experimental peak is approximated by Lorentz function, the dependency of $\beta\cos\theta$ from $4\sin\theta$ is plotted (Figure 3); in case of Gauss approximation - $(\beta\cos\theta)^2$ from $(4\sin\theta)^2$ (Figure 4). Plotted data is fitted by a line equation. In Lorentz case (1) line slope gives 4ε and intercept – λ/D. In Gauss case (2) slope corresponds to $(4\varepsilon)^2$ and intercept to $(\lambda/D)^2$. Slope gives information about lattice strain ε and intercept provides CSR size D and dislocation density $\rho=3(D)^{1/2}$.

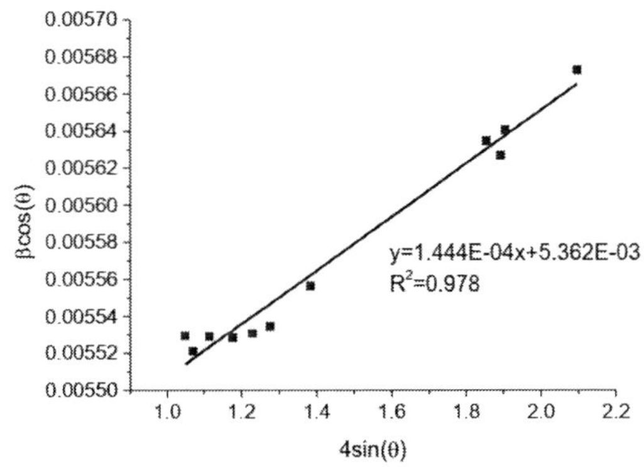

Figure 3. Plots of $\beta\cos(\theta)$ from $4\sin(\theta)$ for Lorentz approximation of initial strontium hexaferrite powder diffraction peaks.

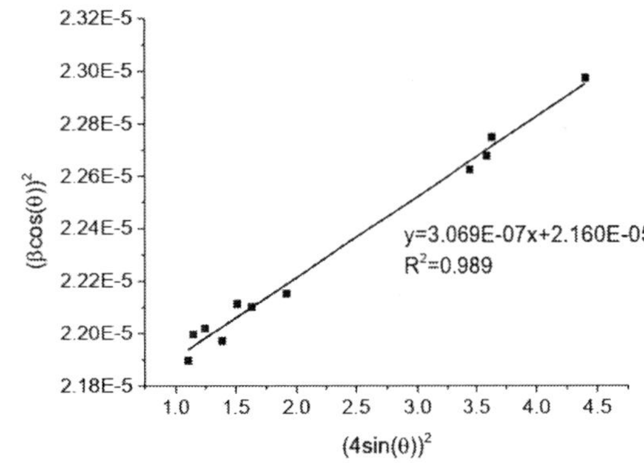

Figure 4. Plots of $\beta^2\cos^2(\theta)$ от $(4\sin(\theta))^2$ for Gauss approximation of initial strontium hexaferrite powder diffraction peaks.

By analysis of Lorentz and Gauss diffraction peaks approximations on Figure 3 and 4, we found out minimal and maximal values of structural parameters, their average and scatter:

$$\frac{\Delta d}{d} = 0.001541 \pm 0.000275\,;\ (17.8\ \%); \tag{3}$$

$$D = (360.91 \pm 4.08)\,\mathring{A}\,;\ (1.1\ \%); \tag{4}$$

$$\rho = (2.3040 \pm 0.0520)\cdot 10^{-5}\,\mathring{A}^{-2}\,;\ (2.2\ \%). \tag{5}$$

By the same Willamson-Hall method we found out minimal, maximal and average values of D, ε and ρ for powders obtained by milling in beater mill with magneto fluidized bed. Dependencies of lattice strain $\varepsilon = \Delta d/d$, CSR size D and dislocation density ρ from milling time are presented on Figure 5, 6 and 7 for strontium hexaferrite milled with effect of alternating magnetic field with 90 mT/m gradient.

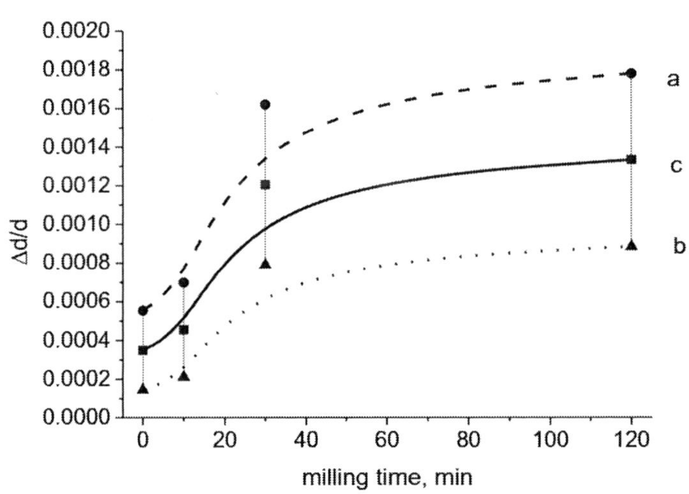

Figure 5. Dependencies of lattice strain from time of milling of initial strontium hexaferrite in beater mill: Lorentz approximation (a), Gauss approximation (b), average (c).

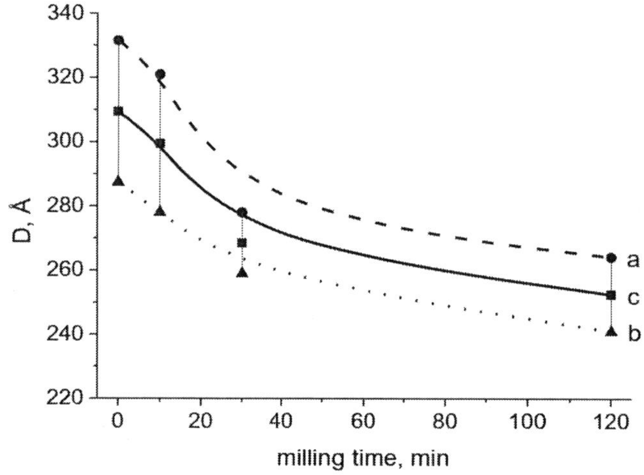

Figure 6. Dependencies of CSR size D from time of milling of initial strontium hexaferrite in beater mill: Lorentz approximation (a), Gauss approximation (b), average (c).

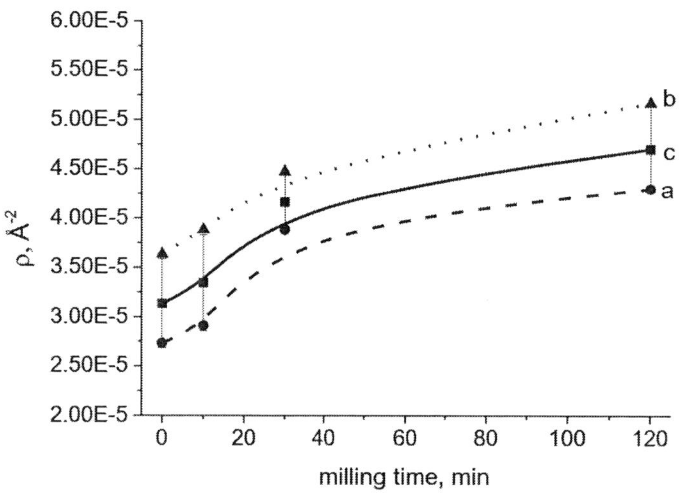

Figure 7. Dependencies of dislocation density ρ from time of milling of initial strontium hexaferrite in beater mill: Lorentz approximation (a), Gauss approximation (b), average (c).

Processing of initial particulate medium with magneto fluidized bed during 10 minutes leads to increase of average $\Delta d/d$ from 0.00035 to 0.00045 (Figure 5), D decreased from 310 to 300 Å (Figure 6) and value of ρ increased from $3.13 \cdot 10^{-5}$ Å^{-2} to $3.34 \cdot 10^{-5}$ Å^{-2} (Figure 7). Subsequent 20 minutes milling gave increase of $\Delta d/d$ to 0.00121, decrease of D to 267 Å and increase of ρ to $4.16 \cdot 10^{-5}$ Å^{-2}. Further, milling with magneto fluidized bed from 30 to 120 minutes changed $\Delta d/d$ to 0.00133 (9.9% vs 245.0% in first 30 minutes), D to 252 Å (5.6% vs 13.8%) and ρ to $4.70 \cdot 10^{-5}$ Å^{-2} (13% vs 32.9%). So, after 30 minutes (when average size is several micrometers), speed of change of structural characteristics decreases with powder particle sizes.

4. Conclusion

Application of magneto fluidized bed to milling of strontium hexaferrite in beater mill provides process intensification and obtaining of submicron powder in 120 minutes. The speed of change of structural characteristic depends not only from intensity of milling process, but also from size of powder particles greatly decreasing with their smaller sizes.

5. References

[1] Kaczmarek W A, Ninham B W 1995 Surfactant-assisted ball milling of $BaFe_{12}O_{19}$ ferrite dispersion *Mat. Chem and Phys.* **40** 21
[2] Goldman A 2006 *Modern Ferrite Technology* (Pittsburgh: Springer) p 438
[3] Kostishyn V G, Panina L V, Timofeev A V, Kozhitov L V 2016 Dual ferroic properties of hexagonal ferrite ceramics $BaFe_{12}O_{19}$ and $SrFe_{12}O_{19}$ *J. Magn and Magn. Mat.* **400** 327
[4] Manafi S A, Joughehdoust S, Kiahosseini S R 2019 Optimization of mechanical alloying parameters for synthesis of nanostructure hexagonal $BaFe_{12}O_{19}$. *J. Aust. Ceram. Soc.* **55** 371
[5] Hodaei A, Ataie A, Mostafavi E 2015 Intermediate milling energy optimization to enhance the characteristics of barium hexaferrite magnetic nanoparticles *J. Alloys Compd.* **640** 162
[6] Valverde J M 2013 *Fluidization of fine powders* (Netherlands: Springer) p 136
[7] Chuen-Shii Chou, Ming-Chun Chu 2005 Preparing magnetic powder using a simple attrition method containing a rotating magnetic field *Adv. Powder Tech.* **16** 435
[8] Egorov I N, Egorova S I 2014 Effect of electromagnetic action on dispersed composition on milling ferromagnetic materials in a hammer mill *Rus. J. of NonFer. Metals* **55** 371
[9] Egorov I N, Egorov N Ya 2016 Energy consumption of milling in beater mill in magnetoliquefied layer *Int. Research J.* **50** 44
[10] Egorov I N, Egorov N Ya 2017 Technological processes intensification in devices with magneto-fluidized bed *MATEC Web of Conf.* **132** 1
[11] Suryanarayana C, Grant Norton M 1998 *X-ray Diffraction: A Practical Approach* (New York: Springer) p 273
[12] Krause W, Nolze G Powder Cell 2.3. Online. Available: http://www.ccp14.ac.uk/ccp/web-mirrors/powdcell/a_v/v_1/powder/e_cell.html
[13] Young R A 1995 *The Rietveld Method* (Oxford: Oxford University Press) p 312
[14] Vives S, Gaffet E, Meunier C 2004 X-ray diffraction line profile analysis of iron ball milled powders *Mat. Sci. And Eng.:A* **366** 229
[15] Lileev A S, Yagodkin Yu D, Lyubina Yu V 2003 Structure and magnetic properties of nanocrystalline alloys based on $Nd_2Fe_{14}B$ obtained by various techniques *J. Magn. and Magn. Mater.* **258** 586
[16] Gubicza J, Kassem M, Ribarik G, Ungar T 2004 The microstructure of mechanically alloyed Al-Mg determined by X-ray diffraction peak profile analysis *Mat. science & engineering A, Str. mat.* **372** 115
[17] Williamson G K, Hall W H 1953 X-ray line broadening from filed aluminium and wolfram *Acta Metall.* **1** 22
[18] Ketov S V, Yagodkin Yu D, Menushenkov V P 2011 Structure and magnetic properties of strontium ferrite anisotropic powder with nanocrystalline structure *J. Alloys and Compd.* **510** 1065
[19] Egorov I N, Egorova S I, Kryzhanovsky V P 2019 Particle size distribution and structural state analysis of mechanically milled strontium hexaferrite *Mat. Sci. Forum* **946** 293
[20] Mote V, Purushotham Y, Dole B 2012 Williamson-Hall analysis in estimation of lattice strain in nanometer-sized ZnO particles *J. Theor. Appl. Phys.* **6** 6
[21] Van Berkum J G M, Delhez R, Keijser Th H, Mittemeijer E J 1996 Diffraction-Line broadening due to strain fields in materials: fundamental aspects and methods of analysis *Acta Cryst.* **52** 730

IOP Conf. Series: Materials Science and Engineering **969** (2020) 012007 doi:10.1088/1757-899X/969/1/012007

Comparison of the Strength Properties of Heat-Resistant Titanium Alloys at Elevated Temperatures

L V Davydenko[1], Yu B Egorova[2] and I M Mamonov[2]

[1]Department of Material Science, Moscow Polytechnic University, 38, Bolshaya Semenovskaya St., Moscow, 107023, Russia
[2]Department of System Simulation and Information Technologies, Academic Affairs of the Stupino branch of the Moscow, 4, Pristantsionnaya St., Stupino, 142800, Russia

E-mail: mami-davidenko@mail.ru

Abstract. The results of statistical studies of the strength properties of rods made of heat-resistant titanium alloys at test temperatures of 20÷600 °C are represented. The initial data for the research were the results of experiments published in various sources from the 1960s to 2019.The chemical composition of the alloys has been compared by aluminum and molybdenum equivalents and tensile strength at various temperatures. With an increase in the degree of alloying with α-stabilizers and neutral hardeners, the heat-resistant properties of alloys increase and remain at higher temperatures. However, this may reduce the thermal stability of alloys with an aluminum equivalent of more than 9.0%. The highest strength properties while maintaining thermal stability have alloys with aluminum equivalents of 8.5–9.0% and molybdenum of 1.0–4.5%. Regression dependences of the average values of strength properties on the test temperature, silicon content, and aluminum and molybdenum equivalents of alloying elements and impurities have been determined.

1. Introduction

Currently, hundreds of experimental titanium compositions have been developed in the Russian Federation with more than 50 titanium alloy grades being used in industry for various purposes, of which about 10 alloys are designed for temperatures up to 600 °C [1, 2]. Heat-resistant alloys have been created mainly based on the α-phase with a small amount of β-phase (≤ 10 %), therefore, by the phase composition, they belong to the pseudo α- and α+β-classes. In Table 1, titanium alloys are arranged in the order of increasing the content of β-stabilizers.

The possibility of using titanium alloys at elevated temperatures can first be inferred by the dependence of short-term strength properties on temperature. For a theoretical evaluation of the tensile strength of α-, pseudo α- and α+β-titanium alloys of different classes at room temperature, the below equation can be used [3, 4]:

$$\sigma_в = 380 + 65[\mathrm{Al}]_{eq}^{str} + 45[\mathrm{Mo}]_{eq}^{str} \quad (1)$$

where $[\mathrm{Al}]_{eq}^{str}$ is the structural aluminum equivalent of α-stabilizers and neutral hardeners; $[\mathrm{Mo}]_{eq}^{str}$ is the structural molybdenum equivalent of β-stabilizers [2]. However, for heat-resistant alloys, equation (1) should be adjusted since structural equivalents do not consider the effect of silicon, which is an obligatory alloying component. This is due to the high lattice mismatch between silicon and titanium. These causes blocking of dislocations at sufficiently high temperatures and prevents their climb and cross gliding. The silicon content should be restricted by its solubility limit in the α phase since silicides do not lead to increased heat resistance but greatly reduce the ductility and processability of alloys. In heat-resistant titanium alloys, the silicon content is limited to 0.15÷0.4 %.

Table 1. Nominal Composition and Operating Temperature of Heat-Resistant Titanium Alloys [2, 5–15].

Alloy Class	Alloy	Chemical Composition	Maximum Operating Temperature, °C
Pseudo α	VT18	7.7Al-0.6Mo-11.0Zr-0.75Nb-0.12Si	550÷600
	VT18U	Ti-6.5Al-2.5Sn-4Zr-1Nb-0.7Mo-0.15Si	550÷600
	VT20	Ti-6.5Al-1Mo-1V-2Zr-0.15Si	450÷500
	VT41	Ti-6.2Al-1.2Mo-4.0Sn-3.3Zr-1.2Nb-0.35Si-0.5W-0.1Fe	550÷600
α+β	VT25	Ti-6.5Al-1.5Sn-4Zr-2Mo-1W-0.15Si	500÷550
	VT9	Ti-6.5Al-3Mo-1.5Zr-0.25Si	500÷550
	VT36	Ti-6.2Al-2.1Sn-3.6Zr-0.7Mo-5.0W-0.15Si	550÷600
	VT8	Ti-6.5Al-3.3Mo-0.3Si	450÷500
	VT8-1	Ti-6.5Al-3.5Mo-1.2Sn-1.2Zr-0.2Si	450÷500
	VT8M	Ti-6.3Al-3.3Mo-1.0Sn-1.0Zr-0.15Si	450÷500
	VT8M-1	Ti-5.5Al-3.8Mo-1.2Sn-1.2Zr-0.18Si	400÷450
	VT46	Ti-6.0Al-2.0Mo-3Sn-0.8V-0.25Fe-2.5Zr-0.7Nb-0.2Si-0.05C--0.1O	500÷550
	VT25U	Ti-6.5Al-1.8Sn-3.8Zr-4Mo-1W-0.2Si	500÷550
	VT3-1	Ti-6Al-2.5Mo-1.5Cr-0.5Fe-0.3Si	400÷450

Aluminum equivalent $[\mathrm{Al}]_{eq}^{str}$ characterizes the thermal stability of alloys and is determined by the equation [2]:

$$[\mathrm{Al}]_{eq}^{str} = \%\mathrm{Al} + \%\mathrm{Sn}/3 + \%\mathrm{Zr}/6 + 10[\%\mathrm{O} + \%\mathrm{C} + 2\%\mathrm{N}] \tag{2}$$

The thermal stability criterion is the plasticity of at least $\delta = 10$ % and $\psi = 20$ % preserved by the samples after they are exposed to a voltage of 246 MPa and a temperature of 540 °C [2]. Alloys lose thermal stability due to the release of the α_2-phase (Ti_3Al) at $[\mathrm{Al}]_{eq}^{str} \geq 9$ %; therefore, the structural aluminum equivalent is sometimes called the thermal stability coefficient [5]. The molybdenum equivalent $[\mathrm{Mo}]_{eq}^{str}$ characterizes the stability of the β-phase and is determined by equation [2]:

$$[\mathrm{Mo}]_{eq}^{str} = \%\mathrm{Mo} + \%\mathrm{Ta}/4.5 + \%\mathrm{Nb}/3.3 + \%\mathrm{W}/2 + \%\mathrm{V}/1.4 + \%\mathrm{Cr}/0.6 + \%\mathrm{Mn}/0.6 + \\ + \%\mathrm{Fe}/0.4 + \%\mathrm{Ni}/0.8 \tag{3}$$

Silicon is a β-stabilizer but does not affect the quantity and stability of the β-phase in commercial titanium alloys and therefore is not considered in formula (3).

The study objective was to compare heat-resistant titanium alloys based on an evaluation of the tensile strength depending on the test temperature and chemical composition expressed in aluminum and molybdenum equivalents and considering the silicon content.

2. Source Materials and Research Techniques

In the study, the results of experiments published in various sources from the 1960s to 2019 have been used as source materials. Generally, the VIAM study results published in papers, handbooks, and monographs have been used [5–15, 17–23]. The alloying degree of the alloys has been estimated using aluminum and molybdenum equivalents by the equation (2), (3). Statistical studies have been performed using the STATISTICA software package and included correlation-regression analysis with a confidence probability of 0.95 using standard techniques [16].

3. Research Results and Discussion

The chemical composition of heat-resistant alloys in aluminum and molybdenum equivalents is shown in Figure 1. A tendency towards a decrease in aluminum equivalent and an increase in molybdenum equivalent of alloying elements is observed. Figure 1 shows that the content of alloying elements for alloys operating at temperatures up to 400÷450 °C should not exceed $[\mathrm{Mo}]_{eq}^{str} \approx 3.5 \div 6.5$ % and $[\mathrm{Al}]_{eq}^{str} \approx 7.0 \div 7.5$%; up to 450÷500 °C – $[\mathrm{Mo}]_{eq}^{str} \approx 1.5 \div 4.5$% and $[\mathrm{Al}]_{eq}^{str} \approx 7.5 \div 8.5$%; up to 500÷550 °C – $[\mathrm{Mo}]_{eq}^{str} \approx 2.5 \div 4.5$% and $[\mathrm{Al}]_{eq}^{str} \approx 8.5 \div 9.0$%; up to 550÷600 °C – $[\mathrm{Mo}]_{eq}^{str} \approx 1.0 \div 2.5$% and $[\mathrm{Al}]_{eq}^{str} \approx 8.5 \div 9.0$%. The average equivalent values can be estimated to optimize the alloy composition by the ratio:

$$[\mathrm{Al}]_{eq}^{str} = 9.53 \div 0.4[\mathrm{Mo}]_{eq}^{str} \quad (4)$$

VT18 and VT41 are considered the most heat-resistant alloys at temperatures of 550÷600 °C. However, in practice, VT18 alloy is not used due to low processability, and VT41 alloy is at the stage of industrial development [12]. The high strength properties of the latter are determined by multicomponent alloying, which leads to the formation of a carbide phase based on tungsten and additional hardening of the α-solid solution with iron added in an amount within the solubility range [12, 15, 21, 22]. The most heat-resistant serial alloys are VT18U and VT25U [23].

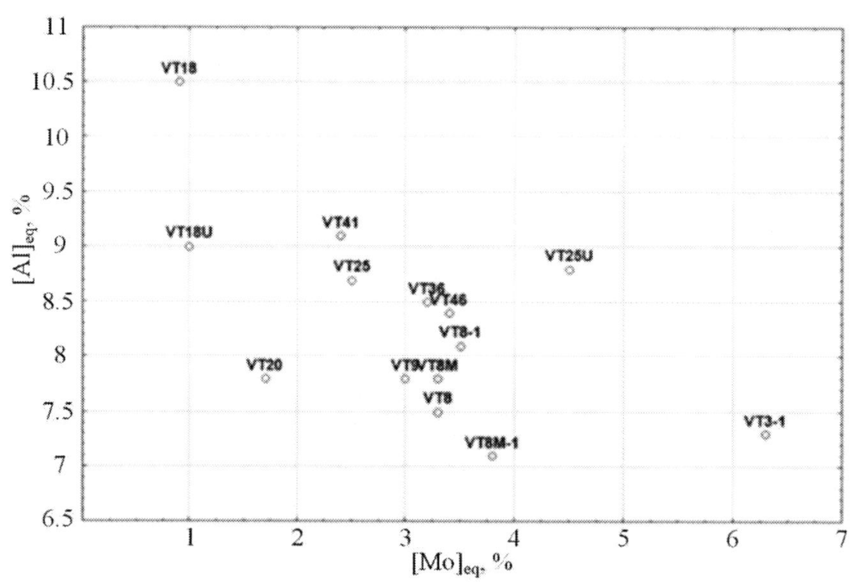

Figure 1. Chemical Composition of Heat-Resistant Titanium Alloys in the Coordinates $[\mathrm{Al}]_{eq}^{str} - [Mo]_{eq}^{str}$.

IOP Conf. Series: Materials Science and Engineering **969** (2020) 012007 doi:10.1088/1757-899X/969/1/012007

An analysis of the published literature shows that the strength and heat-resistant properties of titanium alloys strongly depend on the type of semi-finished product, the process modes of heat treatment, the type and parameters of the structure, and the grade composition fluctuations (Table 2).

Table 2. Mechanical Properties of Various Semi-Finished Products Made of Heat-Resistant Titanium Alloys.

Alloy	Semi-Finished Product (Annealing)	Tensile Strength, MPa, at temperatures, °C				References
		20	500	550	600	
VT18 U	Rod 20 мм. Si = 0.1-0.12 %	935÷1,010	–	–	510÷600	[17]
	Disk stamping	915÷1,015	600÷700	560÷650	–	[15, 18]
	Forging up to 50 kg	910÷1,100	–	–	≥ 560	[20]
VT46	Rods	1,170÷1,250	770÷850	–	–	[13]
	Forging 15-25 кг	1,080÷1,165	760÷830	730÷810	625÷740	[23]
VT41	Rods 45 mm	1030÷1,180	730÷830	630÷750	–	[12, 15, 18, 19]
	Forging up to 220 kg	1,078÷1,192	–	–	675÷755	
	Rods 18-22 mm. Fe = 0.06 %	1,010÷1,035	–	–	635÷665	[21]
	Rods 18-22 mm. Fe = 0.12 %	1,145÷1,190	–	–	680÷715	
	Forging. Fe = 0.05-0.06 %	1,055÷1,115	–	705÷785	655÷725	[22]
	Forging. Fe = 0.11-0.12 %	1,085÷1,165	–	810÷840	735÷780	

E.g., the absolute properties dispersion of the same type of semi-finished products may reach 100÷150 MPa at both room and elevated temperatures (Table 2). Fluctuations in the iron content within the brand composition (0.06÷0.12 % m.) for the VT41 alloy lead to a divergence of $\Delta\sigma_{в}$ = 100÷180 MPa. For some alloys, mechanical properties published in literary sources are extremely contradictory and require clarification. Therefore, at the first stage, the temperature dependences of the tensile strength of various semi-finished products (rods, sheets, forgings) after standard annealing have been analyzed based on an integration of the data published.

For the dependence of strength properties on temperature, the exponential law proposed by N.S. Kurnakov is usually applied [6]:

$$\sigma_{в} = \sigma_0 \cdot \exp(-bt) \tag{5}$$

where σ_0 is the free term characterizing the strain resistance extrapolated to 0 K; b is the temperature coefficient. For each alloy, a regression analysis has been performed using the exponential dependence (4) and a 3rd-degree polynomial (5):

$$\sigma_{в} = \sigma_0 + b_1 t + b_2 t^2 + b_3 t^3 \tag{6}$$

Then, a comparison between actual and regression tensile strength values has been performed, which has shown that ratio (4) is valid only within a rather narrow temperature range of ~ 100÷400 °C. Within the studied temperature range of 20÷600 °C, for most alloys, a 3rd-degree polynomial gives the best approximation. The regression coefficients of the model (5) for rods made of various alloys, the statistical error S of the tensile strength estimate, and the correlation coefficient R are given in Table 3.

At the second stage, the dependence of the tensile strength of various semi-finished products on the aluminum and molybdenum equivalents and the silicon content (0.15÷0.35 %) at various temperatures has been analyzed (Table 4). At room temperature, an increase in the aluminum and molybdenum equivalents by 1.0 % leads to an increase in the tensile strength by an average of ~ 60 and ~ 50 MPa, respectively, which is comparable with relation (1). For silicon, the strength coefficient of 234 MPa/% m. is close to the literature data of 120÷300 MPa/% m. [2]. With increasing temperature, the effect of silicon on strength increases to 500÷600 MPa/% m. at 200÷500 °C. According to [17], for the VT18U alloy hardened with silicon, the strength coefficient values at room temperature and 600 °C are approximately the same and equal to 400÷500 MPa /% m., while for the Ti-8-1-1 alloy, they are much different and equal to 150 MPa/% m. (at 20 °C) and 410 MPa /% m. (at 600 °C). From Table. 4, it follows that alloys containing more α-stabilizers and neutral hardeners have higher strength properties, and the degree of their effect on strength is maintained up to 600 °C. With increasing temperature, the effect of β-stabilizers on the alloy strength decreases significantly, and at a temperature of 600 °C, the tensile strength increases on average by ~ 12 MPa with an increase in the molybdenum equivalent by 1.0 % m. (Table 4).

Table 3. Parameters of a 3rd Degree Polynomial Model (5) for Theoretical Evaluation of the Tensile Strength of Rods Made of Heat-Resistant Titanium Alloys within the Temperature Range of 20÷600 °C.

Alloy	$[Mo]_{eq}^{str}$ %	$[Al]_{eq}^{str}$ %	σ_0, MPa	b_1	b_2	b_3	R	S, MPa
VT18	0.9	10.5	1.135	-1.256	0.0030	$3.234\cdot10^{-6}$	0.97	53
VT18U	1.0	9.0	1.061	-1.887	0.0045	$-4.155\cdot10^{-6}$	0.93	60
VT20	1.7	7.8	1.088	-3.345	0.0105	$-1.105\cdot10^{-5}$	0.98	34
VT41	2.4	9.1	1.161	-2.085	0.0061	$-6.664\cdot10^{-6}$	0.98	46
VT25	2.5	8.7	1.121	-0.789	0.0021	$-3.255\cdot10^{-6}$	0.98	50
VT9	3.0	7.8	1.187	-1.470	0.0034	$-3.881\cdot10^{-6}$	0.99	40
VT36	3.2	8.5	1.102	-0.920	0.0011	$1.453\cdot10^{-6}$	0.99	33
VT8	3.3	7.5	1.129	-1.329	0.0030	$-3.530\cdot10^{-6}$	0.99	33
VT46	3.4	8.4	1.200	-1.172	0.0029	$-4.012\cdot10^{-6}$	0.99	26
VT25U	4.5	8.8	1.190	-1.501	0.0039	$-4.512\cdot10^{-6}$	0.99	49
VT3-1	6.3	7.3	1.097	-1.771	0.0045	$-5.251\cdot10^{-6}$	0.99	28

Table 4. Results of a Regression Analysis of the Dependence between the Tensile Strength of Rods and the Chemical Composition of Heat-Resistant Titanium Alloys at Various Temperatures.

Test Temperature, °C	Regression Model	R	S*
20	$\sigma_в = 386+60[Al]_{eq}^{str} + 50[Mo]_{eq}^{str} + 234Si$	0.92	56
100	$\sigma_в = 299+60[Al]_{eq}^{str} + 45[Mo]_{eq}^{str} + 369Si$	0.94	58
200	$\sigma_в = 215+60[Al]_{eq}^{str} + 45[Mo]_{eq}^{str} + 497Si$	0.94	57
300	$\sigma_в = 167+60[Al]_{eq}^{str} + 40[Mo]_{eq}^{str} + 590Si$	0.95	60
400	$\sigma_в = 120+60[Al]_{eq}^{str} + 40[Mo]_{eq}^{str} + 495Si$	0.95	51
500	$\sigma_в = 86+60[Al]_{eq}^{str} + 40[Mo]_{eq}^{str} + 475Si$	0.96	48
600	$\sigma_в = 52+58[Al]_{eq}^{str} + 12[Mo]_{eq}^{str} + 372Si$	0.95	45

Note: *S is the statistical error of the model.

Thus, with increasing the degree of alloying with α-stabilizers and neutral hardeners, the heat-resistant properties of alloys increase and remain at higher temperatures. However, it should be considered that the thermal stability of alloys with $[\mathrm{Al}]_{eq}^{str} \geq 9$ % may be simultaneously reduced.

4. Conclusions

1. A comparison of heat-resistant titanium alloys by chemical composition expressed in aluminum and molybdenum equivalents has been performed. This allows justifying the optimal composition of new alloys and evaluating their strength properties at various operating temperatures.
2. Based on statistical studies, mathematical dependencies of the most typical tensile strength values on the test temperature have been obtained for the rods made of heat-resistant titanium alloys.
3. Regression models have been determined to estimate (with a confidence probability of 0.95) the dependence of average values of the rod strength properties on the silicon content and aluminum and molybdenum equivalents of alloying elements at test temperatures of 20÷600 °C.

References

[1] Ilyin A A, Egorova Yu B, Skvortsova S V, Mamonov A M 2012 Different Types of Classifying Domestic Titanium Alloys *Titan* **2(36)** 11–18
[2] Ilyin A A, Kolachev B A, Polkin I S 2009 *Titanium Alloys. Composition, Structure, Properties. Handbook* (Moscow: VILS – MATI) p 520
[3] Egorova Yu B, Davydenko L V, Belova S B, Chibisova E V 2015 Statistical Forecasting the Mechanical Properties of Titanium Alloys Depending on Aluminum and Molybdenum Equivalents *Materials Science* **5** 24–30
[4] Ilyin A A, Polkin I S, Egorova Yu B, Davydenko L V 2018 On the Use of Aluminum and Molybdenum Equivalents of Alloying Elements and Impurities in the Production of Titanium Alloys *Titan* **4** 11–19
[5] Nochovnaya N A, Antashev V G, Shiryaev A A, Alekseev E B 2015 Choosing Composition of a New Heat-Resistant Titanium Alloy Using Mathematical Simulation Techniques *Titan* **1** 10–17
[6] Miklyaev P G Mechanical Properties of Light Alloys at the Pressure Processing Temperatures and Speeds (Moscow: Metallurgy) p 280
[7] Tumanov A T 1973 *Aviation Materials. Magnesium and Titanium Alloys* (Moscow: ONTI) p 560
[8] Kablov E N 2010 *Aviation Materials. Titanium Alloys* (Moscow: VIAM) p 96
[9] Solonina O P, Glazunov S G 1976 *Heat Resistant Titanium Alloys* (Moscow: Metallurgy) p 448
[10] Glazunov S G, Moiseev V N 1974 *Structural Titanium Alloys* (Moscow: Metallurgy) p 368
[11] Fridlyander I N, Senatorova O G, Osintsev O E 2001 *Mechanical Engineering. Encyclopedia. Vol. 2-3. Non-Ferrous Metals and Alloys* (Moscow: Mechanical Engineering) p 880
[12] Kablov E N, Kashapov O S, Pavlova T V, Nochovnaya N A 2016 Developing an Experimental Prototype Technology for the Manufacture of Semi-Finished Products from Pseudo-α Titanium Alloy VT41 *Titan* **2** 33–38
[13] Kashapov O S, Pavlova T V, Kalashnikov V S, Kondratyev A R 2016 Investigating the Effect of Alloying Element Contents on the Properties of High-Strength Heat-Resistant Pseudo-α Alloy VT46 *Proceedings of VIAM* **9(45)** 44–50.
[14] Borisova E A 1980 Titanium Alloys. Metallography of Titanium Alloys (Moscow: Metallurgy) p 464
[15] Kashapov O S, Novak A V, Nochovnaya N A, Pavlova T V 2013 Situation, Problems, and Prospects of Creating Heat-Resistant Titanium Alloys for GTE Parts *Proceedings of VIAM* **3** Online. Available: http://viam-works.ru/ru/articles?year=2013&num=3
[16] Borovikov V P 2013 *Popular Introduction to Modern Data Analysis in the STATISTICA System* (Moscow: Hotline – Telecom) p 288
[17] Trubochkin A V 2004 *Effect of Silicon and Educing Intermetallides on the Structure and Properties of Heat-Resistant Pseudo-Alpha-Titanium Alloys* (Ekaterinburg: USTU) p 160

[18] Pavlova T V, Kashapov O S, Nochovnaya N A 2012 Titanium Alloys for Gas Turbine Engines *All materials. Encyclopedic Handbook* **5** 8–14
[19] Nochovnaya N A, Shiryaev A A, Alekseev E B, Antashev V G 2014 Optimizing Heat Treatment Modes for Blade Blanks from an Experimental Heat-Resistant Alloy *MITOM* **12** 22–26
[20] OST1 90197-89 *Titanium Alloy Discs and Shafts Forgings and Stampings*
[21] Kashapov O S, Pavlova T V, Istrakova A R, Kalashnikov V S 2015 Effect of Iron on the Mechanical Properties of Rods Made of Heat-Resistant Titanium Alloy VT41 *Proceedings of VIAM* Online. Available: http://viam-works.ru/plugins/content/journal/uploads/articles/pdf/784.pdf
[22] Kashapov O S, Pavlova T V, Istrakova A R, Kalashnikov V S 2015 Effect of Iron on the Mechanical Properties of Forgings Made of Heat-Resistant Titanium Alloy VT41 *Proceedings of VIAM* Online. Available: http://viam-works.ru/plugins/content/journal/uploads/articles/pdf/869.pdf
[23] Kashapov O S, Pavlova T V, Istrakova A R, Kalashnikov V S 2014 Improving the Strength Characteristics of Heat-Resistant Pseudo-α-Titanium Alloys *Aviation Materials and Technologies* **S5** 73–80

IOP Conf. Series: Materials Science and Engineering **969** (2020) 012008 doi:10.1088/1757-899X/969/1/012008

Wear Resistance Structural Aspects of Materials after Laser Processing

A V Brover, G I Brover and I A Topolskaya

Don State Technical University, 1, Gagarin Sq., Rostov-on-Don, 344000, Russia

e-mail: brover@mail.ru

Abstract. The conducted experiments show that the structures generated by laser processing are able to dissipate friction energy effectively with transformations at different structure levels. Thus, it becomes possible to manage surface strength and wear resistance of materials using the concept of "structural adaptability" of friction pairs that expands the range of normal friction and wear processes. The irradiated steel wear resistance is defined by its original hardness as well as its ability to deformation strengthening during friction. The temperature and strength load of irradiated steel surface layers through friction destabilize austenite towards $\gamma \rightarrow \alpha$ transformation. This results from alloy carbide formation which causes depleted austenite by carbon, and contributing to its transformation into martensitic deformation. It's necessary to regulate the amount and degree of remaining austenite stability in laser-hardened steel and alloys, applied to specific load conditions, that provides that required operation features.

1. Introduction

The tribotechnical material durability is provided when the excessive internal energy accumulated during temperature and strength load manages to dissipate before reaching the critical level which causes the surface damage.

The resilient energy dissipation through loading of laser irradiated surface layers is realized in following ways: by structure dispersing; by appearance and movement of dislocations; by formation of deformation martensite; by turn of structural elements, and others [1–8].

Some ways of increasing metal properties range have been used in laser processing [9–12]:

- increase of dispersion of structure at local plastic deformation as a result of dynamic polygonization in austenitic, formation of microdomain (fragments) with high density of dislocations inherited at accelerated cooling as well as in the result of phase riveting in polymorphic transformation
- nanoprecipitations use (formation of nanoscale precipitations) affected by plastic deformation during thermal stress.
- martensite deformation forming in the irradiated zones during external thermal and strength load in the friction process.

It should be taken into consideration that martensite transformation is a deformation hardening mechanism due to increasing amount of martensite, on the one hand, and microstress relaxation and use of martensite transformation as an extra plastic deformation mechanism, on the other hand. These two factors operate simultaneously, but they are competing

In case the first factor is dominating, high strength is provided along with satisfactory flexibility. If the second one is dominating, the flexibility increases considerably with sufficient strength.

In the case of optimal balance of both processes, martensite transformation during loading provides the best mechanical properties range.

A heterogeneous structure is optimal for friction pairs, when the reconstruction of one structure into another, stable at a higher level, takes place through changing external thermodynamic parameters (temperature, pressure, component concentration), i.e. the phenomenon of structural-energy adaptability is realized. Formed structures after adaption dispense the most effectively the energy introduced into a tribosystem, and minimize the wear of friction pairs. This structure variant is achieved through martensite optimal amount in laser processing zones in two-phase austenitic-martensitic areas of a hardened layer. At the same time, technological and structural plasticity is provided, which results from the transformation of part of austenite into martensite at loading.

There are following possible ways of choosing heat-hardening modes of the surface layers of steel to provide a given resistance of products. First of all, the alloy should experience structural transformations during friction, which ensure an increase in the density of dislocations on the working surfaces of products. Or, the selected material after surface heat treatment should retain the initial density of dislocations at external loading, that is, have a stable indifferent structure. This option is achieved in the case of obtaining martensitic structure mainly in laser treatment zones, with a little amount (less than 20%) of metastable austenite transforming into martensite through loading.

To obtain the considered variants of structural condition in the surface layers of products, laser processing was used at various radiation modes. Structures with the greatest possible stability in relation to external influence, or with the possibility of optimal restructuring and additional hardening in operation by forming secondary structures were created.

2. Research methods

Materials for research were samples of alloyed steel P6M5.

Metal tests were carried out using optical and scanning probe microscopes. The identification of the phase compounds in the laser treatment zones was carried out by the method of X-ray diffraction analysis. The study was carried out on diffractometer "DRON-0,5" in the filtered FeK_{α}- radiation with registration of the intensity by scintillators.

Pulsed laser surface processing was carried out using technological equipment "Kvant-16" with a change of the radiation power density in the range of 80-150 MW/m^2.

3. Results and discussion

To confirm the considerations above on the structural aspects of the wear resistance of materials after laser treatment it's necessary to results of experimental researches of wear resistance of alloyed steel P6M5 in dry friction conditions, as per the "disc- block" schedule on the friction machine.

Samples of steel P6M5 were made in the form of rings with a diameter of 30 mm and a width of 10 mm and were subjected to volumetric standard heat treatment. As counterbody was used samples of the liners of the bearing steel in the form of pads (HB 130-180). The cleanliness of the contact surfaces was estimated at R_a=0,63 μm. The tests were carried out at a load of 500 N and a linear slip speed of 190 m/min., with an average contact area of 100 mm^2. The wear was estimated by weight loss in analytical scales to 0,0001 g. Some rings were irradiated at Kvant-16 with a power density of 100 MW / m^2. As a result of laser hardening, a metal layer of 80-100 μm depth with an average hardness of 10.5-12 HPa was formed on the surface of the rings.

The test results, shown in Figure 1, reveal a clear advantage in the wear resistance of the laser-irradiated layers compared to standard thermal treatment. In particular, during the test period, the weight losses of irradiated rings are 13 times lower than the basic variant.

Examining the surface of the rings after the tests shows that the volumetric-hardened rings have a damaged surface of contact with the grip and drift areas of the metal. On the contrary, the rings subjected to laser treatment had a fairly clean surface after the tests. The features of structural and phase transformations in laser-hardened surface layers during friction should be considered, which lead to increased wear resistance of the material.

IOP Conf. Series: Materials Science and Engineering **969** (2020) 012008 doi:10.1088/1757-899X/969/1/012008

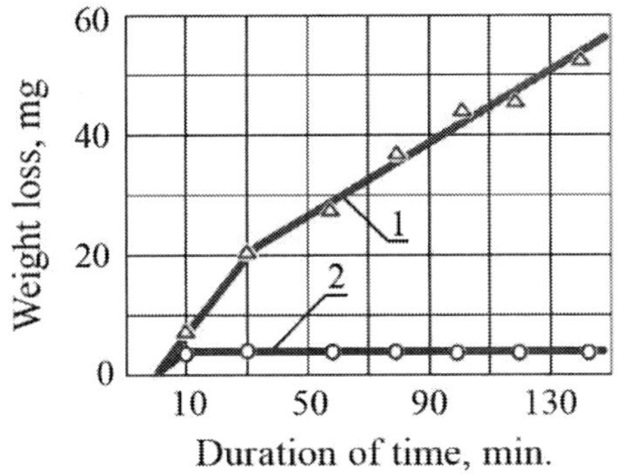

Figure 1. Wear curves of P6M5 steel samples after heating with 1220°C and 3 times leave (1), laser processing (2).

Figure 2, curve 1, shows that with laser hardening of alloyed steels, tetragonal martensite and some residual austenite are formed in the structure of surface layers. A different number of martensite and austenite in laser-hardened zones was reached changing radiation modes in the work, and a connection between the development of phase transformations and the properties of steels is established.

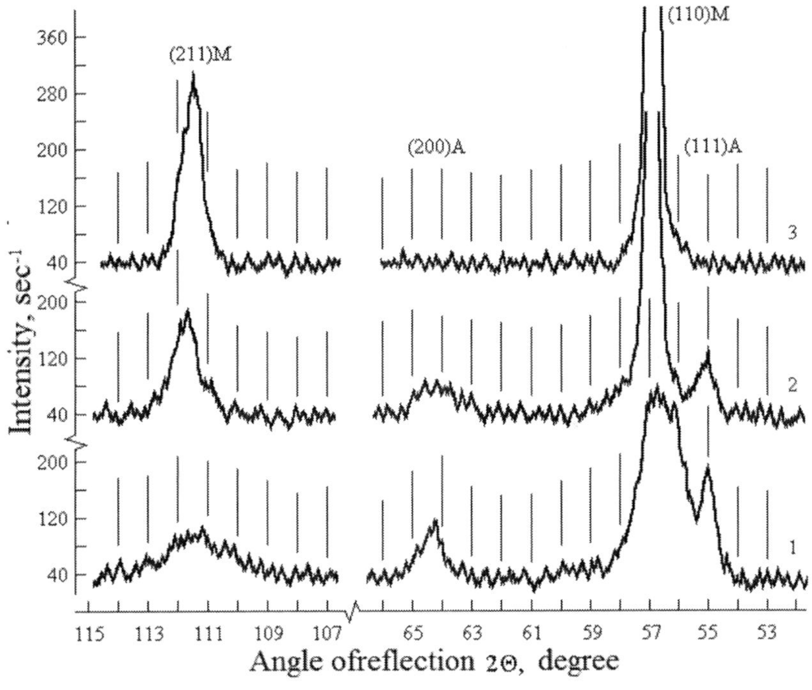

Figure 2. Fragments of diffractogram of steel P6M5: 1 – volumetric and laser hardening; 2 – option 1 and a 15-minute friction test; 3 – option 1 after a friction test for 150 minutes.

The study found that for maximum hardening (10-11,5 HPa) processing must be carried out with a radiation density of 80-100 MW / m^2, which ensures that a certain number of insoluble original carbides are stored in the volume of irradiated areas (30% in the case of laser hardening of tool steels). It is also important that during laser hardening the density of defects in the crystal structure of α-, γ-phases is achieved, exceeding the same for volumetric heat treatment by 10 times.

Further dissolving of carbides, by increasing the energy parameters of laser radiation and the initial stages of melting the surface, is accompanied by reducing the volume of areas with high carbon concentration and alloying elements, due to higher efficiency of diffusion processes and increasing the amount of residual austenite, which results in some reduction in the hardness of the processed metal

(8-10 HPa). Thus, by changing the laser processing modes, you can create a different degree of hardening of the main phases of steels and achieve uniform release of carbide and intermetallic particles due to the heat generated on the wiping faces and surfaces during friction in the operation of products.

It is established that tetragonal martensite along with high hardness is characterized by the ability to intensive deformation hardening during wear that provides it with high strength and wear resistance compared to low-carbon martensite reinforced by a dispersed carbide phase. The described variant of laser hardening and structural adaptability is applicable to metal-working tool on the wiping faces of which high contact stress and temperatures are developing.

Laser processing also increases the stability of the high-temperature γ-phase to martensitic transformation, thereby contributing to its partial or sufficiently complete preservation in the quenched state [13–19]. The ability of the metastable γ-phase of laser hardening to turn into martensite deformation, with the releasing of disperse carbides under the influence of external stresses and deformations, causes the hardening of the irradiated metal, as the martensite of deformation is an untempered martensite, and in its wear resistance it is not inferior to the tetragonal martensite of cooling containing the same amount of carbon. Figure 2, curves 2, 3, shows that the thermal strength load of irradiated metal leads to almost complete disappearance of the austenite reflections on x-ray diffractogram due to the flow of $\gamma \rightarrow \alpha$ transformation in friction conditions.

Figure 3 shows the microstructure of irradiated steel P6M5 after friction tests and a histogram of the distribution of surface profile heights along the reference line marked in Figure 3, a, crossing the measured dispersed separated carbides. Attention has to be drawn to the disperse carbide phase separation, oriented be slip lines in the hardened layer zones after friction tests (see Figure 3, a). The consequence is texture formation of disperse precipitation of 2-10 nm (Figure 3, b). The positive result of texture formation effect in the surface layers of irradiated steels is a reduction in the friction coefficient on the wiping faces of products in operation.

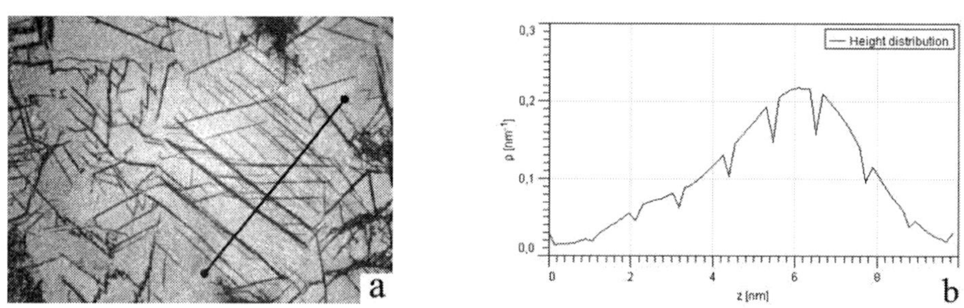

Figure 3. The structure of laser-hardened steel P6M5 after friction tests: a – optical microscopy (×500); b –profile altitude distribution histogram.

It is necessary to note peculiarities of two-phase steels behavior at different types of wear. Particularly, in abrasive wear, heterogene structures have less wear resistance than completely martensitic ones. It can be explained by the intensive wear of the soft structural component. In the conditions of adhesion wear and lubricated friction, two-phase structures are characterized by high wear resistance. In this case, the martensite areas play a part of a strong frame, and the wearable austenite areas form micro-cavities on the working surface, that retain lubrication and improve friction conditions.

It can be concluded that the amount of residual austenite must be regulated depending on the specific load conditions during operation, and its stability in laser hardening structures by changing the radiation modes. This allows to control the mechanical properties of steel and to obtain a high level of those that are most important in the specific conditions of operation.

It's important to pay peculiar attention to the role of strengthening phases in the wear resistance of laser-irradiated metal [20–21]. Inclusions, shots, are one of the factors that can increase their wear resistance. Due to the appearance of external stresses in the laser processing zones, dislocations are able to avoid the carbide particles by cross slip, leaving prismatic dislocation loops behind the particles. Thus, during external load, the shots are the localizers of deformation. Relaxation processes of the translatory-rotational type take place next to them and have a wave-like properties.

Therefore, from the point of view of thermodynamics, far from balance, "strengthening phase – matrix" system behavior in external load is determined by self-regulating processes of dissipative structures that control excessive energy outflow. Inclusions in steel can be considered as a set of relaxation material, which is activated when certain critical stresses are reached upon it.

The increased wear resistance of steel after laser hardening is determined by formation of austenite, quite unusual in properties, along with very hard metastable martensite. Its high original hardness and heat resistance, combined with the ability to turn into martensite of deformation, create conditions for reliable resistance to plastic deformation of metal; reduce the possibility of coalescence and micro-fracture of the surface under friction.

4. Conclusion

1. The structures obtained during laser processing are able to dissipate effectively the energy supplied by friction with transformations at various structural levels. Thus, it becomes possible to control consciously the surface strength and wear resistance of materials, using the concept of "structural adaptability" of friction pairs that extend the range of normal friction and wear processes.
2. The wear resistance of irradiated steels is determined both by their original hardness and by the ability to mechanical hardening during friction.
3. The thermal-strength load of surface irradiated layers of steel at friction destabilizes austenite to $\gamma \rightarrow \alpha$ transformation. It results from the process of alloyed carbide formation, causing the austenite carbon depletion and contributing to its transformation into martensite of deformation.
4. For specific loading conditions, it's necessary to control the quantity and stability of residual austenite in laser-hardened steels and alloys which ensures obtaining required operational properties.

References

[1] Ginberg A M and Ivanov A F 1982 *Wear-Resistant and Antifriction Coatings* (Moscow: Machine building)
[2] Kravchenko G N and Alekseev V V 1986 Influence of plastic deformation by crushing and cyclic loading on properties of surface layer of steel 30XGSN2A *Metal Science and Heat Treatment* **9** 23
[3] Grigoryants A G and Safonov A N 1987 *Methods of Surface Laser Processing* (Moscow: Higher school)
[4] MacClintock F and Argon A 1970 *Deformation and destruction of materials* (Moscow: World)
[5] Panin V T, Likhachev V A and Grinyaev Yu V 1985 *Structural levels of deformation of solid bodies* (Novosibirsk: Science)
[6] Lyubarskiy I V and Palatnik L S 1976 *Friction physics of metals* (Moscow: Metallurgy)
[7] Rybakova L M and Kuksenova L I 1985 Metal science on friction and wear *Metal Science and Heat Treatment* **5** 16
[8] Marchenko E A 1979 *About the nature of metals surface destruction in friction* (Moscow: Science)
[9] Gorbach V G, Sidoruk I V and Izmajlov E A 1988 Martensitic-austenitic steels as an effective tool and structural material *Metal Science and Heat Treatment* **8** 9
[10] Brover A V 2005 Structural features of the process of surface hardening of steel with concentrated energy flows *Materials Science* **9** 18
[11] Brover A V 2008 The complex of mechanisms of metal materials hardening in impulse laser processing *Perspective Materials* **1** 63
[12] Brover A V 2006 Effects of structural-energy adaptation of surface thermal strengthed steel at

friction, Hardening Technologies and Coatings **5** 43
[13] Tushinskij L I 1990 *Theory and technology of metal alloys hardening* (Novosibirsk: Science)
[14] Serebryakov V G and Estrin E I 1992 Influence of deformation on mechanical properties of two-phase austenitic-martensitic steel *The Physics of Metals and Metal Science* **2** 130
[15] Malinov L S and Chejlyah A P 1988 Influence of metastable residual austenite on mechanical properties of steel X12M *Metal Science and Heat Treatment* **8** 12
[16] Bernshtejn M L, Kaputkina L M and Prokoshkin S D 1982 The structure and substructure of austenite formed during the heating of quenched and thermomechanically hardened steels *The Physics of Metals and Metal Science* **54(6)** 150
[17] Bushe N A and Kopyt'ko V V 1981 *Compatibility of friction surfaces* (Moscow: Science)
[18] Belyj A V, Karpenko G D and Myshkin N K 1991 *Structure and methods of formation of wear-resistant surface layers* (Moscow: Machine building)
[19] Bekrenev A N and Bezuglov A Yu 1995 A metallic system self-alignment during nonquasyequilibrium relaxation *Physics and Chemistry of Materials Treatment* **2** 122
[20] Portnoj K I and Babich B N 1974 *Dispersion-strengthened materials* (Moscow: Metallurgy)
[21] Popov A A, Farber V M and Bronfin B M 1974 Influence of deformation on the integration of M23C6 carbide in austenitic steel *The Physics of Metals and Metal Science* **38(2)** 337

IOP Conf. Series: Materials Science and Engineering **969** (2020) 012009 doi:10.1088/1757-899X/969/1/012009

Martensite Nucleation under Conditions of Austenite Superplasticity and External Magnetic Field

V N Pustovoit, Yu V Dolgachev and L P Aref'eva

Department of Physical and Applied Material Science, Don State Technical University, 1, Gagarin Square, Rostov-on-Don, 344000, Russia

E-mail: yuridol@mail.ru

Abstract. The thermodynamic conditions corresponding to the instability of the crystal lattice before the martensitic transformation are analyzed. The instability of the crystal lattice before the martensitic transformation can be crucial for the occurrence of such a phenomenon as superplasticity of the martensitic transformation. Phenomena arising in an instability state are considered, a theoretical analysis of possible structural states detected experimentally is carried out. The phenomenon of superplasticity of the martensitic transformation under tension in the $M_s - M_d$ interval and the effect exerted by the magnetic field in this interval are described. As a result of exposure to an external magnetic field in this temperature range, the resulting magnetostrictive stresses in ferromagnetically ordered austenite nanoclusters under superplastic conditions initiate the formation of martensite.

1. Introduction

The analysis of data on phase transformations in a magnetic field [1–6] leads to the conviction that this problem has not yet been completely solved theoretically. The magnitude of the energy of a constant magnetic field with a strength of up to 2.4 MA/m is 1–2 orders of magnitude less than the driving force of most phase transformations. Only a thermodynamic consideration of the issue can lead to the conclusion that the heat treatment in a magnetic field is futile. However, this conclusion is in contradiction with the actual results [7–11]. It is necessary to take into account the influence of the field on the mechanism of nucleation and the kinetics of phase transformations, which substantially determines the structure and properties of the products of the transformation. Also, it is necessary to take into account the phenomenon of superplasticity of austenite, which can have a significant effect on the development of martensitic transformation when exposed to a magnetic field. When the transformation is thermodynamically possible, even a small energy stimulus can have a significant effect on the kinetics. Thus, a comprehensive study of the effect of the magnetic field on the martensitic transformation seems quite relevant.

The following tasks were posed: to analyze the thermodynamic conditions corresponding to the instabilities of the austenite crystal lattice before the martensitic transformation, to consider the phenomena arising in the instability state, to conduct a theoretical analysis of possible structural states, to consider the phenomenon of superplasticity of the martensitic transformation under tension in the $M_s - M_d$ interval and the effect exerted by the magnetic field in this interval.

2. Experimental Procedures

Mechanical tests were performed on a tensile testing machine 2054 R-5 with a record of the deformation diagram in the load range 0–50 kN. Deformation of the samples was carried out in the temperature range $M_d - M_s$. For testing at low temperatures (up to -70 °C), the tensile testing machine was equipped with a cryostat. The sample was placed in an acetone-filled cryostat; the temperature was reduced by adding dry ice to acetone. To measure the temperature, a copper-constantan thermocouple was used with a working temperature range from -270 to 400 °C. During the research, samples were used from a model alloy 26N30 (0.26% C, 30% Ni, the rest is iron).

3. Results and Discussion

In most cases, structural phase transitions are first-order transitions. However, if in principle the transformation can be continuous and the lattice deformation during the transition is small (first-order phase transitions are close to second-order transitions), then they can also be considered in the representations of L. D. Landau [12]. On the one hand, such transformations are characterized by the presence of hysteresis and a region of coexistence of two phases (as in phase transitions of the first kind). On the other hand, when approaching the transition point, the phase structures become similar, for example, by decreasing the degree of tetragonality of the tetragonal phase during the cube – tetragon transition, and certain elastic constants of both phases decrease ($c' = (c_{11}-c_{12})/2$). The elastic constants of both phases remain finite at the phase equilibrium point, but vanish after the point T_0 (at the points of phase stability loss). The described behavior of the elastic constants is confirmed by the data obtained in [13]. These results suggest that lattice instability occurs prior to the martensitic transformation.

In a number of cases, instabilities of the crystal lattice near the point of martensitic transformation are characterized by short-range order of displacements. They are described using the concept of fluctuation displacement waves [14, 15], which allows one to analyze the diffuse scattering patterns of electrons and x-rays. Electron-diffraction and X-ray diffraction methods in [16] revealed and investigated diffuse scattering in austenite and martensite of Fe-Ni alloys (27–50 %Ni) in the temperature range from -130 to 600 °C. A regular transformation of diffuse scattering is observed, which is associated with a change in the atomic-crystalline structure of austenite and martensite when approaching the temperatures of the onset of direct M_s and reverse A_s martensitic transformations. Diffuse scattering is described by the spectrum of fluctuation displacement waves and indicates the presence of significant atomic displacements due to the "softening" of the crystal lattice near the points M_s and A_s.

From the thermodynamic theory of pre-martensitic states [13–16], a similar theory of solid solution decomposition, it follows that near the point of martensitic transformation there are local regions with short-range order of displacements. These regions, in contrast to Frenkel fluctuations, are characterized by a gradual decrease in order with distance from the fluctuation center and the absence of a clear interphase boundary. In diffraction patterns, they correspond to diffuse scattering regions, the intensity of which is determined by the Fourier transform. The main characteristics of these states can be associated with the elastic modules of the initial phase.

The fluctuations of the displacement waves "prepare" the austenite crystal lattice for martensitic transformation, thus, regions with a short-range order of displacements are the most prepared places for the formation of martensite nuclei. The minimum possible size of such fluctuations (about 1 nm) is comparable with the sizes of the minimum critical size of a ferromagnetically ordered cluster (about 0.66 nm [6]) and with the sizes of the critical martensite nucleus calculated from the dislocation model of the martensitic nucleus [17] (the critical radius of the embryo is 3 nm, its critical half-thickness is 0.125 nm). In such regions, the elastic constants decrease almost to zero and therefore even small stresses are enough to initiate a martensitic transformation. As was shown [4–6], before the martensitic transformation begins, ferromagnetic nanoclusters arise in the paramagnetic matrix of austenite. With a high degree of probability, in the places where ferromagnetic nanoclusters are formed, fluctuations of displacement waves can be expected. Under the influence of an external magnetic field, ferromagnetic nanoclusters, having absorbed the magnetic field energy through magnetostrictive stresses, will change the fields of elastic forces in the crystal lattice, including in regions with short-range order of displacements (where this is much easier to do). A change in the fields of elastic forces in the microvolumes of the matrix lattice contributes to a decrease in the energy barrier for the formation of a critical size nucleus center.

The instability of the crystal lattice before the martensitic transformation can be crucial for the occurrence of such a phenomenon as superplasticity of the martensitic transformation. The martensitic transformation proceeds by a displacement or "shear" process, which at the macroscopic level can be described as a deformation with an invariant plane. In the case where along with the martensitic transformation there is an applied external stress, phenomena such as the martensitic transformation (caused by stress and strain), the formation of new nucleation sites, and also superplasticity can occur. Superplasticity is understood as the ability of a material to plastic deformation many times higher than expected. In some cases, this can be hundreds of percent, but for low-plastic materials, the ability to deform by tens of percent should also be attributed to superplasticity. In another way, superplasticity can be defined as a special state of a material capable of large deformation. Superplasticity due to the state of the metal can be of two types: structural (superplasticity of materials with ultrafine grains) and transformations (superplasticity due to phase transformation).

In the work of Z. Nishiyama [18] it was established that the maximum plasticity is observed in the temperature range from the point M_d to the point M_s. Here and in the text, M_d is the temperature lying above M_s, but below the equilibrium temperature of the martensitic transformation of T_0. Below M_d, martensite may form as a result of deformation of the initial austenite. The phenomenon of superplasticity of austenite before martensitic transformation is used in practice to correct the product. During hardening, an experienced thermist a little earlier than the martensitic transformation begins, removes the tool from the hardening medium and produces deformation. A characteristic of the superplasticity of the martensitic transformation is the presence of a plasticity peak. Below M_s (after the onset of martensite formation caused by cooling), the ductility sharply decreases. Therefore, the mechanical processing of products in the temperature range M_d – M_s (where superplasticity is manifested) initiates the formation of martensite as a result of deformation. This is carried out easily even at low stresses, since the superplastic metal material resembles resins, hot glass, thermoplastics, and other viscous bodies in their behavior during deformation [19].

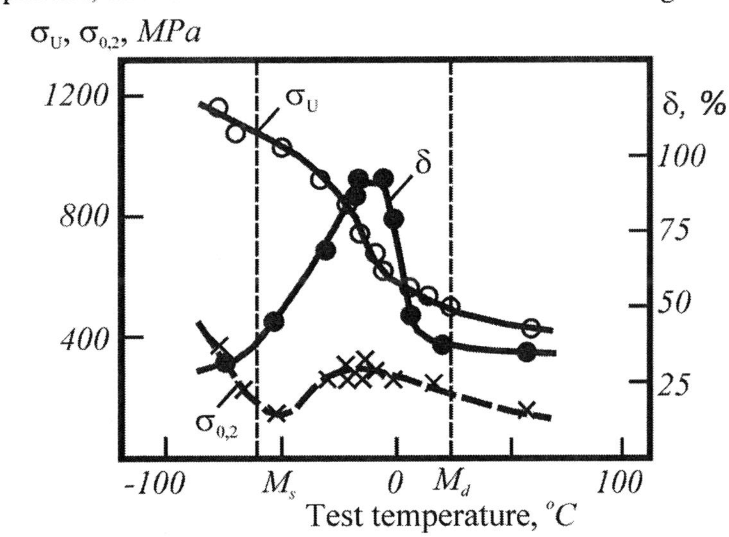

Figure 1. The influence of the test temperature on the mechanical properties of 26N30 alloy.

To observe the phenomenon of superplasticity, an experiment was carried out with a 26N30 metastable austenitic alloy (M_s = -60 °C, M_d = 25 °C), which was tested by stretching in the temperature range M_d – M_s. The dependence of the mechanical properties on the tensile test temperature is shown in figure 1. As can be seen, in the M_d – M_s interval, a peak of superplasticity is observed, where the relative elongation is almost 100%. This elongation is explained by the pre-martensitic instability of the crystal lattice, and also by the fact that when martensitic crystals arise (as a result of deformation), high local stresses are removed and thereby the formation and concentration of the neck are prevented. Immediately before the temperature M_s, an abnormal inverse dependence of the yield strength is observed. The decrease in the yield strength of metastable austenitic steel before M_s is due to the fact that, when tensile near M_s, elastic stresses cause the instantaneous formation of a large number of martensite crystals, which leads to a noticeable macrodeformation (jump in the tensile diagram due to the development of martensitic transformation). An increase in the yield strength occurs after the appearing martensitic crystals begin to play the role of a skeleton when the sample is stretched.

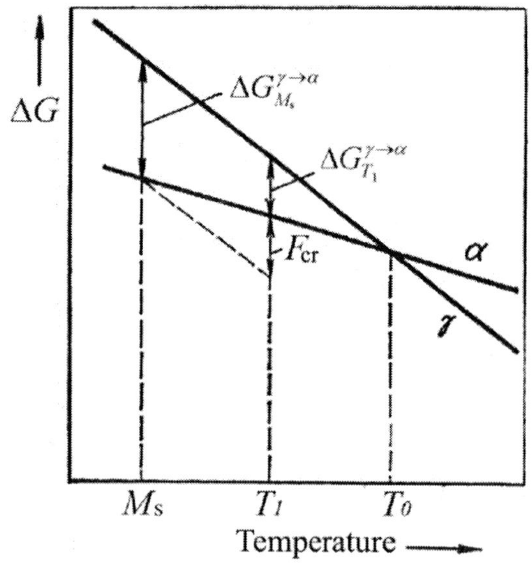

Figure 2. Schematic diagram of the change in the chemical free energy of the martensitic transformation ΔG.

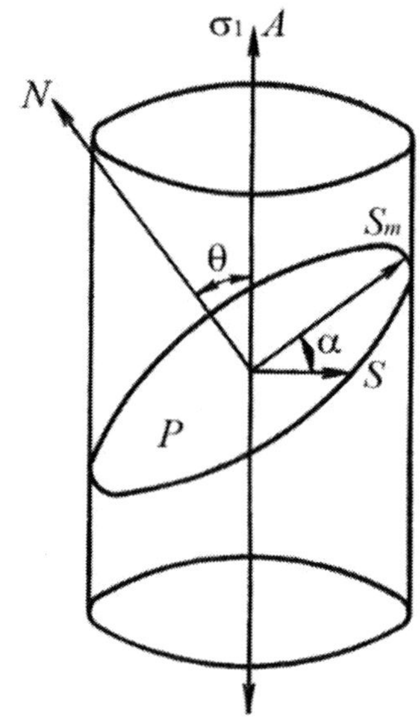

Figure 3. Schmitt factor diagram describing the effect of applied stress.

Since the martensitic transformation is carried out by the cooperative shear motion of atoms, it is easy to imagine that the applied elastic stress promotes the transformation. However, the role of plastic deformation in the martensitic transformation is very complex. To understand the martensitic transformation caused by deformation, it will be more productive to consider not the influence of deformation, but the effect of the applied stress. The figure schematically shows the change in the chemical free energy of martensite and austenite depending on temperature. In this figure, T_0 is the temperature at which austenite and martensite are in thermodynamic equilibrium. The difference between the free energies of austenite (γ) and martensite (α) $\Delta G_{M_s}^{\gamma\to\alpha}$ at a temperature of M_s is a critical driving force for the onset of martensitic transformation. When stresses are applied to austenite at a temperature T_1 between (M_s and T_0), the driving force of the transformation F, due to the stresses, is added to the chemical driving force $\Delta G_{M_s}^{\gamma\to\alpha}$. The martensitic transformation begins at a critical stress, when the total driving force is equivalent to $\Delta G_{M_s}^{\gamma\to\alpha}$. In figure 2, $F_{cr} = \Delta G_{M_s}^{\gamma\to\alpha} - \Delta G_{T_1}^{\gamma\to\alpha}$ it represents the critical driving force necessary for the martensitic transformation caused by stress at T_1. The value of this driving force of the transformation depends on the stress and orientation of the transformed martensitic plate relative to the applied stress (see Figure 3):

$$F_{cr} = \tau\gamma_0 + \sigma\varepsilon_n, \quad (1)$$

where: τ – shear stress acting along the shear direction during the transformation in the habit plane of martensite; γ_0 – shear strain during transformation along the shear direction of the shape strain in the habit plane; σ – normal component of stress acting perpendicular to the habit plane; ε – dilatational component of the shape deformation during transformation.

The figure 3 shows the Schmitt factor diagram describing the effect of the applied stress σ_1 along the axis A when the martensitic transformation is induced in the H-plane P with the normal N, where S is the direction of shape deformation for martensite and S_m is the maximum elongation in the H-plane during shape deformation. When a stress σ_1 is applied to the sample, for any given orientation of the martensitic plate, τ and σ can be expressed as follows:

$$\tau = \frac{1}{2}\sigma_1(\sin 2\Theta)\cos\alpha; \quad (2)$$

$$\sigma = \frac{1}{2}\sigma_1(1+\cos 2\Theta), \quad (3)$$

where: σ_1 – absolute value of the applied stress; Θ – angle between the axis of the applied stress and the normal to the habit plane; α – angle between the direction of shear during the transformation and the direction of maximum shear in the habit plane induced by the applied stress. Based on equations (1), (2) and (3), the driving force of the transformation due to the applied stress σ_1 can be expressed as:

$$F = \frac{1}{2}\sigma_1(\gamma_0(\sin 2\Theta)\cos\alpha \pm \varepsilon_0(1+\cos 2\Theta)). \quad (4)$$

When the martensitic transformation is induced by the stress applied to the polycrystalline austenite, and the grain orientations in the austenite are randomly distributed, the martensitic plate is formed first, the orientation of which provides the maximum value F.

An external magnetic field acts on ferromagnetically ordered clusters in austenite, which leads to local anisotropic deformation of the atomic lattice, which is caused by the forces of exchange and magnetic interactions. Calculations show that a single-domain particle always tends to be elongated, since this form reduces the energy of the demagnetizing field [20]. In crystals with a direction of easy magnetization <100>, these strains can easily interact with the Bane strain or initiate the mechanism of reconstructive martensitic transformation proposed by V. S. Kraposhin [21]. Due to the anisotropy of local spontaneous and forced magnetostrictive deformations in microvolumes with a spin order, the structural equivalence of the octahedral interstitials (in the BTC sublattice of the Bane) is violated and a favorable situation arises for the preferred ordered arrangement of carbon atoms in the interstices of the type [[0, 0, 1/2]] and [[1/2, 1/2, 0]], which is controlled by an energy difference of 10^{-22} J [22] created by a uniform magnetostrictive stress of the order of 10 MPa along the [001] axis. In this case, the probability of formation as a result of the transformation of the ordered phase in those microvolumes of the matrix where short-range order has already taken place can significantly increase. It should be noted that the preferred orientation of carbon under the action of magnetostrictive deformations will lead to greater deformation than magnetostriction itself. This increased deformation creates stresses in the surrounding austenitic matrix, which, under the conditions of austenite superplasticity, become sufficient to initiate shear martensitic transformation (for example, by the mechanism described in [21]) already outside the ferromagnetically ordered nanocluster.

4. Conclusions

Thus, it was shown that before the start of the martensitic transformation, crystal lattice instability is observed, which is accompanied by a decrease in the elastic constants at the transformation point to zero and causes the presence of superplasticity in the $M_s - M_d$ range. As a result of exposure to an external magnetic field in this temperature range, the resulting magnetostrictive stresses in ferromagnetically ordered austenite nanoclusters under superplastic conditions initiate the formation of martensite.

References

[1] Schastlivtsev V M, Kaletina Y V, Fokina E A and Mirzaev D A 2016 *Met. Sci. Heat Treat.* **58** 247
[2] Pustovoit V N, Dolgachev Y V and Aref'eva L P 2019 *Materials Science Forum* **946** 304
[3] Sadovskii V D, Malinen P A and Mel'nikov L A 1972 *Met. Sci. Heat Treat.* **14** 775
[4] Pustovoit V N and Dolgachev Yu V 2012 *Metal Science and Heat Treatment* **53** 515
[5] Dolgachev Y V, Pustovoit V N, Filonenko I O and Ivankov I V 2020 *Vestnik of Don State Technical University* **20(1)** 51
[6] Pustovoit V N and Dolgachev Y V 2017 *Emerging Materials Research* **6(2)** 249
[7] Kaletina Yu V 2008 *Met. Sci. Heat Treat.* **50** 413
[8] Pustovoit V N, Dolgachev Y V and Aref'eva L P 2020 *Materials Science Forum* **989** 79
[9] Sadovskii V D 1966 *Met. Sci. Heat Treat.* **7** 441
[10] Pustovoit V N and Dolgachev Y V 2018 *Materials Performance and Characterization* **8(2)** 320
[11] Pustovoit V N, Dolgachev Y D and Dombrovskii Y M 2017 *Solid State Phenomena* **265** 745
[12] Landau L D and Lifshitz E M 2013 *Statistical Physics* (Oxford: Elsevier) p 544
[13] Enami K, Hasunuma J, Nagasava A and Nenno S 1976 *Scripta Metallurgica* **10** 879
[14] Mori M, Jamada J and Shirane G 1975 *Sol. State Comm.* **17** 127
[15] Teklu A, Ledbetter H, Kim H, Boatner L A, McGuire M and Keppens V 2007 *Metallurgical and Materials Transactions A* **35** 3149
[16] Pushin V G, Romanova R R, Tyapkin Y D, Bujnov N N. and Kondrat'ev V V 1977 *Physics of Metals and Metallurgy* **43(4)** 826
[17] Gridnev V N and Petrov Yu N 1967 *Metal Science and Heat Treatment* **9** 586
[18] Nishiyama Z 1978 *Martensitic Transformation* (New York: Acad. Press) p 471
[19] Novikov I I, Portnoy V K and Levchenko V S 1981 *Acta Metallurgica* **29(6)** 1077
[20] Neel L 1949 *Compt. Rend. Acad. Sci.* **228**(8) 664
[21] Kraposhin V S, Pankova M N, Talis A L and Freiman Y A 2003 *In Journal de Physique IV* **112** 119-122
[22] Mishin D D 1991 *Magnetic materials* (Moscow: Vysshaya Shkola) p 384

IOP Conf. Series: Materials Science and Engineering **969** (2020) 012010 doi:10.1088/1757-899X/969/1/012010

Rolled Stock Structure Preparation for Cold Forging of Pearlite Steel Grades

G V Pachurin[1], A A Fillipov[1] and M V Mukhina[2]

[1]Nizhny Novgorod State Technical University named after R. A. Alekseev, 24, Minin str., Nizhny Novgorod, 603950, Russia
[2]Minin Nizhny Novgorod State Pedagogical University, 1, Ulyanova str., Nizhny Novgorod, 603002, Russia

E-mail: pachuringv@mail.ru

Abstract. Special requirements are applied to the structural-mechanical characteristics of rod bolt products manufactured by cold die forging of pearlite steel class. A very common and productive method of obtaining finished metal products with high mechanical properties is their landing on cold heading (cold-heading) machines made of rolled metal, which must have a certain quality along the entire length: a sufficient ductility, homogeneous structure and parameters of technological properties, absence of any defects. Hardware plants receive the initial blank for the manufacture of various fasteners in the form of hot-rolled steel, which is not possible to use for the landing of high-strength hardware, due to the poor quality structure and surface layer. Therefore, the calibrated rolled stock is subjected to intermediate heat treatment before driving, which should guarantee the quality of the metal products at transitions of the cold heading. Further, the driven products of strength class 8.8 and up, made of alloy steel grades, must be subjected to hardening with subsequent tempering. However, such thermal operation often leads to decarburization of the surface, cracking and warping of finished products, thus increasing labor intensity, energy consumption and, consequently, the final production cost. The paper proposes an alternative method of the rolled stock structural preparation, which includes isothermal treatment of pearlite steel grades 35X and 38X, which excludes from the technological process the recrystallization annealing and quenching with the release of finished bolt products, and which minimizes any appeared scratches, warping and decarburization on the rolled stock and the rolled product. The structure of rolled stock induced in the drawing die, and the surface hardening by reduction, and thread rolling provide the parameters of mechanical properties of M8 bolts and pins, corresponding to the strength class 9.8.

1. Introduction

Reliable operation of metal products significantly depends on the nature of the material [1–3], operating conditions [4–6] and modes of its processing [7, 8], forming the structure and properties of the materials used [9–11], and which are provided by the optimization of technological processes [12, 13].

The rod bolt products manufactured by cold forging of pearlite class steel are subject to special requirements for their structural and mechanical characteristics [14]. A very common and productive way to obtain ready-made metal products with high structural and mechanical properties is to drive these from the rolled stock, which must have a certain quality – a uniform (along the entire length) ductility, structure, manufacturability and absence of defects [15–17].

A wide range of hardware products (bolts, pins and screws) for which specific requirements to strength properties are not regulated, are made, generally of steel grades according to GOST 1050-2013, GOST 4543-2016, and also GOST 380-2005. The above noted standards do not regulate the parameters of the steel structure and surface, which are formed when passing through the drawing die and during the heading. This creates certain risks of signs of flaws, such as deformation cracks and warping [18–20] in the driven rod products.

The chemical composition of steel grades and mechanical properties of rolled metal used for heading of metal rod products of different strength classes for cars, tractors and other special equipment are regulated by GOST 10702-2016. For high strength hardware products the mechanical characteristics are regulated according to effective standard ISO 898-1: 1999. The required strength class is determined by the grade of steel [21, 22] and the heat treatment mode of rolled metal and the driven finished products [4, 23].

The hardware manufacturing enterprises receive blanks for manufacturing of various fasteners in the form of hot-rolled steel, which cannot be used for the high-strength hardware heading due to poor quality structure and the surface layer. Hot-rolled metal from structural steel of pearlite class, which is purchased in non-heat-treated condition from suppliers of metallurgical products, has a "pearlite-ferrite" structure. Before rod product driving, heat treatment and drawing should form the necessary microstructure and the state of the rolled surface [24–26], which would guarantee the quality of the product at all transitions of the cold driving [27–29].

According to GOST 1759.4-87, the fasteners of strength class 8.8 and up, made of alloy steel grades, driven in this way should be subjected to hardening and tempering. Unfortunately, the above-mentioned thermal operation provokes the appearance of decarburization, cracking and warping on the finished products, thus increasing the labor intensity, energy intensity and, consequently, the final production cost of the product.

In this paper we have proposed an alternative method of preparing the rolled stock structure that includes isothermal mechanical treatment of pearlite steels 35X and 38XA with the purpose of eliminating from the technological process the recrystallization annealing, as well as the annulment of hardening with the drawing-back of finished bolt products which would result in a minimization or reduction of risk of the rolled stock and product warping and decarburization. At the same time, this method allows obtaining the mechanical characteristics of the rolled products and the surface quality that fully comply with GOST 10702-2016.

2. Study Methods

The paper has studied samples of rolled steel of pearlite class 35X and 38X, widely used in the manufacture of critical fasteners of motor, body and other groups of automobile, tractor and specialized machinery.

Based on the analysis of experimental data we have proposed the method of isothermal mechanical treatment of the rolled stock made of the investigated steel grades for the manufacture of long length rod products with a low molded head and various pin items corresponding to the strength class 8.8 or up without quenching and tempering of these.

The rolled stock isothermal mechanical treatment pattern for steel grades 35X and 38XA includes as follows:

- initial rolled stock thermal treatment;
- acid solution etching after annealing;
- drawing;
- isothermal treatment (@ 470 °C);
- acid solution etching after isothermal treatment;
- drawing;
- heading of high strength bolts and pins.

To form the microstructure of the rolled steel for driving of bolts with a diameter of 7.8 mm, the process of isothermal treatment - annealing of the 2nd kind was used. The transformation of austenite of these steels in niter bath at temperature of 470 °C occurs in the temperature interval of sorbitic transformation 650–470 °C. The austenite obtained at this temperature of the thermal bath belongs to the eutectoid type. The investigated method of isothermal cooling at a temperature of 470 °C is proposed to obtain a calibrated rolled stock after plastic drawing operation for the heading of short-length, medium-length and long-length bolts with a volume-molded head and pins of various lengths.

Considering certain niter bath length limitations in the case of the wire isothermal treatment, the exposure of the treated calibrated rolled stock does not exceed five or six minutes. To complete the austenitic transformation in metal, the cooling time of the wire with a diameter of 8.2–8.6 mm at a temperature of 470 °C in niter bath is 3.9 minutes (234 s).

The rolled metal samples were prepared by drawing to obtain diameters of 8.1 mm; 8.2 mm; 8.3 mm; 8.4 mm; 8.5 mm and 8.6 mm with the subsequent isothermal treatment in a niter bath at 470 °C. Final drawing was carried out to obtain diameter of 7.8 mm with various degrees of compression in the drawing mill die.

M8 bolts and studs thus obtained were subjected to tensile testing according to GOST 1759.4-87.

3. Results and Discussions

The chemical composition and mechanical properties of 35X and 38X steels grades corresponded to GOST 10702-16. The microstructure of the initial rolled stock is of "pearlite + ferrite" type.

The mechanical properties of the isothermally treated rolled products are shown in table 1.

Table 1. Isothermally treated rolled stock properties.

Grade of steel	Rolled stock diameter, mm	Characteristics			
		strength		ductility	
		σ_v	$\sigma_{0.2}$	δ	ψ
		MPa		%	
35X	8.0; 8.1; 8.2; 8.3; 8.4; 8.5; 8.6	878	630	20.5	57
38XA	8.0; 8.1; 8.2; 8.3; 8.4; 8.5; 8.6	882	637	20.1	57

After the isothermal treatment, the samples of rolled products made of steel 35X and 38XA with diameters 8.1 mm; 8.2 mm; 8.3 mm; 8.4 mm; 8.5 mm and 8.6 mm at the temperature of the salt bath 470 °C have a "niter-like pearlite" pattern. The samples of steel grade 35X after the above mentioned treatment have hardness HB 249, and the samples of steel grade 38XA have same of HB 254.

The rolled stock mechanical characteristics after the isothermal treatment at a temperature of 470 °C and with subsequent drawing with different degrees of drafting are shown in table 2.

The calibrated rolled stock microstructure of both steel grades is of "niter-like pearlite" pattern. It was founded that the strength characteristics of the rolled stock isothermally processed at 470 °C with an increase in the drafting degree during drawing from 4.9 to 17.7 % had been constantly improving. So, the limit tensile strength of the calibrated rolled steel 35X with the degree of drafting of 4.9 to 17.7 % had increased from 860 to 940 MPa, and that of the rolled stock made of steel 38XA - from 930 to 1060 MPa.

The calibrated rolled steel 35X and 38X testing shows that the conventional yield strength after isothermal mechanical process treatment also increases. It was revealed that the rolled stock conventional yield strength before cold heading of steel grade 35X at a drafting ratio of 4.9 to 17.7 % increases from 770 to 880 MPa, and that of steel grade 38XA - from 800 to 950 MPa.

The rolled steel isothermal mechanical treatment at temperature of 470°C with the subsequent drawing with the drafting degree of 4.9 to 17.7 % provides a continuous elongation reduction for steel grade 35X from 16 to 12.2 % that for steel grade 38XA - from 15.7 down to 12.2 %.

The relative transverse contraction after the isothermal mechanical treatment of the rolling stock before cold heading of steel grade 35X with drafting ration of 4.9 to 17.7 % decreases from 54 to 52 %, and that for steel grade 38XA - from 53.1 down to 50.4 %.

Table 2. Strength and ductility characteristics of the rolled stock after the isothermal treatment of steel grades 35X and 38XA @ 470 °C.

Grade of steel	Rolled stock diameter, mm	Drafting degree, %	Characteristics: strength σ_v, MPa	strength $\sigma_{0.2}$, MPa	ductility δ, %	ductility ψ, %
35X	7.8	4.9	878	769	16	54
38XA	7.8	4.9	918	797	15.5	53.1
35X	7.8	7.2	896	791	15.8	53.8
38XA	7.8	7.2	922	807	15	53
35X	7.8	9.5	905	801	14.5	53
38XA	7.8	9.5	931	813	14.1	52.7
35X	7.8	11.6	916	810	14	52.5
38XA	7.8	11.6	959	835	13.5	52.1
35X	7.8	14.2	928	831	13.2	51.8
38XA	7.8	14.2	991	863	13.0	51.0
35X	7.8	15.7	944	862	13.1	51.1
38XA	7.8	15.7	1003	901	12.4	50.4
35X	7.8	17.7	971	887	12.5	50.0
38XA	7.8	17.7	1067	948	12	49.1

Of the calibrated rolled stock made of steel grade 35X and 38XA, after the isothermal mechanical treatment at temperature of 470 °C, some M8 short and long-length bolts were fabricated with a low cut head and M8 pins - by heading in the cold heading machine.

After the cold heading of bolts, these were subjected to tensile testing and the bolt strength and ductile properties were identified. The results of the tensile tests of the bolts are presented in table 3.

Table 3. Tensile tests of bolts made of steel grade 35X and 38XA.

Bolt, M8	Number of bolts, pcs	σ_v, MPa	Ψ, %	δ, %	HB
35X	21	820	44.5	10.7	254
38XA	17	950	44.5	10.7	286

After the pin cold forging and drawing-down, their tensile tests were also carried out and strength and ductile characteristics were identified. The pin tensile test results are presented in table 4.

Table 4. Tensile tests of pins made of steel grade 38X and 38XA.

Pin, M8	Number of pins, pcs	σ_v, MPa	Ψ, %	δ, %	HB
35X	18	912	44.5	10.7	248
38XA	25	938	44.5	10.7	277

4. Conclusions

We have investigated the possibility of isothermal treatment of the rolling stock with diameter of 7.8 mm made of structural steels of pearlite class 35X and 38XA for cold forging of bolt with cut head and pins with strength class 8.8 and up without quenching and tempering.

It is proved that M8 bolts and pins made of calibrated rolled steel mentioned above, and strengthened by subsequent surface ductile deformation during the drawing and thread rolling, have achieved a set of strength and ductile characteristics corresponding to the strength class 9.8 according to the requirements of GOST 52643-2006.

References

[1] Pachurin G V, Goncharova D A, Filippov A A, Shevchenko S M, Mukhina M V and Kuzmin N A 2018 Development of fatigue test technology of sheet automobile materials *Eastern-European journal of enterprise technologies* **5 12(95)** 31–37

[2] Pachurin G V 2012 Life of Plastically Deformed Corrosion-Resistant Steel *Russian Engineering Research* **32 9–10** 661–664

[3] Guslyakova G P, Zhbannikov S I and Pachurin G V 1993 Fatigue failure resistance of deformed structural steels *Materials Science* **28(2)** 182-185

[4] Pachurin G V and Vlasov V A 2014 Mechanical properties of sheet structural steel at operating temperatures *Metallurgy and heat treatment of metals* **4(706)** 48–53

[5] Galkin V V and Pachurin G V 2016 Interrelation of structural and mechanical characteristics and fatigue resistance of hot-deformed material Modern science-intensive technologies **5–3** 435–444

[6] Pachurin G V 2014 The effect of temperature on the mechanical properties of the sheet structural steel *Fundamental research* **1** 18–23

[7] Pachurin G V, Filippov A A and Pachurin V G 2015 The effects of temperature of isothermal treatment with drawing on the mechanical properties of hot-rolled steel grade 40X *Modern science-intensive technologies* **2** 98–106

[8] Pachurin V G, Galkin V V and Pachurin G V 20014 Evaluation of deformation inhomogeneity in rolled products with wedge profile *Fundamental research* **11–4** 765–773

[9] Filippov A A, Pachurin G V, Naumov V I and Kuzmin N A 2016 Low-Cost Treatment of Rolled Products Used to Make Long High-Strength Bolts *Metallurgist* **59 9–10** 810–815

[10] Pachurin G V 1990 Fatigue failure at normal temperature in pre-deformed alloys *Metallurgy and heat treatment of metals* **10** 35–38

[11] Romanovskaya E V 2014 Creation of a new project on the basis of own research and development *Vestnik of Minin University* **1** 9

[12] Romanovskaya E V 2014 Investigation of practical aspects of development of production technologies at machine-building enterprises *Vestnik of Minin University* **4** 9

[13] Pachurin G V and Guschin A N 2007 Enhancing operational durability of metal products by technological methods *Mechanical Engineering Bulletin* **6** 62–65

[14] Pachurin G V and Vlasov V A 2014 Mechanical properties of sheet structural steels at operating temperatures *Metal Science and Heat Treatment* **56 3–4** 219–223

[15] Filippov A A and Pachurin G V 2014 The main trends of development of production of high-strength fasteners *International journal of applied and fundamental research* **8–4** 30–35

[16] Pachurin G V, Shevchenko S M, Filippov A A, Mukhina M V and Kuzmin N A 2018 Defining Role of Metal Performance for cold bolt upsetting (bolt head) *International Journal of Mechanical Engineering and Technology* **9(4)**

[17] Rudskoy A I 2011 *Drawing: study guide* (St.Petersburg: Polytechnical University Publishing House)

[18] Pachurin G V, Kudryavtsev S M, Solovyev D V and Naumov V I 2018 *The body of a modern car: materials, design and production: study guide* (St. Petersburg: Lan)

[19] Pachurin G V and Filippov A A 2008 Resource-saving and environmentally friendly surface treatment of rolled metal before cold heading *Ecology of industry in Russia* 13–15

[20] Filippov A A, Pachurin G V, Kuzmin N A and Matveyev Yu I 2018 Method of formation of structural and mechanical properties of rolled steel for the rod products heading *Ferrous metals* **4** 36–40

[21] Filippov A A, Pachurin G V and Naumov V I 2018 *Getting quality rolled stock and safety issues in hardware production: study guide* (Stary Oskol; TNT)

[22] Filippov A A, Pachurin G V and Kuzmin N A 2018 Evaluation of the quality of rolled steel for cold die forging, Proceedings of higher educational institutions *Ferrous metallurgy* **61(7)** 551–556

[23] Pachurin G V 2008 Ruggedness of structural material and working life of metal components, *Steel in Translation* **38(3)** 217–220

[24] Pachurin G V and Filippov A A 2008 Economical preparation of 40X steel for cold upsetting of bolts *Russian Engineering Research* **28(7)** 670–673

[25] Pachurin G V, Shevchenko S M, Mukhina M V, Kutepova L I and Smirnova J V 2016 The Factor of Structure and Mechanical Properties in the Production of Critical Fixing Hardware 38XA *Tribology in Industry* **38(3)** 385–391

[26] Filippov A A, Pachurin G V, Matveyev Yu I and Kuzmin A N 2016 Comparison of technological methods of the rolled stock surface structural and mechanical properties preparation for the heading of metal products in order to reduce the impact of hazardous and harmful factors on employees *Fundamental research* **10–1** 88–96

[27] Pachurin G V, Filippov A A and Pachurin V G 2015 Quality of surface and structural condition of the rolled stock for metal products made of steel grade 40X *Advances in modern natural science* **1–3** 476–481

[28] Pachurin G V and Filippov A A 2008 Rational reduction of hot-rolled 40X steel before cold upsetting *Steel in Translation* **38(7)** 522–524

[29] Filippov A A, Pachurin V G and Pachurin G V 2015 Obtaining high-quality calibrated rolled stock for high-strength bolts heading *Modern knowledge-intensive technologies* 3 87–92

[30] Filippov A A, Pachurin G V, Naumov V I and Kuzmin N A 2015 The effect of the surface and structural state on the quality of rolled stock used for bolt production *Fundamental research* **10–1** 77–82

IOP Conf. Series: Materials Science and Engineering **969** (2020) 012011 doi:10.1088/1757-899X/969/1/012011

Study of the Features of Damage Accumulation in Steel 12Cr18Ni10Ti at Low Temperatures Using Non-Destructive Testing Methods

A A Khlybov, M S Anosov and D A Ryabov

Nizhny Novgorod State Technical University n.a. R.E. Alekseev, 24, Minin St., Nizhny Novgorod, 603950, Russia

E-mail: hlybov_52@mail.ru

Abstract. The paper considers the features of damage accumulation during fatigue loading of 12Cr18Ni10Ti steel, in particular at low temperatures. The character of changes in the magnetic and acoustic characteristics, as well as the index of the fractal dimension of microstructures during the accumulation of damage to the metal structure, is studied. The correlation of the considered indicators is shown, which indicates the possibility of using indicators such as the acoustic parameter, magnetic characteristics (Hc/Br ratio, as well as the coercive force Hc) and the fractal dimension of the metal microstructure to assess structural damage, such as in laboratory conditions, and during the operation of individual parts and structural elements in a wide range of low temperatures. The use of the considered indicators in the complex will significantly increase the accuracy of forecasting metal damage and its residual life.

1. Introduction

The active development of strategically important territories of the Arctic and the Far North necessitates the reliable functioning of technical facilities under extreme climatic conditions, which significantly reduce the operational reliability and resource of mechanisms and structural elements, which can lead to serious accidents and economic consequences.

During operation, structural elements and mechanisms of the Northern version are subjected to various temperature and power influences, many structural elements operate in conditions of low and high cycle fatigue. Moreover, in the process of operation, cases of nucleation and development of a crack are often observed. The process of damage accumulation begins almost immediately after the application of a variable load, proceeds covertly, practically without visible external signs, and can reach up to 80% of the resource of the sample material. Hidden processes of material degradation lead to a change in physical and mechanical characteristics: elastic moduli, strength and other characteristics. The period of damage accumulation ends with the formation of macrocracks and the destruction of the structure. Such a crack can be detected visually or using flaw detection tools. With the functioning of structures and mechanisms at low temperatures, the appearance of microcracks in the material is unacceptable, because the work of crack propagation with decreasing temperature decreases several times [1, 2].

In this regard, the task is to assess the degree of degradation of the material during its operation even before the appearance of macrocracks, and to ensure constant monitoring of the structural state of the material of construction with the use of reliable diagnostic tools. A comparative analysis of modern methods of non-destructive testing shows that the most promising methods for monitoring the structural state of the material at the stages of operation before macrodefects include physical acoustics, as well as the magnetic method for monitoring and diagnosing the material [3, 4, 5].

In recent years, approaches to nonlinear dynamics and fractal representations developed by V.S. Ivanova, A.G. Kolmakov, V.I. Trefilov, et al. [6–11], which make it possible to quantify the processes of structure degradation based on a quantitative criterion — fractal dimension.

2. Choice of material and design of a stand for tests

Of scientific interest is the study of the mechanisms of deformation and fracture during elastoplastic cyclic deformation of metastable austenitic steels, in particular steel 12Cr18Ni10Ti in connection with the peculiarities of phase transformations during deformation, which significantly affects the physicomechanical characteristics of the material and its resource.

Steel 12Cr18Ni10Ti is widely used for the manufacture of welded structures, vessels and individual parts operating at low temperatures, at high pressures and in aggressive environments.

The chemical composition of steel 12Cr18Ni10Ti is as follows (%): C - 0.11; Si - 0.7; Mn 0.16; Ni - 10.2; S is 0.012; P is 0.008; Cr - 18.16; Ti - 0.6; Fe is the basis.

For fatigue tests, samples were prepared whose geometry is shown in Figure 1 a. For testing, a specialized stand for fatigue testing was developed, the scheme of which is shown in Figure 1 b.

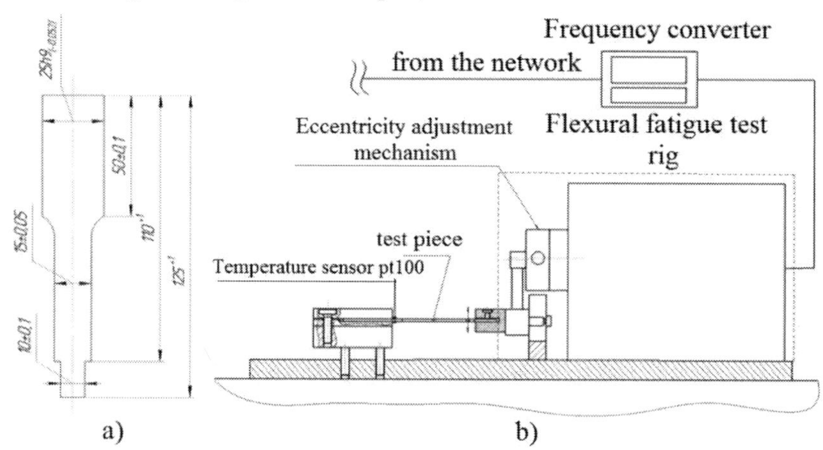

Figure 1. The geometry of the sample (a) and the test bench for fatigue (b).

Test conditions: loading according to the cantilever bending diagram (cycle asymmetry coefficient R = -1) at temperature t = + 20°C and t = -100°C on a fatigue testing machine (Figure 1, b) taking into account the requirements of GOST 25.502-79. The frequency of the elastoplastic cyclic deformation was set using a frequency converter equal to 8.3 Hz (500 cycles/min).

The stress amplitude (σ_{max}) was calculated on the basis of the loading scheme, sample dimensions and mechanical characteristics of the material. During the tests, the following were recorded: the number of cycles N and the amplitude of the stresses in the cycle. The studies were carried out at a stress amplitude σ_{max} of 305 MPa.

To conduct fatigue tests at low temperatures, a cryochamber was used to ensure uniform cooling of the sample and automatic maintenance of the test temperature due to the flow of chilled air from a bath with liquid nitrogen. The temperature in the working zone of the sample was monitored using pt100 temperature sensors (Figure 1, b).

3. Methods of processing acoustic signals and microstructure images

Preliminary tests were carried out to estimate the number of cycles until sample failure (N_Σ). The study of the structures of materials, registration of the parameters of the ultrasonic signal and magnetic

IOP Conf. Series: Materials Science and Engineering **969** (2020) 012011 doi:10.1088/1757-899X/969/1/012011

characteristics was carried out before the start of testing and after operating time with a step of ~ 10% of the limit number of cycles N_Σ.

The microstructure was studied using a KYENCE-VHX 1000 optical microscope. The polished surface was etched in a 10% oxalic acid solution. To assess the fractal dimension of microstructures (D_F), software was developed in the MATLAB environment [6].

The study of material structures, registration of the parameters of the ultrasonic signal and magnetic characteristics was carried out in the zone with the highest stresses, where microcracks were subsequently detected in the material.

For acoustic measurements, the «Astron» measuring and computing acoustic complex (MCC) was used [12]. The device is designed for non-destructive testing of the physicomechanical characteristics of solids by measuring the propagation time of ultrasonic vibrations (ultrasonic testing) and measuring the attenuation coefficient of the received signal. The device allows measuring the propagation time of elastic impulses with an accuracy of 10^{-9} s. The error in measuring the attenuation coefficient (K_{zat}) is 5%.

MCC "Astron" is able to work both in separate and in a separately combined mode with various types of elastic waves. A longitudinal wave sensor was used in the work. The nominal frequency of the sensors is 5 MHz. Glycerin was used as a contact liquid, the viscosity level of which at low temperatures retains its efficiency.

We used pulsed excitation and reception of elastic pulses. As a reference pulse, the first reflected pulse was used, relative to which the velocity and attenuation were measured by comparing its characteristics with the characteristics of subsequent pulses. This allows one to take into account only the effects associated with the path of an elastic impulse in a controlled medium [12–15].

The speed of the ultrasonic wave C when passing through the test sample was determined by the following formula:

$$C = \frac{2L}{t}, \tag{1}$$

where L is the thickness of the sample, m; t is the transit time of the corresponding type of wave through the sample, s. The speed measurement error is ± 2 m / s.

The attenuation coefficient K_{zat} was determined by comparing the amplitude range of the second reflected pulse compared to the first reference:

$$K_{zat} = \frac{1}{L} \cdot 20 \log \frac{A_2}{A_1}, \tag{2}$$

where A_2 is the amplitude of the second reflected pulse, A_1 is the amplitude of the first reflected pulse.

As an acoustic characteristic, materials taking into account the change in the propagation velocity of ultrasonic waves (C_1 and C_2 - transverse with longitudinal and transverse polarization and C_3 - longitudinal, relative to the sample) used the dimensionless parameter D, determined by the formula:

$$D = \frac{C_1 + C_2}{C_3}. \tag{3}$$

Damage (W) of the material was determined on the basis of acoustic measurement data according to the technique shown in [16].

To evaluate the magnetic characteristics, a magnetic metal analyzer/coercimeter MA-412MM was used. The following magnetic characteristics were evaluated: residual magnetization Br, coercive force Hc, maximum induction Bs, ratio Hc/Br.

At each stage, at least 5 measurements, both acoustic and magnetic characteristics, were carried out, followed by statistical processing of the obtained data.

The results obtained at different stages of testing were recorded with a special database of test parameters (DB) and were used later as initial data for training an artificial neural network.

4. Experimental research results

The microstructures of the material under study at different stages of production and with different damage are shown in Figure 2 at test temperatures t = + 20°C and t = -100°C.

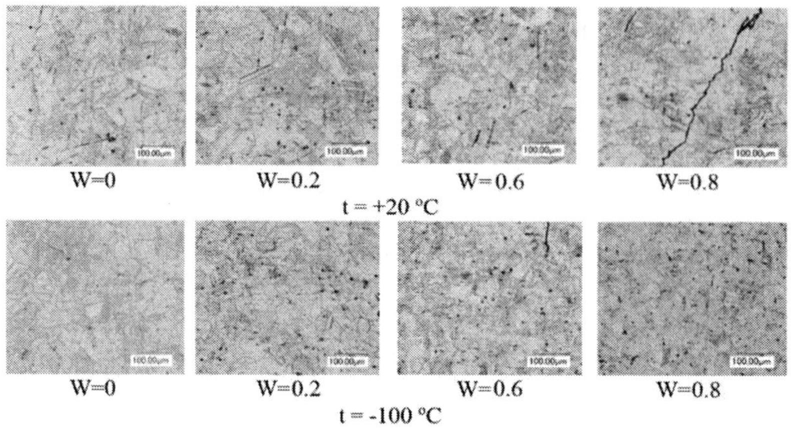

Figure 2. Microstructures (x300) at different stages of operating time and with different damage *W*.

Based on the analysis of microstructures with different operating time and damage (*W*) and the estimation of fractal dimension (D_F) from the images of microstructures, the dependences of the fractal dimension of the microstructure of the working area of the sample (D_F) on the damage of the metal (*W*) at temperatures *t* = + 20°C and *t* = -100°C (Figure 3).

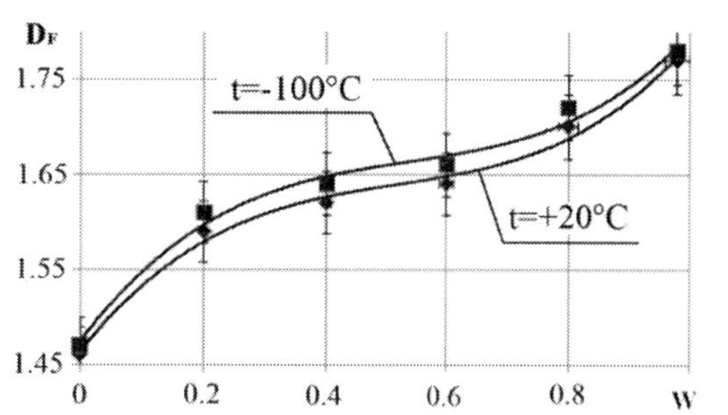

Figure 3. Dependence of the fractal dimension of the microstructure (D_F) on damage (*W*) for steel 12Cr18Ni10Ti for test temperatures.

An analysis of the obtained dependences allows us to talk about the relationship between the fractal dimension of the microstructure in the working zone of the sample and the value of the metal damage. At a low temperature *t* = -100°C, the nature of the dependence in question does not change, however, the values of the fractal dimension of the microstructure are slightly higher. As studies have shown, the values of the fractal dimension of the structure during the test also depend on the initial structure of the material.

When studying the features of changes in the parameters of ultrasonic waves (surface, longitudinal and transverse), it was found that the most informative is the pattern of change in the acoustic parameter D. The obtained dependences of the acoustic parameter D on the damage to the material W for temperatures *t* = + 20°C and *t* = -100°C are shown in Figure 4.

As can be seen from figure 4, the obtained dependence of the acoustic parameter D on the damage value, along with the index of the fractal dimension of the microstructure (Figure 3), has a definite relationship with the damage values of the sample material. On the obtained dependences, 3 characteristic stages can be distinguished, however, with a decrease in temperature, the character of the dependence changes. So, at a temperature of *t* = + 20°C, an increase in the parameter *D* is observed, while at *t* = -100°C its decrease is observed during the accumulation of damage to the material *W*.

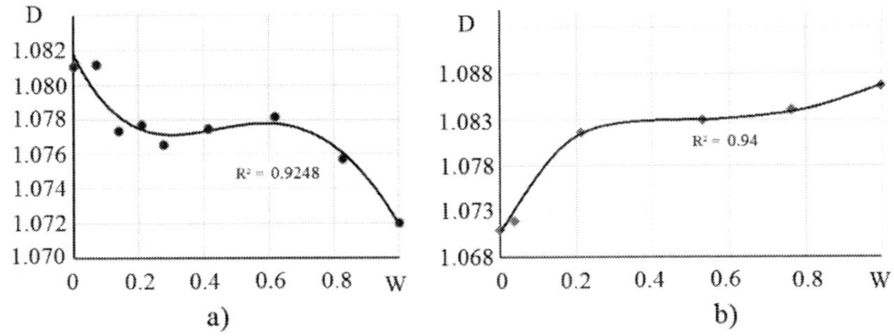

Figure 4. The dependence of the acoustic parameter *D* on the value material damage *W* for temperatures *t* = + 20°C (a) and *t* = -100°C (b).

The results obtained indicate the need to take into account, along with parameter *D*, other acoustic parameters (velocity and attenuation coefficient of waves in the material) to determine the stage of destruction of the material and its damage taking into account the test temperature.

In figure 5 shows the nature of the change in the magnetic characteristics of 12Cr18Ni10Ti steel during fatigue tests at different test temperatures.

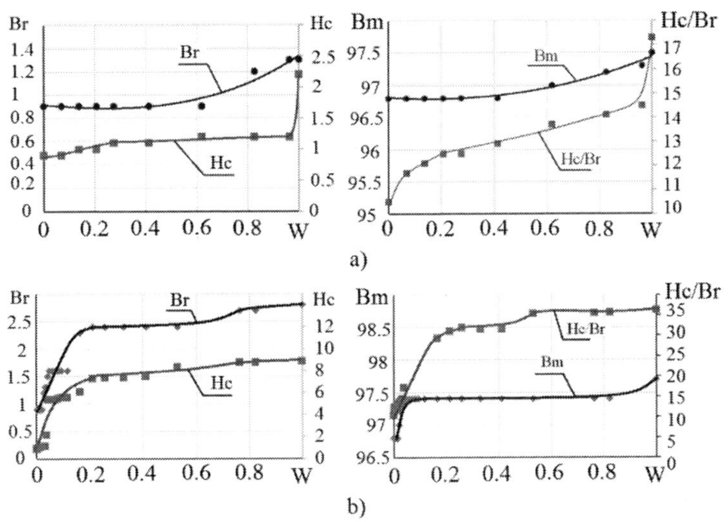

Figure 5. Dependences of the magnetic characteristics of 12Cr18Ni10Ti steel on material damage W at temperatures: a) t = + 20°C and b) t = -100°C

An analysis of the magnetic characteristics indicates their mutual correlation, and the Hc/Br ratio, as well as the coercive force Hc for which the most intense change is observed in the entire temperature range under consideration, can be taken as the most informative. At low temperatures, there is such a feature as an intensive growth of all magnetic characteristics in the initial stages of the test, which also indicates the intensity of phase transformations in the material. At a temperature of *t* = + 20°C, the most intense increase in magnetic characteristics is observed at the stage of prefracture of the sample.

Based on numerous informative parameters (Figures 3–5) that characterize the damage to the material during loading, it is necessary to use tools to assess the damage to the material on the basis of the whole complex of heterogeneous diagnostic information. Taking into account all the identified indicators in the complex will make it possible to identify the stage of degradation and destruction of the material and its residual life.

5. Conclusion

1. In the course of experimental studies, the dependence of the fractal dimension of the metal microstructure (DF) on the parameter of its damage is shown. The nature of the dependence is preserved over the entire temperature range under consideration, namely, a monotonic increase in the fractal dimension of the microstructure during the accumulation of metal damage is observed.
2. To predict the damage to the metal, the acoustic parameter D can be used, which reflects the change in the propagation velocity of elastic waves in the metal. However, in connection with the peculiarities of the change in the considered parameter at room and low temperatures (the nature of the dependence changes), it is additionally necessary to take into account the change in all parameters of the elastic waves in aggregate to increase the accuracy of predicting metal damage.
3. An analysis of the magnetic characteristics showed that the Hc/Br ratio, as well as the coercive force Hc, for which the most intense change is observed in the entire temperature range under consideration, can be taken as the most informative. At low temperatures, there is an intensive growth of all magnetic characteristics at the initial stages of the test, which also indicates the intensity of phase transformations in the material, since at room temperature the change in magnetic characteristics is less intense.
4. Based on numerous informative parameters characterizing the damage to the material during fatigue loading, it is necessary to use tools to assess the damage to the material on the basis of the whole complex of heterogeneous diagnostic information in order to increase the accuracy of the stage of degradation and destruction of the material and its residual life.

References

[1] Terent'ev V F 2015 *Fatigue of metals* (Moscow: Science) p 484
[2] Finkel V M 1970 *Fracture physics: crack growth in solids* (Moscow: Metallurgy) p 376
[3] Klinenko A P 1969 *Cold in mechanical engineering* (Moscow: Mechanical Engineering) p 248
[4] Kabaldin Y G *Russian Engineering Research* **37** 866–873
[5] Semashko N A 2002 *Acoustic emission in experimental materials science* (Moscow: Mechanical Engineering) p 240
[6] Klyuev V V 2006 *Non-destructive testing. Handbook in 8 volumes. Vol.3 Ultrasonic testing* (Moscow: Mechanical Engineering) p 861
[7] Zuev L B 1997 *Technical Physics* **42** 1094–1096
[8] Scherbinin V Ye 1996 *Magnetic control of metal quality* (Yekaterinburg: UrO. RAS) p 263
[9] Kabaldin Y G 2020 *IOP Conf. Series: Materials Science and Engineering* **709 033076**
[10] Savenkov G G 2011 *PMTF* **52(6)** 177–184
[11] Ivanova V S 1994 *Synergetics and fractals in material science* (Moscow: Science) p 384
[12] Krasikova I E 2016 *Electron microscopy and strength of materials. Series: Physical materials science, structure and properties of materials* **22** 3–9
[13] Trefilov V I 1999 *Sat. FIPS-99. Fractals and applied synergetics* 10–11.
[14] Zaiser M 1999 *Acta Mater* **47(8)** 2463–2476
[15] Sagaradze V V 2013 *Hardening and properties of austenitic steels* (Yekaterinburg: UrO. RAS) p 720
[16] Uglov A L 2016 *Defectoscopy* **2** 3–10
[17] Avdushev A A 2007 *Bulletin of St. Petersburg State Electrotechnical University "LETI" SERIES "Instrument-building and information technologies"* **1** 31–36
[18] Muravyov V V 1996 *Sound Speed and Structure of Steels and Alloys* (Novosibirsk: Nauka) p 183
[19] Toozandehjani Meysam 2015 *Journal of Materials Science* **50(7)** 2643–2665
[20] Uglov A L 2015 *Physical Mesomechanics* **18(6)** 111–115.

Acknowledgments

The study was carried out under the grant of the Russian Science Foundation No. 19-19-00332 "Development of scientifically sound approaches and hardware and software for monitoring damage to structural materials based on artificial intelligence approaches to ensure the safe operation of technical facilities in arctic conditions".

IOP Conf. Series: Materials Science and Engineering **969** (2020) 012012 doi:10.1088/1757-899X/969/1/012012

Effect of Special Misorientations – Crystallographic Ordered Boundaries in Recrystallization in Aluminium Alloy of Al-Si-Mg System

P L Reznik[1], A A Redikultsev[1] and M L Lobanov[1, 2]

[1]Ural Federal University named after the first President of Russia B. N. Yeltsin, 19, Mira st., Ekaterinburg, 620002, Russia
[2]M. N. Mikheev Institute of Metal Physics of Ural Branch of the Russian Academy of Sciences, 18, S. Kovalevskaya st., Ekaterinburg, 620990, Russia

E-mail: p.l.reznik@urfu.ru

Abstract. The deformed and recrystallized states of surface regions of hot-rolled aluminium alloy plates of Al-Si-Mg system were investigated by the method of orientation microscopy (EBSD). In the deformed state mainly shear texture components were detected. In the recrystallized structure, there was a significant dispersion of deformation texture due to an increase in the proportion of components that were located at the edges of the deformation texture. It is shown that the crystallites, which are boundary a deformed matrix with the misorientation approximately corresponding to rotation axis <552> at an angle of 52.5°, become the nuclei of recrystallization.

1. Introduction

Interest in texture is, among other things, related to the manifestations of anisotropy of physical properties [1], as well as plasticity, hardness and tendency to resolution of textured material [2–7]. It is more important to note that texture in a deformed material has a noticeable influence on the orientation-dependent properties of the material even after subsequent processing (deformation treatment, heat treatment) [8]. The specified features mean that texture development accounting is a tool to optimize technologies of material production [9] or to create new technologies [10].

Targeted deformation influence on metal is accompanied by texture appearance [11, 12], and also due to dislocations sliding is accompanied by deformation of all its crystallites [13]. Based on the assumption that sliding systems possess maximum tangential stresses, deformation texture development models of Sachs [14], Taylor [15] and some others [16] have been designed.

Understanding texture transformation processes during recrystallization is an important and unresolved task. The inability to predict the texture of annealing, or the degree of its scattering due to insufficient knowledge of some microstructural components [16] of the deformed state does not allow to solve the problem of predictable changes in the characteristics of the final production product.

Modern experimental methods of studying the texture state of the material (based on EBSD [16–18]) allow tracking the movement of boundaries between elements of the mesostructure during recrystallization. Thus, the possibilities to describe boundary motion at the level of sliding and dislocation sliding are opened, which leads to explanation of recrystallization processes using evolution of special boundaries.

It is important to note the role of local texture. The local texture determines the structure of the boundary between two neighboring crystals, and also determines the direction of the recrystallization process (for example, macro texture of recrystallization). In this context, modern research of crystallographic aspects of nuclei "nucleation" at any recrystallization as well as correlation of deformation orientations with orientations of primary recrystallization (PR) and then PR orientations with orientations of either normal or abnormal growth come to the foreground [19–21].

Revealing of a role of special borders at recrystallization processes in a hot-rolled plate from aluminium alloy of system Al-Si-Mg is the purpose of the present work.

2. Materials and methods

Samples of an aluminium alloy AlMg1SiCu (1330) served as research materials. The following sample composition (wt %) was determined by atomic emission analysis: 1.0 Mg, 0.24 Fe, 0.62 Si, 0.14 Zn, 0.1 Mn, 0.19 Cu, 0.06 Cr. The sum of residual impurities is 0.15 wt %. Samples were taken from 23.6 mm thick plates that had been hot-rolled in two different modes [22].

An electron microscope (ZEISS CrossBeam AURIGA) was used to determine the orientation of individual grains and to analyze the local texture (20 kV; Oxford Instruments analysis system; the scanning step – 0.1 μm.). The estimated error limit of the crystal lattice orientation was on average ± 0.6°, but did not exceed ± 1°. Misorientations from 2 to 15° were marked as small angular boundaries between local volumes. High angles were detected when the misorientation was ≥ 15°.

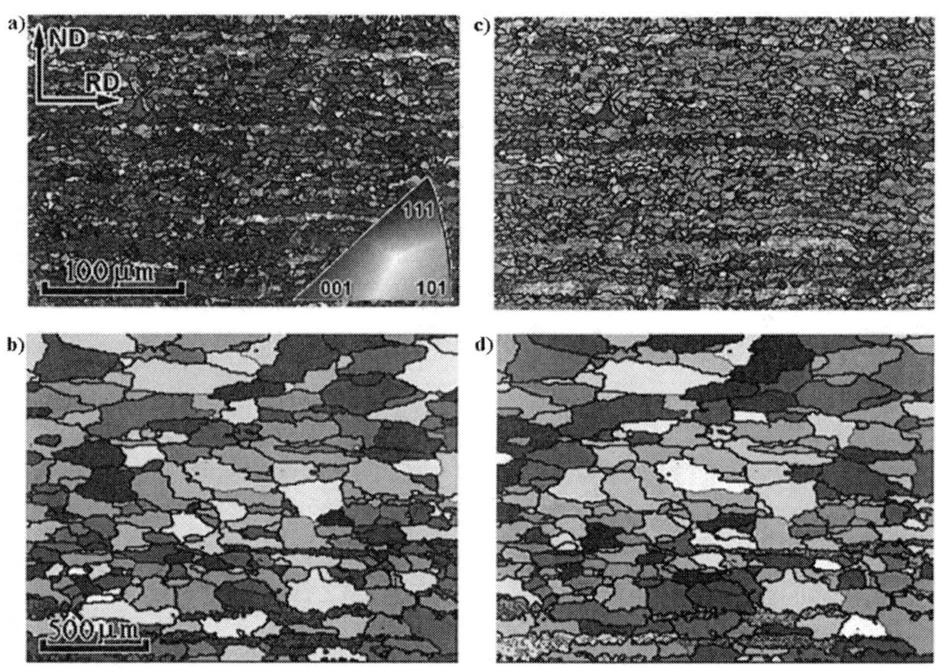

Figure 1. Orientation maps of the surface area aluminum alloy plates (EBSD). (a, b) ND ∥ Y; (c, d) RD ∥ X; (a, c) deformed state; (b, d) recrystallized state.

The orientation analysis was performed in laboratory coordinates. The axes of the coordinate system were connected with the normal to its plane (Y ∥ ND), with hot rolling directions (X ∥ RD), and perpendicular to their direction (Z ∥ TD).

3. Results and discussion

Based on the results of metallographic analysis of specimen structure by thickness depending on the hot rolling mode, it is important to note the following. Based on the results of metallographic analysis of specimen structure by thickness depending on the hot rolling mode, it is important to note the following.

The structure of the sample, which was obtained by deformation mode number 2, consisted of deformed grains stretched in RD (figure 1a, c). Such structure was typical for the entire thickness of the plate.

In the surface layers (up to 1/4 of thickness) of samples from sheets, which were obtained by the rolling regime number 1, the structure consisted mainly of recrystallized grains. The orientation of such grains along the rolled direction was less pronounced (figure 1b, d).

The texture of surface layers at a distance up to about ~ 1/4 of the plate thickness after rolling in deformation state consisted of the set of discrete dispersion orientations: strongly {001}<320>, {114}<110> and weaker {113}<110>, {112}<110>, {223}<110>, (figure 2c, d).

The texture of the surface layers in the recrystallized state consisted of the next set of discrete dispersion orientations: strong {012}<321>, {114}<110> and weaker {001}<110>, {113}<112>, {331}<123 >, (figure 2e, f).

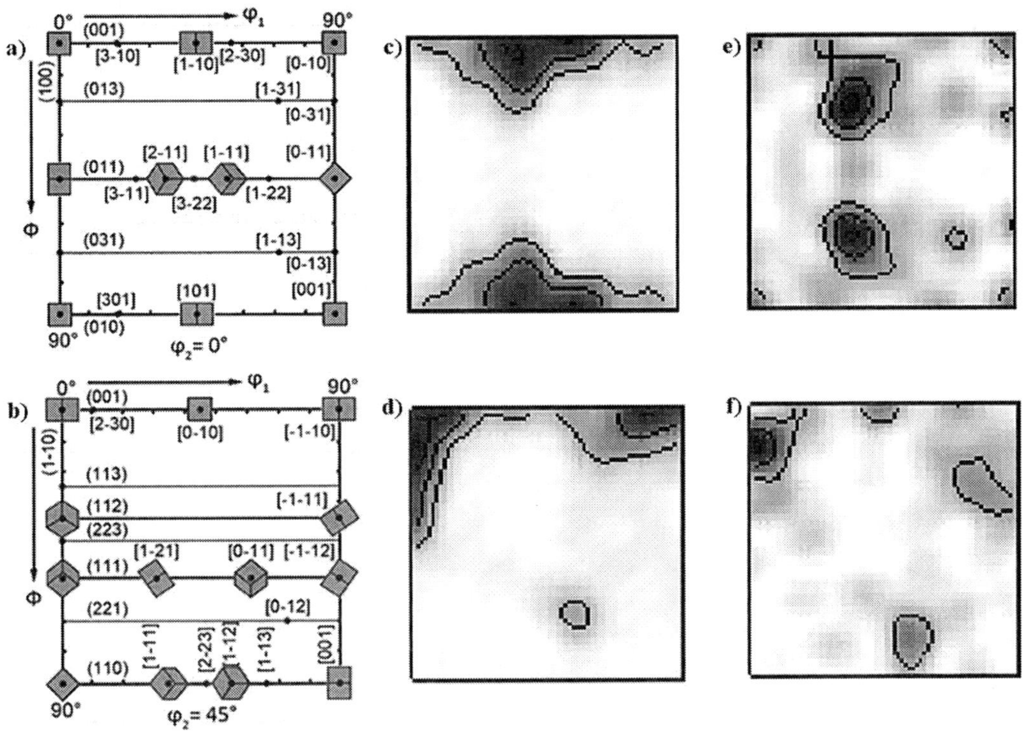

Figure 2. Orientation distribution function. The surface of aluminum alloy plates. (a, b) ideal orientations, view from TD; (a, c, e) – φ = 0° sections; (b, d, f) – φ = 45° sections; (c, d) deformed state; (e, f) recrystallized state.

One may state that in recrystallization caused the growth of grains with orientations that were contained in small quantities in the deformation texture and located at the edges of scattering areas of the main deformation orientations (figure 2). It was often observed that the "carriers" of recrystallization orientations were near-boundary crystallites located inside the deformed grains of one of the main deformation orientations (figure 3).

Vector analysis based on the results of orientation microscopy (EBSD) in the form of Euler angles showed that the grains growing during recrystallization are always in specific misorientation relative to one of the main deformation orientations. This misorientation is formed by turn around one of the crystallographic axes <552> at approximately 52–53°. It is an intermediate misorientation between CSL orientations Σ25b (51.68°, <331>) and Σ45c (53.13°, <221>). EBSD shows both CSL misorientations in the form of CSL boundaries, sometimes transiting into each other between deformed grains and between growing recrystallized and deformed grains as well as in [23]. At the same time, it was observed [23] that recrystallization nuclei were formed in the deformed grains split by boundaries close to CSL Σ3. For the deformation texture components, the CSL Σ3 misorientation is observed between the orientations located at the edges of the main dispersion deformation orientations.

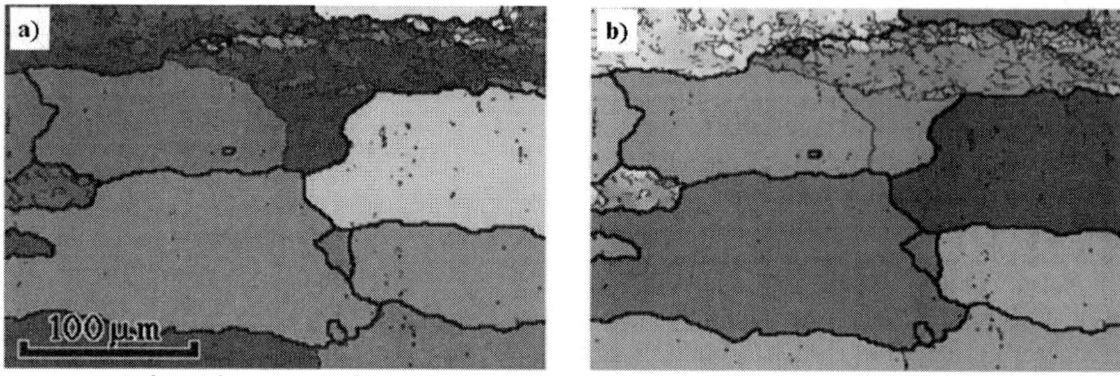

Figure 3. Orientation maps of local recrystallization (EBSD). (*a*) from ND ǁ Y; (*b*) from RD ǁ X.

4. Summary

Thus, the results of this article confirm the results obtained in [23] that the nuclei of recrystallization in an aluminum alloy are crystallites which are with a deformed matrix a high-angle crystallographically ordered border, approximately corresponding to the misorientation: rotation axis <552> at an angle of 52.5°

References

[1] Wassermann G and Grewen J 1962 *Texturen metallischer Werkstoffe* (Berlin: Springer) p 808

[2] Hirsch J 2001 *Proc. Int. Conf. on Advances in the metallurgy of aluminum alloys (Indianapolis)* (Novelty: ASM International) p 431

[3] Engler O, Schafer C and Brinkman H-J 2012 Crystal-plasticity simulation of the correlation of microtexture and roping in AA 6xxx Al–Mg–Si sheet alloys for automotive applications *Acta Materialia* **60** 5217–5232

[4] Mishra S, Kulkarni K and Gurao N P 2015 Effect of crystallographic texture on precipitation induced anisotropy in an Aluminium magnesium silicon alloy *Materials and Design* **87** 507–519

[5] Loginov Yu N, Golovnin M A, Lobanov M L and Doroshenko N M 2016 Investigation of the influence of the technology for the production of flat products from the Al-Mg-Si system alloy on the anisotropy of properties *Light alloy technology* **3** 69–74

[6] Li L, Flores-Johnson E A, Shen L and Proust G 2016 Effects of heat treatment and strain rate on the microstructure and mechanical properties of 6061 Al alloy *International Journal of Damage Mechanics* **25** 26–41

[7] Shen Y, Garnier J, Allais L, Crepin J, Ancelet O and Hiver J-M 2011 Experimental and numerical characterization of anisotropic damage evolution of forged Al6061 T6 alloy *Procedia Engineering* **10** 3429–3434

[8] Rusakov G M, Lobanov M L, Redikul'tsev A A and Belyaevskikh A S 2014 Special misorientations and textural heredity in the commercial alloy Fe-3% Si *Physics of Metals and Metallography* **115** 775–785

[9] Belyaevskikh A S, Lobanov M L, Rusakov G M and Redikul'tsev A A 2015 Improving the production of superthin anisotropic electrical steel *Steel in Translation* **45** 982–986

[10] Redikultsev A A, Rusakov G M and Lobanov M L 2018 The possibility of obtaining electrical steel with texture {100}<001> *Solid State Phenomena* **284** 483–488

[11] Vishnyakov Ya D, Babareko A A, Vladimirov S A and Egiz I V 1979 *The theory of texture formation in materials and alloys* (Moscow: Nauka) p 343

[12] Shtremel' M A 1997 *Strength of alloys. Part II. Deformation* (Moscow: MISiS) p 527
[13] Schmid E 1924 *Proc. Int. Congress on Applied Mechanic* (Delft: J. Waltman, jr.) p 353
[14] Sachs G 1928 Zur Ableitung einer Fliessbedingung *Zeitschrift des Vereines Deutscher Ingenieure* **72** 734–736
[15] Taylor G I 1938 Plastic Strain in Metals *Journal of the Institute of Metals* **62** 307–324
[16] Rollett A, Humphreys F, Rohrer G S and Hatherly M 2004 *Recrystallization and Related Annealing Phenomena: Second Edition* (Oxford: Elsevier) p 574
[17] Maitland T and Sitzman S 2007 *Electron Backscatter Diffraction (EBSD) Technique and Materials Characterization Examples* (Berlin: Springer) p 522
[18] Lobanov M L, Yurovskikh A S, Kardonina N I and Rusakov G M 2014 *Methods of investigations material's texture. Work-book* (Ekaterinburg: Ural University Publishers) p 115
[19] Lobanov M L, Danilov S V, Pastukhov V I, Averin S A, Khrunyk Y Y and Popov A A 2016 The crystallographic relationship of molybdenum textures after hot rolling and recrystallization *Mater. Des.* **109** 251–255
[20] Lobanov M L, Redikul'tsev A A, Rusakov G M and Danilov S V 2015 Interrelation between the orientations of deformation and recrystallization in hot rolling of anisotropic electrical steel *Met. Sci. Heat Treat.* **57** 492–497
[21] Redikul'tsev A A, Lobanov M L, Rusakov G M and Lobanova L V 2013 Secondary recrystallization in Fe – 3 % Si alloy with (110)[001] single-component texture *Phys. Met. Metallogr.* **114** 33–40
[22] Lobanov M L, Loginov Yu N, Danilov S V, Golovin M A and Karabanalov M S 2018 Effect of hot rolling rate on the structure and texture condition of plates of the Al – Si – Mg alloy system *Met. Sci. Heat Treat.* **60** 32–38
[23] Reznik P, Zorina M and Lobanov M 2019 Role of crystallographic misorientations in the evolution of texture in fcc metals *Mater. Today: Proc.* **19** 187–189

Acknowledgments
Express our acknowledgements to the program № 211 of the Government of the Russian Federation № 02.A03.21.0006, which is aimed at supporting the leading universities of the Russian Federation.

Choosing the Rational Heat Treatment Conditions for High-Strength Cold-Resistant Weldable Steel with Yield Strength of more than 600 N/mm^2

P P Poletskov, A S Kuznetsova and O A Nikitenko

Department of Material Processing Technologies, Nosov Magnitogorsk State Technical University, 38, Lenin Avenue, Magnitogorsk, 455000, Russia

E-mail: allakuznetsova.mgtu@mail.ru

Abstract. Rational heat treatment conditions for the new weldable cold-resistant steel grade 20G2SMRA ensuring a guaranteed yield strength $\sigma_{0,2} \geq 600$ N/mm2 combined with the lowtemperature impact strength KCV-60 $\geq$ 50 J/cm2 have been searched. The effect of various quenching and tempering conditions on the steel microstructure parameters, hardness, and impact strength at a temperature of -60 °C has been studied. The dependencies of the microstructure parameters, hardness (HV1), and impact strength (KCV-60) on the quenching and subsequent tempering temperature are represented. Rational heat treatment conditions have been found for a new weldable cold-resistant steel with high yield strength.

1. Introduction

The need for high-strength cold-resistant weldable steels to retrofit the mining and construction equipment has increased with the development of hard-to-reach northern and Arctic regions. These industries experience high demands on reducing metal consumption and manufacturing costs, as well as increasing load-bearing capacity, reliability, and service life. Correspondingly, complex requirements are applied to the mechanical characteristics of materials used for the manufacture of welded supporting system elements, i.e. high strength ($\sigma_{0,2} \geq 600$ N/mm^2, $\sigma_{в} = 700$–1100 N/mm^2) and weldability (Ceqv $\leq$ 0.53), while maintaining allowable ductility and toughness at low temperatures (KCV$^{-60} \geq 50$ J/cm^2) [1–4].

According to the lean alloying concept, the steel quality and the highest chemical composition properties are determined by the heat treatment conditions [5–10]. In this case, choosing the processing scheme and parameters is of great importance.

According to [11–14], in the world and domestic practice, thermally improved high-strength weldable thick steel plates with a high yield strength are mainly supplied, capable of ensuring a yield strength of over 600 N/mm^2 combined with low-temperature toughness.

The main advantage of using quenching technology with high tempering relative to the direct heat treatment, in particular, for the development of new steel products with a thickness within 8 ... 50 mm is the removal of the size assortment restrictions for thicknesses above 20 mm at a minimum steel alloying degree [11].

IOP Conf. Series: Materials Science and Engineering **969** (2020) 012013 doi:10.1088/1757-899X/969/1/012013

In this case, the best combination of strength and toughness may be ensured if the steel hardness after quenching exceeds the minimum permissible values, which, in turn, is possible if quenched steel contains at least 90 % martensite [12, 15–17]. Herewith, the choice of tempering temperature should ensure softening required to achieve a satisfactory strain capacity and machinability of the steel supplied [18].

Thus, the work objective is studying the effect of heat treatment conditions (quenching and tempering) on the structure and properties of new high-strength steel with elevated cold resistance.

2. The Research Material and Technique

The research material was specimens of the newly developed steel grade 20G2SMRA with a below chemical composition (Table 1) [19].

Table 1. Chemical Composition of the New Steel Grade 20G2SMRA.

	Chemical Composition, % wt., no more										
Steel Grade	C	Si	Mn	S	P	Cr+Cu	Ni+Mo	Al	Ti	B	Ceqv
20G2SMRA	0.200	0.550	1.60	0.005	0.012	0.16	0.35	0.050	0.005	0.0050	0.53

The manufacture of experimental flat steel specimens and their subsequent heat treatment was performed at the scientific and production complex of LLC EC Thermodeform-MSTU and the CUC RI NanoSteel under FSBEI HE Nosov Magnitogorsk State Technical University (Magnitogorsk). The quenching and tempering temperature varied within the below ranges:

- quenching temperature gradually increased from Ac_3+30 °C to Tmax = 1000 °C in increments of 50 °C (with a holding of 2.5 min/mm at a quenching temperature);
- tempering temperature increased from 200 to 600 °C in increments of 100 °C (with a tempering duration of 2.5 min/mm + 25 min).

After quenching and tempering, the specimens were cooled by water and air, respectively.

According to the results of tests for micro-hardness and impact bending, for each steel, a heat treatment option has been determined, which provides the best combination of impact strength and the required level of hardness.

Tests for impact bending were performed using an impact tester RKP 450 according to GOST 9454 on transverse specimens 10.0 x 5.0 mm in size with a V-shaped cut. Test specimens were cooled to - 40 °C in a LAUDA master PL1 refrigerator-thermostat.

The Vickers hardness was measured according to GOST 9450-60 by indenting a diamond pyramid with a 136° angle between the opposite faces using a Buehler Micromet Hardness Tester at a load of 1 kg.

Critical points were determined by differential scanning calorimetry (DSC) using an STA Jupiter 449 F3 (NETSCH) simultaneous thermal analyzer at a heating rate of 10 °C/min to 1000 °C in argon of 99.998 % purity, as well as complex dilatometric analysis using the Gleeble 3500 simulator [9, 10].

To study the microstructure, the specimens were ground polished and etched in a 4% nital by immersion of the polished surface in a bath with the reagent.

Qualitative and quantitative analysis was performed using an Axio Observer (Zeiss) optical microscope with a 50-1000-fold increase using the Thixomet PRO metallographic image processing software and a scanning electron microscope (SEM) - JSM 6490 LV.

The volume fraction of phases was determined by separating the fraction of the relevant structural components using the Thixomet PRO software with built-in techniques adapted according to GOST 8233-56 [20]. The data obtained were statistically processed in the Microsoft Office Excel environment.

The steel grain size was determined according to GOST 5639. The average austenite grain size was determined by linear measurements using software with built-in techniques [20], adapted according to GOST 8233 and GOST 5639. The former austenite grain boundaries were determined by etching specimens in a freshly prepared picric acid solution saturated at room temperature with the addition of (1–10) % surfactants while heating the reagent to (50–70) °C [21].

3. The Study Results

Based on the DSC results and dilatometric analysis, the critical points have been determined for the new steel grade 20G2SMRA (Table 2) [20].

Table 2. The Critical Steel Temperature Measuring Results.

Steel Grade	Research Technique	Critical Temperature t, °C	
		A_{c1}	A_{c3}
20G2SMRA	DSC	725	814
	Dilatometer	730	820

The Results of the Quenching Temperature Effect on the Steel Structure and Properties

Using optical microscopy, it has been found that after quenching from 850 to 1000 °C in all cases, the steel grade 20G2SMPA studied has a similar morphological structure of packet (lath) martensite formed according to the diffusion-free shearing mechanism by transverse sliding (Figure 1, *a*, *b*).

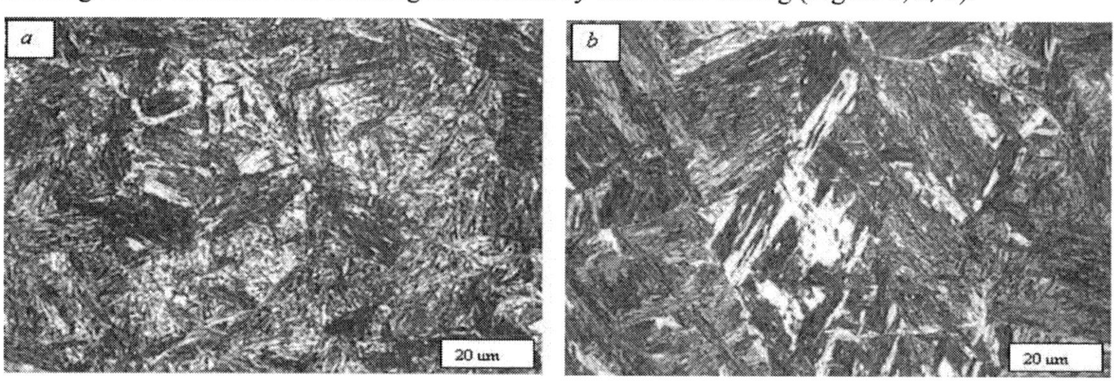

Figure 1. Steel Microstructure after Quenching from: 850 °C (a), 1000 °C (b), x 1000.

Analysis performed using SEM (Figure 2, *a*, *b*) has shown that in the 20G2SMRA steel α-phase structure, the predominant morphological component is lath martensite (Ml), the volume fraction of which exceeds 70%. The area of martensitic packets increases from 20 to 500 μm^2, depending on the quenching temperature. The width of the individual laths making up the packet varies within 150 to 800 nm. The total number of laths in the packet varies from 4–6 to 12–16 pieces.

Also, the structure includes individual plates, the transverse size of which is several times larger than the width of individual laths of martensite packets, and reaches 1–3 μm. The plate length is about 30 μm. This morphology can be identified as plate martensite (Mp), the presence of which causes a decrease in the stress amplitude and, as a result, a decrease in the risk of quenching cracks.

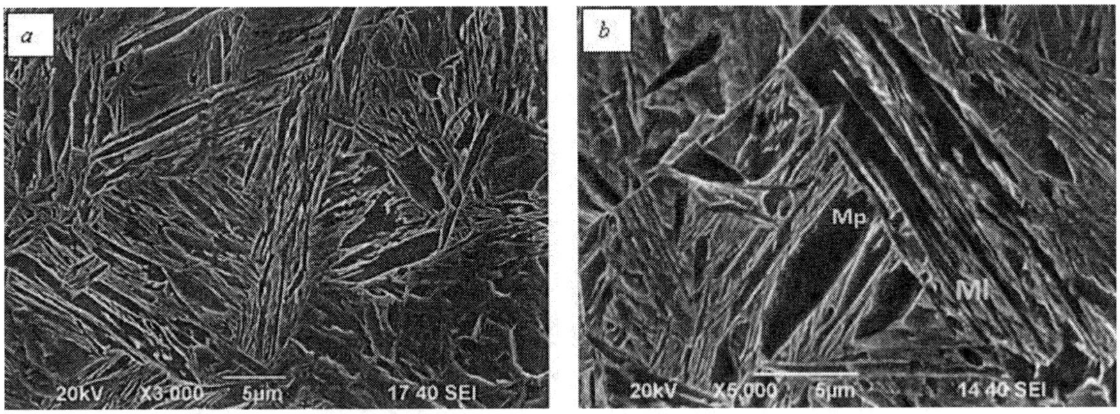

Figure 2. Steel Microstructure after Quenching from 930 °C, SEM.

The results of studying the dependence of the steel microstructure parameters (actual size of austenite grain) on the quenching temperature and its relationship with mechanical properties are shown in Figure 3.

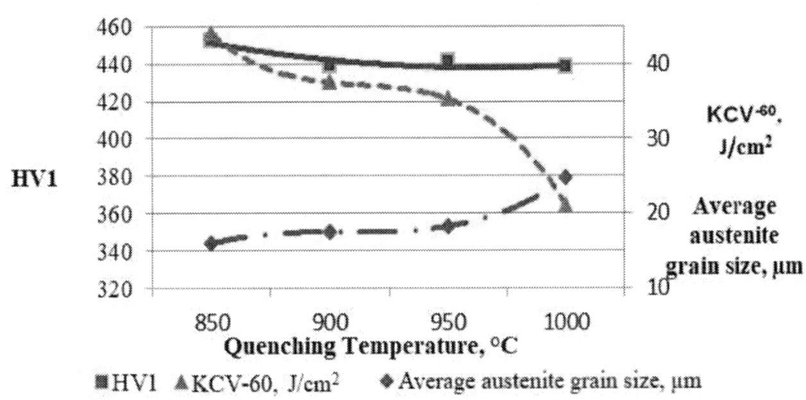

Figure 3. Dependence of the Austenite Grain Size, Impact Strength KCV^{-60}, and Hardness HV1 on the Quenching Temperature.

The data obtained (Figure 3) has shown that the smallest grain size is within 16-18 µm (grain No. 9). The increase in structural parameters observed indicates an increase in the actual austenite grain size (grain No. varies from 8 to 5) caused by an increase in the austenitization temperature, which leads to a decrease in the impact strength of steel within 850–1000 °C.

Within the entire temperature range under study, the hardness HV1 has similar values due to the constant chemical composition of the initial austenite in hypereutectoid steels, which changes when temperature increases above the AC_3 point [22].

According to the experimental study results, the maximum impact strength KCV^{-60} combined with satisfactory hardness HV_1 for steel grade 20G2SMPA is achieved at a quenching temperature of ~ 850 °C.

The Results of the Tempering Temperature Effect on the Steel Structure and Properties

During the study, it was found (Figure 4) that within the tempering temperature from 200 to 500 °C, there are no qualitative changes in the martensite morphology, and the lath α phase structure is preserved up to 500 °C. However, an ever-stronger distortion of the lath boundary shape and structure is observed with an increase in the tempering temperature caused by the formation of a fragmented substructure due to the relaxation of the defective structure formed as a result of the $\gamma \rightarrow \alpha$ transformation and rearrangement of the dislocation structure.

Also, the process is accompanied by the release of the second phase particles; the initial stage of the most intense decomposition of the supersaturated solid solution is fixed at 300 °C, which is accompanied by an increase in the ductile-brittle transition temperature. At a tempering temperature of 400 °C and more, coagulation and spheroidization of carbide particles are observed, contributing to a change in the α phase morphology in the sequence '*tempered martensite* → *tempered troostite* → *tempered troostosorbite*'. At a tempering temperature of 600 °C, *tempered sorbite* becomes the basic structural component.

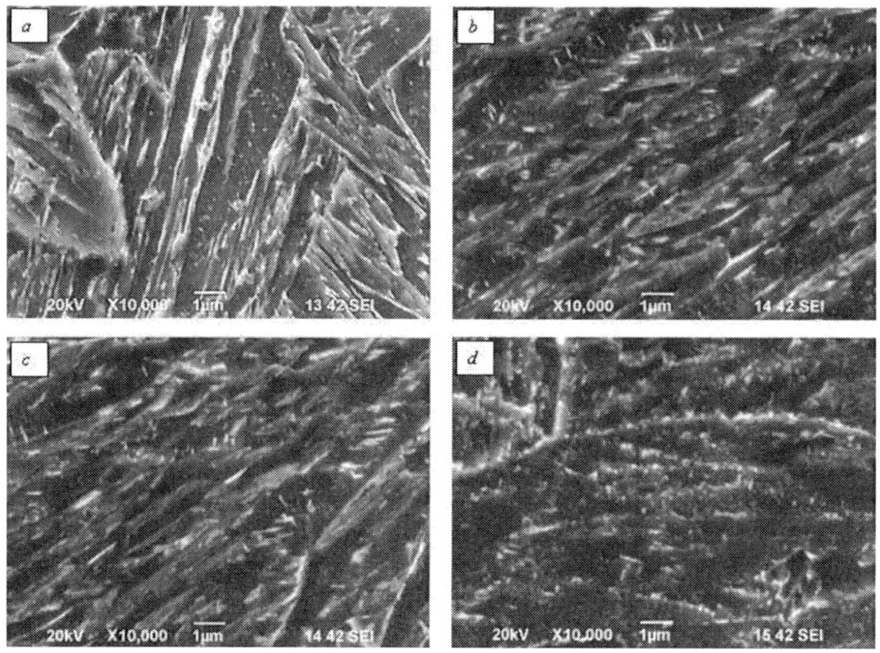

Figure 4. Steel Microstructure after Quenching from 850 °C and Subsequent Tempering at: 300 (a), 400 (b), 500 (c), 600 (d) °C. SEM.

The setup tempering temperature was determined based on its effect on structural changes and, as a result, the hardness HV_1 and impact strength KCV^{-60} (Figure 5).

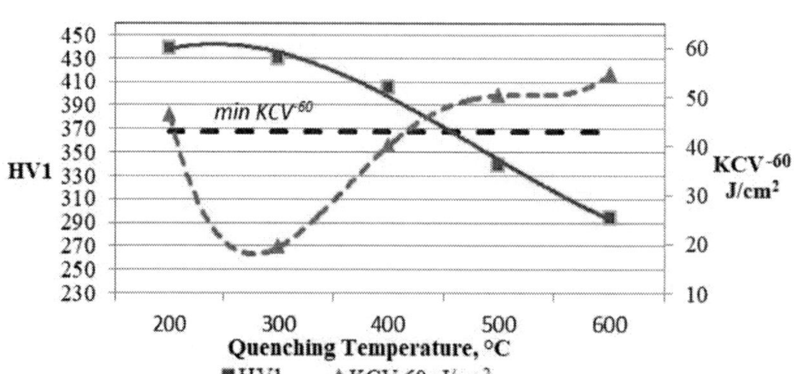

Figure 5. Effect of Tempering Temperature on the Distribution of Hardness HV_1 and Impact Strength KCV^{-60} in Steel Grade 20G2SMRA.

Figure 5 shows that against the background of a monotonous decrease in hardness (HV_1) within 200 to 400 °C, the impact strength (KCV^{-60}) decreases relative to the quenched state, which indicates the development of irreversible temper brittleness. When tempering within 500–600 °C, a drop in impact strength is not observed due to the Mo presence in the steel composition.

A change in the tempering temperature within 400–600 °C leads to a noticeable decrease in the microhardness magnitude by about 1.5 times due to a decrease in the internal shear stresses in lath martensite. Herewith, with a temperature increase, the cold brittleness threshold of steel grade 20G2SMRA is significantly reduced.

Thus, based on the dependencies obtained (Figure 3, 5), it has been found that the processing conditions of quenching from 850 °C followed by high tempering at 600 °C meet the requirements set for new steel grade 20G2SMPA.

4. Conclusion

As a result of studying the new weldable cold-resistant 20G2SMRA steel grade, dependencies of the microstructure parameters, hardness (HV_1), and impact strength (KCV^{-60}) on the quenching and subsequent tempering temperature have been obtained.

It has been established that the best heat treatment conditions to strengthen flat products made of new high-strength steel grade 20G2SMRA are as follows:

- quenching from 850 ± 20 °C for 25 minutes, at which a mixed structure is formed consisting of 70 % lath martensite and about 30 % high-temperature plate martensite, which ensures
- maximum impact strength KCV^{-60}= 40 J/cm² at satisfactory hardness HV_1 452;
- subsequent high tempering at 600 °C for 40 min, at which a tempered sorbitol structure is formed, which ensures guaranteed achievement of the required hardness HV_1 291 combined with impact strength KCV^{-60} = 61 J/cm².

Ensuring such a favorable combination of strength and toughness at a test temperature of up to -60 °C allows using the new steel grade 20G2SMRA in structures operating in the Far North.

Reference

[1] Khlusova E I and Sych O V 2018 Creating Cold-Resistant Structural Materials for the Arctic. History, Experience, Current State *Innovations* **11(241)** 85–92

[2] Stolyarov V I, Nikitin V N, Efron L I and Lazko V G 1993 State and Prospects of Development of Technology and Composition of High-Strength Weldable Steels with a Yield Strength of 700 N/mm² *Steel* **6** 61–67

[3] Chukin M V, Poletskov P P and Nabatchikov D G 2017 Analysis of Technical Requirements for Ultra-Cold-Resistant Flat Products *Bulletin of the South Ural State University. Series: Metallurgy* **17(2)** 52–60

[4] Chukin M V, Poletskov P P and Gushchina M S 2019 Development of Import-Substituting Technology to Manufacture Flat Products from High-Strength Structural Steel of the Northern Version *Manufacture of Flat Products* **4** 5–11

[5] Solntsev Y P and Titova T I 2002 *Steel for the North and Siberia* (St. Petersburg: Khimizdat) p 352

[6] Solntsev Yu P 2005 *Cold Resistant Steels, and Alloys* (St. Petersburg: Khimizdat) p 480

[7] Poletskov P P, Gushchina M S and Berezhnaya G A 2015 Studying the Effect of Heat Treatment on the Mechanical Properties of High-Strength Flat Products *Bulletin of the Nosov Magnitogorsk State Technical University* **4(52)** 88–92

[8] Salganik V M, Poletskov P P and Gushchina M S 2015 Specifics of Obtaining Nanostructured High-Strength Flat Products *Bulletin of the Sukhoy Gomel State Technical University* **1(60)** 27–30

[9] Poletskov P P, Denisov S V and Nikitenko O A 2019 Studying the Decomposition of Supercooled Austenite in Low-Carbon Pipe Steel Using the Gleeble 3500 Complex *News of Higher Educational Institutions. Ferrous Metallurgy* **62(3)** 235–240

[10] Poletskov P P, Nikitenko O A, Kuznetsova A S and Efimova Y Y 2019 Investigation of the Effect of Nickel Content on the Structural and Phase Transformation and Properties of High-Strength Cold-Resistant Complex-Alloyed Steel *Journal of Chemical Technology and Metallurgy* **54(6)** 1291–1297
[11] Nikitin V N, Nastich S Yu and Smirnov L A 2016 High-Strength Steels with Lean Alloying for Quarrying and Mining Equipment *Steel* **10** 57–66
[12] Pontremoli M, Weber L, Dilg K, Schwinn F, Knauf G, Lippe M, Erhardt B, Finger M 2006 HighStrength Steels for Thick Sheets, Pipes, and Profiles *Ferrous Metals* **4** 58–66
[13] Schröter F 2003 Höherfeste Stähle für den Stahlbau – Auswahl und Anwendung *Bauingenieur* **78(9)** 426–432
[14] Kern A and Schriever U 2005 Niobium in Quenched and Tempered HSLA-Steels Recent advances of niobium containing materials in Europe *Proceedings of the symposium of the 30th anniversary of Niobium Products Company* 109–119
[15] Yeh Y 1979 *Heat Treatment of Steel* (Moscow: Metallurgy) p 264
[16] Juna H J, Kanga J S, Seob D H, Kangb K B and Park C G 2006 Effects of Deformation and Boron on Microstructure and Continuous Cooling Transformation in Low Carbon HSLA Steels *Materials Science and Engineering A* **422** 157–162
[17] Luxenburger G, Bockelmann M and Wolf P 2004 High Strength Quenched and Tempered (Q T) Steels for Pressure Vessels *International Journal of Pressure Vessels and Piping* **81(2)** 159–171
[18] Meskin V S 1964 *Steel Alloying Fundamentals* (Moscow: Metallurgy) p 684
[19] Poletskov P P, Nikitenko O A, Kuznetsova A S and Salganik V M 2020 Studying the Kinetics of Phase Transformations during the Decomposition of Supercooled Austenite of New High-Strength Steel of High Cold Resistance *CIS Iron and Steel Review* **19** 56–59
[20] Koptseva N V, Chukin M V and Nikitenko O A 2012 Use of the Thixomet PRO software for Quantitative Analysis of the Ultrafine-Grain Structure of Low-and Medium-Carbon Steels Subjected to Equal Channel Angular Pressing *Metal Science and Heat Treatment* **54(7–8)** 387–392
[21] Poletskov P P, Gushchina M S and Alekseev D Yu 2018 Studying the Effect of Controlled Pipe Steel Rolling on the Structural State of Hot-Deformed Austenite *Bulletin of the Nosov Magnitogorsk State Technical University* **16**(3) 67–77
[22] Sadovsky V D and Fokina E A 1986 *Residual Austenite in Quenched Steel* (Moscow: Nauka) p 113

Acknowledgments
The research was executed under financial support of the RF Ministry of Education and Science within the frame- work of realization of RF President's grant (Agreement No. 075-15-2020-205 dated 17.03.2020 (int. No. MK- 1979.2020.8)).

IOP Conf. Series: Materials Science and Engineering **969** (2020) 012014 doi:10.1088/1757-899X/969/1/012014

Microalloyed Steels for Car Parts

G F Mukhametzyanova[1], T V Shveyova[2] and V I Astaschenko[1]

[1]Department of materials, technologies and quality, Kazan Federal University, 10A, Syuyumbike Avenue, Naberezhnye Chelny, 423812, Russia
[2]50/15, Mira Avenue, Naberezhnye Chelny, 423819, Russia

E-mail: gulnara-ineka@mail.ru

Abstract. The composition and properties of three constructional microalloyed steel with vanadium, niobium and nitrogen is investigated. High strength and ductility steel in rolling mill products and in forgings were revealed, manufactured according to various flow diagram. The resistance of microalloyed steels to grain growth during heating to 1300 °C has been established. The opportunity of realization of the forging heat forgings in their heat treatment is shown. Output testing was carried out of steels in the manufacturing crankshaft and engine connecting rods. For heavily loaded machine components, the use of 38G2AF, AC40H2NMAF and AC30X3NMAFB steels is recommended. High resistance to grain growth is shown when heating steels microalloyed with vanadium, niobium and nitrogen. High properties of the steel AC40H2NMAF, AC30X3NMAFB and 38G2AF make it possible to recommend them for the manufacture of heavily loaded parts - crankshafts and connecting rods. The possibility of realizing forging heat of forgings made microalloyed steels with vanadium, niobium and nitrogen during their heat treatment to ensure a high level properties and structural requirements.

1. Introduction

Steel has been and remains the main structural material used in mechanical engineering, machine tool industry and other industries. Not only its technological and mechanical properties depend on the structural phase state of steel, but also the reliability and durability products in operation. Many indicators steel quality, including its structure, are laid at the stages of metallurgical production. A powerful factor affecting the alloy purity, changes in crystallization conditions, grain structure, homogeneity of the structure, and other property indicators are metallurgical technologies such as deoxidation, refining, modification, and alloying [1–5]. Thanks to alloying, it is possible to control the structure and steel properties in a very wide range. The degree of alloying is always determined by the level of the created properties and the functional belonging of the steel. For heavily loaded case-hardened products, a mandatory requirement is the hardening of the deep layers of the part, combined while maintaining of high plastic properties in these zones. In addition, such details are required to ensure minimal deformation and skellering, as well as the elimination cracks during their thermochemical treatment, which forces the use oil as a quenching medium [6–8]. The solution to these two problems in the complex is achieved by using alloy steels for the manufacture of such parts. The larger the cross section of the hardened product, the greater should be the degree of alloyed steel. At the same time, alloying should still be limited by the level at which the required hardenability and granularity are achieved in the adopted quenching technology. Excessive alloying, that is, exceeding that necessary to achieve the required hardenability, is irrational. This is due to the fact that the addition of all (except nickel) alloying elements leads to a decrease in the reliability of steel under overload conditions and an increase in the cold brittleness threshold. As long as alloying increases hardenability in a given product, it helps to increase reliability, but if the content of alloying elements exceeds that required for through hardening in this section, the reliability of steel decreases [5, 9–11]. Alloying with several elements (complex alloying) more effectively increases hardenability than alloying with one element. It should also take into account the effect of steel alloying on the process diffusion saturation the surface of parts during thermochemical treatment, the formation of carbides and the amount of residual austenite in the hardened layer [12–14]. In selecting the steel, it is necessary to pay attention to the economic side of the issue: the more alloying elements it contains, the higher the cost [6–7].

Thus, it is indisputable that in most steel grades there is an irrational use of alloying elements, since only a certain amount of them is used for effective alloying of the die and the necessary formation compounds in the cross section products of a certain size [15–25]. In this case, the actual direction is the use of scientifically based sparingly alloyed and microalloyed steels for products of various functional purposes.

2. Working purpose

The study of the properties and experience of using microalloyed and sparingly alloyed steels in the manufacture of automobile parts.

3. Materials and research methods

Microalloyed steel vanadium, niobium and nitrogen were used in the work. Preference is given to steel AC40H2NMAF, AC30X3NMAFB, 38G2AF and sparingly alloyed steel 18HGR, in which the present chemical elements contribute to the hardening of alloys due to quench age hardening. Such steels have a reduced tendency to cracking and these elements can be introduced into steel with a wide range of carbon contents [17, 20]. Sample capture for the manufacture samples for mechanical and technological tests and for determining the chemical composition of steel was performed in accordance with GOST 7564 and GOST 7565. The content of chemical elements in steels was determined according to GOST 18895-97 on Spektrolab and AFS–51 spectrographs with specialized software SBP and Next and analyzers AN–7529 and AN–7560. Metallographic studies were performed using Neofot-21 and IM–7200 microscopes with a panoramic image analysis system and the Thixomet–PRO software product. The steel contamination with non-metallic inclusions was assessed according to GOST 1778, grain size was determined by oxidation (steel AC40H2NMAF, AC30X3NMAFB and 38G2AF) and cementation method (steel 18HGR and 20H2N4A) according to GOST 5639-82. Hardenability of steels was determined by end quenching standard samples according to GOST 5657-69. The microhardness was determined on a Durimet instrument at a load of 0.1 N (100 gf) with averaging the results measurements not less than 15 prints of the Vickers pyramid. Heat treatment of the samples was carried out in laboratory conditions using SNOL 1.6–2.5–1/11-V2 and SNOL 1.6–2.0–0.8/9–M1 heating furnaces, and gas furnaces were used in the production heating company "Holcroft" and units design "Teploproekt". To determine the hardness used devices TSH–2M and TK–2M.

Tensile and toughness tests were carried out on a ZD-20 tensile testing machine and a PSW-30 pendulum head in accordance with GOST 1497 and GOST 9454. Heat treatment of billets from steel AC40H2NMAF and AC30X3NMAFB was carried out according to the regime: quenching from a temperature of 930°C to oil and tempering at 630 °C; for steel 18HGR – quenching from 860 °C into water and tempering at 200 °C with air cooling.

Fatigue strength of crankshafts was estimated by loading the cheeks with variable bending moment, and torsion by loading the connecting rod neck with variable torque. The connecting rods were tested according to an alternating loading cycle with reproducing hydrodynamics in the sliding bearing of the crank head with a tensile stress of 4 to 8 and a double amplitude of 14 to 20. Machining by cutting was evaluated in laboratory conditions and in production on automatic lines for the mechanical processing of crankshaft blanks, connecting rods and gears of the engine and bridges of the car.

IOP Conf. Series: Materials Science and Engineering **969** (2020) 012014 doi:10.1088/1757-899X/969/1/012014

4. Working results and their discussion

The content of chemical elements in the studied steels is shown in table 1. Steel AC40H2NMAF smelted in an open-hearth furnace, treated with synthetic slag and rolled into a square of 154 ± 2 mm at the Serov Metallurgical Plant; steel AC30X3NMAFB – melted and rolled to a circle of 70 mm at Izhstal Production Association; 18XГР steel – smelted at the Oskol Electrometallurgical Plant (OEMK) by direct reduction from metallized pellets and rolled into a circle of 90 mm from a continuously cast billet with a section of 360 x 300 mm; 38G2AF steel – smelted and rolled into a circle of 150 mm at OEMK.

Table 1. The chemical composition of steels.

Steel	The content of chemical elements, %.											
	C	Si	Cr	Ni	Mn	Mo	V	P	S	Nb	B	N_2
AC40H2NMAF	0.37	0.20	2.12	0.72	0.45	0.11	0.11	0.025	0.007	-	-	-
AC30X3NMAFB	0.31	0.24	2.54	0.74	0.40	0.20	0.07	0.021	0.011	0.03	-	-
18HGR	0.18	0.28	1.22	0.26	1.21	0.09	-	0.018	0.027	0.002	0.0022	0.010
38G2AF	0.37	0.62	0.13	0.05	1.50	0.01	0.19	0.009	0.053	-	-	0.012

Steel AC30X3NMAFB and 38G2AF are used for the manufacture connecting rods of an internal combustion engine, steel AC40H2NMAF – for crankshafts, and steel 118HGR – toothed parts of the engine and gearbox of the KAMAZ vehicle. In the delivery state, the contamination of steel AC40H2NMAF with oxides and sulfides does not exceed 2.5 points, and steel AC30X3NMAFB with oxides no more than 3 points, sulfides 2 points and nitrides 1 point. The mechanical properties of heat-treated steel in accordance with the recommendations of normative and technical documentation is at a high level and higher than that of commercially used steels 42HMFA and 40HN2MA for crankshafts and engine rods (Table 2).

Table 2. The mechanical properties of steels.

Steel	Property metrics					
	σ_B, MPa	σ_T, MPa	δ, %	ψ, %	KCU, J/cm^2	HB
AC40H2NMAF (connecting rods)	1210	1080	12.5	56	110	302
GOST 4543 40HN2MA (connecting rods)	≥ 980	≥ 835	≥ 12.0	≥ 55	≥ 98	245-285
AC30X3NMAFB (crankshaft)	1110	1030	12.5	54	88	321
Specification 14-1-5520 42HMFA (crankshaft)	≥ 850	≥ 730	≥ 12.0	≥ 42	≥ 80	255-277

Note: Heat treatment of billets with a cross section of 25x25 for steels used for connecting rods - quenching from 850 ° C to oil, tempering at 620 °C, cooling to oil; for steels used for crankshafts - quenching from 860 °C to oil, tempering at 580 °C, cooling to oil.

The manufacturability of steels with nitride-vanadium hardening was evaluated at the stages of forgings manufacturing (induction heating up to 1260 ± 20 °C, hot forging and heat treatment) and during the mechanical processing forgings at various cutting operations (turning, drilling, grinding). In the process of manufacturing forgings, steel showed its adaptability and after heat treatment, including quenching from 930 °C to oil and tempering at 630 °C, high strength and plastic properties were achieved. It should be noted that the heating temperature for hardening of forgings from steel AC30X3NMAFB and AC40H2NMAF is somewhat higher than that of steels 42HMFA and 40HN2MA – 930 °C instead of 860 °C, which is explained by the mechanism of their nitride-vanadium hardening.

When machining 1000 pcs. connecting rods with a hardness in the range of 255–302HB, the high technological effectiveness of the steel used in all cutting operations is established. The test results of connecting rods made of steel AC30X3NMAFB after various types of heat treatment (Table 3) showed that the fatigue limit of all experimental and industrial parts was 6 tons compression with a double cycle amplitude of 24 tons, which is 50% higher than the similar performance parts made from 40HN2MA steels. A slight decrease in hardness and the presence of ferrite (less than 10%) in the steel structure is observed in the part, the heat treatment of which included the implementation of forging heat from the billets (cooling after stamping). To neutralize the decarburized layer, inherited by the parts from the operation of induction heating of billets for stamping and heat treatment in an oxidizing atmosphere of forgings, shot blasting is used. Thanks to this treatment, the endurance of the connecting rods is increased, which is confirmed by metallographic studies, the study surface microhardness and fatigue strength tests [22].

Table 3. Properties of steel AC30X3NMAFB after heat treatment.

Heat treatment schedule				Structure	Hardness, HB
T_{aust},°C	Cooling means	T_{temp}°C	Cooling means		
940	Air	650	Air	Sorbite	285
930	Oil	650	Oil	Sorbite	302
930	Air	705	Oil	Sorbite	278
950 (stamping end temperature)	Air (2-5°C/c) on the line	705	Air	Sorbite + Ferrite	255

The mechanical processing the billet of the crankshaft on an automatic line showed that the deepest drilling operations are the limiting stage of the cutting process. The reason for the low resistance of drills is ⌀5.5÷10.0 mm. the sticking of the billet material to the cutting tool and the formation of poorly removed chips from the cutting zones served as Taking into account the relationship between machinability by cutting and material properties, parameters of the cutting mode and design features of the tool, the possibility of improving machinability in drilling operations billets from complex alloyed heat-optimized steel AC40H2NMAF is investigated. It was found that the optimal parameters when drilling billets from new steel are:

- for drills ⌀ 5.5 mm and ⌀ 5.7 mm – inputting 0.1 mm / rev at a speed of 660 rpm;
- for drills ⌀ 8.5 mm and ⌀ 8.7 mm - inputting 0.17 mm / rev at a speed of 400 rpm;
- for drills ⌀ 9.0 mm and ⌀ 10.0 mm – inputting 0.15 mm / rev at 360 rpm.

The main attention during the pilot industrial testing is given to indicators of the service part properties – their fatigue strength. Bench tests parts made of experimental steel showed that the endurance limits of crankshafts are: bending – 1000 kg · m, torsion - 900 kg · m, which exceeds these properties for currently used 42HMFA steel (900 and 800 kg M, respectively). The advantages of steels microalloyed with such elements include their high resistance to grain growth during heating, as evidenced by the temperature dependence of grain size changes in 38G2AF steel (Figure 1). In comparison with the 20H2N4A and 18HGR steels widely used in automotive industry, 38G2AF steel retains a fine-grained structure over the entire temperature range of alloy deformation (1000–1250 °C), which indicates the possibility of its quenching from temperature after deformation processing. A good example is the AC30X3NMAFB structural steel chilled immediately after completion of hot plastic deformation and having a grain size of 7–8 points (table 3). Similar results were obtained with the forging heat of forgings of a connecting rod made of 38G2AF steel. During controlled forging of the connecting rod forgings, the following indicators took place: the temperature of the onset of deformation –1220 °C, the end of the deformation – 940 °C, cooling with cold compressed air (3.0–3.5 atm.) To a temperature of 450–550 °C and beyond in workshop conditions, to room temperature in calm air, the steel acquired a ferrite-pearlite structure (ferrite is less than 10%), grain size 6–7 points and hardness 285–293 HB.

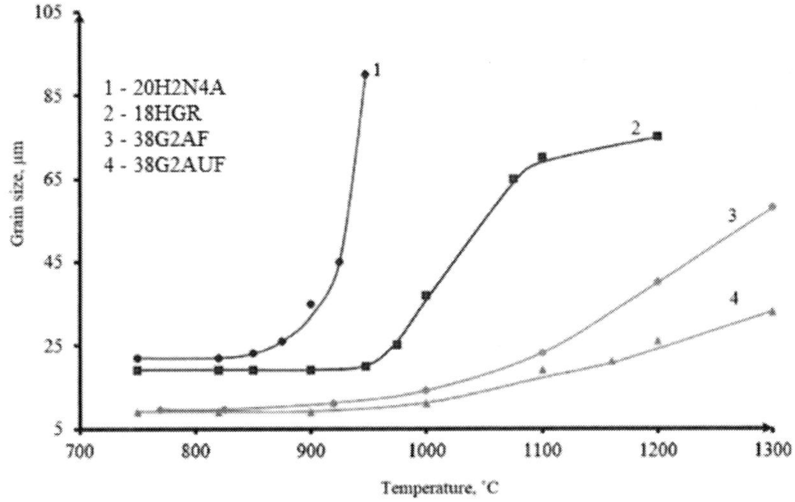

Figure 1. Temperature dependence of grain growth in steel.

On the metal the connecting rod parts, the following strengths were achieved: tensile strength of 980 MPa, yield strength of 840 MPa, elongation of more than 19% and relative narrowing of more than 45%, which indicates the possibility of implementing residual forging heat for heat treatment forgings.

Conclusions

- High resistance to grain growth is shown when heating steels microalloyed with vanadium, niobium and nitrogen.
- High properties of the steel AC40H2NMAF, AC30X3NMAFB and 38G2AF make it possible to recommend them for the manufacture of heavily loaded parts – crankshafts and connecting rods.
- The possibility of realizing forging heat of forgings made of steels microalloyed with vanadium, niobium and nitrogen during their heat treatment to ensure a high level of properties and structural requirements.

References

[1] Kirkhoff H U 2013 The future is steel *Ferrous metallurgy* **5** 71–76
[2] Jopp K 2015 Steel remains the dominant material in the automotive industry Black metals **9** 71–73
[3] Voskoboinikov V G, Kudrin V A and Yakupov A M 1979 *General metallurgy* (Moscow: Metallurgy) p 430
[4] Bicheev A M 1988 *Metallurgy steel* (Moscow: Metallurgy) p 480
[5] Meskin V S 1964 *The basis for the steel alloy* (Moscow: Metallurgy) p 684
[6] Lakhtin U M and Rakhshtadt A G 1980 *Heat treatment in mechanical engineering. Handbook* (Moscow: Machine Building) p 783
[7] Tylkin M A 1981 *Repair Service Thermist Reference* (Moscow: Metallurgy) p 648
[8] Mukhametzyanova G F, Zapadnova N N and Zapadnova E A 2019 Features of technological materials application in thermal manufacturing *Materials Science Forum* **946** 199–204
[9] Brown M P 1982 *Microalloying steel* (Kiev: Science dumka) p 303
[10] Shamari U 2008 High Performance Steels for Automotive *Black metals* **10** 48–51
[11] Black B 2014 Improved high-strength steels for the automotive industry - from microstructure to nano-structure *Black metals* **12** 83–88

[12] Zinchenko V M 2001 *Gear surface engineering using chemical heat treatment methods* (Moscow: Publishing House of MSTU N E Bauman) p 303

[13] Kalner V D 1984 *Steel semi-finished products and parts heat treatment quality control* (Moscow: Mechanical Engineering) p 384

[14] Vildanov A G, Mukhametzyanov I R, Astaschenko V I and Mukhametzyanova G F 2019 The investigation of properties the ball pins of the steering rod of the car *Ad alta-journal of interdisciplinary research* **9** **2** 138–141

[15] Astaschenko V I, Zapadnova E A, Zapadnova N N and Mukhametzyanova G F 2016 Predicting structure micro-alloyed steel products for different purposes *IOP Conference Series: Materials Science and Engineering* **134** **1** 012029

[16] Nagel M, Drissen J, Kovacs S, Pudritz O, Tomitz A, Ratzkov H and Deneke-Arnold H 2019 New developments in the field of high-strength microalloyed automotive steels *Black metals* **3** 43–48

[17] Hulka C and Klinkenberg X 2005 Niobium alloyed steel: state and development trends *Black metals* **12** 45–50

[18] Morozov Yu D, Goli-Oglu E A, S. Nastich Yu and Knyshev A V 2012 Modern approaches to the production of low-carbon micro-alloy pipe steel at Stekkel mills *Black metals* **9** 8–14

[19] Medina S F, Valles P, Calvo J and Cabrera J M 2020 The origin and growth of precipitates in microalloyed vanadium steel: physical theory and experimental results *Physics of metals and metal science* **120** **1** 37–47

[20] *Problems of the production and using steels with vanadium: materials of the international scientific and technical seminar* (Yekaterinburg: State Scientific Center of the Russian Federation OAO UIM) p 436

[21] Shveyova T V, Astaschenko V I and Shveyov A I 2015 Microalloy steels: Energy saving prospect at manufacture of details of cars *Biosciences Biotechnology Research Asia* **12** 297–299

[22] Astaschenko V I, Rudnitsky N M and Ustilovsky S Ya 1984 Hardening of connecting rods of automobile engines during shot cleaning *Automotive industry* **8** 10–11

[23] Chukin M V, Koptseva N V, Nikitenko O A and Efremova Yu Yu 2011 Mechanical properties of carbon structural steel with ultrafine-grained structure *Black metals* **6** 54–59

[24] Polokov S A 2019 Methodological basis for the selection of materials for gears according to the criterion of wear resistance *Mechanical Engineering Issues and Machine Reliability* **4** 54–62

[25] Astaschenko V I, Shveyova T V, Permyakov D N, Shveyov A I and Zapadnova E A 2016 Heat resistance of microalloyed steels with technological metal redistribution *Metal technology* **8** 2–7

IOP Conf. Series: Materials Science and Engineering **969** (2020) 012015 doi:10.1088/1757-899X/969/1/012015

An Analytical Model for Calculation of the Steel Hardness after Continuous Cooling

M V Maisuradze[1], M A Ryzhkov[1] and D I Lebedev[1, 2]

[1]Ural Federal University, 19, Mira Str., Yekaterinburg, 620000, Russia
[2]Institute of Metal Physics, 18, S. Kovalevskoi str., Yekaterinburg, 620000, Russia

E-mail: m.v.maisuradze@urfu.ru

Abstract. A number of modern engineering steel grades were analyzed using the CCT diagrams plotted by means of the dilatometer. The statistical analysis of the experimental data allowed for the determination of the equations connecting the critical cooling rate to achieve fully martensitic microstructure and the martensite hardness with the chemical composition of the steel. The dependence of the steel hardness on the cooling rate in the range of 0.1...30 °C/s was determined. It was shown that this dependence had the logarithmic character for every studied steel grade. The equations connecting the coefficients of the logarithmic function with the chemical composition of steel were obtained.

1. Introduction

Steel is the most widely used material in the industry. By varying the chemical composition of steel and the technological parameters of its heat treatment (heating temperature, cooling rate, etc.), microstructure constituents with different morphology and properties can be obtained: ferrite, pearlite, upper and lower bainite, martensite, residual austenite [1–3]. There are a lot of calculation methods that allow predicting the microstructure and mechanical properties of steel after the implementation of various heat treatment processes with different accuracy [4–8]. In this paper, we analyze the results of dilatometric studies of modern engineering steels [9–14] in order to identify the relations in the microstructure and hardness formation during the continuous cooling.

2. Experimental

The commercial and experimental engineering steels of various alloying systems were studied: Cr-Ni-Mo, Cr-Mn-Mo, Cr-Mn-Si-Ni-Mo, Cr-Mo-V (Table 1).

The transformations of the austenite under continuous cooling conditions were studied by means of Linseis L78 R.I.T.A. dilatometer using cylindrical specimens (diameter 3...6 mm, length 10 mm). The austenitization parameters were Ac_3 + 30...50 °C, 15 minutes. Continuous cooling was carried out at constant cooling rates 0.1...30 °C/s. Analysis of dilatometric data was carried out according to the method described in [15, 16]. The microstructure was studied using MEIJI IM7200 optical microscope. The steel hardness was determined using the Rockwell hardness tester (scale C). At least three hardness measurements were performed on each dilatometric specimen.

Table 1. Chemical composition of the studied steels, wt. %.

Steel	C	Cr	Mn	Si	Ni	Mo	V	W	S	P
20Cr2Ni2Mo	0.18	2.13	0.35	0.07	2.17	0.40	0.010	-	0.024	0.025
20Cr2Ni4	0.17	1.31	0.42	0.23	3.33	0.02	0.007	-	0.011	0.014
25Cr2Ni4Mo	0.24	1.43	0.34	0.28	4.06	0.32	0.000	0.07	0.005	0.005
38CrNi3Mo	0.36	0.87	0.37	0.26	2.80	0.21	0.003	0.02	0.009	0.005
38CrNi3MoV	0.32	1.26	0.33	0.27	3.02	0.34	0.125	0.02	0.001	0.006
31NiCrMo13-4	0.31	1.08	0.53	0.23	3.10	0.23	-	-	0.018	0.009
HY-TUF	0.24	0.34	1.38	1.55	1.69	0.40	0.010	-	0.023	0.019
20Cr2Mn2SiNiMo	0.18	2.35	1.99	1.03	1.17	0.34	0.007	-	0.018	0.013
20Cr2MnSiNi2Mo	0.15	2.55	0.67	0.76	2.31	0.77	-	0,02	0.014	0.013
30Cr2MnSiNi2WMo	0.26	1.76	1.14	0.94	2.15	0.37	0.008	1,11	0.001	0.007
30CrMnSiNi2	0.29	1.08	1.13	0.98	1.51	0.04	0.005	-	0.001	0.005
20CrMn2NiMo	0.21	1.04	1.70	0.01	1.18	0.31	0.005	-	0.019	0.025
25CrMnMo	0.26	0.68	0.87	0.24	0.01	0.21	-	-	0.009	0.009
38CrMnMo	0.37	0.81	0.64	0.24	0.09	0.15	-	-	0.004	0,011
42CrMo4-mod1	0.43	0.97	0.91	0.24	0.13	0.16	0.003	-	0.008	0.007
42CrMo4-mod2	0.42	1.10	1.05	0.24	0.31	0.55	0.001	-	0.007	0.007

3. Results and discussion

The values of the critical cooling rate, v_M, corresponding to the minimum cooling rate at which a completely martensitic microstructure is formed, were determined from the CCT-diagrams of the studied steels. An equation relating the critical cooling rate, v_M, with the chemical composition of steel, wt. %, was obtained using the regression analysis of the experimental data.

$$v_M = 62{,}5 - 89{,}7C - 7{,}3Cr - 5{,}7Mn - 9{,}1Si - 6{,}7Ni - 1{,}0Mo + 16{,}9V + 7{,}2W. \quad (1)$$

A verification of the equation (1) by means of the Fisher criterion showed that the obtained model had a satisfactory adequacy: the ratio of the theoretical value of the Fisher criterion to the calculated one was more than 40. It should be noted that the majority of the alloying elements taken into account in equation (1) reduce the value of the critical cooling rate, i.e., increase the hardenability of steel. The only exceptions are the strong carbide-forming elements – vanadium and tungsten. Their presence in the steel composition, even as the impurities, increases the critical cooling rate. This is explained by the formation of the dispersed and hardly soluble carbide particles in steels alloyed with vanadium and tungsten. These particles serve as substrates for the formation of a new phase during the austenite transformation below Ar_3.

The dependence of the hardness on the cooling rate was determined for each studied steel grade. It was found that for all the steels under consideration the hardness values were in a logarithmic dependence on the cooling rate (until a completely martensitic microstructure was obtained). Figure 1 plots the typical examples of the hardness dependence on the cooling rate for the steels of various alloying systems. The experimental data were satisfactorily described by a logarithmic function.

$$HRC = k \cdot \ln(v_{cool}) + h_1, \quad (2)$$

where HRC – the hardness of steel; v_{cool} – cooling rate, °C/s; k, h_1 – the constant parameters. The parameter k corresponds to the slope of the dependence of hardness on the cooling rate in logarithmic scale. The parameter h_1 corresponds to the hardness of steel at a cooling rate 1 °C/s.

The values of the parameters k and h_1 depend on the chemical composition of the steel. The following equations were obtained using the regression analysis of the experimental data (the content of alloying elements in wt.%):

$$k = 6{,}01 + 13{,}83\mathrm{C} - 0{,}74\mathrm{Cr} - 2{,}74\mathrm{Mn} + 1{,}33\mathrm{Si} - 1{,}13\mathrm{Ni} - 2{,}78\mathrm{Mo} - 1{,}57\mathrm{W}, \quad (3)$$

$$h_1 = -12{,}69 + 89{,}71\mathrm{C} + 2{,}62\mathrm{Cr} + 8{,}34\mathrm{Mn} + 4{,}61\mathrm{Si} + 6{,}73\mathrm{Ni} + 17{,}45\mathrm{Mo}. \quad (4)$$

Note that the carbon content had the highest effect on the values of the parameters k and h_1, and the chromium content had the least effect. An increase of any alloying element content led to an increase in the parameter h_1 value. This is due to the fact that all the alloying elements taken into account increase the hardenability of steel and, consequently, the amount of martensite in the microstructure at a certain cooling rate (particularly at 1 °C/s).

An increase in the content of Cr, Mn, Ni, Mo, W in steel led to a decrease in the parameter k value, which characterized the slope of the dependence of the hardness on the cooling rate in a logarithmic scale. This fact is explained by the effect of the alloying on the increase of the range of the cooling rates at which martensite is formed. As a result, the hardness values are less dependent on the cooling rate. Si and C, on the contrary, increased the value of the parameter k, and, accordingly, the slope of the dependence of hardness on the cooling rate, which is probably due to the influence of these elements on the hardness of the α phase (martensite and bainite).

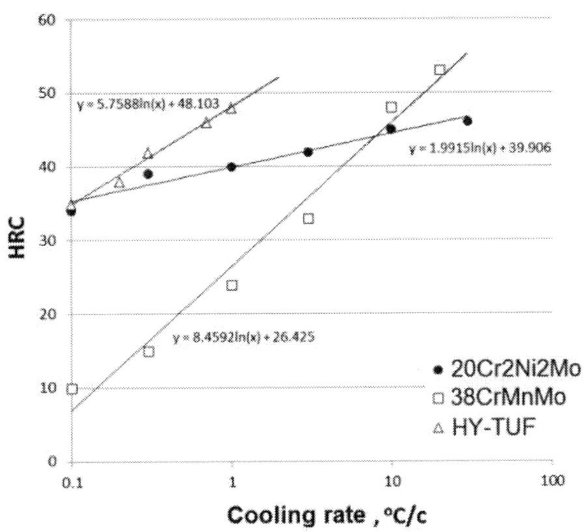

Figure 1. The dependence of hardness on the cooling rate for different steels grades.

The dependence of the hardness of martensite on the content of alloying elements was determined by means of the regression analysis of experimental data:

$$HRC_M = 33{,}40 + 55{,}49\mathrm{C} + 2{,}04\mathrm{Si}, \quad (5)$$

where HRC_M – the hardness of martensite; C, Si – the carbon and silicon content, wt. %.

Thus, knowing the chemical composition of the steel and using equations (1–5), it is possible to determine the hardness of the steel after cooling at any cooling rate in the range of 0.1...30 °C/s with the acceptable accuracy. This, in turn, will make it possible to evaluate the strength characteristics of steel (yield strength and tensile strength) according to the known dependences [17–19].

The following algorithm was proposed for calculating of the hardness of steel obtained after the continuous cooling at different rates:

- the critical cooling rate, v_M, is determined by equation (1);

- the parameters of the logarithmic equation (2), k and h_1, are determined by equations (3) and (4), respectively;
- the hardness of steel is determined according to equation (2) in the range of cooling rates from 0.1 °C/s to v_M;
- the hardness of martensite is determined by equation (5).

Figure 2 shows the examples of the comparison between the experimental dependences of hardness on the cooling rate (both for the steels studied in this work and for the steel from the reference [20]) with the results of calculations by the proposed model. As can be seen, the analytical equations give an adequate prediction of the hardness values.

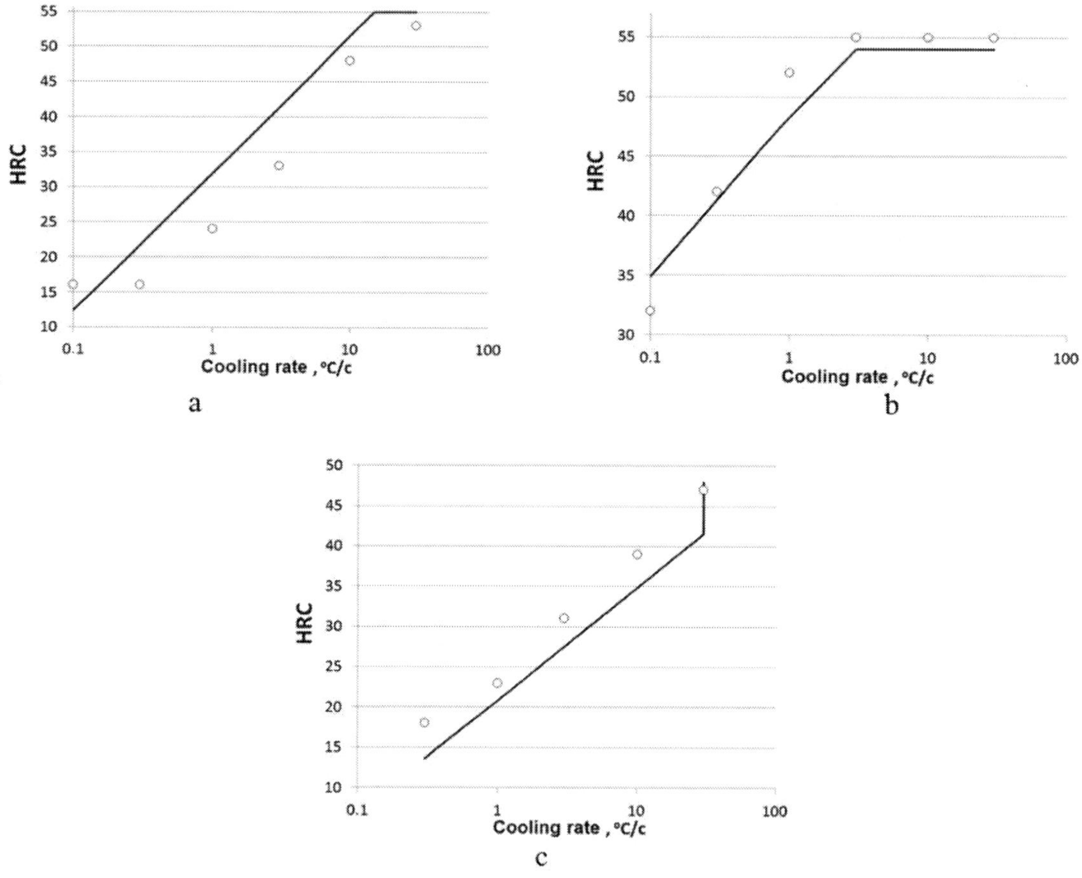

Figure 2. Experimental and calculated dependences of the hardness of steel on the cooling rate: a) 38CrMnMo; b) 38CrNi3Mo; c) 25CrMnV [20].

4. Conclusions

The dependence of the critical cooling rate on the chemical composition of steel was determined. The strong carbide-forming elements (V, W) increased the critical cooling rate, i.e., reduced the hardenability of steel, since these elements formed hardly soluble dispersed carbide particles acting as the substrates for the formation of a new phase.

The logarithmic dependence of the steel hardness on the cooling rate was obtained. The parameter k of the logarithmic equation corresponds to the slope of the dependence of hardness on the cooling rate in logarithmic scale, and the parameter h_1 corresponds to the hardness of steel at a cooling rate 1 °C/s. Equations relating the parameters of the logarithmic equation, k and h_1, with the chemical composition of steel were determined.

An algorithm has been developed that allows to determine the steel hardness obtained after the continuous cooling at constant rates 0.1...30 °C/s using the chemical composition of steel.

References

[1] Gomez M, Rancel L, Escudero E and Medina S F 2014 *J. Mater. Sci. Tech.* **30** **511**
[2] Bhadeshia H and Honeycombe R 2017 *Steels: Microstructure and Properties* (Amsterdam: Elsevier) p 488
[3] Totten G E 2006 *Steel Heat Treatment. Metallurgy and Technologies* (Boca Raton: CRC Press) p 848
[4] Sankaran K K and Mishra R S 2017 *Metallurgy and Design of Alloys with Hierarchical Microstructures* (Amsterdam: Elsevier) p 506
[5] Lin J, Balint D and Pietrzyk M 2012 *Microstructure Evolution in Metal Forming Processes* (London: Woodhead Publishing) p 416
[6] Pereloma E and Edmonds D V 2012 *Phase Transformations in Steels* (London: Woodhead Publishing) p 680
[7] Anufriev N P, Maisuradze M V and Yudin Yu V 2011 *Met. Sci. Heat Treat.* **53** 189
[8] Maisuradze M V, Yudin Yu V and Ryzhkov M A 2015 *Met. Sci. Heat Treat.* **56** 512
[9] Maisuradze M V, Ryzhkov M A, Yudin Yu V and Ershov A A 2014 *Metallurgist* **58** 712
[10] Maisuradze M V, Ryzhkov M A and Kuklina A A 2017 *Solid State Phen.* **265** 712
[11] Maisuradze M V, Ryzhkov M A, Yudin Yu V and Kuklina A A 2017 *Met. Sci. Heat Treat.* **59** 486
[12] Maisuradze M V and Ryzhkov M A 2018 *Solid State Phen.* **284** 351
[13] Maisuradze M V, Ryzhkov M A and Surnaeva O A 2018 *Met. Sci. Heat Treat.* **60** 339
[14] Maisuradze M V and Lebedev D I 2019 *Steel Transl.* **49** 813
[15] Kop T A, Sietsma J and Van Der Zwaag S 2001 *J. Mater. Sci.* **36** 519
[16] Ryzhkov M A and Popov A A 2011 *Met. Sci. Heat Treat.* **52** 612
[17] Datsko J, Hartwig L and McClory B 2001 *J. Mater. Eng. Perf.* **10** 718
[18] Pavlina E J and Van Tyne C J 2008 *J. Mater. Eng. Perf.* **17** 888
[19] Basu P, Acharyya S K and Sahoo P 2018 *Silicon* **10** 1257
[20] Wever F, Rose A, Peter W and Straßburg W 1961 *Atlas der Wärmebehandlung der Stähle* (Berlin: Berichtiger Nachdruck) p 257

Acknowledgments

The work was supported by the Act 211 of Government of the Russian Federation, contract № 02.A03.21.0006, the Ministry of Science and Higher Education of the Russian Federation, project №11.1465.2014/K, and by the state assignment of the M.N. Miheev Institute of Metal Physics of Ural Branch of Russian Academy of Sciences, theme "Laser".

The Features of Bainite Transformation in Alloyed Steels

Yu V Yudin, M V Maisuradze and P D Lebedev

Ural Federal University, 19, Mira Str., Yekaterinburg, 620000, Russia

E-mail: m.v.maisuradze@urfu.ru

Abstract. The main factors affecting the kinetics of isothermal bainite transformation in alloyed steels were determined using the algorithms of the computer modelling of phase transformations in the solid state. The experimental and calculated transformation kinetics was described and compared using well known Kolmogorov-Johnson-Mehl-Avrami (KJMA) equation and Austin-Rickett (AR) equation. The effect of the initial configuration and the nucleation rate of the new phase nuclei on the parameters of the bainite transformation was analysed. The calculated kinetics of phase transformation at different nucleation rates was compared with the experimentally observed for HY-TUF and D6AC steels. The mechanism of the bainite transformation in alloyed steels was explained.

1. Introduction

Discussions about the "mechanism" of the bainite transformation in steels have been going on for a long time, since the discovery of the bainite by Davenport and Bain in the 1930s [1]. However, despite the critical differences, some of the distinguishing features of the bainite transformation in steels are not in doubt. These features consist in the differentiation of the bainite morphology into "upper" and "lower", in the presence of the temperature of the onset of the bainite transformation, B_s, and in the incompleteness of the bainite transformation [2–4].

Currently, there are two main concepts that explain the nature of bainite transformation. The first one considers the nucleation and growth of bainite as a diffusion-controlled process of increment growth of the "ledges" at the interface between α-plates and the surrounding austenite. The second concept is based on a diffusionless transformation, when the bainite is formed by successive autocatalytic nucleation and growth of α-subunits.

The morphological classification of bainite into "upper" and "lower" made it possible to combine two different concepts of the bainite formation: it is believed that the upper bainite is formed mainly due to the diffusion, and the lower bainite is formed by the shear rearrangement of the crystal lattice [5]. However, in the literature there are other varieties of bainite [6–8], for example, granular, columnar, etc., formation of which is not yet explained.

2. Experimental

The commercial alloyed steels HY-TUF and D6AC were studied (the chemical composition is listed in Table 1).

The experimental kinetics of isothermal bainite transformation in the HY-TUF and D6AC steels was studied using LINSEIS L78 "R.I.T.A." dilatometer. The heating temperature of the steel specimens (diameter 4 mm, length 10 mm) was 925 °C, the holding time was 15 minutes. The cooling rate to the isothermal transformation temperature was 30 °C/s in order to prevent the austenite transformation before the required temperature was reached. The isothermal holding time at a temperature in the range of 300...500 °C was 2...4 hours. As a result, the dependences of the bainite fraction on the time were obtained.

Table 1. The chemical composition of the studied steels, wt. %

Steel	C	Cr	Mn	Si	Ni	Mo	V
HY-TUF [9]	0.24	0.31	1.35	1.42	1.71	0.40	0.01
D6AC [10]	0.49	1.11	0.79	0.23	0.46	0.98	0.11

The mathematical description of the experimental kinetics of the isothermal austenite transformation was carried out using Kolmogorov-Johnson-Mehl-Avrami (KJMA) equation (1) [11–15] and Austin-Rickett (AR) equation (2) [16]:

$$P = 1 - e^{-k\tau^n}, \tag{1}$$

$$P = 1 - \frac{1}{1 + k\tau^n}, \tag{2}$$

where P – the austenite transformation fraction; τ – time, s; k – temperature-dependent parameter; n – exponent.

Computer simulation of the phase transformation was implemented using the MATLAB© software. The software package was developed on the basis of the procedures for constructing the wave fronts in an isotropic medium [17]. The simulation method is described in detail in [18].

The results of the analysis of the experimental and simulated transformation kinetics were presented as the kinetic curves in the logarithmic coordinates "ln(–ln(1–P))–ln(τ)" and "ln[P/(1–P)]–ln(τ)" in the case of the equations KJMA (1) and AR (2), respectively.

3. Results and discussion

The experimental kinetics of the isothermal bainite transformation in the studied steels at different temperatures (lower and upper bainite range) are shown in Figure 1 and 2. Figure 2 depicts the transformation kinetics in the logarithmic coordinates of the KJMA (1) and AR (2) equations. Note that in the majority of cases, the KJMA kinetics showed an inflection, characterized by a change in the value of exponent n. The only exception is the formation of upper bainite in D6AC steel (Figure 2, c) where the KJMA kinetics had a linear form. In turn, the AR kinetics had a linear form in the majority of cases except the one for the upper bainite in D6AC steel at 420 °C (Figure 2, d): in this case, an explicit inflection was observed on the AR kinetic curve.

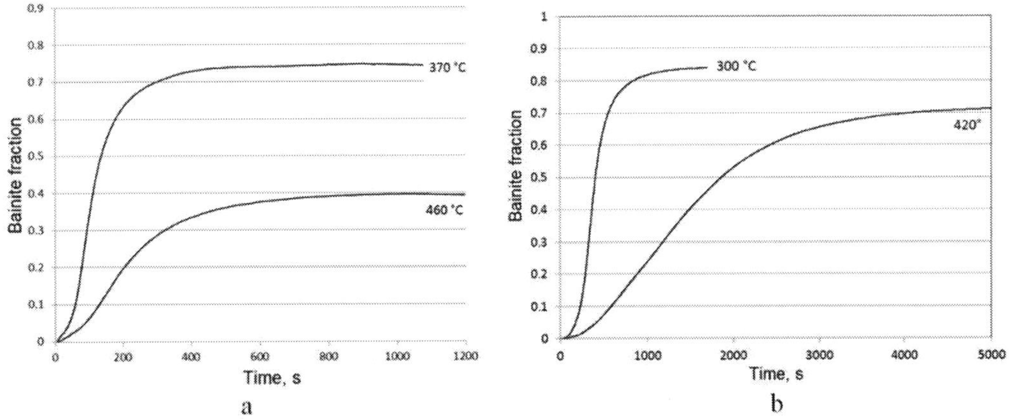

Figure 1. Experimental kinetics of isothermal bainite transformation in steels HY-TUF (a) and D6AC (b).

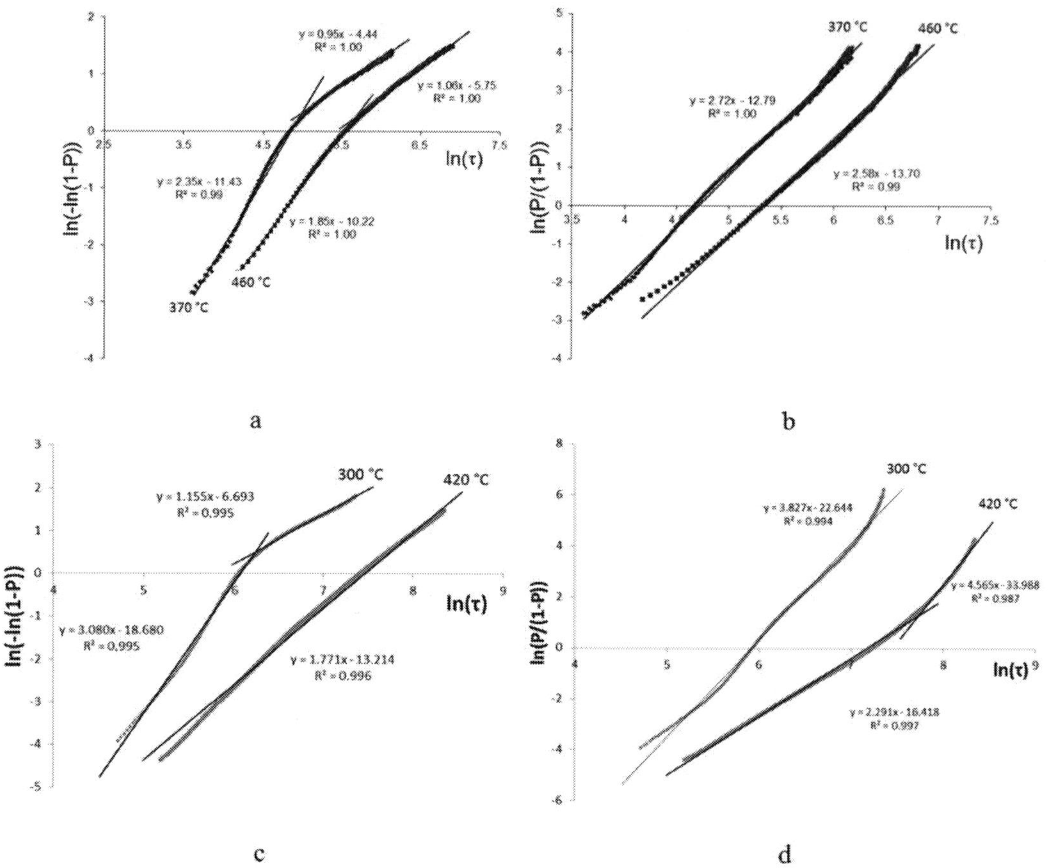

Figure 2. Kinetic curves of isothermal bainite transformation in logarithmic coordinates for steels HY-TUF (a, b), D6AC (c, d): a, c) KJMA equation; b, d) AR equation.

The degree of linearity of the kinetic curve reflects the adequacy of the mathematical description of the entire process of isothermal transformation if the constant coefficients of Equations (1) and (2) are used. To characterize the adequacy of the mathematical description of the observed kinetics of transformation, we used the sum of the squared differences between the experimental and calculated fractions of bainite at each moment of time in the range of the bainite fraction 0.02...0.98. It was found that the AR equation provides up to 40 times better description of the experimental kinetics of bainite transformation compared to the KJMA equation, except for the transformation in D6AC steel at 420 °C. Thus, there are some conditions of the bainite transformation that determine the nature of the process, and these conditions can depend both on the temperature of the transformation and on the chemical composition of the steel. For example, D6AC steel contains more carbide-forming elements (Mo, V) than HY-TUF steel, and the latter contains a significant amount of silicon and nickel (up to 2 wt.%).

To understand the physical phenomena behind the change in the nature of the bainite transformation, a computer simulation of the phase transformation was carried out.

First of all, the calculation was performed for the number of the new phase nuclei randomly distributed in the virtual volume: the average concentration of the new phase nuclei at the beginning of the calculation was 1.25×10^{-2} %, the nucleation rate of the new phase was assumed to be 0. The simulation results showed that the description of the kinetics of the phase transition by the KJMA equation gave a satisfactory result throughout the entire transformation (Figure 3, a). The kinetic curve had a linear form, the exponent of the KJMA equation is $n \approx 3$, and the value $\ln(k) \approx -13.56$. The obtained value of the exponent n is consistent with the classical theory of phase transformations [19, 20], which relates the exponent n of the KJMA equation to the shape of the growing particles of a new phase (in this work, three-dimensional growth of nuclei leads to the growth of quasispherical particles).

When a concentration gradient of nuclei in the virtual volume was applied, a substantially different kinetics of the phase transition was observed (Figure 3, b). From the beginning of the transformation to the fraction 0.60...0.65, the exponent n was 2.8, and ln(k) was −11.0. These values are close to the ones obtained previously for the random distributed nuclei. Upon reaching the transformed fraction 0.60...0.65, an inflection of the kinetic curve was observed, and the transformation proceeded with another parameters: $n = 1.3$; $\ln(k) = -5.2$. The obtained values of the parameters of the KJMA equation satisfactorily coincide with the values obtained for the experimentally observed kinetics of bainite transformation in the steels under study. Note, that the kinetic curves in the AR coordinates have exactly the opposite characteristics: for the uniform nuclei distribution an inflection is observed on the kinetic curve (Figure 4, a), for the concentration gradient of the nuclei the kinetic curve has a linear form (Figure 4, b). Thus, the nature of the bainite transformation in steel of a certain composition depends on the distribution pattern of the critical nuclei able to grow at a specific temperature.

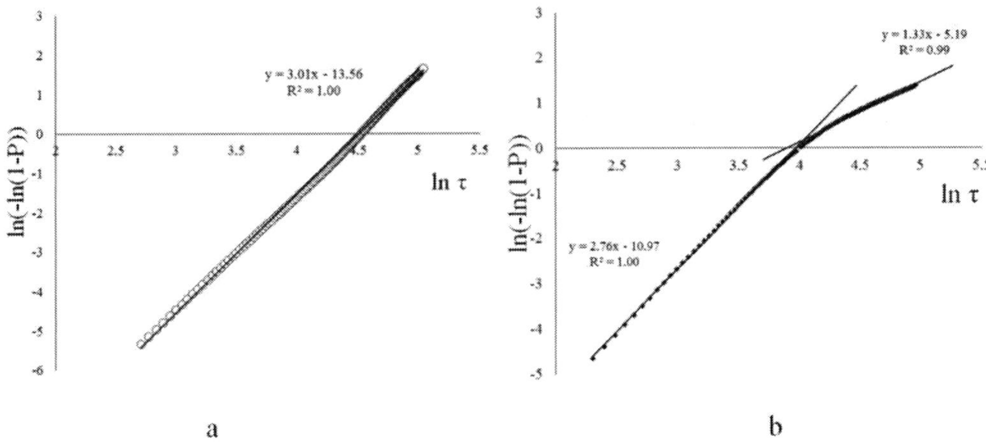

Figure 3. Kinetic curve of the simulated phase transition obtained for the different initial distribution of the new phase nuclei (KJMA equation, zero nucleation rate): a) random; b) concentration gradient.

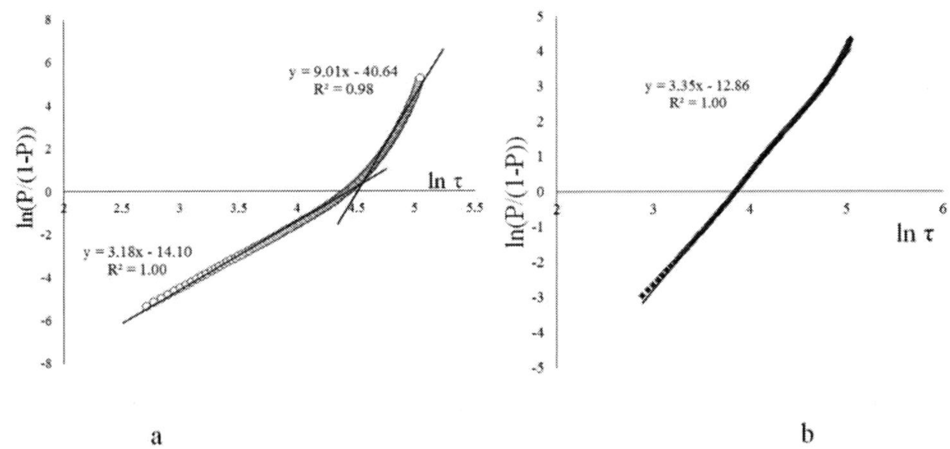

Figure 4. Kinetic curve of the simulated phase transition obtained for the different initial distribution of the new phase nuclei (AR equation, zero nucleation rate): a) random; b) concentration gradient.

The effect of the nucleation rate on the transformation process was evaluated. From 1 to 10 new nuclei of the second phase were added into the virtual volume at every calculation step. It was found that the increase in the nucleation rate led to the linearization of the KJMA kinetics (Figure 5, a), and a characteristic inflection appeared on the AR kinetics (Figure 5, b). So, the adequacy of the description of the transformation kinetics by the AR equation decreased.

Since the AR equation is most suitable for the description of the experimental kinetics of isothermal bainite transformation in the majority of cases, it is most likely that during the bainite transformation in the studied steels there were no nucleation of new nuclei. In this case, the transformation proceeded only due to the growth of the nuclei formed in the volume of the austenite during cooling from the austenization temperature to the temperature of isothermal holding below B_s. Thus, at the initial stage the bainite transformation is similar to the martensite transformation: supercooling below the B_s temperature leads to the formation of a certain number of nuclei depending on the degree of supercooling. Subsequently, these nuclei grow by layers, uniformly, which is characteristic of a diffusion-controlled process, and new nuclei do not form during the transformation.

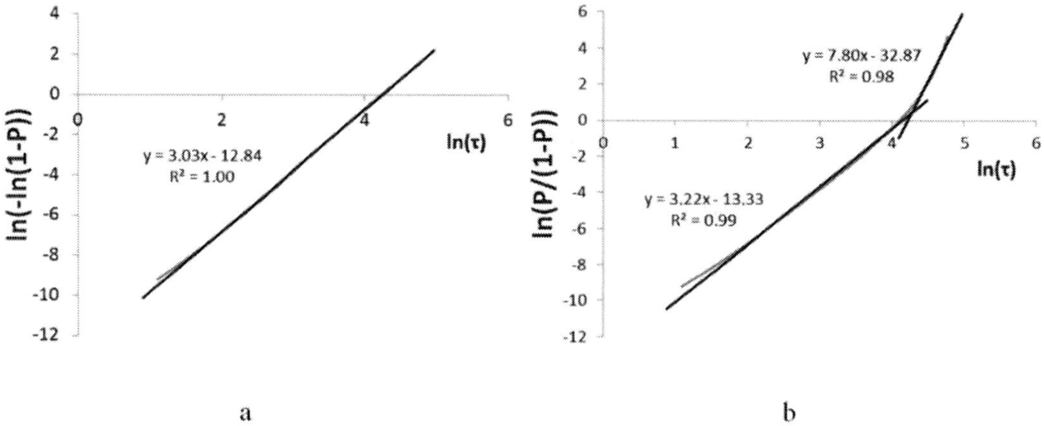

Figure 5. Kinetic curves of the simulated phase transition (concentration gradient of the nuclei, nucleation rate 5 nuclei per cycle): a) KJMA equation; b) AR equation.

In the case of the upper bainite formation in D6AC steel (Figure 2, c, d), the nature of the transformation is similar to the diffusion-controlled growth at a nonzero nucleation rate. This is due to the carbide formation during the isothermal treatment, since D6AC steel contains a large amount of the carbide-forming elements and does not contain a sufficient amount of silicon, which suppresses the precipitation of carbides. Dispersed carbides precipitated from the austenite can serve as the substrates or additional nuclei for the growth of the bainite colonies.

4. Conclusions

The kinetics of isothermal bainite transformation in steels HY-TUF and D6AC at the temperature of lower and upper bainite formation was studied. It was found that the lower and upper bainite transformation in HY-TUF steel is adequately described by the Austin-Rickett equation. For D6AC steel the Austin-Rickett equation was suitable only for the formation of lower bainite. The kinetics of the upper bainite formation in D6AC steel was adequately described by the KJMA equation.

The computer simulation showed that the initial distribution of new phase nuclei in the virtual volume and the nucleation rate had a significant effect on the kinetics of the phase transformation.

It was shown that the bainite transformation in steel occurred under the following conditions: initial uneven distribution of nuclei, layer-by-layer growth, lack of formation of new nuclei during the transformation. Thus, the bainite transformation combines the features of a martensite transformation (the formation of a certain number of nuclei depending on the degree of supercooling below the equilibrium temperature B_s) and diffusion-controlled transformation (uniform layer-by-layer growth of nuclei).

References

[1] Davenport E S and Bain E C 1930 *Trans. AIMME* **90** 117
[2] Bhadeshia H 2015 *Bainite in Steels: Theory and Practice* (Boca Raton: CRC Press) p 616
[3] Fielding L C D 2013 *Mater. Sci. Technol.* **29** 383
[4] Wu H D, Miyamoto G, Yang Z G, Zhang C, Chen H and Furuhara T 2017 *Acta Mater.* **133** 1
[5] Durand-Charre M 2004 The bainite transformation *Microstructure of Steels and Cast Irons* 223–237
[6] Smirnov M A, Pyshmintsev I Yu and Boryakova A N 2010 *Metallurgist* **54** 444
[7] Zajac S, Schwinn V and Tacke K H 2005 *Mater. Sci. Forum* **500** 387
[8] Krauss G and Thomps S W 1995 *ISIJ Int.* **35** 937
[9] Maisuradze M V, Yudin Yu V and Kuklina A A 2019 *Metallurgist* **63** 849
[10] Maisuradze M V, Yudin Yu V and Kuklina A A 2019 *Mater. Perf. Char.* **8** 80
[11] Kolmogorov A N 1937 *Bul. AS USSR Math. Ser.* **1** 355
[12] Johnson W A and Mehl R F 1939 *Trans. AIMME* **135** 416
[13] Avrami M 1939 *J. Chem. Phys.* **7** 1103
[14] Avrami M 1940 *J. Chem. Phys.* **8** 212
[15] Avrami M 1941 *J. Chem. Phys.* **9** 177
[16] Austin J B and Rickett R L 1939 *Trans. AIMME* **964** 1
[17] Lebedev P D and Uspenskii A A 2017 RU Patent **2017662074**
[18] Yudin Yu V, Kuklina A A, Lebedev P D and Maisuradze M V 2018 *Steel Transl.* **48** 684
[19] Cahn R W and Haasen P 1996 *Physical Metallurgy Vol. 2* (Amsterdam: North-Holland) p 940
[20] Christian J W 2002 *The Theory of Transformations in Metal and Alloys* (Amsterdam: Pergamon) p 1200

Acknowledgments

The work was supported by the Act 211 of Government of the Russian Federation, contract № 02.A03.21.0006, the Ministry of Science and Higher Education of the Russian Federation, project №11.1465.2014/K.

IOP Conf. Series: Materials Science and Engineering **969** (2020) 012017 doi:10.1088/1757-899X/969/1/012017

Influence of the Thermal Deformation Treatment on the Structural State and Properties of Ti-Mo Microalloyed Steels of the Ferritic Class

A I Zaitsev[1,2], A V Koldaev[1] and N A Arutyunyan[2]

[1]Bardin Central Research Institute of Ferrous Metallurgy, 23/9, Radio str., Moscow, 105005, Russia
[2]Faculty of Chemistry, Lomonosov Moscow State University, 1-3, Lenin Hills, Moscow, 119991, Russia

E-mail: aizaitsev@chermet.net

Abstract. Investigation on influence of thermal deformation treatment on the special features of structural state, phase precipitates, strengthening mechanisms, and mechanical properties was carried out for low-carbon Ti-Mo microalloyed steel of ferritic class. The following parameters were analyzed: the finishing temperature of rolling and the cooling rate of the steel after rolling end to the coiling temperature. Methods of scanning and transmission electron microscopy, thermodynamic calculation, kinetic analysis, and testing of mechanical properties were used. It was shown that with an increase in the cooling rate from 3 to 22 °C/s, the type of carbide precipitates changes from interphase to ferritic. When fast cooling is applied, mainly mechanisms of dislocation and grain boundary strengthening are realized. When cooling is slowed down, the fraction of precipitation hardening increases. It was established that ferritic precipitates make a smaller contribution to precipitation hardening of steel than interphase ones. To create conditions favorable for the formation of interphase precipitates, the rates of $\gamma \rightarrow \alpha$ phase transformation and cooling should be close.

1. Introduction

High-strength low-carbon microalloyed steels of a ferritic class are promising for widespread use in transport, construction, engineering, and other branches of modern technology and industry due to a remarkable combination of difficultly-to-combine technological and service properties [1–3]. In particular, they possess simultaneously high strength, ductility, formability, and fatigue and corrosion resistance with good weldability. An effective increase in the strength characteristics of the considered steels having a ferritic structure is achieved as a result of a superposition of various strengthening mechanisms [1]. The main ones, grain refinement and dispersion hardening, in many respects, are controlled by the presence of phase precipitates of different types and dispersion. The dependence of the phase precipitate characteristics on the chemical composition and parameters of the thermal deformation treatment of steel, and their effect on the microstructure and mechanical properties of the metal has been the subject of many studies (see, for example, [4–8]). As a result, hot-rolled Ti-Mo microalloyed steels with a yield strength of up to 700 MPa and satisfactory ductility and other properties were produced [9]. However, in order to stably provide and continue the development and increase of strength characteristics, more precise information is needed on the specifics of the influence of metal processing parameters and resulting phase precipitates on the implementation of different strengthening mechanisms and properties of rolled products.

Grain-boundary strengthening is facilitated by the carbide, carbonitride precipitation in austenite during hot deformation [10], which inhibit recrystallization and contribute to the formation of a fine-grained microstructure. For dispersion hardening, precipitates from 1 to 3 nm in size are most effective [11]. As a rule, they are formed during the $\gamma \rightarrow \alpha$ phase transformation (interphase) and upon further cooling of the steel (ferritic). Interphase precipitates are arranged in layers parallel to the moving front of the γ/α transformation [12]. Their presence leads to an increase in strength of steel not only by the dispersion hardening mechanism, but also contribute to the formation of a fine-grained structure due to inhibition of grain growth. The interconnectedness of various strengthening mechanisms significantly complicates making unambiguous conclusions about the influence of technological parameters of metal processing and the characteristics of phase precipitates on mechanical and other properties. In accordance with the results of studies [13–16], it has been unequivocally established that the maximum strength properties are achieved if the temperature of coiling hot-rolled strip into a roll (T_c) is 650 °C. However, the effect of rolling end including finishing temperature (T_f), and especially the cooling rate from T_f to T_c on the features of microstructure formation, nanosized carbide precipitates, and the properties of low-carbon Ti-Mo microalloyed ferritic steels have not yet been determined. Their establishment, using the optimum T_c = 650 °C, has been the subject of this work.

2. Materials and methods

The study was performed using steel of laboratory vacuum induction smelting. The chemical composition was as follows (wt. %): 0.048C, 0.17Si, 1.29Mn, 0.003P, 0.004S, 0.06Cr, 0.06Ni, 0.20Mo, 0.01Cu, 0.09Al, 0.13Ti, 0.0042N. The billets were heated in a resistance furnace to a temperature of 1250 °C, held for 1 hour, and rolled into strips 3 or 8 mm thick. The rolled products were cooled in three ways: in the air-water or air stream created by the compressor, as well as in air to obtain various cooling rates. After cooling to 650 °C, the strips were placed in a furnace heated to the same temperature and held for 30 minutes, then cooled with the furnace to room temperature, thereby simulating cooling of a coiled roll. Samples were prepared from the obtained strips for mechanical tests and microstructure studies. The mechanical properties were determined using HECKERT FP-100/1 tensile testing machine. The results obtained together with the hot-rolling parameters are presented in the table 1.

Table 1. Parameters of hot-rolling modes and mechanical properties of the studied steels.

Lot No.	Rolled thickness, mm	T_f, °C	Cooling rate to T_c, °C/c	$\sigma_{0.2}$, MPa	σ_B, MPa	δ, %
1	8	950	6	670	780	20.0
2	8	950	5	650	770	20.0
3	8	950	3	620	770	20.5
4	3	900	22	690	800	18.0
5	3	900	10	680	780	16.5
6	3	900	7	640	740	15.0

The microstructure was studied by scanning electron microscope using a JSM-6610LV (JEOL) instrument equipped with an INCA Energy Feature XT energy dispersive microanalysis system, INCA Wave 500 wave dispersive spectrometer, and transmission electron microscopy (TEM) using a JEM200CX microscope. The type of nanosized precipitates was determined by the original method based on the analysis of their reflections in microdiffraction images [2].

The thermodynamic calculation of the regions of phases existence in the steel under study was carried out using a computer model [17, 18], implemented on the basis of proprietary software. It consisted in solving the problem of finding the conditions of thermodynamic equilibrium in multicomponent, multiphase systems with given external and internal parameters (temperature, pressure, chemical composition). As a result, the types of equilibrium phases, their quantities and compositions were determined by analysis of the complete possible set for an alloying system of steel. The kinetics of TiC precipitate formation in austenite was simulated using the developed model [19].

3. Results

As can be seen from the table 1, the values of yield ($\sigma_{0.2}$) and tensile (σ_B) strengths of the rolled products are in the ranges of 620–690 MPa and 740–800 MPa, respectively, and depend on both the cooling rate and T_f. Lots of steel subjected to faster cooling (No. 4–6) have higher strength properties. At the same time, in the case of a lower cooling rate, but a higher T_f = 950 °C (No. 1, 2), the values of $\sigma_{0.2}$ and σ_B are higher than those of the rolled steel No. 6. At the maximum cooling rate of steel, the highest values were achieved: $\sigma_{0.2}$ = 690 MPa and σ_B = 800 MPa. The relative elongation of the obtained samples is higher at T_f = 950 °C and does not depend on the cooling rate. The noted circumstances indicate the preference for using T_f equal to 950 °C compared with 900 °C. With an equivalent cooling rate, in this case, a complex of higher properties of rolled products is achieved. For the highest cooling rate (22 °C/s), the ductility remains quite high - 18%.

The results of thermodynamic calculation of the equilibrium phase composition depending on the temperature for the studied steel in the range of 400–1350 °C are presented in figure 1.

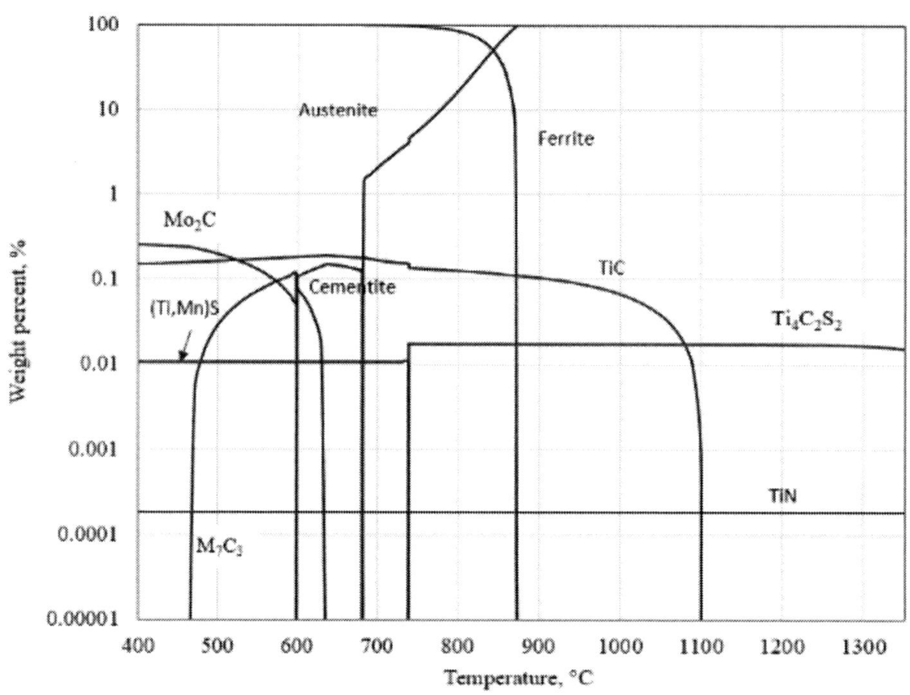

Figure 1. Temperature dependences of equilibrium fractions of phases in the studied steel.

It can be seen that at heating temperatures for rolling, stable phases in austenite are TiN nitride and titanium carbosulfide - $Ti_4C_2S_2$. The titanium content in the studied steel significantly exceeds its required amount for the complete binding of nitrogen. Therefore, nitrogen is completely removed from austenite at high temperatures. As a result, the precipitation of AlN becomes impossible, and titanium carbide formation below 1100 °C is possible in the region of hot deformation temperatures in austenite and in ferrite. This means that formation of austenitic precipitates can be expected. However, from the kinetic dependences (figure 2) it follows that the reaction of titanium carbide precipitation at the used high temperatures of finishing rolling will be far from complete, especially for $T_f = 950$ °C. Therefore, a sufficiently high titanium content will be retained in the steel under study for the further formation of interphase and ferritic precipitates of TiC, including with the participation of Mo [14–17].

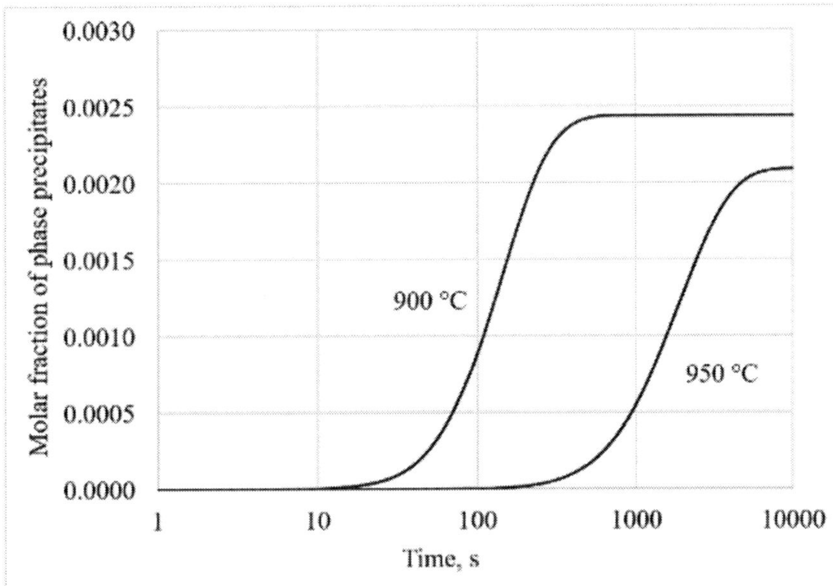

Figure 2. Isothermal precipitation of titanium carbide at 900 °C and 950 °C.

According to the calculation, the temperature of A_{e3} is 873.4 °C. The formation of cementite is possible below 683.4 °C. This temperature can be taken as A_{e1}. The conditions for the formation of Mo_2C and complex iron carbide M_7C_3 occur at temperatures below 637 and 600 °C, respectively. As a result of the significant consumption of carbon on the titanium carbide formation, including Mo containing precipitates [14–17], and diffusion inhibition, the precipitation of cementite, Mo_2C, and M_7C_3 is almost completely suppressed.

The results of the study using scanning electron microscopy show that the metal matrix of rolled products produced by all modes consists mainly of block ferrite, which has mostly equiaxed block shape, less often elongated. The dislocation density in ferrite varies from moderate to high with increasing cooling rate. Cementite or perlite was not detected in any of the rolling products.

Carbide precipitates of submicron sizes in the studied rolled samples are few (lots No. 4–6) or absent (lots No. 1–3). This indicates that their formation in austenite practically does not occur, as follows from the data in figure 2.

In all studied rolled samples, nanosized carbide precipitates are present (figure 3). Basically, these are precipitates formed in ferrite and of a mixed type (nucleated and partly grown by the interphase mechanism, and then increased their volume in the ferrite region), less often - interphase precipitates. In different areas, in variable proportions, as a rule, simultaneously precipitates of different types are present, including within the same ferrite block. However, there are areas in which precipitates of only or almost only one type are observed. Typical sizes of mixed-type precipitates are 2–3 nm (up to ~ 4 nm); in some areas they are larger: 2-5 nm (up to ~ 8 nm). Their significant part (possibly large) has a size of less than 2 nm. It is seen that with an increase in the cooling rate, the precipitation mechanism shifts more and more towards the realization of a mixed and ferritic mechanism. In rolled products cooled at maximum rate (No. 4), fields are observed where interphase precipitates with a very small distance between layers are visible, followed by sections with pure ferritic precipitates (figure 3b). Interphase precipitates are more common in rolled samples produced by a lower cooling rate.

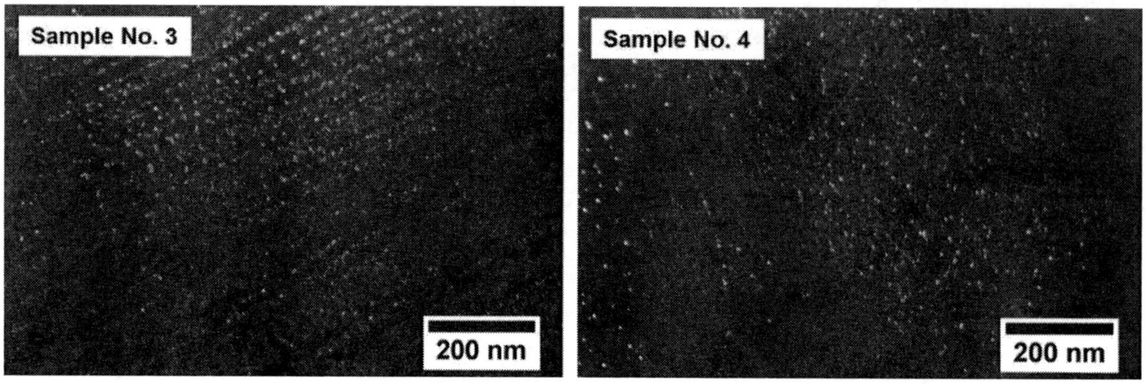

Figure 3. Results of the TEM study (dark-field image in the TiC reflection) of carbide precipitates in the samples: a - No. 3, b - No. 4.

4. Discussion

An analysis of the obtained results of the study including the mechanical properties and microstructure shows that an increase in the strength characteristics of rolled products is observed with a rise of the cooling rate and is accompanied by an increase in the dislocation density without changing the morphology of the ferrite structure. Moreover, the main role of phase precipitates in strengthening during rapid cooling consists in dispersion hardening due to the formation of ferritic precipitates mainly at dislocations, in the form of randomly located separate or small clusters of precipitates. At the same time, austenitic precipitates are practically absent, and the fraction of interphase particles is quite small.

With a decrease in the cooling rate of rolled products, the mechanism of austenite → ferrite transformation changes accompanying by the formation of a much less dislocation structure, which leads to a decrease in strength properties. In addition, the decrease in the rate of phase transformation taking place in this case contributes to the formation of interphase precipitates and to increase, respectively, the contribution of precipitation hardening to the strengthening of steel.

The dependence of realization of the interphase precipitation mechanism on the cooling rate and finishing temperature of rolling can be explained by analogy with the approach [20, 21] on the basis of an analysis of two following processes: phase precipitation and $\gamma \rightarrow \alpha$ phase transformation in steel. The nucleation of a new precipitation layer at the γ/α phase boundary occurs when a certain critical titanium concentration is reached, which is determined from the thermodynamic conditions of TiC nucleation. When the phase boundary moves, titanium diffuses from it to the nucleated precipitates. This diffusing titanium flux is compensated by its transition from the γ phase through the phase boundary. Thus, the titanium concentration at the moving γ/α boundary increases, and the moment of reaching the critical concentration for nucleation determines the distance between the layers. Therefore, the rates of the movement of the interphase boundary and the titanium diffusion, which ensures the nucleation and growth of carbide precipitates, should be comparable. If the metal is cooled too quickly, the rate of the interphase boundary turns out to be too high and the formation of precipitates by the interphase mechanism is violated: after passing through the front of the phase transformation, the ferrite remains supersaturated with respect to the carbide phase, which leads to the disordered formation of precipitates in ferrite. Therefore, there is an optimal combination when the nucleation and growth of interphase precipitates takes place, but the distance between the layers is minimal.

The conclusion made is in good agreement with the results of the analysis of the type and amount of precipitates in the studied steel, depending on the cooling rate. The increase in the cooling rate of rolled products results in the decrease in the distance between the layers of interphase precipitates and their fraction. The maximum cooling rate of rolled lot No. 4, apparently, is almost the limit for the implementation of the mechanism of interphase precipitation at a given T_f.

5. Conclusion

The highest strength properties $\sigma_{0.2} = 690$ MPa, $\sigma_B = 800$ MPa with good ductility $\delta = 18\%$ of rolled products from the studied steel containing wt.%: C-0.048, Mo-0.20, Ti-0.130, were achieved when using a maximum cooling rate of 22 °C/s from the finishing temperature of rolling 900 °C to $T_c = 650$ °C.

Simultaneously higher strength properties and ductility are achieved at a higher finishing temperature of rolling - 950 °C, compared with 900 °C. Strength characteristics also regularly improved with increasing cooling rate of rolled products from T_f to T_c.

With rapid cooling of rolled products, high strength characteristics are achieved as a result of the implementation, mainly, of dislocation and grain-boundary hardening mechanisms. This is due to the formation of a higher dislocation structure of ferrite and ferritic carbide precipitates.

A decrease in the cooling rate slows down the phase transformation, which contributes to the formation of interphase precipitates and to increase, respectively, the fraction of precipitation hardening in the strengthening of steel.

References

[1] Lesch C, Kwiaton N, and Klose F B 2017 *Steel Research Int.* **88** 1700210
[2] Zaitsev A, Koldaev A, Arutyunyan N, Dunaev S and D'yakonov D 2020 *Processes* **8** 646
[3] Deng X, Fu T, Wang Z, Liu G, Wang G and Misra R D K 2017 *Met. Mater. Int.* **23** 175
[4] Bai D Q, Yue S, Sun W P and Jonas J J 1993 *Metall. Mater. Trans.* A **24** 2151
[5] DeArdo A J 2003 *Int. Mater. Rev.* **48** 371
[6] Sanz L, Pereda B and López B 2017 *Mater. Sci. Eng.* A **685** 377
[7] Jiao Z B, Luan J C, Miler M K, Chung Y W and Liu C T 2017 *Mater. Today* **20** 142
[8] Zhang K, Wang H, Sun X J, Sui F L, Li Z D, Pu E X, Zhu Z H, Huang Z Y, Pan H B and Yong Q L 2018 *Acta Metall. Sinica (English Letters)* **31** 997
[9] Funakawa Y, Shiozaki T, Tomita K, Yamamoto T and Maeda E 2004 *ISIJ Int.* **44** 1945
[10] Jonas J J and Weiss I 2013 *Met. Sci.* **13** 238
[11] Gladman T 1999 *Mater. Sci. Technol.* **15** 30
[12] Lagneborg R, Siwecki T, Zajac S and Hutchinson B 1999 *Scand. J. Metall.* **28** 1
[13] Koldaev A V, Zaitsev A I, Krasnyanskaya I A and D'yakonov D L 2019 *Metallurgist* **63** 487
[14] Zhang K, Li Z, Wang Z, Sun X and Yong Q 2016 *J. Mater. Res* **31** 1254
[15] Park D B, Huh M Y, Shim J H, Suh J Y, Lee K H and Jung W S 2013 *Mater. Sci. Eng.* A **560** 528
[16] Funakawa Y and Seto K 2007 *Mater. Sci. Forum.* **539-543** 4813
[17] Shaposhnikov N G, Koldaev A V, Zaitsev A I, Rodionova I G, D'yakonov D L and Arutyunyan N A 2016 *Metallurgist* **60** 810
[18] Shaposhnikov N G , Mogutnov B M, Polonskaya S M, Kolesnichenko A P and Belyavskii P B 2004 *Materials Science* **11** 2
[19] Koldaev A V, Shaposhnikov N G, 2016 *Probl. Chern. Met. Materials Science* **3** 5
[20] Lagneborg R and Zajac S 2001 *Metall. Mater. Trans.* A **32** 39
[21] Zajac S 2005 *Mater. Sci. Forum. Trans. Tech. Publ.* **500** 75

Acknowledgments

This research was supported by the grant of the Russian Science Foundation (Project No. 18-19-00639) and was performed at Bardin Central Research Institute of Ferrous Metallurgy.

IOP Conf. Series: Materials Science and Engineering **969** (2020) 012018 doi:10.1088/1757-899X/969/1/012018

Study of Methods for Increasing Ductility and Formability of Cold-Rolled Ti-Stabilized IF Steels

A I Zaitsev, I G Rodionova and A V Koldaev

Bardin Central Research Institute of Ferrous Metallurgy, 23/9, Radio str., Moscow, 105005, Russia

E-mail: aizaitsev@chermet.net

Abstract. Study of the influence of the chemical composition and parameters of the complex treatment technology on characteristics of the microstructure, phase precipitates, the solid solution state, and mechanical properties was carried out for hot-rolled and cold-rolled Ti-stabilized IF steels. Cold-rolled products were annealed according to the regime of continuous hot-dip galvanizing units. The following methods of investigation were used: light, scanning and transmission electron microscopy, mechanical properties testing, thermodynamic analysis of phase stability conditions, and determination of the interstitial impurities content in a solid solution by internal friction. The possibility and conditions of a significant increase in ductility, formability of cold-rolled steels are established. They are obtained not by the traditional way consisting in reducing the content of interstitial elements and impurities, but by giving them a favorable form of existence. This is achieved by optimizing the chemical composition, first of all, the sulfur content, and parameters of thermo-deformation treatment of steel, which ensure complete carbon binding in $Ti_4C_2S_2$ as a result of TiS transformation already at the hot-rolling stage. To reduce the yield and tensile strength, it is necessary to decrease the content of interstitial elements and to increase the temperature of recrystallization annealing of cold-rolled steel.

1. Introduction

The rapid development of the automotive industry, high-speed transport, instrumentation, and industrial, household, and other technical objects makes it necessary to produce elements and assemblies of an increasingly complex shape by progressive stamping methods. For these purposes, cold-rolled steels, including coated ultralow carbon IF steels, are widely used [1,2]. However, they, as a rule, have insufficiently high indicators of ductility and formability (relative elongation of 44–50%, coefficients of normal plastic anisotropy, r – 2.0–2.2, and strain hardening, n – 0.20–0.24), which makes it difficult or impossible to obtain stamped products of complex shape. Cracks and other defects are observed on the body of the metal that hinder the improvement of the design and service characteristics of new technical means. In world practice, an increase in ductility and formability of rolled products is achieved mainly owing to the use of a high-purity metal charge and technological methods for obtaining low (less than 0.002%) carbon and nitrogen contents in IF steels [3–6]. This is associated with significant costs. On the other hand, it is well known that impurities and interstitial elements participate in the formation of a number of phase precipitates in IF steels, including complex composition [7–16], and not their absolute content is important, but the form of the presence of impurities and interstitial elements [13–17]. Therefore, a considerable problem is to establish the possibilities and conditions for achieving fundamentally improved indicators of ductility and formability of cold-rolled steels using not the traditional way of obtaining an extremely high degree of impurity and interstitial element cleanliness, but by giving them a favorable form of existence. This is the subject of present study.

2. Materials and methods

The most economical cold-rolled Ti-microalloyed (stabilized) IF steels annealed in continuous units were studied. Attention was focused on continuous hot-dip galvanizing unit, CHGU, because obtaining high ductility and formability in unit of this type is difficult due to the limited duration of stay of the strip in it. The steels were smelted in a vacuum induction furnace with a magnesite crucible with a steel capacity of 7–8 kg and casted in a vacuum in one ingot. The results of determining their chemical composition are presented in table 1.

Table 1. The chemical composition of the studied steels.

Melting no.	Content of elements, wt. %													
	C	Si	Mn	P	S	Cr	Ni	Cu	Al	N	Mo	V	Nb	Ti
1	0.004	0.01	0.10	0.007	0.007	0.02	0.02	0.03	0.041	0.003	0.003	0.001	-	0.057
2	0.006	0.01	0.11	0.796	0.067	0.03	0.02	0.02	0.042	0.005	0.004	0.002	-	0.060
3	0.004	0.01	0.11	0.782	0.286	0.02	0.02	0.03	0.036	0.004	0.002	0.002	0.001	0.059
4	0.003	0.01	0.09	0.006	0.007	0.03	0.01	0.02	0.035	0.004	0.001	0.001	-	0.063

The ingots were heated to 1220 °C, held for 1 hour, and rolled on DUO-300 reversible hot rolling laboratory mill with 12–14% reduction in the last pass into strips of 3 mm thickness. After rolling end at various finishing temperatures, Tf, the strip was cooled in a stream of air with a cooling rate of 10–15 °C/s to the temperature of the strip coiling into a roll, Tc. Then it was placed in a furnace heated to this temperature, held for 30 minutes, followed by cooling with the furnace, which simulated the cooling of a roll. The main parameters of the thermo-deformation metal treatment are presented in table 2.

Table 2. Parameters of the thermo-deformation treatment of steels.

Melting no.	Tf, °C	Tc, °C	Cooling rate to Tc, °C/s
1	885	740	10-12
2	920	720	13-15
3	930	720	13-15
4	920	720	13-15

Subsequently, hot-rolled strips were pickled and rolled at the DUO-QUARTO 320 cold rolling laboratory mill with 75% reduction into strips of 0.75 mm thick. The obtained cold-rolled steel was subjected to recrystallization annealing on a universal testing machine according to a regime simulating annealing in the CHGU with a strip moving rate of 100 m/min. The main parameters were as follows: annealing temperature, Ta – 810–830 °C (see below), delayed cooling to 690 °C, accelerated cooling to 475 °C, accelerated cooling after a galvanizing bath to 190–200 °C. After annealing, the bands were subjected to temper rolling with a relative elongation of 0.8%. Samples were made of the obtained hot-rolled and cold-rolled products to study the characteristics of the structural state and the properties of the metal.

The microstructure was studied by methods of light microscopy, scanning electron microscopy, SEM, using a JSM-6610LV (JEOL) instrument equipped with an INCA Energy Feature XT energy dispersive microanalysis system, INCA Wave 500 wave dispersion spectrometer, and transmission electron microscopy, TEM, using a JEM200CX microscope. Herewith, the type of nanosized precipitates was determined by the original method based on the analysis of their reflections in microdiffraction images [18]. The mechanical properties and carbon content of the solid solution were determined by tensile methods on a HECKERT FP-100/1 tensile testing machine and internal friction, respectively. The thermodynamic calculation of the regions of existence of phases in the steel under study was carried out using the approaches presented in detail in [19, 20].

3. Results

By means of light microscopy the average grain diameter was measured in the rolling direction (dx) and in the direction perpendicular to the rolling direction (dy). Using these values the average grain diameter (da) was calculated as the arithmetic mean of dx and dy, as well as the ratio dx/dy, which is in fact, a characteristic of the degree of elongation of the grain shape. The coefficient of variation characterizing the heterogeneity of the grain structure was found as the ratio of the standard deviation to the average grain size. The results of the study of the microstructure and mechanical properties of hot-rolled products are presented in tables 3.

Table 3. Parameters of the microstructure and mechanical properties of hot-rolled steels studied.

Melting No.	d_y, mkm	d_x, mkm	d_x/d_y	d_a, mkm	Coefficient of variation	$\sigma_{0.2}$, MPa	σ_B, MPa	δ, %
1	15.2	16.2	1.07	15.7	0.59	201	291	43.4
2	14.2	16.2	1.14	15.2	0.52	163	289	43.2
3	14.3	16.3	1.14	15.3	0.55	199	287	43.1
4	14.2	16.0	1.13	15.1	0.50	159	283	44.5

The test results of the mechanical properties and parameters of annealing of cold rolled steels studied are presented in table 4.

Table 4. Annealing parameters and mechanical properties of cold rolled steel studied.

Melting No.	T_a, °C	$\sigma_{0.2}$, MPa	σ_B, MPa	δ, %	r	n
1	825	143	294	48.7	2.05	0.27
2	830	148	313	50.9	2.09	0.26
3	810	158	312	46.9	2.06	0.26
4	830	128	279	51.7	2.40	0.26

The results of thermodynamic calculation showed that the main types of phase precipitates in the studied steels are TiN, TiS, $Ti_4C_2S_2$, TiC. Nitrogen at high temperatures is completely bound in TiN. Depending on the chemical composition and processing parameters of the steel, carbon is predominantly involved in the formation of $Ti_4C_2S_2$ or TiC. This was confirmed by the TEM study, which found out a significant amount of such precipitates in rolled metal (figure 1).

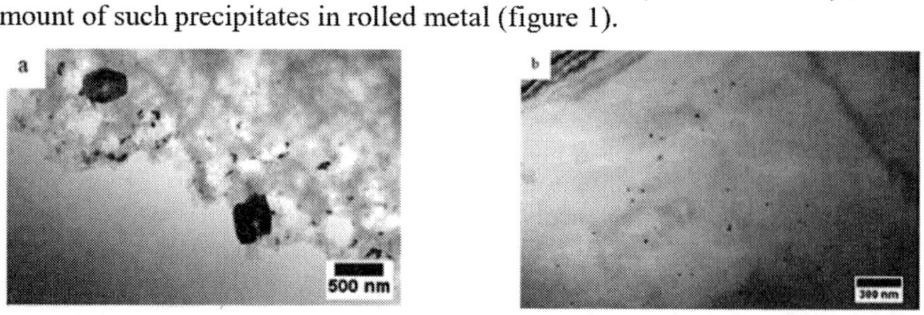

Figure 1. Typical view of precipitates in cold rolled steel: a - titanium carbosulfide (melting No. 2, x20000), b - titanium carbide (melting no. 1, x30000).

4. Discussion

Data of table 3 show that a slightly larger grain size was obtained for hot-rolled steel of melting No.1 produced using a lower finishing temperature of rolling and a higher temperature for coiling the strip into a roll. This may be due to a lower cooling rate of the strip after the rolling end. For rolled steels of melting No. 2, 4 the values of yield strength are lower (table 3). Determination of the carbon content in the solid solution by the method of internal friction showed that the samples No. 1 and 3 contain about 6 and 7 ppm carbon, respectively, while the carbon in the solid solution in steels No. 2 and 4 is absent. The calculation according to the generally accepted ratio of the solid solution hardening of steel caused by the presence of 6-7 ppm carbon in the solid solution showed that this value is about 25-30 MPa. Thus, the increased values of the yield strength of rolled steels of melting No. 1 and 3 are mainly associated with solid solution hardening with carbon. Therefore, the preservation of even minor carbon concentrations in a solid solution can lead to a significant increase in the yield strength.

From the data of table 4 it can be seen that the highest ductility and formability are obtained for cold-rolled steels No. 2 and 4, which do not contain carbon in a solid solution of hot-rolled products. According to the results of the study by SEM, TEM, and local X-ray spectral analysis of rolled products it was shown that this is achieved due to the preferential binding of carbon in $Ti_4C_2S_2$ (figure 1a). There are much more such precipitates in rolled products from steel of melting No. 4 and, especially No. 2, than in the case of melting No.1 and 3. Analysis of the phase stability conditions in the studied steels confirmed the conclusion made. It should be noted that steels of melting No. 2 and 4 are characterized by the highest and the lowest carbon content, respectively, which determines the amount of $Ti_4C_2S_2$ precipitates formed.

The kinetics studies of strain-induced $Ti_4C_2S_2$ precipitation performed for IF steel of model composition close to investigated, (wt.%): 0.004C, 0.02Si, 0.18Mn, 0.006P, 0.014S, 0.09Ti, 0.05Al, 0.004N [12], led to the following conclusions. The "nose" of the C-shaped curve, corresponding to the extreme combination of thermodynamic and kinetic stimuli for the formation of precipitates, is at a temperature of about 940 °C. If TiS precipitates are present in the steel, the formation of $Ti_4C_2S_2$ precipitates occurs at 910–950 °C by heterogeneous nucleation on the surface and transformation of TiS to $Ti_4C_2S_2$. This process is rather fast and ends in 20-30 s. Thus, there are all prerequisites for the complete binding of carbon from a solid solution to $Ti_4C_2S_2$ during hot rolling of steel and its end at 910–950 °C. Studies performed for steels with a low sulfur content up to 0.004 - 0.005 wt.% showed that in the absence or low TiS content, the formation of $Ti_4C_2S_2$ precipitates occurs through self-nucleation and growth of nuclei of this phase. This leads to a significant increase in the time of complete transformation, reaching values of more than 1000 s. As a result, predominant carbon binding occurs not in $Ti_4C_2S_2$, but in TiC and can create the preconditions for its conservation in the solid solution. This is actually observed in the case of steels No. 1 and 3.

The noted features are also confirmed by the results of the thermodynamic calculation of phase stability conditions. In particular, they indicate that upon heating billets of steel of melting No. 1, 2, and 4 for rolling to 1220 °C, titanium sulfide is formed in them. This, during subsequent hot rolling, facilitates the formation of titanium carbosulfide precipitates by transforming TiS precipitates already present and leads to more efficient carbon removal from the solid solution. Moreover, due to the higher finishing temperature of rolling and the mobility of atoms of the phase-forming components, the formation of titanium carbosulfide precipitates in steels No. 2 and 4 is more intense and leads to the complete removal of carbon from the solid solution. To a lesser extent, the transformation under consideration in steel of melting No. 1 leads to a decrease in the fraction of carbon bound to titanium carbosulfide, to incomplete subsequent binding to titanium carbide and to its conservation in solid solution.

Due to the low sulfur content in steel of melting No. 3, during the heating of billets for rolling, TiS formation does not occur at all. Therefore, the precipitation of titanium carbosulfide occurs by the mechanism of independent nucleation and the growth of precipitates of this phase, which has a significantly lower rate compared to the transformation of TiS precipitates. As a result, a significant part of the carbon is bound into finely dispersed titanium carbide precipitates and remains in solid solution.

The aforementioned circumstances lead to lower ductility and formability of cold-rolled steel of No. 1 and, especially, No. 3, compared to melting No. 2.

For cold-rolled products from all the studied steels, the uniform microstructure and rather high ductility and formability indicators were obtained. Nevertheless, for steels No. 2 and 4 they are higher than in the case of steels No. 1 and 3. This is due to the predominant bonding of carbon from the solid solution to $Ti_4C_2S_2$ already at the stage of hot rolling of steel. It is important to note that the content of impurities and interstitial elements in steels No. 2 and 4 is at a level no lower than in steels No. 1 and 3, which indicates the possibility of a significant increase in ductility and formability by giving them a favorable form of existence. Analysis of the data in table 4 shows that to obtain low values of strength characteristics, it is necessary to reduce the content of interstitial elements. An increase in the temperature of recrystallization annealing of rolled products also contributes to this.

5. Conclusions

Thus, the results obtained unambiguously indicate the possibility of a significant increase in the ductility and formability of cold-rolled IF steels annealed according to the CHGU regime by giving a favourable existence form to the impurities and interstitial elements. For this, the sulfur content should be sufficiently high (at least 0.006 wt.%) to ensure the stability of TiS precipitates during heating of the steel for rolling, and the possibility of complete carbon binding in $Ti_4C_2S_2$ already at the stage of hot rolling. This is facilitated by the use of high finishing temperatures of rolling 910-950 °C, at which $Ti_4C_2S_2$ precipitates are formed with a significant rate due to the transformation of TiS. Otherwise, predominant carbon binding occurs not in $Ti_4C_2S_2$, but in TiC, which can create the prerequisites for its conservation in solid solution and lowering the level of properties of rolled products. A decrease in the strength characteristics is favoured by a decrease in the content of interstitial elements and an increase in the temperature of recrystallization annealing of cold-rolled steel.

References

[1] Mendoza R, Huante J, Alanis M, Gonzalez-Rivera C and Juarez-Islas J 2000 Processing of ultra low carbon steels with mechanical properties adequate for automotive applications in the as-annealed condition *Mater. Sci. Eng.* A **276** 203-209

[2] Tsunoyama K 1998 Metallurgy of Ultra-Low-C Interstitial-Free Sheet Steel for Automobile Applications *Phys. Status Solidi* A **167** 427

[3] Elsner A, Kaspar R, Ponge D, Raabe D and Zwaag S 2004 Recrystallisation texture of cold rolled and annealed if steel produced from ferritic rolled hot strip *Mater. Sci. Forum* **467-470** 257

[4] Jian C, Yiyu Z and Lixin Z 2005 *Strategic Study CAE* **7** 21

[5] Zhang L and Thomas B G 2003 State of the art in evaluation and control of steel cleanliness *ISIJ Int.* **43** 271

[6] Zhang L, Zhi J, Mei F, Zhu L, Jiang X, Shen J, Cui J, Cai K and Thomas B G 2006 *Ironmaking Steelmaking* **33** 129

[7] Ray R K and Ghosh P 2010 Thermodynamics of precipitation and textural development in batch-annealed interstitial-free high-strength steels *Mater. Manuf. Processes* **25** 195

[8] Hua M, Garcia C I, Eloot K and DeArdo A J 1997 Identification of Ti-S-C-containing Multi-phase Precipitates in Ultra-low Carbon Steels by Analytical Electron Microscopy *ISIJ Int.* **37** 1129

[9] Carabajar S, Merlin J, Massardier V and Chabanet S 2000 *Mater. Sci. Eng.* A **281** 132

[10] Regon W, Jorge Jr A M and Balancin O 2002 *Acta Microsc.* **11** 71

[11] Regon W, Jorge Jr A M and Balancin O 2003 *Scripta Mater.* **48** 773

[12] Dong F. Xue F, Du L and Liu X 2015 Promoting Ti4C2S2 strain induced precipitation during asymmetrical hot rolling to improve r value and advantaged texture in Ti stabilized IF steel *J. Alloys Comp.* **620** 240

[13] Takechi H 1994 Metallurgical Aspects on Interstitial Free Sheet Steel From Industrial Viewpoints *ISIJ Int.* **34** 1-8

[14] Shin E J, Seong B S, Han Y S, Hong K P, Lee C H and Kang H J 2003 Effect of precipitate size and dispersion on recrystallization behavior in Ti-added ultra low carbon steels *J Appl. Cryst.* **36** 624
[15] Choi J Y, Seong B S, Baik S C and Lee H C 2002 Iron Steel *ISIJ Int.* **42** 889
[16] Zaitsev A I 2017 Prospective directions for development of metallurgy and materials science of steel *Pure Appl. Chem.* **89** 1553
[17] Ray R K, Ghosh P and Bhattacharjee D 2009 *Mater. Sci. Techn.* **25** 1154
[18] Zaitsev A, Koldaev A, Arutyunyan N, Dunaev S and D'yakonov D 2020 Effect of the chemical composition on the structural state and mechanical properties of complex microalloyed steels of the ferritic class *Processes* **8** 646
[19] Shaposhnikov N G, Mogutnov B M, Polonskaya S M, Kolesnichenko A P and Belyavskii P B 2004 Thermodynamic modeling as a tool for improving heating technology for ingots of steel 12Kh18N10T for rolling *Materials Science* **11** 2-9
[20] Shaposhnikov N G, Koldaev A V, Zaitsev A I, Rodionova I G, D'yakonov D L and Arutyunyan N A 2016 Regularities of titanium carbide precipitation in low carbon Ti–Mo-microalloyed high strength steels *Metallurgist* **60** 810-816

Acknowledgments
The study was carried out under agreement on granting the subsidy No. 075-11-2018-170 dated November 26, 2018. Unique identifier of Project - RFMEFI62418X0050.

Electromagnetic Properties of Composite Materials Based on Nd-Compound Modified Fe-Si-Al Alloy

A I Goryachko, S N Ivanin and V Yu Buz'ko

Department of radiophysic and nanotechnologies, Kuban State University, 149, Stavropolskaya, Krasnodar, 350040, Russia

E-mail: Alexandr_g_i@mail.ru

Abstract. The investigation of microstructure, optical and electromagnetic properties of powder prepared from Nd-compound doped Al6Si9Fe85 (Alsifer) alloy. The result of energy-dispersive X-ray spectroscopy is an evidence that the complementary Nd-containing Nd-Fe phase is on the surface of Alsifer particles. The magnetic particles distribution histogram in powder after mechanical treatment was plotted and analyzed. The electromagnetic properties (components magnetic permeability and dielectric permittivity) for produced composites with Nd-compound doped Alsifer powder with 50-90% (wt.) was studied. The nonlinear concentration dependence of magnetic and dielectric permittivity versus concentration Nd-compound doped Alsifer particles was observed. The magnetic resonance in composites was detected that shifted in low frequencies with increased magnetic filler concentration. The resonance field, absorption linewidth, g-factor and microwave absorption ability of Nd-compound doped Alsifer powder were determined from FMR experiment.

1. Introduction

The composite materials with magnetic fillers are frequently used in construction of electromagnetic shields, for microwave emission protection, for electromagnetic compatibility of electronic technical devices, and as component of microwave absorption materials [1−6]. The electromagnetic parameters which are magnetic ($\mu=\mu'+i\mu''$) and dielectric ($\varepsilon=\varepsilon'+i\varepsilon''$) permittivity's are major parameters for estimation of ability of dispersed composite materials for microwave absorption. The magnetic soft materials are suitable for high values of magnetic permittivity for composites in ultra-high frequency (UHF) range. The alloy Fe-Si-Al family is one of the main ferromagnetic soft magnetic materials used as magnetic powder fillers for composite materials [7−12]. Doping or modification of AlSiFe alloy is way to improve electromagnetic characteristic to microwave electromagnetic waves range [13−18].

In this work, for the powder of a commercial $Al_6Si_9Fe_{85}$ alloy modified by the addition of Nd compounds, the microstructural, optical, and electromagnetic properties were investigated. The magnetic and dielectric properties for manufactured composite materials with different concentrations of the investigated magnetic powder filler in paraffin matrix were studied depending on the frequency of the applied electromagnetic field.

2. Materials and measurement

The commercially available highly permeable ferromagnetic Nd-compound modified FeSiAl alloy (China, Chengdu Huarui Industrial Materials Co., Ltd.) was used in this study. The alloy sample was previously mechanically ground, followed by grinding in a ceramic mortar for 1 hour at room temperature in ethanol. Next, the obtained powder was transferred to filter paper and dried in a vacuum oven for 2 hours at 40 ° C.

Micrographs of the Nd-compound modified AlSiFe alloy powder were obtained using a "JEOL JSM – 7500F" scanning electron microscope by mean of the back-reflected electrons detection mode (COMPO). The energy dispersive analysis (EDA) method was used to study the elemental composition of the test sample using a "INCA X-Sight" energy dispersive X-ray spectrometer.

The optical diffuse reflectance spectra of the investigated alloy powder were measured on a "Hitachi U-3900" spectrophotometer with a two-channel integrating sphere in the range 300–900 nm.

The powder EPR spectrum was obtained at room temperature (20 °C) on "JEOL JES FA-300" EPR spectrometer at a frequency of 9.14 GHz with a modulation frequency of 100 kHz.

To determine the magnetic and dielectric properties of the studied Nd-compound modified AlSiFe alloy, AlSiFe/paraffin composites were prepared by means of hot pressing in special brass fixture with a concentration of magnetic filler of 50–90% in increments of 10% by weight. The studied composite materials were made in the form of a toroid with dimensions of an inner diameter of 3.05 mm and an outer diameter of 7 mm. The scattering parameters (S-parameters) of the prepared composites in a coaxial cell were measured using a "Deepace KC901V" vector network analyzer in the transmission line mode in 0.015–7 GHz range.

3. Results

Figure 1 shows the EDA spectra for a particle of the studied magnetic powder. According to the obtained results, it follows that the main fraction of particles contains Fe, Si and Al chemical elements (Figure 1, a) that correspond to the AlSiFe magnetic alloy under study. However, a detailed study of the surface of the powder in the detection mode of back-reflected electrons showed the microparticles whose contrast shade was significantly different from the main powder under study. The additional elemental analysis of the selected particle was carried out, according to the results of which it follows that on the surface of the studied Fe-Si-Al alloy an Nd and Fe elements are present. The fractions of Nd and Fe elements corresponds to Nd2Fe14B chemical compound.

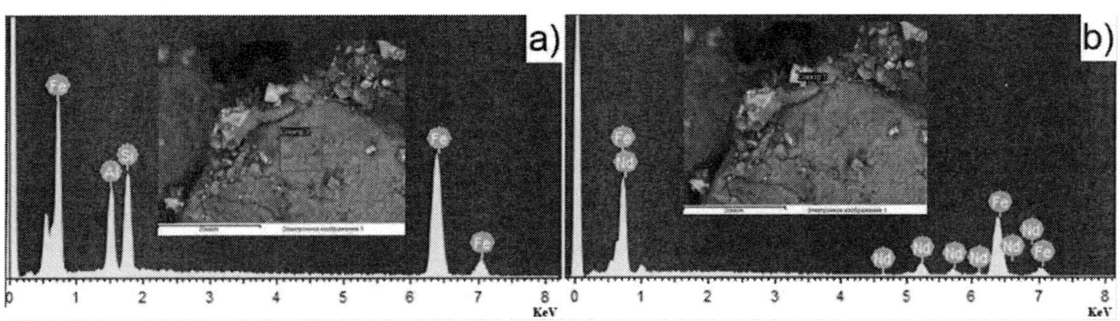

Figure 1. Energy dispersive analysis spectrum with a selected analysis zone for (a) Fe-Si-Al powder and (b) Nd-compound particles.

Microphotograph of the obtained powder of $Nd_2Fe_{14}B$ modified AlSiFe alloy is presented on figure 2a, a. As can be seen from the figure, a large fraction of the particles of the sample under study represents an irregular fragmented and stone-like shape. Based on the obtained photographs the histograms of the particle size distribution of the powder under investigation were calculated depending on their percentage using the "ImageJ" program (figure 2b). The obtained data revealed that the ground sample has a wide particle size distribution in the range from 10 to 115 microns. The main fraction of particles is in the range of 11-46 microns, the total percentage of which is more than 60% of the total.

Figure 3 shows the optical diffuse reflectance spectrum for ground $Nd_2Fe_{14}B$ modified AlSiFe alloy in the range 300–900 nm. The magnetic powder has a light reflection of not more than 20% in investigated spectral range. There is a gradual decrease in light reflection from 17% at $\lambda = 780$ nm to 12.7% at $\lambda = 380$ nm.

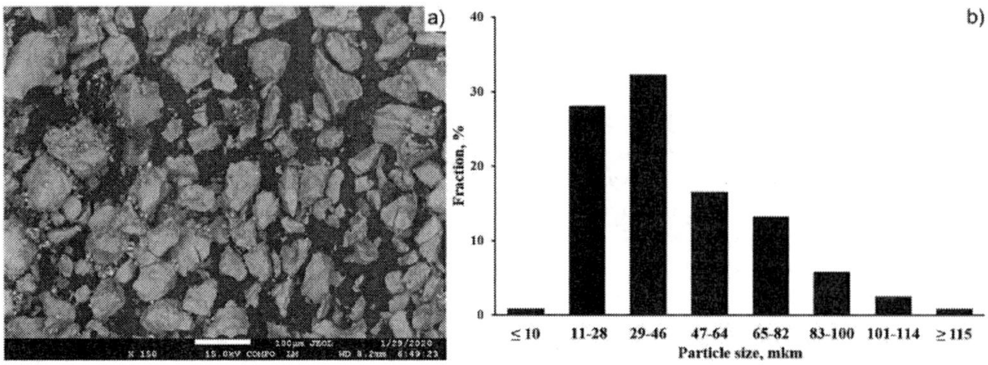

Figure 2. A microphotograph of the studied powder at ×150 magnification (a) and histogram of the particle size distribution (b).

The value of the reflectance minimum for the AlSiFe modified Nd micropowder under study is in the range of 358–418 nm, not exceeding 13%.

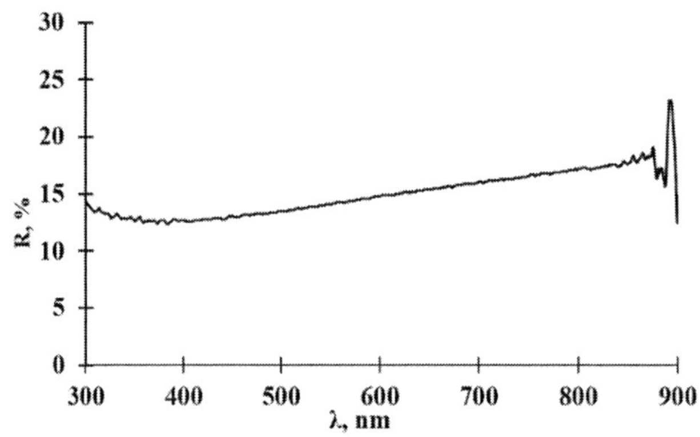

Figure 3. The diffuse reflectance spectrum of $Nd_2Fe_{14}B$ modified AlSiFe powder.

The values of magnetic permeability ($\mu = \mu' + i\mu''$) and dielectric permittivity ($\varepsilon = \varepsilon' + i\varepsilon''$) in a complex form for fabricated composite materials based on the studied powder of $Nd_2Fe_{14}B$ modified AlSiFe alloy were calculated from experimentally measured S-parameters values according to the Nicholson-Ross-Weir algorithm [19–21].

The values of real and imaginary parts of magnetic permeability (μ', μ'') in the studied frequency range are presented in Figure 4. As can be seen from the data in Figure 4, a, in the low-frequency region an increase in μ' is observed with increasing filler concentration ($\mu'_{50\%}$=1.90, $\mu'_{60\%}$=2.41, $\mu'_{70\%}$=2.83, $\mu'_{80\%}$=3.40, $\mu'_{90\%}$=4.71 at 20 MHz). A sharp decrease in μ' value is observed at a frequency of ≈ 1 GHz, for all studied composite materials as the frequency of electromagnetic radiation increases. Based on the data of Fig. 4b, for μ" values there is a peak in magnetic losses, which corresponds to the resonant frequency. A shift of the resonance to the low-frequency region was observed with an increase in the filler concentration ($f_{r,50\%}$≈1.65 ГГц, $f_{r,60\%}$≈1.55 ГГц, $f_{r,70\%}$≈1.1 ГГц, $f_{r,80\%}$≈0.56 ГГц, $f_{r,90\%}$≈0.19 ГГц).

In addition, it should be noted that the authors of [22] investigated the magnetic properties of composite materials based on ferromagnetic alloys. The analysis hen the data of [22] to reveal that magnetic resonance is also observed for composite materials based on plate-like AlSiFe particles in the UHF region. As can be seen from the high-frequency spectra obtained in [22], the μ' and μ'' values are significantly higher than for our values for the composite materials investigated in this work. The main reason may be the high uniformity of the structure of the plate-like particles of AlSiFe alloy used in [22], which entails a uniform distribution of magnetic particles inclusions in the composite material [23].

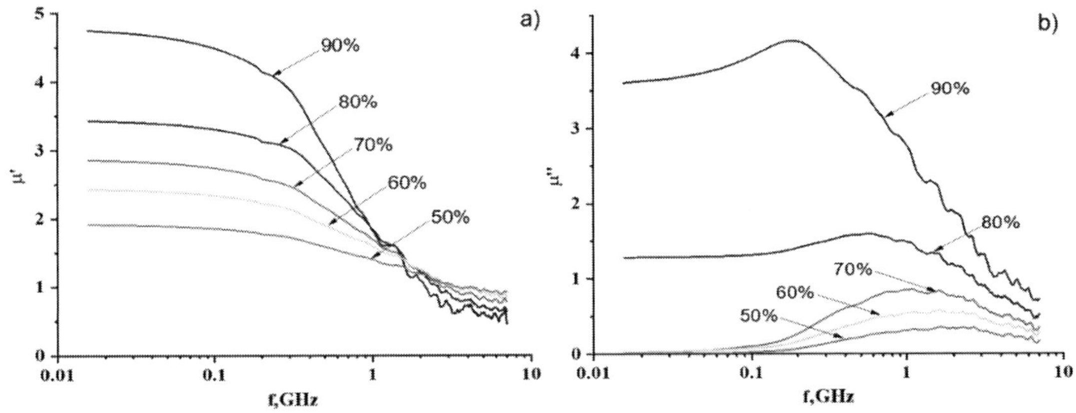

Figure 4. The frequency dependence of magnetic permeability components for composite based on $Nd_2Fe_{14}B$ modified AlSiFe in paraffin.

The dependences of the dielectric constant parts (ε' and ε") for the investigated composites are shown in Fig. 5 (a, b). According to the experimental data, it can be seen that with an increase in the magnetic filler concentration, both ε' and ε" increase for all the samples under study. It can be seen that the ε' values remain almost unchanged in the entire investigated frequency range for composite materials with $Nd_2Fe_{14}B$ modified AlSiFe filler concentrations of 50–80% ($\varepsilon'_{50\%}\approx3.6$, $\varepsilon'_{60\%}\approx4.5$, $\varepsilon'_{70\%}\approx5.95$, $\varepsilon'_{80\%}\approx9.8$). However, for investigated composite material with a concentration 90% of the studied magnetic powder a decrease in ε' is observed from 24.3 to 20.8. Based on the analysis of dielectric permittivity parts data (ε' and ε") in the studied frequency range, we can conclude that there are no relaxation processes for the manufactured composite materials.

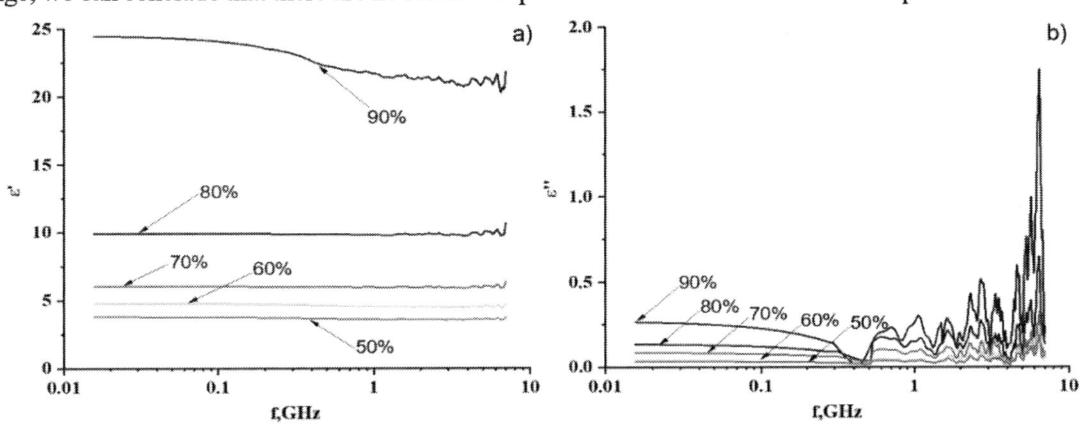

Figure 5. The frequency dependence of dielectric permittivity components for composite based on $Nd_2Fe_{14}B$ modified AlSiFe in paraffin.

Figure 6 shows the EPR spectrum of the studied $Nd_2Fe_{14}B$ modified AlSiFe powder. The resonance field H_r, the absorption line width ΔH (from peak to peak), and the intensity of the first derivative of the absorption signal I were determined from the EPR spectrum. The EPR spectrum has the following parameters: H_r = 115.56 mT, ΔH = 138.18 mT, I = 1662 a.u. The g-factor value was calculated by the formula (1):

$$g = h \cdot \frac{\nu}{\mu_B} \cdot H_r \quad (1)$$

where h – Planck constant (6.626×10^{-34} J·s), ν – EPR spectrometer electromagnetic frequency (9.14 GHz), μ_B – Bohr magneton (9.274×10^{-24} J/T), H_r – resonance field (115.56 mT). The g-factor calculated by formula (1) has a value g = 5.65.

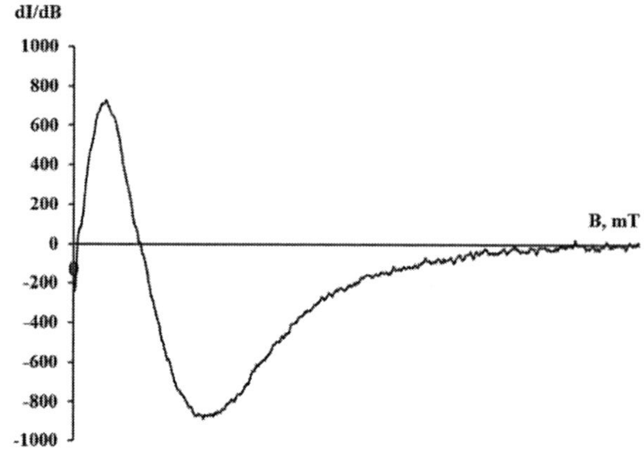

Figure 6. EPR spectrum of the studied $Nd_2Fe_{14}B$ modified AlSiFe powder.

According to the study [24], the g-factor for Fe compounds can vary from 4 to 6. Based on this, we can conclude that the EPR signal for the $Nd_2Fe_{14}B$ modified AlSiFe powder under study gives Fe ions in materials structure. It is also worth noting that composite films based on AlSiFe powder with various concentrations in epoxy resin were studied in [25]. By analyzing the FMR spectra of the magnetic AlSiFe-based composite films from [25], we can conclude that magnetic phase A [25] has a similar EPR spectrum shape as the $Nd_2Fe_{14}B$ modified AlSiFe powder in present study.

4. Conclusion

As a result of the study the data on the microstructure, chemical composition, optical diffuse reflectance properties of the $Nd_2Fe_{14}B$ modified AlSiFe alloy the powder after mechanical machining were obtained. The analysis of the calculated magnetic permeability for composites with $Nd_2Fe_{14}B$ modified AlSiFe alloy powder as magnetic filler showed rather high values in the studied frequency range. Magnetic resonance in frequency range 0.2-3 GHz was detected for $Nd_2Fe_{14}B$ modified AlSiFe powder-based composite materials under study. Data analysis showed the dependence of the microwave absorption resonance frequency on the concentration of the investigated AlSiFe powder filler. Based on the obtained electromagnetic characteristics, it can be concluded that studied $Nd_2Fe_{14}B$ modified AlSiFe powder filler is a promising material for development of microwave absorbing materials in the low frequency range of 0.2–2 GHz.

References

[1] Deng J L and Feng B 2012 Carbonyl Iron. Graphite Double-Layer Structural Absorbing Composite *Advanced Materials Research* 557–559

[2] Pourabdollahi H and Zarei A R 2017 Hydrothermal synthesis of carbonyl iron-carbon nanocomposite: Characterization and electromagnetic performance *Results in Physics* 1978–1986

[3] Vyzulin S A, Buz'ko V Y, Kalikintseva D A, Goryachko A I, and Sarin L I 2019 Magnetic and dielectric properties of composites based on magnetic microspheres *Journal of Physics Conference Series* **1389**

[4] Vyzulin S A, Kalikintseva D A, Sarin L I, Buz'ko V Yu, Goryachko A I and Kolantsov O A 2019 Shielding and microwave absorption characteristics of composites based on graphite and magnetic material ECOM-P *Radiation and Scattering of Electromagnetic Waves* 332–335

[5] Teber A, Cil K, Yilmaz T, Eraslan B, Uysal D, Surucu G 2017 Manganese and Zinc Spinel Ferrites Blended with Multi-Walled Carbon Nanotubes as Microwave Absorbing Materials *Aerospace* **4** 1

[6] Zheng H and Sun H 2018 Study on Microwave Absorbing Properties of CoFeBSiNb/Graphene Composites *Material Sciences* **8 1** 29–36
[7] Cai X, Jiang X, Xie W, Mu J and Yin D 2018 Effect of particle size on the preparation and microwave absorption properties of FeSiAl magnetically soft alloy hollow microspheres *Defence Technology* **14 5** 477–483
[8] Zhou L, Huang J, Wang X, Su G, Qiu J and Dong Y 2019 Mechanical, dielectric and microwave absorption properties of FeSiAl/Al2O3 composites fabricated by hot-pressed sintering *Journal of Alloys and Compounds* **774** 813–819
[9] Li Q, Feng Z, Yan S, Nie Y and Wang X 2015 Comparison of the Magnetic and Absorption Properties of Flaky Super Sendust and Sendust Alloys *Journal of Electronic Materials* **44 10** 3777–3782
[10] Starostenko S N, Rozanov K N, Shiryaev A O, Lagar'kov A N and Shalygin A N 2017 Estimation of the superhigh-frequency magnetic permeability of alsifer from the measured permeability of composites *Physics of the Solid State* **59 11** 2203–2210
[11] Zhou T D, Liang D F, Deng L J and Luan D C 2011 Electron Structure and Microwave Absorbing Ability of Flaky FeSiAl Powders *Journal of Materials Science & Technology* **27 2** 170–174
[12] Sakai K, Wada Y and Yoshikado S 2008 Composite electromagnetic wave absorber made of permalloy or sendust and effect of sendust particle size on absorption characteristics *PIERS Online* **4 8** 846–853
[13] Yun Y W, Kim S W, Kim G Y, Kim Y B, Yun Y C and Lee K S 2006 Electromagnetic shielding properties of soft magnetic metal and ferrite composites for application to suppress noise in a radio frequency range *Journal of Electroceramics* **17** 467–469
[14] Zhou T D, Tang J K and Wang Z Y 2010 Influence of Cr content on structure and magnetic properties of Fe–Si–Al–Cr powders *Journal of Magnetism and Magnetic Materials* **322** 17
[15] Wang X, Gong R, Li P, Liu L and Cheng W 2007 Effects of aspect ratio and particle size on the microwave properties of Fe–Cr–Si–Al alloy flakes *Materials Science and Engineering: A* **466** 178–182
[16] Wang J, Fan X, Wu Z and Li G 2015 Synthesis, microstructure and magnetic properties of $Fe_3Si_{0.7}Al_{0.3}$@SiO_2 core–shell particles and Fe_3Si/Al_2O_3 soft magnetic composite core *Journal of Solid State Chemistry* **231** 152–158
[17] Xu H, Bie S, Jiang J, Yuan W, Chen Q and Xu Y 2016 Electromagnetic and microwave absorbing properties of the composites containing flaky FeSiAl powders mixed with MnO_2 in 1–18 GHz *Journal of Magnetism and Magnetic Materials* **401** 567–571
[18] Liu J, Ma T, Tong H, Luo W and Yan M 2010 Electromagnetic wave absorption properties of flaky Fe–Ti–Si–Al nanocrystalline composites *Journal of Magnetism and Magnetic Materials* **322 8** 940–944
[19] Nicolson A M and Ross G F 1970 Measurement of the Intrinsic Properties of Materials by Time-Domain Techniques *IEEE Transactions on Instrumentation and Measurement* **19 4** 377–382
[20] Ivanin S N, Buz'ko V Y, Goryachko A I and Panushkin V T 2020 Electromagnetic characteristics of heteroligand complexes of gadolinium stearate *Journal of Physical Chemistry* **94 8**
[21] Luukkonen O, Maslovski S I and Tretyakov S A 2011 A stepwise Nicolson-Ross-Weir -based material parameter extraction method *IEEE Antennas and Wireless Propagation Letters* **10** 1295–1298
[22] Rozanov K N, Bobrovskii S Y, Lagarkov A N, Mishin A D 2017 Revealing the effect of interaction between inclusions on the effective microwave permeability of composites *Procedia Engineering* **216** 85–92
[23] Rozanov K N, Osipov A V, Petrov D A, Starostenko S N and Yelsukov E P 2009 The effect of shape distribution of inclusions on the frequency dependence of permeability in composites *Journal of Magnetism and Magnetic Materials* **321 7** 738–741
[24] Bennett J E, Gibson J F, Ingram D J E, Haughton T M 1961 Electron Resonance Studies of Haemoglobin Derivatives. II. Results for Types A, B, C, D and F Myoglobin Crystals

Proceedings of the Royal Society A: Mathematical, Physical and Engineering Sciences **262** 395–408

[25] Vyzulin S A, Kevraletin A L, Syr'ev N E, Buz'ko V Y, Michenko E S and Rozanov K N 2018 Magnetoresonance Properties of an AlSiFe-Based Composite Film System *Bulletin of the Russian Academy of Sciences: Physics* **82** **5** 585–587

IOP Conf. Series: Materials Science and Engineering **969** (2020) 012020 doi:10.1088/1757-899X/969/1/012020

Thermodynamic Analysis of the Deoxidation Ability of Aluminum in the Melt of the Fe–Cr–C System at 1600 °C

G G Mikhailov, L A Makrovets and O V Samoilova

Department of Materials Sciences, Physical and Chemical Properties of Materials, South Ural State University, Chelyabinsk, 454080, Russia

E-mail: mikhailovgg@susu.ru

Abstract. In the course of the present work, we performed a thermodynamic analysis of the deoxidation ability of aluminum in the melt of the Fe–Cr–C system at 1600 °C using the technique of constructing a special type of diagram – the surface of the solubility of components in a metal (SSCM), which binds changes in the composition of the metal melt with changes in the composition of deoxidation products. The calculation of the SSCM was based on the equilibrium constants of the reactions of the elements with oxygen interaction in a metal melt, as well as on the first-order interaction parameters (according to Wagner) of elements in liquid iron. During the calculation, the parameter $e_{Cr}^{Al} = 0.023$ was estimated. Based on the results of the calculation of the SSCM, the concentration regions of the existence of non-metallic phases – deoxidation products in equilibrium with liquid metal were determined.

1. Introduction

In the production of chromium (including stainless) steels, special attention is paid to considering the possibility of oxidation of chromium and its transition from metal to non-metallic inclusions and even to slag. Such a process is highly undesirable and the technology of steelmaking should take into account the thermodynamic aspects of the interaction in the Fe–Cr–O system [1–7]. In order to avoid oxidation of chromium, elements with a sufficiently high affinity for oxygen, in particular aluminum, are introduced into the metal melt. The literature contains data on the thermodynamics of interactions in the Fe–Al–O system [8–15]. However, the literature on the thermodynamics of the interaction of aluminum with oxygen in the liquid metal of the Fe–Cr and Fe–Cr–C systems is very limited [16–18].

The aim of this work is a thermodynamic analysis of the deoxidation ability of aluminum in the melt of the Fe–Cr–C system at 1600 °C.

2. Methodology of thermodynamic modeling

A thermodynamic analysis of the interaction processes in the Fe–Cr–Al–O–C system was carried out using the technique of constructing the solubility surface of components in a metal (SSCM), which is a diagram linking the compositions of a liquid metal with the composition of the formed nonmetallic inclusions [19].

To construct the SSCM, data are needed on the equilibrium constants of high-temperature chemical reactions involving oxygen that occur in the melt of the system under study (table 1). The activity of the components of the oxide melt was calculated according to the theory of subregular ionic solutions, the energy parameters of which are given in table 2. The component activity (CrO) was calculated at zero parameters, which actually corresponds to modeling according to the theory of ideal ionic solutions. The features of this approach to the description of slag melt with the participation of chromium oxides of different valencies are described in detail in [7]. The activities of the components of the solid solution of oxides $|Al_2O_3, Cr_2O_3|_{s.s.}$ were calculated according to the theory of regular ionic solutions, the energy parameter of which is $Q_{12} = +16737$ J/mol. The activity of the components of the solid solution of spinels $|FeAl_2O_4, FeCr_2O_4|_{s.s.}$ was determined by the theory of ideal ionic solutions. The activities of the components of the metal melt were calculated using the first-order interaction parameters (according to Wagner) of elements in liquid iron (table 3). The activity of a pure solid Cr_3O_4 compound was assumed to be unity. The activities of the components of the gas phase were determined through partial pressures.

Table 1. Equilibrium constants of high-temperature chemical reactions involving oxygen according to [7, 15, 20].

Chemical reaction*	log K = $-A/T + B$	
	A	B
$(FeO) = [Fe] + [O]$	6320	4.734
$(CrO) = [Cr] + [O]$	12516	6.495
$(Cr_2O_3) = 2[Cr] + 3[O]$	36332	15.675
$(Al_2O_3) = 2[Al] + 3[O]$	58383	18.063
$\|FeCr_2O_4\|_{s.s.} = [Fe] + 2[Cr] + 4[O]$	50873	22.744
$\|FeAl_2O_4\|_{s.s.} = [Fe] + 2[Al] + 4[O]$	76069	27.365
$\|Cr_2O_3\|_{s.s.} = 2[Cr] + 3[O]$	41800	17.800
$\|Al_2O_3\|_{s.s.} = 2[Al] + 3[O]$	64000	20.480
$\|Cr_3O_4\| = 3[Cr] + 4[O]$	56170	24.300
$\{CO\} = [C] + [O]$	1168	–2.070
$\{CO_2\} = [C] + 2[O]$	9616	2.510

*Note: the components of oxide melt are shown in parentheses, in the metal melt – in square brackets, solids are indicated in line brackets, and the gas phase – in braces.

Table 2. Energy parameters of the theory of subregular ionic solutions according to [21].

System	Q_{ijkl} (J/mole) for oxide melt		
$FeO–Al_2O_3$	$Q_{1112} = +212$	$Q_{1122} = –21502$	$Q_{1222} = –11091$
$FeO–Cr_2O_3$	$Q_{1113} = 0$	$Q_{1133} = +12115$	$Q_{1333} = –60565$
$Al_2O_3–Cr_2O_3$	$Q_{2223} = +17155$	$Q_{2233} = +34310$	$Q_{2333} = +17155$
$FeO–Al_2O_3–Cr_2O_3$	$Q_{1123} = +12715$	$Q_{1223} = +366$	$Q_{1233} = –28293$

Table 3. The first-order interaction parameters (according to Wagner) of elements in liquid iron e_i^j (1600 °C).

Element i	Element j			
	Cr	Al	O	C
Cr	–0.0003 [22]	+0.023*	–0.179 [7]	–0.120 [22]
Al	+0.012 [16]	+0.045 [23]	–1.980 [23]	+0.091 [22]
O	–0.055 [24]	–1.170 [11]	–0.200 [22]	–0.450 [22]
C	–0.024 [22]	+0.043 [22]	–0.340 [22]	+0.140 [22]

*Note: data obtained in the course of this work.

3. Results and discussion

Figure 1 shows the results of the simulation of the SSCM of the Fe–Cr–Al–O system (in the absence of carbon in the metal melt). Data is given in logarithmic coordinates, concentrations are indicated in weight percent. In Figure 1(a) in region I, in equilibrium with a metal melt, an oxide melt (O.m.) of variable composition (FeO, Al_2O_3, Cr_2O_3, CrO) will form as deoxidation products; in region II – a solid solution of spinels $|FeAl_2O_4, FeCr_2O_4|_{s.s.}$; in region III – a solid solution of oxides $|Al_2O_3, Cr_2O_3|_{s.s.}$; in region IV – the Cr_3O_4 compound. According to Figure 1(a) the formation of an oxide melt is possible only at very low concentrations of aluminum (about 10–6 wt. %) and sufficiently low concentrations of chromium (not more than 0.01 wt. %). The Cr_3O_4 compound is formed at high chromium concentrations (about 10 wt. %) and [% Al] ≤ 10–4 wt. %.

In the presence of aluminum in a metal melt (starting from [% Al] = ~ 0.0001 wt. % and more) for any concentration of chromium, the only reaction product will be a solid solution of oxides $|Al_2O_3, Cr_2O_3|_{s.s.}$. The calculation procedure made it possible to determine the composition of the solid solution of oxides. When the oxygen concentration in the liquid metal is 0.01 wt. % (see Figure 1(a)): for the concentration of chromium in the metal [% Cr] = 0.1 wt. % in the composition of the solid solution of oxides the molar fraction of chromium oxide is equal to $x_{|Cr_2O_3|}$ = 0.0061; for [% Cr] = 1 wt. % $x_{|Cr_2O_3|}$ =0.0070; for [% Cr] = 10 wt. % $x_{|Cr_2O_3|}$ = 0.1144. And already at a concentration of oxygen in the metal of 0.001 wt. % in the composition of the solid solution of oxides the molar fraction of chromium oxide does not exceed 0.0026.

With an increase in aluminum concentration, the oxygen concentration in the metal melt decreases and may decrease to 0.0003 wt. %, which indicates a high deoxidizing ability of aluminum in iron-chromium melts. The results are consistent with experimental data from Ohta et al. [17]; there are some deviations for 20 wt. % chromium (see Figure 1(b)), which can be associated with the concentration limitations of the interaction parameters during the calculation.

Figure 2 shows the results of the simulation of the SSCM of the Fe–Cr–Al–O–C system for carbon concentrations of 0.1 wt. % (low carbon steel) and 0.4 wt. % (medium carbon steel). In addition to areas II – IV, which coincide in designations with Figure 1(a), when carbon appears in the metal melt, a region V arises, where the gas phase $\{CO, CO_2\}$ (the product of the interaction of carbon and oxygen dissolved in liquid iron) is in equilibrium with the metal melt.

In the presence of carbon at 1600 °C, the formation of an oxide melt as nonmetallic inclusions is thermodynamically unlikely (region I is not shown in the diagrams), and the formation of a spinel solid solution is possible only at 0.1 wt. % carbon in the metal. For low-carbon steels (see Figure 2(a)), the formation of the gas phase as a deoxidation product is possible only at extremely low concentrations of aluminum in the metal melt. The deoxidation pattern at [% C] = 0.1 wt. % in the presence of aluminum is as follows: starting from [% Al] = ~ 0.0001 wt. % and more for any concentration of chromium, the only interaction product will be a solid solution of oxides $|Al_2O_3, Cr_2O_3|_{s.s.}$. The composition of the solid solution of oxides is as follows: when the oxygen concentration in the liquid metal is 0.01 wt. % (see Figure 2 (a)) depending on the concentration of chromium in the liquid metal, the molar fraction of chromium oxide varies from $x_{|Cr_2O_3|}$ = 0.0005 ([% Cr] = 0.01 wt. %) to $x_{|Cr_2O_3|}$ = 0.0910 ([% Cr] = 10 wt. %). And already at a concentration of oxygen in the metal of 0.001 wt. % in the composition of the solid solution of oxides, the molar fraction of chromium oxide does not exceed 0.0022.

For a carbon concentration of 0.4 wt. % (see Figure 2 (b)) the probability of gas formation increases, but still, at practically significant aluminum concentrations (starting from 0.001 wt. % and more) $|Al_2O_3, Cr_2O_3|_{s.s.}$ will act as nonmetallic inclusions with a predominance of aluminum oxide in the composition of the solid solution (the molar fraction of chromium oxide does not exceed 0.0229).

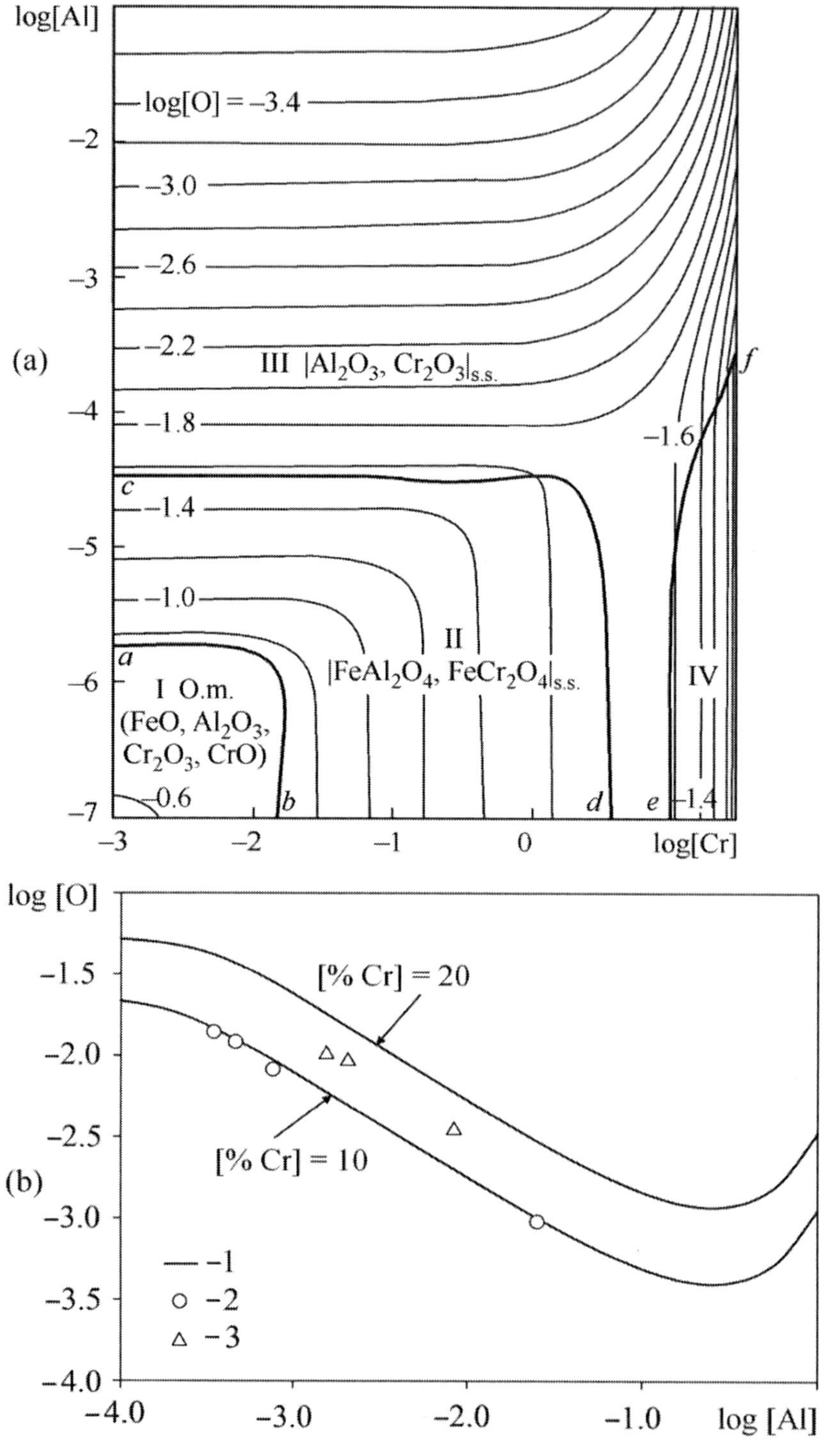

Figure 1. The surface of the solubility of components in a metal of the Fe–Cr–Al–O system at 1600 °C: (a) a general view with iso-oxygen sections; (b) iso-compositional sections for 10 and 20 wt. % chromium. 1 – our simulation results; 2 and 3 – experimental data according to Ohta et al. [17] for 10 and 20 wt. % Cr accordingly.

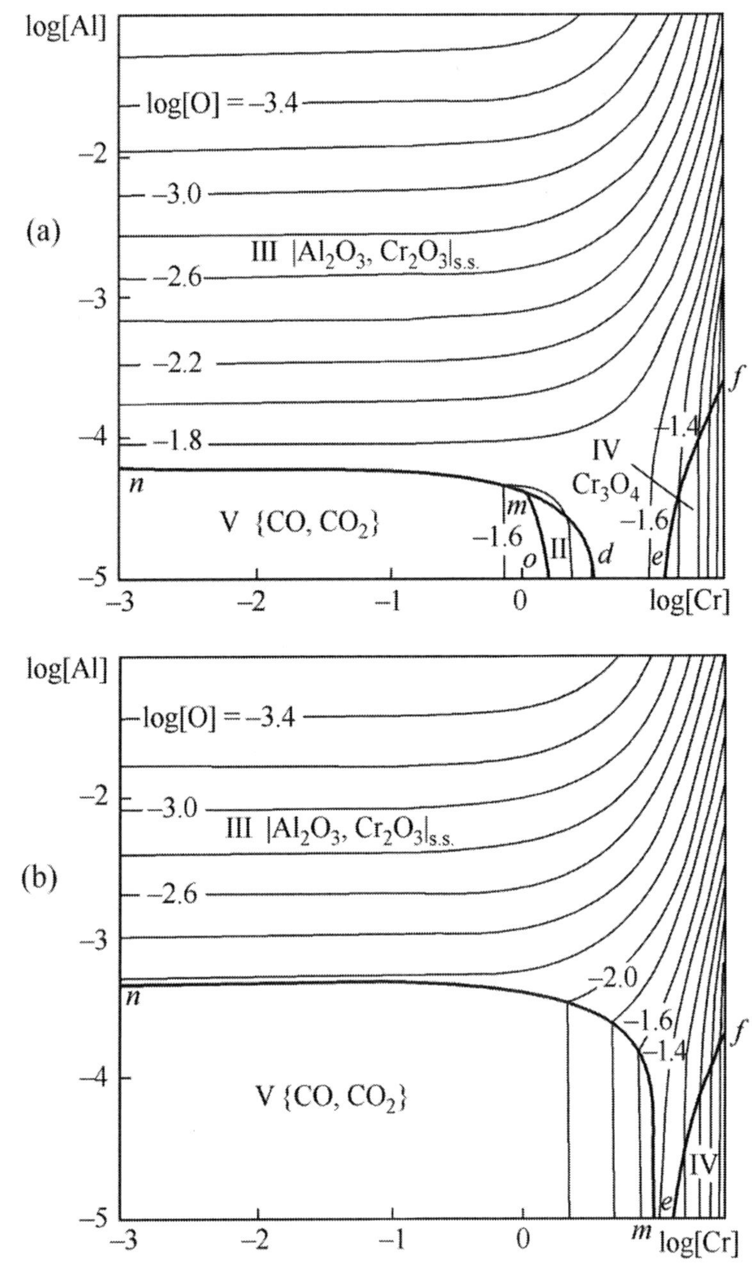

Figure 2. The surface of the solubility of components in a metal of the Fe–Cr–Al–O–C system at 1600 °C: (a) [% C] = 0.1 wt. % and (b) [% C] = 0.4 wt. %.

4. Conclusions

According to the thermodynamic calculation and its graphical representation in the form of SSCM diagrams, it was established that at practically significant aluminum concentrations (corresponding to the actual technological process of deoxidation) in the metal melt of the Fe–Cr–Al–O–C system, the main reaction product is solid solution of oxides $|Al_2O_3, Cr_2O_3|_{s.s.}$ with a predominance of aluminum oxide in the composition of the solid solution. With an increase in aluminum concentration, the oxygen concentration in the metal decreases and may decrease to 0.0003 wt. %, which indicates a high deoxidizing ability of aluminum in iron-chromium melts.

References

[1] Hilty D C, Forgeng W D and Folkman R L 1955 *Trans. AIME. J. Met.* **203** 253
[2] Fruehan R J 1969 *Trans. AIME* **245** 1215
[3] Janke D and Fischer W A 1976 *Arch. Eisenhüttenwes.* **47** 147
[4] Ghosh A and Murthy G V R 1986 *Trans. ISIJ* **26** 629
[5] Itoh T, Nagasaka T and Hino M 2000 *ISIJ Intern.* **40** 1051
[6] Jung I H, Decterov S A and Pelton A D 2004 *Metall. Mater. Trans.* B **35** 493
[7] Samoilova O V and Makrovets L A 2019 *Bull. South Ural State Univ. Ser. Metallurgy* **19(3)** 5
[8] Gokcen N A and Chipman J 1953 *Trans. AIME* **197** 173
[9] Fruehan R J 1970 *Metall. Trans.* **1** 3403
[10] Janke D and Fischer W A 1976 *Arch. Eisenhüttenwes.* **47** 195
[11] Prox H, Hino M and Ban-Ya S 1997 *Tetsu-to-Hagané* **83** 773
[12] Itoh H, Hino M and Ban-Ya S 1997 *Metall. Mater. Trans.* B **28** 953
[13] Kang Y, Thunman M, Sichen D, Morohoshi T, Mizukami K and Morita K 2009 *ISIJ Intern.* **49** 1483
[14] Paek M K, Jang J M, Kang Y B and Pak J J 2015 *Metall. Mater. Trans.* B **46** 1826
[15] Mikhailov G G, Samoilova O V, Makrovets L A and Smirnov L A 2019 *Steel in Translation* **49** 522
[16] Kishi M, Inoue R and Suito H 1994 *ISIJ Intern.* **34** 859
[17] Ohta H and Suito H 2003 *ISIJ Intern.* **43** 1301
[18] Jo J O, Jung M S, Park J H, Lee C O and Pak J J 2011 *ISIJ Intern.* **51** 208
[19] Mikhailov G G, Leonovich B I and Kuznetsov Yu S 2009 *Thermodynamics of Metallurgical Processes and Systems* (Moscow: MISIS Publisher)
[20] Fuwa T and Chipman J 1960 *Trans. AIME* **218** 887
[21] Mikhailov G G, Makrovets L A and Samoilova O V 2020 *Mater. Sci. Forum* **989** 204
[22] Sigworth G K and Elliott J F 1974 *Metal Sci.* **8** 298
[23] Park J H and Todoroki H 2010 *ISIJ Intern.* **50** 1333
[24] Ban-Ya S and Iguchi Y 1988 *The 19th Committee on Steelmaking, Steelmaking Data Sourcebook, Japan Society for the Promotion of Science* (New York: Gordon and Breach) 95

Acknowledgements

The work was supported by Act 211 Government of the Russian Federation, contract № 02.A03.21.0011.

IOP Conf. Series: Materials Science and Engineering **969** (2020) 012021 doi:10.1088/1757-899X/969/1/012021

Application of the Polymeric Material RIMAMID for Production of Machine Parts

A V Shaparev[1], I A Savin[1] and S N Ptichkin[2]

[1]Kazan National Research Technical University named after A. N. Tupolev-KAI, 10, K. Marx St., Kazan, Tatarstan, 420111, Russia

[2]Public Joint Stock Company KAMAZ, 2, Avtozavodskiy Prospect, Naberezhnye Chelny, Republic of Tatarstan, 423827, Russia

E-mail: meganeiii@yandex.ru

Abstract. Modern polymeric materials have properties that, in some respects, exceed the properties of traditionally used materials (steel, aluminum and titanium alloys, wood). The uniqueness of the Rimamide polymer material is that it is possible to pre-design the material in such a way as to give the product from it the properties necessary for a particular application.

The production of blanks and the use of Rimamide material for the manufacture of sliding and rolling bearings, bushings for wheels and rollers, pulleys, blocks, brackets, wheel hubs, gears, sprockets, gears and worm wheels and other parts to reduce weight, noise and vibrations. The characteristics of the material, examples of the manufacture of blanks, the advantages and disadvantages of the RIMAMID material for the manufacture of machine parts, prospects for use in mechanical engineering are shown.

Due to their properties, they can be used in almost all industries. For example, modern aviation and space technology is characterized by the intensive use of new materials, technologies, and promising designs based on them. When using RIMAMID material using antifriction components (molybdenum sulfide, graphite, etc.), the friction coefficient is significantly reduced and the friction units can be operated without lubrication.

The main advantages of this material is a low coefficient of friction for steel, which provides a 3-4-fold increase in the service life of parts in the friction units, as well as low weight, high strength, good machinability on metal cutting machines, resistance to aggressive environments, and environmental safety.

1. Introduction

The expansion of the use of polymers in the production of machine parts is due to their unique properties such as high specific strength (strength to density ratio), wear resistance, resistance to chemical influences, the possibility of varying properties over a wide range due to the modification of polymers and combining them with various ingredients, good dielectric characteristics, etc.

The use of plastics in modern engineering has reached a fairly high level compared to steel, due to a significant decrease in the mass of parts, an increase in durability, reliability, and a reduction in the cost of manufacturing parts. Moreover, due to the reduction of waste during processing, the utilization rate of materials increases significantly (the average utilization of plastics is approximately two times higher than for metals).

2. Statement of the research problem

RIMAMID polymer material [1] has good prospects for use in aircraft and automotive industries, as it allows you to replace heavy and expensive materials in the manufacture of machine parts and assemblies, such as stainless steel, aluminum, antifriction cast iron, porcelain, fluorine-plast, bronze, brass, textolite, wood-fiber plastics and other antifriction and structural materials [2]. The comparative density of materials for the manufacture of machine parts is shown in table 1.

In addition to low weight, the main advantage of this material is the low coefficient of friction on steel, which provides an increase in the service life of parts of the friction unit by 3 to 4 times. When using the RIMAMID material using antifriction components (molybdenum sulfide, graphite, etc.), the friction coefficient is significantly reduced and the friction units can be operated without lubrication [3].

RIMAMID material for a single production is made in the form of blanks (plate, rod, pipe, etc.) of various sizes for further machining into products on turning, milling and other processing machines (figure 1). To produce a batch of parts, casting technologies are used for the manufacture of blanks from RIMAMID material, which are as close as possible to the dimensions of the part being manufactured (figure 2) [4, 5, 6].

Table 1. Comparative density of materials for the manufacture of machine parts compared to RIMAMID [1].

Name of material	Density (kg / m$^{3)}$	Comparative density with RIMAMID material
Copper	8900	7.77
Bronze	8700	7.60
Brass	8500	7.42
Steel, iron	7800	6.81
Cast Iron	7000	6.11
Aluminum	2700	2.36
Ftoroplast	2210	1.93
RIMAMID	1145	1.00
Kapron	1100	0.96
Polyethylene	920	0.80

Due to the high wear resistance of the material (ten times greater than bronze) and a low coefficient of friction, the noise level in the sliding units is reduced, and it is also possible to use parts in dusty installation locations. In addition, the Rimamid material is highly resistant to dynamic shock loads, which can significantly improve the reliability of equipment in difficult conditions. High heat resistance of the material ensures long-term operation of products at temperatures from –40 to +160 ° C while maintaining the properties of the material [7, 8].

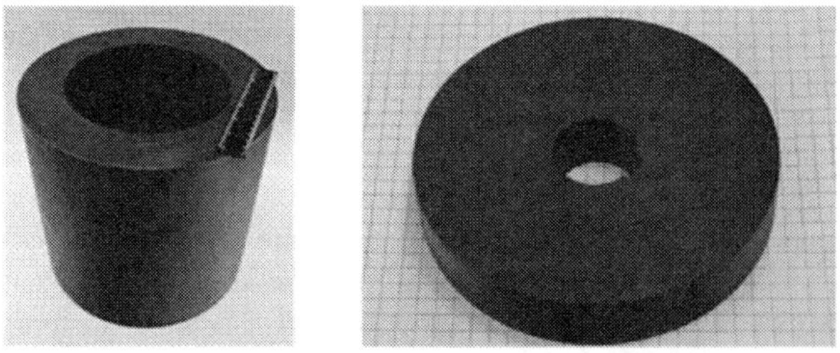

Figure 1. Blanks for bushings and gears.

IOP Conf. Series: Materials Science and Engineering **969** (2020) 012021 doi:10.1088/1757-899X/969/1/012021

The material RIMAMID is manufactured and supplied by the following brands [2]:

- RIMAMID-200, which has high hardness and wear resistance;
- RIMAMID-200-U, which has high impact strength;
- RIMAMID-200-STU, which has an additional high heat resistance and light resistance.

The results of the study of the properties of the material "RIMAMID" produced by NPO "Start-Plast" are presented in table 2.

Table 2. The Properties of the polymer material RIMAMID

No.	Name indicator	RIMAMID 200	RIMAMID 200-U	RIMAMID 200-STU	Test methods
1	Melting point (C°)	220–225	218–223	218–223	GOST 21553
2	Density (kg/m^3)	1145–1150	1135–1145	1135–1145	GOST 15139
3	Tensile stress, tensile strength (MPa)	80–85	not less than 30	not less than 30	GOST 11262
4	Relative elongation at break (%)	25–30	25–40	25–40	GOST 11262
5	Tensile modulus (MPa)	2800–3200	2200–2800	2200–2800	GOST 9550
6	Charpy impact strength of the specimen without notch (20° C) (kJ/m^2)	25–40	without kink	without kink	GOST 4647
7	Water absorption in 24 hours (%)	1.0–2.0	1.0–1.5	1.0–1.5	GOST 4650
8	Maximum water absorption (%)	2.5–3.0	1.5–2.0	1.5–2.0	GOST 4650
9	Thermal conductivity at room temperature (W/(m °C))	0.29	0.29	0.29	GOST 23630.2
10	Average coefficient of linear thermal expansion by 1° C in the temperature range from 0 to 50° C	$9.8 \cdot 10^{-5}$	$9.8 \cdot 10^{-5}$	$9.8 \cdot 10^{-5}$	GOST 15173
11	Shore hardness D	80–85	75–80	75–80	GOST 24621
12	Friction coefficient for steel without lubrication	0.15–0.3	0.2–0.3	0.2–0.3	GOST 11629
13	Friction coefficient for steel with lubricant	0.04–0.08	0.04–0.08	0.04–0.08	GOST 11629
14	Content of extractable substances (%)	1.0–3.0	1.0–1.5	1.0–1.5	GOST 17824

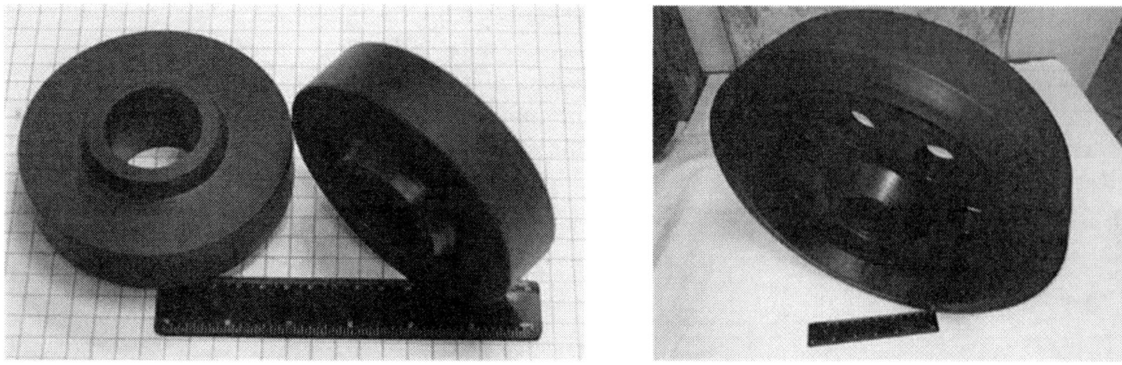

Figure 2. Blanks for gears and track machine rollers, as close as possible to the dimensions of the manufactured parts.

Currently, from the material *"RIMAMID"* are made:

- sliding and rolling bearings, bushings for wheels and rollers, guides and liners of friction units with or without lubrication [9];
- pulleys, blocks, support and guide rollers of hoisting mechanisms [10];
- brackets, wheel hubs and other parts that are subject to increased requirements for impact resistance (figure 3) [11];
- gears, sprockets, gear and worm wheels, screws, screws of various devices and mechanisms in order to reduce noise and vibration (figure 4), etc. [12].

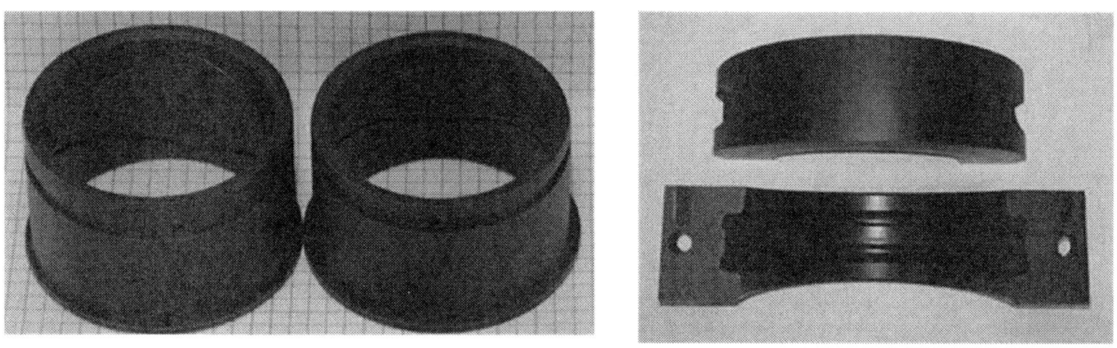

Figure 3. Bushings for wheels and rollers, guides and liners of friction units.

3. Prospects for the use of RIMAMID

Modern polymeric materials possess properties that, in some respects, exceed the properties of traditionally used materials (steel, aluminum and titanium alloys, wood), and the higher the specific characteristics of the composite, the lighter or stronger the construction. The uniqueness of the RIMAMID polymer material is that you can pre-design the material in such a way as to give the product from it the properties necessary for a particular application [13].

Due to their properties, they can be used in almost all industries. For example, modern rocket and space technology is characterized by the intensive use of new materials, technologies, and promising designs based on them [14].

Due to their properties, they can be used in almost all industries. For example, modern rocket and space technology is characterized by the intensive use of new materials, technologies, and promising designs based on them. The development of the commercial sphere of application of new technologies and equipment in space and the global competition for receiving orders for the delivery of payloads to Earth's orbit stimulate a decrease in the cost of launching cargo and force developers of rocket and space technology to actively use composites to optimize the energy-mass characteristics of launch vehicles and increase the volume of payload [15].

Figure 4. Gears from the material RIMAMID.

In the future, it is possible to use RIMAMID material for the manufacture of car radiator grilles with subsequent metallization or surface painting, fan blades, sliding bearings, ventilation pipes, accelerator pedals, steering wheels, control buttons, interior parts, etc. [16].

The use of RIMAMID material a has the following advantages over metals:

- The structure of the RIMAMID material is homogeneous, therefore, unlike metals, there are no phases of different hardness, microcracks do not form and uneven wear occurs [17].
- Gear wheels made of RIMAMID material do not form strong intermolecular bonds at the point of contact, therefore there is no pitting and braking of the wheels [18].
- During operation of parts made of RIMAMID material during wear, abrasive particles do not form, leading to wear of the working surfaces of the teeth and contamination of the lubricant [19].
- RIMAMID does not enter into an electrochemical interaction leading to surface oxidation [20].

4. Conclusion

The experience in the production, implementation and operation of Rimamide material at NPO "Start-Plast" enterprise revealed the following problems:

- large industries and consumers (aircraft, automotive, shipbuilding, energy, etc.) avoid the introduction of new materials, because additional tests, certification, development and approval of new standards are required. These processes are labor intensive and costly. In turn, the absence of large consumers impedes the further development of the production base for the

production of Rimamid material, the improvement and development of new parts and the corresponding technological and design documentation;
- it is necessary to improve the system of technical regulation, the creation of modern standards governing the production and testing methods of new materials. The solution of these problems will create the necessary conditions for a wider use of Rimamide material.

The main advantages of this material is a low coefficient of friction for steel, which provides a 3–4-fold increase in the resource of parts in friction units, as well as low weight, high strength, good machinability on metal cutting machines, resistance to aggressive environments, and environmental safety.

References

[1] Polymer material "Rimamide" Online. Available: https://innokam.pro/catalog/103
[2] About the polymeric material "Rimamide" Online. Available: https://rimamid.ru/
[3] Shaparev A V and Savin I A 2016 Calculation of the Amount of the Reduction Required for the Formation of Compound Layers during Cold Rolling of Bimetals *Materials Science Forum* **870** 328–333
[4] Shaparev A V and Savin I A 2017 Calculation of Joint Plastic Deformation to Form Metal Compound in Cold Condition *Solid State Phenomena* **265** 313–318 DOI: https://doi.org/10.4028/www.scientific.net/ssp.265.313
[5] Gavariev R V, Savin I A and Leushin I O 2016 Impact of the functional coating on service durability of injection molds for the zinc alloys pressure casting *Non-feroous Metals* **1** 66–70 DOI: 10.17580/tsm.2016.01.11
[6] Leushin I O, Subbotin A Yu and Geyko M A 2015 Recycling of galvanized steel scrap for use in cast iron melting in induction melting facilities *CIS Iron and Steel Review* **10** 19–22
[7] Leushina L I 2018 Recycling Cullet from Quartz Ceramic Shells Used in Investment Casting *Steel in Translation* **48 11** 699–703
[8] Leushin I O, Leushina L I and Koshelev O S 2017 Creep of the investment mold material *Russian Metallurgy* **13** 1092–1095
[9] Khusainov R M, Golovko A N, Petrov S M, Yurasov S Y and Balabanov I P 2017 Selecting optimal cutting tools for lathes *Russian Engineering Research* **37(4)** 351–353 DOI: 10.3103/S1068798X17040128
[10] Balabanov I P, Simonova L A and Balabanova O N 2015 Systematization of accuracy indices variance when modelling the forming external cylindrical turning process *IOP Conference Series: Materials Science and Engineering* **86(1)** DOI: 10.1088/1757-899X/86/1/012010
[11] Safronov N N, Mingaleeva L B and Savin I A 2018 Optimization of charge material composition in shs process with ferrosilide fabrication from gaseous wastes of metallurgical production *Ferrous metals* **(2)** 53–59
[12] Shaparev A V and Savin I A 2018 Influence of the State of the Contact Surfaces on the Formation of the Joint of Steel and Brass during Cold Cladding *Solid State Phenomena* **284** 319–325 DOI: https://doi.org/10.4028/www.scientific.net/SSP.284.319
[13] Gavariev R V and Savin I A 2018 Research of the Mechanism of Destruction of Compression Molds for Casting under Pressure of Color Alloys *Solid State Phenomena* **284** 326–331 DOI: https://doi.org/10.4028/www.scientific.net/SSP.284.326
[14] Gavariev R V, Savin I A and Leushin I O 2019 To the Question of Casting Alloys of Non-Ferrous Metals in the Metal Mold *Materials Science Forum* **946** 631–635 DOI: https://doi.org/10.4028/www.scientific.net/MSF.946.631
[15] Gavariev R V, Savin I A and Leushin I O 2017 Increasing the quality of the surface of zinc castings by applying multi-layer protective coatings *Non-feroous Metals* **5** 84–88 DOI: https://doi.org/10.17580/tsm.2017.05.13

[16] Gavariev R V and Savin I A 2017 Improvement of Surface Quality of Casting Produced by Casting under Pressure *Solid State Phenomena* **265** 988–993 DOI: https://doi.org/10.4028/www.scientific.net/ssp.265.988
[17] Savin I A and Akhmedeev M 2020 Connection of the Steel Pipes Having a Polymeric Covering on Internal and External Surfaces *Solid State Phenomena* **299** 766–771 Online. Available: https://doi.org/10.4028/www.scientific.net/SSP.299.766
[18] Gavariev R V, Savin I A and Soldatkina E N 2020 Choice of Protective Coating of Metal Molds for Casting Non-Ferrous Alloys *Solid State Phenomena* **299** 867–871 Online. Available: https://doi.org/10.4028/www.scientific.net/SSP.299.867
[19] Shaparev A V and Savin I A 2017 Calculation of Joint Plastic Deformation to Form Metal Compound in Cold Condition *Solid State Phenomena* **265** 313–318 DOI: https://doi.org/10.4028/www.scientific.net/ssp.265.313

IOP Conf. Series: Materials Science and Engineering **969** (2020) 012022 doi:10.1088/1757-899X/969/1/012022

Action Planning for Improvement of Design and Operating Parameters of Ball-Bearing Steel Tube Annealing

B P Yur'ev, V A Dudko and A V Korelin

Institute of New Materials and Technologies, Federal State Autonomous Educational Institution of Higher Education "Ural Federal University named after the first President of Russia B. N. Yeltsin", 28, Mira str., Yekaterinburg, 620002, Russia

E-mail: yurev-b@mail.ru

Abstract. Annealing process was reviewed for hot rolled 100Cr6 steel tubes in a chamber furnace with subhearth burners. Based on the obtained results of thermotechnical testing at the chamber furnace it was detected that the current annealing curve conceptually do not allow production of satisfactory quality tube in the course of annealing. As a result an adjusted curve corresponding to the annealing requirements for 100Cr6 steel and chamber furnace capabilities was proposed. Causes were determined for poor quality tube annealing and detailed review was carried out for chamber furnace capabilities in terms of required heat treatment duration ensuring sufficiently high quality of annealed tubes. List of actions allowing significant improvement of chamber furnace performance and annealed tube quality was considered. Chamber furnace renovation option was proposed where the total heat treatment time of tubes may be reduced significantly through heat transfer process intensification in the furnace working space, uniformity of tube heating in the batch, annealing quality and therefore furnace performance may be improved.

1. Tube Annealing Process

Annealing is usually used to relief internal stresses, improve metal mechanical properties and machinability, reduce hardness and prepare the structure for further heat treatment [1–14].

The article deals with the issues related to the annealing process review for 100Cr6 steel tubes in one of the chamber furnaces with subhearth gas burners of Pervouralsk Novotrubny Plant (PNTZ) and recommendations development to allow the treated tube quality improvement.

The furnace is intended for high temperature annealing called in practice annealing to granular perlite [15–19]. Requirements for annealed 100Cr6 steel microstructure and hardness are defined by GOST 801–78. The fuel in this furnace is burnt outside the working space in special fireboxes located under the furnace hearth.

The furnace has the following working space dimensions: length – 9550 mm; width – 2700 mm; height – 1460 mm. Number of gas burners – 11. The furnace is heated with natural gas with combustion heat of kJ/m^3. The maximum tube heating temperature is 900°C. With a special vehicle the batch is loaded into the furnace and unloaded from it after annealing. Number of tubes in the batch and their weight depend on the batch length, tube diameter and wall thickness. The batch weight may vary from 10 to 20 t.

100Cr6 steel tube annealing process consists of the batch heating to the temperature of 790 °C and subsequent cooling. Heating and cooling conditions in the course of annealing are determined by the annealing curve shown in the instruction for tube heat treatment. Temperature in the furnace working space is checked by three roof chromel alumel thermocouples, located 100–150 mm above the annealed tube batch. Temperature inside the batch in the course of tube annealing is measured by a side-mounted thermocouple inserted through a furnace wall hole located at the distance of 4.5 m from the furnace front face. Side-mounted thermocouple is installed between the first and second tube layers at a three layer batch and between the second and the third layers if number of layers in the batch is four or more. Thermocouple readings are registered automatically by a potentiometer recorder with the scale of 400 to 900 °C. Annealed tube length is 6–8 m, and diameter is 60–160 mm.

Annealing quality is checked by two methods. About 10 % of all annealed tubes in the batch are inspected on a coercimeter where annealed steel hardness is checked. Samples are taken from reference tube front ends for laboratory microstructure and hardness analysis. Steel with the structure of 1–4 according to scale no. 5 and Brinell hardness of 179–207 what complies to the impression diameter of 4.2–4.5 mm is considered acceptable.

2. Experimental

Testing was focused on research of tube annealing conditions in the furnace central portion where heat treatment conditions are something like that at which the furnace and the batch may be considered as infinitely long. The experience shows that such conditions are met in the furnace portion beginning at the distance of 0,6 m farther the tube front ends and ends at the same distance nearer the rear ends.

The first stage of thermotechnical testing at the chamber furnace was related to regularity study of the temperature conditions for the batch tube annealing. The major method of required information gaining at that was temperature measurement in the selected points of the annealed tube batch using chromel-alumel thermocouples. Temperature distribution was examined throughout the batch length, width and height as well as in the batch cross section vertically and horizontally. The second stage was related to determination of the batch tube optimal heating time at the recommended temperatures ensuring the required annealed tube quality.

The work program provides for action plan development intended for chamber furnace performance improvement. At this investigation stage some conclusions with regard to causes of annealed tube quality deterioration and plans for their elimination became possible. Certain constructive proposals are available for furnace capacity increase. Actions proposed below for tube annealing quality and capacity improvement in the chamber furnace result from the thermotechnical testing and review of chamber furnace capabilities at the current annealing process. These actions may be divided in two groups: those that do not require and those that require revamping of the current equipment.

Actions from the first group require no additional costs or time and may be implemented immediately. Actions from the second group require additional costs, model research, practical development, etc. and may be considered as a plan for further work.

Let's consider actions requiring no revamping of the current equipment. The proposed actions result from the need to adjust the current annealing curve and quality inspection procedure. Review of high quality annealing capabilities for 100Cr6 steel tubes according to the current curve is considered above. Evidence obtained from the experiments demonstrates that the current curve conceptually does not allow obtaining of sufficient annealing quality because of the batch bottom tube underheating.

As experience shows the tube batch section may be divided into three zones (figure 1) with regard to tube heating adequacy in these zones as per annealing temperature condition requirements of 100Cr6 steel. In the first zone standard tube heating is ensured, however it is somewhat more lengthy than required on the batch outside. In the second zone the tubes are underheated although slightly. In the third zone the tubes are significantly underheated what causes annealed tube unacceptability both with regard to microstructure and hardness. The stated deficiency of annealing may be currently avoided only by one method: temperature increase for bottom tubes minimum to 780°C and their holding at this temperature for at least two hours. Increase of total annealing time at that may be partially or completely offset by increased heating and cooling intensity. Current heating temperature margin comprises 25–30 °C. Maximum heating temperature according to the roof thermocouple in the instruction is 810°C. At that the observed maximum temperature on the tube surface is 815–820 °C. Therefore, heating may be carried out at the temperatures of 820–830°C, without fear of overheating as at that the tube surface temperature will be no higher than 835–840°C. Such temperatures of up to 840–850°C are not dangerous with regard to overheating [15]. Thus in one of the experiments the heating temperature was within 825–850°C, sample time in the furnace was four to seven hours. Sample cooling rate equaled 20–30°C/h. Out of six annealed samples no reject was detected. Hardness measurements gave impression diameter of 4.2–4.3 mm. At that samples with the impression of 4.3 were cooled more slowly. No impact was detected for hardness from heating time and rate. Only in case of very massive batches overheating structure may be expected due to lateral tubes held at the temperatures of 820–830 °C, generally not dangerous with regard to overheating, for too long. In such cases heating shall be carried out in two stages, first at the temperature of no higher than 800 °C, and at the temperature of 820 °C at the end. Cases when such complication and additional procedure extension is necessary may be determined only by direct experiments.

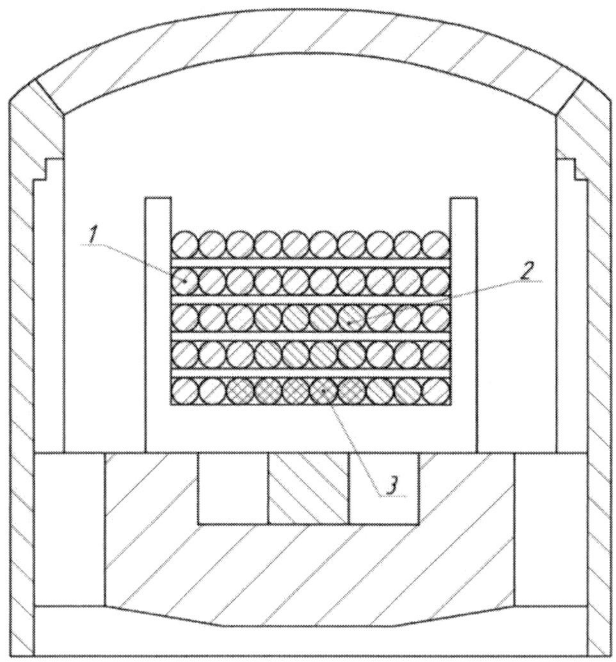

Figure 1. Various tube heating zones of batch cross section: 1 – standard heating zone; 2 – slight underheating zone; 3 – significant underheating zone.

As to cooling it may also be slightly more intensive than in current practice. Cooling rate monitoring is the most complicated. Roof thermocouple readings can give approximate information on the cooling rate. As experience shows roof thermocouple readings decrease about twofold slower as compared to the lateral tube surface temperature. As tube cooling rate shall not exceed 30–60 °C/h, temperature reduction rate according to the roof thermocouple shall not exceed 15–30 °C/h. Practically the risk of too rapid cooling is eliminated if the temperature reduction rate according to the roof thermocouple is 10–20 °C/h. At that initial shock cooling shall not be below 740 °C, and subsequent cooling is preferable with approximately the same intensity ensuring decreased readings of the roof thermocouple with the rate of 10–20 °C/h. It is expedient to reduce breaks in air supply for temperature equalization to 30 minutes and to increase air supply periods to 1.0–1.5 hours (currently the time ratio is reverse). Annealing curve inclusive of all proposed changes is shown in figure 2. It is broken into three intervals: heating, equalizing and cooling. Heating is proposed with the same intensity to the temperature of 800 °C according to the side-mounted thermocouple. Then the burners are turned off and equalizing lasts for 1.5–2.0 hours, during which the temperature in the batch equalizes to the level of 790–800 °C. The last stage – cooling begins from the batch sudden air purging to the temperature of 740 °C according to the roof thermocouple. Then uniform, significantly less intensive cooling with the rate of 10–20 °C/h according to the roof thermocouple continues with intermittent breaks in cooling air supply. At that it shall be considered that more intensive annealing increases requirements to temperature monitoring accuracy in the furnace.

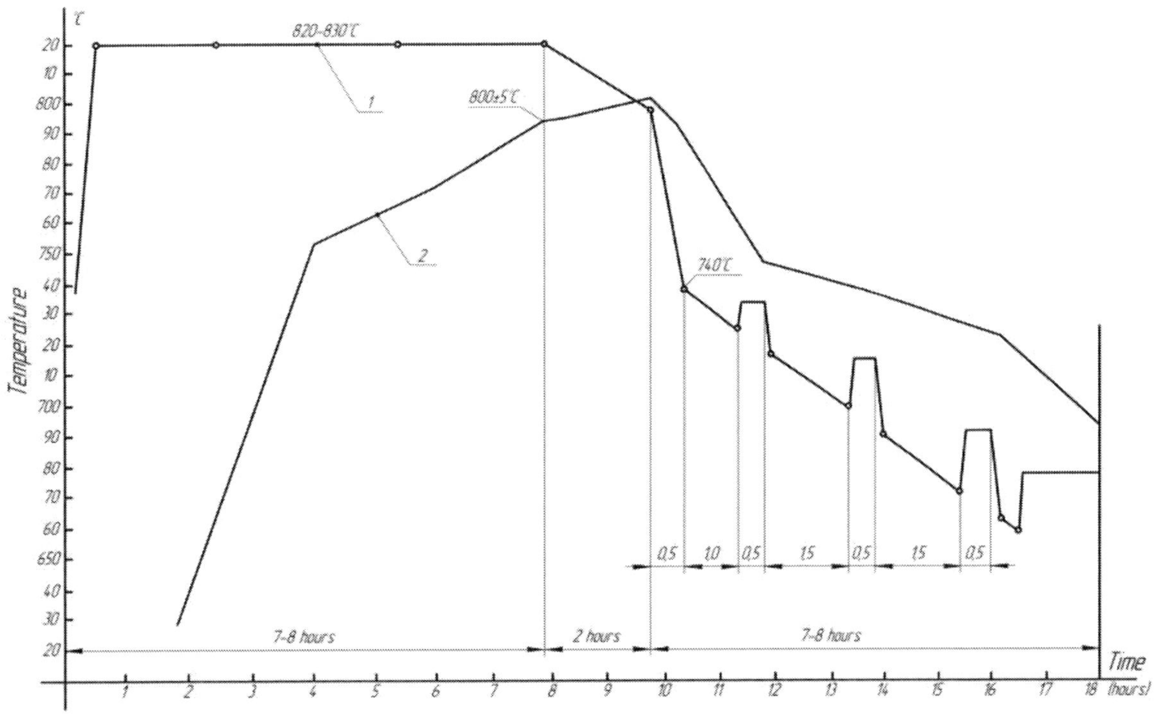

Figure 2. Recommended annealing curve for 100Cr6 steel tubes: 1 – roof thermocouple curve; 2 – side-mounted thermocouple curve.

Currently (as noted above) annealing quality is inspected by two methods complementing each other: tube hardness inspection on a coercimeter and laboratory microstructure and hardness inspection of samples. Disadvantage of the current inspection procedure is its one-sidedness. It is due to the sampling procedure of reference tubes in the annealed lot and specimen cut out only from the tube front ends. At that underheating of batch bottom tubes is neglected. To make inspection a more efficient method of annealing quality improvement it is suggested that two top lateral tubes of the batch and two middle tubes of the bottom layer shall be taken for inspection (figure 3). Specimens from the top tubes shall be cut from front ends and from the bottom tubes – at the distance of 0.9–1.0 m from the tube ends. With such sampling procedure batch tube underheating and overheating are not neglected.

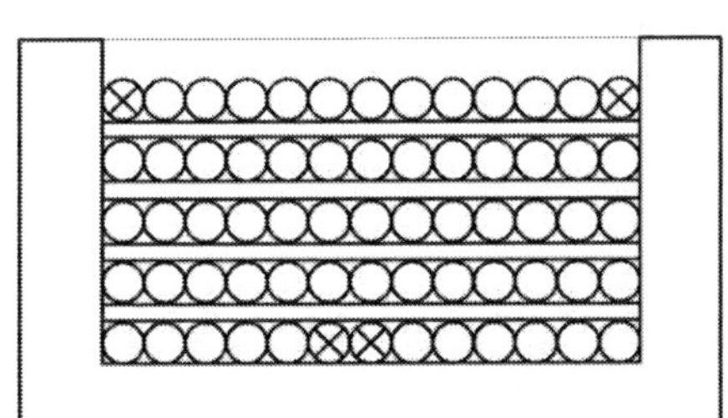

Figure 3. Proposed reference tube sampling procedure, ⊗– reference tube.

The idea of the current furnace revamping is of interest. To increase the convective heat transfer rate between the heating gasses and the batch it is suggested that fans shall be installed in the furnace rear part to ensure intensive gas blending throughout the furnace volume [20]. Heating gasses drawn in from the furnace with the fan through a special gas duct (i.e. formed by a double roof) are fed into the furnace front part. Thus a closed loop is created for heating gas circulation. In case of long-term circulation some gasses draw through the tubes and ensure heating of each tube from two sides – from outside and from inside. Rough estimates show that at that annealing time reduction by 4–5 hours may be expected. Together with the heat transfer process acceleration the tube heating (cooling) uniformity and therefore annealing quality improves. At that tubes may be placed on the cross beams without spacers, and therefore tube distortion may be avoided.

With quite good gas blending with an additional fan the need for very accurate maintaining of steady operation of separate burners disappears and fuel burning process may be controlled via a grandmaster. Besides the temperature monitoring may be significantly simplified and reduced to the temperature inspection in one representative point. Overall annealing process automation conditions improve significantly as control with one gate valve based on one thermocouple readings presents no challenge.

Besides the proposals considered above some other proposals are also of interest. For example, cooling air injection into the furnace not only from the bottom but also from the top. With the corresponding design this would ensure significantly better blending and more homogeneous cooling. It makes sense to preheat cooling air with the flue gas heat, etc. It is only necessary to confirm every time the expedience of changes introduced into the furnace design with the corresponding testing.

3. Summary

The results of the annealing process investigation for 100Cr6 steel tubes in a chamber furnace obtained in the work are of certain interest for specialists dealing with steel heat treatment. Tube annealing curve adjusted according to the requirements for 100Cr6 steel annealing conditions allow as distinct from the current one reasonable choice of the required heating, holding and cooling time of the tube batch at the corresponding temperatures and assurance of steel production with the required hardness and microstructure properties. The proposed furnace revamping allowing tube heating in the batch from two sides undoubtedly will contribute to annealed tube quality and furnace performance improvement.

References

[1] Ovchinnikov V V 2013 *Heat Treatment Process* (Moscow: FORUM Publishing House, Infra R&D Center)

[2] Oskolkova T N 2009 *Steel and Alloy Heat Treatment* (Moscow: Heat Engineer)

[3] Arzamasov B N 2015 *Material Engineering* (Moscow: Mechanic Engineering)

[4] Gulyaev A P 1986 *Metal Science* (Moscow: Mechanic Engineering)

[5] Ovchinnikov V V 2011 *Metal Science* (Moscow: Infra – M)

[6] Novikov I I, Zolotorevsky V S, Portnoy V K 2014 Metal Science *Heat Treatment, Alloys* (Moscow: MISIS)

[7] Teplukhin G N and Gropyanov A V 2011 *Metal Science and Heat Treatment* (St. Petersburg: Saint Petersburg State Technological University of Plant Polymers)

[8] Smirnov M A, Schastlivtsev V M and Zhuravlev L G 1999 *Fundamentals of Steel Heat Treatment* (Ekaterinburg: Russian Academy of Sciences, Ural Branch)

[9] Ning Li, Chunxiang Cui, Yaqi Zhao, Qiangxiong Zhang and Lina Bai 2018 Structure and properties of GCr15 modified by multiphase ceramic nanoparticles *Fe-C composite inoculants, Materials Science and Engineering* **A 738** 63–74

[10] Cao Y J, Sun J Q, Ma F, Chen Y Y and Xie K 2017 Effect of the microstructure and residual stress on tribological behavior of induction hardened GCr15 steel *Tribology International* **115** 108–115

[11] Xie Jianhui, Alpas Ahmet T and Northwood Derek O 2005 The role of heat treatment on the erosion–corrosion behavior of AISI 52100 steel *Materials Science and Engineering* **A 393** 42–50
[12] Oezel M, Janitzky T, Beiss P and Broeckmann C 2019 Influence of steel cleanliness and heat treatment conditions on rolling contact fatigue of 100Cr6 *Wear* **430-431** 272–279
[13] Kan R U and Haazen P T 1987 Physical Metallurgy *Phase Transformations in Metals and Allows and Alloys with Particular Physical Properties* (Moscow: Metallurgy)
[14] Ellermann A and Scholtes B 2015 The strength differential effect in different heat treatment conditions of the steels 42CrMoS4 and 100Cr6 *Materials Science and Engineering* **A 620** 262–272
[15] Rauzin Ya R 1978 *Chromium Steel Heat Treatment* (Moscow: Mechanic Engineering)
[16] Smirnov M A, Pyshmintsev I Yu, Laev K A and Akhmedianov A M 2012 Impact of High Temperature Thermomechanical Treatment on High Chromium Steel Properties *South Ural State University Bulletin, Metallurgy* **39** 85–88
[17] Filippov Yu O, Eremeni E N and Losev A S 2007 Structure and Properties of High-Chromium Steels Modified with Refractory Particles *Omsk Science Bulletin* **2(56)** 101–103
[18] Bityukov S M, Laeb K A, Lefler M N, Zhukova S Y, and Kocheshkova E V 2011 Martensitic steel with 13% Cr for corrosion-resistant oil pipe *Steel in Translation* **41(2)** 171–174
[19] Smirnov M A, Pyshmintsev I Yu and Laev K A 2015 High Chromium, Corrosion Resistant Steel Properties After High Temperature Thermomechanical Treatment *Moscow State Technical University Bulletin* **3** 78–82
[20] Ksenofontov A G 2014 *Heating Unit Analysis and Design* (Moscow: Moscow State Technical University named after N. E. Bauman)

IOP Conf. Series: Materials Science and Engineering **969** (2020) 012023 doi:10.1088/1757-899X/969/1/012023

Graphical Representation Reliability of Experimental Data Concerning to Structural and Surface Properties of Substances

D V Mayorov[1], T T Gorbacheva[2] and Yu O Velyaev[3]

[1]Tananayev Institute of Chemistry and Technology of Rare Elements and Minerals Tananaeva, Kola Science Center Russian Academy of Sciences (ICTREM KSC RAS), 26a, Akademgorodok, Apatity, Murmansk Region, 184209, Russia
[2]Institute of Industrial Ecology of the North, Kola Science Center Russian Academy of Sciences (INEP KSC RAS), 14a, Akademgorodok, Apatity, Murmansk Region, 184209, Russia
[3]Polytechnical Institute, Sevastopol State University, 33, Universitetskaya, Sevastopol, 299053, Russia

E-mail: d.maiorov@ksc.ru

Abstract. Using experimental data on the pore distribution of a sample of magnesium-aluminum layered double hydroxide, various variants of their graphical representation are analyzed. It is shown that graphical interpretation of the obtained data in the coordinates dV/dD – D (dV/dlog(D) – D) or V – D is not of practical (scientific) interest and may lead to incorrect conclusions about the predominance of pores of a certain diameter in the sample and other errors. It is concluded that to avoid possible errors and provide the most complete information about the pore distribution of the studied samples, the presented graphic materials should be supplemented with histograms of the pore distribution.

1. Introduction

In scientific and engineering literature, there are two approaches to presenting experimental porosimetry data. The first one is $dV/dD - D$ or $dV/d\log(D) - D$ [1–15] graphs, where the values of dV/dD ($dV/d\log(D)$) plotted along the Y axis are in $cm^3/(g{\cdot}nm)$ (or $cm^3/(g{\cdot}Å)$). Here, V means pore volume (usually in cm^3/g) in a certain range of diameters D_{pore} ($a < D_{pore} < b$, usually in nm or Å), while D means average pore diameter (D_{av}) in this range, because it's almost impossible to measure the pore volume for a predefined diameter value (for example, $D = 5.629...$ nm).

In the second approach, the particle volume distribution is also presented as $dV/dD - D$, but the dV/dD values are expressed in cm^3/g [3, 6, 9, 16, 17]. This implies that there is a $V - d$ relationship, which is explicitly indicated in [18].

However, we believe that both of these presentations are of little practical (scientific) interest and may lead to incorrect conclusions about the predominance of a certain pore diameter or total pore volume in a specimen.

This paper aims to show the incorrectness of such representations of experimental data and the resulting potentially erroneous conclusions, as well as the incompleteness of the information presented.

2. Experimental procedure and discussion

Table shows the experimental data on the pore diameter distribution in a specimen of magnesium-aluminum layered double hydroxide (Mg-Al LDH). The product was obtained by solid-phase synthesis by mixing crystalline magnesium and aluminum chlorides with ammonium carbonate and its composition can be described by the formula of $Mg_4Al_2(OH)_{12}CO_3{\cdot}3H_2O$ [19]. The pore distribution data for the specimen was collected using a TriStar 302 specific surface area and porosity analyzer based on nitrogen desorption isotherm using the BJH (Barret-Joyner-Halenda) method. Classification into pore size intervals (1÷10) and column names (a÷d) were added by us, as described below.

Table 1. Experimental data on specific volume and pore area distribution in the synthesized Mg-Al LDH specimen by pore diameter.

Pore Diameter Range (nm)		Average Diameter (nm)	Incremental Pore Volume (cm³/g)	Cumulative Pore Volume (cm³/g)
No 10	214.9 ÷ 48.3	55.9	0.022561	0.022561
No 9	48.3 ÷ 24.9	29.6	0.016884	0.039445
No 8	24.9 ÷ 15.1	17.7	0.014623	0.054068
No 7	15.1 ÷ 12.1	13.3	0.007331	0.061400
No 6	12.1 ÷ 9.7	10.6	0.007419	0.068818
	9.7 ÷ 8.1	8.8	0.005555	0.074374
No 5	8.1 ÷ 6.9	7.4	0.005206	0.079580
	6.9 ÷ 6.0	6.4	0.004661	0.084241
No 4	6.0 ÷ 5.2	5.5	0.004206	0.088448
	5.2 ÷ 4.6	4.9	0.003797	0.092244
	4.6 ÷ 4.1	4.3	0.003500	0.095744
No 3	4.1 ÷ 3.7	3.9	0.003327	0.099070
	3.7 ÷ 3.3	3.5	0.013257	0.112327
	3.3 ÷ 3.0	3.2	0.005640	0.117967
No 2	3.0 ÷ 2.8	2.9	0.001478	0.119446
	2.8 ÷ 2.4	2.6	0.000979	0.120424
	2.4 ÷ 2.3	2.3	0.000331	0.120756
No 1	2.3 ÷ 2.0	2.1	0.000294	0.121050
	2.0 ÷ 1.8	2.9	0.000108	0.121157
a		b	c	d

Based on the representation of the collected data in the form of a differential $dV/dD - D$ distribution (Figure 1a), it can be concluded that pores with an average diameter of 3.5 nm predominate in the specimen, i.e. the distribution is monomodal. And that is the only conclusion that can be made. But the average value of a parameter (in this case, pore diameter) characterizes a given value range of the parameter (between a and b). Based on Figure 2a, this range remains unknown for both D_{av} = 3.5 nm and other values of D_{av}.

As can be seen from Table, for pores in the diameter interval 10 (48.3÷214.9 nm), the average diameter is 55.9 nm, which is not even the arithmetic mean value $D_{ar.mean}$.($D_{ar.mean}$ = (48.3 + 214.9) / 2 = 131.6 nm) and can probably be interpreted as a weighted average. Same applies to the interval 7-9 and, given a more precise calculation of *av*, is likely to be true for the other intervals. In addition, if the available hardware (or software) could produce data with a greater resolution (narrower pore diameter ranges), one would probably see a slightly different distribution pattern and, accordingly, could draw completely different conclusions.

Some additional information can be obtained from the integral pore distribution by diameter (Figure 1b). Specifically, pores with an average diameter of 3.2 to 3.5 nm have a volume of 0.018897 cm^3/g, which is ~15.6% of the total pore volume (0.121157 cm^3/g (Table, column d). This distribution pattern is detected without micropores, whose volume in this case is extremely small (less than 1%).This is insufficient for a prevailing value found on the basis of differential distribution data (Figure 1a). In addition, integral distribution gives only the average pore diameter value in a given interval, which does not answer the question about the actual pore diameter intervals.

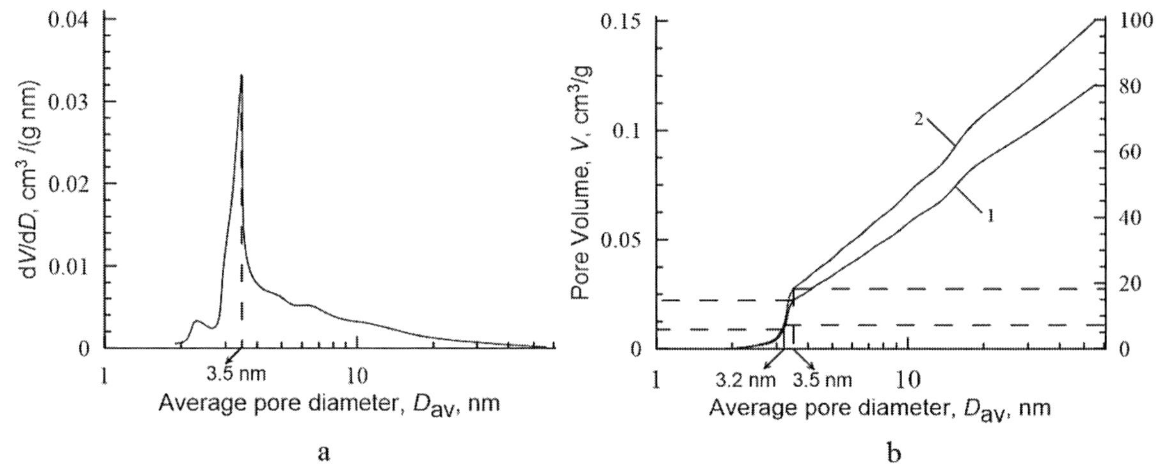

Figure 1. Differential ($dV/dD - D$) (a) and integral (b; 1 – in cm^3/g; 2 – in %) pore distribution in the synthesized Mg-Al LDH specimen (desorption branch).

Visualizing porosimetry data (in this case, Table) as a histogram (Figure 2), which is identical to the data in Table, but easier to interpret, has no such drawbacks. Using such a histogram, it is possible to find the exact volume of the pores in a given pore diameter range. For example, Figure 2a shows detailed histograms for each of the pore volume ranges in Table, while Figure 2b (for convenience of interpretation) shows these for the combined intervals 1÷10 defined by us. In addition, it can be seen that the volume of macropores ($D_{pore} > 50$ nm) equal to 0.022561 cm^3/g is comparable with the volume of pores with $D_{av} = 3.5$ nm and amounts to ~18.6% of the total pore volume of the specimen. Thus, based on the data in Table or Figure 2, completely different conclusions can be drawn and more complete information about the porosity of the specimen can be obtained:

1. The Mg-Al LDH specimen is mesoporous (mesopore volume ~80% of the total pore volume).
2. Mesopore distribution ($2 < D_{pore} < 50$ nm) is polymodal dominated by pores with $D_{av} = 3.5$, 17.7, and 29.6 nm in the diameter ranges 3.3÷3.7, 15.1÷24.9, and 24.9÷48.3 nm, respectively.

Conclusion 2 indicates that the conclusion about the monomodality of the pore distribution and the predominance of pores with an average diameter of 3.5 nm, drawn from the data in Figure 1a, is erroneous.

The second approach to presenting the particle volume distribution ($V - D$) is also not without flaws and leads (upon detailed examination) to unexpected results. Figure 3 shows the pore volume distribution in a $V — d$ space (according to Table). In Figure 3a, the pore distribution is given with the preservation of experimental data points (which somewhat improves the situation), while in Figure 3b, the experimental data points are excluded (which is more common in the literature). By processing the data in Figure 3b (Figure 3c), we find the total (combined) pore volume of the specimen (Figure 3c), which is 0.121157 cm^3/g. The obtained value matches the experimental data, which is not surprising, since we know the position of the experimental data points and the curves were plotted using these. What if experimental data points are missing?

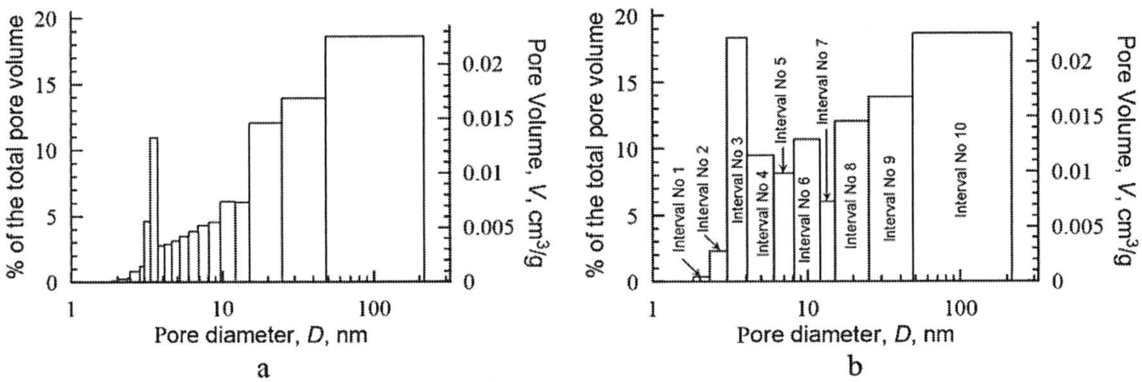

Figure 2. Pore diameter histograms of the Mg-Al LDH specimen (desorption branch).

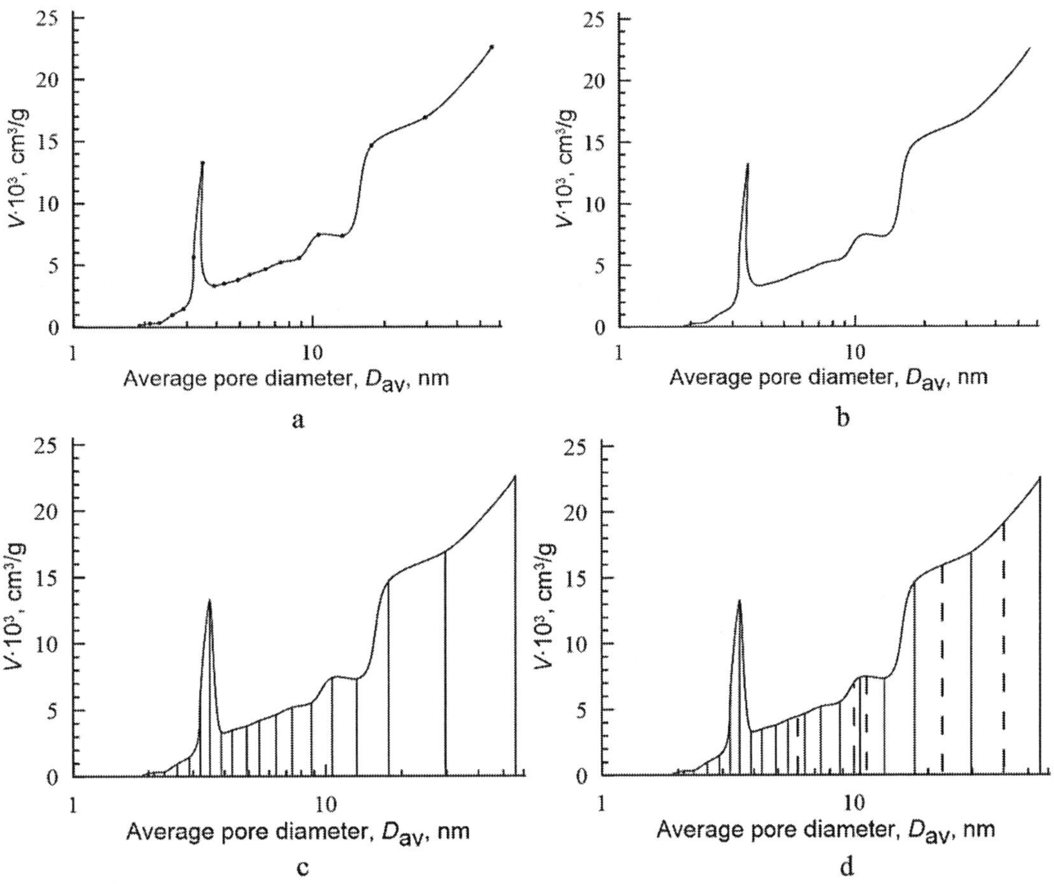

Figure 3. Pore diameter distribution of the Mg-Al LDH specimen (desorption branch and processing results).

In Figure 3d, the total pore volume is found using additional data points (*D* values) (dashed curves) while keeping the data points from Figure 3c. Here we see a total pore volume larger than the actual one (0.121157 cm^3/g) and, accordingly, higher than 100%. What if we find the total pore volume, not knowing the data in Table, based on other values of *D*? It is extremely unlikely that we will succeed and choose (i.e. guess) the actual values of *D*, based on which the curve is plotted, and accordingly, the actual pore volume value (0.121157 cm^3/g). Most likely, the values of *D* will be different, and we will get a total pore volume value that is either lower or higher than the actual one.

In addition, such a distribution does not answer the question raised earlier concerning D_{av} and the pore range corresponding to this value.

3. Conclusion

Thus, we believe that presenting pore diameter distribution in a $dV/dD - D$ or $dV/d\log(D) - D$ and $V - D$ space is not justified and may lead to erroneous conclusions. Most appropriate is presenting porosimetry data either in the form of Table or in the form of histograms (Figure 2 and [20–23]). If presenting the pore diameter distribution in a $dV/dD - D$, $dV/d\log(D) - D$, or $V - D$ space conveys any additional information unknown to us, for the sake of clarity and completeness, the respective findings should be presented together with histograms.

References

[1] Schmitt M S, Fernandes C P, Fabiano G W, Bellini da Cunha Neto J A, Rahner C P and Santiago dos Santos V S 2015 Characterization of Brazilian tight gas sandstones relating permeability and Angstrom-to micron-scale pore structures *Journal of Natural Gas Science and Engineering* **27** 785–807 Online. Available: https://doi.org/10.1016/j.jngse.2015.09.027

[2] Cervantes-Martinez C V, Emo M, Lebeau B, García-Celmad M-J, Stébé M-J and Blin J-L 2019 Insights of the kolliphor/water system for the design of mesostructured silica materials *Microporous and Mesoporous Materials* **285** 231–240 Online. Available: https://doi.org/10.1016/j.micromeso.2019.05.019

[3] Xiao D, Lu Z, Jiang S and Lu S 2016 Comparison and integration of experimental methods to characterize the full-range pore features of tight gas sandstoned. A case study in Songliao Basin of China *Journal of Natural Gas Science and Engineering* **34** 1412–1421 Online. Available: http://dx.doi.org/10.1016/j.jngse.2016.08.029

[4] Trublet M, Maslova M V, Rusanova D and Antzutkin O N 2016 Mild syntheses and surface characterization of amorphous $TiO(OH)(H_2PO_4)\cdot H_2O$ ion-exchanger *Materials Chemistry and Physics* **183** 467–475 Online. Available: http://dx.doi.org/10.1016/j.matchemphys.2016.09.002

[5] Maslova M V, Rusanova D, Naydenov V, Antzutkin O N and Gerasimova L G 2012 Extended study on the synthesis of amorphous titanium phosphates with tailored sorption properties *Journal of Non-Crystalline Solids* **358** 2943–2950 Online. Available: https://doi.org/10.1016/j.jnoncrysol.2012.06.033

[6] Zhang L, Lu S, Xiao D and Gu M 2017 Characterization of full pore size distribution and its significance to macroscopic physical parameters in tight glutenites *Journal of Natural Gas Science and Engineering* **38** 434–449 Online. Available: http://dx.doi.org/10.1016/j.jngse.2016.12.026

[7] Assaker K, Lebeau B, Michelin L, Gaudin P, Carteret C, Vidal L, Bonne M 2015 Blin J L 2015 Zn-TiO_2 mesoporous oxides prepared by mechanical milling *Journal of Alloys and Compounds* **649** 1–10 Online. Available: http://dx.doi.org/10.1016/j.jallcom.2015.07.085

[8] Naboulsi I, Lebeau B, Michelin L, Carteret C, Bonne M and Blin J-L 2018 Influence of crystallization conditions and of gaseous ammonia treatment on mesoporous TiO_2 properties *Microporous and Mesoporous Materials* **262** 1–12 Online. Available: https://doi.org/10.1016/j.micromeso.2017.11.025

[9] Cervantes-Martinez C V, Emo M, García-Celma M-J, Stébé M-J and Blin J-L 2019Morphosynthesis of porous silica from biocompatible templates *Chemical Engineering Research and Design* **151** 179–189 Online. Available: https://doi.org/10.1016/j.cherd.2019.09.006

[10] Kuznetsova T F and Eremenko S I 2014 Synthesis of aerogel-type mesoporous silica *Colloid Journal* **76 3** 327-333 Online. Available: https://doi.org/10.1134/S1061933X14030089

[11] Julve D, Ramos J, Pérez J and Menéndez M 2011 Analysis of mercury porosimetry curves of precipitated silica, as an example of compressible porous solids *Journal of Non-Crystalline Solids* **357** 1319-1327 Online. Available: https://doi:10.1016/j.jnoncrysol.2010.12.042
[12] Kuznetsova T F and Eremenko S I 2013 Properties of sol-gel derived silica membranes *Inorganic Materials* **49** 159–164 Online. Available: https://doi.org/10.1134/S002016851302012X
[13] Kuznetsova T F, Rat'ko A I and Eremenko S I 2012 Textural and adsorption properties of mesoporous silicophosphate *Colloid Journal* **74 1** 78-84 Online. Available: https://doi.org/10.1134/S1061933X11060111
[14] Vorontsova O A, Lebedeva O E and Roessner F 2009 Synthesis of layered hydroxides stable in redox media *Kinetics and Catalysis* **50 6** 863-866 Online. Available: https://doi.org/10.1134/S002315840906010X
[15] Zima T M, Baklanova N I and Lyakhov N Z 2010 Composites of mesoporous Al2O3 and Fe- or Co-containing nanoparticles *Inorganic Materials* **46 8** 852-857 Online. Available: https://doi.org/10.1134/S0020168510080091
[16] Hua J 2015 Synthesis and characterization of bentonite based inorgano–organo-composites and their performances for removing arsenic from water *Applied Clay Science* **114** 239-246 Online. Available: http://dx.doi.org/10.1016/j.clay.2015.06.005
[17] Zaleski R, Kierys A, Grochowicz M, Dziadosz M and Goworek J 2010 Synthesis and characterization of nanostructural polymer–silica composite: Positron annihilation lifetime spectroscopy study *Journal of Colloid and Interface Science* **358** 268-276 Online. Available: http://doi:10.1016/j.jcis.2011.03.008
[18] Maslova M V and Gerasimova L G 2011 The Influence of Chemical Modification on Structure and Sorption Properties of Titanium Phosphates *Russian Journal of Applied Chemistry* **84 1** 1-8 DOI 10.1134/S1070427211010010
[19] Matveev V A and Majorov D V *Method of obtaining layered hydroxide of madnesium and aluminium* RU Patent **2678007**
[20] Zakharov V I, Matveev V A, Mayorov D V, Balbukova A A, Kondratenko T V and Knyazeva A I 2012 A study of the effect of nepheline acid processing conditions on structural-surface properties of silica products formed *Russian Journal of Applied Chemistry* **85** 1643–1648 Online. Available: https://doi.org/10.1134/S1070427212110018
[21] Zhang Z and Yang Z 2013 Theoretical and practical discussion of measurement accuracy for physisorption with micro- and mesoporous materials *Chinese Journal of Catalysis* **34** 1797–1810 Online. Available: https://doi.org/10.1016/S1872-2067(12)60601-9
[22] Fedoročková A, Plešingerová B, Sučik G, Raschman P and Doráková A 2014 Characteristics of amorphous silica prepared from serpentinite using various acidifying agents *International Journal of Mineral Processing* **130** 42–47 Online. Available: http://dx.doi.org/10.1016/j.minpro.2014.05.005
[23] Gerasimova L G, Shchukina E S, Maslova M V, Gladkikh S N, Garaeva G R and Kolobkova V M 2017 Synthesis of Rutile Titanium Dioxide from Russian Raw Materials *Polymer Science, Series D* **10 1** 23–27 Online. Available: https://doi.org/10.1134/S1995421217010099

IOP Conf. Series: Materials Science and Engineering **969** (2020) 012024 doi:10.1088/1757-899X/969/1/012024

Thermal Treatment of Cemented Alloyed Steels for the Cold Plastic Deformation

V I Astashchenko, T V Shveyova and A I Shveyov

Kazan (Volga Region) Federal University, 18, Kremlyovskaya street, Kazan, 420008, Russia

E-mail: astvi-52@mail.ru

Abstract. Disclosed is a method of thermal treatment of blanks from alloyed cement steels for cold volumetric stamping and extrusion. The technology ensures the formation of a ferritic-pearlitic structure in steel with a minimum microhardness of individual structural components. Examples of manufacturing of toothed parts by the method of cold wrapping and thin-walled tubular articles during cold extrusion are given. It is shown that cyclic resistance of parts made by cold plastic deformation and hardened by method of chemical-thermal treatment is increased by three or more times. According to the results of metallographic analysis, the ability of steel to cold plastic deformation and the ability to obtain high-quality machine parts are predicted.

1. Introduction

In the XXI century, the attitude of the machine-building complex to the production of automotive equipment changed significantly. The automotive industry focuses on the production of competitive products that meet the requirements of European and world standards for reliability, durability, ecological properties and deadweight coefficient. An important condition at the stages of technological conversion of metal into a part is the saving of fuel-energy and material resources. The most promising area of simultaneous solution of these problems should include the development and implementation of low-waste and high-efficiency processes for forming parts. Such processes are various methods of cold plastic deformation: extrusion, stamping, landing, reduction, etc., which are increasingly used in mechanical engineering plants. Thanks to these technologies, the metal utilization rate reaches 90% or more [1–3]. It is significantly reduced, and in some cases, and there is no need for subsequent machining of such metal products. In addition, in the process of forming parts by this method, and especially parts of complex geometry, a favorable deformation texture is created in metal products, which contributes to improving their performance. Low carbon steels [4–7] are known to have high processing properties in cold pressure treatment. However, the use of such steels in the automotive industry is limited by their reduced strength and the heavy operating conditions of many parts. For products that simultaneously experience static and dynamic loads in operation, high-strength alloy steels are recommended and often used, which in turn have an insufficient level of ductility at room temperature, which makes them difficult to deform in a cold state. This is especially noticeable in the manufacture serration toothed and thickwalled tubular parts.

The main condition for the implementation of cold plastic deformation technologies in the manufacture of high-tensile-strength steel products is to ensure homogeneity of the structure and properties of the alloy over the entire volume of the semi-finished product [8–9].

The purpose of the work is to create a favorable structure in alloyed cement steels for cold plastic deformation of metal products.

2. Research materials and methods

Studies were carried out on widely used in the automotive industry structural cement alloyed steels of 20HGNMTA and 12HN3A grades used for the manufacture of toothed parts and piston pins of the internal combustion engine (ICE).

For metallographic studies, optical microscopes “Neofot-21” (Germany) and IM-7200 (Japan) with the licensed software product “Trixomet-PRO” of the Video Test-M system were used. The hardness was determined in accordance with the standards on certified instruments TSh-2M and TK-2M. Microhardness of structural components was estimated by means of the Durimet microhardness gage (Germany) at load of indenter 9.8N. Heat treatment of samples and full-scale parts was carried out in laboratory and production conditions in electric furnaces M-12 and in gas-heated mod furnaces. 34898 “Heat Project,” respectively. Plastic deformation of steel semi-finished products was carried out on mod presses. KB 0034.

The bending cyclic durability of the gear parts as part of the KAMAZ car assemblies was evaluated during tests of the driving bridges on the AV-604 model bench. The chemical composition of metal semi-finished products, blanks and parts was determined according to GOST 18895-97 on the SPEKTROLAB emission spectrometer, on the MFS-51 microphotometer and AFS-51 spectrograph, as well as on fashion devices. AN-7529 and AN-7560 for the determination of carbon and sulphur.

3. Results of work

The main criteria of structural state of steels characterizing high ductility of alloy is dispersion of grain structure, quantity and morphology of structural and phase components, degree of contamination and concentration with non-metallic inclusions [4,10].

The first parameters of the structure are formed during the heat treatment of the semi-finished product and primarily as a result of prolonged (40 hours or more) spheroidizing annealing. But even after such long exposure in the structure of alloyed steels, there are areas with traces of non-diffusion decay of austenite (Figure 1).

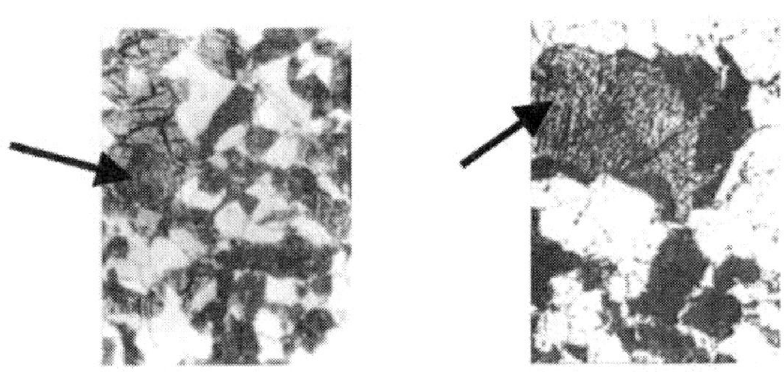

Figure 1. Products of diffusion-free conversion to steel after annealing of blanks.

The presence of microparticles with martensitic (Figure 1) or bainite structure in steel having a ferritic-pearlitic base creates a stress state between gradient structural components and reduces the plastic properties of alloys. In order to increase technological ductility of structural steel it is necessary to have structure of granular, rather than plate pearlite, which increases isotropicity of alloy, reduces its hardening during deformation and improves stamping [11–12].

It was established that the chemical composition of the studied steels for all elements is within the permissible limits of GOST 4543 and technical regulation TU 14-1-5509. At the same time, fluctuations in individual chemical elements in steel 20HGNMTA ranged from 1.19 to 4.0 times, and in steel 12HN3A – from 1.31 to 2.80 times.

Widely used methods of heat treatment of such steels do not provide stable production of the recommended structure and hardness of blanks for cold plastic deformation. This is strongly evidenced by the significant fluctuations in hardness of the steel 12HN3A after annealing, which varied from 176 to 197 HB. Therefore, when developing the steel softening technology, it is necessary to take into account not only the fluctuations of the chemical elements, but also the structural heterogeneity in micro-volumes (Figure 1).

We investigated the influence of temperature-time parameters of heating, cooling and diffusion decay of undercooled austenite on the resulting structure, hardness and size of steel grains. Table 1 shows the results of analysis of such properties in steel 12HN3A after various treatments. The most favorable parameters in terms of structure and hardness for cold plastic deformation are this steel after thermal treatment, which includes austenitization at a temperature of 880°C with subsequent diffusion isothermal γ-α conversion at temperatures of 650°C and 620°C (Table 1 mode No. 4).

Table 1. Steel properties 12XH3A after various types of heat treatment.

Mode number	Scheme of heat treatment	Microstructure	Size of grain, mm	Hardness, HB
1.	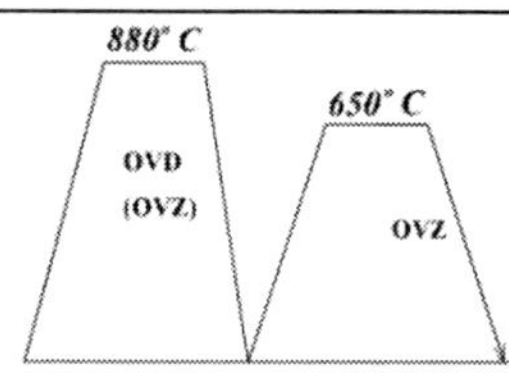	P+F P < 80%	0.017–0.026	187 (159)
2.	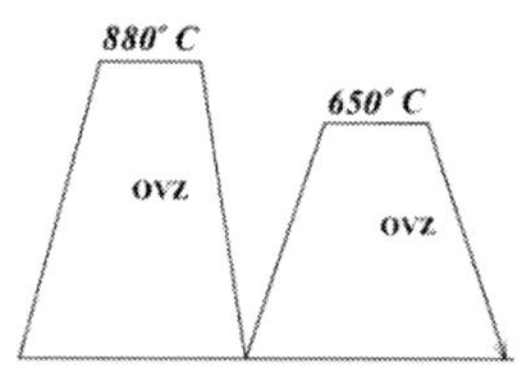	P+F P < 80%	0.017–0.028	159
3.	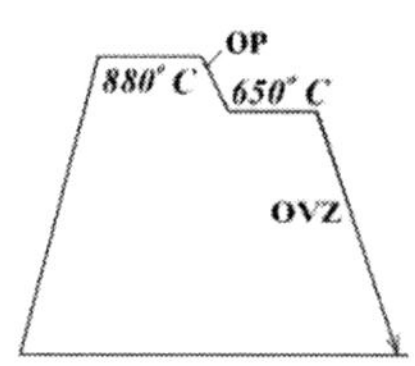	P+F P > 80%	0.016–0.022	149–153
4.	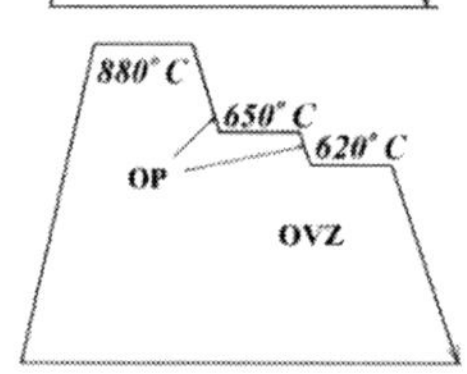	P+F P > 85%	0.014–0.021	146–149
5.	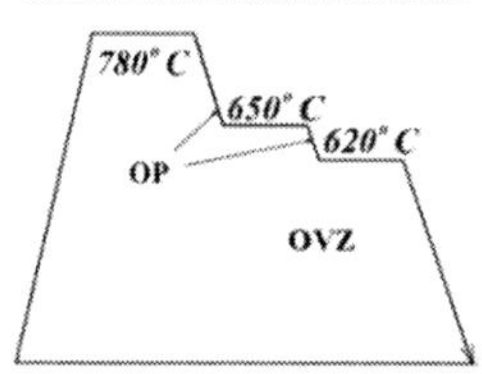	P+F P > 85%	0.017–0.038	146–149

*Designation: Cooling with furnace (OP), in water (OP), in the air (HIA), P – perlite, F – ferrite.

The introduction of this technology of heat treatment of steel 12XH3A in the manufacture of piston fingers ICE has shown its economy and processability. High-quality parts with minimum allowance for mechanical treatment (Figure 2) are obtained from preheated blanks by cold extrusion method.

The developed annealing technology can be used as a softening heat treatment for steel blanks 20HGNMTA. To intensify the process of recrystallization of steel and provide a fine-grained structure in it, the austenization temperature can be increased to 900–920 °C. This is confirmed by the stability and growth of the grain when heating this steel to a temperature of ~ 950°C, which can be explained by the presence of alloying refractory elements in it - molybdenum and titanium [13–22].

Heat treatment according to the proposed technology of steel blanks 20HGNMTA showed that their hardness does not exceed 163 HB, the microhardness of perlite is in the range of 227–286 HV, ferrite – 179–210 HV at the difference between these structural components less than 80 HV. This state of steel made it possible to obtain parts with high tooth accuracy by cold stamping (COS) (Figure 3).

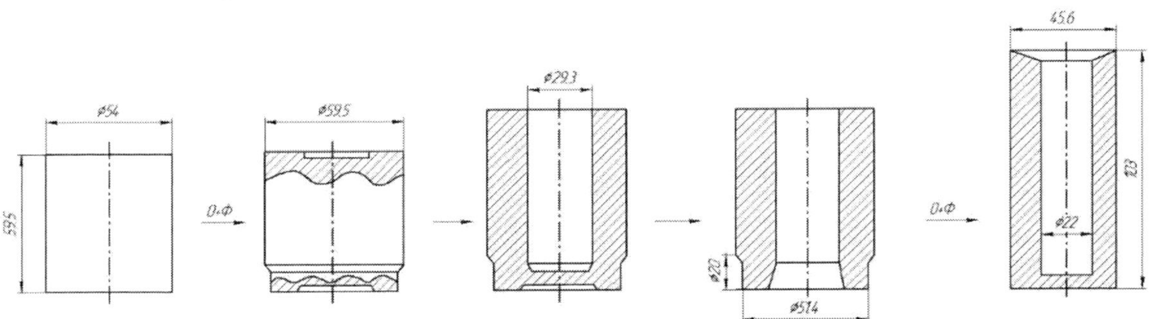

Figure 2. Diagram of production of ICE piston pins and interoperative annealing of billet from steel 12HN3A. O – annealing; F – phosphatizing.

Figure 3. Satellite made by cold stamping (steel 20HGNMTA).

Bench tests of wheel gears of driving axles with satellites manufactured by the XOS method showed the bending durability of parts within 1070–1150 thousand loading cycles, which is about three times higher than the resistance of mating parts included in the mechanism link.

4. Conclusions

1. Disclosed is a method of heat treatment of steel 12HN3A and 20HGNMTA for cold plastic deformation, involving austenizing and decay of supercooled austenite at temperatures of 650 and 620°C.
2. Three-fold increase of cyclic resistance of cemented toothed parts made by method of cold hoop forging was revealed.
3. The technology of annealing blanks is recommended for use at enterprises in the manufacture of parts by cold plastic deformation.

References

[1] Fonstein N M 1988 *Two-phase low-alloy steels for the production of products by various cold deformation methods* (Moscow: TsNIIChermet)
[2] Navrotsky G A 1987 *Forging and stamping* (Moscow: Engineering) p 384

[3] Kuzmin B A 1977 *Metallurgy, metal science and structural materials* (Moscow: Higher School) p 301

[4] Pachurin G V 2012 *Preparation of steel blanks for obtaining high-strength fasteners* (Publishing House LAP LAMBERT Academic Publishing) p 129

[5] GOST 9045 – 95 2003 *Cold-rolled thin-sheet rolled stock made of low-carbon high-quality steel for cold stamping. Technical specifications* (Minsk: Publishing House Interstate Council for Standardization, Metrology and Certification) p 11

[6] GOST 10702 – 78 2006 *Rolled stock of high-quality structural carbon and alloyed steel for cold extrusion and landing. Specifications* (Moscow: Standantinform) p 12

[7] Gliner R E and Astashchenko V I 2015 *Introduction to Metal Quality Management: a textbook* (Kazan: Publishing house of Kazan University) p 351

[8] Astashchenko V I, Galimov E R, Shveev A I, Shveyeva T V, Pankratov D L, Rodkin I M and Shibakov R V 2015 *Method for determining the suitability of steel for cold plastic deformation* RU Patent **2568887**

[9] Astashchenko V I, Shveev A I and Shveeva T V 2016 Diagnostics of the properties of steel articles according to the criterion of microhardness *Metal science and heat treatment* **58 5–6** 303–307 DOI 10/1007/s11041-016-0008-6

[10] Astashchenko V I, Shveev A I and Shveyeva T V 2011 *Quality control and heredity of the steel structure at the technological metal distribution* (Moscow: Academia) p 239

[11] Ilyin S I and Koryagin Yu D 2009 *Steel Heat Treatment Technology: Tutorial* (Chelyabinsk: Publishing Center of SUSU) p 120

[12] Kalner V D 1984 *Quality control of heat treatment of steel semi-finished products and parts* (Moscow: Mechanical Engineering) p 463

[13] Shveyova T V, Astashchenko V I and Shveyov A I. 2015 Microalloy steels: Energy saving prospect at manufacture of details of cars *Biosciences Biotechnology Research Asia* **12** 297–299

[14] Firger I V 1982 *Thermal treatment of alloys: Handbook* (Leningrad: Mechanical engineering) p 304

[15] Astaschenko V I, Zapadnova E A, Zapadnova N N and Mukhametzynova G F 2016 Predicting structure mikro-alloey steel products for different purposes *IOP Conference Series: Materials Science and Engineering* **134 1** 012029

[16] Pesin A M, Shveyova T V and Pustovoytov D O 2020 Features of using cast and deformed steel for machine parts *Ferrous metals* **1** 43–49

[17] Shveyova T V, Pesin A M and Pustovoytov D O 2017 Behavior of micro-alloyed steel at thermal treatment *Solid State Phenomena* **265** 177–180

[18] Smirnova L A 2007 *Problems of production and use of vanadium steels: Materials of the International Scientific and Technical Seminar* (Yekaterinburg: Ural Branch of the Russian Academy of Sciences) p 436

[19] Nagel M, Drissen I, Kovach S and Pudritz O 2019 New developments in the field of high-strength microalloyed autobile steels *Ferrous metals* **3** 41–43

[20] Olshevski G 2016 New steel HSX 710 combining strength and ductility *Ferrous metals* **1** 50–52

[21] Chukin M V, Koptseva N V, Nikitenko O A and Efimova Yu Yu 2011 Mechanical properties of carbon structural steel with ultra-fine grain structure *Ferrous metals* **6** 54–59.

[22] Pesin A, Chukin M, Tandon P, Pustovoitov D, Drigun E, Pesin I and Kozhemyakina A 2018 Development of theory and technologies for hybrid metal forming of steel complex parts *27th International Conference on Metallurgy and Materials, Conference Proceedings* 261–265

IOP Conf. Series: Materials Science and Engineering **969** (2020) 012025 doi:10.1088/1757-899X/969/1/012025

Study of Physical Properties and Physical Processes in Nano-thin Spatial Dissipative Structures

V B Malkov[1], O V Chemezov[2] and A V Malkov[3]

[1]Institute of High-Temperature Electrochemistry, Ural Branch Russian Academy of Sciences, 22, S. Kovalevskoy, Yekaterinburg, 620219, Russia
[2]Ural Federal University named after the first President of Russia B. N. Yeltsin, 19, Mira street, Yekaterinburg, 620002, Russia
[3]ZAO NPTs ROSNA, 9, Studencheskaya, Yekaterinburg, 620026, Russia

E-mail: o.v.chemezov@urfu.ru

Abstract. Two-dimensional geometric object "neutral surface" – successfully used to study thin plates. The use of this two-dimensional geometric object is not possible for the study of physical properties and physical processes in nano-thin SDS (spatial dissipative structures). The SDS lattice undergoes a rotational curvature around three mutually perpendicular directions, and the nano-thin SDS itself remains flat. We have developed a method for the study of nano-thin SDS with rotary lattice curvature. We used as a basis the method of visualizing the rotational curvature of the lattice by constructing a two-dimensional geometric object – the surface of the lattice curvature by nano-thin SDS. This method of investigation involves application of the lattice curvature surface with nano-thin SDS for the selected crystallographic direction as a "neutral surface". The method will find application in nanotechnology.

1. Introduction

For the study of thin plates a two-dimensional geometric object is successfully used – a "neutral surface", that is, the geometric place of the points equally spaced from the upper and lower surfaces of the thin plate [1]. However, the use of the "neutral surface" is not possible for the study of physical properties and physical processes in nano-thin spatial dissipative structures (SDS) [2–8]. The lattice of nano-thin SDS experiences internal bending as opposed to thin plates, and as a whole the lattice of nano-thin SDS remains flat [9].

Really, we will obtain a flat rhomb by constructing a "neutral surface" of nano-thin SDS hexagonal selenium, the lattice of which undergoes elastic rotational curvature around [001]. Thus, the construction the "neutral surface" of the nano-thin SDS with the inner bending of the lattice does not allow visualization of the lattice rotational curvature of the nano-thin SDS.

Visualization can be performed for lattice rotary curvature of nano-thin SDS, namely for nano-thin crystals of hexagonal selenium with elastic rotary curvature of lattice around [001] after stage of hardening of nano-thin SDS [10–13]. We propose a method for visualization of the lattice rotary curvature of nano-thin SDS earlier [10]. The visualization of the rotational curvature of the lattice of nano-thin SDS was achieved by us by constructing a two-dimensional geometric object - the surface of the curvature of the lattice of nano-thin SDS for the selected crystallographic direction (Figure 1).

However, the above method [10] allows visualizing elastic rotational curvature of the lattice of nano-thin SDS, but does not allow investigating physical properties and physical processes in nano-thin SDSs.

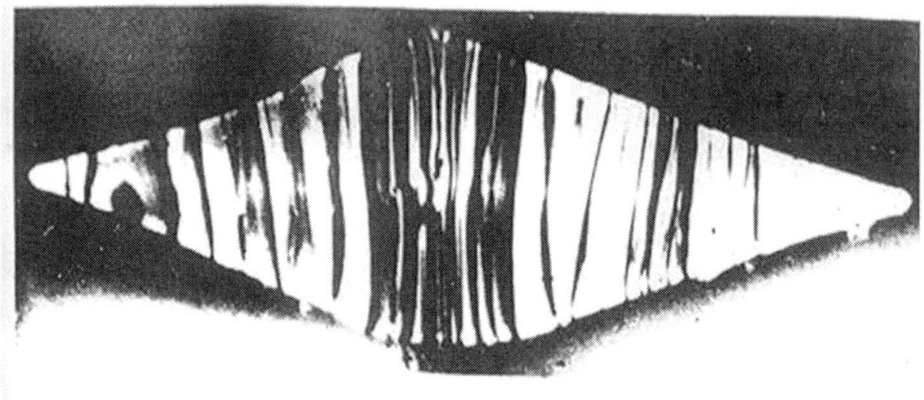

a)

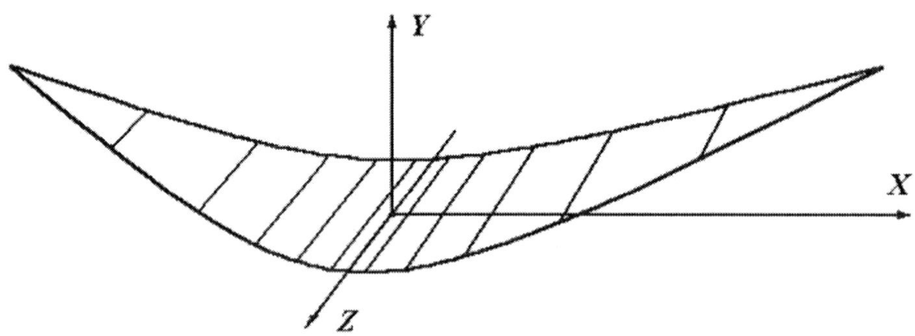

b)

Figure 1. Electron microscopic image of nano-thin diamond-shaped SDS of hexagonal selenium with elastic rotational curvature of lattice around [001] and system of bending extinction contours of parallel [001] coinciding with short diagonal of rhombus, JEM-200CX JEOL Ltd, Japan. a) The lattice curvature surface with a nano-thin SDS for [100] hexagonal selenium, taken as a "neutral surface", the maximum curvature of which coincides with the short diagonal of the rhombus; b) Result of construction the lattice curvature surface with a diamond-shaped nano-thin SDS around [001] for [100].

In view of the above, we have developed a method for investigating nano-thin SDS with rotary lattice curvature. We founded this means of research on the application of the method of visualizing the lattice rotary curvature [10] by constructing a two-dimensional geometric object – the lattice curvature surface with a nano-thin SDS for the selected crystallographic direction. In addition, the above method involves applying the lattice curvature surface of the nano-thin SDS to a selected crystallographic direction as a "neutral surface". Besides, we perform a comparative analysis of the "neutral surfaces" of the nano-thin SDS and the previously studied reference thin plates [1]. This procedure is necessary to identify the bending features of the lattice curvature surface of the nano-thin SDS for the selected crystallographic direction used as the "neutral surface". We are using new information on the bending features of the "neutral surface" of nano-thin SDS to investigate the physical properties and physical processes in nano-thin SDS.

Thus, we propose the following sequence of actions:

1. the two-dimensional geometric object – the lattice curvature surface with nano-thin SDS is created for selected crystallographic direction [10, 11];
2. the lattice curvature surface of the nano-thin SDS for the selected crystallographic direction using as a "neutral surface";
3. a comparative analysis performing of "neutral surfaces" of nano-thin SDS and thin reference plates with previously identified properties and detecting on this basis new information on bending features of the "neutral surface" of nano-thin SDS;
4. new information identified on the features of "neutral surface" bending using to investigate the physical properties and physical processes in nano-thin SDS.

The lattice curvature surface of the nano-thin SDS, used as the "neutral surface" of the nano-thin SDS, allows us to move from solving three-dimensional to solving two-dimensional problems. Indeed, in work [1] it is indicated, "...one can consider the plate as not possessing thickness, that is, as a geometric surface...".

2. Results and discussion

We illustrate the proposed method of investigating physical properties and physical processes in nano-thin SDSs by the following examples.

We investigated the physical property of hexagonal selenium nano-thin SDS using the proposed method. We studied the origin of the interblock torsion boundary at the center of the diamond-shaped nano-thin SDS with elastic rotational lattice curvature. To do this, we constructed the lattice curvature surface with a diamond-shaped nano-thin SDS (figure 1 a) for [100], the lattice of which experiences non-uniform elastic rotational curvature around [001] (figure 1 a) [10–14]. We present the result of the construction the lattice curvature surface with a diamond-shaped nano-thin SDS around [001] for [100] in figure 1 b. We applied the surface lattice curvature of the nano-thin SDS around [001] for [100] (figure 1 b) as the "neutral surface" of the hexagonal selenium nano-thin SDS and established the features of its bending.

We have previously found that the maximum value of the surface lattice curvature of the nano-thin SDS for [100] coincides with the short diagonal of the diamond-shaped nano-thin SDS (figure 1 a) with the inhomogeneous elastic rotational lattice curvature of the nano-thin SDS around [001] [14]. Accordingly, the feature of the "neutral surface" bending of the nano-thin SDS was the maximum curvature of the "neutral surface" of the nano-thin SDS (figure 1 b) in the direction of the short rhombus diagonal.

We performed a comparative analysis of the "neutral surfaces" of the nano-thin diamond-shaped SDS for [100] (figure 1 b), the reference wedge-shaped thin plate and the reference rounded thin disk [1]. As a result, we identified new information on the features of the "neutral surface" bending of nano-thin SDS. Stresses arising from the action of forces on the nano-thin SDS had a maximum value at the center of the long diagonal (Figure 1 b) of the "neutral surface" of the diamond-shaped nano-thin SDS.

The formula of stress distribution in the diamond-shaped nano-thin SDS:

$$\sigma_{rr} = -\frac{2F_1 \cos\phi}{r(\alpha + \frac{1}{2}\sin 2\alpha)} \tag{1}$$

where F_1 is the force acting in the direction of the long diagonal of the diamond-shaped plate; φ is the angle of deviation of the direction of action of the force F_1 from the long diagonal of the diamond shaped plate; α – the angle between the long diagonal of the diamond-shaped plate and the side of the rhombus.

We took into account that the maximum curvature of the lattice curvature surface of the nano-thin SDS for [100] coincides with the short diagonal of the diamond-shaped nano-thin SDS. In addition, we analyzed formula 1 and determined that the maximum value σ_{rr} at the given F_1 force value, α angle value and r value is achieved at $\varphi = 0$, in other words, along the long diagonal of the diamond-shaped nano-thin SDS. As a result, we concluded that the maximum value σ_{rr} at under the above conditions will be at the intersection point of the diagonals of the diamond-shaped nano-thin SDS, that is, at the center of the diamond-shaped nano-thin SDS.

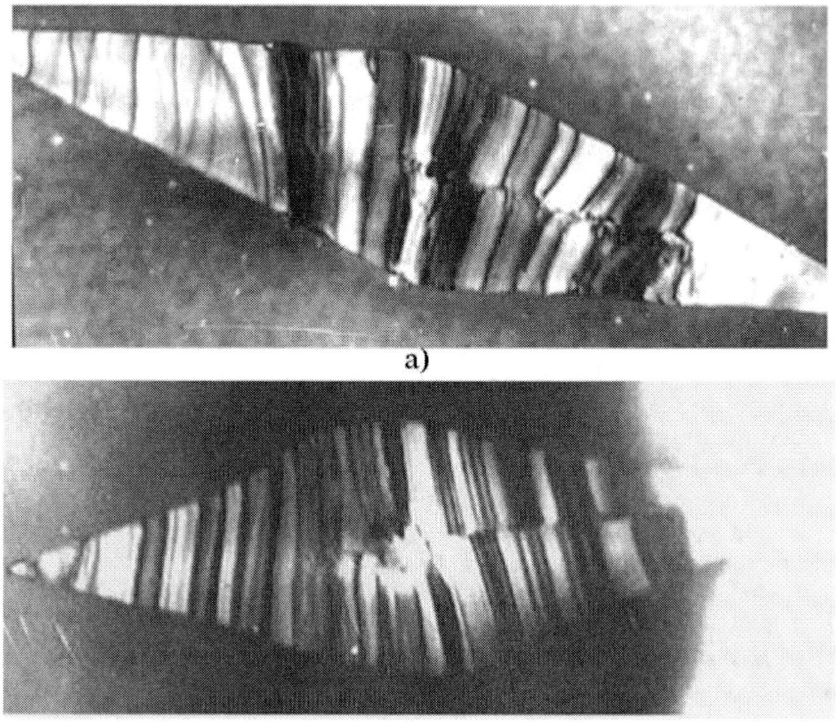

a)

b)

Figure 2. Electron microscopic images of diamond-shaped nano-thin SDS of hexagonal selenium (a, b) whose lattice experiences non-uniform rotational curvature around [001] with interblock torsion boundaries. The above mentioned torsion boundaries originated at the center of the diamond-shaped nano-thin SDS and developed by elongation along the long diagonal of the diamond-shaped nano-thin SDS, up to reaching the growth front of the nano-thin SDS in the amorphous film. JEM-200CX JEOL Ltd, Japan.

New information on the maximum stress value at the intersection point of the neutral surface diagonals of the diamond-shaped nano-thin SDS allowed us to explain the physical property of the nano-thin PDS. Namely, we showed that the emergence of interblock boundaries occurs in the center of diamond-shaped nano-thin SDS [15–17]. We noted that in this experiment, the SDS lattice experiences a non-uniform elastic rotational curvature around [001] (Figure 1 a), (Figure 2 a).

Subsequently, we examined the physical process in nano-thin SDS of hexagonal selenium using the proposed SDS study method. In fact, we studied the process of development of the interblock torsion boundary originating at the point of diagonals intersection of the diamond-shaped nano-thin SDS [18–20], along the long diagonal of the nano-thin diamond-shaped SDS (figure 2 a, figure 2 b). To do this, we constructed a lattice curvature surface with a diamond-shaped nano-thin PDS for [100], the lattice of which experiences non-uniform rotational curvature around [001] [10] (figure 1).

We applied the curvature surface of the nano-thin SDS lattice for [100] (figure 1 b) as the “neutral surface” of the nano-thin SDS and established the features of its bending. We performed a comparative analysis of the "neutral surfaces" of the nano-thin diamond-shaped PDS hexagonal selenium for [100] (figure 1 b), the reference wedge-shaped thin plate and the reference rounded thin disk [1]. As a result, we identified new information on bending features of the "neutral surface" of the nano-thin SDS. Stresses arising from forces on the nano-thin SDS had maximum values along the long diagonal of the "neutral surface" of the diamond-shaped nano-thin PDS of hexagonal selenium.

This new information allowed us to explain the physical process in the nano-thin diamond-shaped SDS of hexagonal selenium. The above process was the development of the interblock torsion boundary originating in the center of the diamond-shaped nano-thin SDS. Indeed, the development of the interblock torsion boundary originating in the center of the diamond-shaped nano-thin SDS (figure 1 a), (figure 2 a), was realized along the long diagonal of the nano-thin diamond – shaped SDS. Therefore, the process took place along the direction where the stresses arising from the action of forces on the nano-thin SDS had maximum values.

3. Conclusion

Thus, using the proposed method, we investigated and got our explanation the physical property – the origin the interblock torsion boundary at the center of the diamond-shaped nano-thin SDS (Figure 1 a), (Figure 2 a).

Also, with the help of the proposed method, we investigated and got our explanation of the physical process - the development along the long diagonal of the nano-thin diamond-shaped SDS of the interblock torsion boundary [15–20] (Figure 2 a, Figure 2 b).

References

[1] Landau L D and Lifshits E M 1987 *Theoretical physics Theory of elasticity* (Moscow: Science Main edition of physical and mathematical literature) p 60
[2] Prigozhin I and Kondepudi D 2002 *Modern thermodynamics. From heat engines to dissipative structures* (Moscow: World) p 461
[3] Nikolis G and Prigozhin I 1979 *Self-organization in non-equilibrium systems* (Moscow: World) p 512
[4] Nikolis G and Prigozhin I 1990 *Knowledge of complex* (Moscow: World) p 342
[5] Haken G 1985 *Synergetrics. Instability hierarchies in self-organizing systems and devices* (Moscow: World) p 419
[6] Klimontovich Yu L 1990 *Turbulent movement and the structure of chaos: A new approach to the statistical theory of open system* (Moscow: Science) p 320
[7] Ivanovsky A L 1999 *Quantum chemistry in materials science. Nanotubular forms of matter* (Yekaterinburg: UrO RAS) p 175
[8] Bazarov I P, Gevorkyan E V and Nikolaev P N 1989 *Non-equilibrium thermodynamics and physical kinetics* (Moscow: Publishing house MSU) p 240
[9] Bolotov I E and Kolosov V Yu 1980 Bend of thin-film crystals of selenium, found along the extinction contours *Izv. Academy of sciences of the USSR Series physical* **44 6** 1194–1197
[10] Malkov V B, Nikolaenko I V, Shveikin G P, Malkov A V, Pushin V G, Malkov O V and Shulgin B V 2015 *Method for visualization of rotary curvature of the lattice of nanofilament crystals* RU Patent **2570106**
[11] Malkov V B, Nikolaenko I V, Shveikin G P, Pushin V G, Malkov A V, Shulgin B V and Malkov O V 2018 Change in the geometry of the lattice of nano-thin, spatial dissipative structures formed in an amorphous film *Reports of the Academy of Sciences* **166 11** 649–651
[12] Malkov V B, Nikolaenko I V, Shveikin G P, Malkov A V, Pushin V G, Shulgin B V, Malkov O V and Plaksin S V 2017 *Method for obtaining dissipative structures* RU Patent **2637396**
[13] Malkov V B, Nikolaenko I V, Shveikin G P, Pushin V G, Malkov A V, Malkov O V and Shulgin B V 2018 Formation of dissipative structures in amorphous film *Reports of the Academy of Sciences* **478 5** 543–545

[14] Malkov V B, Malkov A V, Malkov O V, Pushin V G, Shulgin B V and Agalakov S P 2008 The phenomenon of elastic rotational curvature of the lattice of nano-thin crystals of hexagonal selenium around [001] as part of the asymmetric theory of elasticity *Collection of report Kharkiv nanotechnol. Assembly* **2** 18–23

[15] Bolotov I E, Kolosov V Yu and Malkov V B 1986 Electron Microscopy Investigation of Crystals Based on Bend-Contour Arrangement 3. Formation of Subgrain Boundaries in Dislocation-Free Crystals of Selenium *Phys. Stat. Sol. (a)* **95** 377–383

[16] Bolotov I E, Kolosov V Yu and Malkov V B 1986 A new type of interblock boundaries in thin-film crystals of selenium *Crystallography* **1** 204–206

[17] Bolotov I E, Kolosov V Yu and Malkov V B 1984 Application of Bend-Contour Method the Investigation of Block Formation in Thin-Film Crystals *Budapest Electron Microscopy* **2** 42–45

[18] Malkov V B, Malkov A V, Pushin V G, Strekalovsky V N and Malkov O V 2012 *Method of diagnosing the effect of changing the sign of the disorientation vector along the interblock boundaries in nanofilament crystals* RU Patent **2534719**

[19] Malkov V B, Nikolaenko I V, Shveikin G P, Pushin V G, Malkov A V, Malkov O V and Shulgin B V 2018 The mechanism of formation of interblock boundaries, the effect of changing the sign of the orientation vector along the broken interblock boundaries in nano-thin spatial dissipative structures *Reports of the Academy of Sciences* **481** **1** 30–32

[20] Bolotov I E and Kolosov V Yu 1982 Investigation of Crystals Based on Bend-Contour Arrangement. 1. Relationship between Bend-Contour Arrangement and Bend Geometry *Phys. Stat. Sol (a)* **69** **1** 85–96

Acknowledgments

Experiments on electron microscopic studies of nano-thin SDS of hexagonal selenium by us were carried out in the electron microscopy department of the Collective Use Center "Testing Center for Nanotechnology and Promising Materials" of the Institute of Metal Physics of the Ural Branch of the Russian Academy of Sciences.

Aperiodic Schrodinger Crystals

V B Malkov[1], **O V Chemezov**[2] **and A V Malkov**[3]

[1]Institute of High-Temperature Electrochemistry, Ural Branch Russian Academy of Sciences, 22, S. Kovalevskoy, Yekaterinburg, 620219, Russia
[2]Ural Federal University named after the first President of Russia B. N. Yeltsin, 19, Mira street, Yekaterinburg, 620002, Russia
[3]ZAO NPTs ROSNA, 9, Studencheskaya, Yekaterinburg, 620026, Russia

E-mail: o.v.chemezov@urfu.ru

Abstract. In 1945, E. Schrödinger predicted the possibility of the existence of "aperiodic crystals," that is, more and more increasing crystalline aggregates, but without a simple lattice repetition mechanism. In the present work, such structures are experimentally prepared using selenium as an example. By thermal gradient treatment of the amorphous selenium film, we have obtained nano-thin spatial dissipative structures (SDS) of hexagonal selenium, the lattice of which undergoes non-uniform rotational curvature around, in the general case, three mutually perpendicular directions. The formation of these nano-thin SDS of hexagonal selenium occurs as a result of cooperative rotations of selenium macromolecules around, in general, three mutually perpendicular directions. Since nano-thin SDS after hardening in air have all the features of crystals each of the above nano-thin SDS of hexagonal selenium obtained at different annealing temperatures corresponds to the concept of "aperiodic crystals" Thus, "aperiodic crystals" were obtained experimentally, the possibility of the existence of which was predicted by E. Schrödinger.

1. Introduction

The concept of "aperiodic crystal", first introduced by Erwin Schrödinger in the book "What is life?" [1]. In the first chapter, on the pages of 12 and 13, E. Schrödinger defines an aperiodic crystal: "... periodic crystals are very interesting and complex objects.... However, compared to aperiodic crystals, they seem somewhat elementary and boring. The difference in structure here is the same as between ordinary wallpaper, on which the same pattern is repeated with the correct periodicity, and a masterpiece of embroidery, say, Raphaelev tapestry, which repeats the complex, consecutive and complete design pattern drawn by the great master...". Further, on the page 66 [1], E. Schrödinger gives an example of an aperiodic crystal: "... it is possible to imagine two different ways of building more and more associations. One is a relatively uniform way of repeating again and again the same structure in three directions.... The other way is to build a more and more increasing aggregate without a boring repetition mechanism. This is a case of an increasingly complex organic molecule, in which each atom, each group of atoms plays an individual role that is not quite equivalent to the role of other atoms and groups of atoms. We can definitely call this formation an aperiodic crystal...".

On page 82 [1], E. Schrödinger points to the mechanisms for the formation of ordered structures: "It turns out that there are two different "mechanisms" that can produce ordered phenomena: a

IOP Conf. Series: Materials Science and Engineering **969** (2020) 012026 doi:10.1088/1757-899X/969/1/012026

statistical mechanism that creates "order from disorder"," and a new mechanism that produces "order from order".

Thus, on the one hand, Schrödinger points to the possibility of the existence of "aperiodic crystals" –... "aggregates without a boring repetition mechanism," and, on the other hand, determines the mechanism for their formation –... "a new mechanism that produces "order from order"."

How right was E. Schrödinger, pointing to the possibility of the existence of aperiodic crystals? To answer this question, consider nano-thin spatial dissipative structures (SDS) formed in amorphous films.

2. Results and discussion

It is known that in amorphous films of selenium, during their thermogradient processing, nano-thin spatial dissipative structures of hexagonal selenium are formed [2–6]. In the process of electron-microscopic studies, and the methods of transmission electron microscopy, electronography and bending extinction circuits present on electron-microscopic images of nano-thin SDS were investigating nano-thin SDS of hexagonal selenium, presented in figure 1 [2, 3, 7–17].

Nano-thin diamond-shaped SDS of hexagonal selenium (figure 1, a) are characterized by a system of bending extinction contours parallel to the short diagonal of rhombus [2, 3, 6, 7]. Microdiffraction studies of nano-thin SDS of hexagonal selenium [7–10] and analysis of flexural extinction circuits on electron microscopic images of nano-thin SDS make it possible to conclude: nano-thin diamond-shaped SDS (figure 1, a), formed in an amorphous film with one-sided heating of the lower surface at T = 453 K, are in a non-equilibrium state – their lattice experiences non-uniform elastic rotational curvature around [001], coinciding in direction with the axis OZ (figure 2). The rotation of the lattice of nano-thin diamond-shaped SDS of hexagonal selenium (figure 1, a) around [001] reaches 18^0 [2, 3, 6].

Thus, indeed, the formation of nano-thin diamond-shaped SDS of hexagonal selenium (figure 1, a) occurs "...without a boring mechanism of repetition" [1], and as a result of cooperative progressive rotations of Se macromolecules around [001].

Nano-thin diamond-shaped SDS of hexagonal selenium formed in an amorphous film during its thermal gradient treatment at T = 423 K (figure 1, b), characterized by a linear fan-like system of bending extinction circuits on electron-microscopic images.

Microdifraction studies of nano-thin SDS of hexagonal selenium (figure 1, b) and analysis of the system of bending extinction circuits present on electron-microscopic images of nano-thin SDS make it possible to conclude: – their lattice experiences inhomogeneous elastic rotational curvature around two mutually perpendicular directions – around [001] and around the OX axis (figure 2) [2, 3, 6, 18, 19]. The rotation angles of the lattice of nano-thin diamond-shaped SDS of hexagonal selenium with a linear fan-shaped system of bending extinction circuits on electron-microscopic images reach: around [001] – 18°; about OX axis (figure 2) – 22°.

Formation of nano-thin diamond-shaped SDS of hexagonal selenium (figure 1, b) occurs as a result of cooperative rotations of selenium macromolecules around two mutually perpendicular directions.

Nano-thin STS of hexagonal selenium with curved gabitus are formed in an amorphous film during thermal gradient treatment by heating its lower surface at T = 413 K (figure 1, c).As a result of studies of nano-thin SDS conducted by electron-microscopic and microdifraction methods, it was found that nano-thin STS with curved gabitus (figure 1, c) are in a non-equilibrium state, their lattice experiences non-uniform elastic-plastic rotational curvature around three mutually perpendicular directions: around [001], around the OX axis and about OY axis (figure 2). The angles of rotation of the lattice of nano-thin SDS with the curved gabitus reach: around [001] – 25^0, around the axis OX – 32^0, around the axis OY- 35^0 (figure 2) [2, 5, 19].

Thus, the formation of nano-thin diamond-shaped SDS of hexagonal selenium (figure 1, c) also occurs... "without a boring mechanism of repetition" [1], and as a result of cooperative rotations of selenium macromolecules around three mutually perpendicular directions (figure 2) [2, 5, 19].

IOP Conf. Series: Materials Science and Engineering **969** (2020) 012026 doi:10.1088/1757-899X/969/1/012026

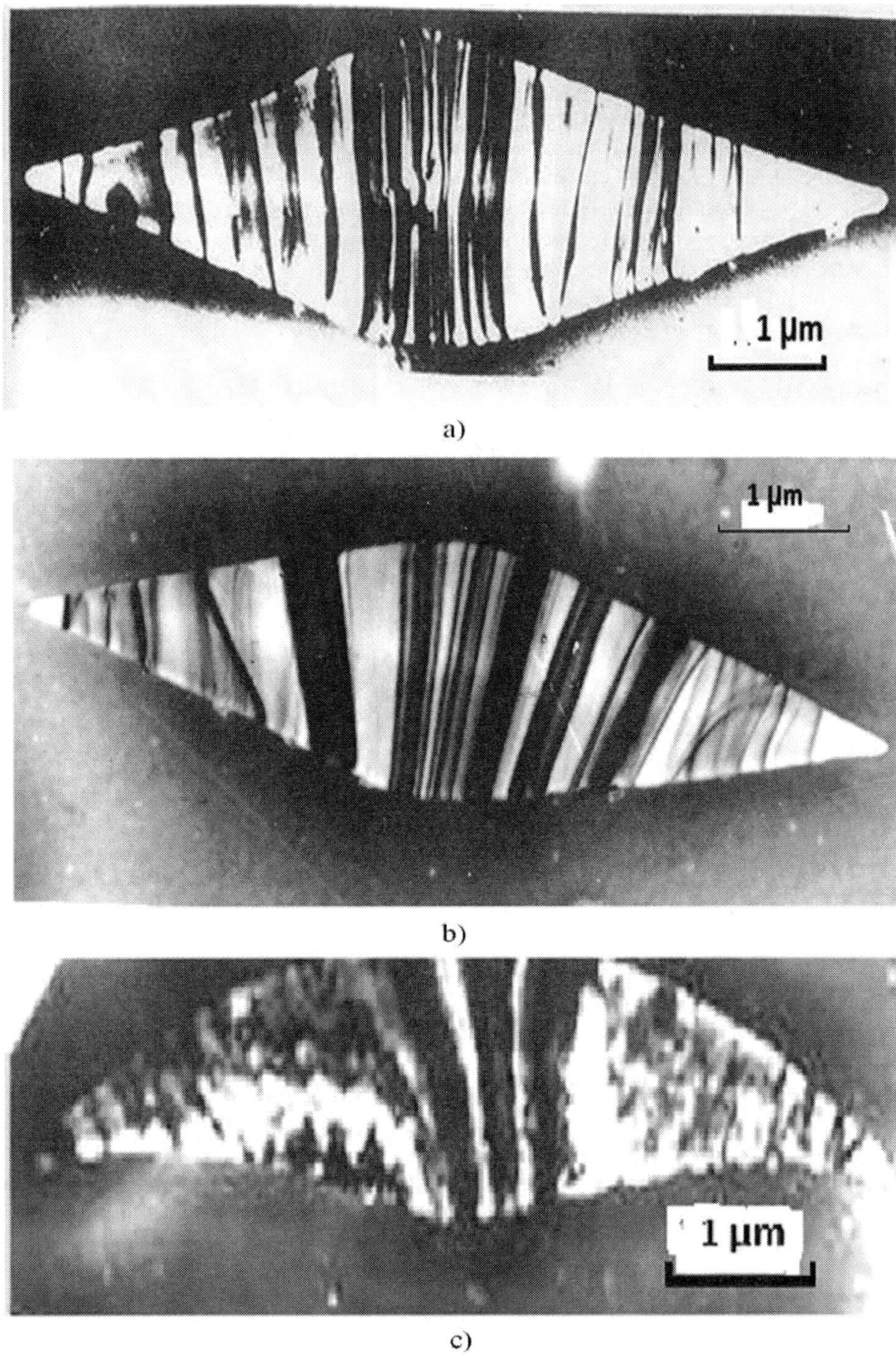

Figure 1. Electron microscopic image of nano-thin SDS hexagonal selenium, JEM-200CX JEOL Ltd, Japan: a) having a diamond-shaped gabitus and a system of bending extinction contours parallel to [001], coinciding with the short diagonal of the rhombus; b) having a diamond-shaped gabitus and a linear fan-shaped system of bending extinction contours; c) with curved habitus.

Since the structural units of the lattices of nano-thin diamond-shaped SDS of hexagonal Se with a parallel short diagonal of the rhombus are a system of bending extinction circuits on electron-microscopic images (figure 1, a); nano-thin diamond-shaped SDS of hexagonal Se with a linear fan-like system of bending extinction circuits on electron microscopic images (figure 1, b); nano-thin SDS of hexagonal Se with curved habitus and a non-linear system of bending circuits on electron microscopic images (figure 1, c) are selenium macromolecules [2, 5, 19–21]. Since it is clear that the rotation of the lattice of nano-thin SDS around one, two or three mutually perpendicular directions is the result of cooperative rotations of selenium macromolecules forming lattices of nano-thin SDS of hexagonal selenium (figure 1, a, figure 1, b, figure 1, c) [2, 5, 19–21], around one, two or three mutually perpendicular directions, respectively (figure 2).

The nano-thin SDS of hexagonal selenium shown in figure 1, a, figure 1, b, figure 1, c differ in complexity. The lattice of nano-thin diamond-shaped SDS (figure 1, a) experiences non-uniform elastic rotational curvature around [001], the lattice of nano-thin diamond-shaped SDS (figure 1, b) experiences non-uniform elastic rotational curvature around two mutually perpendicular directions [11], around – [001] and around the axis OX (figure 2); the lattice of nano-thin diamond-shaped SDS with curved gabitus (figure 1 c) experiences non-uniform elastic-plastic rotational curvature around three mutually perpendicular directions, around [001], around the OX axis and around the OY axis (figure 2).

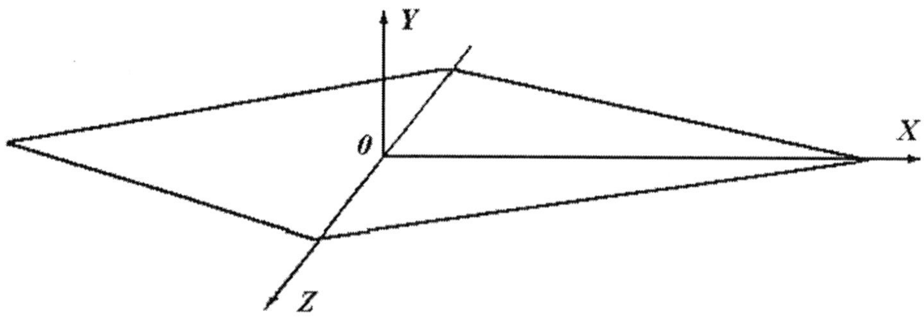

Figure 2. Coordinate system associated with nano-thin SDS of hexagonal selenium.

3. Conclusion

Since nano-thin SDS after hardening in air have all the features of crystals [2, 8, 10, 11], each of the above nano-thin SDS of hexagonal selenium (figure 1, a, figure 1, b, figure 1, c) corresponds to the concept of "aperiodic crystal" [1].

As a result, by thermal gradient treating an amorphous selenium film, nano-thin SDS of hexogonal selenium are obtained, the lattice of which experiences a non-uniform elastic or elastic-plastic rotational curvature around one, two or three mutually perpendicular directions. The formation of these nano-thin SDS of hexagonal selenium occurs as a result of cooperative rotations of selenium macromolecules around one, two or three mutually perpendicular directions. Thus, "aperiodic crystals" were obtained experimentally, the possibility of the existence of which was predicted by E. Schrödinger [1].

References

[1] Shredinger E 2002 *What is life? Physical aspect of a living cell* (Moscow-Izhevsk: N I C "Regular and chaotic dynamics")

[2] Malkov V B, Nikolaenko I V, Shveikin G P, Malkov A V, Pushin V G, Shulgin B V, Malkov O V and Plaksin S V 2017 *Method for obtaining dissipative structures* RU Patent **2637396**

[3] Malkov V B, Nikolaenko I V, Shveikin G P, Pushin V G, Malkov A V, Malkov O V and Shulgin B V 2018 Formation of dissipative structures in amorphous film *Reports of the Academy of Sciences* **478 5** 543–545

[4] Malkov V B, Shveikin G P, Pushin V G, Malkov A V, Shulgin B V and Malkov O V 2018 Nano-thin crystal lattice symmetry, the isomorphism of direct and inverse space and the principle of Curie symmetry *Reports of the Academy of Sciences* **479 1** 18–20
[5] Malkov V B, Shveikin G P, Plaksin S V, Pushin V G, Malkov A V, Malkov O V and Shulgin B V 2019 Nano-thin spatial dissipative structures with rotational lattice curvature around three mutually perpendicular directions *Reports of the Academy of Sciences* **488 6** 619–623
[6] Malkov V B, Nikolaenko I V, Shveikin G P, Pushin V G, Malkov A V, Shulgin B V and Malkov O V 2018 Change in the geometry of the lattice of nano-thin, spatial dissipative structures formed in an amorphous film *Reports of the Academy of Sciences* **166 11** 649–651
[7] Hirsch P B, Howie A, Nicholson R B and Pashley D W *1965 Electron Microscopy of Thin Crystals* (London: Butterworths)
[8] Tomas G and Goringe M J 1979 *Ttransmission electron microscopy of materials* (New York)
[9] Utevsky L M 1973 *Diffraction electron microscopy in metal science* (Moscow: Metallurgy)
[10] Andrews K V, Dyson D J and Keown S R 1968 *Interpretation of Electron Diffraction Patterns* (London)
[11] Bolotov I E and Kolosov V Yu 1980 Bend of thin-film crystals of selenium, found along the extinction contours *Izv. Academy of sciences of the USSR Series physical* **44 6** 1194–1197
[12] Bolotov I E, Kolosov V Yu and Malkov V B 1986 Electron Microscopy Investigation of Crystals Based on Bend-Contour Arrangement 3. Formation of Subgrain Boundaries in Dislocation-Free Crystals of Selenium *Phys. Stat. Sol. (a)* **95** 377–383
[13] Bolotov I E, Kolosov V Yu and Malkov V B 1986 A new type of interblock boundaries in thin-film crystals of selenium *Crystallography* **1** 204–206
[14] Bolotov I E, Kolosov V Yu and Malkov V B 1984 Application of Bend-Contour Method the Investigation of Block Formation in Thin-Film Crystals *Budapest Electron Microscopy* **2** 42–45
[15] Bolotov I E and Kolosov V Yu 1982 New possibilities of the method of bending extinction circuits *Theses of reports 12 All-Union Conf. electron microscopy* 186–187
[16] Bolotov I E, Kolosov V Yu and Kozhin A V 1982 Using the map of extinction contours in the study of the bending of crystals growing in thin amorphous films *Theses of reports 12 All-Union Conf. electron microscopy* 138–139
[17] Bolotov I E and Kolosov V Yu 1982 Possibilities of the method of extinction contours in the study of the bending of thin-film objects *Factory laboratory* **48 11** 53–57
[18] Malkov V B, Malkov A V, Malkov O V, Pushin V G, Shulgin B V and Agalakov S P 2008 The phenomenon of elastic rotational curvature of the lattice of nano-thin crystals of hexagonal selenium around [001] as part of the asymmetric theory of elasticity *Collection of report Kharkiv nanotechnol. Assembly* **2** 18–23
[19] Malkov V B, Nikolaenko I V, Shveikin G P, Malkov A V, Pushin V G, Shulgin B V, Malkov O V and Plaksin S V 2019 *Diagnostic technique for the evolution of nano-thin spatial structures* RU Patent **2687876**
[20] Vunderly B 1976 *Physics of macromolecules* (Moscow: World) p 623
[21] Tager A A 1978 *Physical chemistry of polymers* (Moscow: Chemistry) p 574

Acknowledgments
Experiments on electron-microscopic studies of nano-thin SDS of hexagonal selenium were carried out in the electron microscopy department by the collective use center “Testing Center for Nanotechnology and Promising Materials” of the Institute of Metal Physics of the Ural Branch of the Russian Academy of Sciences.

Kinematic Viscosity of Melt Prepared from an Amorphous $Fe_{72.5}Cu_1Nb_2Mo_{1.5}Si_{14}B_9$ Ribbon

Ye A Kochetkova[1], **Yu N Starodubtsev**[1,2] **and V S Tsepelev**[1]

[1]Reserch Center for Physics of Metal Liquids, Ural Federal University, 19, Mira str., Ekaterinburg, 620002, Russia
[2]Gammamet Research and Production Enterprise, 92, Tatishcheva str., Ekaterinburg, 620028, Russia

E-mail: v.s.tsepelev@urfu.ru

Abstract. In this work, the viscosity of a melt prepared by melting an amorphous ribbon of $Fe_{72.5}Cu_1Nb_2Mo_{1.5}Si_{14}B_9$ alloy was studied. The amorphous ribbon was produced by rapid quenching of the melt, which was preheated at a temperature above (overheated mode) and below (not overheated mode) critical temperature T_k. A hysteresis loop was found in the temperature dependence of viscosity, the widest part of which falls on a temperature from 1670 K to T_k = 1760 K. In this temperature range, the melt has the highest activation energy and the largest cluster size. It has been proven that the melt viscosity increases with decreasing cluster size. In the low-temperature region, an overheated melt has the highest viscosity and the smallest cluster size comparable to the atomic size. It is shown that the heredity of the structure is preserved in the melt from the amorphous ribbon despite a number of structural transitions.

1. Introduction

The physical properties of liquid metals are characterized by various quantities, such as density, viscosity, diffusion, surface tension, thermal conductivity, electrical resistance, and others. Among these quantities, one can distinguish viscosity, which has a high structural sensitivity. The temperature dependences of the viscosity of multicomponent melts have a number of features that are not observed in pure liquid metals. Irreversible changes in the temperature dependence of viscosity at the stages of heating and cooling were detected, and the branch point corresponded to the critical temperature T_k [1]. During a long isothermal exposure of the melt, viscosity fluctuations occurred, and the amplitude of the oscillations significantly exceeded the random measurement error [2–5]. The largest amplitude of viscosity fluctuations was observed in the region of critical temperature [4].

The anomalous behavior of multicomponent melts is associated with a nonequilibrium state that occurs when the temperature changes. For rearrangement of the short range order and for approaching the equilibrium state, a certain time is required [1]. In turn, the relaxation time depends on the structure of the melt. At the melting stage, a short-range structure inherited from the multiphase solid state is retained in the melt. The structural component of the melt are clusters, the size of which depends on temperature [6]. With increasing temperature, the cluster size decreases, and at a temperature above T_k the melt structure becomes the most uniform.

If the melt is heated above the critical temperature and then transferred to the amorphous solid state using rapid quenching, it can be expected that a homogeneous structure will remain in the amorphous state.

Heredity was studied on an amorphous ribbon obtained after the melt was held above and below the critical temperature [7–10]. It was proved that the amorphous ribbon after holding the melt above the critical temperature has a less ordered atomic structure [8], a larger molar volume [7], greater plasticity [8, 9], and the hardness and fracture toughness [9], higher crystallization enthalpy [7, 10]. After nanocrystallization of an amorphous ribbon obtained from $Fe_{72.5}Cu_1Nb_2Mo_{1.5}Si_{14}B_9$ melt with overheating above a critical temperature, an increased proportion of small nanocrystals 1–2 nm in size was discovered [10]. Thus, it was shown that the heredity of the structure is preserved not only in the amorphous solid state after melt quenching, but after crystallization of the amorphous ribbon.

In this work, we determined the depth of structural heredity that the material retains after melt preparation. For this purpose, the viscosity of the melt obtained by melting the amorphous $Fe_{72,5}Cu_1Nb_2Mo_{1,5}Si_{14}B_9$ ribbon was studied, and the amorphous ribbon was obtained by rapidly quenching the melt, which was heated above and below the critical temperature T_k.

2. Experimental

The $Fe_{72,5}Cu_1Nb_2Mo_{1,5}Si_{14}B_9$ alloy was melted in a vacuum induction furnace at temperature 1820 K and cooling in a flat mould. An amorphous ribbon 25 μm thick was produced by the planar flow casting process. When choosing the temperature regime, it was taken into account that in the samples after melting the alloy in a vacuum induction furnace, the critical temperature was 1760 K. Before the casting, the melt was prepared according to two modes. In the first mode, the melt was heated to 1750 K, kept at this temperature for 5 minutes, and then casting was carried out. The melt obtained from this amorphous ribbon will be called "not overheated" melt. In the second mode, the melt was heated to 1795 K with holding at this temperature for 5 minutes, then cooled to 1750, kept for 5 minutes, and casting was carried out. The melt obtained from this amorphous ribbon will be called "overheated" melt.

The crystallization temperature of the amorphous ribbon and the melting temperature were measured using differential thermal analysis. Kinematic viscosity of the melts is determined by torsional vibration method based on measuring the vibration damping factor [11]. The measurement error of the kinematic viscosity ν makes up ± 0.06 $m^2 \cdot s^{-1}$. In the first heating – cooling cycle, the amorphous ribbon was turned into a toroidal core and heated to 1720 K, followed by cooling to 1470 K. Then the second heating – cooling cycle from 1470 to 1870 K and back to 1470 followed, and immediately followed the third cycle, which completely repeated second cycle.

3. Results and discussion

Figure 1 shows the temperature dependences of the kinematic viscosity for the first heating – cooling cycle and for the heating stage of the second cycle in the melt obtained from the amorphous $Fe_{72.5}Cu_1Nb_2Mo_{1.5}Si_{14}B_9$ ribbon. From the figure it follows that immediately after the amorphous ribbon is melted, the melt viscosity is almost the same both in the overheated and not overheated melts. However, as the temperature rises to about 1540 K, the viscosity curves diverge noticeably. The viscosity of the overheated melt goes up sharply, and the viscosity of the not overheated melt changes slightly. At temperatures above 1540 K, the viscosity decreases due to an increase in the free volume of the liquid [12, 13].

Figure 1 also shows that in the overheated melt, the temperature dependence of viscosity goes to a stable trajectory, with the exception of a small loop in the temperature range 1670–1720 K. In the not overheated melt, the trajectory of the temperature dependence of viscosity gradually decreased. Thus, the melt obtained from the amorphous ribbon is nonequilibrium, and the not overheated melt has the greatest deviation from the equilibrium structure. The difference in the viscosity indicates a different structure of atomic groups participating in a viscous flow.

In the model of hard spheres, the melt viscosity can be represented as [14]:

$$\nu \propto a\sqrt{\frac{k_B T}{m}} \tag{1}$$

where a is the atomic size, m, m is the atomic mass, kg, k_B is the Boltzmann constant, $J \cdot K^{-1}$. If clusters are involved in the viscous flow, it must be taken into account that the cluster mass increases proportionally, and the cluster size is not proportional to the number of atoms [15]. Figure 2 shows the dependence of the relative melt viscosity ν_c/ν, where ν_c is the viscosity of a liquid consisting of clusters, on the number of atoms in a cluster n. From the figure it follows that the viscosity of the liquid should decrease with increasing cluster size.

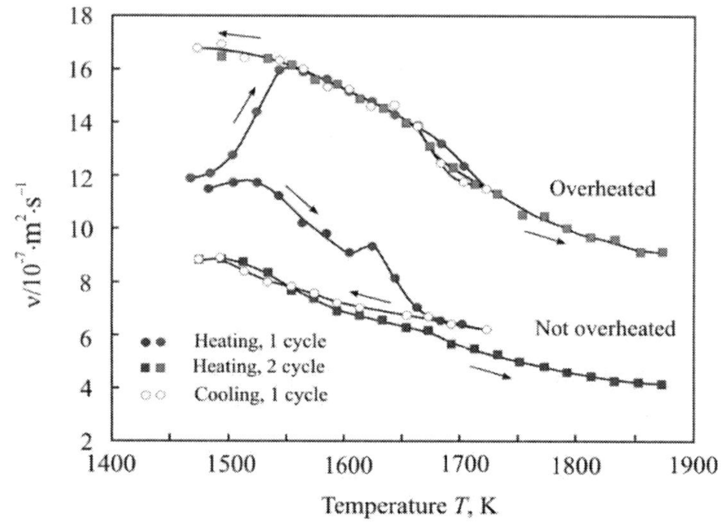

Figure 1. Temperature dependences of kinematic viscosity ν for the first heating – cooling cycle and for the heating stage of the second cycle in the overheated and not overheated melt prepared from the amorphous $Fe_{72.5}Cu_1Nb_2Mo_{1.5}Si_{14}B_9$ ribbon.

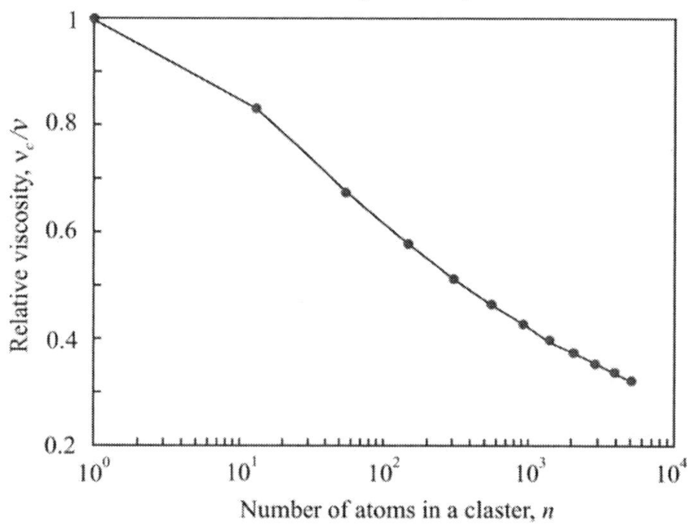

Figure 2. Relative melt viscosity ν_c/ν depending on the number of atoms in a cluster n.

Thus, in an overheated melt, increased viscosity can be associated with the small size of atomic groups in the equilibrium state of the liquid. The same conclusion can be drawn on the basis of general concepts [1], since the interaction energy increases with a decrease in cluster size, and this should lead to an increase in the melt viscosity.

When the amorphous ribbon is heated, crystallization and melting occur with the phase transition temperatures, which are shown in **Table 1**. At the first stage of crystallization (T_{x1} and T_{x2} are the temperatures of the beginning and the end), nanocrystalline grains are formed whose composition is close to $Fe_{80}Si_{20}$ and the average size is 10 nm [16]. It is important that in an amorphous ribbon prepared from a overheated melt, there are noticeably more small grains 1–2 nm in size [10]. At the second stage of crystallization (T_{x3} and T_{x4} are the temperatures of the beginning and end), metastable phases based on Fe-B and Fe-Si are formed [17]. The melting temperature of an amorphous ribbon with overheating (T_l and T_s are the liquidus and solidus temperatures) are slightly lower than in an amorphous ribbon without overheating. The Fe – B, Fe – Nb, and Nb – B clusters dominate in the melt [18]. Apparently,

IOP Conf. Series: Materials Science and Engineering **969** (2020) 012027 doi:10.1088/1757-899X/969/1/012027

the finer nanocrystalline structure of the amorphous ribbon obtained in the overheated mode became the basis for the formation of a finely dispersed melt structure.

Table 1. Phase transition temperature

Melt preparation regime	Crystallization temperature, K				Melting temperature, K	
	T_{x1}	T_{x2}	T_{x3}	T_{x4}	T_l	T_s
Overheated	770	800	940	965	1390	1455
Not overheated	782	805	950	970	1400	1460

Figure 3 shows the dependences of viscosity on temperature in the third heating – cooling cycle in the form convenient for analysis using the Arrhenius equation

$$\nu = \nu_0 \cdot e^{\frac{E_a}{RT}}, \tag{2}$$

where ν_0 is a pre-exponential factor with the dimension of the kinematic viscosity m^2·s^{-1}, E_a is the activation energy of the viscous flow, J·mol^{-1}, R is the gas constant, J·K^{-1}·mol^{-1}. After taking the logarithm we obtain

$$\ln \nu = \ln \nu_0 - \frac{E_a}{RT}. \tag{3}$$

Thus, the logarithm of the kinematic viscosity is a linear function of the inverse absolute temperature.

Figure 3 shows that despite the stabilization of the trajectories of the temperature curves of viscosity in the third heating – cooling cycle, viscosity is irreversible in the critical temperature region, which manifests itself in the form of a hysteresis loop. The branching of the curves begins at a temperature of 1670 K. Moreover, there is also an increased slope of the curves in this region, which indicates an increase in the activation energy of the viscous flow. The results of calculations of the activation energy E_a and pre-exponential factor ν_0 are presented in **Table 2**. The entire temperature region was divided into intervals that vary in the behavior of viscosity. The calculated values are obtained by linear fitting of the experimental points for each temperature range.

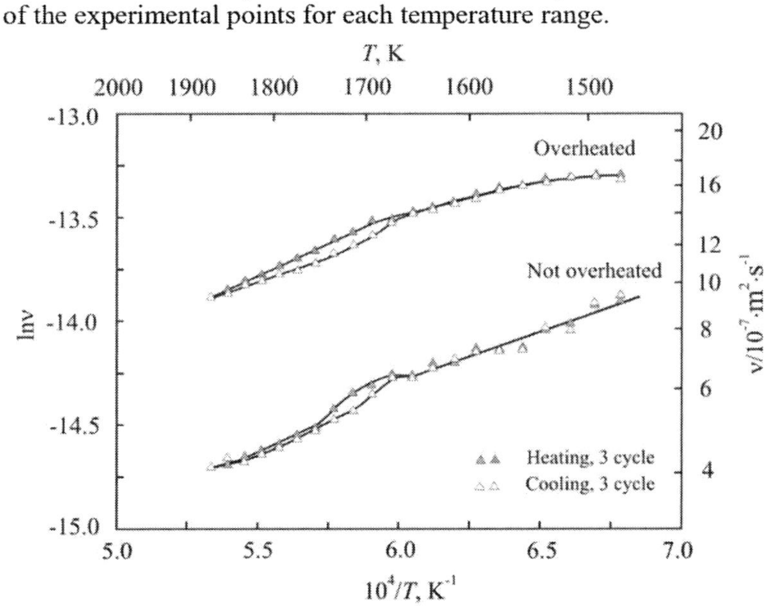

Figure 3. The logarithm of the kinematic viscosity lnν as a function of the reciprocal absolute temperature $10^4/T$ in the third heating – cooling cycle of the overheated and not overheated melt prepared from the amorphous ribbon.

Table 2. The activation energy E_a, cluster size d, and pre-exponential factor ν_0 in the overheated and not overheated melt obtained from the amorphous $Fe_{72.5}Cu_1Nb_2Mo_{1.5}Si_{14}B_9$ ribbon.

Temperature range	Heating			Cooling		
	E_a/kJ·mol^{-1}	$\nu_0/10^{-8}$·m^2·s^{-1}	d/nm	E_a/kJ·mol^{-1}	$\nu_0/10^{-8}$·m^2·s^{-1}	d/nm
Overheated melt 1790 K → 1750 K						
1470 – 1670 K	23.2	26.3	0.05	26.0	21.1	0.05
1670 – 1870 K	51.4	3.47	2.5	45.7	4.8	1.5
Not overheated melt 1750 K						
1470 – 1670 K	37.6	7.6	0.4	38.2	4.0	1.5
1670 – 1750 K	83.1	0.27	380	76.1	0.26	410
1750 – 1870 K	44.1	2.31	6	38.4	3.47	2.5

Table 2 also shows the calculated cluster size d. The calculation was performed using the Eyring model based on the transition state theory [13]. In this model, the pre-exponential factor ν_0 depends on the cluster size clusters d participating in the viscous fluid flow and this relationship can be transformed in the relation [19]:

$$d = \left(\frac{2\pi k_B T}{\rho}\right) \cdot \left(\frac{\mathrm{v}_f}{\mathrm{v}}\right)^{\frac{2}{3}} \cdot \nu_0^{-2}, \tag{4}$$

where ρ is the melt density, kg·m^{-3}, v_f/v is the relative free volume. To calculate the cluster size, typical values of the melt density and the relative free volume of nanocrystalline alloys were used [20]. For a temperature of 1600, 1700, and 1800 K, we used the densities of 6550, 6500, and 6450 kg·m^{-3} and the relative free volumes of 0.035, 0.043, and 0.05, respectively.

From the **Table 2** it follows that the activation energy is higher in the overheated melt. In addition, in the region of the hysteresis loop (see Figure 3), the activation energy is approximately two times higher.

When discussing cluster size data, it must be taken into account that model representations differ from real liquid. For example, atoms and clusters of different composition and size coexist in a real liquid. The model selects the cluster size in a homogeneous fluid, which should correspond to the experimental value of the kinematic viscosity. **Table 2** shows that the cluster size is higher in the melt obtained from the not overheated ribbon. Moreover, the largest cluster size falls on the hysteresis loop of kinematic viscosity, where there is also the highest activation energy of the viscous flow.

For the low-temperature region in the overheated melt, an effective cluster size of 0.05 nm was obtained, which can well be taken as the atomic size, taking into account the accuracy of the model approximation. Thus, in this temperature range, the melt viscosity is associated with the vibration of individual atoms, which are mainly located near their equilibrium positions. With increasing temperature, the mobility of atoms increases, but the strong bonds Fe-B, Fe-Nb(Mo) and Nb(Mo)-B inhibit the decomposition into individual atoms and ensure the formation of clusters.

4. Conclusions

In the work, the viscosity of the melt prepared by melting an amorphous ribbon from an $Fe_{72.5}Cu_1Nb_2Mo_{1.5}Si_{14}B_9$ alloy was studied. An amorphous ribbon was produced by rapid quenching of the melt, which was prepared before casting at a temperature above (overheated mode) and below (not overheated mode) critical temperature T_k. The activation energy E_a and pre-exponential factor ν_0 were calculated from the Arrhenius equation for different temperature ranges. Cluster size was determined from the Eyring model based on the transition state theory. The results of the study allow us to draw the following conclusions.

1. Immediately after melting the amorphous ribbon, the melts have close kinematic viscosity, however, with increasing temperature, the viscosity of the overheated melt becomes much higher than the viscosity of the not overheated melt. The transition to an equilibrium state occurs faster in an overheated melt.
2. After repeated thermal cycling, a hysteresis loop is retained in the temperature dependence of viscosity, the widest part of which is between 1670 K and T_k = 1760 K. In this temperature range, the melt has the highest activation energy and the largest cluster size.
3. In the low-temperature region, an overheated melt has the highest viscosity and the smallest cluster size comparable to the atomic size. The melt viscosity increases with decreasing cluster size.
4. Structural heredity recorded in an amorphous ribbon after rapid quenching of the melt remains in the melt obtained after heating and melting the ribbon, despite a number of structural transitions.

References

[1] Baum B A 1979 *Metal liquids – Problems and hypotheses* (Moscow: Science Press)
[2] Lad'yanov V I, Logunov S V and Pakhomov S V 1998 *Russian Metall* **5** 23
[3] Vasin M G and Lad'yanov V I 2003 *Phys. Rev. E* **68** 051202
[4] Lad'yanov V I, Bel'tyukov V I, Kamaeva L V, Tronin K G and Vasin M G 2003 *Melts* **1** 32
[5] Zamyatin V M, Baum B A, Mezenin A A and Shmakova K Yu 2010 *Melts* **5** 19
[6] Calvo-Dahlborg M, Popel P S, Kramer M J, Besser M, Morris J R and Dahlborg U 2013 *J. Alloys Comp.* **550** 9
[7] Manov V P, Popel S I, Buler P I Manukhin A B and Komlev D G 1991 *Mater. Sci. Eng. A* **133** 535
[8] Starodubtsev Yu N, Son L D, Tsepelev V S, Tyagunov G V, Tishkin A P and Korobka O B 1992 *Melts* **4** 76
[9] Tabachnikova E D, Bengus V Z, Egorov D V, Tsepelev V S and Ocelík V 1997 *Mater Sci Eng A* **226–228** 887
[10] Tsepelev V, Starodubtsev Yu and Konashkov V 2017 *EPJ Web of Conferences* **151** 04006
[11] Tsepelev V, Konashkov V, Starodubtsev Y, Belozerov Y and Gaipisherov D 2012 *IEEE Trans. Magn.* **48** 1327
[12] Batschinski A I 1913 *Z. phys. Chem.* **84** 643
[13] Glasstone S, Laidler K and Eyring H 1941 *The theory of rate processes. The kinetics of chemical reactions, viscosity, diffusion and electrochemical phenomena* (New York – London: McGraw Hill)
[14] Longuet-Higgins H C and Pople J A 1956 *J. Chem. Phys.* **25** 884
[15] Poole C P Jr and Owens F J 2003 *Introduction to nanotechnology* (New Jersey: John Wiley & Sons)
[16] Tsepelev V, Starodubtsev Yu, Konashkov V and Belozerov V 2017 *J. Alloys and Comp.* **707** 210
[17] Lyasotskii I V, Dyakonova N B, Vlasova E N, Dyakonov D L and Yazvitskii M Yu 2006 *Phys. stat. sol. (a)* **203** 259
[18] Pan S P, Qin J Y, Gu T K, Yang L and Bian X F 2009 *J. Appl. Phys.* **105** 013531
[19] Tsepelev V, Starodubtsev Yu, Wu K M and Kochetkova Ye A 2020 *Key Eng. Mater.*
[20] Tsepelev V, Starodubtsev Yu, Konashkov V, Wu K and Wang R 2019 *J. Alloys and Comp.* **790** 547

Acknowledgement
The work was carried out within the framework of the state work No. FEUZ-083-0020.

IOP Conf. Series: Materials Science and Engineering **969** (2020) 012028 doi:10.1088/1757-899X/969/1/012028

Synthesis, Structure and Properties of $Zn_{0.3}Ni_{0.7-x}Co_xFe_2O_4$ Ferrite

D P Sherstyuk, A Yu Starikov and V E Zhivulin

Faculty of Materials Science and Metallurgy Engineering, South Ural State University, 76, Lenin prospect, Chelyabinsk, 454080, Russia

E-mail: sherstiukd@susu.ru

Abstract. The article presents the results of a study of ferrite $Zn_{0.3}Ni_{0.7-x}Co_xFe_2O_4$, where x = 0–0.6 in increments of 0.2. The samples were obtained by solid-phase synthesis in a tube furnace with silicon carbide heaters at a temperature of 1150 °C for 5 hours. X-ray phase analysis showed that all the samples obtained are monophasic and have a spinel structure. X-ray structural analysis revealed that the substitution of cobalt atoms for nickel atoms leads to a monotonic increase in the parameters (a and V) of the unit crystal lattice. The elemental composition of the synthesized samples was monitored using an energy dispersive analyzer installed on a Jeol JSM 7001F scanning electron microscope. The aim of this work was to obtain new monophasic samples using previously developed technology.

1. Introduction

Since 1950, ferrites based on solid solutions of $NiFe_2O_4$ and $ZnFe_2O_4$ found its application in industry [1]. Nowadays with the development of technology, the ferrites under investigation are used everywhere. For example, to suppress electromagnetic interference and incoming noise, to prevent external vibrations using cores made of Ni-Zn ferrite. Particular attention was given to ferrite tubes, which are a core with a hole through which passes current-carrying wire. Ferrites are one of the most important magnetic materials since they have a wide possibility of their technological application. Such materials are used in magnetic cores for high frequency transformers in many microwave components. They are also magnetically resistant materials that are difficult to demagnetize and relatively inexpensive.

Nickel-zinc-cobalt (Ni-Zn-Co) ferrites are affordable materials having the necessary magnetic and electrical properties for high frequency applications in the radio frequency domain. The studied ferrites have high electrical resistivity, low eddy current losses, high magnetic permeability, relatively high Curie temperature and high saturation magnetization. Also, cores of this type are used in visualization devices, audio equipment, telecommunications devices, automotive electronics and various cable systems. Parts in the form of rods are used for radio antennas, chokes, inductors, transformers, as well as power filters and DC signal filters. Flat core can suppress microprocessor radiation noise, prevent resonance and suppress communication when connecting two microprocessors, suppress electromagnetic interference through a cable from electrical equipment when connecting to a flat cable [2, 3].

Depending on the scope and achievement of certain properties, nickel-zinc ferrites get certain method, and if necessary, are doped with ions of other elements that change properties. Currently, several methods for obtaining Ni-Zn ferrites, the most famous of them are methods: solid-phase synthesis, co-precipitation, sol-gel, spontaneous combustion, etc. [4–8].

By analyzing modern scientific literature, you can find many publications in which, in addition to research on Ni-Zn ferrite, they explore ferrite with the addition of alloying elements. The result is substitution of a part of the ions of the starting components by doping ions elements. The most widely studied properties of nickel-zinc ferrites with alloying with Al [9, 10], Cu [11–13], Cr [14–16], Co [4, 17], Nd [18, 19], etc. Also, in addition to one alloying element, the possibility of creating a material in which ions of the initial matrix of 2 or more elements are replaced is actively introduced in research papers [20–22]. After a review of the literature data, a solid-phase synthesis method was chosen to obtain samples, as well as an alloying element for the initial composition of $Zn_{0.3}Ni_{0.7}Fe_2O_4$ ferrite. Such synthesis conditions make it possible to obtain samples of new compositions and properties.

2. Experimental part

The first stage of the study is the synthesis of a series of samples with the general formula $Zn_{0.3}Ni_{0.7-x}Co_xFe_2O_4$, where x – takes values from 0 to 0.6 in increments of 0.2. Solid-phase synthesis was chosen as a method of obtaining the material. The components of the original mixture - iron oxide Fe_2O_3, nickel oxide NiO, zinc oxide ZnO, cobalt oxide CoO were mixed in the mass ratio shown in table 1. After mixing the components in a ball mill to a homogeneous powder, the samples were compacted on a hydraulic press. The final stage of the synthesis is the ferritization of pressed samples in a tube furnace with silicon carbide heaters. The sintering process was carried out at a temperature of 1150 °C for 5 hours of isothermal exposure. Sintering temperature is selected based on the analysis of literature data. The basic composition of $Zn_{0.3}Ni_{0.7}Fe_2O_4$ has a relatively high temperature of the second-order phase transition [1].

Table 1. The composition of the initial charge.

№	Formula	wt. %			
		NiO	ZnO	CoO	Fe_2O_3
1	$Zn_{0,3}Ni_{0,7}Fe_2O_4$	22,1183	10,3279	–	67,5538
2	$Zn_{0,3}Ni_{0,5}Co_{0,2}Fe_2O_4$	18,9566	10,3269	3,1696	67,5469
3	$Zn_{0,3}Ni_{0,3}Co_{0,4}Fe_2O_4$	15,7956	10,3258	6,3385	67,5400
4	$Zn_{0,3}Ni_{0,1}Co_{0,6}Fe_2O_4$	12,6352	10,3248	9,5068	67,5332

The chemical composition and surface morphology of the samples were investigated on a scanning electron microscope Jeol JSM 7001F equipped with an Oxford INCA X-max 80 X-ray spectrometer for elemental analysis.

The data about phase composition and lattice parameters were investigate on powder X-ray diffractometer Rigaku Ultima IV in the angular range 2θ from 15° to 90° with the scanning speed 2 °/min.

The Curie temperature was determined on a Netzsch STA 449 F1 Jupiter differential scanning calorimeter (DSC) in platinum crucibles in an argon atmosphere. The Curie temperature is the temperature corresponding to the temperature of a phase transition of the second kind. Ferromagnetic and paramagnetic modifications have different heat capacities. The peak of the DSC curve corresponds to this transition. Therefore, DSC measurements are sufficient for determination of the Curie temperature [23].

3. Results and discussion

Table 2 shows the elemental composition of the synthesized samples, as well as the gross formulas calculated from it. From table 2 it is seen that the elemental composition of the synthesized samples is in good agreement with the initial charge of the samples. An insignificant mismatch of the given composition with the obtained data is due to the fact that when zinc is heated to the sintering temperature, intense volatilization of zinc and oxygen loss occur [1].

Table 2. The elemental composition recorded in three spectra.

№	Chemical composition, atom. %					Calculated formula
	O	Fe	Co	Ni	Zn	
1	47,66	37,13	-	11,10	4,11	$Zn_{0,27}Ni_{0,73}Fe_2O_4$
2	47,91	36,47	3,75	7,58	4,29	$Zn_{0,27}Ni_{0,49}Co_{0,24}Fe_2O_4$
3	48,40	35,62	7,01	4,68	4,29	$Zn_{0,27}Ni_{0,29}Co_{0,44}Fe_2O_4$
4	48,21	34,82	10,72	1,94	4,31	$Zn_{0,25}Ni_{0,11}Co_{0,63}Fe_2O_4$

The microstructure of the surface of the samples consists of many crystallites of various sizes, sintered together. Crystallites have a characteristic inherent of cubic syngony.

X-ray phase analysis showed that all prepared samples are monophasic and have a spinel structure. In figure 1 presents x-rays of the studied samples.

Due to the fact that Zn and Co atoms have different ionic radii, when Zn atoms (r (Ni^{2+}) = 0.49 Å [24]) are replaced by Co atoms (r (Co^{2+}) = 0.58 Å [24]), the crystal lattices - the growth of unit cell parameters.

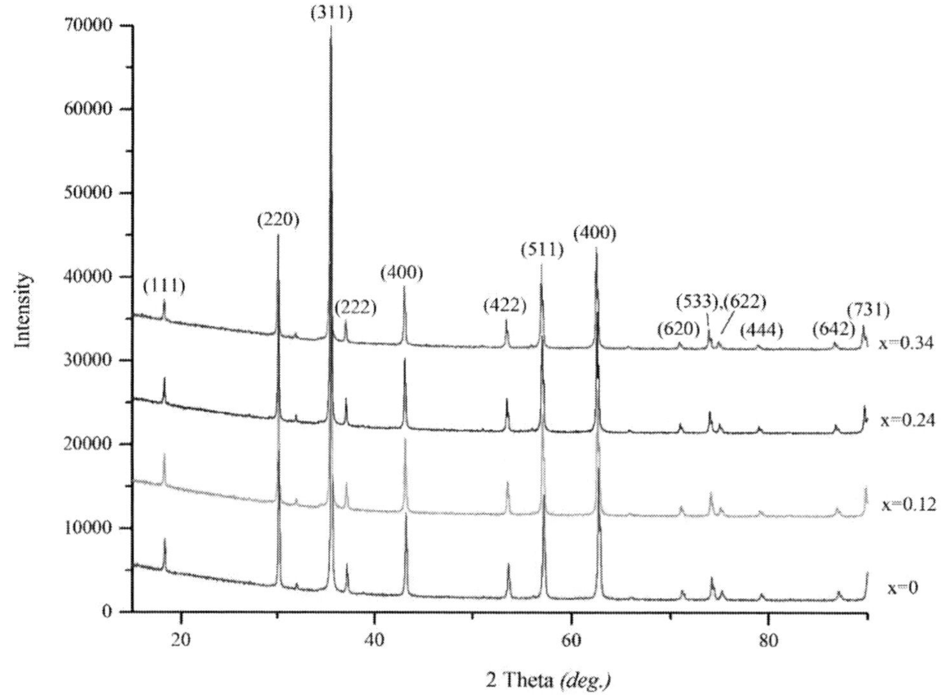

Figure 1. X-ray diffraction patterns of the $Ni_{0.3}Zn_{0.7-x}Co_xFe_2O_4$ system.

Figure 2 shows the dependence of the change in the parameter *a* and the volume *V* of the unit cell on the degree of cobalt substitution. The graph shows that the values monotonically increase with increasing concentration of the substitute element. This is due to the difference in the ionic radii of cobalt and zinc.

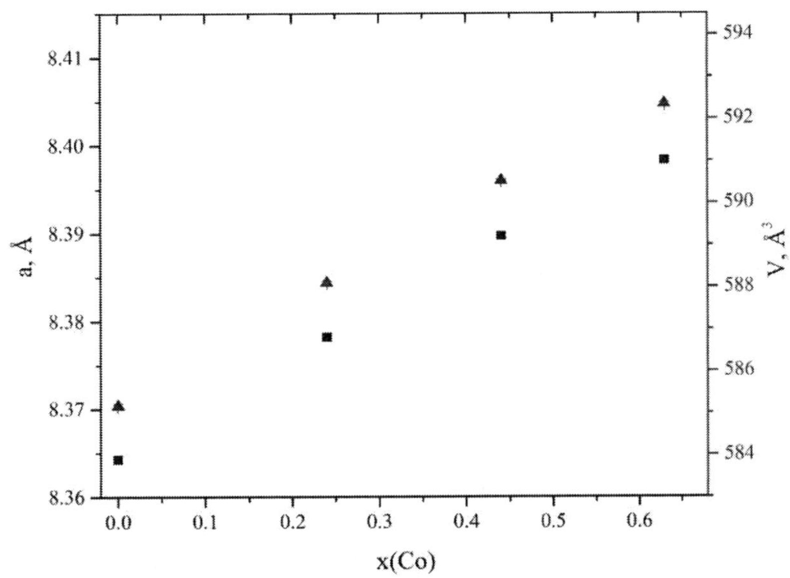

Figure 2. The dependence of the crystal lattice parameters *a, V* on the degree of substitution with cobalt.

DSC curves were obtained at a heating rate of 20 °/min to a temperature of 800 °C. The sample weighed was 150 mg. Figure 3 shows DSC curves of samples. Figure 3 shows that in the temperature range 200–600 °C, all samples have an endothermic process.

The inflection point of the DSC curve corresponds to the temperature of the ferromagnet – paramagnet phase transition (Curie temperature).

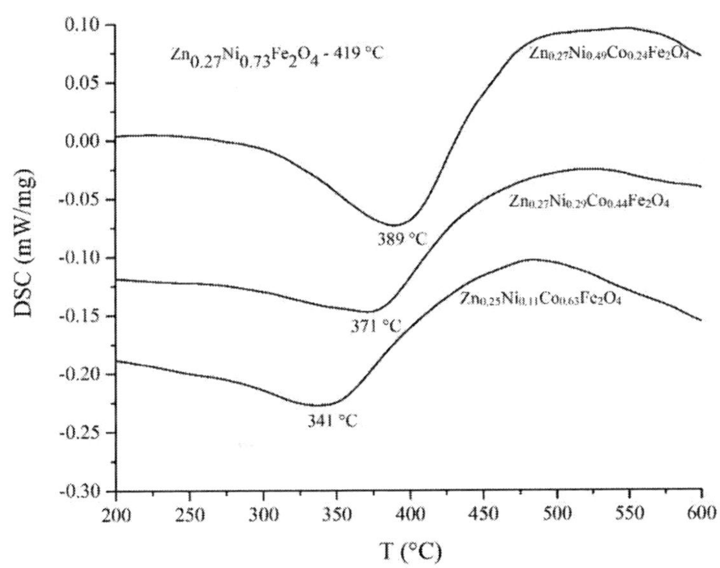

Figure 3. DSC curves of Ni-Zn-Co ferrite.

4. Conclusions

A set of physicochemical parameters has been worked out to ensure the production of nickel-zinc-cobalt ferrite $Zn_{0.3}Ni_{0.7-x}Co_xFe_2O_4$ with a spinel structure. The optimal temperature regime of the solid-phase reaction was determined: temperature $T = 1150$ °C, duration 5 h.

A technology has been developed for producing ferrites of a new chemical composition with spinel structure. Thermal analysis was carried out; samples were studied on a differential scanning calorimeter. The analysis determined the Curie temperature of the samples. Controlling the Curie temperature of the material will allow the use of these ferrites in various temperature conditions.

References

[1] Juravlev G I 1970 *Chemistry and technology of ferrites* (Leningrad: Publ. Chemistry) p 188
[2] Sitidze Yu, Sato Kh 1964 *Ferrites* (Moscow: MIR publ) p 407
[3] Smith Ya 1961 *Ferrites* (Moscow: Foreign publishing house) p 504
[4] Amer M A, Tawfik A, Mostafa A G, El-Shora A F and Zaki S M 2011 *J. Magn. Magn. Mater.* **323** 1445–1452
[5] Dalal M, Mallick A, Mahapatra A S, Mitra A, Das A, Das D, Chakrabarti P K 2016 *Material Res. Bull.* **76** 389–401
[6] Sarveena, Kumar G, Kumar A, Kotnala R K, Batoo K M, Singh M 2016 *Ceram. Int.* **42** 4993–5000
[7] Ramakrishna A, Murali N, Margarette S J, Wegayehu T, Joythi N, Sailaja B, Kumari C C, Samatha K, Veeraiah V 2018 *Adv. Powder. Technol.* **29** 2601–2607
[8] Houshiar M, Jamilpanah L 2018 *Mater. Res. Bull.* **98** 213–218
[9] Paramesh D, Kumar K, Reddy P 2017 *J. Magn. Magn. Mater.* **444** 371–377
[10] Hashim M, Alimuddin A, Kumar S, Ali S, Koo B H, Chung H, Kumar R 2011 *J. Alloy. Compd.* **511** 107–114
[11] Yang W, Kan X, Liu X, Wang Z, Chen Z, Wang Z, Zhu R, Shezad M 2019 *Ceram. Int.* **45** 23328–23332
[12] Ramakrishna K S, Srinivas C, Prajapat C L, Meena S S, Mehar M V K, Potukuchi D M, Sastry D L 2018 *Ceram. Int.* **44** 1193–1200
[13] Houshiar M, Jamilpanah L 2018 *Mater. Res. Bull.* **98** 213–218
[14] Ashtar M, Munir A, Anis-Ur-Rehman M, Maqsood A 2016 *Mater. Res. Bull.* **79** 14–21
[15] Gabal M A, Al Angari Y M, Al-Agel F A. *J 2015 Magn. Magn. Mater.* **391** 108–115
[16] Gabal M A, Bayoumy W A, Saeed A, Al Angari Y M 2015 *J. Mol. Struct.* **1097** 45–51
[17] Ghodake J S, Kambale R C, Shinde T J, Maskar P K, Suryavanshi S S 2016 *J. Magn. Magn. Mater.* **401** 938–942
[18] Qian K, Yao Z, Lin H, Zhou J, Haidry A A, Qi T, Chen W, Guo X 2020 *Ceram. Int.* **46** 227–235
[19] Javed H, Iqbal F, Agboola P O, Khan M A, Warsi M F, Shakir I 2019 *Ceram. Int.* **45** 11125–11130
[20] Slimani Y, Almessiere M A, Sertkol M, Shirsath S E, Baykal A, Nawaz M, Akhtar S, Ozcelik B, Ercan I 2019 *Ultrason. Sonochem.* **57** 203–211
[21] Hossain M D, Khan M N I, Nahar A, Ali M A, Matin M A, Hoque S M, Hakim M A, Jamil A T M K 2020 J. Magn. Magn. Mater. **497** 162978
[22] Rady K E, Elsad R A 2020 *J. Magn. Magn. Mater.* **498** 166195
[23] Gudkova S A, Vinnik D A, Zhivulin V E, Chernukha A S, Zherebtsov D A, Trofimov E A, Trukhanov A V, Trukhanov S V, Kalandija M, Semisalova A S 2019 *J. Magn. Magn. Mater.* **470** 101–104
[24] Shannon R D 1976 *Acta. Crystallogr. Sect.* A **32** 751–767

Acknowledgments

The reported study was funded by RFBR, project number 20-38-70057.

ThermoCalc Modelling of the Effect of Chemical Composition on the Phase Transformations Temperatures of Ti-0.4Al Alloy

A G Illarionov[1,2], A S Ushakova[1] and F V Vodolazsky[1]

[1]Ural Federal University,19, Mira str., Yekaterinburg, 624001, Russia
[2]Institute of Engineering Science, Ural Branch of the Russian Academy of Sciences, 18, S. Kovalevskaya str., Yekaterinburg, 620108, Russia

E-mail: a.g.illarionov@urfu.ru

Abstract. The effect of chemical composition of Ti-0.4Al α-titanium alloy (PT-1M alloy) on the phase transformation temperatures up to melting temperature has been investigated using ThermoCalc software. The possibility of appearance of Ti_2N nitride, $Ti_x(CN)$-type carbonitride, and Ti_2Fe-type intermetallic in the alloy structure in equilibrium state has been proved by thermodynamic calculations. It has been proposed the so-called aluminum equivalent (Al_{eq}) for PT-1M alloy to analyze the effect of chemical composition on the phase transformation temperatures change. The following equivalents have been suggested: $Al_{eq}^{T\beta}$ for $\beta \leftrightarrow (\beta+\alpha)$ – transformation, Al_{eq}^{Tliq} for $L \leftrightarrow \beta$ – transformation, and Al_{eq}^{Tsol} for $(L+\beta) \leftrightarrow \beta$ – transformation. It has been shown the linear correlation between the characteristic temperatures of phase transformations in the investigated alloy and the proposed aluminum equivalents.

1. Introduction

Ti-0.4Al alloy (also known as PT-1M alloy) refers to a low-alloyed α-titanium alloys [1]. This alloy is characterized by good technological plasticity both in hot and cold state, high corrosion resistance, and moderate strength [2]. PT-1M alloy is used for manufacture of deformable semi-finished products (sheets, rods, pipes etc.) for products operating in sea and fresh water with operating temperatures up to 150 °C [3, 4]. The change in the chemical composition varying within the alloy grade of PT-1M alloy should affect the phase transformations temperatures: liquidus temperature (T_{liq}), solidus temperature (T_{sol}), $\beta \leftrightarrow \alpha+\beta$ – polymorphic transformation temperature etc. These phase transformation temperatures are important for a reasonable choice of processing parameters of casting, forming, and heat treatment [5, 6, 7]. One of the promising ways to obtain the phase transformations temperatures is a thermodynamic calculation using ThermoCalc software [8, 9]. However, there is a lack of information about the influence of chemical composition of PT-1M alloy on the phase transformation temperatures in this alloy up to melting temperature. The main purpose of the present study was to calculate the phase transformation temperatures of PT-1M titanium alloy with different chemical composition varying within the alloy grade, as well as to estimate the effect of chemical composition on these temperatures.

2. Experimental

The chemical composition of PT-1M alloy is Ti-0.2…0.7 Al-0.2 Fe-0.3 Zr-0.1 Si-0.12 O-0.07 C-0.04 N-0,006 H (wt.%), where Al is an alloying element, and Fe, Zr, Si, O, C, N, and H are impurities according to GOST 19087-91 standard.

Several chemical compositions varying within the alloy grade were chosen for the investigation. Selected compositions provided different combinations of α-stabilizers (Al, O, N, C) and β-stabilizers (Fe, Si) such as: min α + max β, min α + min β, max α + max β, max α + min β, where min α and max α is minimum and maximum content of α-stabilizers, relatively; min β and max β is minimum and maximum content of β - stabilizers, correspondingly. Minimum and maximum content of Al as well as maximum content of impurities were taken from GOST standard. Minimum content of impurities was taken as 1/10 of maximum content set up by GOST standard. Zirconium was considered as a neutral element. The possible presence of hydrogen was not taken into account due to its low content.

Two additional chemical compositions of PT-1M alloy (average grade composition and composition used in [9,10]) were also studied. The chemical composition of all investigated alloys is given in Table 1.

Table 1. Chemical composition of investigated alloys (wt.%).

Alloy #	Al	Zr	Si	Fe	O	N	C
Alloy 1 – min α + max β	0.2	0.3	0.1	0.2	0.012	0.004	0.007
Alloy 2 – max α + min β	0.7	0.03	0.01	0.02	0.12	0.04	0.07
Alloy 3 – max α + max β	0.7	0.3	0.1	0.2	0.12	0.04	0.07
Alloy 4 – min α + min β	0.2	0.03	0.01	0.02	0.012	0.004	0.007
Alloy 5 – Average grade composition	0.45	0.165	0.055	0.11	0.066	0.022	0.0385
Alloy 6 – Melt composition	0.4	–	–	0.04	0.08	0.01	0.03

Temperatures of phase transitions were calculated using ThermoCalc software. This software is based on model of classical thermodynamics, which deals with systems in equilibrium state [11].

3. Results and discussion

The dependence of change in the number of phases in all investigated alloys in the temperature range from 100°C to 2000°C has been obtained using ThermoCalc software. The calculation results for Alloy 4 (min α + min β) and Alloy 3 (max α + max β) are given in Figure 1.

Based on the calculation results it has been determined the thermodynamically probable temperatures of possible phase transitions such as liquidus temperature ($T_{\beta\leftrightarrow L}$), solidus temperature ($T_{\beta\leftrightarrow L+\beta}$), $\beta\leftrightarrow\beta+\alpha$ – polymorphic transformation temperature (T_{β}), and $\beta+\alpha\rightarrow\alpha$ – polymorphic transformation temperature ($T_{\beta+\alpha\rightarrow\alpha}$), as well as onset temperatures of forming of titanium carbide (T_{TixC}), Ti_2Fe-type intermetallic (T_{Ti2Fe}), and Ti_2N nitride (T_{Ti2N}). The obtained temperatures for all examined alloys are given in Table 2.

Table 2. Temperatures of phase transitions of investigated alloys.

Alloy #	$T_{\alpha+Ti2N\leftrightarrow\alpha}$, °C	$T_{\alpha+TixC\leftrightarrow\alpha}$, °C	$T_{\alpha+Ti2Fe\leftrightarrow\alpha}$, °C	$T_{\alpha+Ti2Fe\leftrightarrow\beta+\alpha}$, °C	$T_{\alpha\leftrightarrow\beta+\alpha}$, °C	$T_{\beta+\alpha\leftrightarrow\beta}$, °C	$T_{\beta\leftrightarrow L+\beta}$, °C	$T_{\beta+L\leftrightarrow L}$, °C
Alloy 1	–	341	–	546	–	886	1635	1677
Alloy 2	151	646	409	–	837	949	1672	1745
Alloy 3	153	663	–	553	–	946	1640	1735
Alloy 4	–	339	375	–	825	892	1665	1681
Alloy 5	–	530	–	546	–	920	1653	1715
Alloy 6	–	478	505	–	724	913	1671	1697

As it can be seen from Table 2 and Figure 1, the calculations point out the possibility of appearance in the alloy structure of such equilibrium phases as Ti_2N nitride with HCP lattice, titanium carbide (Ti_xC) with FCC lattice, and Laves phase on the base of Ti_2Fe intermetallic.

According to the calculations, their equilibrium content does not exceed tenths of a percentage point. The analysis of data given in Table 2 showed that the stability of these phases is directly depends on the content of nitrogen, carbon, and iron for Ti_2N, Ti_xC, and Ti_2Fe intermetallic, correspondingly.

There is a lack of information about the formation of above-mentioned phases in PT-1M alloy. It is reported [12, 13] that formation of Ti_2Fe intermetallic occurs in a commercial pure titanium (which is close to investigated PT-1M alloy) upon long tempering (approximately 500 h) in the temperature range from 300 °C to 600 °C. Formation of Ti_2Fe is a result of the appearance on the grain boundaries the micro segregations enriched with iron [12, 13]. The temperature range of formation of Ti_2Fe intermetallic calculated for PT-1M alloy in the present study is from 375 °C to 553 °C (Table 2) and corresponds well to a previously mentioned data.

According to TermoCalc calculations, Ti_2Fe intermetallic contains mainly titanium and iron, and lesser zirconium. The ratio of these elements gives $(TiZr)_xFe$ compound, where x=2.3, i.e. close to stoichiometry of Ti_2Fe compound. It is reported about the possibility of simultaneous enrichment with zirconium and iron of grain boundary areas in Ti-Al-Zr alloy with Fe impurity [14].

It should be noted that in PT-1M alloy containing more than 0.11% Fe (wt.%) the formation of Ti_2Fe intermetallic occurs at substantially constant temperature of 550±4 °C by α+Ti2Fe$\leftrightarrow\beta$ eutectoid reaction. The calculation results showed that in the course of this reaction the alloy containing more than 0.11% Fe (wt.%) does not turn into single α-phase state during cooling. Therefore, the temperature of $\beta\leftrightarrow(\beta+\alpha)$ – polymorphic transformation ($T_{\beta+\alpha\rightarrow\alpha}$) cannot be calculated (Table 2).

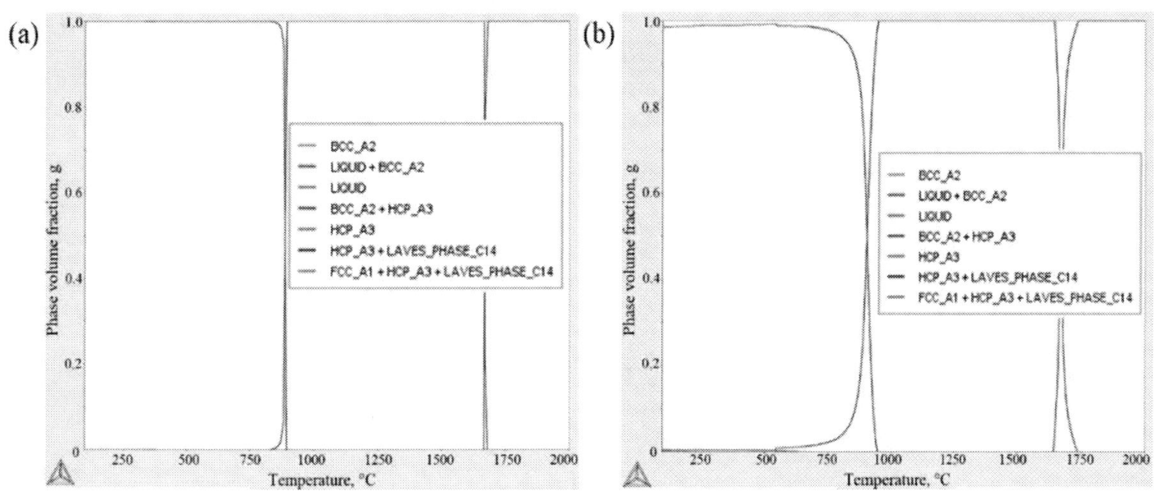

Figure 1. Temperature dependence of volume fraction of phases in Alloy 4 (min α + min β) (a) and Alloy 3 (max α + max β) (b) obtained by use Thermocalc software, where LIQUID is liquid state, BCC_A2 is titanium β-solid solution with BCC lattice, HCP_A3 is titanium α-solid solution with HCP lattice, LAVES_PHASE_C14 is Laves phase on the base of Ti_2Fe intermetallic, FCC_A1 is phase on the base of titanium carbide with FCC lattice, Ti_2N is titanium nitride.

Lowering of iron content from 0.02 wt.% to 0.04 wt.% in PT-1M alloy results in Ti_2Fe-type intermetallic precipitation from α-phase, and decrease in onset temperature of precipitation from 505°C to 375°C. This is consistent with published data [15] where it is shown that in Ti-Fe alloy the precipitation of intermetallic phase is possible at iron content higher than 0.01 wt.%.

It was reported in [16] that carbide phase was observed in a commercial pure titanium. It was noted that usually this phase is not a carbide, but a carbonitride. The calculated chemical composition of the carbide in PT-1M alloy [9,10] confirms the above data. According to ThermoCalc calculations, the obtained equilibrium phase is $Ti_x(CN)$ carbonitride, where x = 1.64, and C/N ratio equals 10/1. The stoichiometry of $Ti_x(CN)$ compound is close to that of Ti_2C. The formation of Ti_2C-type carbide was noted in [17] which considered to the formation of carbide phases in high alloyed titanium β–alloys. It should be noted that that the possibility of the formation of such a carbide in PT-1M alloy at carbon content up to 0.07 wt.% has been confirmed in [18]. According to this study, in Ti-C system at carbon content less than 0.08 wt.% the carbide phase precipitates precisely from from α-phase.

The lack of information about the presence of nitrides and carbides in most industrial titanium alloys is probably due to the insignificant concentration of nitrogen and carbon in these alloys. Therefore, it is hard to reveal the nitrides and carbides in the alloy structure by means of microscopy. The other possible reason is that both nitrogen and carbon are dissolved in α– and β–solid solutions within the solubility limit. The latter can be proved by the fact that calculation using ThermoCalc software does not confirm the formation of nitrides in several of investigated alloys (Table 2).

In order to analyze the effect of chemical composition of PT-1M alloy on the phase transformation temperatures such as liquidus temperature ($T_{L\leftrightarrow\beta}$), solidus temperature ($T_{L+\beta\leftrightarrow\beta}$), $\beta\leftrightarrow\alpha+\beta$- polymorphic transformation temperature ($T_{\beta\leftrightarrow\beta+\alpha}$) it was decided to use a so-called aluminum equivalent (Al_{eq}) similarly to strength equivalent and structural aluminum equivalent [1]. The following aluminum equivalents were suggested to use: $Al_{eq}^{T\beta}$ for $T_{\beta\leftrightarrow\beta+\alpha}$ temperature, Al_{eq}^{Tliq} for $T_{L\leftrightarrow\beta}$ temperature, and Al_{eq}^{Tsol} for $T_{L+\beta\leftrightarrow\beta}$ temperature. The effect of each alloying element in the proposed aluminum equivalents on the corresponding temperature was estimated through equivalent content of aluminum with a certain coefficient. This was proposed based on the binary phase diagrams of Ti-X system, where X is one of alloying elements (Al, Fe, Zr, Si, O, N, C) of the investigated alloy. The coefficient was calculated using the binary phase diagrams [19] and published data [20, 21] proportionally to effect of X element on the corresponding temperature in Ti-X system in comparison to that on aluminum on the same temperature in Ti-Al system.

Data used for calculating $Al_{eq}^{T\beta}$ Al_{eq}^{Tliq}, Al_{eq}^{Tsol} aluminum equvalents are given in Table 3, Table 4.

Table 3. Data for calculating the coefficients of alloying elements for $Al_{eq}^{T\beta}$.

Element	Temperature of transformation, °C	C_β at transformation, wt.%	Change in T_β temperature at adding of 1 wt.% of element	Calculation coefficient for $Al_{eq}^{T\beta}$ (rounded)	Source
Al	–	–	14.5	1	[20]
O	–	–	200	13.8	[20]
N	–	–	550	37.9	[20]
C	–	–	200	13.8	[20]
Zr	605	67.37	–4.12	–0.3	[21]
Fe	595	17.07	–16.84	–1.2	[21]
Si	865	0.65	–26.92	–1.9	[21]

The calculation coefficient of alloying elements may assume both positive and negative values. The coefficient assumes positive value if element increases $T_{\beta\leftrightarrow\beta+\alpha}$, $T_{L\leftrightarrow\beta}$, and $T_{L+\beta\leftrightarrow\beta}$ temperatures; the coefficient assumes negative value if element decreases these temperatures. Data reported in [20, 21] were used to calculate $Al_{eq}^{T\beta}$. According to [20], for alloy containing up to 2% Al (wt.%) adding of 1 wt.% of aluminum increases T_β by 14.5°C. This value was taken for normalization of other values. The coefficients for O, N, C were taken from [21]. The coefficients for Zr, Fe, Si were calculated based on data given in [22], minimum temperature of $\beta\leftrightarrow\beta+\alpha$ – polymorphic transformation and corresponding content of element.

Table 4. Data for calculating the coefficients of alloying elements for Al_{eq}^{Tliq}, Al_{eq}^{Tsol}.

Element	Temperature of liquidus/solidus transformation, °C	Concentration at liquidus/solidus transformation, wt.%	Change of T^{liq}/T^{sol} at adding of 1 wt.% of element	Calculation coefficient for $Al_{eq}^{Tliq}/Al_{eq}^{Tsol}$	Source
Al	1700/1697	3.0/3.0	10.0/9.0	1.0/1.0	[19]
O	1720/1720	1.7/2.8	28.9/17.7	2.9/2.0	[19]
N	2020/2020	1.2/1.9	291.7/184.2	29.2/20.4	[19]
C	1645/1645	0.5/0.2	–54.3/–163.4	–5.5/–18.0	[19]
Zr	1540/1540	50.0/50.0	–2.6/2.6	–0.3/–0.3	[19]
Fe	1085/1085	32.8/24.8	–17.8/–23.6	–1.8/–2.6	[19]
Si	1330/1330	8.6/2.9	–39.8/–113.5	–4.0/–12.6	[19]

Based on the coefficients given in Table 3, Table 4 the following formulas for the calculation of $Al_{eq}^{T\beta}$, Al_{eq}^{Tliq}, and Al_{eq}^{Tsol} aluminum equivalents for PT-1M alloy were suggested:

$$Al_{eq}^{T\beta} = C_{Al} + 37.9C_N + 13.8(C_O + C_C) - 1.9C_{Si} - 1.2C_{Fe} - 0.3C_{Zr}, \quad (1)$$

$$Al_{eq}^{Tliq} = C_{Al} + 29.2C_N + 2.9C_O - 5.5C_C - 4.0C_{Si} - 1.8C_{Fe} - 0.3C_{Zr}, \quad (2)$$

$$Al_{eq}^{Tsol} = C_{Al} + 20.4C_N + 2.0C_O - 18.2C_C - 12.6C_{Si} - 2.6C_{Fe} - 0.3C_{Zr}, \quad (3)$$

where C_{Al}, C_N, C_O , C_C, C_{Si}, C_{Fe}, C_{Zr} is the content of Al, N, O, C, Si, Fe, and Zr, correspondingly (wt.%).

The calculated values of $Al_{eq}^{T\beta}$, Al_{eq}^{Tliq}, and Al_{eq}^{Tsol} aluminum equivalents for all investigated alloys are listed in Table 5.

Table 5. Calculated values of aluminum equivalents for different chemical compositions of PT-1M alloy.

Alloy No	1	2	3	4	5	6
$Al_{eq}^{T\beta}$	0.23	5.3	4.9	0.7	2.8	2.5
Al_{eq}^{Tsol}	–1.66	0.30	–1.36	–0.01	–0.68	0.12
Al_{eq}^{Tliq}	–0.52	1.74	0.99	0.23	0.61	0.69

In Figure 2 are shown the relationships between T_β, T_{liq}, T_{sol} temperatures and corresponding $Al_{eq}^{T\beta}$, Al_{eq}^{Tliq}, Al_{eq}^{Tsol} aluminum equivalents established using ThermoCalc data as well as formulas (1), (2), (3) suggested before. The dependence of crystallization interval $T_{liq} - T_{sol}$ (ΔT) on difference of corresponding aluminum equivalents $Al_{eq}^{Tliq} - Al_{eq}^{Tsol}$ ($\Delta Al_{eq}^{Tliq\text{-}Tsol}$) was also estimated.

As in can be seen in Figure 2a and Figure 2b, a good linear correlation is observed between T_β, T_{sol} temperatures and corresponding $Al_{eq}^{T\beta}$, Al_{eq}^{Tsol} aluminum equivalents. The dependence obtained for T_β, T_{sol} temperatures of PT-1M alloy was approximated on pure titanium taking $Al_{eq}^{T\beta} = 0$ and $Al_{eq}^{Tsol} = 0$. The difference between the approximated and the experimentally determined values ($T_\beta = 882.5$°C, T_{sol} = 1668°C) [22] did not exceed 1°C. Calculated T_β value for Alloy 6 (melt composition of PT-1M alloy) was 913°C, which is consistent with T_β value equal 920°C determined by method of testing quenching reported in [5]. In general, the calculated T_β value range 886…949°C for the investigated chemical compositions of PT-1M alloy is consistent with published data of T_β values 890°C [23] and 925°C [9]. The difference between calculated maximum and minimum T_β values for PT-1M alloy is 63°C and corresponds to that of α-titanium alloys. The statistical analysis of T_β values of industrial titanium alloys ingots showed that the difference between calculated maximum and minimum T_β values for VT20 alloy is 65°C.

The correlation between T_{liq} temperature and Al_{eq}^{Tliq} aluminum equivalent (Figure 2c) is not as linear as that observed between T_{β}, T_{sol} and corresponding aluminum equivalents (Figure 2a, Figure 2b). The dependence obtained for T_{liq} and Al_{eq}^{Tliq} of PT-1M alloy was approximated on pure titanium taking $Al_{eq}^{Tliq} = 0$. The approximated value of T_{liq} was 1687°C which exceeds the values by 19°C (Figure 2c).

The dependence character of crystallization interval (ΔT) on difference of corresponding aluminum equivalents ($\Delta Al_{eq}^{Tliq\text{-}Tsol}$) is close to linear (Figure 2d). The dependence obtained for ΔT and $\Delta Al_{eq}^{Tliq\text{-}Tsol}$ of PT-1M alloy was approximated on pure titanium taking $T_{liq} - T_{sol} = 0$ and $Al_{eq}^{Tliq} - Al_{eq}^{Tsol} = 0$. The approximated value of ΔT exceeds the values by 5°C (see formula on Figure 2d). Thus, it can be concluded that the obtained dependence is quite reasonable.

According to the calculations, the crystallization interval of PT-1M alloy varies from 15°C to 95°C depending on the chemical composition of the alloy. The average value of crystallization interval is 52°C. The calculation results of T_{liq}, T_{sol}, and ΔT values are in a good agreement with that of experimental data reported in [7]. The crystallization interval of titanium alloys may vary from 15°C (for VT1L alloy) to 80°C (for VT21L alloy), and the average value of ΔT of most casting titanium alloys (e.g. VT6L, VT3-1L, VT9L, VT14L) is 60°C, which is consistent the results obtained in the present study.

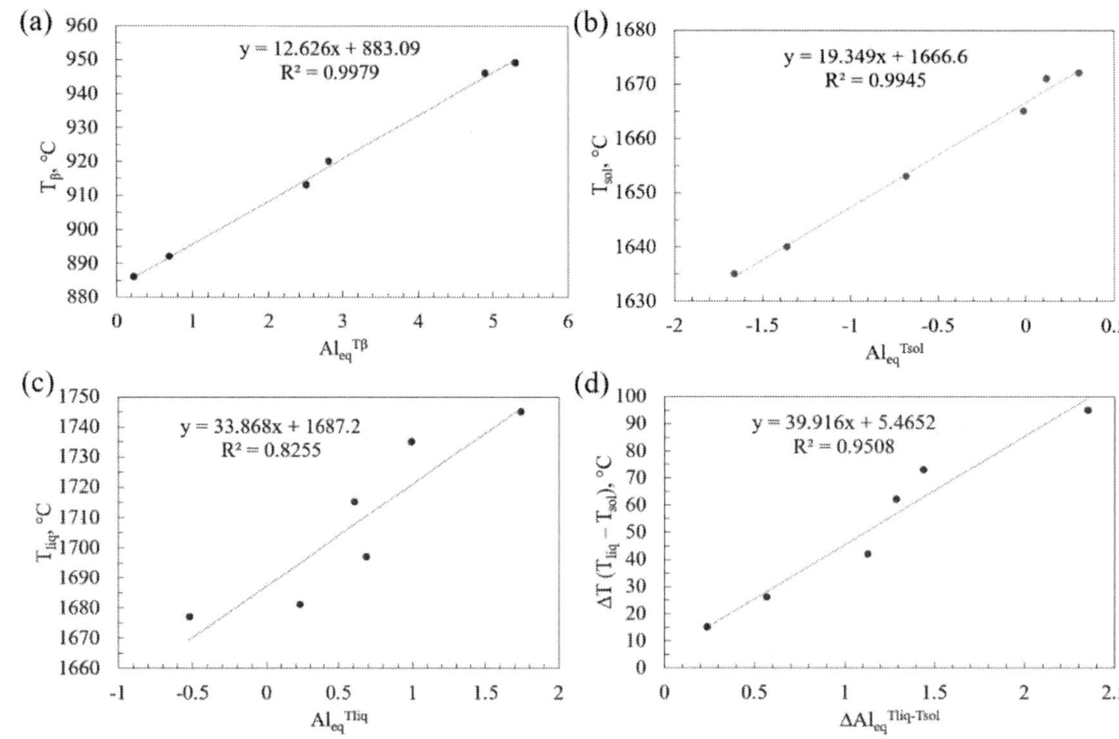

Figure 2. Dependence of T_{β}, T_{sol}, T_{liq} and crystallization interval (ΔT) on $Al_{eq}^{T\beta}$ (a), Al_{eq}^{Tliq} (b), Al_{eq}^{Tsol} (c), $\Delta Al_{eq}^{Tliq\text{-}Tsol}$ (d) for PT-1M alloy.

Thus, the dependence of ΔT on $\Delta Al_{eq}^{Tliq\text{-}Tsol}$ reported in the present paper can be used to developt and optimize the chemical composition of titanium alloys in order to minimize the crystallization interval ensuring [24] higher casting properties.

Conclusions

1. It has been established the effect of chemical composition and temperature on the phase composition of PT-1M titanium alloy. The possibility of appearance of Ti_2N nitride, $Ti_x(CN)$-type carbonitride, and Ti_2Fe-type intermetallic in the alloy structure in equilibrium state has been proved by thermodynamic calculations using ThermoCalc software.

2. It has been proposed using of so-called aluminum equivalents (Al_{eq}) for PT-1M alloy to analyze the influence of chemical composition on the phase transformations temperatures change. It has been suggested $Al_{eq}^{T\beta}$ aluminum equivalent for $\beta \leftrightarrow (\beta+\alpha)$-transformation, Al_{eq}^{Tliq} aluminum equivalent for $L \leftrightarrow \beta$-transformation, and Al_{eq}^{Tsol} aluminum equivalent for $(L+\beta) \leftrightarrow \beta$ – transformation as well as calculation formulas:

$$Al_{eq}^{T\beta} = C_{Al} + 37.9C_N + 13.8(C_O + C_C) - 1.9C_{Si} - 1.2C_{Fe} - 0.3C_{Zr}, \quad (4)$$

$$Al_{eq}^{Tliq} = C_{Al} + 29.2C_N + 2.9C_O - 5.5C_C - 4.0C_{Si} - 1.8C_{Fe} - 0.3C_{Zr}, \quad (5)$$

$$Al_{eq}^{Tsol} = C_{Al} + 20.4C_N + 2.0C_O - 18.2C_C - 12.6C_{Si} - 2.6C_{Fe} - 0.3C_{Zr}, \quad (6)$$

where C_{Al}, C_N, C_O, C_C, C_{Si}, C_{Fe}, C_{Zr} is the content of Al, N, O, C, Si, Fe, and Zr, correspondingly (wt.%).

3. It has been found out the linear correlation dependence between the characteristic temperatures (T_β, T_{liq}, T_{sol}, and ΔT ($T_{liq} - T_{sol}$)) of PT-1M alloy and the proposed aluminum equivalents ($Al_{eq}^{T\beta}$, Al_{eq}^{Tliq}, and Al_{eq}^{Tsol}, and $\Delta Al_{eq}^{Tliq\text{-}Tsol}$ ($Al_{eq}^{Tliq} - Al_{eq}^{Tsol}$).

References

[1] Ilyin A A, Kolachev B A, Polkin I S 2009 *Titanium alloys. Composition, Structure, Properties. Reference book* (VILS-MATI, Moscow) p 520

[2] Ushkov S S, Chechulin B B 1995 Experience in the field of titanium application in shipbuilding and off-shore structures *Titanium* **3–4** 38–40

[3] Briker A S, Zobachev Yu Ye, Molchanova N S, Razuvayeva I N, Khesin Yu D 1995 Prospects of using titanium for improving reliability and service life of various units and items of equipment in civil ships *Titanium* **3–4** 41–43

[4] Ushkov S S and Chechulin B B 1995 The main tasks and steps of introduction of titanium in shipbuilding for production of shipboard power plants and machinery *Titanium* **3–4** 44–47

[5] Aleksandrov V K, Anoshkin N F and Belov A F 1996 *Semi-finished products of titanium alloys* (ONTI-VILS, Moscow) p 584

[6] Kosmatskiy Y I, Fokin N V, Filyaeva E A and Barichko B V 2017 Study on the Deformation Ability Billets of Titanium Alloy Grade PT-1M *Bulletin of the South Ural State University Ser. Metallurgy* **17** 83–91

[7] Illarionov A G and Popov A A 2014 Production and Operating Properties of Titanium Alloys: Textbook (*Ural. Univ.*) p 136

[8] Kuznetsov A V, Sokolovskii V S, Salishchev G A, Belov N A and Nochovnaya N A 2016 Thermodynamic modeling and experimental study of phase transformations in alloys based on γ-tial *Metal Science and Heat Treatment* **58(5–6)** 259–267

[9] Vodolazskiy F V, Gornostaeva E A and Illarionova S M 2020 Calculation of phase composition of α-and near-α-titanium alloys *Solid State Phenomena* **299** 403–408

[10] Vodolazskiy F V, Barannikova N A and Illarionov A G 2020 Structure, texture and mechanical properties through the section of the hot-extruded tube of titanium alloy PT-1M *Materials Science Forum V* **989** 139–144

[11] Pingfang S and Sundman B 2006 Thermo-Calc Software System (*Thermodynamic Framework and Data*) p 64

[12] Nesterova Ye V, Rybin V V, Ushkov S S and Kolodkina G I 1980 Investigation of the nature of the thermal embrittlement of low alpha alloys of titanium *Physics of Metals and Metallography* **50(6)** 83–89

[13] Kryukov I I, Nesterova Ye V, Rybin V V and Rybnikov A I 1981 Boundary precipitates in technically pure titanium *Physics of Metals and Metallography* **52(4)** 172–174

[14] Ushkov S S, Vlasova I G, Kiyevskaya N Kh, Kolodkina G I, Parshikova G Ye and Plamadyala N V 1984 Peculiarities of microsegregations of impurity and alloying elements in alpha-alloys of titanium *Physics of Metals and Metallography* **57(1)** 180–184

[15] Raub E, Raub Ch J, Rosebel E, Compton U B, Geballe T H and Matthias B T 1967 The α-Ti-Fe-solid solution and its superconducting properties *J. of Less Common Metals* **12** 36–40
[16] Kolobov Yu R, Kashin, O A, Sagymbaev E E, Dudarev E F, Bushnev L S, Grabovetskaya G P, Pochivalova G P, Girsova N V and Stolarov V V 2000 Structure and mechanical and electrochemical properties of ultrafine-grained Ti *Russian Physics Journal* **43(1)** 71–78
[17] Zhau D, Ebel T, Yan M and Quian M 2015 Trace carbon in biomedical beta-titanium alloys: recent progress *JOM* **67** 2236–2243
[18] Yan M, Qian M, Kong C and Dargusch M S 2014 Impacts of Trace Carbon on the Microstructure of As-Sintered Biomedical Ti-15Mo Alloy and Reassessment of the Maximum Carbon Limit *Acta Biomater* **10(2)** 1014–1023
[19] Murray J L 1947 *Phase diagrams of binary titanium alloys* (Metals Park, Ohio: ASM International) p 345
[20] Liao Q, Deng C, Qu H, Yang Y, Nan L and Yang J 2012 Effects of Deformation rate on microstructure and tensile properties of cold rolled Ti-3Al-2,5V alloy *Proc. Symp. 12th World Conf. on Titanium* **1** 278–281
[21] Wang Q, Dong C and Liaw P K 2015 Structural stabilities of b-Ti alloys studied using a new Mo equvivalent derived from [β/(α + β)] phase-boundary slopes *Metallurgical and materials transactions* **46** 3440–3447
[22] Zwicker U 1979 *Titan and Its Alloys* (Moscow: Metalurgy) p 512
[23] Anisimov A P, Antipov V, Kopylov V N, Rtishcheva L P, Smirnov V G and Kharin A A 2016 RU Patent **2016068748 A1**
[24] Kolachev B A, Yelagin V I and Livanov V A 2001 *Metal Science and Heat Treatment of Non-Ferrous Metals and Alloys* (Moscow: MISIS) p 416

Acknowledgments
The work was financially supported by Russian Science Foundation grant No. 18-79-10107.

IOP Conf. Series: Materials Science and Engineering **969** (2020) 012030 doi:10.1088/1757-899X/969/1/012030

The Possibility of Using Tyumen Keramzite Clay in the Production of Ceramic Materials

I A Pavlova, A A Getman and E P Farafontova

Institute of new materials and technologies, Ural Federal University named after the First President of Russia B. N. Yeltsin, 19, Mira Str. Ekaterinburg 620002, Russia

E-mail: i.a.pavlova@urfu.ru

Abstract. The properties of red low-melting clay with the aim of its use in the production of building ceramics are investigated. The presented sample of clay raw materials by mineral composition refers to kaolinite clays with mixed-layer formations in the form of illite and ferruginous montmorillonite with Na-, Ca- and Mg-interlayer hydrated exchange cations. It is characterized by an average content of free SiO_2 (10–25%). The clay is medium plastic with a low content of coarse inclusions, it refers to medium raw materials. Clay is medium-drying, medium-sensitive to drying, non-sintering. Mechanical strength during firing at a temperature of 900°C is 27.5 MPa. When firing samples in the temperature range 900–1050°C, a significant black core is formed. At a firing temperature of 1050°C, swelling of the samples occurs. Considering the properties of this clay, it cannot be recommended to produce face and ordinary building bricks. However, the studied clay can be used to produce expanded clay. The introduction of 0.5% of spent technical oil allows to get the expansion coefficient of 3.83. The temperature range of expansion is 1050–1170°C. The density of expanded clay granules in this case is 560 $kg \cdot m^{-3}$.

1. Introduction

Fusible red-burning clays are often overburden or accompanying rocks in mining. Deposits of fusible clays are ubiquitous. The properties of clays of various deposits are not the same and can be used to produce certain types of ceramics [1–5]. Clay properties vary even within the same deposit. Red-burning clays are used on a large scale in the production of building ceramics, and in smaller ones for the production of pottery. Building materials always remain in demand due to the increasing rate of construction in both urban and rural conditions. Fusible clays may be suitable for the production of ceramic bricks, expanded clay, and ceramic tiles. To determine the suitability of clay in the production of a particular type of product, it is necessary to study its ceramic properties and determine the production technology [6–8]. In this regard, research work on determining the suitability of clay in the production of a particular type of product always remains in demand. This project is devoted to the study of the properties of low-melting clay of the Tyumen region in the production of ceramics.

2. Experimental procedure

The chemical composition of the materials is determined according to [9–18]. The content of free SiO_2 in clay was determined according to [19].

Differential thermal analysis was performed by derivatograph (MOM). Heating was carried out in a platinum crucible up to 1000°C with a speed of 10° per min. Al_2O_3 was used as a reference.

Plasticity was tested in accordance with [19]. The plasticity of clay tempered with water is its ability to deform under the influence of a load without breaking the continuity (formation of cracks), take any shape and save it after the removal of the load. The plasticity of clays most of all depends on their disperse and mineral composition. The more dispersed the clay, the higher its plasticity. In ceramic technology, there are many methods for estimation of the plasticity of clays and ceramic masses. All these methods are divided into 2 groups: direct and indirect. Indirect methods cannot simultaneously evaluate the plasticity and molding ability of clays and plastic ceramic masses. As an example of the method of indirect plasticity estimation, the Vasiliev method proposed by [19] is used. When determining clay plasticity using the Vasiliev method, the plasticity number P is determined, which is the difference between the relative humidity of the clay at the lower yield strength and the relative humidity of the same clay determined at the plastic limit [20].

The sensitivity of clays to drying was determined by the method of Z. A. Nosova. The sensitivity of clays to drying is their ability or tendency to crack during drying. The cause of cracks is the different shrinkage over the cross section and on the surface of the product. Shrinkage cannot occur evenly for the entire product, as humidity in the surface and in its central zones varies during the drying process. The nonuniformity of shrinkage causes internal local stresses in the material, which can exceed its mechanical strength and cause deformation of the raw material and the formation of cracks. It is the nonuniformity of shrinkage and internal local stresses caused by shrinkage that cause cracking and defects during drying. The coefficient of sensitivity to drying is the ratio of the volumetric shrinkage during drying to the real porosity of the material in the air-dry state. The smaller this coefficient, i.e. the less clay shrinkage during drying, the easier and safer the drying of products. Clays highly sensitive to drying have coefficient of sensitivity of more 1.5; medium-sensitive clays have this index of 1–1.5; insensitive clays have coefficient of sensitivity less than 1.

The ratio of materials to sintering is determined according to [19]. To determine the sintering ability of clay from the mass of optimal molding moisture content (W = 18%), cubes of size 50×50×50 mm were formed. Samples are dried and fired at a temperature of 800–1050°C.

3. Results and discussion

Clay samples with a career humidity of 20% in the form of pieces from 5 to 15 cm were ground and dried under room conditions. The color of clay in its raw form is blue-gray. The dried clay was ground on laboratory runners to a grain size of less than 3 mm. The chemical composition of clay is presented in Table 1. According to the content in the clay Al_2O_3 in the calcined state (16.47%), the clay belongs to the semi-acid raw (Al_2O_3 – 28–14%). According to the content of coloring oxides (TiO_2+Fe_2O_3 is 8.85%), clay belongs to clays with a high content of coloring oxides. Clay is characterized by a low content of carbonate inclusions, with a low content of water-soluble salts (1.05 mg·eq per 100 g of clay). The content of free quartz is 25%, which allows to attribute the studied clay to clays with an average content of free quartz (10–25%). The predominant size of quartz particles is from 0.5 to 0.063 mm.

Table 1. The chemical composition of clay.

Oxide	SiO_2	Al_2O_3	CaO	Fe_2O_3	MgO	Na_2O	K_2O	TiO_2	LOI
Content in natural clay (%)	61.15	14.80	1.00	7.12	1.62	1.10	1.90	0.84	10.12
Content in calcined matter (%)	68.06	16.47	1.11	7.92	1.80	1.22	2.11	0.93	–

The results of the differential thermal analysis are presented in Figure 1. On the DTA curve, endothermic effects are observed at low temperatures at 136°C, while the loss on ignition is 4.3%. The endothermic effect at 153.4 °C is associated with the removal of interlayer molecular water from illite. A slight bend at 180 °C against the background of the main endothermic effect (on the DTG curve) associated with the removal of water is characteristic of montmorillonite with Ca- and

Mg-interlayer hydrated exchange cations. The next endothermic effect at 394 °C is associated with the release of constitutional water and partial restructuring of the illite structure, while the mass loss is 0.8%. The endothermic effect at 560 °C is associated with the dehydration of kaolinite and the formation of an amorphized metakaolinite phase. Loss on ignition is 4.5%. The exothermic effect at 911°C is associated with the crystallization of mullite. The total loss on ignition by heating the clay is 10.3%.

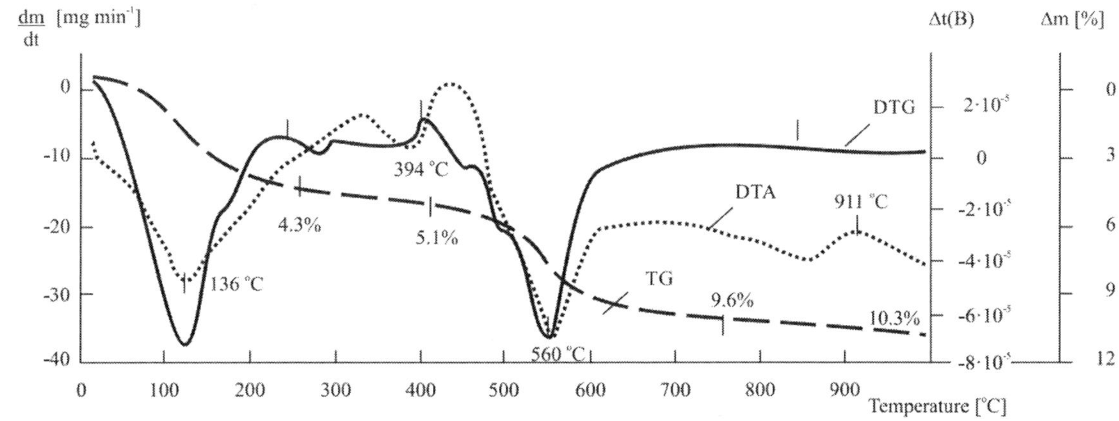

Figure 1. DTA results of clay.

The content of coarse fractions in the clay (particles larger than 0.5 mm) is 0.66%. These are raw materials with a low content of coarse inclusions (less than 1%).

By the size of the predominant coarse inclusions, clay refers to raw materials with small inclusions.

In terms of material composition, inclusions in clay are mainly pyrolusitic with a low content of quartz particles and rare inclusions of limonite.

Fine fractions in clay (Table 2) were determined by the sedimentation method according to [19].

Table 2. The distribution of fine fractions in clay.

Particle size (mm)	> 0.06	0.06–0.01	0.01–0.005	0.005–0.001	< 0.001
Fraction content (%)	8.53	16.39	7.72	13.68	53.68

The total content in the clay of particles less than 10 microns in size is 75.08%, which makes it possible to attribute it to medium-sized raw materials.

The plasticity of clay according to the method of Vasiliev is 22.1. Humidity at the lower yield strength is 35.9%, humidity at the limit of rolling into a tow is 13.8%. The plasticity number of white clay is 22.1. Clay is a medium plastic raw material.

The sensitivity coefficient of clay to drying by the method of Z. A. Nosova is 1.26. This is a medium drying clay. Clay air shrinkage is 7.9%, i.e. clay is medium drying.

To determine the technological properties of clay from the mass of optimal molding moisture content (W = 19%), samples were formed – tiles in size 50×50×10 mm and cubes in size 50×50×50 mm. Samples were dried and calcined at a temperature of 900, 950, 1050°C. Data on the amount of shrinkage, water absorption, porosity, density and mechanical compressive strength are shown in Table 3.

At a firing temperature of 900 and 950°C, the clay has a water absorption of 7.1–7.3%. At a firing temperature of 1050°C, the water absorption is only 2%, but the samples swell. Thus, blue clay is a non-sintering raw material.

At all firing temperatures, clay samples have a significant black core, which is associated with the dispersion of clay (particle content less than 1 μm in size is 53.68%) and the presence in the clay of a high content of iron oxides and organic substances. The dispersion of clay contributes to the formation of an impermeable crust on the surface of the samples during firing.

Table 3. Sintering degree and mechanical compressive strength of blue clay.

Firing temperature (°C)	Air shrinkage (%)	Fire shrinkage (%)	Total shrinkage (%)	Water absorption (%)	Open porosity (%)	Apparent density (kg·m^{-3})	Compressive strength (MPa)
900	7.9	1.1	9.0	5.8	11.0	1910	41.9
	7.5	0.8	8.3	5.3	9.7	1840	34.6
	7.4	0.5	7.9	7.1	13.4	1900	54.3
	7.9	1.5	9.4	8.4	16.3	1930	52.2
	7.7	0.8	8.5	8.7	16.7	1920	–
950	7.8	0.3	8.1	7.8	14.8	1.90	30.5
	8.1	0.6	8.7	7.3	13.9	1.90	37.4
	7.7	0.8	8.5	7.2	13.7	1.89	31.9
	8.1	0.4	8.5	7.2	13.6	1.90	–
	8.1	0.7	8.8	7.2	13.5	1.88	–
1050	8.1	3.5	11.6	2.1	4.2	1.95	
	8.1	3.6	11.7	2.4	4.7	1.92	swollen
	8.2	3.2	11.4	2.5	4.7	1.88	samples
	7.7	3.3	11.0	1.4	2.5	1.83	
	7.9	3.7	11.6	1.5	2.7	1.77	

Access of oxygen to the inner layers is limited, while a reducing medium appears inside the samples, which leads to the transition of Fe^{3+} to Fe^{2+}. When dispersed organic impurities burn out, sooty carbon forms, which settles in the pores of the products. After drying of water-saturated images of calcined clay, part of the soot goes to the surface of the products, painting them gray. The low gas permeability of the samples during firing also leads to the destruction of some samples, which makes it impossible to obtain not only face, but also ordinary building bricks.

To determine the suitability of clay in the production of expanded clay, a series of samples were made by plastic molding with a molding moisture content of 18–19%. As intumescent component used technical waste oil in an amount of 0.5%. Sample cylinders were molded with a size of 15×15 mm. Dried at room temperature during the day. Heat treatment was carried out at 200°C and fired at a temperature of 1050, 1090, 1130, 1170°C. Samples during firing were installed in a preheated oven to a predetermined temperature. It was kept for 7 min and sharply cooled in air. The average density of the obtained granules and the coefficients of expansion of the mass are given in Table 4.

Table 4. Expanded clay density and expansion coefficient.

Firing temperature (°C)	1050	1090	1130	1170
Average density of granules (kg·m^{-3})	1280	800	630	560
Expansion coefficient	1,51	2,31	2,85	3,83

The temperature range of expansion is 1050–1170°C with the introduction of technical waste oil. The minimum density of expanded clay samples was obtained by firing at a temperature of 1170°C. Thus, clay is suitable for producing artificial porous aggregates for concrete [20–22]. With the introduction of waste technical oil, the expansion coefficient is up to 3.83.

4. Conclusion

The presented sample of clay raw materials by mineral composition refers to kaolinite clays with mixed-layer formations in the form of illite and ferruginous montmorillonite with Na-, Ca- and Mg-interlayer hydrated exchange cations. It is characterized by an average content of free SiO_2 (10–25%). It is medium plastic, with a low content of coarse inclusions and refers to medium-dispersed raw materials. Clay is medium-drying, medium-sensitive to drying, non-sintering. The mechanical strength during firing of sample cubes at a temperature of 900°C is 27.5 MPa. When firing samples in the temperature range 900–1050°C, a significant black core is formed. At a firing temperature of 1050°C, swelling of the samples occurs. Considering the properties of this clay, it cannot be recommended for obtaining face and ordinary building bricks. However, the studied clay can be used for the production of expanded clay. The introduction of 0.5% of spent technical oil allows to get the expansion coefficient of clay 3.83. The temperature range of expansion is 1050–1170°C, while it is possible to obtain expanded clay with a density of granules of 560 $kg \cdot m^{-3}$.

References

[1] Moreno-Maroto J M, Rodrigiuez M U and Ceacero C J C 2020 *Construction and Building Materials* **237** 117583
[2] Galán-Arboledas R-J, García A M and Bueno S 2016 *Key Engineering Materials* **663** 105–114
[3] Galán-Arboledas R-J and Bueno S 2016 *Key Engineering Materials* **663** 62–71
[4] Harrati A, Manni A and El Bouari A 2020 *Materials today: proceedings* **04** 344
[5] Voišnienė V, Kizinievič O, Kizinievič V 2019 *IOP Conf. Series: Materials Science and Engineering* **660** 012065
[6] Murray H H 2007 *Applied Clay Mineralogy* (Amsterdam: Elsevier) p 180
[7] Barton C D and Karathanasis A D 2002 *Encyclopedia of Soil Science* **74** (New-York: Marcel Dekker)
[8] Tushar Kanti Sen 2017 *Clay Minerals: Properties, Occurrence and Uses* (New-York: Nova Science Publushers, Inc.)
[9] GOST 2642.2-2014 *Refractories and refractory raw materials. Determination of loss on ignition*
[10] GOST 2642.3-2014 *Refractories and refractory raw materials. Methods for determination of silicon (IV) dioxide*
[11] GOST 2642.4-2016 *Refractories and refractory raw materials. Methods for determination of aluminium oxide*
[12] GOST 2642.5-2016 *Refractories and refractory raw materials. Methods for determination of iron (III) oxide*
[13] GOST 2642.6-2017 *Refractories and refractory raw materials. Methods for determination of titanium (IV) oxide*
[14] GOST 2642.7-2017 *Refractories and refractory raw materials. Methods for determination of calcium oxide*
[15] GOST 2642.8-2017 *Refractories and refractory raw materials. Methods for determination of magnesium oxide*
[16] GOST 2642.9-2017 *Refractories and refractory raw materials. Methods for determination of chrome (III) oxide*
[17] GOST 2642.10-2018 *Refractories and refractory raw materials. Methods for determination of phosphorus (V) oxide*
[18] GOST 2642.11-2018 *Refractories and refractory raw materials. Method for determination of potassium and sodium oxides*
[19] GOST 21216-2014 *Clay raw materials. Test methods*
[20] Sapozhnikova M A, Pavlova I A 2020 *Solid State Phenomena* **299** 3–7
[21] Kashcheev I D, Zemlyanoi K G, Shul'meister T M 2016 *Glass and Ceramics* **72** 341–344
[22] Pavlova I A 2017 *Glass and Ceramics* **74** 104–106

Technogenic Raw Materials in High-Alumina Chamotte Production

I A Pavlova, A A Getman and E P Farafontova

Institute of new materials and technologies, Ural Federal University named after the First President of Russia B. N. Yeltsin, 19, Mira Str., Ekaterinburg 620002, Russia

E-mail: i.a.pavlova@urfu.ru

Abstract. The paper presents studies on the production of a high-alumina aggregate with an Al_2O_3 content of 75% based on fine-grained corundum dust generated by grinding fused corundum. Finely dispersed corundum powder is a by-product and due to the low Al_2O_3 content (93–95%) is not used in further production. It is proposed to obtain high-alumina aggregate without joint grinding of the components. Aggregate was obtained by pressing through a "false grain" from a mixture of kaolin and corundum dust in a certain ratio. The samples firing was carried out at a temperature of 1700°C. The phase composition of the obtained aggregate is represented by 46–51% corundum, 45–49% mullite. The water absorption of such chamotte is 14–16%, open porosity is 33–35%, apparent density is 2230–2290 kg·m^{-3}. Now, it was not possible to obtain a high-density filler grade ZML and ZMK that meets the requirements of GOST 23037–78. The production flow chart of high-alumina chamotte based on kaolin and corundum dust should include a joint fine grinding of the components, or a separate grinding of corundum dust until it passes completely through the 0.063 mm mesh.

1. Introduction

A promising direction in the development of the production of aluminosilicate refractories is the production of materials with increased operational characteristics [1]. Despite the very wide spread of chamotte in metallurgy, its use is being reduced. It is replaced by new refractory materials with higher service properties, the use of which is due to the continuous intensification of the thermal service of furnaces [2–7]. The growth of their application is limited by a relatively high cost.

Aluminosilicate materials have aluminum and silicon oxides as the refractory base. The structure, properties and application field of these materials depend on the content of Al_2O_3 and SiO_2. On this basis, they are divided into semi-acidic (15–30% Al_2O_3), chamotte (30–46% Al_2O_3) and high-alumina (over 46% Al_2O_3). In turn, high-alumina refractories are divided into mullite-siliceous (45–62% Al_2O_3), mullite (62–72% Al_2O_3), mullite-corundum (72–90% Al_2O_3) and corundum refractories (over 90% Al_2O_3) [8–10].

Mullite refractories include materials with 62–72% Al_2O_3 content. The phase composition of mullite refractories is mainly represented by mullite, also corundum and glass phase (6–12 wt.%). Mullite refractories are usually obtained in two stages: first, mullite synthesis, and then producing products from it. As starting materials for the mullite synthesis, mainly technical alumina, pure refractory clays, kaolins and quartz are used [11–13]. To transfer the silica formed during the silicates firing into the mullite, technical alumina is introduced in the mass composition. Mullite formed during the interaction of silica with technical alumina is called secondary. Production technology in most cases is determined by the properties of raw materials. The main troubles in the production of high-alumina products are associated with a large shrinkage of technical alumina during firing as a result of the conversion of γ-alumina to α-alumina, with low sintering ability of α-alumina and with the difficulty of obtaining a specified structure. To reduce shrinkage during the firing of products, a chamotte flow chart is used, i.e. part of the material is preliminarily fired on chamotte. The main difference between high-alumina refractory technology and multi-chamotte technology is the production of high-alumina aggregate, which is prepared in various ways. Upon receipt of the briquette from technical alumina and refractory clay, alumina is preliminary ground to a particle size of 3–5 μm. Briquettes are prepared by plastic or semi-dry molding and fired at 1700°C. Chamotte with water absorption of 1–2% is obtained by firing the briquette in a rotary kiln. Then a granular product is prepared by crushing, coarse grinding and fine grinding.

When fused alumina is crushed, dust is formed [14]. It is captured by electrostatic precipitators and stored in the enterprise [15]. This by-product is substandard material to produce corundum refractories, as there is an insufficient content of Al_2O_3 (93–95%). In addition, electrocorundum is used as an abrasive material [16–17].

To produce chamotte refractories, kaolins with a content of 30–39% Al_2O_3 are used. Due to the supplanting of chamotte refractories by high-alumina refractory materials, it is important to produce materials in demand with high content of Al_2O_3 based on kaolins. Thus, the aim of this project is to develop a technology for the production of high-alumina chamotte based on enriched kaolin and corundum dust generated during crushing of electrocorundum.

2. Experimental procedure

Phase composition was determined by an X-ray phase method in a Miniflex 600 (Cu Kα-radiation, λ = 1.541862 Å, recording interval 3.00 – 90.00 deg, scanning step 0.02 deg), Rigaku, Carl Zeiss, Japan, with MiniFlex Guidance and PDXL Basic data treatment package program control and data collection. Identification of diffraction maxima was carried out using a JSPDS data bank. The chemical composition of quartz sand was determined by emission spectral analysis with inductively coupled plasma in an Optima 4300 DV (Perkin Elmer, USA) optical emission spectrometer.

The water absorption, open and total porosity and apparent density were determined in accordance with [18]. The grain size composition of finely milled powders was determined by sedimentation analysis in a SLAD 2201 (Shimadzu Corp.) particle laser diffraction analyzer according to [19]. Sieve analysis was carried out in accordance with [20]. The plasticity number of clay raw materials was determined according to [21]. The real density of the materials was determined by the pycnometric method according to [22]. The specific surface of the materials was determined according to [23].

3. Results and discussion

To solve the problem of obtaining high alumina material in this project, the introduction of finely dispersed corundum dust in the process of enrichment of kaolins without joint grinding of raw materials is considered. Thus, the modernization of the existing production of chamotte will consist in the installation of a thermal unit providing a firing temperature of chamotte 1700°C.

Enriched kaolin of Kyshtymskoye deposit and finely dispersed corundum dust were used as raw materials. The chemical composition is presented in Table 1. Kaolin is the basic raw material. The mineral composition of kaolin is as follows: muscovite (up to 5 wt.%), quartz (up to 2 wt.%), kaolinite (95–98 wt.%). The content of Al_2O_3 in corundum dust is 93–95 wt.%. According to the XRD analysis corundum content is 90%. As impurities, thialyte and rutile are present in an insignificant amount, 1.2 and 0.16 wt.% respectively. According to the results of chemical and X-ray phase analysis, it is possible to determine the content of the glass phase in corundum dust, it is about 10%.

IOP Conf. Series: Materials Science and Engineering **969** (2020) 012031 doi:10.1088/1757-899X/969/1/012031

Table 1. The chemical composition of materials.

Material	SiO_2	Al_2O_3	CaO	Fe_2O_3	MgO	R_2O	TiO_2	LOI
Kaolin (%)	43.66	40.59	0.16	0.98	0.16	0.56	0.62	13.28
Fine substandard corundum powder (%)	0.05-0.3	93.0-95.0	0.2-0.24	0.13-0.5	0.44-0.62	0.3-0.55	2.2-5.0	–

The particle size distribution of kaolin is shown in Figure 1. Particles with a size of less than 2 μm in kaolin is 8–15%. Particles smaller than 63 μm in size more than 95%. Particles 2–55 μm in size are distributed almost uniformly and make up about 85%.

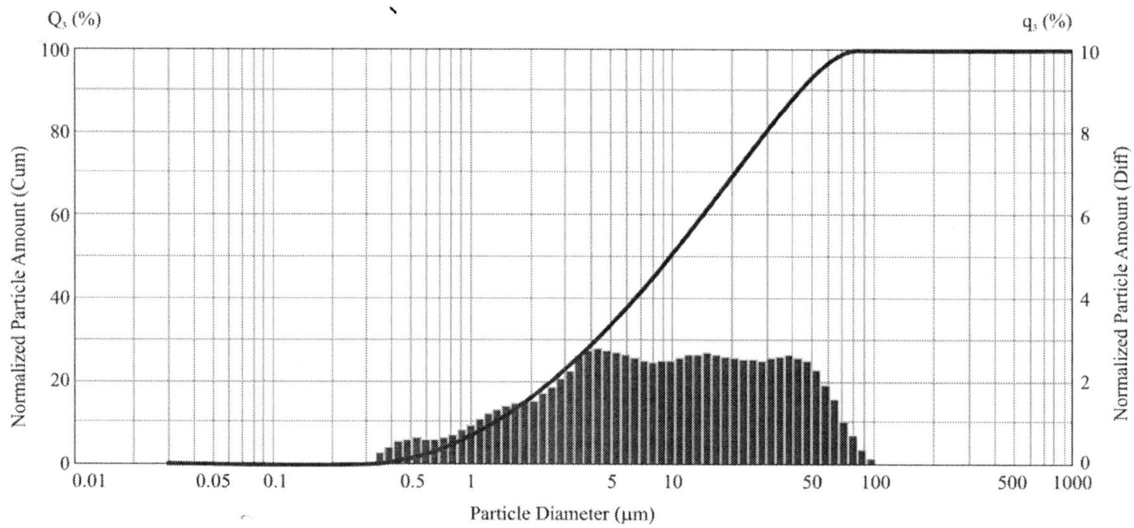

Figure 1. The dispersed composition of kaolin.

The granulometric composition of Kyshtymskoe kaolin, determined in accordance with [21], is given in Table 2. According to the particle content of less than 1 μm (43.2%), kaolin in accordance with [24] is a medium dispersed raw material.

Table 2. The granulometric composition of Kyshtymskoe kaolin.

Particle size (mm)	> 0.06	0.06–0.01	0.01–0.005	0.005–0.001	< 0.001
Fraction content (%)	3.6	16.9	17.5	31.7	30.3

The specific surface area of kaolin in the delivery state is 16035 $cm^2 \cdot g^{-1}$, real density is 2.65 $g \cdot cm^{-3}$. The plasticity number of kaolin is 9.1, kaolin in accordance with [24] refers to moderately soft raw materials.

The particle size distribution of finely dispersed corundum is presented in Table 3. The main fractions of the powder are particles 0.16–0.063 mm in size and less than 0.063 mm.

Table 3. The distribution of fine fractions in corundum.

Particle size (mm)	> 0.5	0.5–0.16	0.16–0.063	< 0.063
Fraction content (%)	3.07	4.52	51.45	40.96

The composition of the mixture to obtain high-alumina chamotte was calculated in view of 75% Al_2O_3 content in chamotte. Plastic cakes of kaolin with a moisture content of 20–25%, obtained after the filter press, were carefully mixed with corundum dust. The plastic mass was dried at 105°C. It was ground to a grain size of less than 3 mm. To reduce dust during the chamotte firing, the briquette must have a transport strength of at least 4 MPa. To determine the briquette forming parameters (humidity and pressing pressure) in order to obtain a solid raw briquette, the obtained powder was moistened with a technical lignosulfonate solution of density 1.05 g·cm^{-3} to a moisture content of 5, 7, 9%. Then, cylinder samples of 20×20 mm size were molded at a specific pressure of 5, 10, 15, 20, 25 MPa. After drying at a temperature of 105°C, the compressive strength was determined. Figure 2 shows the dependences of the briquette strength on humidity and pressing pressure.

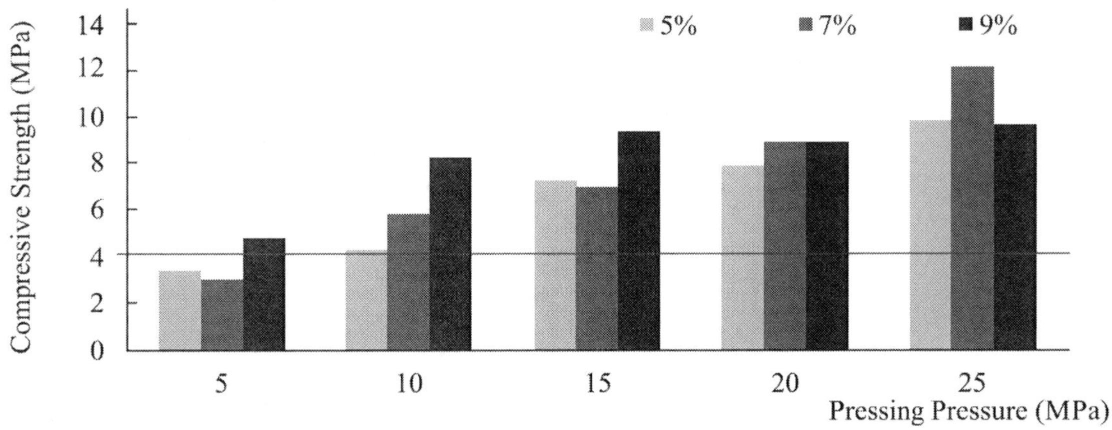

Figure 2. Dependence of compressive strength of unfired briquette on specific pressing pressure and relative humidity.

Strength of 4 MPa is possessed by samples formed at a moisture content of 9% at a pressure of 5 to 25 MPa [26–27]. Samples formed at lower humidity at a compression pressure of 5 MPa will not be strong enough and will crumble during firing, increasing dust extraction. At a pressing pressure from 10 to 25 MPa with molding humidity of 5, 7, 9%, briquettes with a strength of more than 4 MPa were obtained. The most favorable conditions for briquette molding are the lowest humidity and pressing pressure. Redundant moisture causes the formation of increased porosity after firing. The use of high values of molding pressure suggests the presence of expensive equipment, and the use of low ones - insufficient contact between the particles of corundum and kaolin. Thus, samples formed at a humidity of 5 and 7% at a molding pressure of 10 MPa were fired at a temperature of 1700°C. The water absorption, porosity, apparent density, and strength of fired samples were determined. The results are presented in Table 4.

Table 4. Post-firing properties of the samples.

Molding humidity, %	Total shrinkage (%)	Water absorption (%)	Open porosity (%)	Apparent density (kg·m^{-3})	Compressive strength (MPa)
5	1.0	15.6 ± 0.7	34.8 ± 1.4	2230 ± 30	30.5 ± 5
7	1.1	14.5 ± 0.8	33.2 ± 1.3	2290 ± 50	39.3 ± 6

Complete shrinkage is 1% at molding humidity of 5%. With an increase in molding humidity to 7%, it leads to an increase in shrinkage of 1.1%. Water absorption and porosity of samples molded at a humidity of 7% is less than that of samples molded at a humidity of 5%. This effect can be explained by the fact that a more moistened mass provides better molding conditions and a denser contact between the particles of kaolin and corundum. In addition, the density and strength of samples molded at 7% are greater than those molded at 5%.

The XRD results are shown in Figure 3. The following phases were found in the samples of high-alumina chamotte are corundum, mullite, thialite, rutile, quartz. The presence of quartz can be explained by rubbing it in the process of sample preparation on the XRD. Samples were ground in an agate mortar. The high hardness of corundum causes grinding of quartz from a mortar and pestle. The absence of cristobalite indicates its complete interaction with corundum with the formation of secondary mullite.

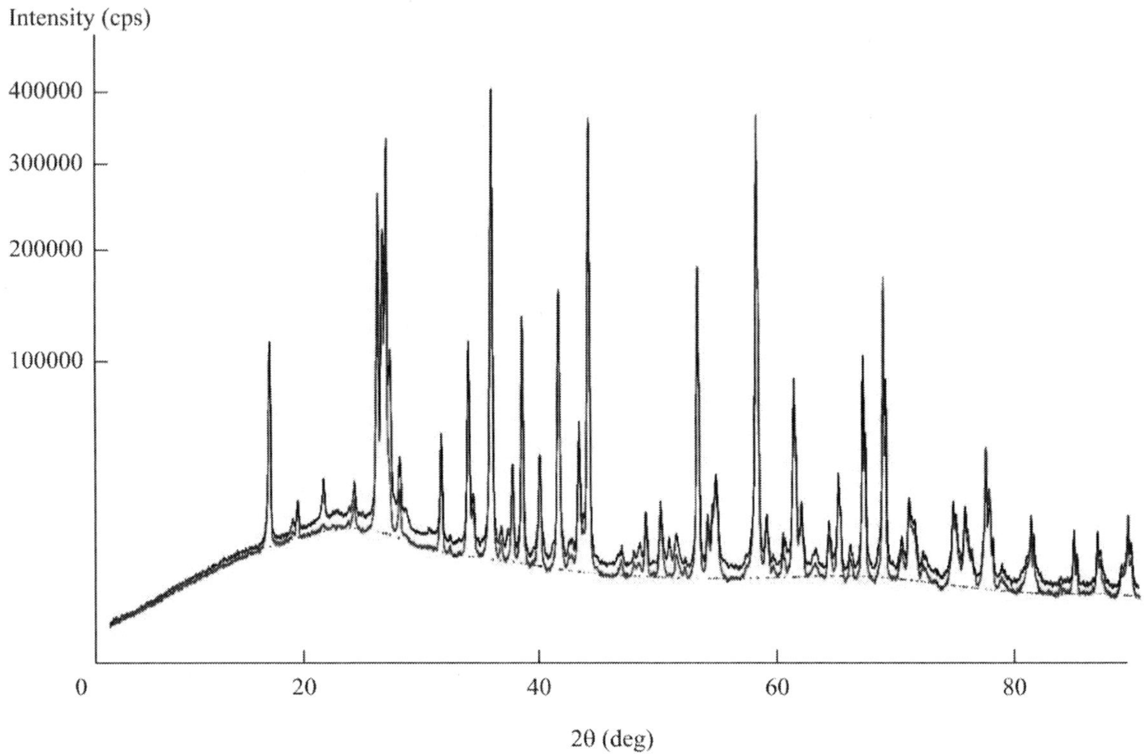

Figure 3. XRD results of high-alumina chamotte.

The results of X-ray phase analysis are presented in Table 5.

Table 5. Phase composition of high-alumina chamotte.

Molding humidity, %	Corundum	Mullite	Thialite	Rutile
5	46	49	3	2
7	51	45	3	1

The results of a semi-quantitative analysis according to the results of XRD are within the error of the method. The composition of high-alumina chamotte is as follows: 46–51% corundum, 45–49% mullite, rutile ~ 1%, thialite ~ 3%.

4. Conclusion

As a result of the research, samples of high-alumina aggregate with an Al_2O_3 content of 75% were obtained on the basis of finely dispersed corundum dust generated by grinding fused corundum. The phase composition of chamotte is represented by 46–51% corundum, 45–49% mullite. The water absorption of such chamotte is 14–16%, open porosity 33–35%, apparent density 2230–2290 $kg \cdot m^{-3}$. In accordance with [25], aluminosilicate aggregates of the ZML and ZMK brands should have a water absorption of not more than 3%. Thus, as a result of the studies, a high-density aggregate of the ZML and ZMK brands was not obtained in accordance with [25]. Further research should be aimed at reducing water absorption, for example, by using a joint grinding of kaolin and corundum, or by replacing kaolin with refractory clay. The flow chart for the production of high-alumina chamotte should include joint

IOP Conf. Series: Materials Science and Engineering **969** (2020) 012031 doi:10.1088/1757-899X/969/1/012031

fine grinding of corundum dust with a plastic component, or separate grinding of corundum dust until it passes completely through a 0.063 mm mesh, followed by mixing with a plastic component.

References

[1] Pivinskii Yu E and Dyakin P V 2011 *Refractories and Industrial Ceramics* **52** 264–271
[2] Andrews A, Adam J and Gawu K Y 2013 *Ceram. Int.* **39** 779–783
[3] Weinberg A V, Varona C and Chaucherie X 2018 *Refractories WorldForum* **10** 57–62
[4] Zubashchenko R V 2011 *Refractories and Industrial Ceramics* **52** 143–145
[5] Shun Yao, Shengli Wu and Heng Zhou 2019 *Ironmaking & Steelmaking* **47** 1–8
[6] Chudíková D, Hirjak R and Mišaneková V 2018 *Interceram* **67** 50–57
[7] Primachenko V V, Martynenko V V and Ustichenko V A 2007 *Refractories and Industrial Ceramics* **48** 421–424
[8] Sengupta P 2020 *Refractories for the Cement Industry* (Switzerland: Springer Nature) p 37
[9] Biswas S and Sarkar D 2020 *Introduction to Refractories for Iron- and Steelmaking* (Switzerland: Springer Nature) p 30
[10] Soltan A M, Pöllmann H and Kaden R 2015 *J. Eur. Ceram. Soc.* **8** 18–38
[11] Skurikhin V V and Ermakov I N 2004 *Glass and Ceramics* **61** 346–351
[12] Prutskov D V, Troyan V D and Malyshev I P 2000 *Refractories and Industrial Ceramics* **41** 56–59
[13] Sharapova V V 2014 *Refractories and Industrial Ceramics* **55** 45–48
[14] Lang K, Tvrdík L and Kovář P 2019 *Refractories Manual* **68** 40–45
[15] Tukhareli V D, Pushkarskaya O Y and Tukhareli A V 2018 *Solid State Phenomena* **284** 1030–1035
[16] Shumyacher V M, Danilova E S and Pushkarev I O 2011 *Russian Engineering Research* **31** 901–902
[17] Mikhailov G G, Morozova A G and Bamburov V G 2018 *Materials Science Forum* **946** 192–198
[18] GOST 2409–2014 *Refractories. Method for determination of bulk density, apparent and true porosity, water absorption*
[19] GOST 8.777–2011 *State system for providing measurement unification. Dispersed composition of aerosols and suspensions. Determination of particle size by diffraction laser radiation*
[20] GOST 27707–2007 *Unshaped refractories. Methods for determination of grain composition*
[21] GOST 21216–2014 *Clay raw materials. Test methods*
[22] GOST R 56300–2014 *Refractories and refractory raw materials. Methods of true density determination*
[23] GOST 21043–87 *Iron ores and concentrates. Method for determination of external specific surface*
[24] GOST 9169–75 *Clayish materials for ceramic industry. Classification*
[25] GOST 23037–78 *Unmoulded refractories. Aggregates for concrete articles, masses, mixtures, coatings and mortars. Specifications*
[26] Kashcheev I D and Pavlova I A 2006 *Glass and Ceramics* **63** 86–88
[27] Kashcheev I D, Glyzina A É and Finkel'shtein A B 2019 *Refractories and Industrial Ceramics* **60** 362–364

Features of Shear Banding in Grains with Orientation {111}<110> in the Fe–3%Si Alloy

M L Lobanov[1, 2], P L Reznik[1] and A A Redikultsev[1]

[1]Ural Federal University named after the first President of Russia B. N. Yeltsin, 19, Mira st., Ekaterinburg, 620002, Russia
[2]M. N. Mikheev Institute of Metal Physics of Ural Branch of the Russian Academy of Sciences, 18, S. Kovalevskaya st., Ekaterinburg, 620990, Russia

E-mail: m.l.lobanov@urfu.ru

Abstract. EBSD method were used to study the texture of BCC crystals {111}<110> Fe-3%Si. It is shown that in the state after cold rolling shear bands with habit plane at a slope of ~ 20–40° to the RD were formed as a result of 70% deformation. Theoretical analysis was performed and experimental data were analyzed which showed that the matrix reorientation into orientation {110}<001> occurs when grain stripping with orientation {111}<110>. Such effects are observed either due to rotation around the <100> axis or due to rotation around the <110> axis. By means of dislocation sliding, most of the SBs blocks are reoriented to orientations close to {111}<112>.

1. Introduction

Transformations, which determine the structure and texture of grain electrotechnical steel or Fe – 3% Si alloy (GOES) are caused by formation of sharp texture {110} <001> at one of the final stages of steel processing [1, 2], are not fully understood. One of the most important tasks in the process of texture formation in GOES is to create orientation {110} <001>. This can be achieved due to operations preceding secondary recrystallization. After significant cold deformation, the areas of orientation {110}<001> are formed in shear bands (SBs). Shift bands appear in crystallites {111}<112>, such areas then develop at primary recrystallization [3–7].

From the research point of view, the most mysterious element of the deformation mesostructure is SB. To date, a large amount of experimental data [8–17] on their morphology, the correlation between the matrix and crystallite orientation in the stripes, and their location in the matrix have been accumulated. Theoretical views of their formation are presented in detail in [18–23]. However, not all aspects of this issue are considered in them.

The main prerequisite for the formation of shear bands in crystals, i.e., loss of stability of homogeneous deformation, is the "hardness" of its orientation, characterized by a large Taylor coefficient [14]. The Dilmore theory assumes that strong localization of deformation in SB promotes macroscopic deformation of a crystal even without changing its orientation as a whole [18]. When rolling metal, regardless of the type of its crystal lattice, the SB should be located at angles close to ± 35 ° to the rolling plane. Formation of SB has another important aspect that connects the reorientation of the crystal lattice inside them with the geometric softening inside the plasticity model of Taylor [24]. SBs are treated as areas where the shift deformation is localized. However, little is known about their internal structure and formation during cold rolling. Thus, it is important to study the basic laws of shear band formation to understand texture formation in GOES.

However, little is known about the internal structure of SPs when they are formed during cold rolling. Hence, it is necessary to study the basic laws of shear band formation. This approach can be useful for understanding texture formation in GOES.

Observation of texture evolution during secondary recrystallization shows that orientation {111}<112> is the main one, which is "absorbed" in the first place. This process is also related to SBs crystallography. Establishing this connection is also an important part of the problem to be solved. Orientation {111}<112> is dominant in the surface layers - the result of primary recrystallization GOES. It is likely that in the dominant after deformation orientations {111}<110>, {112}<110> SBs are the main "candidate" for the sources of grain formation with the orientation {111}<112>.

Presentation of results of research of Fe-3% Si alloy in the state of cold deformation, in which shear bands were formed in grains with orientation {111}<110> is the purpose of this paper.

2. Materials and methods

Samples were taken from a 0.65 mm thick Fe – 3% Si alloy in cold rolled condition. The total deformation was approximately 70%. The study of microstructure and texture was carried out with a scanning step of 0.1 and 0.025 μm using the JEOL JSM6490LV raster electron microscope at an accelerating voltage of 20 kV. The estimated error limit of the crystal lattice orientation was on average ± 0.6°, but did not exceed ± 1°. Misorientations from 2 to 15° were marked as small angular boundaries between local volumes. High angles were detected when the misorientation was ≥ 15°.

A coordinate system was adopted with the following axes: cold rolling direction (RD), normal to its plane (ND), and the direction perpendicular to both (TD).

Using the matrix of rotation *R*, which was obtained from the ratio $F_1=RF_2$, the misorientations between the crystal lattices of different orientations were determined. Where the matrix *F*, calculated from experimentally determined EBSD Euler angles, describes the orientation of the ferrite lattice. In the analysis to determine the closeness to the special misorientations of the CSL, the obtained matrices of rotation *R* were compared with the matrices corresponding to certain special misorientations from [25], by the method used in [26, 27]:

$$H_{ij} = C_j R^{-1} C_i T, \tag{1}$$

where matrix T_1 is used as *T*, describing the variant of special misorientation of CSL from [25], $C(C_J)$ – matrices of type $\begin{pmatrix} 1 & 0 & 0 \\ 0 & 1 & 0 \\ 0 & 0 & 1 \end{pmatrix}$ describing in general 24 symmetrical turns in a cubic lattice:

$$i(j)= 1, 2, \ldots, 24. \tag{2}$$

For each matrix of misorientation *H*, taking into account the full range of its symmetric variations, we calculated the values of angles Θ and pivot axes [*uvw*], using known mathematical formulas:

$$\Theta = \arccos\{[h_{11} + h_{22} + h_{33} - 1)/2\}, \tag{3}$$

$$u:v:w=[h_{12} - h_{21}]:[h_{23} - h_{32}]:[h_{31} - h_{13}], \tag{4}$$

where h_{ij} are elements of *H* matrix. In the obtained sets of versions of deviation parameter values (angle and pivot axis) related to one *H* matrix, only one variant with the lowest angle value Θ was chosen. Matrix *T* corresponding to the given version was accepted as the closest special misorientation of CSL, and corner Θ was accepted as a deviation of experimentally defined orientation from orientation of CSL.

3. Results and discussion

The different propensity of different grain orientations to the formation of shear bands was established. Using orientation analysis of the microstructure with a large pitch (figure 1, a, b, c, grain I) it is shown that the banding does not occur in grains with an orientation close to {001}<110>.

Grains with orientation close to {111}<112> (figure 1, a, d, grain II) and {111}<110>, {112}<110> (figure 1 a, b, c, grain III and IV) contain stripes. Only in some areas was it possible to determine the orientation within SBs. For example, in grain II, the orientation inside SB corresponds to {110}<001> (figure 1, d). This corresponds to the misorientation ∑9 (or ∑27) according to [21–23].

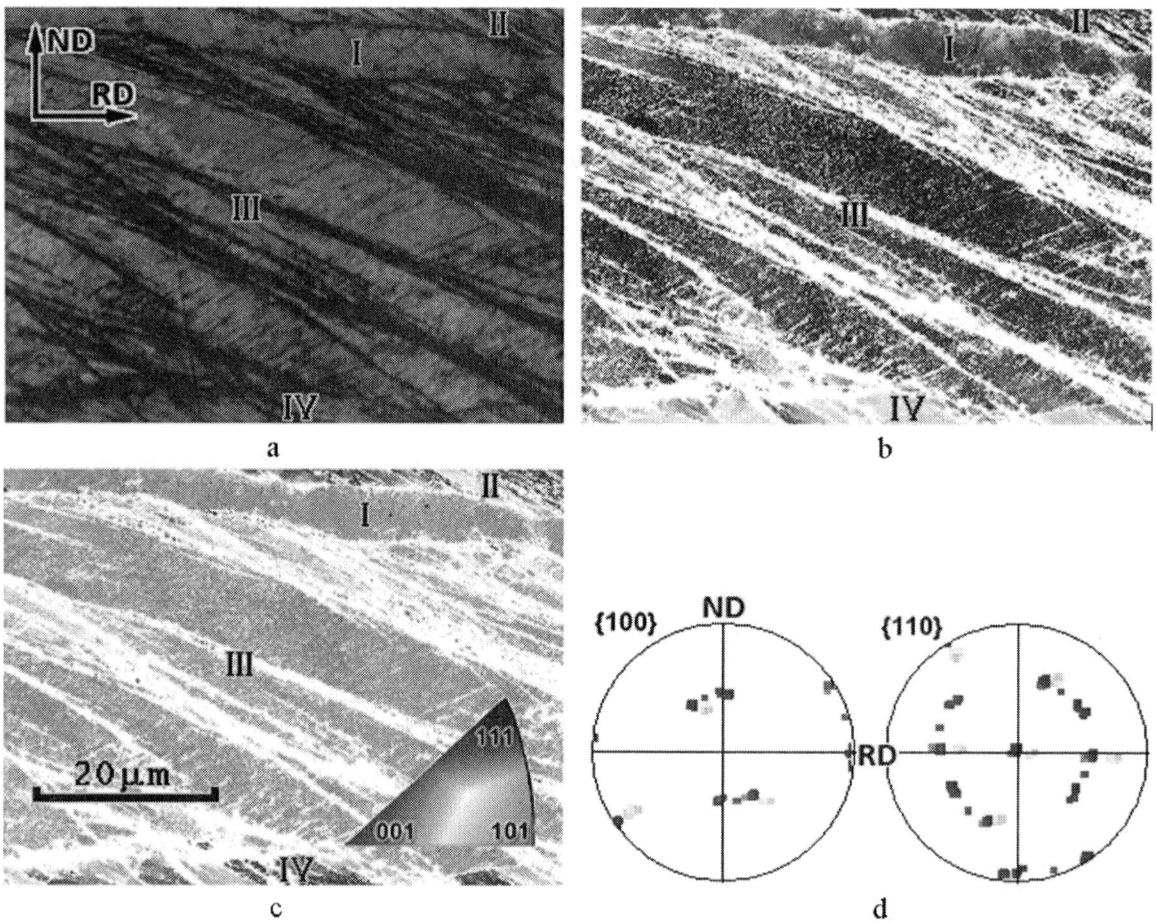

Figure 1. Microstructure of Fe-3%Si alloy polycrystal (4 grains: I, II, III, IV). Cold rolled state; deformation ε ~ 70%; EBSD: a – orientation map band contrast; b - orientation map from ND; c – orientation map from RD; d – DPF {100} and {110} of grain II.

The establishment of morphological and crystallographic features of SBs in grain with orientation close to {111}<110> (figure 2) was possible due to the increasing density of scanning points (step 0.025 µm). The fragment on figure 2 (a, b, c, d) contains three SBs characterized by close orientations in the central area, thus essentially differing in thickness (from 0.5 to 1.5 µm) and angles of a slope to a rolling plane ~ 20°, 35 and ~ 38°. Presumably, SBs were formed at different points in time, i.e. at different values of deformation. The strips have a block structure. The blocks are stretched along the plane of habitus. The length of the blocks reaches 2 µm. The number of blocks within the cross-section of wide SB varies from one to five. The misorientation between individual blocks can reach several tens of degrees. Central blocks of SBs have orientation close to {110}<001> with a significant magnitude of scattering (up to ~15° around the axis <uv0>). On both sides of wide SBs the orientations of peripheral blocks are close. The orientations of these blocks are related by rotation at angles of 15-25° around the <110> axis close to the TD direction concerning the matrix orientation.

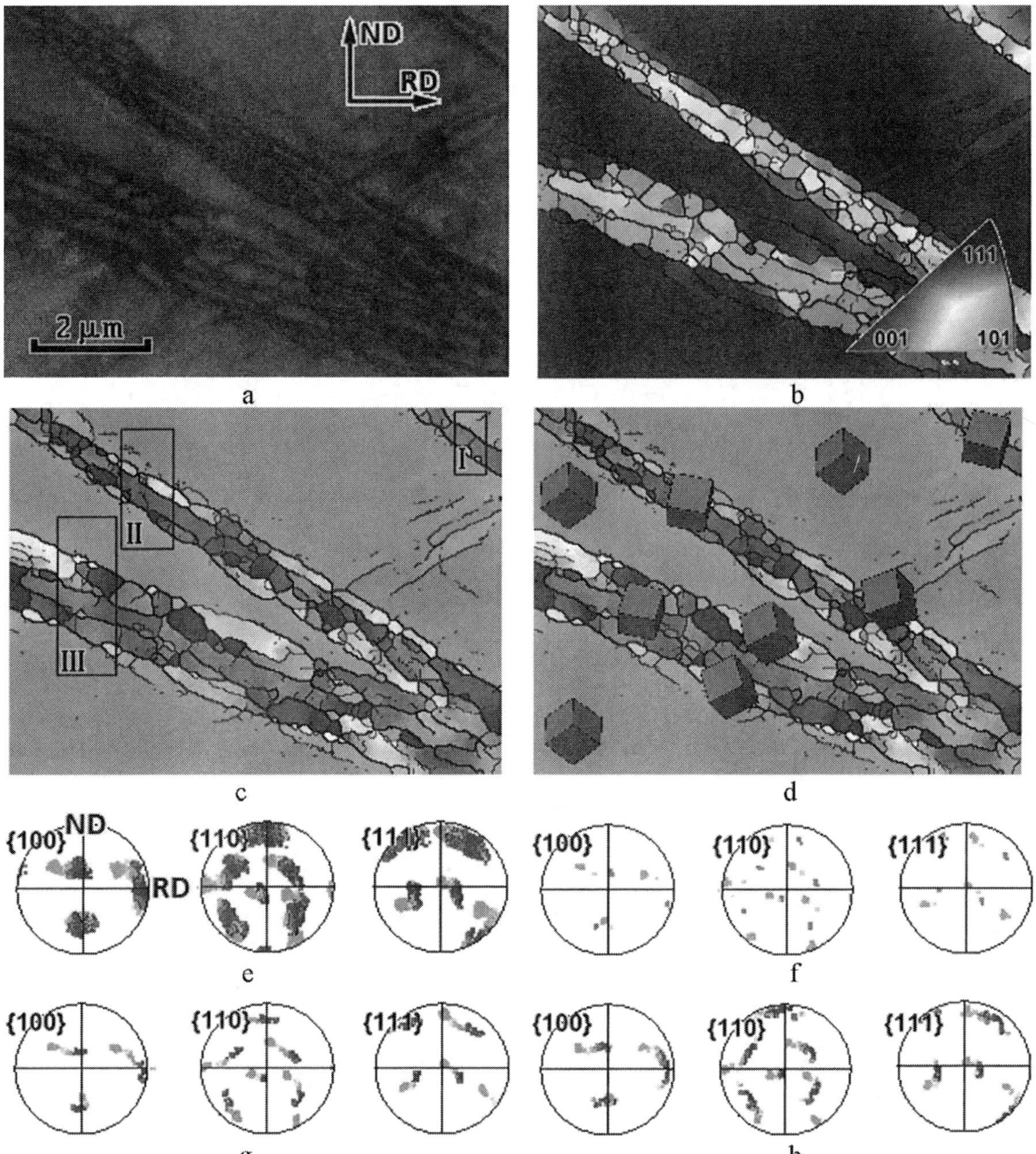

Figure 2. Microstructure and texture of Fe-3% Si single crystal {111}<110>. Cold rolled state; ε ~ 70%; EBSD: a – orientation map band contrast; b – orientation map from ND; c – orientation map from RD; d – orientation map from RD, indicating orientations of individual structural elements in the form of elementary crystalline cells; e-h – DPF {100}, {110}, {110} corresponding to the micro areas marked with rectangles in "c"; e – areas shown in "c"; f – I; g – II; h – III.

The vector analysis allows us to describe the misorientation between the central units of all SBs and the matrix orientation {111}<110> as CSL ∑15±(4-10°) (rotation angle 48.19° around the axis <210>), which does not coincide both with the misorientation of ∑9 (or ∑27) in deformed grains {111}<112> and with the misorientation of ∑5 (or ∑29a) in deformed grains {110}<110> [17], but in some sense is intermediate between them. It should also be noted that the correctness of the analysis is significantly affected by the orientation scattering, which continues due to the deformation of both the SBs and the matrix orientation and is reflected in the DPF (figure 2, e, f, g, h).

4. Summary

Thus, it can be assumed that when the grain is banding with the orientation {111}<110>, the matrix undergoes a sharp reorientation into the orientation {110}<001> either due to the rotation around the <100> axis (less likely) or due to the rotation around the <110> axis (more likely). During further deformation, most of the SBs blocks are reoriented by dislocation sliding into orientations close to {111}<112>, which play a significant role in subsequent recrystallization processes.

References

[1] Lobanov M L, Redikultsev A A and Rusakov G M 2011 Electrotechnical anisotropic steel. Part I. history of development *Metal Science and Heat Treatment* **53** 326–332
[2] Lobanov M L, Redikultsev A A and Rusakov G M 2011 Electrotechnical anisotropic steel. Part II. State-of-the-art *Metal Science and Heat Treatment* **53** 355–359
[3] Ushioda K and Hutchinson W B 1989 *ISIJ Int.* **29** 862–867
[4] Cruz-Gandarilla F, Penelle R and Mendoza-Leon H 2005 *Mater. Sci. Forum* **495–497** 483–488
[5] Dorner D, Zaefferer S and Raabe D 2007 *Acta Mater.* **55** 2519–2530
[6] Dorner D, Adachi Y, Tsuzaki K and Zaefferer S 2007 *Mater. Sci. Forum* **550** 485–490
[7] SunMi Shin, SamKyu Chang and de Cooman B C 2008 *ISIJ Int.* **48** 1788–1794
[8] Sokolov B K, Gubernatorov V V, Gervasyeva I V, Sbitnev A K and Vladimirov L R 1999 The Deformation and Shear Bands In The Fe-3%Si Alloy *Textures and Microstructures* **32** 21–39
[9] Paul H, Driver J H, Maurice C and Jasien′ski Z 2003 Shear band microtexture formation in twinned face centred cubic single crystals *Materials Science and Engineering. A* **359** 178–191
[10] Xue Q and Gray G T 2006 III Development of Adiabatic Shear Bands in Annealed 316L Stainless Steel: Part I. Correlation between Evolving Microstructure and Mechanical Behavior *Metallurgical and Materials Transactions A* **37A** 2436–2446
[11] Dorner D, Zaefferer S and Raabe D 2007 Retention of the Goss orientation between microbands during cold rolling of an Fe3%Si single crystal *Acta Mater.* **55** 2519–2530
[12] Dorner D, Adachi Y and Tsuzaki K 2007 Periodic crystal lattice rotation in microband groups in a bcc metal *Scripta Materialia* **57** 775–778
[13] Peirs J, Tirry W and Amin-Ahmadi 2012 Microstructure of adiabatic shear bands in Ti6Al4V *Materials Characterization* **75** 79–92
[14] Nguyen-Minh T, Sidor J J, Petrov R H and Kestens L A I 2012 Occurrence of shear bands in rotated Goss ({110}<110>) orientations of metals with bcc crystal structure *Scripta Materialia* **67** 935–938
[15] Lobanov M L, Rusakov G M, Redikul'tsev A A, Karabanalov M S and Lobanova L V 2011 Shear Bands in Fe–3% Si–0.5% Cu Alloy *Steel in Translation* **41** 559–564
[16] Rusakov G M, Lobanov M L and Redikul'tsev A A 2014 Lattice Reorientation in the Shear Bands of {112}⟨131⟩ Crystallites in an Fe–3%Si Alloy *Technical Physics* **59** 1242–1244
[17] Redikultsev A A, Rusakov G M and Lobanov M L 2018 The Possibility of Obtaining Electrical Steel with Texture {100}<001> *Solid State Phenomena* **284** 483–488
[18] Dillamore I L, Roberts J G and Bush A C 1979 Occurrence of shear bands in heavily rolled cubic metals *Mater. Sci.* **13** 73–77
[19] Hutchinson J W and Tvergaard V 1981 Shear band formation in plane strain *International Journal of Solids and Structures* **17** 451–470
[20] Mahesh S 2006 Deformation banding and shear banding in single crystals *Acta Materialia* **54** 4565–4574

[21] Rusakov G M, Lobanov M L, Redikultsev A A and Kagan I V 2009 Model of {110}<001> Texture Formation in Shear Bands during Cold Rolling of Fe-3 Pct Si Alloy *Metallurgical and materials transactions A* **40** 1023–1025

[22] Rusakov G M, Redikul'tsev A A, Kagan I V and Lobanov M L 2010 Mechanism of Formation of Shear Bands upon Cold Deformation of a Commercial Fe-3% Si Alloy *The Physics of Metals and Metallography* **109** 662–669

[23] Lobanov M L, Rusakov G M, Redikul'tsev A A and Lobanova L V 2013 Formation of Special Misorientations Related to Transition Bands in Structure of Deformed and Annealed Single Crystal (110)[001] of Fe-3% Si Alloy *The Physics of Metals and Metallography* **114** 27–32

[24] Tailor G I 1938 Plastic strain in metals *Journ. Inst. Met.* **62** 307–324

[25] Kaybyshev O A and Valiyev R Z 1987 *Grain boundaries and properties of metals* (Moscow: Metallurgiya) p 213

[26] Kabanova I G and Sagaradze V V 1999 Statistical analysis of mutual misorientations of austenite (martensite) crystals after martensitic $\gamma\rightarrow\alpha\rightarrow\gamma$ ($\alpha\rightarrow\gamma\rightarrow\alpha$) transformations *The Physics of Metals and Metallography* **88** 44–52

[27] Tereshchenko N A, Yakovleva I L, Kabanova I G and Mirzaev D A 2019 Special Misorientations in Low-Temperature Isothermal Bainite of High-Carbon Manganese–Silicon Steel *The Physics of Metals and Metallography* **120** 874–880

Acknowledgments
The authors are grateful for assistance in the program of leading universities of the RF with the aim or improving their competitiveness No. 211 of the RF government No. 02.A03.21.0006.

The Influence of the Initial Billet on the Mechanical Properties of Pipes

A T Zhakupova[1], A V Bogomolov[2] and A N Zhakupov[2]

[1]Faculty of Earth and Environmental Sciences, D. Serikbayev East Kazakhstan State Technical University, 69, Protozanov Street, Ust-Kamenogorsk, 070004, Kazakhstan
[2]Department of Metallurgy, S. Toraighyrov Pavlodar State University, 64, Lomov Street, Pavlodar, 140008, Kazakhstan

E-mail: bogomolov71@mail.ru

Abstract. The article researched the influence on the pipes mechanical properties initial billet obtained by the traditional method and the proposed method, in which the piercing process is excluded. The proposed method involves casting the initial billet for the production of seamless pipes with a cavity obtained on a continuous casting machine. The production of hot-rolled seamless pipes from a hollow billet is an urgent issue today, since central porosity and axial segregation after casting are excluded, affecting the quality of the pipes obtained in the subsequent stages of production: rolling on a mandrel mill and reducing. The result is a minimal deviation of the mechanical properties of steel samples obtained by double deformation of a hollow continuously cast billet from samples obtained according to the traditional scheme, including the piercing process.

1. Introduction

At the present stage of steel production development, the most rational way to obtain steel billets for the production of seamless pipes is the continuous casting process, due to its technical and economic advantages: productivity, product yield, etc. Solid billets are characterized by axial segregation heterogeneity with any manufacturing methods, therefore, removing metal from the axial zone of the billet is advisable [1–4]. The authors of [5–7] propose to use a hollow continuously cast billet instead of a sleeve pierced on a cross-helical mill as the initial billet for pipe production. However, the pipe obtained by rolling from a hollow cast billet undergoes lesser degrees of deformation and in this regard, experimental researches and mechanical properties analysis of the obtained products from a hollow cast billet and a pierced sleeve are necessary [8–11].

2. Object and research methods

For the research, templates with a length of 150 mm were selected from a continuously cast billet with a diameter of 210 mm from steel grade 09Mn2Si and from a sleeve obtained after piercing from this billet. The experimental scheme is shown in Figure 1. The casting of a continuously cast billet was carried out in the electric steel melt shop of the LLP "KSP Steel", and the billet was pierced in the pipe rolling shop of the same plant on a two-roll cross-helical mill with mushroom-shaped rolls.

Upset of samples with a diameter of 30 mm and a length of 75 mm was carried out on a hydraulic press at the laboratory of S. Toraighyrov PSU to ensure a compression equal to 30%, which is similar to the total compression of the rolling process on a continuous mill and reducing on a reduction mill. Heat treatment of the upsetted samples was carried out in the furnace SNOL 7.2/1300 at the laboratory of S. Toraighyrov PSU.

The mechanical properties determination of the initial billets, upsetted and heat-treated samples was carried out according to GOST 1497-84 “Metals. Tensile test methods” in the laboratory of LLP “KSP Steel”.

Samples for testing were taken in the transverse *T* and longitudinal *L* directions, three units from each template, as shown in Figure 2 for a solid billet and in Figure 3 for a hollow billet.

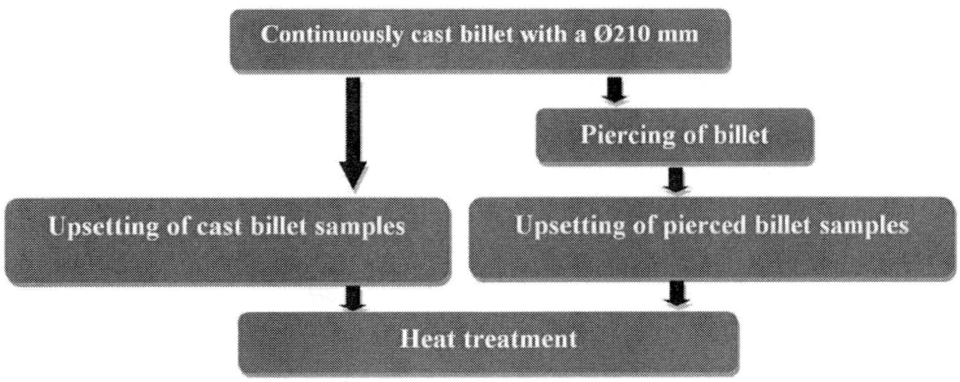

Figure 1. Experimental scheme.

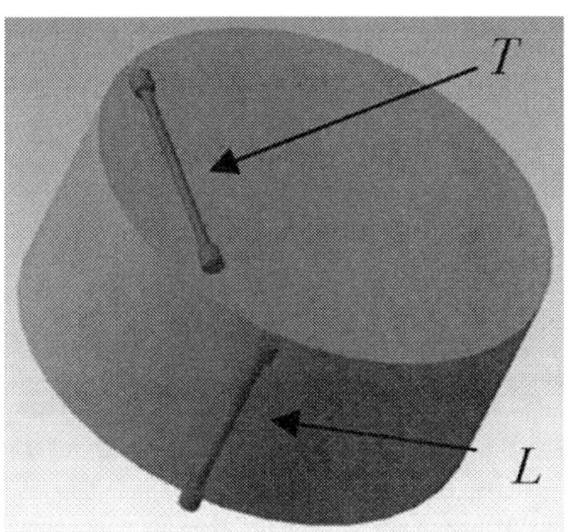

Figure 2. Cutting of transverse and longitudinal samples for mechanical tests from a solid billet.

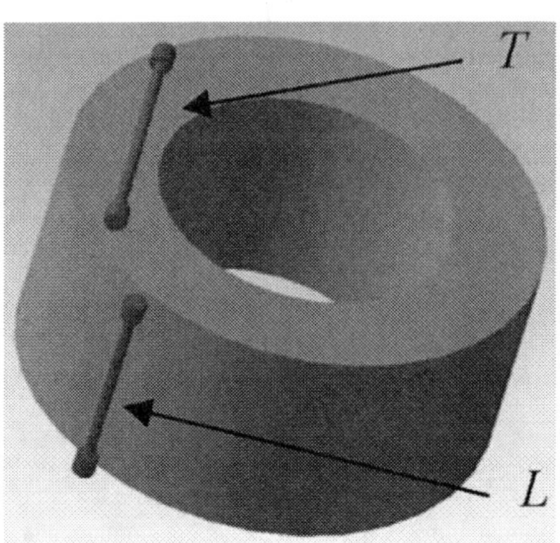

Figure 3. Cutting of transverse and longitudinal samples for mechanical tests from a hollow billet.

3. Research results and discussion

In the production of seamless hot-rolled pipes according to GOST 632-80 "Casing pipes and couplings to them", GOST 633-80 "Tubing and couplings to them" and similar foreign API 5CT "Casing and tubing" only mechanical properties are standardizing indicators, namely, tensile strength, yield strength and elongation.

The initial mechanical properties and density of the cast hollow billet are similar to the properties of a solid continuously cast billet, since at this stage the product does not yet undergo deformation. In this regard, the conditional initial values of the mechanical properties of the hollow billet are taken from the test results of tensile tests of samples taken from solid continuously cast billets. Table 1 shows the test results in comparison with samples cut from a pierced sleeve [12–15].

Table 1. The mechanical properties of the initial billets.

Type of billet	Test sample	Tensile strength, MPa	Yield strength, MPa	Elongation, %
Cast	transverse	437;445;414 432	415;435;368 406	6.2;5.2;6.4 5.9
	longitudinal	478;455;457 463	424;448;391 421	5.5;6.9;7.6 6.7
Pierced	transverse	596;538;634 589	517;485;504 502	9.1;10.5;8.1 9.2
	longitudinal	689;666;539 631	457;525;547 510	11.5;10.5;10.5 10.8

As can be seen from the data in table 1, the values for all indicators of the pierced billet are significantly higher than that of cast.

From the standpoint of the theory of metal forming, longitudinal rolling is now presented as a "regular upsetting", with a continuous supply of metal under the peens. The processes of upset and longitudinal rolling are studied in detail in the works of S. I. Gubkin, I. M. Pavlov, A. I. Tselikov, Ya. M. Okhrimenko and A. A. Presnyakov [16–20]. On the basis of these research and work, the rolling and reducing processes were modeled by another type of metal forming – upsetting. To determine the mechanical properties after rolling and reduction processes, cylindrical samples were taken with pierced and solid cast billets. To provide 30% compression samples 75 mm long upsetting to achieve equal 52.5 mm in length. Regarding the size of the compression, a pipe with a wall of ~ 10 mm, rolled from a sleeve with a wall thickness of 33 mm, was taken as an example, and a total compression of 30% was achieved. Before upsetting, the samples were heated in an electric furnace to 1000 °C, providing better deformation conditions and simulating factory technological parameters. Time in the furnace at this temperature was calculated on the basis of 1.0–1.5 minutes per 1 mm section, which, respectively, for samples with a diameter of 30 mm was 30–45 minutes.

After upsetting, standard specimens for testing were made from these samples. The test results are shown in table 2.

Analyzing the results of table 2, it is seen that the deviation of the mechanical properties values practically decreased to 3%.

To ensure a smaller discrepancy between the values for each sample and to obtain a more accurate averaged value, all samples were heat-treated in an SNOL 7.2/1300 electric furnace. The heat treatment mode, based on the chemical composition of the steel and the resulting strength group:

- heating temperature – 880–890°C;
- heating rate ~10 °C/min;
- heating time – 1.0 min per 1 mm section;
- cooling rate ~ 100 °C/c;
- tempering temperature – 600–620 °C;
- holding time at tempering – 2.0 min per 1 mm section.

Table 2. The mechanical properties of the upsetted samples.

Type of billet	Test sample	Tensile strength, MPa	Yield strength, MPa	Elongation, %
Cast	transverse	767;808;789 / 788	643;622;620 / 628	14.3;14.9;14.8 / 14.7
	longitudinal	802;809;781 / 797	623;635;659 / 639	14.8;14.8;15.1 / 14.9
Pierced	transverse	817;792;799 / 803	625;651;646 / 641	14.9;15.0;14.9 / 14.9
	longitudinal	821;814;798 / 811	640;635;665 / 647	15.3;15.1;14.8 / 15.1

Table 3 shows the results of the final mechanical properties obtained after heat treatment.

Table 3. Mechanical properties of heat-treated samples.

Type of billet	Test sample	Tensile strength, MPa	Yield strength, MPa	Elongation, %
Cast	transverse	790;801;786 / 792	633;642;629 / 635	14.8;14.9;15.1 / 14.9
	longitudinal	793;799;794 / 795	636;640;645 / 640	15.5;14.8;15.0 / 15.1
Pierced	transverse	807;793;796 / 799	650;641;649 / 647	14.4;15.3;15.3 / 15.0
	longitudinal	809;810;797 / 805	651;653;644 / 649	15.0;15.4;15.6 / 15.3
GOST 632-80, GOST 633-80		not less than 689	552-758	not less than13
API 5CT		not less than 689	552-758	not less than13 (for a sample with a cross section of 70-80 mm^2)

According to the results of heat treatment, the discrepancy between the values decreased, which indicates the alignment of the mechanical properties and structure along the cross section of the samples. The values of the indicators are within the permissible limits of mechanical properties according to GOST 632-80, GOST 633-80 and API 5CT, and corresponds to strength group E (according to GOST) or N80 (according to API).

4. Conclusion

The possibility of producing seamless hot-rolled pipes from a hollow billet is substantiated, providing the necessary set of mechanical properties that are regulated by the standards GOST 632-80, GOST 633-80 and API 5CT. The deviation of the mechanical properties of products obtained from cast and piercedbillets is less than 3%. The results of casting, deformation and subsequent heat treatment of the researched pipe samples correspond to the standardized indicators of the standards GOST 632-80, GOST 633-80 and API 5CT, which confirms the feasibility of producing pipes from a hollow billet, considering improving the quality of the structure.

References

[1] Hutny A, Warzecha M, Derda W and Wieczorek P 2016 Segregation of Elements in Continuous Cast Carbon Steel Billets Designated for Long Products *Arch. Metall. Mater.* **61** **4** 2037–2042

[2] Zou L, Zhang J, Liu Q, Zeng F, Chen J and Guan M 2019 Prediction of central carbon segregation in continuous casting billet using a regularized extreme learning machine model *Metals* **9** 1312

[3] Kanaev A, Bykov P, Bogomolov A and Reshotkina E 2012 Reducing the Central Porosity of Continuous_Cast Billet by Modification of the Solidification Process *Steel in Translation* **42** **8** 643–645

[4] Kanaev A, Bogomolov A and Reshotkina E 2010 Defects and Thermal Hardening of Reinforcement Rolled from Continuous Cast Billet Process *Steel in Translation* **40** **6** 586–589

[5] Hress O, Isayev O, Chebotaryova O and Wu K 2019 Numerical simulation of macrosegregation impurities in the solidifying continuous casting steel billet *Materials Physics and Chemistry* **1** **1** 1–13

[6] Pickering E 2013 Macrosegregation in Steel Ingots: The Applicability of Modelling and Characterisation Techniques *ISIJ International* **53** 935–949

[7] Luchkin V, Vorobej S and Levchenko G 2005 Continuous casting billet with different sections quality specificity *Metal and casting of Ukraine* **5** 30–33

[8] Krayushkin N, Pribytkov I and Shatokhin K 2019 Formation of temperature fields and thermal stresses arising during solidification of cylindrical continuously cast steel billets *Izvestiya. Ferrous Metallurgy* **62** **1** 57–61

[9] Gabelaya D and Kabakov Z 2019 Optimization of the Gap Parameters of Supporting Rollers of Curvilinear Continuous Casting Machine *Metallurgist* **63** 823–828

[10] Stulov V 2017 Cooling of a mold at preforming cylindrical continuous cast steel billets *Journal of Machinery Manufacture and Reliability* **46** **1** 57–62

[11] Rajora S, Butola M and Khare K 2018 Calculating Numerical Derivatives using Fourier Transform: some pitfalls and how to avoid them *European Journal of Physics* **39** **6** 143–149

[12] Morozova O, Sivak A and Smolyakov A 2015 Dorn with a controlled profile of the working surface for the mold of the continuous casting machine of hollow billets *RU Patent* **2543627**

[13] Krayushkin N, Pribytkov K, Shatokhin S 2019 Temperature Fields and Thermal Stress in the Solidification of Cylindrical Continuous-Cast Steel Billet *Steel in Translation* **49** **1** 20–23

[14] Nicholas Grundy A, Münch S, Feldhaus S and Bratberg J 2019 Continuous Casting of High Carbon Steel: How Does Hard Cooling Influence Solidification, Micro- and Macro Segregation? *IOP Conference Series: Materials Science and Engineering* **529** **1** 012069 1–7

[15] Kabakov Z and Tsurko V 2019 The Determination of effectiveheat capacity of carbon *Proceedings of Higher Educational Institutions. Ferrous Metallurgy* **57** **2** 15–19
[16] Presnyakov A 1988 *Deformation zone in metal forming* (Alma-Ata: Nauka) p 108
[17] Gabelaya D, Mashchenko M and Kabakov Z 2019 Calculation of changes in specific volumes of Fe-C system alloys depending on carbon content and temperature *Proceedings of Higher Educational Institutions. Ferrous Metallurgy* **62** **8** 627–631
[18] Gubkin S 1939 *Forging and pressing of nonferrous metalsand alloys* (Moscow: Metallurgy) p 325
[19] Galkin V, Mikhalenko F and Ulanov P 2001 Analysis of the performance of the leafsprings sheets obtained by treatment of rolled strip *KSP OMD* **8** 3–7
[20] Galkin V, Ulanov P and Pachurin G 2006 Simulation of rolling billets springs with variable cross section *Mechanics and Mechanical Engineering* **8** **4** 911–914

IOP Conf. Series: Materials Science and Engineering **969** (2020) 012034 doi:10.1088/1757-899X/969/1/012034

Ablative Laser Processing of Metals and Dielectrics in an Electrostatic Field

D A Kochuev, A S Raznoschikov and R V Chkalov

Vladimir State University, 87, Gorky Street, Vladimir, 600000, Russia

E-mail: j.larenax@gmail.com

Abstract. The work considers the task of synthesis of highly dispersed spherical powder granules by laser ablation method. To improve the geometry of synthesized particles and increase the process productivity, an electrostatic system for collecting ablation products was used, which allows the formed particles to leave the area of laser irradiation without being subjected to excessive exposure to a heat source. Using an external field makes particle motion predetermined, ensuring their adhesion to the electrodes. The paper shows the experimental results on the removal of ablated particles, as well as an assessment of the effectiveness of electrostatic field for accelerating particle motion.

1. Introduction

The synthesis of fine powders by the method of fem to second laser ablation of material provides high-purity spherical particles with a controlled dispersion of sizes [1–3]. During laser ablation treatment, the resulting ablation products can make a significant contribution to the process. Ablated particles can repeatedly fall under the next laser pulse, thereby causing reflection, scattering, and absorption of a significant part of the radiation [4–6]. Laser ablation products, in addition to direct interaction with laser radiation, significantly increase the lifetime and temperature of a laser-induced plasma plume [7–9], which decrease the energy entering processing area.

As the most relevant way to solve the described problem is to use an electrostatic field to remove ablated particles. Using an electrostatic field allows ablation products to be removed from the laser radiation propagation region without deteriorating surface quality and causing dispersion of synthesized micro- and nanoparticles [10–14].

2. The study of electrostatic field influence on the mass of ablated material

The experiment was carried out in accordance with the scheme shown in figure 1, under the following conditions: the processed materials are titanium VT1-0, and aluminum oxide, pulse repetition rate – 10000 Hz, full pulse duration – 300 fs, energy of each pulse – 55–145 μJ, lens focal length – 200 mm, processing area – 8x1 mm, scanning density – 30 lines / mm, scanning speed – 10 mm / s, number of pulses – $1.2 \cdot 10^7$ pulses, electrostatic field strength – 30 kV. The processing took place under normal conditions, argon was used as the working medium.

To study the efficiency of using the electrostatic field, a series of experiments was carried out with different energies in a laser pulse. The experiments revealed a positive effect of electrostatic field on the overall laser ablation dynamics. As can be seen from the graphs presented in figure 2, with a sequential increase in the average radiation power to 950 mW, an increase in fraction of removed material is observed both when using an electrostatic field and without it. Further, however, for a treatment mode without an electrostatic field, a significant decrease in function is observed. In turn, for ablation in an electrostatic field at a power of 1450 mW for titanium, the efficiency increase is about 85%, for alumina 70%.

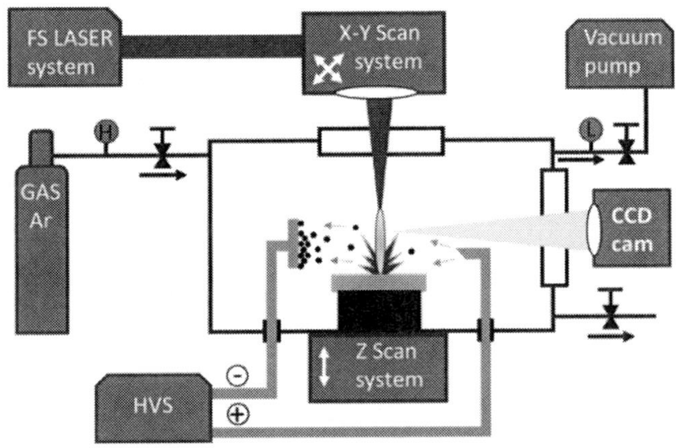

Figure 1. Experimental setup of laser ablation processing.

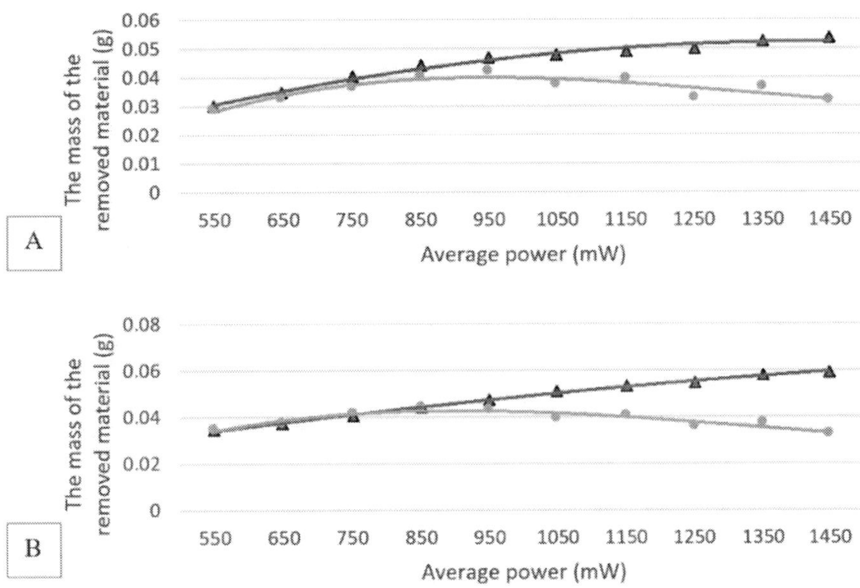

Figure 2. Dependence of removed material mass on laser radiation average power in conditions of an electrostatic field (red lines) and without it (green lines). A – for alumina, B – for titanium.

3. Evaluation of the effectiveness of using the electrostatic field for the removal of ablation products

When using an electrostatic field of high intensity, the mechanics of the expansion of ablated particles becomes predetermined. Ablation products, acquiring a charge, are deposited on high-voltage electrodes [15–17]. To confirm the assumption of increasing the efficiency of laser ablation, we estimate the rate of particle removal from the laser radiation propagation region. To assess, first of all, it is necessary to determine a number of conditions: the distance that a particle must travel to exit the probable region of laser radiation propagation is taken to be 0.5 mm; the time for which a particle must leave a given region is 100 μs (the time interval between pulses based on the laser repetition rate). Particles characteristic diameter synthesized as a result of ablation by ultrashort pulses ranges from 5 nm to 1 μm and depends on the material and the mode of exposure (figures 3, 4) [18–20].

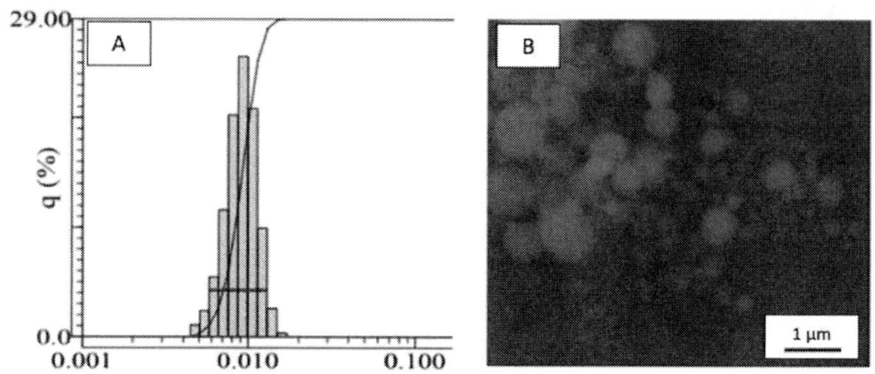

Figure 3. The obtained nanopowder of oxide ceramics based on Al_2O_3: A – histogram of nanoparticle sizes distribution, B – SEM image of obtained nanoparticles.

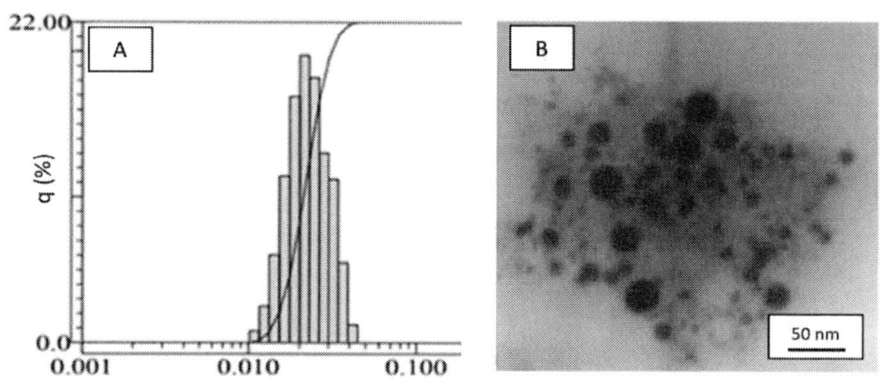

Figure 4. Titanium nanopowder obtained in argon medium (granule dispersion 10-40 nm): A – histogram of nanoparticle sizes distribution, B – SEM image of obtained nanoparticles.

To determine the force necessary for a particle to overcome the region of propagation of laser radiation in the interpulse time interval, we write the following relation:

$$F_{el} - F_{st} = m \cdot a, \qquad (1)$$

where F_{el} – electrostatic field strength, F_{st} – Stokes drag force for spherical particles, m – particle mass, a – acceleration.

$$F_{el} = q \cdot E \qquad (2)$$

where q – particle charge, E – field strength.

Particle charge is determined from the relation $q=c \cdot \varphi$, c – particle capacity, φ – electric potential. Particle capacity is defined as $4\pi\varepsilon\varepsilon_0 R$, ε – dielectric constant of argon (taken equal to one), ε_0 – $8{,}85 \cdot 10^{-12}$ F/m (electric constant), R – radius of a spherical particle. The potential is determined from the relation $\varphi=U/2$ the field strength at the interelectrode gap is 30 kV, ablated particles are located in the center of interelectrode gap, thus the voltage is 15 kV.

$$F_{st} = 3\pi \cdot \mu \cdot D \cdot v, \tag{3}$$

where μ – dynamic viscosity coefficient, D – particle diameter, v – particle velocity.

$$q \cdot E - 3\pi \cdot \mu \cdot D \cdot v = \frac{mdv}{dt}, \tag{4}$$

$$\frac{dv}{dt} + 3\pi \cdot \mu \cdot D \cdot \frac{v}{m} = \frac{qE}{m}. \tag{5}$$

Define the constants:

$$k_1 = \frac{3\pi\eta D}{m}, \tag{6}$$

$$k_2 = \frac{qe}{m}. \tag{7}$$

Consider a first-order differential equation with respect to the function v(t) with the initial condition:

$$\frac{dv}{dt} + k_1 v = k_2, \quad v(0) = 0. \tag{8}$$

A particular solution to the problem has the following form:

$$v(t) = \frac{k_2}{k_2}(1 - e^{-k_1 t}), \tag{9}$$

$$v(t) = \frac{qe}{3\pi\eta D}\left(1 - \exp\left(-\frac{3\pi\eta D}{m}t\right)\right). \tag{10}$$

The obtained velocity distribution for spherical titanium particles depending on the size, presented in graphical form, is shown in figure 5. Similar results are obtained for alumina powders. The estimation method shows the acquisition by ablated particles of considerable speed as a result of exposure to an electrostatic field.

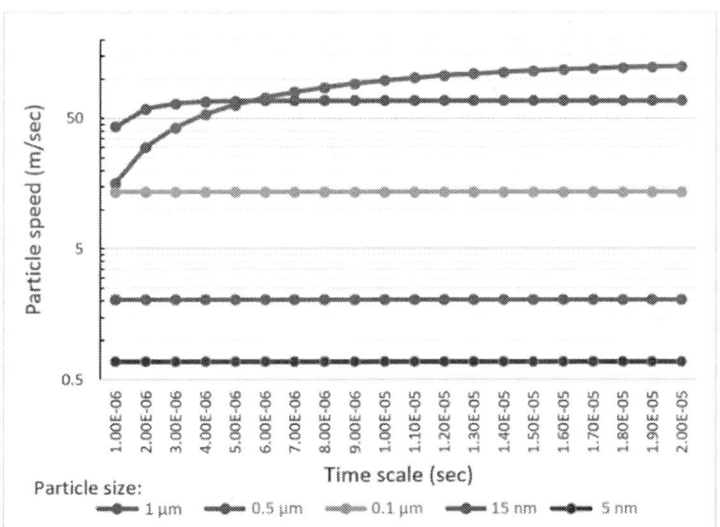

Figure 5. Dependence of movement speed of spherical titanium particles in an electrostatic field of 30 kV on diameter.

References

[1] Leshchik S D, Shupan P I, Sergienko I G and Linnik D A 2016 Synthesis of nanoparticles by method of a laser ablation of metal materials in liquid in the mode of nanosecond impulses *Vesnik of Yanka Kupala State University of Grodno*

[2] Stuchebryukov S D and Sukhov V M 2020 Spectral dependence of the maximum efficiency of the near field for spherical gold nanoparticles *Protection of Metals and Physical Chemistry of Surfaces*

[3] Gerke M N, Khorkov K S, Kutrovskaya S V, Nogtev D V, Prokoshev V G and Arakelyan S M 2011 Formation of nanostructures during femtosecond laser ablation in vacuum *The Journal Perspective Materials*

[4] Kavetskaya I, Voloshina T, Karavanskii V and Krasovskii V 2009 Optical properties of the gold nanoparticle *Condensed media and interfaces*

[5] Leshchik S D, Shupan P I, Sergienko I G and Linnik D A 2016 Synthesis of nanoparticles by method of a laser ablation of metal materials in liquid in the mode of nanosecond impulses *Vesnik of Yanka Kupala State University of Grodno*

[6] Zazhogin A P and Reza F A 2008 Dynamics of ablation processes and the development of a near-surface laser plasma of aluminum alloys by double laser pulses *Bulletin of the Belarusian State University*

[7] Chuchman M P, Shuaibov A K and Mesarosh L V 2011 Spatial and radiative characteristics of the plasma of a laser aluminum torch *High Temperature*

[8] Uspensky S A, Bykovsky D P and Petrovsky V N 2014 Steam plasma plume for titanium welding with powerful fiber laser *Lasers in science, technology, medicine*

[9] Osipov V V, Shitov V A, Lisenkov V V and Zolotonosha A V 2008 Dynamics of evaporation of a fast-moving target under action of high power laser radiation *Atmospheric and oceanic optics*

[10] Kochuev D A, Chernikov A S, Khorkov K S, Chkalov R V, Voźnesenskaya A A and Prokoshev V G 2019 Surface processing of titanium in the medium of n-hexane by ultrashort laser pulses *Journal of Physics: Conference Series*

[11] Kochuev D A, Chkalov R V, Prokoshev V G and Khor'kov K S 2020 Effect of Laser Radiation on the Surface of a Solid Body and the Formation of Micro- and Nanostructures *Bull. Russ. Acad. Sci. Phys.*

[12] Kochuev D, Chkalov R, Prokoshev V and Khorkov K 2019 Laser radiation impact on a solid surface and formation of micro- and nanostructures *EPJ Web Conf.*

[13] Kozadaev K V, Goncharov V K, Mikitchuk E P and Prokhorovich M A 2014 Laser synthesis and optical diagnostics of silver plasmon nanostructures on the surface of glass *Physico-chemical aspects of studying clusters, nanostructures and nanomaterials*

[14] Ivanov A M, Myslitskaya N A and Kapelevich M S 2011 Production of nanoparticles by laser ablation of metal surfaces in a liquid and study of the obtained structures by optical methods *Izvestia KSTU*

[15] Ismail R A and Fadhil F A 2014 Effect of electric field on the properties of bismuth oxide nanoparticles prepared by laser ablation in water *J. Mater. Sci. Mater. Electron.*

[16] Kochuev D 2019 Formation of powder materials under the combined influence of pulse laser radiation and electric arc discharge *EPJ Web Conf.*

[17] Kazakevich V S, Kazakevich P V, Yaresko P S and Kamynina D A 2016 Laser ablation of gold in liquid argon *Physical education in universities*

[18] Grigorieva A M 2019 Laser synthesis and modeling of ceramic materials spherical nanopowders formation *Days of science of vlsu students*

[19] Ivashchenko A, Kochuev D and Chkalov R 2019 Laser-Induced-Electro-Explosion Synthesis of Powder Materials (LIEES) *Materials Today: Proceedings*

[20] Kochuev D A, Khorkov K S, Ivashchenko A V, Prokoshev V G and Arakelian S M 2018 Formation of microspheres under the action of femtosecond laser radiation on titanium samples in hydrocarbons *Journal of Physics: Conference Series*

IOP Conf. Series: Materials Science and Engineering **969** (2020) 012035 doi:10.1088/1757-899X/969/1/012035

Correlation between Iron Substitution Degree in the M-Type Hexaferrites Based Crystal Structures and Dopant Content in the Crystal Growth Feeding Melt

V Zhivulin, O Zaitseva and D Vinnik

Material science & physics and chemistry of materials department, South Ural State University (national research university), 76, Lenin prospekt, Chelyabinsk, 454080, Russia

E-mail: vinnikda@susu.ru

Abstract. In this paper the results of the investigation of the properties of the obtained crystals of hexaferrites grown from the sodium carbonate flux doped with 1–12 mass % of metal oxide (aluminum, titanium, manganese, cobalt, and copper) are presentd. The ratio of concentration of doping elements in the crystal and in the feeding melt has strong dependence on nature of dopant. Doping metals can be ordered in a line Ti > Al > Mn > Co > Cu where Ti has maximal ratio of 0.5 and Cu of 0.005. Hexagonal M type ferrites attract the reseachers interest due to great capacity for the accommodation of different metals at high doping levels that provides a possibility for wide tuning of physical properties by cation substitution.

1. Introduction

M-type hexaferrites with a magnetoplumbite structure have found application in various branches of science and technology as single magnetic materials and magnetic composites fillers for information storage devices, magneto-optical devices [1, 2]. In recent years, hexaferrites are increasingly used in devices in the millimeter and subterahertz frequency ranges. Such devices are, for example, tunable resonators [3], insulators [4] and radio-absorbing coatings, in which hexaferrites (primarily barium) are used both in the form of films and in the form of bulk materials [1].

A promising area of research is the creation of materials with a minimum reflectivity of electromagnetic radiation to protect large objects [5, 6]. The main requirement for materials is absorption in a wide frequency range [7]. In the microwave (GHz) region, hexaferrites are of greatest interest. Substitution (i.e. substitution iron atoms with other elements atoms) can tune the of ferromagnetic resonance frequency. At the same time, hexaferrites can be used as emitters in information and telecommunication systems, where, on the one hand, a low level of losses is important, and on the other hand, an increase in the resonance frequency (which can be achieved by partial substitution of iron atoms in the hexaferrite matrix with dopant element) expands the capabilities of devices and devices based on such material. In particular, the transition to frequencies up to 100 GHz and higher will provide an increase in data transmission between satellites. This will lead to the development of new generation telecommunication services based on space systems [8, 9].

By virtue of all the above, the study of the structure and properties of these materials is of both scientific [10, 11] and practical interest. In the last 15 years, the direction in which the researchers interest is focused on the preparation and study of M-type hexaferrites doped with different elements

(Al /Ti / Mn / Co / Ni / Cu / W / Zn / Cr) has acquired great relevance [12]. Atomic substitutions in ferrites are commonly studied to manipulate the magnetic and electrical properties of the resulting materials.

It has been found that the functional properties of ferrites can be significantly changed by partial or complete substitution of iron cations with dopants. Such changes in the structure allows one to significantly vary the ferrites properties, in particular, to change the uniaxial magnetic anisotropy level, the magnitude of the coercivity, and the Curie temperature [13, 14].

In the literature, there are quite a few data on the substitution nature of atoms in the lattice of ferrites by dopants. There is a series of works related to the hexaferrites production in the form of a fine-crystalline material doped with La, La-Co, La-Nd, Er-Zn-Co. But the authors of the works disagree about the nature of the replacement of REE cations and cations of codopants. Some authors argue that REE^{3+} are embedded in the Fe^{2+} position [15], others claim that bivalent codopants replace the iron cation, and REE^{3+} – barium atoms [16] (when it comes to barium hexaferrite). Still others argue that both REE and bivalent codopant substitution barium positions [11, 17–21].

At the same time, no serious X-ray diffraction studies were performed in any of the above works. The study of the mechanism of substitution of REE and codopant cations is an important task for understanding the effect of impurities on the magnetic properties of hexaferrites.

Obviously, the powder or nano-scaled materials do not allow one to study their crystal structure with the required degree of accuracy. This requires large enough single crystals.

Summarizing the brief analysis of publications, the following should be said:

Work on the creation of crystalline substituted hexaferrites (in particular, in the form of single crystals) is relevant, and their results will be in demand by modern high-tech industries. Of particular relevance are works aimed at the development and improvement of methods for alloying hexaferrites in order to purposefully modify their properties, as well as to obtain and study new types of ferrite structures, stabilized due to the partial replacement of iron atoms with atoms of other elements. The development of this direction is hindered by the lack of sufficient information on the mechanisms of substitution of iron atoms, on the effect of the composition and number of substituents on the crystal structure of substituted ferrites, as well as on the structure and properties of doped ferrites.

The complex of work has the task to start the next step in the preparation of substituted crystalline materials with magnetoplumbite structure – to prepare the samples and to study the structure and properties of this type crystalline phases with a gradually varying degree of substitution. The results that are planned to be obtained in the future will be aimed at managing the crystallization process - i.e. the possibility of targeted production of single crystals of hexaferrites with gradient substitution or targeted prevention of the formation of crystals with inhomogeneous substitution. In the published literature, there is no information on works aimed at obtaining such a result.

One of the obvious ways of producing structures with a gradient substitution is the growing of relatively large crystals in a relatively small amount of melt thereby, that the proportion of dopant in the process of growing the crystal is gradually regularly varied, which will lead to a natural change in the degree of iron substitution in the growing crystal ... To understand the possibilities and limitations of this approach, as well as to carry out preliminary calculations of the charge and determine other conditions for growing crystals with, it is useful to know how the degree of iron substitution is related to the concentration of the dopant in the melt feeding the crystal growth. Analysis of experimental data of this kind is a main part of the study.

2. Experimental technique

Single crystals of barium hexaferrite partially substituted with aluminum, titanium, manganese, cobalt, and copper, up to 12 mm in size, were grown from a flux based on sodium carbonate by spontaneous crystallization. The growth of single crystals was carried out in platinum crucibles with a volume of 60–300 ml. Exposure at a maximum temperature of 1300–1350 °C was carried out for 24 hours, after which it was slowly (at a rate of 4 °C/h) cooled to the lower limit of the temperature range (900 °C). In this case, iron oxide (Fe_2O_3), barium carbonate, dopant oxide and sodium carbonate were used as starting components. The use of a flux allows the crystallization process to be carried out at temperatures below 1300 °C, while other methods require heating to significantly higher temperatures. After crystallization, the content of the crucibles was washed in boiling nitric acid in order to extract crystals from flux residues.

Since in the implementation of the method of spontaneous crystallization, the growth rate of crystals located in different parts of the crucible can differ significantly, which leads to a different concentration of the dopant, then for averaging the composition, 3–4 of large well-faceted crystals were selected from each experiment. After that, the composition, structure and properties of the samples were studied. The composition of the samples was investigated using a scanning electron microscope with an energy dispersive spectrometer.

3. Results and discussion

On the whole, it is obvious that an dopant content increasing in the melt should lead to an increase in the substitution degree. But whether this process occurs gradually or abruptly, is there a threshold after which an increase in the dopant content in the melt ceases to affect the substitution degree in the crystal and the threshold value up to which the dopant present in the melt does not integrate into the lattice of the growing crystal – the answers to such questions are extremely important for realizing the process of growing a crystal with gradient substitution.

The obtained dependences of the degree of substitution x in structures of the $BaFe_{12-x}M_xO_{19}$ type on the mass fraction of M in the melt are shown in Figure 1–2.

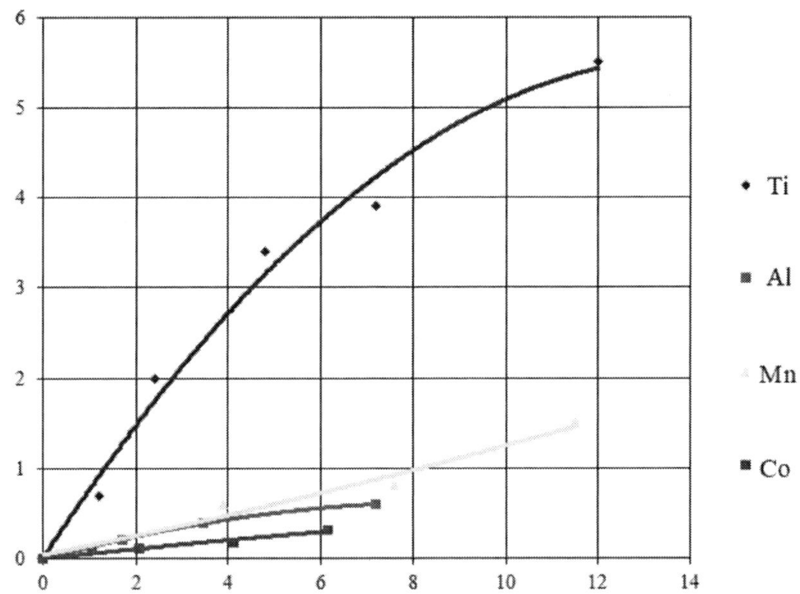

Figure 1. Dependence of x in the $BaFe_{12-x}M_xO_{19}$ on the mass fraction of M (Ti, Al, Mn, Co) in the melt when growing crystals by the method of spontaneous crystallization.

The results allow us to conclude that the proportion of the dopant gradually increases depending on its presence in the melt. In addition, they make it possible to select from the total number of studied systems those that are most promising from the point of view of growing larger crystals with a noticeable gradient in the dopant distribution. A first systems in which the crystals are formed were $BaFe_{12-x}Ti_xO_{19}$, $BaFe_{12-x}Al_xO_{19}$ and $BaFe_{12-x}Mn_xO_{19}$. It is in them the relatively small changes in the dopant concentration in the melt lead to the maximum change in the dopant content in the crystal.

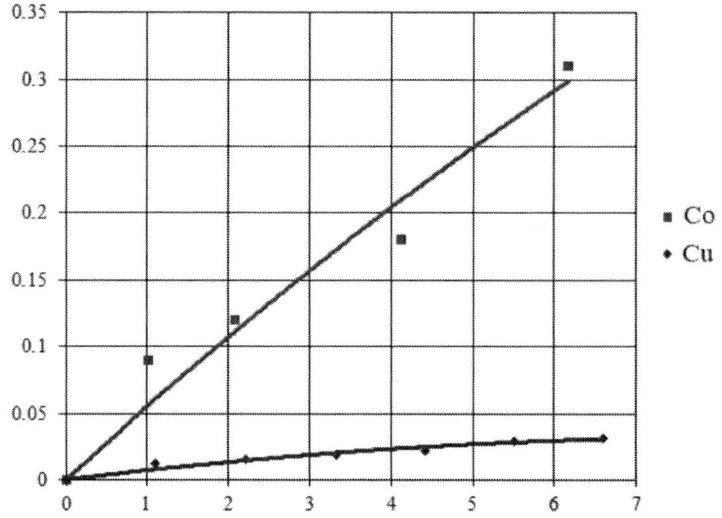

Figure 2. Dependence of x in the $BaFe_{12-x}M_xO_{19}$ on the mass fraction of M (for cobalt and copper) in the melt when growing crystals by the method of spontaneous crystallization.

The results allow us to conclude that the proportion of the dopant gradually increases depending on its presence in the melt. In addition, they make it possible to select from the total number of studied systems those that are most promising from the point of view of growing larger crystals with a noticeable gradient in the dopant distribution. A first systems in which the crystals are formed were $BaFe_{12-x}Ti_xO_{19}$, $BaFe_{12-x}Al_xO_{19}$ and $BaFe_{12-x}Mn_xO_{19}$. It is in them the relatively small changes in the dopant concentration in the melt lead to the maximum change in the dopant content in the crystal.

4. Conclusions

The results of experiments accompanied by the formation of substituted crystal structures based on hexaferrites are analyzed. Crystals of substituted hexaferrites investigated (primarily by EPMA method) in order to study the relationship between degree of iron substitution in forming crystalline structures based on M-type hexaferrite with a dopant concentration in the feed growth melt. From the total number of studied systems, we selected those that are most promising from the point of view of growing large crystals with a noticeable gradient in the dopant distribution. Such systems are primarily the system in which $BaFe_{12-x}Ti_xO_{19}$ crystals are formed, as well as the $BaFe_{12-x}Al_xO_{19}$ and $BaFe_{12-x}Mn_xO_{19}$ systems. It is in them the relatively small changes in the dopant content in the melt lead to the maximum change in the crystal dopant content. Please follow these instructions as carefully as possible so all articles within a conference have the same style to the title page. This paragraph follows a section title so it should not be indented.

References

[1] Zhang W, Peng B, Zhang W, Zhou S and Schmidt H 2010 Ultra large coercivity in barium ferrite thin films prepared by magnetron sputtering *J. Magn. Magn. Mater.* **322** 1859–1862
[2] Xu H, Zhang W, Peng B and Zhang W 2011 Properties of barium hexa-ferrite thin films dependent on sputtering pressure *Appl. Surf. Sci.* **257** 2689–2693
[3] Song Y-Y, Ordóñez-Romero C L and Wu M 2009 Millimeter wave notch filters based on ferromagnetic resonance in hexagonal barium ferrites *Appl. Phys. Lett.* **95** 142506
[4] Harris V G 2012 Modern Microwave Ferrites *IEEE Trans. Magn.* **48** 1075–1104
[5] Cho H-S and Kim S-S 1999 M-hexaferrites with planar magnetic anisotropy and their application to high-frequency microwave absorbers *IEEE Trans. Magn.* **35** 3151–3153

[6] Matsumoto M and Miyata Y 1996 A gigahertz-range electromagnetic wave absorber with wide bandwidth made of hexagonal ferrite *J. Appl. Phys.* **79** 5486–5488
[7] Ghasemi A, Hossienpour A, Morisako A, Liu X and Ashrafizadeh A 2008 Investigation of the microwave absorptive behavior of doped barium ferrites *Mater. Des.* **29** 112–117
[8] Vishnevsky K O, Grebenyuk A Yu, Karasev O I, Sokolov A V, Chulok A A and Yaroslavtsev A B 2014 *Forecast of scientific and technological development of Russia: 2030. Transport and space systems* (Moscow: Ministry of Education and Science of the Russian Federation, National Research University Higher School of Economics) p 40
[9] Vishnevsky K O, Grebenyuk A Yu, Karasev O I, Sokolov A V, Chulok A A and Yaroslavtsev A B 2014 *Forecast of scientific and technological development of Russia: 2030. New materials and nanotechnology* (Moscow: Ministry of Education and Science of the Russian Federation, National Research University Higher School of Economics) p 52
[10] Denisov V M, Denisova L T, Irtyugo L A, Patrin G S, Volkov N V and Chumilina L G 2013 High-temperature heat capacity of $BaFe_{12}O_{19}$ and $BaSc_{0.5}Fe_{11.5}O_{19}$ *Phys. Solid State* **55** 240–242
[11] Trukhanov A V, Kostishyn V G, .Panina L V, Jabarov S H, Korovushkin V V, Trukhanov S V and Trukhanova E L 2017 Magnetic properties and Mössbauer study of gallium doped M-type barium hexaferrites *Ceram. Int.* 12822-12827
[12] Pullar R C 2012 Hexagonal ferrites: A review of the synthesis, properties and applications of hexaferrite ceramics *Prog. Mater. Sci.* **57** 1191–1334
[13] Mariño-Castellanos P A, Anglada-Rivera J, Cruz-Fuentes A and Lora-Serrano R 2004 Magnetic and microstructural properties of the Ti^{4+}-doped Barium hexaferrite *J. Magn. Magn. Mater.* **280** 214–220
[14] Teh G, Nagalingam S and Jefferson D 2007 Preparation and studies of Co(II) and Co(III)-substituted barium ferrite prepared by sol–gel method *Mater. Chem. Phys.* **101** 158–162
[15] Xie Y, Hong X, Wang X, Zhao J, Gao Y, Ling Y, Yan S, Shi L and Zhang K 2012 Preparation and electromagnetic properties of La-doped barium-ferrite/polythiophene composites *Synth. Met.* **162** 1643–1647
[16] Li W, Qiao X, Li M, Liu T and Peng H-X 2013 La and Co substituted M-type barium ferrites processed by sol–gel combustion synthesis *Mater. Res. Bull.* **48** 4449–4453
[17] Kakizaki K, Taguchi H and Hiratsuka N 2004 Magnetic properties of La–Co substituted barium ferrite thin films with large magnetic anisotropy *J. Magn. Magn. Mater.* **272–276** 2241–2243
[18] Huang X, Zhang J, Wang H, Yan S, Wang L and Zhang Q 2010 Er^{3+}-substituted W-type barium ferrite: preparation and electromagnetic properties *J. Rare Earths* **28** 940–943
[19] Li J, Zhang H, Li Q, Li Y and Yu G 2013 Influence of La-Co substitution on the structure and magnetic properties of low-temperature sintered M-type barium ferrites *J. Rare Earths* **31** 983–987
[20] Almessiere M A, Sözeri H, Trukhanov A V, Slimani Y, Ali S, Acar E, Baykal A 2019 Tb^{3+} ion substituted Sr-hexaferrites as high quality microwave absorbers *J. Magn. Magn. Mater.* **491** 165595
[21] Slimani Y, Almessiere M A, Guner S, Alahmari F S, Yasin Ghulam, Trukhanov A V, Baykal A 2020 Influence of Tm–Tb substitution on magnetic and optical properties of Ba–Sr hexaferrites prepared by ultrasonic assisted citrate sol-gel approach *Mat. Chem. Phys.* **253**123324

Acknowledgments
The work was supported by Russian science foundation, project No. 19-73-00144.

IOP Conf. Series: Materials Science and Engineering **969** (2020) 012036 doi:10.1088/1757-899X/969/1/012036

Investigation of Microwave Characteristics of a Composite Based on Copper-Substituted Nickel-Zinc Ferrite

V Zhivulin[1], P Zezyulina[2] and A Vasijeva[1]

[1]Material science & physics and chemistry of materials department, South Ural State University (national research university), 76, Lenin prospekt, Chelyabinsk, 454080, Russia

[2]Laboratory of composite materials electrophysics, Institute for theoretical and applied electromagnetics, 13, Izhorskaya street, Moscow ,13125412, Russia

E-mail: vinnikda@susu.ru

Abstract. The results of ferrite synthesis and composite preparation and investigation are presented. The Cu substituted Ni-Zn ferrite was prepared in single phase form. The morphology and chemical composition were obtained. The composites on the base of experimental $Ni_{0.28}Zn_{0.62}Cu_{0.1}Fe_2O_4$ and commercially available one were prepared. The microwave properties of all samples were investigated. It was checked that the used synthesis method (temperature and time range) can be used for ferrite producing for industry. The microwave properties of composites made of synthesized ferrites are very close to those of composites based on commercially available one. All samples exhibit broad magnetic loss peaks associated with a natural ferrimagnetic resonance.

1. Introduction

Ferrite materials are widely used today. Since 1950, nickel-zinc ferrites with a spinel structure have been actively used as a material for creating various components. The properties that they possess make it possible to use them, for example, in electronics, as choke coils, inductors, transformers [1-3], as well as as a material for absorbing electromagnetic radiation [4-7], which must have high electrical resistance, low dielectric loss, high hardness, high Curie temperature and chemical stability.

The production of such materials is possible using various synthesis methods, from which the final properties of the material are very strong. The most popular methods are considered: solid-phase synthesis [8], coprecipitation [9], sol-gel [10-11], auto-combustion [12-13], etc. In this work, as the most promising from the point of view of ease of preparation, method of solid-phase synthesis.

The creation of new materials is one of the main tasks for the development of science and technology. Based on this principle, scientists are faced with the task of producing materials whose properties can be manipulated. By changing the composition of the initial mixture, the properties of the material change dramatically [14-17]. There are a huge number of publications describing certain results of obtaining Ni-Zn ferrites by various methods and with the addition of one or more additional elements. Spinel ferrites with varying chemical substitution and strong correlation between functional properties attract much attention due to fundamental and practical importance [18-21]. The authors of [22] showed that the introduction of Dy^{3+} cations into Ni-Cu-Zn-ferrite leads to an increase in the magnetic properties of the product. Sm^{3+} cations, according to [23], increase the magnetic permeability of Cu-Zn-ferrite, La^{3+} cations decrease the magnetization and coercive force of Zn-Cu-Ni-ferrite [24], and also increase the activation energy of conduction. An increase in the magnetic and electrical characteristics of spinel ferrites is associated with the magnetocrystalline anisotropy of rare earth elements and with the exchange interaction between Fe^{3+} and Ln^{3+} cations.

This work presents the results of obtaining and studying nickel-zinc ferrite doped with copper.

2. Experimental

The initial components for preparing the samples were powders of iron oxides (Fe2O3), nickel (NiO), zinc (ZnO), and copper (CuO). All the chemicals used were of analytical grade.

The starting materials were measured in the specified stoichiometric ratios, mixed and ground for 30 minutes in an agate mortar. Table 1 shows the calculation of the mass fraction of the starting materials required for the synthesis of samples in the total mass of the charge.

Table 1. Initial weight ratios of charge compositions (wt.%).

Target composition	Charge composition (wt.%)			
	Ni	Zn	Cu	Fe
$Ni_{0.24}ZnC_{0.65}Cu_{0.1}Fe_2O_4$	0.08	0.22	0.03	0.67

After grinding, the resulting mixture was compressed into a tablet. The pressing was carried out in a metal mold with a diameter of 20 mm using a laboratory hydraulic press. The pressing force was 3 t/cm^2.

The resulting tablets were placed on a platinum sheet in a high-temperature electric furnace and sintered at temperatures of 1150 °C for 5 hours. The platinum backing is required to prevent the sample from interacting with the furnace lining elements. The furnace heating rate was 400 °C/hour. The furnace cooling rate to 900 °C was 100 °C/hour, at lower temperatures the cooling rate was not controlled.

The phase composition and structure of the obtained samples were studied using a Rigaku diffractometer, model Optima IV (Cu radiation). The elemental composition was studied using a Jeol JSM7001F scanning electron microscope equipped with an INCAX-max 80 X-ray energy dispersive spectrometer (Oxford Instruments). Figure 1 shows SEM image of obtained ferrite sample. The crystals are typically shaped for spinel structure. Figure 2 shows the X-ray diffraction spectra of the synthesized sample. Figure 1 shows that the obtained sample is monophasic and have a spinel structure. The dashes indicate the literature data [25]. Table 2 presents the chemical composition and average sample formula.

Figure 1. Electronic image obtained from a sample cleavage Ni0.24ZnC0.65Cu0.1Fe2O4.

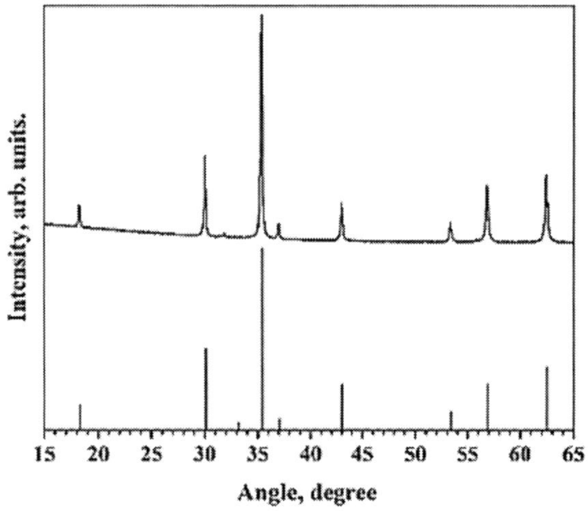

Figure 2. X-ray diffraction pattern of experimental Ni0.24ZnC0.65 Cu0.1Fe2O4 sample. The dashes indicate the literature data [25].

Table 2. The chemical composition and average sample formulas.

Target composition	Charge composition (wt.%)					Sample formula
	O	Fe	Ni	Cu	Zn	
$Ni_{0.24}ZnC_{0.65}Cu_{0.1}Fe_2O_4$	53.11	31.81	4.26	1.49	9.34	$Ni_{0.28}Zn_{0.62}Cu_{0.1}Fe_2O_4$

The frequency dependences of complex permeability and permittivity are measured in the frequency range of 0.1 to 20 GHz by the transmission-reflection (Nicolson-Ross) technique [26] in a 7/3 mm coaxial line. The sintered ceramics and composite samples for microwave measurements are shaped to fit the coaxial line.

Composite samples for microwave study are prepared by manual mixing of the ferrite powder with molten paraffin wax. The volume fractions of the powder in composite samples are 10% and 30%. The thickness of the samples is about 2 mm. Figure 3 presents the measured frequency dependences of permeability and permittivity of composite samples.

The microwave properties of composites made of synthesized ferrites are very close to those of composites based on commercially available one. All samples exhibit broad magnetic loss peaks associated with a natural ferrimagnetic resonance. A small discrepancy in the permittivity may be due to the presence of inhomogeneities in the samples or to a gap in the line during measurement.

The magnetic loss peak shifts toward low frequencies, from about 1 GHz to about 0.6 GHz with volume fraction increase. This can be attributed to the interaction between ferrite particles in composites.

The measured frequency dependences of permeability and permittivity of sintered sample are shown in Figure 4.

The frequency of the magnetic peak loss is beyond the region if the measurement. The shift of the magnetic peak loss of composites toward high frequency compared to sintered samples may be due to the influence of a demagnetizing field on the ferrite particles [27].

3. Conclusion

In the current paper the results of ferrite synthesis and composite preparation and investigation are presented. The Cu substituted Ni-Zn ferrite was prepared in single phase form. The morphology and chemical composition were obtained. The composites on the base of experimental $Ni_{0.28}Zn_{0.62}Cu_{0.1}Fe_2O_4$ and commercially available one were prepared. The microwave properties of all samples were investigated. It was checked that the used synthesis method (temperature and time range) can be used for ferrite producing for industry.

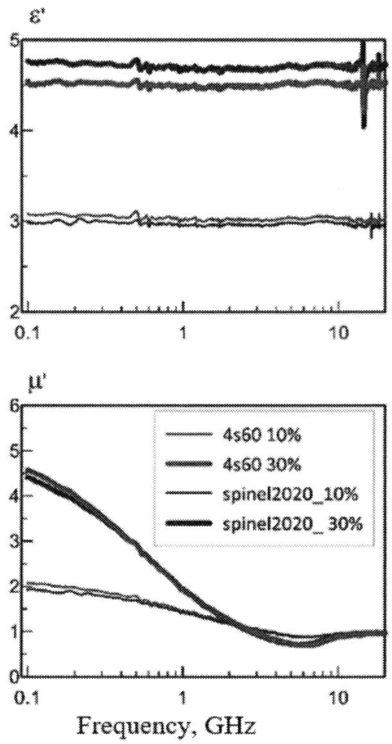

Figure 3. The frequency dependences of permeability and permittivity of composite samples.

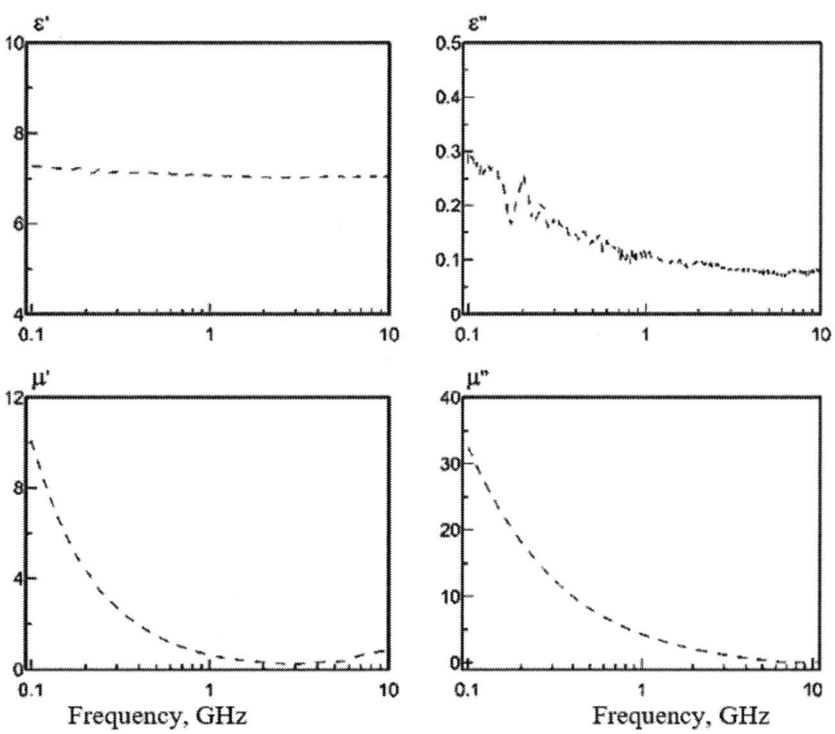

Figure 4. The frequency dependences of permeability and permittivity of ferrite samples.

IOP Conf. Series: Materials Science and Engineering **969** (2020) 012036 doi:10.1088/1757-899X/969/1/012036

References

[1] Kumar Zhuravlev G I 1970 *Chemistry and Technology of Ferrites* (Leningrad: Chemistry publ.) p 192
[2] Kumar Sitidze Yu and Sato Kh 1964 *Ferrites* (Moscow: MIR publ.) p 407
[3] Kumar Smith Ya and Vein Kh 1961 *Ferrites* (Moscow: Foreign publishing house) p 504
[4] Kostishin V G, Vergazov R M, Andreev V G, Bibikov S B, Morchenko A T, Kaneva I I and Mayorov V R 2011 Influence of technological factors on the dielectric constant and radio-absorbing characteristics of nickel-zinc ferrites *Materials of electronic technology* **2** 42–46
[5] Kumar A M, Varma M C, Dube C L, Rao K H and Kashyap S C 2008 Development of Ni–Zn nanoferrite core material with improved saturation magnetization and DC resistivity *J. Magn. Magn. Mater.* **320** 1995–2000
[6] Hwang Y 2006 Microwave absorbing properties of NiZn-ferrite synthesized from waste iron oxide catalyst *Mater. Lett.* **60** 3277–3280
[7] Kuznetsov A, Mel'nikov V, Gyrdasova O, Bazuev G and Novikov S 2011 Synthesis of spinel $Ni_{0.75}Zn_{0.25}Fe_2O_4$ and the properties of a coating obtained by gas-flame spraying *Theor. Found. Chem. Eng.* **45** 455–460
[8] Amer M A, Tawfik A, Mostafa A G, El-Shora A F and Zaki S M 2011 Spectral studies of Co substituted Ni–Zn ferrites *J. Magn. Magn. Mater.* **323** 1445–1452
[9] Dalal M, Mallick A, Mahapatra A S, Mitra A, Das A, Das D and Chakrabarti P K 2016 Effect of cation distribution on the magnetic and hyperfine behaviour of nanocrystalline Co doped Ni–Zn ferrite ($Ni_{0.4}Zn_{0.4}Co_{0.2}Fe_2O_4$) *Mater. Res. Bull.* **76** 389–401
[10] Sarveena, Kumar G, Kumar A, Kotnala R K, Batoo K M and Singh M 2016 Investigation of structural, magnetic and Mössbauer properties of Co^{2+} and Cu^{2+} substituted Ni–Zn nanoferrites *Ceram. Int.* **42** 4993–5000
[11] Ramakrishna A, Murali N, Margarette S J, Wegayehu Mammo T, Krishna Joythi N, Sailaja B, Sailaja Kumari C C, Samatha K and Veeraiah V 2018 Studies on structural, magnetic, and DC electrical resistivity properties of $Co_{0.5}M_{0.37}Cu_{0.13}Fe_2O_4$ (M = Ni, Zn and Mg) ferrite nanoparticle systems *Adv. Powder Technol.* **29** 2601–2607
[12] Houshiar M and Jamilpanah L 2018 Effect of Cu dopant on the structural, magnetic and electrical properties of Ni-Zn ferrites *Mater. Res. Bull.* **98** 213–218
[13] Paramesh D, Kumar K and Reddy P 2017 Effect of Aluminium substitution on the electrical properties of Ni-Zn nanoferrites *J. Magn. Magn. Mater.* **444**
[14] Sugimoto T, Shimotsuma Y and Itoh H 1998 Synthesis of uniform cobalt ferrite particles from a highly condensed suspension of β-FeOOH and β-$Co(OH)_2$ particles *Powder Technol.* **96** 85–89
[15] Pannaparayil T, Marande R and Komarneni S 1991 Magnetic properties of high-density Mn-Zn ferrites *J. Appl. Phys.* **69** 349–351
[16] López Pérez J A, López Quintela M A, Mira J, Rivas J and Charles S W 1997 Advances in the Preparation of Magnetic Nanoparticles by the Microemulsion Method J. Phys. *Chem. B* **101** 8045–8047
[17] Shafi K V P M, Gedanken A, Prozorov R and Balogh J 1998 Sonochemical Preparation and Size-Dependent Properties of Nanostructured $CoFe_2O_4$ Particles *Chem. Mater.* **10** 3445–3450
[18] Trukhanov A V, Astapovich K A, Turchenko V A, Almessiere M A, Slimani Y, Baykal A, Sombra A S B, Zhou D, Jotania R B, Singh C, Zubar T I, Tishkevich D I and Trukhanov S V 2020 Influence of the dysprosium ions on structure, magnetic characteristics and origin of the reflection losses in the Ni–Co spinels *J. Alloys Compd.* **841** 155667
[19] Almessiere M A, Slimani Y, Gungunes H, Kostishyn V G, Trukhanov S, Trukhanov A and Baykal A 2020 Impact of Eu3+ ion substitution on structural, magnetic and microwave traits of Ni–Cu–Zn spinel ferrites *Ceram. Int.* **46**
[20] Slimani Y, Unal B, Almessiere M A, Korkmaz A D, Shirsath S E, Yasin G, Trukhanov A V and Baykal A 2020 Investigation of structural and physical properties of Eu3+ ions substituted

$Ni_{0.4}Cu_{0.2}Zn_{0.4}Fe_2O_4$ spinel ferrite nanoparticles prepared via sonochemical approach *Results Phys.* **17** 103061

[21] Almessiere M A, Trukhanov A, Khan Phd F A, Slimani Y, Tashkandi N, Turchenko V A, Zubar T, Chushkova D, Trukhanov S, Panina L and Baykal A 2019 Correlation between microstructure parameters and anti-cancer activity of the $[Mn_{0.5}Zn_{0.5}]$(EuxNdxFe2-2x)O4 nanoferrites produced by modified sol-gel and ultrasonic methods *Ceram. Int.* **46** 7346–7354

[22] Shirsath S E, Kadam R H, Patange S M, Mane M L, Ghasemi A and Morisako A 2012 Enhanced magnetic properties of Dy^{3+} substituted Ni-Cu-Zn ferrite nanoparticles *Appl. Phys. Lett.* **100** 42407

[23] Sattar A A, Wafik A H, El-Shokrofy K M and El-Tabby M M 1999 Magnetic Properties of Cu–Zn Ferrites Doped with Rare Earth Oxides *Phys. status solidi* **171** 563–569

[24] Roy P K, Nayak B B and Bera J 2008 Study on electro-magnetic properties of La substituted Ni–Cu–Zn ferrite synthesized by auto-combustion method *J. Magn. Magn. Mater.* **320** 1128–1132

[25] Kedesky, Katz 1953 *Ceramic Age* **62** 29–34

[26] Nicolson A M and Ross G F 1970 Measurement of the Intrinsic Properties of Materials by Time-Domain Techniques *IEEE Trans. Instrum. Meas.* **19** 377–382

[27] Salem M M, Morchenko A T, Panina L V, Kostishyn V G, Andreev V G, Bibikov S B and Nikolaev A N 2015 Dielectric and Magnetic Properties of Two-Phase Composite System: Mn-Zn or Ni-Zn ferrites in Dielectric Matrices *Phys. Procedia* **75** 1360–1369

Acknowledgments
This research was funded by RFBR, project no. 20-52-53020. In part of the ferrites preparation the work was supported by RFBR, project no. 20-08-00716.

Determination of Mechanical Characteristics Plasma Hardened Wheel Steel

A V Bogomolov[1], A T Kanaev[2] and T E Sarsembaeva[2]

[1]S. Toraighyrov Pavlodar State University, 64, Lomov Street, Pavlodar, 140008, Kazakhstan
[2]L. N. Gumilyov Eurasia National University, 2, Satpayev Street, Nur-Sultan, 010008, Kazakhstan

E-mail: bogomolov71@mail.ru

Abstract. A technique for measuring the mechanical properties of plasma-hardened wheel steel with nanoindentation is applied. The purpose of the work is to demonstrate and substantiate the serious differences between the mechanical properties and behavior of materials in nano-volumes from those obtained in traditional macroscopic tests. The features of measuring mechanical properties – hardness, Young's modulus, elastic recovery – affecting the wear resistance of the surface layers of the material are given. Measurement of the physic mechanical properties of the material makes it possible to evaluate and select the optimal technology for surface modification by surface plasma hardening. It is noticed that the objectivity of determining the hardness, elastic modulus, elastic recovery and flow stress depends on the parameters of the measuring equipment used and strict adherence to the requirements for the imprint depth depending on the thickness of the hardened layer. The studies carried out on wheel steel samples from the rim and from the ridge of a railway wheel subjected to surface plasma hardening under the same conditions. The HV and H of the wheel rim are greater than the corresponding values of the flange, and the young modulus of the wheel flange, on the contrary, is greater for the flange.

1. Introduction

Reliability and durability of machine parts and mechanisms operating under conditions of friction and wear, in many cases, is ensured by increasing the hardness of the surface layer. Under real operating conditions, minimization of wear depends on the elasticity and resistance to deformation of the surface layer not less than on hardness [1–5]. These properties are determined by the basic physical and mechanical characteristics of the surface layer, namely, hardness, elastic modulus and elastic recovery. These parameters are measured using instrumental indentation (nanoindentation) according to GOST R 8.748-2011 (ISO 14577-1: 2015), and their change is possible by surface modification, for example, by plasma hardening.

With the advent and development of the continuous measurement nanoindentation method, it became possible to quantify some important characteristics of metallic materials within individual submicroscopic zones, in particular, Vickers hardness, Young's modulus, flow stress, etc. Thus, in [6], nanoindentation aluminum alloys measured the submicrocrystalline hardness.

Speaking about the physical nature of nanohardness, it should be said that, despite the increased amount of information obtained by nanoindentation, the physical justification of micromechanisms of hardness remains poorly satisfactory. So, an example of the insufficiency of existing ideas about the nature of hardness is the problem of explaining the reasons for the large-scale size factor, which manifests itself in an increase in the number of hardness with a decrease in the load and dimensions of the print, especially at depths h <1 μm. An explanation of the reasons for the size factor by the dislocation plasticity mechanism by introducing the required number of dislocations did not lead to a resolution of the problem [7–9]. Formally, it is possible to introduce the required number of dislocations into the material under study, however, their density at h <100 nm becomes unrealistically large (>10^{14} sm^{-2}). In addition, this contradicts the results of microstructural studies, which do not detect such a number of dislocations. Obviously, the mechanical properties and behavior of materials in nano-volumes can, for a number of reasons, differ greatly from those obtained in traditional macroscopic tests, since with the decrease in the size of the loaded region by many orders of magnitude, many new factors begin to affect the material properties. The localization of the load leads to a strong hardening of the material in the deformation zone and the resulting large stress gradients can strongly affect the mechanisms of plastic flow. Therefore, it is not clear in which direction the material moves from under the indenter. It is believed that in plastic materials the material flows from under the indenter towards the free surface. In fact, mass transfer is directed into the volume, which compacts the material in the zone of local deformation [10–12].

Currently, there is a lot of direct evidence of the growing role of nonequilibrium point defects in mass transfer with a decrease in the size of the contact spot. There is also convincing evidence of a significant change in the structure of materials under the indenter because of amorphization, phase transformations, and the formation of a nanocrystalline structure. However, in existing theories of hardness proceeding from dislocation mechanisms of plastic deformation, these circumstances are not taken into account and are not explained [13–16].

In addition to all this, due to the small size of the deformed region (~ h), large relative strain rates $\varepsilon \sim v/h$ is realized even at small absolute penetration rates - v. As a result, the nanoscale hardness of a material can be two or more orders of magnitude higher than the yield strength. This implies the need to justify physical ideas about the nature of hardness, in general, and nanohardness, in particular.

Experimental studies show that when working out the optimal regimes of plasma hardening, it is necessary to ensure a smooth change in structure and microhardness in depth. The solution to this problem is facilitated by the use of the nanoindentation method when studying the physic mechanical properties of plasma-hardened steel wheels.

2. Object and research methods

In this work, the measurement of some mechanical characteristics of wheel steel subjected to surface plasma hardening was carried out using the UPNN-170 installation of the PlazmaCenter scientific and production company (St. Petersburg).

Technical characteristics of the UPNN-170 installation: rated current – 120 A, rated operating voltage – not more than 42 V, argon consumption 5 l/min, cooling water consumption 180–220 l/h. Hardness was measured by nanoindentation at the Center for the Study of Material Properties of Tomsk Polytechnic University.

Wheel steel is represented by two samples: sample No. 1 is cut from the rim, and sample No. 2 is from the crest of a railway wheel subjected to surface plasma hardening under the same conditions.

Samples for research with dimensions of 20x30 mm were subjected to grinding and polishing on a LaboPol-5 machine. Because of electrochemical polishing, the height of surface irregularities did not exceed 10 nm. Processing and analysis of the obtained surface images was carried out using Nova software. During the experiment, a probe sensor of the SPMProdeNSC 15/AIBS type was used. The Vickers hardness thus determined is equal to the average pressure on the contact surface of the "indenter-sample". The hardness was measured at room temperature under continuous loading with a linearly increasing in time load up to 150 mH.

The loading and unloading of the indenter, as well as the recording of the P-h diagram (applied load and indenter penetration depth) were carried out automatically. This method of measuring hardness, called the method of kinetic hardness (continuous indentation of an indenter), allows one loading-unloading cycle to determine the depth of unreduced h_{max} and recovered (plastic) h prints, Young's modulus, as well as the work of plastic and elastic deformation during indentation. Figure 1 shows the kinetics of the change in the load P and the penetration depth h of the indenter in the loading – unloading cycle (a) and the diagram of the dependence of the load P on the penetration depth h (b), where P_{max} is the maximum load applied to the sample, h_{max} is the penetration depth of the indenter; hf–residual area after unloading. *Apl* and *Aerabot* of plastic and elastic deformation, tgα is the slope of the linear part of the unloading part.

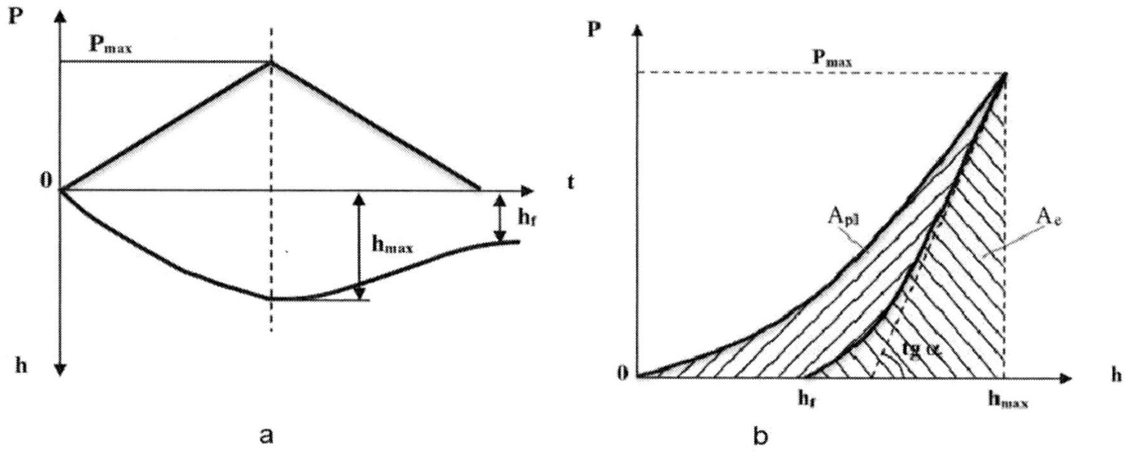

Figure 1. Dependences of the load P and the penetration depth h of the indenter on time t in the loading - unloading cycle (a) and a diagram of the dependence of P on h (b).

The maximum immersion depth of the indenter h_{max} using data from a tapping scanning probe microscopy measured the indenter footprint. The speed of loading and unloading the indenter was 300 mH/min. To process the test results used the method of Oliver and Farr [10].

The nanoindentation method consists in pressing an indenter with a diamond tip, with a load acting on it, into the surface layer of the material under study and determining the thickness of this layer using the software of a nanohardness meter. Processing of experimental data is carried out according to the results of measurements of at least 3 prints obtained under the same experimental conditions. Figure 2 shows the indenter prints in the surface layer of the material at a certain distance from the surface.

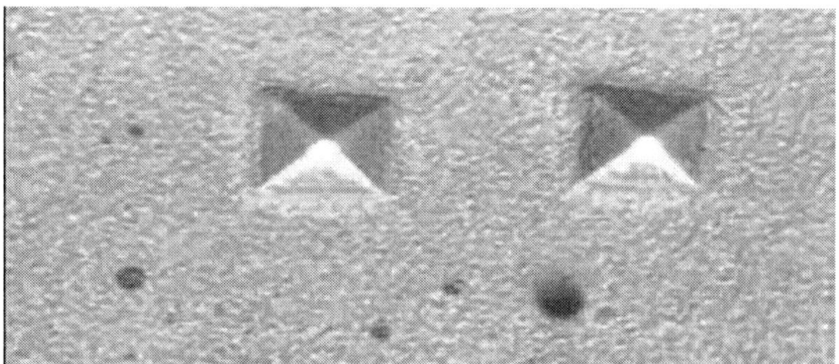

Figure 2. Imprint indenter in the near-surface layer of the material.

The work of the NanoHardness Tester nanosolid hardness tester is that because of passing a current pulse through the coils located in the magnetic field of a permanent magnet, an electric field is created that presses on the indenter with a diamond tip. The indenter drops to a point on the surface of the sample with a predetermined load. After the load on it has reached its maximum value, and the direction of the current of the coils is changing in the opposite direction, it returns to its original position.

After the load reaches its maximum value, the indenter begins to unload, the load acting on it gradually decreases to zero, and he returns to his original position. In this case, the unloading curve is produced, which shows that upon indentation the sample is deformed elastically elastic, i.e. the material from under the indenter does not return to its former position to the end. The mismatch between the load and unload lines is probably due to inelasticity.

3. Results and discussion

The results of determining the mechanical characteristics of plasma-hardened wheel steel according to are presented respectively in Table 1 for the wheel rim and in Table 2 for the wheel flange. Figure 3 shows one of the experimental loading and unloading curves showing the indentation process on sample No. 1 from the rim of a railway wheel. Similar curves were obtained for sample No. 2 wheel flange.

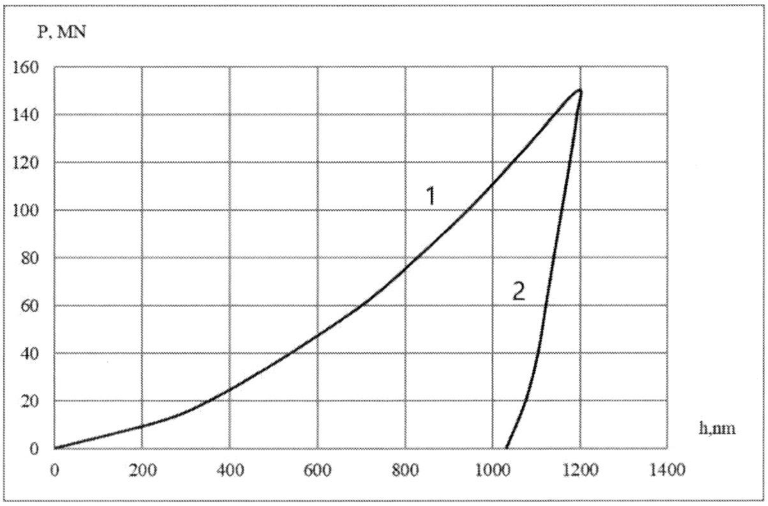

Figure 3. Loading (1) and unloading (2) curves for sample 1.

The plot of diagram 2 (Figure 3) gives information related to the elastic properties of the material; it is caused by the relaxation of energy accumulated during elastoplastic deformation during indenter penetration. The mismatch of curves 1 and 2 indicates the presence of permanent deformation after completion of the indentation procedure.

A comparison of the data in Tables 1 and 2 shows that the HV and H of the wheel rim are larger than the corresponding flange values, and the Young's modulus of the wheel flange, on the contrary, is larger for the flange compared to the rim (231.7–246.8).

The results obtained complement well the data presented by other researchers [17–20].

4. Conclusion

The objectivity of determining hardness, modulus of elasticity and elastic recovery depends on the parameters of the measuring equipment used and strict adherence to the requirements for the depth of the imprint depending on the depth of plasma hardening.

The mechanical properties and behavior of materials in nano-volumes can be very different from those obtained in traditional macroscopic tests, since with a decrease in the size of the loaded region (by many orders of magnitude), many new factors begin to affect the material properties.

The localization of the load leads to a strong hardening of the material in the deformation zone, the large stress gradients arising from this can strongly affect the mechanisms of plastic flow, and the structure of the materials under the indenter can significantly change because of amorphization, phase transformations, and the formation of a nanocrystalline structure.

Table 1. Mechanical characteristics of the wheel rim.

N of experiment	Mechanical characteristics of the wheel rim at 20 μm deviation from the edge of the sample deep into the metal			Mechanical characteristics of the wheel rim at 1700 μm deviation from the edge of the sample deep into the metal		
	HV	H, MPa	E, GPa	HV	H, MPa	E, GPa
1	398.00	4222.86	217.32	466.22	4939.52	254.11
2	385.32	4082.40	226.33	378.08	4005.66	245.44
3	361.22	3827.01	211.70	373.25	3954.54	254.95
4	389.10	4122.39	221.29	354.25	3753.25	245.98
5	358.02	3793.15	229.06	383.95	4067.88	230.76
6	395.26	4187.66	252.30	397.99	4216.62	248.90
7	366.76	3885.80	243.07	385.83	4087.81	230.62
8	386.15	4191.21	235.22	369.62	3916.01	261.58
9	406.59	4307.70	233.89	398.50	4222.04	253.07
10	463.92	3855.66	236.25	386.23	4092.04	251.74
11	366.71	3885.21	242.59	–	–	–
Aver.	388.80	4032.80	231.70	384.38	4072.39	241.41

Table 2. Mechanical characteristics of the wheel flange.

N of experiment	Mechanical characteristics of the wheel flange at 40 μm deviation from the edge of the sample into the metal			Mechanical properties of the wheel flange at 100 μm deviation from the edge of the specimen deep into the metal		
	HV	H, MPa	E, GPa	HV	H, MPa	E, GPa
1	350.74	3715.99	287.72	392.01	4153.16	210.41
2	354.93	3760.39	254.69	356.29	3774.81	210.65
3	334.65	3545.52	219.53	387.87	4109.37	230.99
4	365.16	3568.82	232.74	370.25	3922.27	241.84
5	367.65	3895.16	239.30	375.79	3981.44	247.86
6	–	–	–	366.42	3882.17	337.77
Aver.	354.6	3697.2	246.8	374.77	3323.51	246.59

References

[1] Firstov S 2008 Establishment of limit values of hardness, elastic deformation and the corresponding stress of materials by the method of automatic indentation *Materials Science* **61 8** 15–21

[2] Makarov A, Pozdeeva N and Savrai R 2012 Improving the wear resistance of hardened structural steel by nanostructural friction treatment *Friction and wear* **22 6** 444–445

[3] Gogolinsky K, Lvov N and Useinov A 2017 The use of scanning probe miroscopes and nanosolid testers to study the mechanical properties of solid materials at the nanoscale *Factory laboratory, Diagnostics of materials* **73 6** 28–36

[4] Tsui T, Pharr G and Oliver W 1995 Nanoindentation and Nanoscratching of Hard Carbon Coatings for Magnetic *Materials Research Society* **383**

[5] Leyland A and Matthews A 2019 On the significance of the H/E ratio in wear control: a nanocomposite coatings approach to optimized tribological behavior *Wear* **246 1** 1–11
[6] Chikova O, Shishkina E and Petrova A 2014 Measurement by hard nanoindentation of hardness of submicrocrystalline industrial aluminum alloys obtained by dynamic pressing *Physics of Metals and Metallurgy* **115 5** 555–560
[7] GolovinYu, Ivolgin V and Korenkov V 2017 Determination of the complex of mechanical properties of materials in nano volumes by nanoindentation methods *Condensed Matter and Interphase Boundaries* **3 2** 122–135
[8] Goldstein M, Litvinov V and Bronfin B 1986 *Metallophysics of High - Durable Alloys* (Moscow: Metallurgy) p 310
[9] Sosnin N, Ermakov S and Topolyansky P 2013 *Plasma technology* (St. Petersburg: Publishing House of the Polytechnic University) p 403
[10] Oliver W and Pharr G 2019 An improved technique for determining hardness and elastic modulus using load and displacement sensing indentation experiments *Journal Materials Research* **7 6** 1564–1583
[11] Golovin Yu 2008 Nanoindentation and mechanical properties of materials in a nanoscale (review) *FTT* **50 12** 1113–1142
[12] Kanayev A and Bogomolov A 2018 Increase of wear resistance and contact-fatigue strength of wheel steel by plasma hardening *Solid State Phenomena* **284** 1144–1150
[13] Kanayev A, Bogomolov A and Sarsembayeva T 2017 Overall hardening of solid-rolled wagon wheels by volume quenching and surface plasma processing *Solid State Phenomena* **265** 706–711
[14] Rybin V, Malyshevsky V and Khlusova E 2019 Technology for creating structural nanostructured steels *Metallurgy and heat treatment of metals* **648 6** 3–7
[15] Kozlov E, Popova N and Koneva N 2014 Fragmented structure formed in bcc steels during deformation *Izv. RAS, Physical Series* **68 10** 1419–1427
[16] Tushinsky L 2004 *Structural theory of structural strength of materials* (Novosibirsk, NSTU Publishing House) p 400
[17] Urtsev V, Gornostyrev Yu and Kornilov V 2012 Nanoengineering in the steel industry *Steel in translation* **42 2** 130–131
[18] Cui D, Wang H and Li L 2015 Optimal design of wheel profiles for high-speed trains *Proceedings of the institution of mechanical engineers Part F-Journal of rail and rapid transit* **229** 248–261
[19] Garnham J, Franklin F and Fletcher D 2007 Predicting the life of steel rails *Proceedings of the institution of mechanical engineers, Part F-Journal of rail and rapid transit* **221** 45–58
[20] Pandkar A, Arakere N and Subhash G 2015 Ratcheting-based microstructure-sensitive modeling of the cyclic hardening response of case-hardened bearing steels subject to Rolling Contact Fatigue *International journal of fatigue* **73** 119–131

IOP Conf. Series: Materials Science and Engineering **969** (2020) 012038 doi:10.1088/1757-899X/969/1/012038

Formation of Impurity Inclusions in Silicon when Smelting in Ore-Thermal Furnaces

N V Nemchinova, V V Hoang and A A Tyutrin

Irkutsk National Research Technical University, 83, Lermontova Street, Irkutsk, 664074, Russia

E-mail: ninavn@yandex.ru

Abstract. Silicon production by carbon reduction of silica-containing raw materials is performed in ore-thermal furnaces. The process proceeds at high temperatures (above 2,000°C). Consumers place high demands on the chemical purity of the silicon smelted, but the finished product contains a negligible amount of impurities. We have analyzed the causes for the impurity inclusions in silicon forming in smelting based on the data on the behavior of various elements penetrating the process with quartzite and reducing agent ashes. The chemical composition of furnace slag and silicon samples from the Silicon JSC of RUSAL (Russia) has been analyzed using various techniques, i.e. metallographic, X-ray fluorescence, X-ray diffraction, and X-ray spectral microanalysis. It has been established that the main source of non-metallic inclusions in silicon are furnace slag particles enmeshed in the melt when it is released from the tap. Due to the high degree of iron passing into the melt when smelting, the main metal inclusions are metal aluminum solutions based on iron, silicon, and other elements: $FeSi_2Al_{0,3}$, $FeSi_{1,5}Al_{0,3}$, and $FeSi_2Ti$. The main way to improve the smelted silicon quality is to select feed with a minimum amount of impurities and conduct the process at optimal temperatures in the quartzite melting zone.

1. Introduction

The metallurgical industry is an integral part of the developed world countries, including Russia [1, 2]. The production of metallurgical silicon (Si_{met}), which is used as a deoxidizer in the cast iron manufacture, a modifier of the alloy properties, an alloying element in alloys, and a feedstock to produce polycrystalline silicon, organosilicon materials, and silanes [3–7], is developing due to expanding the raw material base by involving new ore and man-made feeds and improving the process [8–14]. The global Si_{met} production reaches ~ 1 million tons per year; the largest producing country is China. The Norwegian Elkem Concern is the second-largest silicon manufacturer. In Russia, the largest metallurgical silicon manufacturer is Silicon JSC of RUSAL.

2. Research Objective

Silicon is obtained in ore-thermal furnaces (OTF) from silica-containing raw materials by reduction with carbon in the main reaction [4, 8, 15, 16]:

$$SiO_2 + 2C = Si + 2CO.$$

Quartzite is mixed with a mixture of carbonaceous reducing agents, and the resulting charge is loaded into an OTF. In the furnace, silica is reduced in the following stages: the SiO formation due to the silica evaporation in a reducing atmosphere, the interaction of SiO_2 with carbon to form SiC, and the interaction of SiC with silica and SiO to form elemental silicon. Due to the inevitably forming intermediate compound - gaseous silicon monoxide - the target product extraction is 70–74%.

To obtain a marketable product meeting the consumer requirements, the silicon smelted should contain minimum impurities (Fe, Ca, Al, Ti, etc.). At Silicon JSC, the charge for silicon production consists of quartzite obtained from the Cheremshansk deposit, being the enterprise's ore base, and a mixture of carbonaceous reducing agents (CRA): charcoal, petroleum coke, and coal from various deposits. The research objective was studying the behavior of the charge components when melting in OTF to analyze the mechanism of forming impurity inclusions in smelting products - silicon and slag from Silicon JSC.

3. Impurities Penetrating the Carbothermic Process

The choice of silica-containing feed for silicon smelting has very stringent requirements for chemical purity since most of the impurity elements penetrate the smelting process with the ore part of the charge (55% Fe, 9% Ca, 60% Al). The Cheremshansk quartzite, the deposit of which is represented by a bed of mono-mineral quartz sandstones, has the following composition, % wt.: SiO_2 – 98.75–99.46 (average 99.2), Fe_2O_3 – 0.14; CaO – 0.03 and Al_2O_3 – 0.37. Due to the stable crystal lattice, quartz usually does not contain a large amount of structural impurities. According to [17], the main share of impurities falls on monovalent (K, Na, Li) and divalent (Ca, Mg, Ni, Cu, Cr) alkaline earth cations, but additionally, they also include trivalent (Fe, Al) and tetravalent (Mn, Ti, Ge, etc.) atoms. Also, impurities penetrate the process with CRA ash. Along with chemical purity, CRA should have satisfactory reactivity to ensure the reduction in OTF. Depending on the origin, each carbon material differs in ash content and reactivity (Figure 1). In this regard, in industrial production, a mixture of active but expensive charcoal, low-ash low-reactivity petroleum coke, and hard coals produced in Russia, Kazakhstan, and Colombia (these materials are characterized by average ash content and reducing properties) is used.

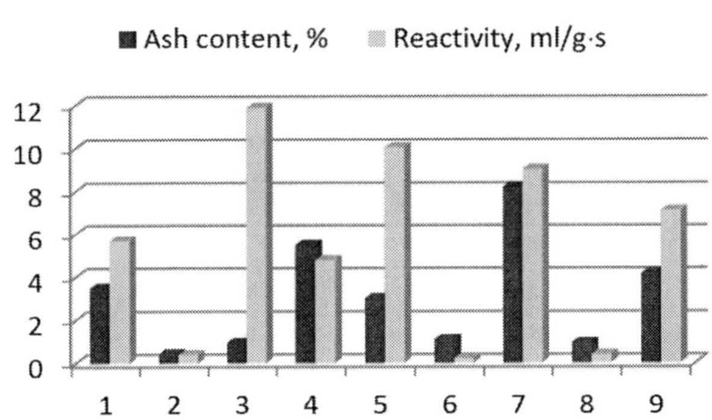

Figure 1. Comparison of the Carbonaceous Reducing Agent Properties: 1 – Charcoal, 2 – Petroleum coke, 3 – Wood chips, 4 – Coal, 5 – Lignin, 6 – Graphite, 7 – Peat coke, 8 – Pitch coke, 9 – Semi-coke.

4. Results and Discussion

We have studied the silicon and furnace slag samples from Silicon JSC. The Simet samples have been taken under the normal operating mode of OTF No. 6 (25 MVA capacity), and the slag samples from the taphole during the scheduled repair of the furnace. The chemical composition of samples has been studied by metallographic, X-ray fluorescence (XRF), X-ray diffraction, and electron microprobe analysis (EMPA) techniques.

Silicon production is considered slag-free, unlike similar ferroalloy production processes, but a small amount of slag is formed during carbothermic smelting (up to 5 % in high-power smelting and 2–3% in small and medium-sized furnaces), which reduces silicon extraction by an average of 2.2 and 1.2%, respectively).

The source of non-metallic inclusions in Si_{met} is slag particles enmeshed in the melt when it is released from the tap, that further crystallize reducing the finished product quality. Metallographic analysis has been performed using an Olympus GX-51 microscope (Olympus, Japan) equipped with an Altera20 digital camera. XRF has been performed using an S4 Pioneer spectrometer (Bruker AXS, Germany).

According to the studies, furnace slag contains intermetallic compounds and silicon inclusions, but to a greater extent - refractory oxygen-containing phases with a complex composition (as a result of the under-reduction of their oxide elements, including crystalline and amorphous silica). Also, the samples contain silicon carbide as a result of intermediate high-temperature interactions between silica and a reducing agent (according to the X-ray diffraction analysis performed using a DRON-7.0 diffractometer with Cu-K_α radiation on a nickel filter, the average Si content in slags is 44.1% wt).

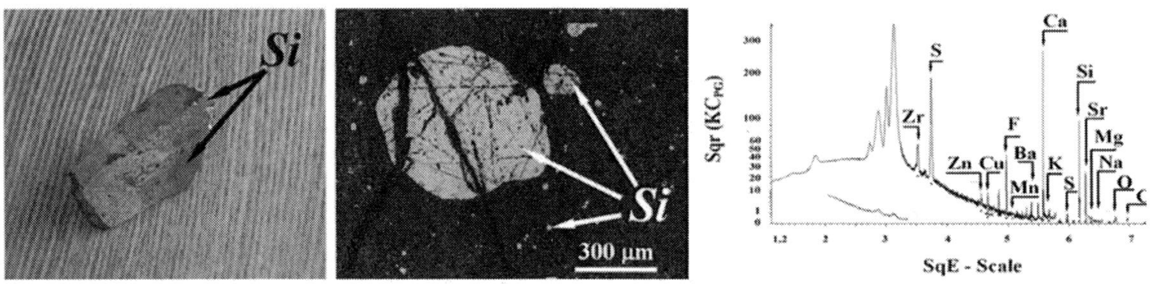

a b c

Figure 2. Furnace Slag of Silicon Production: a – general view of the sample; b – metallographic analysis (dark area); c – the XRF result.

We have also performed EMPA of Si_{met} samples using a Superprobe JXA-8200 electron probe analyzer (Jeol Ltd., Japan) (Figure 3, Table). Photos of the backscattered electron images of the samples under study (COMPO mode - atomic number contrast, TOPO mode - topographic contrast) have been taken using a scanning electron microscope, and the phase chemical composition determined using an energy-dispersive spectrometer. Measures to prepare a sample for metallographic studies included cutting samples using a Labotom-5 cutting machine (Denmark), pressing samples into the DuroCit resin to prepare samples for grinding, grinding and polishing the sample surface using a Tegramin – 25 grinding and polishing machine (Denmark), and etching the sample surface with an etchant of the following composition: 90 ml H_2O+5 ml HNO_3+5 ml HF.

According to the studies, the main inclusions in Si_{met} after smelting are intermetallic compounds, mainly silicides – $FeSi_2Al_{0,3}$, $FeSi_{1,5}Al_{0,3}$ (with Ti impurity), and $FeSi_2Ti$ (with Al impurity). The presence of Zr in inclusions is also confirmed by XRF since as a lining component, this element can penetrate the melt and intermetallic compounds. Non-metallic inclusions are oxygen-containing phases.

5. Analysis of Forming Impurity Inclusions in the Finished Product

The main aluminum supplies are related to corundum Al_2O_3 contained in quartzite (sometimes the OTF is fed with kaolinite $Al_2O_3 \cdot 2SiO_2 \cdot 2H_2O$, which decomposes when heated to 1,250°C with the formation of mullite $3Al_2O_3 \cdot 2SiO_2$). The reduction of aluminum from its oxide occurs through the stage of forming stable aluminum oxycarbides. Mono-oxycarbide Al_2OC decomposes in a peritectic reaction with the formation of aluminum metal ($Al_2OC \leftrightarrow 2\,Al + CO$), and in the high-temperature zone (above 2,000°C), aluminum mono-oxycarbide can be reduced to form aluminum carbide ($Al_2OC + 3C = Al_4C_3 + 2CO$). Aluminum carbide is reduced to silica at a temperature above 2,000°C ($2Al_4C_3 + 3SiO_2 = 8Al + 3Si + 6CO$).

In the presence of iron, silicon, copper, nickel, or their oxides, Al_4C_3 can interact with them to form metallic aluminum solutions based on iron, silicon, and other metals [19]. This can explain the high content of this impurity in silicon.

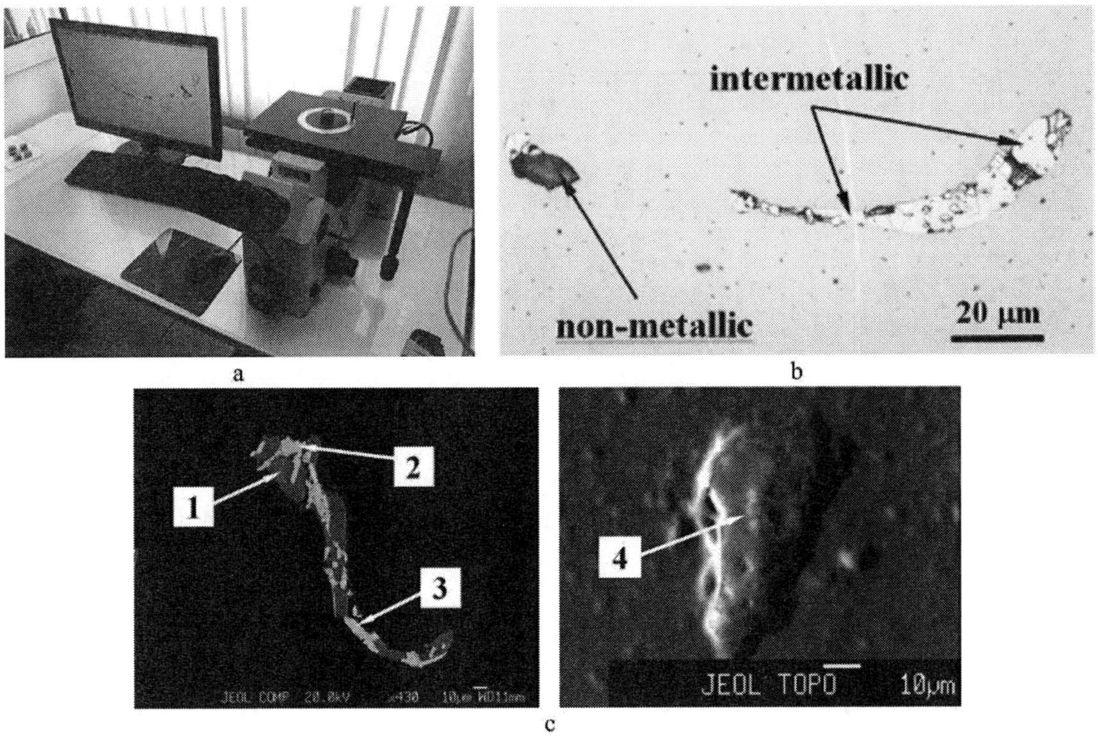

Figure 3. Silicon Sample Study Results: a – general view of the sample under study, b – metallographic analysis data (light area), c – EMPA (points 1–4 – see Table 1).

Table 1. EMPA Results for a Silicon Sample.

Probe point No. (Figure 3 b)	Si	Fe	Ti	Al	Zr	C	Total
1	51.661	42.721	0.000	5.618	0.000	-	100.000
2	41.125	34.883	2.548	9.291	12.154	-	100.000
3	35.725	32.491	24.842	1.196	5.745	-	100.000
4	24.250	-	-	-	-	75.750	100.000

According to the results of summarizing numerous analytical production data, various authors believe that ~ 50–88.7 % of aluminum passes into silicon smelted [4]. According to the latest data published, this figure is taken equal to 89% [20].

From the presence in the finished product point of view, one of the most problematic impurities is iron, which penetrates the OTF as hematite Fe_2O_3. In the presence of a gaseous reducing agent CO, it transforms into FeO while forming elemental Fe and CO_2. Iron carbide formed at 700–800°C ($3Fe + C = Fe_3C$) is spent to form aluminum oxycarbide and elemental iron by interaction with alumina. Elemental iron dissolves at high temperatures and therefore, is accumulated in the finished product. From iron oxide, the metal is reduced even at a temperature above 950°C ($FeO + CO = Fe + CO_2$).

The iron has a beneficial effect as it easily destroys SiC and interacts with SiO, thereby contributing to a shift of the reaction toward silicon formation but forming silicides ($Fe + nSiC = FeSi_n + nC$; $2Fe + SiO = FeSi + FeO$) [21].

When using materials with high iron content in the charge, the charge melting, rather than reduction processes develop first, which leads to the under-reduction of silica [22]. Therefore, ingress of feed with a significant content of iron-containing minerals into the process should be avoided; from this point of view, one of the ways to prepare feed for smelting is the magnetic separation of quartz grains.

The degree of Ca passing into the melt varies within a wide range (36–84% [4]) and depends mainly on the charge material composition and the operating mode. Calcium can penetrate the process with quartzite minerals: diopside, sphene, and limestone. In OTF, along with the reduction of silica, partial reduction of impurities from the reducing agent ash CaO occurs. During the smelting, part of calcium evaporates, passes into the reaction gas, is blown out of the furnace, and released as CaO dust. Part of gaseous CaO reacts with under-reduced silica to form silicates at 2,130°C ($2CaO + SiO_2 = 2CaO \cdot SiO_2$). Under-reduced CaO together with alumina and silica forms complex refractory composite oxides (slag phase), which deteriorate the Si_{met} grade. Thus, to reduce the Ca passing into silicon, it should be transferred into the gaseous state to a maximum extent, and hardly decomposable complex silicates should not be allowed to penetrate smelting.

Titanium penetrates the process as sphene, rutile, and ilmenite. According to [21], the ilmenite reduction sequence in the Si–O–C system is as follows:

$$FeTiO_3 \rightarrow TiO_2 + Fe \rightarrow Ti \rightarrow TiC.$$

Magnesium penetrates OTF as oxide (less commonly – magnesioferrite $MgFe_2O_4$). Magnesium forms stable silicide with silicon Mg_2Si. The degree of this impurity passing into silicon is taken equal to 10–35% [20].

6. Conclusions

According to studies, the main impurity inclusions in Si_{met} are represented by intermetallic compounds based on metal solutions of aluminum with iron silicides. The presence of aluminum carbide indicates the incompleteness of the recovery process. Silicon production slag is mainly composed of refractory compounds based on under-reduced metal oxides penetrating the process with a charge. When silicon is released from OTF, the slag is enmeshed in the melt, which causes the presence of non-metallic inclusions in the finished product. The main way to improve the quality of silicon smelted is to select feed with minimum impurities and conduct the process at optimal temperatures in the silica melting zone.

References

[1] Sizyakov V M, Vlasov A A and Bazhin V Y 2016 Strategy tasks of the Russian metallurgical complex *Non-ferrous metals* **1** 32–37

[2] Holappa L 2010 Toward sustainability in ferroalloys production *Proc. of the 12th Int. Ferroalloys Congress: Sustainable Future* 1–10

[3] Katkov O M 1999 *Smelting of Technical Silicon* (Irkutsk: ZAO Kremniy) p 244

[4] Gasik M I and Gasik M M 2011 *Electrothermics of Silicon* (Dnepropetrovsk: National Metallurgical Academy of Ukraine) p 487

[5] Schei A, Tuset JKr and Tveit H 1998 *Production of High Silicon Alloys* (Trondheim: Tapir Academic Press) p 363

[6] Andresen B 2010 The metallurgical silicon process revisited *Proc. Int. Conf. Silicon for the Chemical and Solar Industry X* 11–23

[7] Nemchinova N V 2017 *Physicochemistry and silicon carbothermy* (Irkutsk; Irkutsk State Technical University) p 288

[8] Ringdalen E 2015 Changes in Quartz During Heating and the Possible Effects on Si Production *JOM* **67** 484–492

[9] Zobnin N N, Baisanov S O, Baisanov A S and Musin A M 2020 *Proceedings of Irkutsk State Technical University* **24(2)** 444–459

[10] Alf T H, Espen D, Therese V B, Havard M, Per C, Trond A, Judit S and Per E W 2016 Carbon Neutral Metal Production – a new environmental friendly method for production of silicon *Proc. Int. Conf. Silicon for the Chemical and Solar Industry XIII* (Kristiansand) pp 1–8

[11] Nemchinova N V, Leonova M S and Tyutrin A A 2017 Experimental works on melting the blinded charge in silicon production *Proceedings of Irkutsk State Technical University* **21(1)** 209–217
[12] Tesfahunegn Y A, Magnusson T, Tangstad M and Saevarsdottir G 2018 Effect of electrode shape on the current distribution in submerged arc furnaces for silicon production - A modelling approach *Journal of the Southern African Institute of Mining and Metallurgy* **118(6)** 595–600
[13] Børset M T, Kolbeinsen L, Tveit H and Kjelstrup S 2015 *Energy* **90(2)** 1916–1921
[14] Nemchinova N V, Tyutrin A A, Korepina N A and Belskii S S 2018 On the possibility of carbonaceous dust waste use of prebaked anode production in silicon metallurgy *IOP Conf. Series Materials Science and Engineering* **411(1)** 012052
[15] Ringdalen E and Tangstad M 2012 Reaction mechanisms in carbothermic production of silicon, study of selected reactions *Int. Smelting Technology Symposium (The Minerals, Metals & Materials Society)* 195–203
[16] Vangskåsen J 2012 *Metal-producing Mechanisms in the Carbothermic Silicon Process* (NTNU-Trondheim) p 83
[17] Kuzmina V G and Kravets B N 1990 *Gangue quartz minerallurgy* (Moscow: Nedra) p 293
[18] Gasik M 2013 *Handbook of Ferroalloys: Theory and Technology* (Butterworth-Heinemann) p 536
[19] Schedrovitsky Ya S 1975 *Ferroalloy production in closed furnaces* (Moscow: Metallurgy) p 312
[20] Eryomin V P 2008 *Technology and Equipment of Ore Thermal Production "Electrothermy-2008"* 204–212
[21] Katkov O M 1993 *Russian Journal of Non-Ferrous Metalls* **3–4** 37–40
[22] Ragulina R I and Emlin B I 1972 *Electrothermy of silicon and silumin* (Moscow: Metallurgy) p 239

Acknowledgment
This work has been performed under research work 0667-2020-0037 within the scope of the State task of the Ministry of Education and Science of the Russian Federation for the performance of academic studies for 2020 and the planning period of 2021 and 2022.

IOP Conf. Series: Materials Science and Engineering **969** (2020) 012039 doi:10.1088/1757-899X/969/1/012039

Resource Saving Design of High Resistible Lining of a Teeming Ladle

L S Molchanov[1], B M Boichenko[2] and Ye V Synehin[2]

[1]Department of Steel Metallurgy, Institute of Ferrous Metallurgy named after Z. I. Nekrasov National Academy of Science of Ukraine, 1, Akademika Starodubova Square, Dnipro, 49050, Ukraine
[2]Department of Steel Metallurgy, National Metallurgical Academy of Ukraine, 4, Gagarina Ave., Dnipro, 49600, Ukraine

E-mail: Sinegin.ev@gmail.com

Abstract. It is described in the article a process of erosive destruction of teeming ladle lining. Special attention is paid to erosion by metal jet during tapping from melting furnace, for example BOF or EAF. This phenomena lead to steel contamination by non-metallic inclusions (NI). In case the steelmaking plant does not have any opportunity to remove NI from steel, for example argon blowing or electromagnetic stirring, a significant part of them remains in steel. Authors have proposed a lining design of teeming ladle, which alter a flow pattern in a way to move NI up. To confirm this hypothesis authors have developed a methodology of "water" modelling and high-temperature modelling with use of fusible metal. As a result of experiment topography of lining destruction for traditional and developed lining design has been achieved. An efficiency of developed lining design has been confirmed.

1. Introduction

Modern metallurgical production provides for the proceeding of complex technological operations to obtain the necessary properties of metal products. Refining of liquid steel proceeds under various physical and chemical conditions, that significantly increases the role of teeming ladle. The most significant design of metallurgical ladles affect the quality of metal melt directly after smelting, especially if operations of blowing with neutral gas in the ladle are excluded from the production cycle. That worsens significantly the conditions for NI removing from the melt, especially at the finishing stages of steel production [1–5].

The interaction of the metal melt with the lining in the casting ladle is based on two physical processes: hydrodynamic destruction of the lining by fluid flows and the ascent of NI in a liquid medium.

It is well known that NI significantly affect the quality of metal products. The removal of NI from steel occurs due to the action of Archimedes force and the convective metal flows. For NI of small size, the influence of convection prevails, and for large ones – Archimedes force [6]. This process can be intensified by increasing the speed of the upward flows of the melt using blowing or electromagnetic stirring. However, the use of these technological operations for ladles of small capacity is practical because of the significant heat losses and the risk of drawing slag inclusions from the coating slag back into the metal when argon is blown [7–9]. Electromagnetic processing requires significant capital expenditures for its implementation [10, 11]. Therefore, an alternative technologically expedient method of refining steel from NI is the exposure of the molten metal in the ladle.

The floating of a single NI in a liquid metal melt is described by the Stokes law. However, for a number of reasons, including the intensive movement of flows caused by the tapping of steel from a furnace, the Stokes law is violated [12, 13]. The study of the floating of a NI ensemble in the melt by computer and mathematical modeling is extremely difficult. The best option for this purpose is physical modeling. The aim of the study is to carry out a physical simulation to determine the characteristics of the floating of an ensemble of NI in liquid steel and define the design of lining to intensify removal of NI.

2. Goals of research

At previous stages of research, it was defined that a shape change of the working space of the ladle significantly affects the removal efficiency of NI [14]. At the same time, the influence of the defined rational parameters of the working space of the teeming ladle on the hydrodynamic destruction are not established. Thus, the goal of this study is to determine the influence of geometrical parameters of the teeming ladles lining on the hydrodynamic features of the interaction of refractory materials with high-temperature metallurgical melts (slags and metals).

3. Methodology of research

The study of the hydrodynamic features of the interaction of the teeming ladle lining with metallurgical melts was based on the use of the π-theorem [15–19] and included two stages: cold modeling and high-temperature modeling using liquid tin (99.5 % Sn).

Mechanical or erosive destruction of the lining by a jet of liquid metal is the most intensive in the bottom area of the ladle at the beginning of steel tapping from a furnace. It is obvious that the jet impulse plays a significant role in the destruction, the value of which depends on the acceleration of gravity g, the volumetric flow rate q, the fluid density ρ and the length of the jet h. As the parameter that characterizes the resistance of the refractory to erosion destruction, its strength σ was assumed. Destruction rate of the refractory is described by the mass flow m, which indicates the weight of the refractory material separated from the lining surface during its contact with the jet. Generalized formula is follow:

$$m = f(g, q, \rho, h, \sigma). \tag{1}$$

The similarity numbers obtained by the Markov method [15] are given in Table 1. Magnitude of most of them is close to zero, which testifies of their adequacy.

Table 1. Similarity numbers for describing erosive wear of refractories.

Parameter	Magnitude	Unit	Base unit	Similarity number	Magnitude
m	10^{-2}	[kg·m^{-2}·sec^{-1}]	-	$\pi_1 = \frac{m}{\rho \cdot \sqrt{g \cdot h}}$	10^{-7}
σ	10^{4}	[kg·m^{-1}·sec^{-2}]	-	$\pi_2 = \frac{\sigma}{\rho \cdot g \cdot h}$	10^{-1}
ρ	10^{3}	[kg·m^{-3}]	[kg]	-	-
g	10^{1}	[m·sec^{-2}]	[sec]	-	-
h	10^{0}	[m]	[m]	-	-
q	10^{-1}	[m^{3}·sec^{-1}]	-	$\pi_3 = \frac{q}{\sqrt{g \cdot h^5}}$	10^{-3}

In Table 2 the calculated scales of the selected parameters were shown. The experiment included a series of 5 consecutive fillings for a ladle model with a traditional lining and a developed one. These designs of ladle lining design are presented in Figure 1.

The operating the lining of teeming ladles during the experiment was as close as possible to industrial conditions (according to the research conditions, the drop height of the metal jet and the duration of the liquid flow were controlled).

Table 2. The values of the parameters for the model and nature, their scale and the domain of selected similarity numbers.

Parameters and similarity numbers	Original	Model	Scale
Mass flow m [$kg\cdot m^{-2}\cdot sec^{-1}$]	$1,5\cdot 10^{-2}$	$4,15\cdot 10^{-4}$	1:36
Strength of refractory σ, [$kg\cdot m^{-1}\cdot sec^{-2}$]	40000	221	1:181
Fluid density ρ, [$kg\cdot m^{-3}$]	7200	1000	1:7,2
Length of the jet h, [m]	3…4	0,12…0,16	1:25
Volumetric flow rate q, [$m^3\cdot sec^{-1}$]	58…77	0,018…0,024	1:3182
π_1	$3,3\cdot 10^{-7}…3,8\cdot 10^{-7}$		-
π_2	0,1…0,2		-
π_3	$5,8\cdot 10^{-4}…1,2\cdot 10^{-3}$		-

In order to simulate the lining of teeming ladle with calculated strength (see Table 2), a sandy-clay mixture was used for low-temperature modeling in a ratio of 21:5 (at high temperature, mortar MMK-72 [20] after drying and thermal strengthening and heating to 400 °C), which was formed into a ladle in a wet state. Lining drying was carried out naturally by holding for 10 days at a temperature of 28…30 °C.

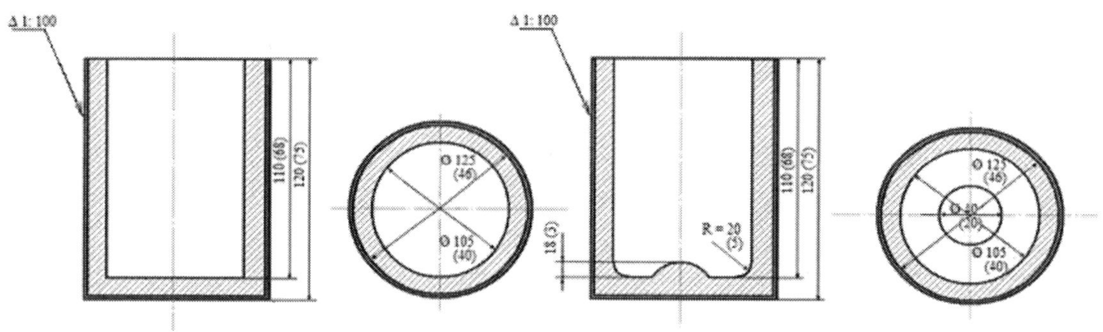

a) b)

Figure 1. Sketch of teeming ladles lining: a – traditional design; b – developed design; without brackets, the values for "water" modeling are given, and in brackets for high-temperature modeling.

The study of the process of erosion destruction of the teeming ladle lining was carried out according to the following methodology. After filling the ladle with liquid, it was held for 2…3 minutes, after which the water was drained through the upper edge of the ladle. Each campaign of the lining of the teeming ladle model assumed a consistent acceptance of 5 heats. In order to eliminate errors in the implementation of the experiment, each experiment was repeated 5 times.

At the end of each experiment in the series, a mechanical scanning of the surface of the lining of the teeming ladle model was performed. The measurement accuracy was 0.1 mm.

In order to obtain the detailed map of lining wear during the operation of teeming ladle, its surface was cut by four planes (Figure 2) passing through the vertical axis of the ladle model (planes

I-I, II-II; III-III; IV-IV). There were also 5 points in the bottom part of the bucket along the line of intersection of each of the planes: 0 is a center of the bottom part (a place where the jet falls) 1 is 25% of the bottom part radius; 2 is 50% of the bottom part radius; 3 is 75% of the radius of the bottom part and 4 is 100% of the radius of the bottom part. Along the projection of each of the planes on the side surface of the walls of the lining 5 points were distinguished: 0' is the junction of the bottom and the walls of the ladle; 1' is 25% of the height of the working space of the ladle; 2' is 50% of the height of the working space of the ladle; 3' is 75% of the height of the ladle working space and 4' is 100% of the height of the ladle working space.

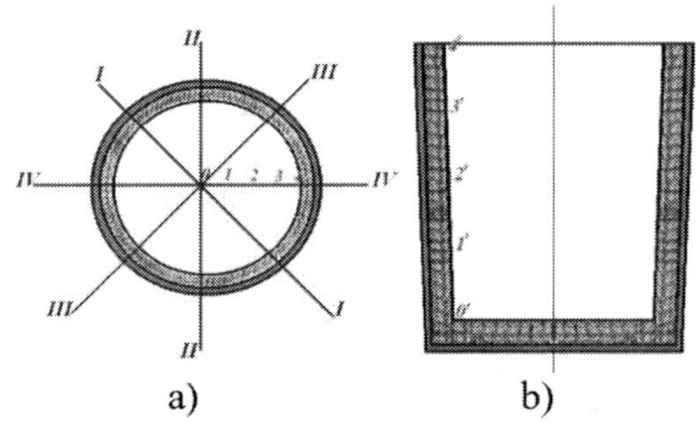

Figure 2. Scheme of topography of the wear of the teeming ladle lining during the "water" simulation: a) the horizontal projection; b) vertical projection on the half-plane I-0.

The high-temperature modeling of destruction of the teeming ladles lining a contamination level of the metal melt with NI were also determined, for which samples of the melt after melting and filling and holding of tin were taken in each series of experiments. The obtained samples were analyzed by using metallographic methods of analysis. Determining the nature of the origin of NI was carried out by X-ray analysis, which allowed determining their mineralogical compound.

4. Results of studies

The results of studies of the hydrodynamic destruction of the teeming ladle lining during operation and the influence of the lining design on this process are presented in Figure 3. During the research, a detailed topography of the erosional destruction of the steel-teeming ladles in the process of operation for the standard design and optimized was obtained (Figure 3). Thus, in the case of the use of the linings of both structures, considerable wear is observed along the O-II line, due to the peculiarities of the experiment (liquid was drained along this line simulated melt after filling).

It can be seen in the Figure 3 a, c, the significant erosion destruction of traditional design of lining to almost ½ the height of the working space of the ladle, which is caused by considerable turbulization of melt flows. Due to this, the energy of the fluid flow is significantly increased, destroying the lining. The average rate of the bottom destruction when "water" modeling was 0.73 mm per melt, and when high-temperature modeling was 1.41 mm per melt.

In the case of using the developed lining design (Figure 3b, d), the turbulence of the flows decreases, which leads to a decrease in the destruction rate. On the topography of the destruction, it can be observed as a decrease in the total destruction area of the lining by ¼ of the ladle height. The average destruction rate for "water" modeling was 0.59 mm per melt, and for high-temperature modeling it was 1.14 mm per melt.

As a result of research, it was established that the application developed lining design makes it possible to decrease the destruction rate of the lining by 18.1% (according to "water" modeling) and by 17.2% (according to high-temperature modeling).

A study of samples of tin taken during high-temperature modeling made it possible to determine that contamination of the melt with NI decreases by 35.3% when using the developed design of the ladle lining in comparison with the traditional. The X-ray analysis of the samples let to determine the presence of NI of the $Al_2O_3 - SiO_2$ system.

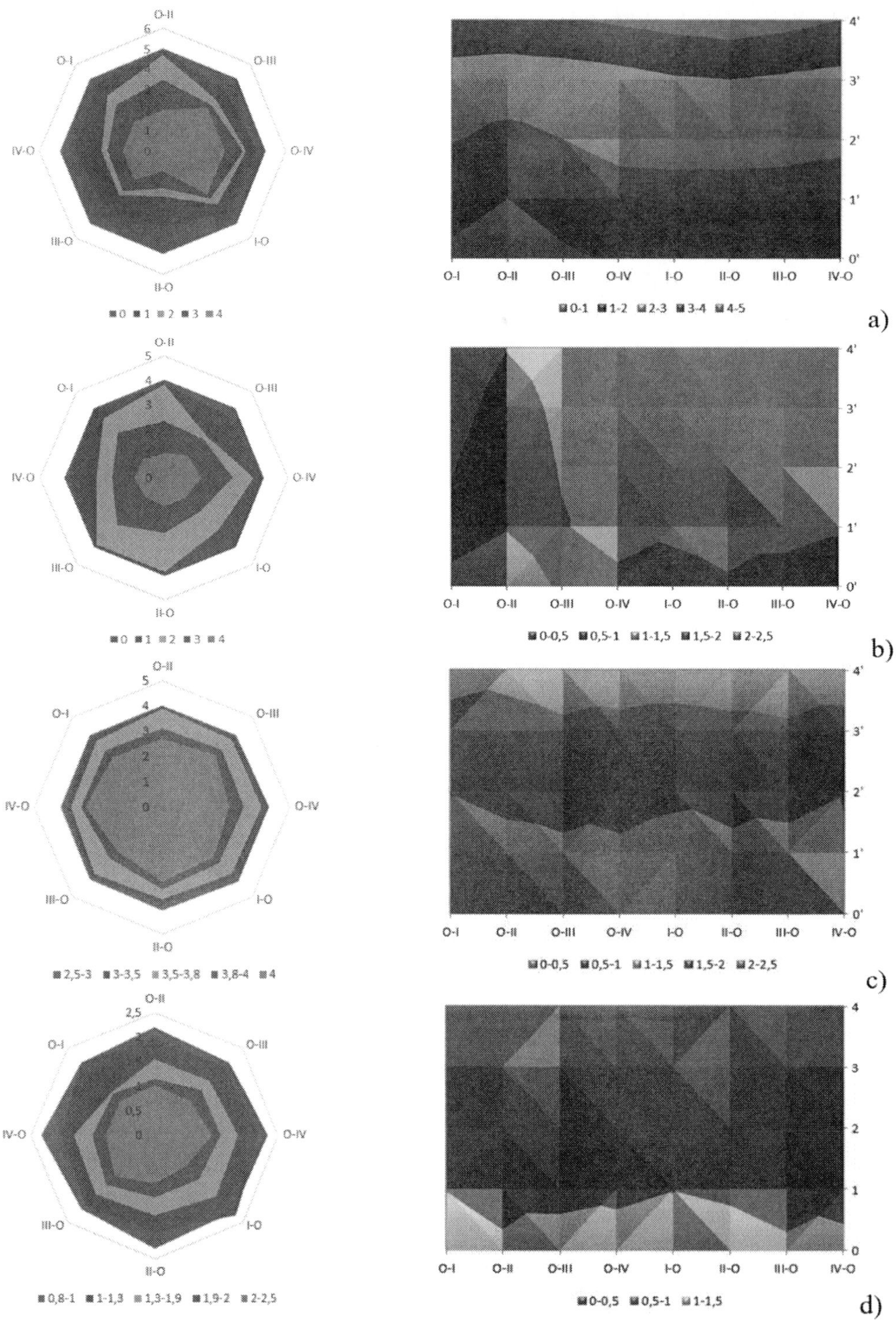

Figure 3. Topography of lining wear on the ladle model (legend's colored markers show the depth of wear in millimeters): a) the traditional design of lining after "water" modeling; b) developed design of lining after "water" modeling; c) the traditional design of lining after high-temperature modeling; d) developed design of lining after high-temperature modeling.

5. Summary

A complex physical modeling of the influence of the geometrical parameters of the working space of the teeming ladle on the destruction of the lining was carried out. The efficiency of the developed lining design for teeming ladles has been confirmed. The use of the proposed lining design allows increasing the life of refractories by 17.2%, and to increase the purity of the metal by 35.3%.

References

[1] Abdrakhmanov R I, Shevchenko S V and Igonin A Iu 2012 Exploitation of teeming ladles with various lining *Steel* **2** 30–33
[2] Iakushev E V, Alekseev E M, Kutcevalov V M and Mikheev A E 2010 Improving of facing technology of work layer of teeming ladle *Metallurgist* **2** 49–52
[3] Koleda I V 2004 Use of magnesium-carbon refractories in work layer of teeming ladle *Ferrous metallurgy* **10** 67–69
[4] Primachenko V V, Martynenko V V, Babkin L A, Kasian G I and Mintc A Ia 2002 Experience of application a domestic low-cement concrete as a lining of teeming ladles *Steel* **12** 19–20
[5] Stoliarovskii O A, Sadovskii V G and Danskov V A 2010 Experience of application heat-resistant concretes in a lining of metallurgical aggregates *Steel* **6** 83–85
[6] Okhotskyi, V B, Kostolov O L, Cimonov V K, Kovalov D A, Tarakanov A K and Kucher A H 1997 *Theory of metallurgical processes* (Kyiv: IZMN)
[7] Velychko O H, Stoianov O M, Boichenko B M and Niziaev K H 2016 *Technology of steel quality improvement* (Dnipropetrovsk)
[8] Ziborov A V, Zyranov V V, Kuznetcov M S and Kulagin S A 2008 Influence of technology of ladle treatment on steel pollution by non-metallic inclusions *Metallurgist* **3** 39–40
[9] Siddiki F, Kodak A V, Iavtushenko P M and Trotcan A I 2008 Influence of parameters of ladle treatment of steel grade St10 on pollution of continuous cast billet (round 180 mm) by non-metallic inclusions *Casting processes* **1** 38–40
[10] Sang-Ik Chung, Young-Ho Shin and Jong-Kyu Yoon 1992 Flow Characteristics by Induction and Gas Stirring in ASEA-SKF Ladle *ISIJ International* **32 12** 1287–1296.
[11] Lidong Teng 2017 Effect of EMS on Inclusion Removal in Ladle Furnace for Specialty Steel Production *Proceedings of AISTech* **20171395–1402**
[12] Mastriukov B S 1996 *Thermal physic of metallurgical processes* (Moscow: MISIS)
[13] Frolov J G 1988 Course of colloid chemistry. Surface phenomena and dispersed systems (Moscow: Chemistry)
[14] Molchanov L S, Synehin Y V, Lantukh O S and Ryshkova I S 2018 Research of Non-metallic Inclusions Removal in Teeming Ladles of Various Design *Journal of Engineering Sciences* **5** **2** C1–C4
[15] Markov B L 1984 *Physical modeling in metallurgy* (Moscow: Metallurgy)
[16] Gukhman A A 1974 *Application of similarity theory for study of heat and mass transfer processes* (Moscow: Vysshaia shkola)
[17] Gukhman A A 1973 *Introduction in similarity theory* (Moscow: Vysshaia shkola)
[18] Rumiantcev V D 2006 *Theory of heat and mass transfer* (Dnipropetrovsk: Porogi)
[19] Baptizmanskii V I, Iakovlev Iu N and Paniotov Iu S 1992 *Organizing of experiment* (Kyiv: UMKVO)
[20] GOST 6137-2015, Refractory alumosilicate mortars. Specifications

IOP Conf. Series: Materials Science and Engineering **969** (2020) 012040 doi:10.1088/1757-899X/969/1/012040

The Effect of Temperature and Roasting Time on the Conversion of Zinc Ferrite to Zinc Oxide in the Electric Arc Furnace Dust

A G Ryazanov, A V Senin and N A Kornilov

South Ural State University, 76, Lenin Prospect, Chelyabinsk, 454080, Russia

E-mail: ryazanov_andrei@mail.ru

Abstract. The dust of electric arc furnaces (EAF-dust) contains significant amounts of zinc in the form of various chemical compounds. Therefore, dust can be used as a raw material for zinc extraction. EAF-dust containing 20 wt.% Zn has been studied. The effect of dust roasting with Ca O and Na_2CO_3 additives on the degree of conversion of $ZnFe_2O_4$ to Zn O was studied experimentally. The effect of the roasting temperature of 700, 850 and 1000 °C and the roasting duration of 1, 2 and 3 hours were established. Ca O and Na_2CO_3 additives in dust contribute to the destruction of $ZnFe_2O_4$ due to the formation of $Ca_2Fe_2O_5$ and $NaFeO_2$. After treatment, zinc is presented in the form of acid soluble zinc oxide. The content of acid-soluble zinc increases from 25% in the initial dust to 97–98% in the roasting product ("cinder") under optimal roasting conditions.

1. Introduction

Zinc is widely used for hot and electrolytic galvanizing of steel products. Up to 80 % of zinc produced in the Russian Federation is used in galvanizing processes [1]. World consumption of zinc increases annually by 1.1–1.5 % [3]. A part of galvanized metal scrap is melted in electric arc furnaces. During smelting, zinc evaporates and is oxidized by atmospheric oxygen to form various solid oxide compounds of zinc. 15–25 kg of zinc-containing dust is formed per ton of steel produced [8, 9], up to 90 % of the initial zinc goes into dust [4]. EAF-dust contains up to 20 wt.% zinc [5–7]. Recycling of dust by remelting method does not apply due to the harmful effect of zinc on the furnace lining [18]. In different countries, including the Russian Federation, some of the dust is stored at specialized landfills, and some of the dust is processed through using zinc sublimation technology using carbon, for example, the Waelz-process [10–15]. Products of the Waelz-process (waelz-oxide) are processed by leaching in sulfuric acid, after the removal of halogens [17, 25, 26]. Zinc is extracted from the sulphate solution by electrolysis. Direct dust leaching is often not carried out due to the presence of zinc ferrite $ZnFe_2O_4$ in dust, about 60 % of zinc in dust is part of zinc ferrite [5–7, 16]. Alkaline treatment of dust has an advantage over reduction technology, since the use of expensive carbon materials (for example, coke) is not required. Leaching EAF-dust in alkaline solutions is possible, but the limiting factor is the presence of zinc ferrite in dust [18–23].

High-temperature roasting of dust with CaO and Na_2CO_3 additives is one of the options for increasing the proportion of acid-soluble zinc (in the form of ZnO) in dust. CaO and Na_2CO_3 contribute to the destruction of $ZnFe_2O_4$ due to the formation of $Ca_2Fe_2O_5$ and $NaFeO_2$. Previously, we found the optimal amount of CaO and Na_2CO_3 additives to the initial dust for roasting [2].

In this study, the effect of temperature and roasting time on the conversion of zinc ferrite to zinc oxide was studied overviwed.

2. Experimental

2.1. Materials and apparatuses

The chemical composition of EAF-dust is given in Table 1. The following reagents were used as additives: calcined calcium oxide GOST 8677-79, soda ash GOST 5100-85. The roasting was carried out in a muffle furnace in aerial environment. The phase composition of the samples was determined by the method of X-ray analysis (diffractometer Rigaku Ultima IV, Nanotechnology R&E Centre, South Ural State University [24]).

Table 1. The chemical composition of EAF-dust, [wt.%].

Element	Content	Element	Content	Element	Content
Zn (total)	20.41	Al	0.5	Na	1.98
Zn (in ZnO)	5.04	Fe	24.18	Cl	1.8
Pb	1.43	Ca	7.87	F	0.078

2.2. Procedures

The composition of the mixture for roasting, in grams, is 50.0 dust; 6.28 CaO and 17.79 Na_2CO_3. The components were thoroughly mixed in a mechanical mixer. The mixture in the form of a pyramid was placed on the surface of the refractory substrate. Periclase-chromite brick was used as a substrate. The roasting of the mixture was carried out at a temperature of 700, 850 and 1000 °C for three hours. The pyramids retained their shape at a temperature of 700 °C. The pyramids softened and turned into an oblate cylinder form at 850 and 1000 °C. The change in the shape of the pyramids is associated with the melting of Na_2CO_3. The softening effect of the charge must be considered in the technological process. At a constant temperature of 1000 °C, roasting was performed using an exposure of 1, 2, and 3 hours.

The roasting products ("cinder") were investigated for the content of total zinc and acid-soluble zinc in the form of ZnO. The fraction of soluble zinc ε_{Zn} (in %) was calculated from equation (Eq. 1):

$$\varepsilon_{Zn} = \frac{(\%Zn)_{ZnO}^{cinder}}{(\%Zn)_{total}^{cinder}} 100 , \qquad (1)$$

where $(\%Zn)_{ZnO}^{cinder}$ is the zinc concentration in zinc oxide, wt.%; $(\%Zn)_{total}^{cinder}$ is the total concentration of zinc in cinder, wt.%. The transition degree of zinc γ_{Zn} (in %) from zinc ferrite to zinc oxide was calculated from equation (Eq. 2):

$$\gamma_{Zn} = \left(1 - \frac{\left((\%Zn)_{total}^{cinder} - (\%Zn)_{ZnO}^{cinder}\right) m_{cinder}}{\left((\%Zn)_{total}^{dust} - (\%Zn)_{ZnO}^{dust}\right) m_{dust}}\right) 100 , \qquad (2)$$

where $(\%Zn)_{total}^{dust}$ is the total concentration of zinc in the initial dust, wt.%; $(\%Zn)_{ZnO}^{dust}$ is the zinc concentration in zinc oxide in the initial dust, wt.%; m_{dust} is the mass of initial dust, g; m_{cinder} is the mass of cinder, g.

3. Results and Discussion

The X-ray diffraction patterns of the four samples (initial EAF-dust and cinder after treatment at 700, 850 and 1000 °C for 3 hours) are given as an example in Figure.

With increasing temperature, the intensity of the $ZnFe_2O_4$ reflections consistently decreases until disappearing, and the reflex intensity of ZnO, $Ca_2Fe_2O_5$ and $NaFeO_2$ increases. The experimental results are in complete agreement with the results of thermodynamic analysis on cinder composition changes during heating, which were established earlier [2].

The effect of temperature and roasting duration on the increase in the content of acid-soluble zinc are given in the Table 2. The content of acid-soluble zinc increases from 25 % in the initial dust to 97–98 % in the cinder under optimal roasting conditions (1000 °C, the roasting time of 3 hours).

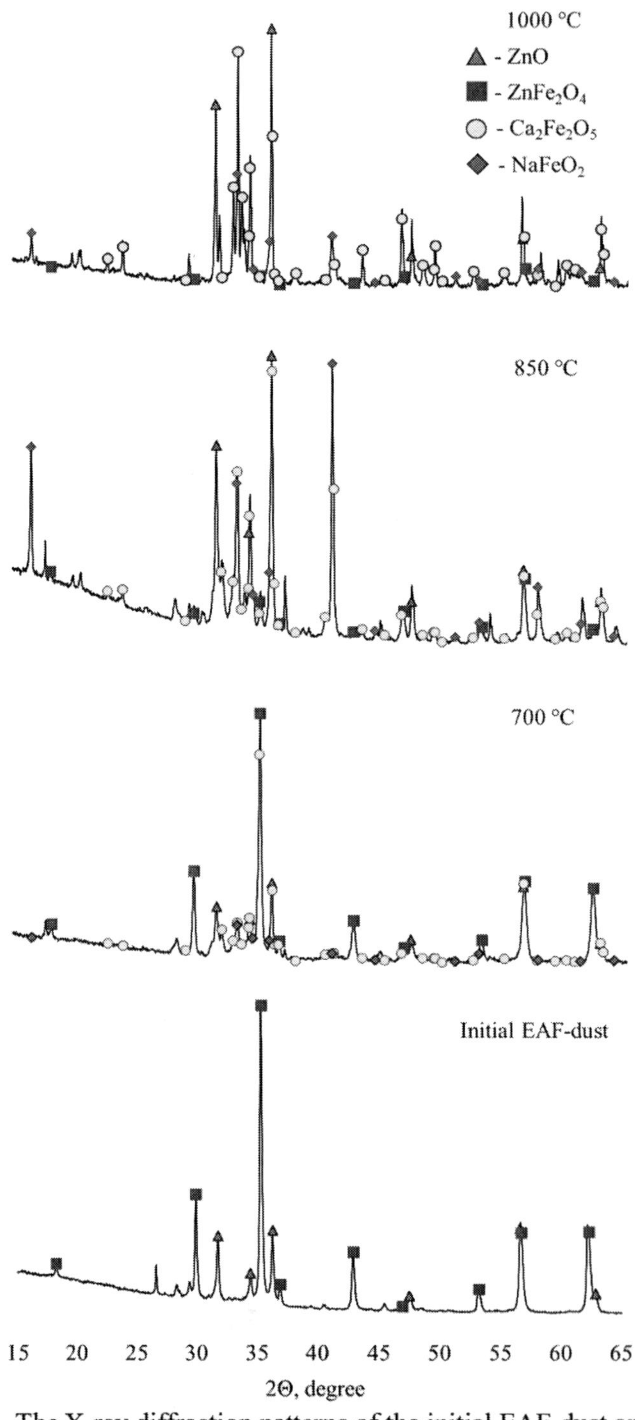

Figure 1. The X-ray diffraction patterns of the initial EAF-dust and cinder.

Table 2. The influence of temperature and the roasting time on the transition degree of zinc γ_{Zn} and on the fraction of soluble zinc ε_{Zn}.

Temperature, [°C]	Roasting time, [hour]	γ_{Zn}, [%]	ε_{Zn}, [%]
Initial EAF-dust	–	–	24.7
700	3	42.4	50.6
850	3	91.6	92.8
1000	**3**	**97.5**	**97.9**
1000	2	89.2	90.8
1000	1	83.9	86.2

4. Summary

The effect of dust roasting with CaO and Na2CO3 additives on the degree of conversion of ZnFe2O4 to ZnO was studied experimentally. The effect of the roasting temperature of 700, 850 and 1000 °C and the roasting duration of 1, 2 and 3 hours were established. The content of acid-soluble zinc increases from 25 % in the initial dust to 97–98 % in the cinder under optimal roasting conditions (1000 °C, the roasting time of 3 hours).

The roasting product ("cinder") with a zinc as zinc oxide can be used in alkaline leaching to produce zinc pregnant solution.

References

[1] Online. Available: https://www.ugmk.com/analytics/surveys_major_markets/tsink/

[2] Ryazanov A G, Senin A V and Khmeleva O V 2019 Zinc Ferrite Decomposition by Sodium Carbonate and Calcium Oxide at Roasting of Steelmaking Dust *Materials Science Forum* **946** 474–479

[3] Online Available: http://www ilzsg org/static/

[4] Senin A V and Chumanov I V 2013 Thermodynamic analysis of refining the iron melt from zinc *Bulletin of the South Ural State University Series Metallurgy* **13 1** 176–178

[5] Pereskoka V V, Kamkina L V, Projdak Y S, Stovpchenko A P and Kvichanska M I 2010 Reducing and thermal treatment of electric filter dust of electric arc furnace *Bulletin of the Priazov State Technical University* **21** 13–16

[6] Shawabkeh R A 2010 Hydrometallurgical extraction of zinc from Jordanian electric arc furnace dust *Hydrometallurgy* **104 2** 61–65

[7] Pickles C A 2009 Thermodynamic analysis of the selective chlorination of electric arc furnace dust *J. Hazard Mater.* **166 2** 1030–1042

[8] Guézennec A G, Huber J C, Patisson F, Sessiecq P, Birat J P and Ablitzer D 2005 Dust formation in electric arc furnace: birth of the particles *Powder Technology* **157** 2–11

[9] Huber J-C, Patisson F, Rocabois P, Birat J-C and Ablitzer D 1999 Some means to reduce emissions and improve the recovery of electric arc furnace dust by controlling the formation mechanisms *Global Symposium on Recycling Waste Treatment and Clean Technology* **II** 1483–1492

[10] Panshin A M, Leontiev L I, Kozlov P A, Dubanov V G, Zatonsky A V and Ivakin D A 2012 Technology of dust processing of electric arc furnaces of OJSC Severstal in the Welz complex of JSC CZP *Ecology and Industry of Russia* **11** 4–6

[11] Roth J L, Frieden R, Hansmann T, Monai J and Solvi M 2001 PRIMUS a new process for recycling byproducts and producing virgin iron *Rev. Metall* **98 11** 987–996

[12] Antrekowitsch J and Antrekowitsch H 2001 Hydrometallurgically recovering zinc from electric arc furnace dusts *JOM* **53 12** 26–28

[13] Pickles C A 2008 Thermodynamic analysis of the selective carbothermic reduction of electric arc furnace dust *J. Hazard Mater.* **150** **2** 265–278
[14] Wu C C, Chang F C, Chen W S, Tsai M S and Wang Y N 2014 Reduction behavior of zinc ferrite in EAF-dust recycling with CO gas as a reducing agent *J. Environ. Manage* **143** 208–213
[15] Suetens T, Klaasen B, Van Acker K and Blanpain B 2014 Comparison of electric arc furnace dust treatment technologies using exergy efficiency *Journal of cleaner production* **65** 152–167
[16] Montenegro V, Agatzini-Leonardou S, Oustadakis P and Tsakiridis P 2016 Hydrometallurgical treatment of EAF dust by direct sulphuric acid leaching at atmospheric pressure *Waste and Biomass Valorization* 7 6 1531–1548
[17] Morcali M H, Yucel O, Aydin A and Derin B 2012 Carbothermic reduction of electric arc furnace dust and calcination of waelz oxide by semi-pilot scale rotary furnace *Journal of Mining and Metallurgy Section B Metallurgy* **48** **2** 173–184
[1] Dutra J B, Paiva P R P and Tavares L M 2006 Alkaline leaching of zinc from electric arc furnace steel dust *Minerals engineering* **19** **5** 478–485
[18] Chairaksa-Fujimoto R, Inoue Y, Umeda N, Itoh S and Nagasaka T 2015 New pyrometallurgical process of EAF dust treatment with CaO addition *International Journal of Minerals Metallurgy and Materials* **22** **8** 788–797
[19] Yakornov S A, Pan'shin A M, Grudinsky P I, Dyubanov V G, Leont'ev L I, Kozlov P A and Ivakin D A 2017 Thermodynamic analysis of the zinc ferrite decomposition by lime in the dust of the electric arc furnace *News of higher educational institutions Non-ferrous metallurgy* **5** 28–33
[20] Moridera H, Kotoni S, Takeuchi M, Shigezumi T, Miyahara S and Matauoka S 1995 Status of environment friendly process and study on zinc recovery from stell plant dust in Japan *Galvatech'95 Conference Proceeding* 549–551
[21] Sloop J D 2000 EAF Dust recycling at Ameristeel *Proceedings of Fourth International Symposium on Recycling of Metals and Engineered Materials* 421–426
[22] Daiga V R and Horne D A 2000 Production of crude zinc oxide from steel mill waste oxides using a rotary hearth furnace *Proceedings of Fourth International Symposium on Recycling of Metals and Engineered Materials* 361–368
[23] Online. Available: Information on http://nano susu ru
[24] Menad N, Ayala J N, Garcia-Carcedo F, Ruiz-Ayúcar E and Hernandez 2003 A Study of the presence of fluorine in the recycled fractions during carbothermal treatment of EAF dust *Waste Management* **23** **6** 483–491
[25] Morcali M H, Yucel O, Aydin A and Derin B 2012 Carbothermic reduction of electric arc furnace dust and calcination of waelz oxide by semi-pilot scale rotary furnace *Journal of Mining and Metallurgy B: Metallurgy* **48** **2** 173–184

IOP Conf. Series: Materials Science and Engineering **969** (2020) 012041 doi:10.1088/1757-899X/969/1/012041

On the Efficiency of Nitrogen-Containing Gaseous Waste Plasma Afterburning

A V Matushkin[1], S V Anakhov[2] and G V Kharina[2]

[1]Welding technology chamber, Ural Federal University, 19, Mira str., Yekaterinburg, 620078, Russia

[2]Physics and mathematics chamber, Russian State Vocational Professional University, 11, Mashinostroiteley, Yekaterinburg, 620012, Russia,

E-mail: 227433@rambler.ru

Abstract. The paper is devoted to the study of the plasma neutralization technology for toxic waste in the gaseous phase. It is noted that for effective implementation of plasma processes in waste processing technologies, it is necessary to search the optimal solutions taking into account the criteria of quality, productivity, cost and safety. In order to determine the feasibility of using plasma incineration technology, attention is paid to the definition of the appropriate nomenclature for gaseous toxic waste. The technology of plasma neutralization for thermal waste processing products – superoxicants (dioxins) and nitrogen-containing gas emissions (ammonia and nitrogen oxides) is studied. A modernized design of the plasma torch for the gaseous waste utilization is proposed. The known methods of thermal destruction for ammonia and nitrogen oxides are considered. Temperature approximations of their decomposition duration and efficiency in the range of plasma heating temperatures are found. The gas-dynamic parameters of the air-plasma flow in the process of nitrogen-containing gases thermal heating by plasma jet are determined using machine modelling methods. The effectiveness of the considered plasma neutralization technology is proved. The results of these studies should allow us to further formulate commonly used in engineering practice principles and methods for designing high-performance plasma heating technology.

1. Introduction

One of the most promising areas of the electroplasma technologies application [1–3] is their implementation in environmental projects related to the waste thermal processing [4]. The use of plasma torches in environmental projects is based on the effect of high-energy (T_{pl} > 2000 K) impact on materials. This effect allows to subject the compounds that make up this substance to deep decomposition – plasma incineration ("combustion"). However, for effective implementation of plasma processes in waste processing technologies, it is necessary to search the optimal solutions taking into account the criteria of the achieved result quality, productivity, cost and safety.

This article presents a number of results obtained in the framework of R&d under the RFBR grant "Development of fundamental scientific bases for the application of plasma incineration processes in waste recycling technologies". In the course of this grant, the authors conduct research related to the elucidation of new physical laws about a plasma-arc discharge in a gas – dynamic flow, the relationship of these laws with the plasmatron properties, the physical concepts development about the processes of heat transfer accumulation, mass and charge flows in plasma-chemical systems and the features of polymorphic transformations in them. The results of these studies allow us to formulate commonly used in engineering practice principles and methods for designing high-performance plasma heating technology [5]. The application of such developments, in turn, will be a new impetus for further fundamental research in the field of plasma and environmental technologies – the development of new plasma torches for heating materials, obtaining a new quality of waste disposal, switching to new algorithms of low-waste environmental technologies, solving numerous material science problems, etc.

One of the tasks solved in the study is to determine the nomenclature of gaseous toxic waste, for which it is advisable to use plasma incineration technology. It is known that the main costs for the implementation of environmental projects fall mainly on the gas treatment system [6], which includes several stages of passing gas emissions through cyclones, scrubbers, adsorbers, electromagnetic and mechanical filters. In this regard, the use of plasma technologies at the afterburning stage of waste gases can significantly improve the quality, simplify and reduce the cost of waste treatment and neutralization technologies, as well as solve the problem of neutralization of gaseous supertoxicants (dioxins, furans, etc.) formed during combustion.

2. Technique of researches

During the past period of project implementation (2019 – early 2020), studies were conducted on the processes of plasma incineration of supertoxicants (dioxins) in waste disposal and neutralization technologies [7]. The known methods of thermal neutralization of dioxins are considered and analyzed in detail. Based on a small amount of known information [8, 9] on their high-temperature neutralization (at temperatures of 15.000 and 50.000), the authors made approximations of their decomposition duration in the range of plasma heating temperatures based on the Arrhenius equation for the reaction rate constant. On the basis of the obtained approximating dependencies, the criteria for the effectiveness of dioxins high-temperature neutralization were obtained and the applicability of plasma incineration technology was justified.

As part of the study, it was noted that the assessment of the effectiveness for neutralizing toxic gases using arc-type plasma torches is a multi-parametric task, since in addition to the design parameters, it is necessary to take into account the gas-dynamic and heat-energy parameters of the technology. For this purpose, the gas-dynamic parameters of the air-plasma flow in the mixing chamber (MC) developed by the authors for the environmental technologies plasmatron [10] were determined using mathematical modelling methods. In the MC interaction of two gas streams – forming a plasma arc (jet) and a secondary flow of toxic gas is organized by constructive and technological methods. Various configurations of the MC (cylindrical and confusor type), as well as different ways of the recycled gas supply into the plasma jet (2- and 4-channel feed at different angles to the axis of the plasma jet) were studied. Two ways of heating the volume of the disposed gas in the MC with the use of plasma jets of different configurations are considered. According to the research results, the optimal design features of a plasma torch for eco-technologies was determined. This technology use a plasma arc of at least 170 mm in length and a temperature on the axis of the cylindrical MC of at least 7000 K, which provides heating of the neutralized gas mixture in the range of 2–10 ms at average temperatures of 2000–4000 K. Supply of toxic gases – through 4 pipes with a diameter of 4 mm, located perpendicular to the MC axis at a distance from the nozzle of 11 mm.

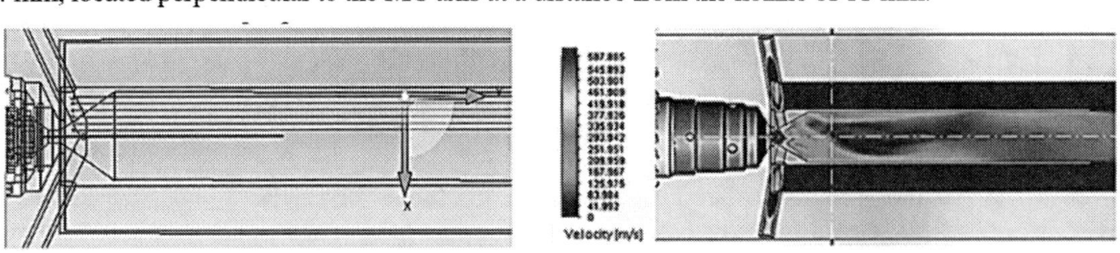

a b

Figure 1. Calculation of gas dynamic parameters in the MC of plasma heating system: (a) linear trajectories for velocities and temperatures calculating; (b) velocity distribution.

In the course of the work, methods for calculating and analyzing the trajectory of gases (linear and spiral), their velocities, time and heating temperatures were proposed. Using these methods, the characteristic temperatures, rates and heating times of the disposed gas in various areas of the mixing chamber are calculated (Figure 1). The results correlate with the orders of the dioxins decomposition duration at such temperatures, thus indicating the effectiveness of the technology. This conclusion is also confirmed when evaluating the effectiveness of dioxins detoxification in accordance with the proposed criteria [11]. The obtained data allow us to evaluate the effectiveness of neutralization for dangerous types of nitrogen-containing gases formed in various technological processes (for example, metallurgical processes using nitric acid), as well as during thermal waste processing. These compounds should primarily include ammonia and a number of nitrogen oxides.

As you know, ammonia gas (NH_3) is a toxic compound, dangerous (up to fatal) if inhaled. A mixture of ammonia and air is explosive, and if there is a constant source of fire, it is flammable. Among the known classic nitrogen oxides – N_2O (dinitrogen oxide, also known as laughing gas), NO (nitric oxide), N_2O_3 (dinitrogen tetroxide), NO_2 (nitrogen dioxide), N_2O_5 (dinitrogen pentoxide) – hazardous chemicals include nitrogen oxide and dioxide. Nitric oxide irritates the eyes, skin and mucous membranes, and when inhaled causes serious poisoning, up to loss of consciousness and convulsions. Inhalation of nitric oxide, even at concentrations below the threshold, can lead to pulmonary edema. Nitrogen dioxide also irritates the skin and mucous membranes, and its inhalation can lead to chronic lung diseases and poisoning. In addition, with chronic exposure in low concentrations, nitrogen dioxide affects the immune system, reducing the body's resistance to diseases, causing genetic changes in people. The threshold of sensitivity to odors is 10 mg/m^3, which is higher than the MPC (table.1). Symptoms of nitrogen dioxide poisoning differ from those of nitrogen oxide poisoning. However, it should be borne in mind that in some cases, people may be exposed to both of these substances simultaneously. In addition to production waste associated with the use of nitric acid, the formation of nitrogen oxides NO_x occurs during the combustion of all fuel types due to the content of nitrogen-containing organic substances in them. For example, the combustion products of 1 kg diesel fuel contain 20-40 g nitrogen oxides [12]. Nitrogen oxides, as sources of environmental problems such as acid precipitation, ozone depletion and photochemical smog, contribute to the degradation of ecosystems and human health. In this regard, an intensive search for ways to minimize the content of nitrogen oxides in emissions from various sources is quite natural.

Table 1. MPC of dangerous nitrogen-containing gaseous compounds.

n/n	Substances	MPC, mg/m^3		
		Maximum one-time	Daily average	Hazard class
1	Nitrogen dioxide (NO_2)	0.085	0.04	2
2	Nitric oxide (NO)	0.6	0.06	3
3	Ammonia (NH_3)	0.2	0.04	4

3. Results of research and their discussion

Almost all currently known works devoted to the neutralization of ammonia are studies of its decomposition using various catalytic systems made with transition metals (Fe, Pt, Ni, etc.) [13, 14]. Decomposition of ammonia begins at 270 ^{0}C, and at a temperature above 900 ^{0}C, ammonia is almost completely decomposed to nitrogen and hydrogen:

$$2NH_3 \rightarrow N_2 + 3H_2 \quad (1)$$

Decomposition of ammonia usually occurs with the formation of intermediate compounds. In this case, the rate of ammonia decomposition is largely determined by the rate of nitrogen

IOP Conf. Series: Materials Science and Engineering **969** (2020) 012041 doi:10.1088/1757-899X/969/1/012041

desorption. It is believed that the following variants of ammonia thermal conversion are possible – without a catalyst, at an elevated temperature (1200–1300 ^{0}C):

$$4NH_3+3O_2 \rightarrow 2N_2+6H_2O \quad (2)$$

in the presence of a catalyst, at elevated temperature:

$$4NH_3+5O_2 \rightarrow 4NO+6H_2O \quad (3)$$

The authors [15] gave the time of decomposition of ammonia by reaction (1) in the temperature range of 700–1500K at a pressure of 3.92 MPa (table 2).

Table 2. Estimated time to establish chemical equilibrium t and characteristic completeness of completion of the decomposition reaction of ammonia X as a function of temperature T [15].

T, K	700	1000	1200	1500
t, c	$3.54{\cdot}10^{12}$	$1.11{\cdot}10^{4}$	6.31	$5.02{\cdot}10^{-3}$
X	0.822	0.9856	0.9951	0.9999

The Figures 2–4 show the results of numerical simulation of ammonia heating by an air-plasma jet in the mixing chamber of plasma torch for eco-technologies (the design and technological parameters correspond to the optimal values given below). Due to the fact that the efficiency of high-temperature decomposition of most substances is proportional to both temperature and heating duration, the product of these characteristics can be taken as an integral parameter. Analysis of the presented data shows that at the plasma heating temperatures in the MC, virtually complete decomposition of ammonia occurs. In this case, the most effective decomposition occurs at a distance of 3–4 mm from the MC axis, but the rest of the volume provides the necessary conditions for decomposition (Figure 3 a). Taking into account that heating occurs in a gas-dynamic non-uniform medium (in a rotating flow), it is advisable to pay attention to the results of calculation along the screw trajectories (with a pitch and diameter characteristic of the distribution of the gas flow in the MC). These results indicate a more effective (1.5–3 times) neutralization of the main mass of gas moving in the MC along a spiral trajectory. You should also note that under the conditions of heating the air plasma jet may be a partial oxidation of ammonia to nitrogen oxide NO (with excess oxygen according to reaction (3)), although given the flow rate of plasma (0.005 kg/s) and recyclable (0.004 kg/s) gases have to go mainly according to the scheme of reactions (1) and (2).

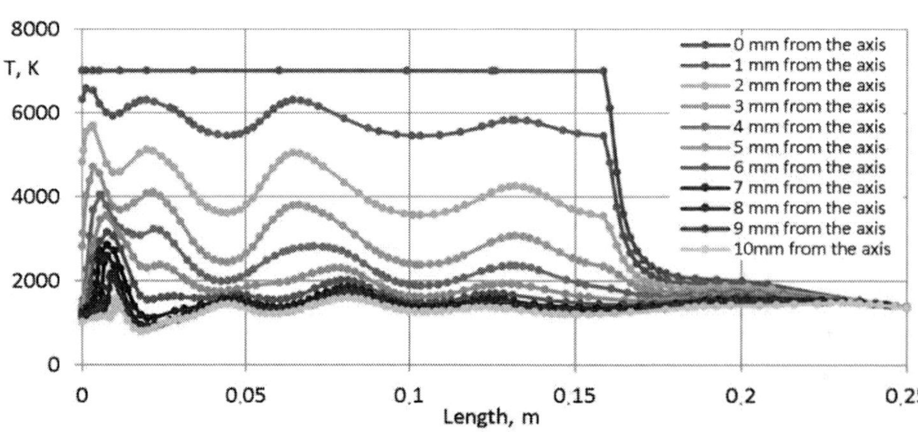

Figure 2. Temperature distribution in the plasma torch MC (calculation by linear trajectories).

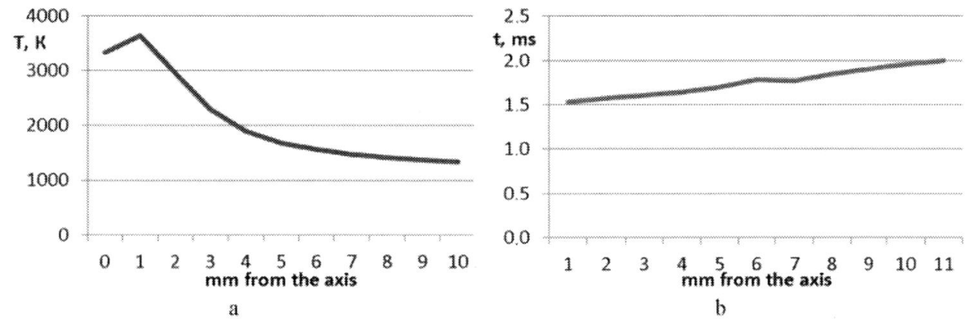

Figure 3. Average value of temperatures (a) and heating duration (b) of ammonia in the cross section of the plasma torch MC.

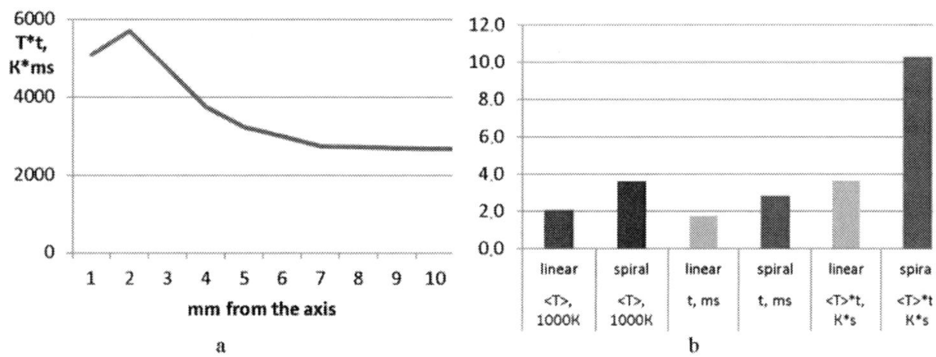

Figure 4. Ammonia heating efficiency (a – integral parameter, b – average parameters by trajectories) in the cross section of the plasma torch MC.

Special attention should be paid to the possibility of plasma neutralization for toxic nitrogen oxides. To date, gas purification from nitrogen oxides is based on the use of a variety of methods: catalytic, biochemical, non-catalytic, thermal, etc. Given the fact that catalytic methods have a number of known disadvantages (high cost of materials, the need for regular regeneration, etc.), you should pay attention to other methods of neutralization. Non-catalytic methods are based on the use of certain reducing agents, such as the above-mentioned ammonia, which is quite effective (up to 90%) in reducing nitrogen oxides [16]. The ammonia is stored in the container as ammonia water, which is injected into the gas stream at a high temperature, thus causing the reduction of nitrogen oxides according to the following schemes:

$$4NO+4NH_3+O_2 \rightarrow 4N_2+6H_2O \tag{4}$$

$$6NO_2+8NH_3 \rightarrow 7N_2+12H_2O \tag{5}$$

As can be seen from diagrams (4) and (5), this conversion of nitrogen oxides produces completely non-toxic compounds.

Thermal methods are practically not suitable for utilization of nitrogen oxides, since at typical temperatures (700–800 ^{0}C), the probability of nitrogen oxidation entering the combustion zone together with air and gas mixture is high [17]. In this regard, it is advisable to use step-by-step cleaning of gas mixtures, using methods aimed at neutralizing nitrogen oxides or step-by-step combustion of fuel. In recent years, plasma-chemical methods based on the destruction of recycled gases in a plasma jet with a temperature of several thousand degrees have become increasingly popular. The most effective is two-stage thermal decomposition of gaseous waste: the first stage is incomplete incineration of waste, and the second stage is the afterburning of incomplete combustion products, such as nitrogen oxides, carbon, and sulfur. In [18], it was shown that at temperatures of

2000–4000 °C, the equilibrium concentration of nitrogen oxide at the outlet remains quite high, about 2.4 % (at the initial concentration of nitrogen oxide 5 %), which corresponded to the degree of decomposition equal to 52 %. The process of high temperature method of nitric oxide decomposition can be described by the scheme:

$$2NO \rightarrow N_2 + O_2 \tag{6}$$

Due to the fact that the amount of nitrogen oxide produced in the system (4) depends directly on the oxygen concentration, the authors [18] proposed using reducing agents (solid, liquid, and gaseous) for oxygen binding. It was found that the introduction of reducing agents into the system shifts the equilibrium of the system (4) towards products, i.e. towards the formation of nitrogen and oxygen, and this shift does not depend on the aggregate state of the reducing agent. Therefore, for the complete decomposition of nitrogen oxides, conducting reducing agents is preferable when using the plasma-chemical method.

To assess the effectiveness of neutralization, the authors calculated the equilibrium concentrations of nitrogen monoxide during its decomposition by plasma-chemical method. The experimental equations obtained in [19] on the plasma-chemical synthesis of nitric oxide were used as a basis. The rate of nitric oxide (II) formation at temperatures from 3500 K and above is determined according to the equation:

$$\frac{dC_{NO}}{d\tau} = 1,5 \cdot 10^{13} \cdot e^{-\frac{360340}{RT}} \cdot \frac{(C'_{NO})^2 - (C_{NO})^2}{\sqrt{C_{O_2}}}, \tag{7}$$

where C_{NO} and C_{O2} are the initial concentrations of nitrogen oxide and oxygen; C_{NO}' is the equilibrium concentration of nitrogen oxide. Since the process of nitric oxide (II) obtaining is reversible, the time τ_p to establish its equilibrium concentration can be found by the formula:

$$\tau_p = \frac{5 \cdot 10^{11} \cdot e^{-\frac{360340}{RT}}}{\sqrt{C_{O_2}}} \cdot C'_{NO}. \tag{8}$$

Consequently, the equilibrium concentration of nitric oxide (II) of the reversible process of its decomposition (6) can be calculated by the equation:

$$C_{NO} = \frac{\tau \cdot \sqrt{C_{O_2}}}{5 \cdot 10^{11} \cdot e^{-\frac{360340}{RT}}}. \tag{9}$$

The choice of the initial oxygen concentration is based on the fact that the content of atomic oxygen in the fuel combustion products varies from 0.4 to 8.0 wt.%. The dependence of the equilibrium time in the system (6) on the temperature is presented in [19]. The calculated concentrations of nitric oxide (II) are shown in table 3. This table shows that as the temperature increases, the equilibrium concentration of nitrogen oxide in the system decreases. In all cases, the content of nitric oxide (II) in the system is less than the MPC value, which indicates a high degree of decomposition.

Figure 5 show the results of numerical modelling of gas-dynamic and thermo-physical processes of nitrogen heating in the plasma torch MC, which can be approximated with a small error to the processes of dangerous nitrogen oxides heating. The results of the analysis show that heating occurs at temperatures above 2000 K (the average temperature is 2700 K when calculated using a linear and 4000 K for a spiral trajectory) within a time of 2–3 ms. Taking into account the data presented in the table 3 it is possible to conclude that the NO content at characteristic temperatures and times of plasma heating in MC is significantly lower than the MPC, which indicates the effectiveness of the proposed method for nitrogen oxides neutralization.

Table 3. Typical decomposition time and NO concentration depending on heating temperature and oxygen concentration.

Oxygen concentration	**Heating temperature**	Time of complete decomposition	NO concentration	
C_{O2}, **mol/dm³**	T, **K**	t, ms	C_{NO}, mol/dm³	C_{NO}, mg/m³
0.0625	**3000**	1	$9.6 \cdot 10^{-10}$	0.029
0.0937	**3000**	1	$1.2 \cdot 10^{-9}$	0.035
0.125	**3000**	1	$1.4 \cdot 10^{-9}$	0.041
0.0625	**4000**	0.001	$2.6 \cdot 10^{-14}$	$0.78 \cdot 10^{-6}$
0.0937	**4000**	0.001	$3.2 \cdot 10^{-14}$	$0.95 \cdot 10^{-6}$
0.125	**4000**	0.001	$3.7 \cdot 10^{-14}$	$1.09 \cdot 10^{-6}$

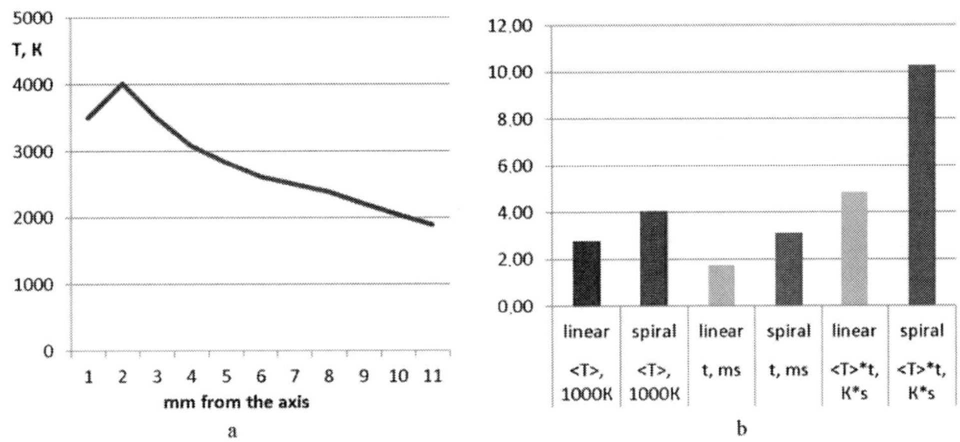

Figure 5. Average value of nitrogen heating temperature (a) and efficiency parameters (b) in the cross section of the plasma torch MC.

Figure 5 show the results of numerical modelling of gas-dynamic and thermo-physical processes of nitrogen heating in the plasma torch MC, which can be approximated with a small error to the processes of dangerous nitrogen oxides heating. The results of the analysis show that heating occurs at temperatures above 2000 K (the average temperature is 2700 K when calculated using a linear and 4000 K for a spiral trajectory) within a time of 2–3 ms. Taking into account the data presented in the table 3 it is possible to conclude that the NO content at characteristic temperatures and times of plasma heating in MC is significantly lower than the MPC, which indicates the effectiveness of the proposed method for nitrogen oxides neutralization.

4. Conclusions

Based on the results of this study, we can conclude that plasma neutralization method for waste containing toxic nitrogen of gaseous phase composition is justified by the example of ammonia, nitrogen monoxide and dioxide. The advantages of this method in comparison with the known technologies of high-temperature incineration and waste disposal are the speed and efficiency of the process. However, we should continue to develop and analyze this eco-technology in order to find optimal parameters for its application, as well as the development of new principles and methods of their design [20].

IOP Conf. Series: Materials Science and Engineering **969** (2020) 012041 doi:10.1088/1757-899X/969/1/012041

References

[1] Cherednichenko V S, Anshakov A S and Kuzmin M G 2011 *Plasma electrotechnological installations* (Novosibirsk: NGTU)

[2] Chernets I 2011 Development of high-power plasma reformer and power supply for large scale applications *20th International Symposium on Plasma Chemistry*

[3] Fridman A 2008 *Plasma Chemistry* (Cambridge: Cambridge University Press)

[4] Anakhov S V and Pyckin Yu A 2012 *Ecological designing: strategy and technologies* (Saarbrucken: LAP LAMBERT Academic Publishing)

[5] Anakhov S V 2014 *Principles and methods of design in electroplasma and welding technologies* (Ekaterinburg: RSVPU)

[6] Leontyev L I 2019 Prospects for solid waste disposal in Russia *Proceedings of the Congress "TEHNOGEN-2019"* 18–24

[7] Anakhov S V, Pyckin Yu A and Matushkin A V 2019 Modeling of the plasma incineration technology of waste utilization and neutralization *Technosphere safety* **1** 129–141

[8] Disinfection technologies Online. Available http://www seu ru/cci/lib/books/dioksiny/8/ 02 htm

[9] Petrov V G and Trubachev A V 2012 Neutralization of organochlorine industrial waste without formation of dioxins *Bulletin of Udmurt University Physics and chemistry series* **3** 64–68

[10] Pyckin Yu A, Anakhov S V and Shakurov S A 2007 *Plasma Torch* RU patent **67909**

[11] Anakhov S V, Matushkin A V, Dorozhkin E M, Lyzhin A I and Pyckin Yu A 2019 Study of plasma incineration processes in ecological waste recycling technologies *EurAsian Journal of BioSciences* **13** 1785–1789

[12] Turkin A V 2014 Investigation of the possibility of practical application of a method for cleaning exhaust gases of a ship's engine by adsorption of a solid substance in the presence of ozone *Bulletin AGTU Ser.: Marine engineering and technology* **3** 89–96

[13] Atroshchenko V I 1949 *Technology of nitric acid* (Moscow: State science and technology publishing house of chemical lit.)

[14] Moiseev M M, Leonov V T and Moiseeva I D 2014 Research of catalysts of decomposition of ammonia *Bulletin of TulSU Natural science* **1/2** 27–33

[15] Dorofeev A A and Yagodnikov D A 2016 Thermodynamic modeling of the equilibrium composition of combustion products and decomposition of liquid rocket fuels based on hydrazine, Science and education *Bauman Moscow state technical University* **5** 29–40

[16] Frolov S G and Roslyakov A D 2009 Reducing harmful emissions of transport diesels by neutralizing nitrogen oxides Bulletin of the Samara state aerospace University **3/19** 138–142

[17] Bratchikova I G 2011 *Physical and chemical bases of engineering ecology, P I Protection of the atmosphere* (Moscow: RUDN) Online. Available: https://lib rucont ru/efd/221313

[18] Mishchenko A V and Kuznetsov S I 2018 Thermal method of neutralization of nitrogen oxides *Bulletin of KhNTU Engineering science* **2/65** 35–40

[19] Fixation of atmospheric nitrogen in RF flare plasma torch Online. Available https://portal tpu ru/SHARED/i/INOVOSELOV/training/Tab2/Fiksatsiya_ptpv pdf]

[20] Anakhov S V 2018 *Principles and methods of plasma torches designing* (Ekaterinburg: RSVPU)

Acknowledgements

The work is executed at support of RFBR grant 19-08-00190

IOP Conf. Series: Materials Science and Engineering **969** (2020) 012042 doi:10.1088/1757-899X/969/1/012042

Analysis of Steel Slag Composition and Properties to Facilitate the Search for Rational Slag Recycling Methods

A A Tyutrin, N V Nemchinova and A N Baranov

Department of Metallurgy of Non-Ferrous Metals, Irkutsk National Research Technical University, 83, Lermontova Street, Irkutsk, 664074, Russia

E-mail: an.tu@inbox.ru

Abstract. This paper is devoted to the challenging problem of steel slag recycling. This study considers the composition and properties of steel slag, namely, slag of electric and pneumatic steelmaking. Slag of the pneumatic and electric steelmaking was found to be composed of calcium silicate Ca_2SiO_4, while phosphorus is represented by a complex calcium silicophosphate. Iron is in oxidized forms of FeO, Fe_2O_3 and Fe_3O_4 with the following ratios to the total mass of iron oxides, %: 51.51:37.16:11.33 for pneumatic steelmaking slag, and 75.92:20.90:3.18 for electric steelmaking slag, respectively. Chemical composition study showed the following content of the main elements for electric steelmaking slag, mass %: 31.6–33.2 Fe; 0.330–0.383 P; 19.15–19.2 Ca; 4.62–5.39 Si; for pneumatic steelmaking slag, mass %: 18.7–20.5 Fe; 0.630–0.710 P; 27.8–28.0 Ca; 4.73–5.04 Si. Electron microprobe analysis of slag samples showed the presence of Fe, C, O, P, Ca, Si, Mn, Ti, Cr, and Mg heterogeneously distributed over the sample crosscut surface. Samples had low content of phosphorus, however, this element was found to be accumulated in clusters together with Si and Ca.

1. Introduction

Ferrous metallurgy enterprises generate a significant amount of iron-containing waste [1–8], e.g. several thousand tons per month, which is unsuitable for reuse due to impurities, which significantly reduces the technical and economic performance of the metallurgical processing. The development of steelmaking is directly linked to innovative technologies for recycling secondary resources and industrial waste. Efficient recycling of non-reclaimable waste and intermediate products will increase the competitiveness and environmental safety of the production.

Due to both internal and external reasons, one of the main strategic goals of the ferrous metallurgy development in the Russian Federation for 2014–2020 and in the long term until 2030 is resource and energy saving, as well as reduction of negative environmental impact [9]. Therefore, recently, increasing attention is paid to research related to ferrous slag waste recycling technologies development.

Nowadays, up to 190 kg/t and 280 kg/t of slag is created in the pneumatic and electric steelmaking processes, respectively [1, 9]. The resulting solid and liquid slag is reused in steelmaking process to purify metal of unfavourable impurities (desulfurization and dephosphorization), to partially replace certain materials (e.g., lime and limestone, fluorspar), to increase the product yield via aluminothermic reduction processes of metal from oxides, as well as to reduce the metal product rejection rate. Also, the resulting slag is used in other industries: in general construction [10–18], road construction [14, 17, 19], railway construction, and agriculture [10]. Fine slag fractions are used as adsorbents. Slag is also recycled in blast furnaces to extract additional metal [20]. Blast furnace slag is used as backfill mixture in mines. However, high volumes of waste can not be recycled in full yet.

One of the main constraints on the reuse of slag from electric and pneumatic steelmaking processes for metal purification is phosphorus accumulation in the slag, which, if reused, returns back to the metal, thereby polluting the final product. Therefore, phosphorus shall be extracted into a separate product to enable slag to be used to purify metal from unfavourable impurities (desulfurization and dephosphorylation). However, rational technology can be selected only after the preliminary study of slag's chemical composition allowing to determine the phase and elemental composition.

2. Study of the chemical composition of the research subject

The subject under research were samples of slag from electric and pneumatic steelmaking – dark grey agglomerates of various sizes up to 300–400 mm. Slag samples contain clusters of metallic iron, which is most specific for electric steelmaking samples.

To study the elemental and phase composition, slag samples were preliminarily crushed to obtain a particle size of 3 mm. To assess the particle size distribution in samples, particle sizes were determined by a sieve method and laser diffraction measurement using the Analysette 22 NanoTec plus particle size analyzer (Fritsch, Germany).

The phase composition of samples was determined by X-ray diffraction analysis (XRD) using the XRD-7000 X-ray diffractometer (Shimadzu, Japan) with a vertical θ–θ goniometer. Metallic iron was removed from the samples prior to analysis. Filming was performed in a step mode for 2θ angle range from 3 to 80° using Cu_{α} radiation. The experimental conditions for the analysis were as follows: 40 kV, 50 mA, exposure – 6 s, step size – 0.05° 2θ. X-ray diffraction patterns were calculated using diffractometer software. PDF-2 powder diffractogram base was used to identify the phases. The analysis results are presented in Table 1.

Table 1. Qualitative and quantitative composition of samples (XRD data).

Slag sample	Content, mass %					
	Ca_2SiO_4	FeO	$Ca_2(SiO_4)_6(Ca_3(PO_4)_2$	Fe_2O_3	Fe_3O_4	Total
Pneumatic steelmaking	53,58	11,77	23,57	8,49	2,59	100
Electric steelmaking	45,52	28,63	16,77	7,88	1,20	100

XRD results for an average sample show that slag samples from pneumatic and electric steelmaking processes consist ~50% of calcium silicate Ca_2SiO_4, while phosphorus is represented by a complex calcium silicophosphate. Iron is in oxidized forms of FeO, Fe_2O_3 and Fe_3O_4 with the following ratios to the total mass of iron oxides, %: 51.51:37.16:11.33 for pneumatic steelmaking slag, and 75.92:20.90:3.18 for electric steelmaking slag, respectively.

To study the elemental composition of slag samples, chemical analysis was performed using atomic absorption spectrometry (AAS) and optical emission spectrometry (OES) utilizing the VARIAN AA 240, VARIAN AA 240 FS, and VARIAN 730-ES spectrometers, as well as methods of x-ray fluorescence spectrometry (XRD) and inductively coupled plasma optical emission spectrometry (ICP-AES) (Table 2).

Spectrometry showed that the samples contain up to 34% iron and 30% calcium. The average phosphorus content was 0.35% in electric steelmaking slag and 0.63% in pneumatic steelmaking slag.

After particle size analysis, the content of basic elements was also determined for slag samples. Element distribution is shown in the diagram (Figure 1), which shows that the content of phosphorus and other elements varies slightly for different particle sizes.

Table 2. Chemical composition of steel slag.

Element	Content, mass %										
	Fe	Ca	P	Si	Mn	Mg	Al	Ti	Cr	Cu	Other
Electric steelmaking slag											
AAS and OES	33,2	19,15	0,33	5,39	4,68	2,11	1,68	0,38	0,8	0,0125	0,1093
XRD and ICP–AES	31,6	19,2	0,383	4,62	n/a*	n/a	n/a	n/a	n/a	n/a	n/a
Pneumatic steelmaking slag											
AAS and OES	20,5	27,8	0,63	4,73	3,84	5,29	1,26	0,61	0,32	0,0028	0,06
XRD and ICP–AES	18,7	28,0	0,71	5,04	n/a	n/a	n/a	n/a	n/a	n/a	n/a

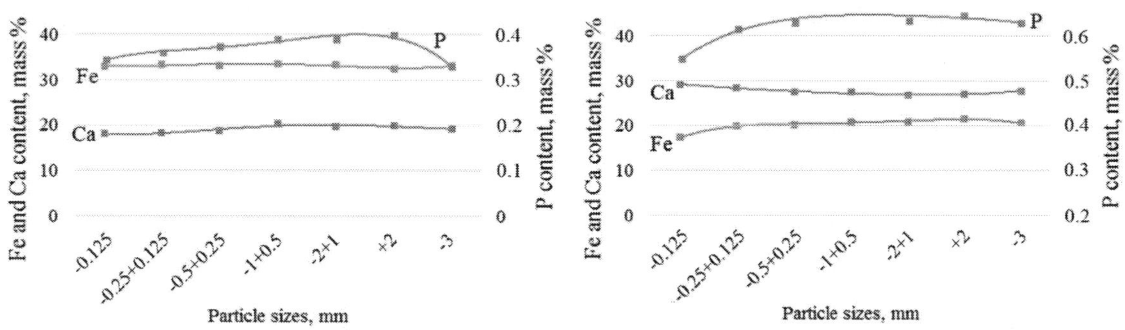

a b

Figure 1. Element content depending on size electric steelmaking (a) and pneumatic steelmaking (b) slag.

3. Slag microstructural analysis

To study the structure of electric and pneumatic steelmaking slag samples, samples were prepared for electron microprobe analysis (EMPA) by sputtering a thin gold film (of nanometer thickness) on their surface with accelerated gold ions in an Ar medium to create a conductive layer. The film was sputtered using SPI Module Sputter/Carbon Coater (Structure Probe Inc., USA). The prepared samples were analysed using JIB-Z4500 scanning electron microscope (JEOL, Japan).

Fe, C, O, P, Ca, Si, Mn, Ti, Cr, and Mg were found in the studied sample of electric steelmaking slag and were heterogeneously distributed over the crosscut surface of the sample (Figure 2a, Table 3). Region 4 corresponds to carbidizable metallic iron (84.89% Fe, 15.11% C). X-ray diffraction results show that iron oxide, mainly FeO, is concentrated in the region 5 (38.29% Fe, 40.09% O), which also contains insignificant amounts of Mn, Mg, Ca, Cr, and C. Region 6 is mainly composed of carbon (62.61%) and a smaller amount of iron oxide. Region 7 is represented mainly by calcium and silicon oxides, and to a lesser extent by phosphates of these elements, which also confirms the XRD results.

EMPA results of pneumatic steelmaking slag element distribution (Figure 2b, Table 4) show that the material has almost the same elemental composition as the electric steelmaking slag. Five microstructure regions (spectrums) of the sample were investigated. Region 2 is mainly represented by calcium and silicon oxides, and also most likely by phosphates of these elements (mainly calcium phosphate). Region 3 corresponds to carbidizable metallic iron (79.81% Fe, 20.19% C). Iron oxide is concentrated in region 4. This region has an increased content of magnesium (11.66%), as well as a small amount of Mn, Ca and C. Region 5 has a heterogeneous composition, mainly represented by calcium oxide (17.40% Ca) and a smaller amount of Fe, Al, C, Ti, Cr, and Mg. Region 6 has a carbon inclusion.

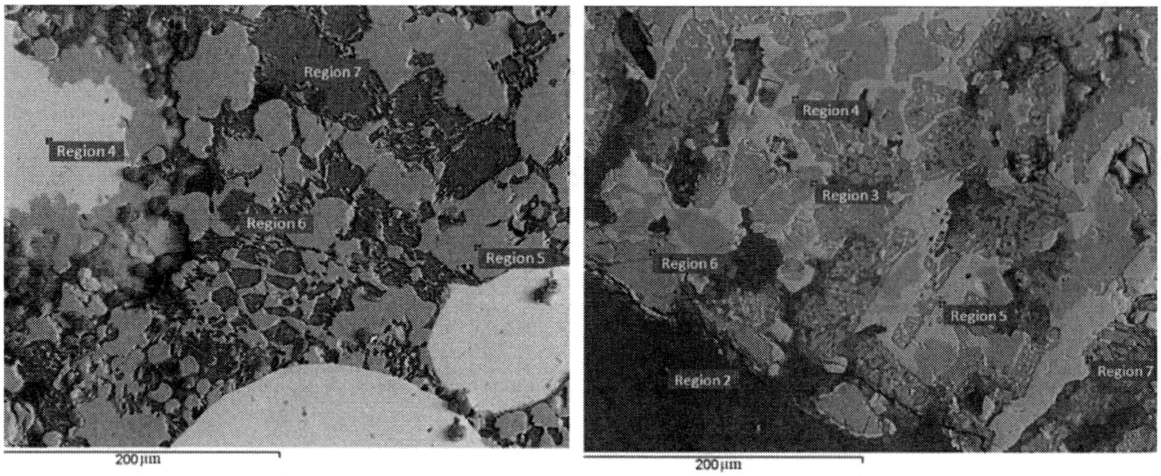

a b

Figure 2. EMPA results for the electric steelmaking (a) and pneumatic steelmaking (b) slag samples by regions.

Table 3. EMPA results for the electric steelmaking slag sample by regions.

Region	Content, mass %											
	C	O	Mg	Al	Si	P	Ca	Ti	Cr	Mn	Fe	Total
Region 4	15.11										84.89	100
Region 5	7.99	40.09	3.18				1.51		1.05	7.90	38.29	100
Region 6	62.61	18.44	0.83	0.16			0.59		0.55	3.07	13.74	100
Region 7	7.82		0.27		12.01	1.67	29.22			0.43	0.94	100
Total	12.52	40.13	1.03	1.62	2.41	0.14	7.11	0.18	0.56	2.98	31.32	100
Max	62.61	49.33	3.18	1.62	12.01	1.67	29.22	0.18	1.05	7.90	84.89	
Min	7.82	18.44	0.24	0.16	2.41	0.14	0.59	0.18	0.55	0.43	0.94	

Table 4. EMPA results for the pneumatic steelmaking slag sample by regions.

Region	Content, mass %												
	C	O	Mg	Al	Si	P	Ca	Ti	V	Cr	Mn	Fe	Total
Region 2	16.6	43.74		0.86	8.57	1.27	27.80	0.18	0.58			0.85	100
Region 3	20.19											79.81	100
Region 4	12.04	38.57	11.66				2.09				6.13	29.50	100
Region 5	13.25	50.41	0.71	5.37	0.34		17.40	2.039	0.44	1.55	0.42	7.73	100
Region 6	46.39	44.49		0.17	1.99		6.14					0.82	100
Total	19.13	38.61	2.01	0.72	5.13	0.52	16.57	0.34		0.19	1.38	15.08	100
Max	46.39	50.41	11.66	5.37	8.57	1.27	27.80	2.39	0.58	1.55	6.13	79.81	
Min	12.04	38.57	0.71	0.17	0.34	0.52	2.09	0.18	0.44	0.19	0.42	0.82	

The sample surface microstructure of electric and pneumatic steelmaking slag was also studied using Olympus GX-51 inverted metallurgical microscope (Japan). Electric and pneumatic steelmaking slag surface microstructure has a heterogeneous composition. Thus, electric steelmaking slag samples have a finer-grained structure (Figure 3a) in comparison with the microstructure of the pneumatic steelmaking slag (Figure 3b), which has needle-shaped grains. Presumably, bright regions of the samples have mainly iron oxide (FeO, Fe_2O_3, Fe_3O_4) and insignificant amounts of Mn, Mg, Cr, which is also confirmed by XRD and EMPA results. Dark regions have high calcium silicate content and a lesser amount of calcium silicophosphate. Also, samples have pores and metallic iron inclusions.

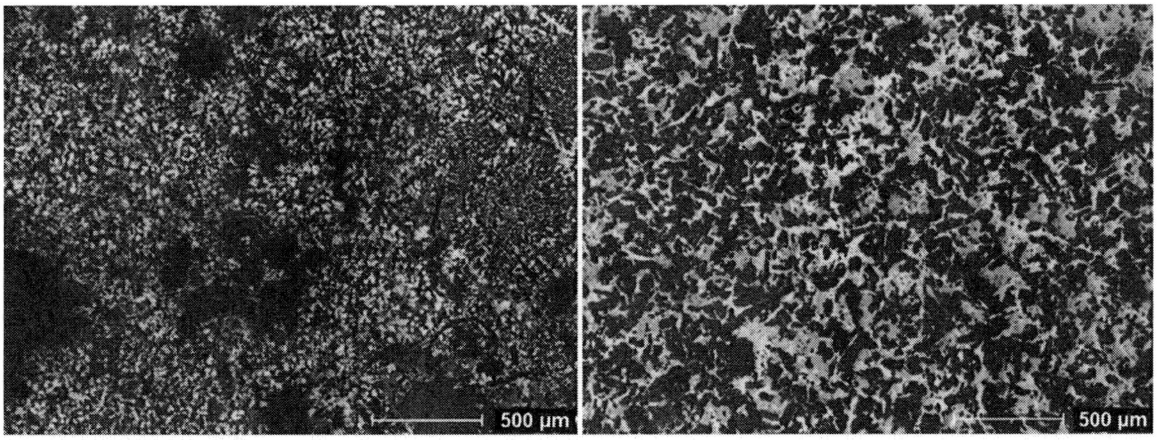

a b

Figure 3. Slag surface microstructure (50x magnification): a – electric steelmaking; b – pneumatic steelmaking.

4. Conclusions

Phosphorus content has negative effects on steel, and reducing its content is an important task. Phosphorus comes into furnaces with minerals of agglomerate and ore rock, and also with fluxes. The main component for binding phosphorus transferring it into slag is lime. To prevent phosphorus reduction from slag to metal (repeated phosphorization), the necessary slag alkalinity and oxidation shall be maintained. The following optimal ratios shall be considered for the slag: (CaO)/(SiO2) = 2.8÷3.1 and (CaO)/(FeO) = 4.5÷5.

Phase composition study of the samples by X-ray diffraction spectrometry showed that electric and pneumatic steelmaking slag is composed 50% of calcium silicate Ca_2SiO_4, while phosphorus is represented by a complex calcium silicophosphate. Iron is in oxidized forms of FeO, Fe_2O_3 and Fe_3O_4 with the following ratios to the total mass of iron oxides, %: 51.51:37.16:11.33 for pneumatic steelmaking slag, and 75.92:20.90:3.18 for electric steelmaking slag, respectively. Chemical composition study using atomic absorption spectrometry, optical emission spectrometry, X-ray fluorescence spectrometry, as well as inductively coupled plasma optical emission spectrometry showed the following content of the main elements for electric steelmaking slag, mass %: 31.6–33.2 Fe; 0.330–0.383 P; 19.15–19.2 Ca; 4.62–5.39 Si; for pneumatic steelmaking slag, mass %: 18.7–20.5 Fe; 0.630–0.710 P; 27.8–28.0 Ca; 4.73–5.04 Si.

EMPA and microscopic metallography results for showed the presence of Fe, C, O, P, Ca, Si, Mn, Ti, Cr, and Mg heterogeneously distributed over the sample crosscut surface. The samples have regions with metallic iron and its oxide. In addition to iron oxide, magnesium and manganese are also detected in this region. Samples had low content of phosphorus, however, this element was found to be accumulated in clusters together with Si and Ca. Also, inclusions of elemental carbon and small amounts of Al, Cr, V, and Ti were detected in the studied samples. Electric steelmaking slag samples have a finer-grained structure in comparison with the microstructure of the pneumatic steelmaking slag, which has needle-shaped grains.

To reincorporate slag into metal purification (desulfurization and dephosphorization), the phosphorus content in the slag shall be reduced. Based on the concluded research, technological solutions for the extraction of fluorine using physical, pyro- and hydrometallurgical methods shall be considered.

References

[1] Nemchinova N V, Shumilova L V, Salkhofer S P, Razmakhnin K K and Chernova O A 2016 *Integrated sustainable waste management. Metallurgical industry* (Moscow: Academy of Natural Sciences Publ.) p 494
[2] Dyudkin D A, Kisilenko V V and Smirnov A N 2009 *Steelmaking. Continuous casting* (Moscow: Teplotechnik) p 528
[3] Patrushov A E, Nemchinova N V, Chernykh V E and Tyutrin A A 2018 *Proceedings of Irkutsk State Technical University* **22(4)** 183–190
[4] Baranov A P, Chernykh V E and Patrushov A E 2017 *Proc. Int. Congress Basic research and applied development of processes for processing and utilization of technogenic formations* 185–189
[5] Patrushov A E 2020 *Proceedings of Irkutsk State Technical University* **24(3)** 672–683
[6] Allibert M 1995 *Slag atlas* p 616
[7] Nemchinova N V, Chernykh V E, Tyutrin A A and Patrushov A E 2016 *Steel in Translation* **46(5)** 368–372
[8] Nemchinova N V, Mineev G G, Tyutrin A A and Yakovleva A A 2017 *Steel in Translation* **47(12)** 763–767
[9] Piatak N M., Parsons M B and Seal R R 2015 *Applied Geochemistry* **57** 236-266
[10] Shi C and Qian J 2000 *Resour. Conserv. Recycl.* **29 (3)** 195–207
[11] Lesovik V S, Zagorodniuk L Kh and Shakhova L D 2011 *Industrial byproducts in the production of dry construction mixtures* (Belgorod: Publishing House of BSTU) p 196
[12] Rekus I G and Shorina O S 2001 *Basics of environmental safety and conservation* (Moscow: Publishing House of MGUP) p 146
[13] Royak S M, Pyachev A V and Shkolnik Ya Sh 1978 *Cement* **8** 4–5
[14] Jung V N 1952 *Binders technology* (Moscow: State publishing house on building materials) p 560
[15] Khamatova A R and Khokhryakov O V 2016 *News of the KSUAE* **36(2)** 221–227
[16] Tsakiridis P E, Papadimitrioua G D, Tsivilisb S and Koroneos C 2008 *Journal of Hazardous Materials* **152** 805–811
[17] Ruzavin A A 2018 *Proceedings of SUSU. Series "Construction and architecture"* **18(3)** 68–72
[18] Yevtushenko Ye I 1996 *Integrated recycling of metal-containing waste* (Belgorod: BelGTASM) p 60
[19] Pankovets A I and Miroyevsky S V 2013 *Casting and metallurgy* **69(1)** 26–27
[20] Kabanov Yu A, Stolyarskii O A and Agapeev E N 2006 *Metallurgist* **50** 39–41

Acknowledgment

This work has been performed under research work 0667-2020-0037 within the scope of the State task of the Ministry of Education and Science of the Russian Federation for the performance of academic studies for 2020 and the planning period of 2021 and 2022.

IOP Conf. Series: Materials Science and Engineering **969** (2020) 012043 doi:10.1088/1757-899X/969/1/012043

Thermal and Physical Properties of Methane Family Hydrocarbon and Oxygen Combustion Products in State-of-the-Art Arc Steel Furnace

G V Voronov and I V Glukhov

Institute of New Materials and Technologies, First President of Russia B. Yeltsin Ural Federal University, 19, Mira street, Ekaterinburg, 620002, Russia

E-mail: ermia12@yandex.ru

Abstract. The work deals with the particular combustion characteristics of methane family hydrocarbons (methane, ethane, propane, butane) and wet natural gas with process oxygen at carbon dioxide and water steam dissociation in a state-of-the-art arc steelmaking furnace. An algorithm is developed to calculate chemistry, the amount and concentration of combustion products at carbon dioxide and hydrogen dissociation, their physical and thermophysical parameters; heating power, balance and actual temperature, heat and pyrometric factors are evaluated considering heat transfer by radiation into unbounded medium. Based on the calculation results the recommendations are given for development of cold charge material heating conditions in order to minimize dusting, carbon oxide and hydrogen and charge material loss.

1. Introduction

Finished and semi-finished steel making in a state-of-the-art arc steel furnace is related to employment of practically all iron containing materials [1], including scrap of various form and dimensions, frequently containing non-ferrous metal additions (copper, tin, lead, stibium, etc.). Charge material composition and structure changes require development of heating conditions with the use of various thermal energy sources, ensuring rapid heating of cold charge materials. Charging is generally arranged into two baskets. The second basket may be fed only after the charge materials of the first basket settle. It is expedient to heat the first basket charge materials to the temperature of 1300 °C with the minimum power consumption, if possible. The heating conditions corresponding to this melting period shall be developed considering the minimum formation of carbon oxide (CO) and hydrogen, charge material wastes and dust. It is expedient to manufacture semi-finished steel in state-of-the-art furnaces following scrap and scrap and coke method with the use of an additional cheap energy carrier – natural gas (NG). However, in some cases natural gas combustion with process oxygen results in local charge material overheating and scars on the lining. Design analysis of natural gas and oxygen combustion conditions allows mitigation of possible negative results of the furnace operation. NG combustion analysis is related to three objectives: determination of the oxidizer consumption, in our case it is the process oxygen [2, 3, 4, 5, 6], component concentration in the combustion products, their quantity and temperature.

Let's consider combustion of natural gas with the chemistry containing methane family hydrocarbons C_nH_{2n+2} (methane CH_4, ethane C_2H_6, propane C_3H_8, butane C_4H_{10}).

Boundary parameters include the following combustion conditions.

1. Initial stoichiometric mixture of fuel and oxidizer thoroughly mixed beforehand with the oxygen flow coefficient of $\alpha_o = 1.0$. Initial temperature and sensible heat of natural gas are $t_g = 0$ °C, $Q_h^g = 0$ kJ and of oxygen – $t_o = 0$ °C, $Q_h^o = 0$ kJ. The volume of the initial mixture and combustion products has a shape of a sphere. The mixture ignites in its center at the pressure of $P = 101325$ Pa.
2. The mixture kinetic combustion takes place by a chain mechanism in the static volume.
3. Internal heat exchange is related to transition of ionized atoms (particles) to combustion product molecules without changing their weight.
4. External heat exchange of combustion products takes place only due to CO_2 and H_2O radiation into the free and unlimited open air volume.

Stoichiometric combustion equation of wet NG is as follows:

- in pure oxygen by stoichiometric ratio:

$$C_nH^w_{(2n+2)} + (3n+1)/2O_2 = n \cdot CO_2 + (n+1)H_2O, \tag{1}$$

where n – number of carbon atoms in hydrocarbon, integral;

- in process oxygen:

$$C_nH^w_{2n+2} + ((3n+1)/2)O_2 + k^{po}((3n+1)/2)N_2 = n \cdot CO_2 + (n+1)H_2O + k^{po}((3n+1)/2)N_2, \tag{2}$$

where $k^{po} = N_2^o/O_2^o = (100 - O_2^o)/O_2^o$ – nitrogen factor of process oxygen; N_2^o, O_2^o – nitrogen and oxygen concentrations in process oxygen, correspondingly, vol. %.

Stoichiometric (pure) oxygen consumption

$$V_{O_2} = 0.01 \cdot \left[\sum \left(\frac{3n+1}{2} \right) \cdot C_nH^w_{(2n+2)} - O_2^w \right], \text{m}^3/\text{m}^3 \tag{3}$$

where O_2^w - oxygen concentration in wet NG, vol.%.

Process oxygen consumption:

$$V_{O_2}^{po} = \left(1 + k^{po}\right) \cdot V_{O_2}, \text{m}^3/\text{m}^3 \tag{4}$$

Actual process oxygen consumption at $\alpha_o > 1.0$:

$$V_\alpha^{po} = \alpha_o \cdot V_{O_2}^{po}, \text{m}^3/\text{m}^3 \tag{5}$$

where α_o – process oxygen consumption factor.

The amount and concentration of the NG combustion product components at $\alpha_o > 1.0$ exclusively of CO_2 and H_2O dissociation:

$$V_{CO_2} = 0.01\left(CO_2{}^w + nC_nH\sum{}^w_{2n+2}\right), \text{m}^3/\text{m}^3 \tag{6}$$

$$V_{H_2O} = 0.01\left[H_2O^w + (n+1)C_nH\sum{}^w_{2n+2}\right], \text{m}^3/\text{m}^3 \tag{7}$$

$$V_{N_2} = 0.01 \cdot N_2{}^w + \alpha_o k^{po} V_{O_2}, \text{m}^3/\text{m}^3 \tag{8}$$

$$V_{O_2^e} = (\alpha_o - 1) \cdot V_{O_2}, \text{m}^3/\text{m}^3 \tag{9}$$

where O_2^e – excess oxygen.

Total amount of NG combustion products:

$$V_\alpha = V_{CO_2} + V_{H_2O} + V_{N_2} + V_{O_2^e}, \text{m}^3/\text{m}^3 \tag{10}$$

Component concentration in NG combustion products:

$$x_i = V_i / V_\alpha \cdot 100 \text{ vol. \%}; \ \sum x_i = 100 \text{vol. \%} \quad (11)$$

Let's review temperature computation for C_nH_{2n+2} hydrocarbon combustion products with initial and boundary conditions introduced so that the obtained results could be used to compare thermal and physical characteristics of the fuel components, generation of heat energy (enthalpy), quality of fuel combustion (Q_3 incomplete combustion), process (H_2) and environmental (CO) parameters when developing temperature and heating conditions of the furnace operation.

To compare heat energy generation of fuel and various composition natural gas combustible components the heating power parameter is used which describes heat content (enthalpy) of combustion products.

Balance equation of hydrocarbon combustion with oxygen will take the following form when estimating the heating power:

$$t^h_{\alpha=1} = Q^{op.}_{lowest} / (0.01 \cdot V^{h\sum c_i}_{\alpha=1} \cdot c_i^{t^h_{\alpha=1}}), °C \quad (12)$$

where $t^h_{\alpha=1}$ – combustion product heating power, °C; $V^h_{\alpha=1}$ – stoichiometric volume of hydrocarbon combustion products, m³/m³; ci – i-th component concentration, vol. %; $c_i^{t^h_{\alpha=1}}$ – i-th component heat capacity at the temperature of $t^h_{\alpha=1}$, kJ/m³·°C.

Physical and thermophysical parameter calculation results are shown in Table 1.

Table 1. C_nH_{2n+2} Hydrocarbon Combustion Product Properties at $\alpha_o = 1.0$, $t_y = 0$ °C, $t_o = 0$ °C and $P = 101325$Pa exclusive of CO_2 и H_2O dissociation.

Substance	$Q^{op.}_{lowest}$, kJ/m³	$V^h_{\alpha=1}$, m³/m³	CO_2, m³/m³/vol.%.	H_2O, m³/m³/vol.%.	$c^{t^h_\alpha}_{CO_2}$, kJ/m³·°C	$c^{t^h_\alpha}_{H_2O}$, kJ/m³·°C	$t^h_{\alpha=1}$, °C
CH_4	35797	3.356	1/29.79	2/59.69	3.534	3.027	3553
C_2H_6	64351	5.624	2/35.56	3/53.45	3.593	3.092	3697
C_3H_8	93573	7.891	3/38.02	4/50.8	3.624	3.127	3774
C_4H_{10}	123552	10.337	4/39.38	5/49.33	3.648	3.142	3833

It is known [7, 8, 9, 10] that CO_2 and H_2O dissociation occurs at the temperature of over 1500 °C. Thermal dissociation refers to internal heat exchanger related to structural change of the combustion product component molecules and occurrence of new molecules (substances) with the initial weight retained. Dissociation degree is the ratio of the dissociated molecule weight to the total molecule weight of the substance (CO_2 and H_2O); it depends on the temperature and concentration of the concerned components [11].

Water steam dissociation in NG combustion products results in additional oxygen and hydrogen:

$$V^{H_2O}_{O_2} = 4.775 \cdot 10^{-5} \cdot a_d \cdot H_2O^d \cdot V_{\alpha_o}, \text{m}^3/\text{m}^3 \quad (13)$$

$$V^{H_2O}_{H_2} = 9.57 \cdot 10^{-5} \cdot a_d \cdot H_2O^d \cdot V_{\alpha_o}, \text{m}^3/\text{m}^3 \quad (14)$$

Total volume of the combustion products including the non-dissociated water steam, hydrogen and oxygen is

$$V'_{H_2O} = (1 - 0.01 \cdot a) \cdot V_\alpha \cdot H_2O^d \cdot 0.01 + V^{H_2O}_{H_2} + V^{H_2O}_{O_2}, \text{m}^3/\text{m}^3 \quad (11)$$

CO_2 dissociation in the combustion products results in free oxygen and carbon oxide formation. Oxygen and carbon oxide volumes are:

$$V^{CO_2}_{O_2} = 5.03 \cdot 10^{-5} \cdot b_d \cdot CO_2^d \cdot V_\alpha, \text{m}^3/\text{m}^3 \quad (16)$$

$$V_{\mathrm{CO}}^{\mathrm{CO_2}} = 10.06 \cdot 10^{-5} \cdot b_d \cdot \mathrm{CO}_2^{\mathrm{d}} \cdot V_\alpha,\ \mathrm{m}^3/\mathrm{m}^3 \qquad (17)$$

The volume of the combustion products including the non-dissociated carbon dioxide, carbon oxide and oxygen, is:

$$V_{\mathrm{CO_2}}^{'} = (1 - 0.01 \cdot b) \cdot V_\alpha \cdot \mathrm{CO}_2^d \cdot 0.01 + V_{\mathrm{O_2}}^{\mathrm{CO_2}} + V_{\mathrm{CO}}^{\mathrm{CO_2}},\ \mathrm{m}^3/\mathrm{m}^3 \qquad (18)$$

where b and a are the dissociation degrees of carbon dioxide and water steam, correspondingly.

The total volume and composition of NG combustion products with due account for water steam (H_2O) and carbon dioxide (CO_2) dissociation is:

$$V_\alpha^{dis.} = V_{\mathrm{H_2O}}^{'} + V_{\mathrm{H_2}}^{\mathrm{H_2O}} + V_{\mathrm{O_2}}^{\mathrm{H_2O}} + V_{\mathrm{CO_2}}^{'} + V_{\mathrm{O_2}}^{\mathrm{CO_2}} + V_{\mathrm{CO}}^{\mathrm{CO_2}} + V_{\mathrm{N_2}} + V_{\mathrm{O_2}},\ \mathrm{m}^3/\mathrm{m}^3 \qquad (19)$$

Component concentrations of natural gas and process oxygen combustion products with due account for H_2O and CO_2 dissociation are:

$$\mathrm{CO}_2^{'} = (1 - 0.01 \cdot b) \cdot V_{\alpha_o}^{w} \cdot \mathrm{CO}_{2d} / V_\alpha^{dis.},\ \mathrm{vol.\%} \qquad (20)$$

$$\mathrm{H_2O}^{'} = (1 - 0.01 \cdot a) \cdot V_{\alpha_o}^{w} \cdot \mathrm{H_2O}_d / V_\alpha^{dis.},\ \mathrm{vol.\%} \qquad (21)$$

$$\mathrm{N}_2^{'} = (0.01 \cdot \mathrm{N}_2^{w} + k^{\mathrm{po}} \cdot \alpha_o \cdot V_{\mathrm{O_2}}^{o}) / V_\alpha^{dis.},\ \mathrm{vol.\%} \qquad (22)$$

where N_2^w – nitrogen concentration in wet NG, vol.%; α_o – process oxygen consumption factor;

$$\mathrm{O}_2^{'} = ((\alpha_o - 1) V_{\mathrm{O_2}}^{o} + V_{\mathrm{O_2}}^{\mathrm{CO_2}} + V_{\mathrm{O_2}}^{\mathrm{H_2O}}) \cdot 100 / V_\alpha^{dis.},\ \mathrm{vol.\%} \qquad (23)$$

$$\mathrm{CO}' = V_{\mathrm{CO}}^{\mathrm{CO_2}} \cdot 100 / V_\alpha^{dis.},\ \mathrm{vol.\%} \qquad (24)$$

$$\mathrm{H}_2^{'} = V_{\mathrm{H_2}}^{\mathrm{H_2O}} \cdot 100 / V_\alpha^{dis.},\ \mathrm{vol.\%} \qquad (25)$$

Balance equation for combustion product temperature computation with due account for incomplete combustion caused by H_2O and CO_2 dissociation will take the form of:

$$t_{\alpha_o=1}^{b} = (Q_{lowest}^{op} - Q_3^{dis.}) / (0.01 \cdot V_{\alpha=1}^{dis.} \cdot \Sigma c_i \cdot c_i^{t_{\alpha=1}^{b}}),\ ^\circ\mathrm{C} \qquad (26)$$

where $Q_3^{\mathrm{dis.}} = 0.01 \cdot V_{\alpha_o=1}^{\mathrm{dis.}} (Q_{CO} \cdot CO' + Q_{H_2} \cdot H_2^{'})$ is incomplete combustion resulting from H_2O and CO_2 dissociation at the temperature of $t_{\alpha=1}^{\mathrm{h}}$

Table 2 shows the computation results of physical and thermophysical properties of methane family hydrocarbon combustion products at $\alpha_o = 1.0$.

It is known [14, 15], that diatomic gasses with symmetrical molecules, such as N_2, O_2, H_2, do not absorb or emit heat energy. The absorption and emission properties are characteristic of triatomic gasses, such as $CO_2$and H_2O [4, 7, 12, 15, 16]. Emissivity factor of combustion product layer depends on the combustion product temperature and on product of the radiating gas concentration and the effective path length:

$$\varepsilon_{cpr} = \varepsilon_{\mathrm{CO_2}} \cdot \left(T_\alpha^b \cdot 0.01 \cdot S_{ef}^{sph} \cdot \mathrm{CO}_2^{'}\right) + \beta \cdot \varepsilon_{\mathrm{H_2O}} \cdot \left(T_\alpha^b \cdot 0.01 \cdot S_{ef}^{sph} \cdot \mathrm{H_2O}^{'}\right), \qquad (28)$$

where $S_{ef}^{sph} = \eta_{ef} \cdot 4V_{sph} / F_{sph}$ is effective path length in a spherical volume of combustion products provided that $V_{sph} = 4/3 \pi R_{sph}^3 = V_{\alpha=1}^{dis.}$ m^3; T_α^b is balance temperature of combustion products, K; β – correction factor, inclusive of higher impact of water steam concentration on H_2O emissivity factor [4, 7] as compared to $S_{\_ef}$; η_{ef} is a gas volume radiation efficiency factor corresponding to the heat energy share reaching the sphere reference surface [8].

Table 2. C_nH_{2n+2} Hydrocarbon Combustion Product Properties with Due Account for CO_2 and H_2O Dissociation at $\alpha_o = 1.0$ and Temperature of $t^h_{\alpha=1.0}$.

Sub stance	a, % $H_2O^{dis.}$	b, % $CO_2^{dis.}$	$V^{H_2O}_{H_2}$ m³/m³ /vol. %	$V^{H_2O}_{O_2}$ m³/m³ /vol. %	$V^{non\ dis.}_{H_2O}$ m³/m³ /vol. %	$V^{CO_2}_{CO}$ m³/m³ /vol. %	$V^{CO_2}_{O_2}$ m³/m³ /vol. %	$V^{non\ dis.}_{CO_2}$ m³/m³ /vol. %	$V^{dis.}_{O_2}$, m³/m³ /vol. %	$Q^{dis.}_3$ kJ/m³ /%	$t^b_{\alpha_o=1}$ °C
CH_4	19.84	62.50	0.38/ 9.89	0.204/ 5.3	1.61/ 41.75	0.63/ 16.34	0.31/ 8.05	0.37/ 9.75	0.514/ 13.4	12051 /33.67	2692
C_2H_6	25.44	62.34	0.73/ 11.1	0.394/ 5.98	2.24/ 34.01	1.25/ 19.03	0.63/ 9.56	0.75 /11.43	1.024/ 15.5	23751 /36.91	2703
C_3H_8	27.06	63.54	1.04/ 11.12	0.562/ 6.02	2.92/ 31.33	1.92/ 20.55	0.96/ 10.29	1.09/ 11.72	1.522/ 16.3	35439 /37.87	2728
C_4H_{10}	28.3	65.13	1.36/ 11.22	0.68/ 5.62	3.59/ 29.67	2.62/ 21.65	1.31/ 10.82	1.39/ 11.53	1.99/ 16.4	47771 /38.66	2746

Thermal ray efficient length in the spherical volume is:

$$S^{sph.}_{ef.} = \eta_{ef.} \cdot 4V_{sph.} / F_{sph.} = 1.333 \cdot \eta_{ef.} \cdot R_{sph}, \text{ m.} \quad (29)$$

Sphere radius equivalent to the stoichiometric volume of combustion products is:

$$R_{sph.} = \sqrt[3]{V^{dis.}_{\alpha=1} \cdot 3/(4\pi)} = 0.621 \cdot \sqrt[3]{V^{dis.}_{\alpha=1}}, \text{ m.} \quad (30)$$

$$S^{sph.}_{ef.} = 1.333 \cdot 0.621 \cdot \eta_{ef.} \cdot \sqrt[3]{V^{dis.}_{\alpha=1}} = 0,828 \cdot \eta_{ef.} \cdot \sqrt[3]{V^{dis.}_{\alpha=1}} \text{ m.} \quad (31)$$

The emissivity factor of the combustion product stoichiometric volume components can be determined using the interpolatory equation and table data [7].

Actual emissivity factor of the combustion product volume is $\varepsilon^a_{cpr.} = \varepsilon_{CO_2} + \beta\varepsilon_{H_2O}$. Based on this amount a correction $\delta = \varepsilon_{CO_2} \cdot \varepsilon_{H_2O}$, related to coincidence of some radiation bands within CO_2 and H_2 spectrum, is calculated, then $\varepsilon'_{cpr.} = \varepsilon^d_{cpr.} - \delta = \varepsilon^d_{cpr.} - \varepsilon_{CO_2} \cdot \varepsilon_{H_2O}$.

Actual balance temperature of hydrocarbon combustion products, considering heat energy output from radiation into the unbounded open space is:

$$t^d_{\alpha=1} = (Q^{op.}_{lowest} - Q^{dis.}_3 - Q_l) / (0.01 \cdot V^{dis.}_{\alpha=1} \cdot \Sigma c_i \cdot c_i^{t^d_{\alpha=1}}) \text{ °C,} \quad (32)$$

where Q_1 – the amount of heat from combustion products released to atmosphere due to CO_2' and H_2O' radiation at the temperature of $t^d_{\alpha=1}$, kJ; $c_i^{t^d_{\alpha=1}}$ – CO_2' and H_2O' heat capacity at the actual temperature of $t^d_{\alpha=1}$, kJ/(m·°C).

Following the recommendations of S. S. Kutateladze [15]

$$Q_l = 4.9\left[\varepsilon'_{cpr.}\left(T^b_{\alpha=1}/100\right)^4 - a_w\left(T_0/100\right)^4\right] \text{ kJ,} \quad (33)$$

where a_w is a unit value atmosphere absorption degree; T_0 is thermodynamic temperature of 273.15K.

Heat pyrometric factor is

$$\eta^{heat}_{pyr.} = (1 - Q_l / (V_g \cdot Q^{op.}_{lowest}))100\%, \quad (34)$$

where V_g – the amount of initial gas, m³.

Temperature pyrometric factor of combustion products may be estimated for the selected initial and boundary conditions:

$$\eta_{pyr.}^{heat} = T_d / T_{\alpha=1}^{b} = \sqrt[4]{1-\varepsilon_{cpr.}^{'}} \quad (35)$$

Table 3 shows computation results of hydrocarbon combustion product properties at $t_{\alpha=1}^{b}$ and radiation heat transfer to unbounded medium.

Table 3. C_nH_{2n+2} Hydrocarbon Combustion Product Properties at the Temperature of $t_{\alpha=1}^{d}$.

Sub stance	$V_{\alpha_o=1}^{dis.}$, m^3/m^3	$V_{spr.}^{dis}$ m^3	$F_{sph.}^{dis.}$ m^2	$R_{sph.}^{dis.}$ m	$S_{ef.}$ m	$0.01S_{ef.}$ $H_2O^{dis.}$, м	$0.01S_{ef.}$ $CO_2^{dis.}$, м	Q_l, kJ	$\eta_{pyr.}^{heat}$ %	t_d, °C	$\eta_{pyr.}^{heat}$, %
CH_4	3.81	3.81	1.23	0.97	1.23	0.53	0.13	1277.38	96.43	2654	98.61
C_2H_6	6.55	6.55	1.47	1.16	1.47	0.51	0.18	1559.05	97.58	2678	99.09
C_3H_8	9.28	9.28	1.65	1.31	1.65	0.53	0.20	1793.11	98.08	2708	99.28
C_4H_{10}	12.2	12.2	1.81	1.43	1.81	0.55	0.22	1987.85	98.39	2729	99.65

2. Conclusion

Using the proposed procedure a chemistry and actual temperature computation program was developed for combustion products of methane family hydrocarbons (methane, ethane, propane, butane) with industrial oxygen in Microsoft Excel with due account for dissociation (of water steam and carbon dioxide) and radiation heat transfer to unbounded medium.

Hydrogen in the amount of 9.89 vol.% to 11.22 vol.% and carbon oxide in the amount of 16.34 vol.% to 21.65 vol.% are generated in combustion products with due account for the water steam and carbon dioxide dissociation what results in incomplete combustion [17, 18, 19, 20] of 33.67% to 38.66% based on these parameters. Thus preliminary kinetic combustion of natural gas with oxygen causes combustion in a layer of charge materials of not initial fuel mixture with oxygen but H_2O and CO_2 dissociation products with free oxygen (O'_2 vol. %), which content varies within 13.4 vol. % – 16.4 vol. %.

Design heat output, balance and actual temperatures describe the possible (maximum) level and shall be used to compare gaseous medium temperature condition in the furnace work space.

Heat and temperature pyrometric factors significantly dependent on the fuel combustion conditions and boundary heat exchange conditions shall be used in computation as a first approximation.

References

[1] GOST 2787-75 Ferrous Secondary Metals. General Technical Requirements
[2] Startsev V A, Voronov V G and Lobanov V I 2004 *The Scrap-Carburetor Process in the Production of Steel in Open-Hearth Furnaces* (Ekaterinburg: UGTU-UPI) p 225
[3] GOST 5583-78 (ISO 2046-73) Technical and Medical Oxygen Gas. Specification
[4] Kazantsev E I 1975 *Industrial furnaces Reference Manual for Calculations and Design* (Moscow: Metallurgy) p 368
[5] Lisienko V G, Shleimovich E M, Ladygichev M G, Sannikov S P and Shchelokov Ya M 2008 *Temperature: Theory, Practice, Experiment: reference book in 3 volumes, V. 1, book 1: Temperature Inspection Methods* (Moscow: Teplotechnik) p 549
[6] Metallurgy. Premium Engineering Online. Available: http://www/premen.ru/ru/content/air/apply/metal/ (28.03.18).
[7] Kitaev B I, Zobnin B F and Ratnikov V F 1970 *Thermotechnical Calculations for Metallurgical Furnaces* (Moscow: Metallurgy Publishing House) p 528
[8] Baum V A and Budrin D V 1951 *Metallurgical Furnaces* (Moscow: Metallurgizdat) p 976
[9] Budrin D V 1963 *Metallurgical Furnaces v. 1* (Moscow: Metallurgy Publishing House) p 436

[10] Lebedev P D and Shchukin A A 1956 *Heat Process Engineering* (Moscow: Gosenergoizdat) p 384
[11] Prokhorov A M 1983 *Physics Encyclopedia* (Moscow: Soviet Encyclopedia) p 928
[12] Timofeev V N and Karasina E S 1948 *Bulletin of All-Union Thermal Engin. Institute* **9**
[13] Krivandin V A 1977 *Metallurgical Furnaces: workbook for universities* (Moscow:Metallurgy) p 463
[14] Kutateladze S S 1958 *Heat Transfer Reference Book* (Leningrad: Gosenergoizdat), p 413
[15] Hottel H and Mangelsdorf H 1935 *Trans. Am. Inst. Chem. Eng.* **31**
[16] Hottel H and Egbert R 1942 *Trans. Am. Inst. Chem. Eng.* **3**
[17] Gudim Yu A, Zinurov I Yu and Kiselev A D 2010 *Production of Steel in Electric-Arc Furnaces. Designs, Technologies, and Materials: Monograph* (Novosibirsk: NGTU)
[18] Belkovskii A G, Kats Ya L and Krasnyanskii M V 2013 Status of and trends in EAF steelmaking and design *Chern. Metall* **3** 72–78
[19] Alam M, Naser J, Brooks G and Fontana A 2010 Computational fluid dynamics modeling of supersonic coherent jets for electric arc furnace steelmaking process *Metallurgical and materials transactions* **41B** 1354–1367
[20] Voronov G V, Goltsev V A and Glukhov I V 2016 Aerodynamics and Thermal State of State-of-the-Art Arc Steel Furnace *Ferrous Metallurgy and Material Engineering Issues* **1** 28–34

IOP Conf. Series: Materials Science and Engineering **969** (2020) 012044 doi:10.1088/1757-899X/969/1/012044

Mathematical Modelling of the Thermal State of a Ladle During Arc Heating of the Melt

E B Agapitov, M S Sokolova and M A Lemeshko

Department of Heat and Power Engineering Systems, Nosov Magnitogorsk State Technical University, 38, Lenin Street, Magnitogorsk, 455000, Russia

E-mail: margo88k2017@gmail.com

Abstract. The paper analyzes energy losses in the ladle furnace. In particular, the authors focus on electrical losses, namely total energy losses of the electric arc, conditionally including radiation losses of a part of the arc uncovered by slag. Managing these flows creates conditions for increasing the efficiency of heating. The calculations given in the paper showed that electrical losses decreased, during prolonged heating (about 8-12 minutes) with a sufficient amount of slag. This mode provides a reduction in heat loss by 4-5 %. By selecting options of changes in a sequence of stages and their duration, the proposed mathematical model makes it possible to apply a new technology to process molten steel in the ladle furnace characterized by lower power consumption.

1. Introduction

A higher quality and better technical and economic performance of steelmaking units and continuous casting machines are mostly provided by secondary metallurgy. Currently, existing ladle treatment facilities are characterized by relatively low thermal efficiency and, at the same time, a great potential for saving energy and electricity.

A ladle furnace is an intermediate link between a steelmaking unit and a continuous casting machine and ensures their coordinated operation. There are frequent situations in production when neither the thermal state nor the chemical composition of steel in terms of the ladle volume is known at the starting time of treatment [1, 2]. At the same time, the uneven temperature of the melt reaches 40-50 °C by weight and local measurement data with a single immersion thermocouple are not indicative of bulk temperature of the melt, resulting in errors in temperature setpoints for casting the melt on a continuous casting machine [3].

The efficiency of heating the melt is determined by power of the electric arc unit, which is, in its turn, determined by transformer power. When applying alternating current, power should be optimal for a given volume of steel and the geometric dimensions of the ladle. When transformer power is not sufficient, the treatment cycle is extended; and when transformer power is more than it is required, capital expenditure and power overconsumption increase and the ladle lining lifetime decreases [4, 5]. The analysis of Figure 1 shows that the increase in transformer power leads to an increase in the heating rate. However, the same heating rate can be obtained at various specific active power and, consequently, different power consumption (Figure 2) [6].

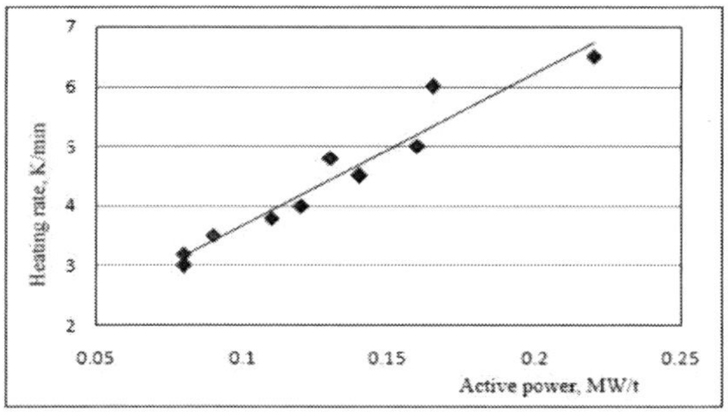

Figure 1. The dependence of the steel heating rate on active power supplied from the transformer.

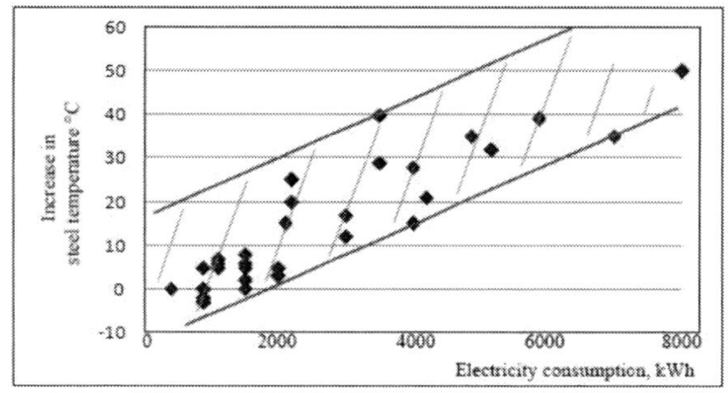

Figure 2. The dependence of the steel temperature increase (heating rate) on the power consumption.

At the same power consumption, a temperature increase is different, since the efficiency of the unit, in addition to the heating rate, is determined by the heating efficiency or energy utilization factor [7, 8]. The dependence of the installed capacity of transformers on the ladle furnace on the weight of the processed steel shows a significant variation in values (Figure 3). By approximating the data by power dependence, we get a rather reliable result ($R^2 = 0.8$), but it is unlikely to accurately calculate the required transformer power for specific technological tasks. Therefore, suppliers equip the facilities with transformers with an overrated installed capacity, which is partially not used.

At the same time, power supplied to the ladle furnace is limited to the lining wear rate, and excessively supplied power inevitably leads to an increase in the wear rate of refractories in the slag belt zone and in "hot spots" of the cover. The so-called "hot spots" are associated with overheating of refractories in the immediate vicinity of the arc, which increases the consumption of refractory materials and leads to damages on an inner surface of the ladle. According to the data provided by Danieli company, maximum allowable specific power of the arc should be about 1.8–2.2 MW/m^2 of the liquid metal surface.

The most important technical and economic indicator of the ladle furnace operation is thermal efficiency of heating the heat. The value of efficiency is determined by both design features of the unit and the secondary treatment technology. Statistical processing of the available data [9] shows a significant variation in values, and for estimated calculations with sufficiently high confidence ($R^2 = 0.89$), the power dependence shown in Figure 4 can be used only for small capacity ladles.

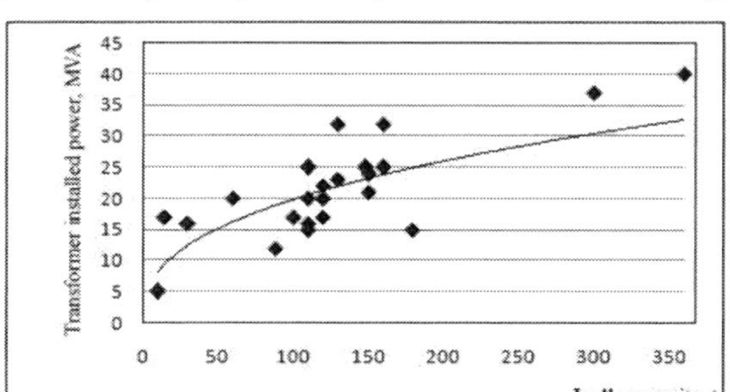

Figure 3. Dependence of the transformer power of the ladle furnace on the ladle capacity.

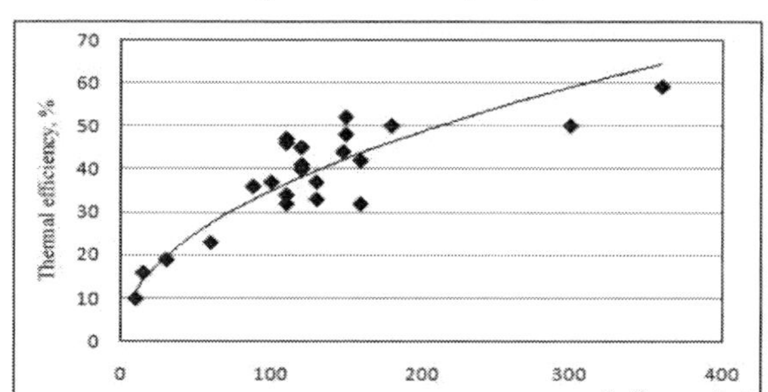

Figure 4. A dependence diagram of the thermal efficiency of the ladle furnace on its capacity.

Thus, these data can be used only for reference, and for the conditions under study we should calculate the relevant heat balance of the unit. The heat radiated from the furnace melting volume is dissipated through the lining of the roof, walls and bottom, escapes with cooling water and gases, and it is also radiated through the charging doors, etc.

Inert gas purging of steel in the ladle furnace, comparing favourably with other secondary treatment methods due to relatively low costs, allows successfully solving such tasks as reducing the heterogeneity of steel in its temperature and chemical composition, improving the conditions for removing non-metallic inclusions, and partially degassing the melt [9]. Heat loss with waste gases is one of the expenditure items of the heat balance. It is usually assumed in calculations that gas temperature has temperature of the melt, when leaving the ladle furnace.

When calculating the heat balance of the ladle furnace, it is difficult to estimate the heat loss through the ladle lining, as well as its accumulation of heat during processing, although this is necessary for an accurate prediction of steel temperature changes. Therefore, this balance item is determined with a known inaccuracy, although it amounts to about one-third of the total heat loss of the furnace on average [10–12].

Heat transfer in the ladle furnace is determined by the conditions of electric arc burning. The immersion depth of the arc, depending on slag thickness, influences the efficiency of heat exchange between electric arcs and the melt and, consequently, the power consumption of the arc device of the unit. Paper [13] describes the dependence of slag thickness in the ladle on the specific energy consumption (Figure 5).

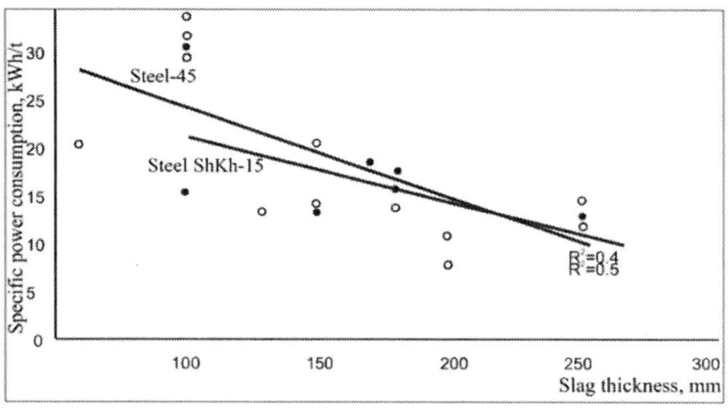

Figure 5. The influence of slag thickness in the ladle on the specific power consumption.

The slag layer thickness should cover at least 2/3 of the arc length in those cases when steel is not desulfurized, but adjusted by temperature only. The calculations [5] show that if 10 mm of the arc is not covered by slag, electric power losses amount to 2–3 % of supplied power.

To determine the heat loss with cooling water, we should know input and output water temperatures at the cover of the ladle furnace, as well as the water flow rate [14]. The ladle furnace is equipped with temperature sensors that determine the water temperature discharged from each circuit and measure the water temperature supplied to the cover. Measurements carried out on the ladle furnace of the basic oxygen furnace shop at PJSC Magnitogorsk Iron and Steel Works (MMK) show that when the heating stage is turned on, the water temperature in the cover at a flow rate of 245 m^3/h increases in proportion to the heating time:

$$t_1 = 0.015 \cdot \tau + 37.09, \tag{1}$$

where t_1 is water temperature at the outlet of the cover, °C; and τ is heating time, s.

Heat for heating cooling water in the ladle cover is calculated as follows:

$$Q_w = \begin{cases} P \cdot \tau_0 \cdot (t_{01} - t_2) \cdot c_w, Q_E = 0 \\ P \cdot c \cdot \int_{\tau_1}^{\tau_2} (0.015 \cdot \tau + 37.09 - t_2) dt, Q_E \to 0 \end{cases} \tag{2}$$

where P is water consumption for the cover, m^3/s; c_w is heat capacity of water at actual temperature, J/kg·°C; t_{01} is water output temperature at the cover, when there is no heating, t_{01} = 37 °C; t_2 is water

IOP Conf. Series: Materials Science and Engineering **969** (2020) 012044 doi:10.1088/1757-899X/969/1/012044

input temperature at the cover, t_2 = 30 °C; τ_0 is the duration of the period without heating, s; τ_1, τ_2 is end and starting time of heating, respectively, s; τ is current heating time, s.

According to the data obtained as a result of the research, the heat loss with cooling water during heating can be 12–18 % of energy supplied to heating.

2. Mathematical description

The system of circulating water supply, which allows fixing water consumption and temperature, has certain inertia. Accounting for heat losses with water over the entire treatment period on the ladle furnace may give an idea of a total level of such losses; however, when preparing operational balances, for example, for heating operations that last 5–7 minutes, may also give an error in the calculations [15].

Heating in the ladle furnace is a sequence of operations, each of them differs in the special features of the processes performed (heating, desulfurization, alloying) and their combination [16, 17]. The calculation of the heat balance at every stage is associated with a certain error, which influences the estimated average mass temperature of the melt by the end of the stage. When temperature is measured selectively, a true value of molten steel temperature at the beginning of the operation remains unknown and can only be estimated by calculation. Accuracy of the calculations was corrected by comparing the predicted calculated and measured values of temperature during processing and, as a result, according to temperature of the ladle transferred to the downstream operations [18].

To predict accurately the temperature change of steel during processing in the ladle furnace, it is necessary to calculate the heat balance items of the unit, when many of these items cannot be correctly estimated. For example, we cannot calculate accurately the heat loss through the ladle lining and the accumulation of heat by the lining during processing. To eliminate temperature gradients along the molten steel depth, the melt is mixed with argon supplied through porous plugs in the ladle bottom. Efficiency of blowing depends not only on the amount of supplied gas, but also on the condition of the plugs, slag on the steel surface and its physical state [19, 20].

As part of the study, we developed a software product adaptable to the conditions of a specific ladle furnace based on the classical equations of heat and mass transfer and a multivariate analysis of the database of heat reports for the unit over a long period. First, we analysed the database of 20,000 heat reports taken from the ladle furnace in the basic oxygen furnace shop at PJSC MMK to evaluate the influence of these factors on the processes of thermodynamics, heat and mass transfer and kinetics, and to integrate the analysis results in the form of correction factors into separate program blocks.

The program allows you to calculate the heat balance of the ladle furnace for a short period, the average mass temperature of the melt, the energy characteristics of heating and processing by entering various initial data. The model is executed in MATLAB with the Simulink software module.

As an example of organizing individual calculation blocks, we describe the block "Heat input from oxidation of electrodes" (Figure 6). The block calculates heat input from oxidation of electrodes during a total heating. During the combustion of 1 kg of graphite, 33520 kJ of heat is released, and the electrode consumption is 0.008 kg/kW·h of electric arc spent on burning. During the operation of the electric arc facility, the electrodes are oxidized with heat input coefficient Q_4=0.074·Q_1.

The heat input at every point of time is calculated according to this equation depending on the electrical power of the unit related to the transformer tap used during heating. Figure 6 shows the trend in heat flux at different taps (columns). As a result of the calculation, the window "Heat input from oxidation of electrodes" contains the total amount of heat for this item for a total period of treatment.

By using this program, we can evaluate the influence of melt heating efficiency at different stages of technological operations (Figure 7) in the ladle furnace and give recommendations for improving the heating technology.

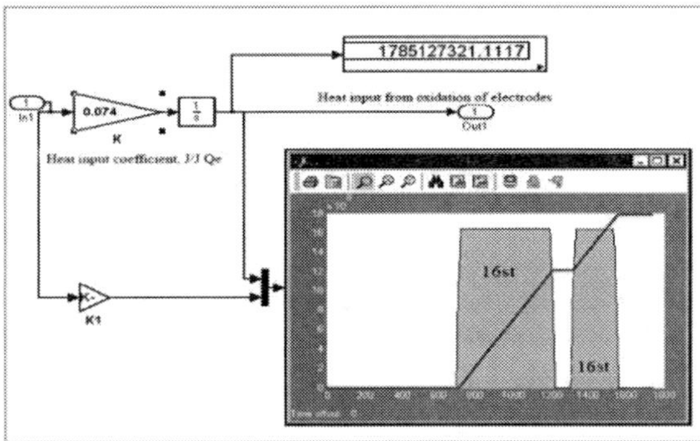

Figure 6. Block “Heat input from oxidation of electrodes”.

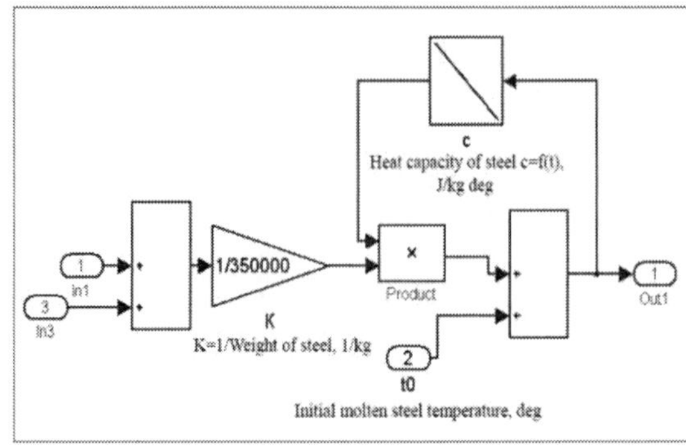

Figure 7. Block “Heating steel”.

Thus, the mathematical model of the power supply system of the ladle furnace simulated the system operation at a level close enough to the actual behaviour studied in field tests. A criterion for the model accuracy was considered to be an insignificant error (most often the root mean square one) of the simulation results. At the model adjustment stage, the heat balance residual error was attributed to electrical losses. Model adequacy was proved by conducting control heats (10 heats of steel grade 09G2S) on the ladle furnace in the basic oxygen furnace shop of PJSC MMK. Inaccuracy in estimating the current temperature of the melt did not exceed 5 °C.

Studies on the database for the ladle furnace of PJSC MMK using the created model of heat balance allowed us to estimate a range of changes in heat loss items (Table 1).

Table 1. Ranges of changes in heat loss items on the ladle furnace during a melting process (melting steel 09G2S, ID 19135).

Heat loss items	Range of changes (%)
Outgoing gases	3-15
Water cooling the cover	15-45
Slag induction	8-10
Adding alloying elements	0.5-1
Electrical loss	10-20

3. Conclusion

Thus, the conducted studies on the mathematical model showed that maximum energy loss in the ladle furnace were attributed to losses with cooling water and in the electrical circuit. It should be noted that electrical losses are total energy losses of the electric arc, conditionally including radiation losses of a part of the arc uncovered by slag. By managing these flows, we create conditions for increasing heating efficiency. The calculations showed that a decrease in these losses is observed during prolonged heating (about 8–12 minutes) with a sufficient amount of slag. This mode provides a reduction in heat loss by 4–5 %. By selecting options of changes in a sequence of stages and their duration, the proposed mathematical model makes it possible to apply a new technology to process molten steel in the ladle furnace characterized by lower power consumption.

References

[1] Agapitov E B, Nikolaev A A and Lemeshko M A 2018 Complex research of energy efficiency of electric arc furnace ladle installations *International Multi-Conference on Industrial Engineering and Modern Technologies* **8602500**

[2] Agapitov E B, Kornilov G P, Khramshin T R, Erofeev M M and Nikolaev A A 2006 Thermal and electrical control of the ladle furnace *Electrometallurgy* **6** 11–16
[3] Dyudkin D A, Bat S Yu, Greenberg S E and Marintsev S N 2003 *Steel production on the ladle furnace* (Donetsk: South-East, Ltd) p 300
[4] Kramarov A D and Sokolov A N 1976 *Electrometallurgy of steel and ferroalloys* (Moscow: Metallurgy) p 440
[5] Erofeev M M and Agapitov E B 2009 Mathematical simulation of nonstationary mixing of steel in a ladle furnace *Russian Metallurgy (Metally)* **7** 571–575
[6] Steinberg L S, Goldberg L A, Kuzina V I, Vladimirov V A and Chaykin B S 1995 The software package for calculating the parameters of heat and mass transfer processes of after-treatment metal *Steel* **11** 14–17
[7] Kats Ya L, Sinelnikov V A and Brodov A A 2004 Ladle furnaces for increasing the efficiency of electric steel melting *Proceedings of the 8th Congress of Steelmakers* 18–24
[8] Nikolsky L E, Smolyarenko V D and Kuznetsov L N 1981 *Thermal work of electric arc furnaces* (Moscow: Metallurgy) p 320
[9] Beitsun S V, Zhadanos A V, Mikhailovskiy N V and Shatalyuk S V 2004 *Prediction of heat losses through the lining of the ladle during out-of-furnace processing of steel* (Zaporozhye: Metallurgy collection of scientific papers) p 124
[10] Volkova Ye N, Bakhman S M and Sheller P R 2006 The influence of various lining materials on the thermal state of the after-treatment steel ladle *Metallurgy and mining industry* **7** 314–320
[11] Kharlamov D A and Merker E E 2004 Development of a control algorithm for slag mode of out-of-furnace steel processing *Energy Saving and Energy Efficient Technologies. Collection of reports of the All-Russian Scientific and Technical Conference* **2** 238–241
[12] Vikhlevshchuk V A, Kharakhulakh V S and Brodsky S S 2000 Ladle steel finishing *(Dnepropetrovsk: System technologies)* p 190
[13] Agapitov E B, Bigeev V A and Erofeev M M 2007 The results of the treatment of the steel melt on the ladle furnace with hollow electrodes *Vestnik of Nosov Magnitogorsk State Technical University* **1(17)** 36–38
[14] Yakimov I A and Gasiyarova O A 2018 The energy-efficient control of the electrical regime of high-power electric arc furnaces: the case of EAF-180 MMK, PJSC *International Conference on Industrial Engineering, Applications and Manufacturing* 1–5
[15] Agapitov E B 2011 Energy saving at jet-plasma bucket processing of steel melt *Vestnik of Nosov Magnitogorsk State Technical University* **4(36)** 88–90
[16] Agapitov E B, Bigeev V A and Lemeshko M A 2011 Assessment of the effect of indistinctly controlled parameters on desulfurization in a ladle furnace *Electrometallurgy* **11** 7–9
[17] Agapitov E B, Lemeshko M A and Sokolova M S Prospects for the use of hollow electrodes for deep desulfurization of steel in the ladle furnace *Materials Science Forum* **989** 474–479
[18] Kartavtsev S V, Nikolaev A A, Neshporenko E G, Demin Yu K and Matveev S V 2016 Analysis of its own energy independence electric steelmaking process *Proceedings of the 2016 IEEE North West Russia Section Young Researchers in Electrical and Electronic Engineering Conference* 587–589
[19] Nikolaev A A, Tulupov P G and Antropova L I 2018 Heating stage diagnostics of the electric arc furnace based on the data about harmonic composition of the arc voltage *Proceedings of the 2018 IEEE Conference of Russian Young Researchers in Electrical and Electronic Engineering* 742–747
[20] Khramshin V R, Evdokimov S A, Nikolaev A A and Karandaev A S 2015 Monitoring technical state of the Smart-Grid technology introduction within the industrial electric networks *Proceedings of the 2015 IEEE North West Russia Section Young Researchers in Electrical and Electronic Engineering Conference* 214–220

IOP Conf. Series: Materials Science and Engineering **969** (2020) 012045 doi:10.1088/1757-899X/969/1/012045

High Temperature Treatment of Castings from Steel Grade 150KhNM

A V Efimov, M G Potapov and N A Kuts

Department of foundry processes and materials science, Nosov Magnitogorsk State Technical University, 38, Lenin Street, Magnitogorsk, 455000, Russia

E-mail: potapovmg@mail.ru

Abstract. High-temperature processing of steel cast is an effective way to obtain a high range of properties of finish product. The main purpose of the study was to determine the critical temperatures and modes of processing of molten steel 150KhNM. The thermal calculation was carried out according to the additivity rule. The holding time of the melt was determined experimentally. Laboratory melts were performed with varying temperatures of maximum heating of the melt above and below tcr. The optimal modes of high-temperature processing of the melt were determined by analyzing the structure of solid samples. The influence of high-temperature treatment on the structure and properties of castings from steel grade 150KhNM has been studied. The main processing parameters of this alloy have been determined. The dependences of the metal overheating influence on the castings mechanical properties improvement are obtained. Examples of improving the quality of products with using optimal modes of high-temperature processing are given. The investigated modes have showed that holding at a critical temperature leads to change in the type of casting-box (raw sand, dry sand, chill mold) and uniformity of distribution. This fact leads to a change in properties and a more homogeneous structure is formed.

1. Introduction

High temperature treatment (HTT) is a highly efficient and successfully developed method of treating molten steel. This is one of the methods of thermal effect on steel. It includes heating steel up to a set temperature, soaking at such temperature within a definite period and cooling down to casting temperature or soaking steel near casting temperature. Molten steel is refined and homogenized, resulting in an increase in a broad range of various performance properties and the quality of the castings [1–3].

The current machine industry, metallurgy and other industries of the metalworking industry are characterized by high requirements for properties of materials. This is attributed to an increasing rate of loading machines, including a trend in decreasing their weight. In many cases, weight is a decisive factor.

Quality and operational reliability of industrial products – machines, pieces of machinery and devices – directly depend on relevant properties of their elements – parts, joints and assemblies. In their turn, structural properties of the latter depend on the properties of materials. Static and dynamic strength, brittle fracture resistance, scuff and wear resistance, resistance to fatigue fracture, physical, corrosion, heat-resisting and other properties of materials are attributed to main factors determining reliability and lifetime of machines.

Main structural materials are metals and alloys. Production of metals and alloys in the countries with a high technical production level significantly increases, especially with an increase in melting steel that undergoes heat treatment. A main objective of steelmaking is production of molten steel with a set chemical composition, definite physical, chemical and casting properties with minimum power consumption, burden materials and melting time. Production of high-purity steels is widely spread. Development of heat treatment operations includes names and parameters of such operations. This includes setting the heating temperature, heating time, soaking time, cooling time, heating and cooling media.

When producing ingots, billets and castings, to control and prevent defects, manufacturers apply many techniques that can be divided by a method of treatment to the specified degree of convention. Dynamic methods are based on a principle of the forced physical effect on a liquid phase during solidification. Such methods usually ensure an active influence not only on heat– and mass transfer in a liquid phase, but also significantly change the nature of the processes in a two-phase zone [2, 3].

Transfer of the system to the equilibrium state or the state near to equilibrium is resulted from heating up to such high critical temperatures. The higher degree of equilibrium and uniformity of alloying component atoms distributed in molten steel, the weaker inherited influence of burden materials and the higher quality of the castings. By heating up to a set temperature, liquid eliminates some burden inheritance according to a trend in a phase transfer, not by a chemical reaction; therefore, molecular formations cannot be a main basis for this process [1, 4–6].

2. Objective

This medium-carbon steel grade for rolls undergoes very complex and power-consuming heat treatment. Hypereutectoid steel grade 150KhNM is used for manufacturing forged and cast rolls for hot rolling and rolls for cold rolling. Despite complexity and power consumption of relevant operations, steel that underwent these types of treatment often did not show significant improvement of mechanical properties.

One of main disadvantages of this steel grade is its tendency to crack. This steel grade also has lower wear and thermal resistance. This paper is aimed at studying the structural factors influencing impact strength and structural homogeneity of the castings from steel grade 150KhNM [7, 8].

These guidelines, written in the style of a submission to J. Phys.: Conf. Ser., show the best layout for your paper using Microsoft Word. If you don't wish to use the Word template provided, please use the following page setup measurements.

3. Materials and methods of studies

To determine main parameters of high temperature treatment, such as critical temperature, we applied a calculation method depending on a chemical composition of the alloy according to the rule of mixtures. Critical temperature is the temperature when molten steel is homogeneous. A homogeneous state contributes to the lack of hysteresis in molten steel and cluster groups. Such high temperature structure of molten steel is well overcooled; we managed to decrease concentrations of detrimental impurities, such as oxygen and nitrogen, directly in molten steel. Thus, the amount of added reducing constituents and non-metallic inclusions decreases due to favorable conditions for their removal [1, 8–10].

Melting heat, as well as other properties, follows the rule of mixtures. Liquidus and solidus temperatures were calculated according to the equation:

$$\frac{TL}{S} = T_0 - \Delta t_{L/S}, \tag{1}$$

where T_0 is melting temperature of the solubilizing agent (iron); $\Delta t_{L/S}$ is a decrease in liquidus/(solidus) temperature resulting from various chemical elements contained in steel.

$$\Delta t = \frac{\Delta t_l}{s(a)} + \frac{\Delta t_l}{s(b)} + \frac{\Delta t_l}{s(c)} \ldots \tag{2}$$

where $\Delta t_l / s(a) + \Delta t_l / s(b) + \Delta t_l / s(c)\ldots$ is liquidus/(solidus) temperature of molten steel with the concentrations of a; b; $c\ldots$ elements according to the binary phase diagrams.

Melting temperatures are given in Table 1.

Table 1. Melting temperature of steel grade 150KhNM.

Elements	Percentage, %	T liquidus, °C	Decrease in Tliquidus, Δ°C	T solidus, °C	Decrease in T solidus, Δ°C
Fe-C	1.45	$t_l = 1470$	$\Delta t_l = 69$	$t_s = 1240$	$\Delta t_s = 319$
Fe-Si	0.29	$t_l = 1537$	$\Delta t_l = 2$	$t_s = 1535$	$\Delta t_s = 4$
Fe-Mn	0.65	$t_l = 1534$	$\Delta t_l = 5$	$t_s = 1532$	$\Delta t_s = 7$
Fe-Cr	0.995	$t_l = 1538$	$\Delta t_l = 1$	$t_s = 1538$	$\Delta t_s = 1$
Fe-Ni	1.05	$t_l = 1530$	$\Delta t_l = 9$	$t_s = 1528$	$\Delta t_s = 11$
			$\sum \Delta t_l = 86$		$\sum \Delta t_s = 332$
$t_l = 1539–86 = 1453$ °C			$t_s = 1539–332 = 1217$ °C		

Mole fractions were used to determine molar melting heat. To calculate a mole fraction of every component, we determined a number of moles of every component (Table 2).

$$n_i = \frac{C_i}{M_i}, \tag{3}$$

where n_i is a number of moles of the component in 100 g of the alloy; C_i is a concentration of the component in the alloy, % (g/100 g); M_i is molar mass of the component.

The sum of moles:

$$\sum n_i = n_1 + n_2 + n_3 + \ldots + n_n. \tag{4}$$

A mole fraction is calculated as a ratio of the number of moles of the component to a total number of moles:

$$m_i = \frac{n_i}{\sum n_i}. \tag{5}$$

Melting heat for every chemical element is taken from reference data. By summing all the components, we calculate a mole fraction of melting heat in kJ/mol (Table 3).

A share of a disordered zone at melting temperature is calculated by the relation:

$$\psi_{melt} = \frac{\Delta H_{melt}}{\Delta H_{dis}} \tag{6}$$

where Ψ_{melt} is a share of the disordered zone at melting temperature; ΔH_{melt} is melting heat, kJ/mol; ΔH_{dis} is thermal effect of full thermal disorder of clusters in liquid, kJ/mol.

In our case, liquidus temperature is taken as melting temperature. Thermal effect (Table 1) of full thermal disorder is calculated by the below empirical dependence:

$$\Delta H_{dis} = -1.26 + 2.95 \cdot 10^{-2} \cdot T_{melt}. \tag{7}$$

Temperature dependence of the share of the disordered zone is expressed by the equation:

$$\psi_{dis} = A\exp\left[\frac{-\Delta H_{dis}}{RT}\right]. \tag{8}$$

Knowing the share of the disordered zone at liquidus temperature, we calculate coefficient A:

$$A = \frac{\psi_{melt}}{\exp\left[\frac{-\Delta H_{dis}}{RT}\right]}. \tag{9}$$

Temperature of disorder (Table 2) is expressed by the share of disorder:

$$T_{dis} = \frac{-\Delta H_{dis}}{(R\ln A)}. \tag{10}$$

Table 2. Mole fractions.

Elements		C	Si	Mn	S	P	Cr	Ni	V	W	Fe	Total
Moleweight		12	28	55	32	31	52	59	61	184	56	
Steel grade 150 KhN M	%	1.45	0.29	0.65	0.018	0.015	0.995	1.05	0	0	95.53	100
	wt	0.064	0.0053	0.006	0.0003	0.0003	0.01	0.0095	0	0	0.904	1.00
ΔH_{melt}		103.2	47600	14630	393.6	25960	21008	17523	23050	46000	14120	
Share of ΔH_{melt}		6.605	252.3	175.6	0.118	7.788	210.1	166.5	0	0	12764	15583

Table 3. Temperature of disorder.

Steel grade	ΔH_{dis}, kJ/mol	ϕ_{dis}	A, coefficient	T_{dis} at ϕ_{dis}= 0.5
150KhNM	49717	0.273214	8.7476	1817 ^{0}C

During steel melting, after adding and melting all burden components, molten steel is heated up to 1800–18200 ^{0}C. Soaking time at these temperatures for all molten steels was 10 minutes, then they were cooled down to steel pouring temperatures. Steel was poured into dry green sand molds. The HTT parameters are based on previous studies on the selection and evaluation of HTT schedules for molten steel to improve its structure and properties.

To analyze the chemical composition of steel, we applied a spectral method.

4. Effect of high temperature treatment on the structure and properties of the castings

The structure of cast steel 150KhNM (Figure 1) consists of granular pearlite and thin cementite network, as well as carbide inclusions. It consists of a pearlite matrix and secondary cementite precipitated on boundaries of dendritic branches and primary grains of austenite as a network and coarse laminae growing from a boundary network inside dendritic branches. Eutectic carbide areas are formed between branches of dendrites. A ferrite skin is formed along the network and laminae of secondary carbide. Depending on the cooling rate below point A1, eutectoid austenite is transformed into lamellar pearlite or partially into lamellar pearlite and partially into granular pearlite. Steel is hypereutectoid [5, 11–21].

At overheating a volume of cementite decreases (Figure 2). A cementite network was almost completely dissolved (the matrix was broken) and became thin and broken. Grains became significantly coarser. Granular pearlite transformed to lamellar pearlite. Overheating with nanosecond electromagnetic pulses: the structure consists of eutectic (a mechanical mixture of pearlite and carbides). Pearlite is ordinary. Dendrites grew in the direction to the anode, when treated with nanosecond electromagnetic pulses; therefore, solidification is directed to the anode. Alloyed carbides of a regular square type were formed due to the direction of crystallization.

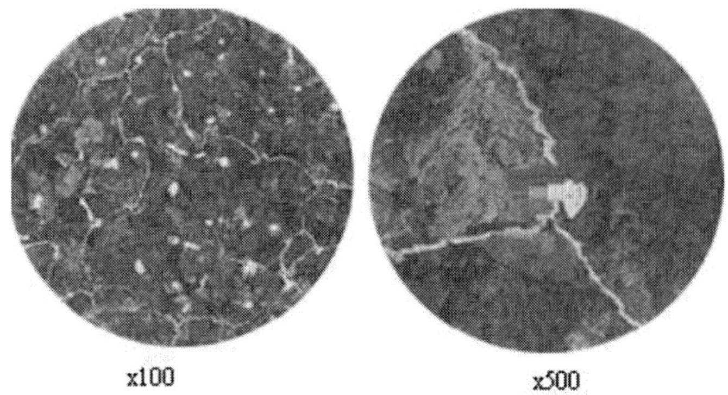

Figure 1. Structure of cast steel 150KhNM that did not undergo HTT.

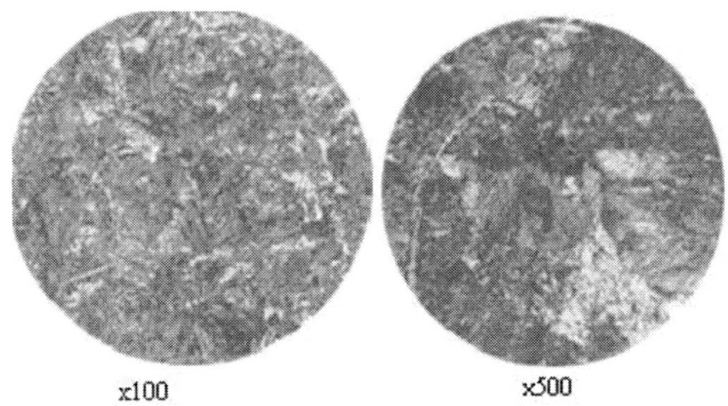

Figure 2. Structure of cast steel 150KhNM that underwent HTT.

Steel that underwent HTT has higher ductile properties, hardness decreased from 45 HRC to 31 HRC, by 1.4 times. Impact resistance increased by 1.4 times, from 4 J/cm2 to 5.6 J/cm2. Overheating contributes to the increase in ductile properties of steel.

5. Conclusion

This paper describes how high temperature treatment influences the structure and properties of the castings from steel grade 150KhNM. Optimum treatment schedules were developed to improve steel properties. It was found that the changed structure had influenced the mechanical properties of the castings under such treatment.

High temperature treatment contributes to the significant improvement of physical, chemical and casting properties of steel, as well as mechanical and strength parameters. It promotes lower inhomogeneity and better ordering of the structure in the castings.

References

[1] Potapov M G, Berezova S V, Ivanova I V and Koshcheev P V 2012 An experimental study of the influence of high temperature treatment of molten sodium thiosulfate on the formation of its macrostructure *Foundry processes* **11** 80–85

[2] Efimov A V, Potapov M G and Chernov V P 2018 Properties of castings from steel 150khnm after high-temperature processing *Foundry* **4** 17–18

[3] Sangho Jeon, Dong-Hee Kang and Shin Hwan Kang 2016 Non-Contact Measurement of Thermophysical Properties of Fe, Fe–C, and Fe–C–Mn Alloys in Solid, Supercooled, and Stable Liquid Phases *ISIJ International* **56 4**
[4] Mikhalkina I V 2015 *Selection and justification of the high temperature treatment schedule for the melt of mold die steel to improve its structure and properties* (Magnitogorsk: Nosov Magnitogorsk State Technical University) p 130
[5] Grigoriev A K and Kodzhaspirov G E 1985 *Thermomechanical hardening of steel in blank production* (Leningrad: Mechanical Engineering) p 143
[6] Bernshtein M L 1968 Thermomechanical treatment of metals and alloys (Moscow: Metallurgy) p 1172
[7] Thu Hoai L and Lee J 2011 Density of Liquid Fe-Mn-C Alloys *Metall and Mater. Trans. B* **42 925** Online. Available: https://doi.org/10.1007/s11663-011-9555-9
[8] Baum B D and Tyagunov G V 1996 Thermal treatment of liquid alloys and steel *Steel* **6**
[9] Chernov V P 2016 *Theory of molten metals: Textbook* (Magnitogorsk: Nosov Magnitogorsk State Technical University) p 79
[10] Baryshev E E, Savina L G and Baum B A 2002 On the role of clusters in the crystallization of metals *Physical properties of metals and alloys* 172–178
[11] Efimov A V and Chernov V P 2017 Time and temperature treatment of castings from steel grade 150KhNM *Vestnik of NMSTU* **1** 79–84
[12] Bazin Yu A and Klimenkov E A 1979 Short-range order of liquid iron alloys with carbon *Ukr. physical Journal* **24 7** 1052-1055
[13] Ivakhnenko I S 1985 Structural Features of Metal Melts *Proceedings of universities. Ferrous metallurgy* **5**
[14] Chernov V P 2012 *Theory of molten metals: Textbook* (Magnitogorsk: Nosov Magnitogorsk State Technical University) p 143
[15] Savina L G 2003 *Effect of high temperature melt processing on the structure and properties of high-carbon ferroalloys* (Yekaterinburg) p129
[16] Lee J, Thu Hoai L and Shin M 2011 Density and Surface Tension of Liquid Fe-Mn Alloys *Metall and Mater. Trans B* **42** 546–549 Online. Available https://doi.org/10.1007/s11663-011-9490-9
[17] Borisov G P 1991 Heredity of the heat-effect on the melt in the region of a quasi-two-phase state *Foundry* **4** 6–8
[18] Filippov K S 1998 Investigation of the state of clusters in metal melts using the equilibrium of free energy between the surface and volume *University News. Ferrous metallurgy* **5**
[19] Spassky A G 1950 *Fundamentals of foundry. Textbook for students of universities of metallurgy* (Moscow: Metallurgizdat) p 319
[20] Kaniski D R 1984 An Update on U.S. Ductile Iron Practicles *Modem Casting* **74 11** 25–27
[21] S Mathur and A Barnawal The melting process *Casting world* **1 3**

IOP Conf. Series: Materials Science and Engineering **969** (2020) 012046 doi:10.1088/1757-899X/969/1/012046

Effect of External Integrated Treatment on the Structure and Properties of the Castings from Steel Grade 35L

A V Efimov, M G Potapov and N A Kuts

Department of foundry processes and materials science, Nosov Magnitogorsk State Technical University, 38, Lenin Street, Magnitogorsk, 455000, Russia

E-mail: potapovmg@mail.ru

Abstract. The methods of influence of external physical and thermophysical factors on the cast steel structure formation are considered in the paper. High-temperature processing of the melt is an effective method of heat treatment. It improves a wide range of mechanical and operational properties of finished castings. The parameters of high-temperature treatment were established experimentallyin the course of this study. Its influence on the structure and properties of 35L steel castings has been shown. The usage of a nanosecond electromagnetic pulse during the crystallization of a casting in the mold is a method of dynamic action on the metal liquid phase. The parameters of nanosecond electromagnetic pulse processing during crystallization were determined by the technical capabilities of the electromagnetic pulse generator.

The study of the properties and structure of samples cast from a melt processed by electromagnetic pulses confirms the improvement of the mechanical and operational properties of castings by improving the primary cast metal structure. The influence of metal overheating, nanosecond electromagnetic pulse, and temperature-time holding on the structure and mechanical properties of castings is revealed. The modes of thermal holding (final technological heat treatment of the metal) were determined experimentally using the method of differential scanning calorimetry. The paper presents the results of changes in improving the properties of cast products due to the use of optimal modes of external complex influences.

1. Introduction

Current development of the machine building industry is attributed to a higher quality of finished products and lower production costs. In current foundries an original structure of steel rarely satisfies the set requirements without additional treatment or any additional effects on manufacturing processes.

Presently, little attention is paid to improving the quality of steel by applying external treatment without using additional alloys or modifying elements.

Now, steelmakers use a variety of methods to improve the quality of steel by treating it at the solidification stage. Properties of finished steel may be early controlled at initial stages by external treatment of molten steel. By varying molten steel treatment schedules and methods, the structure and properties of steel may be changed within a broad range. Conventionally, researchers pay more attention particularly to the second stage, namely a search for optimum solidification conditions. Attempts of treating steel at the melting stage are limited to additional alloying, modifying and refining processes only to optimize the composition, while little attention is paid to applying additional external treatment without using additional elements. Alloying and modifying elements introduced into molten steel often entail not only positive, but also negative side effects: many elements adversely affect steel, while increasing the cost. External physical and thermophysical treatment of molten steel contributes to a significant increase in properties of steel, avoiding impurities, characteristic of alloying and modifying. At this stage external treatment is very efficient, but less often used. It does not contain the elements producing detrimental effect and has low cost.

A method of dynamic effect on a liquid phase of molten steel at the solidification stage was applied in practice to produce castings. One of the treatment methods of solidifying molten steel is electromagnetic pulse. The application of high voltage nano- and subnanosecond pulses constantly expands due to development of, and reduction of prices for tools for their generation [1–6].

Many processing methods applied on cast steel in a solidification process are widely used to control and prevent defects in a steelmaking process. Dynamic methods are based on a principle of the forced physical effect on a liquid phase during solidification. Usually these methods not only actively effect heat and mass transfer in a liquid phase, but also considerably change the nature of processes in a two-phase zone [2, 3–9].

As for the well-known methods of external heat, physical and thermophysical treatment, steelmakers use high temperature treatment of molten steel, treatment with electromagnetic pulse and thermal effect on solid steel [4].

2. Research procedure

High temperature treatment (HTT) includes heating steel up to the specified temperature, soaking at such temperature within a definite period and cooling down to casting temperature. The structure of molten steel is homogenized entailing an increase in a range of various parameters and quality of castings [5, 6].

In this paper, HTT is used as preliminary treatment of steel to prepare it for solidification driven by nanosecond electromagnetic pulse (NEMP).

Molten steel heating temperature is determined by studying the ways of changing physical properties of liquid alloys under heat effect and the relations between the structure formation and a liquid state of molten steel. Critical temperature is the temperature when molten steel is in a homogeneous state. The homogeneous state is responsible for the lack of hysteresis in molten steel, when measuring structure-sensitive properties, and the lack of large cluster groups. Such high temperature structure of molten steel is well overcooled. We manage to decrease concentrations of impurities, such as oxygen and nitrogen, directly in molten steel. Consequently, the amount of added reducing agents and non-metallic inclusions decrease due to favorable conditions for their removal. Critical temperature for steel grade 35L, when molten steel is in a disordered state, is calculated by applying a rule of mixtures and determining relative shares of structural components of molten steel [7, 8].

A share of a disordered zone is calculated by the equation:

$$\psi_{dis} = A \cdot \exp\left[\frac{\Delta H_{dis}}{RT}\right] = 0.5, \qquad (1)$$

where Ψ_{dis} is a share of a disordered zone in molten steel; A is an empirical coefficient, which is individual for every molten steel; ΔH_{dis} is thermal effect of complete thermal disorder of clusters in liquid, kJ/mol; R is absolute gas constant, J/(mol·K); T is actual temperature, K.

Critical heating temperature, when a disorder level is 0.5 is calculated by the equation:

$$T_{dis} = \frac{\Delta H_{dis}}{R(\ln A - \ln 0.5)} = 1870, {}^{\circ}C. \qquad (2)$$

Critical heating temperature of steel grade 35L is 1870°C.

An alloy melting process was adjusted in the Foundry Laboratory of the NMSTU Department of Casting Processes and Materials Science, applying a 60 kg medium-frequency induction crucible furnace with a basic magnesite lining. This furnace was used to melt steel grade 35L, having a chemical composition under GOST 977–88, wt. %: C is 0.32 – 0.4; Si is 0.2 – 0.52; Mn is 0.45 – 0.9; S is up to 0.06; P is up to 0.06.

When melting steel, after adding and melting of all burden components, molten steel was heated up to 1870–1880 ^{0}C. Soaking time at such temperatures during all the experiments was 10 minutes, then molten steel was cooled down to pouring temperature. Soaking time at maximum temperature was calculated based on the results of previous tests. Steel was poured into dry green sand molds.

Molten steel was treated with NEMP directly in green sand molds. Having cast molten steel into the mold, we supplied NEMP through graphite electrodes contacting molten steel. Electromagnetic pulse effected until complete solidification of steel in the mold. The diagram of the unit is given in Figure 1. Then the castings underwent temperature-time treatment.

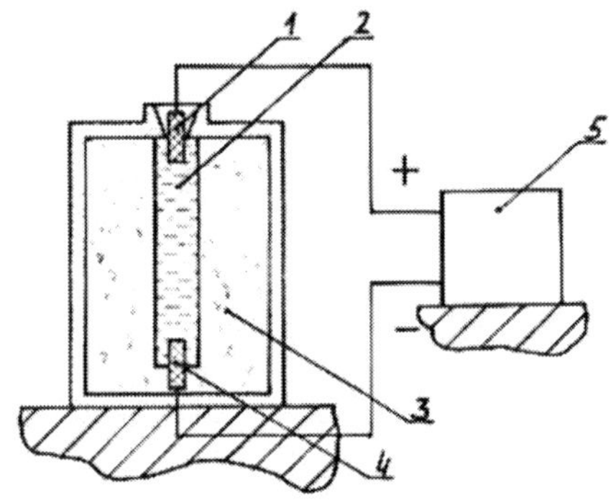

Figure 1. Diagram of the testing unit: 1 is a graphite electrode with an active anode; 2 is a casting; 3 is a green sand mold; 4 is a graphite electrode with a passive cathode; 5 is a source of NEMP.

NEMP parameters: pulse time is 0.5 ns; amplitude is 10 kV; pulse frequency is 1000 Hz.

Temperature-time soaking (TTS) of steel is heat treatment, when short thermal effect is produced on steel in a solid state. Steel was heated up to a set temperature, when structural phase transformations began. To avoid complete phase transformation processes and to fix a definite modified structure after TTS, we applied multiple repeated short cycles of thermal effect under equal conditions. A number of repeated similar operations of thermal effect on the alloy and cooling conditions were determined by experiment.

Temperature-time soaking is based on the pearlite-austenite phase transformation. Pearlite grain sizes depend on the sizes of inherited austenite grains. The larger austenite grains, the larger pearlite grains, as a rule (Figure 2). Austenite grains grow during heating only (they do not become smaller during subsequent cooling); therefore, maximum heating temperature of steel in an austenite state and its inherited grain size determines a final grain size. TTS is aimed at forming a highly dispersed structure due to grain refinement in an austenite area [9, 10–17].

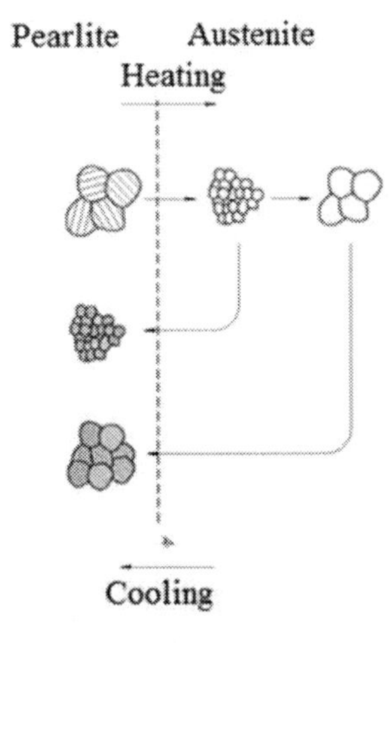

Figure 2. Diagram showing changes in the pearlite grain size depending on heating in the austenite area.

To determine and fix temperatures of beginning phase transitions, we applied a thermal analysis representing a differential scanning calorimetry (DSC) method.

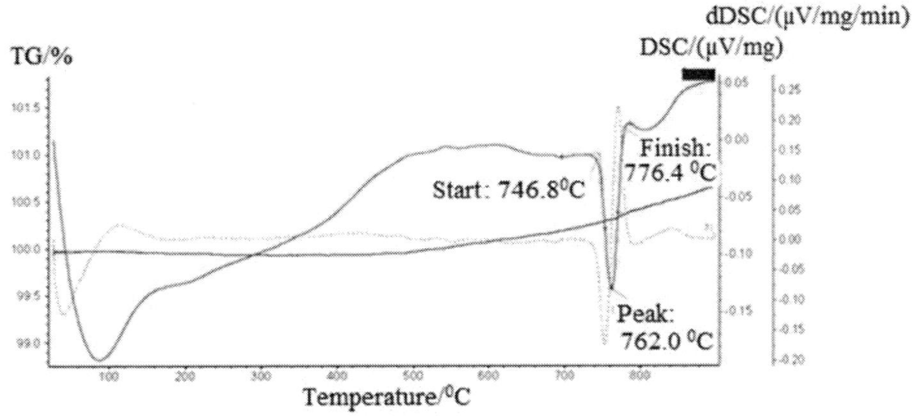

Figure 3. DSC curve of steel 35L.

Within a temperature range of 747 °C–776 °C there is an area characterized by a nucleation period of a new phase. The existing carbides are dissolved and crystallization centers of a new phase (austenite) nucleate. Minimum peak at 760 °C is attributed to maximum formation speed of new crystallization centers.

Steel samples were placed in the furnace preliminary treated up to 760 °C. Casting heating temperature was calculated according to the method suggested by A. P. Gulyaev to determine heating time for a total volume of various products in various media subject to their location in the furnace. Heating time for all the samples is 20 minutes. The samples were cooled in water. The number of treatment cycles, being three under equal conditions, is determined by experiment [10–11, 18–23].

3. Effect of integrated treatment on steel 35L

Medium-carbon cast steel 35L, which did not undergo heat treatment, has a ferrite-pearlite structure, including a Widmanstätten (oriented) distribution of ferrite and a ferrite network along boundaries of former austenite grains. A pearlite to ferrite ratio is 35/65 %. There are minor (insignificant) dendritic areas. The structure has 4–5 points of the Widmanstätten pattern under GOST 5640–68. Minimum and maximum sizes of structural components for the casting samples are from 0.1 μm to 130 μm. Pearlite points is 8–10 (Figure 4).

A structure of sample b that underwent NEMP and HTT also has a ferrite and pearlite structure. It is characterized by the near absence of the ferrite network. Grain number changed from 6–7 to 4–5. Highly-dispersive lamellar pearlite (3–4 points) changed to granular pearlite (1 point). Minimum and maximum sizes of structural components for the steel samples that underwent HTT are from 0.1 μm to 9 μm. (Table 1).

Table 1. Structural relation of the samples from steel 35L.

Sample	Pearlite points	Pearlite to ferrite ratio	Grain number	Interlamellar spacing	Structural inclusion sizes, μm
Cast	6–7	75/25–85/15	over 1	1.0–1.2 μm	0.1–30
Integrated treatment	granular, 1	85/15	4–5	grain diameter is up to 0.25 μm	0.1–9

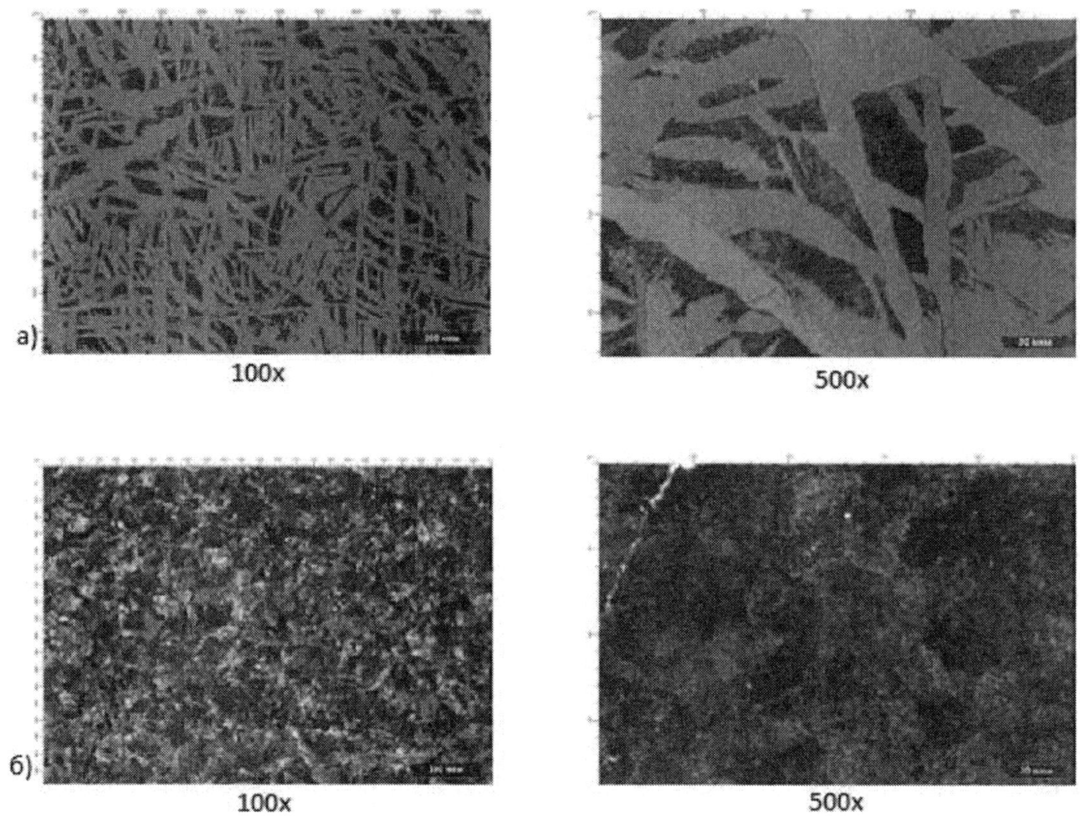

Figure 4. a) cast sample; b) sample that underwent integrated treatment.

Table 2 shows mechanical properties of the samples.

Table 2. Mechanical properties of the samples from steel 35L

Samples	Cast	Integrated treatment
Hardness, HB	185	285
Tensile strength σ_{uts}, MPa	620	950
Wear resistance, Kw	1.32	1.68
Impact strength, KCV	11	11

4. Findings

The steel structure changes significantly under the integrated effect. We noted a significant recrystallization of the structural components. A new finely dispersed and uniform structure is characterized by a total absence of the Widmanstätten pattern; the ferrite network is almost completely dissolved. Lamellar pearlite was fully transformed to granular pearlite. Consequently, hardness of the castings increases by 54 %. Due to a new uniform, fine-grain structure hardness increases by 53 %. When applying temperature-time soaking, wear resistance in the castings increases by 27 %, impact strength does not show a significant change (Figures 5, 6).

High temperature treatment applied during melting and nanosecond electromagnetic pulse at the crystallization stage of molten steel and subsequent time and temperature soaking contribute to a better microstructure of the castings in general, resulting in a higher quality of the castings without additional alloying elements and complex high temperature heat treatment in a solid phase.

Figure 5. Hardness and tensile strength parameters after treatment.

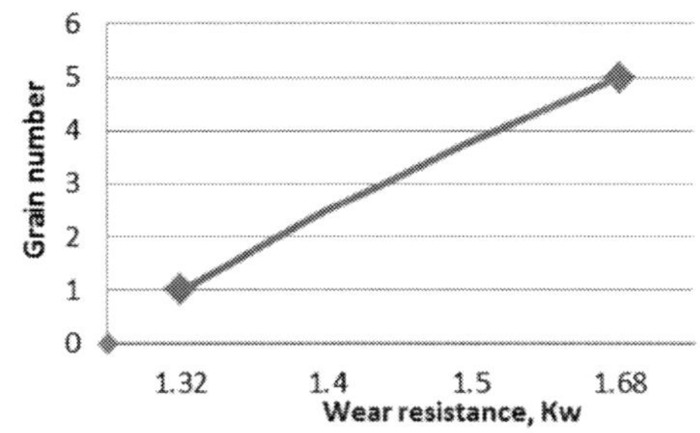

Figure 6. Grain number and wear resistance parameters after treatment.

High temperature treatment applied during melting and nanosecond electromagnetic pulse at the crystallization stage of molten steel and subsequent time and temperature soaking contribute to a better microstructure of the castings in general, resulting in a higher quality of the castings without additional alloying elements and complex high temperature heat treatment in a solid phase.

5. Conclusion

This paper:

- determines main parameters of molten steel 35L and critical temperature of its heating during HTT, and soaking time,
- describes the effect of high temperature treatment on the structure and properties of the castings from steel grade 35L and the relevant process,
- studies the effect of NEMP on the casting structure and properties; describes the developed process of its application, while molten steel is crystallizing, and
- sets schedules of time and temperature treatment as an option of high temperature tempering of steel.

References

[1] Potapov M G, Berezova S V, Ivanova I V and Koshcheev P V 2012 An experimental study of the influence of high temperature treatment of molten sodium thiosulfate on the formation of its macrostructure *Foundry processes* **11** 80–85

[2] Balakirev V F, Krymsky V V and Shaburova N A 1996 *Irradiation of molten metal by powerful electromagnetic pulses* (Chelyabinsk: South Ural State University) p 5

[3] Mikhalkina I V 2015 *Selection and justification of the high temperature treatment schedule for the melt of mold die steel to improve its structure and properties* (Magnitogorsk: Nosov Magnitogorsk State Technical University) p 23

[4] Spassky A G 1950 *Fundamentals of foundry. Textbook for students of universities of metallurgy* (Moscow: Metallurgizdat) p 319

[5] Grigoriev A K and Kodzhaspirov G E 1985 *Thermomechanical hardening of steel in blank production* (Leningrad: Mechanical Engineering) p 143

[6] Bernshtein M L 1968 *Thermomechanical treatment of metals and alloys* (Moscow: Metallurgy) p 1172

[7] Chernov V P 2016 *Theory of molten metals: Textbook* (Magnitogorsk: Nosov Magnitogorsk State Technical University) p 79

[8] Chernov V P 2012 *Theory of molten metals: Textbook* (Magnitogorsk: Nosov Magnitogorsk State Technical University) p 143

[9] Savina L G 2003 *Effect of high temperature melt processing on the structure and properties of high-carbon ferroalloys* (Yekaterinburg) p 129

[10] Efimov A V and Chernov V P 2017 Time and temperature treatment of castings from steel grade 150KhNM *Vestnik of NMSTU* **1** 79–84
[11] Efimov A V and Chernov V P 2018 Influence of external treatment on the structure and properties of steel castings *Current problems of modern science, engineering and education* 11–17
[12] Ivakhnenko I S 1985 Structural Features of Metal Melts *Proceedings of universities. Ferrous metallurgy* **5**
[13] Efimov A V, Potapov M G and Chernov V P 2018 Properties of castings from steel 150khnm after high-temperature processing *Foundry* **4** 17–18
[14] Kaniski D R 1984 An Update on U.S. Ductile Iron Practicles *Modem Casting* **74** **11** 25–27
[15] Lee J, Thu Hoai L and Shin M 2011 Density and Surface Tension of Liquid Fe-Mn Alloys *Metall and Mater. Trans B* **42** 546–549 Online. Available https://doi.org/10.1007/s11663-011-9490-9
[16] Thu Hoai L and Lee J 2011 Density of Liquid Fe-Mn-C Alloys *Metall and Mater. Trans. B* **42** **925** Online. Available: https://doi.org/10.1007/s11663-011-9555-9
[17] S Mathur and A Barnawal The melting process *Casting world* **1** **3**
[18] Baryshev E E, Savina L G and Baum B A 2002 On the role of clusters in the crystallization of metals *Physical properties of metals and alloys* 172–178
[19] Sangho Jeon, Dong-Hee Kang and Shin Hwan Kang 2016 Non-Contact Measurement of Thermophysical Properties of Fe, Fe–C, and Fe–C–Mn Alloys in Solid, Supercooled, and Stable Liquid Phases *ISIJ International* **56** **4**
[20] Bazin Yu A and Klimenkov E A 1979 Short-range order of liquid iron alloys with carbon *Ukr. physical Journal* **24** **7** 1052-1055
[21] Baum B D and Tyagunov G V 1996 Thermal treatment of liquid alloys and steel *Steel* **6**
[22] Borisov G P 1991 Heredity of the heat-effect on the melt in the region of a quasi-two-phase state *Foundry* **4** 6–8
[23] Filippov K S 1998 Investigation of the state of clusters in metal melts using the equilibrium of free energy between the surface and volume *University News. Ferrous metallurgy* **5**

IOP Conf. Series: Materials Science and Engineering **969** (2020) 012047 doi:10.1088/1757-899X/969/1/012047

Methods for Evaluating the Effectiveness of Gas Vortex Stabilization in Plasma Torches for Metal Cutting

I Yu Matushkina[1], S V Anakhov[2] and Yu A Pyckin[3]

[1]Welding technology chamber, Ural Federal University, 19, Mira str., Yekaterinburg, 620078, Russia

[2]Physics and mathematics chamber, Russian State Vocational Professional University, 11, Mashinostroiteley, Yekaterinburg, 620012, Russia,

[3]Physical and chemical technologies of biosphere protection chamber, Ural State Forest Engineering University, 37, Siberian tract, Yekaterinburg, 620038, Russia

E-mail: iri-matushkina@yandex.ru

Abstract. The well-known recommendations for the design of gas-vortex stabilization systems in metal-cutting plasma torches are considered. Various designs of such systems used in the series PMVR plasma torches developed by the authors are presented. It is noted that the method of efficiency assessment developed by the authors should be based on the calculation of the uniformity distribution of the gas flow velocity. The velocities are calculated in the control sections of the gas-air path in the plasma torch. Various simplified and precise estimation methods are proposed. The velocity distribution in control sections for various modifications of plasma torches are presented. Calculations are made based on the "cold" model of gas flow and its heating by a plasma arc. Recommendations are made on the choice of parameters for evaluating the effectiveness of gas-vortex stabilization in plasma torches. The advantages of the new modernized plasma torches, including those using narrow-jet plasma technology, in terms of the gas-vortex stabilization efficiency are demonstrated. The necessity of using several criteria for evaluating the effectiveness of plasma torches design is noted.

1. Introduction

The development of plasma torches for air-plasma cutting, as is known [1, 2], was carried out in the 60s in the USSR (at the Avtogenmash, Institute of electric welding named after E. O. Paton, VNIIEESO), the USA, Japan, Germany and France. At the same time, the main parameters of metal cutting quality were determined. By the mid-70s, the basic principles of designing plasma torches were formulated and the first classifications were created, including various ways to stabilize the arc discharge [3, 4]. A characteristic feature of most designs used in DC arc plasma torches for metal cutting is the use of a gas-vortex stabilization (GVS) system of the plasma arc [1]. In the case of gas-vortex arc stabilization, gas is introduced into the electrode zone through the channels located tangentially to the walls of the arc chamber. Gas supply can be carried out using either one or two gas flows (working fluid) with different designs of swirlers. In a special device – a “swirler” – a spiral vortex flow is created that compresses the arc stream in the open part of the nozzle and in the arc channel and isolates it from the walls. The use of a vortex flow ensures gas mixing in the arc stream, intensifies plasma formation, and increases the voltage of the plasma jet. The tangentially swirled flow of plasma-forming gas (PFG) stabilizes the cathode spot, prevents shunting of the plasma arc and isolates the initial section of the plasma jet from the walls of the nozzle channel. Also it forms the geometry and kinetic properties of the jet at the nozzle outlet, contributing, in addition to improving the above parameters and reducing the acoustic emission [5]. As noted in [6], the stabilization system must be linked to the shape of the electrode, and the stability of the cathodes used for air-plasma cutting strongly depends on the effect of the cathode spot stabilization within the thermochemical cathode insert created by the vortex system [1, 3].

To form a vortex flow in an arc chamber, swirling systems of various designs are used. They can take the form of cylindrical cartridge covering the cathode with threaded screw channels or collar with a hole whose diameter is close to the diameter of the nozzle forming channel. The slots of swirling system are forming the direction of the gas flow tangent to the circumference of the channel. According to the results of experiments conducted in the late 60s [7, 8], it was noted that to ensure the longevity of the electrodes and high stabilization of the arc position in the nozzle channel, it is necessary that the ratio of the gas velocity tangential component V_t to the axial V_0, which is an indicator of the degree of gas spin, was within 7–12. In this case, if $V_t/V_0<5$, the electrode quickly burns out due to the effects of shunting (closing) the arc on the nozzle surface. However, many questions related to the dependence of the arc spot movement on the speed of the swirling gas flow were still insufficiently studied in those years.

A large amount of experimental data on the study of the operation modes for various plasmatron designs was generalized and optimized by the end of the 70s at The Institute of Thermophysics under the leadership of academician M. F. Zhukov [9, 10]. Two design solutions were proposed for linear plasma torches with GVS: one- and two-chamber arc stabilization zone. It was noted that there is a radial gradient of gas density in the electric arc chamber, which leads to the appearance of an Archimedean force. This force pushes the arc stream into the axial zone of the channel when it is deflected from the axis by disturbing forces. The gas vortex flow stabilizes the arc until the wall-mounted turbulent layer penetrates the axial zone of the chamber. The use of two vortex chambers – the central (main) and butt-end (additional) allows avoiding the limitations inherent for a single-chamber plasma torch on the type of gas and current. This effect is due to varying gas flow rates supplied through the two chambers. Generalization of experimental data on cathode erosion made it possible to determine the best values of the circumferential gas velocity at the entrance to the vortex chamber at 150–200 m/s.

In these works, there were presented the calculation of the electric arc chamber in the plasma torch. These calculations allow to define the flow geometry and parameters, to ensure reliable ignition and stable burning of the electric arc, excluding a heat lock channel during operation of the plasma torch. Based on the results of studies on the vortex chamber aerodynamics, conducted in a wide range of changes in the determining parameters, the following recommendations were made for the design of the plasma torch:

1. Ratio of the cross-section area for the twist ring to the cross-section area of the output electrode diameter should be 3.5–5.
2. Number of holes in the twist ring – at least four with a uniform location around the circle; the total area of the passage sections must be such that at a given full pressure and temperature of the supplied gas, its flow rate is about 0.3–0.5 the speed of sound.
3. Length of tangential channels in the twist ring (to obtain a sharply directed gas jet) must be at least 3–4 calibers.
4. Value of the interelectrode gap is selected based on the following conditions:
   - possibility of its breakdown by the oscillator voltage;
   - ensuring that there is no double arcing when cutting;
   - absence of gas-dynamic locking of the gas flow in the gap section.

Compliance with these requirements ensures good GVS of the electric arc on the axis of the discharge chamber and reliable oscillator start-up of the device. It should be noted that these recommendations remain as the main principles of GVS designing in modern publications [11], but they require, in our opinion, analysis for the loss of pressure in the system and the uniformity of distribution of gas dynamic parameters (the flow rate and pressure of the PFG, etc.) in the vortex flow at the output from it with subsequent design corrections. Dynamic analysis also showed that the electric arc chamber of the plasma torch could be regarded as a heat nozzle, in which the gas is heated by the arc discharge [9], and therefore in the design of arc chamber by gas-dynamic criteria, you must enter the correlation for thermal heating of gas flow.

2. Technique of researches

The above arguments and design recommendations, as already noted, were formulated as a result of generalization of experimental data in the study of plasmatrons. At the same time, the above-mentioned works noted the absence of a strict theory of vortex flow in the electric arc chamber of the plasma torch. In this regard, it should be noted that by the end of the 90-ies of the XX century, the emergence of opportunities for automating numerical solutions to complex gas-dynamic problems allowed us to conduct a number of such developments and generalize the results [12–15]. These studies confirmed the effect of energy separation in a vortex tube due to high- and low-frequency instabilities associated with the formation of large-scale coherent shear vortex structures in the tube. It was also noted that in plasma torches with GVS, the average integral temperature of the jet increases with an increase in the degree of spin. In addition [14] using numerical analysis, it was investigated the effect of the ratio for inlet diameter tangential swirl gas injection nozzle to the diameter of swirl nozzles on the flow in the channel of the plasma torch and found the dimensions of the entrance holes at their various numbers (six and eight). However, a very simplified geometric model of the electric arc chamber was analyzed, and the solutions were in a two-dimensional approximation.

Summing up the consideration of various well-known principles and methods of GVS designing, it should be noted that the problem of the plasma arc stabilization in plasma torches for metal cutting is multi-factorial. In our opinion, there is a gap in the wide variety of research results that affect the efficiency of such plasma torches, related to the study of the gas flow uniform distribution factor across the plasma torches section, both in vortex and electric arc chambers. The research conducted earlier by the authors [3, 16] showed that the applied methods of GVS in such plasmatrons do not ensure the proper uniformity of the PFG flow distribution along the cross section of the gas-air path (GAP), thereby reducing the efficiency of the plasmatron. As a result, the problem arises of studying this factor in standard and upgraded plasma torches, determining its physical and mathematical equivalent, which allows its use for the search for new design solutions that improve the operation of the plasma torches, the efficiency and application breadth of plasma cutting technologies.

For this purpose, one- and two-flow plasma torches of the PMVR series with various systems of GVS are considered. As the base model for gas dynamic analysis, the PMVR -M plasmatron produced by company Polygon (Yekaterinburg) was selected (Figure 1 a), which has proven itself at many metallurgical and pipe production enterprises in the Ural region of the Russian Federation.

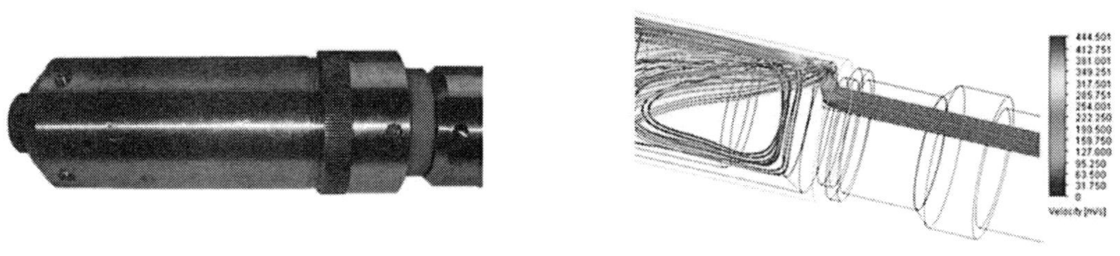

a b

Figure 1. PMVR-M plasma torch: a – working model, b – speed distribution by GAP.

The uniform distribution of the PFG flow rate across the cross-section of the plasma torch channels is an important parameter that determines the efficiency of stabilization and shielding of the plasma arc, and ultimately affects the performance and quality of cutting, as well as the reliability of the plasma torch. This distribution was determined by numerical methods in the FlowWorks application of the SolidWorks software in the process of calculating the gas dynamic and thermo-physical parameters of the gas flow along the GAP of the plasma torch. When calculating the velocity distribution over the PMVR-M plasma arc (Figure 1 b), it is seen that the uneven distribution of the gas flow through the swirler channels is due to an asymmetric gas supply to it through a small expansion chamber. This unevenness persists in the nozzle of the plasma torch, thereby affecting the efficiency and quality of its operation.

Using symmetrical supply of plasma gas flow into GVS, two expansion chambers and swirlers with a stepped distribution of gas through the channels of GAP and a tangential flow of gas in the nozzle allows to increase the distribution uniformity in the nozzle of the plasma torch, to improve the kinematic and energy parameters of the arc plasma (jet) for the single-stream torches series PMVR-5. The PMVR-5.1 plasma torch (PMVR-2M in the former nomenclature) uses a system of gas dynamic filters (additional walls with perforations in the expansion chamber) and increases the size of the expansion chamber (Figure 2). In the PMVR-5.2 plasma torch (PMVR-3M in the former nomenclature), two swirlers and two expansion chambers (forming and stabilizing) are used – Figure 3, the PMVR-5.3 plasma torch is supplemented with a symmetrical gas supply to the expansion chamber (via 2 channels) and changing the geometry of the forming swirler – Figure 4.

Figure 2. PMVR -5.1 plasma torch with the GVS.

Figure 3. GVS of the PMVR-5.2 plasma torch.

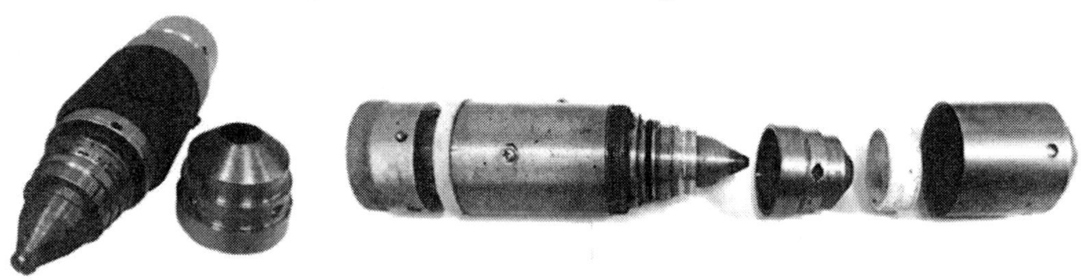

a b

Figure 4. Experimental model of the PMVR-5.3 plasma torch (design elements).

Additional compression of the plasma jet due to the flow of secondary gas in two-stream plasma torches ("narrow-jet plasma" technology) of the PMVR-9 series (Figure 5) allows to further increase the uniformity and kinematic characteristics of the plasma flow at a small distance over the nozzle of the plasma torch. Developed plasma torches of this type can significantly increase the productivity and energy efficiency of cutting technology, as well as improve the quality of cutting seams [17].

Figure 5. Experimental model of the PMVR-9.1 plasma torch (design elements).

When evaluating the GVS efficiency in metal-cutting plasma torches, you can rely on several key indicators. First, you should pay attention to the implementation of the above recommendations on the ranges and ratios of gas flow rates in the nozzle of plasma torches. However, the width of the declared speed ranges does not allow us to make a qualitative comparative analysis of various GVS designs, and, consequently, to identify their advantages and disadvantages. In this regard, as already noted, it is advisable to take as a key criterion the distribution of speeds and mass expenditures across the GAP section in the plasma torch. It is desirable that these distributions are sufficiently uniform already at the input to the 1st swirler of GAP. However, due to the fact that the main functions of the GVS system are performed in the nozzle, it is advisable to take the section at the input to the nozzle (near the plane of the butt-end surface of the cathode) and the section at the input to the cylindrical nozzle channel (the output section) as control sections (CS) for calculating. A similar principle is advisable to use in the system of two-stream plasma torches PMVR-9, using a double nozzle system (Figure 6). In addition to determining the gas-dynamic parameters, it is also advisable to calculate the temperature distributions in the CS, as well as to see the change in these parameters outside the plasma torch nozzle (within the interaction of the plasma jet with the metal surface).

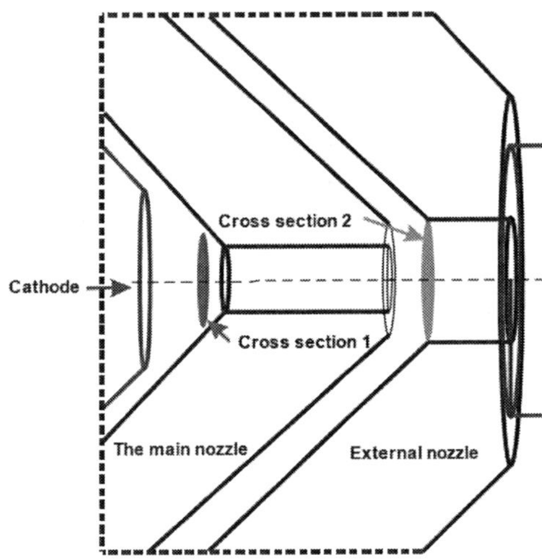

Figure 6. Location of the ring control sections (CS) in the PMVR-9.1 plasma torch.

When calculating the uniformity factors, there are several problems associated with determining the geometry of the calculated trajectory, the number of calculated points, and selecting the uniformity criterion. It has been experimentally established that it is advisable to calculate along the middle line of the CS, using reasonable machine modeling capabilities (50–200 calculated points along the perimeter of the trajectory). If there is a shortage of time and machine capacity, it is possible to limit the calculation to 4 symmetrically located points (relative to the supply line of the PFG to the expansion chamber [16]), but due to the undulating nature of the speed distribution (Figure 7), the accuracy of determining the efficiency of GVS will be low.

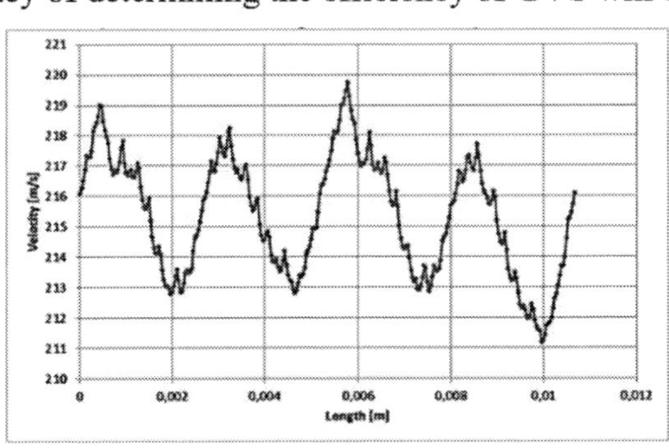

a

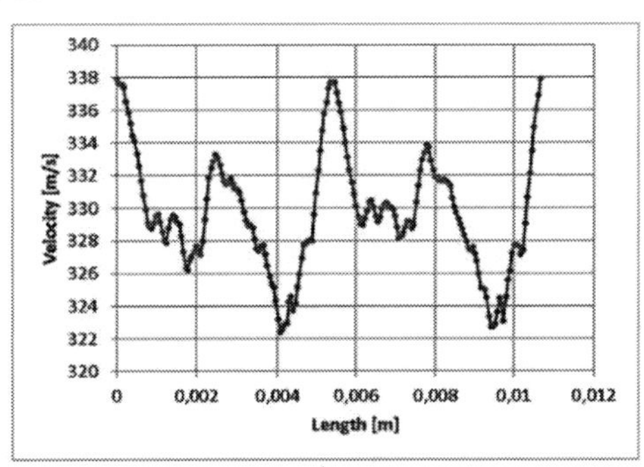

b

Figure 7. Velocity distribution in the control sections (CS) of the PMVR-9.1 plasma torch (Figure 6):a – in section 1, b – in section 2.

IOP Conf. Series: Materials Science and Engineering **969** (2020) 012047 doi:10.1088/1757-899X/969/1/012047

In order to find a criterion for evaluating the degree of the flow distribution uniformity when analyzing a large number of control points, the obtained velocity distributions for various plasma torches were analyzed using statistical methods. A statistical hypothesis was made that the presented velocity dependencies obey either the normal or uniform distribution law of random variables. However, testing the hypothesis using the Pearson χ^2 criterion did not provide statistically significant confirmation. It was also not confirmed when analyzing other velocity distributions and when changing the frequency calculation intervals. For this reason, the parameters that are often used when processing large arrays of random variables were considered as criteria:

1) Scope of variation $R = V_{max} - V_{min}$,

2) Ratio of variation $L = \dfrac{V_{max}}{V_{min}}$,

3) Average linear deviation (ALD): $a = \sum_{i=1}^{n} \left| V_i - \bar{V} \right| / n$,

4) Mean square deviation (MSD): $S = \sqrt{\sum_{i=1}^{n} (V_i - \bar{V})^2 / n}$,

5) Coefficient of variation $F = \dfrac{S}{\bar{V}}$.

For large sample sizes (in our case n>40), $S \approx \sigma$, where σ is the dispersion of the random variable.

It is obvious that the values of productivity, quality, energy efficiency and safety of metal cutting in comparison with the values achieved when working with other plasma torches will also be valid and experimentally confirmed parameters from the point of view of GVS efficiency.

3. Results of research and their discussion

The results of calculations for a number of the GVS efficiency criteria for various plasma torches of the PMVR series are shown in Figure 8–10. The calculations considered 2 modes – "cold" (without arc heating) and "hot" (with arc heating), which allow to identify the influence of temperature on the efficiency of GVS.

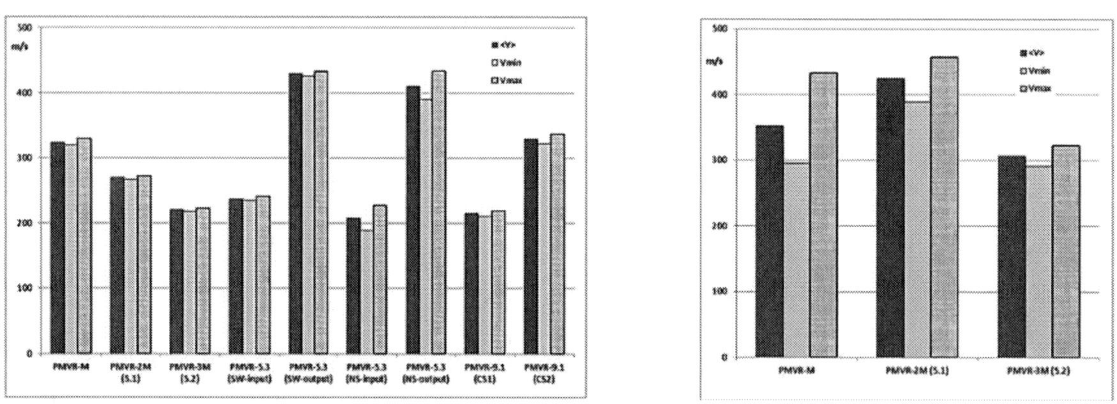

a b

Figure 8. Average, maximum and minimum velocities in control sections (CS) of plasma torches (SW-serial swirler, NW-new (modernized) swirler): a – without heating, b – with arc heating.

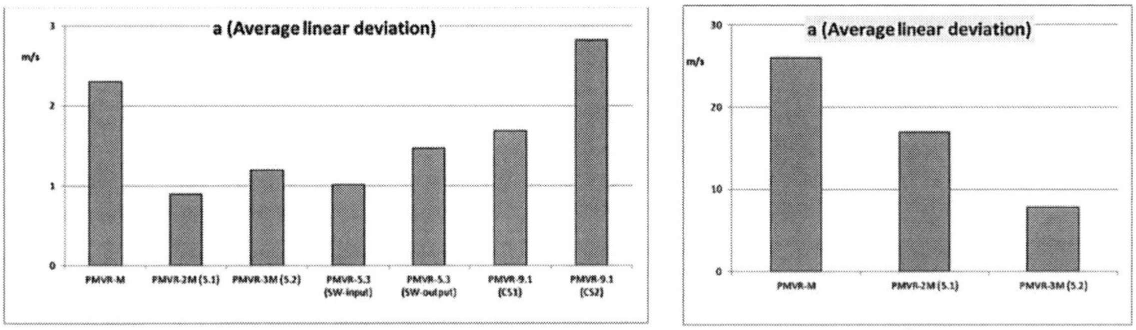

a b

Figure 9. Average linear deviations (ALD) of velocities in the control cross sections (CS) of plasma torches (SW-serial swirler): a-without heating, b-with arc heating.

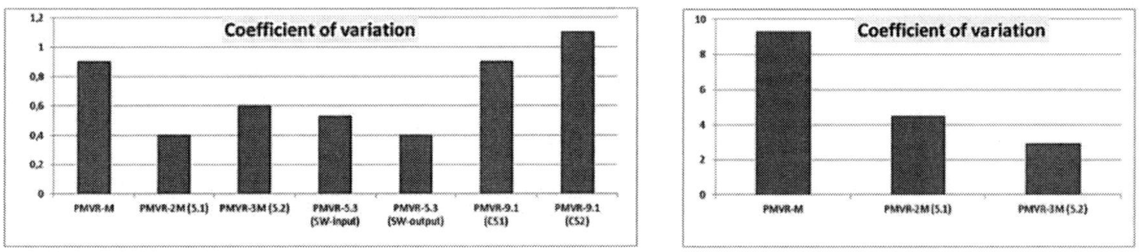

a b

Figure 10. Coefficients of variation of velocity distributions in control sections (CS) of plasma torches (SW-serial swirler): a – without heating, b – with arc heating.

Analyzing the presented results, as well as the design of the corresponding GVS systems, we can conclude that in fact all the models presented correspond to the recommendations stated at the beginning of this article. At the same time, the introduction of the changes described above in the GVS design allows to increase the flow rate of PFG in the plasma torch nozzle (Figure 8) for a number of plasma torch designs, thereby increasing the kinetic energy of the plasma jet and cutting performance. In addition, it can be concluded that a mandatory analysis of the GVS efficiency in the "hot" mode in which the advantages of modernized GVS systems are most significant (2–3 times lower than the values of ALD and F criteria – see Figure 9 and 10). Similar estimates should be made, obviously, for the two-stream plasma torch PMVR-9.1 in order to justify the effectiveness of its operation when heated by a plasma arc. A comparative analysis of other performance criteria generally confirms the conclusions presented here, however, in order to fully justify the effectiveness of a particular GVS design, it is likely to rely on several criteria at the same time, using, among other things, the experimentally determined values of productivity, quality, energy efficiency and safety of metal cutting mentioned above. Additional information about the effectiveness and applicability of a particular plasma torch can also be obtained by analyzing the distributions of plasma jet parameters outside the plasma torch nozzle (see the results obtained by the authors in [18]).

4. Conclusions

It is obvious that the presented results are a justification for the use of machine modelling methods to determine the efficiency of the GVS system in metal-cutting plasma torches and the effectiveness of their design, in general. At the same time, it should be noted that to develop a full-fledged methodology for evaluating the GVS efficiency and design, a broader analysis of both various plasma torch designs and a wide number of parameters of their operation in various ranges is required. For example, to assess the quality of the cathode spot stabilization and focusing, a stabilization criterion can be used, consisting of the determining mode parameters: gas flow, the degree of confusability of the vortex chamber, its diameter, and so on [19]. We should also pay attention to the relationship between GVS and the effect of small- and large-scale turbulent pulsations of the gas flow, which largely determine the probability of arc shunting and the stability for the plasma torch [10]. In this regard, we should also pay attention to the effect of high-frequency acoustic radiation [2], which is largely due to the appearance of this type pulsations. Gas consumption is also an important stabilizing factor that reduces the amplitude of arc vibrations and heat loss [20]. A full-fledged analysis of the GVS efficiency for metal-cutting plasma torches requires taking into account a large number of parameters and criteria mentioned in this article.

References

[1] Cherednichenko V S, Anshakov A S and Kuzmin M G 2011 *Plasma electrotechnological installations* (Novosibirsk: NGTU)
[2] Shirshov I G and Kotikov V N 1987 *Plasma cutting* (Leningrad: Mashinostroenie)
[3] Anakhov S V 2018 *Principles and methods of plasma torches designing* (Ekaterinburg: RSVPU)
[4] Vasiliev K V 1974 *Plasma-arc cutting* (Moscow: Mashinostroenie)
[5] Anakhov S V and Pykin Yu A 2010 *Plasma torches: the problem of acoustic safety Thermophysical and gas-dynamic design principles of low noise plasma torches* (Ekaterinburg: RIO Uro RAN)
[6] Klimenko A A and Lyapin G K 2010 *Designs of electric arc plasma torches* (Moscow: MSTU)
[7] Kostylev A M 1969 *Installation of electric arc (plasma torches)* (Moscow: MVTU)
[8] Dautov G Yu, Zhukov M F and Koroteev A C 1967 *Features of plasma generators with vortex arc stabilization and generalization of research results. Low temperature plasma* (Moscow: Mir)
[9] Zhukov M F and Anshakov A S 1979 *The basis of the calculation of the linear plasma torches schema* (Novosibirsk: ITP SB AS USSR)
[10] Zhukov M F 1977 *Experimental studies of plasmatrons* (Siberian branch)
[11] Malakhovsky V A 1988 *Plasma processes in welding production* (Moscow: Higher school)
[12] Alekseenko V P, Stukalov A S and Yakimov P Y 2009 Calculation of the characteristics of vortex temperature control systems *Proceedings of the Samara scientific center of the Russian Academy of Sciences* **11/5** 170–176
[13] Piralishvili S A, Polyaev V M and Sergeev M N 2000 *Vortex effect Experiment, theory, technical solutions* (Moscow: Energomash)
[14] Butyrev A E, Galanin M P and Gnedenko V G 2007 Mathematical modeling of gas flow in vortex chambers with tangential injection *Preprint of IPM named after M V Keldysh RAS* **85**
[15] Savchenko N V 2009 Regulation of a vortex tube at a constant flow rate of the cooled stream *Proceedings of the Samara SC of the RAS* **11/5(2)** 424–427
[16] Anakhov S V, Pykin Yu A and Matushkin A V 2016 Investigation of systems for gas vortex stabilization of plasma torches *Welding International* **30/4** 310–314
[17] Anakhov S V, Pykin Yu A and Matushkin A V 2016 Narrow jet plasma as the energy efficient and safe technology for metal cutting *Solid State Phenomena* **870** 523–527
[18] Anakhov S V, Matushkin A V and Pyckin Yu A 2019 Effect of a gas-vortex stabilization method on the kinematic properties of the jet in metal-cutting plasmatrons *IOP Conf Series: Materials Science and Engineering* **681**
[19] Mikhailov B I 2001 Influence of the arc chamber geometry, pressure, flow rate and type of plasma-forming gas on the axial stabilization of the arc in gas-vortex plasma torches *Thermophysics and Aeromechanics* **8/1** 133–141
[20] Donskoy A V and Klubnikin V S 1979 *Electroplasma processes and installations in mechanical engineering* (Leningrad: Mashinostroenie)

IOP Conf. Series: Materials Science and Engineering **969** (2020) 012048 doi:10.1088/1757-899X/969/1/012048

Thermodynamic Simulation of Reduction Processes at the Production of Ferrochrome

V A Salina, V I Zhuchkov and O V Zayakin

Laboratory of Steel and Ferroalloys, Institute of Metallurgy of the Ural Branch of the Russian Academy of Sciences, 101, Amundsen str., Ekaterinburg 620016, Russia

E-mail: valentina_salina@.mail.ru

Abstract. The possibility of reduction process of chromium from the Cr_2O_3–FeO–CaO–SiO_2–MgO–Al_2O_3 system by carbon using the method of thermodynamic simulation was showed in the article. The compositions of chromium ore of the deposit Rai-Iz and dust of cyclones of the low-carbon ferrochrome in an amount of 0; 5; 10; 20% were used as a components of the system. The carbon was used a reducing agent. The thermodynamic simulation was performed using the HSC Chemistry 6.12 software package developed from Outokumpu, Finland. The thermodynamic characteristics of CrO (II) compound were entered into the software package's database. The thermochemical data of compounds $CaCr_2O_4$, Fe_3C, Cr_3C_2, Cr_7C_3, $Cr_{23}C_6$, SiC were refined. The calculations were performed using the "Equilibrium Compositions" module in the temperature range 1500–1700°C at a temperature increment of 100°C, at a total pressure of 0.1 MPa in the system and 2.24 $m^3$of N_2. It was determined that the increase in the dust amount from 5 to 20% did not affect practically at the reduction degree of chromium which is equal 90.2; 93.4; 94.6% respectively at the temperatures of 1500, 1600 and 1700°C. The chemical composition of the metal is, %: 65.6–65.1 Cr; 22.7–23.2 Fe; 11.5–11.6 C; 0.2–0.26 Si at the temperature of 1700°C. The thermodynamic simulation results can be used to develop technology for producing ferrochrome using wastes of metallurgical production.

1. Introduction

Significant volumes of industrial wastes (slag, dust, sludge, scale, gas) containing iron, manganese, chromium, nickel, calcium and magnesium are formed during various types of metal products (cast iron, steel, ferroalloys). The consumption of mineral raw materials, production costs, environmental pollution will be decrease at involving of wastes in the metallurgical processing [1–4].

The literature contains data about thermodynamic simulation of the reduction process of elements [5–8], a kinetic model of the process [9], using of chromium-containing slags, dusts in the metallurgical process to obtain commercial product are given [10–14].

The authors of [5] were determined possibly products of the reduction carbothermal process elements of the Fe–Cr–O system thermodynamically. It was found that more Fe–Cr–C solution and less residual carbon content are obtained at high temperatures and low nC: nO ratios (the initial molar ratio of C to O in the sample). Metal carbides are formed at the initial stage of the reduction process and Fe–Cr–C solution at the degree of reduction is quite high.

The results of the thermodynamic analysis of the reduction reactions of Cr from Cr_2O_3 (cored wire filler) by carbon in the standard state and under conditions different from the standard ones were carried out in [6]. It was determined that the products of the reduction reactions are carbides chromium and chromium is formed as a component of the melt in furnace at the temperature of 2227 °C. Chromium oxide Cr_2O_3 has the highest reactivity in the liquid state. Most feasibly direct reduction process by carbon.

The thermodynamics of the interaction of the dust components of the gas purification electric furnace with carbon was studied by the authors of [7] using the Terra software package. It was determined that the metals Fe, Zn, Pb, Na, K, Cu, Cd reduced to the temperature of 1200 °C. Iron has been interacted with carbon at temperatures above 650 °C, forming cementite, and at higher temperatures is in the solid phase in the form of a metal.

In [8], the results of thermodynamic simulation of the reduction process of elements of the composition system, is %: 40 Cr_2O_3; 21 FeO; 15 Al_2O_3; 6 SiO_2; 16 MgO; 2 CaO; 0.006 P_2O_5, silicon ferrosiliconickel (28% Fe, 65% Si, 7% Ni) are given. The HSC Chemistry 6.12 software package (Outokumpu, Finland) was used for simulation. It was found that the increase in the reduction degree of chromium by 31.1% (from 64 to 95.1%) at an increase in the basicity of slag from 0.07 to 1.86 and when the using degree of silicon at reduction of the elements is 94.5%. The study increasing in the reducing agent consumption from 0.5 to 1.05 m_{red}. at the reduction degree of chromium. The obtained value of 95.1% is close to the industrial indicators of low-carbon ferrochrome production.

The kinetic model of the reduction carbothermal process of chromium oxide from slag during the production of stainless steel was developed and presented in [9]. The minimum values temperatures 1600–1645 °C were determined to obtain a final concentration of chromium oxide in the slag of 0.8–1.2 % The model can be used to determine the treatment time and temperature.

The technology of durable complex chromium pellets with specified physicochemical characteristics for the smelting of chromium ferroalloys has been developed by the authors of [10]. Charge materials is chrome ore, fine concentrate, slags of refined ferrochrome, coke. The content of 3 % coke in the composition of the charge decrease the firing temperature of pellets from 1350 to 1200 °C for receiving non-metallized pellets without significantly changing the hardware and technological design of the firing process. The chemical composition of the pellets, %: 44 Cr_2O_3; 6.08 Al_2O_3; 18.65 SiO_2; 12.48 FeO; 1.57 CaO; 9.95 MgO; 7.27 – rest. Average strength of the pellets at the firing temperature 1200 °C is 5325 N/pellet.

The technology producing high-carbon ferrochrome using up to 30-70 % of anthracite in the composition of charge materials (P^d–0.016%, S^d–0.15%) as a partial replacement for traditional coke was developed by authors of [11]. Charge materials is chrome ore, metal concentrate, ferrosilicon slag, chrome waste, quartzite fines, chrome briquettes. The extraction coefficient of chromium is 84.4% when using coke and anthracite (30–50 % per serving portion) and 84.7% when using coke and coal.

The manufacture of flux-cored wires based on dust of gas purification of ferrochrome production was considered by authors of [12]. Powders of silicon, aluminum and dust of gas purification aluminum production are used as reducing agents in an amount of at least 27 %. It was determined that the absorption of chromium in the melt is completely dependent on the fill factor of the cored wire.

A overview of solutions for the implementation resource-saving technologies and utilizing technogenic raw materials is given of [13, 14]. The technologies for processing waste of ferroalloy production and substandard raw materials at developments of Russia were described by authors.

The data on measuring the slag viscosity of the CaO–SiO_2–CrO system are presented by authors of [15]; extracting chromium from high-carbon ferrochrome slag by magnetic separation is in [16]; it is using in the preparation of refractory materials, ceramics, and construction is in [17-19], as well as the effect of boron-containing additives on the slag composition of low-carbon ferrochrome is in [20].

The scope of wastes of chrome production at the production of metal products (pellets, ferrochrome, wire) was described in the literature overview and insufficient knowledge of reduction physicochemical processes of chromium in the Cr_2O_3–FeO–CaO–SiO_2–MgO–Al_2O_3 system was showed.

2. Materials and method

Existing methods of processing waste of the metallurgical industry can reduce the consumption of mineral raw materials, decrease the cost of finished products and improve the environment. Thermodynamic simulation can be to determine the possibility of using waste during obtaining products.

This work is aimed at studying the influence of the cyclone dust additives from the production of low-carbon ferrochrome to chrome ore on the reduction degree of chromium (η_{Cr}) from the Cr_2O_3–FeO–CaO–SiO_2–MgO–Al_2O_3 system within the temperature range of 1500–1700 °C by the method of thermodynamic simulation.

The chemical oxide compositions of the deposit Rai-Iz and dust cyclones of production low-carbon ferrochrome were used as the initial oxide system for thermodynamic simulation from at various ratios of system components (Table 1). The carbon was used as a reducing agent; its consumption was increased by 10 % of the stoichiometrically necessary for the complete reduction of iron and chromium for the formation of iron and chromium carbides.

Table 1. The chemical composition of components system, %.

Component of system	Cr_2O_3	FeO	CaO	SiO_2	MgO	Al_2O_3	(CaO+MgO)/(SiO_2)
Chrome ore	38.0	11.0	0.2	15.0	29.8	6.0	2.0
Dust of cyclones	26.5	9.1	21.2	16.5	19.2	7.5	2.5
Ratio ore : dust							
95 : 5	37.3	11.0	1.3	15.1	29.3	6.0	2.0
90 : 10	36.9	10.9	2.3	15.2	28.6	6.1	2.0
80 : 20	35.7	10.6	4.4	15.3	27.7	6.3	2.1

The thermodynamic simulation was performed using the HSC 6.12 Chemistry software package (Outokumpu, Finland). The software, based on minimization of the Gibbs energy and variation principles of thermodynamics [21]. The equilibrium composition of the Cr_2O_3–FeO–CaO–SiO_2–MgO–Al_2O_3 multicomponent oxide system was determined using the “Equilibrium Compositions” software. Initial parameters of thermodynamic simulation are temperature 1500–1700 °C, total pressure of 0.1 MPa and 2.24 m^3 of $N_{2(gas)}$ as a neutral additive for the acceleration of the computational procedure of searching for the equilibrium composition. The thermodynamic characteristics of CrO (II) were entered into the database of the HSC Chemistry 6.12 software package and the thermochemical data of the compounds $CaCr_2O_4$, Fe_3C, Cr_3C_2, Cr_7C_3, $Cr_{23}C_6$, SiC were specified [22].

3. Result and discussion

The dependence of the reduction degree of chromium (η_{Cr}) on amount of dust cyclones at the temperatures of 1500–1700 °C is shown in figure 1. It was determined that increase the amount of dust in the system to 5% insignificantly decreases the reduction degree of chromium. For example, η_{Cr} at temperatures of 1500, 1600 and 1700 °C decreased by 0.4; 0.3 and 0.2 %, respectively. This is due to a decreasing in the total Cr_2O_3 content in the system from 38 to 37.3 %.

An increasing in the amount of dust from 5 to 20 % in the system had practically no effect on the reduction degree of chromium. It was amounted to 90.2; 93.4; 94.6 % for temperatures of 1500, 1600 and 1700 °C, respectively. The charge the reduction degree of chromium in the system is caused on the one hand by the increasing slag basicity of (CaO+MgO)/(SiO_2) from 2 to 2.1, and with another – decreasing concentration of Cr_2O_3 in the system.

IOP Conf. Series: Materials Science and Engineering **969** (2020) 012048 doi:10.1088/1757-899X/969/1/012048

The reduction degree of chromium was changed slightly (by 0.2–0.4%) at the decreasing of Cr_2O_3 by 2.3% in the system as a result of the combined effect of these oppositely influencing factors.

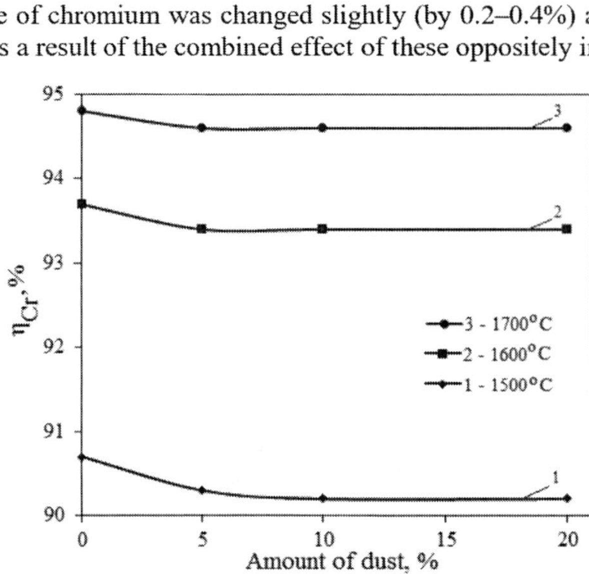

Figure 1. Dependence of the reduction degree of chromium (η_{Cr}) from the amount of dust at the temperatures of 1500, 1600 and 1700 °C.

The content of chromium in the metal [Cr] decreased slightly from 65.6 to 65.3 % at the temperature of 1700 °C and increasing the amount of dust to 5% (Table 2). Moreover, an increasing the temperature of reduction process of chromium from 1500 to 1700°C increases the chromium content in the metal by 1.4–1.6 % an average for all system compositions.

Table 2. The chemical composition of the metal, %.

Element	Ratio ore : dust			
	0	5	10	20
Cr	65.6	65.3	65.2	65.1
Fe	22.7	23.1	23.1	23.2
C	11.5	11.4	11.4	11.6
Si	0.26	0.22	0.21	0.2

The values of the thermodynamic strength of iron and chromium oxides ($\Delta G^0_{298.15\ (FeO)} = -244.299$ kJ/mol; $\Delta G^0_{298.15\ (Cr2O3)} = -1058.966$ kJ/mol) [22] are shown that chromium reduction process after iron, that corresponds to the data obtained in of [23–25]. The amount of iron, silicon and carbon in the metal changes slightly.

The melt containing oxides of magnesium, silicon, aluminium, chromium, iron and calcium is formed during reduction carbothermal of chromium. Phases of calcium silicates are formed at an increasing the amount of dust in the system to 20% and its content increases. The amount of magnesium silicates in the system is decreased. The phase composition of the slag is also represented by magnesia spinel, oxides of chromium, magnesium, aluminium and silicon.

4. Conclusions

It was shown that adding to the chromium ore up to 20% more poorer dust Cr_2O_3 content slightly decreases the reduction degree and the content of chromium in ferrochrome. The results of thermodynamic simulation can be used to develop a technology for producing ferrochrome from chrome ore using wastes of metallurgical production.

References

[1] Temnikov V V, Kalimulina E G and Tleugabulov B S 2018 Analysis of the generation and processing of metallurgical waste at EVRAZ NTMK JSC *Ferrous metals* **(7)** 32–37

[2] Butorina I V and Butorina M V 2018 Review of waste disposal technology for mining and metallurgical industry *Ferrous metals* **(12)** 44–49

[3] Nikitin L D, Dyachok N G, Vashchenko A V, Shentsov A I and Kutran' V I 2019 Employing industrial waste in sinter and hot-metal production *Steel in Translation* **49(7)** 472–477

[4] State report "On the state and protection of the environment of the Russian Federation in 2015". Official site of the Ministry of Natural Resources and Environment of the Russian Federation Online. Available: http://www.mnr.gov.ru/docs/o_sostoyanii_i_ob_okhrane_okruzhayushchey_sredy_rossiyskoy_federatsii/gosudarstvennyy_doklad_o_sostoyanii_i_ob_okhrane_okruzhayushchey_sredy_rossiyskoy_federatsii_v_2016_/

[5] Zhang Yang-ling, Liu Yang and Wei Wen-jie 2013 Carbothermal reduction process of the Fe-Cr-O system *International journal of minerals, metallurgy and materials* **20(10)** 931–940

[6] Kryukov R E, Goryushkin V F, Bendre Yu V, Bashchenko L P and Kozyrev N A 2019 Thermodynamic aspects of Cr_2O_3 reduction of carbon *Steel in Translation* **49(12)** 843–847

[7] Doronin I E and Svyazhin A G 2013 Thermodynamic study of carbon reaction with steel-smeltig dust components *Metallurgist* **57(1–2)** 41–48

[8] Salina, V A, Zayakin O V and Zhuchkov V I 2020 Thermodynamic simulation of the silicothermic reduction of chromium ore elements *Russian Metallurgy* **(1)** 11–16

[9] Leuchtenmüller M, Antrekowitsch J and Steinlechner S 2019 A kinetic study investigating the carbothermic recovery of chromium from a stainless steel slag *Metall. Mater. Trans. B* **50(5)** 2221–2228

[10] Bondarenko I V and Tastanov E A 2019 Obtaining multi-component pellets from finely dispersed chromium concentrates, refined ferrochrome slags and diatomite raw materials of Kazakhstan *Metallurgist* **62(11–12)** 1213–18

[11] Kaliakparov A G, Baltabaev S R, Strakhov V M and Mukhtar A A 2017 Manufacture of high-carbon ferrochrome using anthracite *Metallurgist* **61(9–10)** 765–769

[12] Kozyrev N A, Usol'tsev A A, Prudnikov A N, Kryukov R E and Mikhno A R 2019 Izuchenie svoistv poroshkovoi provoloki na osnove pyli gazoochistki ferrokhroma *Ferrous metallurgy. Bulletin of scientific, technical and economic information* **75(3)** 365–372

[13] Zhuchkov V I, Leontyev L I and Zayakin O V 2019 Promising measures to reduce the harmful effects of man-made waste from ferroalloy production *Theoretical and practical conference with international participation and School for young scientists «Ferroalloys: Development prospects of metallurgy and machine building based on completed research and development». Knowledge E Materials Science* 263–270

[14] Zhdanov A V, Zhuchkov V I, Dashevskii V Ya and Leont'ev L I 2014 Use of waste from ferroalloy production *Steel* **(3)** 92–98

[15] Wu T, Yuan F and Zhang Ya 2018 Viscosity measurements of $CaO\text{-}SiO_2\text{-}CrO$ slag *ISIJ International* **58(2)** 367–369

[16] Bai Z T, Zhang Z A, Guo M, Hou X M and Zhang M 2015 Magnetic separation and extraction chrome from high carbon ferrochrome slag *Mater. Res. Innovations* **5** 113–118

[17] Sahu N, Biswas A and Gajanan U Kapure 2016 A short review on utilization of ferrochromium slag *Miner. Process. Extr. Metall. Rev.* **37(4)** 211–219

[18] Zhuchkov V I, Zayakin O V and Sychev A V 2020 Slags and dusts of ferroalloy production *Russian Metallurgy* **(6)** 662–666

[19] Kascheev I D, Zhuchkov V I and Zayakin O V 2020 Forming and utilizing of ferrochromium production waste *Materials Science Forum* **989** 492–497
[20] Akberdin A A, Kim A S, Ilmaz O Y, Eti Maden Isletmeleri G M and Zhuchkov V I 2013 Boron in ferroalloy production *Proceeding of the XIII INFACON* **1** 325–333
[21] Roine A 2002 *Outokumpu HSC Chemistry for Windows. Chemical reactions and Equilibrium software with extensive thermochemical database* (Pori: Outokumpu research OY)
[22] Glushko V P 1979 *Thermal constants of substances* (Moscow: Akademiya nauk SSSR) p 574
[23] Ryss M A 1985 *Ferroalloy production* (Moscow: Metallurgiya) p 344
[24] Gasik M I, Lyakishev N P and Emlin B I 1988 *Theory and technology of production of ferroalloys* (Moscow: Metallurgiya) p 784
[25] Ryabchikov I V, Mizin V G and Yarovoi K I 2013 Reduction of iron and chromium from oxides by carbon *Steel in Translation* **43(6)** 379–382

Acknowledgments
The work was carried out according to the project no. 18-29-24027 of Russian Found Based Researches.

IOP Conf. Series: Materials Science and Engineering **969** (2020) 012049 doi:10.1088/1757-899X/969/1/012049

Research of Iron-Rich Pellet Oxide Reduction by Carbon

B P Yur'ev and V A Dudko

Institute of New Materials and Technologies, Ural Federal University named after the first President of Russia B. N. Yeltsin, 28, Mira str., Yekaterinburg, 620002, Russia

E-mail: yurev-b@mail.ru

Abstract. Advantages of iron-rich pellet use in steelmaking in electric furnaces as compared to steelmaking scrap were reviewed. Decomposition of carbon and hydrocarbon oxides with carbon release was observed at the Kachkanar iron-rich pellet cooling in shaft furnaces, where they were reduced. The released carbon interacts with iron oxides and alloying elements. Derivatorgraphic research was carried out, which allowed establishing the temperature ranges of the iron oxide and alloying element reduction process as well as reaction mechanism depending on the carbon amount. To investigate the nature of alloying element distribution among the metal and slag phases a mineralogical analysis was carried out for carbon containing iron-rich Kachkanar pellets.
The carried out research on alloying element (vanadium, titanium, manganese) distribution in Kachkanar iron-rich pellets after their post-reduction allowed process flow development for preparation of charge materials for steelmaking using spongy iron. A process was developed for vanadium and titanium containing alloying element production from iron-rich pellets containing carbon.

1. Iron-Rich Pellet Properties

A currently steelmaking in electrical furnaces is increasingly using iron-rich pellets produced by direct reduction methods [1–11]. The pellet basis is iron with carbon content of 0.2 to 2.0–2.4 %. They also contain some amount of unreduced iron oxides (3–12 %) and lean material which amount shall be no more than 3–7 % of the pellet weight. An important pellet property is metallization degree usually comprising 88–97 %. A distinctive feature of this raw material is low content of sulphur, phosphor, copper, nickel, chrome and other additions usually contained in steelmaking scrap in fairly large amounts. This simplifies the melting process and ensures production of high purity steel. Pellet content in the charge materials comprises 60–70 % of its weight. The reduction process in the course of iron-rich pellet production is carried out in a shaft furnace [12–14], where hydrogen and carbon monoxide coming from natural gas as a result of conversion are used as reducing gas.

When iron-rich pellets are cooled in shaft furnaces carbon and hydrocarbon oxides are decomposed with carbon release. Hence, the reduction products contain soot and carbide carbons and carbon diluted in iron. In Kachkanar iron-rich pellets oxides of iron and vanadium, titanium and manganese alloying elements interact with carbon in steelmaking units when making steel; at that iron and alloying elements are distributed among the metal and oxide phases determining reducing element consumption and carbon formation in the pool. Iron oxide reduction by carbon is accompanied by a significant endothermic effect and substantial energy costs at steelmaking in arc electric furnaces [15–17].

The aim of this work is to study the impact of the initial carbon content in iron-rich pellets on the oxide phase reduction process in Kachkanar pellets with the metallization degree of 88 %. The average chemistry of the pellets under research is shown in Table 1. The carried out derivatographic research showed that the bulk of iron-rich pellet oxides is reduced at 1000–1300 °C by carbon, diluted in iron; at that the reaction mechanism significantly depends on the carbon amount. Thus, iron-rich pellet use in steelmaking units requires knowledge of the quantitative dependence between the carbon consumption degree, time and temperature in the stated temperature range.

2. Iron Rich Pellet Reduction

The reduction process of residual oxides in iron-rich pellets by carbon may be conceptually described by a generalized kinetic equation [18]. However, the obtained formulas will be of empirical nature rather than theoretical [19]. They will provide no answer to the question of which oxides are recovered by carbon. At the same time the distribution pattern of alloying elements among the metal and slag phases determines the mechanism of the secondary reduced iron-rich pellet use in steelmaking units.

The obtained results are indicative of the fact, that iron, titanium and vanadium oxides which may interact with carbon are not in pure form, but are binded into complex compounds and glass. The latter amount depends on the melt content at the pellet firing at the Kachkanar Mining and Processing Works, as the metallization temperature in shaft furnaces is below the liquid phase temperature. Study of the mineralogical composition changing in the course of Kachkanar iron-rich pellet reduction by carbon showed that non-metallic phase in such pellets is extremely fine what complicates the mineralogical analysis of samples subjected to thermal reduction treatment. It was also detected that alloying and slag-forming elements were redistributed particularly in the samples with the initial content of 2.63 and 5.20 % C, the slag phase contained Ca, Mg, Al, Ti, V, Si. Besides, calcium titanate and magnesium vanadate inclusions were noted. Titanium-vanadium spinel cemented with dolomite glass was noted in the samples with 2.68 and 2.09 % C at 1320–1350 °C.

It may be concluded that some titanium and aluminium from spinel compounds in initial iron-rich pellets transfer to iron-free compositions. It is possible only in case of iron oxide reduction from metal spinels, besides the noted inclusions of unbound silica in the reduced samples, absent in the initial samples, are indicative of metallic iron formation from iron-calcium-silicate glass. Review of vanadium distribution in the reduced pellets demonstrated that the attempts to distinguish particular mineralogical forms with its presence (except for insignificant amounts of titanium-vanadium spinel) failed. Its uniform inclusions in the metallic iron matrix are noted.

The carried out mineralogical analysis showed that carbon at heating of iron-rich pellets from Kachkanar concentrate is predominantly consumed for Fe and V reduction to the metallic state from glass and spinels, at that the remaining Ti and Al oxides are bound with Ca and Si. The mineralogical analysis is of qualitative character and provides no data on alloying element content in the metallic and oxide phases. At the same time this data is required to develop the process of metallurgical Kachkanar pellet use in steelmaking.

Table 1. Average Chemistry of Kachkanar Pellets.

Pellet Type	Content, %							
	SiO_2	C	Al_2O_3	CaO	MgO	FeO	$Fe_{tot.}$	Fe_2O_3
Initial, fired	3.54	0.10	2.39	2.78	2.59	2.86	61.41	84.64
Iron-rich	3.34	2.53	2.53	2.16	2.75	11.89	80.84	–
	Content, % (continued)							
	Fe_{met}	V_2O_5	P_2O_5	TiO_2	Cr_2O_3	MnO	S	$\frac{CaO}{SiO_2}$
Initial, fired	0.18	0.57	0.04	2.94	0.10	0.23	0.003	0.79
Reduced	71.60	0.72	0.05	3.30	0.11	0.29	0.004	0.65

Please find below the research results on alloying element distribution in Kachkanar metallurgical pellets after their post reduction by carbon. From the previous mineralogical analysis one shall expect the predominant titanium accumulation in the slag phase of iron-rich pellets after their post reduction and partial vanadium transfer to iron. To evaluate Ti, V and Mn reduction degree their content in the slag metallic fractions obtained by sieving of crushed iron-rich pellets after their reduction was determined by chemical analysis. The pellets reduced in a pilot furnace of the Beloretsk Metallurgical Works containing in %: 3.7 of C and 12.4 of FeO; 1.8 of C and 12.1 of FeO were used for the experiments. The metallized iron products were heat treated at 1100 – 1350oC in an open coal heater furnace (weakly oxidizing atmosphere with 1 % of O2 in the mixture with CO, CO2 and N2) and in a controlled-atmosphere furnace, filled with argon and pumped down to 10-8 Pa. Pellet batch (50 – 100 g) in an alumina crucible was placed into a preliminary heated furnace and held at the preset temperature for 1 h. After cooling the pellets were weighted, crushed and sieved into fractions, mm: coarse of over 2.5, medium of 1.0 – 2.5 and fine of less than 1.0.

As per derivatographic analysis in the course of heat treatment the sample weight reduced as follows, %: in weakly oxidizing medium – by 1.8–3.5, in argon atmosphere – by 2.0–3.7, in vacuum – by 3.4–6.6. Together with reduction the metallic iron sintered and coagulated; melt drips showed up on crack and pore surface. The reduced pellet observation results showed that in atmosphere containing CO the pellets were welded already at 1250 °C, moreover the process was accompanied by liquid metal medium release. Argon and vacuum welding temperature is 50 °C higher. The stated phenomena are related to various degrees of carbon consumption for Fe, V, Mn, Ti oxide reduction within the research temperature range by carbon, diluted in iron, and carbides in reactions:

$$[C]_{met} + (Me_nO_m) \rightarrow [Me]_{Fe} + \{CO\}; \quad (1)$$

$$[C]_{Fe} + (Me_nO_m) \rightarrow [Me_xC] + \{CO\}; \quad (2)$$

$$[Fe_3C] + (Me_nO_m) \rightarrow [Me]_{Fe} + [Me_kC_1] + \{CO\}. \quad (3)$$

As the product of these reactions is CO, its presence in gas phase shifts the balance to the left and ensures larger amount of carbon residue. The latter reduces the iron melting temperature and the obtained melt causes pellet adhesion. Absence of gaseous reaction products (1)–(3) in argon and particularly in vacuum results in rapid carbon consumption and iron melting temperature increase. This allows pellet maintaining without adhesion within the range of 1250–1300 °C, where manganese and vanadium carbide solubility in austenite is comparatively high [20], and these elements may to a significant extent transfer to the metallurgical phase. Titanium carbide solubility in iron-rhodic alloys is negligible, thus ferrous titanium spinel reduction causes Ti binding with other pellet components, such as lime, with calcium titanate formation. This is confirmed by data on V, Ti, Mn alloying element dependence on iron content in slag metallic fractions of various size. It was determined that in case of comparatively high C concentration in the pellets (~3.7 %) at 1200–1300 °C the V content is the same in high ferriferous and low ferriferous fractions. At the same time with low C content (~1.5 %) the low ferriferous fraction is enriched with vanadium. These results correlate with mineralogical research data where it was stated that in initial iron-rich pellets vanadium is predominantly bound with slag, and at high temperature reduction by carbon it partially transfers to iron.

The similar picture is observed with regard to manganese but only for atmospheres containing no CO, where reduction is the most intensive. Titanium behaves differently: low ferriferous fractions pellets with high C content (~3,7 %) contain 5–6 times more Ti, what expressly speaks of its concentration in slag inclusions.

Thus, in the course of nonoxidation heating of carbon containing iron-rich pellets vanadium reduction is the most intensive, at that high degree of its transfer into metal is observed for all studied gaseous media. Mn is reduced to a lesser extent and Ti reduction is absolutely insignificant. However, all elements are reduced in vacuum.

Assuming that alloying elements are completely absent in initial iron-rich pellets, V, Mn, Ti reduction degree was calculated. The amount of the reduced elements was determined based on their content in the most ferriferous (90 % Fetot) fraction with the size of over 2,5 mm. As its yield is quite high and comprises 70–80 %, thus error due to alloying element losses in slag metallic fractions with low iron content is insignificant.

Research on post reduction of iron-rich Kachkanar pellets allows recommending of the process flows using spongy iron. It is difficult to reach high metallization degree of Kachkanar pellets in shaft furnaces because of the occurred dense layer of metallurgical iron, preventing diffusion of gaseous reducing agents. Production of pellets containing over 70–75 % of Femet, is related to significant losses of shaft furnace performance. At the same time soot carbon evolved at cooling is highly soluble in iron, allowing active post reduction of practically all oxide inclusions.

Use of carbon reduction potential allows reduction of power consumption in electric steelmaking furnaces due to smaller amount of heat consumed for reduction and slag yield. Preliminary post reduction also allows homogenising of the pellet composition with regard to carbon and stabilizing the melt process conditions.

Based on the obtained results it is clear that iron-rich pellets shall be preliminary heated in atmosphere precluding oxidation of spongy iron with the minimum amount of CO, what will allow intensification of iron and alloying element reduction.

Technically the most real is the atmosphere, containing 0.1–1.0 % of O2 and less than 1 % of CO. It may be obtained in drum furnaces when burning gas with excess air factor of about 1.02–1.05. The stated atmosphere composition will allow pellet heating temperature increase to 1250–1300 °C without their fusion and ensure free movement of the material from the furnace.

Heating to the stated temperatures with the rate of 10–15 °C/min will allow complete post reduction. At that the pellets will get into the temperature zone of possible melt formation (over 1250 °C), the carbon consumption degree will comprise 0.8–0.9. This will preclude iron carbonization and liquid phase release. The higher heating rates of the pellets are hardly allowable as they increase the carbon residue content at the temperature of over 1250 °C and cause material fusion. Heating rates of over 10 °C/min. will result in reduced drum furnace performance and increased conversion costs. Preliminary reduction of charge materials containing spongy iron is more efficient in drum furnaces than in steelmaking furnaces.

It shall be noted that the considered method is acceptable for pellets, containing alloying elements V, Mn, which oxides are easily reducible and carbides are freely soluble in iron, Ti remains in slag and drains with it. Processing of these slags with low Ti is hardly profitable. At the same time the obtained results show the actual way of richer titanium master alloy production particularly for pellets with high TiO_2 content.

The research results show that after heat treatment of iron-rich pellets containing C and V and Ti oxides in weakly oxidizing medium at 1250–1300 °C, vanadium was predominantly in the metallic phase and titanium – in the slag phase, thus after crushing of such pellets the metallic magnetic phase was enriched with vanadium, and less magnetic phase with large slag amount – with titanium. Separation of these phases and the corresponding master alloy production by magnetic separation is quite acceptable. Fineness, type and operating conditions of magnetic separators shall be selected to develop the corresponding process.

3. Summary

The reduction process of Kachkanar iron-rich pellets by carbon released at their cooling in gasses containing carbon oxides and hydrocarbons was studied. Reduction temperatures of the bulk oxides in iron-rich pellets were determined by the derivatorgraphic method. Mineralogical analysis of carbon bearing iron-rich pellets was carried out with a view of studying the nature of alloying element distribution between the metallic and slag phases. To use the reduction potential of soot carbon post reduction of oxide inclusions in iron-rich pellets was suggested in rotating furnaces prior to their getting

IOP Conf. Series: Materials Science and Engineering **969** (2020) 012049 doi:10.1088/1757-899X/969/1/012049

into electric steelmaking units. A process was developed for vanadium and titanium containing master alloy production from carbon containing iron-rich pellets.

References

[1] Shnabel S 2004 From Direct Reduction Iron to High Quality Steel *Ferrous Metals* **6** 20–23
[2] Kudryavtsev V S and Pchelkin S A 1974 *Iron-Rich Pellets* (Moscow: Metallurgy)
[3] Yusfin Yu S, Danshin V V, Pashkov N F and Pitatel V A 1982 *Iron Ore Raw Material Metallization Theory* (Moscow: Metallurgy)
[4] Shvydky V S, Yaroshenko Yu G, Spirin N A and Lavrov V V 2018 Simulation of Iron-Rich Pellet Firing Process with Due Account for Their Physicochemical Transformations *Bulletin of Sibir State Industrial University* **4(26)** 36–40
[5] Kozhukhov A A 2012 Research of Iron-Rich Pellet Role in Slag Frothing in State-of-the-Art Arc Steelmaking Furnaces *Electrometallurgy* **11** 11–16
[6] Kozhukhov A A, Merker E E, Sazonov A V 2010 Review of Arc Furnace Heat Exchange at Iron-Rich Pellet Melting *University Bulletin. Ferrous Metallurgy* **5** 37–40
[7] Grishin A A 2006 Particular Features of Metal Decarbonisation at Continuous Feeding of Iron-Rich Pellets to Arc Furnace Pool *University Bulletin. Ferrous Metallurgy* **11** 21–23
[8] Sazonov A V, Merker E E, Kochetov A I 2008 Particular Features of Metal Decarbonisation at Pellet Melting in Arc Furnace *University Bulletin. Ferrous Metallurgy* **12** 27–29
[9] Kraht A N, Merker E E, Kem A U and Stepanov V A 2015 Reduced Iron Ore Pellets and Efficiency Increase for Their Use in Electric Steelmaking in Arc Furnace *Don State Technical University Bulletin* **3(82)** 35–40
[10] Merker E E and Stepanov V A 2014 Electric Steelmaking Efficiency Increase by Metallized Raw Material Feeding to High Temperature Zone through Channels in Arc Furnace Electrodes *Ferrous Metallurgy* **1** 41–49
[11] Amdur A M 1979 *Research on Thermotechnical Features of Iron-Rich Pellet Melting in Arc Steelmaking Furnaces* (Sverdlovsk: Ural Polytechnic University named after S. M. Kirov)
[12] Povolotsky V U, Lantsov U V, Plotnikova S U 2014 Kinetic and Dynamic Model of Direct Iron Production in Shaft Furnace *Steel* **8** 35–38
[13] Zinyagin G A, Sushkov P N 2004 Research and Process Development of Direct Iron Production in Shaft Furnaces at Charge Material Increased Sulphur Content *University Bulletin. Ferrous Metallurgy* **1** 13–16
[14] Grebennikov V S 1982 *Optimization of Iron-Rich Pellet Production Process in Shaft Furnaces* (Moscow: National University of Science and Technology MISIS)
[15] Merker E E, Sazonov A V and Grishin A A 2008 Particular Features of Iron-Rich Pellet Electric Melting in Arc Furnace *University Bulletin. Ferrous Metallurgy* **2** 31–33
[16] Fedina V V, Merker E E, Kharlamov D A 2004 Improvement of Metallized Iron Ore Raw Material Electric Melting Process in Arc Steelmaking Furnace *Ferrous Metals* **2** 16–19
[17] Nokhrina O I, Rozhikhina I D, Rybenko I A and Khodosov I E 2016 Fundamental Development of High Efficiency Metallization Processes Using Thermodynamic Simulation *University Bulletin. Ferrous Metallurgy* **4(59)** 237–244
[18] Uendlandt U 1978 *Thermal Analysis Methods* (Moscow: Mir)
[19] Delman B 1972 *Heterogenous Reaction Kinetics* (Moscow: Mir)
[20] Golikov I N, Goldshtein M I, Murzin I P 1968 *Vanadium in Steel* (Moscow: Metallurgy)

IOP Conf. Series: Materials Science and Engineering **969** (2020) 012050 doi:10.1088/1757-899X/969/1/012050

Research of Possibility of Processing of Oxidized Nickel Ore by Chloride Sublimation Roasting Technology

O B Kolmachikhina, S E Polygalov and V G Lobanov

Department of Non-Ferrous Metallurgy, Ural Federal University, 17, Mira St., Ekaterinburg, 620002, Russia

E-mail: o.b.kolmachikhina@urfu.ru

Abstract. Oxidized nickel ores (ONO) contain most of the world's nickel reserves. However, choice of a cost-effe nickel ore processing technology is difficult due to the dispersed distribution of nickel compounds in refractory minerals, low nickel content, presence of a large amounts of consuming host rocks, and impossibility of beneficiation by conventional physicochemical methods. The paper presents studies on the ONO processing technology by means of chloride sublimation roasting. Sodium and calcium chlorides were used as chlorinating agents both individually and in various ratios. It has been shown that roasting allows for nickel and cobalt conversion into chloride form followed by selective distillation of the resulting compounds into the gas phase by 89–90% at a roasting temperature of 1100 °C.

1. Introduction

Ural ONO deposits are characterized by a high content of magnesium and silicate components, as well as high content of iron oxides. Pyrometallurgical technologies traditionally being used at Ural plants – matte and ferronickel smelting – are currently unprofitable due to high cost of energy carriers, which led to a halt in production. At the same time, the reserves of oxidized nickel ores in the region are large, which makes it reasonable to search for new technologies that would significantly reduce the cost of ore processing. The most low-cost technologies are geotechnological heap and in-situ leaching, but the problem in this case is that oxidized nickel ores contain significant amounts of slimy particles and clay minerals, which complicate the process of solution percolation through the rock layer, a large solvent consumption, a long leaching time, low nickel recovery due to the impossibility of minerals digestion, etc. [1–12]. Therefore, the application of the technologies mentioned has not been industrially implemented in the region. The traditional hydrometallurgical technologies used for oxidized nickel ores processing abroad, such as autoclave sulfuric acid leaching and the Caron process [13–20] are not applicable for the Ural ores due to the high content of magnesium oxide in the ores and high cost of the processes.

A combined technology, including in the first stage, chloride sublimation roasting of the ore, and in the second, leaching of the obtained fumes might be one of the promising options for the processing of ONO. The essence of chloride sublimation roasting is the formation and distillation into gas phase of volatile chlorides of recoverable metals during the interaction of their initial compounds with a chlorinating agent. In this case, all compounds of the extracted metals are transformed into one class of chlorides having a high vapor pressure at temperatures of 900–1100 °C, which allows the process to be carried out without melting the entire mass of processed raw materials. As applied to the processing of oxidized nickel ores, the process is based on selective chlorination of nickel and cobalt compounds, host rocks remain in their original state,

IOP Conf. Series: Materials Science and Engineering **969** (2020) 012050 doi:10.1088/1757-899X/969/1/012050

and selective distillation of nickel and cobalt chlorides into gas phase. The parameters that determine the course of the process are properties of metal chlorides, composition of gas phase, temperature and duration of roasting. Recovery of metals from chlorides after their condensation is carried out by water leaching under standard conditions.

The temperature of the material during chloride sublimation roasting of oxidized nickel ores should not exceed 900–1100 °C in order to exclude the formation of low melting eutectics. The source material must maintain the porosity necessary for the access of chlorine to nickel and cobalt compounds, which are included in the minerals of iron and silicates.

For relatively poor raw materials such as oxidized nickel ores, solid compounds $CaCl_2$ and NaCl can be used as a chlorinator. In this case, chlorination can occur additionally due to the formation of gaseous chlorine:

$$2NaCl + SiO_2 + 0{,}5O_2 = Cl_2(gas) + Na_2SiO_3. \quad (1)$$

The choice of such compounds stems from their availability and low price. The cost of an alternative reagent, hydrochloric acid, is significantly higher, which negatively affects the expenditures for the process and can be a decisive factor in the choice of processing technology.

2. Research methods

As the object of research, we selected oxidized nickel ore from the Serovskoye deposit, one of the largest in the Urals. The ore composition is shown in Table 1. The ore is typical of the Ural deposits – low nickel and cobalt content with a high content of silica and iron oxides.

Table 1. Composition of nickel ore by major components, %.

Material	Ni	Co	MgO	SiO_2	CaO	Al_2O_3	FeO	Others
Oxidized nickel ore	1.15	0.08	6.25	32.57	1.64	7.13	43.56	11.62

When conducting experiments on chloride sublimation roasting, the residual metal content in the cinder was chosen as the main criterion for the process. Preliminary preparation of raw materials (pelletizing, tableting, etc.) before chloride sublimation roasting was not carried out. Limestone was used as a fluxing agent that prevents sintering and melting of ore. The limestone consumption was established at the stage of preliminary studies, the optimal limestone consumption prevented melting at a calcination temperature of 1100 °C was 28% of the mass of the initial ore. A portion of the ore was thoroughly mixed with a chlorinating agent and limestone and loaded into a crucible with a thin layer.

The initial ore moisture was 12%; no additional wetting of the sample was carried out. Roasting was carried out in a tube furnace with the capture of sublimates. The crucibles were not sealed, oxygen was not supplied during roasting. The consumption of chlorinators ($CaCl_2$ and NaCl) was varied – 10–60% by weight of dry ore, roasting temperature – 900–1100 °C, roasting duration 10–90 min. We studied both the effect of each chlorinating agent separately, and their mixtures in various ratios on the distillation of nickel and cobalt chlorides into the gas phase.

3. Results and discussion

The experiments showed that the transition of nickel and cobalt chlorides to the gas phase begins 10–15 minutes yet after experiment starts, and completely stops after 50 minutes at a process temperature of 1100 °C. The cinder yield is 57.5–62.7% of the mass of the loaded charge and increases with increasing content of the mixture CaCl2. The main cinder compounds according to phase analysis are calcium compounds $3CaO * SiO_2$; $3CaO * Al_2O_3$; $2CaO * Fe_2O_3$; $Ca_3Mg(SiO_4)_2$; CaO, which are formed due to the introduction of limestone into the mixture.

The addition of calcium chloride allows for deeper extraction of both nickel and cobalt into gas phase at a lower consumption of chlorinator and the same roasting conditions - temperature 1100 °C, duration – 40 min. Tests results are presented on Figure 1, 2.

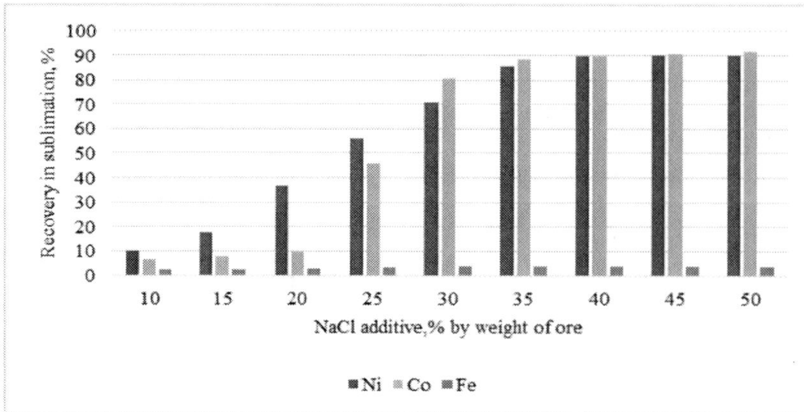

Figure 1. Dependence of the transition of metals into gas phase on the consumption of the chlorinating agent.

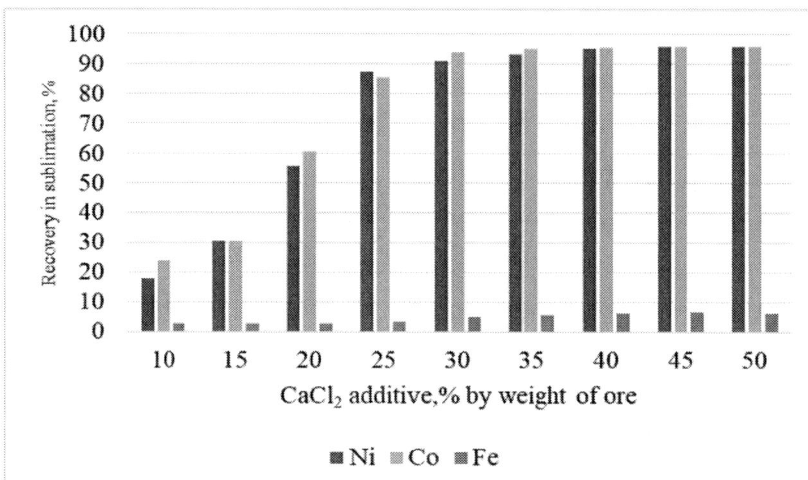

Figure 2. Dependence of the transition of metals into gas phase on the consumption of the chlorinating agent.

The most interesting results were obtained in experiments with a mixture of chlorinating agents NaCl and $CaCl_2$ at a roasting temperature of 1100 °C and a duration of 40 minutes (table 2, figure 3).

Table 2. Consumption of chlorinating agents when using NaCl and $CaCl_2$ together.

Test	The addition of a chlorinating agent,% by weight of dry ore		
	$CaCl_2$	NaCl	Total additive
1	5	30	35
2	10	25	35
3	15	20	35
4	17.5	17.5	35
5	20	15	35
6	25	10	35
7	30	5	35

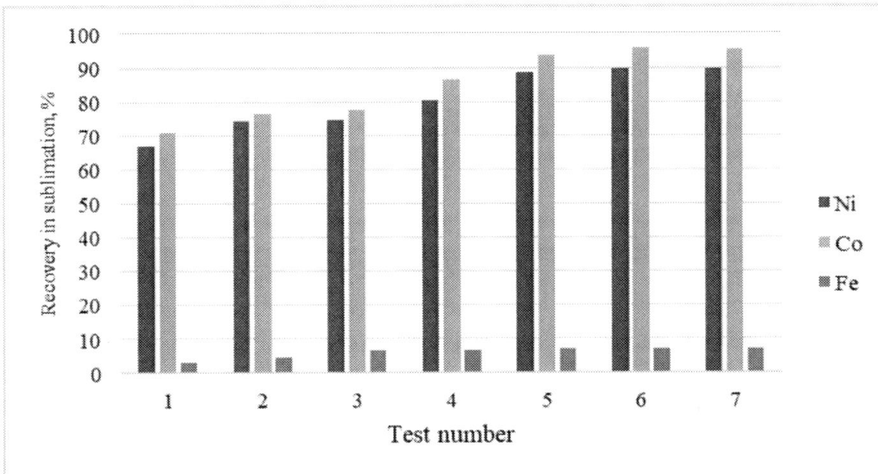

Figure 3. Dependence of the transition of metals into gas phase on the consumption of the chlorinating agent.

4. Conclusion

The addition of reagents in the ratio of 10% NaCl and 25% CaCl2 by weight of dry ore allowed for about 90% nickel and 95% cobalt transfer into sublimates. The transition to sublimatesof iron was about 7%. Despite the low extraction of iron into sublimates, which is a positive factor, since it allows one of the main problems of oxidized nickel ores processing to be solved - the selective separation of nickel and cobalt, iron chlorides make up the bulk of the sublimates. The average composition of sublimates in a series of experiments with a mixture of chlorinators, %: 37–40 NiCl2; 2.7–3 CoCl2; 57.7–58 FeCl2.The percentage of cobalt was very low in the cinder, which suggests an almost complete transition of cobalt to the gas phase. Nevertheless, for a sufficiently complete transition of nickel and cobalt to the gas phase, a temperature of 1100 °C and a duration of at least 40 minutes are required.

References

[1] Kalashnikova M I, Tsymbulov L B, Naboychenko S S and Kolmachikhina O B 2019 Innovative processing applicable to the oxidized nickel ores found in the urals region *J. Non-ferrous metals* **8** 4–12

[2] Khalezov B D, Gavrilov A S, Petrova S A and Mel'chakov S Yu 2019 Investigation of Solid Residues Obtained After Oxidized Nickel Ore Leaching *J. Metallurgist* **63** 860–866

[3] McDonald R G and Whittington B I 2008 Atmospheric acid leaching of nickel laterites review. Part I. Sulphuric acid technologies *J. Hydrometallurgy* **91** pp 35–55

[4] McDonald R G and Whittington B I 2008 Atmospheric acid leaching of nickel laterites review. Part II. Sulphuric acid technologies J. *Hydrometallurgy* **91** 56–69

[5] Watling H R, Elliot A D, Perrot F A and Shiers D W 2009 Impacts of mineralogy on the chemistry and microbiology of heap bioleaching *Conf. 18th International Biohydrometallurgy Symposium (Bariloche)* **71–73** 369–372

[6] MacCarthy J, Nosrati A, Skinner W and Addai-Mensah J 2016 Atmospheric acid leaching mechanisms and kinetics and rheological studies of a low grade saprolitic nickel laterite ore *J. Hydrometallurgy* **160** 26–37

[7] Elfimova L G Korol Yu A and Naboychenko S S 2016 Possibilities of hydrometallurgical processing of oxidized cobalt-nickel ores of Belininskoe deposit *J. Non-ferrous metals* **3** 23–30

[8] Khalezov B D, Gavrilov A S, Petrova S A and Ovchinnikova L. A. 2019 Nickel extraction from solutions using sodium hydrosulfide *J. Non-ferrous metals* **3** 33–38

[9] Oxley A, Smith M E and Caceres O 2016 Why heap leach nickel laterites? *J Minerals Engineering* **88** 53–60

[10] Zablotskaya Yu V, Sadykhov G B, Khasanov M Sh and Smirnova V B 2018 Kinetics of sulphuric acid leaching of nickel from reduced limonite ore of the buruktal deposit *J. Non-ferrous metals* **12** 27–31
[11] Ayanda O S, Adekola A F, Baba A A, Fatoki O S and Ximba B J 2011 Comparative Study of the Kinetics of Dissolution of Laterite in some Acidic Media. *J. Journal of Minerals & Materials Characterization & Engineering* **10** 1457–1472
[12] Sosnovskiy M G, Gulyaev S V and Zarkov A. V.2019 Possible processing of nickel-cobalt ores of the buruktal deposit at Southern Urals nickel plant *J. Non-ferrous metals* **3** 21–27
[13] Loveday B K 2007 The use of oxygen in high pressure acid leaching of nickel laterites *J. Minerals engineering* **21** 531–538
[14] Thompson G and Senanayake G 2018 Effect of iron(II) and manganese(II) on oxidation and co-precipitation of cobalt(II) in ammonia/ammonium carbonate solutions during aeration - An update and insight to cobalt losses in the Caron process for laterite ores *J Hydrometallurgy* **181** 53–63
[15] Senaputra A and Nicol M J 2010 Effect of thiosulfate, sulfide, copper(II), cobalt(II)/(III) and iron oxides on the ammoniacal carbonate leaching of nickel and ferronickel in the Caron process *J. Hydrometallurgy* **105** 60–68
[16] Saykova S V, Panteleeva M V and Saykova D I 2019 Cation exchange processing of buruktal oxidized nickel ore autoclave solutions *J. Non-ferrous metals* **9** 16–21
[17] Ilyas S., Srivastava R R, Kim H, Ilyas N and Sattar R 2020 Extraction of nickel and cobalt from a laterite ore using the carbothermic reduction roasting-ammoniacal leaching process *J Separation and Purification Technology* **232 115971**
[18] Mano E S, Caner L, Petit S, Chaves A P and Mexias A S 2019 Ni-smectitic ore behaviour during the Caron process *J. Hydrometallurgy* **186** 200–206
[19] Whittington B I and Muir D 2000 Pressure acid leaching of nickel laterites: a *review J Mineral Processing and Extractive Metallurgy Review* **21** 527–599
[20] Chen J, Jak E and Hayes P C 2019 Investigation of the reduction roasting of saprolite ores in the Caron process: microstructure evolution and phase transformations *J. Mineral Processing and Extractive Metallurgy: Transactions of the Institute of Mining and Metallurgy*

IOP Conf. Series: Materials Science and Engineering **969** (2020) 012051 doi:10.1088/1757-899X/969/1/012051

Simulation of Physical Fields in Induction Hardening Technology - from Digital Twin to New Technology

Yu Perevalov[1] and V Demidovich[2]

[1]Department of Electrotechnology and Converter Engineering, St. Petersburg Electrotechnical University (LETI), 5, prof. Popov, St. Petersburg, 197376, Russia
[2]Russian Technologies of Induction Heating (RTIH) Ltd., 14/12, Iskrovsky prospect, St. Petersburg, 193313, Russia

E-mail: yyperevalov@yandex.ru

Abstract. A new technology for the final induction heat treatment of large mill rolls has been developed. When designing this technology, installation and control system, a digital twin of the system was created on the basis of combined numerical models of electromagnetic and temperature fields in cylindrical axially symmetrical systems. The final heat treatment of mill rolls consists in a two-stage, multi-frequency induction hardening of body of large rolls. The developed digital twin of the system allows one to evaluate the key features arising during induction hardening, such as temperature distribution, hardness of the hardened layer, stresses that arise during hardening. The design of induction systems through digitalization allows one to improve the design, equipment composition and heat treatment mode to achieve maximum quality and minimize energy costs.

1. Introduction

Rolls' induction hardening has been used in industry for many years, but the technological process is not always designed optimally, just like induction hardening installations are designed without taking into account key production features. Today, it is advisable to create new technological complexes that can reduce energy consumption and increase labor productivity, without reducing the quality of the product. For optimal design, it is advisable to develop a digital twin of a technological complex with numerical models embedded in it for calculating electromagnetic fields in induction block, temperature fields in a workpiece, taking into account stress fields, as well as simultaneously taking into account the levels of active and reactive power. It is also necessary to observe the mechanical vibrations of structural elements that occur during operation and the temperature characteristics of all devices within the induction system.

Digital twins are used by a very large induction heat treatment plants, such as the induction hardening plant for large mill rolls. The power level of this plant reaches the level of 3 MW, therefore, when designing, it is advisable to use computer modeling and numerical models built in specialized software packages.

2. Problem

Induction hardening of large mill rolls was first used in the 1920s in the USSR and the USA [1]. Even then, it was noticed that it was difficult to achieve a constant high quality of the hardened surface of the rolls, the rolls did not always have a high hardness of the hardened layer, and there were also cases when the rolls cracked after hardening due to high residual stresses.

Mill rolls are made of steels with a high chromium content - 8X3SMF, 9X, 9XF, 9X2, 9X2V, 9X2SF, 9X2MF, 9X5MF, 9X2SVF, 60X2SMF, 65X5MF and others. The hardness of the work roll body is in the range of 90-102 HSD. In addition to the high hardness value, one of the important parameters of the hardened layer is uniformity. Hardness differential along the length of the roll body is no more than 2 HSD. The depth of the working layer for alloys 9X1, 9X2, 9X2MF is at least 17 mm, and for alloys 8X3SMF, 65X5MF – at least 30 and 45 mm, respectively. In this case, the mill rolls must withstand high contact pressure and, at the same time, the formation of cracks on the surface of the roll body is not allowed.

To develop the technology for the final heat treatment of mill rolls, a digital twin of the induction system was developed, which includes numerical models of technological heat treatment, two-dimensional models of electromagnetic and thermal fields, models of thermal stresses arising in the process of heating and cooling the rolls. All numerical models are required to predict hardness, grain size, stresses and other characteristics of rolls and the hardened layer [2, 3].

As a result of computer simulation, a two-stage technology for poly-frequency hardening of large mill rolls was developed. The technology consists in induction heating of the roll before hardening at the base frequency (100 Hz) at the first stage, and then there is a hardening stage, which is performed at a different frequency. The temperature distribution and parameters of the power source during the multi-frequency hardening process are shown in Figures 1 and 2. The frequency of the hardening stage is selected based on the depth of the working layer.

Developed technology takes into account the dangerous temperature and stress gradients arising during the hardening process, which allows avoiding the formation of cracks and obtaining a uniform distribution of hardness on the surface of the roll body.

Developed digital twin turned out to be extremely effective for digital control of induction heat treatment complexes (Figure 1, 2). Numerical modeling has shown that it is advisable to use modular induction systems for hardening the mill rolls.

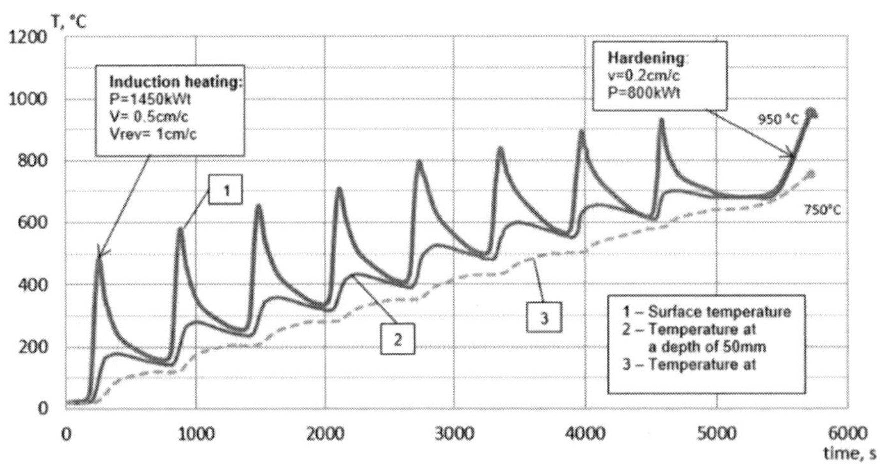

Figure 1. The process of induction heating and hardening of large-sized rolls of the rolling mill.

The use induction heaters modular structure has a significant advantages in induction heating technology, since it allows to optimize the operation modes of equipment in relation to: energy consumption; scale formation; ability to maintain a constant final temperature [4, 5, 6].

The advanced digital control allows to effectively process possible disturbances, such as: changes of speed, diameter and initial temperature of the workpiece, and, therefore, to minimize the amount of defects during heating. Digital program control based on modern intelligent systems is the key to the quality of hardening of large-sized rolls of rolling mills.

Figure 1 shows the process of induction heating and hardening of the roll of the rolling mill. In the first heat treatment pass, the induction heating is performed at the base frequency (100 Hz), and a quenching pass is performed at a different frequency (70 Hz). The frequency of quenching should be determined from the technical requirements for the depth of the working layer, the frequency of heating is selected from the mass and size parameters of the equipment.

$$\sigma_e = (((\sigma_1 - \sigma_2)^2 + (\sigma_2 - \sigma_3)^2 + (\sigma_3 - \sigma_1)^2)/2)^{1/2} \qquad (1)$$

where σ_e – equivalent von Mises stress, σ_1, σ_2, σ_3 – principal stresses. Principal stresses are normal stresses on three mutual perpendicular planes of elemental parallelepiped; planes are chosen so that tangential stresses on them equals zero.

The stress with biggest value defines σ_1, smallest defines σ_3 and σ_2 lies between them. During mill rolls quenching they experience three-dimensional, volumetric stresses. Complex three-dimensional state of tension defines single positive value of equivalent stress, which can be used to predict yield development in plastic material. Von Mises criteria is independent from third deviatory stress invariant [3] (according to Mises criteria for each stressed state of deformable body, both tension and compression, three principal stresses defines single positive value of equivalent stresse, which can be used to predict development of dangerous deformations).

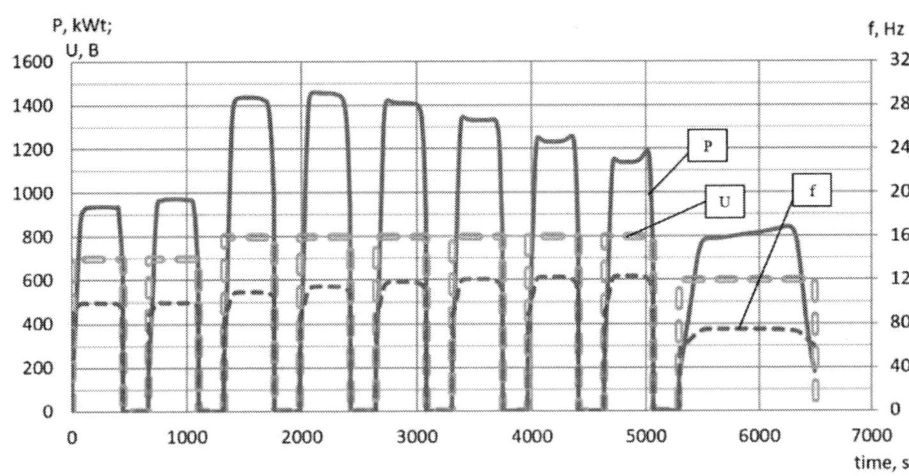

Figure 2. The parameters of the induction system during hardening.

When calculating mechanical stresses developing during quenching of large-dimensioned mill rolls using finite element analysis equivalent stress can be calculated for each element of the model. So one can find places of stress concentration and based on stress distribution quenching setting can be corrected [7, 8, 9].

Obviously, that whole heat treatment process time (pre-heating, heating up to quenching temperature, quenching and cooling) all of the roll material should be in elastic zone of stress-strain curve for roll steels, so for estimation of stress-strain behavior based on the equivalent von Mises stress assurance coefficient should be used to guarantee quenching without specimen destruction (hardened layer delamination, surface cracking) [10, 11, 12].

Figure 3 represent distribution of temperature and equivalent von Mises stress along the roll radius with and without pre-heating. One can see that mill roll pre-heating to 500°C before quenching reduces maximal equivalent stress greatly, so the roll will not deformated critically or fractured with such quenching setting.

IOP Conf. Series: Materials Science and Engineering **969** (2020) 012051 doi:10.1088/1757-899X/969/1/012051

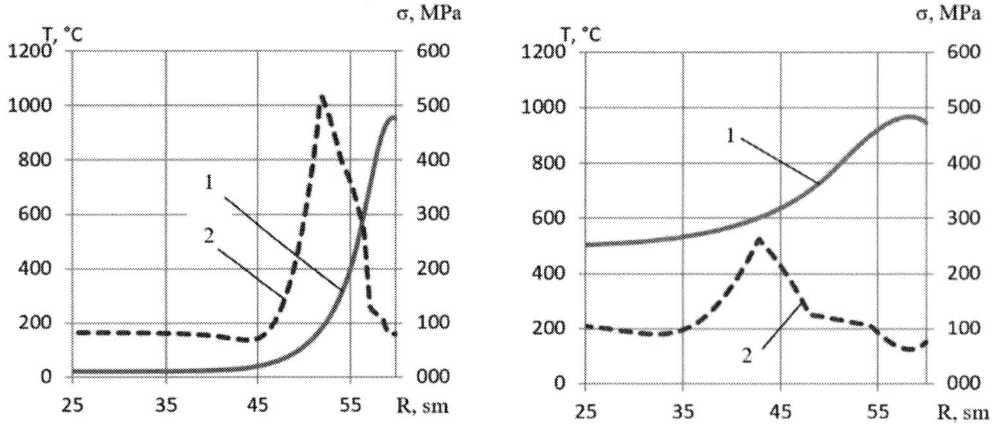

(a) (b)

Figure 3. 1200 mm mill roll quenching (a) single-shot without pre-heating, (b) with 500°C pre-heating. Marked: 1 – temperature distribution behind the inductor vs. roll radius, 2 – equivalent von Mises stress distribution.

Special program should be developed to keep quality level high during large-dimensioned mill rolls quenching. This program should digitally control the quenching process in real-time environment (Figure 4). This digital software should define induction system parameters during pre-heating and quenching. The input for it are the temperature needed behind the inductor, on the roll surface and on the depth defined by technical specification, the output, generated by the program, are values of voltage and current, power level and inductor scanning rate. The software should be able generate the setting of quenching process, ensuring that roll temperature after pre-heating is 500°C, and on the second, quenching inductor run, temperature after inductor on the roll surface and 100 mm deep is exactly 950°C [13, 14, 15].

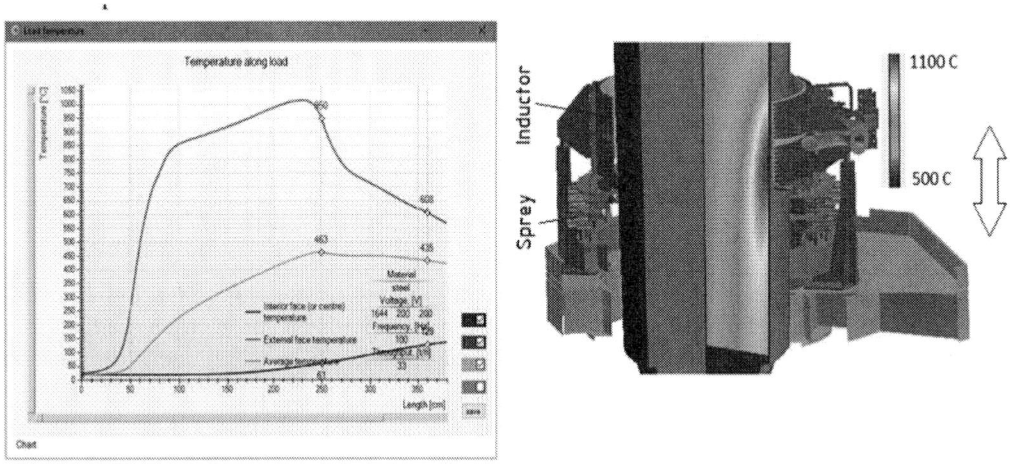

(a) (b)

Figure 4. Sample of calculation window of digital software for roll quenching simulation (a), induction system model (b).

3. Conclusions

New technologies for induction heat treatment of large diameter rolls of rolling mills with a barrel diameter of up to 65" are developed. It was found that the digital model of the induction system must necessarily include the calculation of electromagnetic and thermal fields, the calculation of stress and strain fields, as well as the calculation of power source parameters. This digital model can be used to create an intelligent control system for the installation of induction heat treatment of large-sized rolls of rolling mills, which provide high-quality heat treatment and minimize residual stresses [16, 17, 18, 19, 20].

References

[1] Muehlbauer A 2008 *History of Induction Heating and Melting* (Vulkan Verlag)
[2] Demidovich V, Andrushkevich V and Perevalov Yu 2018 Induction heat treatment of large diameter pipes and coils *9th International Symposium on Electromagnetic Processing of Materials* 1–4 DOI: 10.1088/1757-899X/424/1/012064
[3] Timoshenko S 1976 Strength of Materials, 3rd edition. Part 1. Elementary Theory and Problems (Melbourne (Florida): Krieger Publishing Company) p 456
[4] Demidovich V B, Perevalov Y Y and Andrushkevich V V 2018 Induction heat treatment of large diameter pipes and coils *IOP Conf. Series: Materials Science and Engineering* **424** Online. Available: https://doi.org/10.1088/1757-899X/424/1/012064
[5] Baldan M, Nikanorov A and Nacke B 2018 Hierarchical optimization approaches in designing surface hardening induction systems *IOP Conf. Series: Materials Science and Engineering* **424** Online. Available: https://doi:10.1088/1757-899X/424/1/012067
[6] Barglik J 2018 Induction contour hardening of gear wheels made of steel 300M *IOP Conf. Series: Materials Science and Engineering* 424 Online. Available: https://doi:10.1088/1757-899X/424/1/012062.
[7] Nacke B and Dietrich A 2018 Potentials of single stage induction heating for press hardening of steel blanks *IOP Conf. Series: Materials Science and Engineering* **424** DOI: 012058. 10.1088/1757-899X/424/1/012058
[8] Huu Kien Bui, Abdoulaye B, Berthiau G and Trichet D 2018 Fast 3D simulation of various applications of induction heating *IOP Conf. Series: Materials Science and Engineering* **424** Online. Available: https://doi: 012048. 10.1088/1757-899X/424/1/012048
[9] Gagnoud A. Du Terrail Yves and Budenkova O 2018 3D Numerical Modeling for Inductive Processes *IOP Conf. Series: Materials Science and Engineering* **424** Online. Available: https://doi:012045.10.1088/1757-899X/424/1/012045
[10] Ando T, Katayama D, Hirota N, Koike O, Tatsumi R and Yamato M Structure Formation of Magnetic Particles under Magnetic Fields toward Anisotropic Materials *IOP Conference Series: Materials Science and Engineering* Online. Available: https://doi:424.012076.10.1088/1757-899X/424/1/012076
[11] Demidovich V B, Tchmilenko F V, Perevalov Y Yu and Rastvorova I I 2017 Electromagnetic processing of metal as coupling of multi-physics phenomena *Fundamentals and Applications*
[12] Rudnev V 2019 Trends in Induction Hardening *Gear Technology*
[13] Hájek Jiří, Rot D and Jirinec J 2019 Distortion in Induction-Hardened Cylindrical Part *Defect and Diffusion Forum* **395** 30–44 DOI: 10.4028/www.scientific.net/DDF.395.30
[14] Bukanin V, Ivanov A, Zenkov A and Vologdin V 2018 Induction Hardening of External Gear *IOP Conference Series: Materials Science and Engineering* **327** **022016** DOI: 10.1088/1757-899X/327/2/022016
[15] Aliferov A, Lupi S and Forzan M 2018 Milliseconds pulse induction hardening *International Journal of Microstructure and Materials Properties* **13** **73** DOI: 10.1504/IJMMP.2018.10014735
[16] Hirao M and Ogi H 2017 Measurement of Induction Hardening Depth *Electromagnetic Acoustic Transducers. Springer Series in Measurement Science and Technology* Online. Available: https://doi.org/10.1007/978-4-431-56036-4_13
[17] Dawidowicz M and Dappen S 2017 Quality control in induction hardening *Elektrowaerme International* **75** 75–80
[18] Zheng C, Wu Z and Wang S 2013 Induction hardening process control of wind turbine gear ring **38** 100–103

[19] Yermekova M and Galunin S 2018 Numerical simulation of induction through heating for forging *IEEE Conference of Russian Young Researchers in Electrical and Electronic Engineering* 524-527 DOI: 10.1109/EIConRus.2018.8317149.

[20] Bondar A, Streltsova O, Chmilenko F and Bondarenko D 2019 Prognostification of Specific Steel Resistivity for Modeling Electrothermal Processes *XXI International Conference Complex Systems: Control and Modeling Problems* 370–373 DOI: 10.1109/CSCMP45713.2019.8976724.

IOP Conf. Series: Materials Science and Engineering **969** (2020) 012052 doi:10.1088/1757-899X/969/1/012052

Features of Technogenic Iron Oxide Recovery

O Yu Sheshukov[1,2], M M Mikheenkov[1,2] and L B Vedmid[1,2]

[1]Institute of New Materials and Technologies (INMT), Ural Federal University named after the first President of Russia B. N. Yeltsin, 28, Mira str., Ekaterinburg, 620002, Russia

[2]Institute of Metallurgy of the Ural Branch of Russian Academy of Sciences, 101, Amundsen str., Ekaterinburg, 620016, Russia

E-mail: silast@mail.ru

Abstract. The overview of existing models for iron oxides reducing and an analysis of their applicability to technogenic formations is presented. The thermodynamic analysis data of iron oxides reducing reactions through solid and gas phases and experiment results presenting features of technogenic iron oxide reducing under conditions of solid-phase diffusion (limited diffusion) and partially liquid-phase diffusion (easier diffusion) of iron oxide ions. It is shown that under conditions of iron oxide ions solid-phase diffusion the reduction processes rate is significantly affected by the briquetting pressure, which increases the contact area of reacting substances. The briquetting pressure does not affect the reduction processes rate under conditions of partially liquid-phase diffusion of iron oxide ions. An ion-diffusion catalytic mechanism is proposed to describe the observed effects of technogenic iron oxide reducing.

1. Introduction

A huge number of technogenic formations that accumulated by metallurgical plants is a source of environmental pollution. At the same time, it is also a promising stock of raw materials, of which iron is the most valuable for ferrous metallurgy. Technogenic formation of the iron oxide presented mostly by sludge and scale. In such technogenic formations all iron oxides can be present simultaneously, and they are arranged in layers in the scale. The layered structure of the scale is completely consistent with the diagram of the iron-oxygen state. Above the temperature of 650 °C, the inner layer of scale consists of vustite (FeO), the middle layer of magnetite (Fe_3O_4), and the outer layer of hematite (Fe_2O_3) [1].

Vustite (FeO) is a thermodynamically unstable compound, below the temperature of 560 °C, it decays into α-Fe and magnetite (Fe_3O_4). The crystal lattice is cubic, a = 4.29 Å. The vustite crystal lattice always contains an excess of oxygen, so its composition corresponds more precisely to the non-stoichiometric formula $Fe_{0.91}O$ or $FeO_{1.09}$. As a result of this, the vustite crystal lattice contains a significant number of non-stoichiometry defects in the form of Fe^{2+} ion vacancies [2]. From a structural point of view, vustite refers to subtraction solid solutions: in its lattice, oxygen atoms occupy all the nodes in the anionic sublattice, and some nodes in the cationic sublattice are not occupied by iron ions. If there is a lack of Fe^{2+} cations in the lattice, its electroneutrality is maintained due to the convertation of part of Fe^{2+} cations to Fe^{3+}, so this non-stoichiometric compound can be considered as a substitution solid solution of Fe_2O_3 in FeO [3].

Iron oxide α-Fe_2O_3 (hematite) is characterized by a rhombohedral corundum-type lattice with parameters a = 5.42 Å and α = 55°17'. The unit cell of hematite is formed by four Fe^{3+} and six O^{2-} ions, which form a dense hexagonal package. One-third of the octahedral voids are vacant, and two-thirds of them are filled with Fe^{3+} [2].

Magnetite (Fe_3O_4) has a cubic lattice of "reversed" spinel with a = 8.38 Å. $8Fe^{2+}$, $16Fe^{3+}$ and $32O^{2-}$are present in the unit cell of magnetite. Oxygen ions form a dense cubic lattice containing 32 octahedral and 64 tetrahedral voids in one unit cell. In this case, eight Fe^{2+} and the same amount of Fe^{3+} are distributed (statistically) in octahedral, and the remaining eight Fe^{3+} – in tetrahedral internodes. The presence of Fe^{2+} and Fe^{3+} in the same crystallographic positions (octahedral) facilitates the exchange of electrons and makes magnetite highly conductive [2].

Decomposition of iron oxides occurs due to the influence of gaseous or solid reducing agents. The most important reducing agent of iron, widely used in metallurgical practice, is carbon. Several models have been proposed to describe the mechanism of metal oxides reduction by carbon [4–8] S. I. Popel and his colleagues proposed 4 recovery schemes [9]. The first scheme is based on the most frequently used two-stage recovery scheme based on the adsorption-catalytic mechanism of G. I. Chufarov [10].

According to the first scheme, it is assumed that the process is carried out as a result of two simultaneous reactions:

$$MeO+CO=Me+CO_2\uparrow \tag{1}$$

$$CO_2+C=2CO\uparrow \tag{2}$$

The reduction of iron oxide in this scheme occurs through CO, and the role of solid carbon is reduced to the regeneration of carbon oxide by the Bell-Boudoir reaction.

According to the second (dissociative) scheme, the reduction of metals with a sufficiently high dissociation elasticity ($-\Delta\rho°_{O2} \geq$ 1Pa) can proceed by dissociation of the oxide and subsequent interaction of molecular oxygen with carbon:

$$MeO=Me+1/2O_2\uparrow \tag{3}$$

$$C+1/2O_2=CO\uparrow \tag{4}$$

According to the third scheme, if the reduction and sublimation temperatures of the oxide are close, the interaction of the oxide and the reducing agent is possible due to the evaporation of the oxide with subsequent condensation (adsorption) of its vapors on the carbon surface, where the reduction process occurs:

$$MeO_s=MeO\uparrow \tag{5}$$

$$MeO\uparrow+C_s=MeO_{ads}\cdot C_s \tag{6}$$

$$MeO_{ads}\cdot C_s=Me_s\cdot CO_{ads} \tag{7}$$

$$Me_s\cdot CO_{ads}=Me_s+CO\uparrow \tag{8}$$

In this way, it is possible to reduce the volatile oxides of MoO_3, WO_3, Nb_2O_3, etc.

According to the fourth scheme, the reduction occurs directly during contact of solid phases - the interaction of solid carbon and oxide. This is one of limiting factors in the scheme implementation. At the beginning, this contact is direct, and after formation of separating layer of metal or lower oxide, it is carried out by the diffusion of reagents through this layer. According to the authors [9], the flow of the 4 scheme is very possible, but its contribution to the overall kinetics of metals direct reduction from oxides is negligible, since the actual particles contact surface is an insignificant fraction of their total surface. The second limiting factor of the fourth scheme possibility is the diffusion of reagents, which is quite slow at low temperatures.

According to [11], all the described schemes are based on outdated provisions of the atomic-molecular theory of reduction, which represents atoms as solid indivisible balls, and do not take into account advances in the study of the fine structure of matter and physical theories describing the behavior of elementary particles that make up atoms.

2. Thermodynamic analysis of the reactions of reduction of iron oxides

The total reaction of all four schemes is the reaction (9):

$$MeO+C=Me+CO\uparrow \quad (9)$$

In relation to a specific iron oxide, this interaction is possible by the following reactions (10–12):

$$2Fe2O3+6C= 4Fe+6CO\uparrow \quad (10)$$

$$Fe3O4+4C= 3Fe+4CO\uparrow \quad (11)$$

$$FeO+C= Fe+CO\uparrow \quad (12)$$

Reduction of specific iron oxides through the gas phase is possible by reactions (13-15:)

$$Fe2O3+3CO= 2Fe+3CO2\uparrow \quad (13)$$

$$Fe3O4+4CO= 3Fe+4CO2\uparrow \quad (14)$$

$$FeO+CO= Fe+CO2\uparrow \quad (15)$$

According to [11], when cementite is formed in the system by reaction (16):

$$6FeO+5C=2Fe3C+3CO2\uparrow \quad (16)$$

it is also possible to reduce iron through the gas phase by reaction (17)

$$Fe_3C+CO_2\uparrow=3Fe+2CO\uparrow \quad (17)$$

Thermodynamic calculations were performed to determine the temperature range in which the reduction reactions of specific iron oxides can occur through the solid and gas phase. The results of calculating the Gibbs free energy for reactions (10–17) are shown in table 1.

Table 1. Value of Gibbs free energy of iron oxide reduction reactions.

Reaction №	ΔG^0_T, J/mol, at temperature, °C					
	500	600	700	800	900	1000
10	151.2	47.3	-55.6	-158.0	-260.4	-362.5
11	136.9	71.1	6.0	-58.9	-123.9	-188.8
12	33.7	18.3	2.9	-12.5	-27.9	-43.3
13	-30.3	-29.2	-27.7	-26.5	-25.4	-24.5
14	-1.0	0.2	1.4	2.6	3.8	4.8
15	-1.6	0.7	2.9	5.0	7.0	8.9
16	26.3	15.6	5.0	-5.4	-15.7	-25.8
17	6.8	3.2	-0.4	-4.1	-7.9	-11.9

According to the data presented in table 1, reduction of Fe_2O_3 oxide (reaction 10) by direct interaction of solid phases is thermodynamically possible starting from the temperature of 700 °C, and Fe_3O_4 and FeO (reactions 11 and 12) – starting from the temperature of 800 °C. Through the gas phase, reduction of Fe_2O_3 (reaction 13) is possible in the entire temperature range (from room temperature to 1000 °C). Above the temperature of 600 °C, recovery through the gas phase of Fe_3O_4 and FeO (reactions 14 and 15) is thermodynamically impossible. Since in accordance with the principle of sequence of transformations formulated by A. A. Baikov [12] above the temperature of 570 °C, the transition from higher iron oxides to lower ones proceeds according to the scheme:

$$Fe_2O_3 \xrightarrow{C(CO)} Fe_3O_4 \xrightarrow{C(CO)} FeO \xrightarrow{C(CO)} Fe \tag{18}$$

And according to the data [13] given in table 2, the content of Fe_2O_3 in the scale at a temperature above 700 °C is insignificant, so it is very likely that above this temperature the reduction of Fe_3O_4 and FeO occurs only with the participation of solid carbon in the reactions (11) and (12). Above the temperature of 700 °C through the gas phase, it is also possible to restore iron by reaction (17), but this will only happen if cementite is formed in the system by reaction (16).

Table 2. The content of iron oxides in the slag at elevated temperatures.

Temperature, °C	Scale oxide content, mass. %		
	FeO	Fe_3O_4	Fe_2O_3
600	50	50	0
700	98	2	0
800	96	2	2
900	90	9	1
1000	95	4	1
1090	94	5	1

Since above the temperature of 700 °C for Fe_2O_3, and for Fe_3O_4 and FeO above 800 °C, the reduction reaction is thermodynamically possible through the solid phase, it is difficult to explain the high rate of iron oxides reduction under conditions of practically no diffusion of reagents in the solid phase using the provisions of the classical atomic-molecular theory of reduction.

Under these conditions, there are only two ways to accelerate the reduction of iron oxides with carbon:

1-in the absence of diffusion of reagents, increase the area and number of their contacts;

2-create conditions for favorable diffusion of reagents.

According to [14, 15], the area and number of contacts of reacting solid phases can be increased by their joint grinding and subsequent pressing. In [16], same approach was used in the example of the synthesis of nickel ferrite, zincaluminates and magnesium aluminates from oxides under pressure from atmospheric to 60 kbar. The observed strong acceleration of these reactions is explained by an increase in the contact area of the reacting particles and, thus, an improvement in the conditions for diffusion along the interphase surfaces.

The diffusion rate of reactants can be increased by partial destruction of their crystal lattice. This is possible due to the fluxes introduction into the mixture of scale and carbon, which reduce the melting point of the scale, lead to the formation of local liquid-phase zones, in which the diffusion rate of reacting substances increases by several orders.

3. Experimental method

The rolling scale was used in this work. According to the phase analysis results, the scale contained, mass. %: 43.4 FeO, 27.0 Fe_3O_4 and 29.6 Fe_2O_3, with a Fe_{all} content of 73.4. The coke was used as a reducing agent with a carbon content of 92mass.%. Its amount was 15% over the stoichiometric ratio. To assess the iron oxides reduction rate, in the absence of reactants diffusion,

IOP Conf. Series: Materials Science and Engineering **969** (2020) 012052 doi:10.1088/1757-899X/969/1/012052

the raw mixture was prepared only from scale and coke. To assess the iron oxides reduction rate, under conditions of easier diffusion of reacting substances, a flux based on CaO, MgO, SiO_2, Al_2O_3, Na_2O and CaF_2 was added to the scale and coke mix in an amount of 10% of the mass of the scale. Scale, coke and flux were dried to a constant mass and joint grinded to a specific surface area by Blain≈400 m^2/kg. The material mixture of scale, reducing agent and flux prepared in this way was pressed at pressures of 100, 200 and 300 MPa.

Heat treatment of the pressed raw material mixture was carried out in an argon atmosphere with the removal of gas reaction products. Heating was performed to a temperature of 1200 °C. The thermal characteristics of the raw material mixture were determined during the heating process, and the phase composition of the heating products was determined. Qualitative x-ray phase analysis was performed using an XRD-7000 diffractometer (Shimadzu). Quantitative x-ray phase analysis was performed using a STADI-P diffractometer (STOE, Germany). The survey was performed in CuKα-radiation (40 kV, 30 mA), graphite monochromator, in the range of scattering angles $2\Theta = 10 - 70$ degree, with a step of 0.02 degree. The results were analyzed using the PDF-2 database (Release 2008 RDB 2.0804). Thermal analysis (TA) was performed using the DSC method (differential scanning calorimetry) on the STA 449 F3 Jupiter thermal analyzer (Netzsch-Geratebau GmbH). The experiment was conducted in an argon atmosphere (Ar 99.998 %), under the condition of linear heating at a speed of 10 °/min, the temperature measurement error is no more than ± 1.5 degree. Al_2O_3 crucibles with lids were used for the study. The value of the endothermic effect was calculated in the software module of the device in j/g and converted to kJ/mol by multiplying the data of the device by the mole mass of metallic iron.

4. Results

Figure 1 shows the TA data of the material mixture without the flux addition. Presented data shows, that starting from a temperature of about 300 °C, the DSC curves are separated strictly according to the pressure during pressing. The line for not pressed scale is the upper, and the line for scale, pressed at 300 MPa, is the low. This indicates the pressing limiting influence on the process of studied mixture thermodynamic state converting. Due to a more dense contact of the reacting phases, provided by increasing pressing pressure, the thermodynamic state of the mixture changes at the expense of less heat. At a temperature of about 600°C, a small endothermic effect is observed. It indicates the beginning of the Fe_2O_3 reduction process, according to reaction 19, which is the first stage of the proposed scale reduction mechanism described below. According to thermodynamic calculations, this reaction is possible starting from the temperature of 539 °C.

$$Fe_2O_3+C=2FeO+CO\uparrow \tag{19}$$

Upon further heating at a temperature of about 1000 °C, a small endothermic effect is observed in the non-pressed and pressed scale at 100 MPa, indicating the end of the reducing process of $Fe_2O_3$to Fe_3O_4 by reaction 20 and reducing of Fe_3O_4 to FeO by reaction 21.

$$2FeO+2Fe_2O_3=2Fe_3O_4 \tag{20}$$

$$2Fe_3O_4+2C=6FeO+2CO\uparrow \tag{21}$$

Reactions 21 and 22 are the second stage of the proposed scale reduction mechanism described below. The scale pressed at 200 and 300 MPa has no small endothermic effect, due to the increasing in pressing pressure accelerate the recovery processes, and all hematite is reduced at the first stage. That is confirmed by the data of the phase analysis described below. Figure 2 shows the effect of pressing pressure on the temperature of metallic iron formation beginning from a mixture without flux.

In accordance with Figure 2 when the pressing pressure increases, the metallization beginning shifts to the lower temperature range from 1021.7 °C at 0 MPa to 989.8 °C at 300 MPa.

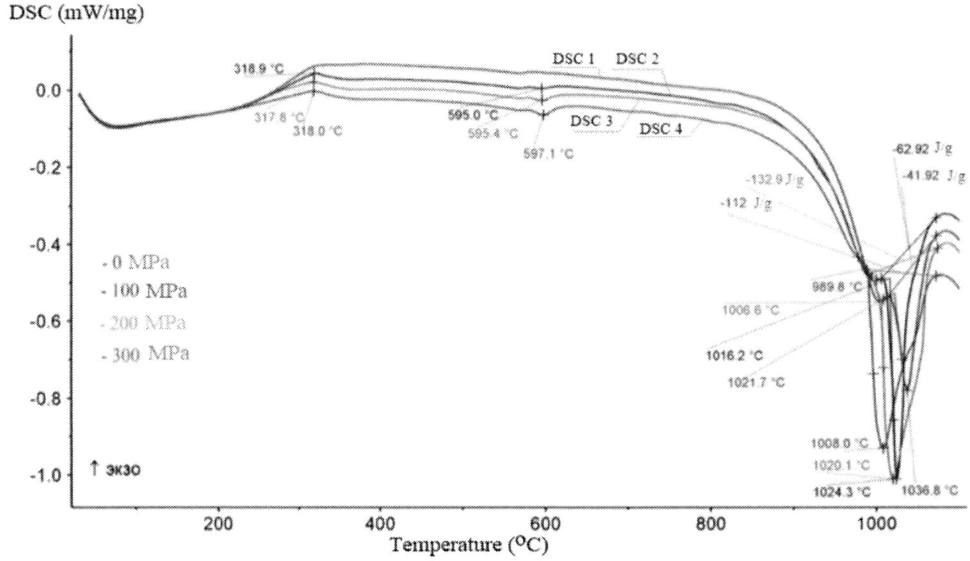

Figure 1. DSC data from of the pressed material mixture of rolled scale and coke without the flux.

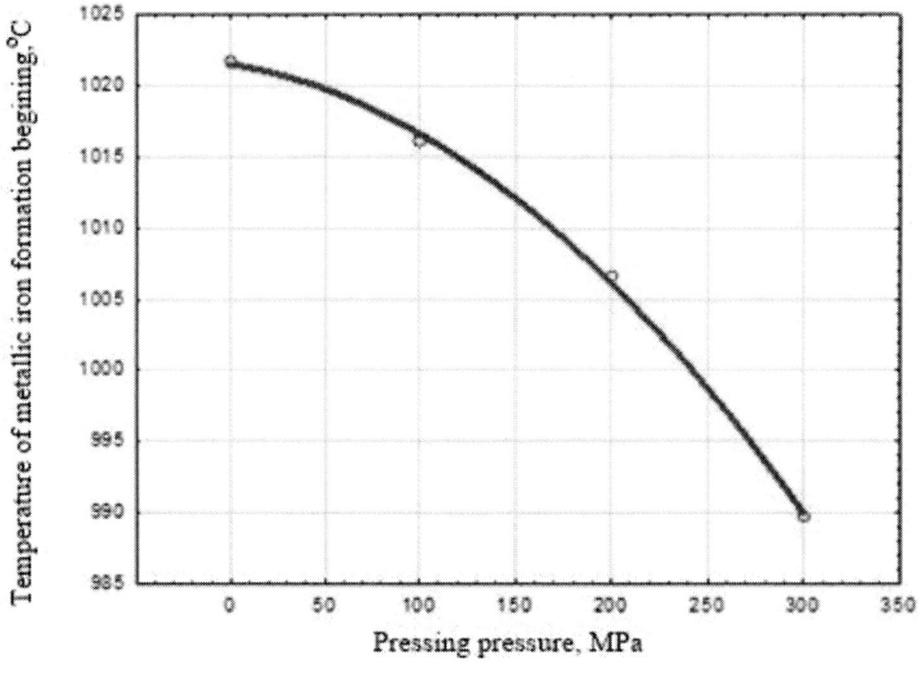

Figure 2. Influence of pressing pressure on the temperature of metallic iron formation beginning from a mixture without flux.

As the pressing pressure increases, the enthalpy of metallic iron formation and the degree of metallization also increases. The change of these parameters depending on the pressing pressure is shown in figure 3. The enthalpy of iron metal formation increases from 2.3 kJ/mol at 0 MPa to 6.6 kJ at 300 MPa. The metallization degree increases from 34.0 % to 72 %, respectively. The data of comparative qualitative phase analysis of the pressed mixture without flux, heated in an argon atmosphere, is shown in Figure 4. Figure 5 shows the change in the heating products phase composition of the mixture without flux, depending on the pressing pressure according to quantitative phase analysis.

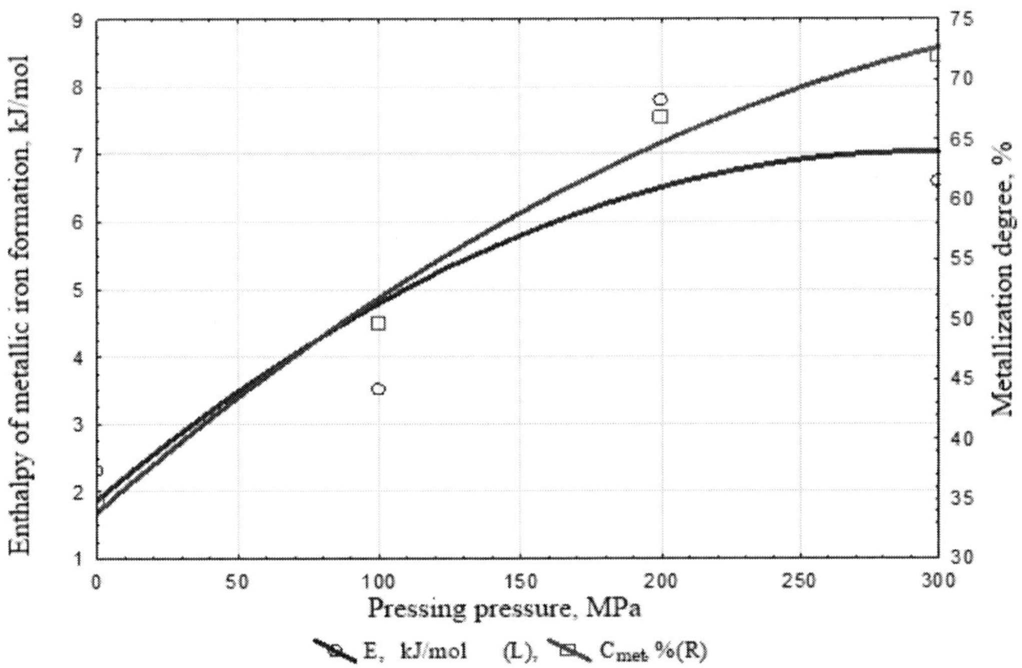

Figure 3. Influence of pressing pressure on the metal iron formation enthalpy and on the mixture without flux metallization degree.

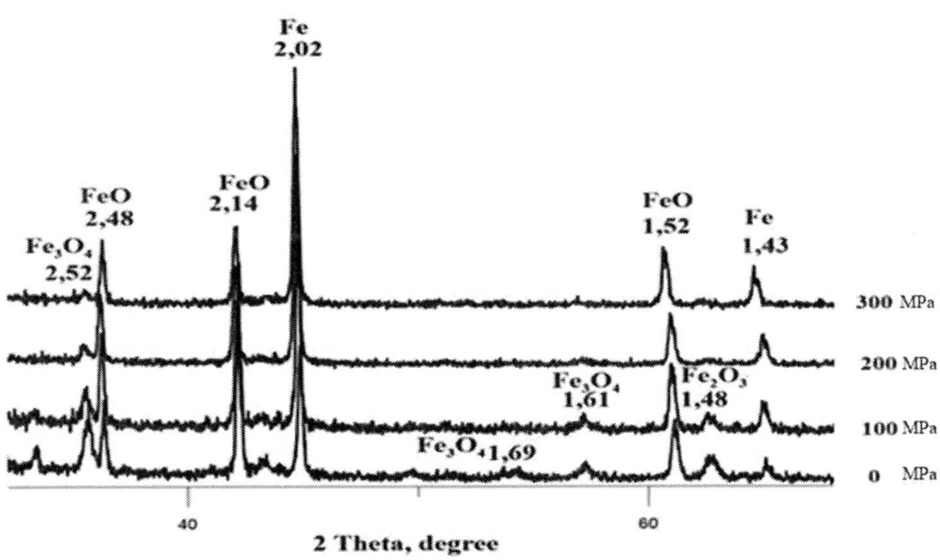

Figure 4. Influence of the pressing pressure of a raw mixture without flux heated in an argon medium on the change in its phase composition.

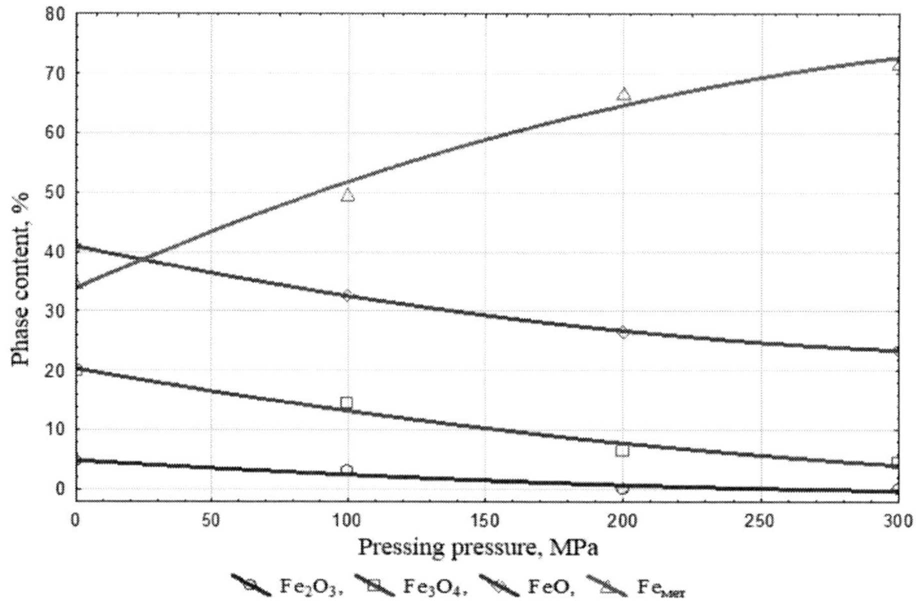

Figure 5. Influence of the pressing pressure of the raw material mixture without flux on the change in the phase composition of heating products according to quantitative phase analysis.

According to quantitative phase analysis, the residual content of Fe_2O_3 in the heating products of non-pressed scale and pressed at 100 MPa is 4.6 and 3.2 %, respectively. There is no Fe_2O_3 in the heating products of scale pressed at 200 and 300 MPa. Thus, the phase data confirm the conclusions made after DSC analysis, that a small endothermic effect at a temperature of 1000 °C indicates the completion of the reducing process of Fe_2O_3 to Fe_3O_4. In addition, from the phase analysis data, it follows that under the same conditions of heat treatment, with an increase in the pressing pressure, the content of iron oxide phases decreases from 65.4 % for non-pressed scale to 28.0% for scale pressed at 300 MPa. The content of metallic iron, on the contrary, increases.

Figure 6 presents TA data of pressed mixture with 10 % flux.

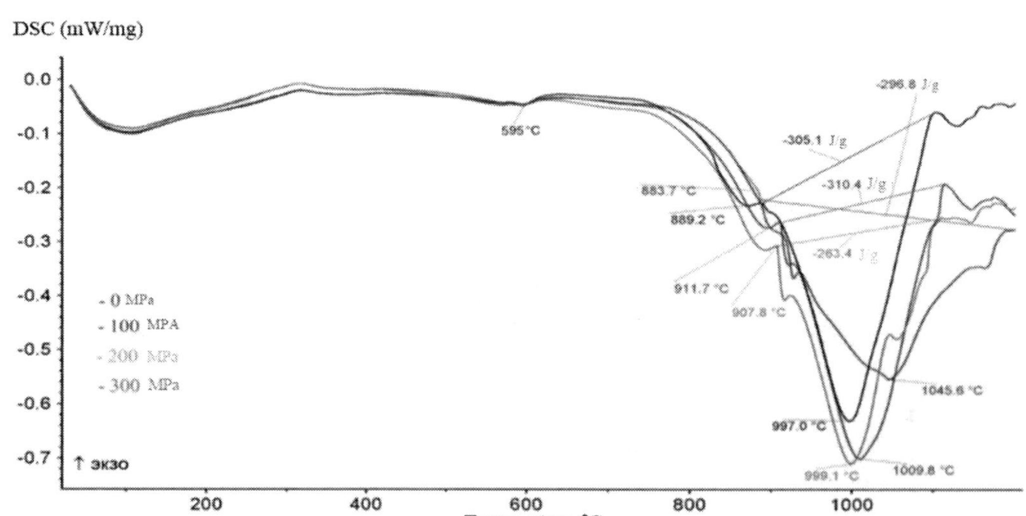

Figure 6. DSC data of the pressed mixture of scale and coke with the addition of 10% flux.

The data presented in Figure 6 show that, at a temperature of 600 °C, all four compositions have a small endothermic effect, and the separation of the DSC curves depending on the pressing pressure also begins at this temperature. Small endothermic effects at a temperature of 1000 °C are observed in all compositions. Figure 7 shows the effect of pressing pressure on the metallic iron formation beginning temperature for the mixture with the flux.

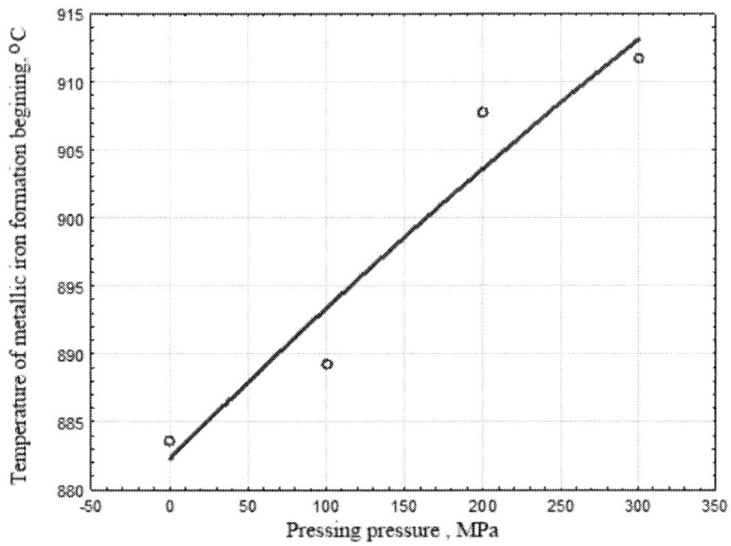

Figure 7. Influence of pressing pressure on the temperature metallic iron formation beginning temperature for the mixture with the flux.

In accordance with figure 7, the material mixture with the flux has the temperature of the metallic iron formation beginning shifted to the higher temperatures from 883.7°C at 0 MPa to 911.7°C at 300 MPa.

Figure 8 shows the pressing pressure effect on the change in the metallic iron formation enthalpy and the metallization degree of the mixture with the flux.

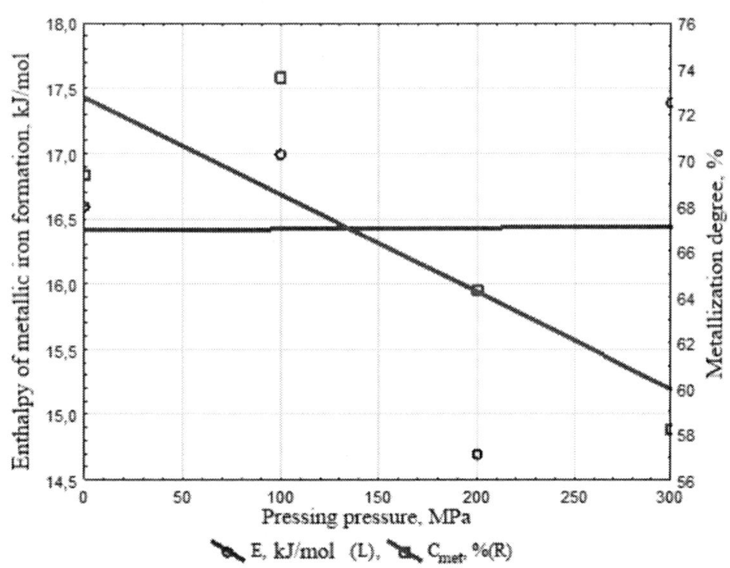

Figure 8. The pressing pressure effect on the change in the metallic iron formation enthalpy and the metallization degree of the mixture with the flux.

According to the figure 8, it can be seen that in the mixture containing flux, with increasing pressing pressure, the metal iron formation enthalpy increases slightly from 16.6 kJ/mol at 0 MPa to 17.4 kJ/mol at 300 MPa, and the metallization degree even slightly decreases, respectively, from 69.3 % to 58.2 %. The data of comparative qualitative phase analysis of the pressed material mixture with 10 % flux, during heating in an argon atmosphere, are shown in figure 9.

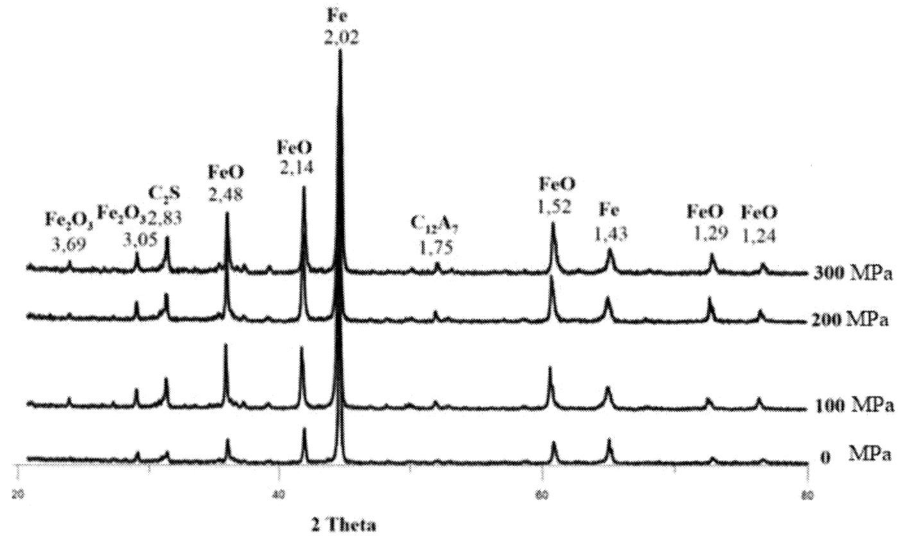

Figure 9. Pressing pressure of the materials mixture containing flux effect on its phase composition change, during heating in an argon atmosphere.

Figure 10 shows the firing products phase composition change depending on the pressing pressure of the mixture containing flux according to quantitative phase analysis.

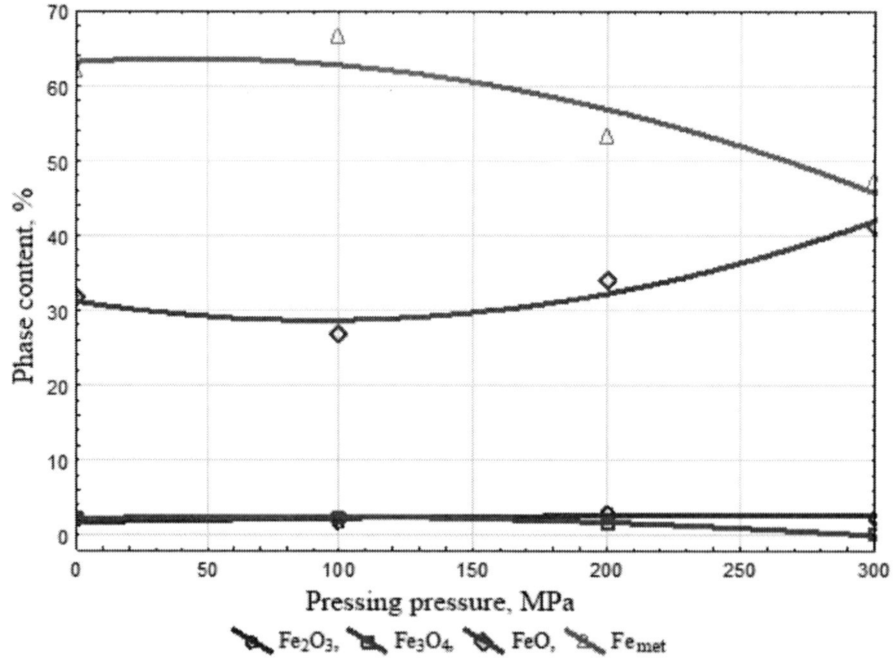

Figure 10. Pressing pressure of a mixture containing flux effect on the heating products phase composition change according to quantitative phase analysis.

The phase analysis data shown in figure 9 and 10 show a small content of hematite and magnetite in the heating products of the mixture containing flux, and it practically does not change with increasing pressing pressure. The FeO content increases with increasing pressing pressure from 31.8 % for non-pressed to 41.4 % for pressed at 300 MPa scale, and the content of metallic iron, on the contrary, decreases from 69.3 % to 46.9 %, respectively.

According to the data presented in figure 7, 8 and 10, it can be seen that the pressing pressure does not significantly affect the measured parameters when the material mixture contains 10% of flux, in contrast to the material mixture that does not contain flux.

5. Discussion

The results of thermodynamic analysis show that the recovery of hematite, magnetite and vustite above the temperature of 700 °C is possible through solid-phase interaction of reagents. Roasting of the mixture based on coke and scale in an argon atmosphere with the gaseous products removing from the reaction zone, i.e. in conditions that exclude the participation of gaseous products in the recovery process, showed that the recovery processes are successfully implemented with a high degree of metallization through the solid phase.

It is hard to explain the observed effects from the classical atomic-molecular theory of reduction point of view. That theory determine that the reacting substances atoms difficult diffusion in the solid phase cannot provide a high rate of the reduction processes. Observed effects can be explained by the theory presented by V. E. Roshchin and colleagues about the role of electronic processes in the iron oxides reduction [17–21]. To explain the observed effects, we propose an ion-diffusion-catalytic mechanism for iron oxides solid-phase reduction, which details statements of V. E. Roshchin's theory in relation to iron oxides of technogenic origin.

According to this mechanism, at the first stage, the solid-phase interaction of the hematite located on the surface of the scale with the solid carbon located in the intergranular space occurs, with the formation of a CO by reaction (19). As indicated above, this reaction is possible starting from the temperature of 539 °C. The endothermic effect observed on the DSC curves at a temperature of $\approx$ 600 °C corresponds to the beginning of the recovery first stage. In accordance with the statements of Roshchin's theory the metallic iron crystal lattice is a crystal structure with cations at its nodes, and collectivized valence electrons fill the space between the cations and bind them to the backbone of the crystal lattice. When metal iron is oxidized, oxygen dissolves in it and adds its electronic shell with valence electrons, becoming an anion. The metal is transformed into a higher oxide after all collectivized electrons becomes binded with anions. During reducing, reverse processes occur. Oxygen does not carry away the previously captured valence electrons when extracted from the scale surface due to inter-grain carbon interaction, since this would disrupt the charges balance in the oxide. In our opinion, it transfers electrons to neighboring iron cations. As a result, an oxygen vacuum is formed on the scale surface, two atoms of divalent iron and, respectively, two molecules of vustit.

Resulting vustite molecules are thermodynamically unstable, since they are thermodynamically predisposed to interact with the below hematite layers to form two magnetite molecules by the reaction (20), which is the beginning of the recovery second stage. At 600 °C, the Gibbs free energy of this reaction is -58.4 kJ/mol. The end of the recovery second stage is the interaction of oxygen of newly formed magnetite molecules with solid intergranular carbon, with the formation of 6 FeO and CO molecules by reaction (21). Small endothermic effects of $\approx$1000 °C correspond to the curves of this reaction, and they are observed only in mixtures with a sufficient amount of residual Fe_2O_3. The reaction (21) is thermodynamically possible starting from a temperature of 700 °C. As a result of this reaction, oxygen will transfer electrons to neighboring iron atoms with the formation of 6 divalent iron atoms (6 vustite molecules, respectively) and the formation of 2 oxygen vacancies. The newly formed vustite molecules also will be associated with the remaining neighboring hematite molecules with the formation of magnetite molecules. Due to the formation of a vustite molecules significant number, the catalytic acceleration of the scale reduction processes will occur with the promotion of divalent iron ions deep into the scale, and oxygen anions to the surface, in the direction of oxygen vacancies. Due to the catalytic acceleration, provided by the vustite significant amount formation, at the third stage of reducing process, it is restored to metallic iron. At the final stage, only vustite molecules and oxygen vacancies will be located on the scale surface. The vustite oxygen interaction with solid carbon will lead to the metallic iron and CO formation by reaction (12). According to thermodynamic calculations (table 1), this reaction is possible from the temperature of 800 °C, and a large endothermic effect in the phase of the highest reaction speed corresponds to this reaction on all DSC curves. During the third stage, oxygen anions will move to the surface, and the cations of metallic iron formed during this process will form an apparent flow of movement into the crystal.

The test results show that without the introduction of flux into the material mixture, the pressing pressure has a significant effect on the firing products phase composition and the recovery process parameters. When the material mixture contains 10 % flux, the pressing pressure effect on the firing products phase composition and the recovery process parameters is minimal. These data indicate that when a sufficient amount of flux is introduced into mixture, the diffusion processes begin to prevail over the dense packing processes associated with pressing pressure. In other words, a introduced flux sufficient mass eliminates a diffusion difficulties significant part in the process mechanism, ensuring that the metallization beginning temperature is close to the minimum limit (due to thermodynamics). Also, we can say that the introduction of flux and pressing to some extent have an equivalent effect associated with an increase in the mobility of ions and the growth of the grains active surface.

6. Conclusions

1. 1. Data of thermodynamic calculations and experimental data are presented. Results can be explained by the solid-phase interaction of scale iron oxides and carbon.
2. 2. To describe observed effects, an ion-diffusion-catalytic mechanism for scale iron oxides solid-phase reduction is proposed, which details and clarifies the statements of the V. E. Roshchin and colleagues theory in relation to iron oxides of technogenic origin.

Refernces

[1] Rodzevich A P 2010 *Physical and Chemical Bases of Metallurgical Processes: Educational Textbook* (Tomsk: Yurginsk Institute of technology)
[2] Gerasimenko A A 1987 *Protection Against Corrosion, Aging and Bio-Damage of Machinery, Equipment And Structures: Reference Book 1* (Moscow: Machine engineering)
[3] Gorshkov V S, Saveliev V G and Fedorov N F 1988 *Physical Chemistry of Silicates and Other Refractory Compounds: Educational Textbook* (Moscow: Higher school)
[4] Gruner L1872 Etudes sur les hauts-formeause *Annales des Mines* 1–14
[5] Rostovtsev S T, Simonov V K and others 1970 *Mechanism of carbon-thermal reduction of metal oxides. Mechanism and kinetics of metal reduction* (Moscow: Science)
[6] Senin A V, Pashkeev I Y and Mikhalov G G 2018 Gas-phase-solid-phase mechanism for recovery of ore materials *Prospects for the development of metallurgy and mechanical engineering using completed basic research and R&D: FERROALLOYS Conf. Proc.* 72–80
[7] Tleugabulov S M, Abikov S B and others 2018 Fundamentals and prospects for the development of reduction steel melting *Metals* **2** 72–77
[8] Averin V V, Korneev V P and Dybanov V G 2010 Solid-phase carbon reduction of electric furnace sludges *University Transactions. Ferrous metallurgy* **9** 10–13
[9] Popel S I, Sotnikov A I and Boronenkov V N 1986 *Theory of Metallurgical Processes: Educational Textbook for Universites* (Moscow: Metallurgy)
[10] Chufarov G I, Men A N and Balakirev V F 1970 *Thermodynamics of Metal Oxide Reduction Processes* (Moscow: Metallurgy)
[11] Roshchin V E and Roshchin A V 2020 Electronic processes during metals recovery and extracting from ores *Electrometallurgy* **3** 14–24
[12] Komatsu W 1964 *Proc. of the 5th Intern. Symp. Reactivity of Solids*

[13] Baikov A A 1948 *Collection of selected works in 2 books Book 2* (Moscow-Leningrad: AS USSR)
[14] Rodzevich A P 2010 *Physical and chemical bases of metallurgical processes: textbook* (Tomsk: Yurginsk Institute of technology)
[15] Levitas V I 2004 Continuum mechanical fundamentals of mechanochemistry *High Pressure Surface Science and Engineering. Section* **3** 159–292
[16] Drickamer H G and Frank C 1973 *Electronic Transitions and the High Pressure Chemistry and Physics of Solids* (London: Chapman and Hall)
[17] Roshchin V E and Roshchin A V 2019 Electron mechanism of reduction process in blast and ferroalloy furnaces *CIS Iron and Steel Review* **17** 14–24
[18] Roshchin V E and Roshchin A V 2015 Physics of metals oxidation and reduction in the solid phase *Metalls* **3** 19–24
[19] Roshchin V E, Gamov P A and Roshchin A V 2019 Electronic recovery theory: implications for the theory and practice of metals extracting from ores *University Transactions. Ferrous metallurgy* **62** **5** 407–417
[20] Roshchin V E, Roshchin A V, Gamov P A and Bilgenov A S 2020 Electro-and mass transfer during the metals reduction by solid carbon in solid complex oxides *Metalls* **1** 33–37
[21] Roshchin V E, Roshchin A V, Berdnikov A A and Goihenberg U N 2008 Formation and sublimation of intermediate products of silicon reduction from its oxide *Metalls* **4** 14–18

IOP Conf. Series: Materials Science and Engineering **969** (2020) 012053 doi:10.1088/1757-899X/969/1/012053

Oxidized Nickel Ore and Galvanic Sludge Combined Treatment Investigation

M V Boshnyak, O B Kolmachikhina and E B Kolmachikhina

Department of Non-Ferrous Metallurgy, Ural Federal University, 17, Mira St., Ekaterinburg, 620002, Russia

E-mail: o.b.kolmachikhina@urfu.ru

Abstract. Galvanic sludges have been accumulated in large quantities in Ural region. These materials contain low concentration of non-ferrous metals (nickel, zinc, chromium) and iron, and are not being processed. Another difficult for treatment nickel containing materials are oxidized nickel ores, deposits of which are located in Ural. However, the nickel ores aren't processed due to low economic efficiency. In this article simultaneous treatment of oxidized nickel ore and galvanic sludge was investigated. The proposed technology includes nickel ore sulfatization at high temperature in presence of sulfuric acid and further water leaching. Therefore, maximum nickel extraction in solution at leaching of oxidized nickel ore after sulfatization was 82 %. Solutions after leaching are neutralized by nickel containing galvanic sludge. Due to partial nickel chromite dissolution in sulfuric acid a total nickel concentration has been substantially increased. Selective nickel extraction from such solutions is possible by sorption methods, and further metal nickel obtaining – by electrowinning.

1. Introduction

Significant quantity of technogenic wastes (electrodeposition sludges in particular) have been accumulated in Ural region. The reason is that machine-building industry enterprises overcoat metallic details by electrodeposited coating from aqua solutions or molten metals salts for increasing corrosive resistance and visual improvement. For example, nickel plating in sulfuric acid electrolytes, which contain various brightening additions, including chlorine and sulfated salts of cobalt and cadmium, is being used widely [1]. Covering by this technology leads to formation of toxic waste waters in large quantities with acids and heavy metals salts containment [2, 3].

Widely applied method of such wastes detoxification is neutralization by lime. This method is cheap and accessible because it doesn't require sophisticated equipment, expensive reagents and allows to achieve high neutralization and purification levels for one stage. Produced sludges contain metals hydroxides (nickel, zinc, iron, etc.) and being industrial wastes. Treatment of galvanic sludges is complicated due to complex composition and low concentration of non-ferrous metals (nickel content in different sludges is estimated at the level of 2–6 %) [4, 5]. At the present time these wastes are disposed on special waste landfill deposits. This compromised solution damages the environment and demands significant investments for construction and operation of toxic wastes landfill [6–12].

At the same time, Ural holds large oxidized nickel ore deposits, which were a raw material during long time for Ural plants (JSC Ufaleynickel, JFC Rezhnickel). Nickel in such ores presents as complex oxidized compound, nickel content in which is about 1 % [13, 14]. Pyrometallurgical methods of oxidized nickel ores treatment used by Ural plants were unprofitable. This is connected, primarily, with

IOP Conf. Series: Materials Science and Engineering **969** (2020) 012053 doi:10.1088/1757-899X/969/1/012053

oxidized nickel ore beneficiation impossibility and melting necessity of poor material in large quantities. Using of such technologies have led to stoppage of all Ural nickel plants in period 2012–2018. Nevertheless, presence of ore materials, large quantity of nickel containing wastes and industrial facilities make finding of new technologies relevant objective from ecological and economical points of view [15–19].

2. Research methods

There is a strong need for a technology which allows to include oxidized nickel ores in treatment, along with nickel-containing galvanic sludges. Investigation of oxidized nickel ore and galvanic sludge simultaneous treatment were carried out on ore of the Buruktal deposit and galvanic sludges which were collected from abandoned sladge landfill.

The Buruktal deposit, placed in Orenburg area, is one of the biggest deposits of oxidized nickel ore in Russia. The Buruktal deposit consists of 7 sections and nickel total reserves are 1.5 million tons. At the present time, this deposit isn't developed due to shutdown of nickel metallurgical plants at Urals.

The main aim of this investigation was opportunities estimation for combined hydrometallurgical treatment of these materials. Chemical analysis of sludge and ore were carried out using atomic absorption spectrophotometer Vario 6 "Analytik Jena AG" (Table 1).

Table 1. Raw materials composition, %.

Material	Fe	Si	Ni	Cr	Ca	Mn	Cu
Galvanic sludge	2.46	16.49	2.7	6.76	5.79	0.15	2.39
Oxidized nickel ore	37.69	21.48	1.35	1.43	0.85	0.61	-

Effectiveness of a hydrometallurgical technology depends on chemical composition of raw materials. The oxidized nickel ore and the galvanic sludge contain iron, chromium, copper, silicon and other elements that makes their hydrometallurgical treatment difficult, in particular, due to high consumption of leaching reagents which are spent not to transfer valuable metals into solution but to neutralize the gangue and other components. Cost of leaching agents at oxidized nickel ores treatment is a significant economic factor. At the moment, sulfuric acid is the most available and cheap reagent for leaching in Russia. Furthermore, sulfuric acid is produced abundantly at Urals copper smelting plants as a minor and an inevitable product.

3. Results and discussion

On the first stage of the study, we carried out investigation of oxidized nickel ore acid consumption. Results have showed that specific sulfuric acid consumption leveled at 0.67–0.78 g of acid/g of ore, it is high order, especially with account of low nickel extraction in solution – maximum value was about 76 %. Such efficiency is characteristic of oxidized nickel ores from Ural deposits.

For the purpose of nickel extraction enhancing preliminary ore sulfatization were carried out by sintering with sulfuric acid in muffle furnace at 270–420 °C. Sulfuric acid consumption was varied in a range of 0.37–1.1 g/g of ore. Physical-chemical basis of this process is formation of water-soluble sulfates of non-ferrous metals and iron during ore heat treatment in presence of sulfuric acid. High temperature level of oxidized ores and acids mixture sulfatization is close to 700 °C. At the further temperature increasing nickel and cobalt sulfates decomposition occurs with oxide formation, wherefore extraction efficiency of these metals in solution is decreased at following leaching [20]. In our opinion, preliminary sulfatization allows to transfer valuable metals into solution in shorter period and at more mild conditions, without application of high temperature conditions and concentrated solutions.

After sintering obtained byproduct has been repulped by water with liquid and solid ratio on a level of 6:1 and has been leached by water for several hours at 50 °C. Obtained suspensions were filtered, cakes were washed. Filtrates and washing waters were analyzed on nickel concentration. At calculation of nickel extraction, we took into account total nickel extraction to filtrates and washing solutions. Such accounting is necessary for possibility estimation of washing solutions including in a technological scheme for the ore treatment. Tests results are presented on Figure 1.

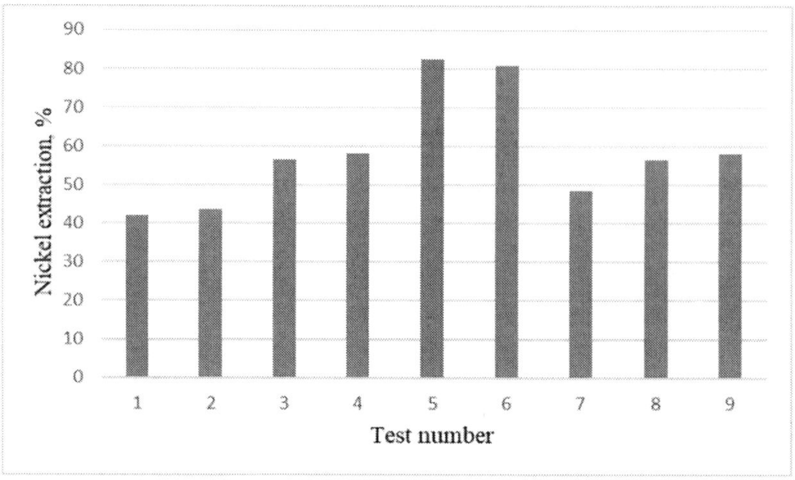

Figure 1. Nickel extraction in solution.

Therefore, maximum nickel extraction in solution at leaching of oxidized nickel ore after sulfatization was 82 %. Conditions of sulfatization were sulfuric acid consumption 0.73 g/g of ore, sintering temperature – 420 °C, sintering duration – 60 minutes. Nickel concentration in such solution was 5,5 $g \cdot l^{-1}$. Also, it has been noted that at sulfuric acid consumption increasing a nickel extraction on solution was dramatically decreased. Obtained solutions pH level was in a rage 3–4. For nickel precipitation from such solution, it is necessary to neutralize solutions. We recommend nickel galvanized sludges to be used as neutralizator due to it consist of calcite which reacts with sulfuric acid according to the equation (1):

$$CaCO_3 + H_2SO_4 = CaSO_4 + CO_2 + H_2O. \quad (1)$$

Value of solutions pH after neutralization must be in range of 6–7. At lower values a nickel hydroxide starts their precipitation from solution. Neutralization was carried out by small dosing of galvanic sludge in solution at continuous agitation and pH monitoring. Obtained data is presented in Table 2.

Table 2. Test results of solution neutralization.

Test	Initial solution			Galvanic sludge weighed portion (g)	Final solution		
	Volume (l^{-1})	pH	Nickel concentration ($g \cdot l^{-1}$)		Volume (l^{-1})	pH	Nickel concentration ($g \cdot l^{-1}$)
1	0.302	4.2	4.5	35	0.295	5.8	8.9
2	0.302	4.2	4.5	45	0.295	6.1	11.4
3	0.302	4.2	4.5	50	0.295	6.4	12.7
4	0.302	4.2	4.5	60	0.295	6.9	15.2

Galvanic sludge addition has allowed to neutralize initial solution. Therewith, due to partial nickel chromite dissolution in sulfuric acid total nickel concentration has been substantially increased from 5.5 $g \cdot l^{-1}$ to 15.2 $g \cdot l^{-1}$. Such low nickel concentrations in solutions don't allow to process it by electrowinning. Selective nickel extraction from such solutions is possible by sorption methods, and further metal nickel obtaining – by electrowinning. It is way forward for this work. In accordance with the findings scheme of oxidized nickel ore and galvanic sludge hydrometallurgical treatment has been proposed (Figure 2).

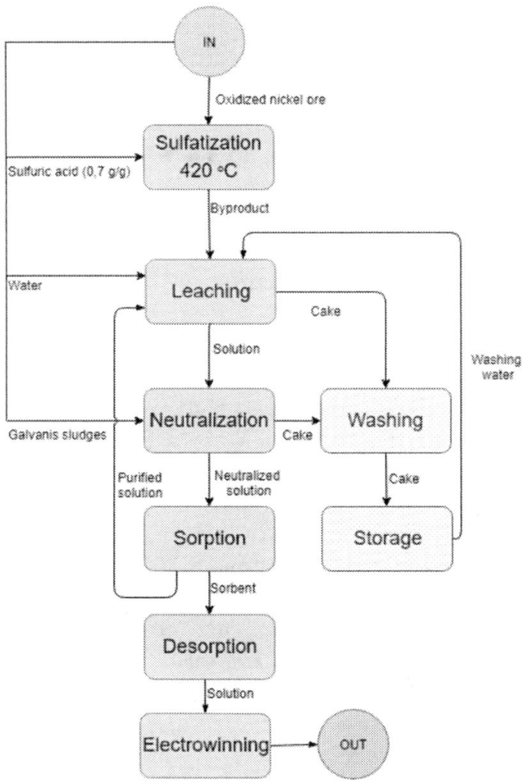

Figure 2. Proposed scheme of oxidized nickel ore and galvanic sludge combined treatment.

Industrial implementation of this technology will lead to release of galvanic sludges landfill deposits, liquidation of emissions which are formed at conventional pyrometallurgical processing of nickel oxidized ores, extracted nickel quality improvement, organization of non-waste processing.

4. Conclusion

The part of proposed technology of simultaneous treatment of oxidized nickel ore and galvanic sludge was tested in this work. Preliminary sulfatization allowed to increase nickel extraction at leaching up to 82 %. Neutralization of solutions after leaching led to increasing of nickel concentration up to 9–15 $g \cdot l^{-1}$. These solutions could be processed by sorption using nickel selective ion exchange resins and further nickel electrowinning.

References

[1] Takuma Y, Sugimori H, Ando A, Mizumoto K and Tahara K 2018 Comparison of the environmental impact of the conventional nickel electroplating and the new nickel electroplating *The International Journal of Life Cycle Assessment* **23** 1609–1623

[2] Bykovsky N, Kantor E, Rahman P, Puchkova L and Fanakova N 2019 Electrochemical treatment of waste water from nickel in galvanic production *IOP Conf. Series: Earth and Environmental Science* **350** 012029

[3] Benvenuti T, Krapf R, Rodrigues M, Bernardes A and Zoppas-Ferreira J 2014 Recovery of nickel and water from nickel electroplating wastewater by electrodialysis *Separation and Purification Technology* **129** 106–112

[4] Tsai L, Fang H, Lin J, Chen C and Tsai F 2009 Recovery and stabilization of heavy metal sludge (Cu and Ni) from etching and electroplating plants by electrolysis and sintering *Science in China Series B: Chemistry* **52** 644–651

[5] Tian L, Chen L, Gong A, Wu X, Cao C, Liu D and Liu Y 2020 Separation and Extraction of Valuable Metals from Electroplating Sludge by Carbothermal Reduction and Low-Carbon Reduction Refining *JOM* **72** 782–789
[6] Vinogradov S S 1998 *Environmentally safe plating industry* (Moscow: Globus)
[7] Surzhko A A 2015 Analysis of galvanic sludge treatment technologies using best available technology criteria *Topical Problems of the Humanities and Natural Sciences* **4** 180–181
[8] Ol'shanskaya L N 2011 Investigation of galvanic processing sludges obtained from consumer-packaged goods production *Polzunovsky Bulletin* **4** 203–206
[9] Makarov V M 2001 *Integrated treatment of galvanic processing wastewater* (Ivanovo: Yaroslavsk State Technical University)
[10] Zainullin Kh, Babkov V, Zakirov D, Chulkov A and Iskanova E 2003 *Utilization of galvanic wastewaters residues* (Moscow: Ore and Metals)
[11] Rubanov Yu, Tokach Yu and Ognev M 2009 Galvanic processes sludges and wastewaters treatment with non-ferrous metals extraction *Modern high technologies* **3** 82–83
[12] Rubanov Y and Tokach Y 2011 *Method of Galvanic Process Sludges Treatment* RU Patent **2422543 C1**
[13] Mikhailov B 2000 Nickel ores in the Urals *Lithology and Mineral Resources* **35** 351–364
[14] Lazarenkov V, Talovina I, Vorontsova N, Mezentseva O and Ryzhkova S 2011 Nickel chlorites in the oxide-silicate nickel deposits of the Urals *Lithology and Mineral Resources* **46** 312–320
[15] Selivanov E, Sergeeva S and Gulyaeva R 2018 Urals oxidized nickel ores: composition, thermal properties, treatment technologies *The role of technological mineralogy in rational subsoil use*
[16] Sadykhov G, Anisonyan K, Kop'ev D, Olyunina T, Zablotskaya Yu, Gonsharov K and Khasanov M 2018 New integrated methods of oxidized nickel ores usage *Institute of Metallurgy and Materials Science named after A.A. Baikov RAS 80 years*
[17] Kyle J 2010 Nickel laterite processing technologies – where to next? *ALTA 2010 Nickel/Cobalt/Copper Conference*
[18] Leont'ev L, Zhuchkov V, Zhdanov A and Dashevskii V 2015 Ferroalloy production in Russia *Steel in Translation* **10** 773–777
[19] Mudd G 2010 Global trends and environmental issues in nickel mining: Sulfides versus laterites *Ore Geol. Rev.* **38** 6–9
[20] Senchenko A, Rybkin S and Aksenov A 2016 *Method of stubborn uranium, pyrite and noble metals materials treatment for uranium extraction and production of noble metals concentrates* RU Patent **2398903 C1**

IOP Conf. Series: Materials Science and Engineering **969** (2020) 012054 doi:10.1088/1757-899X/969/1/012054

Effect of Organic Nutrients and Elevated Temperature on Bioleaching of Sulfide Copper Concentrate

A G Bulaev and V S Melamud

Winogradsky Institute of Microbiology, Research Center of Biotechnology of the Russian Academy of Sciences, 33, Leninsky prospect, Moscow, 119071, Russia

E-mail: bulaev.inmi@yandex.ru

Abstract. The goal of the work was to study the effect of addition of organic carbon source required for microorganisms oxidizing sulfide minerals in the nutrient medium on the rate of copper concentrate bioleaching. The concentrate containing 31.6% Fe, 19.2% Cu, and 2.9% Zn was bioleached in a batch mode for 30 days using mixed culture of acidophilic microorganisms. In the first reactor, biooxidation was performed at 45°C without additional carbon sources. In the second and in the third reactors, biooxidation was performed at 45°C and 55°C and nutrient medium was supplemented with 0.02% of yeast extract. Biooxidation without yeast extract allowed extracting 47 and 92% of copper and zinc. During biooxidation at 45°C using the medium supplemented with yeast extract, 54 and 94% of copper and zinc were extracted. During bioleaching at 55°C, 48% and 84% of copper and zinc were extracted. Thus, yeast extract allowed increasing copper bioleaching rate at 45°C, but temperature increase up to 55°C led to decrease in bioleaching efficiency. It may be explained by the fact that yeast extract increased activity of acidophilic microorganisms, which may consume organic carbon sources, while high temperature may inhibit some microorganisms and, therefore, does not allow increasing bioleaching rate.

1. Introduction

Biohydrometallurgical treatment (bioleaching) of sulfide copper ores with acidophilic microorganisms has been successfully used for copper extraction worldwide and currently provides about 5% of world copper production [1–3]. Main reason to use bioleaching for copper extraction is possibility to treat copper ores, which are inappropriate for pyrometallurgical processing due to the low copper content, specificity of mineral composition, which impedes obtaining high-grade copper concentrates, and high content of different impurities, which decrease copper extraction rate and quality of the copper produced [1, 4].

As biohydrometallurgical technologies are mainly process low-grade copper ores, heap bioleaching is usually used in copper industry due to the low capital and operating expenses [4–6]. The role of heap bioleaching in copper production may increase in the future due to the decline in average copper content in ores and decrease in the resources of conventional copper ores [4–6].

Thus, the main advantages of heap bioleaching are low requirements for the quality of raw materials as well as low costs of metal production [3, 7]. In the same time, heap bioleaching provide low rate of metal extraction and does not provide effective copper extraction from some refractory minerals, for example, chalcopyrite ($CuFeS_2$), the most abundant copper sulfide mineral [2, 8, 9].

Therefore, the development and commercialization of alternative approaches is required to increase the rate and efficiency of biohydrometallurgical treatment of copper ores. Tank bioleaching has been successfully applied for processing sulfide concentrates and provides higher rate of metal extraction in comparison to heap bioleaching. In the same time, this technology is mainly used for treatment of gold-bearing sulfide concentrates due to relatively high costs related to use complex equipment for stirring, temperature maintaining, and aeration [10–12]. It was shown that tank bioleaching may be successfully used for processing cobalt, nickel, and low grade copper-zinc concentrates [1, 5, 12–14]. As copper minerals, including chalcopyrite, are often refractory to biooxidation with acidophilic microorganisms, different methods to increase copper bioleaching rate in tank bioleaching have been proposed. As parameters of tank bioleaching processes in contrast to heap bioleaching may be regulated, different approaches to increase their efficiency may be used. For example, it was shown that copper extraction rate in tank bioleaching processes may be increased by temperature increase or using high chloride concentrations [15, 16].

In our previous work we studied possibility to increase the efficiency of tank biooxidation of sulfide gold-bearing concentrate containing pyrite and arsenopyrite by addition of organic carbon source (yeast extract) required for the constructive metabolism of acidophilic microorganisms oxidizing sulfide minerals at different temperatures [17]. As many of microorganisms involved in biooxidation of sulfide concentrates gaining energy by oxidizing ferrous iron and sulfur compounds may also consume organic nutrients as carbon sources [1, 18, 19], addition of organic compounds in nutrient medium used for bioleaching made it possible to increase both pyrite and arsenopyrite biooxidation rate, as well as gold extraction rate [17]. Thus, it was shown that use of organic nutrients may be a promising approach to increase the efficiency of tank bioleaching of pyrite-arsenopyrite concentrate; therefore, it also may be used to increase the efficiency of chalcopyrite concentrate bioleaching as it may enhance the activity of microorganisms performing bioleaching.

The goal of the present work was to study the possibility to increase the rate of copper concentrate bioleaching by addition of organic carbon source required for acidophilic microorganisms oxidizing sulfide minerals and temperature regulation.

The experiments were performed using mixed culture of moderately thermophilic microorganisms, as moderate thermophiles are usually predominate in industrial bioleaching processes [12, 13, 20]. In the same time, it should be noted that moderately thermophilic acidophiles used in biohydrometallurgical processes are usually able to utilize organic compounds as carbon source and addition of organic nutrients are usually added in the media during their laboratory cultivation [21]. In mixed microbial populations performing bioleaching in the reactors, organic compounds required for some microorganisms are synthesized by autotrophic members of microbial populations that makes it possible to maintain activity of mixo- and heterotrophic acidophiles. Nevertheless, use of additional organic carbon sources may provide higher oxidizing activity of microorganisms [17–19].

2. Materials and methods

Tables 1 and 2 show chemical and mineral composition of the concentrate, which was used in the study.

Table 1. Chemical composition of the concentrate.

Content, %						
Fe	Cu	Zn	S_{total}	$S_{sulfide}$	$S_{sulfate}$	S^0
31.6	19.2	2.9	34	32.3	1.7	< 0.01

Table 2. Mineral composition of the concentrate (based on XRD data).

Content, %				
Pyrite (FeS_2)	Pyrrhotite (FeS)	Chalcopyrite ($CuFeS_2$)	Sphalerite (ZnS)	Quartz (SiO_2)
14	14	67	<5	<5

Biooxidation was performed in a batch mode in stirred tank reactors that contained 1 L of pulp. Mineral nutrient medium used for the experiments with initial pH of 1.5 containing (g/L) $(NH_4)_2SO_4$ – 0.75; KCl – 0.05; $MgSO_4 \times 7H_2O$ – 0.125; K_2HPO_4 – 0.125. The experiments were conducted at aeration of 4 min^{-1} and a stirring speed of 500 rpm. The duration of the experiments was 30 days. Pulp density was 20% (200 g of the concentrate and 1 L of mineral nutrient medium). Mixed culture of moderately thermophilic acidophiles, properties of which are shown in table 3, was used for the bioleaching. The initial number of each strain in the liquid phase of the pulp was about 1×10^8 cells/mL. In the first reactor (control), biooxidation was performed at 45°C without organic carbon sources (i.e. under autotrophic conditions, where carbon dioxide contained in air, was the sole carbon source). In the second and in the third reactors, nutrient medium was supplemented with 0.02% of the yeast extract (YE), which is used for the cultivation of mixo- and heterotrophic acidophiles as organic carbon source, in the beginning of the experiments as well as on the 10th and 20th days, and bioleaching was performed at 45°C and 55°C, respectively.

Table 3. Properties of the strains used for the bioleaching.

Microorganism	Oxidation		Optimum temperature / upper limit, °C	Carbon source
	Fe^{2+}	S^0		
Acidithiobacillus caldus MBC-1	–	+	45/53	CO_2
Sulfobacillus thermosulfidooxidans VKMV 1269^T	+	+	50/60	CO_2, organic compounds
Acidiplasma sp. MBA-1	+	+	50-55/63	Organic compounds

The pH, Eh of the pulp as well as metal (Fe^{3+}, Fe^{2+}, Cu, Zn) concentrations were monitored. The pH of the pulp was adjusted using $CaCO_3$ or H_2SO_4 to maintain the pH value in the range from 1.0 to 1.7. The concentrations of Fe^{2+} and Fe^{3+} ions were determined using the rhodanide method based on formation of a purple coordination complex of ferric iron ions Fe^{3+} and rhodanide SCN^- spectrophotometrically (at 475 nm). The concentration of copper and zinc were determined using a Perkin Elmer 3100 flame atomic absorption spectrometer (Perkin Elmer, USA). The rates of copper and zinc leaching from concentrate were calculated by the concentration of Cu and Zn ions in liquid phase. Sulfur content in the solid residues was measured using the phase method.

3. Results and discussion

Table 4 and figures 1 and 2 show the results of the experiments. It was shown that under different conditions, some differences in liquid phase parameters and, therefore, in metal extraction rates were observed. Despite this, it should be noted that differences in bioleaching conditions affected mainly iron leaching, while copper and zinc bioleaching depended on experimental conditions to a lesser extent.

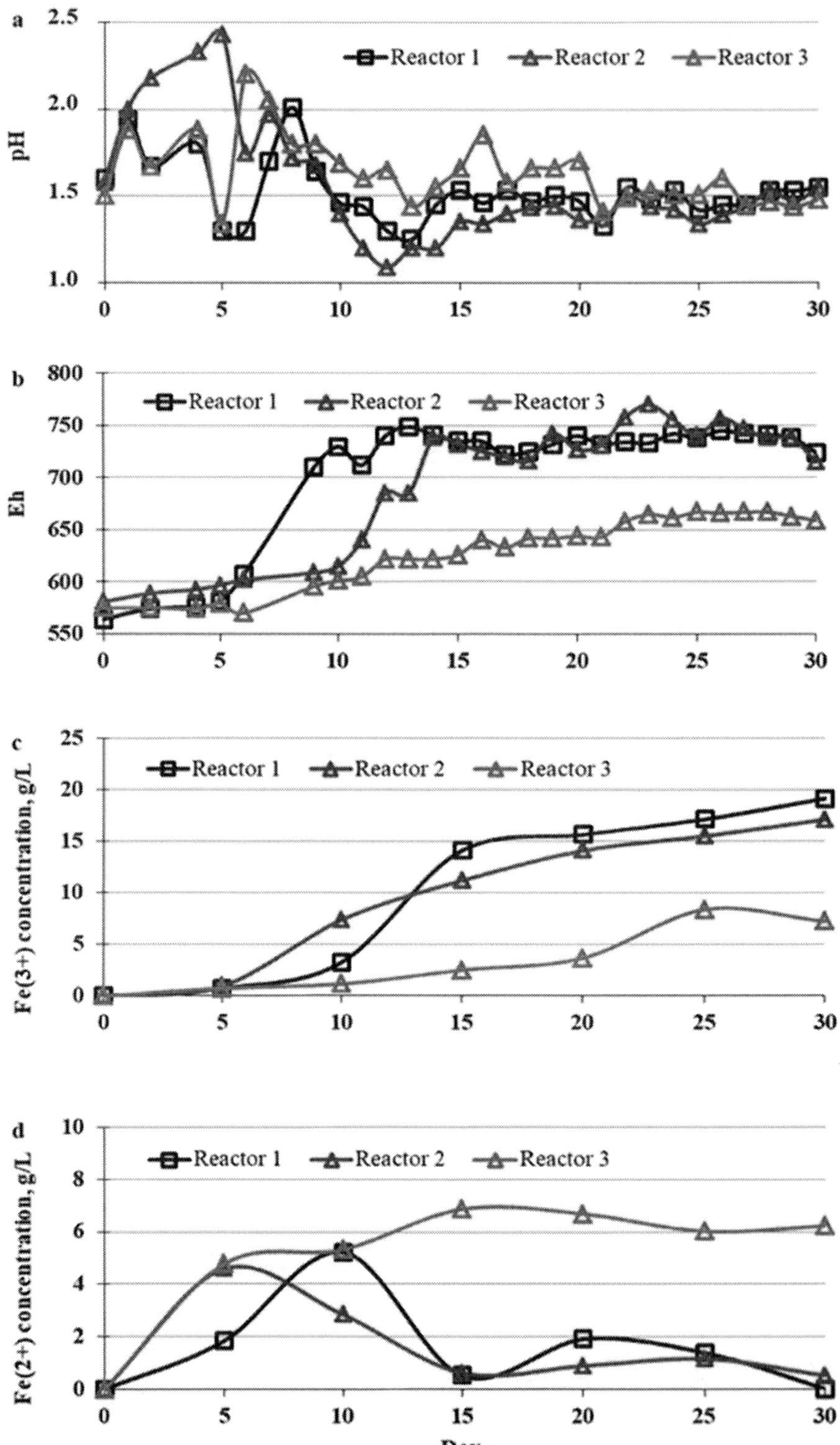

Figure 1. Changes in liquid phase parameters during the bioleaching: a – pH, b – Eh, c – Fe^{3+} concentration, d – Fe^{2+} concentration.

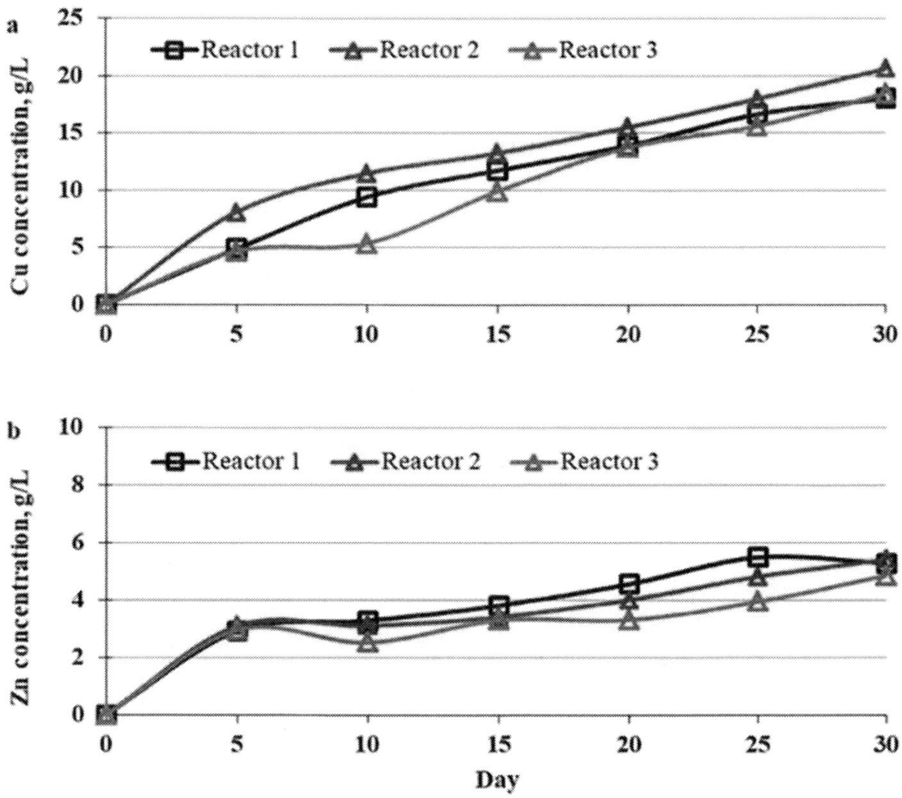

Figure 2. Changes in non-ferrous metals concentrations during the bioleaching: a – Cu, b – Zn.

Table 4. Results of bioleaching.

Conditions			Cu extraction, %	Zn extraction, %	$S_{sulfide}$ oxidation, %	H_2SO_4 consumption, kg/t	$CaCO_3$ consumption, kg/t
Reactor	T, °C	YE addition					
1	45	-	47	92	49	5.2	67.5
2	45	+	53	94	55	10.3	112.5
3	55	+	48	84	23	18.7	5

During bioleaching of the concentrate, iron, copper, and zinc concentrations in the liquid phase increased due to the biooxidation of sulfide minerals, while pH decreases due to the production of sulfuric acid due to the oxidation of sulfur compounds. The Eh value affected mainly by the ratio of Fe^{3+}/Fe^{2+} ions reflects iron-oxidizing activity during the bioleaching of sulfide concentrates. As sulfuric acid and calcium carbonate are used to adjust the pH level, their consumption also reflects biooxidation activity. Thus, these parameters are used to evaluate microbial activity during sulfide concentrate bioleaching.

During the bioleaching, pH values fluctuated due to the biooxidation of sulfur compounds as well as due to sulfuric acid and calcium carbonate addition. To maintain pH level favorable for acidophilic microorganisms, sulfuric acid was added to the pulp of reactors 1 and 2 from the beginning of the experiment to the day 10. Then, the pH levels were low and maintained by calcium carbonate addition (figure 1). In the same time, in the reactor 3, sulfuric acid was added to the pulp up to the 20th day. Nevertheless, the pH was not stable and periodically increased (figure 1a). Sulfuric acid and calcium carbonate consumption (table 4) demonstrates clear difference between the reactors. Both sulfuric acid and calcium carbonate consumptions were the highest in the reactor 2. While in the reactor 3, calcium carbonate consumption was lower by an order of magnitude than in the reactors 1 and 3, while sulfuric acid consumption was several times higher. Thus, it may be concluded that biooxidation of sulfur compounds that leads to decrease in the pH in the reactor 2 was more intensive than in the reactors 1 and 3, while in the reactor 3, biooxidation intensity was probably low. In the same time, higher sulfuric acid consumption in the reactor 2 in comparison to reactor 1 may be explained by more active iron biooxidation in the reactor 2 (figures 1 c and d) that results in H^+ consumption and increase in pH:

$$4Fe^{2+} + O_2 + 4H^+ \rightarrow 4Fe^{3+} + 2H_2O. \quad (1)$$

Eh values in the reactors increased during the bioleaching and its changes reflected the changes in Fe^{3+}/Fe^{2+} ratio (figures 1b, c, and d). Thus, in the reactor 3, Eh was comparatively low (figure 1 b) as Fe^{3+} concentration was lower and Fe^{2+} concentration was higher than in the reactors 1 and 2 (figures 1 c and d). In the reactors 1 and 2, Eh values were significantly higher and exceeded 700 mV after 9 and 14 days of the bioleaching. Ferrous iron oxidation rates in the reactors 1 and 2 differed, but were comparable (figures 1 c and d).

Sulfide sulfur oxidation levels (table 4) confirmed the results of liquid phase parameters, as $S_{sulfide}$ oxidation level was the highest in the reactor 2 and the lowest in the reactor 3.

Thus, differences in bioleaching conditions led to the differences in biooxidation rate. However, copper and zinc extraction rates depended on experimental conditions to a lesser extent (figure 2 and table 4). In the reactor 3, zinc extraction level was lower in comparison to the reactors 1 and 2, while copper extraction levels in the reactors 1 and 3 were almost equal. In the reactor 2, copper extraction was the highest, but the difference with the reactors 1 and 3 was about 5%.

Thus, the addition of organic nutrients into the medium made it possible to increase the intensity of sulfide concentration bioleaching and rate of sulfide sulfur biooxidation. In the same time, the increase in copper and zinc extraction levels was not significant.

4. Conclusions

Thus, organic carbon source made it possible to increase copper bioleaching rate at 45°C, but the increase in temperature up to 55°C led to decrease in bioleaching efficiency. It may be explained by the fact that addition of yeast extract led to the increase in activity of acidophilic microorganisms, which may consume both carbon dioxide and organic substances as carbon source, but high temperature may inhibit some microorganisms in microbial populations and, therefore, does not allow to increase bioleaching rates by increasing the temperature. Despite this, it should be noted that differences in bioleaching conditions affected mainly iron leaching, while copper and zinc bioleaching depended on experimental conditions to a lesser extent. These results may be explained by the fact that sphalerite (ZnS) is comparatively easily leached [22] and, therefore, may be actively bioleached regardless the conditions. In the same time, chalcopyrite bioleaching is determined by different factors, including temperature and Eh, but not only by the high activity of iron- and sulfur-oxidizing activity [2].

In our previous work, we studied the biooxidation of gold-bearing pyrite-arsenopyrite concentrate under the same conditions [17]. It was shown, that organic nutrients addition significantly affected both pyrite and arsenopyrite biooxidation and made it possible to increase gold extraction rate. Thus, the comparison of the results obtained in the present work with those obtained under the same conditions with other sulfide concentrate demonstrated that peculiarities of mineral biooxidation mechanism have a crucial role in the efficiency of different approaches, which may be proposed to improve bioleaching processes.

References

[1] Johnson D B 2014 Biomining - biotechnologies for extracting and recovering metals from ores and waste materials *Cur. Opinion in Biotechnology* **30** 24–31
[2] Watling H R 2006 The bioleaching of sulphide minerals with emphasis on copper sulphides e a review *Hydromet* **84** 81−108

[3] Gentina J C and Acevedo F 2013 Application of bioleaching to copper mining in Chile *Electronic Journal of Biotechnology* **16(3)**
[4] The World Copper Factbook 2017 Online. Available: http://www.icsg.org
[5] Gericke M, Neale J W and van Staden P C 2009 A Mintek perspective of the past 25 years in minerals bioleaching *J. South. Afr. Inst. Mining Metallurgy* **109** 567–585
[6] Northey S, Mohr S, Mudd G, Weng Z and Giurco D 2014 Modelling future copper ore grade decline based on a detailed assessment of copper resources and mining *Resour. Conserv. Recycl.* **83** 190–201
[7] Yin S, Wang L, Kabwe E, Chen X, Yan R, An K, Zhang L and Wu A 2018 Copper bioleaching in China: review and prospect *Minerals* **8** 32
[8] Dreisinger D 2006 Copper leaching from primary sulfides: Options for biological and chemical extraction of copper *Hydromet* **83** 10–20
[9] Schlesinger M E, King M J, Sole K C and Davenport W G 2011 Extractive metallurgy of copper (London: Elsevier) p 481
[10] Sovmen V K, Belyi A V, Danneker M Y, Gish A A, Teleutov A N 2009 Biooxidation of refractory gold sulfide concentrate of Olympiada deposit *Adv. Mater. Res* **71–73** 477–480
[11] Van Aswegen B D, van Niekerk J, Olivier W 2007 *Biomining eds D E Rawlings and B D Johnson* (Berlin: Springer)
[12] Mahmoud A, Cezac P, Hoadley A F A, Contaminea F, and D'Hugues P 2017 A review of sulfide minerals microbially assisted leaching in stirredtank reactors *Int. Biodeterioration & Biodegradation* **119** 118–146
[13] Morin D H R and d'Hugues P 2007 *Biomining* (Berlin: Springer)
[14] Neale J, Seppala J, Laukka A, van Aswegen P, Barnett S and Gericke M 2017 The MONDO minerals nickel sulfide bioleach project: from test work to early plant operation *Solid State Phenomena* **262** 28–32
[15] Batty J D and Rorke G V 2006 Development and commercial demonstration of the BioCOP™ thermophile process *Hydromet* **83** 83–89
[16] Bevilaqua D, Lahti H, Suegama P H, Garcia O, Benedetti A V, Puhakka J A and Tuovinen O H 2013 Effect of Na-chloride on the bioleaching of a chalcopyriteconcentrate in shake flasks and stirred tank bioreactors *Hydromet* **138** 1–13
[17] Bulaev A, Valiev Kh, Bugubayeva A, Amandikova A and Melamud V 2019 Effect of organic nutrients on biooxidation of gold-bearing refractory concentrate *SGEM 2019 Conference proceedings* **19 (1.3)** 793–798
[18] Li Q, Tian Y, Fu X, Yin H, Zhou Z, Liang Y, Qiu G, Liu J, Liu H, Liang Y, Shen L, Cong J and Liu X 2011 The community dynamics of major bioleaching microorganisms during chalcopyrite leaching under the effect of organics *Curr. Microbiol.* 63 164–72
[19] Muravyov M I and Bulaev A G 2013 Two-step oxidation of a refractory gold-bearing sulfidic concentrate and the effect of organic nutrients on its biooxidation *Min. Eng.* **45** 108–14
[20] van Hille R P, van Wyk N, Froneman T and Harrison S T L 2013 Dynamic evolution of the microbial community in BIOX leaching tanks *Adv. Mater. Res.* **825** 331–34
[21] Johnson D B and Hallberg K B 2009 Carbon, iron and sulfur metabolism in acidophilic microorganisms *Adv Microb Physiol* **54** 201–255
[22] Jyothi N, Sudha K N and Natarajan K A 1989 Electrochemical aspects of selective bioleaching of sphalerite and chalcopyrite from mixed sulphides *Int. J. miner Process.* **27** 189–203

Acknowledgments
The work was performed within the framework of the government task of Research Center of Biotechnology of the Russian Academy of Sciences and supported by The Ministry of Science and Higher Education of the Russian Federation.

IOP Conf. Series: Materials Science and Engineering **969** (2020) 012055 doi:10.1088/1757-899X/969/1/012055

Hydrometallurgical Technology for Processing of Galvanic Sludges

O Yu Makoskaya, K S Kostromin and N I Bryantseva

Institute of New Materials and Technologies, Ural Federal University named after the first President of Russia B. N. Yeltsin, 19, Mira Street, Ekaterinburg, 620002, Russia

E-mail: o.i.makovskaia@urfu.ru

Abstract. The problem of processing galvanic sludges, formed as a result of neutralization of technological solutions and wastewater containing heavy non-ferrous metals is considered. At present, sludges are transported to disposal area and are not used in any way. Typically, such sludges contain significant amounts of chromium and nickel, which creates environmental hazard. Investigated sludge contains up to 6.6% Ni and up to 7.4% Cr. The hydrometallurgical technology for processing of these sludges has been developed. Solutions of sulfuric acid and ammonia were used as lixiviants. It is shown that when using a solution of sulfuric acid with pH=1.5, extraction of up to 93.3% Cu, 70.2 Ni, 90.3 Zn is achieved. For selective recovery of nickel from leachates the process of sorption concentration by Lewatit TP207 is proposed.

1. Introduction

Galvanic workshops produce large amounts of waste waters. Solid waste, called galvanic sludge is formed as result of the physicochemical treatment of that waste water. Galvanic sludge is considered to be one of the most highly toxic types of waste and stored mainly in sludge pits. Such practice has not any economic or environmental advantages [1, 2].

In order to reduce the environmental hazard of galvanic production wastes deposited in the sludge, methods of chemical fixation by means of ferritization of solid phase of wastes, silicification, solidification of wastes with the use of inorganic and organic binders, as well as sintering are used [3–7]. One of the ways of processing of galvanic sludges is their application as filling agents and plasticizers in concretes, asphalt concretes and building mortars [8–10]. It is possible to obtain pigments from sludge components [11]. The work [12] describes production of nickel hydroxide suitable for use as an active component of cathode mass for manufacture of $Ni(OH)_2$ electrode of a nickel-cadmium (iron) battery.

At the same time, the sludge can be considered as a significant source of heavy non-ferrous metals and chromium. Content of copper, chromium, nickel or zinc can reach 30 % of dry weight Development of alternative and effective ways reducing the impact on the environment and providing extraction of valuable metals contained in these sludges are of paramount importance.

Considerable variations in sludge composition create additional difficulties for the development of a comprehensive sludge treatment scheme. The most expedient way to process such complex raw material is hydrometallurgical, providing selective extraction of non-ferrous metals.

Mineral acids, ammonia, chelators are used as a solvent. There is information about the use of NaOH as a solvent [13]. In [14] a combined method of recycling of galvanic sludges was proposed, which includes galvanic sludge mixing with additives containing chlorine ions in the ratio Cl^{-} : Me^{n+}= 1.1:1, mechanochemical activation of mixture obtained by dry grinding in the ball mill to a size of 0.5-5 microns, subsequent heat treatment in a muffle furnace at the temperature of 450-500 °C, the resulting cake leaching by waste water own galvanic production at pH = 3.

Further processing of the obtained solutions is also possible by several methods. The most widely used liquid extraction [15, 16] and ion exchange [17]. These methods make it possible to selectively isolate valuable components from solution.

In any case, the choice of processing technology is carried out after studying the chemical and phase composition of each individual sludge sample.

2. Experimental

Sludge produced by nickel-plating/chrome-plating units has been studied. Sludge samples were taken from different points of the sludge storage facility and have noticeable variations in the composition (table 1), so the samples taken for the study were averaged. The phase composition of the sludge is represented by the following compounds: $Ni(OH)_2$, $Zn(OH)_2$, $Cu(OH)_2$, $Ca(OH)_2$, Ca_2SiO_4, SiO_2, $Cr(OH)_3$. The sludge is mixed with sand due to the conditions of its placement in the sludge storage.

Table 1. Chemical composition of sludge.

	Content (%)						
Sample	Ni	Zn	Cu	Pb	Fe	Ca	Cr
1	3.97	1.26	0.82	0.035	0.58	0.01	4.71
2	6.98	1.43	1.14	0.11	0.64	1.71	8.02
3	3.88	1.73	1.03	0.25	0.81	0.84	2.11
4	3.46	1.65	0.93	0.21	0.74	2.19	1.81
5	7.06	1.94	1.55	0.23	0.98	1.21	7.11
Average sample	6.64	1.33	1.06	0.13	0.58	1.26	7.41

Ammonia leaching of the initial sludge was carried out at pH 8–9, temperature 25 °C, sludge mass 20 g, the volume of the solution 150 cm^3, the duration of the process 1.5 hours. The results of the experiment are shown in figure 1.

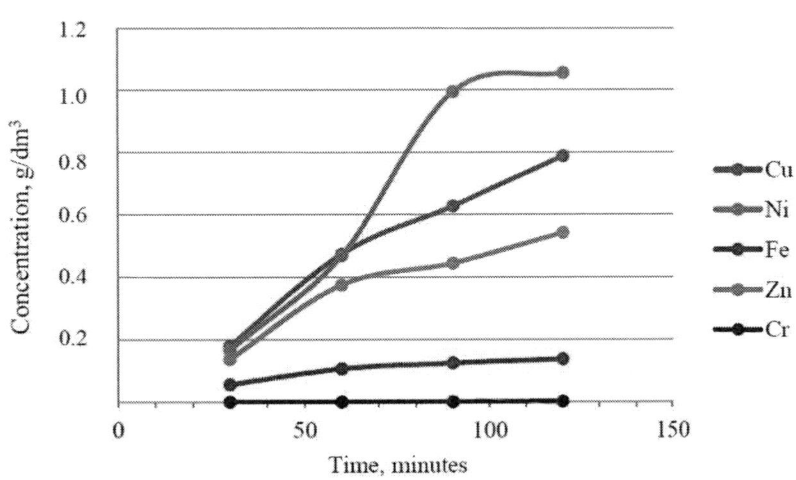

Figure 1. Kinetic of sludge leaching by NH_4OH.

IOP Conf. Series: Materials Science and Engineering **969** (2020) 012055 doi:10.1088/1757-899X/969/1/012055

As a result of ammonia leaching within 90 minutes, solutions containing about 1 g/dm^3 of nickel and 0.628 g/dm^3 of copper were obtained. The rate of nickel transfer to the solution reduces at further leaching. The advantage of ammonia leaching is minimal transition of iron to solution and complete absence of chromium in leachate. However, recovery of nickel in the solution does not exceed 50 % of its initial content in the sludge.

Sulphuric acid has proved to be one of the best reagents from previously tested solvents [18, 19]. Therefore, subsequent leaching experiments were carried out using sulfuric acid.

Sulphuric acid leaching of the initial sludge was carried out at pH 1.5; temperature 25 °C, sludge mass 10 g, solution volume 100 cm^3, pH was regulated by the droplet adding of H_2SO_4 (1 M). Dependence of metals concentration in solution and extent of metals extraction in solution from duration of leaching is presented in figure 2.

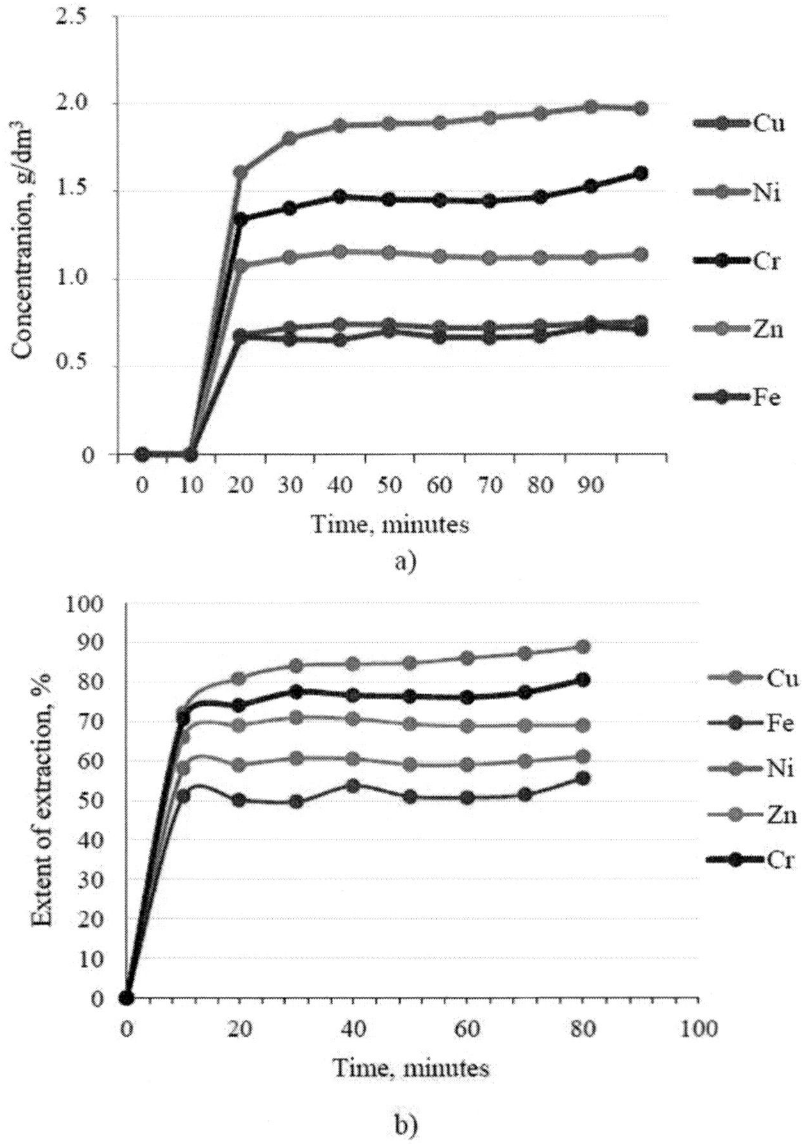

Figure 2. Dependence of metals concentration in solution (a) and extent of metals extraction in solution (b) from duration of leaching.

Nickel recovery reaches 90 % at leaching by solution of sulfuric acid during 90 minutes. The main disadvantage of sulfuric acid leaching is the lack of selectivity. Up to 80 % of chromium and up to 50 % of the iron contained in the sludge passes into the solution. It is necessary to solve the problem of separation of these elements at further processing of solutions.

Sorption on ion exchange resins was used to extract nickel from sulphuric acid leaching solutions. It is advisable to use a sorbent that preferable absorbs nickel at its joint presence with chromium in the solution. Based on published data, the following sorbents were selected for the tests: KU-2×8, Lewatit TP207, Purolite C104, Purolite S930.

To determine the static exchange capacity, all the ion exchangers were converted to the H^+-form by washing with 1H H_2SO_4. A sample of ion exchanger (0.5 g) was brought into contact with 50 cm^3 of nickel (II) solution (1 g/dm^3). To determine the capacity for chromium (III), a solution with concentration of 0.63 g/dm^3 was used. A static exchange capacity was calculated for each ion exchanger; the results are shown in table 2.

Table 2. Ionites static exchange capacity for nickel and chromium.

Ionite	SEC for Ni(II) (g/g)	SEC for Cr(III) (g/g)
Purolite S930	0.039245	0.0002
Lewatit TP-207	0.040525	0.0001
KU 2×8	0.043695	0.0266
Purolite C104	0.03046	0.0077

Based on the obtained data, Lewatit TP-207 was selected as having the best indices for metal selection. Ionite Lewatit TP-207 is a weakly basic macroporous cation exchange resin with chelate groups of iminodiacetic acid. Ionite is intended for the selective removal of heavy non-ferrous metal ions from weakly acidic and slightly alkaline solutions. This ion exchanger showed a high sorption capacity with respect to nickel, but insignificant sorption of chromium when they collectively presence in solution.

The effectiveness of the sorption process strongly depends on the pH value; therefore, an experiment to study the effect of pH on the nickel sorption process was conducted. At pH=2, the concentration of H^+ ions in the functional groups is high and pushes metal cations into the solution due to the high degree of protonation of the functional groups, i.e. competition between H^+ ions and metal cations is observed. At pH 2-4, H^+ ions and metal cations still completely fill the functional groups, but the latter begin to be deprotonated. At pH 7, the carboxylic acid residue of the functional group is deprotonated, and at pH 12 the iminodiacetate group is completely deprotonated (figure 3).

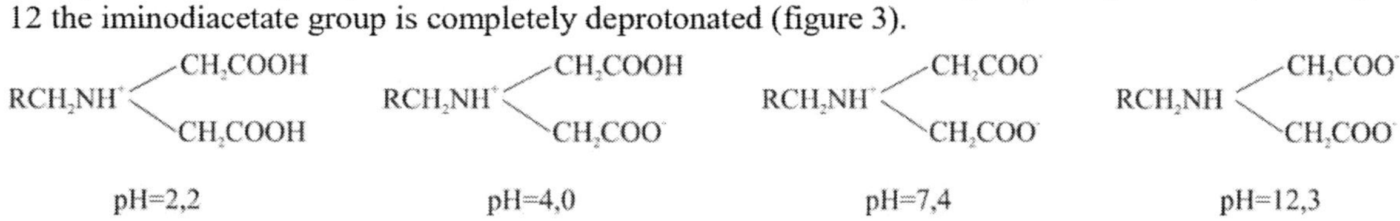

Figure 3. Influence of pH on iminodiacetate functional groups.

Although the latter case is more favorable for the sorption of cations due to the absence of hydrogen ions in the functional groups, maximum of metal cations sorption occurs at a pH value about 5.

To study the effect of pH on the sorption of nickel (II), model solutions of nickel with different pH were prepared. Samples of ion exchanger weighing 1 g in H^+-form were brought into contact with the solution. Duration of sorption was 24 hours; the results are shown in figure 4.

Obtained data shows that ion-exchange resin Lewatit TP207 has the maximum capacity for Ni (II) at pH=4-4.5. Thus, during the leaching operation, obtaining of solutions with a pH close to this value is preferable or the pH of the leachate should be adjusted.

Sorption of nickel from leach solution was carried out in a column with a diameter of 1 cm and a length of 50 cm. Lewatit TP-207 (25 cm^3) was loaded into the sorption column. The solution was passed through a sorption column at a constant rate of 4 specific volumes per hour and 25 cm^3 samples were taken.

During nickel sorption in the column, competitive sorption of calcium ions was detected. The concentration of calcium ions in the leach solution is comparable to the concentration of nickel ions. When the solution was passed through the column, nickel was first sorbed, the ion exchanger turned blue. The colored front moved along the height of the column. Further nickel displacement was observed. Analysis of the filtrates showed the absence of calcium ions in them. With an increase in the total length of the ion exchanger layer by a factor of 10, displacement of sorbed calcium ions and repeated sorption of nickel ions were observed. A similar sorption behavior of calcium is described in [20].

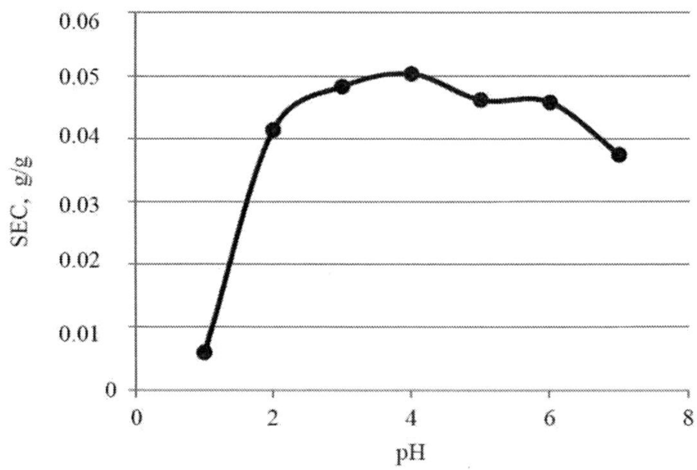

Figure 4. Dependence of SEC for from pH value.

3. Conclusions

Thus, for the processing of nickel-chromium galvanic sludges hydrometallurgical technology can be used. The proposed scheme includes steps of sludge leaching with sulfuric acid solutions, pH correction and selective sorption of nickel by synthetic ion-exchange resin Lewatit TP207. Competitive sorption of calcium ions inhibiting the sorption of nickel ions was detected. To reduce this effect, it is recommended to increase the total height of the ion exchanger layer.

References

[1] Twidwell L G and Dahnke D R 2001 Treatment of metal finishing sludge for detoxification and metal value *The European Journal of Mineral Processing and Environmental Protection* **1(2)** 76–88

[2] Ødegaard H, Paulsrud B and Karlsson I 2002 Wastewater sludge as a resource: Sludge disposal strategies and corresponding treatment technologies aimed at sustainable handling of wastewater sludge *Water Science & Technology* **46(10)** 295–303

[3] Pérez-Villarejo L, Martínez-Martínez S, Carrasco-Hurtado B et al 2015 Valorization and inertization of galvanic sludge waste in clay bricks *Applied Clay Science* **105–106** 89–99

[4] Stanisavljevic M, Krstic I and Zec S 2010 Eco-technological process of glass-ceramic production from galvanic sludge and aluminium slag *Science of Sintering* **42(1)** 125–130

[5] Ozdemir O D, Figen A K and Piskin S 2011 Utilization of galvanic sludge as raw material for production of glass *Proc. of the Int. Conf. on Chemistry and Chemical Process* **10** 45–49
[6] Bocanegra J J C, Mora E E and González G I C 2019 Galvanic sludges: Effectiveness of red clay ceramics in the retention of heavy metals and effects on their technical properties *Environmental Technology & Innovation* **16** 100459
[7] Orescanin V, Mikulic N, Mikelic I L, Posedi M, Kampic S and Medunic G 2009 The bulk composition and leaching properties of electroplating sludge prior/following the solidification/stabilization by calcium oxide *Journal of Environmental Science and Health* **A 44** 1–7
[8] Stepanov S, Morozov N and Morozova N 2016 Efficiency of use of galvanic sludge in cement systems *Procedia Engineering* **165** 1112–1117
[9] Bednarik V, Vondruska M and Koutny M 2005 Stabilization/solidification of galvanic sludges by asphalt emulsions *Journal of Hazardous Materials* **122(1–2)** 139–145
[10] Arsenovic M, Radojevic Z and Stankovic S 2012 Removal of toxic metals from industrial sludge by fixing in brick structure *Construction and Building Materials* **37** 7–14
[11] Andreola F, Barbieri L, Bondioli F, Cannio M, Ferrari A M and Lancellotti I 2008 Synthesis of chromium containing pigments from chromium galvanic sludges *Journal of Hazardous Materials* **156** 466–471
[12] Olshanskaya L, Danilova E and Lazareva E 2018 Extraction of Nickel Hydroxide from Galvanic Waste Production for the Manufacture of the Nickel-Cadmium (Iron) Battery *Ecology and Industry of Russia* **22(6)** 4–9
[13] Świerk K, Bielicka A, Bojanowska I and Maćkiewicz Z 2007 Investigation of Heavy Metals Leaching from Industrial Wastewater Sludge *Polish J. of Environ. Stud.* **16(3)** 447–451
[14] Tokach Y E, Rubanov Y K Pivovarova N A and Balyatinskaya L N 2013 Galvanic Sludge Recycling with the Extraction of Valuable Components *Middle-East J. Sci. Res.* **18(11)** 1646–1655
[15] Silva J E, Paiva A P, Soares D, Labrincha A and Castro F 2005 Solvent extraction applied to the recovery of heavy metals from galvanic sludge *Journal of Hazardous Materials* **B120** 113–118
[16] Shi T, Zou S, Liu J, Zuo L and Lv C 2011 Separation and Recovery of Nickel and Copper from Multi-Metal Electroplating Sludge by Co-Extracting and Selective Stripping *J Civil Environment Eng* **1(1)**
[17] Evaristo A A A, Santos K C R, Costa L C and da Marques M R 2013 Evaluation of ion exchange resins for recovery of metals from electroplating sludge *Polymer Bulletin* **70(8)** 2239–2255
[18] Makovskaya O Yu, Kostromin K S 2019 Leaching of non-ferrous metals from galvanic sludges *Materials Science Forum* **946** 591–595
[19] Silva J E, Soares D, Paiva A P et al 2005 Leaching behaviour of a galvanic sludge in sulphuric acid and ammoniacal media *Journal of Hazardous Materials* **B 121** 195–202
[20] Timofeev K L, Naboychenko S S, Lebed A B and Akulich L F 2013 Kinetics of zinc, calcium and magnesium sorption from water solutions, using the aminediacetic ionite Lewatit TP-207 *Non-ferrous metals* **8** 59–64

IOP Conf. Series: Materials Science and Engineering **969** (2020) 012056 doi:10.1088/1757-899X/969/1/012056

Obtaining of Pigment-Quality Magnetite from Sintering Process Red Mud

A A Shoppert, I V Loginova and J A Napol'skikh

Metallurgy of non-ferrous metals, Ural Federal University, 17, Mira St., Yekaterinburg, 620002, Russia

E-mail: a.a.shoppert@urfu.ru

Abstract. The limited availability of high-quality bauxite and, at the same time, the formation of a considerable amount of harmful red mud (bauxite residue) causes to develop cost-effective technologies for bauxite processing. Selective separation of iron from red mud allows solving this problem and producing marketable products in the form of pigment-quality magnetite. In this context, the possibility of the pigment-quality magnetite obtaining by the interaction of freshly precipitated iron hydroxide (III) of the sintering process red mud with ferrous iron ions in an alkaline medium is studied. It was found that the optimal parameters of the process are: temperature = 90 °C; duration = 4 hours; a ratio of ferrous iron to stoichiometric quantity = 70 %. The yield of the magnetic concentrate under optimal conditions was 69.3 %, and the content of magnetite in concentrate was 47.7 %.

1. Introduction

The most important problem of the alumina industry is the limited availability of high-quality bauxite, so it is necessary to develop cost-effective technologies for processing alternative raw materials. Even though the sintering process or the combined method can hardly compete with the Bayer process due to high-energy costs, this technology is still the main method of low-quality raw materials processing. The obtaining of high quality products and the use of certain advantages of sintering can significantly reduce the cost of alumina.

Another problem is the formation of a considerable amount of red mud (bauxite residue). At the same time, the amount of residue becomes more substantial when the quality of bauxite deteriorates. Therefore, the solution to the problem of processing of aluminum-containing raw materials, including the extraction of useful components from man-made waste – red mud, has not received its logical conclusion.

Bauxite residue forms a kind of man-made deposits, the contents of which have been repeatedly proposed to be processed into building materials for the manufacture of raw and fine ceramics, such as tiles, floor tiles [1, 2], for the manufacture of bricks [3, 4], materials in road construction [5], components in the production of cement and its special brands [5, 6], pigments [7], coagulants for wastewater treatment [8–11], iron concentrate [13, 14], raw materials for the extraction of titanium [15], rare and rare earth metals [16–19].

Many of the developed technologies have not gone beyond laboratory and pilot tests, and as a result, millions of tons of red mud, which poses a threat to the environment and surrounding cities, are thrown into the red mud fields near alumina refinery every year.

To suggest one of the possible solutions to the specified problem, the possibility of selective separation of iron from red mud to produce marketable products in the form of pigment-quality magnetite, iron concentrate, and aluminosilicate residue with a high content of titanium was studied.

2. Experimental

All chemicals (except sinter) used in this study were of analytical grade. The Ural-aluminum-plant two-component bauxite-sintering product with composition presented in Table 1 served as the raw material for the experiments. Figure 1 shows an XRD pattern of the original sintering product.

Table 1. Chemical composition of the bauxite-sintering product.

Component	Al_2O_3	Fe_2O_3	Na_2O	SO_3	TiO_2	CaO	MgO	SiO_2
Content, wt. %	34.8	16.6	33.0	3.2	1.8	4.8	0.8	4.0

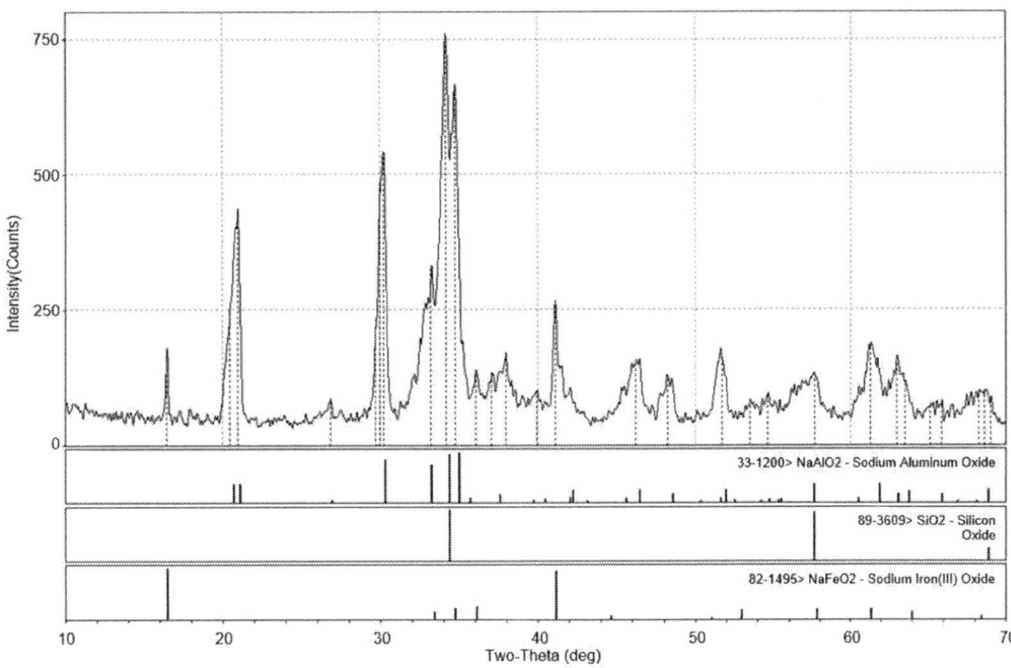

Figure 1. XRD pattern of the original sintering product.

Studies were conducted on the chemical treatment of red mud, as a result of which, due to the reaction (1) in an alkaline medium, iron hydroxide obtained after the interaction of sodium ferrite with water by reaction (2) is converted into magnetite. The source of ferrous iron was $FeSO_4 \times 7H_2O$.

$$2Fe(OH)_3 + Fe^{2+} + 2OH^- = Fe_3O_4 + 2H_2O \tag{1}$$

$$NaFeO_2 + 2H_2O = Fe(OH)_3 + NaOH \tag{2}$$

The reaction (1) was proposed in [20, 21] to obtain a pigment-quality magnetite from jarosite – a waste of zinc solutions purification, which also forms an amorphous ferric hydroxide in an alkaline medium.

In [20] it is shown that reaction (1) proceeds so quickly, that after 5–10 minutes the amount of ferrous ions in the solution remains less than 500 mg/l. However, to obtain a well-crystallized product, the authors carried out the process for 7 hours.

The method of our experiments was as follows. The sample of 100 g of the sintering product was leached in an aluminate solution with a concentration of Na_2O_k = 130 g/l and α_k = 1.65 at 90 °C for 30 minutes. The obtained pulp is then was subjected to filtration, followed by washing of red mud. The red mud was repulped with 170 ml of distilled water and 40 ml of 25 % ammonium hydroxide or 70 ml of 25 % sodium hydroxide solutions, and an iron sulfate solution (60 g $FeSO_4 \times 7H_2O$ for 200 ml H_2O) was added. The resulted pulp was heated to the required temperature and kept for 4 hours.

After exposure, the resulting pulp was filtered, the red mud was washed and dried at 80 °C during the day. Further, the magnetic concentrate was recovered from the red mud by wet magnetic separation on a laboratory magnetic separator.

3. Results and Discussion

3.1. Influence of the type of sintering product

During the experiments, it was found (Figure 2) that the reaction (1) is less effective if the old sintering product was used.

This phenomenon is the most likely connected to the hydrolysis of sodium ferrite by the air water, resulting in the formation of iron hydroxide (III), which by the time of interaction with the bivalent iron will not be as active as the newly deposited. Therefore, in order to obtain the high yield of magnetite by reaction (1), it is necessary to use as fresh sintering product as possible.

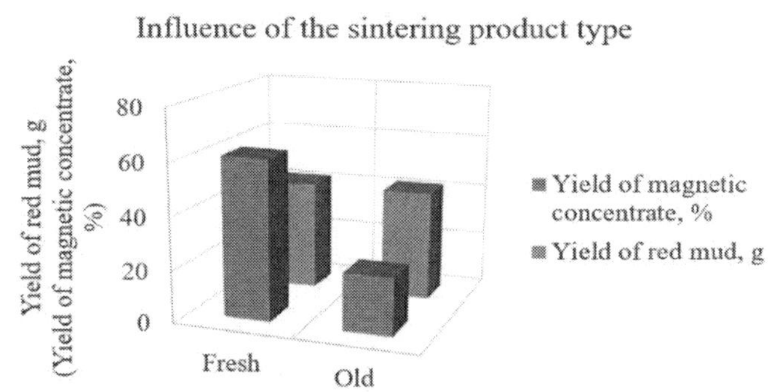

Figure 2. The influence of the type of sintering product on the yields of the magnetic concentrate and the red mud.

3.2. Influence of temperature

Further, the effect of temperature on the formation of magnetite was studied.

In figure 3 the results of experiments at different temperatures of magnetite formation in alkaline medium in the presence of iron (II) sulfate are shown.

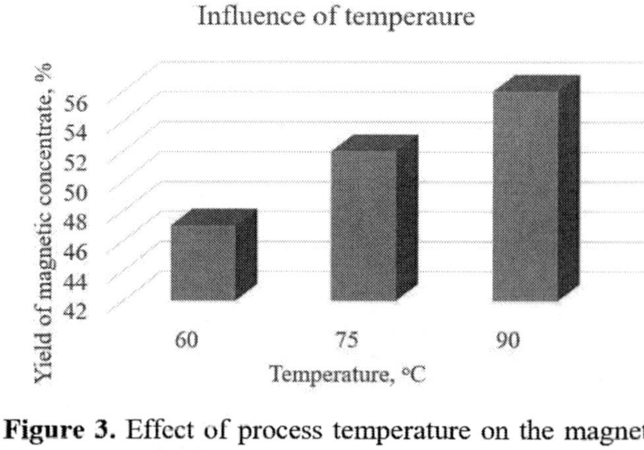

Figure 3. Effect of process temperature on the magnetic concentrate yield.

It is obvious that the temperature increase helps to enhance reaction (1) velocity. Therefore, the optimum temperature of 90 °C was chosen.

3.3. Influence of iron (II) sulfate amount

In figure 4 the influence of ratio of ferrous iron to stoichiometric quantity required for the reaction (1) on the yields of the magnetic concentrate and the red mud is shown.

Excess of iron sulfate leads to an increased yield of red mud and a relatively low yield of the magnetic concentrate, although the degree of reaction (1) is greater in this case.

In addition, as shown in figure 4, the greatest yield of the magnetic concentrate is achieved at 70% ratio of the amount of iron sulfate to stoichiometry, since the degree of reaction in this case is closest to the amount of added ferrous iron, thus the excess of the latter is undesirable, since it leads to an overspending of the reagents.

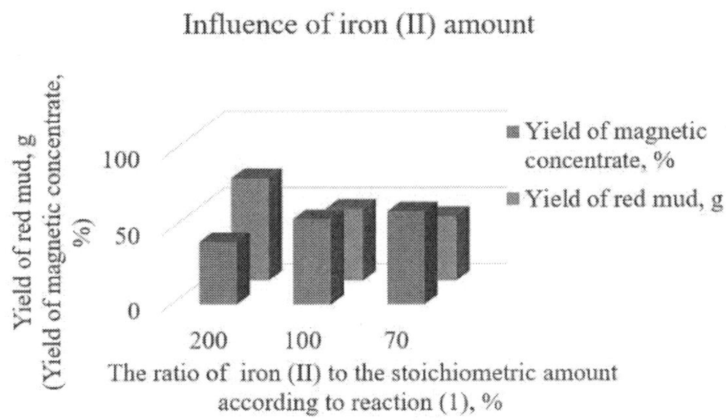

Figure 4. The effect of the ratio of the amount of iron (II) sulfate to the stoichiometric amount required for the reaction (1) on the degree of iron hydroxide conversion.

Incomplete conversion of iron (III) hydroxide to magnetite may be due to the "aging" of the raw material in contact with air, as sintering product was used in the study a few weeks after its receipt.

Therefore, the optimal ratio of the amount of iron (II) sulfate to the stoichiometric amount required for the reaction (1) is 70 %.

When 70 % ratio of the amount of iron (II) sulfate to the stoichiometric amount required for the reaction (1) was taken, the content of magnetite in the magnetic concentrate was 47.7 % (figure 4), indicating insufficient recovery of iron in the concentrate and the need of finding another method of the magnetite separating from the resulting red mud.

4. Summary

The possibility of pigment-quality magnetite obtaining in the processing of bauxite-sintering product was studied.

The optimal parameters of the interaction of iron (III) hydroxide of the red mud with ferrous iron ions with the formation of magnetite in an alkaline medium, in our opinion, are:
temperature = 90 °C; duration = 4 hours; the ratio of bivalent iron to the stoichiometric amount according to reaction (1) = 70 %; the method of addition of ferric sulfate = 4, also, the sintering product must be as fresh as possible.

The yield of the magnetic concentrate under the optimal conditions was 69.3 %, and the content of magnetite in it was 47.7 %. Therefore, it is necessary to study other ways of magnetite recovering from red mud. For example, the use of differences in the density of magnetite and other components of red mud in centrifugal gravity enrichment allowed to increase the content of magnetite to 97 %, but the concentrate yield was only 11 %.

Obtaining a better concentrate with a high yield according to the method proposed in the work will help to increase the complexity of bauxite processing with obtaining of cost-effective products.

References

[1] Guo Y X, Zhang Y H, Huang H W and Meng K 2014 Novel glass ceramic foams materials based on red mud *Ceram. Int.* **40** 6677–6683

[2] Pérez-Villarejo L, Corpas-Iglesias F A, Martínez-Martínez S, Artiaga R and Pascual-Cosp J 2012 Manufacturing new ceramic materials from clay and red mud derived from the aluminium industry *Constr Build Mater.* **35** 656–665

[3] Liu W C, Yang J K and Xiao B 2008 Application of Bayer Red Mud for Iron Recovery and Building Material Production from Alumosilicate Residues *J. Hazard. Mater.* **161** 474–478

[4] Dimas D D, Giannopoulou I P and Panias D 2009 Utilization of alumina red mud for the synthesis of inorganic polymeric materials *Miner. Process. Extr. Metall. Rev.* **30** 211–239

[5] Liu W, Yang J and Xiao B 2009 Review on treatment and utilization of bauxite residues in China *Int. J. Miner. Process* **93** 220–231

[6] Senff L, Hotza D and Labrincha J A 2011 Effect of red mud addition on the rheological behaviour and on hardened state characteristics of cement mortars *Constr Build Mater.* **25** 163–170

[7] Somlai J, Jobbágy V, Kovács J, Tarján S and Kovács T_2008 Radiological aspects of the usability of red mud as building material additive *J. Hazard. Mater.* **150** 541–545

[8] Poulin É, Blais J-F and Mercier G 2008 Transformation of red mud from aluminium industry into a coagulant for wastewater treatment *Hydrometallurgy* **92** 16–25

[9] Rai S, Wasewar K L, Lataye D H, Mukhopadhyay J and Yoo C K 2013 Feasibility of red mud neutralization with seawater using Taguchi's methodology *Int. J. Environ. Sci. Technol.* **10** 305–314

[10] Ahmed M D J K and Ahmaruzzaman M A 2016 Review on potential usage of industrial waste materials for binding heavy metal ions from aqueous solutions *J. Water Process Eng.* **10** 39–47

[11] Altundogan H S, Altundogan S, Tümen F and Bildik M 2000 Arsenic removal from aqueous solutions by adsorption on red mud *Waste Manag.* **20** 761–767

[12] Shoppert A A, Loginova I V, Rogozhnikov D A, Karimov K A and Chaikin L I 2019 Increased As Adsorption on Maghemite-Containing Red Mud Prepared by the Alkali Fusion-Leaching Method *Minerals* **9** 60

[13] Valeev D, Zinoveev D, Kondratiev A, Lubyanoi D and Pankratov D 2020 Reductive Smelting of Neutralized Red Mud for Iron Recovery and Produced Pig Iron for Heat-Resistant Castings *Metals* **10** 32

[14] Zinoveev D, Grudinsky P, Zakunov A, Semenov A, Panova M, Valeev D, Kondratiev A, Dyubanov V and Petelin A 2019 Influence of Na_2CO_3 and K_2CO_3 Addition on Iron Grain Growth during Carbothermic Reduction of Red Mud *Metals* **9** 1313

[15] Shoppert A and Loginova I 2017 Red Mud as an Additional Source of Titanium Raw Materials *KnE Materials Science* **2** 150–157

[16] Chaikin L, Shoppert A, Valeev D, Loginova I and Napol'skikh J 2020 Concentration of Rare Earth Elements (Sc, Y, La, Ce, Nd, Sm) in Bauxite Residue (Red Mud) Obtained by Water and Alkali Leaching of Bauxite Sintering Dust *Minerals* **10** 500

[17] Liu Y Y, Zhao B C, Tang Y and Wan P Y 2014 Recycling of iron from red mud by magnetic separation after co-roasting with pyrite *Thermochim. Acta.* **588** 11–15

[18] Paramguru R K, Rath P C and Misra V N 2005 Trends in red mud utilization – a review *Miner. Process. Extr. Metall. Rev.* **26** 1–29

IOP Conf. Series: Materials Science and Engineering **969** (2020) 012056 doi:10.1088/1757-899X/969/1/012056

[19] Loginova I V, Shoppert A A and Chaikin L I 2016 Extraction of Rare-Earth Metals During the Systematic Processing of Diaspore Boehmite Bauxites *Metallurgist* **60** 198–203
[20] Vu H, Jandová J and Hron T 2010 Recovery of pigment-quality magnetite from jarosite precipitates *Hydrometallurgy* **101** 1–6
[21] Piskunov V M, Matveev A F and Yaroslavtsev A S 1988 Utilizing Iron Residues from Zinc Production in the U.S.S.R. *J. of Metals* **8** 36–39

IOP Conf. Series: Materials Science and Engineering **969** (2020) 012057 doi:10.1088/1757-899X/969/1/012057

Anodic Dissolution of Non-Ferrous Metals in a Glycerate-Alkali Electrolyte

V A Valnev, V G Lobanov and D I Bludova

Department of Non-Ferrous Metallurgy, Ural Federal University, 17 Mira street, Ekaterinburg city, 620002, Russia

E-mail: va.valnev@urfu.ru

Abstract. Studies have been conducted to assess the possibility of refining a lead collector containing precious metals, with the aim of obtaining lead using an economically viable and environmentally safe technology. A method of electrochemical selection in an alkaline glycerate electrolyte is proposed. The behavior of lead, silver and "silver crust" during anodic polarization in glycerate electrolyte was studied using the potentiodynamic method. "Silver crust" – type alloy of the following composition were used in this study, (wt.) %: 77.9 Pb, 12.76 Zn, 3.53 Ag, 0.50 Cu, Sb 0.90. The studies were carried out using the method of a rotating disk. Studied the patterns of anodic oxidation of lead and impurities in alkaline-water-glycerate electrolytes, depending on the concentration of sodium hydroxide and glycerol in the electrolyte, the potential sweep rate.

1. Introduction

Anodic dissolution of non-ferrous metals and alloys is accompanied by the formation of a film of poorly soluble products on the surface of the anodes. The result of this process is the passivation of the anode, further dissolution slows down or stops completely. Metals such as lead, antimony, tin show a greater tendency to passivation. For this reason, electrochemical refining of alloys containing these metals in appreciable amounts is almost impossible.

The refinement of argentiferous lead by the pyrometallurgical method is accompanied by the production of silver crust, where precious metals are concentrated. Silver crust mostly contains lead, zinc, copper, bismuth, iron, tellurium and some other metals [1–4]. Improving the technology of complex processing of this multicomponent ingoing material is relevant.

2. Current technologies

The main purpose in the processing of silver crust is the selection of the extraction of precious metals. Traditional technologies are based on the distillation of crust in electrode furnaces with the aim of rundown to 90% zinc. The resulting residue is additionally cupelled, gold and silver remain in the bath of the furnace. The resulting alloy is poured into molds and sent for refining [5–8].

An alternative method was engineered in 1956 [3]. In an electrolyte containing 80 g/l of Pb and 100 g/l of sulfamic acid, 95% of Pb is recovered into the cathode sediment and 100% of all precious metals are extracted into the slime. Energy consumption was 150 kWh on ton of lead. The slime yield was approximately 17% by weight of the dissolved part of the anode, the sludge contained 82% Ag; 0.1

IOP Conf. Series: Materials Science and Engineering **969** (2020) 012057 doi:10.1088/1757-899X/969/1/012057

Au; 3.5% Pb; 6.5% Cu; 3.5% Zn; 0.6% Bi and 11.3% Sb. Metals are more electropositive than lead, (almost all impurities), remain in the anode slime, only tin is deposited with lead.

The electrolyte based on fluorosilicic acid and silicofluoride lead has become widespread in the electrorefining of raw lead. The lead to be refined, after debondering, was cast into anodes in the form of plates. The resulting sludge contained, %: 12–20 Pb; 4–5 Ag; 6–15 Cu; 25–30 Sb; 9–15 As [3].

Known electroextraction of lead from nitric-acid solutions with obtaining cathodic sediment – metallic lead and anodic - lead dioxide. The anode is titanium, the cathode is lead. When the concentration in the electrolyte, mol/l: Pb 0.5 and HNO_3 0.5 reached a current output of 94% without releasing nitrous gases. To reduce dendritic formation at the cathode, formic acid was added to the electrolyte in the amount of ~ 440 g on 1 ton of cathode lead [9, 20].

Anodic dissolution of alloys containing in addition to lead, antimony and stannic in significant quantities - more than 10%. Using these electrolytes is not possible, since stannic, even at roam temperature, forms an oxide film on the alloy surface, which causes irreversible passivation, reduces the rate of anodic dissolution.

With the joint processing of some lead-containing products at JSC "Uralelektromed", when refining lead, silver crust of the following composition is obtained, in mass %: 78.58 Pb, 15.95 Zn, 0.47 Bi, 0.22 Cu, 0.36 Fe, 0.63 Te, 0.005 As, 0.008 Sb, 0.003 Sn, 0.006 Ni, 0.002 Al, 0.02 Ca, 0.01 Mg, 4.31 Ag [2]. Currently, measures are being taken to select a rational technology for the complex processing of such raw materials.

The purpose of this work is to create a cost-effective and environmentally friendly technology for the integrated extraction of valuable components from polymetallic alloys of reduced composition. Certain prospects in solving this problem have electrochemical selection in alkali-glycerate electrolyte [10–12]. According to [13], when using anodes made of an alloy of lead and antimony, anodic dissolution of both metals occurs under optimal conditions [14–15], but in some cases the surface of the anode is oxidized to lead dioxide, on which the reaction actually takes place.

3. Research methods

The specific task of the experiments was to study the laws of anodic polarization of the alloy and individual components in the electrolyte using the potentiodynamic method and the installation of a rotating disk electrode. This method was used to research specific action of polymetallic alloy and separate components in conditions of anodian polarization.

Electrodes from following metals were used in research: crust alloy, lead, antimony, tin, silver, copper and zinc. Chemical pure reagents were melted in an induction furnace and poured into graphite molds. The obtained cylindrical samples were processed to a diameter of 7 mm and pressed into fluoroplastic forms. The measurements were carried out by the method of a rotating disk using the Volta "EM-04" attachment, the rotation speed of 800 rpm at room temperature is 25 °C.

Anodic dissolution of metals was studied using an IPC-Pro galvanostat-potentiostat, setting the potential change interval from -0.2 to 2 Volts with a sweep rate 100 mV/s, since a dense passivation layer is formed at low speeds. Concentrations of alkali and glycerol in equal proportions were used: 25, 50, 100, 150 g/l. Silver chloride reference electrode was used. The measurements were carried out in an open cell without separation of the auxiliary (platinum) electrode and the reference half-cell of porous glass or membrane partitions [16–19].

4. Discussion of results

The working surfaces of the disk electrodes were carefully processed before the experiment. Figures 1–4 show the potentiodynamic curves of the alloy, lead and silver, obtained under these conditions.

It is most likely that for electrodes made of lead and an alloy based on it, a significant anodic current near the potential of 0 mV during polarization is associated with the formation of lead oxide, which begins in the cathode potential region (-800 mV). Oxygen evolution becomes visually noticeable from 900–1200 mV. With increasing alkali concentration, the potential of this reaction shifts to the negative region. In the potential range of 600–800 mV, a peak is observed which is characteristic of a deeper

IOP Conf. Series: Materials Science and Engineering **969** (2020) 012057 doi:10.1088/1757-899X/969/1/012057

oxidation of lead to dioxide. Under dynamic conditions of potential sweep, lead oxidation occurs at a satisfactory rate in the range of 200–800 mV at a speed of 100–200 mA/cm².

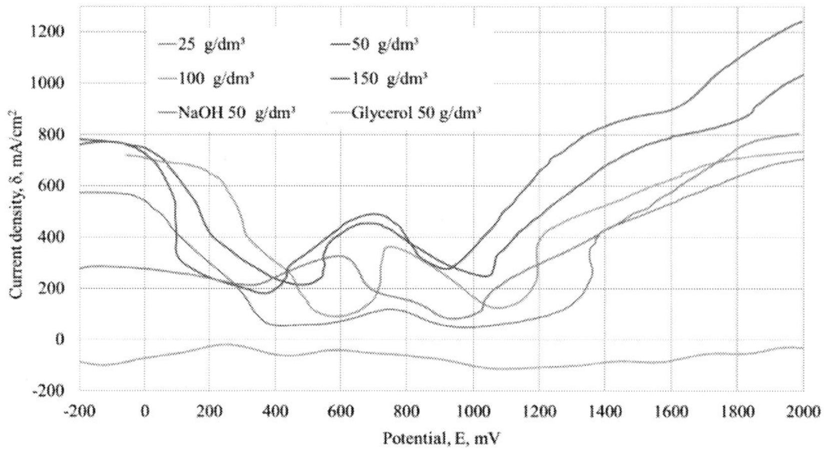

Figure 1. Polarization graphs of foam alloy.

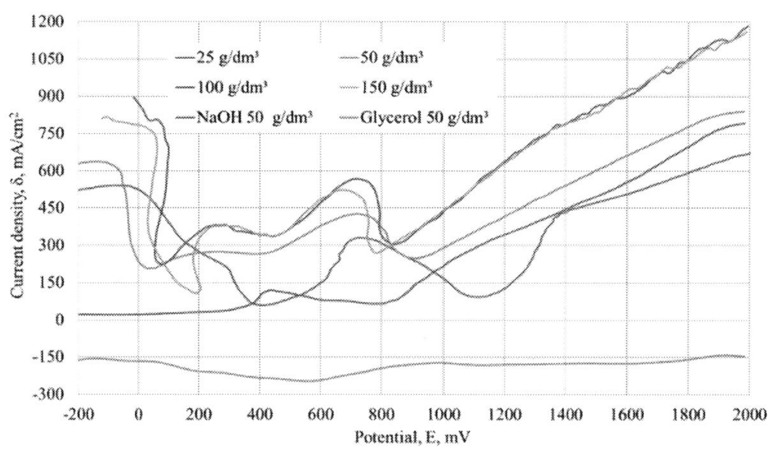

Figure 2. Polarization graphs of lead.

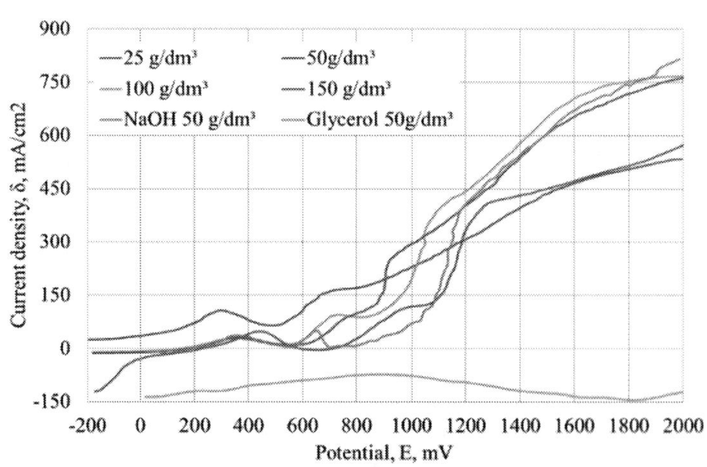

Figure 3. Polarization graphs of silver.

Silver in electrolytes alkali-glycerin is passivated, when measuring the disk was covered with loose black sediment, at the potentials of 700–900 mV, the beginning of oxygen evolution was observed. The current density of anodic oxidation of silver does not exceed 10–20 mA/cm^2. According to the XRD, the precipitate is represented by silver oxide.

With an increase in the velocity of the potential sweep, the beginning of the noted processes shifts to the region of more negative potentials. The optimum scanning speed, allowing to identify confidently differences in the behavior of the alloy components lies in the range from 50 mV/s to 150 mV/s.

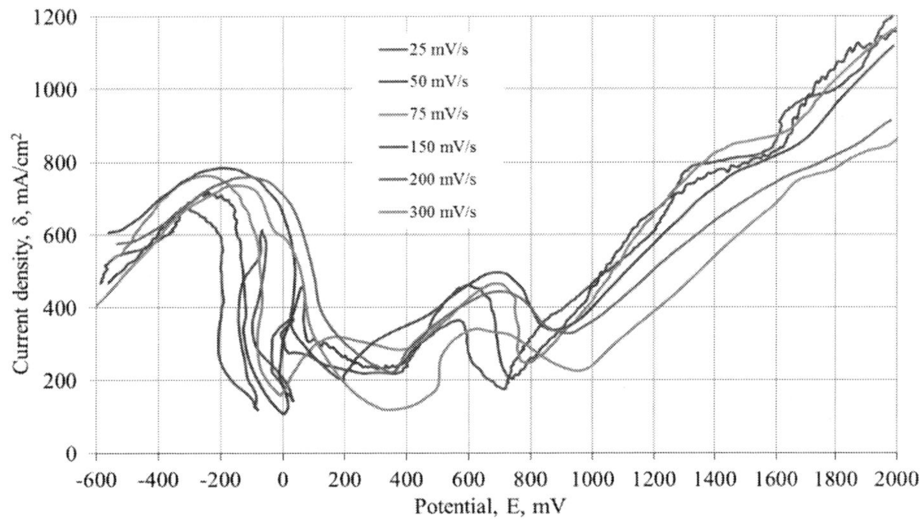

Figure 4. Polarization graphs of lead with different speed sweep at concentration 150 g/l alkali/glycerol.

The graph shows that increasing the sweep rate leads the current density to lower range. However, too low scanning speeds result in significant increase of experiment time and distortion of the curves. According to the graphic optimal scanning speed is situated between 75 mV/s and 200 mV/s, cause a scanning speed of experiments was 100 mV/s.

5. Conclusion

The area of confident oxidation of lead in the alkaline-glycerate electrolyte corresponds to the potentials of 300–800 mV.

For the lead-based polymetallic alloy (silver crust) area of anodic oxidation corresponds to the potentials of 700-800 mV.

The totality of the revealed regularities of the anodic polarization of polymetallic alloys gives reason to assume a confident transition to a solution of lead, antimony and tin, while silver, gold, tellurium are concentrated in the anode sludge.

Silver is oxidized to form an insoluble oxide, which forms the anode sludge. The issue of further research is determination of passivation process, influence on glycerol, electrolyte components, temperature, potential sweep and rotational disc speed.

Reference

[1] Ekilik V V 2004 *Theory of corrosion and protection of metals* (Rostov-on-the-Don) p 67
[2] Korolev A A, Krayukhin S A, Maltsev G I and Filatov E S 2004 Treatment of silver foam by vacuum distillation, *J. Universities' Proceedings Non-Ferrous Metallurgy* **4** 21–29
[3] Baimakov Yu V and Zhurin V I 1963 *Electrolysis in hydrometallurgy* (Moscow: Metallurgizdat)
[4] Lebedev V A 1991 *Theory of electrometallurgy of non-ferrous metals* (Yekaterinburg: Ural Polytechnic Institute)

[5] Loskutov F M 1965 *Metallurgy of lead* (Moscow: Metallurgy)
[6] Zaitsev V Ya and Margulis E V 1985 *Lead and zinc metallurgy* (Moscow: Metallurgy)
[7] Smirnov N P 1977 *Lead refining and processing of industrial products* (Moscow: Metallurgy)
[8] Abdeev M A, Getskin L S and Zaplavniy A Ya 1964 M*odern methods of processing of lead-zinc ores and concentrates* (Moscow: Metallurgy)
[9] Ostanova S V, Pashkov G L, Drozdov S V and Chubarov A V 2002 Study of the process of electroextraction of lead from nitric acid milieu *Intellect–2002* 327–328
[10] Bugenov E S, Nurlybaev R E, Shaidarbekova J K and Murzagulova A A 2013 Electrolysis of antimony from alkaline glycerate solutions *J. Vestnik KazNTU* **1** 112–114
[11] Baiborodov P P and Ezhkov A B 1967 *Glycerate method of lead raw material processing* RU Author's Certificate **195105**
[12] Ishankhodzhaev S 1997 *Chemistry and technology of production of antimony, lead and their compounds in the presence of polyatomic alcohols* (Tashkent)
[13] Baizer M M and Lund H 1983 *Organic electrochemistry* (New York) p 285.
[14] Abdurakhmonov S 1997 *The development and improvement of technology of extraction of gold and antimony from a refractory ores* (Navoii)
[15] Kurkchi E U, Fedorenko A M and Kurkchi U M 2000 Solubility of antimony (III) oxides in aqueous solutions of polyatomic alcohols and calcium hydroxides, barium *Scientific notes of Taurida state University* **2** **13** 185–188
[16] Kogan M S 1994 *Processing of sulphate-oxide fraction of lead accumulator scrap in conditions of small-scale production* (St. Petersburg)
[17] Ostanina T N 2013 *Theoretical electrochemistry: a laboratory course* (Yekaterinburg: Ural Federal University)
[18] Kiselev M G, Mrochek J A and Drozdov A V 2014 *Electrophysical and electrochemical methods of materials processing* (Moscow: INFRA-M).
[19] Kholkina A S 2018 *Electrochemical separation of Pb-Sb-Bi alloys in a mixture of potassium and lead chlorides* (Yekaterinburg: Ural Federal University)
[20] Astapchik S V 2007 *Electroextraction of lead from nitric acid electrolytes* (Krasnoyarsk)

IOP Conf. Series: Materials Science and Engineering **969** (2020) 012058 doi:10.1088/1757-899X/969/1/012058

Study of the Mechanism for Forming a Passivating Layer under Electrochemical Oxidation of Copper (I) Sulfide

O V Nechvoglod and S M Pikalov

Institute of Metallurgy, Ural Branch of the Russian Academy of Science, 101, Amundsen st., Yekaterinburg, 620016, Russia

E-mail: nechvoglodov@mail.ru

Abstract. In this study the patterns of anodic oxidation of copper (I) sulfide were considered. It was shown that during the crystallization of copper sulfide, the formation of both the chalcosine (Cu_2S) phase and the jarleite ($Cu_{31}S_{16}$) phase is possible, which is characterized by a deficiency of copper in the crystal lattice and deviation from the stoichiometric composition. The electrochemical oxidation of the sample in a solution of sulfuric acid was carried out. During oxidation, intermediate non-stoichiometric sulfides ($Cu_{1.74}S$, $Cu_{1.8}S$) were formed in the following sequence: Cu_2S ($Cu_{31}S_{16}$) $\rightarrow$ $Cu_{1.81}S$ $\rightarrow$ $Cu_{1.74}S$ $\rightarrow$ $Cu_{1.6}S$ $\rightarrow$ CuS $\rightarrow$ $S + Cu^{2+}$. The process was accompanied by the transition of copper cations into solution. As elemental sulfur and copper sulfides accumulated on the reaction surface of the sample and, as a result, the rate of electrochemical oxidation decreased due to the difficulty of the product removal and the reagent supply to the reaction zone.

1. Introduction

Sulfide ores are the main raw material for the production of metallic copper [1–3]. Recently, hydrometallurgical technologies have been more widely used for their processing [4–8]. Existing leaching methods include the use of various reagents - oxidizing agents (including bioleaching) [9–11], high pressure (autoclaves) [12] and electric current [13–16]. Due to the fact that electrochemical processes are becoming more widespread in copper metallurgy, the relevance of research in this area is increasing. Of particular interest is the anodic dissolution of sulfide copper materials with the transfer of copper into solution, and elemental sulfur - into sludge [17]. A new alternative direction is the electrolysis of granular materials, in particular, sulfide copper, nickel and copper-nickel intermediates [18, 19]. For the physicochemical substantiation of the electrolysis technology of sulfide copper-containing materials, it is of interest to study the patterns of anodic oxidation of copper (I) sulfide.

Under the influence of an electric current, in a solution of sulfuric acid, Cu_2S is oxidized with the transition of sulfide sulfur (S^{2-}) to the elemental state (S^0) [20], and monovalent copper (Cu^+) to divalent (Cu^{2+}) one. The process has a stepwise mechanism and is accompanied by the formation of intermediate sparingly-soluble copper sulfides [20–22]. Copper cations pass into the solution and insoluble products – sulfides and elemental sulfur – accumulate on the reaction surface of the sample forming a sulfur-sulfide layer. This layer prevents product withdrawal and reagent supply to the reaction zone, causing diffusion limitations. A decrease in the rate of removal of copper cations from the anodic oxidation zone into the electrolyte volume leads to a decrease in the oxidation rate of the initial sample. The passivating layer (film) has a porous structure due to the removal of copper cations into the electrolyte volume, which ensures the further occurrence of electrochemical processes at the Cu_2S/film interface. The rate of electrochemical oxidation of copper (I) sulfide is determined by the dissolution of the passivating layer and the diffusion of reagents and reaction products through it. The formation of a passivating layer occurs due to the phase transformations of Cu_2S under the influence of an electric current. As a result of the electrochemical oxidation of sulfide sulfur, CuS is formed through the series of transformations of sulfides of non-stoichiometric compositions.

The purpose of this study is to study the process of formation of a passivating layer and its structure formed on the surface of copper (I) sulfide as a result of electrochemical oxidation in a solution of sulfuric acid, as well as to study the sequence of phase transformations of copper (I) sulfide under the influence of an electric current.

2. Experimental

For research, a copper (I) sulfide sample was synthesized and its anodic oxidation was carried out to form a sulfur sulfide layer. The synthesis of copper sulfide was carried out by fusing copper (99.99% Cu) with sulfur in an electric resistance furnace at a temperature of 1250 °C in a reducing atmosphere. The melt was drawn into a quartz tube and cooled in air. The resulting rod of copper sulfide was pressed into the fluoroplastic. The working surface area on which the passivation layer was subsequently formed was 1 cm^2. A stainless steel rod was used as a cathode. The electrolyte was a solution of sulfuric acid with a concentration of 100 g/dm^3 H_2SO_4 with a volume of 30 ml. The obtained copper sulfide anode was subjected to electrochemical oxidation for 15 minutes at a current density of 1000 A/m^2 in the current stabilization mode.

The initial sample and products of electrochemical oxidation were analyzed by various methods. X-ray diffraction method was applied to determine phase compositions of the initial copper (I) sulfide sample and the layer of its electrochemical oxidation products using a XRD-7000C diffractometer (Shimadzu Corporation, Japan). Processing of the results of XRD was performed on the basis of diffraction data from The International Centre for Diffraction Data (ICDD). The quantitative phase compositions were determined by the method of local scanning analysis using a scanning electron microscope-microanalysis system JSM-5900LV, equipped with a wave dispersive microanalyzer system INCA Energy 250 (Japan).

3. Results and Discussion

According to chemical analysis, the initial copper (I) sulfide sample contains, %: 77.4 – Cu and 22.6 – S. According to the XRD data (Figure 1, a), it is a non-stoichiometric solid solution of copper and sulfur with a predominant monoclinic structure (sp.gr. P21/n (14)) close in composition to copper sulfide $Cu_{31}S_{16}$ (jarleite). The sample contains a phase of metallic copper which is confirmed by metallographic analysis. Crystallization of the metal phase leads to the formation of a sulfide phase with a copper deficiency – $Cu_{31}S_{16}$, which explains the deviation from the stoichiometric composition. Also, a stoichiometric Cu_2S phase is present in the sample.

As a result of the electrochemical oxidation of copper (I) sulfide, a layer of sparingly soluble products is formed. Under conditions of anodic polarization [20–22], the oxidation of copper sulfide proceeds through a series of intermediate sulfide compounds according to the total reactions:

$$Cu_2S - e^- = CuS + Cu^{2+}, \quad (1)$$

$$CuS - 2e^- = Cu^{2+} + S^0 \quad (2)$$

According to the data presented in [20–22], under the influence of an electric current, copper (I) sulfide can decompose into copper and copper (II) sulfide:

$$Cu_2S = Cu + CuS. \quad (3)$$

According to the XRD data (Figure 1, b), the main phase components of the layer of products of electrochemical oxidation were CuS (covellite) and $Cu_{1.74}S$ (roxbiite), and Cu_9S_5 was in a smaller amount (which corresponds to $Cu_{1.81}S$ - digenite). A weak peaks characteristic of elemental sulfur were also observed on the XRD pattern. The lattice parameters of CuS and $Cu_{1.74}S$ calculated were in a good agreement with the international database. According to scanning electron microscope (Table 1, Figure 2), the $Cu_{1.6}S$ phase (jitite) is formed on the surface of the sample. The reaction product layer had a dense uniform structure formed by fine particles of Cu_xS_y sulfides of 4-6 microns in size.

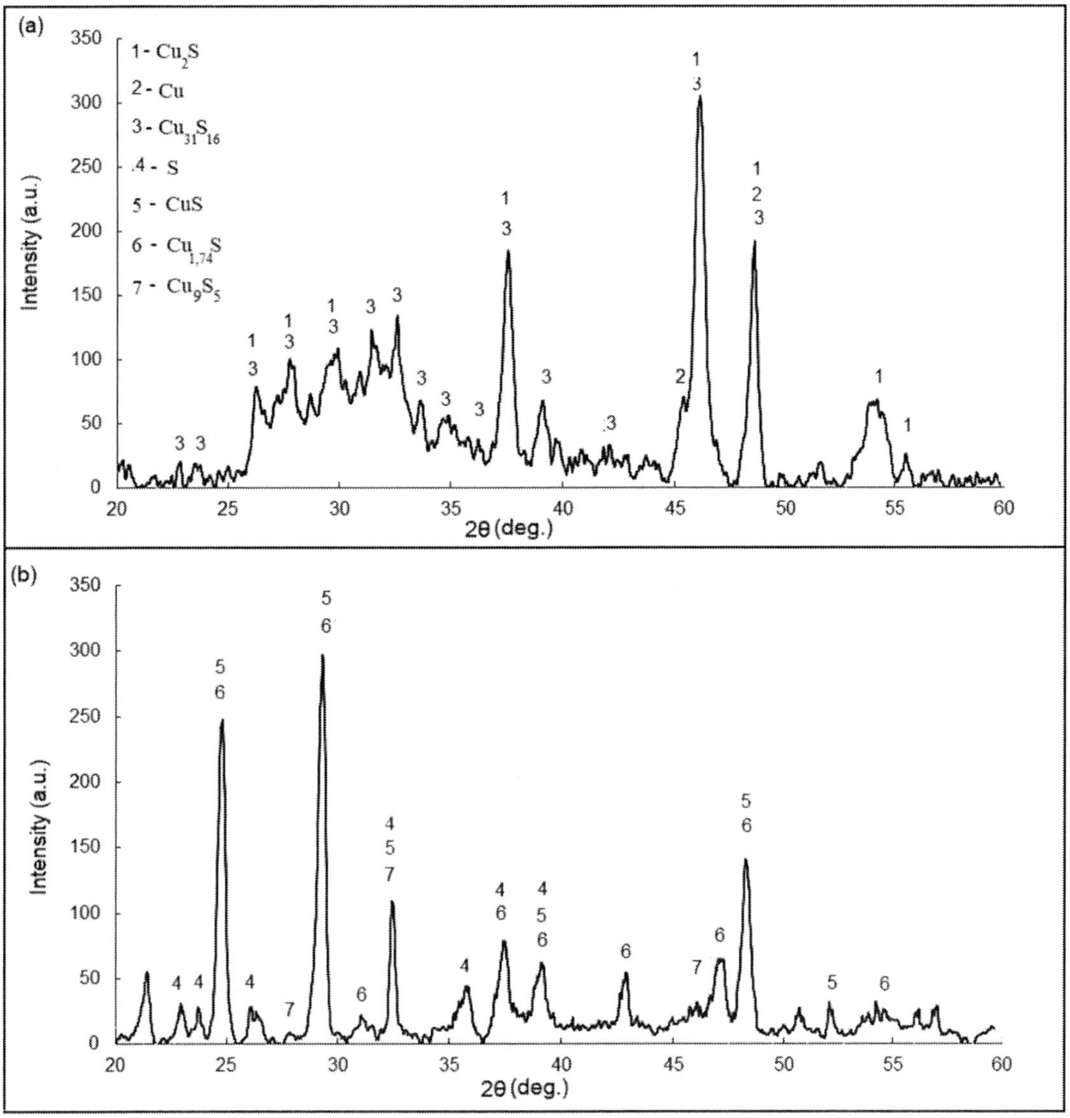

Figure 1. A) XRD pattern of the initial sample of copper sulfide; b) XRD pattern after electrochemical oxidation.

Table 1 Chemical composition at the sensing points (according to Figure 1).

№	Content, at.%		Phase
	S	Cu	
1	37.6	61.5	$Cu_{1,6}S$
2	36.8	61.9	$Cu_{1,6}S$
3	11.7	88.1	Cu-CuS
4	30.0	69.9	Cu-CuS
5	38.9	60.4	$Cu_{1.6}S$

According to the XRD data, the responses characteristic of elemental sulfur have been identified. However, they have a low intensity, which can be explained by sulfur amorphousness, an insignificant amount and dispersion of particles. The film formed comprises an inner sublayer with a dense uniform structure and an outer layer with dense homogeneous and loose inhomogeneous areas. The electrochemical oxidation of copper (I) sulfide (Cu_2S, $Cu_{31}S_{16}$) in a sulfuric acid electrolyte proceeds, at the first stage, with the formation of non-stoichiometric $Cu_{1.81}S$, $Cu_{1.74}S$, $Cu_{1.6}S$ up to CuS. At the second stage, CuS is oxidized to elemental sulfur. The process is accompanied by the transfer of copper cations into the solution. In the interaction of copper cations with a sulfate electrolyte in the presence of oxygen released at the anode, the formation of oxide-sulfate inclusions of the $mCuO \cdot nCuSO_4$ type is possible. The sequence of phase transformations of the test sample of copper (I) sulfide can be represented in a general form as follows: Cu_2S ($Cu_{31}S_{16}$) → $Cu_{1.81}S$→ $Cu_{1.74}S$→ $Cu_{1.6}S$→CuS → S + Cu^{2+}.

Figure 2. A layer of copper sulfides formed on the surface of a Cu_2S sample after electrochemical oxidation.

The decomposition mechanism of copper (I) sulfide by reaction (3) explains the formation of areas of metallic copper on the surface of the electrodes.

In this study, the mechanism of electrochemical oxidation of copper (I) sulfide has been validated. The data on the chemical and phase composition of sparingly soluble products have been specified and expanded. The sequence of phase transformations has been considered.

4. Conclusion

1. The electrochemical dissolution of copper (I) sulfide proceeds with the formation of a film of insoluble products due to the oxidation of sulfide sulfur to copper sulfide and sulfur in an elemental state. As oxidation progresses, intermediate non-stoichiometric sulfides are formed.
2. The sequence of phase transformations of copper (I) sulfide can be represented in a general form as follows: Cu_2S ($Cu_{31}S_{16}$) → $Cu_{1.81}S$ → $Cu_{1.74}S$ → $Cu_{1.6}S$ → CuS → S + Cu^{2+}.

References

[1] Moskalyk R R and Alfantazi A M 2003 *Miner. Eng.* **16** 893
[2] Pedersen K B, Jensen P E, Ottosen L M, Evenset A, Christensen G N and Frantzen M 2017 *Miner. Eng.* **107** 100
[3] Wang M, Chen W and Li X 2015 *Resour. Conserv. Recy.* A **105** 36
[4] Dreisinger D 2006 *Hydrometallurgy* **83** 10
[5] Mizoguchi T and Habashi F 1981 *Int. J. Miner. Process* **8** 177
[6] Letowski F, Kolodziej B, Czernecki M, Jedrczak A and Adamski Z 1979 *Hydrometallurgy* **4** 169
[7] Xu B, Zhong H and Jiang T 2011 *Miner. Eng.* **24** 1082
[8] Hyvarinen O and Hamalainen M 2005 *Hydrometallurgy* **77** 61
[9] Dreisinger D 2006 *Hydrometallurgy* **83** 10
[10] Sheng-hua Y, Ai-xiang W and Guan-zhou Q 2008 *Trans. Nonferrous Met. Soc. China* **18** 707
[11] Wu A X, Xi Y, Yang B H, Chen X S and Jiang H C 2007 *Acta Metall. Sin. (Engl. Lett.)* **20** 117
[12] Jin B, Yang X and Shen Q 2009 *Hydrometallurgy* **96(1)** 57
[13] Abd El Haleem S M and Din Abd El Aal E E 2007*Alloys Compd. J.* **432(1)** 205
[14] Kartal L and Timur S 2018 *Trans. Nonferrous Met. Soc. China* **28** 2143

[15] Lu Z, Tang J, Mendosa M L, Chang D, Cai L and Zhang L J 2015 *Electroanal. Chem.* **745** 37
[16] Fang Z and Chen J 1989 *Acta Metall. Sin. (Engl. Lett.)* B **2** 159
[17] Selivanov E N 2011 Patent RU **2 434 065 C1 20**
[18] Nechvoglod O V, Selivanov E N and Mamyachenkov S V *2012 Defect Diffus. Forum 326–328 383*
[19] Selivanov E N, Nechvoglod O V and Lobanov V G 2013 *IFAC Proc.* **46** 259
[20] Chanturiya V A and Vigdergauz V E 2009 *Electrochemistry of Sulfides Theory and practice of flotation* (Moscow: Ore and Metals Publishing House) p 280
[21] Deng T, Lu Y, Wen Z and Liu D 2001 *Hydrometallurgy* **62** 23
[22] Sabauri G N, Bryukvin V A, Blokhina L I and Levin A M 1993 *Rus. Metallurgy* **6** 24

Acknowledgements
The reported study was funded by RFBR and Sverdlovsk Region, project number 20-48-660020.

IOP Conf. Series: Materials Science and Engineering **969** (2020) 012059 doi:10.1088/1757-899X/969/1/012059

Heat Exchange in High-Power Arc Steel Furnace when the Height of the Slag Layer Changes

A N Makarov and A V Krupnov

Electric Supply and Electrical Engineering Faculty, Tver State Technical University, Tver, Russia

E-mail: AV.Krupnov@yandex.ru

Abstract. Calculation results and analysis of changes in thermal radiation density of arcs by wall height for hundred-ton arc furnaces at different slag layer height and at equal arc power are presented. It was found that with an increase in the height of the slag layer from 70 to 300 mm, that is, with an increase in the depth of the arcs in the slag, the density of thermal radiation flows of the arcs in the most heat-loaded part of the walls decreases by 2 times. The efficiency of arcs increases with the increase in the depth of arcs in the slag and increases by 1.61 times when fully buried.

1. Introduction

The influence of the height of the slag layer on the density of thermal radiation from electric arcs on the walls of an electric arc furnace (EAF) is a topical issue, despite the fact that the stability of the wall lining is affected by many factors [1–10]: melting modes, chemical composition and temperature of the slag, the quality of service of the lining, the materials used, and others. This is due to the fact that the main share in the thermal load of walls is from the thermal radiation of electric arcs.

Therefore, we performed calculations of the heat flux density for the height and perimeter of the walls for a hundred-ton DSP-100 furnace to determine the effect of the height of the slag layer. The power of the arcs in all calculations was assumed unchanged.

2. Method determination of densities of thermal radiation flows from three arcs

Determining the densities of thermal radiation flows from arcs on the wall surface is based on the method described in [11–13]. Figure 1 shows the necessary constructions in the AutoCAD program for calculations performed at scale in MS Excel.

Symbols used in Figure 1: h_W – wall height of EAF, m; d_E – diameter of the electrode, m; h_T, h_M, h_{Sl} – the depth of the arc, respectively, the total, in the metal, in the slag, m; D_{Bath}, $d_{Dec.}$ – accordingly, the diameter of the bath, the decay of the electrodes, m; γ – the angle formed by the horizontal plane of the end of the electrode and the inclined plane that occurs on the electrode due to the electrodynamic movement of the arc from the axis of the electrode to its periphery, deg.; 0…5 – calculation points on the walls of furnace; l_{Arc} – arc length, m; l_{Open} – length of the open part of the arc radiating heat flow to the calculated site, m; r – beam, distance from the arc to the calculated point on the walls, m; α – the angle between the beam r from the middle of the open part of the arc and the perpendicular $N1$ - to the axis of the arc, drawn to the beginning of the beam r on the arc, deg.; β – the angle between the normal $N2$ to the wall surface at the calculated point and the beam r, deg.; Θ – the angle of the electrodynamic deviation of the arc axis from the electrode axis is determined by the method presented in [11–13].

The density of the arc thermal radiation flux falling on the calculated site located on the walls of the EAF was determined by the expression [11]:

$$q = \frac{\alpha_{\text{Arc}} \cdot P_{\text{Arc}} \cdot \cos\alpha \cdot \cos\beta \cdot l_{\text{Open}}}{\pi^2 \cdot r^2 \cdot l_{\text{Arc}}} \cdot e^{-k \cdot r}, \text{kW/m}^2, \tag{1}$$

where α_{Arc} – the percentage of arc power allocated in the arc pillar; P_{Arc} – arc power, kW; k – the coefficient of absorption of the gas atmosphere of the furnace, varies in the range of EAF from 0.1 to 1.3.

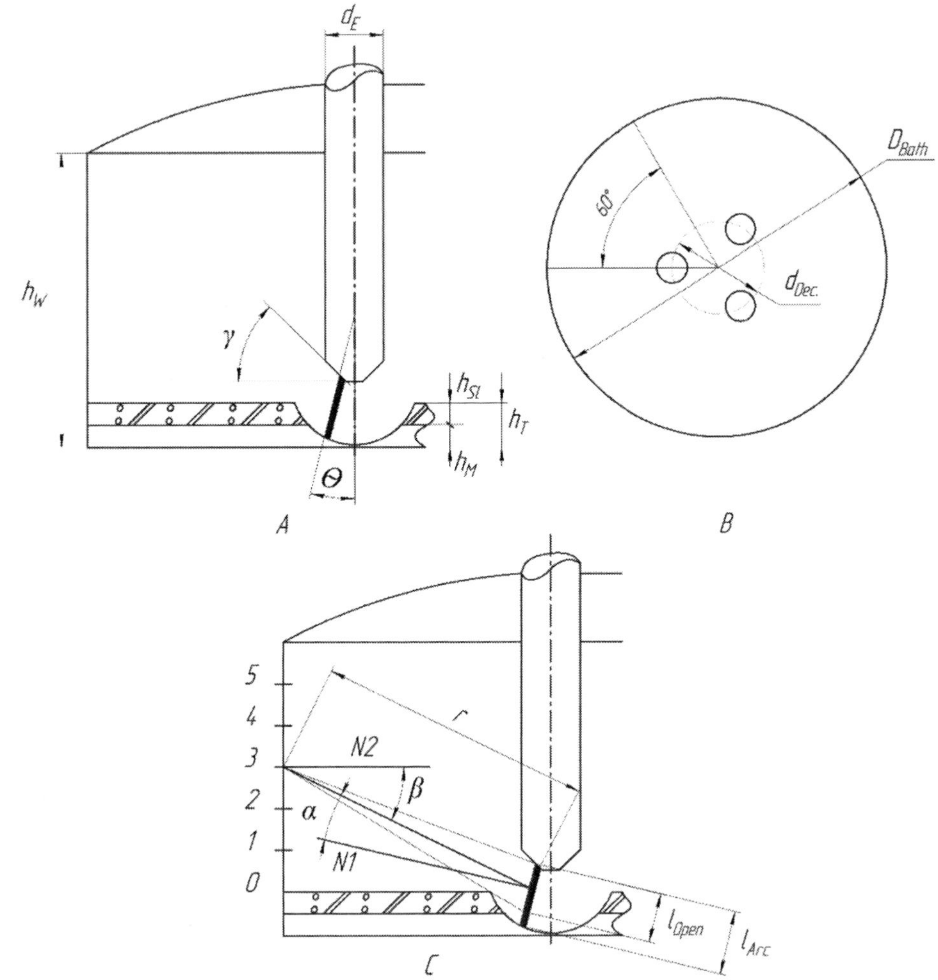

Figure 1. Diagrams of the mutual location of the arc, slag, walls (A); electrodes, metal bath, the scale is reduced (B); for calculating the fluxes of thermal radiation of arcs on the walls (C).

3. Initial parameters for calculations

The furnace atmosphere contains suspensions of solid and liquid particles, and the density of the dust-gas mixture varies from 5 to 50 g/m^3 [14–19]. Therefore, for a complete picture, it is necessary to analyze not only the ray-transparent medium, but also the gas-polluted one. The gas content in the calculation is taken into account in (1) by the absorption coefficient of the furnace's dust and gas atmosphere, which can vary from 0 to 1.4 [11–13]. In calculations, we take the average value of the coefficient $k = 0.7$.

The parameters of the DSP-100 furnace required for building the calculated model in AutoCAD (Figure 1,C) are shown in table 1[11].

Table 1. Parameters of the arc and furnace.

Parameter	D_{Bath}, m	$d_{Dec.}$, m	h_{cm}, m	d_E, mm	γ, deg.	U_{Arc}, V	I_{Arc}, kA	P_{Arc}, MW	l_{Arc}, mm	α_{Arc}
Furnace DSP-100	5,4	1,4	2,5	600	35	260	69,2	18	300	0,92

Calculation of fluxes of thermal radiation from three arcs incident to the settlement site 0…5, which are located on the vertical wall surfaces (Figure 1, C) opposite arcs and between arcs by rotating the vertical plane crossing the electrode, the arc and the wall in front of the arc, at an angle of 60° (Figure 1,B), performed by (1).

To determine the effect of the height of the slag layer on the density of thermal radiation flows on the walls of the DSP-100 furnace, the following values of the arc depth in the metal and slag are taken:

I – h_T = 70 mm – the arc is buried only in metal;
II – h_T = 160 mm – the arc is buried in metal and slag half its height;
III – h_T = 300 mm – the arc is buried at full height in the slag.

4. Result of calculation

At Figure 2 the obtained results of calculating the distribution of heat flux densities from three arcs to the sections of the furnace walls opposite one of the arcs in a dusty environment at k = 0.7 (Figure 2,a) and a ray-transparent atmosphere at k = 0 (Figure 2,b) are presented. Figure 3 shows similar calculation results, but only for the sections of the furnace walls between the arcs. And for simplicity of comparison of the received values, we will reduce them to table 2 [4].

The results of calculating the densities of thermal radiation flows of arcs on the walls of the DSP-100 furnace are confirmed by experimental data on the operation of water-cooled wall panels in the EAF [4].

In all the cases considered, the peak heat flux density of radiation observed on the height of the walls of 0.5–1.0 m. Maximum density of the heat flow, as seen from the results obtained in front of the arc at depth h_T = 70 mm at a height of 0.5–1.0 m and it is 600 kW/m^2 in a ray-transparent environment of furnace. With an increase in the height of the slag layer, the maximum density of thermal radiation flows decreases by 1.3 times when the arc depth increases from 70 mm to 160 mm and by 2 times when the arc depth increases to 300 mm.

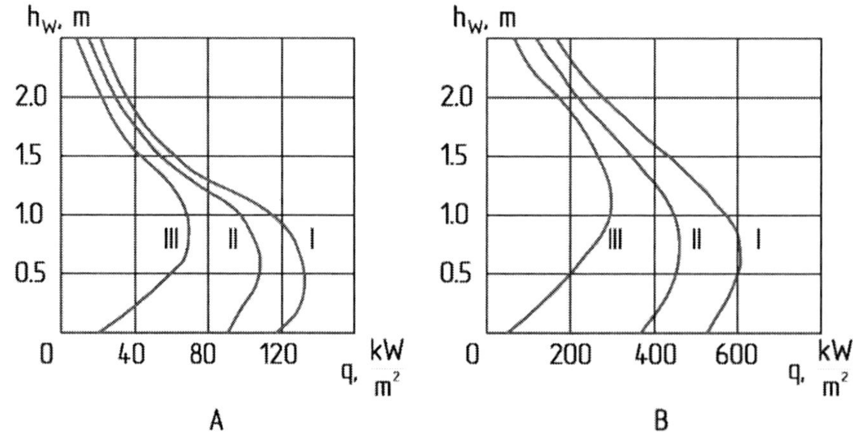

Figure 2. Distribution of heat radiation flux densities of arcs on the sections of walls located opposite the arc: when the arc depth is h_T=70 mm (I), h_T=160 mm (II), h_T=300 mm (III) in the gas-filled atmosphere of the furnace (A), in the ray-transparent atmosphere of the furnace (B).

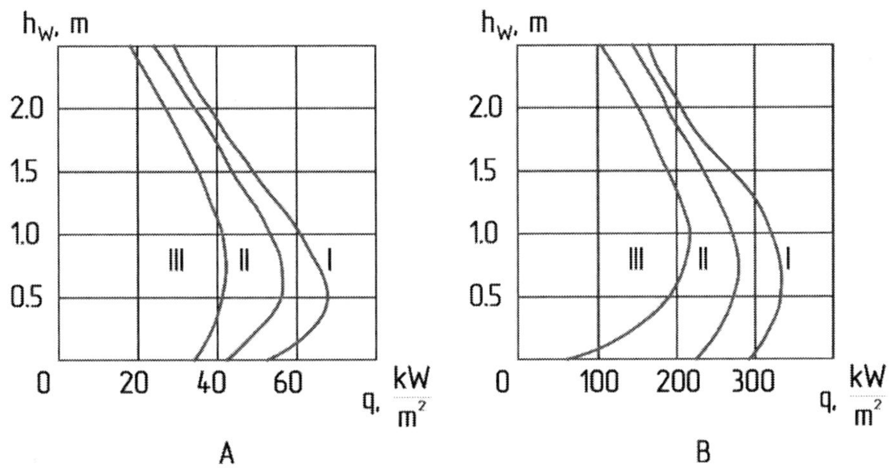

Figure 3. Distribution of heat radiation flux densities of arcs along the sections of walls located between arcs: when the arc depth is h_T=70 mm (I), h_T=160 mm (II), h_T=300 mm (III) in the gas-filled atmosphere of the furnace (A), in the ray-transparent atmosphere of the furnace (B).

When the height of the slag layer increases, the density of heat fluxes of arc radiation is redistributed, at the height of the walls from 0 to 0.5 m it decreases, and at the height of 1.0–1.5 m it increases (see Figures 2, B and 3, B).

This phenomenon is due to the fact that the open part of the arc decreases (Figure 1, C). The density of thermal radiation flows of arcs along the height of the walls decreases by an average of 1.6 times with the increase in the depth of arcs in the slag.

As can be seen from Figure 2 (and table 2) in a dusty environment of the furnace, when the absorption coefficient k = 0,7, density heat flux radiation arcs decrease in 4–5 times in the area of the maximum is from 600 kW/m^2 to 132 kW/m^2, and up to 7–8 times at the height of the walls 1.5–2.0 m. These results correspond to the data given in [3], that in the case of a ray-transparent environment, the heat flow on the wall of the DSP-100 is 350-534 kW/m^2, with maximum dustiness it is reduced by 4–4.5 times to 80–115 kW/m^2.

The obtained results of calculating the densities of thermal radiation flows from three arcs to the wall sections located along the perimeter of the walls between the arcs (Figure 3), allow us to note that the densities of thermal radiation flows are 1.5–2 times less both in a ray-transparent environment (Figure 3, B) and in a dusty environment (Figure 3, A) compared to the densities of thermal radiation flows from arcs to the wall sections located opposite the arcs (Figure 2).

With an increase in the depth of arcs in the slag from 70 mm to 300 mm, the density of thermal radiation flows of arcs to the sections of walls located between the arcs decreases by 1.2–1.6 times in the height of the walls.

According to the method described in [11, 20], the efficiency of arcs was calculated depending on the height of the arc sinking into the slag and the following relations were obtained:

when h_T = 70 mm (h_T / l_{Arc} = 70 / 300 = 0,23) – η_d = 0,46;

when h_T = 160 mm (h_T / l_{Arc} = 160 / 300 = 0,53) – η_d = 0,53;

when h_T = 300 mm (h_T / l_{Arc} = 300 / 300 = 1,0) – η_d = 0,74.

It can be seen that when the arcs are buried in the slag from 70 to 300 mm, that is, from h_T / l_{Arc} = 0.23 to 1.0, the efficiency of the arcs increases by 1.61 times.

The experience of increasing the height of the slag layer on DSP-120 from 238 to 325 mm has shown that the efficiency also increases and the specific power consumption decreases by 22% [21].

Table 2. Values of heat flux densities of arc radiation along the wall height for the main points.

Wall height	The depth of the arcs in the slag and metal bath h_T, mm					
	70 (I)	160 (II)	300 (III)	70 (I)	160 (II)	300 (III)
	Heat flux density of radiation q, kW/m^2					
	in front of the arc					
	in a dusty environment			ray-transparent environment		
0.5	132	106	56	600	450	200
1	115	96	64	572	450	290
1.5	60	54	43	440	344	264
2	37	30	22	278	230	172
	between arcs					
	in a dusty environment			ray-transparent environment		
0.5	68	56	42	330	270	184
1	62	54	41	316	275	217
1.5	50	46	35	270	234	188
2	38	34	28	212	192	154

5. Conclusion

It was found that with an increase in the height of the slag layer and with an increase in the depth of the arcs of the DSP-100 furnace from 70 to 300 mm, the density of thermal radiation flows of the arcs decreases by an average of 1.6 times over the entire height of the furnace walls. With the growth of the slag layer, the share of thermal radiation of arcs on the walls decreases by 1.6 times. The densities of thermal radiation flows of arcs are unevenly distributed both along the height and along the perimeter of the walls.

Depending on the dustiness of the furnace atmosphere, the density of thermal radiation flows of arcs on the walls of the furnace DSP-100 may differ by 4–5 times from the maximum value of 600–650 kW/m^2 in a transparent environment to 130–160 kW/m^2 with a high dustiness of atmosphere of a furnace.

The height of the slag layer affects the efficiency of the arc, so increasing the height of the slag layer increases the efficiency and when the arc is completely buried in the slag reaches 74% for the DSP-100 furnace.

References

[1] Korostelev A A, Renters A A and Semin A E 2018 Increasing the durability of the EAF lining when using hot-briquetted iron in the filling *New refractories* **3** 3–10
[2] Aksel'rod L M 2017 Iron and steel industry, refractory materials: realyty and estimates *New refractories* 11 3–14
[3] Sheshukov O Y, Nekrasov I V and Mikheenkov M A 2016 Stabilization of refining slags by correction of their phase composition *Steel* **46(5)** 339–342
[4] Kuznetsov A V, Dolgopolov V F and Derevyanchenko I V 2010 Improving the lining of electric steelmaking shop units *Steel* **1** 47–52
[5] Kucherenko O L, Repin I V and Yakubovskaya I N 2010 Experience in using direct reduced iron at Moldavian metallurgical plant *Steel* **1** 40–43
[6] Ashraf H and Zetti K M 2015 Modern electric arc furnace processes and their requirements for improved lining concepts *RHI Bulletin* **1** 17–24

[7] Watanabe I 2009 Refractory resources in the world *Taikabytsu Refractories* **61(9)** 447–452
[8] Bodyaev Yu A, Zhuravlev Yu P and Kopcev L A 2010 Optimization of energy balance and choice of operating modes for arc steel furnaces *Steel* **2** 29–31
[9] Lomakin V I, Kazman Ts D and Ustyugov A A 1977 Melting of the lining EAF *Production of electric steel* **6** 56–64
[10] Shalimov A G 2013 Formation of foamed slags in an arc furnace *Electrometallurgy* **6** 8–13
[11] Makarov A N 2014 *Heat transfer in electric arc and torch metallurgy furnaces and energy plants* (St. Petersburg: Lan`)
[12] Blokh A G, Zhuravlev Yu A and Ryzhkov L N 1991 *Radiant Heat Transfer* (Moscow: Energoatomizdat)
[13] Siegel R and Howell Y 1975 *Thermal radiation heat transfer* (Moscow: Mir)
[14] Samokhvalov G V, Chernysh G I 1984 *Electric furnaces of ferrous metallurgy* (Moscow: Metallurgy)
[15] Nikol'skii L E, Smolyarenko V D and Kuznetsov L N 1981 *Thermal Operation of Electric Arc Furnaces* (Moscow: Metallurgiya).
[16] Al'tgausen A P *Electrothermal equipment* (Moscow: Energy)
[17] Krivandin V A and Egorov A V 1989 *Thermal performance and constructions of iron and steel furnaces* (Moscow, Metallurgy)
[18] Kuzmin M G, Cherednichenko V S and Bikeev R A 2014 Water-cooled units of heavy-duty arc steelmaking furnaces *Electrometallurgy* **7** 8–16
[19] Gonzalez O J P Effect of arc length on fluid flow and mixing phenomena in AC arc furnaces *ISIJ Int.* **52(5)** 804–813
[20] Makarov A N 2019 Calculation and analysis relationship between efficiency and position of electric arcs with power consumption in EAF of small and large capacity. Part II. Calculation and analysis of the relationship between the location of arcs, walls and power consumption *Metallurgist* **5** 21–27
[21] Kuznetsov M S, Yakushev E V and Kulagin S A 2010 Influence of the mass of the metal charge and the thickness of the slag on the technology of steel smelting in an arc furnace *Electrometallurgy* **2** 2–6

Effusion Mechanism of α-Layer Formation in Vacuum Casting of Titanium Alloys

D V Saulin, V Z Poylov and N P Uglev

The Department of chemical technologies, Perm National Research Polytechnic University, 29, Komsomolskii av., Perm, 614000, Russia

E-mail: Ouglev@mail.ru

Abstract. The formation of the so-called α-layer when casting titanium alloys significantly impairs the mechanical properties of the metal and leads to the need for additional product surface treatment. It is known that simultaneously with the change in mechanical properties, the composition of the casting surface layer is often changed, and the composition change depth, as a rule, does not coincide with the depth of α-layer formed. The reason for α-layer formation, most often, is considered to be the introduction of foreign elements into the crystal lattice of a metal casting. According to our assumption, the mechanism of α-layer formation is two-stage, and it is associated with an intense diffusive effect on the melt structure in the casting outer layer, resulting from its boiling in vacuum during pouring into porous ceramic shells (CS). The paper suggests possible mechanism for composition changing of the surface α-layer in titanium castings due to the secondary processes of effusion saturation of the metal surface with reverse vapor flows coming from the porous ceramic shell after solidification of the casting. In this case, the primary process is the saturation of the CS pores due to the effusion of metal vapors during the boiling of the poured melt due to the high intrinsic vapor pressure exceeding the pressure in the foundry furnace. Chemical interaction of the alloy component vapors with the CS material leads to the volatile components: silicon, silicon II oxide, zirconium II oxide, saturating the metal surface layer due to reverse secondary effusion. The paper presents experimental results on α-layer formation when casting BT5Л alloy in CS having different composition and porosity. The research results confirm the assumption about both the diffusion mechanism of α-layer formation and the effusion mechanism of changes in its composition.

1. Introduction

When casting products made of titanium alloys, the surface layer is often formed that differs in mechanical properties and composition from the material in the casting volume (core). It is difficult to machine, has high hardness and brittleness, and is a source of cracks during solidification and operation of the product. The condition of the casting outer layer is controlled on a cross-section of the sample to a depth of at least 500 microns by Vickers microhardness determination, followed by removal of the detected α-layer by chemical etching or mechanical treatment [1, 2].

A lot of papers have been devoted to study of α-layer formation causes [3, 4]. It is believed that the casting process is accompanied by oxidation and saturation of the casting surface layer with oxygen due to interaction with residual gases, or due to chemical reactions with the shell material. The influence of many factors: material of the mold face layer; temperature of the form before pouring;

duration of exposure at high temperature; cooling rate; distribution of the components in the surface layer of castings; size and position CS in the centrifugal field during the pouring of the melt, etc. However, it is currently impossible to say that the mechanism of α-layer formation of the during titanium alloys casting in CS is clear to the end, although a number of semi-empirical recommendations have been formulated that significantly reduce the probability of its formation.

In this regard, this paper analyzes the possible causes of α-layer formation and formulates a new hypothesis of such formation, based on experimental data.

2. A brief overview of possible mechanisms of α-layer formation

Let's consider the most probable mechanisms of this effect:

1. Inheritance of the structure when passing the state diagram of the alloy from top to bottom at the temperature when cooling the castingFormatting author affiliations.
2. Influence of the second component on polymorphic transformations in the alloy.
3. Chemical interaction of alloy components with residual oxygen in the furnace volume.
4. Interaction of the melt with gases released during desorption from CS.
5. Release of gases dissolved in the melt during the crystallization.
6. Chemical interaction of the alloy components with CS material.
7. The influence of microstructure and syngony of the CS front layer components on the formation of surface crystal seed during solidification of the alloy.

Mechanism 1: in accordance with Ti-Al diagram, in a case of slow cooling and equilibrium crystallization of the BT5Л melt containing 5% w/w of Al, in the composition of the first equilibrium crystals at the "cold" wall of the ceramic form, the aluminum content in the casting should be approximately 7%, and until the Figure urative point of the alloy passes the solidus temperature, the composition of the wall layer should have increased concentrations of aluminum, while in the subsequent layers of the wall, the melt will be enriched with titanium [5]. For alloys containing 20% w/w of Al, on the contrary, the wall layer should be enriched with titanium. When the melt is further cooled below the solidus temperature, the phase compositions will be located first in the β-Ti region, and then in the α-Ti region. In practice, the composition of the casting wall layer does not correspond to this rule, which indicates the non-equilibrium cooling of the entire alloy volume. When the melt is rapidly cooled, which is typical for the process of industrial casting of titanium alloys in cold ceramic shells, the BT5Л alloy seems to immediately fall into the α-Ti zone, while it may partially remain "trapped" β-Ti zones that affect the casting mechanical properties. But this state should be characteristic of the entire casting volume, and not just for the surface layer. In addition, there are no correlations with the initial temperature and thermal conductivity of the shell. When using CS with high thermal conductivity, for example, from graphite, the degree of non-equilibrium of the crystallization process, and α-layer size in this case should increase. This should also occur when reducing the size of the casting - the smaller mass of the casting, the faster it cools. The latter two dependencies are contrary to experimental data. It should also be noted that the maximum content of aluminum in the surface layer often exceeds the theoretically possible value by almost twice (Figure 1). Given that aluminum and other components are not always contained in the casting surface layer, it can be assumed that they are formed due to a secondary, external process.

Mechanism 2: It is known that impurities of the second component affect the rate of transition of a substance from one to another allotropic state [5]. When changing the parameters of the crystal lattice, the sample may crack. It should be noted that cracking is also found in the surface layers of titanium castings, which are most enriched with impurities of foreign elements in comparison with deeper layers of metal. Such effect allows to link the properties of the casting surface layer with changes in its structure due to changes in composition. However, cracking does not always occur, and it is not clear why α-layer is often present, but there is no extraneous component. For example, from the analysis of curves in Figure 1, 2, it follows that the main drop in the concentration of aluminum occurs at the first 40 microns of the casting depth, after which it does not change, while the microhardness of the same sample continues to decrease up to 500 microns. Strange is the fact that the hardness increases in the direction of the casting wall, where the content of a relatively soft component is greater:

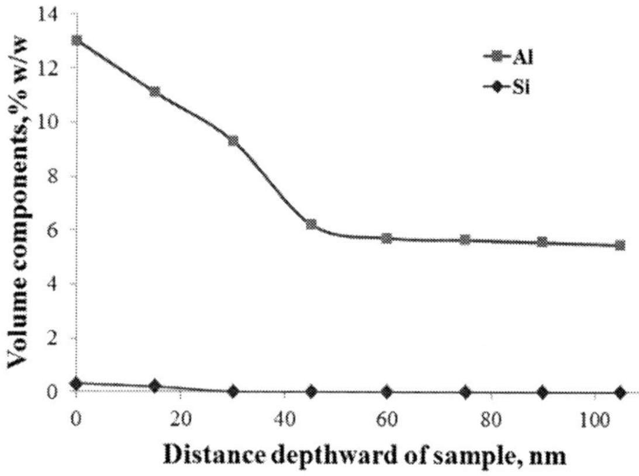

Figure 1. Distribution of aluminum (upper curve) and silicon in surface layers of BT5Л titanium alloy.

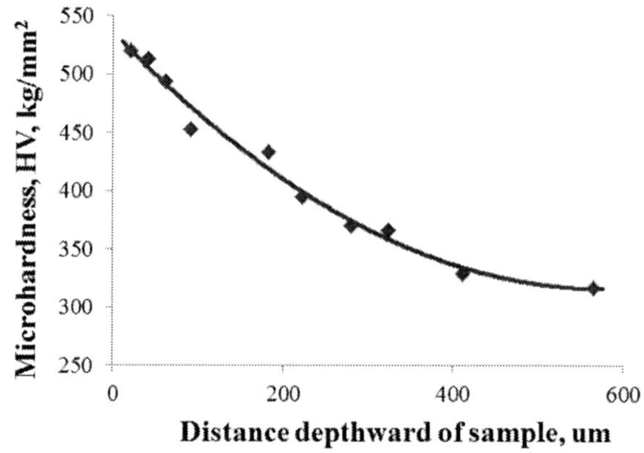

Figure 2. Hardness in α-layer (Microhardness Tester FM-300, USA).

Thus, the hardness of α-layer of titanium castings samples does not correlate with the concentration of aluminum. Therefore, the excess aluminum and silicon contained in the alloy are not the cause of the formation of α-layer in this case.

This result suggests that the reason for the change in hardness is the variable structure of the cast alloy acquired when the sample solidifies, the characteristics of which are functionally and independently related to the distance from the external surface of the sample, and not to the content of impurities of foreign elements.

Mechanism 3: when holding the solid sample of titanium alloy in air at an elevated temperature (700-800°C), it is gradually saturated with oxygen, and the saturation rate increases significantly with increasing temperature, which is associated with an increase in the rate of diffusion of oxygen atoms into the metal volume. The external gas-saturated α-layer has an increased microhardness, which decreases with a decrease in the oxygen content. The thickness of α-layer also gradually increases, reaching 100 microns in 500–1000 hours. A clear logarithmic relationship between microhardness and oxygen content in the titanium alloy has been revealed, which was shown in [6]. When applied to the actual casting conditions of titanium alloys, an unresolved issue in the interpretation of these results is the source of oxygen, possibly diffusing into the melt from the interface. If this question does not arise during heat treatment of solid castings in an oxygen-containing atmosphere, then in the case of casting the alloy in ceramic shells in a vacuum (10^{-3} mm Hg) or purified argon, the source of oxygen is not clear. The temperature of the molten metal is within 1700 °C, aluminum and titanium are intensively vaporized during melting and are chemically very active chemical elements, so they perform the functions of a getter inside the vacuum furnace, completely absorbing residual oxygen, water vapor, and probably also residual nitrogen. At the above operating pressure, no more than 3–4 mg of oxygen can remain in the volume of the centrifugal vacuum furnace (≈ 10 m^3), which is sufficient to form α-layer with a thickness of no more than 1 μm, distributed over the total surface of all the loaded workpieces, and even then, provided that all the residual oxygen passes into the melt.

Mechanism 4. Gases that can be adsorbed on the surface of the pores of ceramic shells include oxygen and nitrogen in the air. Since they have relatively low critical condensation temperatures well below the operating temperature of the CS, capillary condensation is completely excluded. Thus, oxygen can be adsorbed practically only in the form of a monomolecular layer. We will make a rough, overestimated assessment of the amount of residual oxygen introduced into the furnace volume in the adsorbed form. Let's assume that adsorption at one atmosphere leads to the formation of a monomolecular layer of oxygen on the entire inner surface of the ceramic with a density corresponding to the liquid state. When applying the vacuum, the adsorption capacity is proportionally reduced by 5-6 orders of magnitude. Research of foundry ceramics ("Sorbi" MS device for determining the surface) showed that its specific surface, with the porosity of 5 %, corresponds to 1.9 m^2/g, which, in terms of the entire volume of ceramics at full load of the furnace, is estimated to be no more than $0.5 \cdot 10^6$ m^2. The average pore diameter is 26 nm. After vacuuming the furnace volume, no more than 1.2 mg of oxygen can remain on the ceramic surface. At the time of casting metal into molds, the vacuum in the furnace, indeed, for a short time deteriorates due to the desorption of gases, but then quickly recovers. However, it is most likely that this is due to the desorption of water vapor rather than oxygen, since ceramics have some hygroscopicity. When the mold is heated at the time of filling, desorbed gases will mostly move in the pores as a Knudsen flow towards the evacuated volume of the furnace, rather than towards the dead end created by the melt. In any case, the oxygen pressure at the melt surface will be in the range of 10^{-3}–10^{-4} mm Hg, which will not allow the mechanism 4 to have any noticeable significance in the formation of α-layer.

Mechanism 5. The release of dissolved gases during the crystallization of the melt is quite likely, but it should be taken into account that during electron-arc metal remelting in vacuum, the melt enters the storage crucible in fairly small drops with a developed surface and has the high temperature that leads to complete evaporation of foreign volatile components. Therefore, in our opinion, mechanism 5 cannot make a significant contribution to the formation of α-layer.

Mechanism 6: it is assumed that the metal oxides that make up the ceramic shell of the mold, depending on the composition, can partially physically dissolve in the molten titanium in molecular form (with subsequent dissociation), or chemically interact with the melt to form both metal titanites and pure components, such as silicon. For example, in [3] we present the results of the influence study of the ceramic face layer composition on the characteristics of α-layer. It is shown that the oxygen content in the metal boundary layer increases sharply with simultaneous drop in the concentration of titanium, and fluctuations in the content of aluminum and zirconium are insignificant. The results are interpreted as oxygen insertion into the metal layer, accompanied by distortion of the alloy crystal lattice, leading to a significant change in the mechanical characteristics of the metal in the boundary layer. The depth of penetration of both oxygen and other impurities is determined by the diffusion coefficients and the duration of phase contact, and most often, at a significantly shorter length than α-layer (see mechanism 2). It should be noted that solid reaction products formed at the first moment at the interface, especially those that are not soluble in the metal melt, will block the further flow of the reaction due to a sharp decrease in the rate of diffusion supply of components to the contact zone. The latter is confirmed experimentally - for example, it is shown [7] that when a pure titanium melt contacts a ceramic material made of TiO_2 and $TiAl_3$, simple dissolution or chemical interaction does not occur, which is confirmed by the absence of α-layer, although titanium oxides are completely soluble in liquid titanium (see the Ti-O diagram [5]).

Thermodynamic analysis [8, 9] showed that a very limited number of oxides remain chemically inert to titanium at casting temperatures, among which ZrO_2 and Y_2O_3 are of the greatest interest, which are recommended in foreign literature as materials that guarantee a reduction in the thickness of α-layer (Figure 3).

However, despite the significant inertia of the materials, α-layer is formed in this case, although on a small scale. Thus, it is independently shown that the mechanism of formation is not related to chemical interactions in the system. You should notice the fact that during casting in a vacuum furnace, or in a furnace filled with clean argon gas, the cooling rate of the melt poured into cold form is quite high, causing the duration of contact of the melt with the mold material estimated by the equation of Newton-Richman, subject to delays due to crystallization is 6 to 10 minutes, after which the metal solidifies. With further cooling of the solid casting, taking into account the difference in the expansion coefficients of metal and ceramics, we should expect the loss of physical contact between the metal and the wall and termination of possible chemical reactions between them [10]. Even when using components capable of chemical reactions with titanium for CS formation, chemical reactions seem to be strongly hindered not only by a lack of contact time, but also by mass transfer difficulties. So, as an example, it should be pointed out that when casting titanium alloys in chemically active graphite shells (or shells coated with graphite), α-layer is minimal.

Mechanism 7. The influence of the seed on the crystallization process in melts is not in doubt. In this case, the syngony of the introduced seed crystal must be at least isomorphic to the resulting phase. A decrease in the number of seeds usually leads to an increase in the average size of new crystals formed. Given that CS wall material uses granular materials with the grain size of up to 100 microns, we should expect a large-crystal structure directly from the surface of the resulting casting. In fact, the coarse-grained structure of the metal appears in the casting not from the surface layer, but somewhat deeper. It is significant that the microhardness of the sample decreases smoothly both within a single crystal and in the case of a polycrystalline structure of the surface layer of the casting (Figure 4). Thus, we can conclude that the dependence of the hardness on the layer depth is the result of an external general physical factor that does not depend on the macrostructure of the condensed matter.

The crystallization front at the boundary of the liquid and solid phases usually has a sharp break and does not allow an exponential distribution of the substance properties over the depth, which occurs when studying the hardness within α-layer (Figure 3, 4, 6). Strange, also, is the fact that the positive effect on α-layer is provided by materials with different syngony – ZrO2, Y2O3, pyrocarbon (graphite), from which we can conclude that there is no specific effect of these components on the embryo formation during the crystallization of the titanium alloy. Therefore, mechanism 7 also cannot claim to be the general physical cause of the formation of α-layer.

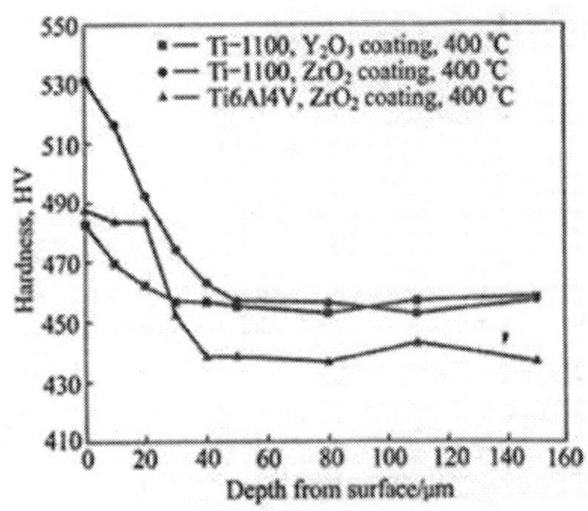

Figure 3. Changes in the micro-hardness of titanium alloys in casting α-layer if casting into ceramic molds made of yttrium and zirconium oxides (the temperature of the ceramic mold before casting is 400°C [3]).

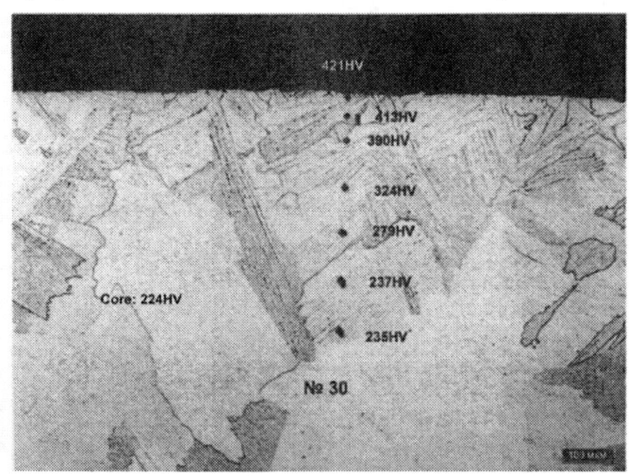

Figure 4. Microstructure and distribution of hardness HV, kg/mm^2 by the depth of BT5Л alloy sample.

Thus, it should be recognized that none of the mechanisms discussed above can claim to fully explain the effect of α-layer, which is the result of some external general physical phenomenon that appears in the liquid during the crystallization.

3. Proposed mechanism of formation of α-layer

One of the general factors that were not previously considered related to the casting conditions of titanium alloys in centrifugal machines is the intrinsic vapor pressure of the molten metal. The condensing vapor of metals and oxides thereof is collected on the chamber walls and in vacuum filters in the form of dust in sufficiently large quantities, which makes it possible to consider the evaporation of the melt during casting a fairly significant factor in the casting process. Calculation of natural vapor pressure of ВТ5Л alloy from temperature based on reference data [11] at temperatures close to 1700 °C leads to the value of 1.2 mm Hg, while the additive contribution of titanium is approximately $5 \cdot 10^{-3}$ mm Hg. In any case, the melt poured into a ceramic shell with open pores, and is under mechanical pressure (in the lower part of the gating system up to 10 atm) due to centripetal acceleration, is under the external pressure of the surrounding gas $1 \cdot 10^{-3}$–10^{-4} mm Hg, which is significantly lower than its own vapor pressure over the metal. The latter is just the condition for boiling the liquid. Under the influence of high mechanical pressure, the metal partially enters the pores of the ceramic, forming convex menisci, with a curvature less than critical, which does not allow it to flow into the pores, since it does not wet the CS material. The metal "boils in the pores", while powerful diffusion flows are formed in the conjugated layers of the casting, which violate the structure of the melt near the surface [12]. It is clear that when the mechanical pressure of the metal increases in the zones most remote from the center of rotation, the curvature of the meniscus, which provides a counterbalancing Laplace counter-pressure, increases. It is known from the general thermodynamic laws that the pressure in the liquid also affects the intensity of evaporation of the melt. When boiling, the metal is further cooled and the outer layer of the melt passes in the quenching mode into a solid non-equilibrium state with a disturbed structure. Thus, in our opinion, α-layer is formed having the properties significantly different from the normal crystalline metal.

Figure 5 shows typical results of the study of the cross-section composition of used CSs after ВТ5Л alloy casting in them in industrial conditions. The results confirm the effusion of titanium vapors into the ceramics pores, and effusion intensity depends on the density. It should be noted that the metal vapor flow is very significant - during the period of filling and cooling of the part, the amount of effusive titanium was up to 1.1 mg per 1 cm^2 of the shell. We have not identified the condensed aluminum on the background of ceramic corundum, but its flow is much larger, and it leads to a fairly fast blocking of the through pores of CS. The depth of this visually noticeable "aliting" of CS is 2–3 mm. Different intensity of steam flows from the melt in CS with different density (porosity) is accompanied by quantitatively different violations in the structure of the surface metal layers of castings, which appear as α-layer through the dependence of microhardness on the distance to the casting surface (Figure 6).

The fact of structure violation of the substance under the influence of diffusion flows is known [13].

Additional experiments on the influence of the porosity of the shell on the regularities of the formation of α-layer were carried out.

4. Experimental methods and results

Experimental samples were made from a model mass in the form of cylinders with a rounded end having diameter of 10 mm, a length of 60 mm, which were soldered to a common gate with a diameter of 50 mm. Previously, the samples were coated with the two-layer protective coating up to 400 microns thick, containing various components that lead to changes in the characteristics of α-layer. After that, 8–10 layers of ordinary ceramic suspension of corundum powder of different sizes were applied to the prepared block. Own binder of the protective layer was prepared on the basis of silica sol and alumina sol. After standard drying procedures, model melting and CS firing, ВТ5Л titanium alloy was poured into the finished ceramic molds on the ДВЛ-200Р centrifugal unit. Experimental samples were cut approximately in the middle and the microhardness and composition of the surface layer were determined using Microhardness Tester FM-300 (USA), and a scanning electron microscope S-3400N, Hitachi, (Japan) with the Brooker unit for quantitative X-ray spectral analysis. Figure 7–18 below shows the results of synchronous analysis

IOP Conf. Series: Materials Science and Engineering **969** (2020) 012060 doi:10.1088/1757-899X/969/1/012060

of Vickers microhardness and composition of a number of samples made of BT5Л alloy. The value of α-layer (the microhardness drop at distance of 400 microns and the length of this layer) is considered by us as an indicator of the intensity of the boiling of the melt associated with the amount of metal trapped in the form of vapors into the ceramic shell.

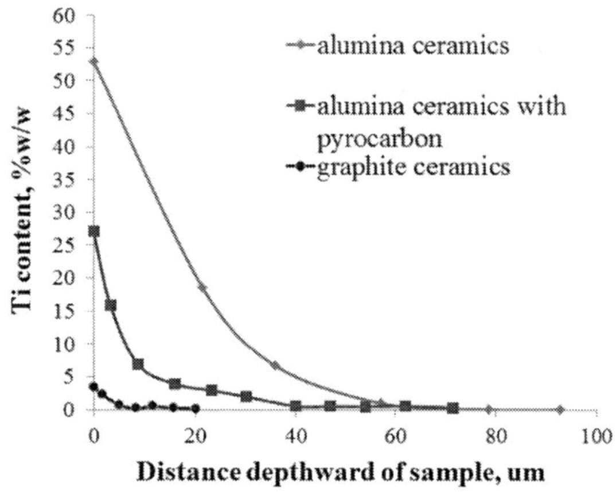

Figure 5. Titanium content in the surface layers of used ceramic shells of various densities.

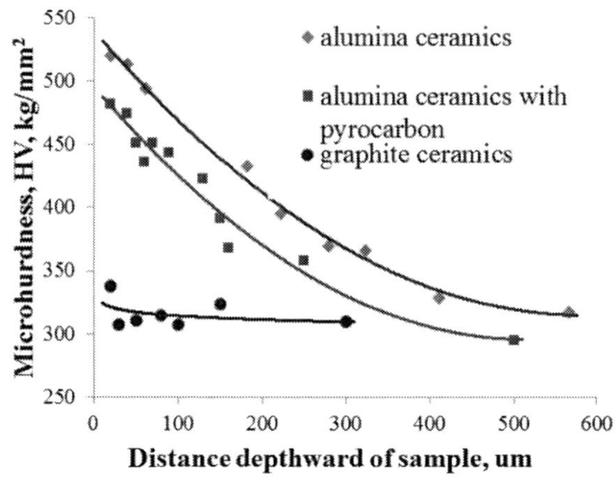

Figure 6. HV microhardness in surface layers of BT5Л parts cast in ceramic shells of different densities.

In some of the drawings below, the content of aluminum and other components is shifted up by 60-70%, which allows extend the scope and present the results more clearly. A slight change in the composition of the surface layer of the metal (Figure 8) is visible only to a depth of 20 microns. At the same time, the length of α-layer (Figure7) is much greater. This pattern is observed for all experimental data, including those presented in this paper, which further confirms the assumption that α-layer is not directly related to the composition of the casting in this area.

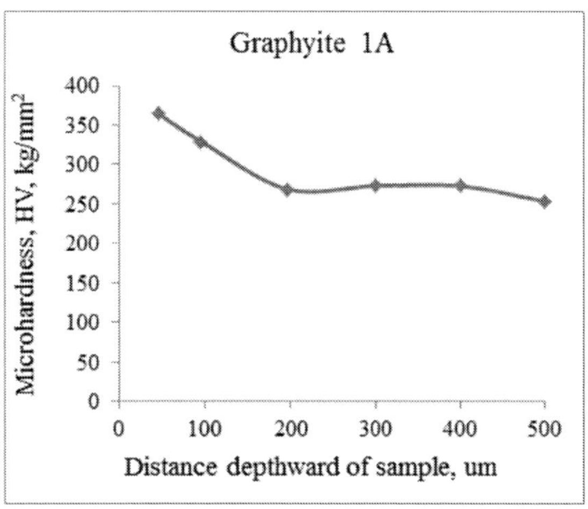

Figure 7. Change in microhardness with the distance to the casting boundary.

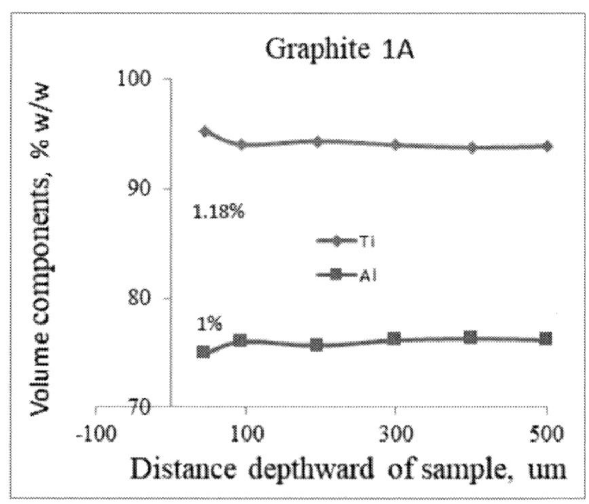

Figure 8. The composition of the surface layer (Al) +70%.

Attention is drawn to the fact of the multidirectional deviation of the concentration of aluminum in the range from 0 to 100 microns in the experiments in Figure 8 and 10 (synchronously with aluminum, the same is seen for titanium). In accordance with the previously described idea of the formation mechanism of α-layer, it can be imagined that in the case of intensive effusion of aluminum into the pores of ceramics, which exceeds the effusion of titanium by several hundred times, their concentration in the adjacent layer decreases faster than that of titanium. In accordance with the balance equation, the part of titanium increases, which is shown in Figure 8. In the next experiment (Figure 9, 10), smaller graphite particles were used, which led to better sealing of the pores and condensation of liquid aluminum in them, as a result of which it accumulated in the wall layers of the CS, restored some of the silicon that is part of the binder, and together with it returned as vapors to the pores of the casting (Figure 10). It is clear that in the case of (Figure 8), a much smaller amount of aluminum vapor returned back, since the pores were open both ways.

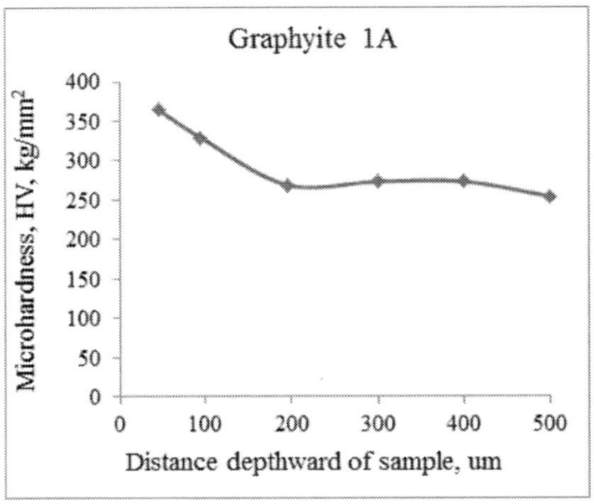

Figure 9. Change in microhardness with the distance to the casting boundary.

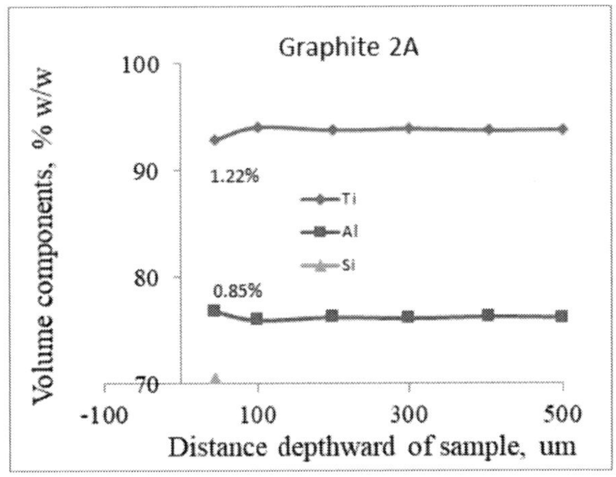

Figure 10. The composition of the surface layer (Al, Si) +70%.

Similar processes are observed when using zirconium oxide as a facial protective layer, both in a mixture with aluminum oxide, and in pure form (Figure 11–14). However, in this case, alumina sol was used as a binder [14–17], so in this experiment series, silicon was not detected in α-layer. It should be noted that in this case, the protective ceramic layer had a significant porosity, so that the amount of evaporated aluminum (and titanium) in the pores of the ceramic is significantly higher than what was observed in the first two experiments (the estimated proportions of displaced aluminum in the layer up to 100 microns are indicated in the Figures in %). Sample 4 (Figure 12) was about twice as close to the axis of rotation as sample 5 (Figure 14), so the melt pressure in the latter case was about twice as high. Perhaps an increase in pressure in the liquid metal can be explained by the decrease in evaporation in this case.

When conducting an experiment using pure zirconium oxide as the front layer (Figure 16), the sample was placed in a low-pressure zone. Almost unimpeded access of aluminum vapors to zirconium oxide led to an intensive chemical reaction with almost complete exhaustion of aluminum. Instead, an almost balanced amount of volatile chemical reaction products returned to the casting. Thus, we can conclude that the second condition for the formation of an intense steam flow from a liquid metal into a ceramic shell is the possibility of chemical interaction of aluminum and ceramic components with the formation of volatile products [18–21].

Additional verification of the possibility of chemical reactions with inert fluorite (Figure 18) and magnesium oxide, when using a silica-based binder, confirmed the above regularities.

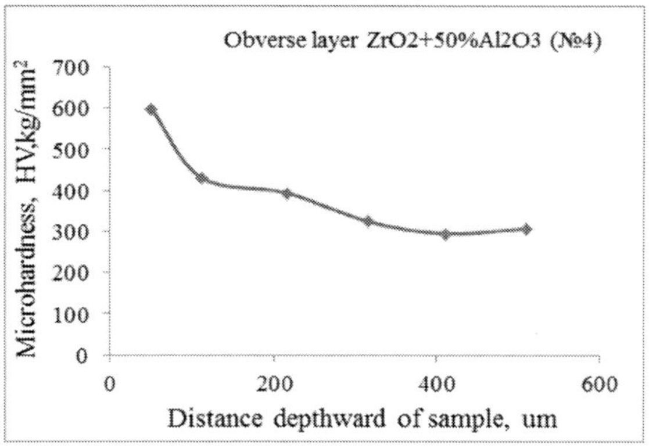

Figure 11. Change in microhardness with the distance to the casting boundary.

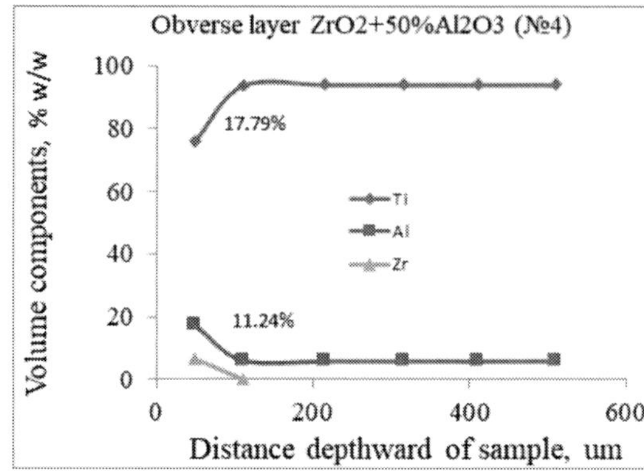

Figure 12. The composition of the surface layer.

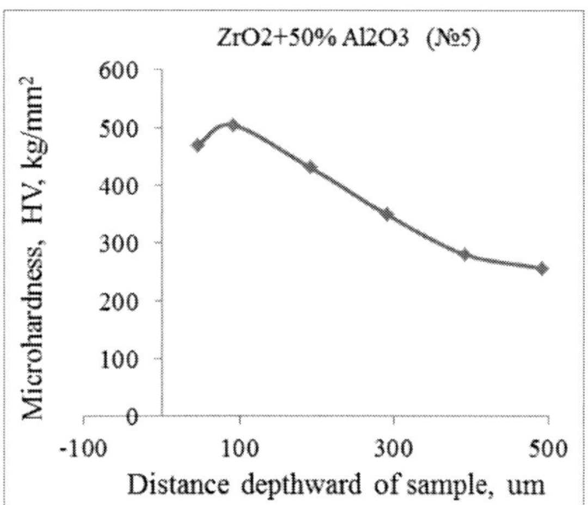

Figure 13. Change in microhardness with the distance to the casting boundary.

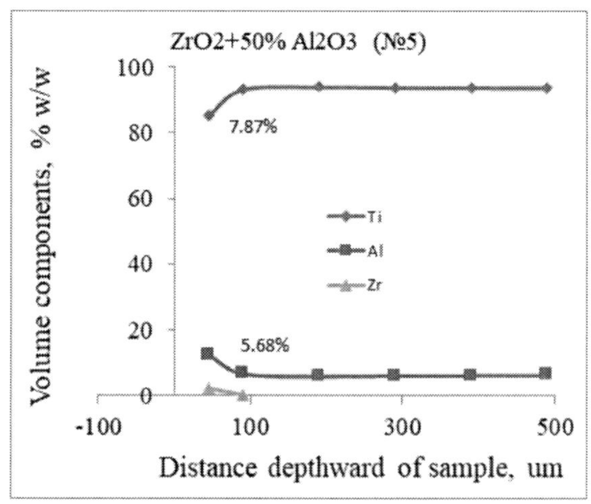

Figure 14. The composition of the surface layer.

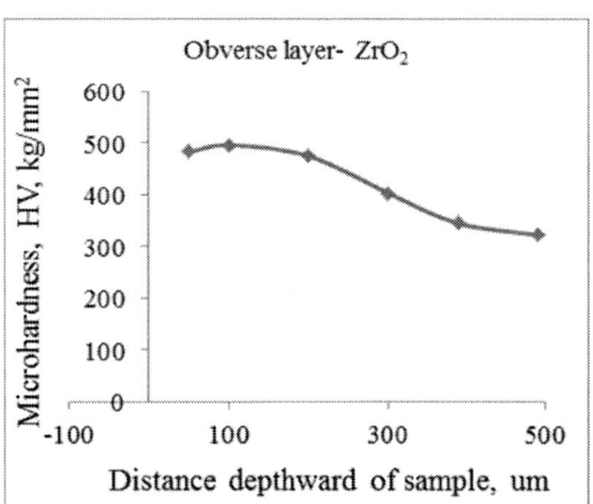

Figure 15. Change in microhardness with the distance to the casting boundary.

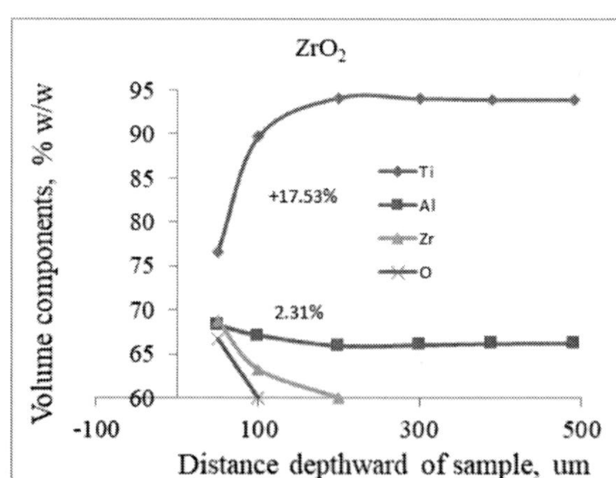

Figure 16. The composition of the surface layer (Al, Zr, O) + 60%.

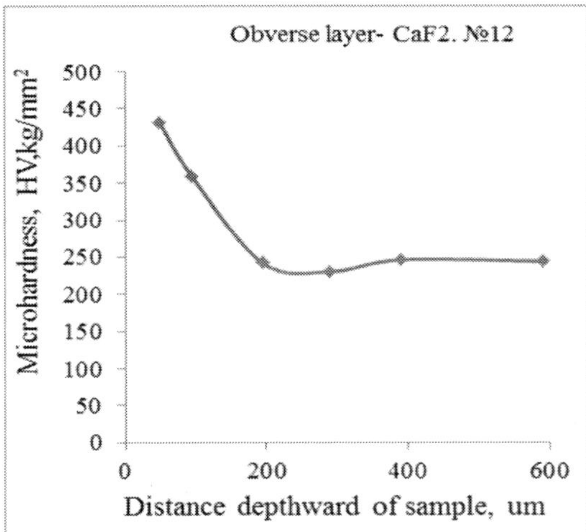

Figure 17. Change in microhardness with the distance to the casting boundary.

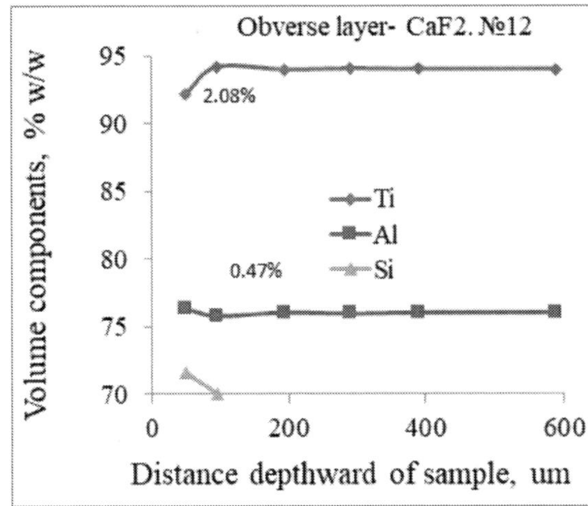

Figure 18. The composition of the surface layer (Al, Si) +70%.

5. Discussion of experiment results

Based on the mechanism of formation of α-layer proposed in section 2, we can assume the following mechanism for changing the composition of the surface layer of the casting: as a result of the boiling process, the aluminum contained in the alloy begins to evaporate into the pores of the shell, taking part of the titanium with it [21]. Further, in the volume of the porous ceramic shell, aluminum condensation occurs, which, under vacuum conditions, accumulates and seals the through pores of the shell. Due to high temperatures, all condensed aluminum in the pores is in a liquid state. The process of casting solidification takes place when the pores of the CS are completely blocked from the side of the poured melt. With a further decrease in temperature, due to different expansion coefficients, the solid part is separated from the surface of the ceramic to form a gap. At the same time, Al, which is located in the pores of CS, retains its liquid state for a long time. The pressure of aluminum vapor at these temperatures is significant, and significantly exceeds the pressure in the casting furnace [11], so it can move for a considerable time (1-2 hours) in the form of a molecular (atomic) flow through the gap between the CS and the metal casting to the low-pressure zone of the furnace. Thus, an increased concentration of Al vapors is formed on the outer surface of the metal, leading to the effusion of its atoms into the pores of the surface layer of the solid casting. In other words, a secondary alitization process is taking place, which causes an increase in the concentration of aluminum within some external part of α-layer. In the case of CS containing silicon oxide (when used as a silica binder), a chemical reaction takes place inside the ceramic [9]:

$$SiO_2 + Al(g) = Si(g) + Al_2O_3. \quad (1)$$

It is clear that in this case, the amount of free aluminum capable of secondary alitization decreases in proportion to the reaction loss (1), which is observed in Figure 16, in contrast to the results of the experiment in Figure10, where the decrease in the proportion of titanium almost completely corresponds to an increase in the proportion of aluminum in the surface layer. It is also possible that a reaction occurs with the formation of a volatile SiO incomplete oxide, but it should be expected that oxygen appears in the outer layers of the casting. The pressure of silicon vapors at high temperatures is much higher than that of aluminum, so the process of silicification of the metal part with silicon vapors occurs by the same mechanism. A similar situation occurs in the case of using zirconium dioxide in CS together with a binder based on an almost inert alumina sol.

Despite the exceptionally low vapor pressure of pure zirconium, it enters the casting by the same mechanism as a gaseous incomplete oxide formed by the reaction:

$$3ZrO_2 + 2Al(g) = ZrO(g) + Al_2O_3. \quad (2)$$

It should be noted that aluminum can also form volatile oxides at high temperatures that can carry oxygen into the outer layer of the casting [21].

6. Conclusion on the work

Thus, the main way to reduce the intensity of the formation of α-layer is to reduce the flow of metal vapors in the CS during casting by reducing the porosity of ceramics, which corresponds to the empirical recommendations known from the literature for casting titanium alloys. This rule is automatically triggered when using fine-grained materials (grain size less than 1 micron) for CS, such as ZrO2, Y2O3, MgO, pyrolytic carbon (graphite), which show the best results with respect to α-layer.

We tested this assumption in a special experiment with a different method of applying a two-layer coating of aluminum and yttrium oxides (Figure 19):

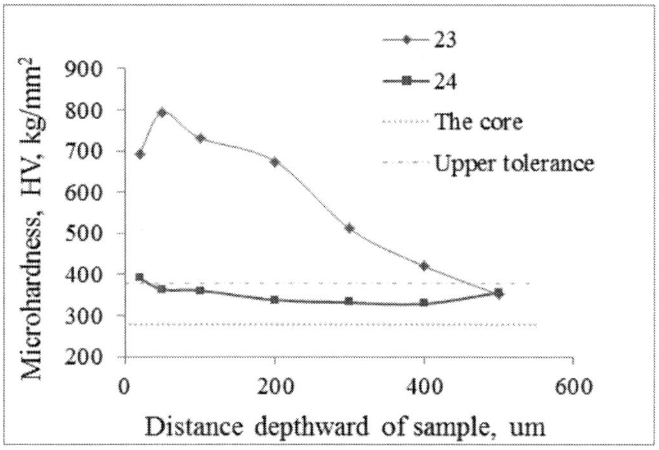

Figure 19. The value of α-layer when using different combinations of oxides in both protective layers applied to the CS: 23 - First layer Al_2O_3, second - Y_2O_3; 24- First - Y_2O_3, second - Al_2O_3.

Of greatest interest are the results for sample 24, which demonstrate almost no α-layer. It is significant that changes in the order of alternating layers of a porous sublayer with thickness of 200 microns made of corundum F1000 (sample 23) were sufficient for the effusive flow of aluminum vapors into this layer from the surface of the boiling melt to cause an intense conjugate diffusion flow in the near-surface layer of the casting with a violation of the metal structure and the formation of unacceptable α-layer.

References

[1] Prokhorov A M 1995 *Chemical encyclopedia* (Moscow: Great Russian Encyclopedia)
[2] Leyen C and Peters M 2003 *Titanium and Titanium Alloys. Fundamentals and Applications* (Weinheim: Wiley-VCH Verlag)
[3] Zhao E, Kong F, Chen Yu and Li B 2011 *Trans.of Nonfer. Met. Soc. of China* **B 21** 348–352
[4] Saha R L, Nandy T K, Misra R D K and Jacob K 1989 *Bull. Mater. Sci* **B 12** 481–493
[5] Lyakishev N P 1996 *State diagrams of double metal systems* (Moscow: Mashinostroenie)
[6] Leyens C 2014 *Oxidation and Alpha–Case Phenomena in Titanium Alloys used in Aerospace Industry* (Luleå University of Technology SE–971 87 Luleå)
[7] Si-Young S, Bong-Jae C, Beom-Suck H, Han-Jun O and Young-Jig K 2008 *J. Mater. Sci. Technol* **B 24** 70–74
[8] Ganeev A A, Demenok A O, Bakerin S V, Kulakov B A, Mukhamadeev I R and Garipov A R 2016 *Bulletin of the South Ural State University (Metallurgy)* **B 16** 70–78

[9] Nikitchenko M N, Semukov A S, Saulin D V and Jaburov A Yu 2017 *Bulletin of PNRPU. Chemical technology and biotechnology* **B 4** 249–263
[10] Chretien J, King M, Proia W and Rudolf S 2010 *Titanium Alpha Case Prevention* (Worcester Polytec. Inst)
[11] Kikoin I K 1976 *Tables of physical quantities. Handbook* (Moscow: Atomizdat)
[12] Ouglev N P, Poilov V Z and Smirnov S A 2017 *Foundry* **B 8** 2–8
[13] Beyer W J 2000 *Non-Cryst. Solids* **B 2** 845-849
[14] Si-Young S, Bong-Jae C, Sang-Ho N, Chang-Su H, Yong-Mun R, Beom-Suck H and Young-Jig K 2007 *Advanced Materials Research Online* **B 26–28** 519–522
[15] Chennakesava R A 2016 *Int. J. of Advanced Research* **B 4** 101–106
[16] Mukhamadeev I R, Demenok O B, Ganeev A A, Pavlynich S P and Alikin P V 2015 *Bulletin of the South Ural State University (Metallurgy)* **B 15** 95–104
[17] Myoung-Gyun K, Shae K K and Young-Jig K 2002 *The Jap. Inst. of Met. Material Transactions* **B 43** 745–750
[18] Bauristhene A M, Mutombo K, Stumpf W E 2013 *J. of The Southern African Institute of Mining and Metallurgy* **B 113** 357–361
[19] Mutombo K, Mazibuko N E, Rossow P 2011 *ITA Conference* 1–25
[20] Branscomb T 2015 *ITA Conference* 1–10
[21] Kostomarov D V 2011 Melts **B 4** 89–96

Acknowkedgments
The research was carried out using scientific equipment of the "Center for science-intensive chemical technologies and physical and chemical research" of PNRPU in accordance with the project of the Perm REC (RFMEFI62120X0038) "Rational subsoil use".

IOP Conf. Series: Materials Science and Engineering **969** (2020) 012061 doi:10.1088/1757-899X/969/1/012061

Study of the Principles for Ensuring the Accuracy of CCM Design Parameters

A L Sotnikov[1], A A Sholomitskii[2] and S M Strichenko[3]

[1]Donetsk National Technical University, 58, Artema St., Donetsk, 83001, Ukraine
[2]Siberian State University of Geosystems and Technologies, 10, Plakhotny St., Novosibirsk, 630108, Russia
[3]Enakievsky Metallurgical Plant, 9, Metallurgov ave., Enakievo, 86429, Ukraine

E-mail: sholomitskij@mail.ru

Abstract. It is established that the defining factor of accuracy of the designing parameters of S-type billet continuous casting machine (CCM) is relative difference of curvature of the adjacent arc-like sections, which form process axis of strand. An analysis of the local radius change of CCM ingot path curvature is carried out. In the process of CCM equipment alignment, it is necessary to ensure alignment of its curvature centers under condition of curvature value equality for minimization of its relative difference. Developed technology of position control and CCM equipment alignment allows meeting this requirement. The developed technology is based on the use of "Vizir 3D" a high-precision measurement system built on the base of electronic total station. Practical experience of alignment showed the efficiency of geodesic and combined method of equipment alignment on both billet and slab CCM.

1. Introduction

Casting of steel at billet continuous casting machines (CCM) gets widespread use among iron and steel plants of small size and volumes of steel products manufacturing [1].

In the process of CCM equipment alignment there occur deviations (errors) of its position in relation to the position stipulated by design and engineering documentation. Position errors depend on many factors attributed to both accuracy of assembly and erection works carried out, and conditions and modes of CCM [2-4] equipment operation, in that condition of refractory materials [5, 6]. Meaning and mode of indicated deviations and errors could be presented as random ones, which cause dispersion of accuracy of the CCM design parameters [7]. As a rule, study and analysis of random deviations are carried out using statistical techniques, based on theory of probability. The techniques allow establish a certain connection between the possible values of random variable and probability of its occurring.

2. Status of issue

Total deviation (error) of CCM equipment position can be determined by using statistical techniques based on experimental data, since numeric values of random variables are grouped around its average value. The measure of its values dispersion near the center of grouping is a standard deviation.

In order to carry out the statistic analysis there were taken experimental data of distribution of the deviation of actual position coordinates of the secondary cooling zone rollers and withdrawal and straightening unit of Danieli billet CCM, illustrated in Figure 1. Since a sufficient number of independent similar close-variant parameters is considered, the law of random dispersion approaches normal distribution law. This law is graphically expressed as a symmetric curve asymptotically approaching axis of abscissa. The area limited by normal distribution curve and axis of abscissa expresses probability of random value entering the given interval. It is proved that with a probability of 99.73 %, i.e. close to 1, the field of random values dispersion relating to average value is within the $\pm 3a$ range, where a is a random variable [8]. This interval is to be compared to permissible deviation of CCM equipment position. If the condition that the permissible deviations of position $[\delta] \geq \pm 3\sigma_\delta$, where σ_δ is a standard deviation of coordinates of CCM rollers position, is fulfilled, then setting accuracy is achieved, i.e. the tolerance is technologically provided.

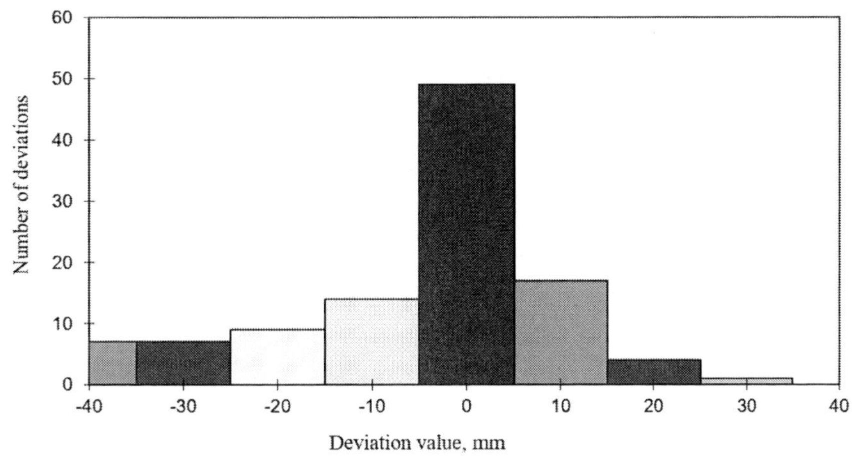

Figure 1. Bar graph of the actual coordinates deviation distribution of the Danieli billet CCM rollers position.

The parameters of position deviations distribution are set out in table 1. Statistical analysis has shown that permissible limit deviation of CCM equipment position during its installation with application of conventional methods and means, is as follows:

$$\sigma_\delta = \sqrt{\frac{1}{n}\sum_{i=1}^{k}\left(\delta_i - \bar{\delta}\right)^2 m_i} = \sqrt{\frac{1}{108}12427.55} = 10.73\,\text{mm}, \tag{1}$$

$$\bar{\delta} = \frac{1195}{108} = 11.06\text{ mm}, \tag{2}$$

where n=108 – number of measurements; δ_i – random variable (roller position deviation), mm; k – number of different values (intervals) of random variable; m_i – rate of random variable occurrence.

Then permissible limit deviation of the billet CCM rollers position coordinates at its control and alignment using conventional methods and means is as follows:

$$[\delta] \geq \pm 3 \cdot 11.06 = \pm 33.18\,\text{mm}, \tag{3}$$

which clearly does not conform to the requirement of design and engineering documentation of CCM manufacturing plant – tolerance limit is ±0.2…0.5mm.

Table 1. Statistical distribution parameters of the position coordinates deviation of secondary cooling zone rollers and withdrawal and straightening unit of Danieli CCM.

Interval δ_i, mm	Interval mid-point, δ'_i, mm	Rate of occurrence, m_i	$m_i\delta'_i$	$\delta'_i-\bar{\delta}$	$(\delta'_i-\bar{\delta})^2$	$m_i(\delta'_i-\bar{\delta})^2$
0…5	2,5	49	122.50	-8.56	73.36	3594.45
5…15	10	31	310.00	-1.06	1.13	35.15
15…25	20	13	260.00	8.94	79.84	1037.89
25…35	30	8	240.00	18.94	358.54	2868.33
35…40	37,5	7	262.50	26.44	698.82	4891.73
Total:	–	108	1195.00	–	–	12427.55

3. A new method for solving the problem

If we consider, for example, only deviation of secondary cooling zone rollers position with respect to "ideal" process axis, i.e. axis radially set in accordance with design and engineering documentation, then, according to the error theory, we can calculate an absolute error in the whole roller guide system of the machine:

$$\sum_{i=1}^{n}\Delta_i\,, \tag{4}$$

where n – number of rollers; Δ_i – deviation of i-roller from the ideal process axis. We can also determine an absolute error of each secondary cooling zone (SCZ) roller:

$$\frac{\sum_{i=1}^{n}\Delta_i}{n}\,. \tag{5}$$

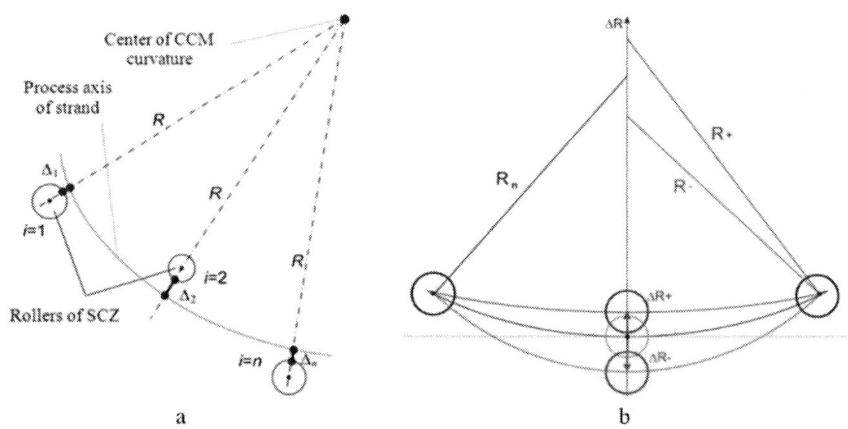

Figure 2. Computational scheme for determination of absolute error and standard deviation of secondary cooling zone rollers position (*a*) and scheme of local curvature radius change of the CCM process axis (*b*).

Standard deviation of roller position with respect to "ideal" process axis will be equal to the following:

$$\sigma=\sqrt{\frac{\sum_{i=1}^{n}\Delta_i^{\,2}}{n-1}}\,. \tag{6}$$

IOP Conf. Series: Materials Science and Engineering **969** (2020) 012061 doi:10.1088/1757-899X/969/1/012061

Table 2 shows the calculation results for secondary cooling zone rollers of Danieli 6-strand billet CCM according to experimental data provided in this work [9]. Analysis of given data shows that the least absolute error and standard deviation of roller position are observed on strand #5, and the most – on strand #6. The present analysis itself yields little information since it does not correlate to any data, presented by the authors of study.

Table 2. Absolute error and standard deviation of the secondary cooling zone roller position of Danieli CCM with respect to "ideal" process axis.

Strand	Rollers of secondary cooling zone					$\sum_{i=1}^{n}\Delta_i$	$\frac{\sum_{i=1}^{n}\Delta_i}{n}$	$\sigma=\sqrt{\frac{\sum_{i=1}^{n}\Delta_i^2}{n-1}}$
	Δ_1	Δ_2	Δ_3	Δ_4	Δ_5			
#1	38.32	9.98	7.89	8.49	0.19	64.49	12.90	20.63
#2	32.87	20.65	1.59	6.86	1.92	60.05	12.01	19.75
#3	38.40	22.42	7.75	5.95	4.78	79.30	15.86	22.89
#4	22.93	28.38	3.36	1.76	0.07	56.36	11.27	18.34
#5	25.10	19.59	0.04	1.59	5.34	40.90	8.18	16.16
#6	38.66	27.19	12.60	10.42	0.87	88.00	17.60	25.01

The scheme of CCM rollers position, shown in figure 2a, allows introducing a concept of "local radius change of process axis curvature". A simulation of the curvature radius change at the change of the area middle roller position was performed for three adjacent rollers of radial area (figure 2b).

If rollers are placed on the circumference of preset radius (*Rn*), then it corresponds to design position (it is shown in figure 2b by middle black line). If the middle roller is shifted by ΔR+ value toward the curvature center, it will lead to the increase of curvature radius to $Rn+\Delta R$ value and, consequently, to the decrease of the process axis curvature in this area. If the roller is shifted towards the opposite side of curvature center by ΔR – value, it will lead to the decrease of curvature radius by $Rn–\Delta R$ value and, consequently, to the increase of the process axis curvature in this area. The results of simulation for some range of ΔR values and different types of CCM are shown in figures 3 and 4, respectively, by means of graphs of the curvature radius changes and process axis curvature.

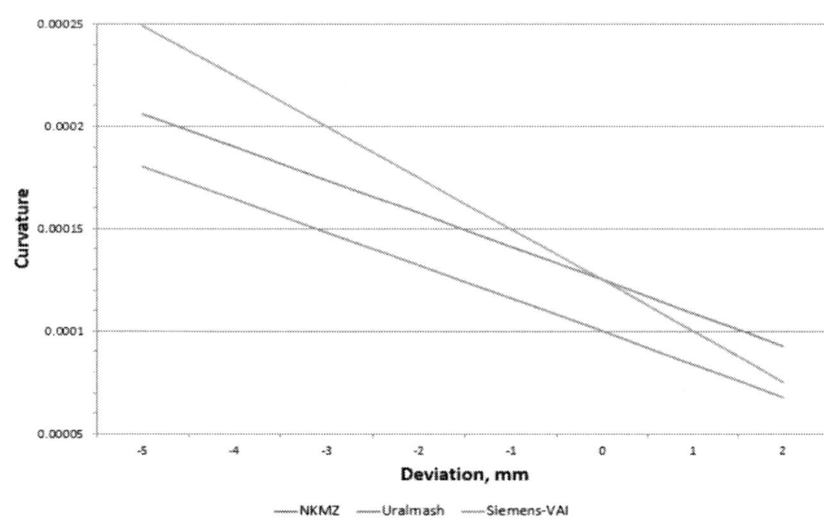

Figure 3. Graphs of the curvature radius change of process axis of the different CCM types.

IOP Conf. Series: Materials Science and Engineering **969** (2020) 012061 doi:10.1088/1757-899X/969/1/012061

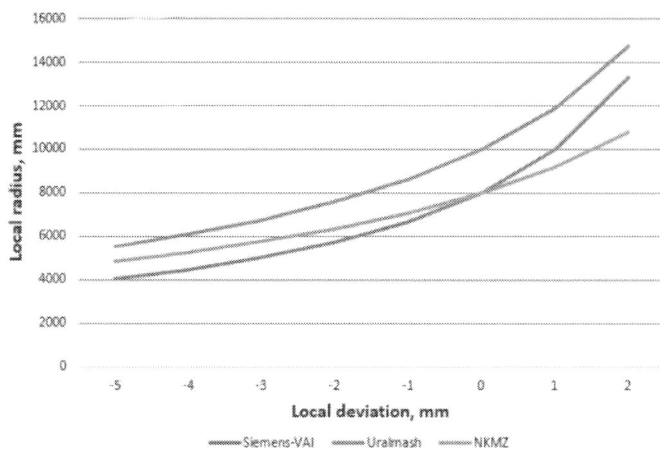

Figure 4. Graphs of the curvature radius change of process axis of the different CCM types.

Many scientific efforts, dedicated to designing of CCM strand process channel and selection of its main parameters [10–12], indicate that deformation stress of ingot guide σ_n for its placing on level depends on the curvature radius R of the process axis, for example, at its simplest [12]:

$$\sigma_n = \frac{Eh}{2R}, \tag{7}$$

where E – module of metal elasticity of the concast ingot at melt temperature; h – ingot thickness.

Indicated dependance shows that the radius change of the process axis curvature leads to the change of guide stress, and this dependence is linear. Curvature radius increase by 10 % will lead to proportional increase of deformation stress by 10% as well, and vice versa. At this, if we consider the possibility of the curvature radius change (for Danieli CCM R=8000 mm) by ±0.2…0.5 mm, it is apparent that it will have little effect to the value of deformation stress.

In case of multi-point unbend of concast ingot, a process axis of CCM strand is formed by a number of adjacent arcs of circumferences having different radius. In this respect the works [10,11] indicate that process axis curvature and degree of deformation are to be constant for the whole area of ingot formation. Curvature function, as well as its derivatives #1 and #2 should not have jumps. And that could be possible if curvature of each arc-like area of process axis, characterized by its radius of arc (curvature radius) will be equal both to each other and to some preset value. Generally, curvature is equal to the following:

$$K = \frac{1}{R}. \tag{8}$$

4. Practical application

As shown by experimental data relating Danieli billet CCM [9], in secondary cooling zone process axis has two areas with different radius of circumference arcs, which form it. Each circumference arc is adjacent to three sequentially located rollers. Using mathematical model of process axis of the CCM strand, developed in the work [9], both actual radius of each circumference arc and coordinates of curvature center were calculated, as well as absolute deviation from CCM curvature center specified by design and engineering documentation. Based on these data it was established that there is a point of concast ingot bending at a roller of section #3 of the secondary cooling zone of each strand. Using these data, we can also give quantitative assessment of the process axis shape deformation in each CCM strand, presented in table 3 and in figure 5.

Comparison of presented data and results of carried out cause-quantity analysis of failures and malfunction of the roller guide [13, 14] of Danieli 6-strand radial billet CCM in operation [9] allows drawing a conclusion that it is the difference of curvature value of the adjacent arc-like sections, which form process axis of strand, that determines failure-free performance of the concast ingot formation area equipment, as well as stability and safety processes of steel continuous casting at CCM. Diagram, presented in figure 5, shows that process axis of strand #3, which, during the considered period, is characterized by minimum number of failures and malfunctions [9], has a transition from one arc-like section to another with curvature radius change by 0.19%, and standard deviation of actual curvature center coordinates is 54.94 mm. As for strands #2, #5 and #6 we can observe that relative difference in curvature radius value tends towards minimum value, and standard deviation of curvature center coordinates – towards maximum value, which can also be associated with low failure-free performance of CCM equipment. As for strands #1 and #4, the situation is also obvious, i.e. increase in deviation of the curvature centers position of adjacent arc-like process axis sections results in difference increase of its curvature radius.

Table 3. Parameters of process axis in secondary cooling zone of Danieli billet CCM.

Strand	Radius of axis curvature in the area between rollers of		Standard deviation of the actual curvature center coordinates	Relative difference in the curvature radius value, %
	section #1 and #3	section №3 and withdrawal and straightening unit		
#1	8325.44	7911.29	334.16	4.97
#2	7928.74	7855.08	186.09	0.93
#3	8029.30	8013.87	54.94	0.19
#4	7672.07	8001.56	358.25	4.12
#5	7846.10	7909.24	201.86	0.80
#6	7983.72	7916.88	112.06	0.93

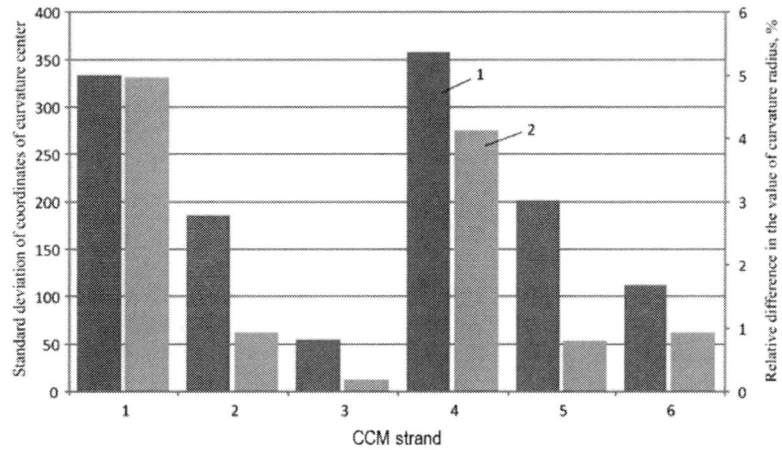

Figure 5. Standard deviation of the actual curvature center coordinates (1) and relative difference in the curvature radius value (2) in the process axis area of secondary cooling zone of Danieli billet.

Similar results of the study were received with using the example of slab CCM, designed by Uralmash. It is equipped with 69 rollers, and 54 of them are located on radial-curved area. Figure 6 shows data on deviation of CCM rollers position from design position. Bar graph of deviations rate in corresponding values ranges shows that deviations follow the normal law of distribution, and the majority of deviations is within ±2 mm range.

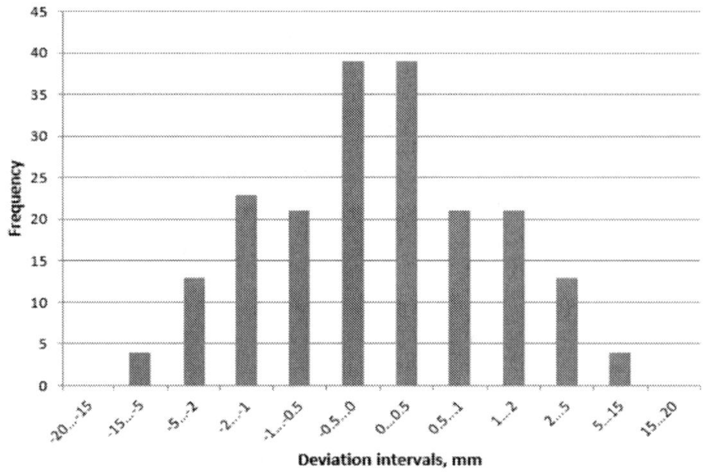

Figure 6. Bar graph of deviation distribution of the actual roller position coordinates of the slab CCM, designed by Uralmash.

The analysis of absolute CCM roller position deviation showed that there are only two sections of process axis, where local values of curvature radius change are equal to over 10 mm. As it was shown before, it leads to increase of curvature radius and corresponding decrease of process axis curvature of the machine.

Working experience of authors in CCM alignment using geodesic methods [15–17], shows that good results in ensuring preset values of the process axis curvature radius of both billet and slab CCM can be achieved with using combined method of position control and alignment of roller guide. The first and the last roller of a certain area (of secondary cooling zone) are installed according to geodesic alignment data, and intermediate rollers – using an arc-like template [18–20]. Such a technology would not result in accumulation of errors, which is usual, when CCM alignment is carried out with using only templates. It must be understood that geodesic measurements are carried out only after machine total shut down, during major overhaul or routine maintenance. However, the efficiency of geodesic methods is very high, as level of rejected concast billets at slab CCM decreased twice, and at billet CCM – decreased by 30–40%. Experience in carrying out geodesic alignment of CCM at the enterprises of the Ukraine and Russia showed that such alignment is to be carried out once a year, and always during carrying out major overhauls.

5. Conclusion

Thus, in the process of CCM equipment alignment in the concast ingot formation area with regard to process axis, in the first instance one should be guided by deviations of equipment position coordinates regarding data preset by design and engineering documentation. In the second instance, by the value of relative curvature difference of the adjacent arc-like sections, which form process axis. Minimum value of difference can be achieved by matching their curvature centers under condition of curvature value equality. Deviation of CCM equipment from process axis in radial direction within the 0.2…0.5 mm range was established on the basis of ensuring minimum deviation of the actual centers coordinates of the adjacent arc-like sections curvature of process axis.

References

[1] Sotnikov A L, Kireev V N and Orobtsev A Y 2013 Continuous casting of steel at bar-casting machines in Ukraine and Moldova *Steel in Translation* **43(12)** 808–814

[2] Gharaibeh Nabeel S, Matarneh Mohammed I and Artyukh V G 2014 Loading Decrease in Metallurgical Machines, *Research Journal of Applied Sciences, Engineering and Technology* **8(12)** 1461–1464

[3] Tsuprun A Yu, Fedosov A V, Kislitsa V V and Pashchuk D V 2012 Study of the effect of the characteristics of harmonic mold-oscillation regimes on the surface quality of continuous-cast ingots *Metallurgist* **56(7-8)**
[4] Yan Xiaobo, Jia Boran, Wang Qiangqiang, He Shengping and Wang Qian 2019 Mold Nonsinusoidal Oscillation Mode and Its Effect on Slag Infiltration for Lubrication and Initial Shell Growth during Steel Continuous Casting *Metals* **9** Online. Available: https://doi.org/10.3390/met9040418
[5] Tochilkin V V, Filatova O A and Umnov V I 2018 Analysis of the Process of Metal Casting and the Structure of Refractory Equipment of the "Tundish Ladle – Submerged Nozzle – Crystallizer" System of a Section CBCM, *Refractories and Industrial Ceramics* **59(1)** 6–9 Online. Available: https://doi.org/10.1007/s11148-018-0172-x
[6] Vdovin K N, Tochilkin V V and Yachikov I M 2016 Designing refractories for the tundish of a continuous caster *Refractories and Industrial Ceramics* **56(6)** 569–573 Online. Available: https://doi.org/10.1007/s11148-016-9889-6
[7] Nikitchenko A, Artiukh V, Shevchenko D, Larionov A and Zubareva I 2019 Application of Nonlinear Dynamic Analysis for Calculation of Dynamics and Strength of Mechanical Systems *Advances in Intelligent Systems and Computing* **983** 496–510 Online. Available: https://doi.org/10.1007/978-3-030-19868-8_49
[8] Kravchenko V S 1975 *Installtion of ship power plants* (Shipbuilding)
[9] Sotnikov A L 2016 *Development of scientific basis and practice of ensuring accuracy of design and process parameters of billet continuous casting machines* (Donetsk National Technical University)
[10] Brovman M Y 2007 *Comtinuous casting of metals* (Ecomet)
[11] Bulanov L V 2004 *Billet continuous casting machines. Theory and calculation* ("Marat" Ural PR and advertising center)
[12] Timokhin A O 2000 Features of CCM process axis calculation and control *Steel* **2** 16–21
[13] Brezinová J, Viňáš J, Brezina J, Guzanová A and Maruschak P 2020 Possibilities for renovation of functional surfaces of backup rolls used during steel making *Metals* **10** 164 Online. Available: https://doi.org/10.3390/met10020164
[14] Maruschak P O, Panin S V, Zakiev I M, Poltaranin M A and Sotnikov A L 2016 Scale levels of damage to the raceway of a spherical roller bearing *Engineering Failure Analysis* **59** 69–78
[15] Mogilny S G, Sholomitskii A A, Revutskii V N and Prigarov V A 2009 "Vizir 3D" measurement system on enterprises of Ukraine: Geodesic control and alignment of process equipment *Geoprofile* **3(6)** 12–19
[16] Mogilny S G, Sholomitskii A A and Sotnikov A L 2009 Geodetic works at examination of alignment of billet CCM equipment *Metallurgical processes and equipment* **2(16)** 19–27
[17] Mogilny S G, Sholomitskii A A, Lunev A A, Sotnikov A L, Vatralik E M and Frolov I S 2012 Impovement of technology of the position control and alignment of CCM equipment *Metallurgical processes and equipment* **3(30)** 12–25
[18] Sotnikov A L and Sholomitskii A A 2017 Monitoring Alignment of Mold Oscillatory Motion with CCM Process Stream Axis *Metallurgist* **60(9–10)** 1046–1053
[19] Sholomitskii A and Sotnikov A 2019 Position Control and Alignment of the CCM Equipment *Materials Science Forum* **946** 644–649
[20] Mogilny S G, Sholomitskii A A and Sotnikov A L 2020 Technical Audit of Rotary Aggregates *Lecture Notes in Mechanical Engineering* 541–549

IOP Conf. Series: Materials Science and Engineering **969** (2020) 012062 doi:10.1088/1757-899X/969/1/012062

Investigation of Influence of Silicon Carbide Phase and Chemical Composition on its Recovery

D A Boldyrev[1], L I Popova[1] and A A Tokarev[2]

[1]Togliatti State University, 14, st. Belarusian, Togliatti, 445020, Russia
[2]South Ural State University, 76, Lenin av., Chelyabinsk, 454080, Russia.

E-mail: 151@nppgroup.ru

Abstract. When employing the cast iron manufacturing technology involving the use of silicon carbide, it was discovered that recovery of silicon carbide and that of its individual elements, viz., carbon and silicon varied. Based on the results obtained from a diffraction analysis and electron microscope investigation with electron microprobe analysis, the genuine structural and chemical features of industrial-grade silicon carbide were described. The phase composition of silicon carbide is inhomogeneous and consists of particles of the α-SiC crystalline silicon carbide and oxidized non-crystalline silicon carbide whose basis is a solid solution "carbon-silicon". Recovery of silicon carbide by molten cast iron determined by the increase in the contents of carbon and silicon during the melting process is inversely on the content of the α-SiC phase.

1. Introduction

There is currently a tendency among cast iron foundries to use steel scrap as a replacement for expensive pig iron as part of charge materials to produce compacted graphite iron (CGI) products. In fact, pig iron being melted promotes formation of a larger amount of nucleation centers, or sites, in the base iron than molten steel scrap [1]. Therefore, when using steel scrap instead of pig iron, it is essential to monitor and adjust the chemical composition of the melt [2, 3]. For this purpose, metallurgical grade silicon carbide is used. It enriches the melt with silicon and carbon (it is believed that silicon carbide favors formation of a pure synthetic melt with nucleation sites (solidification centers)). Silicon carbide contributes to the increase in the amount of solidification centers as a result of which the amount of graphite nodules in the melt also increases, and the tendency of the cast iron to form cementite and ledeburite within its structure (also known as the chilling tendency of cast iron) decreases [4, 5]. The traditional charge materials used for manufacturing CGI products include fine-grain graphite and, normally, ferrosilicon of 45 per cent silicon (ФС/FS45 grade). These materials are primary products as they are manufactured from source materials and are used for the purpose specified which makes them relatively expensive. Notably, the graphite used for its intended purpose is delivered without being subjected preliminarily to activating treatment [6, 7] which explains its low recovery and the ferrosilicon has uncontrollable impurities transferring from the quartzite. Use of silicon carbide, as a source of carbon and silicon, in the iron foundry industry may constitute an advantageous alternative in terms of economy and technology to use of the traditional charge materials [8, 9]. In this case, carbon and silicon are decomposition pure products free of uncontrollable impurities. An additional technical effect resulting from its application is that the total weight and volume of the charge materials used with each ton of silicon carbide is less than those used with each ton of fine-grain graphite and ferrosilicon of 45 per cent silicon.

IOP Conf. Series: Materials Science and Engineering **969** (2020) 012062 doi:10.1088/1757-899X/969/1/012062

However, as silicon carbide is being tried out commercially, its recovery is found inconsistent which manifests itself case by case in the predominant recovery of either carbon or silicon [10, 11]. In order to establish the cause of such peculiar recovery of silicon carbide, one sample of fine-grain silicon carbide with a size of 0-10 mm and the content of SiC being 92–95% was taken from three batches of silicon carbide: 1. conventionally "good" sample; 2. conventionally "bad" sample (with poor recovery of carbon; 3. conventionally "bad" sample (with poor recovery of silicon).

The investigation methods used are presented in Table 1.

Table 1. Investigation methods.

Methods	Equipment
Electron microscope investigation	LEO1455VP scanning electron microscope
Electron microprobe analysis	EDX X-MAX80 energy dispersive spectrometer
X-ray phase analysis	ARL X'Tra X-ray diffractometer

2. Materials and experimental procedure

For an electron microprobe analysis to be performed, certain quantities of the samples have been deposited on a carbon-based substrate [12, 13]. SEM images are shown in figures 1–3, the element contents in the particle X-ray spectra are presented in Table 2.

Table 2. Contents of silicon carbide sample elements in particle X-ray spectra, Wt%.

Sample No.	Spectrum No.	Element, Wt%		
		C	Si	O
1	16	59.2	36.2	4.6
	17	71.5	24.4	4.2
	18	52.0	44.4	3.6
	19	52.3	28.6	19.2
	20	34.7	33.0	32.3
	21	68.9	18.8	12.3
	22	81.8	13.4	4.8
2	1	35.1	33.0	31.8
	2	58.2	36.2	5.6
	3	69.8	18.0	12.2
3	4	55.8	32.2	12.0
	5	76.7	17.8	5.5
	6	51.1	35.8	13.0
	7	48.5	38.6	12.8
	8	35.8	34.7	29.5
	10	53.5	40.5	6.0
	11	51.3	32.9	15.9
	13	42.9	35.0	22.1
	14	92.3	4.2	3.5
	15	62.2	27.6	10.2

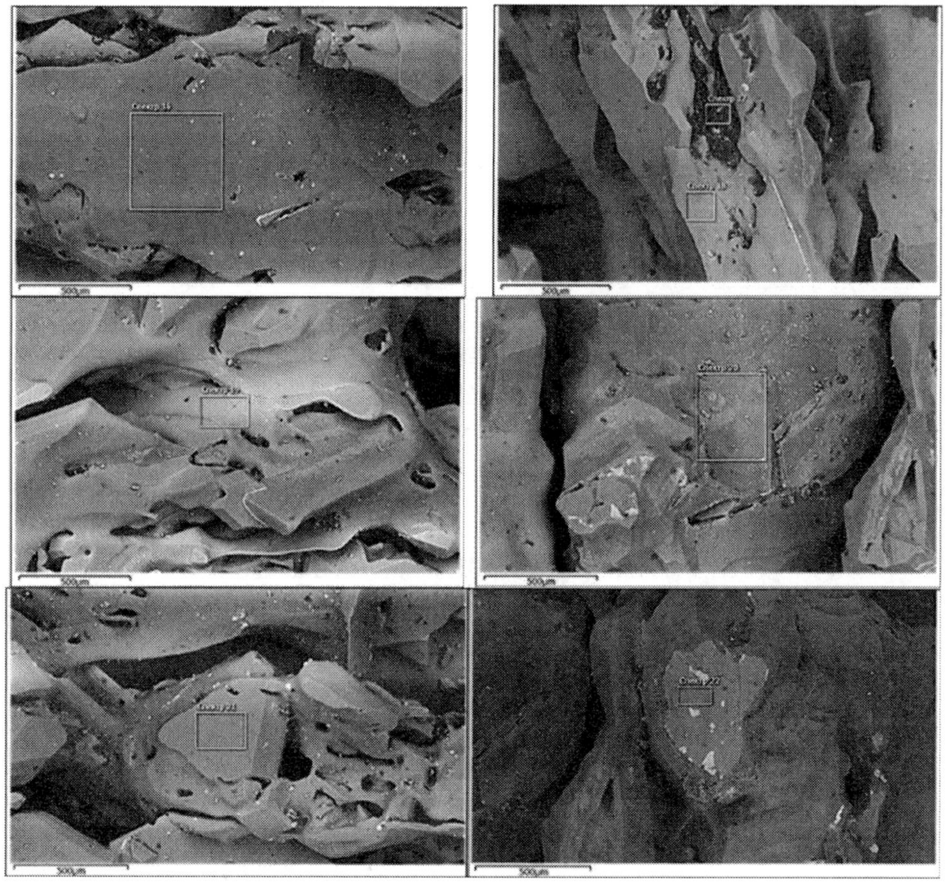

Figure 1. (Scanning) electron microscopic images of a fragment of sample No.1; marked with rectangle is an area subjected to a local electron microprobe analysis.

Figure 2. (Scanning) electron microscopic images of a fragment of sample No.2; marked with rectangle is an area subjected to a local electron microprobe analysis.

Only carbon, silicon and oxygen have been analyzed, the impurities have been left out. The SEM images were obtained with the help of a back-scatter detector. The intensity of the grey color in these images indicates the positions of the above elements in the periodic table (the lighter the grey color, the higher number in the periodic table the element has). The weight percentage values of the elements in the above table are qualitative [14, 15]. The electron microprobe analysis has shown that the samples being investigated have the following four phases: hexagonal silicon carbide (its amount was so insignificant that it was not taken into account further in the analysis process), rhombohedral silicon carbide (α-SiC), graphite, and silicon. Diagrams showing the contents of carbon, silicon and rhombohedral silicon carbide (α-SiC) are depicted in Figure 4.

The morphological types of the particles in each of the three samples analyzed are identical. They are represented by two major types. The first type is characterized by monolithic (solid), homogeneous and less oxidized particles with a higher content of silicon relative to carbon; the second type is distinct for a predominant content of carbon, more oxidized small particles forming agglomerations being bonded by loose and porous carbon compound. The weight stoichiometric 7:3 silicon-to-carbon ratio for both types of the particles is not observed [16]. The actual ratio values are either higher or lower than the above value. The samples differ from each other in terms of the quantitative relationship between the particles of both types and their dimensions. Particles of the second type and particles of the petcoke that have failed to react mainly find themselves in the pores and cavities of the particles of the first type.

The morphological types of the particles in each of the three samples analyzed are identical. They are represented by two major types. The first type is characterized by monolithic (solid), homogeneous and less oxidized particles with a higher content of silicon relative to carbon; the second type is distinct for a predominant content of carbon, more oxidized small particles forming agglomerations being bonded by loose and porous carbon compound. The weight stoichiometric 7:3 silicon-to-carbon ratio for both types of the particles is not observed [16]. The actual ratio values are either higher or lower than the above value. The samples differ from each other in terms of the quantitative relationship between the particles of both types and their dimensions. Particles of the second type and particles of the petcoke that have failed to react mainly find themselves in the pores and cavities of the particles of the first type.

3. Conclusion

Sample No. 1 is predominantly represented by the largest particles of the first type. The particles of the first type are oxidized less as compared to the similar particles of the other samples. In sample No. 2 more oxidized particles of the second type prevail over the rest of the particles with the content of carbon being higher than that of silicon. Conversely, in sample No.3 more oxidized particles of the second type prevail over the rest of the particles with the content of silicon being higher than that of carbon. Based on the results obtained from a combination of qualitative electron microprobe analysis and x-ray phase analysis, it is safe to conclude that industrial-grade silicon carbide is an inhomogeneous material made up of particles of crystalline silicon carbide α-SiC and amorphous silicon oxycarbide. This is corroborated by the absence of an identified crystalline lattice of the silicon-carbon phase of the permanent composition discovered when the x-ray phase analysis was performed. An x-ray phase analysis visualizes it not as a unified phase but as an individual combination of carbon and silicon. The overall deterioration of recovery of the samples of silicon carbide Nos. 2 and 3 by the molten metal as compared to that of sample No. 1 directly results from the fact that the content of the crystalline silicon carbide α-SiC in them increased by 2.5–4.5 times as compared to the amorphous silicon carbide[17, 18]. The analysis of the contents of the carbon and silicon and their recovery by the molten metal from samples Nos. 2 and 3 indicates that despite the highest contents of the carbon and silicon in sample No.3 only the carbon is recovered from it, and consequently, it is mainly the silicon that is oxidized which prevents its content from increasing; the decline in the growth of the carbon (recovery of the carbon) when sample No. 2 was used was caused by its combined minimum content in this sample.

Based on the results obtained from the analyses performed, augmented industrial tests were carried out in the cast iron foundry of AO Avtovaz where grey and compacted graphite iron products were cast in 40-ton AC electric arc furnaces. Silicon carbide in the 0 to 2.5 mm size range with a composition from 88% to 92% SiC was used for this purpose [19, 20]. The tests have shown that molten metal recovered smaller-sized silicon carbide with lower contents of the crystalline phase α-SiC. The above-mentioned sizes and SiC contents have been accepted and approved as the optimum requirements to be met by manufacturers.

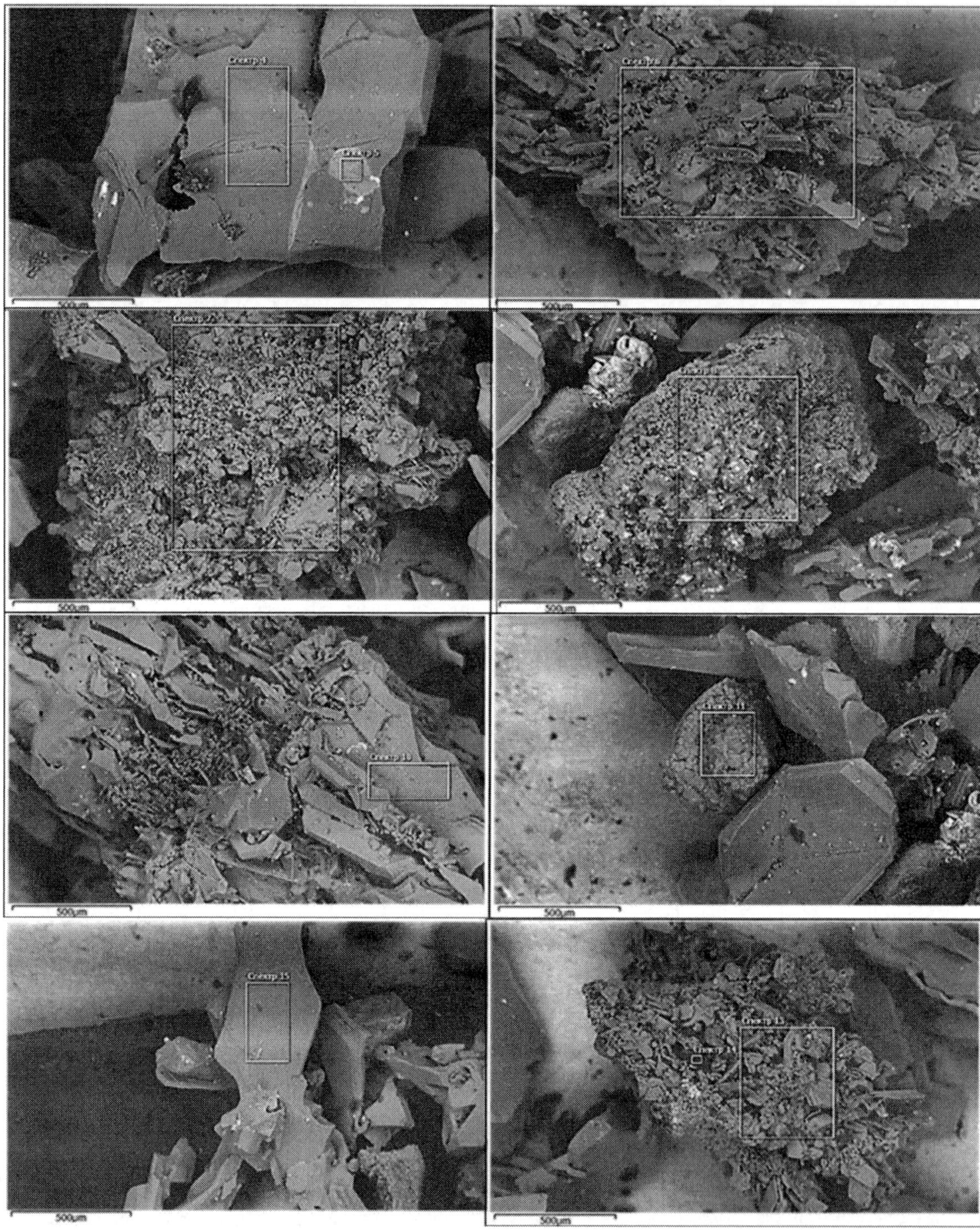

Figure 3. (Scanning) electron microscopic images of a fragment of sample No.3; marked with rectangle is an area subjected to a local electron microprobe analysis.

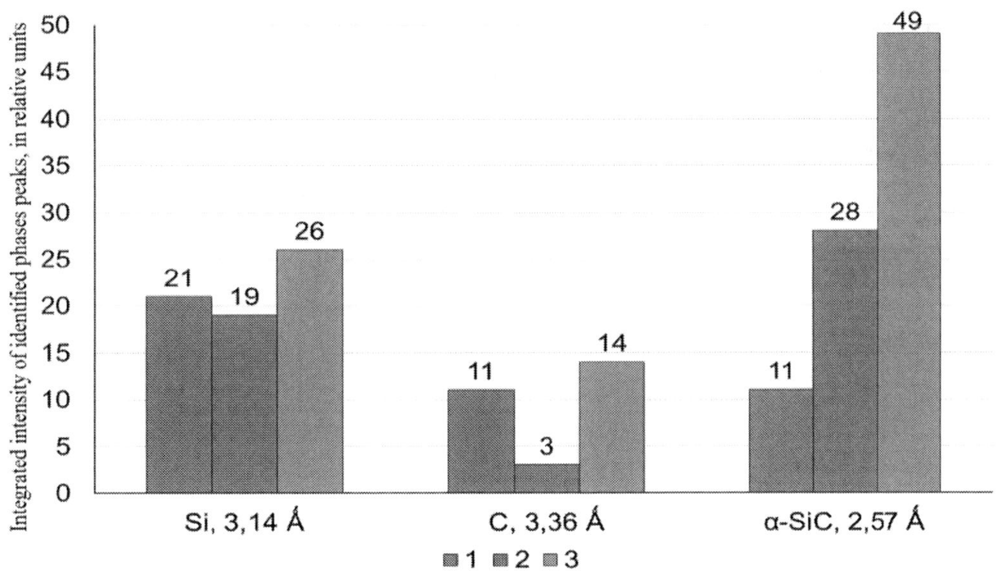

Figure 4. Integral intensities of three discovered phases expressed in relative units.

References

[1] Greenwood N 2008 *Chemistry of the elements* (Moscow: BINOM) p 607
[2] Gasik M I 2006 Smelting steel in arc furnaces of a machine-building complex with replacement of cast iron with carbon-carbide-silicon briquettes *Electrometallurgy* **6** 2–13
[3] Gasik M I, Ovcharuk A N, Semenov I A and Derevianko I V 2004 Progressive technologies of electric steel smelting with replacement of cast iron by CSiC briquettes *Steel* 4 26–30
[4] Zubov V A and Gasik M I 2002 *Electrometallurgy ferrosilicon* (Dnepropetrovsk: State Scientific and Production Enterprise "System Technologies") p 704
[5] Ryabchikov I V 1966 Thermodynamic studies of the Si-O-C system at high temperatures *Izvestia of the USSR Academy of Sciences. Metals* **2** 14–19
[6] Derevianko I V 2005 Development of technology for extracting silicon carbide and graphite from secondary materials of electrothermal production *Metallurgical and mining industry* **3** 22–27
[7] Urazlina O Yu, Gasik M M and Gasik M I 2004 Investigation of the physical properties of thermoanthracite during high-temperature calcination *Metallurgical and mining industry* **4** 17–20
[8] Tolstoguzov N V 1992 *Theoretical foundations and technology of smelting silicon and manganese ferroalloys* (Moscow: Metallurgy) p 238
[9] Porada A N and Gasik M I 1990 *Electrothermia of inorganic materials* (Moscow: Metallurgy) p 232
[10] Volynkina B P, Strakhov V M and Litvin E M 1998 Anthracite-based briquettes for steelmaking *Coke and Chemistry* **9** 36–39
[11] Chalykh E F 1972 Technology and equipment for electrode and electric coal enterprises (Moscow: Metallurgy) p 432
[12] Bestuzhev N I, Mikhailovsky V M, Bestuzhev A N, Chepyzhov B A, Grudnitsky I P and Pashkevich V F 2003 Complex inoculating treatment of liquid cast iron *Foundry* **10** 6–8
[13] Lekakh S N 223 *Resource-saving technologies for obtaining high-quality cast iron for machine-building castings* (Minsk: Science and Technology) p 223
[14] Lekakh S N and Bestuzhev N I 1992 *Out-of-furnace treatment of high-quality cast iron in mechanical engineering* (Minsk: Science and Technology) p 269

[15] Bestuzhev N I, Mikhailovsky V M, Bestuzhev A N, Chepyzhov B A, Grudnitsky I P and Pashkevich V F 2003 Complex inoculating treatment of liquid cast iron *Foundry* **10** 6–8
[16] Peregudov L V, Malashin M M and Druzhbina T V 1982 Interaction of melt and siliceous additives when modifying cast iron *Foundry production* **4** 6–7
[17] Cirilli V and Appendino P 1980 The mechanism of action of inoculants for gray iron Atti *Accord. Nas Lincein. Mem clsei fis Mate Natir.* **2** 16–20
[18] Decrop M and Masere C 1969 Etude de la formation graphite dans le fotes Mecanisme de l'inoculation *Fonderie* **276** 105–120
[19] Brandtzag S R, Linda H and Oze H A 1983 Structural changes in graphite by heat treatment *Lighz Metals* 711–725
[20] Ollesinki R W and Abbaschian G J 1984 The C-Si (cfrbonsilicon) system *Vull Alloy Phase Diagr.* **5 5** 486–489

IOP Conf. Series: Materials Science and Engineering **969** (2020) 012063 doi:10.1088/1757-899X/969/1/012063

Increasing the Strength Properties of Gray Iron Castings without Introducing Additional Modifying and Alloying Additives into Its Composition

N I Gabelchenko, N A Kidalov and A A Belov

Departament of Machines and technology of foundry, Volgograd state technical university, 28, Lenin Avenue, Volgograd, 400005, Russia

E-mail: mitlp@vstu.ru

Abstract. The effect of the differential cooling regimes on the structure and strength properties of gray iron castings in sand molds was investigated. The experimental foundry equipment was developed and assembled in order to produce experimental and control castings. This equipment allows to investigate the crystallization of two castings under initially equal pouring conditions: chemical composition, temperature and speed of mold filling with molten metal. It was shown that due to the using of the differential cooling regimes the experimental castings tensile strength was increased by 13.8% and hardness reduced by 12.3% via excluding the formation of ledeburite in the structure, which leads to increase in the quality index of cast iron by 33%.

1. Introduction

Foundry is the procurement base of engineering, so obtaining high-quality castings with improved structure and mechanical properties is an urgent task. Due to ever-increasing demands of the consumer, the existing technologies for producing castings of high-carbon alloys using modification and alloying have mostly exhausted their capabilities. Therefore, the possibility of increasing the strength of cast iron without changing its chemical composition is of great interest. It is known that the formation of the structure, and hence the properties of cast iron during crystallization and subsequent cooling process occurs in several local structurally sensitive intervals, such as dendritic crystallization, eutectic transformation and eutectoid transformation. Many researchers believe that the strength characteristics of cast iron are directly related only to the secondary recrystallization structure [1–9]. However, the secondary recrystallization structure inherit the properties from structure of primary crystallization and determined by the parameters of the primary structure, enhancing their strengthening effect in some cases and weakening in others [10–12]. Depending on the cooling rate, cast iron can crystallize both in stable (with the formation of austenite-cementite eutectic) and metastable (with the formation of austenite-graphite eutectic) systems. At high cooling rates, common for thin-walled castings, in gray cast iron usually occur ledeburite formation which leads to chilling. At low cooling rates, cast iron loses its strength properties due to the presence of free carbon in the form of graphite in its structure and an increase in the proportion of ferrite component in the metal base [13–18]. Thus, unidirectional change in the cooling rate alone does not solve the problem of increasing the strength properties with preserving the nature of gray cast iron. Therefore, it is necessary to increase the cooling rate in the interval of dendritic crystallization for dispersed dendritic structure formation and then decrease the cooling rate in the interval of eutectic transformation for exclude ledeburite formation, thereby allowing reduce the degree of supercooling of cast iron and crystallization in according to stable system. The use of differential cooling regimes can solve this problem.

2. Materials and methods

To implement the application of differential cooling regimes of gray iron castings an experimental foundry equipment that allows to investigate the crystallization of castings, with the possibility of multidirectional exposure to external factors under the same pouring conditions (such as temperature and chemical composition) was developed and assembled at the "Casting Machines and Technology" department (Figure 1).

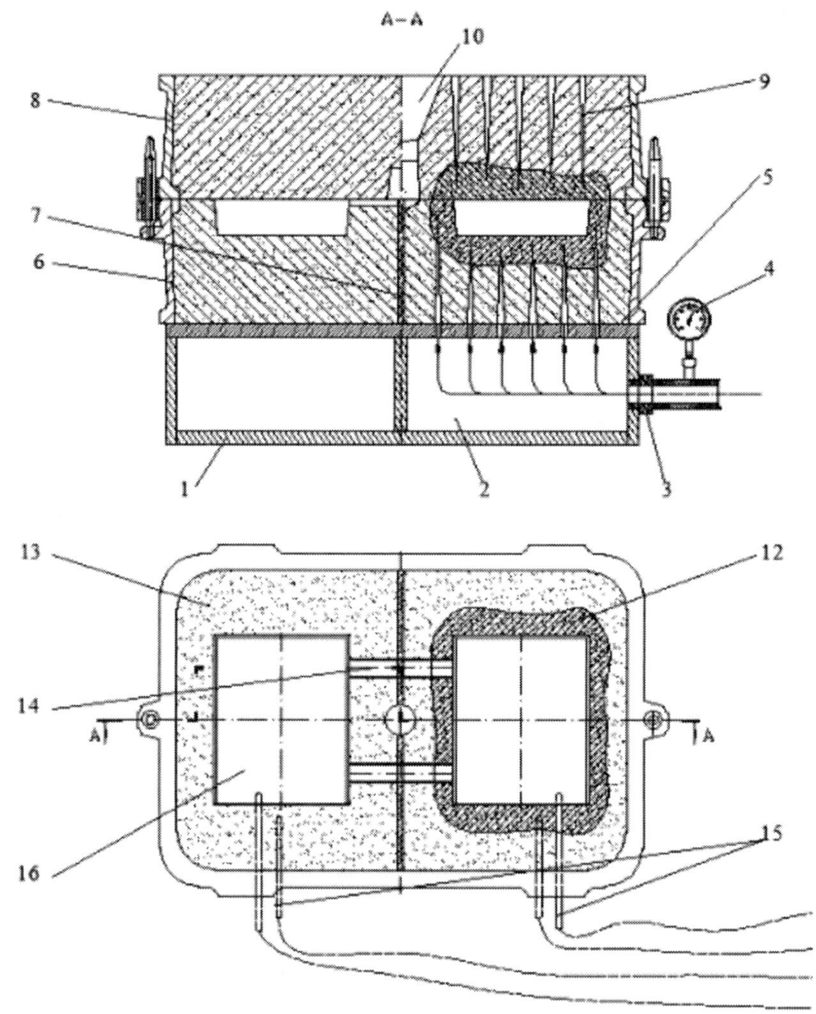

Figure 1. Equipment for the study of castings crystallization.

Foundry equipment for the study of castings crystallization is divided into two independent segments. The cooling of the studied castings in these segments takes place according to individual regimes. A dividing wall in the pallet 1 is installed for cooling gas supply to each of the segments. The supply and control of gas flow into the pallet is carried out through the fitting 3 with a built-in reducer 4. For uniform distribution the accumulated in the pallet gas rises under pressure into the mold through a perforated plate 5. The division of the lower half of the mold 6 by a heat-insulating partition 7 makes it possible to use various compositions of filling and facing sands to change the thermal regime of the mold. Having two independent segments 2 for supplying with various gases allows to adjust the cooling rate of the castings in the desired directions (increase or decrease the cooling rate). To increase the efficiency of gas blowing of lower half of the mold 6 and to improve the conditions of heat removal from the casting in the upper half of the mold 8, the application of prickings 9 is provided. For the castings temperature and cooling rate control the thermocouples 15 was placed both in the mixture 12, 13 and the body of the castings 16 via technological holes. As a result, each of the castings are provided with individual cooling regimes under initially same pouring conditions (such as chemical composition, temperature and speed of mold filling with molten metal) due to the common gating and feeding system 10, 14. This allows you to obtain correct data for further comparison of experimental results.

To investigate the effect of the cooling rate on the structure and properties two castings were poured: one crystallized without additional impact of the cooling rate, and the second with cooling rate control in the intervals of dendritic crystallization and eutectic transformation. Pre-eutectic cast iron of composition: C – 3.47; Si – 1.73; Mn – 0.59; P – 0.072; S – 0.112; Cr – 0.174; Ni – 0.188; Cu – 0.162; Mo – 0.01; V – 0.013; S - 0.9; Fe – the rest (%, wt) was chosen for research. The degree of eutecticity (Se) is – 0.9 [19]. Melting cast iron performed in the "IST-0,06" induction crucible furnace. The pouring temperature was recorded by a type A ("VR5/VR20") tungsten-rhenium thermocouple display unit. Metal crystallization temperature was controlled by a "KSP-4" potentiometer. Metallographic studies of the samples were carried out on Olympus BX51M modular optical microscope with Olympus SC30 digital camera for microstructure photographing. Samples were etched in a 4% HNO3 solution. Etching of samples was carried out by wiping with a cotton swab or dipping into a container with the etching reagent. Static tension tests were performed on cylindrical samples (working part diameter of 6 mm and length of 36 mm) at room temperature in accordance with GOST 1497-84 on a "TsD-40" tensile testing machine. For each option 3-5 samples were tested. Obtained data was averaged. The determination of Brinell hardness (GOST 9012-59) was performed using a "TB 5004" hardness tester. The hardness was determined at a load of 3000 kg with the indenter diameter of 10 mm. The measurement error was not greater than 1% in both tests.

3. Results and discussion

A molten metal was poured at a given temperature into a common gating and feeding system, from which it simultaneously at the same speed entered two cavitys of the lower half of the mold, which are isolated from each other by a heat-insulating partition. Cylindrical specimens with a diameter of 12 mm and a length of 125 mm were cast. The filling and facing sands of the experimental mold consisted of 6% (wt) bentonite clay and 94% (wt) quartz sand. Water was added in an amount of 3.5% (wt). An exothermic carbon-containing additive was additionally introduced into the composition of the facing sand in an amount of 2.5% (wt). At the initial moment of pouring, the mold was blown with air under a pressure of 4 atm. Upon reaching the eutectic transformation temperature, the blowing was ceased and the cooling rate was slowed down due to destruction of the exothermic carbon-containing additive introduced into the facing sand. The control casting was cooled in the normal regime (without blowing with air and without regulating the cooling rate).

The cooling curves of the control and experimental castings are shown in Figure 2.

It can be seen from Figure 2 that, by blowing the mold of experimental casting with air under excessive pressure it was possible to increase the cooling rate by 1.17 times during the primary austenite dendrites precipitation. An increase in the cooling rate led to the formation of a strong reinforcing framework in the form of numerous dendrites of primary austenite in cast iron structure. Next, a mechanism for slowing down the cooling rate in the interval of eutectic transformation was triggered due to the release of heat by previously introduced into the composition of the facing molding sand of the exothermic carbon-containing additive. This led to a decrease in the degree of supercooling of the experimental casting and further crystallization of cast iron according to a stable system with the formation of austenite-graphite eutectic (Figure 3).

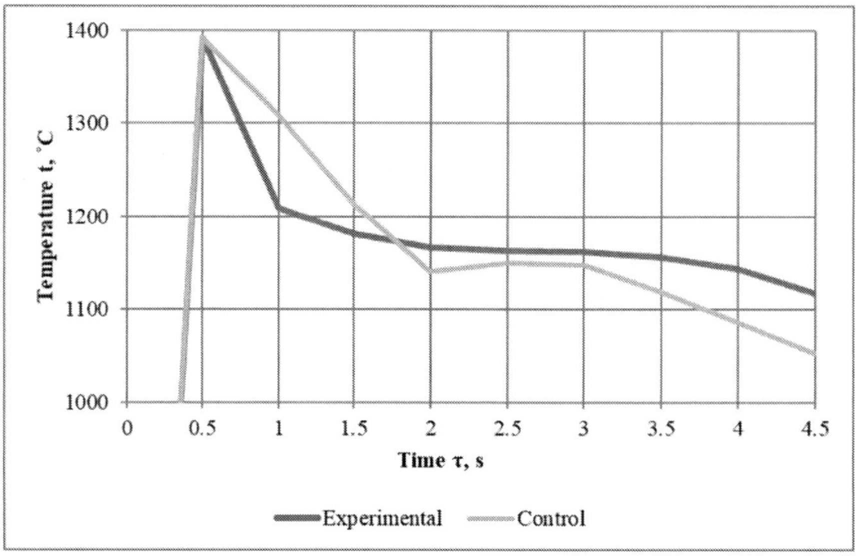

Figure 2. The cooling curves of the control and experimental castings.

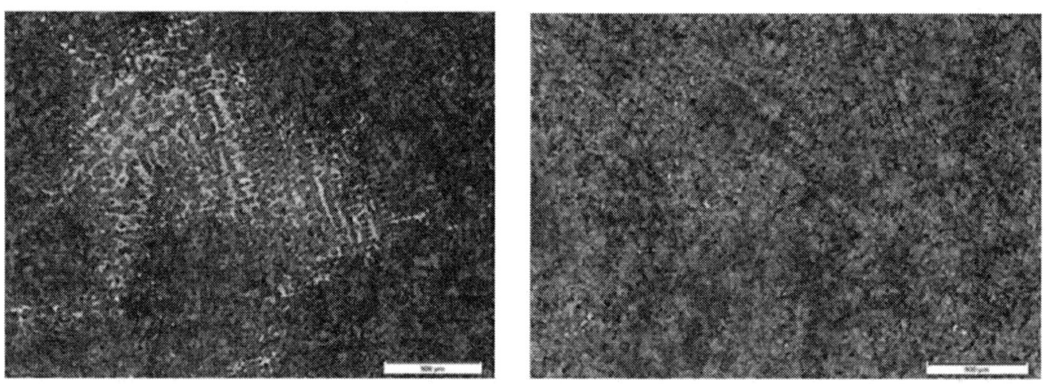

a) b)

Figure 3. Microstructures of the experimental (a) and control castings (b).

The presence of ledeburite in the structure of the control casting led to the chilling which common for mottled cast iron. The mechanical properties of the control and experimental castings are shown in table 1.

The mechanical tests showed that the tensile strength of the experimental castings is 13.8% higher than the strength of the control one. The hardness of the experimental castings was reduced by 12.3%, which led to an increase in the quality index of cast iron by 33%.

Table 1. Mechanical properties of cast iron castings.

casting	tensile strength, $\sigma_в$, N mm^{-2}	hardness, HB	Index of quality K*
Control	232	246	0.9
Experimental	264	219	1.2

*ratio between relative strength and relative hardness [20].

4. Conclusions

- An experimental foundry equipment has been developed and assembled in order to study the simultaneous crystalllization of two castings under the same pouring conditions with obtaining different structure and properties due to an individual cooling regimes.
- It was shown that the use of a multidirectional cooling regime for experimental castings in the interval of dendritic crystallization and eutectic transformation led to an increase in the strength properties of gray cast iron without introducing additional modifying and alloying additives into its composition.

References

[1] Borsato T, Ferro P, Berto F and Carollo C 2017 Mechanical and fatigue properties of pearlitic ductile iron castings characterized by long solidification times *Engineering Failure Analysis* **79** 902–12

[2] Doroshenko V S and Klimenko S I 2016 Intensification of heat removing during castings cooling by fluidization of granular filler in the mold *Metal and casting of Ukraine* **6** 22–26

[3] Doroshenko V S 2013 Control cooling of the casting in sand mold evacuated by filtration of gaseous, liquid refrigerant and movement of sand particles *Foundry Technologies and Equipment* **10** 32–37

[4] Jabbari Behnam M, Davami P and Varahram N 2010 Effect of cooling rate on microstructure and mechanical properties of gray cast iron *Materials Science and Engineering A* **528** 583–588

[5] Muumbo A 2003 Processing of Semi-Solid Gray Cast Iron Using the Cooling Plate *Technique Materials Transactions* **44** 893–900

[6] Shim S, Seong H, Jeong J and Lim S 2007 Application of Inclined Cooling Plate to Fabrication of SemiSolid Cast Iron *Materials Science Forum-MATER SCI FORUM* **544–5** 195–198

[7] Korovin V A and Leushin I O 2010 Relation between modification and graphitization processes in cast iron Stahl und Eisen *Black metals* **7** 30–32

[8] Pivovarskiy E 1965 *High quality cast iron* (Moscow: Metallurgy) p 650

[9] Górny M and Tyrała E 2013 Effect of cooling rate on microstructure and mechanical properties of thin-walled ductile iron castings *J. Mater. Eng. Perform.* **22**

[10] Il'inskii V A, Kostyleva L V, Gabelchenko N I and Santalova E A 2000 Steels and alloys crystallization study *Foundry Technologies and Equipment* **4** 5–7

[11] Katarzhin A I, Sukhanova E A, Kostyleva L V and Il'inskii V A 2003 Thermokinetic features of structure formation in the intervals of local crystallization of steel and cast iron castings *Metallurgy of Machinery Building* **5** 23–26

[12] Il'inskii V A, Zhukov A A, Kostyleva L V and Abdullaev E V 1988 New in the theory of graphitization. The relationship between primary and secondary crystallization of graphitized iron-carbon alloys *Metal Science and Heat Treatment* **10** 10–16

[13] Christian J W 2002 *The theory of transformations in metals and alloys* (Oxford: Elsevier Science Ltd) p 1200

[14] Bogachev L N, Dubinin N P, Egorenkov I P 1961 *Handbook of a caster: Iron casting* (Moscow: Mashgiz.) p 774

[15] Girshovich N G 1979 *Crystallization and properties of cast iron in castings* (Moscow: Metallurgy) p 562

[16] Bunin K P and Taran Yu N 1972 *The structure of cast iron* (Moscow: Metallurgy) p 160

[17] Gabelchenko N I, Belov A A and Meshcheryakova O A 2019 Production of High-Quality Castings from Gray Iron by Differentiated Cooling in Iron Mould *Int. Sci. and Techn. Conf. on Engineering – Innovation Technol. in Eng.: From Design to Production of Competitive Products* **973** 15–20

[18] Gabelchenko N I, Kidalov N A, Belov A A, Savchenko A I and Grigoreva N V 2020 Study of the effect of differentiated cooling modes on the structure and mechanical properties of gray cast iron *IOP Conference Series: Materials Science and Engineering. International Conference of*

Young Scientists and Students. Topical Problems of Mechanical Engineering **747**
[19] Sofroni L, Riposan I, Brabie V and Chisamera M 1985 *Cast iron* (Bucharest, Romania: Editura Didactica si Pedagogica)
[20] Zhukov A A 1978 *Handbook of iron casting* (Leningrad: Mechanical engineering) p 758

IOP Conf. Series: Materials Science and Engineering **969** (2020) 012064 doi:10.1088/1757-899X/969/1/012064

The Impact of the Production Stages of Grade D Steel on its Contamination and the Chemical Composition of Nonmetallic Inclusions

D V Rutskii, N A Zyuban and G V Babin

Department of Materials Technology, Volgograd State Technical University, 28, Lenin avenue, Volgograd, 400005, Russia

E-mail: tecmat@vstu.ru

Abstract. Optical and electronic microscopy methods are used to evaluate the contamination of grade D carbon steel with nonmetallic inclusions in steel samples taken at various stages of steelmaking (electric arc furnace (EAF) discharge → ladle furnace (LF) → vacuum deoxidation (VD) → continuous casting) at an electric furnace melting shop. Steel contamination with nonmetallic inclusions and inclusion chemical composition are assessed. It is demonstrated that the aluminum deoxidation of the semiproduct leads to the formation of corundum nonmetallic inclusions (Al_2O_3) in molten steel. The total amount of corundum reaches 52% for all the stages of steelmaking process. The identification and evaluation of nonmetallic inclusions demonstrates that deoxidation performed at all the stages of secondary steelmaking reduce the amount of inclusions. After vacuum deoxidation and adding Al and SiCa, corundum inclusions become globular in shape with a particle size less than 6 μm. When the continuously cast billet has solidified, the total amount of nonmetallic inclusions does not change, whereas silicate inclusion content decreases and corundum contamination grows, with corundum inclusions being of an irregular shape. A heavy corundum contamination is caused by aluminum secondary deoxidation during steel casting as well as by the fact that the molten residue in the pouring nozzle is carried into the solidifying continuously cast billet.

1. Introduction

Nowadays the advent of modern technologies for the production and transportation of hydrocarbons at locations with harsh operational environment calls for the making of steel grades with high mechanical, physical and chemical properties. A steel manufacturing process is the major factor which affects steel quality and determines chemical, physical and structural heterogeneity of finished products as well as the presence of nonmetallic inclusions. The inclusion cleanliness of steel is an important parameter which determines steel quality, and existing steelmaking processes may effectively reduce the amount of inclusions [1–5]. Somehow, even small quantities of impurities may reduce impact strength; besides, corrosive nonmetallic inclusions (NMI) worsen the corrosion resistance of steel. Hence, the production of steel grades with a low content of NMI is a topical and challenging task of metallurgy.

Controlling the formation, shape, distribution as well as phase composition of NMI has a major effect on steel quality [6]. It has also been demonstrated that the inclusion cleanliness of steel has a significant impact on the formation of contact fatigue defects in rail steel [3].

2. Main part

NMI composition and final oxygen content in steel are conditioned by deoxidizers used, their composition, quantity, addition sequence and methods of adding into molten steel [7–9].

Currently aluminum has been most widely applied as a deoxidizer at the initial stages of ladle treatment (when molten steel has been discharged from the furnace). Aluminum has the highest deoxidation power compared to other commonly used deoxidizing agents, such as Si, Mn, etc. When aluminum is used for steel killing, corundum inclusions as well as Al_2O_3 inclusions emerge. Aluminum residual content in steel as well as aluminate inclusions content heavily affect the stability of the continuous casting process [10, 11]. Moreover, Al_2O_3 content determines the degree of inclusion deformation at rolling temperatures and triggers the formation of contact fatigue defects [9, 12, 13]. In order to reduce the negative impact of aluminum inclusions on the teeming of steel and steel quality, molten steel may be treated with calcium compounds. Thus, calcium carbide CaC_2 was added as a deoxidizer at the initial stages of steel production (when a semiproduct was discharged from the furnace). It has been established that calcium affinity to oxygen is much higher than that of aluminum [14–16]. Somehow. calcium carbide has not been widely used for killing steel. It has been revealed that carbide calcium may reduce the total amount of inclusions in a continuously cast billet. The authors also demonstrate that if calcium carbide, and not aluminum, is added at the initial stages of steel ladle treatment, the NMI content in samples grows. Somehow, the phase composition of inclusions alters because of the decrease in aluminum oxide content (Al_2O_3) [7, 9].

Nowadays aluminum and calcium based compounds, silicocalcium being the most common one, are added to molten steel at the final stages of ladle treatment, namely, before casting. This is done in order to improve steel castability, reduce the impact of corundum inclusions on steel quality and decrease the amount of nonmetallic inclusions. Consequently, the chemical composition and shape of aluminate inclusions may alter. Adding calcium compounds does not always have a positive effect and may trigger the formation of new types of corrosive nonmetallic inclusions, like Type I and Type II.

Such inclusions heavily deteriorate steel corrosion resistance [4, 5], impact strength [2, 3] and fatigue strength [12, 13]. It has been demonstrated that when calcium is added, $(CaO \cdot Al_2O_3)+CaS$ complex oxysulfide inclusions of dfferent shapes emerge and may trigger the build up of molten metal in the pouring nozzle. Consequently, it deteriorates steel castability and contaminates the continuously cast billet with nonmetallic inclusions [10, 17].

The paper is focused on the assessment of the effect of steel production stages at an electric furnace melting workshop (electric arc furnace (EAF) discharge → ladle furnace (LF) → vacuum deoxdation (VD) → continuous casting) on the chemical composition and NMI contamination of steel at different stages of steel production.

Steel samples of two heats of grade D steel were investigated; their chemical composition is given in Table 1.

Upon the discharge from the furnace, molten steel was deoxidized with aluminum AB 87 at the amount of 0.9 kg/ton. The amount of the aluminum added was calculated to achieve oxygen content not greater than 4 ppm after killing.

Then steel was ladle treated and vacuum deoxidized in order to obtain the final chemical composition; steel composition and the amount of the materials consumed are given in Tables 1 and 2. The weight of steel in the steel-teeming ladle was 150 tons.

Inclusions were identified with optical (using Leica 8 C) and electronic (using FEI Versa 3D) microscopy methods. NMI phase composition was assessed by recalculating the TDS analysis data with the help of the Thermocalc application. The evaluation of NMI contamination was performed automatically with the Axalit Soft application in compliance with the GOST 1778–70 method P.

Table 1. Chemical composition of the samples, taken in different stages of D steel treatment.

Element, %	Heat №1			Heat №2		
	EAF smelting	After vacuum deoxdation	Continuous cast billet	EAF smelting	After vacuum deoxdation	Continuous cast billet
Ca	–	0.0001	0.0011	–	0.0001	0.0011
Al	0.0217	0.0140	0.025	0.0237	0.0154	0.018
C	0.3643	0.4342	0.465	0.3743	0.4442	0.475
Si	0.2288	0.2082	0.226	0.2008	0.2062	0.216
Cr	0.1337	0.142	0.153	0.1237	0.132	0.143
Ni	0.0858	0.0868	0.087	0.0838	0.0848	0.0865
Ti	0.0017	0.0013	0.001	0.0016	0.0014	0.0012
Cu	0.1755	0.1754	0.175	0.1645	0.1654	0.1760
Mo	0.018	0.0169	0.017	0.019	0.0179	0.017
Sn	0.009	0.0092	0.009	0.008	0.0082	0.008
P	0.0089	0.0099	0.010	0.0115	0.0128	0.0130
S	0.0444	0.0149	0.007	0.0468	0.0037	0.006

Table 2. Amount of introduced charge materials during deoxidation of steel with aluminium.

Materials, kg	Number of sample									
	LF–1	LF–2	LF–3	LF–4	VC	LF–1	LF–2	LF–3	LF–4	VC
	heat 1					heat 2				
Upon the discharge from the furnace, molten steel was deoxidized with aluminum AB 87 at the amount of 0.9 kg/ton										
ALBRA	87	–	–	–	10	40	–	–	–	10
Alsiflux	366	–	110	–	–	213	–	100	–	–
CaO	313	740	245	414	–	798	328	500	167	–
Al	30	30	–	13.2	36.5	6.8	0.4	–	13.2	27
FeSiMn18	–	238	–	85	–	–	373	–	85	–
Anthracite	–	210	–	66	–	–	28	–	27	–
FeSi65	–	–	–	75	–	–	–	–	–	–
Rice husk	–	–	–	–	180	–	–	–	–	–
SiCa	–	–	–	–	98.5	–	–	–	–	84.4

The identification of nonmetallic inclusions with optical and electronic microscopy methods reveals that oxide NMI are presented by the following compounds (Figure 1 and 2):

- corundum (Al_2O_3; T_{melt} = 2050 °C; ρ = 3.96 g/cm^3) with a content of not less than 50% of the total amount of NMI (see Figure 2a),
- calcium aluminate compounds containing ($xCaO \cdot yAl_2O_3$) with alumomagnesian spinel ($MgO \cdot Al_2O_3$) at the centre of the inclusion particle (see Figure 2b);

- silicates containing $MnO \cdot SiO_2 \cdot Al_2O_3$, (spessartite $Mn_3Al_2Si_3O_{12}$, T_{melt} =1195 °C; ρ = 4.18 g/cm^3) (see Figure 2c).

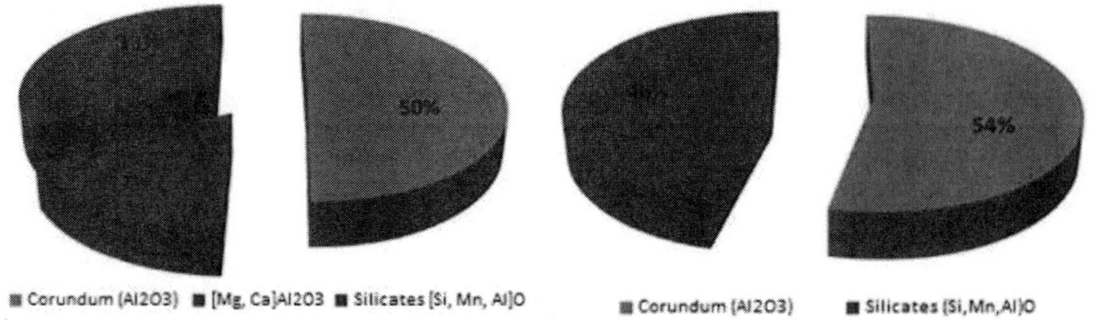

Figure 1. Relative fraction of oxide non-metallic inclusions detected at all stages.

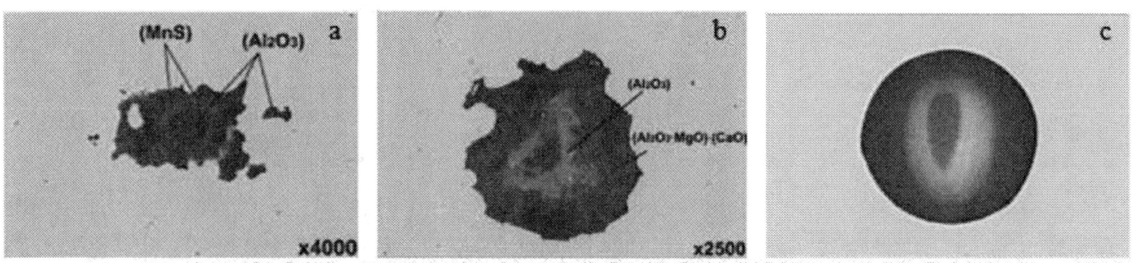

a – corundum (Al_2O_3); b – calcium aluminate ($xCaO \cdot yAl_2O_3$) based on spinel ($MgO \cdot Al_2O_3$); c – silicate (spessartite $Mn_3Al_2Si_3O_{12}$, $T_{пл}$=1195 °C; ρ = 4,18 g/sm^3)

Figure 2. The characteristic types of oxide non-metallic inclusions.

The evaluation of NMI contamination at different stages of metallurgical treatment at an electric furnace melting shop is presented in Figure 3.

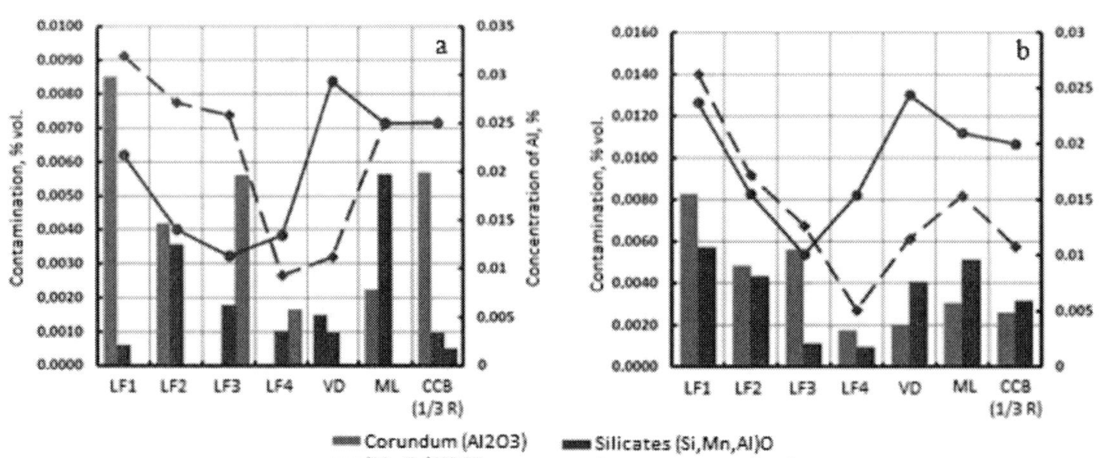

a – heat 1; b – heat 2

Figure 3. Changes of steel contamination with different type of non-metallic inclusions on smelting stages.

The analysis reveals that the samples taken after killing steel at the initial stages of secondary metallurgy are most heavily contaminated with NMI (see Figure 3).

The LF treatment of steel reduces overall NMI content to approximately 0.0019% vol., with the prevalence of corundum based (Al_2O_3) inclusions (see Figure 3b); silicate ($Mn_3Al_2Si_3O_{12}$) content reaches its lowest value of 0.0010% vol. The addition of 85 kg of FeSiMn18 and 75 kg of FeSi 65 results in the formation of globular-shaped corundum inclusions. Combined with argon blowing, it facilitates the transport of inclusions into slag.

The amount of NMI with a high content of corundum increases in the samples taken after VD and adding 36.5 kg (81 m) of aluminum wire and 98.2 kg (204 m) of SiCa (see Figure 3); these corundum inclusions not always being of globular shape (see Figure 2b). Unlike calcium aluminates, corundum inclusions are hard to remove from molten metal and their residual amount is high in the samples taken both from the tundish as well as from the continuously cast billet. Moreover, corundum inclusion contamination increases after teeming into the tundish to 0.0022% ÷ 0.0030% vol. (see Figure 3). Killing steel with SiCa after VD alters the shape and composition of corundum inclusions; besides, silicate inclusion content increases to 0.0055% ÷ 0.0058% vol. (see Figure 3).

The increased contamination of steel with corundum NMI may be explained by the secondary oxidation of aluminum during casting, a partial erosion of the inner surface of the corundum pouring nozzle with molten metal flow, and an alteration of aluminum content after VD. In heat No. 2 corundum inclusion content remains unchanged after the continuously cast billet has solidified (see Figure 3b). The fact can be explained by adding a smaller amount of aluminum (see Figure 2) and, consequently, a lower content of aluminum in the molten metal (see Table 1, Figure 3b).

The data presented in Figure 3 prove that a reduced amount of aluminum results in a lower NMI contamination (see the left hand side of the curves in Figure 3), while an increased amount of aluminum leads to an increase in NMI contamination (see the right hand side of the curves in Figure 3).

In order to determine the efficiency of the current aluminum deoxidation methods, an aluminum deoxidation curve for steel was calculated using data available [18–23]. With this aim the K_p values of the equilibrium constant for aluminum deoxidation as a function of temperature were evaluated.

The data calculated for the degree of molten steel deoxidation are presented in Figure 3. A comparison of the theoretical and empirical data reveals that the current method of killing steel with aluminum at a rate of 0.93 kg per ton can result in soluble oxygen content [O] of 3.7913 ppm (in the sample taken after killing steel in the ladle, heat No. 2).

The calculation of the deoxidation curve demonstrates that in order to keep oxygen content below 4 ppm, the amount of aluminum for killing steel in the ladle should be reduced from 0.93 to 0.70 ÷ 0.80 kg per ton (see Figure 4).

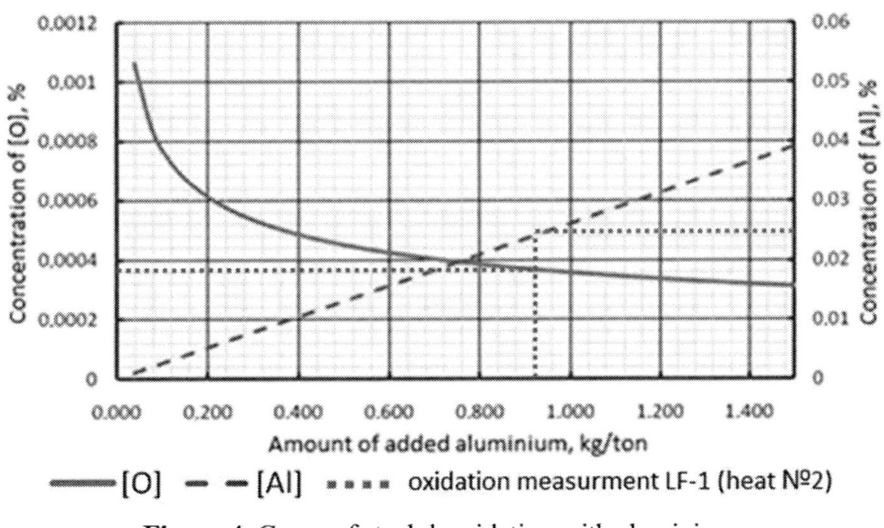

Figure 4. Curve of steel deoxidation with aluminium.

3. Conclusion

The analysis of NMI chemical and phase composition as well as the evaluation of NMI contamination prove that the amount of aluminum deoxidant needs to be decreased in order to reduce NMI contamination of the continuously cast billet. A theoretical analysis demonstrates that if dissolved oxygen content [O] in killed steel is not to exceed 4 ppm, it is sufficient to reduce the amount of the deoxidizing agent from 0.93 to 0.7 ÷ 0.8 kg per ton.

An increased amount of dissolved aluminum leads to the growth of NMI content, with the inclusions being predominantly presented by hard melting non-deforming corundum particles (Al_2O_3) with the shape which makes the further pressure shaping process more difficult.

Secondary metallurgy decreases NMI contamination of steel and alters corundum inclusions, modifying their chemical composition and shaping them like spheres. It needs to be noted that following VD, during steel casting and solidification, corundum inclusions of unfavorable shape emerge.

Alongside with corundum inclusions, globular silicate inclusions containing the oxides ($MnO{\cdot}SiO_2{\cdot}Al_2O_3$) were identified. The latter emerge after killing molten steel with Si additions.

Steel semiproduct deoxidation should be performed at an early stage of steelmaking (when discharging steel from the EAF) using less aluminum: the amount of aluminum deoxidizer should be reduced 0.70 ÷ 0.80 kg per ton.

Aluminum content is as follows:

[Al] 0.020 ÷ 0.030% (for steels with restricted Al content)

[Al] 0.020 ÷ 0.030% (for marketable workpieces).

Aluminum content should be adjusted after VD and not greater than by 0.005%. In case if aluminum content needs to be adjusted by a greater value, calcium silicon wire needs to be added at the amount of 1 kg per ton of steel immediately before discharge with molten steel killed in the ladle for 10 ÷ 15 minutes.

References

[1] Kuznetsov A A, Sumin S N and Kozlov G S 2014 Change in oxygen activity during out-of-furnace steel treatment upon deoxidation with aluminum or calcium carbide *Sci-Tech. Progress in metallurgy* 30–33

[2] Morozova T V 2012 *The influence of steel production technology on the uniformity of the structure and contamination of the metal with non-metallic inclusions in order to increase the reliability of main pipelines* (Moscow: abstract) p 23

[3] Dub A V, Barulenkova N V and Morozova T V 2005 Non-metallic inclusions in low alloy pipe steel *Metallurgist* **4** 67–73

[4] Zajcev A I, Rodionova I G annd Semerin G B 2011 New types of unfavorable non-metallic inclusions based on Mg-AL_2O_3 and metallurgical factors determining their content in the metal. Part I *Metallurgist* **2** 50–55

[5] Zajcev A I, Rodionova I G and Semerin G B 2011 New types of unfavorable non-metallic inclusions based on Mg-AL_2O_3 and metallurgical factors determining their content in the metal. Part II *Metallurgist* **3** 28–33

[6] Gamanyuk, S B, Zyuban N A, Ruckij D V and Ananeva A N 2017 Investigation of the influence of deoxidation modes on the formation and arrangement of sulfides in medium-carbon structural steels *Steel* **2** 15–19

[7] Ruckij D V, Zyuban N A, Neklyudov I V and Agarkov A Yu 2019 Influence of calcium carbide as a deoxidizing agent on the chemical composition and contamination with non-metallic inclusions *Black metals* **5(1049)** 15–21

[8] Grigorovich K V, Krasovskij P V and Trushnikova A S 2002 Analysis of non-metallic inclusions – the basis for quality control of steel and alloys *Analytics and control* **6** **2** 133–142

[9] Ruckij D V, Babin G V, Gamanyuk S B, Morozov V V and Kornev Yu L 2019 Optimization of the composition and amount of non-metallic inclusions in a continuously cast billet based on the deoxidation of the melt with aluminum and calcium carbide *Metallurgist* **4** 41–48

[10] Agarkov A Yu, Ruckij D V, Zyuban N A, Babin G V, Kirilichev M V and Morozov V V 2020 Revealing the nature of the formed "build-ups" on the inner wall of the pouring nozzle during continuous casting of steel grade C45E *Theory and technology of metallurgical production* **1(32)** 11–17

[11] Bozheskov A N, Kazakov V V and Korostelev A A 2015 Use of argon-purged dosing nozzles to increase the castability of steels *Steel* **7** 3–15

[12] Grigorovich K V 2006 Investigation of the structure and metallurgical quality of rail steels from various manufacturers *Metals* **5** 1–16

[13] Grigorovich K V and Garber A K 2011 Analysis of the processes of secondary treatment of carbon steels *Perspective materials* **13** 13–25

[14] Takashi Kimura and Hideaki Suito 1994 Calcium deoxidation equilibrium in liquid iron *MMTB* **25 b** 33–42

[15] Kotelnikov G I, Zubarev K A and Movenko D A 2016 Calcium iron deoxidation curve plotting *Electrometallurgy* **6** 10–18

[16] Fujiwara H, Tano M, Yamamoto K and Ichise E 1996 Solubility and activity of calcium in molten iron in equilibrium with lime and thermodynamics of calcium containing iron melts *ISIJ int* **35 9** 1063–1071

[17] Jae H S and Joo H P 2017 Formation mechanism of oxide-sulfide complex inclusions in high-sulfur-containing steel melts *Metallurgical and Materials Transactions B* **49** 311–324

[18] Bigeev A M and Bigeev V A 2000 *Steel metallurgy. Theory and technology of steel melting. Textbook for universities* (Magnitogorsk: MGTU) p 544

[19] Sponseller D A and Flim R A 1964 The solubility of calcium in liquid iron and third element interaction affects *Tans. Aime* **239** 876–887

[20] Ajzatulov R S, Xarlashin P S, Protopopov E V and Nazyuta L Yu 2004 *Theoretical foundations of steelmaking processes. Textbook for universities* (Moscow: Misis) p 320

[21] Laptev D M 1965 *Solution thermodynamics tasks and exercises* (Moscow: Metallurgy)

[22] Kudrin V A 2003 *Theory and technology of steel production: a textbook for universities* (Moscow: Mir) p 528

[23] Bornackij I I 1974 *Physical and chemical foundations of steelmaking processes* (Moscow: Metallurgy) p 320

IOP Conf. Series: Materials Science and Engineering **969** (2020) 012065 doi:10.1088/1757-899X/969/1/012065

Acoustic Methods for Testing Gas Turbine Engine Blades

V N Trofimov[1], I N Pyankov[1,2] and V A Pyankov[2]

[1]Dynamics and Strength of Machines Department, Perm National Research Polytechnic University, 29, Komsomolsky prospect, Perm, 614990, Russia
[2]United Engine Corporation - Perm Motors, Perm, Russia

E-mail: tvn_perm@mail.ru

Abstract. In this paper the application technology of the ultrasonic testing method in the process of the PS-90A aircraft engine manufacturing is considered. It is known that the development and implementation of monocrystalline turbine blades for aircraft gas turbine engines are aimed at ensuring the material effectiveness or increasing the blade service life in comparison with polycrystalline blades. It is achieved at the cost of the increased long-term material strength in the same temperature range. Based on the results of the process and metallographic studies analysis, it has been established that during the structure formation under the internal stress action, hot crystallization cracks with an opening from 3 to 8 μm or more and a length of 3 to 35 mm along the grain boundaries can appear. Grains formation in most blades happens with their subsequent exit to the lock end, a bridge. The formation of hot cracks most likely occurs at a large grain disorientation angle (more than 25°). Grain boundaries on blades with hot crystallization cracks are located, as a rule, on the bridge at the junction of the blade root with the blade shank ("bottleneck") with orientation along the blade axis. The method of contact laser-ultrasonic flaw detection (CLUFD) was tested on blades in the region of bridges using both pilot samples and standard parts in order to solve an urgent production and technical problem, that is timely detection of hot crystallization cracks in rotating turbine engine blades at the manufacturing stage. The use of the phased antenna array technology (PAR) in combination with a portable X-32 flaw detector represents a new level of blade ultrasonic testing. The flaw detector operates in a wide frequency range, which allows solving a very large range of various ultrasonic testing problems. The advantages and possibilities of the PAR technology with the portable X-32 flaw detector can be considered on the example of the turbine blade lock testing. S-scan (sector-scan) makes it possible to interpret the received signals quickly and easily.

1. Introduction

It is known that the development and implementation of monocrystalline turbine blades for aircraft gas turbine engines are aimed at ensuring the material effectiveness or increasing the blade service life in comparison with polycrystalline blades. It is achieved at the cost of the increased long-term material strength in the same temperature range.

High-temperature destruction of polycrystalline blades, as is known, occurs in the weakest places, which are grain boundaries. A single crystal blade usually consists of one or several macrograins, whose boundaries are transverse to the external load. And since there are few or no such boundaries in single-crystal blades, this circumstance ensures a significantly longer blade service life.

Vacuum precise casting is widely applied for manufacturing high-pressure turbine rotor blades for the 1st and 2nd stages of the PS-90A engine using the models cast from heat-resistant nickel alloy ZhS32-VI by means of the high-speed directional solidification method.

Blades have a complex structure with cooling channels in the inner cavity, divided by 1.5 mm thick bridge edges. Cooled turbine blades are hollow structures with various air-cooling channels (radial channel, deflector channel, etc.) made of polycrystalline equiaxed, directional and monocrystalline structures by the method of investment casting.

Many studies have established and experimentally confirmed that, according to a set of turbine blade strength properties the [001] direction is an optimal axial orientation. Monocrystalline rotating turbine blades of modern gas turbine engines are cast with such an axial orientation [1].

A blade structurally consists of several macrograins, whose boundaries are located mainly parallel to the blade axis. The growth of one grain relative to another occurs in crystallographic directions. Due to the arising internal stresses, the grain boundary may go unparallel to the crystal growth axis in the [001] direction. This can be indicated by the internal grain growth and the rotation of the grains relative to each other at a certain angle. Hot and cold cracks occur in castings due to the internal stresses that are greater or comparable to the tensile strength of the material. Both grain boundaries and the grains themselves are destroyed in the process of their formation.

Large-angle grain boundaries have a significant negative effect on the parameters of the material strength properties. This fact necessitates the rejection of blades. Particularly stringent requirements for this quality parameter are imposed on blades crystallized from alloys with a low carbon content, where the strength of high-angle boundaries is low.

Research in the crystal structure perfection of a single crystal heat-resistant alloy [2] shows that the volume of a single crystal is divided by several level boundaries into large blocks (with disorientation angles in several degrees). Each block is divided into smaller blocks with disorientation angles in less than 1°, and they, in turn, into even smaller ones.

Based on the results of the process and metallographic studies analysis, it has been established that during the structure formation under the internal stress action, hot crystallization cracks with an opening from 3 to 8 μm or more and a length of 3 to 35 mm along the grain boundaries can appear. The formation of grains in most blades occurs with their subsequent exit to the lock end, a bridge. The formation of hot cracks most likely occurs at a large grain disorientation angle (more than 25°). The grain boundaries on blades with hot crystallization cracks are located, as a rule, on the bridge at the junction of the blade root with the shank ("bottleneck"), oriented along the blade axis.

2. Materials and blade defects

Turbine blades (TBs) are the main parts of a gas turbine engine (GTE) which operate at high temperatures and are exposed to high static, vibration and temperature stresses. They are exposed to both corrosive and erosive gases. The strength characteristics of operational TBs determine the engine reliability and its service life. In particular, the breakdown of the operational TB can lead to a non-localized engine destruction. JSC "UEC-Perm Motors" widely uses non-destructive testing methods in the manufacture of the PS-90A gas turbine engines for medium-range and long-range aircraft Tu-204, Il 96-300, etc.

Possible defects (Figure 1) of the turbine blades made by casting methods, in particular by the method of high-speed directional solidification (HSDS), are hot crystallization cracks with an opening from 3.0 to 8.0 μm and more and a length from 3.0 to 35.0 mm along the macrograins boundaries. Hot crystallization cracks appear when the structure is formed under the influence of internal stresses. The formation of grains in most blades occurs with their subsequent exit to the lock end, a bridge.

Cracks are defects of the TB locking part after the gas turbine engine operation. The central and edge regions of the first spline groove of the TB locking part is the area of the most probable crack initiation. Cracks parallel to the splined groove have a length of 0.5 to 3.0 mm, a depth of 1.0 mm and a width of 0.2 mm.

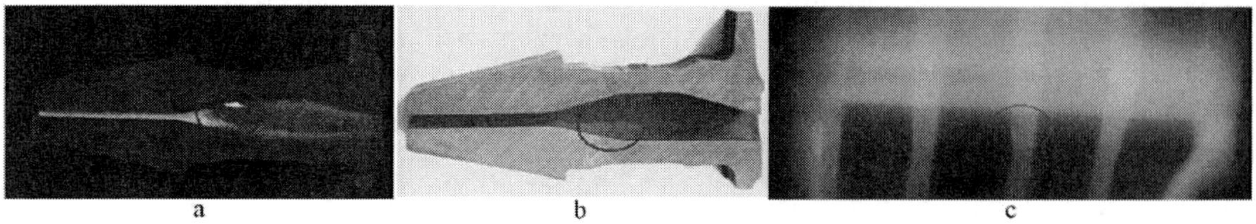

a b c

Figure 1. Bridge of the blade locking part: with the crystallization crack (a); with the macrograin boundary (b); with the slag clogging (c).

3. Advanced methods for non-destructive cast blade testing

At present, in order to prevent the usage of blades with hot cracks in production, serial bridge macrostructure testing in the region of the blade roots and shank is carried out, taking into account additional requirements for grain boundaries with the intersection of specially defined zones and the introduction of a maximum disorientation angle, which should not be more than 25° (The angle was determined by statistical data). Inspection of the turbine rotating blades is performed by two testing methods: visual-optical and X-ray (the Laue method).

The first stage of cast aircraft engine blade inspection is visual-optical (visual). Testing is carried out with the use of lighting and magnifying optical devices that help to see the crystal boundary better and determine the crystal disorientation. Re-inspection of the blades that do not meet the disorientation angle standards is carried out by X-ray diffraction analysis (the Laue method) using the KROS PRDU installation.

The installation includes an X-ray protective camera, a workstation, a system control panel and a Digora Optime laser reader. The software solves the problem of automatic recognition of the Laue reflexes, determination of their coordinates and transformation of the epigram into a stereographic projection. The accuracy of orientation measurements should be determined by parallel surveys of specially selected samples using a diffractometer. The surface of the inspected object must be prepared in advance. The surface of the investigated blades is prepared by etching, just as for the macrostructure testing in accordance with the serial part production technology. If necessary, in order to improve the quality of the epigrams, the inspection site can be etched locally in accordance with the manufacturing technology of the part.

It is allowed to produce blades with a grain disorientation angle of up to 25°. The problem appears when grains are disoriented at 25°–26°. Cast blades with such a disorientation of macrograins are classified as defective. Defective blades are very expensive for production operation. The total number of discarded blades reaches 20–40 %.

However, the measurements made do not always show the actual state of the bridge macrostructure along the entire length of the blade shank and root. The boundaries of two grains on the surface and in the bridge of the part may not coincide. The assessment of a grain boundary transition from one zone into another (on the blade back and pressure side) is inaccurate.

The analysis of the blade defectiveness showed the need to improve the technological process of non-destructive blade testing, taking into account their design features and elastic properties, in particular, cast nickel alloy ZhS32-VI.

The method of contact laser-ultrasonic flaw detection (CLUFD) was tested in the bridge region using both pilot samples and standard parts in order to solve an urgent production and technical problem, that is timely detection of hot crystallization cracks in rotating turbine engine blades at the manufacturing stage [3].

To carry out blade flaw detection, a UDL-2M flaw detector and a specialized transducer PLU-6P-01 were used (Figure 2). The PLU-6P-01 model is a broadband optical-acoustic transducer, whose installation in the rotating blade root pocket of the second stage turbine engine for bridge testing is shown in Figure 2c. This method is based on the use of the thermoelastic effect [4].

Elastic waves (longitudinal waves from 0.1 to 100 MHz) are excited in the blade when a sufficiently intense light beam of pulsed laser radiation is exposed to its surface. The laser radiation absorbed by the metal heats the thin surface layer of the blade. The specific radiation power does not exceed the threshold value at which ablation (evaporation of the metal surface layer) occurs. The amplitude of the generated pulses increases when, for example, the metal surface is coated with a thin layer of water or oil, which prevents the expansion of the heating zone. The latter is associated with a change in the longitudinal waves diagram. Their maximum coincides with the normal to the blade surface.

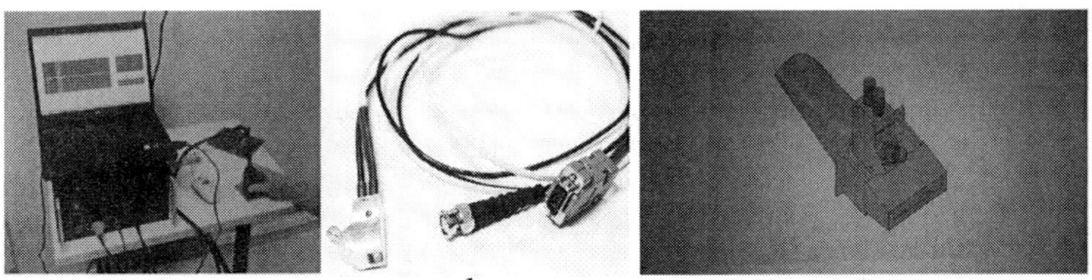

a b c

Figure 2. The exterior view of the UDL-2M flaw detector (a), the exterior view of specialized sensor PLU-12U-02 (b) and installation of specialized converter PLU-12U-02 on the blade surface (c).

Laser emits light pulses at a duration of 10 nanoseconds with a pulse energy of 0.1 J with a repetition rate of up to 1 kHz. A short laser pulse is transmitted through an optical fiber to an optical-acoustic generator. The subsequent thermal expansion causes the ultrasonic pulses excitation. An excited short broadband ultrasonic pulse propagates deep into the blade under study.

The pulse scatters along the blade structure inhomogeneity. Further, a broadband piezoelectric contact receiver with the PVDF film as a piezoelectric element registers reflected and backscattered longitudinal ultrasonic waves. Reflected and scattered waves carry information about the acoustic properties of the blade metal structure and defects. To obtain a two-dimensional image, the PLU-12U-02 sensor is manually moved along the blade face perpendicular to the bridge under study.

The signals from the transducer are subjected to time and frequency filtering with subsequent formation in the matrix.

The final result of the signal processing is the bridge cross-section image on B scan.

A small diameter of the probe pulse increases the sensitivity to defects detection, whose size is equal to or less than the ultrasound wavelength. What is more, with a short ultrasonic pulse duration, high accuracy of ultrasound velocity measurement equal to 1% is achieved. It becomes possible to measure the cross-section of ultrasonic waves back-scattering with a high spatial resolution of up to 0.05 mm, as well as measure the attenuation of ultrasonic waves in a wide frequency range of 0.1–100 MHz. In fact, there is no uncontrolled ("dead") zone with one-sided access to the part.

Testing is carried out in two main directions with the introduction of longitudinal L-waves: perpendicular to the shank platform and the blade root.

We tested model samples with artificial defects in the form of a 1.5–2.0 mm long and 1.5 mm wide groove, equal to the bridge width. The defects were grooved by the electroerosion method in different bridge regions from the side of the leading (entering) and exit edges of the blades (Figure 3). Bridge testing was carried out in two places: on the shank closer to the root and in the root pocket. The results of model samples testing are presented in Figure 3. The revealed defects are marked with a circle.

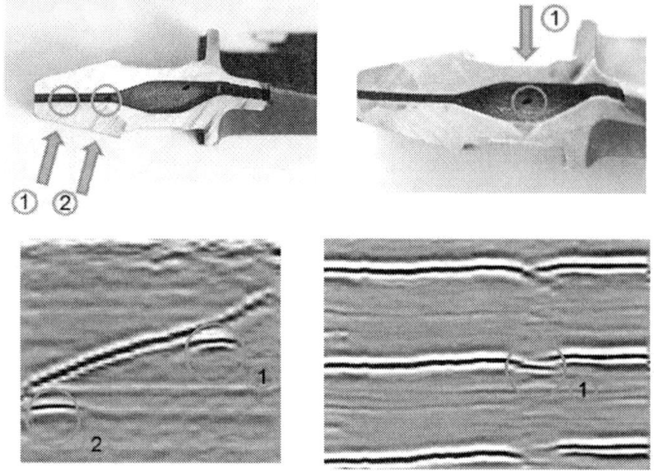

Figure 3. Results of revealed artificial defects in model samples.

10 castings of the second stage HPTBs were tested by the method of contact laser-ultrasonic flaw detection. Those blades were rejected according to the results of X-ray diffraction analysis done with the KROS PRDU device to determine grain disorientation on the shank. The grain disorientation angle on different blades ranged from 29.5° to 55.9°. Testing was carried out from the tapered part of the shank. Six blades were found relatively acceptable. Four blades were selected according to the availability of ultrasonic waves reflection from the bridge region. After that the samples cut along the bridges were tested by the LUM1-OV1 method followed by the metallography.

Metallographic studies in the course of layer-by-layer metal removal in the bridge cross-section in places with fixed coordinates of the reflected signals showed the presence of grain boundaries.

It was found that ultrasound reflection on three out of four blades was due to a pronounced boundary between two adjacent grains. A hot crystallization crack with a length of about 2.0 mm and a maximum opening of up to 0.8 mm was found on one blade from the side of the exit bridge. Metal discontinuity flaw was found in the same blade, in the bridge from the side of the leading (entering) edge.

Taking into account the requirements of serial testing, namely: the introduction of the maximum permissible angle of grain disorientation equal to 25°, as well as having the coefficient value of the ultrasonic wave reflection from the interface, it is possible to introduce the cutoff level of the signal amplitude scattered back from the controlled bridge volume. This will allow increasing the blade testing efficiency by the laser-ultrasonic method. The possibility of using the CLUFD method for the timely detection of hot crystallization cracks in rotating turbine blades is shown experimentally (on the example of model samples and standard blades).

Due to the complex geometry of the blade airfoil and locking piece, it is rather difficult to carry out ultrasonic testing of the blade locking piece using traditional ultrasonic testing methods. The blade geometry limits access to the testing regions. The use of phased array technology (Figure 4) in combination with the portable X32 flaw detector represents a new level of ultrasonic blade testing [5–9]. The flaw detector operates in a wide frequency range, which allows solving a very large range of various ultrasonic testing problems [10–21]. The advantage and capabilities of the PA technology together with the portable X-32 flaw detector can be considered on the example of the TB locking piece testing. When using the X-32 flaw detector and a PAR-transducer, it becomes possible to detect defects in the blade locking piece accurately and determine their location and dimensions. The area of the platform from which the sounding of the blade locking piece was carried out was (10x5) mm2.

S-scan (sector-scan) makes it possible to interpret the received signals quickly and easily.

To confirm the validity of the inspection results, an image obtained on the S-scan of the turbine blade locking piece with detected 0.5 mm and 1.5 mm cracks is superimposed on the TB photograph.

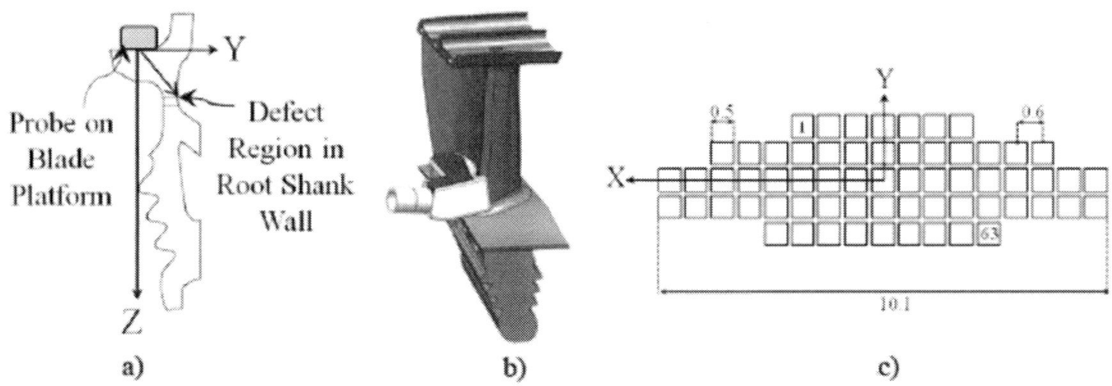

Figure 4. Schematic drawings show the probe on the turbine blade and the component layout. All dimensions are given in mm.

The optimal angles of ultrasonic vibrations input are within the sector from 0o to 55o. Signals become less clear at angles greater than 45°.

The inspection results show that an X-32 flaw detector and PAR-transducers allow testing turbine blades more efficiently in comparison with traditional technologies based on the use of a single-channel ultrasonic flaw detector [22].

4. Conclusion

1. Methodologies and techniques for ultrasonic testing of various-sized blades with different purposes for the PS-90A engine and its family were developed and introduced into production.
2. The CLUFD method together with the existing X-ray control method was carried out experimentally using standard and model blades, which made possible to increase the diagnostic informativity of the results when assessing cast blades quality.
3. Testing of the turbine blade locking piece by the PAR technology was carried out experimentally using standard and model blades.

References

[1] Yelenevsky D S and Pavlinich S P 2011 Effect of the azimuthal crystallographic orientation of single-crystal turbine blades of ground turbine engines on their resistance to cracking in operation *Gas turbine technologies* 34–38

[2] Luttsau V G, Kostyukova E P, Toloraya V N and Kornilova O M 1978 Study of the degree of perfection of nickel heat-resistant alloy crystals *News of USSR Academy of Sciences. Metals* **3** 166–170

[3] Gusev V E and Karabutov A A 1991 *Laser optics* (Moscow: Nauka) p 304

[4] Kulakov M A and Morozov A I 1982 Visualization of the subsurface structure of objects using an acoustic microscope *Letters to the ZhTF* **8** **12** 719–721

[5] Portnov O V, Zotov K V and Prokhorenko A A 2007 Experience in using the X-32 flaw detector on a phased array *Control Diagnostics* **11** 61–63

[6] Lane C J L, Dunhill A, Drinkwater B W and Wilcox P D 2010 Inspection of single crystal aerospace components with ultrasonic arrays *AIP Conference Proceedings* **1211** 2060–2067

[7] Lane C J L, Dunhill A, Drinkwater B W and Wilcox P D 2012 The development of a 2D ultrasonic array system for the in situ inspection of single crystal turbine blades *AIP Conference Proceedings* **1430** 865–872

[8] Lane C J L and Wilcox P D 2012 The effect of beam directivity on the inspection of anisotropic materials using ultrasonic arrays *AIP Conference Proceedings* **1430** 889–896
[9] Lane C J L, Dunhill A, Drinkwater B W and Wilcox P D 2011 The ultrasonic measurement of crystallographic orientation for imaging anisotropic components with 2D arrays *AIP Conference Proceedings* **1335** 803–810
[10] Cumpsty N 2003 *Jet Propulsion: A simple guide to the aerodynamic and thermodynamic design and performance of jet engines* (Cambridge: Cambridge University Press)
[11] Carter P, Cox D C, Gandin C A and Reed R C 2000 *Materials Science and Engineering* **A280** 233–246
[12] Ford T 1997 *Aircraft Engineering and Aerospace Technology* **69** 564–566
[13] Auld B A 1973 *Acoustic Fields and Waves in Solids* (New York: John Wiley & Sons)
[14] Červený V 2001 *Seismic Ray Theory* (Cambridge: Cambridge University Press)
[15] Musgrave M J P 1970 *Crystal Acoustics* (London: Holden-Day)
[16] Rose J L 1999 *Ultrasonic Waves in Solid Media* (Cambridge: Cambridge University Press)
[17] Frasier D, Whetstone J, Harris K, Erickson G and Schwer R 1990 Process and Alloy Optimization for CMSX-4 Superalloy Single Crystal Airfoils *High Temperature Materials for Power Engineering* 1281–1300
[18] Kline R A 1999 "SAFT Imaging in Anisotropic Media *Review of Progress in Quantitative Nondestructive Evaluation* **18** 913–919
[19] Spies M and Jager W 2003 *Ultrasonics* **41** 125–131
[20] Holmes C, Drinkwater B W and Wilcox P D 2005 *NDT&E International* **38** 701–711
[21] Jones A T and Baxter C 1995 *Meas. Sci. Technol.* **6** 131–133
[22] Dumas P, Clossen-Von Lanken Schulz M, Opheys M and Rauschenbach H 2006 Nondestructive Inspection of Blade Roots and Blade Attachment Grooves *Ideal Application of Phased Array Inspection Technique* **2 8 2**

IOP Conf. Series: Materials Science and Engineering **969** (2020) 012066 doi:10.1088/1757-899X/969/1/012066

Increasing the Service Lifetime of a Working Elements of Mixer-Pneumosuperchargers for the Building Mixtures Preparation

N I Gabelchenko[1], A A Belov[1] and A I Gabelchenko[2]

[1]Departament of Machines and technology of foundry, Volgograd state technical university, 28, Lenin Avenue, Volgograd, 400005, Russia
[2]Limited liability company "GA group" (LLC "GA group"), 38, Michurinsky prospekt, 277, Moscow, 119192, Russia

E-mail: mitlp@vstu.ru

Abstract. Comparative studies of the blade's materials for mixer-pneumosuperchargers operated under abrasive conditions were carried out. For this purpose in the production environment at the construction site the service lifetime of the blades offered by various manufacturers was measured on four "SO–241" mixer-pneumosuperchargers. Production of building mixture in quantity declared by the manufacturer was used as a control period. It is shown that blades made of 110G13L steel retained their service properties almost until the completion of the control period. Also, the composition and technology for producing the working elements of mixer-pneumosuperchargers from wear-resistant white cast iron alloyed with chromium and manganese was developed. The test results showed that the blades made of the proposed material exceeded the service lifetime of the blades made of 110G13L steel.

1. Introduction

Currently, the wear problem of consumables for mixer-pneumosuperchargers become quite urgent due to the lack of centralized supplies and the presence of uncontrolled suppliers on the Russian market of metal products.

The principle of mixer operation is mixing of building mixture (semi-dry mortar), which includes: cement, sand, water, chopped basalt thread and supplying the finished mixture to the place of laying. The mixing element is the blades operating under abrasive conditions. Alloys such as cast iron and wear-resistant steel alloys are used as material for blades. The technology for producing blades is also different – this is either using of rolled metal or casting. In recent years, the leader in the production of sheet metal blades was the Swedish wear-resistant Hardox 450 steel with a hardness of 450 HB and designed to meet special demands on wear resistance [1–6]. The foundry technology of manufacturing the working elements is characterized by the use of various alloys based on iron, starting from steel 15, reinforced by depositing, to high-carbon alloys, such as cast irons [7–16]. However, according to consumers, the blades often do not withstand the declared service lifetime. Therefore, the task was to obtain the blades for mixer-pneumosuperchargers with a high and guaranteed service lifetime.

In this regard, comparative studies of the service lifetime of the working elements of mixer-pneumosuperchargers from various materials were carried out along with the selection of the optimal material, considering the price-quality ratio, and technology development for producing blades with stable operational properties.

2. Results and discussion

To compare the service lifetime of the working elements of mixer-pneumosuperchargers at production conditions with the service lifetime declared by the suppliers, the blades made of the most commonly used materials (SCh 25 cast iron, steel 15 reinforced with H10 steel, 110G13L steel and Swedish wear-resistant Hardox 450 steel) were selected. Each set of four blades from the selected materials was installed in the "SO–241" mixer-pneumosuperchargers: two blades ensure the movement of the mixture components to the center and two blades for axial movement of the components in the opposite direction (figure 1). The experiment involved four mixer-pneumosuperchargers.

(a) (b)

Figure 1. Solution mixing blades: a – for moving components of mixture to the center, b – for axial movement of components in the opposite direction.

The amount of semi-dry mortar (cement, sand and water in cubic meters) produced by the mixer-pneumosuperchargers was measured until the failure of mixing blades or reaching the declared service lifetime (figure 2).

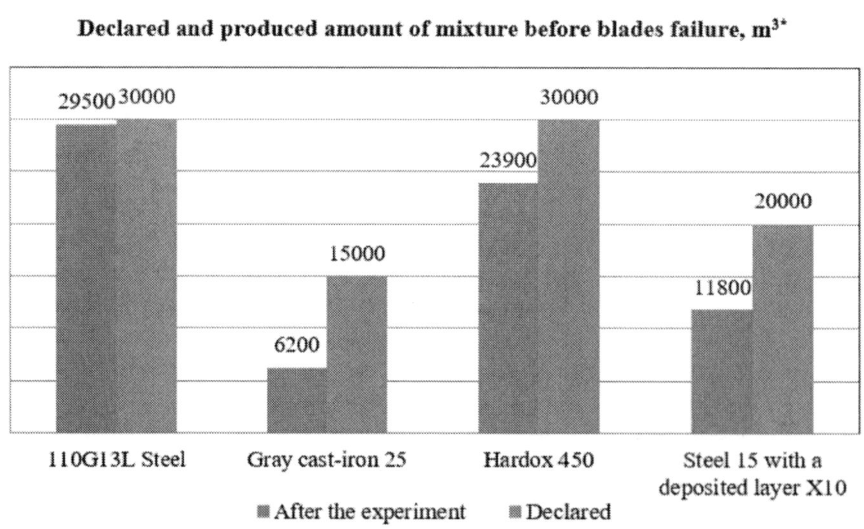

Figure 2. The results of a comparative study conducted at production conditions.
*The values of the produced amount of the mixture were rounded to hundreds.

Figure 2 shows that by blades made of gray cast iron was produced 58.7% less mixture before replacement, due to intensive wear of working surfaces than was declared by the suppliers, for Hardox 450 steel reduction was 20.4% and for 15 steel reinforced with X10 steel reduction was 41%. Blades made of 110G13L steel retained their service properties almost until the completion of the control period.

However, given the high cost and certain technological difficulties of the production of castings from 110G13L steel a composition of white cast iron alloyed with chromium and manganese was developed by request of a construction organizations using mixer-pneumosuperchargers and showed results for wear resistance tests comparable to 110G13L steel at an affordable price [17, 18]. At the same time, the task was set to increase the service lifetime of such blades.

Alloying with chromium and manganese was chosen, since they inhibit the graphitization process, which is guarantees production of white cast iron with high hardness.

According to the sizes of the two blades, different in configuration, provided by the construction organization, wooden patterns with a thickness of 10 mm, pattern allowances and two holes with a diameter of 10 mm were made (Figure 3).

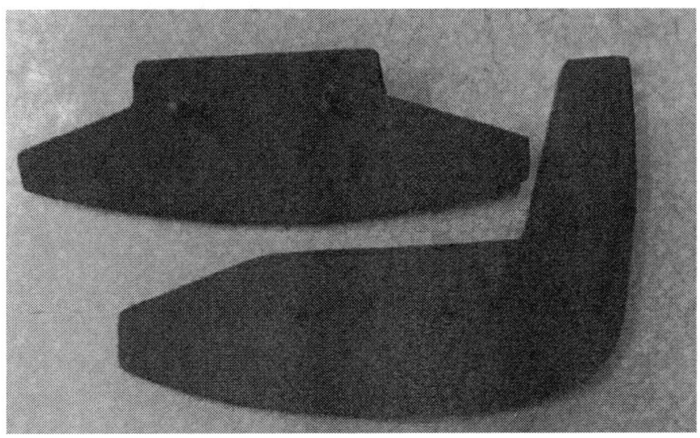

Figure 3. Patterns of the mixer-pneumosuperchargers blade.

To obtain holes in the castings, rods made of a mixture on an organic binder material with good knockability and providing the necessary surface cleanliness were used [19, 20]. The composition of the rod's mixture is presented in table 1.

Table 1. The composition of the rod's mixture.

Component	Weight fraction, %
Sand brand $2K_2O_202$	95
Technical lignosulfonates	3
Mustard filter cake	2

The rods were dried in an "EKPS–10" resistance furnace at 240–260 °C for 30 minutes. For molding a sand-clay mixture was used. The composition of the charge included purchased steel scrap (GOST 2787–78), cast iron (GOST 4832–95), ferrochrome (GOST 4757–91) and ferromanganese (GOST 4755–91). The charge was melted in the "IST–0,06" induction furnace. Low tempering was carried out to relieve internal stresses of castings. For heat treatment of castings, the "SNOL 80/12" electric furnace was used. Simultaneously with the molding and filling of the blades, additional samples were made to determine the chemical composition, conduct metallographic studies and determine the mechanical properties of the obtained castings. The chemical composition of the obtained castings is presented in table 2.

Table 2. The chemical composition of white cast iron (wt. %).

C	Mn	Si	Cr	Ni	S	P	Fe
2.53	2.09	1.36	5.15	0.093	0.030	0.0050	rest

To carry out metallographic studies, metallographic sample were made from a metal of additional sample. The presence and shape of graphite in cast iron was assessed. To study the microstructure, etching was performed with a 4% solution of nitric acid in ethanol. The microstructures of samples of white cast iron are presented in Figure 3.

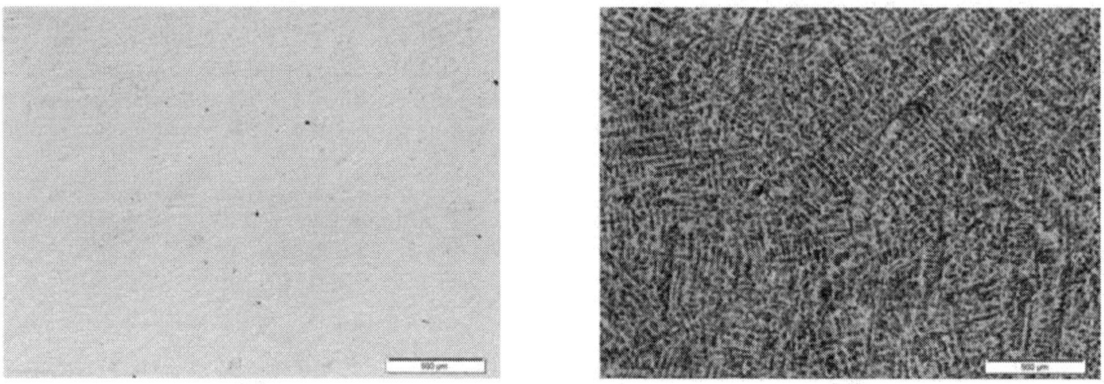

(a) (b)

Figure 4. Microstructures of white cast iron samples: a – without etching; b – with etching.

As can be seen from Figure 4, alloying of cast iron with manganese leads to inhibiting the release of graphite, perlite grain refining and cast iron bleaching. The introduction of chromium increased the strength and hardness of castings and inhibited carbon graphitization process. The ratio of carbon, manganese and chromium obtained in the composition of cast iron when interacting with sulfur neutralized its harmful effect. Minor amount of fine graphite inclusions had a flaky shape (Figure 4a). Figure 4b shows that the microstructure of cast iron consists of perlite and ledeburite. The hardness of the test cast iron was 470 HB, which exceeds the hardness of Hardox 450 steel.

The experimented batch of alloyed white cast iron was manufactured and tested under production conditions. The test results showed that the blades made of the proposed material failed only after the production of 43,000 m^3 of the building mixture, which exceeded the service lifetime of the blades made of 110G13L steel.

3. Summary

As a result of the work, the following results were obtained:

- the composition of white cast iron was developed for the manufacture of working elements of mixer-pneumosuperchargers;
- An experimental technology was developed for the production of working elements of mixers-pneumosuperchargers by casting in sand-clay molds;
- the service lifetime of the blades was significantly increased, which made it possible to recommend the proposed cast iron composition and technology for producing castings for the manufacture of working elements of mixer-pneumosuperchargers.

References

[1] Wei L, Bodnar R, Amsden C, Ecklund J and Baer B 2013 Comparison of abrasion and impact wear properties of chromium carbide hardfacing alloys on HARDOX® 450 and mild steel base plates *Materials Science and Technology Conference and Exhibition* **3** 2194–2204

[2] Ivanov Y, Gromov V, Konovalov S, Kormyshev V, Aksenova K and Teresov A 2017 Structure

and properties of strengthening layer on Hardox 450 steel *Materials Science and Technology* **33(17)** 2040–2045

[3] Konovalov S V, Kormyshev V E, Kapralov E V, Ivanov Y F, Zavatski O V and Gromov V E 2016 Formation of the Increased Wear-Resistant Properties of Hardox 450 Steel by Deposited Coatings *IOP Conference Series: Materials Science and Engineering*

[4] Konovalov S, Kormyshev V, Gromov V and Ivanov Y 2016 Metallographic Examination of Forming Improved Mechanical Properties Via Surfacing of Steel HARDOX 450 with Flux Cored Wire *Materials Science Forum* **870** 159–162

[5] Białobrzeska B, Konat Ł and Jasiński R 2017 The Influence of Austenite Grain Size on the Mechanical Properties of Low-Alloy Steel with Boron *Metals* **7**

[6] Ulewicz R, Mazur M and Bokůvka O 2013 Structure and Mechanical Properties of Fine-Grained Steels *Periodica Polytechnica Transportation Engineering* **41(2)** 111–115

[7] Kritskij D Y, Tyurin S I, Kovaleva A A and Gil'Manshina T R 2018 Manganese steel structure effect on the performance characteristics of large-size cast product parts *Coal* **7** 9–12

[8] Bolobov V I and Bochkov V S 2017 Effect of strengthening treatment on durability of lining in apatite-nepheline ore ball mills *Mining Journal* **1** 57–60

[9] Bolobov V, Chupin S and Bochkov V 2017 About increasing wear resistance of rock-breaking tool to abrasion by using mechanical and thermo-mechanical treatment *International Review of Mechanical Engineering* **11(5)** 301–304

[10] Borovikov E M, Yakovlev B N and Tsentsiper B M 1998 Production of special castings from high-manganese steel *Foundry production* **9** 6–7

[11] Kolokoltsev V M, Vdovin K N, Gorlenko D A and Gulin A E 2016 Calculation of stacking fault energy and its influence on abrasive wear resistance of hadfield cast steel cooled at different rates *CIS Iron and Steel Review* **11** 35–40

[12] Vdovin K N, Feoktistov N A, Gorlenko D A and Koptseva N V 2018 Wear of high-manganese steel castings *Ferrous metals* **9** 48–53

[13] Vdovin K N, Feoktistov N A, Gorlenko D A, Chernov V P and Khrenov I B 2017 Influence of Alloying and Heat Treatment on the Abrasive and Impact–Abrasive Wear Resistance of High-Manganese Steel *Steel in Translation* **47(11)** 705–709

[14] Kostyleva L V, Gapich D S, Motorin V A and Novikov A E 2019 Microstructure and abarizive wear resistance of heavy duty parts from high-strength cast iron in chisel plows *Ferrous metals* **3** 37–42

[15] Kostyleva L V, Gapich D S, Novikov A E and Motorin V A 2020 Wear-resistant cast iron containing spheroidal graphite with a two-layer Ledeburitic–Martensitic shell *Russian Metallurgy* **3** 231–237

[16] Takhirov A A, Gushchin N S, Nuraliyev F A 2013 Production of Grinding Elements of Asphalt Preparation Equipment. Foundry *Technologies and Equipment* **7** 2–7

[17] Gabelchenko N I, Belov A A, Kidalov N A, Gabelchenko A I and Volkova N V 2016 The study of wear-resistant materials of the blades for mixers – pneumosuperchargers *Bulletin of the Samara Scientific Center of the Russian Academy of Sciences* **18** 175–178

[18] Gabelchenko N I, Belov A A and Skorobogatova A V 2017 Investigation of metal consumables working in conditions of abrasive wear *Bulletin of the Volgograd State Technical University. Ser. Problems of materials science, welding and strength in mechanical engineering* **10** 136–140

[19] Kidalov N A, Osipova N A, Chertova O O, Potashova I E, Aliev D O and Kiselev I V 2012 *The binder for heat hardening foundry rods* RU Patent **2449855**

[20] Kidalov N A, Osipova N A, Chertova O O and Potashova I E 2012 Research of the properties of thermal cured core mixtures using residual products of processing mustard seeds *Russian Caster* **1** 43–46

IOP Conf. Series: Materials Science and Engineering **969** (2020) 012067 doi:10.1088/1757-899X/969/1/012067

Effect of a Carbonaceous Additive on the Structure of an Aqueous Sodium Silicate Film on the Surface of a Refractory Chromite Filler

N A Kidalov, N V Grigoreva and A S Adamova

Departament of Machines and technology of foundry, Volgograd state technical university, 28, Lenin Avenue, Volgograd, 400005, Russia

E-mail: mitlp@vstu.ru

Abstract. The results of studying the effect of a carbonaceous additive on the structure formation of aqueous sodium silicate film on the surface of a refractory chromite filler was shown. Ultradispersed pyrolytic carbon (pyrocarbon) was used as a carbonaceous additive. The structure of liquid glass films on the surface of a refractory filler was investigated after curing at temperatures of 180-400-600-900 °C, which are comparable to the temperature of convective drying of liquid glass molds and cores, as well as molds heating temperature during pouring with liquid metal. The structure of liquid glass films on a refractory filler was investigated using a Versa 3D two-beam scanning electron microscope, which made it possible to show the effect of pyrocarbon on the structure of a silicate binder film in each temperature range.

1. Introduction

Currently, liquid glass is one of the common binders in the producing of molding and core sands for foundry, due to a number of its positive qualities, such as non-toxicity, relatively low cost, high binding capacity, etc. [1–3]. The study of the curing processes of liquid glass mixtures is reflected by many researchers, and the study of liquid glass films formation on the surface of a refractory filler is of particular interest. The quality of the silicate binder film's structure and its adhesion to the refractory filler directly affect the strength characteristics of molding and core sands. But when working with liquid glass, it is also necessary to take into account the residual strength, which affects the difficulty of knocking out cores and castings from the mold. The works devoted to solving the problems of poor knocking-out ability propose and consider various methods such as modifying sodium silicate solution and liquid glass mixture [4–7], adding organic hardeners, optimizing compositions and using softening additives that affect residual strength [8–11].

Of the above methods, the introduction of organic, inorganic and complex additives into the mixture is most universal. The softening process with the introduction of these additives is realized due to the release of CO and CO_2 under high-temperature exposure. At the same time, with gas exit the internal stresses inside the mold are created and accompaned with binder's film integrity violation [12–14].

It is known that in the 70s the question of the aqueous sodium silicate film structure formation starts to attract attention [15]. Research methods and technologies of that time did not allow maintain fully detailed structural studies. The physical and mechanical properties are largely influenced by the quality of the liquid glass film, therefore, studies that allow investigate the effect of curing methods on films properties will lead to possibility of giving recommendations for control properties of liquid glass molds and cores.

In this work, the chromite sand was used as a filler, which is widely used in the manufacture of steel castings using core and facing mixtures, and also provides a cheaper alternative to special fillers such as zircon, while reducing or eliminating casting defects associated with using quartz sand [16]. Ultradispersed pyrolytic carbon (pyrocarbon) was used as a carbonaceous additive, which is carbon films formed by crystallization from gas phase at high pressure and temperature (1400–1600 °C) [17–19].

The purpose of this work is to study the effect of a carbonaceous additive on the structure formation of aqueous sodium silicate film on the surface of a refractory chromite filler.

2. Materials and methods

To determine the behavior of pyrocarbon at elevated temperatures, thermogravimetric (TGA) and differential thermal (DTA) analyzes were carried out simultaneously using a Q-1500D derivatograph manufactured by MOM, Hungary. A sample with pyrocarbon was loaded in the device and heated up to a temperature of 950 °C at 10 °/min rate.

To study the effect of a carbonaceous additive on the structure of aqueous sodium silicate film, mixtures were prepared from chromite sand (TU 0741-001-23081308) as a filler and liquid glass (GOST 13078-81) as a binder. For comparative analysis, the mixture 1 did not contain a carbonaceous additive. Pyrocarbon was added to the mixture 2 in an amount of 2%. The samples were treated in a furnace at temperatures of 180-400-600-900 °C, which is comparable to the temperature of convective drying of liquid glass molds and cores, as well as molds heating temperature during pouring with liquid metal. The mixtures compositions for samples are presented in Table 1.

Table 1. Mixtures compositions, %.

Mixtures №	Chromite sand AFS 45-55	Pyrocarbon	Liquid glass
1	94	–	6
2	92	2	6

After heat treatment of the samples, the structure of liquid glass films on a refractory filler was investigated using a Versa 3D two-beam scanning electron microscope, which made it possible to show the effect of pyrocarbon on the structure of a silicate binder film in each temperature range.

3. Results and discussion

Analysis of the derivatogram (Figure 1) showed that heating up to 400 °C does not cause any significant changes in pyrocarbon, which is confirmed by the DTA and DTG data.

The weight loss of the sample in this temperature range is no more than 5% via the burnout of impurities in pyrocarbon. At temperatures above 400 °C, a sample's weight loss with a small endothermic effect is observed due to the oxidation of the carbonaceous additive with the release of CO and CO_2. At the same time weight loss continuous throughout the experiment, which is confirmed by the direction of the TGA curve.

Figure 2 shows the structure of liquid glass films sample, in which no carbon additive was introduced.

At sample temperature of 180 °C (Figure 2 a), the places of contact of the filler grains (cuffs) are swollen due to the evaporation of water (from liquid glass), but no film defects such as tears were found, the surface is solid in comparison with similar samples using quartz sand [20]. This is due to the high heat-storage ability of chromite (2380 compared to 1628 $W \cdot c^{1/2}/(m^2 \cdot K)$ for quartz sand), which leads to formation of moisture condensation zone at greater depth than in quartz sand mixtures. At the same time, on samples cured at 400 and 600 °C (Figure 2. b, c), defects formed by water evaporation are observed, the structure of the film is disturbed, which is clearly visible both on the chromite grains and on the cuffs.

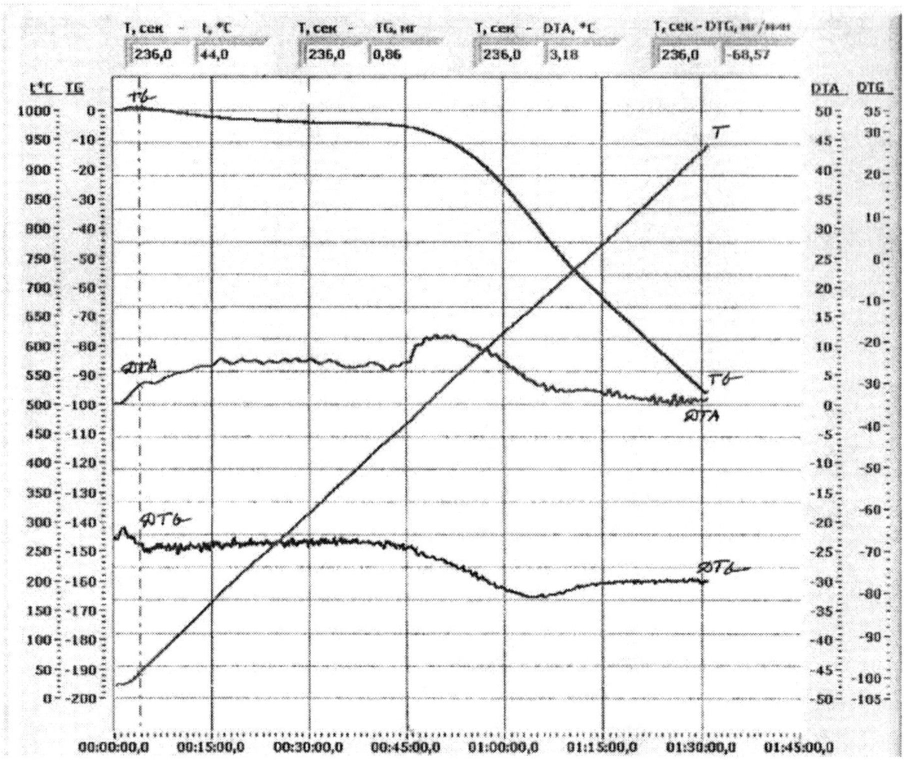

Figure 1. Differential thermal curves of pyrocarbon.

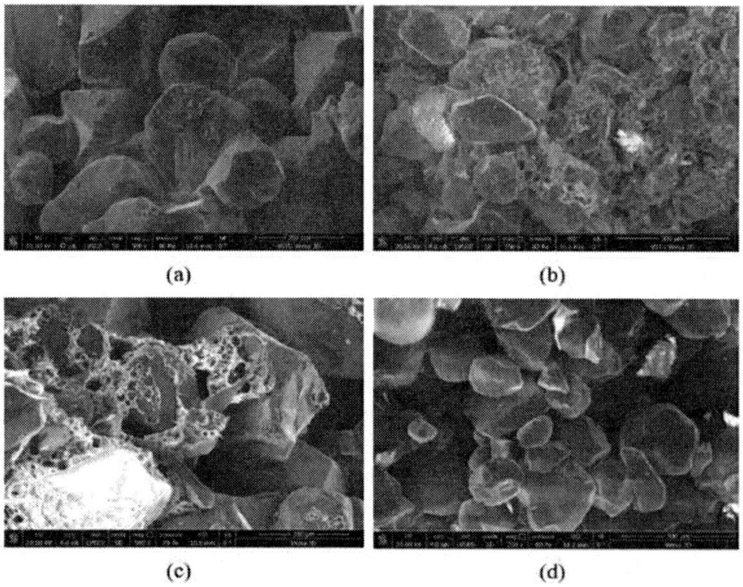

Figure 2. Structure of liquid glass films on refractory chromite filler grains without pyrocarbon addition at heating temperature of: a – 180 °C (×500); b – 400 °C (×250), c – 600 °C (×500), d – 900 °C (×250).

At sample temperature of 900 °C (Figure 2 d), the structure of the film has no cracks and chips, which is associated with the ability of liquid glass to melt, passing into a liquid state at temperatures above 793 °C [3].

Figure 3 shows the structure of liquid glass films sample in which carbonaceous additive was introduced.

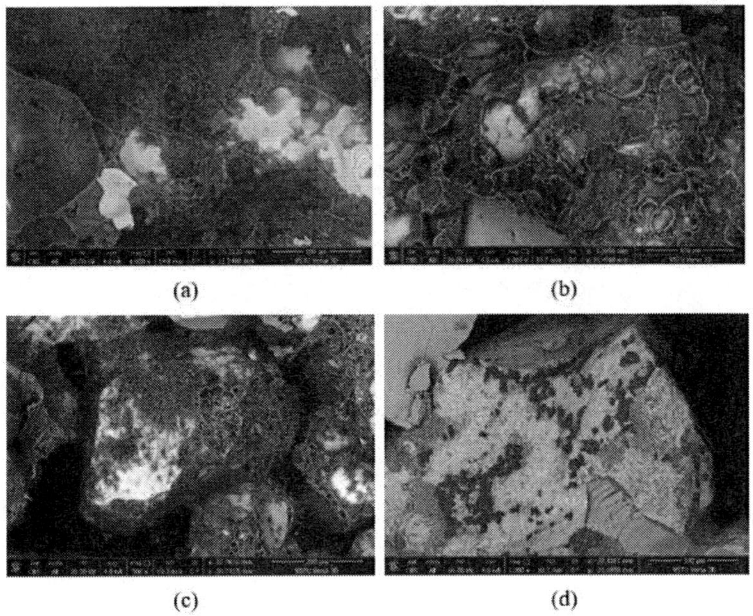

Figure 3. Structure of liquid glass films on refractory chromite filler grains with pyrocarbon addition at heating temperature of: a – 180 °C (×1000); b – 400 °C (×1000), c – 600 °C (×500), d – 900 °C (×1000).

Defects associated with the removal of water from the composition of liquid glass, as well as the release of CO and CO_2 and impurities of the carbonaceous additive, are observed on the binder's film in the samples in which pyrocarbon was introduced.

At 180 °C (Figure 3 a) the film is smoother, with visible cracks formed during shrinkage of the binder. There are also traces of gas escaping in the form of small circular holes. The films structure of samples after heat treatment at temperatures of 400 and 600 °C (Figure 3 b, c), contains defects associated with the removal of water (globular structure of the film), and more film tears formed by release of CO and CO_2 due to pyrocarbon oxidation in this temperature range, which is confirmed by the endothermic effect on differential thermal curves (Figure 1). At temperature of 600 and 900 °C (Figure 3 c, d), via carbonization process during the release of CO_2 a formation of crystals is occurred on the aqueous sodium silicate film, which provokes tearing of the film and its delamination from the refractory filler.

4. Conclusion

It has been experimentally established that the addition of ultradispersed pyrocarbon affects the structure formation of the aqueous sodium silicate film on the refractory filler grains.

A film of aqueous sodium silicate on sample that did not contain pyrocarbon has defects associated with the removal of water (component of liquid glass) at temperatures of 400 and 600 °C. At temperature of 180 and 900 °C, the structure of the film is continuous with no tears, which is associated with the heat-storage ability of chromite and the ability of the binder to melt at temperatures above 793 °C.

On the liquid glass films of the sample in which the carbon additive was introduced, observed defects are associated with the removal of water, the release of CO and CO_2 during heating of the samples, as well as the formation of carbonization crystals, which provokes tearing of the film and its delamination from the refractory filler.

The study of the effect of various additives on the structure formation of aqueous sodium silicate film on the surface of refractory fillers will allow in the future to predict the process of structure formation of liquid glass mixtures.

References

[1] Zhukovsky S S and Kvasha F S 2012 Production of rods from liquid-glass mixtures in the mass production of castings *Foundry worker of Russia* **7** 41

[2] Zhukovsky S S and Boldin A N 2002 *Foundry technology: molding and core mixtures* (Bryansk: BSTU) p 470

[3] Korneev V I and Danilov V V 1996 *Liquid and soluble glass Production Edition* (St. Petersburg: Stroyizdat) p 216

[4] Krutilin A N, Guminsky Yu Yu, Rusevich O A and Kulbitskaya L V 2018 Improving the use of liquid glass mixtures. overview information. Part 1. Modification *Casting and metallurgy* **1(90)** 47

[5] Krutilin A N, Guminsky Yu Yu, Rusevich O A and Kulbitskaya L V 2018 Improving the use of liquid glass mixtures. 4. Combined hardening *Casting and metallurgy* **4(93)** 38

[6] Kidalov N A, Osipova N A and Potashova I E 2012 The use of waste oil extraction production to improve the knockability of liquid glass mixtures *Procurement in engineering* **11** 5

[7] Nefedov K Y 2004 Easily knock-out water glass molding and core sands *Foundry* **1** 18

[8] Kidalov N A, Osipova N A and Potashova I E 2015 Filtration products of vegetable oils to improve the knockability of liquid-glass mixtures *Foundry* **10** 29

[9] Kidalov N A and Osipova N A 2005 The effect of metal oxides on the properties of liquid glass mixtures for the manufacture of foundry molds and cores *Mechanical engineer* **3** 46

[10] Stechman M, Rözycka D and Baliński A 2003 Modification of aqueous sodium silicate solutions with morphoactive agents IV Conference on Chemical Technology *Polish Journal of Chemical Technology* **5(3)** 47

[11] Al-Saraireh F M 2018 An assessment of the efficiency of utilizing complex modifiers for softening the liquid-glass mixtures to improve iron and steel casting *ARPN Journal of Engineering and Applied Sciences* **13(9)** 3231

[12] Kidalov N A, Volchkov V M and Knyazeva A S 2015 Optimization of mixture composition for foundry *Conference on Creativity in Intelligent Technologies and Data Science* **535** 341.

[13] Kidalov N A, Volchkov V M and Osipova N A 2005 Optimization of sodium silicate bonded sand composition *Foundry* **6** 11

[14] Akulova M V, Slizneva T E, Sokolova Yu A and Sokolova A G 2018 Nanostructuring sodium silicate solutions applied as binding substance of molding sands in foundry *Mechanical Engineering, Automation and Control Systems* **560(1)** 012136

[15] Ryzhkov I V and Tolstoy V S 1975 *Physicochemical basis for the formation of the properties of mixtures with liquid glass* (Kharkov: Vishka school.) p 139

[16] Ravi S, Thiel J, Bryant N, Giese S and Schneiter T J 2018 Use of dilatometry to evalute the high temperature characteristics of silica in chromite sand *73 World Foundry Congress* 447

[17] Mamaev K V 2004 Core sand with antipickup ultradispersed pyrocarbon additive *Foundry* **1** 20

[18] Aleksandrov V M, Vasin Yu P, Loginovskii A N and Ovechkin I V 1974 Pyrocarbon as a binder for moulds and cores *Russ. Cast. Prod.* **4** 145

[19] Aleksandrov V M, Barvinskii V A, Vasin Yu P, Eremin A I, Ivanov B G and Korobov G V

1974 Pyrocarbon coating of finely divided refractory moulding materials *Russ. Cast. Prod.* **3** 89

[20] Kidalov N A, Grigorieva N V, Adamova A S and Zatyamin D A 2019 Influence of methods for curing a liquid glass mixture on strength characteristics and structure of the casting form *Bulletin of the Volgograd State Technical University. Ser. Problems of materials science, welding and strength in mechanical engineering* **10(233)** 78

Acknowledgments
The reported study was funded by RFBR, project number 19-33-90111.

IOP Conf. Series: Materials Science and Engineering **969** (2020) 012068 doi:10.1088/1757-899X/969/1/012068

Features of Design of Chill Molds for Casting of Non-Ferrous Metal

R V Gavariev[1], K N Gavarieva[2] and E N Soldatkina[1]

[1]Kazan national research technical university named after A. N. Tupolev-KAI, 10, K. Marx St., Kazan, Tatarstan, 420111, Russia
[2]Kazan Federal University, Naberezhnye Chelny Institute, 10A, Prospekt Syuyumbike, Naberezhnye Chelny, 423812, Russia

E-mail: Gavarievr@mail.ru

Abstract. The article discusses methods for calculating the main parameters of the chill mold in the design process. Formulas for calculating the thickness of the working wall, as well as recommendations for choosing the material of the chill mold, depending on the size and shape of the resulting castings, are specified. Disadvantages of existing methods for calculating the chill mold parameters are indicated. Solutions for the use of multi-layer coatings applied by physical vapor deposition on the forming surface of the chill mold are proposed, which increases the efficiency of the casting process in the chill mold, regardless of the shape of the casting and the design of the chill mold.

1. Introduction

Compared to traditional casting methods, the chill casting process is an effective way to obtain blanks by reducing the consumption of molding materials, reducing sources of waste, improving the quality and performance of castings, high productivity, the ability to automate and mechanize the process, reducing capital costs, reducing the cost of castings. These advantages are due to the repeated use of metal foundries in comparison with sand and clay forms. However, the effectiveness of using the chill mold will depend on its design and material, which makes the design stage one of the most difficult at the pre-production stage [1].

2. Results

Based on the research [2], we can say that when designing chill molds, one of the main issues is the choice of the thickness and material of the working walls, since these parameters affect the operational stability of the chill mold. There are various methods for determining the wall thickness of the chill mold S. The most widely used method is based on the theoretical analysis of the stress-strain state of the working walls of chill molds of various designs (Figure 1a) [3, 4].

For single-layer and double-layer steel and cast-iron chill molds with liquid cooling, the wall thickness must be selected according to the schedule (Figure 1a), with the upper limit values being the most preferred options. An exception is the situation when the chill mold is used for casting castings of complex configuration with closely spaced protrusions. In this case, the values for the lower limit are used. In addition, in the case of casting castings using sand rods, it is necessary to choose a double thickness of the casting.

When choosing a cylindrical design of chill molds for hollow castings, it is necessary to use a graph (Figure 1b), in which the values of the parameters N and K are determined by the formulas [5]:

$$K = \frac{S_1}{2R}; \tag{1}$$

$$S = 2NR, \tag{2}$$

where S_1 – the thickness of the casting; R – outer radius of the casting.

When casting solid cylindrical castings, the thickness of the profile walls can be determined with the ratio $S = 1.4\,R$.

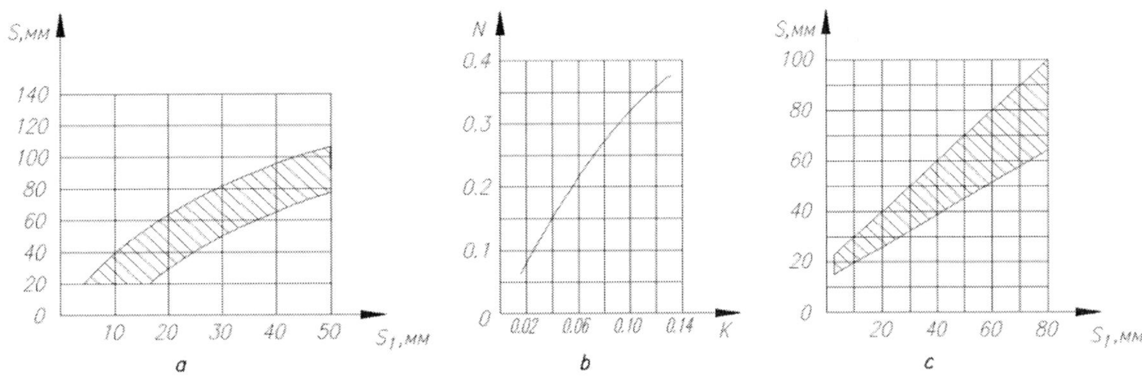

Figure 1. Interdependence of parameters of the chill molds.

According to the method [6], the wall thickness of the chill mold S when casting castings from ferrous and non-ferrous alloys must be determined according to the schedule (Figure 1c). In this case, values are chosen depending on the melting temperature of the molten melt for high temperature alloys is preferred upper limit, and for low temperature – lower limit.

Despite the fact that these methods have been tested for a long time in practice, they have a disadvantage due to the fact that the proposed graphs (Figure 1) are generalized for various designs of chill molds and therefore may have an error associated with averaging the operating loads. In this regard, additional methods for determining the thickness of the working walls are applicable for specific cases.

In [7], the wall thickness of flat chill molds is proposed to be calculated using the formula:

$$S = \frac{k}{2} A \left(1 + \sqrt{1 + \frac{8\lambda_2}{A\alpha_1}} \right), \tag{3}$$

where k – empirical coefficient; λ – coefficient of thermal conductivity (indices "1" and "2" refer to the casting and chill mold, respectively),

$$A = \frac{3Q_1}{c_2\rho_2\left(T_{cr} - T_{2\mathrm{p}}\right)}; \tag{4}$$

$$\alpha_1 = \frac{\lambda_{\mathrm{cr}}}{X_{cr}}; \tag{5}$$

$$Q = X_1\rho_1\left[c_1'\left(T_{\mathrm{p}} - T_{cr}\right) + r_1\right], \tag{6}$$

where X_1 – half the wall thickness of a flat casting; c – specific heat (the «'» symbol indicates a liquid state); ρ – density; r_1– specific heat of solidification of the casting; T_p, T_{cr} – pour and crystallization temperatures; X_{cov} – thickness of the paint layer (coating).

This method of calculating the wall thickness has passed laboratory and factory tests and can be successfully applied to enterprises that produce mainly flat castings. However, there are significant disadvantages associated with the restriction on the scope of use only for flat chill molds, as well as the need to determine a large amount of empirical and theoretical data depending on the material of the chill mold and casting, which makes the calculation quite complex and suitable only for experienced design engineers.

The calculation method is relatively simple [8]:

$$S = 11\sqrt{2S_1} \tag{7}$$

This method has shown good results when casting cast iron castings, but if it is necessary to obtain castings from other alloys, its use becomes impossible.

The method developed on the basis of theoretical research is quite accurate [9]:

$$S = \frac{3B}{2K_P} X_1 \left(K_B + \frac{1 + K_B r_1}{T_p - T_{2in} c_1} \right), \tag{8}$$

where B is the coefficient that takes into account the propensity of the chill mold to warping, and its material to oxidation;

$$K_P = \frac{2X_1}{R_{th}}; \tag{9}$$

$$K_B = \frac{b_2}{b_1}, \tag{10}$$

where R_{th} – given the wall thickness of the casting; b_1, b_2 – the coefficients of accumulation of heat to the casting material and form; $T_{2\text{in}}$ – the initial temperature of the chill mould; r_1– specific heat of solidification of the casting, c_1 – specific heat of the casting material.

Calculations using formula (8) are fairly accurate, but they require a large amount of preparatory work related to the definition of specific values of coefficients and parameters for each specific case, so its widespread use in practice is unlikely.

In addition to difficulties in calculating the size of the working wall, there is a problem of choosing the material of the chill mold, the method of its manufacture [10]. Gray cast iron, high-strength cast iron, cast steel, structural carbon steel, aluminum and copper alloys, as well as alloy steels and alloys with special properties can be selected as the material of the coil.

The choice of a specific brand of metal as a chill mold material is due to two factors – physical and mechanical properties that affect the strength, and economic efficiency. Thus, the option of expensive copper alloys or special alloyed steels is justified under the influence of intense thermohydrodynamic loads or with a mass volume of castings production. The use of cast iron or steel is due to the size of the casting, as well as the version of the coil and the type of cooling (air or water). In special cases, chill molds made of aluminum alloys are used with the use of anodizing, most often for small castings [11, 12]. At the same time, there is no unambiguous method for evaluating the suitability of materials for manufacturing the working walls of the chill mold. There are various design parameters of the special assessment.

For brittle materials, depending on the presence of a heat-protective coating, the following formulas are used [13]:

$$Z_0 = \frac{\sigma_B \lambda}{\alpha_T E}; \tag{11}$$

$$Z_0' = \frac{\sigma_B}{\alpha_T E}, \tag{12}$$

where σ_B – the limit of the tensile strength; λ – thermal conductivity; α_T – thermal expansion coefficient; E – elastic modulus.

For malleable materials, the parameters K and K' are used, depending on the failure mechanism under alternating loads:

$$K = \frac{\lambda\delta}{\alpha_T E}; \tag{13}$$

$$K' = \frac{\lambda a_{IS}}{\alpha_T E}; \tag{14}$$

where δ – elongation; a_{IS} – impact strength.

$$\mathrm{P} = \frac{a_{IS}}{\sigma_{YS}\left(\alpha_T \theta_c - \dfrac{2\sigma_T}{E}\right)}; \tag{15}$$

where σ_T – yield strength; $\theta_c = \theta_p \dfrac{b_1}{b_1 + b_2}$ – chill mold temperature at the time of filling; θ_p – the temperature of the metal at the time of casting, calculated from the initial temperature of the coil, as from zero.

The higher the values of the parameters specified in formulas 11–15, the higher, under equal conditions, the operational stability of the chill molds [14, 15].

The considered theoretical and practical methods of calculation and assessment of suitability show the complexity of the design process of chill molds, so it is important for designers to use universal methods to improve the performance of the chill mold in order to increase the number of castings obtained.

Based on the results of the study [16, 17], it was found that coating the forming surface by physical vapor deposition will reduce the coefficient of friction between the form and the interacting metal, reduce the pressure on the forming parts and reduce the temperature on the forming surface at the time of pouring the molten metal, which should improve the quality of the resulting castings. These factors will reduce the magnitude and amplitude of vibrations of tensile stresses in the forming parts both in the closed and open state of the form, which will increase the period of operation until the formation of cracks of the 1st type. In addition, the formation of compressive stresses in the coating applied by the physical vapor deposition method can contribute to slowing the growth of type 1 cracks in the material of forming parts and the formation of cracks in the coating itself. Coating by physical vapor deposition will slow down the growth of cracks, as well as increase wear resistance, which will increase the operational life of chill molds and the quality of the resulting castings [18].

It is possible to use simple single-layer coatings, such as titanium nitride (TiN) or molybdenum nitride (MoN), as well as complex multi – layer coatings (titanium carbonitride (TiCN) – titanium nitride (TiN) – molybdenum nitride (MoN) [19, 20]. The proposed coatings have positive properties, namely: a small thickness (up to 10 microns), which will not affect the accuracy of the resulting castings; high adhesive ability, which contributes to a strong adhesion to the chill mold material; the ability to slow down the growth of cracks formed both on the forming surface and in the chill mold material; the chemical inertness of the coating after application, which will provide protection from sticking of the poured melt to the forming surface, especially when casting aluminum alloys; relative ease of coating application; high microhardness of the coating, providing a long service period of the coating.

3. Conclusions

Thus, there are a large number of methods for selecting the material and calculating the thickness of the chill mold walls, obtained as a result of generalization of experimental data, taking into account various specific conditions of the casting process. However, each method has disadvantages that limit the scope of application. Designers face a difficult practical task to determine all the conditions of the developed casting process that affect the specific values of the chill mold parameters. In this regard, given the relative complexity of these methods, there are promising universal ways to improve the performance properties of the chill mold, which will allow for a wider use of most methods for selecting materials and calculating the wall thickness of the chill mold using any of the methods. One of these solutions is a method for applying nitride and carbonitride coatings by physical vapor deposition on the forming surface of a coil or metal rod.

References:

[1] Burakov S L, Veinik A I and Dubinin N P 1980 Die-cast (Moscow: Mashinostroenie) p 415
[2] Gavarieva K N, Simonova L A, Pankratov D L and Gavariev R V 2017 Development of expert systems for modeling of technological process of pressure casting on the basis of artificial intelligence *IOP Conf. Series: Materials Science and Engineering* **240 012019** DOI: 10.1088/1757-899x/240/1/012019
[3] Yan Q S, Yu H, Lu G, Xiong B W, Lu B P and Zou X 2014 Effect of crystallization pressure on secondary dendrite arm spacing of vacuum counter-pressure casting aluminum alloy *School of Aeronautic Manufacturing Engineering Nanchang Hangkong University* **24 5** 1194–1199
[4] Leushin I O and Chystyakov D G 2016 Technologies for Use in the Formation of a Differentiated Structure in Iron Billets Used in Glass Molds *Metal Science and Heat Treatment* **5–6 58** 299–302
[5] Gavariev R V, Savin I A and Leushin I O 2017 Improvement of zinc castings surface quality by laminated protective coating *Non-ferrous metals* **5** 84–88 DOI: 10.17580 / tsm.2017.05.13
[6] Hu X P, Zhao G Q and Wang W M 2010 Solidifying pressure and microstructure of alsi10cu3 in die sleeve in high pressure die casting *International Journal of Cast Metals Research* **23 5** 289–295
[7] Leushin I O, Leushina L I and Koshelev O S 2017 Creep of the investment mold material *Russian Metallurgy* **13** 1092–1095
[8] Gavariev R V and Savin I A 2017 Improvement of surface quality of casting produced by casting under pressure *Solid State Phenomena* **265** 988–993 DOI: 10.4028/www.scientific.net/SSP.265.988
[9] Gavariev R V and Savin I A 2018 Research of the mechanism of destruction of compression molds for casting under pressure of color alloys *Solid State Phenomena* **284** 326–331 Online. Available: https://doi.org/10.4028/www.scientific.net/SSP.284.326
[10] Pankratov D L and Gavariev R V 2019 Improving the quality of castings made of non-ferrous metal alloys when casting in metal molds *IOP Conf. Series: Materials Science and Engineering* **570 012072** DOI: 10.1088/1757-899X/570/1/012072
[11] Pankratov D L, Gavariev R V and Gavarieva K N 2016 Influence of multilayer coatings on the operational stability of molds for injection molding *IOP Conference Series: Materials Science and Engineering* **134 1 012031** DOI: 10.1088/1757-899X/134/1/012031
[12] Leushina L I, Leushin I O, Plokhov S V and Deev V B 2018 Recycling Cullet from Quartz Ceramic Shells Used in Investment Casting Steel in Translation **48 11** 699–703
[13] Gavarieva K N, Simonova L A, Pankratov D L, Shibakov V G and Gavariev R V 2018 Application of multi-agent system for control of parameters of precision stamping process of bevel gears *IOP Conf. Series: Materials Science and Engineering* **412 012020** DOI: 10.1088/1757-899X/412/1/012020
[14] Gavariev R V, Savin I A and Leushin I O 2019 To the question of casting alloys of non-ferrous metals in the metal mold *Materials Science Forum* **946** 631–635 Online. Available: https://doi.org/10.4028/www.scientific.net/msf.946.631
[15] Safronov N N, Mingaleeva L B and Savin I A 2018 Optimization of charge material composition in shs process with ferrosilide fabrication from gaseous wastes of metallurgical production *Ferrous metals* **2** 53–59

[16] Gavariev R V, Savin I A and Leushin I O 2016 Impact of the functional coating on service durability of injection molds for the zinc alloys pressure casting *Non-ferrous metals* **1** 66–70 DOI: 10.17580/tsm.2016.01.11

[17] Gavariev R V, Savin I A and Soldatkina E N 2020 Choice of Protective Coating of Metal Molds for Casting Non-Ferrous Alloys *Solid State Phenomena* **299** 867–871 Online. Available: https://doi.org/10.4028/www.scientific.net/ssp.299.867

[18] Leushin I O, Subbotin A Yu and Geyko M A 2015 Recycling of galvanized steel scrap for use in cast iron melting in induction melting facilities *CIS Iron and Steel Review* **10** 19–22

[19] Gilman V N, Balabanov I P and Faskhutdinov A I 2019 Improving the efficiency of shaving through the use of wear-resistant coatings *IOP Conference Series Materials Science and Engineering* **570:012024** DOI: 10.1088/1757-899X/570/1/012024

[20] Raab G I, Podrezov Y M, Danylenko M I, Borysovska K M, Aleshin G N and Shafigullin L N 2016 Dislocation structure evolution during plastic deformation of low-carbon steel *Mathematics Education* **11 6** 1563–1576

IOP Conf. Series: Materials Science and Engineering **969** (2020) 012069 doi:10.1088/1757-899X/969/1/012069

On the Issue of Heat Balance of Molds for Injection Molding of Non-Ferrous Metal Alloys

R V Gavariev[1] and K N Gavarieva[2]

[1]Kazan national research technical university named after A. N. Tupolev-KAI, 10, K. Marx St., Kazan, Tatarstan, 420111, Russia
[2]Kazan Federal University, Naberezhnye Chelny Institute, 10A, Prospekt Syuyumbike, Naberezhnye Chelny, 423812, Russia

E-mail: Gavarievr@mail.ru

Abstract. The article presents a study of the thermal balance of injection molding molds. The main components of the balance equation are indicated, as well as the factors that affect them. It is revealed that the most important factor affecting the thermal balance is the material of the mold. The use of coatings applied by the method of physical vapor deposition on the forming surfaces of molds is proposed. These coatings provide a reduction in friction between the forming surface and the poured melt at the time of filling, which reduces the temperature of the forming surface by 10-12°C. Among the considered options, the most optimal indicators were for titanium and molybdenum metal nitride (Ti, Mo) N.

1. Introduction

When designing a mold for injection molding, an important issue is to ensure the pre-operation of heating or cooling the mold. This action allows you to obtain the best initial conditions for the interaction of molten metal with the mold, which ultimately improves the quality of the resulting castings, and also increases the service life of the mold. The solution to this problem is possible only by accurately determining the components of the heat balance equation. At the same time, it is necessary to take into account all factors that can directly or indirectly affect the components of the heat balance [1, 2].

2. Results

In General, there are two components of the heat balance of the mold under a steady thermal regime, one of them Q_{in} – input heat and Q_{out} – output heat. The equality of these values determines the thermal balance between the mold and the casting. The solution of this equation allows you to determine the need for pre-cooling or heating the mold. The heat input can be calculated with sufficient accuracy using a well-known method [3]:

$$Q_{in} = Q_{oh} + Q_{cr} + Q_c + Q_{fr}, \tag{1}$$

where Q_{oh} – heat of metal overheating, Q_{cr} – heat of crystallization, Q_c – heat of solid casting, Q_{fr} – frictional heat.

In this case, the heat of friction and the heat of overheating together make up about 12% of the total value of the amount of heat supplied [4, 5].

IOP Conf. Series: Materials Science and Engineering **969** (2020) 012069 doi:10.1088/1757-899X/969/1/012069

Calculating the heat output is not an easy task, since there are many gaps in the mold. However, the heat released by the mold to the environment consists of two main components. One of them is Q_{cl} – the heat withdrawn from the closed mold through the side surfaces, the other is Q_{op} – the heat withdrawn from the forming surface in the open state of the mold. That is, in the end, the heat output can be calculated using the following formula [6]:

$$Q_{out} = Q_{cl} + Q_{op}, \tag{2}$$

It is also worth noting that the heat released to the environment is about 3% of the total heat released in the mold [7].

The main factor determining the value of the components of the input and output heat is the material of the mold, which is almost impossible to significantly change the effect. However, there are ways that allow you to partially change the components of the heat balance. One of them is the use of various lubricants that can significantly reduce the temperature of the working surface. These lubricants can be liquid, powdered, or a mixture of liquid and powder. Their application is that after a certain number of cycles, the lubricant is applied to the work surface [8]. However, a significant disadvantage of using such lubricants is that they are washed off from sharp edges and corners, in addition, the destruction of molds begins with sharp edges and corners on the forming surfaces. In addition, for these lubricants, even on even flat areas, the temperature distribution will be uneven, this moment will depend on the distance to the gating system. Another important point is that the application of liquid lubricants must be carried out cyclically through a small number of cycles, depending on the size of the resulting casting [9, 10].

Thus, given the fact that in modern engineering almost all working parts have a complex configuration, it is necessary to better solve the problem of reducing the contact temperature on the working surfaces. Coatings applied by physical vapor deposition (PVD) can serve as a solution to this issue [11, 12].

A distinctive feature of this method is the ability to apply a coating with almost any metal as a base. This makes it possible to purposefully improve the pre-set performance characteristics of the surface layer, which ultimately increases the durability of the tooling, as well as the quality of the resulting castings. Another important feature of coatings applied by the PVD method is the presence of these coatings of a fine-grained densely packed structure, which provides increased resistance to destruction in conditions of high temperatures and stresses [13].

The thermal interaction between the mold and the molten metal after the PVD coating is applied changes for several reasons. One of them is due to the physical properties of the coating, the other is to reduce the coefficient of friction [14, 15].

Despite the possibility of using refractory metals as the basis of a multifunctional coating, their thickness of 3–10 microns is so small that the use of even such a metal as molybdenum, which has increased refractory properties, does not give any significant changes in the terms of the heat balance and is no more than 0,3% of the heat input [16, 17].

Another important point for studying the effect of coating properties on the thermal balance is the cooling of the mold between injections of molten metal. It can be assumed that between cycles, the coating should reduce the interaction rate between the mold and the environment. In turn, the preservation of heat inside the mold should reduce the influence on the magnitude of the amplitude change in the surface layer, which, ultimately, will increase the resistance index of the molds [18, 19]. However, experimental data disproved this hypothesis. The temperature difference for molds in the open state with different coatings in comparison with uncoated ones gave a difference of no more than 0.2 %.

Thus, the only moment associated with coatings that can significantly affect the thermal balance is the change in the value of the heat flow from the friction between the molten metal and the surface layer of the forming surfaces by reducing the coefficient of friction between these surfaces. The value of this coefficient depends entirely on the composition of the coating, and a study was conducted to determine the optimal one. For comparison, the following compositions of single-layer coatings applied by the PVD method were selected: TiN, MoN, TiCN, (Ti, Mo) N, and as a comparison with the classical technology of manufacturing molds, inserts with a nitrided forming surface were studied.

Determining the temperature of the mold TM in the molten metal casting zone was performed using thermocouples, one end of which was connected to the temperature indicator, the other end was fixed in blind holes located next to the forming surfaces. The temperature was measured at several points and at different heights from the forming surface. According to the tests, it was found that the highest temperature values correspond to a layer 8–10 mm from the forming surfaces (figure 1), which is confirmed by the existing theory [20].

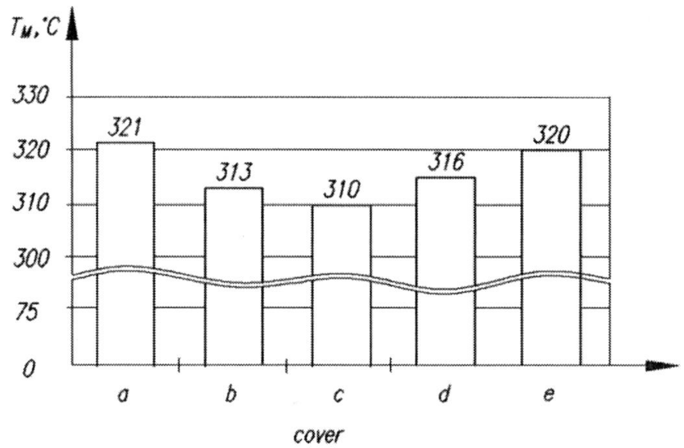

Figure 1. Mold temperature depending on the coating composition, a – nitrided surface, b – TiN (titanium nitride), c – MoN (nitride of molybdenum), d – TiCN (the titanium carbonitride), e – (Ti,Mo)N (the nitride of the metals titanium and molybdenum).

The best values correspond to MoN coating, which is due to the fact that molybdenum is a good base for lubricants. Thus, MoN coating is able to reduce the operating temperature of the forming surfaces of the mold by 10–12 °C in comparison with the nitrided surface. This means that the preheating temperature of the mold before pouring molten metal will be lower by the specified 10–12 °C. It is also worth noting that among the studied coatings, the worst results were shown by coatings complex in composition, but better than the nitrided surface.

Despite the fact that different grades of steel can be used as the mold material, as well as the fact that the melting temperature of the poured alloys differs depending on their compositions, the curve of the dependence of the number of thermal cycles on the temperature difference in the mold material has approximately the same nonlinear character. Despite the possible different initial conditions of the injection molding process, it can be said that the smaller the temperature difference in the mold material, the greater the number of casting cycles until the destruction of the forming surfaces. Thus, even a slight decrease in the temperature drop can have a positive effect on the resistance of casting equipment. And in particular for the injection molding process of zinc alloys, the distinctive feature of which is a relatively small melting point of about 380°C, even such a small decrease in the maximum operating temperature will give a much greater effect than for aluminum or brass alloys.

3. Conclusions

The use of wear-resistant coatings applied by the PVD method, which consists in the condensation of a substance from the plasma phase in a vacuum with ion bombardment of the surface on a special installation, allows you to reduce the temperature on the forming surfaces of the mold. The supplied heat flows are reduced due to an abnormally low coefficient of friction on the forming surfaces, which reduces the maximum heating temperature of the mold material. This leads to an increase in the operational stability of injection molding molds, as well as to an increase in the quality of the resulting castings.

References

[1] Goryunov I I 1974 *Injection molds* (Leningrad: Mashinostroenie) p 255
[2] Leushin I O, Leushina L I and Koshelev O S 2017 Creep of the investment mold material *Russian Metallurgy* **13** 1092–1095
[3] Safronov N N, Mingaleeva L B and Savin I A 2018 Optimization of charge material composition in shs process with ferrosilide fabrication from gaseous wastes of metallurgical production *Ferrous metals* **2** 53–59
[4] Leushin I O and Chystyakov D G 2016 Technologies for Use in the Formation of a Differentiated Structure in Iron Billets Used in Glass Molds *Metal Science and Heat Treatment* **5–6** **58** 299–302
[5] Gavariev R V, Savin I A and Leushin I O 2017 Improvement of zinc castings surface quality by laminated protective coating *Non-ferrous metals* **5** 84–88 DOI: 10.17580/tsm.2017.05.13
[6] Hu X P, Zhao G Q and Wang W M 2010 Solidifying pressure and microstructure of alsi10cu3 in die sleeve in high pressure die casting *International Journal of Cast Metals Research* **23** **5** 289–295
[7] Gavarieva K N, Simonova L A, Pankratov D L and Gavariev R V 2017 Development of expert systems for modeling of technological process of pressure casting on the basis of artificial intelligence *IOP Conf. Series: Materials Science and Engineering* **240** **012019** DOI: 10.1088/1757-899x/240/1/012019
[8] Gavariev R V and Savin I A 2017 Improvement of surface quality of casting produced by casting under pressure *Solid State Phenomena* **265** 988–993 DOI: 10.4028/www.scientific.net/SSP.265.988
[9] Gavariev R V and Savin I A 2018 Research of the mechanism of destruction of compression molds for casting under pressure of color alloys *Solid State Phenomena* **284** 326–331 Online. Available: https://doi.org/10.4028/www.scientific.net/SSP.284.326
[10] Pankratov D L and Gavariev R V 2019 Improving the quality of castings made of non-ferrous metal alloys when casting in metal molds *IOP Conf. Series: Materials Science and Engineering* **570** **012072** DOI:10.1088/1757-899X/570/1/012072
[11] Pankratov D L, Gavariev R V and Gavarieva K N 2016 Influence of multilayer coatings on the operational stability of molds for injection molding *IOP Conference Series: Materials Science and Engineering* **134** **1** **012031** DOI: 10.1088/1757-899X/134/1/012031
[12] Kapranos P, Brabazon D, Midson S P , Naher S and Haga T 2014 Advanced casting methodologies: inert environment vacuum casting and solidification, die casting, compocasting, and roll casting, casting, semi-solid forming and hot metal forming *Comprehensive Materials Processing* **5** 3–37
[13] Gavarieva K N, Simonova L A, Pankratov D L, Shibakov V G and Gavariev R V 2018 Application of multi-agent system for control of parameters of precision stamping process of bevel gears *IOP Conf. Series: Materials Science and Engineering* **412** **012020** DOI: 10.1088/1757-899X/412/1/012020
[14] Gavariev R V, Savin I A and Leushin I O 2019 To the question of casting alloys of non-ferrous metals in the metal mold *Materials Science Forum* **946** 631–635 Online. Available: https://doi.org/10.4028/www.scientific.net/msf.946.631

[15] Yan Q S, Yu H, Lu G, Xiong B W, Lu B P and Zou X 2014 Effect of crystallization pressure on secondary dendrite arm spacing of vacuum counter-pressure casting aluminum alloy *School of Aeronautic Manufacturing Engineering Nanchang Hangkong University* **330063 24 5** 1194–1199
[16] Gavariev R V, Savin I A and Leushin I O 2016 Impact of the functional coating on service durability of injection molds for the zinc alloys pressure casting *Non-ferrous Metal* **1** 66–70 DOI: 10.17580/tsm.2016.01.11
[17] Gavariev R V, Savin I A and Soldatkina E N 2020 Choice of Protective Coating of Metal Molds for Casting Non-Ferrous Alloys *Solid State Phenomena* **299** 867–871 Online. Available: https://doi.org/10.4028/www.scientific.net/ssp.299.867
[18] Jayal A D, Badurdeen F, Dillon Jr O W and Jawahir I S 2010 Sustainable manufacturing: Modeling and optimization challenges at the product, process and system levels CIRP *Journal of Manufacturing Science and Technology* **2(3)** 144–152 DOI: 10.1016/j.cirpj.2010.03.006
[19] Gilman V N, Balabanov I P and Faskhutdinov A I 2019 Improving the efficiency of shaving through the use of wear-resistant coatings *IOP Conference Series Materials Science and Engineering* **570:012024** DOI: 10.1088/1757-899X/570/1/012024
[20] Raab G I, Podrezov Y M, Danylenko M I, Borysovska K M, Aleshin G N and Shafigullin L N 2016 Dislocation structure evolution during plastic deformation of low-carbon steel *Mathematics Education* **11 6** 1563–1576

Manufacture of Ceramic Molds and Cores from Inorganic Materials Using Permanent

L G Znamenskii, O V Ivochkina and T V Stepanova

South Ural State University, 76, Lenin prospekt, Chelyabinsk, 454080, Russia

E-mail: ivochkinaov@susu.ru

Abstract. The paper describes the technology for the manufacture of ceramic molds and cores from inorganic materials using permanent masters. A technique for reinforcing molds and cores using ceramic tubes is proposed. The effect of pulse currents of different power on the slurry has been studied. In the experimental part, the performed experiments and the data obtained are described. In contrast to the conventional Show-process based on ethyl silicate binder (basic option), the technology developed for the manufacture of ceramic molds from inorganic materials ensures improving the quality and economic and environmental efficiency of investment castings. The quality improvement is determined by improved physical and mechanical characteristics of ceramic molds and the stable ceramic cracking. The improvement in environmental indicators is ensured by using safe inorganic molding materials and eliminating the solvent burning. Improving economic indicators is determined by reducing power consumption for calcining ceramic molds, casting defects, and molding material costs. Thus, the technology developed allows improving the quality of investment castings manufactured for the country's oil and gas complex.

1. Introduction

The development of heavy engineering industries such as oil and gas, power engineering, shipbuilding, and aerospace engineering leads to the year-by-year growing demand for high-quality and high-precision castings. Conventional casting does not allow satisfying the customers' demand due to the limitations of these techniques.

One of the promising areas is the manufacture of castings in ceramic molds [1, 2, 3]. Well-known techniques for casting in fixed-volume ceramic molds have a lot of advantages:

- lack of limitations on the cast alloy, high chemical resistance,
- lack of limitations on the casting weight,
- high casting surface quality,
- short mold and core manufacture time as compared to shell ceramic molds in investment casting,
- low gas content in the molding sand.

However, the conventional technology for the production of ceramic molds and cores has some shortcomings:

- the high cost of source materials for the preparation of molding sands,
- the high consumption of expensive binder,
- the high toxicity of the materials used,
- the mold tendency to forming large cracks,

- the need for precise execution of process; any minimum violations cause irreversible defects in the mold elements and often even their destruction [4, 5, 6].

The main research objective is to develop the optimal composition and technology for the manufacture of ceramic molds and cores using permanent masters with improved properties. The technique should ensure the below technical advantages:

- low molding duration,
- high physical and mechanical characteristics of ceramic molds and cores,
- increased gas permeability,
- replacement of large cracks with small ones,
- environmental safety at the binder and gelling agent preparation and the solvent removal stages.

2. Theoretical

The technology for the manufacture of ceramic molds and cores using permanent masters comprises the below stages:

- preparing a ceramic slurry based on a refractory filler and a silica binder [7, 8, 9],
- preparing a gelling agent,
- introducing the gelling agent into the slurry while continuously stirring,
- pouring the finished slurry into the forming tools,
- curing molds and cores in the tooling,
- removing masters or cores from the molds or tooling,
- calcining mold halves and cores,
- assembling molds,
- pouring metal [10].

In each specific case, choosing the optimal slurry composition requires a comprehensive analysis to ensure the required process properties [11, 12, 13]. The manufacture of castings with thin walls and complex relief requires a high content of fine fractions in the filler; the manufacture of large thick-walled castings requires using a large amount of medium and coarse filler fractions to prevent mold buckling. The proper distribution of the solid and liquid phases in the slurry ensures the required strength of the mold elements at a sufficiently dense filler grain packing, the minimum slurry shrinkage when curing, and a low tendency to large cracking [10].

Currently, the most widespread molds are those based on a hydrolyzed ethyl silicate binder solution. Ethyl silicate (ETS) is the product of the reaction between ethyl alcohol and silicon tetrachloride. To use ethyl silicate as a binder, it should be hydrolyzed, i.e. ethoxy groups C_2H_5O in ethers should be replaced with hydroxyl OH ones. In the course of hydrolysis, silicic acid is converted to a silica sol. Fused quartz, corundum, chamotte sand, and mullite are most often used as fillers [10].

As part of the study, the manufacture of a slurry based on a colloidal silica sol binder instead of a hydrolyzed ethyl silicate solution has been proposed, which improves the environmental situation, eliminates the fire hazard of the production process, and ensures high physical and mechanical characteristics of ceramic molds and cores based on permanent masters [14].

The slurry fluidity is a very important property since it ensures the composition mobility, it's spreading over the tooling, and the ability to qualitatively fill the entire cavity and reproduce the tool contour [15, 16].

The slurry is prepared by stirring the ingredients in a mixer at an impeller speed of 1,500 ... 1,800 rpm for 30 ... 40 minutes.

The gelling agent is introduced into the slurry just before it is molded. Thorough mixing of the slurry while feeding the gelling agent is a very important factor for its uniform distribution and subsequent high-quality curing throughout the entire mold or core volume [17].

The slurry to gel conversion time is one of the most important parameters of the entire ceramic mold manufacture process. The gelling time depends on the binder quality and the type and amount of gelling agent. On the one hand, the gelling time should be minimum, and on the other hand, it should ensure the possibility to perform the required operations, i.e. preparing the slurry for molding, processing it (if necessary), and directly, pouring into the mold-core tooling; a time reserve (about 20 sec) should also be provided. Too short gelling does not allow performing high-quality preparatory operations, and too long one leads to stratification of the slurry in the mold, which may cause the mold defects (buckling, macrocracking, etc.) [18].

Alkalis (NaOH), urotropine ($(CH_2)_6N_4$), ferrochrome slag, etc. are conventionally widely used as gelling agents. In the technology developed, an aluminoborophosphate concentrate is proposed to gelatinate the slurry. To ensure uniform distribution of the gelling agent throughout the ceramic slurry, the ABPC requires preliminary preparation. To do this, water is poured into the initial aqueous aluminoborophosphate concentrate solution to a density of 1150 ... 1300 kg/m^3.

The aqueous aluminoborophosphate concentrate solution density of less than 1150 kg/m^3 does not allow ensuring stable and uniform slurry curing in the mold. The gelling agent density of more than 1300 kg/m^3 causes premature gel precipitation in separate slurry microvolumes. This significantly reduces the ceramic strength.

Fast and uniform gelling agent distribution leads to accelerated molding and eliminates large cracking of the ceramics. In this case, small cracking occurs, which is required to ensure the gas permeability of the ceramics.

Immediately after the gelling agent preparation, it is introduced into the previously prepared slurry while continuously stirring. Then, the finished molding sand is molded into the tooling.

To avoid defects in ceramic molds, before pouring the metal, the liquid composition should be almost completely removed. To improve this process, ceramic tubes are installed in the mold immediately after it is filled with the slurry; they contribute to the complete and stable solvent removal under the effect of pulse currents. In this case, solvent removal is a volumetric process rather than a superficial one, like in most electric furnaces. As a result, the manufacture of ceramic molds and cores using permanent masters is accelerated, and their high crack resistance is ensured, as well as conditions are created to reduce the temperature and duration of subsequent calcination [10].

Depending on the gelling agent amount, the slurry is cured in the molds for 5 ... 15 minutes. The mold or core is placed in a microwave oven to remove the solvent under the effect of pulse current at a frequency of 2,450 MHz and a power of 600 ... 2,000 W, depending on the mold or core weight.

A power of less than 600 W does not ensure high-quality solvent removal, which causes defects in ceramic molds, cores, and investment castings. The pulse electric current power of more than 2000 W is impractical due to unjustified energy consumption.

The next process stage is mold calcination.

Thermal treatment of molds depends on several factors:

- type of alloy poured into a ceramic mold,
- overall mold dimensions,
- configuration of the casting,
- the prevailing wall thickness of the casting,
- the weight of the molded metal.

Depending on the combination of the above factors, three calcination modes are conventionally used:

- for large castings - calcination at 900 °C for 2 hours,
- for medium castings with a wall thickness of up to 20 mm, the mold is calcined at 500 °C, or the surface is simply dried using gas burners,
- for small castings with a wall thickness of up to 4 mm, molding is performed without calcining immediately after the solvent burns out.

The use of the technique developed allows a significant decrease in the calcination temperature and duration. The temperature regime is 350 ... 600 °C; the duration is limited to 30 ... 90 minutes, depending on the mold and core dimensions. Reducing the thermal treatment temperature and duration makes the technology more energy-efficient, especially when manufacturing large complex-shaped castings.

IOP Conf. Series: Materials Science and Engineering **969** (2020) 012070 doi:10.1088/1757-899X/969/1/012070

3. Experimental

The ceramic slurry is made of silica sol base (20 % wt.) and refractory filler (80 % wt.). A mix of flint and granular fused quartz at a ratio by weight of 1:1 is used as a refractory filler. The slurry is prepared by stirring the ingredients in a mixer at an impeller speed of 1,500 rpm for 30 minutes. Then, the gelling agent is prepared. Water is poured into the initial solution of aluminoborophosphate concentrate to obtain three solutions of different densities: 1,150; 1,250; 1,300 kg/m3. These three gelling agents with different densities are introduced, respectively, into three containers with prepared ceramic slurry at a rate of 10 % wt. of the gelling agent in the slurry while continuously stirring. Then, three slurries obtained are poured into the tooling (three master molds). Ceramic tubes with a diameter of 5 ... 8 mm and a thickness of 2 ... 3 mm are installed in the curing slurry in the mold with an interval of 3 ... 5 cm. They reinforce the ceramics, increase its strength and crack resistance, and ensure stable accelerated removal of the solvent under subsequent exposure to pulse currents. The slurry is cured in the mold, after which the permanent masters are removed. The molds are placed in a microwave oven, where the solvent is removed at a frequency of 2,450 MHz and a power of 800 W for 2 minutes. The final stage is calcining the molds and pouring molten metal into them to obtain 20GL steel castings. The characteristics of the molds and castings obtained are given in table 1.

Table 1. Characteristics of Molds and Castings.

Indicator	The Technique Claimed at a Gelling Agent Density, kg/m^3		
	1,150	1,200	1,300
1. Molding duration, min.	60	50	45
2. Bending strength of samples, MPa:			
a) after curing	3.5	3.8	2.5
b) after calcination	6.0	6.3	5.5
3. Large cracks	no	no	no
4. Gas permeability, units	6.0	5.0	9.0
5. Calcination temperature, ° C	450	500	600
6. Calcination duration, min	70	60	90
7. Flashes on 20GL steel castings	no	no	no

In the next experiment, the microwave oven current power to remove the solvent from ceramics has been varied: 600; 1,200; 2,000 watts

Ceramic cores are made according to the above technique. The gelling agent is diluted to a density of 1,250 kg/m^3. Removing the solvent from ceramic cores by exposure to pulse currents allows improving the ceramics strength and reducing the calcination temperature and duration. This improves the quality of manufacturing investment castings, especially the complex-shaped ones made of alloy steels.

The effect of the current power on the ceramic core properties is given in table 2.

Table 2. Effect of the Current Power on the Ceramic Core Properties.

Property	Current Power, W		
	600	1,200	2,000
1. Bending strength of samples, MPa:			
a) after removing the solvent	3.6	4.8	5.3
b) after calcination	6.5	7.4	7.0
2. Calcination temperature, ° C	500	400	350
3. Calcination duration, min	45	40	30

IOP Conf. Series: Materials Science and Engineering **969** (2020) 012070 doi:10.1088/1757-899X/969/1/012070

4. Conclusion

Analysis of the data obtained has shown that the technology and composition developed allow shortening the mold manufacture cycle and improving the physical and mechanical characteristics of ceramic molds while reducing the power consumption. This ensures reducing defects in the investment castings and improving their quality.

The use of cheaper materials significantly reduces the product costs, and the elimination of toxic and harmful conventional ceramic mix components leads to improving the environmental situation and working conditions in the workshop.

The further research area will be testing the technology on a pilot batch of complex-shaped castings for the oil and gas industry and adjusting the process parameters to obtain consistently high results.

Reference

[1] Mozhaytsev V V and Pirainen V Yu 2015 Import Substitution of Castings and Modernization of Foundry Production in Russia *Foundry of Russia* **7** 36–43

[2] Leushina L I, Leushin I O and Geiko I V 2015 Possible Risks of Low-Temperature Calcination of Shell Molds *Foundry* **4** 27–29

[3] Leushina L I, Leushin I O and Plokhov S V 2014 Modelling View of Calcining Quartz-Based Shell Molds *Foundry of Russia* **12** 28–31

[4] Uglev N P 2014 Thermomechanical Properties of Ceramics for Investment Casting *Foundry* **5** 16–20

[5] Tkachenko S S, Emelyanov V O and Martynov K V 2019 On Ensuring the Strength of Ceramic Molds for Artistic Casting *Foundry* **6** 29–31

[6] Ordin D A 2016 Properties of Ceramic Slurries Based on Water-Colloidal Binders *Foundry* **3** 21–25

[7] Kwon St S, Kulikov V Yu, Shcherbakova E P and Arinova S K 2019 Studying the Effect of Shell Composition on its Properties during Investment Casting *Foundry* **5** 19–22

[8] Murkina A S 2013 Binding Solutions in Investment Casting *Foundry of Russia* **2** 14–17

[9] Emelyanov V O, Martynov K V, Naumova A K and Ekimova E S 2013 Process Properties of Water-Based Binders for Investment Casting *Foundry of Russia* **6** 23–25

[10] Lakeev A S *Techniques for Investment Casting* (Kiev: Technics, 1984) p 160

[11] Dyachkov V N, Nikitin K V and Barinov A Yu 2015 Technology for Preparing Ceramic Molds for Pouring during Investment Casting *Foundry of Russia* **12** 27–30

[12] Chernov V P and Safonova E A 2015 Ash of TPPs as a Refractory Filler for the Investment Casting Slurries *Foundry of Russia* **6** 32–35

[13] Chernov V P and Safonova E A 2014 Studying the Survivability of Refractory Slurries Based on ETS-40 *Foundry of Russia* **11** 29–32

[14] Ivochkina O V et al. Theoretical Fundamentals of Electron-Ion Processing of Crystalline Hydrate Binders in Investment Casting *Foundry of Russia* **8** 29–32

[15] Dyachkov V N, Paramonov A M, Nikitin K V and Nuzhdin G S. 2013 Improving the Technology for the Manufacture of Critical Steel Castings by Investment Casting to Improve their Quality *Foundry of Russia* **10** 36–38

[16] Bryukhanova E V and Golotenkov O N 2013 Improving the Technology for the Manufacture of Ceramic Molds to Cast Corrosion-Resistant Steels *Foundry of Russia* **7** 23–24

[17] Doshkarzh I, Gabriel J, Gousht M and Pavelka M 1979 *Investment Casting* (Moscow: Mashinostroenie) p 296

[18] Stryuchenko A A and Zakharchenko E V 1988 *Ceramic Molds in Investment Casting Using Permanent Masters* (Moscow: Mashinostroenie) p 128

IOP Conf. Series: Materials Science and Engineering **969** (2020) 012071 doi:10.1088/1757-899X/969/1/012071

The Simplifying the Calculation of the Strip Profile during Rolling in a Four-High Mill Stand

M I Rumyantsev, A N Kolybanov and O K Yesipova

Department of Materials Processing Technologies, Nosov Magnitogorsk State Technical University, 38, Lenin Street, Magnitogorsk, 455000, Russia

E-mail: mikhail.rumyantsev54@bk.ru

Abstract. The task of simplifying the calculation of deformation of the rolls and the strip profile during rolling in a four-high (quatro) stand has been formulated and stated in the article. The design and computer experiment results on the basis of which nonlinear approximations have been obtained for simplified calculation of the transverse rigidity moduli of the roller unit of the quarto finishing stand of the wide - strip hot mill 2500 have been described. The calculation of the transverse rigidity moduli of the roller unit of the quarto finishing stand is performed depending on the rolling force and the regulating force.

1. Introduction

The main element of the system "Four-High Mill Stand" is a roll unit (roll system). Elastic deformations of the roll system affect the cross-sectional shape (profile) of the strip and can cause "transverse thickness variation" and "nonflatness" defects [1–4]. The task of calculating the deformation of the rolls for prediction of strip profile is important for controlling the rolling process. In the most general formulation, it is necessary to consider the joint description of the stress-strain state (SSS) of both the roll system and the rolled strip using the three-dimensional theory of elasticity and plasticity. There are various methods for calculating roll deformation [5–11]. But the calculations by simple methods [5–6] give a noticeable error, and the calculations by more accurate methods [7–11] are, as a rule, labor-consuming.

A model of deformation of a quarto roller unit under symmetric loading was created and developed by the scientists: Salganik V.M., Melzer V.V., Poletskov P.P. and Omelchenko B.Ya. at the Metal Forming Department of Nosov MSTU in 1970–2009 [12–14]. Good results were obtained with its use in solving a number of problems [14–16]. The essence of the task of simplifying the calculation is to replace the complex of analytical dependencies that make up this model with two approximations, which are obtained from the results of computer simulation performed according to the experimental designing procedure [17].

2. Features of the roll unit deformation model

The transverse thickness variation (profile) of the strip at the exit of the rolls:

$$\delta h = \frac{K_c - K_e}{K_c} \cdot (w_{p1} + w_{p2} - z_{p1} - z_{p2}) + \frac{K_e}{K_c} \cdot \delta H, \tag{1}$$

where w_{p1} and w_{p2} are the deflections of the axes of the upper and lower work rolls; z_{p1} and z_{p2} are the current profiles of the upper and lower work rolls; δH is transverse thickness variation of the strip at the entrance to the rolls; K_c is coefficient of elastic compliance of the work rolls:

$$K_c = \frac{\pi}{2 \cdot \frac{1-v_p^2}{E_p} \cdot \ln \frac{3.04 \cdot D_{WR} \cdot E_p}{P_{1W} \cdot (1-v_p^2)}}, \quad (2)$$

where v_p and E_p are Poisson's ratio and Young's modulus for compressing the material of the work roll. For example, for a steel forged roll $v_p = 0.3$ and $E_p = 20.59 \cdot 10^4$ MPa. For the alloyed cast-iron roll of a two-layer casting$E_p = 14.71 \cdot 10^4$ MPa and $v_p = 0.27$; DWR is the diameter of the barrel of the work roll; $P_{1W}=P/W$ is the rolling force related to the strip width; K_e is the effective coefficient:

$$K_e = \frac{K_{\Delta h} \cdot K_c}{K_{\Delta h} + K_c}. \quad (3)$$

$K_{\Delta h}=P/\Delta h$ is the coefficient of the strip compliance.

The profiles of the work rolls $z_{p1}$and z_{p2}, and also necessary for other calculations, the profiles of the support rolls (z_{o2} and z_{o2}) are the initial data. The deflections of the work rolls w_{p1} and w_{p2} are determined taking into account the deflections of the support rolls (w_{o1} and w_{o1}) and the sinking of the work rolls as the beams on an elastic base. And the loading of the roll unit not only by the rolling force P, but also by the regulating force of the anti – bending (+Q) or the additional bending (-Q) of the work rolls is displayed. In general, the model includes 11 equations and is implemented by using the matrix method, which requires the application of a special computer program.

3. A possible approach to simplify the calculation of roll deformation and strip profile

In order to simplify the calculation of the transverse strip thickness variation, the method of approximation of the roll deflection dependence on the dimensions features of the roll unit and the loading conditions can be applied. With this approach, a multiple regression analysis will produce an equation that, given the current characteristics of the roll unit, the strip width, the effort values P and Q, will allow the value δh to be determined by a simple calculation using the formula. Such approach was used, for example, in [6] for calculating the deformation of the roll unit of the cold rolling quarto-mill 1700.

In a more general form, the indicated simplifying approach can be applied if the dependence for calculating the modulus of the transverse rigidity of the roll system will be constructed. The transverse strip thickness variation is expressed in terms of the moduli of the transverse rigidity of the roller system from the rolling force (M_W^P) and the regulating force (M_W^Q) as follows [18–19]:

$$\delta h = \frac{P}{M_W^P} - \frac{Q}{M_W^Q}. \quad (4)$$

The stiffness moduli are numerically equal to the inverse of the change in the transverse strip thickness variation from a single increment of the rolling force (δh_P) and the regulating force (δh_Q) according to the following formulas:

$$M_W^P = \frac{1}{\delta h_P} = \frac{P' - P''}{\delta h(P') - \delta h(P'')}, \quad (5)$$

$$M_W^Q = \frac{1}{\delta h_Q} = \frac{Q' - Q''}{\delta h(Q') - \delta h(Q'')}, \quad (6)$$

where, $\delta h(P')$, $\delta h(P'')$ and $\delta h(Q')$, $\delta h(Q'')$, are the values of the transverse thickness variation of the strip at different values of the rolling force ($P'>P''$) and the regulating force $Q'>Q''$, respectively, and all other conditions being equal.

Thus, to simplify the calculation of the deformation of the roll unit and the profile of the strip during rolling in the quarto stand, it is necessary to find the dependence of the modulus of the transverse rigidity of the roller system on the influencing factors x_1; x_2; x_3 …x_n:

$$M_W = f(x_1; x_2; x_3 ... x_4). \quad (7)$$

4. Experiment design

The experiment was carried out by means of a computer program in MS EXCEL, and a mathematical model [12–14] was implemented. The finishing stand of the wide – strip hot mill 2500 of the PJSC "MMK" was selected as the subject of the study. The goal of the experiment was to replace the above model with a polynomial dependency.

According to the experimental designing technique [17], a second-order symmetric composite rotatable design was chosen. And since four influencing factors had been chosen, the design was four-factor. In total, 25 experiments were conducted, 16 of which correspond to the full factorial experiment ("core" of the design), 8 experiments with "star" points and one experiment in the middle of the design. The range of the factors change was chosen according to the possible change of these factors at the wide - strip hot mill 2500. The rolling force at the points of the design was assumed to be 16 MN, and the regulating force was 0.2 MN. The levels of variation of the factors were presented in table 1.

Table 1. Levels of factors variation.

Factors	W, [mm]	l_{OR}, [mm]	D_{WR}, [mm]	$K_{\Delta h}$, [MPa]
Main level, (0)	1625	2200	690	20460
Variation interval	187	100	10	8120
Upper level, (+1)	1812	2300	700	28580
Lower level, (-1)	1438	2100	680	12340
Star point, (+2)	2000	2400	710	36700
Star point, (-2)	1250	2000	670	4220

5. Experimental results

After the regression equations had been obtained according to the method [17], and the transition from them to polynomials with natural factors had been made, two equations for the roll stiffness moduli from the rolling force M_W^P and the regulating force M_W^Q had been got:

$$\begin{aligned} M_W^P = & -26465 + 3.227 \cdot W + 2.918 \cdot l_{OR} + 62.39 \cdot D_{WR} + 1.45 \cdot K_{\Delta h} - 0.018405 \cdot W \cdot l_{OR} + \\ & + 0.00666 \cdot W \cdot D_{WR} + 0.0011846 \cdot W \cdot K_{\Delta h} - 0.02412 \cdot l_{OR} \cdot D_{WR} - 0.001767 \cdot l_{OR} \cdot K_{\Delta h} + \\ & + 0.002833 \cdot D_{WR} \cdot K_{\Delta h} + 0.010365 \cdot W^2 + 0.009477 \cdot l_{OR}^2 - 0.00975 \cdot D_{WR}^2 - 0.000007 \cdot K_{\Delta h}^2 \end{aligned} \quad (8)$$

$$\begin{aligned} M_W^Q = & 1990 + 0.732 \cdot W - 1.3359 \cdot l_{OR} - 3.6758 \cdot D_{WR} + 0.387 \cdot K_{\Delta h} - 0.001243 \cdot W \cdot l_{OR} - \\ & - 0.0030986 \cdot W \cdot D_{WR} - 0.0002147 \cdot W \cdot K_{\Delta h} + 0.00081 \cdot l_{OR} \cdot D_{WR} - 0.000261 \cdot l_{OR} \cdot K_{\Delta h} + \\ & + 0.0006084 \cdot D_{WR} \cdot K_{\Delta h} + 0.0011014 \cdot W^2 + 0.000747 \cdot l_{OR}^2 + 0.006 \cdot D_{WR}^2 - 0.00000023 \cdot K_{\Delta h}^2 \end{aligned} \quad (9)$$

After the calculation by equations (equation 1, equation 2), the value $K_{\Delta h}$ should have the dimension kg/mm^2.

Figure 1–2 presents the diagrams of correspondence of the rolls stiffness modules from the rolling force and the regulating force, respectively, calculated with the use of the model [12–14] and the obtained equations (8–9). Table 2 shows the relative δ and absolute errors Δ, the coefficient of determination (degree of conformity) R^2 and the effectiveness [20] E_δ of the obtained mathematical model with $|\delta|$ no more than 5%. The indicated quality estimates of the model show that its compliance with the experimental results can be considered excellent.

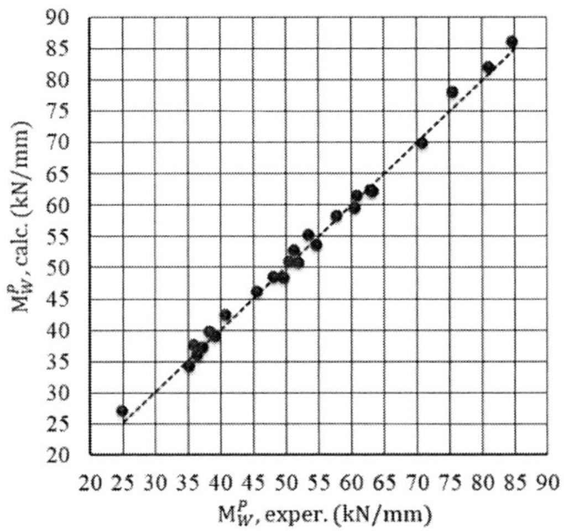

Figure 1. Correspondence diagram of the stiffness modulus from the rolling force.

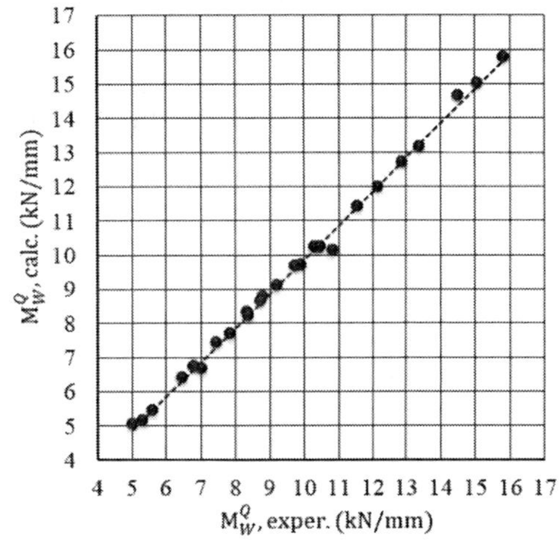

Figure 2. Correspondence diagram of the stiffness modulus from the regulatory force.

Table 2. Estimated characteristics of the mathematical model.

Roller stiffness modulus from the rolling force					
δ, [%]	Δ_{av}, [%]	Δ, [kN/mm]	Δ_{av}, [kN/mm]	R^2	E_5, [%]
-3.74…6.55	0.16	-1.80…2.32	0.06	0.99	96
Roll stiffness modulus from the control force					
δ, [%]	Δ_{av}, [%]	Δ, [kN/mm]	Δ_{av}, [kN/mm]	R^2	E_5, [%]
-6.98…0.79	-1.73	-0.71…0.12	-0.15	0.99	92

The dependences of the rolls stiffness moduli on the width of the strip under the operating rolling force and the regulating force at the maximum, average and minimum values of the influencing factors are shown in figure 3. From figure 3 it can be seen that the greater the length of the inter-roll contact l_{OR} is the greater the stiffness moduli are. But if the modulus M_W^Q is monotonously decreasing with the strip width increase, then the modulus M_W^P has the minimum, the position of which depends on the width. With an increase in the length of the inter-roll contact, the indicated minimum shifts to the region of large widths.

6. Conclusion

The calculation task of the rolls elastic deformation of the strip rolling mill is important not only for assessing the rigidity of the working stand, but also for calculating the transverse profile of the strip. The solution of this problem with the use of simple methods has a noticeable error, and the calculations by more accurate methods are laborious. In order to simplify the calculation of the transverse strip thickness variation, it is possible to use the method of approximation of the calculation results of roll deformations by using exact methods in a wide range of variation of the roll unit features and loading conditions. In this case, it is advisable to take the moduli of the transverse rigidity of the roller unit from the rolling force and from the regulating force as the approximated value, and the actual approximation should be non-linear.

When the method of the experiment designing is used, second-degree polynomials are obtained, which makes it possible to simplify the calculation of the transverse rigidity moduli from the rolling effort and from the regulating force for the roll unit of the finishing stand of the wide - strip hot mill 2500 of the PJSC MMK. The effectiveness of the mathematical model in the form of these polynomials with an admissible error of 5% is 96% for the modulus of rigidity from the rolling force and 92% from the regulating force. The calculation according to these expressions is made in a short time, which is a significant advantage for production conditions.

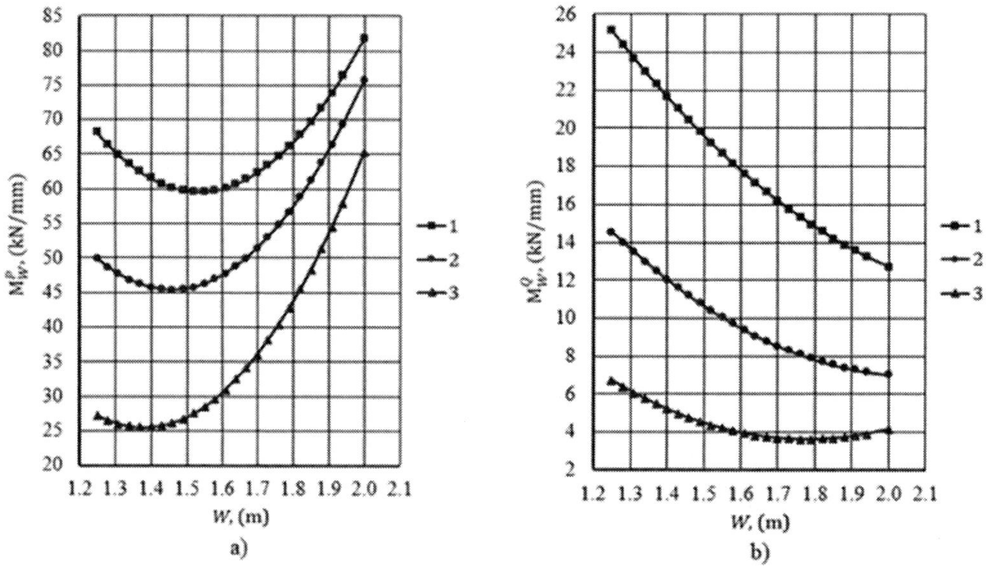

Figure 3. Dependence of the rolls stiffness moduli on the width of the strip with the variation of the rolling force (a) and the regulating force (b): 1- l_{OR} = 2400 mm, D_{WR} =710 mm, $K_{\Delta h}$ =36700 MPa; 2- l_{OR} =2200 mm, D_{WR} =690 mm, $K_{\Delta h}$ =20460 MPa; 3 - l_{OR} =2000 mm, D_{WR} =670 mm, $K_{\Delta h}$ = 4220 MPa.

References

[1] Shatalov R L 2004 Control of strip quality and deformability in rolling *Steel in Translation* **34(9)** 43–47

[2] Shatalov R L, Maksimov E A and Babkin A G 2011 Adjustment of rolling mills to produce flatter steel strip *Steel in Translation* **41(10)** 845–848

[3] Shatalov R and Genkin A 2015 Sheet mill control in steel strip hot rolling *Journal of Chemical Technology and Metallurgy* **50(6)** 624–628

[4] Belsky S M, Mazur I P, Lezhnev S N and Panin E A 2016 *Shaping strip for sheet rolling* (Temirtau) p 161

[5] Tretyakov A, Garber E and Davletbaev G 1976 *Calculation and study of rolling rolls* (Moscow: Metallurgy) p 256

[6] Grigoryan G G, Zheleznov Yu D, Cherny V A, Kuznetsov L A and Zhuravsky A G 1975 *Setting, stabilization and control of the sheet rolling process* (Moscow: Metallurgy) p 368

[7] Polukhin V P 1972 *Mathematical modeling and computer calculation of sheet rolling mills* (Moscow: Metallurgy) p 512

[8] Budakva A A, Konovalov Yu V and Tkalich K N 1986 *Profiling rolls sheet mills* (Moscow: Technique) p 190

[9] Bolobanova N L 2015 *Development of methods for modeling profiling and elastic deformations of rolls of sheet mills with the aim of improving the technology of rolling wide lanes* (Cherepovets) p 123

[10] Zhengwen Yuan, Hong Xiao, Hongbiao Xie 2014 Practice of Improving Roll Deformation Theory in Strip Rolling Process Based on Boundary Integral Equation Method *Metallurgical and Materials Transactions* **45(2)** 1019–1026
[11] Shao Jian, Li Bo, He Anrui, Sun Wenquan, Liu Zhubin and Zhou Jie 2015 Roll Deformation Model of Four High Hot Rolling Mills Based on Improved Influence Function Method *The Open Automation and Control Systems Journal* **7** 93–99
[12] Salganik V M, Melzer V V and Omelchenko B Ya 1977 Calculation of profiling rolls of sheet rolling mills (Sverdlovsk) p 57
[13] Salganik V M 2002 Mathematical modeling of roll load and deformation in a four-high strip mill. Metal Forming *The University of Birmingham*
[14] Salganik V M 2009 *Modeling deformations and loads of the quarto roll system and improving the quality of sheet products along the profile* (Magnitogorsk) p 133
[15] Salganik V M, Poletskov P P, Sinitsky O V, Soloviev A G and Lytyaykina Yu B 2003 *Increasing the flatness of wide bands on the hot rolling mill 2500 of MMK in the conditions of preferential production of narrow bands* (Cherepovets)
[16] Omelchenko B Ya 2006 *Model of the formation of the strip profile in the finishing group of the wide-strip hot rolling mill. Materials of the 64th scientific and technical conference on the results of scientific research for 2004–2005* (Magnitogorsk)
[17] Novik F S and Arsov Ya B 1980 *Optimization of metal technology processes by experiment planning methods* (Moscow: Mechanical Engineering) p 304
[18] Barkaya V F, Rokotyan S E and Ruzanov F I 1976 Sheet metal change (Moscow: Metallurgy) p 264
[19] Rokotyan S 1991 Theory of rolling and the quality of the metal (Moscow: Metallurgy) p 223
[20] Rumyantsev M I and Tulupov O N 2018 Further developments in simulation of metal forming processes *CIS Iron and Steel Review* **16** 21–24.

IOP Conf. Series: Materials Science and Engineering **969** (2020) 012072 doi:10.1088/1757-899X/969/1/012072

On Issue of Strip's Break during Cold Rolling

S M Belskiy and I I Shopin

Lipetsk State Technical University, 30, Moscow st., Lipetsk, 398055, Russia

E-mail: Belsky-55@yandex.ru

Abstract. The strip breaks during the cold rolling lead to significant losses in a continuous mill: downtime, associated with the elimination of consequences, and additional consumption of metal. Strip break occurs at a specific strip location, which further complicates the identification and analysis of its root causes, since a critical deviation of the parameters in only one small section of the strip may already be the cause of the subsequent break during cold rolling. In this work, the methods for checking hypotheses and regression modeling are used to identify the key parameters of hot-rolled strips that affect the strip break. Then, on the basis of the developed mathematical models, threshold values for key parameters are determined, exceeding which increases the probability of strip break.

1. Introduction

The cold rolling of steel strips on a continuous wideband mill is often accompanied by incidents such as the strip's shifting along the work rolls' barrel, tension throws and flatness's violations, which can lead to the emergency conditions or break of the strip. The elimination of the consequences of break is associated with an additional production's costs in the form of unplanned downtime, increased of consumption coefficient and breakdowns of rolling rolls [1–30].

The defects in the cross-section profile of hot-rolled rolling are the main cause of the violation of strips' flatness during the cold rolling, at that the internal stresses, arising during the rolling process and distributed over the strip width, can exceed the strength limit, which inevitably leads to the strip's break. But even if the strips aren't breaked, the resulting non-flatness of cold-rolled strips leads to rolls' defects [1–6] and the need for additional technological correction operations [7–22], which increase the cost price and thereby reduce the competitiveness of the finished product [23–30].

2. Binary logistic regression of cold rolled strip's break forecast

The parameters of hot-rolled strip, such as the flatness and cross-section profile, are usually described through the integral estimates within the entire length of the strip. The integral estimates of parameters show the overall quality level of the strips, produced in the hot rolling mill, but don't provide information about its variation. In this case, the cold rolled strip's break occurs at a specific coordinate along the length, so in the cause-and-effect analysis it is better to use distributed parameter estimates, rather than integral parameters.

The distribution of breaks of the transformer steel grade E3A during the first cold rolling was analyzed in relation to the coordinate along the length of the hot-rolled strip. In figure 1 it can be seen that the largest number of breaks occurs on the head section of the hot-rolled strip.

On the first 5 % of the length, the probability of break is 6–7% and gradually falls to almost zero by 24 % of the strip length. In the central part of the strip (from 25 % to 83 % of the length), there are almost no breaks. On the tail section of the strip (starting from 84 % of the length) there is an increase in the part of breaks from 0 % to 2.5 %. Taking into account the above, the analysis of the influence of profile parameters on the strip breakage during the first cold rolling should be conducted in three zones separately: the head section (from 0 % to 24 %), the central section (from 25 % to 83 %) and the tail section (from 84 % to 100 %).

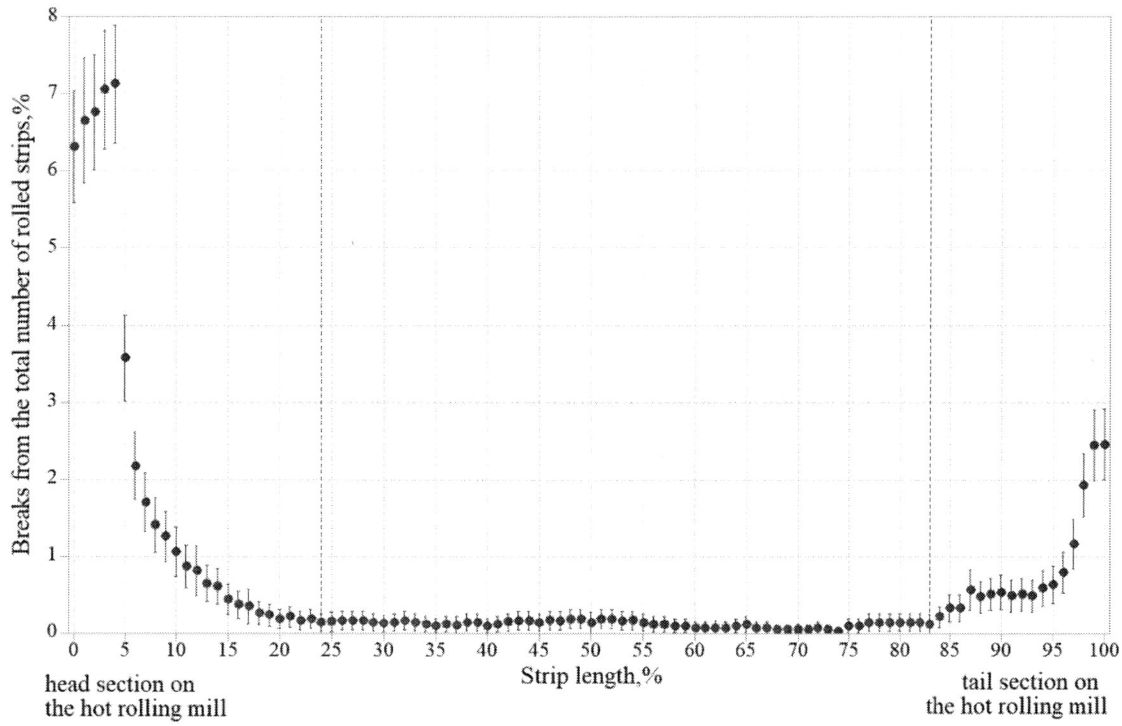

Figure 1. Distribution of breaks in the E3A assortment along the hot-rolled strips' length with an indication of the 95 % confidence interval.

For the subsequent regression analysis potentially significant factors were selected by based on a statistical hypothesis checking method similar to the work [23]. The parameter was selecting as potentially affecting on the breaks of strips during the cold rolling, if there was a statistically significant difference between medians or dispersions on strips with and without breaks.

Taking into account that for all the analyzed parameters of the flatness and cross-section profile the observed distribution does not correspond to the normal distribution, the *P*-criterion for the medians was calculated using the Kruskel – Wallis test, and for the dispersions – according to the Levene test [24, 25].

Let F_a be an asymmetric component of flatness, I-Unit (1 I-Unit = 10 mkm/m = 10^{-5}); F_s be a symmetrical component of flatness, I-Unit; C be a cross-section profile's convexity, mkm; TV be a cross-section profile thickness, mkm; D be a cross-section profile's convexity displacement from the center of strip, mkm; W be a cross-section profile's wedge, mkm; W_{co}^{rs} be a strip edge wedge from operator-side, mkm; W_{cd}^{rs} be a strip edge wedge from drive-side, mkm; R_{sq} be a determination coefficient of cross-section profile.

Let's build a mathematical model of the probability of breaks during the cold rolling on the head section of the hot-rolled strip (within the framework of this article, the mathematical models for the central and end sections are not given). To do this, we use a sampling for one steel grade (E3A) and one

IOP Conf. Series: Materials Science and Engineering **969** (2020) 012072 doi:10.1088/1757-899X/969/1/012072

standard size (2.5 × 1055 mm) for the period from May 2018 to February 2019. The total number of rolled strips, used in the analysis, is 4546, 485 strips of them were with breaks.

Based on the results of calculations, we obtain the following binary logistic regression [26] for the forecast of the strip break:

$$P_{break} = \frac{e^Y}{1+e^Y}, \quad (1)$$

$$\begin{aligned} Y = a_0 + a_1F_a + a_2F_s + a_3C + a_4D + a_5W_{co}^{rs} + a_6W_{cd}^{rs} + a_7W + a_8R_{sq} + +a_9F_a^2 + a_{10}F_s^2 + \\ +a_{11}C^2 + a_{12}D^2 + a_{13}W^2 + a_{14}CD + a_{15}CW_{cd}^{rs} + +a_{16}CR_{sq} + a_{17}DR_{sq} + a_{18}W_{cd}^{rs}R_{sq} + \\ +a_{19}F_s^3 + a_{20}C^2W_{cd}^{rs} + a_{21}CW_{cd}^{rs}R_{sq}, \end{aligned} \quad (2)$$

where P_{break} is the probability of strip's break.

The presented mathematical model is adequate. First, the model explains more than a half of all breaks in the cold rolling mill (the determination coefficient of the resulting regression model is 0.6304). Second, the acceptable accuracy of the regression model is provided by a small number of significant factors, because the value of the adjusted determination coefficient (equal to 0.6292) is comparable to the value of the determination coefficient (equal to 0.6304) [27]. All factors used in this model are statistically significant, because the *P*-criterion for them is less than 0.001. These properties allow us to use the developed regression model in the technological process for predicting breaks during the cold rolling on the head section of the strip.

3. Influence of parameters of hot-rolled rolling on breaks of cold-rolled strips

We'll study the influence of the hot-rolled rolling flatness and cross-section profile parameters on the breaks during the cold rolling using the developed binary regression model. For the convenience of displaying the analysis results, we'll plot the graphs of the strip break probability dependence during cold rolling on of related factors' pairs.

An increase of the edge's wedge values on both sides of the strip leads to an increase in the probability of breaks.

Exceeding by the symmetrical and asymmetric components of flatness the optimal diapason's boundaries also leads to a sharp increase of the breaks during the cold rolling. The determination coefficient's decrease of the cross-section profile below 0.92 increases the probability of breaks during the cold rolling on the head section of the strip. At the same time, the cross-section profile convexity's increasing reduces the impact of the determination coefficient on the strips' breaks. Thus, increasing of the strip cross-section profile's convexity from 30 microns to 40 microns shifts the lower boundary of the region with a low probability of breakage for the determination coefficient from 0.92 to 0.86. Increasing of the cross-section profile's wedge leads to increasing of the breaks' probability, while increasing of the strip cross-section profile's convexity above 35 microns radically reduces the influence degree of wedge on the breaks. The deflection of the strip cross-section profile's convexity on the drive-side is more critical than on the operator-side, which may be due to the asymmetry of the temperature field along the width of the hot-rolled strip. On the basis of the developed regression model, the recommendations for the flatness parameters and cross-section profile parameters of the hot-rolled strips which provides minimum number of breaks during the cold rolling were worked out. The optimal values of the strip cross-section profile's convexity couldn't be determined unambiguously. This is due to the fact that: firstly, the system for regulating of the strip cross-section profile's convexity works out a single setting value 30 mkm; secondly, if the other factors are maintained in the optimal diapasons, the influence of the strip cross-section profile's convexity on the strips' breaks disappears. However, the modelling shows that increasing of the strip cross-section profile's convexity extends the optimal diapasons for other parameters. Using the regression modeling, we also obtained the share estimate of a specific factor influence on the breaks probability, when the factor leaves the optimal diapason.

As a result, we became able to classify sections of the hot-rolled strip according to the degree of criticality in terms of breaks during the cold rolling (figure 2) and "highlight" the most problematic areas for the operator of the cold rolling mill. 80 % of breaks falls on the most critical zones (5 or more considered parameters are out of the optimal diapasons), despite the fact that only 11 % of measurements are analyzed.

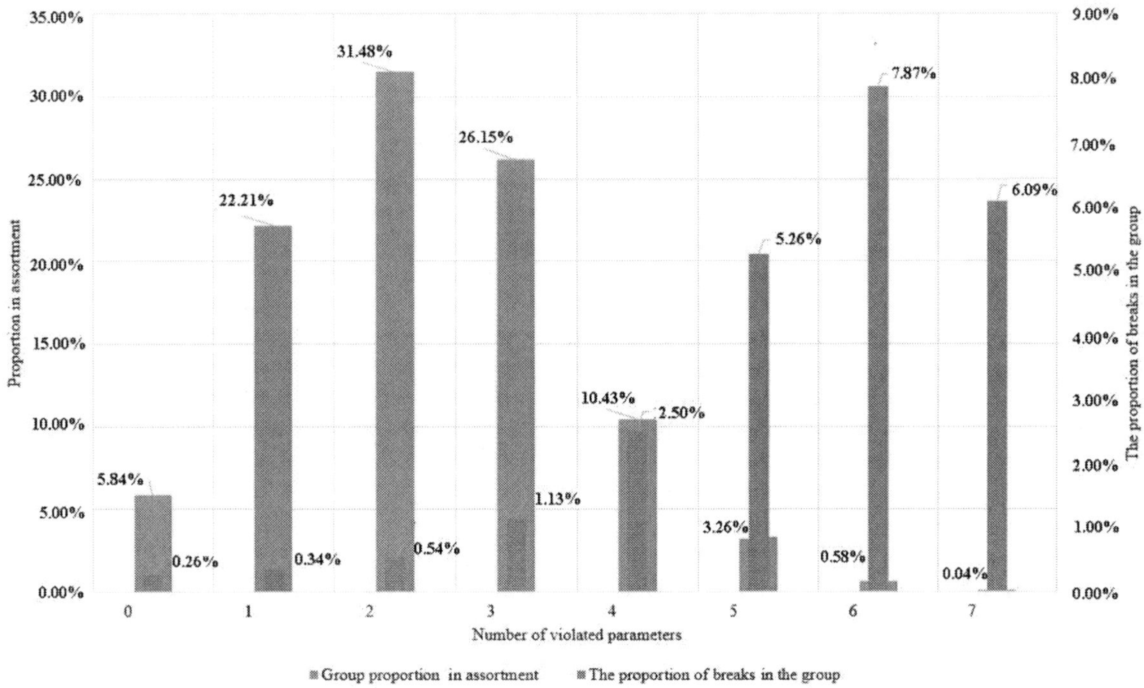

Figure 2. Influence of the flatness and cross-sectional profile parameters exceeding the optimal boundaries on strip breaks during cold rolling.

4. Conclusions

The flatness and cross-section profile parameters of a hot-rolled strip have a significant effect on the breaks during the first cold rolling. When we analyze the influence of the flatness and cross-section profile on the breaks of strips, it is best to use parameters, distributed along the length of the strip, rather than their average value for the entire roll. The developed binary regression model of the strip breaks' probability during cold rolling of hot-rolled strips can be used for a visualization of the most critical zones for the operator of a cold rolling mill. The optimal values of the flatness and cross-section profile parameters were determined using a regression analysis, which ensuring a breaks minimal number during cold rolling of hot-rolled strips.

References

[1] Korolev A A 1987 *Mechanical Equipment of Rolling and Pipe Shops* (Moscow: Metallurgy)

[2] Shopin I I and Belskiy S M 2016 Simplified model of deflected mode of coil on coilwinder *Rolling* **5** 13–17

[3] Shopin I I and Belskiy S M 2016 Lamellar model of deflected mode of coil on winder *Rolling* **8** 3–7

[4] Bel'skii S M, Mukhin Yu A, Mazur S I and Goncharov A I 2013 Influence of the cross section of hot-rolled steel on the flatness of cold-rolled strip *Steel in Translation* **43(5)** 313–316

[5] Maindonald J and Braun W J 2010 *Data Analysis and Graphics Using R: An Example-Based Approach* (Cambridge University Press)

[6] Belsky S M, Kotsar S L and Polyakov B A 1990 Calculation of the distribution of rolling force across the width of the strip and residual stresses in the strip by the variational method *Proceedings of higher educational institutions. Ferrous metallurgy* **10** 32–34
[7] Predeleanu M and Gilormini P 1997 *Advanced Methods in Materials Processing Defects* **45** (Elsevier Science)
[8] Shinkin V N 2017 Mathematical model of technological parameters' calculation of flanging press and the formation criterion of corrugation defect of steel sheet's edge *CIS Iron and Steel Review* **13** 44–47
[9] Shinkin V N 2016 The mathematical model of the thick steel sheet flattening on the twelve-roller sheet-straightening machine. Message 1. Curvature of sheet *CIS Iron and Steel Review* **12** 37–40
[10] Shinkin V N 2016 The mathematical model of the thick steel sheet flattening on the twelve-roller sheet-straightening machine. Message 2. Forces and moments *CIS Iron and Steel Review* **12** 40–44
[11] Shinkin V N 2017 Arithmetical method of calculation of power parameters of 2N-roller straightening machine under flattening of steel sheet *CIS Iron and Steel Review* **14** 22–27
[12] Lim Y, Venugopal R and Ulsoy A G 2014 *Process Control for Sheet-Metal Stamping Process Modeling, Controller Design and Stop-Floor Implementation* (Springer)
[13] Lin J, Balint D and Pietrzyk M 2012 *Microstructure Evolution in Metal Forming Processes* (Woodhead Publishing)
[14] Banabic D 2016 *Multiscale Modeling in Sheet Metal Forming* (Springer)
[15] Shinkin V N 2016 Preliminary straightening of thick steel sheet in a seven-roller machine *Steel in Translation* **46(12)** 836–840
[16] Shinkin V N 2017 Failure of large-diameter steel pipe with rolling scabs *Steel in Translation* **47(6)** 363–368
[17] Shinkin V N and Kolikov A P 2011 Simulation of the shaping of blanks for large-diameter pipe *Steel in Translation* **41(1)** 61–66
[18] Shinkin V N and Kolikov A P 2011 Elastoplastic shaping of metal in an edge-bending press in the manufacture of large-diameter pipe *Steel in Translation* **41(6)** 528–531
[19] Hingole R S 2015 *Advances in Metal Forming: Expert System for Metal Forming* (Springer)
[20] Chakrabarty J 2010 *Applied Plasticity* (Springer)
[21] Shinkin V N 2016 Geometry of steel sheet in a seven-roller straightening machine *Steel in Translation* **46(11)** 776–780
[22] Shinkin V N 2017 Calculation of bending moments of steel sheet and support reactions under flattening on the eight-roller straightening machine *Ferrous metals* **4** 49–53
[23] Belskiy S M and Shopin I I 2019 Application of the saddle coefficient for estimating the quality of a hot-rolled steel strips *Ferrous metals* **9** 9–13
[24] Lvovskij E N 1982 *Statistical Methods for Construction of Empirical Formulas* (High school)
[25] Kobzar A I 2006 *Applied Mathematical Statistics: For Engineers and Researchers* (Fizmatlit)
[26] Hastie T, Tibshirani R and Friedman J 2017 *The Elements of Statistical Learning: Data Mining, Inference and Prediction* (Springer)
[27] Draper N R and Smith H 1986 *Applied Regression Analysis* **1** (Finance and Statistics)
[28] Shinkin V N 2017 Simplified calculation of the bending torques of steel sheet and the roller reaction in a straightening machine *Steel in Translation* **47(10)** 639–644
[29] Freedman D A 2009 *Statistical Models: Theory and Practice* (Cambridge University Press)
[30] Venables W N and Ripley B D 2002 *Modern Applied Statistic with S* (Springer)

IOP Conf. Series: Materials Science and Engineering **969** (2020) 012073 doi:10.1088/1757-899X/969/1/012073

Simulaling the Faceted Tube Drawing

R A Okulov[1,2] and N V Semenova[1]

[1]Institute of Fundamental Education, The Ural Federal University named after the first President of Russia B. N. Yeltsin, 19, Mira street, Yekaterinburg, 620002, Russia
[2]Institute of Metallurgy, Ural Branch of the Russian Academy of Sciences, 101, Amundsen street, Yekaterinburg, 620016, Russia

E-mail: n.v.semenova@urfu.ru

Abstract. The paper studies cold-sinking of four-, six-, and eight-faceted steel tubes from round billets in a single pass. The built mathematical model of the process objectively describes the drawing and allows forecasting the geometric and deformation parameters of the pipes processed, as well as the power parameters of the process. The model accuracy has been confirmed by a series of parallel calculations and a comparative calculation in two software packages. The results have shown satisfactory convergence. The theoretical experiment results have been verified by performing a full-scale one. The results have also been consistent. The research results are of practical use to the process equipment manufacturers and consumers.

1. Introduction

The manufacture of shaped tubes with high dimensional accuracy while reducing production costs is a relevant scientific and technical problem. Cold sinking is one of the widespread ways to produce shaped precision tubes. The modern packages implementing the finite element technique have proven themselves as a reliable tool in searching for the production problem solutions. Mathematical models allow forecasting the processing result with high accuracy. The process equipment developers and consumers need recommendations allowing them to forecast the processing results at the process equipment design stage and improve it during operation. The research objective is to build a mathematical model objectively describing the shaping, the analysis of which allows developing recommendations for improving this process. The study has been supported by the Russian Foundation for Basic Research (project No. 20-21-00063/20 Rosatom) according to the State Assignment of IMET UB RAS under the Program of Fundamental Research of State Academies. Works on the simulation of form drawing of shaped tubes using the finite element technique are known. In [1–4], the stress-strain state was studied based on a numerical method for the form drawing of shaped tubes. In the work [5], two simulation techniques were applied to study the tube wall deformations. Simulating the effect of friction and the lubricant behavior is described in detail in [6]; a full-scale experiment has been performed. The work [7] studies the drawing of heat-exchange tubes using a mathematical model. In [8], the stress-strain state was analyzed using the *ANSYS* software package for the form drawing of oval tubes. In [9], the cold drawing of precision seamless tubes was considered using the *DEFORM* software package. The work [10] is devoted to studying the tube drawing accuracy using finite element simulation. In [11], a similar process was considered for thin-walled steel tubes. In [12], a numerical method was applied to improve the shape of the working tool used in the tube drawing. A similar problem for square tubes was solved in [13]. In the listed and similar works, the problem raised herein has not been solved; this determines the scientific novelty of the study.

2. Description of the Mathematical Model and Discussion of the Numerical Experiment Results

The finite element technique was chosen as a tool to solve the problem raised herein. The *DEFORM* software package was chosen to generate a solution using the finite element technique. When choosing the size of the finite element mesh partition elements, a series of preliminary calculations were made, and the accuracy of their results analyzed. The rationale for choosing the element size and the estimation of its impact on the calculation accuracy is described in [14–16]. To improve the calculation accuracy, five parallel computer experiments were conducted, and the results averaged. To estimate the accuracy of the theoretical numerical experiment results, a comparative calculation has been performed in the *ANSYS* software package. The results have shown satisfactory convergence.

For the tubular product consumers, the section shape is of particular importance. Figure 1 shows a tube cross-section diagram specifying the dimensions. Under this research, the change of the below parameters depending on the face number of the tubes being processed has been studied: the wall deflection in the face center L, the actual wall thickness Sa, the radii of the face conjugation on the outer and inner surfaces R and r, and the elongation μ.

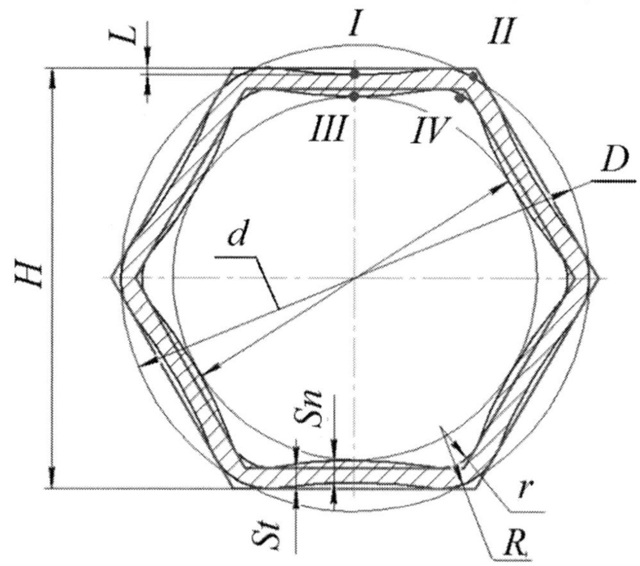

Figure 1. Six-Faceted Tube Cross-Section Diagram: L – the wall deflection in the face center; D and d – the diameters of the circumscribed and inscribed circles; St and Sa – theoretical and actual wall thickness; R and r – radii of the face conjugation on the outer and inner surfaces; H – the polyhedron height. Characteristic points: A, B, C, and D.

When implementing the drawing, the tubular product manufacturers consider the power indicators [17], i.e. the drawing force P, the drawing force work W, the energy consumption of the process E, and the displaced material amount q. Herein, changes in these parameters depending on the face number of the tubes shaped have been determined.

To estimate the quality of deformation parameters, the deformation intensity has been studied. The changes in the deformation intensity at the characteristic points depending on the product face number have been studied. 4 characteristic points were chosen: A, B, C, and D located at the conjunctions of the outer and inner surfaces of the tubes shaped. The characteristic point locations are specified in Figure 1.

As part of a theoretical study, three shaping cases have been considered: four-, six- and eight-faceted tubes. The cross-sectional profile on the outer contour of the tubes under consideration is set by the diameter of the circumscribed circle of the calibrating die section Dsc, which is the same for all three faceted tube processing cases considered.

The shaping has been performed on a die, the cross-section of which is shown in Figure 2. The inner die channel has a crimping section in the form of a truncated pyramid with flat faces. The calibrating section size is 16.4 mm. The working tool shape is described in detail in [18]. The crimping section to the die axis inclination angle is 10^0; there was no rounding radius between the crimping and calibrating working tool sections. Steel 20 GOST 1050 was chosen as the billet material. The billet was a round tube with a wall thickness of 1.5 mm and a diameter of 19 mm. The Amontons-Coulomb friction coefficient was taken equal to 0.15.

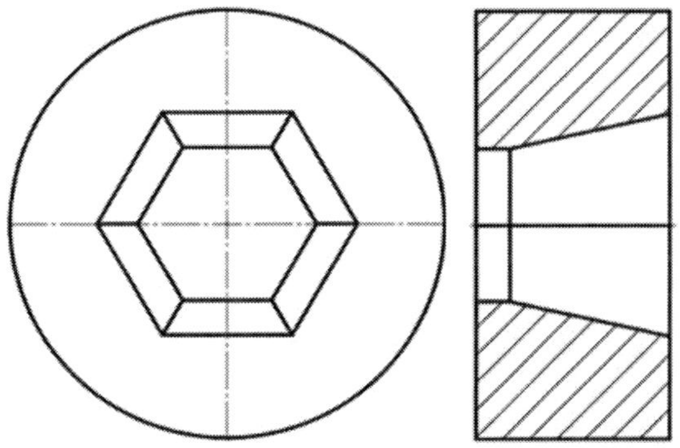

Figure 2. Working Tool Shape Diagram.

Based on the numerical experiment results, a comparative analysis of the parameters studied has been performed for the cases of shaping tubes with a different face number. The data are given in Table 1.

Table 1. Numerical Experiment Data.

Parameter	Tube Face Number		
	4	6	8
Actual wall thickness in the face center *Sa* (mm)	1.53	1.61	1.65
Deflection in the face center *L* (mm)	0.71	1.39	1.01
Elongation μ	1.032	1.099	1.131
The radius of face conjugation on the outer surface *R* (mm)	2.50	2.55	2.74
The radius of conjugation of the faces on the inner surface *r* (mm)	1.34	2.02	2.38
Drawing force *P* (H)	8515	15456	18893
Draw force work *W* (J)	8784	16992	21371
Displaced material mass *q* (kg)	2.92	11.40	16.11
Energy consumption of the process *E* (J / kg)	3009	1491	1327
Deformation intensity at point *I*	0.210	0.408	0.467
Deformation intensity at point *II*	0.255	0.184	0.172
Deformation intensity at point *III*	0.163	0.221	0.230
Deformation intensity at point Point *IV*	0.552	0.620	0.546

As can be seen from these data, the wall thickness *Sa* increases with an increase in the face number. The elongation μ and the conjugation radii r and R demonstrate the same behavior, i.e. increase with increasing the face number. The face deflection *L* has divergent dynamics.

In turn, with an increase in the fact number, the drawing force *P* increases, like the drawing force work *W* and the displaced material mass unit *q*, while the processing energy consumption *E* decreases.

The shape of the tubes to be processed also significantly affects the deformation intensity. With an increase in the face number, the deformation intensity at points I and III increases and at point II decreases; at point IV, the tendency to increase first prevails, but when processing tubes with six or more faces, it decreases: the deformation intensity changes nonlinearly and shows a divergent tendency when the face number changes. The greatest deformation intensity value is achieved at point IV.

3. Verifying the Theoretical Research Results and Describing the Full-Scale Experiment

To verify the computer experiment results, a full-scale one has been performed exemplified by shaping a six-faceted tube from a billet, the shape and mechanical properties of which, as well as the working tool and process parameters, correspond to those adopted in the theoretical experiment. To reduce the statistical error in processing the results, the full-scale experiment was performed on three groups of specimens. To ensure identical mechanical properties, the specimens were made of the same tube billet.

The tubes were shaped on a tensile testing machine MI-40U, for which tooling was prepared, which allowed reproducing the drawing and measuring the power indicators of the process. Instruments and process equipment are described in [19]. The geometrical parameters of the tube processed were measured using coordinate measuring devices with optomechanical conversion UIM-23 and DIP3 with a measurement accuracy of up to 5 μm using *Global Performance Power SHAPE 7080* software [20]. The deformation intensity was studied according to the technique [21] at characteristic point A. The study is described in more detail in [22, 23].

Comparative analysis of the theoretical and full-scale experiment results is represented in Table 2. The results have shown satisfactory convergence.

Table 2. Theoretical and Full-Scale Experiment Results.

Parameter	Theoretical	Full-scale.
Actual wall thickness in the face center Sn (mm)	1.53	1.52
Deflection in the face center L (mm)	0.71	0.72
Elongation μ	1.032	1.04
The radius of face conjugation on the outer surface R (mm)	2.50	2.47
The radius of face conjugation on the inner surface r (mm)	1.34	1.33
Drawing force P (H)	8515	8328
Draw force work W (J)	8784	895
Displaced material mass q (kg)	2.92	3.16
Energy consumption of the process E (J / kg)	3009	3295
Strain intensity. Point I	0.210	0.21

4. Conclusions and Practical Significance of the Results

As a result of the research, a mathematical model of cold-sinking of four-, six-, and eight-faceted tubes from seamless steel tubes in a single pass has been developed. The model accuracy has been confirmed by a series of parallel calculations and a comparative calculation in two software packages. The results have shown satisfactory convergence. The theoretical experiment results have been verified by performing a full-scale one. The results have also been consistent.

The mathematical model developed objectively describes the process and allows forecasting the shaping result. The model represented allows determining the power indicators of shaping, as well as the geometric and deformation parameters of the tube processed. The data obtained based on this model can be used in developing recommendations for improving the process equipment.

References

[1] Gonzalez G M, Sousa J R M and Sagrilo L V S 2019 A modal finite element approach to predict the lateral buckling failure of the tensile armors in flexible pipes *Marine Structures* **67** 102628

[2] Musrati W, Medjo B, Gubeljak N, Štefane P, Veljić D, Sedmak A and Rakin M 2019 Fracture assessment of seam and seamless steel pipes by application of the ring-shaped bending specimens *Theoretical and Applied Fracture Mechanics* **103** 102302
[3] Hatala M, Botko F, Peterka J, Bella P and Radic P 2020 Evaluation of strain in cold drawing of tubes with internally shaped surface *Materials Today: Proceedings* **22** 287–292
[4] Barrett T J and Knezevic M 2019 Deep drawing simulations using the finite element method embedding a multi-level crystal plasticity constitutive law: Experimental verification and sensitivity analysis *Computer Methods in Applied Mechanics and Engineering* **354** 245–270
[5] Dong J, Bao X and Zheng X 2012 Comparison of Two Methods of Finite Element Modeling for Elbows with Unequal Wall Thickness *Physics Procedia* **33** 498–504
[6] Carretta Y, Bech J, Legrand N,Laugier M, Ponthot J-P and Boman R 2017 Numerical modelling of micro-plasto-hydrodynamic lubrication in plane strip drawing *Tribology International* **110** 378–391
[7] Bella P, Durcik R, Ridzon M and Ludovit P 2018 Numerical simulation of cold drawing of steel tubes with straight internal rifling *Procedia Manufacturing* **15** 320–326
[8] Khademi-Zahedi R 2019 Application of the finite element method for evaluating the stress distribution in buried damaged polyethylene gas pipes *Underground Space* **4** 59–71
[9] Bella P, Bucek P, Ridzon M, Mojzis M and Parilak L 2017 On numerical modelling of multi-rifled tube drawing *Procedia Engineering* **207** 2382–2387
[10] Boutenel F, Delhomme M, Velay V and Boman R 2018 Finite element modelling of cold drawing for high-precision tubes *C. R. Mecanique* **346** 665–677
[11] Palengat M, Chagnon G, Favier D, Louche H, Linardon C and Plaideau C 2013 Cold drawing of 316L stainless steel thin-walled tubes: Experiments and finite element analysis *International Journal of Mechanical Sciences* **70** 69–78
[12] Béland J-F, Fafard M, Rahem A, D'Amours G and Côté T 2011 Optimization on the cold drawing process of 6063 aluminium tubes *Applied Mathematical Modelling* **35** 5302–5313
[13] Shu G, Jin X, Zhang Y, Gu Y, Zheng B and Jiang Q 2019 Experimental and numerical study of cold-drawn duplex stainless steel square tube columns *Journal of Constructional Steel Research* **156** 155–166
[14] Okulov R A, Parshin S V and Spiridonov V A 2014 *Bulletin of mechanical engineering* **5** 82–84
[15] Okulov R A and Parshin S V 2015 *Metallurgist* **2** 55–57
[16] Okulov R A, Parshin S V and Spiridonov V A 2014 *Rolled metal production* **6** 4–7
[17] Okulov R A, Parshin V S and Spiridonov V A 2012 *Bulletin of mechanical engineering* **9** 80–81
[18] Okulov R A and Parshin S V 2015 *Metallurgist* **1–2** 156–158
[19] Okulov R A and Semenova N V 2019 *Don's Engineering Gazette* **5** 21–25
[20] Okulov R A and Parshin S V 2013 *Bulletin of Magnitogorsk State Technical University G. I. Nosova* **4** 40–44
[21] Smirnov-Alyaev G A and Chikidovsky V P 1972 *An experimental study in the processing of metals by pressure* (Moscow: Mechanical Engineering) p 360
[22] Okulov R A, Parshin S V and Spiridonov V A 2014 *Russian Engineering Research* **8** 509–511
[23] Okulov R A 2015 *Improving the processes of profiling faceted pipes based on modeling of the deformation zone in order to increase their accuracy* (Ekaterinburg: Ural Federal University named after the first President of Russia B.N.Yeltsin) p 152

Acknowledgments
The work was performed according to the State task of the Institute of Metallurgy of the Ural Branch of the Russian Academy of Sciences in the framework of the Program of Basic Research of State Academies. The work was supported by the Russian Foundation for Basic Research (project No. 20-21-00063 Rosatom).

IOP Conf. Series: Materials Science and Engineering **969** (2020) 012074 doi:10.1088/1757-899X/969/1/012074

Statistical Regression Analysis of Breakages in Cold Rolling

S M Belskiy and I I Shopin

Lipetsk State Technical University, 30, Moskovskaya street, Lipetsk, 398055, Russia

E-mail: Belsky-55@yandex.ru

Abstract: The maximum probability of strip's breakage during the cold rolling is observed when the values of edge wedge are greater than 150 microns and the saddle coefficient is less than –0.075 and greater than 0.050. It is also shown that there is a pronounced synergistic effect of influence of combination of the saddle coefficient and the edge wedge on the breakage of the strip. When the edge wedge is reduced below 50 microns, the saddle coefficient ceases to affect the breakage of the strip during the cold rolling. At the same time, when the edge wedge is higher than 160 microns, the breakage is minimal with the saddle coefficient in the range from 0 to 0.025.

1. Introduction

The strip breakage during processing on the continuous cold rolling mill leads to significant losses: unplanned downtime, associated with the elimination of the consequences of breaks, the increased consumption of metal of the working and support rolls due to the injuries.

One of the main reasons for the strips' breaks during the rolling is the presence of the defects on the hot-rolled blank. These can be either the edge's defects or defects in the shape of the cross-section profile of the hot-rolled strip. If the edge's defects can be detected either visually or by the specialized automated surface quality control systems for the rolled strips, then the defects in the cross-section profile cause the strip to lose its flat shape during the subsequent cold rolling. In turn, the flatness of the cold-rolled strip leads to the need for the additional technological leveling operations [1–33], which increase the production cost.

2. Criteria for evaluation of defects of strip's profile

To detect defects in the cross-section profile of a hot-rolled strip, a criterion for evaluating their criticality for the subsequent cold rolling is required.

In order to formalize the estimation of the parameters of the cross-section profile of the hot-rolled strips, their parabolic approximation is usually used [12] and such parameters as, for example, the convexity P and the wedge shape W are calculated (figure 1).

The parameters, that characterize the cross-section profile, include the determination coefficient:

$$R^2 = 1 - \frac{\sum_j \left[H_{sc}^j - f(x_j)\right]^2}{\sum_j \left[H_{sc}^j - H_m\right]^2}, \tag{1}$$

where R^2 is the coefficient of determination, H_m is averaged across the width of the thickness of the strip, H_{sc} is the measured value of strip thickness at j-th point in width, x_j is the coordinate of the measured point (the distance between the measured point of the profile and middle of the band), $f(x) = ax^2 + bx + c$ is the approximating parabola of the measured cross-sectional profile, whose coefficients are determined by least squares.

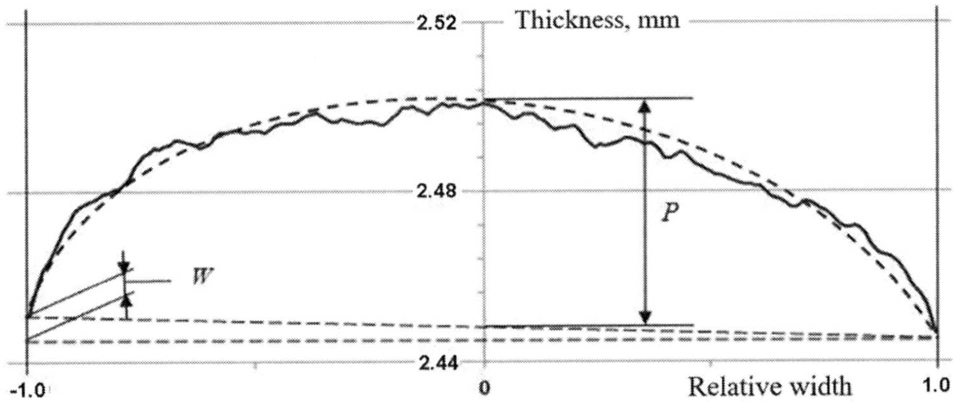

Figure 1. Real cross-section profile of hot-rolled strip and parameters of its estimation.

The determination coefficient of the cross-section profile is a value that characterizes the share of dispersion of the actual profile, that is due to its parabolic approximation; in other words, it shows how close the actual cross-section profile is to the parabolic approximation. Quite often, there are the cases when the cross-section profiles of the hot-rolled strips with different deviations from the approximating parabola have similar, and sometimes identical, values of the determination coefficient.

One of the alternative criteria, for evaluating the quality of the strip's cross-section profile, to the determination coefficient is the saddle coefficient *Ks* [13]. This value is intended to characterize the cross-section profile with the edge thickenings.

The saddle coefficient, as well as the determination coefficient, characterizes the deviation of the real cross-section profile from the approximating parabola, describing an increase in the contribution to its value of deviations with an increase in the distance from the strip's center to its edges:

$$K_s = \frac{\sum_j \left[H_{sc}^j - f(x_j)\right] \cdot |x_j|}{\sum_j \left[H_{sc}^j - H_m\right] \cdot |x_j|}. \tag{2}$$

The integral character of the saddle coefficient allows us to characterize the stability of other parameters of the cross-section profile. For example, the deviation of the saddle coefficient from zero is associated with an increase in the mean square deviation of the convexity of the cross-section profile. This important feature of the saddle coefficient allows you to reduce the number of factors and thus to simplify the regression mathematical model.

Another parameter for the quality of the cross-section profile of the strip is the value of the edge-boundary wedge W_{rs}, which characterizes the difference in thickness within 40 mm from the edge of the strip.

3. Model for predicting the probability of strip's break

We apply the above parameters of the strip's cross-section profile in probability forecast model of breakage at the cold rolling.

Based on the results of experiments, we obtain the following binary logistic regression [23] of the forecast probability of breakage as a function of the saddle coefficient and the edge-boundary wedge:

$$P_{break} = \frac{e^Y}{1+e^Y}, \tag{3}$$

$$\begin{aligned} Y = &-3,402 - 6,91K_s + 0,0282W_{rs} - 411K_s^2 - 0,000429W_{rs}^2 - 0,0483K_sW_{rs} + \\ &+4093K_s^3 + 0,000002W_{rs}^3 + 4,13K_s^2W_{rs} + 56866K_s^4, \end{aligned} \tag{4}$$

where P_{break} is the probability of strip's break.

The presented model has the adequacy property, since: first, the model explains half of all breaks in the cold rolling mill (the determination coefficient of the obtained regression model is R_m^2 = 48.9 %); secondly, the acceptable accuracy of the regression model is provided by a small number of the significant factors, since the value of the adjusted determination coefficient ($R_{m,\ adj}^2$ = 43.1 %) is comparable to the value of the determination coefficient (48.9%) [24].

The determination coefficient of model is calculated using the formula:

$$R_m^2 = 1 - \frac{\sum_j \left(P_{fact}^j - P_{break}^j\right)^2}{\sum_j \left(P_{fact}^j - P_{fact}\right)^2}. \tag{5}$$

The adjusted determination coefficient is calculated as follows:

$$R_{m,\,adj}^{\;\;2} = 1 - \left(1 - R_m^2\right)\frac{n-1}{n-k}, \tag{6}$$

where P_{break}^j is the probability of breakage of the j-strip in model, P_{fact}^j is the indication of breakage of the j-th strip (0 – no break, 1 – break), P_{fact} is the share strips with the breaks from the total number of analyzed strips, n is the total number of analyzed strips, k is number of factors in the regression model.

These properties allow us to use the developed regression model in the technological process to predict the probability of strip's breakage during the cold rolling.

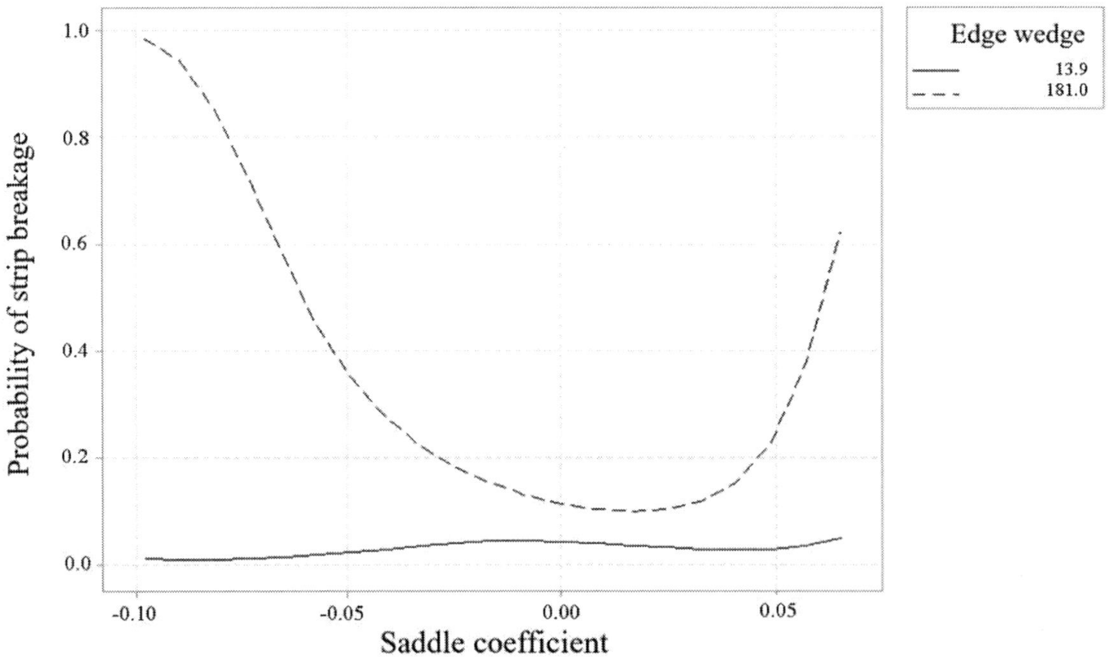

Figure 2. Dependence of probability of strip breakage on saddle coefficient of the maximum and minimum values of edge wedge.

4. Statistical testing of hypotheses

To analyze the significance of factors in the regression model, we use the method of statistical hypothesis testing [25]. Null hypothesis: the factor does not have a significant impact. An alternative hypothesis: the analyzed factor in the regression model has a significant impact on the probability of breakage during the cold rolling of the hot-rolled strip. The *P*-criterion is the probability that the null hypothesis is correct. We accept that the null hypothesis is rejected, when the *P*-criterion is less than 0.05. The sense of the *P*-criterion is the probability that the regression model is the result of coincidence of the random fluctuations of factors and has no the real physical regularities:

$$P_{value} = p_1 + p_2, \tag{7}$$

where P_{value} is the *P*-criterion, p_1 is the probability that the average of the first sample is part of the second sample, p_2 is the probability that the average of the second sample is part of the first sample.

5. Conclusions

The parameters of the cross-section profile of the hot-rolled strip (the saddle coefficient and the edge wedge) can be used to predict the probability of the strip's breakage during the cold rolling. At the values of the edge wedge below 50 microns, the probability of breaking the hot-rolled strip during the cold rolling is significantly reduced (see figure 2). At the values of the saddle coefficient in the range from 0 to 0.025, the probability of breakage of the hot-rolled strip during the cold rolling is significantly reduced.

The breakage of the hot-rolled strip during the cold rolling is influenced by the saddle coefficient and the edge wedge in an interrelation to each other. The maximum probability of breakage is observed when the edge wedge values are greater than 150 microns and the saddle coefficient is less than –0.075 and greater than 0.050.

References

[1] Korolev A A 1987 *Mechanical Equipment of Rolling and Pipe Shops* (Metallurgy)
[2] Belskiy S M and Shopin I I 2019 Application of the saddle coefficient for estimating the quality of a hot-rolled steel strips *Ferrous metals* **9** 9–13
[3] Bel'skii S M, Mukhin Yu A, Mazur S I and Goncharov A I 2013 Influence of the cross section of hot-rolled steel on the flatness of cold-rolled strip *Steel in Translation* **43(5)** 313–316
[4] Bel'skii S M and Mukhin Yu A 2009 Classification of regulation principles for strip flatness *Steel in Translation* **39(11)** 1012–1015
[5] Predeleanu M and Gilormini P 1997 *Advanced Methods in Materials Processing Defects* (Elsevier Science)
[6] Shinkin V N 2016 The mathematical model of the thick steel sheet flattening on the twelve-roller sheet-straightening machine. Message 1. Curvature of sheet *CIS Iron and Steel Review* **12** 37–40
[7] Shinkin V N 2016 The mathematical model of the thick steel sheet flattening on the twelve-roller sheet-straightening machine. Message 2. Forces and moments *CIS Iron and Steel Review* **12** 40–44
[8] Shinkin V N 2017 Mathematical model of technological parameters' calculation of flanging press and the formation criterion of corrugation defect of steel sheet's edge *CIS Iron and Steel Review* **13** 44–47
[9] Shinkin V N 2017 Arithmetical method of calculation of power parameters of 2N-roller straightening machine under flattening of steel sheet *CIS Iron and Steel Review* **14** 22–27
[10] Lim Y, Venugopal R and Ulsoy A G 2014 *Process Control for Sheet-Metal Stamping Process Modeling, Controller Design and Stop-Floor Implementation* (Springer)

[11] Sheinin V I, Levin B V, Blokhin D I and Favorov A V 2003 Identification of nonstationary changes in stress state of geomaterials by infrared radiometry data *Journal of Mining Science* **39(5)** 431–437
[12] Belskiy S M 2017 Parameters of evaluation of shape cross section of hot-rolled steel strips. Message 1. The determination coefficient *Ferrous metals* **10** 65–70
[13] Belskiy S M 2017 Parameters of evaluation of shape cross section of hot-rolled steel strips. Message 2. The saddle coefficient *Ferrous metals* **11** 42–47
[14] Banabic D 2016 *Multiscale Modeling in Sheet Metal Forming* (Springer)
[15] Frank V 2013 *Lecture notes in production engineering* (Springer)
[16] Shinkin V N 2016 Geometry of steel sheet in a seven-roller straightening machine *Steel in Translation* **46(11)** 776–780
[17] Shinkin V N 2016 Preliminary straightening of thick steel sheet in a seven-roller machine *Steel in Translation* **46(12)** 836–840
[18] Shinkin V N 2017 Failure of large-diameter steel pipe with rolling scabs *Steel in Translation* **47(6)** 363–368
[19] Shinkin V N 2017 Simplified calculation of the bending torques of steel sheet and the roller reaction in a straightening machine *Steel in Translation* **47(10)** 639–644
[20] Chakrabarty J 2010 *Applied Plasticity* (Springer)
[21] Sheinin V I, Blokhin D I, Gaisin R M, Maksimovich I B, Maksimovich Il B and Khodarev V V 2014 Complex diagnostics of the technical condition of a monolithic "diaphragm-wall" after long-term conservation *Soil Mechanics and Foundation Engineering* **51(4)** 196–203
[22] Hingole R S 2015 *Advances in Metal Forming: Expert System for Metal Forming* (Springer)
[23] Hastie T, Tibshirani R and Friedman J 2017 *The Elements of Statistical Learning: Data Mining, Inference and Prediction* (Springer)
[24] Draper N R and Smith H 1986 *Applied Regression Analysis* **1** (Finance and Statistics)
[25] Lvovskij E N 1982 Statistical Methods for Construction of Empirical Formulas (High school)
[26] Shinkin V N and Kolikov A P 2011 Simulation of the shaping of blanks for large-diameter pipe *Steel in Translation* **41(1)** 61–66
[27] Shinkin V N and Kolikov A P 2011 Elastoplastic shaping of metal in an edge-bending press in the manufacture of large-diameter pipe *Steel in Translation* **41(6)** 528–531
[28] Shinkin V N 2017 Calculation of bending moments of steel sheet and support reactions under flattening on the eight-roller straightening machine *Ferrous metals* **4** 49–53
[29] Belsky S M, Kotsar S L and Polyakov B A 1990 Calculation of the distribution of rolling force across the width of the strip and residual stresses in the strip by the variational method *Proceedings of higher educational institutions. Ferrous metallurgy* **10** 32–34
[30] Shopin I I and Belskiy S M 2016 Simplified model of deflected mode of coil on coilwinder *Rolling* **5** 13–17
[31] Shopin I I and Belskiy S M 2016 Lamellar model of deflected mode of coil on winder *Rolling* **8** 3–7
[32] Balueva M A, Blokhin D I, Savatorova V L, Talonov A V and Sheinin V I 2009 Simulation of microcrack influence on temperature variations within geomaterials under deformation *Journal of Mining Science* **45(6)** 576–580
[33] Maindonald J and Braun W J 2010 *Data Analysis and Graphics Using R: An Example-Based Approach* (Cambridge University Press)

Acknowledgments
The reported study was funded by RFBR, project number 19-38-90257.

IOP Conf. Series: Materials Science and Engineering **969** (2020) 012075 doi:10.1088/1757-899X/969/1/012075

Study of the Manufacturing Options for an Asymmetric Plate Made of 12X18N10T Steel

G V Shimov and N I Bushueva

Department of Metallurgy and Materials Science, Ural Federal University named after the First President of Russia B. N. Yeltsin, 19, Mira Str., Ekaterinburg, 620049, Russia

E-mail: g.v.shimov@urfu.ru

Abstract. The analysis of possible options for manufacturing an asymmetric plate of 12X18N10T steel is presented. A phenomenological model of damage accumulation was used to analyse the probability of metal micro – and macro – destruction. The results of FEM modelling of technological processes of plate manufacturing are presented. Due to these results were made the conclusions about feasibility of their application based on the metal damage calculation. The choice of a rational technology to produce a plate is justified, in which the greatest metal ductility reserve takes place at a dangerous point. According to the proposed scheme, an experimental batch of plates have been made.

1. Introduction

The production technological scheme of any shaped profile is developed depending on the material, quality requirements, volume of the produced batch, profile configuration and specific production conditions. It is important to say that determining the optimal scheme for a group profiles manufacturing or a specific shaped profile is a demanding task requiring simultaneous consideration of variety factors. There are more and more scientific studies of technologies to produce metal products using the method of damage assessment [1–4] considered in the article due to the development of old criteria and widespread use of new criteria for destruction [5–15]. This paper uses the phenomenological theory of damage to predict the probability of damage at a point.

2. Research material and methodology

The theory and practice of manufacturing shaped profiles show [16] that with increasing axes of symmetry in the cross section of the profile, the manufacturing technology becomes easier. The monosymmetric profiles containing relatively thin webs are non-technical in most of the known schemes due to the complexity of editing the finished profiles in the cold state. The analysis and selection of the most suitable technological scheme for small-scale production of a monosymmetric thin-walled highly accurate profile made of corrosion-resisting steel type 18–10 for the plate is performed (Figure 1). The length ranged from 50 to 100 mm. The work's aim was to offer a rational technology for manufacturing this plate according to the requirements of the technical conditions.

The complexity of manufacturing is that the plate must have the increased head strength section of transition from the plate to the head in the "danger zone". The reliability of the plate under reactor conditions is evaluated by both the full strength and metal ductility. The metal must remain ductile throughout the operation cycle in the reactor. That is why the metal ductility reserve is very important performance in this case.

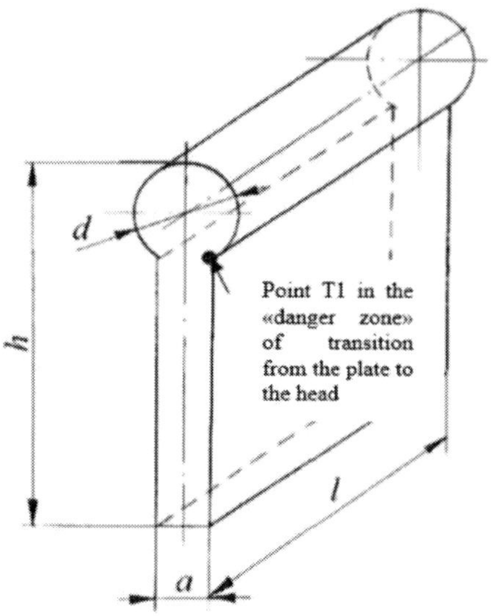

Figure 1. Shape and dimensions of the plate.

The possibility of obtaining a finished plate was evaluated during computer and physical modeling according to the methods mentioned above. Also, important to take into account the change of the accumulated degree of shear deformation, damage and metal ductility reserve in the "danger zone".

3. Option for manufacturing the plate by rolling

Flat rolling is the traditional method of producing these types of profiles. This method provides the best performance in comparison with other one. The case of manufacturing a plate by rolling from a blank with a thickness of 3 mm and a length of 50 mm, the rolls diameter $D = 130$ mm was considered. The use of a double gauge is necessary to prevent the plate from bending (Figure 2). There is a need to separate the plates after rolling, which leads to inferior surface quality of the plate head.

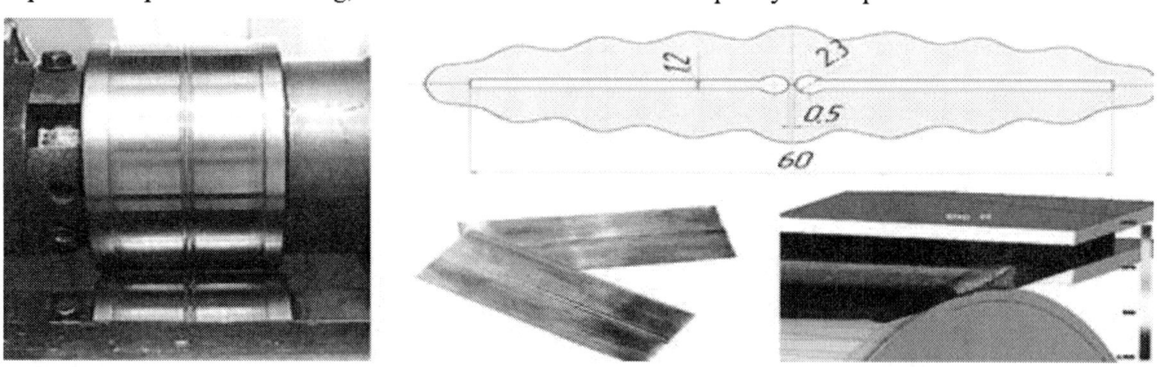

Figure 2. Gauge and sectors for rolling double plates and rolled plates on the laboratory mill Duo-130.

The FEM-modeling process problem in DEFORM-3D for the purpose of analyzing stress-strain state during rolling is solved. The 12H18N10T steel hardening curve was set tabular according to the literature data [17]. The elongation ratio across the plate λ_p is comprised 2.5. The elongation ratio across the head λ_g – 1.78. The rolling process modeling has shown that it is possible to produce a plate in one pass. Energy-power parameters of rolling are obtained as follows: rolling force $P = 1.06$ MN, pressure metal on the rolls $p = 1.89$ kN, mill torque $M = 8.59$ kN·m.

The resulting plates in shape and size fully met the requirements of the technical conditions. The difference between the rolling-out on the profile elements did not lead to a crescent-shaped strip due to the use of a double gauge. It can be concluded that the method is generally suitable for industrial use.

A phenomenological model of damage accumulation is used to analyze the probability of metal micro-and macro-destruction. The stress state indicator in the phenomenological theory of damage is used as integral characteristics σ/T, where $\sigma = (\sigma_{11} + \sigma_{22} + \sigma_{33})/3$; σ_{11}, σ_{22}, σ_{33} are principal normal stress, $T=(0.5s_{ij}s_{ij})^{1/2}$ is the intensity of the tangential stresses, s_{ij} are deviator stress tensor components). The degree of shear deformation Λ is used as a generalized characteristic of the accumulated deformation of a material particle. This parameter is calculated along the trajectory of the particle moving [18].

The metal damage at the point has the additivity property ($\omega = \sum_{i=1}^{n} \omega_i$) and is determined at the time by the formula:

$$\omega = \int_0^{\Lambda} \frac{a \cdot \Lambda^{a-1}}{\Lambda_p^a} d\Lambda, \qquad (1)$$

where a is the functional coefficient that characterizes the rate of metal damage accumulation, depending on the metal properties, Λ_p is the maximum permissible degree of shear deformation at the specified indicators of the stress state σ/T and μ_σ.

The functional coefficient a depends on the index σ/T and is calculated using the formula:

$$a = a_0^{1+\alpha\sigma/T} \qquad (2)$$

where a_0 is the coefficient that depends on the grade of steel and corresponds to the a value when $\sigma/T = 0$, α is the power factor, $\alpha \approx 0{,}238$ for all steel grades [19].

The coefficient a_0 is determined experimentally when testing for alternating torsion at different pressures. For cold rolled steel sheet 12H18N10T $a_0 = 1{,}63$ [20].

The rolling process was divided into 10 time stages and in each time point was calculated values Λ_i, ω_i и Λ_Σ, ω_Σ accumulated over the period of the deformation point. An unfavorable stress-strain state was found in the "danger zone" when analyzing the results of rolling simulation. The components of stress and strain tensors were determined to estimate the metal ductility reserve in the transition zone from the plate to the head. The stress state index σ/T and the accumulated degree of shear strain Λ at T1 are also calculated.

The metal ductility reserve due to the presence of significant tensile stresses in the "danger zone" is relatively small after rolling ($\Lambda_\Sigma = 1.92$ when $\sigma/T = 0.32$). The reserve of metal ductility is found in the corresponding diagram $\Lambda_p = f(\sigma/T)$ [18], where Λ_p is the maximum permissible degree of shear deformation Λ_p. It is composing 2.31 at $\sigma/T = 0.32$ for 12H18N10T steel. The diagram of the dependence of the accumulated degree of shear deformation on the coefficient σ/T at point T1 and the ductility diagram for 12H18N10T steel [19] are presented in Figure5. Curve 1 in Figure. 5 characterizes the dependence of Λ_p on σ/T and μ_σ, which is determined when testing samples on a high-pressure bench. Curve 2 in Figure 3 illustrates the process of accumulation in time of the degree of shear deformation Λ_i when the point T1 passes the deformation zone.

Damage at a danger point after rolling is composing $\omega_\Sigma = 0.28$. The damage value of 0,28 exceeds the first micro-fracture criterion ω^* (for steel 12H18N10T $\omega^* = 0.25$), which indicates the probability of micropores and micro-cracks in the "danger zone" [19]. The indicator σ/T is shifted to the positive region $\sigma/T = 0.32$ due to the resulting tensile stresses at the last stages of deformation (the moment of final head formation in the rolls). The metal ductility reserve is so small that there is a possibility of microcracks at the point of transition of the plate into the head of the plate in this stress-strain state scheme.

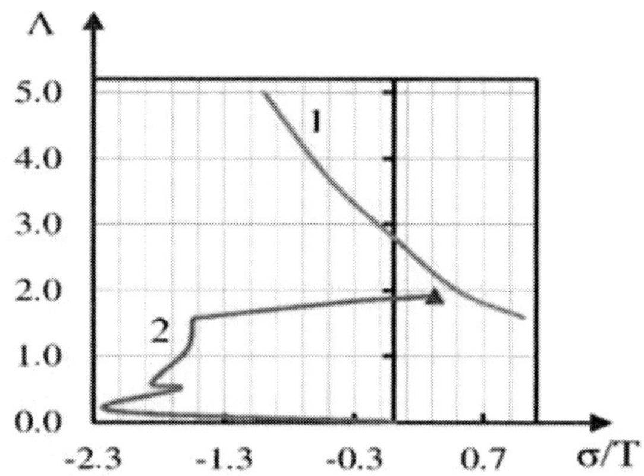

Figure 3. Dependence diagram of Λ on σ/T during rolling a plate in a double gauge: 1 – ductility diagram $\Lambda_p = \Lambda_p(\sigma/T)$ for 12H18N10T steel; 2 – dependence curve $\Lambda_i = \Lambda_i(\sigma/T)$ for point T_1.

4. Option of manufacturing the plate by heading

An alternative production technology is heading due to the shape of the plate. The FEM-modeling task of the process in DEFORM-3D was set and solved for the purpose of analyzing stress-strain state during heading. The heading process was also divided into 10-time stages (Figure 4).

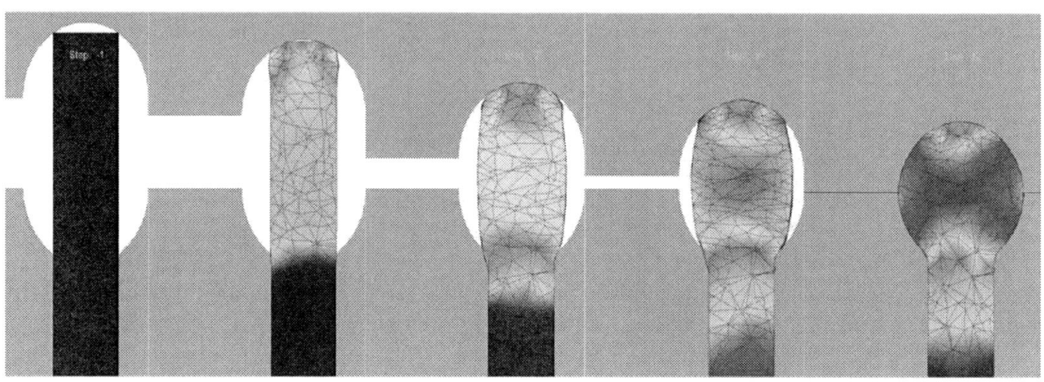

Figure 4. The stages of heading.

The diagram of the dependence of the accumulated degree of shear deformation on the indicator σ/T at point T1 and the metal ductility diagram for 12H18N10T steel [19] are presented in Figure 5. The maximum permissible degree of shear deformation Λ_p is composing 4.33 when $\sigma/T = -0.77$ for 12H18N10T steel. The metal ductility reserve in the "danger zone" is relatively large ($\Lambda_\Sigma = 0.26$ when $\sigma/T = -0.77$). This phenomenon is explained by a favorable stress-strain state scheme during sinking strain, which does not provoke an increase in tensile stresses in the "danger zone".

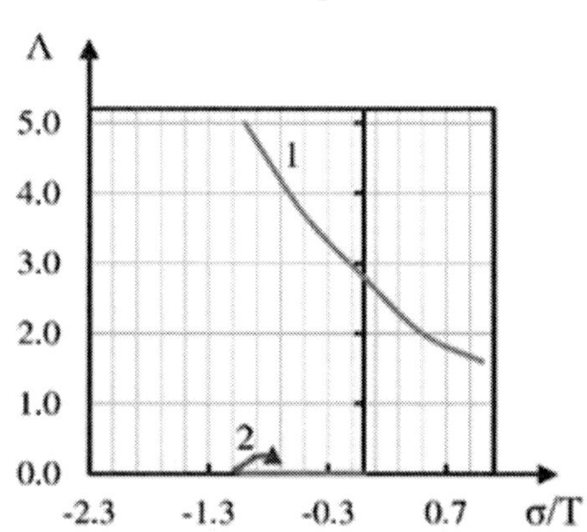

Figure 5. Dependence diagram of Λ on σ/T during heading: 1 – ductility diagram Λp = $\Lambda_p(\sigma/T)$ for 12H18N10T steel; 2 – dependence curve $\Lambda_i = \Lambda_i(\sigma/T)$ for point T_1.

The damage at the dangerous point after heading is $\omega_\Sigma = 0.029$, which is significantly less than the first micro-destruction criterion ω^*. A favorable stress-strain state scheme occurs in the hearth during almost the entire heading cycle. This scheme is close to the all-round compression scheme. The indicator $\sigma/T < 0$ varies in the range from −0.77 to −1.09. This is the stress-strain state scheme in accordance with the diagram in Figure 5 assumes "unlimited" ductility of the metal under specified conditions [18]. It is ensuring that there is no metal destruction at the T1 point and vicinity of this point during the heading process. It can be concluded that there is no damage in the form of micropores and microcracks in the "danger zone" after heading. A pilot batch of three plates was made using the proposed technology.

5. Conclusion

The phenomenological model of damage accumulation allowed to estimate the probability of micro-and macro-destruction of metal in the "danger zone" as well as the resource of the ductility for various stress-strain state schemes. It was possible to choose a rational option for making a plate based on the results of damage calculation.

The most suitable for small-scale production of the internal combustion engine mounting plate was recognized as the technological version of the heading on the finished plate. In this case, only one operation is provided – abrasive cutting of the end parts of the plate after heading.

References

[1] Botkin A V 2012 *News of Ufa: State aviation technical University* **8** 98–103
[2] Bogatov A A, Ovchinnikov A S and Salikhyanov D R 2018 *Materials Science Forum* **918** 140–144
[3] Akchurin R Z, Bogatov A A, Salikhyanov D R and Ovchinnikov A S 2018 *Metallurgist* **10** 1039–1047
[4] Bogatov A A, Lezhnev S N and Tolkushkin A O 2018 IOP *Conf. Ser.: Mater. Sci. Eng.* **461** 012086
[5] Oh S I, Chen C C and Kobayashi S 1979 ASME *j. Eng. Ind.* **101** 36–44
[6] Hambli R, and Reszka M 2002 *Int. J. Mech. Sci.* **44** 1349–1361
[7] Ogawa N, Shiomi M and Osakada K 2002 *Int. J. Mach. Tools Manuf.* **42** 607–614
[8] Komischke T and Hora P 2018 IOP *Conf. Ser.: Mater. Sci. Eng.* **461** 012041
[9] Dunand M and Mohr D 2014 *Journal of the Mechanics and Physics of Solids* **66** 133–153
[10] Mohr D and Marcadet S 2015 *Int. Journal of Solids and Structures* **67–68** 40–55
[11] Komischke T, Hora P, Domani G, Plamondon M and Kaufmann R 2018 *Procedia Manufacturing* **15** 176–184
[12] Amaral R, Teixeira P, Azinpour E, Abel D 2016 MATEC *Web of Conferences* **80** 03004
[13] Hu Q 2018 *Acta Mech. Solida Sin.* **31** 259
[14] Pesin A M, Pustovoitov D O and Lokotunina N M 2015 *Quality in the materials forming* **2** 63–67
[15] Botkin A V and Valiev R Z 2011 *Deformation and destruction of materials* **7** 17–22
[16] Vydrin V N 1977 *The Production of Shaped Profiles of High Precision* (Moscow: Metallurgy) p 184
[17] Smirnov S V and Veretennikova I A 2015 *Diagnostics, resource and mechanics of materials and structures* **4** 6–17
[18] Bogatov A A, Mijiritskii O I and Smirnov S V 1984 *Plastic Resource of Metals in Forming* (Moscow: Metallurgy) p 144
[19] Bogatov A A 2002 *Mechanical Properties and Models of Metal Fracture* (Ekaterinburg: UPI) p 329
[20] Bogatov A A 2017 *Metallurgist* **1** 70–74

Acknowledgments
The study was made within the base part of state job in the field of scientific activity № 0836-2020-002.

IOP Conf. Series: Materials Science and Engineering **969** (2020) 012076 doi:10.1088/1757-899X/969/1/012076

Investigation of Lateral Surface Contour at Forging of Billets with Upsetting Ratio Above 2.5

O Anishchenko[1], V Kukhar[1] and E Klimov[2]

[1]Pryazovskyi State Technical University, 7, Universytetska str., Mariupol, 87555, Ukraine
[2]Kremenchuk Mykhailo Ostrohradskyi National University, 20, Pershotravneva str., Kremenchuk, 39600, Ukraine

E-mail: kvv.mariupol@gmail.com

Abstract. List of the most used functions (circle, parabola, ellipse, hyperbola, cosine, hyperbolic cosine), which are applied to approximate of billets contours at upsetting and pointed to the lack of a universal equation for these purposes, was analyzed. The advantages and disadvantages of each type of equations are pointed. The paper presents the results of upsetting of lead cylindrical specimens with a upsetting ratio 2.0–4.0. The experiment shows the stages of a "double barrel" forming on the lateral surface of a billet and its subsequent transition to a single one. The authors propose to use the "superformula" recently invented by Johan Gielis as a universal equation that allows to approximate the contour of the lateral surface of a billet at all stages of upsetting. The paper presents comparative initial data necessaries to determine upsetting force and pressure, taking into account the barrel-shape calculated by the superformula, and the coefficient of friction determined by Gubkin's method. The equations coefficients have find, which give approximately the same results comparables to upsetting of specimens from lead, and these equations proposed for the calculation of the forming at upsetting from double-barrel to single-barrel workpiece.

1. Introduction

Upsetting of a billet is a common forging operation, intended primarily for the manufacture of forgings with large cross-sections from billets of smaller cross section, refining the cast structure of the ingot, pre-forming of the billet for further piercing or closed die forging. In addition, upsetting helps to receive a grain flow in the deformable metal to improve the performance properties of the finished parts, to weld hydrogen flakes, to remove scale from the surface and align the mechanical properties along the axis of the forgings [1, 2]. A limitation of the upsetting process is the limiting height H_0 to diameter D_0 ratio (upsetting ratio or aspect ratio), the excess of which leads to a bulk buckling. In the literature from the beginning of the last century to the present is copied the ratio that ensures the sustainability of the upsetting [1, 2]:

$$H_0 / D_0 \leq 2.5. \quad (1)$$

Due to limitation (1), it is often necessary to use cogged forgings as primary billets instead of cheaper rolled stocks product, because the latter with a cross-sectional size of more than 250 mm is produced in a limited amount.

2. State of the art review

The wide application of upsetting in the forming processes makes it necessary to constantly research and improving. Studies are carried out in the field of choosing the optimal conditions of contact friction between billet and dies, in particular, lubricants for upsetting [3–6], identifying the influence on the process of geometric dimensions, temperature, upsetting reduction ratio (engineering strain) and strain rate [7, 8], chemical composition of billet material [5, 9], stress state of billet [10, 11], predicting the curvature of its lateral surface and choosing the optimal preform for subsequent die forging [7, 9, 12–15]. Primary materials for research are increasingly used billets in shells, powder and bimetallic specimens [5, 16, 17].

Modern technologies, equipment and tools have long allowed to upset billets outside the inequality (1). For example, PStPÖ1Z-4000 hydraulic press of 40 MN force provides precise centering of the billet and the main hydraulic cylinder axes due to a centering device located between the press columns, similar to the interaction of a piston, engine cylinders, etc. [18–20]. As a result, the probability of buckling for billets with upsetting ratio $H_0/D_0 \leq 3.3$ is eliminated at forging. Upsetting of ingots and pre-rolled stocks with the ratio $H_0/D_0 = 2.0$–2.5 is recommended to be carried out between open dies with a concave spherical face, which centres the axis of the preformed workpiece coaxially with the press axis [1, 2]. Billeted ingots with $H_0/D_0 > 3.0$ are upset with the location of the shank in the hole of the lower die [21].

Increasing the limiting ratio H_0/D_0 allows to reduce the pressure and force at upsetting, to reduce the cost of the initial billet (the ingot is cheaper than rolled product), to increase the forging ratio and improve physical and mechanical properties of forgings. But in the literature there are not enough researches about upsetting of billets with the ratio $H_0/D_0 \leq 2.5$ [4, 13, 14]. This is due to the limited technological capabilities of processing and lack of forging theory of such ultrahigh bodies, namely the justification for the appearance and subsequent disappearance of the contour of the "double barrel" on the lateral surface of the upset billets. The evolution of the contour of billets affects the change in pressure and force during their upsetting. To calculate these parameters formulas of two types are used use (Table 1).

The upsetting force according to the first type of formulas (2)–(9) depends on the contact area of the billet with the tool F and the upsetting pressure p, which, in turn, is determined by the yield strength of the deformed material y_s multiplied by the correction coefficient C, that takes into account the diameter to height ratio H_0/D_0 and the coefficient of friction μ between the end face of the billet and the upsetting dies. In this case, for hot upsetting, the coefficient of friction a priori is taken $\mu = 0.5$. The coefficient of friction is not a constant value in the process of deformation. The spread in values of the coefficient of friction, determined by many different methods, is significant for the same material with the same geometric shape. For upsetting operations, most often used values of the friction coefficient are those that are determined by the ring compression test (upsetting of ring specimens), which does not provide a geometric similarity of the method to the considered process of upsetting of cylindrical billets and, therefore, casts doubt on the correspondence of the calculated value of the coefficient to its real value during the upsetting of a cylindrical billet.

Assignment to the coefficient of friction of its maximum value equal to 0.5 is the upper estimate of the upsetting force and is justified in the upsetting of ingots without lubrication. In the technological process of hot forging, in which one of the transitions can be, for example, upsetting of a billet with a layer of glass lubricant [21], the values of the friction coefficient can be much lower than 0.5. In this case, the calculated values of the upsetting force will be significantly higher than the real values. The end surface area of the billets F in the first type of equations (2)–(9) is overestimated, since it is determined by the diameter calculated without taking into account barrel formation, i.e., exceeding the value of the end diameter under conditions of real upsetting. The yield strength y_s is usually selected from reference books in which its value is fixed according to the results of tensile tests, while it would be more correctly to apply the conditional yield stress $y_{0.2}$ determined by compression tests. In addition, hardening of the metal during the upsetting process, especially in the cold state, is not always taken into account. At the same time, the friction forces arising between the billet and the die at upsetting determine one or another barrelling shape of the upset billet, characterized primarily by the radius of curvature R, which is determined by the diameters of the barrel D_b and the ends D_m of the billet. These values are quite accurately measured experimentally.

Table 1. Equations for calculating the C coefficient [22–26].

Equations that take into account the coefficient of friction μ, the friction factor m (μ=m) and the geometric dimensions of the billets	
$C=1+\dfrac{\mu D}{3H}$ (2)	$C=\left(1+\dfrac{\mu D}{3H}\right)^2$ (3)
$C=\dfrac{1}{2}\left(\dfrac{\mu D}{H}\right)^2\left[\exp\left(\dfrac{\mu D}{H}\right)-\left(\dfrac{\mu D}{H}\right)-1\right]$ (4)	$C=\left[1+\dfrac{\mu D_0}{H_0}\exp\left(\dfrac{3\varepsilon}{2}\right)\right]^{-1}$ (5)
$C=1+\dfrac{\mu D}{H3\sqrt{3}}$ (6)	$C=\dfrac{\mu D_0\sqrt{H_0}}{3H^{3/2}}$ (7)
$C=\dfrac{H_0}{H}\left(1+\dfrac{m}{3\sqrt{3}}-\dfrac{2R_0}{H}\sqrt{\dfrac{H_0}{H}}\right)$ (8)	$C=\dfrac{8bR_{ideal}}{H}\left\{\left[\dfrac{1}{12}+\left(\dfrac{H}{bR_{ideal}}\right)^2\right]^{3/2}-\left(\dfrac{H}{bR_{ideal}}\right)^3+\dfrac{m}{24\sqrt{3}\left(e^{b/2}-1\right)}\right\};$ $b=\dfrac{4mH}{R_{ideal}\sqrt{3}}\left(1+\dfrac{2mH}{R_{ideal}3\sqrt{3}}\right)$ (9) R_{ideal} – radius of the specimen upsetted without a barrel ($\mu=0$)
Equations that take into account the diameters of the ends, "barrels" and the radius of curvature of the contour of billets lateral surface	
$C=\left[\left(1-\dfrac{2R}{R_b}\right)\ln\left(1-\dfrac{R_b}{2R}\right)\right]^{-1}$ (10)	$C=\left[\left(1-\dfrac{2R}{D_b}\right)\ln\left(1-\dfrac{D_b}{2R}\right)\right]^{-1}$ (11)
$C=\left[\left(1-\dfrac{4R}{D_b}\right)\ln\left(1-\dfrac{D_b}{4R}\right)\right]^{-1}$ (12)	$C=\left[\dfrac{\left(D_b^2-4RD_m\right)\ln\left(1-\dfrac{D_b}{4R}\right)}{D_b^2}\right]^{-1}$ (13)

In the equations of the second type (10)–(13) the use of parameters D_b and D_m obtained by direct measurements, instead of the value of friction coefficient calculated from the results of indirect measurements without full compliance with scale and geometric similarity, will increase the accuracy of calculations parameter C and, accordingly, upsetting force and pressure. However, when calculating the radius of curvature R according to the equation:

$$R = \frac{\left[1+(y')^2\right]^{3/2}}{y''} \tag{14}$$

where y', y'' – the first and second derivative functions approximating the contour of the meridional generating lateral surface of the upset billet, it is necessary to know the type of function that approximates the contour of the lateral surface of the "barrel". Small inaccuracies in the choice of function can lead to large spread in value of R, since the first and second derivatives of the approximating function are part of the equations (10)–(13).

Today, researchers intuitively assume in advance one or another type of contour of the lateral surface of the upset billet, and hence the type of approximating function (Table 2), and then select the appropriate values of the coefficients included in this function. That is, for the upsetting of billets there is no universal function, which is able to describe any of the possible contours of the lateral surface of the deformed billets (in particular, contour of the "double barrel").

Table 2. Some equations for approximating the barrel-shaped lateral surface of the billet after upsetting.

Equation	Designation
Parabola: $y = \frac{2(D_m - D_b)}{H^2}x^2 + \frac{D_b}{2}$	D_b, D_m, H – diameters of a barrel, an end face and height of the billet
Ellipse: $\frac{x^2}{a^2} + \frac{y^2}{b^2} = 1$	$a = \frac{H}{2\sqrt{1-\left(\frac{D_m}{D_b}\right)^2}}$, $b = \frac{D_b}{2}$
Circle: $x^2 + (y-b)^2 = R_{circle}^2$	$b = \frac{D_b^2 - D_m^2 - H^2}{4(D_b - D_m)}$, $R_{circle} = \frac{(D_b - D_m)^2 + H^2}{4(D_b - D_m)}$
Cosine: $y = a\cos(bx)$	$a = \frac{D_b}{2}$, $b = \frac{2}{H}\arccos\left(\frac{D_m}{D_b}\right)$
Hyperbolic cosine: $y = a\cosh(bx)$	$a = \frac{D_m}{2}$, $b = \frac{2}{H}\operatorname{arg\,cosh}\left(\frac{D_b}{D_m}\right)$
Hyperbola: $-\frac{x^2}{a^2} + \frac{y^2}{b^2} = 1$	$a = \frac{H}{2\sqrt{\left(\frac{D_b}{D_m}\right)^2 - 1}}$, $b = \frac{D_b}{2}$

Purpose of the research is the process of "barrelling" during upsetting of billets with upsetting ratio $H_0/D_0 = 2.0$–4.0, identification of a universal equation for approximating the barrel-shaped contours of upset billets, and the most acceptable formulas for calculating the upsetting force and pressure.

IOP Conf. Series: Materials Science and Engineering **969** (2020) 012076 doi:10.1088/1757-899X/969/1/012076

3. Materials and methodology of research

The research was performed on lead billets with a diameter of D_0 = 30 mm with upsetting ratio H_0/D_0 = 2.0–4.0. The upsetting was carried out without lubrication between flat dies with a stroke rate of 2 mm/s. The coefficient of friction between the billet and dies (by the Gubkin method) was $\mu = 0.41$. The upsetting reduction ratio determined by the equation:

$$\varepsilon = (H_0 - H)/H_0. \tag{15}$$

On the lateral surface of the billets at a distance along the height equal to 0.1 from the original height H_0, performed circular lines. The curvature of the contour of the lateral surface at upsetting was studied by the results of changing the diameter of these lines during the gradual upsetting of billets with upsetting reduction ratio of 0.1 to 0.5 with a step of 0.1.

In the paper [27] there is information that a huge number of different contours are quite accurately described by the superformula of J. Gielis. In the polar coordinates for the radius vector r and the angle φ the Gielis superformula has the form:

$$r = \frac{1}{\sqrt[n_1]{\left[\left(\left|\frac{1}{\alpha}\cos\left(\frac{k}{4}\phi\right)\right|\right)\right]^{n_2} + \left[\left(\left|\frac{1}{b}\sin\left(\frac{k}{4}\phi\right)\right|\right)\right]^{n_3}}} \tag{16}$$

where n_1, n_2, n_3 – parameters that determine the shape of the curve; a, b – dimensions (values of semi-axes); k – coefficient that characterizes the number of repeating fragments.

In forming processes, superformula (16), as well as its simplified variant – superellipse Lame – has already been successfully used as universal equations to approximate the contours of deformable sheet billets and further analysis of their stress-strain state [28–30]. Regarding the process of billet upsetting, previous attempts to approximate the superformula have shown that the selection of coordinate system is important for the possibility and accuracy of the approximation.

Figure 1 shows a diagram of the selection of the coordinate system of the approximating function relative to the geometric dimensions of the deformed barrel-shaped billet. According to this scheme, the approximation of experimental data can be performed in a system of dimensional and dimensionless coordinates. If the contour of the lateral surface of the upset billets is a "single barrel", then a system of dimensionless coordinates $y0x$ with measurement limits on both axes from zero to one is chosen. The x-axis is parallel to the axis of the billet and coincides with the generatrix, which forms a cylinder with a diameter equal to the diameter of the end face D_m. The y-axis intersects the x-axis in the plane of the maximum barrel diameter of billet D_b. The dimensionlessness of the coordinates Xi and Yi was ensured by dividing according to the measured real size h_i (mm) by the maximum possible size, which is equal to half the height of the upset billet $H/2$, i.e. $X_i = h_i/(H/2)$, as well as the measured real size:

$$d_i = (D_b^i - D_m)/2 \tag{17}$$

when $x = h_i$ at the maximum possible size equal to the half-difference between the diameters of the "barrel" and the end face of the billet, i.e.

$$Y_i = \frac{(D_b^i - D_m)/2}{(D_b^{\max} - D_m)/2}. \tag{18}$$

In this coordinate system ($y0x$) the contour of the "single barrel" is circumscribed by the radius vector r of the superformula when the angle φ changes within [0; 90°]. If the contour of the lateral surface of billets at upsetting evolves from "double" to "single barrel", the coordinate system $y0_1x_1$ is chosen, in which the x_1 axis coincides with the central axis of the billet, and the dimensionlessness of the Y_i

IOP Conf. Series: Materials Science and Engineering **969** (2020) 012076 doi:10.1088/1757-899X/969/1/012076

coordinate is provided by dividing the measured real radius of the barrel along the axis of the billet by the maximum possible one, i.e.:

$$Y_i = \frac{D_b^i / 2}{D_b^{\max} / 2} = D_b^i / D_b^{\max} . \quad (19)$$

4. Analysis of results

Figure 2 shows the billets with a "double barrel" on the lateral surface after upsetting, and in Figure 3 shows the results of increasing the diameter of the circular lines D relative to the original diameter D_0 depending on the height of the billets H, which decreases at upsetting. The location of the circular lines was measured by the parameter f – the distance from the lower end to the circular line, in particular, ratio $f/H = 0.5$ for the line bisecting the specimen height H.

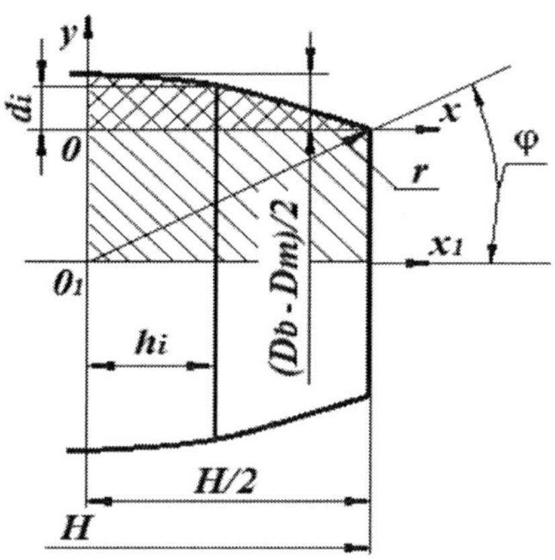

Figure 1. Diagram for selecting the coordinate system of the approximating function.

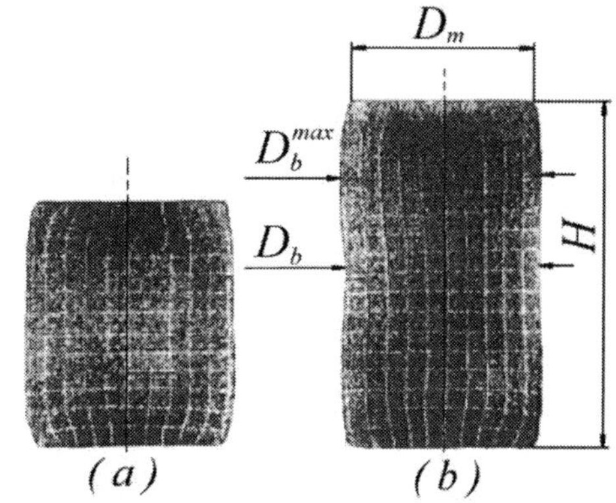

Figure 2. Appearance of deformed semi-finished products with a double-barrel on the lateral surface upsetted from the billet with initial dimensions ratio: $H_0/D_0 = 2.5$ (*a*); $H_0/D_0 = 3.5$ (*b*).

Figure 3(*b-d*) indicates that "double barrelling" occurs when billets with a ratio $H_0/D_0 > 2$ are upset. In addition, the "double barrel" is fixed at upsetting reduction ratio of the specimens not more than 0.3. When $H_0/D_0 = 2$ upsetting of billets increase the ratio D/D_0 (at $f/H = 0.5$) in 1.05–1.47 times with increasing ε within 0.1…0.5. Increasing the upsetting ratio to $H_0/D_0 = 2.5$ also increases the maximum diameter of the "barrel" ($D/D_0 = 1.49$). In addition, in the upsetting reduction ratio $\varepsilon = 0.1$–0.3, a "double barrel" appears on the lateral surface of the billets. With increasing ε from 0.1 to 0.4, the maximum diameter ratio of the "barrel" is shifted in the area from the ends to the center of the billet and is fixed at ratio $f/H = 0.78$, 0.77, 0.73 and 0.68, respectively, for billets with upsetting ratio $H_0/D_0 = 2$; 2.5; 3; 3.5. The difference between the maximum diameters ratio of the "double barrel" and the diameter in the central part of the specimen (at ratio $f/H = 0.5$) is insignificant and equal to $\Delta(D/D_0) = 0.01$–0.04. The "double barrel" for upset billets with $H_0/D_0 = 3$ is already almost indistinguishable at $\varepsilon \geq 0.3$. The maximum diameters ratio of the "barrel" is fixed in the range $f/H = 0.76$–0.74, and the maximum difference $\Delta(D/D_0) = 0.02$ at $\varepsilon = 0.2$. It is noticeable that the circular zones, which stand from the end at a distance of $0.1N$ with increasing ε become the end surface of billets. Thus, if lines 1-4 have five experimental measurements of diameters D, then line 5 has only four measurements. The increase in the upsetting ratio of the billets to $H_0/D_0 = 3.5$ (Figure 3(*d*)) with values D/D_0at upsetting almost does not differ from the graphs (Figure 2(*c*)) for upsetting of billets with $H_0/D_0 = 3$. In particular, "double barrel" is fixed in the range of upsetting reduction ratio $\varepsilon = 0.1$–0.3 and by the parameter $\Delta(D/D_0)$ is also quite insignificant. As for upsetting of billets with $H_0/D_0 = 4.0$, then at $\varepsilon = 0.08$ they lost resistance and buckled so that their further deformation was impossible.

The superformula that approximates the contour with a "double barrel" in the coordinates $y0_1x_1$, reproduces the same contour instead of a straight line on the end surface of the billet. Since the end is flat, the approximate values of the superformula coefficients when changing the angle φ from 0 to arctg[$(D_m/2)/(H/2)$] are not valid, and the calculation of the contour of the "double barrel" should be performed when changing the angle φ within {arctg[$(D_m/2)/(H/2)$]; $\pi/2$}.

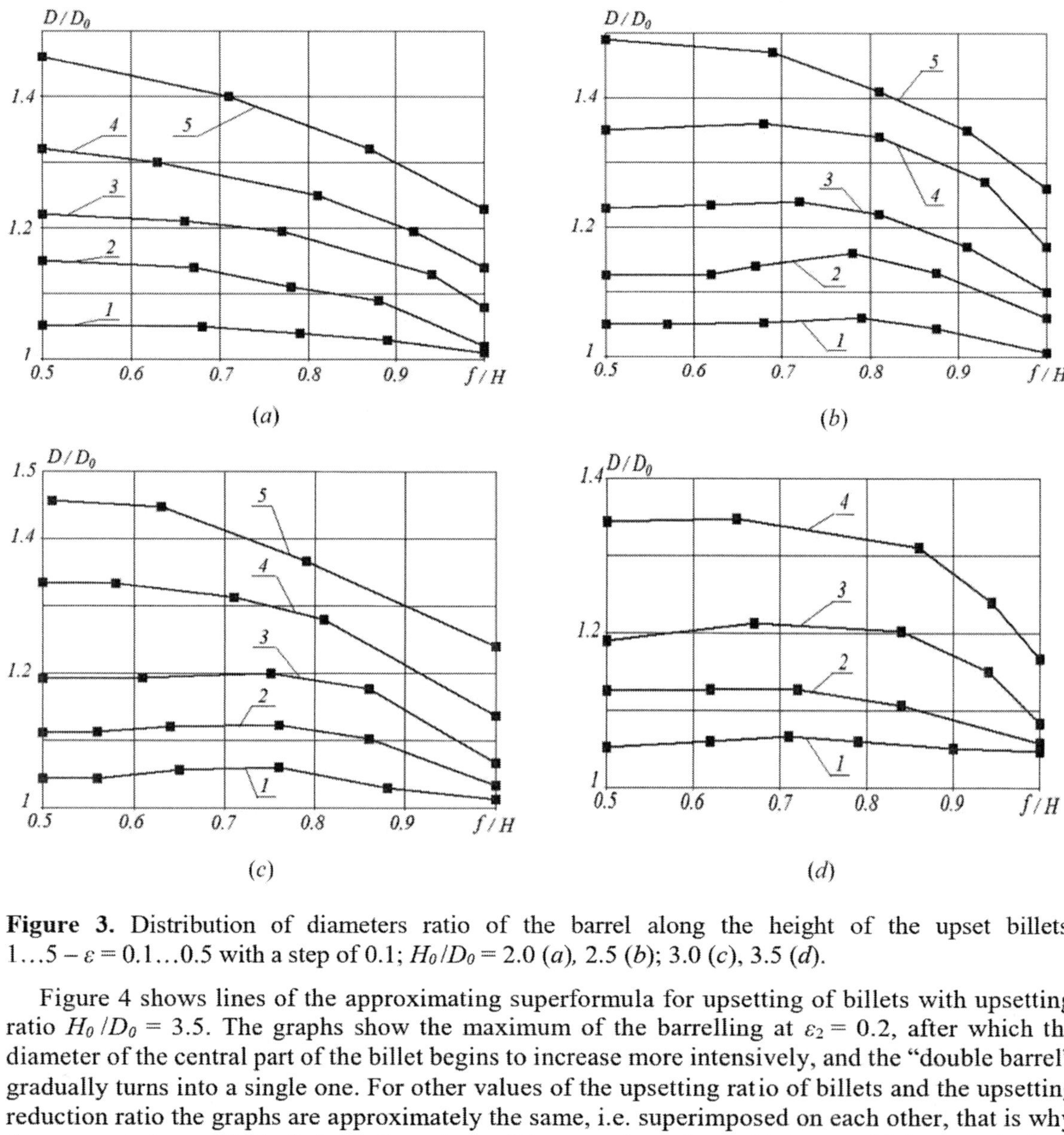

Figure 3. Distribution of diameters ratio of the barrel along the height of the upset billets: 1…5 – ε = 0.1…0.5 with a step of 0.1; H_0/D_0 = 2.0 (*a*), 2.5 (*b*); 3.0 (*c*), 3.5 (*d*).

Figure 4 shows lines of the approximating superformula for upsetting of billets with upsetting ratio $H_0/D_0 = 3.5$. The graphs show the maximum of the barrelling at $\varepsilon_2 = 0.2$, after which the diameter of the central part of the billet begins to increase more intensively, and the "double barrel" gradually turns into a single one. For other values of the upsetting ratio of billets and the upsetting reduction ratio the graphs are approximately the same, i.e. superimposed on each other, that is why they are not shown. Table 3 shows the values of the superformula coefficients, which were obtained by approximating the generating lateral surfaces of the upset billets, including "single" and "double barrel". The calculations followed the recommendations of Johan Gielis [27] to assign (for cases similar to ours) the value of the coefficient $k = 4$. However, later our calculations showed that the change in the values of k in the range from 3 to 5, which already determined other values of the coefficients $n_1 \dots n_3$, can provide an approximation of experimental data with less error. But such a result can be guaranteed only in the first quadrant of the Cartesian coordinate system.

Table 3. Values of the superformula coefficients.

Upsetting ratio of billets, H_0/D_0	Upsetting reduction ratio, ε	Value of the superformula coefficients		
		n_1	n_2	n_3
2.0	0.1	1.4 (0.7)	13.5 (2.1)	1.3 (1.6)
	0.2	33 (2.1)	34 (2.1)	25.4 (1.4)
	0.3	2.1 (8.9)	12.3 (0.5)	1.6 (0.7)
	0.4	1.6 (2.2)	15.4 (1.2)	1.1 (1.0)
	0.5	6.1 (1.2)	66.2 (1.4)	3.7 (1.2)
2.5	0.1	2.7	10	2.6
	0.2	2.2	13.8	2
	0.3	1.8	8.1	1.7
	0.4	1.8	7.5	1.7
	0.5	2.3	11.8	1.6
3.0	0.1	7.9	11.9	7.8
	0.2	8	10.5	7.9
	0.3	3.1	7.5	3
	0.4	7.6	11.4	5.8
	0.5	3.4	15.9	2.2
3.5	0.1	3.7	19.9	3.5
	0.2	1.5	10.7	1.4
	0.3	3.4	7.3	3.4
	0.4	2.4	7.6	2.2
	0.5	3.3	11.5	2.3

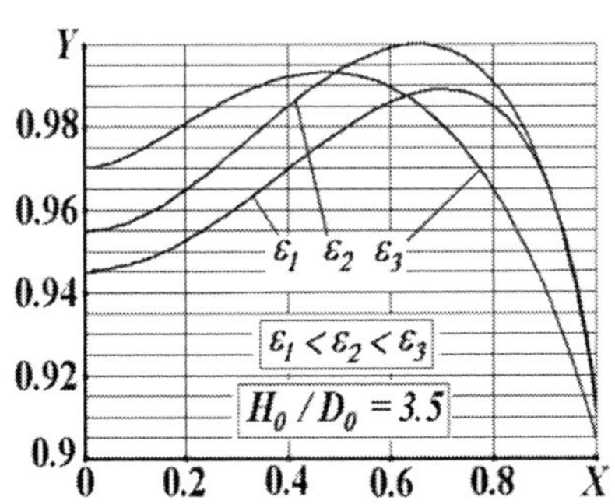

Figure 4. Approximation of the contour of the double barrel on the lateral surface of the upset billet with $H_0/D_0 = 3.5$; $\varepsilon_1 = 0.1$; $\varepsilon_2 = 0.2$, $\varepsilon_3 = 0.3$.

The coefficients $n_1 \dots n_3$ of the superformula were used to determine the parameters C according to equations (10)–(13), as well as their comparison with the same parameters in equations (2), (3). As a result, it was found that equation (2) coincides with formulas (12), (13), if μ = 0.25 and 0.43, and equation (3) – if μ = 0.20 and 0.34. Therefore, equation (13) is not suitable to determine C, because the values of $\mu = 0.34$, 0.43, as well as $\mu = 0.41$, are overrated. Probably the most suitable for the calculation of upsetting pressure and force are formulas (2) and (12). The values of the approximating coefficients $n_1 \ldots n_3$ in quotation marks shown in table 3 for billets with a upsetting ratio $H_0/D_0 = 2$, i.e. upset without a "double barrel", obtained for the case when the coordinates $y0x$ were selected, and the ordinate Y_i was calculated by equation (18).

5. Conclusions

Upsetting of cylindrical billets with the ratio $H_0/D_0 > 2$, at the initial stage creates the contour of the "double barrel" on their lateral surface, which at a upsetting reduction ratio $\varepsilon > 0.3$ turns into a "single barrel". The difference between the maximum diameters ratio of the "double barrel" and the diameter in the central part of the specimen is quite insignificant and equal to 0.01–0.04. To approximate all possible contours of the lateral surfaces of upset billets, in particular, "single" and "double barrel", it is advisable to use the universal superformula of J. Gielis. The most acceptable equation for calculating the correction parameter C in the equation to determine upsetting pressure and force, which uses the radius of curvature of the lateral surface of a billet.

References

[1] Dineshbabu C, Arivazhagan R, Venkatesh R, Balasubramani K and Periyasamy R 2020 *Materials Today: Proceedings* **21** 601–611
[2] Petrov M, Kalpin Y and Petrov P 2020 *Procedia Manufacturing* **47** 1504–1511
[3] Ragab A R 2002 *Materials Science and Engineering* **A334** 114–119
[4] Kulkarni K M and Kalpakjian S 1969 *Journal of Engineering for Industry* 743–754
[5] Kumar P, Ranjan R K, Singh D and Singh V 2019 *AIP Conference Proceedings* **2142** 170023
[6] Egea A J S, Martynenko V, Abate G, Deferrari N, Krahmer D M and López de Lacalle L N 2019 *The International Journal of Advanced Manufacturing Technology* **102** 1623–1633
[7] Wang X, Li H and Sandrashekhara K 2017 *Journal of Materials Processing Technology* 1–24
[8] Anishchenko A, Kukhar V, Artiukh V and Arkhipova O 2018 *MATEC Web of Conferences* **239** 06006
[9] Mahesh E S, Jyothi K, Rao K V and Murty M S 2015 *International Journal of Scientific and Research Publications* **5(10)** 1–9
[10] Sahu M K, Valarmathi A, Baskaran S, Anandakrishnan V and Pandey R K 2014 *Proceedings of the Institution of Mechanical Engineers, Part B: Journal of Engineering Manufacture* **228(11)** 1501–1507
[11] Krallics G, Ree A, Bobor K and Szombathelyi V 2013 *Materials Science Forum* **752** 85–94
[12] Ashtiani H R R and Shahsavari P 2018 *Prediction Journal of Advanced Materials and Processings* **6(1)** 29–45
[13] Kukhar V, Balalayeva E, Prysiazhnyi A, Vasylevskyi O and Marchenko I 2018 *MATEC Web of Conferences* **178** 02003
[14] Markov O E, Gerasimenko O V, Shapoval A A, Abdulov O R and Zhytnikov R U 2019 *International Journal of Advanced Manufacturing Technology* **103(5-8)** 3057-3065.
[15] Usami M and Oya T 2014 *Procedia Engineering* **81** 371–376
[16] Essa K, Kacmarcik I, Hartley P, Plancak M and Vilotic D 2012 *Journal of Materials Processing Technology* 212 (4) 817–824
[17] Egorova R V, Egorov M S and Skorikov A V 2013 *Metallurgist* **57(5-6)** 442–448
[18] Korol S, Moroz M, Yelistratov V and Moroz O 2019 *IEEE International Conference on Modern Electrical and Energy Systems (MEES)* 30–33
[19] Korol S O, Moroz M M, Korol S S and Serhiienko S A 2017 *Scientific Bulletin of the National University University* **5** 56–61
[20] Chernenko S, Klimov E, Chernish A, Pavlenko O and Kukhar V 2018 *International Journal of Engineering and Technology(UAE)* **7(4)** 120–124

[21] Anishchenko A S, Feofanov Yu V and Bogun A B 1992 *Chemical and Oil and Gas Engineering* **11** 33–35
[22] Altan T, Ngaile G and Shen G 2005 *Cold and hot forging: Fundamentals and Applications* (Ohio: Materials Park, ASM International) p 333
[23] Torrente G 2018 *Materials Research* **21(4)** 1–7
[24] Zhou X and Li Z 2006 *Journal of Engineering Mechanics* **132(2)** 149–157
[25] Aluko O and Adeyemi M B 1998 *Journal of Materials Engineering and Performance* **7(4)** 474–478
[26] Christiansen P, Martins P A F and Bay N 2019 *Manufacturing Engineering Education* **4** 85–104
[27] Gielis J A 2003 *American Journal of Botany* **90(3)** 333–338
[28] Anishchenko A, Kukhar V, Artiukh V and Arkhipova O 2018 *MATEC Web of Conferences* **239** 06007
[29] Anishchenko O S, Kukhar V V, Grushko A V, Vishtak I V, Prysiazhnyi A H and Balalayeva E Yu 2019 *Materials Science Forum* **945** 531–537
[30] Anishchenko O, Kukhar V, Artiukh V, Trebukhin A and Zotkina N 2020 *Advances in Intelligent Systems and Computing* **982** 809–817

IOP Conf. Series: Materials Science and Engineering **969** (2020) 012077 doi:10.1088/1757-899X/969/1/012077

Integrated Diagnostic Method Analysis of Use in the Inspection of Rolling Mills

S N Rednikov[1], E N Akhmedyanova[2] and K T Akhmedyanova[3]

[1]South Ural State University, 76, Lenin prospekt, Chelyabinsk, 454080, Russia
[2]South Ural State Agrarian University, 13, Gagarin street, Troitsk, 457100, Russia
[3]Kazan State Power Engineering University, 51, Krasnoselskaya str., Kazan, 420066, Russia

E-mail: srednikov@mail.ru

Abstract. Organizing maintenance of equipment in accordance with its technical condition is a way to reduce production costs in the metallurgical industry. Equipment condition assessment is possible by various methods, but one of the most promising is the method of monitoring the thermal fields of devices. The method of monitoring the equipment temperature fields is not new and is often used for diagnostics of metallurgical equipment. This article analyzes the economic efficiency of the applied method in comparison with the method of vibration diagnostics, control current signal, a method of condition monitoring of hydraulic fluids. The authors describe their experience in equipment diagnostics.

1. Introduction

Creating modern technological units, designers strive to increase the unit capacity of machines in an effort to reduce the cost of producing a production unit. But the growth of specific indicators leads to the creation of products with limiting characteristics and a slight change in operating parameters increases the number of equipment failures, which means that equipment downtime and reduces production efficiency. Thus, an increase in pressure in hydraulic control systems leads to a reduction in their service life and requires improving the quality of cleaning the working fluid. The use of advanced technologies does not guarantee itself an increase in production efficiency. You can significantly increase efficiency by properly organizing the operation of systems and using the optimal maintenance system of metallurgical production. Under the performance indicator use methods of diagnosis in this work is the ratio of the costs of diagnostic procedures using existing certified methods, the costs for applying the developed methods of complex diagnostics primary with a medium probability of defects to be detected rolling mills.

The method of thermal fields analysis has a high efficiency index at a low cost of diagnostics[1, 2, 3, 4, 5, 6, 7, 8]. Its application in a complex of methods of vibration diagnostics, acoustic vibration assessment, analysis of test signals, analysis of wear products increases the level of forecast reliability.

2. Problem

Considering the accidents that occur at metallurgical plants in our country and abroad, we can see that the distribution of accidents is uneven by type of equipment and by primary causes. An analysis of accidents at metallurgical enterprises in our country shows that they are mainly associated with a violation of the functioning of cooling and control systems. Since the problem of equipment failures exists and to increase production efficiency, it is necessary to introduce advanced technologies and improve maintenance, so it is necessary to consider the types of maintenance and their use. Since the 18th century, the main thing in the metallurgical industry was repair by failure, which was typical for small enterprises of a periodic cycle or for equipment with a small and chaotic load.

In some cases, this approach is still found today. The maintenance method, in which equipment is repaired or replaced only if it fails or runs out of service, is often referred to as reactive maintenance.

Reactive maintenance has a number of disadvantages: the possibility of unscheduled downtime due to sudden equipment failures, and expensive and lengthy repairs due to the severity and extent of defects. At the same time, this approach did not allow to plan repair activities and optimize the repair personnel, to plan the spare parts fleet with high efficiency, since there is a possibility of sudden failure of several different units at the same time, so the need for repair work may exceed the capabilities of the repair service of the enterprise. In addition, an emergency failure of one element in metallurgical equipment is often accompanied by an avalanche destruction of serviceable elements in the process chain, and this leads to significant production costs.

3. Solution approach

The thermal type of non-destructive testing is based on the registration of changes in the thermal or temperature fields of controlled objects. It is applicable to objects made of any materials. According to the nature of the interaction of fields, there are methods of passive or natural radiation (the object is not affected by an external energy source) and active (the object is heated or cooled from an external source). The measured informative parameter is the temperature or heat flow.

When monitoring by the passive method, heat flows or temperature fields of working objects are measured in order to determine malfunctions that occur in the form of places of increased heating.

Thermal imaging diagnostics consists of two stages. The first is thermal imaging. The second part-image processing and analysis of anomalies and defects-is the most important part and requires high qualification. At this stage, images are interpreted, object temperatures are determined, and the reasons for temperature deviations from the norm are analyzed. To do this, determine the temperature of objects based on the results of thermal imaging and compare with the standard temperatures or thermal fields, determine the cause of losses, temperature deviations from the norm.

The solution of two and three-dimensional heat conduction equations by modern computing systems has been repeatedly described in [1, 8, 9, 10, 11, 12, 13]. For an axisymmetric problem, a solution in a cylindrical coordinate system is often sufficient. The system of equations in this case can be presented in the classical version:

$$(1-s)\rho_k c_k \frac{\partial t_k}{\partial \tau} = \alpha_v (T-t) + \lambda_e \left(\frac{\partial^2 t}{\partial R^2} + \frac{1}{R}\frac{\partial t}{\partial R}\right) \tag{1}$$

$$(-s)uc_g \frac{\partial T}{\partial x} = \alpha_v (T-t) \tag{2}$$

boundary condition: $\tau = 0; t = t(x,R)$, $R = 0; \frac{\partial t}{\partial R} = 0; T = T_g$, $x = H; \frac{\partial t}{\partial x} = 0$, $R = R; k(t-t_b) = \lambda_e \frac{\partial t}{\partial R}$.

where: α_v — given heat transfer coefficient; s - parameter of the heat flow direction; $\rho_k c_k$— the material density and heat capacity; λ_e — the material thermal conductivity; c_g — heat capacity of the gas at the border; k — heat transfer coefficient to the environment; R, R_n — respectively, current and external radii; H —height; t —material temperature; u — coolant velocity; t_b — ambient temperature; T, T_g— coolant temperature at the border; τ — time.

IOP Conf. Series: Materials Science and Engineering **969** (2020) 012077 doi:10.1088/1757-899X/969/1/012077

Methodology of integrated primary diagnostics was successfully applied to the inspection of the equipment in the rolling mill at Ural steel (Novotroitsk). We used complex diagnostics to assess thermal fields with the calculation of temperature distribution in the volume of the most important equipment, vibration diagnostics, current diagnostics, assessment of contamination and chemical stability of working environments.

The cost of work on metallurgical equipment technical diagnostics includes: the cost of materials used; salary of diagnosticians; deductions from the payroll; the electricity cost; overhead costs; other expenses. When determining the cost of materials and electricity used in the diagnosis process, the current prices are applied and reasonable amounts of expenditure are laid down. The payroll is calculated based on the analysis of real time expenditures or according to the time standards for carrying out diagnostic activities, taking into account the hourly rate of diagnosticians. Thus, according to [14, 15, 16, 17, 18, 19, 20, 21], the standard time for analyzing documentation per unit of equipment is 0.8 man-hours, 1.8 man-hours are allocated for analyzing accidents, 8 man-hours are allocated for evaluating the remaining resource, and 10 man-hours are allocated for preparing a technical conclusion. These time expenditures are common both for complex diagnostics and for diagnostics using vibration analysis methods, thermal fields, current methods, acoustic methods, and when using equipment condition assessment by analyzing the state of working fluids.

The analysis of actual time spent directly on carrying out measures for diagnostics of metallurgical equipment by methods of analysis of external thermal fields showed the average time spent per unit of equipment in the amount of 0.75 man-hours. Moreover, the work included preparing and configuring the device, removing the distribution of external temperature fields, fixing the results of measurements, and analyzing the results of thermography.

Thermal imaging systems showed the lowest diagnostics cost with defects, high detection probability (no less than 0.67) and minimal time expenditures. High time costs and higher diagnostics cost, but with a higher (0.72) detection probability of defects, were demonstrated by vibration diagnostics methods. Methods for assessing the state of units for working fluids, analysis and lubricants using the spectrometric method when studying the maximum cost gave a reliable minimum time forecast.

The applied control by the current signal, by the reaction mechanisms, methods, ultrasonic control were used selectively and gave intermediate results. When using the greatest efficiency was achieved at the preliminary stage with vibroacoustic and current diagnostics by methods of further refinement.

An analysis of the time spent on diagnosing a unit of metallurgical equipment by various methods of assessing the state of aggregates is given in table 1 .

To determine the efficiency coefficient, it is necessary to find the cost of conducting diagnostic measures. The typical scope of activities performed during diagnostics includes analysis of technical documentation for the object being diagnosed, visual monitoring of the state of aggregates, directly carrying out diagnostic activities, analysis of the results of diagnostics and evaluation of the residual resource of the object, preparation of a technical conclusion. Analysis of the tables shows a twofold reduction in the required time spent on diagnostics using thermal fields compared to other types of diagnostic measures at a lower cost of diagnostic measures.

To assess the effectiveness of the applied methods of technical condition assessment, it is necessary to consider the probability of detecting defects during diagnostic measures. When carrying out measures to diagnose the state of metallurgical equipment, statistics of confirmed and not confirmed by control disassembly of equipment and failures of defects were recorded. Based on the ratio of detected anomalies and confirmed defects during monitoring, a conclusion was made about the probability of establishing a correct diagnosis of the equipment condition assessment. Diagnostic measures were performed by a single group of diagnosticians, which eliminated the need to introduce an amendment to the different probability of detecting defects with different qualifications of personnel. Diagnostic measures were performed at monthly intervals.

Table 1. Time spent on diagnostic measures.

	Vibration diagnostics	Analysis of external thermal fields	Current signal analysis	Analysis of working fluids
Time spent working with technical documentation	20.6	20.6	20.6	20.6
Time spent on performing diagnostic measurements	10		7	4
The time needed for processing of measurement results	10	0.75	5	10
Time spent on analyzing the received data	8.1		8.1	8.1
Subtotal	48.7	21.35	40.7	42.7
Standard time	56.98	24.98	47.62	49.96

Table 2. Probability of detecting defects.

Months before the audit of mechanisms	Vibration diagnostics	Analysis of external thermal fields	Current signal analysis	Analysis of working fluids
5	0.2	0.1	0	0.03
4	0.49	0.35	0.1	0.05
3	0.72	0.67	0.4	0.2
2	0.92	0.86	0.605	0.41
1	0.93	0.9	0.74	0.8

4. Method of solution

The diagnostic interventions efficiency ratio was determined by the study in progress. A primary complex external field analysis method was applied. The efficiency is compared with the most effective of the certified methods - the vibration diagnostic method. During the research, the coefficient of effectiveness of diagnostic measures was determined by methods of complex primary diagnostics using the method of analysis of external thermal fields. The resulting efficiency coefficient is compared with the most effective of the certified methods, the method of vibration diagnostics. Calculation results

The diagnostic effectiveness ratio of measures from the moment of diagnosis is shown in Figure 1.

The results of using the complex diagnostics method. The efficiency coefficient was determined three months before equipment failure. Combined diagnostics efficiency exceeded vibration diagnostics efficiency by 34% (K_{epp} = 1.41, K_{ev} = 1.06).

Diagnostics combined external thermal analysis fields time cuts troubleshooting in half. Failure of operating time forecast for 2-3 months before the critical state is received. Failure detection combined diagnostics reduces costs. Failure detection reduces costs. Procedure planning and maintenance and repair combined unit diagnostics simplifies.

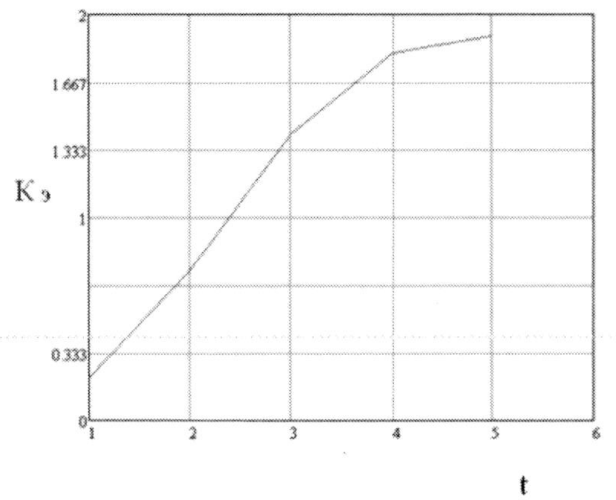

Figure 1. Results of calculating the coefficient of effectiveness of diagnostic measures by analyzing external thermal fields.

References

[1] Brovman M J 2007 *Continuous casting of metals* (Jekomet) p 484

[2] Krivandin V A, Filimonov Ju P 1986 *Theory of construction and calculations of metallurgical furnaces* (Moscow: Metallurgy) p 479

[3] Lisienko V G, Lobanov V I, Kitaev B I 1995 *Thermophysics of metallurgical processes* (Moscow) p 240

[4] Mastrjukov B S 1978 *Theory, designs and calculations of metallurgical furnaces. Volume 2. Calculations of metallurgical furnaces* (Moscow) p 272

[5] Plahtin V D 1983 *Reliability, repair and installation of metallurgical machines* (Moscow) p 415

[6] Alekseeva T V, Babanskaja V D, Bashta T M 1989 *Technical diagnostics of hydraulic actuators* (Moscow) p 264

[7] Rednikov S N, Akhmedyanova E N, Zakirov D M 2018 Experience in Using Combined Diagnostic Systems for Assessing State of Metallurgical Equipment *Global Smart Industry Conference* Online. Available: https://ieeexplore.ieee.org/document/8570148

[8] Sherkunov V G, Rednikov S N, Vlasov A E 2016 Mathematical modeling of galvanic coating deposition processes at different speed modes *Bulletin of the Magnitogorsk State Technical University. G.I. Nosova* **14** **2** 101–106.

[9] Rednikov S N, Muromcev N N 2011 Determination of the temperature of the medium under study at high pressures *Bulletin of the Samara Scientific Center of the Russian Academy of Sciences* **13** **1–3** 620–622

[10] Rednikov S N 2015 Viscosity characteristics of hydrocarbons at high pressures *Procedia Engineering* **129** 839–844

[11] Bataev V A 2007 *Methods of structural analysis of materials and quality control of parts* (Moscow) p 224

[12] Zedginidze I G 1976 Planning an experiment in the study of multicomponent systems (Moscow) p 390

[13] Kashheev I D 2004 *Properties and application of refractories* (Moscow: Teplotehnik) p 352

[14] Kazancev E I 1975 *Industrial furnaces. Reference Guide for Calculation and Design* (Moscow) p 368

[15] Smirnov A N, Piljushenko V L, Minaev A A 2002 *Continuous Casting Processes* (Doneck) p 536

[16] Grebenik V M, Ivanchenko F K, Pavlenko B A 1991 *Mechanical equipment of metallurgical plants. Mechanical equipment of converter and open hearth shops* (Kiev) p 287

[17] Pavlova G A 2011 Statistical analysis of accidents and injuries at metallurgical plants *Technosphere safety technologies* Online. Available: http://ipb.mos.ru/ttb

[18] Rednikov S N, Rahmatullin B B 2012 The method of express diagnostics of hydraulic system units *Dynamics of machines and work processes*

[19] Fominyh A V, Chinjaev I R, Shanaurin A L 2016 Pipe fittings as the basis of passive protection systems *Valve construction* 4 58–63
[20] Rednikov S N, Zakirov D M, Platov S I 2018 *Complex diagnostics of metallurgical equipment* (Magnitogorsk) p 75
[21] Sherkunov V G, Rednikov S N, Vlasov A E 2016 Influence of the dynamics of the electrolyte movement in a galvanic bath on the uniformity of the applied coating *Bulletin of the Magnitogorsk State Technical University. G.I. Nosova 14 3* 32–38

IOP Conf. Series: Materials Science and Engineering **969** (2020) 012078 doi:10.1088/1757-899X/969/1/012078

Improvement of the Device for Continuous Pressing of Non-Ferrous Metals and Alloys and Computer Simulation of the Deformation Process in This Device

S Lezhnev[1], A Naizabekov[1] and E Panin[2]

[1]Rudny Industrial Institute, 38, 50 let Oktyabrya str., Rudny, 111500, Kazakhstan
[2]Karaganda State Industrial University, 30, Republic av., Temirtau, 101400, Kazakhstan

E-mail: sergey_legnev@mail.ru

Abstract. A new design of the device for continuous pressing is proposed, which combines the principles of the ECAP and Linex processes. To assess the possibility of the process and the efficiency of metal processing, computer modeling was performed. It is found that the greatest load is received by the mobile segments that ensure the movement of metal through the matrix. The distribution of stress-strain state parameters in the longitudinal section of the workpiece is extremely uneven, which is due to the transverse action of the pulling force. In the first two channels, due to a backpressure the compressive stresses occur. In the output channel, the metal experiences tensile stresses.

1. Introduction

The development of the modern world has been going on at an ever-increasing pace for a long time. The reality around us is constantly being modernized and supplemented with new elements, and first of all, various modern and in many cases innovative technology. At the same time, the constantly occurring development of technology simultaneously imposes higher requirements for the quality of non-ferrous metals and semi-finished products used for its production. Non-ferrous metals and alloys have been of great importance for more than a decade. They are used in almost all industries, including aircraft construction, chemical industry, mechanical engineering, and others. And in the manufacture of many parts for household appliances, cars and aircraft, etc. non-ferrous metals and alloys have no competitors at all.

The increased requirements for the quality and properties of non-ferrous metals and alloys dictate the conditions for the development of new innovative technologies for processing non-ferrous metals and alloys by pressure or improving existing technologies, but in any case aimed at obtaining after pressure treatment of non-ferrous metals and alloys with an ultrafine-grained (UFG) structure. This structure is characterized by an average grain size from 0.1 to 1 microns [1], and the increased interest in UFG materials is due to their improved physical and mechanical properties [2].

One of the most promising and economically justified methods for obtaining an ultrafine-grained structure, both in ferrous and non-ferrous metals and alloys, is the severe plastic deformation (SPD) method [3]. In many works of scientists from different countries of the world [4–10], it has been proved more than once that the processing of various metals and alloys by the SPD method makes it possible to significantly reduce the initial structure of the metal without significantly changing the initial size of the workpiece.

Currently, numerous schemes have been developed not only for the implementation of severe plastic deformation: high-pressure torsion (HPT) [11]; equal-channel angular pressing (ECAP) [12]; comprehensive forging [13]; multi-axis deformation [14] and others, but also tools for their implementation. For example, numerous matrix designs for equal-channel angular pressing have been developed [15–20].

Further development of the scientific direction related to severe plastic deformation is associated with the development of new innovative methods for producing long-length products from ferrous and non-ferrous metals with an ultra-fine-grained structure. These methods include: the method of deforming the Conform [21]; equal-channel angular drawing [22]; combined processes "ECAP – drawing" [23], "rolling – ECAP" [24] and many others. A method of continuous pressing of non-ferrous metals and alloys called "Linex" is also known [25]. This method of deformation, although it allows you to obtain long products from non-ferrous metals and alloys, but most often from soft, but has one significant drawback: it provides an increase in the quality of metal products only when the size of the source material is repeatedly reduced, which leads to significant energy and labor costs.

2. Concept of new device for pressing

Based on this method of deformation and the tool for its implementation, it is proposed to create a new device for continuous pressing of non-ferrous metals and alloys without significantly changing the initial dimensions of the workpiece. This will be achieved by the fact that the device for continuous pressing of non-ferrous metals and alloys will include a movable part made in the form of two tape gripping blocks with rigid inserts, as well as a fixed part in the form of an equal-channel step matrix made without side walls (Figure 1).

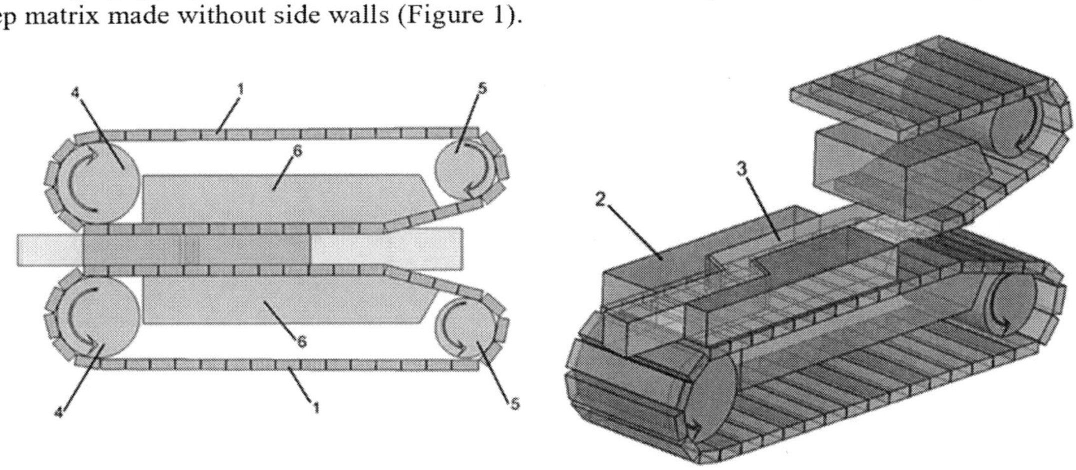

Figure 1. Device for continuous pressing of non-ferrous metals and alloys: 1 - movable belt blocks, 2 - fixed matrix with three channels of the same cross-section, two of which are parallel to each other, and the middle channel is located at an angle to the input and output channels, 3 - blank, 4 - idler pulleys, 5 - drive pulleys, 6 - fixed locking blocks.

Pressing in this device is performed as follows. The workpiece is fed to the device, where movable tape blocks capture the workpiece and push it through the channels of the fixed matrix. Each belt gripper block is clad on two pulleys, one of which is idle, and the other is driven by an electric motor. This is why the tape gripping blocks are set in motion. The horizontal forming of tape gripping blocks is created by their movement along the workpiece and fixed locking blocks that perform a clamping role. To reduce the deformation force, grease is applied to the walls of the fixed matrix, while there is no grease supply to the moving parts in order to increase the gripping capacity.

To assess the possibility of implementing the continuous pressing process in the proposed device and study it, a computer simulation of this process was carried out in the Deform software package. The initial dimensions of the workpiece were 10x15x100 mm. The material of the billet is AL-1100 alloy. The deformation was performed at room temperature. The angle of the channel junction in the matrix is 145°, to reduce the back pressure at the channel junctions, rounding with a radius of 5 mm was made, and the coefficient of friction with fixed side segments was assumed to be 0.1. For a stable process, it is necessary to ensure a high level of adhesion of the workpiece with the moving segments, so the coefficient of friction was set to 0.8.

3. Results and discussion

The simulation results were used to study the load force on all segments of the device, as well as the stress-strain state in the workpiece (figure 2).

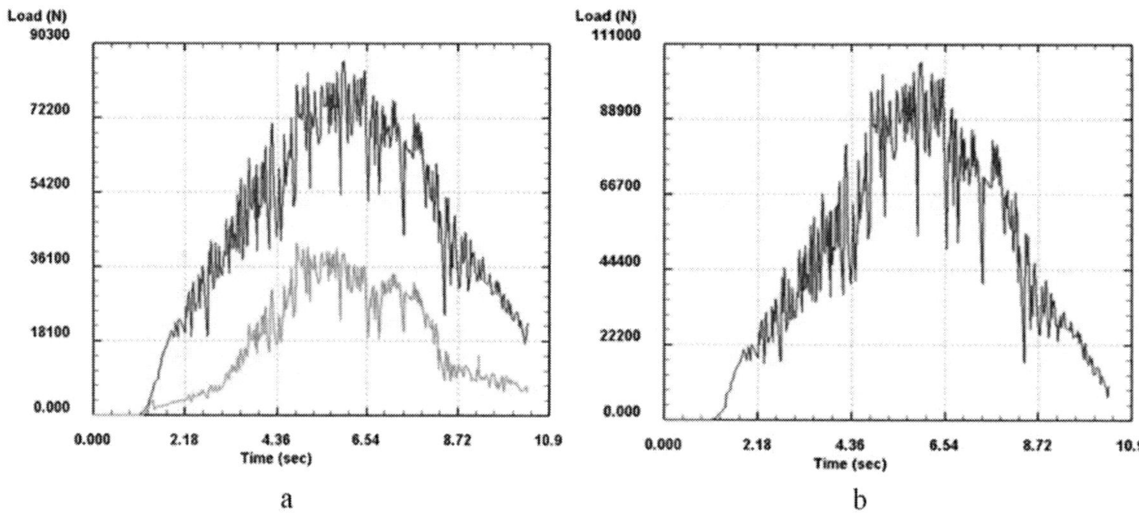

Figure 2. Load force on fixed (a) and movable (b) segments.

On fixed segments, the amount of force varies, as in the case of a conventional ECAP – on a large segment, where there is a large backpressure, the force reaches 85 kN, on a small segment, the force is much lower, about 35 kN. On movable segments, the force value is the same, about 107 kN.

Due to the fact that the main force that causes the movement of metal through the channels of the matrix acts not in the longitudinal but in the transverse direction, the distribution of equivalent strain over the width of the workpiece is extremely uneven. This, in turn, is due to the nature of filling the matrix channels with deformable metal. Figure 3a clearly shows that due to the transverse action of the force, the second joint in the matrix is not filled, since the metal, after passing the inclined channel, moves in the direction of movement of the moving segments. As a result, the width of the workpiece metal gets a deformation from 0.4 to 6.2.

The reason for this phenomenon can be easily understood if we consider the stress state in the plane of the longitudinal section of the workpiece (figure 3b). When deforming in this device, all the metal gets a movement in the longitudinal direction. However, due to the construction of the matrix, when moving in the first channel, the metal experiences a backpressure from the joint angle, and when moving in the second channel, it experiences a backpressure from the inclined face of a large segment. Therefore, in these zones, compressive stresses of up to -20 MPa prevail in the metal. Everything changes after the metal passes the inclined channel. Here, there are no more obstacles in its path, and there is no backpressure. Therefore, in the output channel, the metal experiences tensile stresses up to +60 MPa.

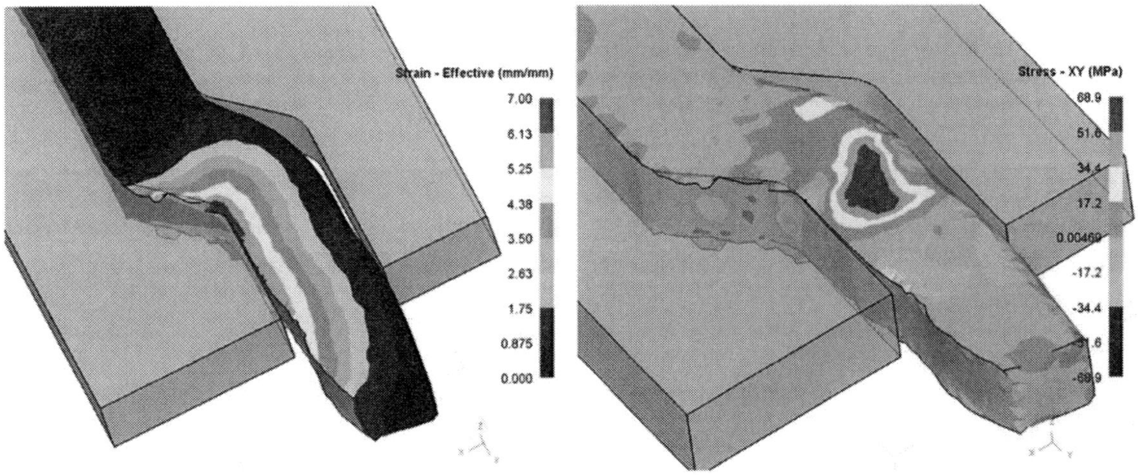

a b

Figure 3. Equivalent strain (a) and stress in the plane of the longitudinal section of the workpiece (b).

4. Conclusion

A new design of the device for continuous pressing is proposed, which combines the principles of the ECAP and Linex processes. To assess the possibility of the process and the efficiency of metal processing, computer modeling was performed. It is found that the greatest load is received by the mobile segments that ensure the movement of metal through the matrix. The distribution of stress-strain state parameters in the longitudinal section of the workpiece is extremely uneven, which is due to the transverse action of the pulling force. Therefore, to align the stress-strain state parameters on the section of the workpiece, it is recommended to perform an even number of deformation cycles with the workpiece turning 180 degrees after each cycle.

References

[1] Valiev R Z and Alexandrov I V 2000 Nanostructured materials obtained by intensive plastic deformation (Moscow: Logos)
[2] Valiev R Z and Alexandrov I V 2007 Volume nanostructured metal materials. Obtaining, structure and properties (Moscow: Akademkniga)
[3] Utyashev F Z 2008 Modern methods of intensive plastic deformation (Ufa: UGATU)
[4] Utyashev F Z 2013 Scientific and technical Bulletin of Spbpu. Physical and mathematical Sciences **4–1** 204
[5] Naizabekov A B, Lezhnev S N and Volokitina I E 2015 Metal Science and Heat Treatment **57 5–6** 254
[6] Langdon T G 2020 Advanced Engineering Materials **22 1** 1900442
[7] Zhao S L, Zhang H M, Cui Z S, Chen D and Chen Z 2020 Journal of Alloys and Compounds **834** 155136
[8] Nakashima K, Horita Z, Nemoto M and Langdon T G 2000 Materials Science and Engineering A **281 1–2** 82
[9] Zhilyaev A P, Ringot G, Huang Y, Cabrera J M and Langdon T G 2017 Materials Science and Engineering A **688** 498
[10] Kral P, Dvorak J, Sklenicka V, Horita Z, Takizawa Y, Tang Y, Kvapilova M and Svobodova M 2020 Materials Science and Engineering A **785** 139383

[11] Valiev R R, Murashkin M Yu, Sabirov I N, Salimgareev H Sh, Smirnov I V and Valiev R Z 2015 RU Patent **2547984**
[12] Raab G I 2009 Development of scientific bases of technologies of intensive plastic deformation and creation of equipment according to the scheme of equal-channel angular pressing for obtaining ultra-fine-grained metal semi-finished products (Ufa: UGATU)
[13] Salishchev G A, Valiakhmetov O R, Galeev R M and Malysheva S P 1996 Metals **4** 86
[14] Martynenko N S, Lukyanova E A, Serebryany V N, Anisimova N Yu, Kiselevsky M V, Yurchenko N Yu, Salishchev G A, Dobatkin S V and Estrin Yu Z 2018 Technology of light alloys **2** 51
[15] Polovnikov V M, Kenderov V V, Kandaurov I V and Latysh V V 2009 RU Patent **2356679**
[16] Naizabekov A B, Ashkeev Zh A, Lezhnev S N and Toleuova A R 2005 Steel in Translation **35** **2** 37
[17] Kuchkin V V, Osokin E P, Rybin V V, Rybin Yu I and Shcherbel R D 2008 RU Patent **2333062**
[18] Ivanov A M and Alexandrova E S 2013 RU Patent **2492950**
[19] Ivanov A M 2013 RU Patent **2477662**
[20] Raab G I, Kulyasov G V, Polozovsky V A and Valiev R Z 2002 RU Patent **2181314**
[21] Ghaforian Nosrati H, Khalili K, Gerdooei M 2020 Metallurgical and Materials Transactions B **51** **2** 519
[22] Zhao H, Ren Y, Yang B and Qin G 2019 Journal of Alloys and Compounds **81** 152002
[23] Naizabekov A, Volokitina I, Panin E, Volokitin A, Lezhnev S, Garstka T, Knapinski M, Latypova M and Zhumagaliev D 2020 Journal of Chemical Technology and Metallurgy **55** **3** 499
[24] Naizabekov A, Lezhnev S, Panin E, Volokitina I, Arbuz A, Koinov T and Mazur I 2019 Journal of Materials Engineering and Performance **28** 200
[25] Voorhes W G 1975 USA Patent **3922898**

Modeling Dynamic Parameters of Hard Alloys during Shock Wave Regeneration

I Savchenko[1], A Shapoval[2] and A Gurenko[3]

[1]Department of Cybersecurity and Information Technology, University of Customs and Finance, 2/4, Vladimir Vernadsky Street, Dnipro, 49000, Ukraine
[2]Department of Manufacturing Engineering, Kremenchuk Mykhailo Ostrohradskyi National University, 20, Pershotravneva Street, Kremenchuk, 39600, Ukraine
[3]Department of Analytical Economics and Management, Dnipro State University of Internal Affairs, 26, Gagarina, Dnipro 49005, Ukraine

E-mail: alexshap.as@gmail.com

Abstract. Construction of the shock adiabat of a porous multi-component mixture and the parameters of compression of this medium by shock waves. As a research technique for the present work, we considered the description of a solid by the equations of hydrodynamics when a shock wave propagates through a substance or mixture of substances when the shock compression pressure exceeds the yield strength of the medium components. When shock-wave compression of solids is determined, the parameters of the medium during its multiple compression using the equations of state of multicomponent systems, which are represented by continuous analytical dependencies. The fundamental possibility of using shock-wave processing as a factor stimulating the processes of destruction of powder products of any configuration to obtain high-quality powder for further formation, sintering, and production of tools for various purposes is shown, using developed dependencies. Technology for the regeneration of hard alloys is developed, an environmentally friendly processing technology is proposed super hard materials, cermet components of obsolete military equipment, and different types of ammunition.

1. Introduction

Striving for dominating positions in the market, encourages the leadership of the countries and large companies, as well as other influential stakeholders of today's world, to develop an innovative approach to business processes engineering reengineering to increase the economic and overall business performance leading to higher revenues. The implementation of innovative approaches will allow to increase business efficiency, which calls for traditional tried and tested approaches and methods to be adapted for the development of modern society [1].

It is necessary to determine multiple compressed medium parameters for the process of shock wave compression in solids. This task can be made relatively easy and performed in the case of single-component systems, but the task gets complicated in case of the dynamic systems compression, which consists of several components. In this case, it is necessary to determine the equation of multicomponent systems state, which are to be represented with continuous analytical dependencies or, Experimental data on shock compressibility for many metals, rocks, minerals, and other substances was published [2–7].

IOP Conf. Series: Materials Science and Engineering **969** (2020) 012079 doi:10.1088/1757-899X/969/1/012079

2. The main part of the article

There are several methods of tungsten cobalt hard alloys scrap regeneration: pyro and hydrometallurgical, chemical (chlorine, zinc), thermochemical, and using explosion energy of blasting explosives. The first two methods are very complicated, multi-staged, and laborious, requiring equipment that is stable in aggressive mediums and consuming a lot of electricity. The final products of these methods are composite compounds of tungsten, which require additional processing. Besides, all of these imply harmful work conditions and contamination of the environment [8–13].

The purpose of these studies is to form the shock adiabat of a porous multi-component mixture along with the shock wave compression of the medium. The equation of the shock wave adiabatic of the material $P=P(\rho)$ is widely used for the calculation purpose, which is based on the experimental dependency between the speed of the shock wave D and the mass speed at its front $D=c_0+\lambda u$, where c_0 – hydrodynamic sound speed, λ – material constant. Basis this dependency and the mass and impulse conservation laws at the front of the shock wave, we shall have the following result [14–17].

$$P=\frac{\rho_0 c_0^2\left(1-\rho_0/\rho\right)}{\left[1-\lambda\left(1-\rho_0/\rho\right)^2\right]^2}; \tag{1}$$

$$E=E_0+\frac{1}{2}\left[\frac{c_0\left(1-\rho_0/\rho\right)}{1-\lambda\left(1-\rho_0/\rho\right)}\right]^2 \tag{2}$$

where P and E – pressure and energy at Hugonio curve; ρ – front shock wave density; ρ_0 – initial density of the material. Boundary compression value is determined, provided: $1-\lambda\left(1-\rho_0/\rho\right)=0$. Resulting

$$\frac{\rho}{\rho_0}=\frac{\lambda}{1-\lambda}=\frac{1}{1-1/\lambda}. \tag{3}$$

When plotting shock adiabat of multicomponent mediums at 10 GPa and higher pressure, the hydrodynamic model is used since the process of the shock wave propagation only slightly depends on the initial conditions. Usually, multicomponent media are considered which, likewise mixtures, have no volume defect:

$$V=\frac{1}{\rho}=\sum_{i=1}^{n}\alpha_i V_i \ ;\sum_i \alpha_i=1 \tag{4}$$

where Vi – the specific volume of the ith component of the medium; α_i – weight fraction; V – specific volume of the whole medium.

In addition to this, it is considered that a multicomponent medium is a homogeneous medium. This assumption is a consequence of the fact that the shock adiabat of the medium at high pressures is weakly dependent on the structure of the medium, starting with some grain size. The single-phase model of the medium, in turn, assumes single temperature, single-speed model of the medium, whose thermodynamic properties can be described based on the Mi-Gruneisen type equations with the thermal part of pressure and energy, determined by the Debye theory.

The study [18] demonstrates the possibility of constructing a substance shock adiabat using only the value of the volumetric (hydrodynamic) sound velocity in this substance and the following dependence $P(\mu)$, where $\mu=u/c_0$. Dimensionless coordinate $P = Du / c_0^2$; $\mu=u/c_0$.

Experimental data on the dynamic compressibility of metals and minerals data represented by generalized shock adiabatic [19]. For a mixture of two substances (solid or liquid) in the absence of experimental data, sound velocity can be calculated using the formula recommended in the study [20].

$$\frac{1}{\rho_0 c_0} = \frac{\alpha}{\rho_{01} c_{01}} + \frac{1-\alpha}{\rho_{02} c_{02}}, \tag{5}$$

where α – the weigh share of the first component; ρ_{01}, ρ_{02}, c_{01}, c_{02} – density and sound velocity of the first and second components respectively.

The construction of the shock adiabat of the non-porous mixture of two solids is based on the following assumptions: provided the equality of mass velocities of the components along the front of the shock wave, the distribution of energies by the components is proportional to their mass content.

When constructing the shock adiabat [21-22] the following assumption was made – when the shock wave propagates on the sample of the non-porous mixture, the pressure in both components on the shock wave front is leveled and no heat exchange occurs. In this case, the mass velocity of the two materials mixture particles is determined by the formula:

$$u_{см}^2 = \alpha \cdot u_1^2 + (1-\alpha) u_2^2 . \tag{6}$$

The specific volume of the mixture in the initial state and the final state is calculated using the following ratios:

$$V_{см} = \alpha V_{01} + (1-\alpha) V_{02} \tag{7}$$

where V_{01}, V_{02} – specific volumes of the first and the second components in the initial stage and at the shock wave front respectively.

The basic assumption is used in the construction of the shock adiabat of a porous medium: a solid is a non-porous mixture of two solids for which the relation (5 – 7) is true. This assumption was based on the fact that the constants c_0 and λ, which characterize the dynamic compressibility of solids, do not differ much, which allowed building the equation of the porous medium, where hydrodynamic sound velocity c_0 and the λ_{0n} parameter in the equation $D= c_0+ \lambda_{0n} u$ are generalized and dependent on the corresponding c_{0i} and λ_i components.

The specific volume of the medium in the initial stage, before the shock wave front, is also calculated: $V_{0см} = \sum_{i=1}^{n} \alpha_i V_{0i}$. Substituting ρ_i we shall establish the final dependency $\rho_{cm} = \rho_{cm}(P)$:

$$\rho_{см} = \left[\sum \left\{ \frac{1}{\rho_{0i}} \alpha_i \left[1 - \left(\sqrt{\frac{\rho_{0i} c_{0i}^2}{4\lambda_i^2 P} + \frac{1}{\lambda_i}} - \frac{c_{0i}}{2\lambda_i} \sqrt{\frac{\rho_{0i}}{P}} \right)^2 \right] \right\} \right]^{-1} \tag{8}$$

Therefore, knowing the shock compressibility adiabats of each i-th component (c_{0i}, λ_i), the dependency $P = P(\rho_{см})$ can be constructed (8).

The $\rho_{0см}$ shall be calculated using expression (5):

$$\rho_{0см} = \frac{1}{V_{0см}} = \frac{1}{\sum_{i=1}^{n} \alpha_i V_{0i}} = \frac{1}{\sum_{i=1}^{n} \frac{\alpha_i}{\rho_{0i}}} \tag{9}$$

Now it will be easy to calculate λ_{cm}:

$$\lambda_{CM} = \left(\frac{\tilde{\rho}_{CM}}{\tilde{\rho}_{CM}-\rho_{CM}}\right) = \left[1-\frac{\sum_{i=1}^{n}\alpha_i V_{0i}\left(1-{}^{1}\!/_{\lambda_i}\right)}{\sum_{i=1}^{n}\alpha_i V_{0i}}\right]^{-1} \tag{10}$$

Feeding λ_{cm} and c_{0cm}, the shock wave adiabatic of the mixture in (P, ρ_{CM}) coordinates, will be presented as follows:

$$P = \frac{\sum_{i=1}^{n}\frac{\alpha_i}{\rho_{0i}}\left(1-\left[\rho_{CM}\sum_{i=1}^{n}\frac{\alpha_i}{\rho_{0i}}\right]^{-1}\right)}{\left(1-\frac{\sum_{i=1}^{n}\frac{\alpha_i}{\rho_{0i}}}{\sum_{i=1}^{n}\frac{\alpha_i}{\rho_{0i}\lambda_i}}\left(1-\frac{1}{\rho_{CM}\sum_{i=1}^{n}\frac{\alpha_i}{\rho_{0i}}}\right)\right)^2\sum_{i=1}^{n}\frac{\alpha_i}{\rho_{0i}^2 c_{0i}^2}}. \tag{11}$$

Figure 1 illustrates the calculated shock adiabats in $(P - u)$ coordinates for tungsten, containing hard alloys: VK8 – 1, carbon, which contains – 5,642 % weight, cobalt – 8 % weight, and tungsten – rest (curve 1); VK15 – 2, carbon, which contains, – 5,213 % weigh, cobalt – 15 % weight and tungsten – rest (curve 2), VN8 – 3, carbon, which contains, – 5,642 % weight, nickel – 8 % weight and tungsten - rest (curve 3); VN15 – 4, carbon, which contains, – 5,213 % weight, cobalt – 15 % weight and tungsten – rest (curve 4). Using the established dependencies, the technology of hard alloys regeneration was developed [1].

Marketing strategy implies offering the international quality standard product at 5–10% less than the market price [1].

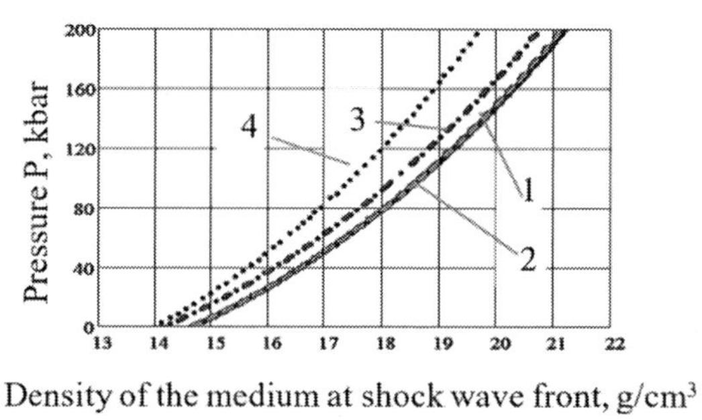

Figure 1. Impact of adiabat.

New technology is resistant to negative factors. The center can operate at no loss, selling over 11 tons of products a year. In the case of a price drop by 10–20% or material cost increase by 20–50%, product profitability will reach 60–70%, which is still rather high.

Economical indices of the project are as follows:

- net present value (NPV) for the settlement period of project realization (5 years) is over 1265 thousand USD, which corresponds to the effectiveness condition, where NPV should be >0;
- investments profitability index (JR) – 6,0 (JR>1);
- recoupment of capital investments is 8 months from the issuance of the loan.

Industrial development of the above-described technology, within a short timeframe, will allow solving several important issues:– hard alloys scrap utilization, its further reduction and consequent significant cut down of raw material usage;– making use of new technology which omits low

IOP Conf. Series: Materials Science and Engineering **969** (2020) 012079 doi:10.1088/1757-899X/969/1/012079

productive ways of deficit raw material utilization and manufacturing products of international quality standard.

3. Conclusions

The obtained equation of the shock adiabat of hard alloy allowed to optimize, in terms of defect formation, the load parameters of the processed materials. The propagation of the shock wave through a substance or mixture of substances when the compression pressure exceeds the flux limits of the medium components is described by the hydrodynamic equations regardless of the aggregate state of the substance, that is, the solid, as well as the fluid, is described by the variables pf pressure P and volume V. Fundamental potential of shock–wave treatment as a stimulating factor for powder products destruction of any configuration, to obtain high-quality powder for further formation, caking and production of tools of multiple uses was demonstrated.

References

[1] Savchenko Iu, Gurenko A and Naumenko O 2016 The cutting-edge industrial technology of mining tool manufacturing *Mining of Mineral Deposits* **10 4** 105–110

[2] Markov O, Gerasimenko O, Aliieva L, Shapoval A and Kosilov M 2019 Development of a new process for expanding stepped tapered rings *Eastern-European Journal of Enterprise Technologies* **2(1–98)** 39–46

[3] Dragobetskii V, Zagirnyak V, Shlyk S, Shapoval A and Naumova O 2019 Application of explosion treatment methods for production items of powder materials *Przeglad Elektrotechniczny* **95(5)** 39–42

[4] Bast J, Gorbatyuk S M, and Kryukov I Yu 2011 Study of the temperature fields in the mold of a horizontal continuous caster *Metallurgist* **55(3–4)** 163–166

[5] Gorbatyuk S, Kondratenko V and Sedykh L 2019 Investigation of the Deep Hole Drill Stability When Using a Steady Rest *Materials Today: Proceedings* 11 258–264

[6] Zagirnyak M, Zagirnyak V, Moloshtan D, Drahobetskyi V and Shapoval A 2019 A search for technologies implementing a high fighting efficiency of the multilayered elements of military equipment *Eastern-European Journal of Enterprise Technologies* **6/1(102)** 33–40

[7] Markov O, Gerasimenko O, Aliieva L and Shapoval A 2019 Development of the metal rheology model of high-temperature deformation for modeling by finite element method *EUREKA, Physics and Engineering* **(2)** 52–60

[8] Glukhov L M, Gorbatyuk S M, Morozova I G and Naumova M G 2016 Effective laser technology for making metal products and tools *Metallurgist* **60(3–4)** 306–312

[9] Shapoval A, Mos'pan D and Dragobetskii V 2016 Ensuring High-Performance Characteristics For Explosion-Welded Bimetals *Metallurgist* **60 3–4** 313–317

[10] Gorbatyuk S M, Pashkov A N, Zarapin A Y and Bardovskii A D 2019 Development of Hot-Pressing Technology for Production of Aluminum-Based Metal-Matrix Composite Materials *Metallurgist* **62(11–12)** 1261–1266

[11] Lutsenko I, Fomovskaya E, Oksanych I, Koval S and Serdiuk O 2017 Development of a verification method of estimated indicators for their use as an optimization criterion, Eastern-*European Journal of Enterprise Technologies* **2 486** 17–23, 2017

[12] Kukhar V, Balalayeva E, Prysiazhnyi A, Vasylevskyi O and Marchenko I 2018 Analysis of the relation between edging ratio and deformation work done in pre-forming of the workpiece by bulk buckling *MATEC Web of Conferences* **178**

[13] Keropyan A M, Gorbatyuk S M, Bibikov P Y and Bardovski A D 2019 Influence of Roughness of Working Surfaces of the Wheel–Rail System of Open-Pit Locomotives with an implementable Adhesion Coefficient *J of Friction and Wear* **40(1)** 73–79

[14] Lutsenko I, Fomovskaya E, Koval S and Serdiuk O 2017 Development of the method of quasi-optimal robust control for periodic operational processes *Eastern-European Journal of Enterprise Technologies* **4 2–88** 52–60

IOP Conf. Series: Materials Science and Engineering **969** (2020) 012079 doi:10.1088/1757-899X/969/1/012079

[15] Shapoval A N and Shapoval A A 2002 Development of The Unit for Multi-Stage Vibration Drawing of Metal Product *Non-ferrous metals* **4** 77–82

[16] Lutsenko I, Vihrova E, Fomovskaya E and Serdiuk O 2016 Development of the method for testing of efficiency criterion of models of simple target operations *Eastern-European Journal of Enterprise Technologies* **2 480**

[17] Lutsenko I 2015 Identification of target system operations Development of global efficiency criterion of target operations *Eastern-European Journal of Enterprise Technologies* **2 2** 35–40

[18] Lutsenko I 2016 Definition of efficiency indicator and study of its main function as an optimization criterion *Eastern-European J of Enterprise Technologies* **6 2** 24–32

[19] Anishchenko A, Kukhar V, Artiukh V and Arkhipova O 2018 Superplastic forming of shells from sheet blanks with thermally unstable coatings *MATEC Web of Conferences* **238**

[20] Markov O E, Gerasimenko O V, Shapoval A A, Abdulov O R and Zhytnikov R U 2019 Computerized simulation of shortened ingots with a controlled crystallization for manufacturing of high-quality forgings *Int J of Advanced Manufacturing Technology* **103(5–8)** 3057–3065

[21] Dragobetskii V, Zagirnyak M, Naumova O, Shlyk S and Shapoval A 2018 Method of determination of technological durability of plastically deformed sheet parts of vehicles *Int J of Engineering and Technology* **7 4** 92–99

[22] Anishchenko O S, Kukhar V V, Grushko A V, Vishtak I V, Prysiazhnyi A H and Balalayeva E Yu 2019 Analysis of the Sheet Shell's Curvature with Lame's Superellipse Method during Superplastic Forming *Materials Science Forum* **945** 531–537

[23] Dragobetskii V, Naumova E, Shapoval A, Shlyk S and Moloshtan D 2019 Improving the Operational Reliability of Stamped Parts of Electrical Engineering Machines and Electrical Products *Pr of the Int Con on Modern Electrical and Energy Systems* **8896532** 506–509

[24] Gorbatyuk S, Shapoval A, Mos'pan D and Dragobetskii V 2016 Production of periodic bars by vibrational drawing *Steel in Translation* **46 7** 474–478

[25] Belas E, Bogoboyashchyy V, Grill R, Izhnin I, Vlasov A and Yudenkov V 2003 Time relaxation of point defects in p- and n-(HgCd)Te after ion milling *Journal of Electronic Materials* 32 7 698–702

[26] Bogoboyashchii V, Vlasov A and Izhnin I 2001 Mechanism for Conversion of the Conductivity Type in Arsenic-Doped p-Cd x Hg1– x Te Subject to Ionic Etching *Russian Physics Journal* **44 1** 61–70

[27] Bogoboyashchii V V and Inzhin I I 2000 Mechanism for Conversion of the Type of Conductivity in p-Hg1–xCdxTe Crystals upon Bombardment by Low-Energy Ions s *Russian Physics Journal* **43** 627–636

[28] Hrudkina N, Aliieva L, Abhari P, Markov O and Sukhovirska L 2019 Investigating the process of shrinkage depression formation at the combined radial-backward extrusion of parts with a flange *Eastern-European Journal of Enterprise Technologies* **2 5/1(101)** 49–57

[29] Hrudkina N, Aliieva L, Markov O, Marchenko I, Shapoval A, Abhari P and Kordenko M Predicting the shape formation of hollow parts with a flange in the process of combined radial-reverse extrusion *Eastern-European Journal of Enterprise Technologies* **4/1(106)** DOI: 10 15587/1729-4061 2020 203988

[30] Babak V P and Shchepetov V V 2018 Wear Resistance of Amorphous-Crystalline Coatings with Lubricants *Journal of Friction and Wear* **39(1)** 38–43

IOP Conf. Series: Materials Science and Engineering **969** (2020) 012080 doi:10.1088/1757-899X/969/1/012080

Surface Texture of Deformed Copper Wire

M A Zorina, M S Karabanalov and Yu N Loginov

Department of Heat Treatment and Physics of Metals, Institute of New Materials and Technologies, Ural Federal University named after the first President of Russia B. N. Yeltsin, 19, Mira Street, Ekaterinburg, 620002, Russia

E-mail: m.a.zorina@urfu.ru

Abstract. The texture of the cold-drawn copper wire was investigated along the radius and on surface by means of EBSD. The wire fiber texture after drawing has been shown to consist of a set of a main orientations analogically to the rolling texture. These texture orientations are associated to the stress state in drawing. The central area of the wire was characterized by two main orientations: faint <001> and powerful <111>-fibers. Main orientations {110} <001> and {112} <111> were distinguished within these fibers. The peripheral region is represented by a shear texture that rotates 90° to the center texture and scattered at ± 10°. The surface texture scattered close to axial and also has deviated preferred components: strong {110} <001> and weak {112} <111> orientations.

1. Introduction

Crystallographic texture is one of the ways to obtain products with an improved set of orientation-dependent physicochemical and / or mechanical properties [1–2]. The emergence and development of texture occurs due to directed deformation effects, realized in cramped conditions [3, 4], which the product is subjected to at the stages of processing. Drawing is one of the main technological operations in the manufacture of a great number of metallic materials and products characterized by advanced electrical conductivity, such as Cu, Ag, Au, and Al. The electrically-conductive elements made of copper are subjected to the strong deformation impacts (rolling, drawing) during the processing, which result in the formation of intense crystallographic texture. The research of the features of texture transformations in copper with the use of modern techniques of structural analysis allows to develop approaches to controlling the operational and functional properties of the face-centered cubic (FCC) metals. Note that considering, the texture shaping on a specific treatment path allows you to optimize present technologies [5] or even develop new ones.

The location of deformation textures in metals can be explained from the point of view of the classical theories [6–8]. An important consequence of almost all present models of the development of principal grain orientations during deformation is the emergence of sustained orientations at comparatively high degrees of deformation [9]. These orientations, characterized by precise crystallographic indices, do not change their spatial position in the further deformation, which is associated with the equilibrium of slip systems acting in inverse directions. For a given deformation pattern, the stress state forms a set of distinct orientations of grains, located with a certain misorientation relative to each other and forming a complex limited texture. It is worth noting that the large amount (but well-defined) of stable orientations generated by deformation of metallic materials with a cubic lattice determines their symmetrical location relative to the state of external stresses [4, 9]. Crystal plasticity models are used to model high strain plasticity and texture evolvement. The main disadvantage of most present models is that in some cases they are unable to predict all types of experimentally obtained textures, as well as texture conversions.

It is known from the mechanics of a plastically deformable body that the spreading of deformation during drawing is influenced by boundary conditions such as a tool configuration (the half-angle of die taper), a friction stress, a back-pull stress, etc. [11, 12]. Shear textures are often found at the surface of rolled products. It needs to be investigated because it is immediately related to the anisotropy of the metal. Surface texture influences friction and transfer layer formation during sliding. Crystallographic reorientation (i.e. texture evolution) under the contact surface may be a possible cause of the decrease in the coefficient of friction during sliding wear. Crystallographic reorientation can lead to a decrease in the effective shear stress in the near-surface areas and a decrease in the frictional force required to maintain relative motion [12].

Drawing is known [13–15] to bring about the formation of the complex axial texture of the wire. The main characteristic of drawing texture of FCC-metals is that it contains two basic axial components: <111> and <100>. The authors of [16] report that during drawing of the copper wire the <111> texture component was observed along with the <112>. Namely the last has been considered to be metastable. The <112> axial texture is formed from the parent orientation <100> at high strains in the surface layer of the wire [17].

Studies by Lee, Park et al. [18, 19] have shown that the shear stresses can be ignored in the center of the wire and that they increase from the center to the surface of the wire. Differences in shear stresses might lead to heterogeneous allocation of the components of the axial texture along the radial direction of the drawn wire. Thus, the allocation of the components of the drawing texture in the radial direction of the deformed wire can differ significantly [11].

This paper examines the drawing texture of FCC metals in terms of copper wire and the accordance between the main orientations of the texture on center, peripheral areas and surface of the cold-drawn wire.

2. Experimental procedure

The semi-products of M001 copper (analogue of ETP copper) were used in the study. Wire 1.65 mm in diameter was obtained from copper rod 8 mm in diameter on a multi-pass drawing machine MCM 85. Drawing was carried out in 10 passes at a drawing speed on the last drawing block of 20 m/s. Deformation was carried out without intermediate annealing. Tandem drawing was carried out using a set of polycrystalline dies. The half angle of the taper of the drawing die was 10°. Stable boundary conditions for heat transfer and friction were ensured by a system for supplying liquid lubricant and devices for its regeneration and temperature control, which was equipped with a drawing mill. For this, the cutting fluid Unopol G 570 was used. The total deformation as a result of 10 drawing passes was 23.51, which corresponds to a logarithmic deformation of 3.16 and a reduction area (RA) of 95.7%.

The analysis of the texture state was carried out by the method of electron backscattering diffraction (EBSD) using a ZEISS CrossBeam AURIGA field emission gun scanning electron microscope (FEG SEM) equipped with Nordlys HKL Channel 5 hardware and software for recording and analyzing diffraction patterns. The scanning step was 0.2 μm, the error in determining the orientation of the crystal lattice was ± 1°. Orientation maps and pole figures (PF) of two types: 1) as a projection of the pole outputs (all existing in the analyzed region are shown); 2) as a distributed polar density (showing the main orientations and the degree of their scattering) were used in texture analysis. A laboratory coordinate system was taken to denote local crystallographic orientations, the axes of which refer to the axial direction (AD // X), normal to the horizontal plane (ND // Y) and directly transverse to them (TD // Z). All selected directions give the right triplet of vectors.

In order to simulate the stress-strain state under drawing of copper wire the boundary-value problem has been solved using RAPID-2D software. This calculation is based on the finite element method. The program unit makes it possible to transfer a coldworked workpiece from tool to tool and, thus, to simulate multistage drawing. The boundary-value problem of determining the angles of inclination of the main deformation axes along the radius of a copper wire is solved using a lightweight version of the

ABAQUS program with 1000 finite elements by the FEA. The die cone angle was set at 10° in accordance with the technological process. The reciprocity of the metal with the drawing tool was specified using the Coulomb friction law. Friction coefficient 0.05. Low and constant values of the coefficient of friction were taken from the average values of the experimental data for drawing copper products to simplify the calculation [20]. The copper wire was indicated as a nonlinear strain-hardening medium, namely, the power dependence between stress and strain was taken. The artificial diamond drawing die was regarded as an elastic medium in the simulation.

3. Results and discussion

The texture of the wire after 10 draw passes was a complex fiber (figure 1). Previous studies [11, 21] demonstrate uneven allocation of the main texture orientations along the radius of the cold-drawn copper wire. The texture of the metal after drawing within the entire product was complex axial, which was related by the symmetry of the deformation pattern during drawing. Though, a layer-by-layer (local) analysis of the crystallographic orientation of grains along the radius of the wire showed noticeable distinctions between the central, peripheral areas and the surface even in the type of the formed texture of deformation (figure 1).

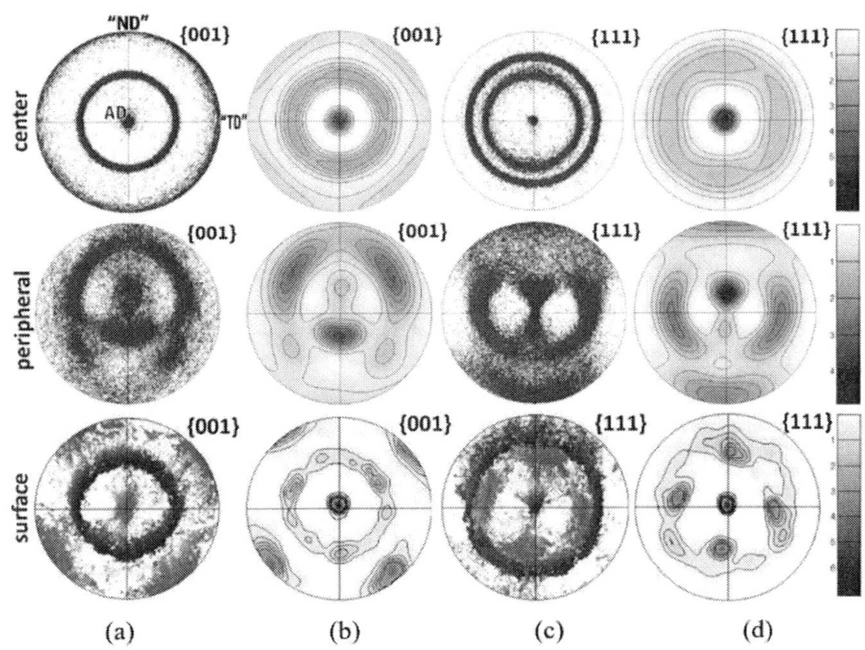

Figure 1. The EBSD analysis in various areas along the radius of the copper wire: {001} and {111} PF: (a), (c) – separate poles (scattered data); (b), (d) – pole density distribution (contouring data).

The central area of the wire was characterized by two main orientations: <001> and <111> fibers (figure 1). Main orientations {110} <001> and {112} <111> were distinguished within these fibers (figure 2).

A bounded texture with two lines symmetric with respect to the AD, unequal in intensity of components, close to {332} <113> (in the used coordinate system) was formed in the peripheral area as a result of deformation (figure 1). Apparently, the inequality of the {332} <113> components (asymmetry of the deformation texture) is associated with the asymmetry of the stress state, which is implemented at drawing in the peripheral area. The scatter of both orientations is of a complicated nature, mostly rotating around the axes belonging in the TD – AD plane.

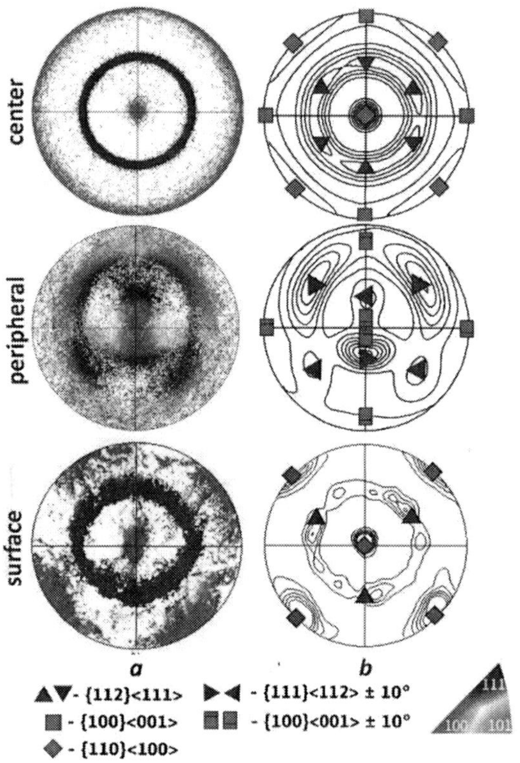

Figure 2. {100} PF of the copper cold-drawn wire texture; (a) – scattered data; (b) – contouring data + ideal orientations.

An analysis of the relative location of the main components of the peripheral and central areas of the crystal geometry indicates that they are associated with spinning around the TD by an angle of nearly 80°, or 90° ± 10°, where ± 10° is the drawing conicity angle. Analogous distinction in the center and on the surface were noted during cold rolling of aluminum with colossal friction between the rolls and the strip [9, 22] or during hot rolling of electrical steel [23].

Thus, in the peripheral area (figure 1) texture was submitted by deviated main components, that are faint, scattered {100} <001>, {112} <111> – orientations and powerful {111} <112>, {332} <113>. Importantly that this texture can be viewed not as axial, but as a complex bounded texture. Main orientations in the peripheral area can be viewed as a shear texture component that are connected with the main orientations of the central area by the rotation at an angle of 90° about the axis perpendicular to AD (figure 2). Therefore, orientation {332} <113> corresponds to {112} <111> rotated at 90°, scattered at ±10° and {100} <001> scatters from AD at ±10 °.

It is considered [9] that the shear texture components develop in the surface layer. The texture of directly to the surface is studied in this paper. The surface texture of the cold-drawn wire is different from the one of the peripheral areas. The surface texture scattered close to axial (figures 1, 2), but also has deviated main components: powerful {110} <001> and faint {112} <111> orientations. Moreover, the {112} <111> component is blurred around the AD, and the {110} <001> component has a limited character. Such a difference in the surface texture could be associated with the dynamic recrystallization processes occurring in the surface layers as a result of heating. But it was shown in [24] that the precise orientation {100} <001> is the main component of recrystallization in a copper wire. This orientation is absent in the surface layer. This orientation is not present in the surface layer. Apparently, this textural state is associated with the stress-strain characteristics of surface layers of the wire.

The rotation of texture in drawing is related to the specific features of the state of stress in this processing method. To estimate this rotation, a number of boundary value problems were solved for drawing of a copper wire. The distribution of the principal elongation directions (PED) calculated for each finite element is shown by arrows in Fig. 3. Here, the term PED is used as a special case of designating the principal elongation directions during drawing: one main deformation of elongation and two main deformations of shortening. The inclination of the PED was measured at the level of metal exit from the calibrating die strip. As is seen from this PED distribution, the inclination of PEDs increases with the coordinate along a radius. Its nonlinear character is clearly visible, and the rate of change of the angle of inclination is seen to decrease when the periphery (i.e., the wire surface) is approached (figure 3).

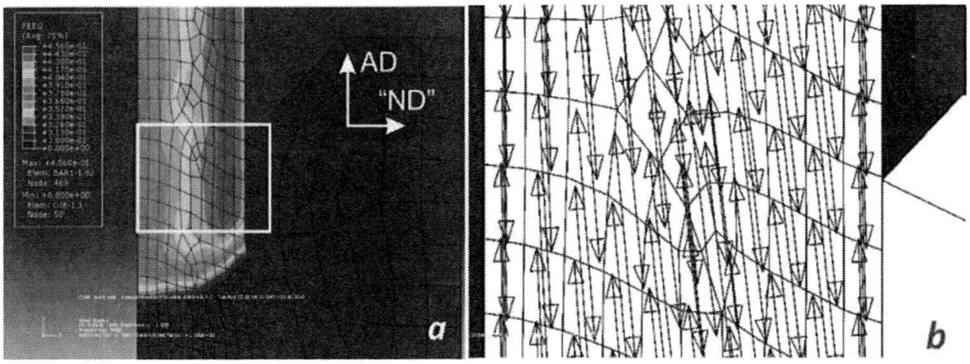

Figure 3. The formulation of the drawing problem in the ABAQUS software package with a finite element mesh: (a) – half of the deformation zone with the distribution of the deformation degree (levels); (b) – coordinate grid in half the deformation zone with indicated principal elongation directions (PED) along the radius.

It follows from the changes in the coordinate grid that its rectangular cells elongate without line distortions at the center of the workpiece; that is, tensile deformation takes place (figure 3(b)). Tensile deformation is principal, since additional shear are absent. Rectangular grid cells transform into parallelograms near the periphery (i.e., the workpiece surface). The long diagonals of the parallelograms in peripheral regions are inclined to the AD. Thus, shown that the deformation zone is nonuniform in drawing: the strain rates are different in this zone, and principal elongation axes have different angles of inclination along a workpiece radius. If the texture forms due to the elongation of material fibers, the directions of the principal strain axes indicate the deviation of the texture from the workpiece axis. Note, that the inclination of PEDs from AD on the surface becomes minimal, and it follows that the surface texture should rotate, as in the center of the wire. There is no complete coincidence of the texture of the center and the surface, because obviously, the real (3D) stress state is significantly more complicated than the modeled (2D) state.

4. Conclusion

Identity of texture of cold-drawn wire are associated with the stress state during drawing. Thus, the drawing texture is more correctly considered as a complicated rolling texture rather than a complicated fiber one.

The paper shows that the texture of deformed copper wire is a series of main orientations analogically to rolling texture. The texture orientations are associated with the deformation state in the drawing. The subsequent fibers develop locally in the central part of the wire: {112} <111> is referred to overall <111>-fiber and {110} <001> and {100} <001> is referred to overall <100>-fiber. Texture analogically to the shear one was generated in the periphery area of the wire. But it was revolved to the texture of the central area at an angle of 90±10°. The surface texture scattered close to axial and also has deviated preferred components: strong {110} <001> and weak {112} <111> orientations.

The results of the study indicate the opportunity and need of progress numerical models of the evolution of texture and physical and mechanical properties depending on the stress-strain state. This contribute to the forecast of the products properties both at the stage of manufacturing and during operating.

References

[1] Belyaevskikh A S, Lobanov M L, Rusakov G M, and Redikul'tsev A A 2015 Improving the production of superthin anisotropic electrical steel *Steel Transl.* **45(12)** 982–986
[2] Lobanov M L, Danilov S V and Urtsev V N 2020 Effect of structure and texture on failure of pipe steel sheets produced by TMCP *IOP Conference Series: Materials Science and Engineering* **709(4)** 044010
[3] Gottstein G 2004 *Physical Foundation of Materials Science* (Berlin: Springer-Verlag)
[4] Vusnyakov Ya D, Babareko A A, Vladimirov S A, and Égiz I V 1979 *Theory of Texture Formation in Metals and Alloys* (Moscow: Nauka)
[5] Wright R N 2011 *Wire technology: process engineering and metallurgy* (Elsevier Inc.)
[6] Sachs G 1928 *Trans. Faraday Soc.* **24** 84–92
[7] Taylor G I 1938 *J. Inst. Met.* **62** 307–324
[8] Bishop J F W and Hill R 1951 *Philos. Mag.* **42** 414–427
[9] Hölscher M, Raabe D, and Lücke K 1994 Relationship between rolling textures and shear textures in f.c.c. and b.c.c. metals *Acta Metall. Mater.* **42** 879–886
[10] Kraft F F, Chakkingal U, Baker G and Wright R N 1996 *Proc. 6th Int. Conf. Met. Form.* **60** 171–178
[11] Zorina M A, Karabanalov M S, Stepanov S I, Demakov S L, Loginov Yu N and Lobanov M L 2018 Fiber vs Rolling Texture: Stress State Dependence for Cold-Drawn Wire *Metall. Mater. Trans.* **A 49** 427–433
[12] Kuhlmann-Wilsdorf D 1980 *Dislocation concepts in friction and wear Fundamentals of Friction and Wear of Materials* (Metals Park Ohio) p 119
[13] Linben G, Mengelberg H D and Stuwe H P 1964 *Z. Met.* **55** 600–604
[14] Schlafer U and Bunge H J 1972 *Texture* **1** 31–49
[15] Montesin T and Heizmann J J 1992 *J. Appl. Crystallogr.* **25** 665–673
[16] Wang Y, Huang H Y, and Xie JX 2011 *Mater. Sci. Eng.* **A 530** 418–425
[17] Chen J, Yan W, Liu C X, Ding R G and Fan X H 2011 *Mater. Charact.* **62** 237–242
[18] Park H and Lee D N 2003 *Metall. Mater. Trans. Phys. Metall. Mater. Sci.* **A 34** 531–541
[19] Park H and Lee D N 2002 *Mater. Sci. Forum* **408–412** 637–642
[20] Komori K 2003 *Int. J. Mech. Sci.* **45** 141–160
[21] Demakov S L, Loginov Y N, Illarionov A G, Ivanova M A and Karabanalov M S 2012 *Phys. Met. Metallogr.* **113** 681–686
[22] Lobanov M L, Loginov Yu N, Danilov S V, Golovin M A and Karabanalov M S 2018 Effect of Hot Rolling Rate on the Structure and Texture Condition of Plates of the Al-Si-Mg Alloy System *Met. Sci. Heat Treat.* **60** 322–328
[23] Lobanov M L, Redikul'tsev A A, Rusakov G M and Danilov S V 2015 Interrelation between the orientations of deformation and recrystallization in hot rolling of anisotropic electrical steel *Met. Sci. Heat Treat.* **57** 492–497
[24] Zorina M A, Lobanov M L, Makarova E A and Rusakov G M 2018 Primary recrystallization texture in FCC-Metal with low packing defect energy *Met. Sci. Heat Treat.* **60** 329–336

Acknowledgments

The authors are grateful for cooperation of the support program of leading RF universities with the aim of increasing their competitiveness No. 211 of the RF government No. 02.A03,21.0006.

IOP Conf. Series: Materials Science and Engineering **969** (2020) 012081 doi:10.1088/1757-899X/969/1/012081

Evaluating the Range of Applicability of Existing Models of Stress Distribution in the Neck Forming on Cylindrical Specimens during Tensile Testing

M V Erpalov

Metal forming department, Ural Federal University named after the first President of Russia B. N. Yeltsin, 19, Mira Street, Ekaterinburg, 620002, Russia

E-mail: m.v.erpalov@urfu.ru

Abstract. This paper is devoted to the post-processing the results of testing the cylindrical specimens for tension. The tensile test are considered from the point of view of studying the rheological properties of metals and alloys. The main problem in identifying hardening curves from tensile test results is the formation of a neck, inside which the stress state of the material differs from uniaxial. The paper considers theoretical solutions to the problem of stress distribution in the neck region, namely, the models of Bridgman, Davidenkov-Spiridonova and Ostsemin. These models make it possible to calculate equivalent stress values even after specimen deformation localizes in the neck. Using the finite element method, the range of applicability of these models was evaluated. It was shown in the paper that the Ostsemin stress distribution model allows to achieve the best results of the material hardening curves identification regardless of its rheological properties. However, the range of applicability of this model is limited by the maximum strain value of 1.5-2.0. With further deformation of the specimen, an overestimation of the calculated values of equivalent stresses with respect to the proper values is observed.

1. Introduction

The rheological properties of metals and alloys are the dependence of the flow stress on the strain and strain rate, the temperature conditions of a metal forming process and the structural state of a workpiece. Taking into account the principle of minimum energy supplied on the deformation, these properties determine the uniformity of the strain distribution in the volume of the workpiece and, as a result, characterize the operational properties of the finished product [1]. The rheological properties of materials also determine the force, torque and power parameters of any metal forming process, which are important technological factors in the production [2].

There are various types of tests to study the rheological properties of metals and alloys [3]. Among them are tensile tests [4–6], compression [7–10] and torsion tests [11–15], a combined process of compression with torsion, and a number of others [16]. At the same time, one of the simplest testing method from the point of view of the implementation is the method of the tensile testing of cylindrical specimens. However, despite the deceptive simplicity and extremely wide distribution, this testing method is rarely applied to study the materials hardening curves. This is due to the complexity of post-processing the experimental data corresponding to the stage of concentrated deformation of the specimen in the forming neck. In the neck a simple uniaxial stress state of the material changes into a more complex triaxial state, which corresponds to the inhomogeneous distribution of axial tensile stress and other ones along the diameter d in the minimum cross-section of the specimen (figure 1). This leads to the fact that the equivalent flow stress cannot be found by simply dividing the tensile force P by the cross-sectional area of the specimen. Doing it, you can calculate only the true stress in the neck, which is distributed uniformly along any cross-section of the specimen:

$$\sigma_{true} = \frac{4P}{\pi d^2}. \tag{1}$$

But in order to determine the equivalent (effective) stress, you need to use the correction coefficient K, which takes into account the heterogeneity of the stress distribution in the neck [17]:

$$\sigma_{eq} = \frac{\sigma_{true}}{K}. \tag{2}$$

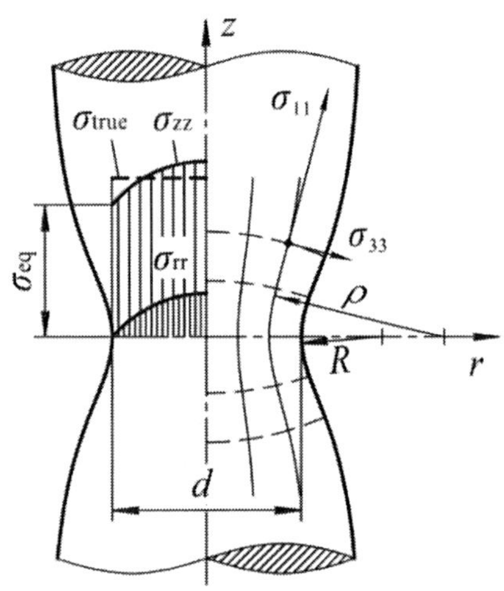

Figure 1. Inhomogeneous stress distribution in the minimum cross-section of the neck:
σ_{zz} – axial tensile stress;
σ_{rr} – radial stress;
σ_{11} – maximum principal stress;
σ_{33} – minimum principal stress.

There is no exact solution to the problem of tensile stress distribution in the neck of the specimen under tension. However, there are various theoretical models of stress distribution. They make it possible to calculate the value of the correction coefficient K based on the geometric parameters of the neck. Among these parameters are the specimen diameter d and the radius of curvature of the neck profile R (figure 1). Bridgman [17], as well as by Davidenkov and Spiridonova [18] developed such models as far back as in the middle of the last century. However, relatively recently Ostsemin proposed new model of stress distribution in the works [19, 20].

It should be noted that the use of any of mentioned above theoretical models represents a certain difficulty associated with assessing the reliability of the hardening curves obtained. In addition, there are no works devoted to the solution of this problem in the research literature. Therefore, the aim of this study is to evaluate the effectiveness of the application of these theoretical models for identifying the hardening curves of metals and alloys according to the results of tensile tests of cylindrical specimens. The addition aim of this study is to determine their applicability depending on the type of the hardening law of a material and its ductility.

2. The models of stress distribution in the neck

The well-known models of stress distribution in the neck of the specimen are based on a number of assumptions. Among the general ones is the assumption that the specimen has rotational symmetry about the z axis. In addition, the specimen is symmetrical in the direction of positive and negative z values relatively to the plane perpendicular to the axis and passing through the minimum cross-section of the neck. The flow stresses satisfy the von Mises condition and at each instant of a time can be determined by the accumulated strain value. Nevertheless, it is assumed that in the neck section strain changes insignificantly and is uniformly distributed along the specimen radius at each instant of a time. The specimen material is isotropic.

In [17], Bridgman suggested that in the immediate vicinity of the neck the contour of the specimen can be approximated by a tangent circle, and the surface of the principal stresses σ_{33} (dashed lines in figure 1) can be approximated by a sphere. Given these assumptions, he found the stress distribution in the neck section of the specimen:

$$\begin{cases} \sigma_{rr} = \sigma_{\varphi\varphi} = \sigma_{eq} \cdot \ln \dfrac{a^2 + 2aR - r^2}{2aR}; \\ \sigma_{zz} = \sigma_{eq} + \sigma_{eq} \cdot \ln \dfrac{a^2 + 2aR - r^2}{2aR}, \end{cases} \tag{3}$$

where r is the radial coordinate and a is the specimen radius in the minimum cross-section of the neck, $a = d/2$.

Based on the equation for calculating tensile stress:

$$P = 2\pi \int_0^a \sigma_{zz} r \mathrm{d}r , \tag{4}$$

the correction coefficient K according to the Bridgman model (3) can be found by the formula:

$$K_{Bridgman} = \left(1 + 2\frac{R}{a}\right) \cdot \ln\left(1 + \frac{a}{2R}\right). \tag{5}$$

Another model of stress distribution proposed by Davidenkov and Spiridonova [18] is based on the assumption that in the minimum neck cross-section the radius of curvature of the maximum principal stress lines σ_{11} varies along the radius r according to the function:

$$\rho = R\frac{a}{r}. \tag{6}$$

Taking into account equation (6), the stress distribution in the neck can be written in the form:

$$\begin{cases} \sigma_{rr} = \sigma_{\varphi\varphi} = \sigma_{eq} \cdot \dfrac{a^2 - r^2}{2aR}; \\ \sigma_{zz} = \sigma_{eq} + \sigma_{eq} \cdot \dfrac{a^2 - r^2}{2aR}, \end{cases} \tag{7}$$

and the correction coefficient K can be calculated as follows:

$$K_{Davidenkov} = 1 + \frac{a}{4R}. \tag{8}$$

Ostsemin [19, 20] put forward new assumptions regarding the lines of principal stresses. He considered that the lines of the maximum principal stress σ_{11} are hyperbolas, and the lines of the minimum principal stress σ_{33} are ellipses. Thus, the stress distribution in the neck was obtained:

$$\begin{cases} \sigma_{rr} = \sigma_{\varphi\varphi} = \sigma_{eq} \cdot \ln \dfrac{a^2 + aR - r^2}{aR}; \\ \sigma_{zz} = \sigma_{eq} + \sigma_{eq} \cdot \ln \dfrac{a^2 + aR - r^2}{aR}. \end{cases} \tag{9}$$

According to the Ostsemin model, the correction coefficient K can be calculated by the formula:

$$K_{Ostsemin} = \left(1 + \frac{R}{a}\right) \cdot \ln\left(1 + \frac{a}{R}\right). \tag{10}$$

3. Research methodology

The solution of formulated problem is possible only with the use of computer simulation methods. This is because you have ability to compare the material hardening curves calculated using different theoretical models of stress distribution in the neck with the hardening curve that is set in the software during the formulation of the simulation problem. Another advantage of the computer simulation is that the solution to the problem is not limited to the type of the hardening law of a material and its ductility.

Figure 2 presents the hardening curves of the test materials used to simulate the tension of specimens in the Deform-2D software. We examined two materials with hardening, one material had a constant flow stress, and the last one had the effect of softening. We used standard cylindrical specimens (figure 3) corresponding to the Russian standard GOST 1497.

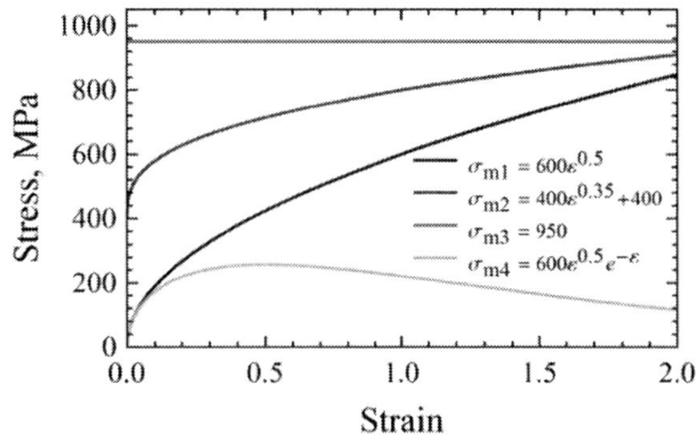

Figure 2. Stress-strain curves of test materials.

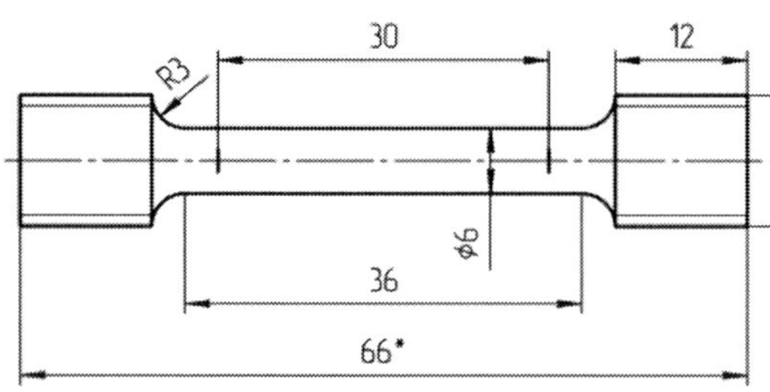

Figure 3. Specimen dimensions.

In order to measure the profile of the forming neck, we used an approach based on the analytical description of all points belonging to the surface of the specimen by the following function:

$$r = r_1 - (r_1 - a) \cdot \left(1 + \frac{z^2}{c}\right)^{-1}, \tag{11}$$

where z and r are the axial and radial coordinates of points on the specimen surface, r_1 is the radius of the gauge length of the specimen corresponding to the beginning of the neck formation and c is the parameter sensitive to differences in material properties. At each simulation step, the coefficients of equation (11) were found by the least-squares deviation method, while the coordinates of all points on the gauge length of the specimen were used for fitting.

From equation (11), we obtained the formula for calculating the radius of curvature of the neck:

$$R = \frac{c}{2 \cdot (r_1 - a)}. \tag{12}$$

It should be noted that described approach is not effective at all simulation steps. For example, for the test material m2 (you can see its hardening law on figure 2) with the total specimen extension of 7.5 mm, the accuracy of the neck profile fitting is 98.3% (figure 4a). And with the extension of 10 mm corresponding to the strain value of 2.44, the fitting accuracy decreases to 88.3% (figure 4b). As you can see, the largest error appears in the center of the neck. This leads to inaccurate calculation of the radius of curvature of the neck R. To increase the accuracy of the post-processing, we applied the procedure in which the number of taken into account points on the surface was limited iteratively. In each iteration, starting from the second one, in order to fit the neck profile we took only those points corresponding to the specimen surface that were located between the inflection points of the function (11) with the coefficients found in the previous iteration. Practice has shown the convergence of this algorithm at 3-4 iterations. Figure 4c shows an example of using an iterative procedure for the same specimen with the extension of 10 mm. As you can see, the fitting accuracy increases up to 96.0%.

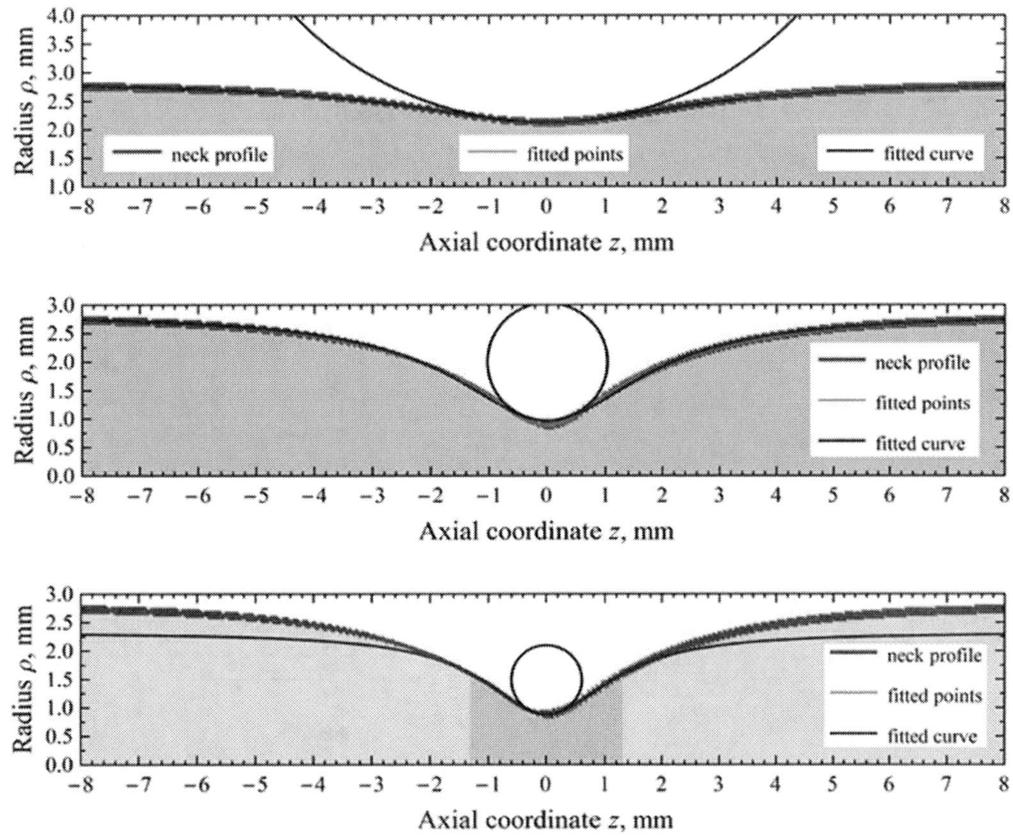

Figure 4. The results of fitting the neck profile for the test material m2: (a) extension of 7.5 mm; (b) extension of 10 mm; (c) extension of 10 mm, iteration procedure was used.

Having determined at each simulation step the values of the radius of the neck curvature R, the radius of the specimen a, and also the tensile force P, using the equations (5), (8) and (10), we calculated the values of the correction coefficients K according to the Bridgman, Davidenkov-Spiridonova and Ostsemin models. Then, using equations (1) and (2), the true and equivalent stresses were determined depending on the strain values. We calculated the last ones according to the equations:

$$\varepsilon_{eq} \approx \varepsilon_{true} = 2\ln\frac{d_0}{2a}, \tag{13}$$

where d_0 is the initial diameter of the gauge length of the specimen, $d_0 = 6$ mm.

4. Results

Figure 5 presents the results of the calculation of the hardening curves of the test materials in comparison with the corresponding hardening curves that were set during formulation of the finite element simulation problems. A comparative analysis of the simulation results shows that taking into account the inhomogeneity of the stress distribution in the neck can significantly increase the accuracy of the hardening curves identification, while post-processing the results of real tensile tests. As you can see from figure 5, Ostsemin stress distribution model provides the best result. However, the hardening curve identification remains its high accuracy only up to the strain values of approximately 1.5-2.0. With values of the strain greater than 2, a noticeable overestimation of equivalent stresses relative to a proper hardening curve is observed. The type of the material hardening law has no significant effect on the study of rheological properties.

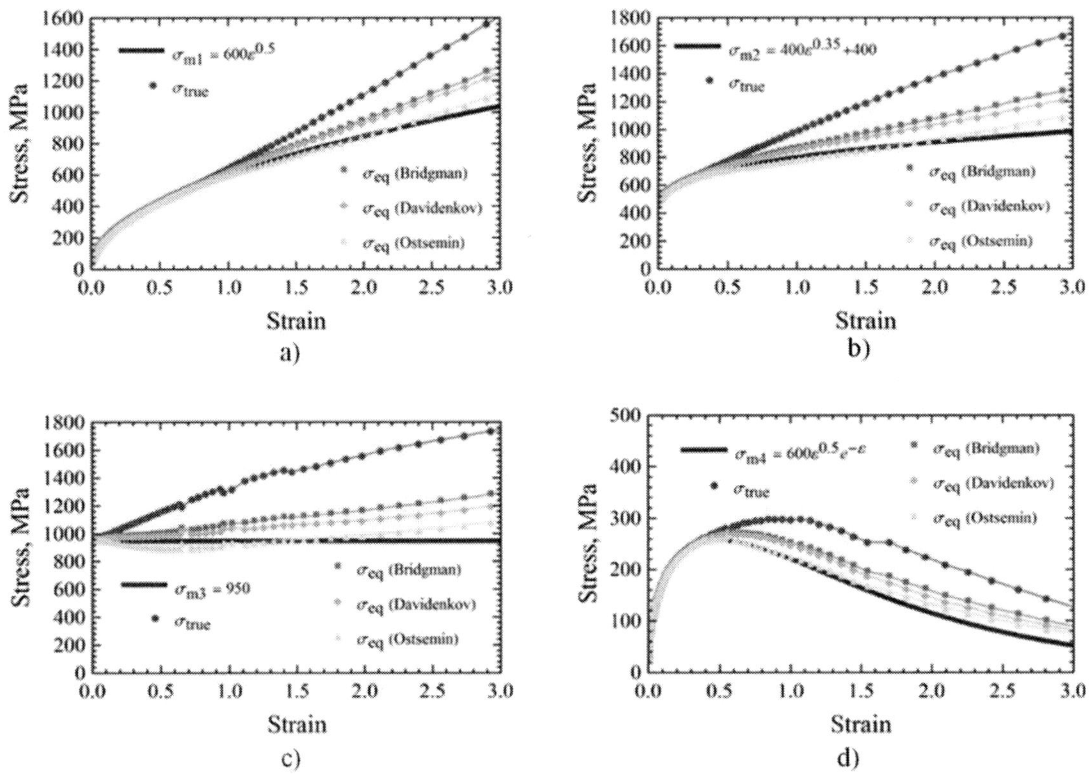

Figure 5. The comparison of the stress-strain curves of the test materials m1 (a), m2 (b), m3 (c) and m4 (d) obtained using different models of stress distribution in the minimum cross-section of the neck.

5. Conclusions

In the work, known models of stress distribution in the neck of a cylindrical specimen under tension are considered. The computer simulation of tensile tests showed that the Ostsemin model allows to achieve the best results of the material hardening curves identification regardless of its rheological properties. However, the range of applicability of this model is limited by the maximum strain value of 1.5-2.0. With further deformation of the specimen, an overestimation of the calculated values of equivalent stresses with respect to the proper values is observed. This phenomenon may be associated with an inhomogeneous distribution of strain over the neck cross-section and should be studied in the course of further researches.

References

[1] Loginov Yu, Potapov A and Shalaev N 2012 *Titan* **37** 36–42
[2] Nadai A 1963 *Theory of flow and fracture of solids* (New York: McGraw-Hill Book Company) p 840
[3] Rabotnov Ju N 1962 *Strength of materials* (Moscow: Fizmatgiz) p 456
[4] Farber V M, Polukhina O N, Vichuzhanin D I, Khotinov V A and Smirnov S V 2019 *Metal Science and Heat Treatment* **61** 274–279
[5] Panov D, Pertsev A, Smirnov A, Khotinov V and Simonov Y 2019 *Materials* **12** 2058

[6] Khotinov V A, Polukhina O N, Selivanova O V and Farber V M 2019 *Inorganic Materials: Applied Research* **10** 939–944
[7] Gladkovsky S V, Potapov A I and Lepikhin S V 2015 *Diagnostics, Resource and Mechanics of materials and structures* **4** 16–26
[8] Smirnov A S and Konovalov A V 2015 *Diagnostics, Resource and Mechanics of Materials and Structures* **1** 61–72
[9] Smirnov A, Belozerov G, Konovalov A, Shveikin V and Muizemnek O 2016 *AIP Conf. Proc.* **1785** 040068
[10] Khoddam S and Hodgson D 2018 *Mech. Mater.* **122** 1–8
[11] Erpalov M V and Pavlov D A 2018 *Ferrous metals* **12** 72–76
[12] Erpalov M V and Pavlov D A 2018 *CIS Iron and Steel Review* **16** 71–75
[13] Fields D S and Backofen W A 1957 *ASTM Proceeding* **57** 1259–1272
[14] Laber K, Kawalek A, Sawicki S, Dyja H, Borowski J, Lesniak D and Jurczak H 2016 *Key Eng. Mater.* **682** 356–361
[15] Khoddam S, Hodgson P D and Bahramabadi M J 2011 *Mater. Des.* **32** 1903-1909
[16] Edalati K and Hortia Z 2016 *Mater. Sci. Eng.* A **652** 325–352
[17] Bridgman P W 1952 *Studies in Large Plastic Flow and Fracture, with Special Emphasis on the Effect of Hydrostatic Pressure* (New York: McGraw-Hill) p 362
[18] Davidenkov N N and Spiridonova N I 1946 *Proc. ASTM* **46** 1147–1159
[19] Ostsemin A A 2009 *Strength of Materials* **41** 356–362
[20] Dilman V L and Ostsemin A A 1998 *Factory Laboratories* **5** 47–49

Acknowledgments
This work was supported by a grant MK–1878.2020.8 from the President of the Russian Federation for young scientists – candidates of sciences.

Hot Isostatic Pressing of AISI 6150 Steel

D G Chkalova and A S Raznoschikov

Vladimir State University, 87, Gorky Street, Vladimir, 600000, Russia

E-mail: j.larenax@gmail.com

Abstract. Hot isostatic pressing (HIP) is a promising way to improve the quality of structural materials, and therefore the structures themselves. The article presents the results of experimental studies on the application of technology for processing critical products made of AISI 6150 steel in order to reduce the number of pores in material and formed weld. The influence of HIP process on the structure of AISI 6150 steel samples was investigated using x-ray computed tomography. The experimental results show a significant improvement in structure of material, a decrease in the number and size of pores relative to the initial values. The experiments presented in this work were carried out on the basis of a HIP setup located at Vladimir State University.

1. Introduction

At present, several methods are known for reducing the number of discontinuities formed in critical details. These include reductive heat treatment (HT) [1–4], mechanical-thermal treatment (MTT) [5–7] and electric pulse treatment (EPT) [8–10]. The action of HT and MTT is based on the fact that constant or cyclic heating or, in combination with it, mechanical and electro-impulse action (deformation) accelerate diffusion processes in the material of the part, which leads to the "healing" of some part of micropores and microcracks. However, these processes are characterized by a very long duration and low efficiency, only those discontinuities with a size of less than 1 μm are eliminated. At the same time, if tensile stresses arise in the discontinuity region, this can lead to an increase in these defects. Gas pores are not accessible for mechanical removal (stripping) and are stress concentrators, centers of nucleation of fatigue and thermal fatigue cracks.

The most effective way to eliminate internal pores and shells in full-crystalline castings of various alloys is hot isostatic pressing, which has been used for many years to process castings for critical purposes [11–14]. HIP has a number of advantages that determine its practical attractiveness. It is a complex technological process, the essence of which is the simultaneous exposure of a metal to high temperature and pressure obtained by pumping inert gas, more often argon, into a vessel. Under these conditions, processes characteristic of pressure sintering develop in a powder billet. In a monolithic workpiece with internal isolated defects such as pores, friability, cracks in the defect zone, plastic deformations occur, as a result of which the defect walls are closed. The development of diffusion processes leads to a gradual blurring of the interface and the disappearance of the defect [15, 16].

The combined effect on the workpiece of high pressures and temperatures can effectively eliminate (heal) micro- and macrodefects that occur in monolithic workpieces when they are obtained by traditional methods (casting, forging, stamping, rolling, welding) [17–20].

2. HIP processing of AISI 6150 steel

For the experiments, an automated hot isostatic press was used, designed for heat treatment of metal and ceramic products. The main technological parameters of the HIP setup used for the experiments are presented in table 1.

Table 1. HIP technological parameters.

Parameter	Value
Useful diameter	250 mm
Useful height	350 mm
Maximum filling mass	up to 30kg
Camera type	cylindrical
Working pressure	up to 2500 bar
Working temperature	up to 2000 °C
Temperature gradient	less than 3 °C
Heating / cooling speed	controlled
Pressure medium	argon
Furnace type	graphite
Work mode	manual, automatic

As samples for investigating the effectiveness of hot isostatic pressing operation, was used a blank of a hydraulic equipment part operating at a pressure of up to 50 MPa. Sample material – AISI 6150 steel (after welding operation). Parts containing the maximum number of defects were used as samples. The aim of this approach was to study the effectiveness of eliminating various types of defects by hot isostatic pressing. In accordance with the literature, the application of HIP processing effectively eliminates pores, internal cracks in material, which do not have an exit to surface, increase material isotropy, which is widely used for processing parts obtained by SLM method [21–23]. The elimination of pores and cracks emerging on surface is possible when the part is encapsulated in a shell that prevents pores from filling with the medium (argon) used to increase pressure in vessel, however, this operation was not considered in this work.

The HIP process took place with the following parameters: working temperature – 1000 °C, working pressure 1500 bar, rate of rise and decrease in temperature – 6 °C/min, holding time – 7 hours. The graph of HIP processing is shown in figure 1.

Figure 2 shows the surface of the processed sample formed by a section of plane of the sleeve welded joint. The image was obtained by x-ray tomography. The same area was considered before and after HIP, orientation of the part is saved. The figure 2a shows many pores in the weld area.

As a result of HIP treatment, a significant improvement in the structure of material is observed. The pore content drops significantly, 5.39% before treatment and 0.21% after HIP. The total percentage of pores after the hot isostatic pressing operation in the weld area is comparable to the porosity of the base material. The pore content in the initial sample outside the weld pool before and after treatment in HIP is shown in figure 2. The nature of pores distribution in weld is different from the porosity of base material. Round pores with a diameter of up to 2 mm are recorded, which were probably formed as a result of the union of gas cavities. The presence of lack of fusion, uneven penetration corresponding to codes 403, 4013 [24] was not detected.

Despite the extremely positive result of processing, the gas-conditioning of welded joints must be treated with caution. With HIP, pores that have access to the seam surface or lie near it do not close and can even open [25]. Non-discontinuities filled with air are also not welded. As a result of processing at high pressure, they only flatten. Subsequent heat treatment, due to expansion during heating of gas contained in discontinuities, they increase in volume and can break the metal. The latter, in particular, is characteristic of castings obtained by injection molding. In the case of filling defects with hydrogen (hydrogen porosity in castings, welds, etc.), it is dissolved in a solid solution and partial extraction into the atmosphere of the gas bath with successful welding of the defect itself.

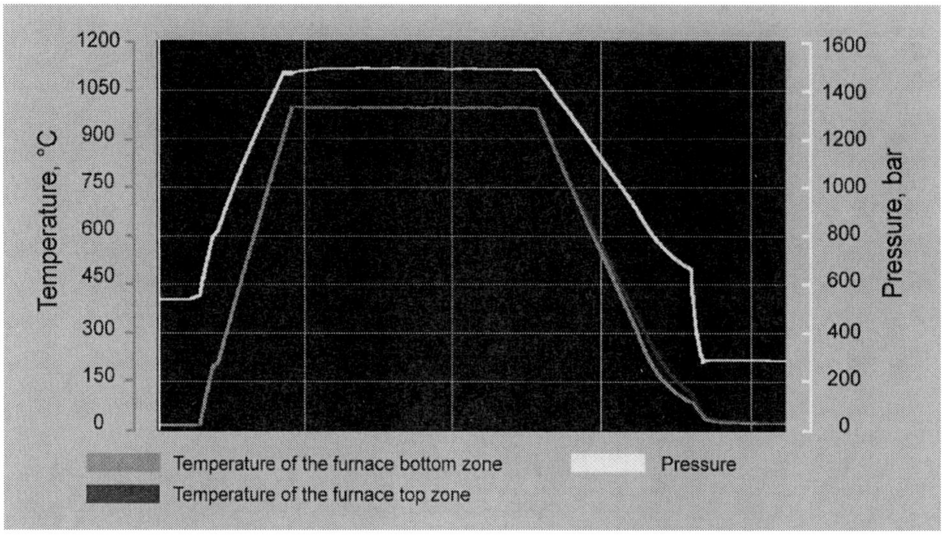

Figure 1. Graph of HIP processing cycle.

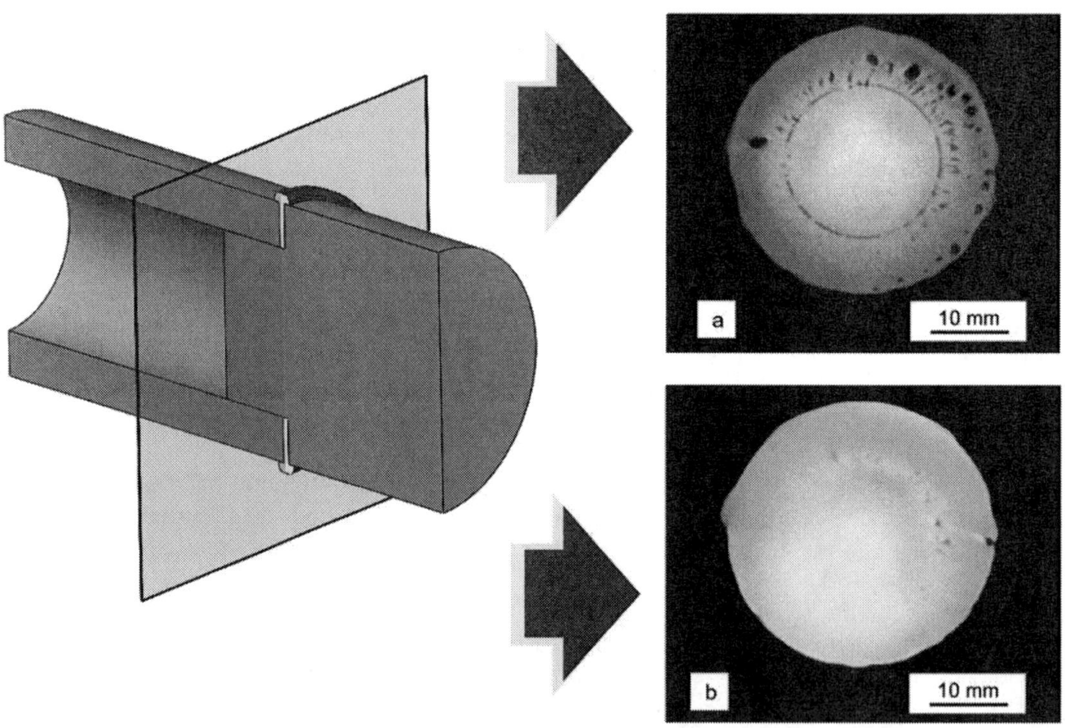

Figure 2. Image of the seam cross-section: a - before HIP, b - after HIP.

3. Conclusion

Hot isostatic pressing is a relatively long and complex technological process associated with the use of expensive equipment. In this regard, the use of such technology was limited only to those branches of engineering – aerospace technology and nuclear energy, where high capital and operating costs are justified [26–30]. Analysis and generalization of the results achieved abroad allow to consider that HIP is an extremely effective and promising technological process. Intensively conducted research works are constantly expanding the scope of gas-static treatment and increase its efficiency in production conditions. However, hot isostatic pressing is not a tool that can and should be used for all materials and all types of products. Ultimately, the possibility of using this process is determined by the cost of the resulting product.

References

[1] Pakdee U, Chiangga S, Suwannatus S, Duangsawat B, Oopathump C and Klumdoung P 2015 Reductive heat treatment of austenitic stainless steel substrate for growth of hybrid carbon nanostructures *Proceedings 2015 International Conference on Science and Technology*
[2] Rowell R M, Andersone I and Andersons B 2012 *Heat treatment. Handbook of Wood Chemistry and Wood Composites, Second Edition*
[3] Totten G E 2006 *Steel heat treatment: Metallurgy and technologies*
[4] Vrancken B, Thijs L, Kruth J P and Van Humbeeck J 2012 Heat treatment of Ti6Al4V produced by Selective Laser Melting: Microstructure and mechanical properties *J. Alloys Compd.*
[5] Lausmaa J 2001 *Mechanical, Thermal, Chemical and Electrochemical Surface Treatment of Titanium*
[6] Durai T G, Das K and Das S 2008 Corrosion behavior of Al-Zn/Al2O3 and Al-Zn-X/Al2O3 (X = Cu, Mn) composites synthesized by mechanical-thermal treatment *J. Alloys Compd.*
[7] McElroy R J and Szkopiak Z C 1972 Dislocation-substructure-strengthening and mechanical-thermal treatment of metals *Int. Metall. Rev.*
[8] Shi X D, Xue Q G, Wang J S, Cang D Q, Liu S H and Niu S Z 2005 Effect of electric pulse treating time on solidification structure of Al-18%Si alloy *Jinshu Rechuli/Heat Treat. Met.*
[9] Shi X, Wang J, Xue Q, Cang D, Wang L and Wu S 2005 Effect of electric pulse treatment on solidification structure of an Al-15% Si alloy *Rare Met.*
[10] Vakulenko I A, Nadezhdin Y L, Sokirko V A, Volchok I P and Mitiaev A A 2013 Electric pulse treatment of welded joint of aluminum alloy *Sci. Transp. Progress. Bull. Dnipropetr. Natl. Univ. Railw. Transp.*
[11] Bocanegra-Bernal M H 2004 Hot isostatic pressing (HIP) technology and its applications to metals and ceramics *J. Mater. Sci.*
[12] Tammas-Williams S, Withers P J, Todd I and Prangnell P B 2016 Porosity regrowth during heat treatment of hot isostatically pressed additively manufactured titanium components *Scr. Mater.*
[13] Kumar A, Bai Y, Eklund A and Williams C B 2017 Effects of Hot Isostatic Pressing on Copper Parts Fabricated via Binder Jetting *Procedia Manuf.*
[14] Atkinson H V. and Davies S 2000 Fundamental aspects of hot isostatic pressing: An overview *Metall. Mater. Trans. A Phys. Metall. Mater. Sci.*
[15] Essa K, Jamshidi P, Zou J, Attallah M M and Hassanin H 2018 Porosity control in 316L stainless steel using cold and hot isostatic pressing *Mater. Des.*
[16] Hsu K-T, Wang H-S, Chen H-G and Chen P-C 2016 Effects of the Hot Isostatic Pressing Process on Crack Healing of the Laser Repair-Welded CM247LC Superalloy *Metals (Basel)*
[17] Kruk S I, Sagalevich V M, Gorokhov V Y and Lukin V I 1991 Application of gas-static treatment for improvement of the welded joints quality *Welding International* **11** 8–11
[18] Lavery N P, Cherry J, Mehmood S, Davies H, Girling B, Sackett E, Brown S G R and Sienz J 2017 Effects of hot isostatic pressing on the elastic modulus and tensile properties of

316L parts made by powder bed laser fusion *Mater. Sci. Eng. A*

[19] Elrakayby H, Kim H K, Hong S S and Kim K T 2015 An investigation of densification behavior of nickel alloy powder during hot isostatic pressing *Adv. Powder Technol.*

[20] Tammas-Williams S, Withers P J, Todd I and Prangnell P B 2016 The Effectiveness of Hot Isostatic Pressing for Closing Porosity in Titanium Parts Manufactured by Selective Electron Beam Melting *Metall. Mater. Trans. A Phys. Metall. Mater. Sci.*

[21] Voznesenskaya A A, Kochuev D A, Chkalov R V, Kireev A V and Morozov V V 2020 Research of post-processing approaches for parts obtained by the method of selective laser melting *Journal of Physics: Conference Series*

[22] Finfrock C B, Exil A, Carroll J D and Deibler L 2018 Effect of Hot Isostatic Pressing and Powder Feedstock on Porosity, Microstructure, and Mechanical Properties of Selective Laser Melted AlSi10Mg *Metallogr. Microstruct. Anal.*

[23] Tillmann W, Schaak C, Nellesen J, Schaper M, Aydinöz M E and Hoyer K P 2017 Hot isostatic pressing of IN718 components manufactured by selective laser melting *Addit. Manuf.*

[24] Ryabtsev I A, Rozert R, Turyk E and Ryabtsev I I 2017 Classification and characteristic of defects of deposited layers according to the international standard ISO 6520-1:2007 *The Paton Welding Journal* **9** 44–49

[25] Geenen K, Röttger A and Theisen W 2017 Corrosion behavior of 316L austenitic steel processed by selective laser melting, hot-isostatic pressing, and casting *Mater. Corros.*

[26] Rao G A, Sankaranarayana M and Balasubramaniam S 2012 Hot isostatic pressing technology for defence and space applications *Def. Sci. J.*

[27] Park D J, Kim H G, Park J Y, Jung Y I, Park J H and Koo Y H 2016 FeCrAl and Zr alloys joined using hot isostatic pressing for fusion energy applications *Fusion Eng. Des.*

[28] Busom Descarrega J, Calviani M, Hutsch T, López Sola E, Pérez Fontenla A T, Perillo Marcone A, Sgobba S and Weißgärber T 2020 Application of hot isostatic pressing (HIP) technology to diffusion bond refractory metals for proton beam targets and absorbers at CERN *Mater. Des. Process. Commun.*

[29] Tejado E, Carvalho P A, Munoz A, Dias M, Correia J B, Mardolcar U V. and Pastor J Y 2015 The effects of tantalum addition on the microtexture and mechanical behaviour of tungsten for ITER applications *Journal of Nuclear Materials*

[30] Crapps J, Clarke K, Katz J, Alexander D J, Aikin B, Vargas V D, Montalvo J D, Dombrowski D E and Mihaila B 2013 Development of the hot isostatic press manufacturing process for monolithic nuclear fuel *Nucl. Eng. Des.*

IOP Conf. Series: Materials Science and Engineering **969** (2020) 012083 doi:10.1088/1757-899X/969/1/012083

The Model for Predicting the Radius of Curvature of the Neck Forming on Cylindrical Specimens during Tensile Testing

M V Erpalov and D A Pavlov

Metal forming department, Ural Federal University named after the first President of Russia B. N. Yeltsin, 19, Mira Street, Ekaterinburg, 620002, Russia

E-mail: m.v.erpalov@urfu.ru

Abstract. The article is devoted to the testing of cylindrical specimens for tension. The main tendency of tensile tests is the improvement of methods for determining equivalent stresses acting in specimens while deformation localizes in the neck, i.e. methods for determining the hardening curves of materials up to the fracture. The most reliable way to study the rheological properties of materials is the direct measurement of the neck profile, namely the determination of the specimen diameter in the minimum cross-section of the neck and the radius of its curvature. However, direct measurements of the neck are associated with certain difficulties. They include a small number of measurements corresponding to different instants of the test, as well as inaccurate determination of the radius of the neck curvature due to the subjective approach to the measurement process. The measurements become more accurate if you use some equation for the analytical description of the neck profile and perform the measurements using optical methods during the test. However, such processing algorithms are often inaccessible to researchers. In this regard, this work proposes a model for predicting the radius of the neck curvature that is formed during tensile testing the cylindrical specimens. This model relies on the volume conservation law and the analytical description of the neck profile using the one-parameter equation. To calculate the radius of the neck curvature at each instant of the test it is necessary to know only the current values of elongation and the minimum specimen diameter. The effectiveness of the proposed mathematical model is evaluated using the computer simulation. The simulation results show that the use of predicted radius of the neck curvature allows calculating the equivalent stresses close to proper ones without the need of the direct neck profile measurement.

1. Introduction

One of the most common methods for studying the mechanical properties of materials is the tension of cylindrical specimens [1, 2]. During the test, it is possible to determine the mechanical characteristics of materials such as yield strength, tensile strength, percentage extension and percentage reduction of area, as well as a number of other characteristics [3–7]. However, during standard tensile tests it is not possible to achieve large degrees of uniform deformation because the softening of the specimen caused by a decrease in its cross-sectional area begins to prevail over the strain hardening of the material. As a result, the deformation gradually localizing in a narrow area called a neck. Due to the complexity of processing the experimental data corresponding to the stage of concentrated deformation of the specimen, tensile tests have not been widely used to study the rheological properties of materials. However, tensile tests are very promising for their wider use.

To define hardening curves of the materials based on the results of the tensile tests, it is necessary to calculate the equivalent stresses taking into account the inhomogeneous distribution of the axial tensile stresses in the neck area [8–11]:

$$\sigma_{eq} = \frac{P}{F \cdot K}, \tag{1}$$

where P is the tensile force, F is the cross-sectional area of the specimen and K is the correction coefficient that takes into account the inhomogeneous stress state of the material. The values of equivalent stresses have to be associated with the values of equivalent strains determined by the formula:

$$\varepsilon_{eq} = 2\ln\frac{d_0}{d}, \tag{2}$$

where d_0 is the original diameter of the specimen gauge length and d is the current value of the diameter corresponding to minimum cross-section of the specimen.

The main difficulty in defining the hardening curves of the material is the determination of the correction coefficient K that may correspond to the Bridgman [8], Davidenkov-Spiridonova [9] or Ostsemin [10, 11] models. The Ostsemin model is the most advanced among them. According to it, the formula to calculate the correction coefficient is as follows:

$$K = \left(1 + 2\frac{R}{d}\right) \cdot \ln\left(1 + \frac{d}{2R}\right), \tag{3}$$

where R is the radius of the neck curvature in the minimum section of the specimen.

The measurements of the radius R are associated with certain difficulties. Usually to measure the neck profile the specimen is removed from the grips of a test setup. Then the radius of the neck curvature is determined using an instrumental microscope [12], digital image correlation method [13] or using a set of templates and enlarged specimen images obtained with a projector [8]. In this case, the number of neck measurements is very limited and the test procedure is long-time and hard. Moreover, the neck profile measurements are subjective because a researcher visually selects the circle to approximate the neck profile.

The use of some equation to describe the neck profile is the most advanced measuring procedure. Knowing such an equation, one can calculate the desired radius of the neck curvature based on methods of mathematical analysis [14, 15]. At the same time, the specimen can stay in the grips of the test setup during the neck profile measurement that is performed with the use of the photography [16] or the video recording of the testing process [17, 18]. However, the algorithms for the neck images processing are closed and inaccessible to a wide circle of researchers.

A radically different approach is proposed in [15, 19], where the authors claim that the radius of the neck curvature is not depended on the properties of the material. They propose the statistical models for calculating the equivalent stresses according to a large number of experiments. However, the results of another study [20] show that the material properties affect the radius of the neck curvature and the assumption proposed in [15, 19] is incorrect.

According to [20, 21] the one-parameter equation may approximate the neck profile for any hardening curve of the material and the instant of the test:

$$\rho = \frac{d_1}{2} - \frac{d_1 - d}{2} \cdot \left(1 + \frac{z^2}{c}\right)^{-1}, \tag{4}$$

where ρ and z are the radial and axial coordinates of points on the specimen surface, d_1 is the diameter of specimen when the neck starts forming and c is the material coefficient. The universal equation (4) makes it possible to develop a mathematical model for predicting the radius of curvature R without performing direct measurements. This work is devoted to this problem.

2. The model of necking the specimen

Let consider the specimen having the original gauge length L_0, the parallel length L_c and the diameter of the original gauge length d_0 (figure 1). We divide the tension of specimen into two stages. The first stage corresponds to the uniform specimen deformation (figure 2), and the second stage – to the neck formation (figure 3).

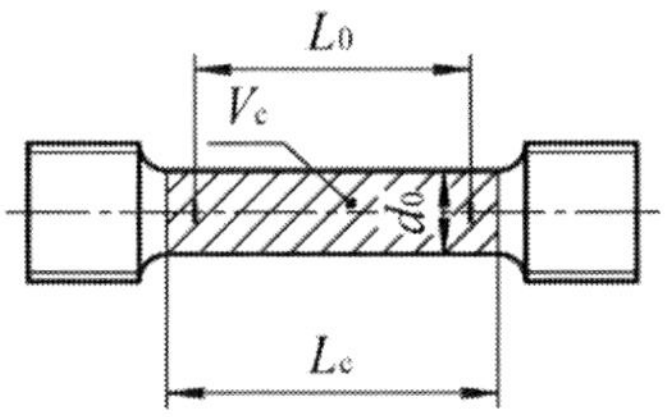

Figure 1. The specimen taken for the model development.

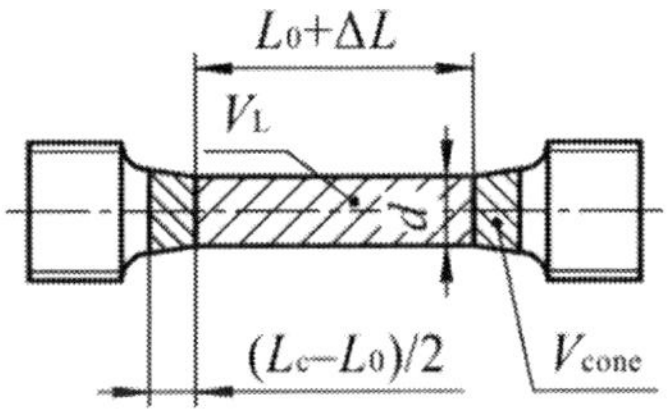

Figure 2. The stage of uniform deformation.

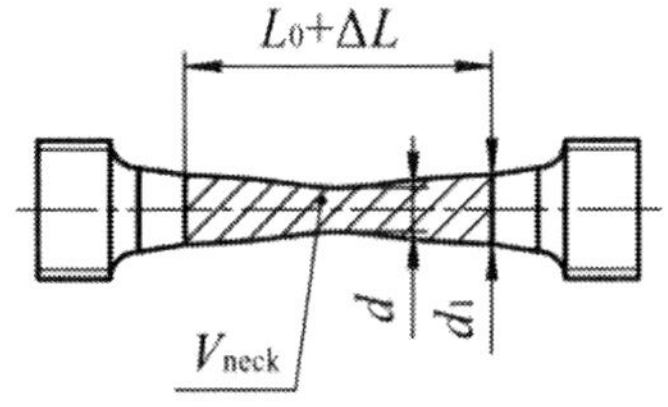

Figure 3. The stage of the neck formation.

We assume that at the first stage the part of deformation is uniformly developed along the gauge length:

$$L = L_0 + \Delta L, \tag{5}$$

where ΔL is the specimen elongation. The other part of deformation is developed into two conical parts having the length

$$L_{\text{cone}} = \frac{L_c - L_0}{2}. \tag{6}$$

We also assume that the conical parts are deformed with the constant length and the cross-sections attached to the specimen grip sections do not change the size. This assumption makes it possible to establish the law of the diameter change d at the first stage of testing the specimen. Thus, we consider the volume constancy equation:

$$V_c = V_L + 2V_{\text{cone}}, \tag{7}$$

where V_c is the volume of specimen corresponding to the original parallel length, V_L is the volume corresponding to the current gauge length L and V_{cone} is the volume of specimen conical part. In accordance with the equation (7), the diameter of the specimen at each instant of the uniform deformation satisfies the formula:

$$d = \frac{d_0\left[L_0 - L_c + \sqrt{9(L_0 + L_c)^2 + 12(L_0 + 2L_c)\Delta L}\,\right]}{4L_0 + 2L_c + 6\Delta L}. \tag{8}$$

The second stage of the test occurs when the tensile force assumes the maximum value, i.e. the deformation of the specimen starts to localize in the neck. We consider that the whole deformation is concentrated within the gauge length during the neck formation and the shape of the neck corresponds to the equation (4). Thus, knowing the analytical description of the neck shape we can write the equation of the volume balance in the following form:

$$\Delta V = V_L - V_{\text{neck}}, \tag{9}$$

where ΔV is the material volume displaced along the specimen during tension and V_{neck} is the volume of the specimen into the neck. We represent equation (9) in the following form:

$$\frac{\pi d_1^2}{4}(\Delta L - \Delta L_1) = \frac{\pi d_1^2}{4}(L_0 + \Delta L) - \pi \int_{-(L_0+\Delta L)/2}^{+(L_0+\Delta L)/2} \left[\frac{d_1}{2} - \frac{d_1 - d}{2} \cdot \left(1 + \frac{z^2}{c}\right)^{-1} \right]^2 dz, \quad (10)$$

where ΔL_1 is the elongation of the specimen corresponding to the maximum tensile force and d_1 is the diameter of specimen when the neck starts forming. The diameter d_1 is determined by equation (8) for the specimen elongation equal to ΔL_1.

The numerical solving of the equation (10) allows to find the values of parameter c corresponding to each specific elongation ΔL. We should note that it is necessary to measure the diameter d over the entire stage of the neck formation in order to take into account the influence of the material properties on the neck dimensions. Based on the values of the parameter c, it is possible to predict the radius of curvature of the resulting neck [20, 21]:

$$R = \frac{2c}{d_1 - d}. \quad (11)$$

The predicted radius of the neck curvature R allows calculating the values of the correction coefficient K in accordance with the equation (3) at each instant of the test. Then the material hardening curve can be defined in accordance with equations (1) and (2).

3. The validation of the proposed necking model

Evaluating the accuracy of the proposed mathematical model that allows identifying hardening curve of the material without the need of the direct neck profile measuring was performed by simulating the tensile process in the Deform-2D software. The use of the computer simulation for the research is able to consider any hardening curve of the material and to compare the hardening curve obtained using the proposed mathematical model with it.

We carried out the simulation experiment on five-fold specimens corresponding to the Russian standard GOST 1497. Specimens had the original diameter of the gauge length equal to 6 mm and the parallel length equal to 36 mm. For the simulation, we considered four types of hardening laws: two materials m1 and m2 had a hardening effect only, one material m3 had a constant flow stress and the material m4 had a hardening effect for small strain values and a softening effect for larger strains. The hardening laws of the test materials are the following:

$$\sigma_{m1} = 600\varepsilon^{0.5}; \quad (12)$$

$$\sigma_{m2} = 400\varepsilon^{0.35} + 400; \quad (13)$$

$$\sigma_{m3} = 950; \quad (14)$$

$$\sigma_{m4} = 600\varepsilon^{0.5}\exp(-\varepsilon). \quad (15)$$

Figure 4 presents a comparison of the actual specimen profiles obtained by the results of computer simulation with the predicted profiles obtained on a basis of the proposed mathematical model of necking the specimen. The comparison is available for all considered stress-strain curves. For each test material, we considered four different instants of the test. They corresponded to the change in the diameter d of approximately 0.5 mm. Specimen profiles were defined throughout the parallel length section.

As you can see from figure 4, the accuracy of predicting the whole specimen profile is quite high. You can observe the greatest discrepancy between the actual and predicted profiles in the transition area from the parallel length to the gauge length for the material m1 and the material m4. This is due to the strong hardening effect of these materials when even the grips sections of the specimen begin to deform. In addition you can see from figure 4 that the predicted specimen profiles lie a little wider than the actual profiles for the strongly hardening material m1 and a little narrower for the material m3 having no hardening effect at all. Thus, the proposed mathematical model of necking the specimen gives an average result for the entire range of the materials hardening curves. For the hardening law of the material m2 that is most common in real tests, the model gives the best result for predicting the specimen profile.

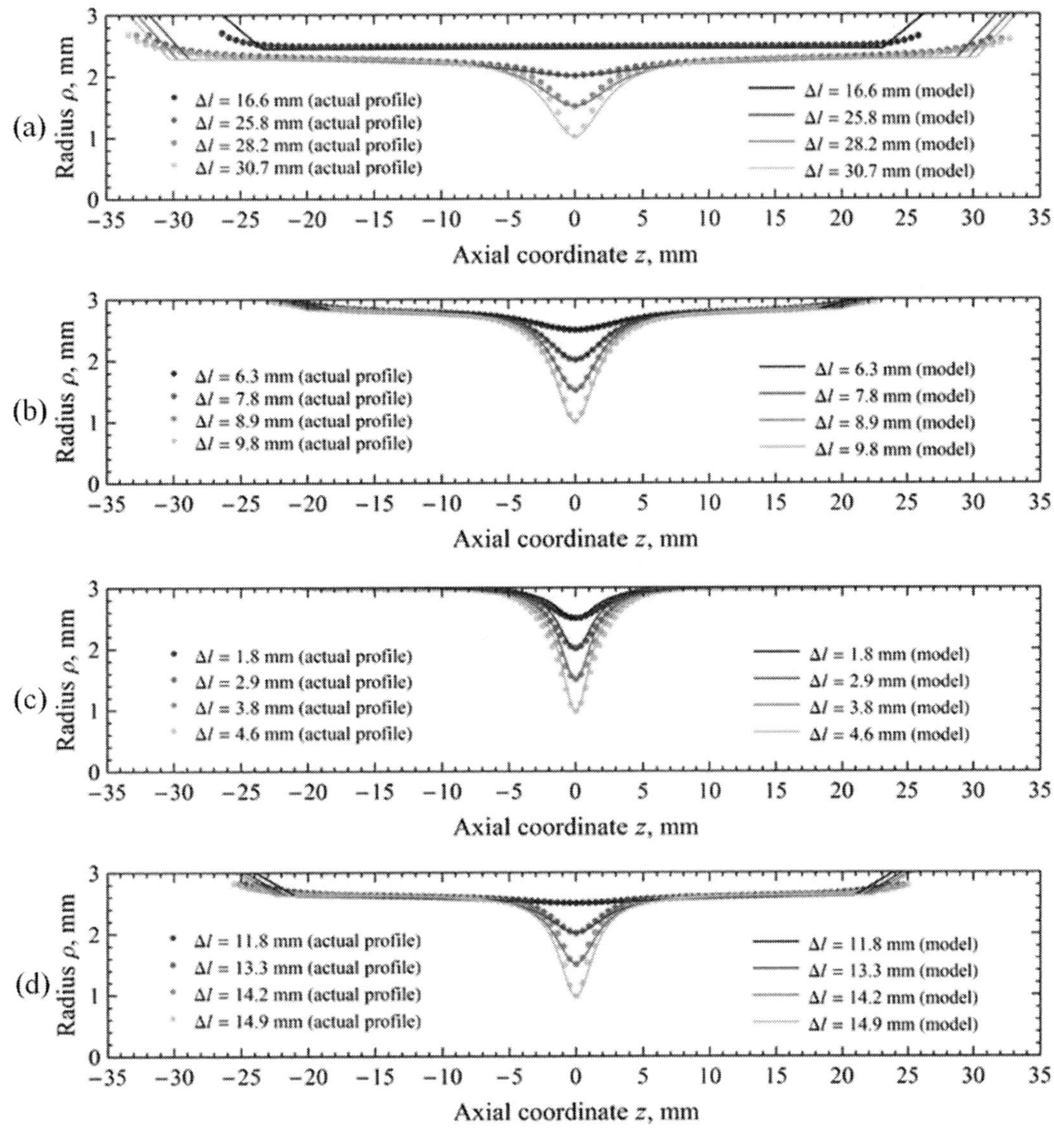

Figure 4. Actual specimen profiles (dots) in comparison with the predicted ones (lines): (a) – the test material m1, (b) – m2, (c) – m3 and (d) – m4.

Based on the described above mathematical model of the neck formation, we defined the hardening curves of the test materials in comparison with the hardening curves that were earlier inputted in the software Deform-2D when the simulation problems were formulated. To calculate the equivalent stresses we used the model of stress distribution in the neck according to Ostsemin. Figure 5 shows the results of calculations of the equivalent stresses as well as the true ones. We determined the true stresses acting in the specimens by equation (1) assuming the correction coefficient K to be constant and equal to 1.

As you can see from figure 5, the equivalent stresses calculated in accordance to the proposed mathematical model are closer to the proper stress values than the true stresses calculated without taking into account the influence of an inhomogeneous stress distribution in the neck while the methods of direct measuring the neck profile are not available. The maximum deviations of the true and equivalent stresses from the actual ones are respectively: 30.9 and 12.1% for the material m1, 52.8 and 9.6% for the material m2, 64.1 and 14.9% for the material m3 and 89.5 and 29.9% for the material m4. You can observe the greatest deviations for the material m3 and the material m4 at the softening stage. This is most likely because the deformation localizes in the neck more intensively while there is no material hardening. In the case of hardening materials, the model gives more good convergence.

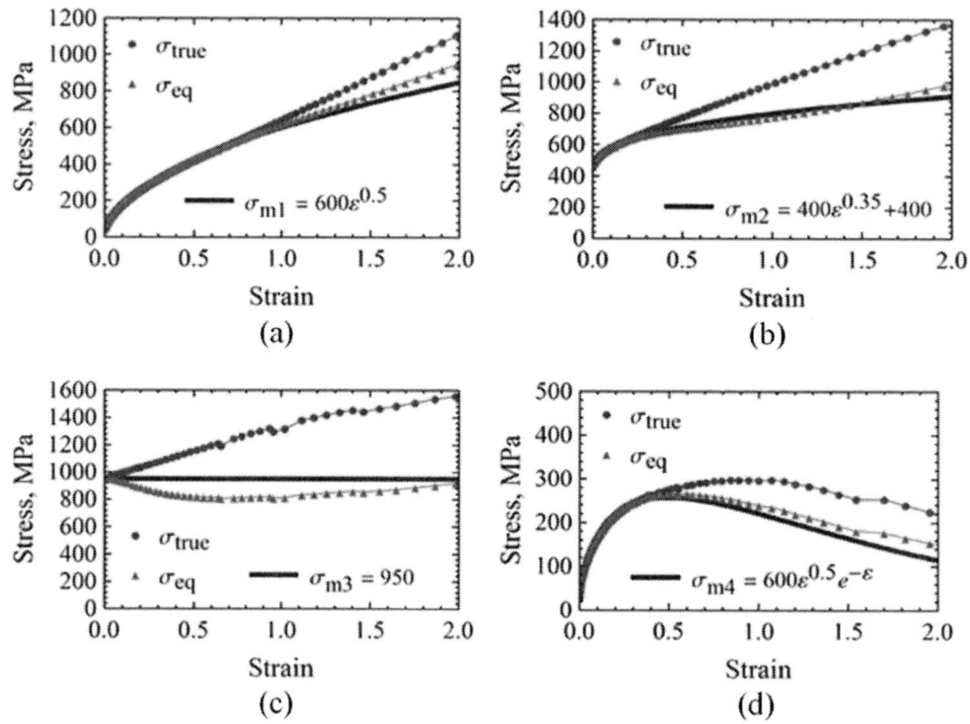

Figure 5. The stress-strain curves in equivalent and true coordinates in comparison with the proper hardening curves: (a) – the test material m1, (b) – m2, (c) – m3 and (d) – m4.

4. Conclusions

The mathematical model of necking the specimen taking into account the effect of grips sections allows predicting the radius of the neck curvature and calculating the equivalent stresses acting in the specimen. The results of computer simulations show that identifying the stress-strain curves on a basis of the proposed model is more preferable than the calculation of the true stresses used in the absence of the possibility to perform the direct measurements of the neck profile.

References

[1] Nadai A 1963 *Theory of flow and fracture of solids* (New York: McGraw-Hill Book Company) p 840
[2] Rabotnov Ju N 1962 *Strength of materials* (Moscow: Fizmatgiz) p 456
[3] Farber V M, Polukhina O N, Vichuzhanin D I, Khotinov V A and Smirnov S V 2019 *Metal Science and Heat Treatment* **61** 274–279
[4] Farber V M, Polukhina O N, Vichuzhanin D I, Khotinov V A and Smirnov S V 2019 *Metal Science and Heat Treatment* **61** 405–410
[5] Panov D, Pertsev A, Smirnov A, Khotinov V and Simonov Y 2019 *Materials* **12** 2058
[6] Khotinov V A, Polukhina O N, Selivanova O V and Farber V M 2019 *Inorganic Materials: Applied Research* **10** 939–944

[7] Farber V M, Arabey A B, Khotinov V A, Morozova A N and Karabanalov M S 2018 *Metal Science and Heat Treatment* **59** 735–740
[8] Bridgman P W 1952 *Studies in Large Plastic Flow and Fracture, with Special Emphasis on the Effect of Hydrostatic Pressure* (New York: McGraw-Hill) p 362
[9] Davidenkov N N and Spiridonova N I 1946 *Proc. ASTM* **46** 1147–1159
[10] Ostsemin A A 2009 *Strength of Materials* **41** 356–362
[11] Dilman V L and Ostsemin A A 1998 *Factory Laboratory* **5** 47–9 (in Russian)
[12] Smirnov S V, Vichuzhanin D I and Nesterenko A V 2015 *PNRPU Mechanics Bulletin* **3** 146–164
[13] Vichuzhanin D I, Khotinov V A and Smirnov S V 2015 *Diagnostics, Resource and Mechanics of materials and structures* **1** 73–89
[14] Vorobyov E V 2008 *J. Problems of strength* **3** 92–99
[15] Mirone G 2004 *Int. J. Solids Struct.* **41** 3545–3564
[16] Sancho A, Cox M J, Cartwright T, Davies C M, Hooper P A and Dear J P 2019 *Int. J. Solids Struct.* **176–177** 191–206
[17] Sancho A, Cox M J, Aldrich-Smith G, Cartwright T, Davies C M, Hooper P A and Dear J P 2018 *EPJ Web Conf.* **183** 02063
[18] G'Sell C, Hiver J M, Dahoun A and Souahi A 1992 *J. Mater. Sci.* **27** 5031–5039
[19] La Rosa G Mirone G and Risitano A 2003 *Metall. Mater. Trans.* A **34A** 615–624
[20] Erpalov M and Nukhov D 2020 *Mater. Sci. Forum* **989** 372–377
[21] Erpalov M V, Nuchov D Sh and Pavlov D A 2020 *J. Chem. Technol. Metall.* **55** 544–549

Acknowledgments
This work was supported by a grant MK–1878.2020.8 from the President of the Russian Federation for young scientists – candidates of sciences.

IOP Conf. Series: Materials Science and Engineering **969** (2020) 012084 doi:10.1088/1757-899X/969/1/012084

The Use of Finite Mixture Models and EM-Algorithm to Analyze Grain Structure in HPT-Nanostructured Metallic Materials

A Stolbovsky[1,2]

[1] Ural Federal University named after the first President of Russia B. N. Yeltsin, 19, Mira St., Ekaterinburg, 620002, Russia
[2]M. N. Miheev Institute of Metal Physics, UB RAS, 18, S. Kovalevskoy St., Ekaterinburg, 620108, Russia

E-mail: stolbovsky@imp.uran.ru

Abstract. Analysis of grain size distribution's histograms of Nb and Ni subjected to high-pressure torsion at cryogenic temperatures and Nb_3Sn layers formed in Nb/Cu–Sn composite wires after the diffusion annealing has been carried out using statistical analysis method based on the application of finite mixture models and using an expectation–maximization algorithm with the estimation of fitting accuracy by the Bayesian information criterion. It has been established that the approximation by the model with a single component of logarithmic standard distribution is the most suitable for all examined experimental distributions in contrast to the model with two components. Besides, the use of the proposed approach allows to practically eliminate an influence of the additional errors in the experimental data which seem to be introduced at transmission electron microscopy image processing and constructing histograms of grain size distributions.

1. Introduction

In the last decades materials with ultrafine-grained (UFG) structure attract great attention of the researchers due to their unique physical and mechanical properties. UFG materials possess a grain size below 1000 nm and exhibit significantly higher strength, unusual mechanical behavior and a complex of unique properties in comparison to their coarse-grained counterparts [1–3].

The application of severe plastic deformation (SPD) enables obtaining bulk UFG materials. A number of various techniques of SPD have been developed by now [4], and one of the earliest and most common among them is high-pressure torsion (HPT) [5,6]. This method makes it possible to produce a structure close to nanocrystalline [7–14] at room temperatures, and HPT at cryogenic temperatures allowed the authors of [15–19] to obtain a true nanocrystalline structure in pure metals.

It was shown [20–22] that grain boundaries as an element of structure playing a determinative role in UFG materials markedly differ from those in coarse-grained polycrystals and are referred to as "non-equilibrium" or deformation-modified boundaries. Besides, such grain boundaries are characterized by high density of defects and, accordingly, enhanced energy.

Since the grain structure is the most important factor responsible for the special properties of such materials, in the studies devoted to investigations of structure in nanostructured materials [7–10,13–16,18,19] transmission electron microscopy (TEM) with a subsequent processing of TEM images is used. However, in a lot of these studies, the statistical analysis of structure usually deals with the grain sizes distributions described by a logarithmic standard (normal) distribution. At the same time, it was shown [23] that the histograms of the grain size distribution in the nanostructured materials both of the HPT and of the nanostructures of recrystallization origin are not always described only by the logarithmic standard distribution. Besides, the different models can be used to fit the experimental distributions. That may not lead to the same results because of different accuracy of fitting techniques and models [24, 25]. Since experimental distributions may be a mixture of several log-normal distributions with different values of the mean, standard deviation and volume fraction, a small amount of one of distributions may be estimated as contamination. Moreover, a logarithmic normal distribution is not symmetrical that additionally complicates the process of fitting.

Therefore, the aim of this work is to propose a robust statistical approach to analyze experimental grain sizes distributions.

2. Results and discussion

The results of studies [15, 18] devoted to nanostructuring of pure metals by HPT at cryogenic temperatures had been used. It is assumed that an effect of recrystallization both during deformation and after that was completely eliminated. The data on the nanostructure formed during the annealing of highly deformed composites is used [26] as well.

The analysis of grain sizes distributions was performed in terms of that the experimental distributions of grains are described by lognormal function [27]. Pearson's mixture model-based approach known as finite mixture models was chosen to describe the experimental distributions [26]. An expectation-maximization (EM) algorithm [28, 29] has been used for fitting the experimental distributions.

In addition, it is suggested that the experimental distribution may be both single distribution and a sum of separated logarithmic standard distributions, so the expression of the model is as follows:

$$f(x)=\sum_{i=1}^{N} A_i \cdot \frac{1}{x} \cdot \frac{1}{\sqrt{2\pi} \cdot \sigma_i} \cdot e^{-\frac{(\ln x-\mu_i)^2}{2\sigma_i^2}}, \quad (1)$$

where μ_i and σ_i are the mean and the standard deviation of the variable of the natural logarithm distribution; A_i is a scale factor, which is the estimation of volume fraction of the grain group in the experimental distribution; i – grain group number; N – the amount of distributions in the model.

EM-algorithm is common technique used to determine the parameters of a mixture. At the same time, EM-algorithm is of particular appeal for finite normal mixtures with an a priori given number of components [28, 29]. Therefore, in order to use EM-algorithm with the most efficiently it is necessary to describe experimental distribution as normal distribution's mixture.

Since, a log-normal distribution is a continuous probability distribution of a random variable whose logarithm is normally distributed, the model (1) can be represented in logarithmic scale as:

$$f(x)=\sum_{i=1}^{N} A_i \cdot \frac{1}{\sqrt{2\pi} \cdot \sigma_i} \cdot e^{-\frac{(\ln x-\mu_i)^2}{2\sigma_i^2}}, \quad (2)$$

Due to μ_i and σ_i have logarithmic scale, these parameters should be turned into normal scale by follows expressions:

$$M_i = e^{\mu_i+\frac{\sigma_i^2}{2}}, \quad (3)$$

$$S_i = \sqrt{\left(e^{\sigma_i^2}-1\right) \cdot e^{2\cdot\mu_i+\sigma_i^2}}, \quad (4)$$

where M_i and S_i are the mean and the standard deviation of the standard distribution in the normal scale, respectively.

IOP Conf. Series: Materials Science and Engineering **969** (2020) 012084 doi:10.1088/1757-899X/969/1/012084

It should be noted that the approach assumes that initial data for processing is a raw set of grain sizes. In order to process experimental grain sizes distributions, the distributions should be decomposed into a raw set of grain sizes in contrast to fitting of distributions by curves in accordance with the technique of work [23–25]. The next step of processing is to represent the set of grain sizes on the logarithmic scale. Then the data was processed by EM-algorithm with different the amount of distributions (N) in accordance with the model. The advantage of the technique is the use of EM-algorithm with enough amount of random start points to avoid incorrect results. Moreover, modern implementations of EM-algorithm for fitting of a Gaussian mixture model allow to compute the Bayesian information criterion (BIC) [30] to assess the number of clusters in the data. That can be used to select the number of components (N) in a model in an efficient way.

In Table 1, the results of the calculation are summarized, as well. It can be seen that the approximation by the model with a single component ($N = 1$), in accordance with the calculated BIC values, is the most suitable for both the grain structure of Nb and Ni subjected to HPT at cryogenic temperatures and Nb_3Sn structure (Composite) formed in Nb/Cu–Sn composite wires after the diffusion annealing in contrast to the model with two components ($N = 2$).

Table 1. Results of approximation.

Material	M (nm)	S (nm)	A_i	BIC
Approximation by the model with a single component ($N = 1$)				
Nb (HPT 80K)	77.42	22.16	1	95.81
Ni (HPT 80K)	92.62	29.17	1	53.69
Composite	72.87	20.19	1	343.59
Approximation by the model with two components ($N = 2$)				
Nb (HPT 80K)	76.06	25.21	0.62	107.71
	79.55	14.47	0.38	
Ni (HPT 80K)	91.10	31.96	0.78	60.59
	97.62	7.78	0.22	
Composite	71.97	20.59	0.89	359.85
	79.85	11.39	0.11	

In addition, the data processing results of experimental grain sizes histograms from works [15,18,26] with using the most suitable model are represented in Figure 1 where the experimental histograms and calculated curves in both normal and logarithmic scale are shown.

Earlier [23] it was shown that the fitting of the experimental grain sizes histograms by only single log-normal distribution gives significant imprecision for all these materials. Whereas, use of additional normal distribution in the fitting model allowed to describe the experimental data with much greater accuracy in accordance with the calculated chi-square values. Besides, the fraction of the additionally introduced standard distribution in all cases was considerable and even exceeded the fraction of the logarithmic distribution in the case of Nb.

It can be concluded that the common technique based on the analysis of TEM image with a subsequent processing and constructing histograms of grain size distribution introduce additional normally distribution errors. This fact seems to be the main reason for appearing the additional normal distribution in experimental histograms. At the same time, the use of statistical analysis to describe the experimental distributions in terms of finite mixture models with using EM-algorithm allowed to practically eliminate an influence of the additional errors in the experimental data.

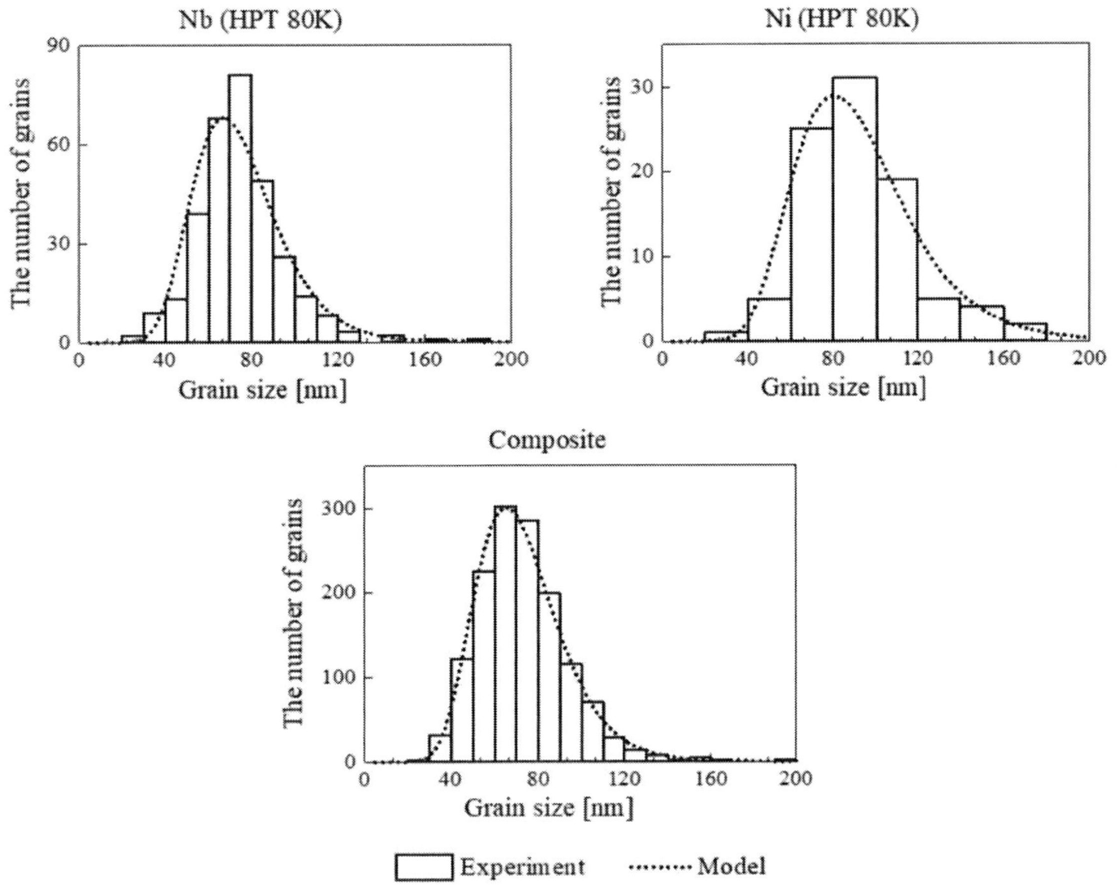

Figure 1. Experimental histograms of grain size distribution [15,18,26] and approximation results.

3. Summary

The analysis of grain sizes distributions was performed in terms of that the experimental distributions of grains are described by the lognormal function. Pearson's mixture model-based was chosen to describe the experimental distributions. An expectation-maximization algorithm has been used for fitting the experimental distributions of the grain structure of Nb and Ni subjected to HPT at cryogenic temperatures and Nb_3Sn structure formed in Nb/Cu–Sn composite wires after the diffusion annealing.

The analysis of fitting accuracy results was shown that the approximation by the model with a single component is the most suitable for all examined experimental distributions in contrast to the model with two components. Besides, the use of the proposed approach allows to practically eliminate an influence of the additional errors in the experimental data which are introduced at TEM image processing and constructing histograms of grain size distributions.

References

[1] Gleiter H 2000 *Acta Materialia* **48** 1–29
[2] Valiev R, Zhilyaev A and Langdon T 2014 *Bulk Nanostructured Materials: Fundamentals and Applications* (Hoboken, New Jersey, USA: TMS, Wiley)
[3] Wilde G and Divinski S 2019 *Materials Transactions* **60**(7) 1302–1315

[4] Valiev R, Estrin Y, Horita Z, Langdon T, Zehetbauer M and Zhu Y 2006 *The Journal of The Minerals, Metals & Materials Society* **58(4)** 33–39
[5] Zhilyaev A and Langdon T 2008 *Progress in Materials Science* **53(6)** 893–979
[6] Pippan R, Scheriau S, Hohenwarte A and Hafok M 2008 *Material Science Forum* **584–586** 16–21
[7] Popova E, Popov V, Romanov E and Pilyugin V 2006 *Physics of Metals and Metallography* **101(1)** 52–57
[8] Popova E, Popov V, Romanov E and. Pilyugin V 2007 *Physics of Metals and Metallography* **103(4)** 407–413
[9] Popov V, Popova E and Stolbovskiy A 2012 *Materials Science & Engineering: A* **539** 22–29
[10] Gapontseva T, Degtyarev M, Pilyugin V, Chashchukhina T, Voronova L and Patselov A 2016 *Physics of Metals and Metallography* **117(4)** 336–347
[11] Zhang H, Huang X and Hansen N 2008 *Acta Materialia* **56** 5451–5465
[12] Zhang H, Huang X, Pippan R and Hansen N 2010 *Acta Materialia* **58** 1698–1707
[13] Krasnoperova Yu,Voronova L, Degtyarev M, Chashchukhina T and Resnina N 2015 *Physics of Metals and Metallography* **116(1)** 79–86
[14] Krasnoperova Yu, Degtyarev M, Voronova L, and Chashchukhina T 2016 *Physics of Metals and Metallography* **117(3)** 267–274
[15] Popov V, Popova E, Stolbovskii A, Pilyugin V and Arkhipova N 2012 *Physics of Metals and Metallography* **113** 295–301
[16] Popov V, Popova E, Stolbovskiy A and Pilyugin V 2011 *Materials Science & Engineering: A* **528** 1491–1496
[17] Rathmayr G and Pippan R 2011 *Acta Materialia* **59** 7228–7240
[18] Popov V, Popova E, Kuznetsov D, Stolbovskii A and Pilyugin V 2015 *Physics of Metals and Metallography* **115** 682–691
[19] Popov V, Stolbovsky A, Popova E and Pilyugin V 2010 *Defect and Diffusion Forum* **297–301** 1312–1321
[20] Valiev R and Musalimov R 1994 Physics of Metals and Metallography **78** 666–670
[21] Kolobov Yu, Grabovetskaya G, Ivanov M, Zhilyaev A and Valiev R 2001 *Scripta Materialia* **44** 873–878
[22] Wilde G, Ribbe J, Reglitz G, Wegner M, Rösner H, Estrin Y, Zehetbauer M, Setman D and Divinski S 2010 *Advanced Engineering Materials* **12** 758–764
[23] Stolbovsky A and Farafontova E 2018 *Solid State Phenomena* **284** 431–435
[24] Stolbovsky A and Farafontova E 2018 *Solid State Phenomena* **284** 425–430
[25] Stolbovsky A 2020 *Solid State Phenomena* **299** 376–380
[26] Popova E, Deryagina I, Romanov E, Dergunova E, Vorobyova A and Balaev S 2011 *Journal of Nano Research* **16** 69–75
[27] Vaz Fatima M and Fortes M A 1988 *Scr. Metall.* **22** 35–40
[28] McLachlan G and Peel D 2000 *Finite Mixture Models* (USA: John Wiley & Sons, Inc.)
[29] Dempster A, Laird N and Rubin D 1977 *Journal of the Royal Statistical Society, Series B* **39**(1) 1–38
[30] Schwarz G 1978 *Annals of Statistics* **6**(2) 461–464

Acknowledgments
The research was carried out within the state assignment of Minobrnauki of Russia (theme "Function" No. AAAA-A19-119012990095-0).

IOP Conf. Series: Materials Science and Engineering **969** (2020) 012085 doi:10.1088/1757-899X/969/1/012085

Effect of Thermomechanical Treatment Parameters on Structure, Phase Composition and Mechanical Properties of Ti-3Al-5Mo-4.5V Titanium Alloy

A G Illarionov, S I Stepanov and S L Demakov

Heat Treatment and Physics of Metals Department, Ural Federal University, 19, Mira Str., Ekaterinburg, 620002, Russia

E-mail: s.i.stepanov@urfu.ru

Abstract. The structure, phase composition and mechanical properties of (α+β) – titanium alloy solution treated at 850 °C, cold-rolled at the reduction ratio in the range of 0…45%, followed by ageing at 450, 500, 550 °C for 0.5, 1.5, 3 hours was studied using XRD, microindentation and tensile testing. The influence of strain level at cold rolling and time-temperature parameters of ageing on the formation of structure and phase composition of solution treated and water quenched Ti-3Al-5Mo-4.5V alloy was investigated and discussed in terms of tensile properties and microhardness. The parameters of low temperature thermomechanical treatment (LTMT) of the (α+β) – alloy were proposed to obtain a high-strength state.

1. Introduction

The (α+β) – titanium alloy Ti-3Al-5Mo-4.5V (Russian Grade –VT16) was developed to manufacture fasteners in aircraft industry employing cold deformation [1–4] as well as the other high strength product, including medical implants [5]. Strain-induced martensitic transformations and twinning can occur during cold deformation of the solution treated and water quenched alloy depending on the stability of the metastable phases fixed in Ti-3Al-5Mo-4.5V, [1, 6–10]. This is reflected in a staged nature of the decay of metastable phases, which were previously studied during continuous heating in [9, 11–12]. At the same time, strengthening treatment in industrial conditions of include solution treatment and cold deformation with the isothermal ageing as a final stage of the processing [13]. However, data on the development of structure and mechanical properties of the cold-rolled VT16 alloy during isothermal ageing are limited [14–15]. The purpose of the study was to investigate the effect of reduction ratio at cold rolling of the Ti-3Al-5Mo-4.5V on structure and mechanical properties, which are formed during ageing with different time-temperature parameters.

2. Material and Methods

The commercial hot-rolled 12 mm rod of VT16 alloy of the following composition – Ti-2.9Al-4.8Mo-5.1V-0.12O (wt.%) with β-transus of 860 °C was manufactured by VSMPO-AVISMA Corporation. The processing of the rods included solution treatment (ST) at 850 °C followed by water quenching (WQ). The next step was cold rolling (CR) with reduction ratios of 9, 16, 31, 45%. In accordance with [9] a two-phase structure consisting of a small fraction of the primary α-phase in a matrix of α″-orthorhombic martensite obtained as a result of β → α″ – transformation is fixed in the alloy during water quenching from 850 °C. An increase in the reduction ratio at CR promotes the development of a strain-induced

$\alpha'' \rightarrow \alpha' + \beta_a$ transformation in the ST alloy. β_a – is a BCC phase with an anomalously high lattice spacing which is detected starting from the reduction ratio of 16%. The ST and CR rods were aged at temperatures of 450, 500, 550 °C for 0.5, 1.5 and 3 hours. The phase composition of the alloy was studied using XRD on a Bruker D8 Advance in Cu K_α-radiation. Vickers microhardness values were determined using an attachment to Neophot-21 microscope at a load of 1 N. Tensile tests were carried out on an Instron 3382 in accordance with ISO 6892.

3. Results and discussion

A comparative analysis of the diffraction patterns of the Ti-3Al-5Mo-4.5V alloy in the ST and aged state revealed that α'' – martensite decomposed according to the $\alpha'' \rightarrow \alpha + \beta$ scheme, which was accompanied by the appearance of diffraction peaks of the α and β phases (figure 1a). A similar character of transformations upon heating is characteristic of the alloy, which was ST and CR at reduction ratio of 9%. An increase in the ageing temperature from 450 °C to 550 °C contributes to a deeper development of decomposition in the alloy after ST and CR at 9%. This is evidenced by a decrease in the intensity of martensite peaks and by a corresponding increase in the intensities of the α and β-peaks (figure 1a) with increasing the ageing temperature at the same dwell times. However, the VT16 alloy did not reach a completely equilibrium state. The peaks of the α'' – martensite (figure 1a) preserved in the diffraction patterns at a maximum ageing temperature of 550 °C and dwell time of 3 hours.

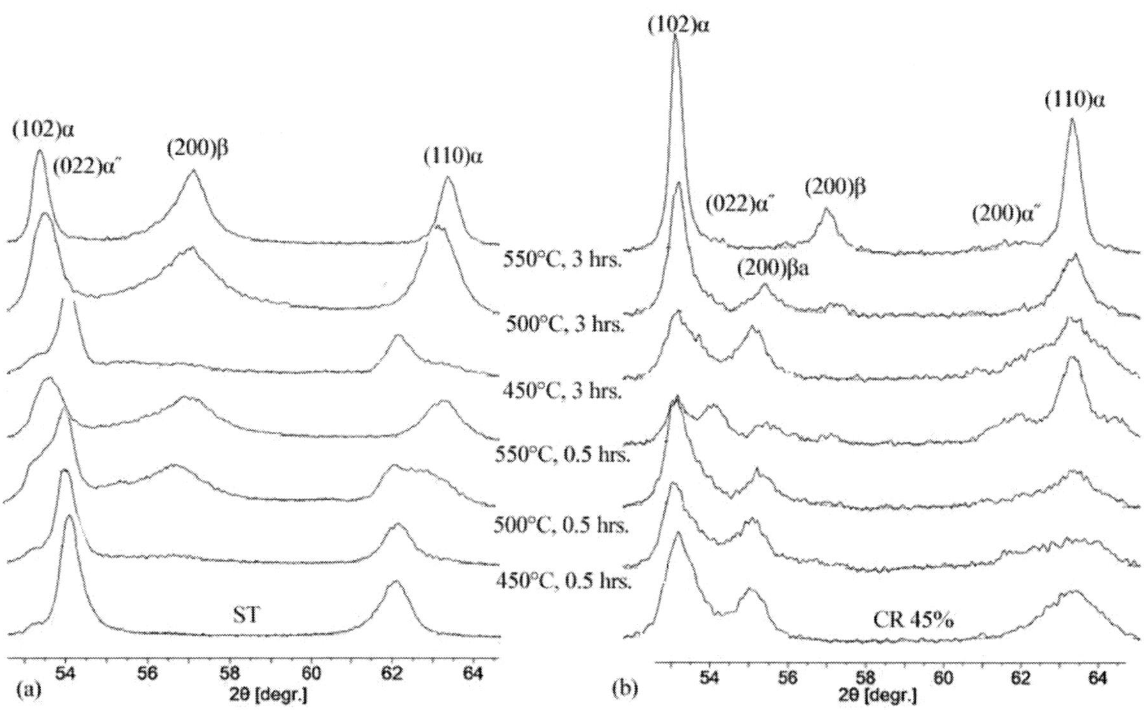

Figure 1. Diffraction patterns of ST alloy (a) and after CR at reduction ratio of 45% (b) before and after ageing at 450, 500, 550 °C for 0.5 and 3 hours.

The decay scheme during ageing changes for the alloy with the reduction ratios of 16, 31, 45%, in which as a result of strain-induced transformation α'' – martensite transforms into α' – martensite and βa-phase peaks appear in the diffraction pattern (figure 1b) and becomes the following:

1. $\alpha'(\alpha'') \rightarrow \alpha'(\alpha'')_{rich} + \alpha'(\alpha'')_{lean} \rightarrow \beta + \alpha$
2. $\beta_a \rightarrow \alpha + \beta$

where $\alpha'(\alpha'')_{rich}$ and $\alpha'(\alpha'')_{lean}$ - $\alpha'(\alpha'')$ are phases respectively enriched and depleted with β-stabilizers as a result of diffusion, which takes place during ageing of $\alpha'(\alpha'')$ – martensite obtained by WQ and subsequent CR. β – BCC-phase formed of $\alpha'(\alpha'')_{rich}$ and β_a-phases due to their diffusion enrichment with β-stabilizers (Mo, V) as a result of $\alpha'(\alpha'')_{rich} \rightarrow \beta$ and $\beta_a \rightarrow \beta$ – transitions.

Analysis of the diffraction patterns of the aged alloy after CR at the reduction ratios in the range of 16–45% demonstrated that the decay of α′(α″) – martensite starts earlier than of the β_a-phase. The decay accelerates with increasing ageing temperature from 450 to 550 °C (figure 1b) similar to ST samples.

Based on XRD data the lattices parameters of α and β-phases (c/a_α, $FWHM_\alpha$ and a_β) formed during ageing (figure 2) were calculated. The influence of the reduction ratio on the intensity of the decay process during ageing was analyzed.

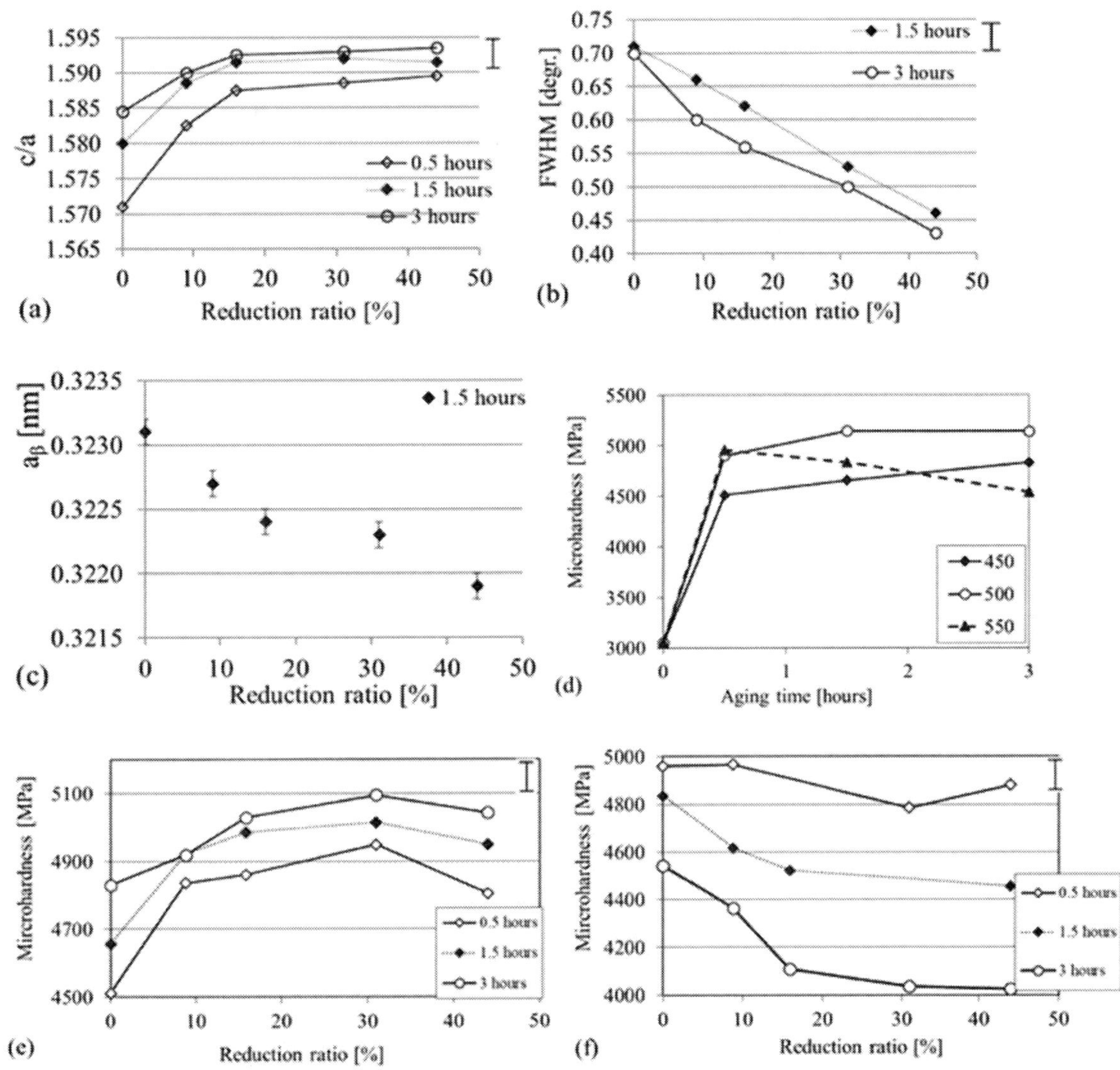

Figure 2. Lattice parameters c/a (a), FWHM of (100)α diffraction peak (b) and a_β (c) depending on the CR reduction ratio after ageing at 500 (b) and 550 °C (a, c). Microhardness depending on the ageing time of solution treated Ti-3Al-5Mo-4.5V (d) and on the reduction ratio of the cold-rolled alloy aged at 450 °C (e) and 550 °C (f).

It was demonstrated that an increase in the reduction ratio results is the acceleration of decay. This was confirmed by a regular growth c/a of the α-phase with an increase in the reduction ratio upon CR as was shown for ageing at 550 °C (figure 2a). According to [16], the parameter c/a of the α phase increases with increase in the content of α-stabilizing elements such as Al in VT16 and with a decrease in the concentration of β-stabilizers such as V and Mo.

The development of a more equilibrium state also leads to a regular decrease in the full width at half maximum (FWHM) of the diffraction peaks of α-phase formed during the ageing of the rods with higher reduction ratio (figure 2b). In addition, an increase in the reduction ratio at the same time-temperature parameters of ageing is accompanied by a decrease in the lattice spacing of the β-solid solution (figure 2c) as the result of enrichment with β-stabilizers (Mo, V) which are characterized by a smaller atomic radius than titanium according to [17].

The acceleration of decay processes during ageing with a higher reduction ratio is due to the characteristic increse in a number of crystal structure defects such as dislocations, nonequilibrium point defects such as vacancies and interphase boundaries [9], which serve as the sites for heterogeneous nucleation during decay and accelerate the diffusion [18]. The observed acceleration of decay processes in the ST alloy with an increase in the reduction ratio affects its microhardness in the aged state (figure 2 d, e, f).

Ageing in the temperature range of 450 ... 550 °C for 0.5, 1.5 and 3 hours contributes to an increase in microhardness values both after ST (3100 MPa) and after CR with the average microhardness values: 3300 MPa (ε=9%) 3600-3700 MPa (ε=16, 31%) and 4200 MPa (ε=45%). The influence of the temperature-time parameters of ageing is fundamentally close for ST and CR states, only the level of the values changes. Therefore, the further analysis was based on the ST sample.

The decay in the ST alloy after ageing even for 30 minutes provided an increase in hardness by 1.5...1.6 times up to 4500 MPa (T_a = 450 °C), 4900 MPa (T_a = 500 °C), 4950 MPa (T_a = 550 °C). (figure 2d). The increase in microhardness with an increase in the ageing temperature is attributed to the acceleration of α″-martensite decomposition in accordance with the above-mentioned XRD data (figure 1, 2). An increase in ageing time from 0.5 to 3 hours had a different effect on microhardness for ageing temperatures (figure 2d). A permanent growth of microhardness is characteristic of ageing at 450 °C. Constant values starting from dwell time of 1.5 hours were observed for 500 °C. A decrease in microhardness starting from dwell time of 1.5 hours took place at 550 °C. This is in a good agreement with the XRD data (figure 1), as well as with the general laws of decay of metastable phases during ageing of light alloys [19].

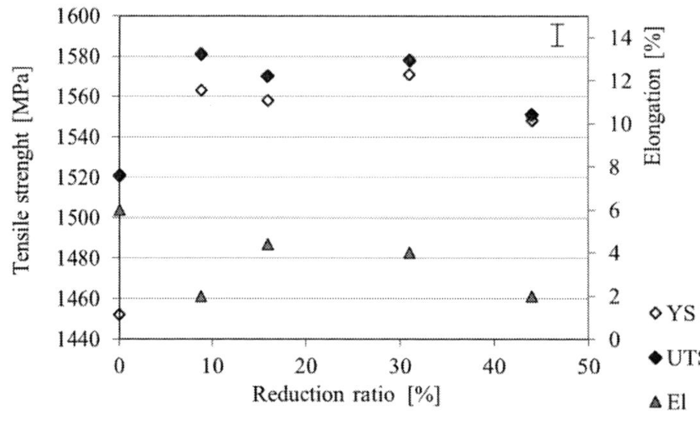

Figure 3. Mechanical properties vs. reduction ratio at cold rolling for rods of Ti-3Al-5Mo-4.5V aged at 500 °C for 3 hours.

According to the XRD data (figure 1), the decay of α″ – martensite during ageing at 450 °C up to a 3 hour dwell did not provide the equilibrium (α + β)-structure. Therefore, hardness increased with increasing exposure time (figure 3d) in accordance with [20]. Structure close to equilibrium was formed at holding for 1.5...3 hours at an ageing temperature of 500 °C and consisted mainly of α and β phases. Therefore, the growth of microhardness stopped and reached a constant level at these dwells. An increase in the ageing temperature up to 550 °C resulted in (α+β) – structure that was close to equilibrium starting from 0.5 hour exposure (figure 1). The coagulation of decomposition products and stress relaxation processes took place at exposures for 1.5 and 3 hours, which led to a decrease in microhardness.

An increase in the reduction ratio during CR, accelerated the decomposition of metastable phases and led to higher microhardness values after ageing at 450 °C compared to the ST sample (figure 2e). Active development of softening processes occurred with increasing dwell time during ageing at 550 °C (figure 2f). Therefore, ageing of ST and CR samples at 500 °C for 3 hours was chosen for mechanical testing (figure 3), which provided high strength state along with the equilibrium structure according to the microhardness values (figure 2d) and XRD data (figure 1).

Thus, the LTMT including ST at 850 °C with WQ followed by CR at reduction ratio of 16% and ageing at 500 °C for 3 hours provided the most significant combination of mechanical properties of Ti-3Al-5Mo-4.5V (VT16) titanium alloy (YS > 1550 МПа, UTS > 1560 МПа, EL > 4 %).

References

[1] Moiseev V N 2001 *Metal Science and Heat Treatment* **43** 73–77
[2] Dunaev V V and Shirshov A A 2009 *Russian Eng. Research* **29** 864–870
[3] Xingwu L, Masyue S and Junpeng C 2009 *Met.Science and Heat Treatment* **51** 594–598
[4] Ivanov A, Orlov A and Golubovskii E 2019 *Mat. Today: Proc.* **19** 2163–2166
[5] Illarionov A G, Grib S V and Huppeev A V 2019 *Materials Science Forum* **946** 309–314
[6] Moiseev V N 1972 *Metal Science and Heat Treatment* **14** 391–395
[7] Maltsev M 1990 *Fiz. Met. Metalloved* **12** 97–103
[8] Popov A A, Illarionov A G, Stepanov S I and Ivasishin O M 2014 *Phys. of Metals and Metallography* **115** 517–522
[9] Illarionov A G, Demakov S L, Stepanov S I and Illarionova S M 2015 *Phys. of Metals and Metallography* **116** 267–273
[10] Xue Q, Ma Y J, Lei J F, Yang R and Wang C 2018 *J. of Mater. Science & Tech* **34** 2507–2514
[11] Maltsev M V 1976 *Fiz. Met. Metalloved* **41** 1225–31
[12] Maltsev M V 1977 *Tsvetn. Met.* 68–70
[13] Khorev A I 2011 *Russian Eng. Research* **31** 1227–32
[14] Stepanov S I, Illarionov A G, Demakov S L and Stepanova E D 2016 *AIP Conf. Proc.* **1785** 040084
[15] Grabovetskaya G P, Ratochka I V, Mishin I P, Lykova O N and Zabudchenko O V 2019 *Met. Science and Heat Treatment* **60** 580–588
[16] Zwicker U, 1974 *Titan und Titanlegierungen* (Berlin: Springer)
[17] Miracle D B and Senkov O N 2017 *Acta Mater.* **122** 448–511
[18] Collings E W 1984 *The Physical Metallurgy of Titanium All*oys (Ohio, Metals Park: Am. Soc. for Metals)
[19] Polmear I 2006 *Light Alloys: From Traditional Alloys to Nanocrystals* (Butterworth-Heinemann: Elsevier)
[20] Kolachev B A, Elagin V I and Livanov V A 2005 *Metal Science and Heat Treatment of Non-Ferrous Metals and Alloys* (Moscow: MISIS)

Acknowledgement
This study was financially supported by the State Assignment, grant number 0836-2020-0020. The research equipment was purchased with the support from Act 211 of the Government of the Russian Federation, contract No. 02.A03.21.0006.

IOP Conf. Series: Materials Science and Engineering **969** (2020) 012086 doi:10.1088/1757-899X/969/1/012086

Computer Simulation of the Metal Deformation Process in an Equal Channel Step Matrix with Additional Backpressure

A Naizabekov[1], S Lezhnev[1] and E Panin[2]

[1]Rudny industrial institute, 38, 50 let Oktyabrya st., Rudny, 111500, Kazakhstan
[2]Karaganda industrial university, 30, Republic av., Temirtau, 101400, Kazakhstan

E-mail: cooper802@mail.ru

Abstract. Theoretical studies of the process of round cross-section billets pressing in equal-channel step matrices of various designs by modeling are carried out. It was found that the backpressure created by the narrowing of the matrix channels has a favorable effect on the stress-strain state of the metal, which in turn allows for better processing of the structure of the deformable metal. It is also found that when using an equal-channel step matrix with narrowing channels, the deformation force increases, which cannot be attributed to the positive sides of these matrix designs. It was found that the most optimal results were obtained when using an equal-channel step matrix with a narrowing of the intermediate channel.

1. Introduction
The development of technologies for the production of ferrous and non-ferrous metals and alloys with ultrafine-grained (UFG) structure by severe plastic deformation methods has been an actual direction of modern materials science for more than a decade. It is metals and alloys with an ultrafine-grained structure that are considered as promising structural and functional materials of the next generation, since most of these materials in this structural state have unique properties.

One of the most effective methods for obtaining UFG materials with severe plastic deformation is equal-channel angular pressing [1–6], and, in particular, pressing in an equal-channel step matrix [7–11]. At the same time, it is known from [11–19] that one of the main favorable factors affecting the possibility of obtaining an ultrafine-grained structure with a smaller number of pressing cycles is backpressure. One of the ways to create a backpressure during metal pressing in an equal-channel angular matrix is to narrow the output channel of the matrix [20–21]. In the process of metal pressing in an equal-channel step matrix, in contrast to the angular matrix, the backpressure is created due to the presence of a second junction of the matrix channels. Additional backpressure during compression in an equal-channel step matrix can be achieved, as well as in angular compression, by narrowing the channels of the matrix. There are several possible options:

- narrowing only the output channel of the matrix;
- narrowing of the intermediate (inclined) channel of the matrix;
- gradual narrowing of the intermediate channel and then the output channel of the matrix.

2. FEM-models
This work is devoted to the theoretical study of the deformation in an equal-channel step matrix with additional backpressure based on the simulation of these processes in the Deform software

package, in order to determine the optimal design of an equal-channel step matrix, which allows creating the most favorable stress-strain state for obtaining metal with an ultrafine structure at lower energy costs.

For these studies, drawings of equal-channel step matrices with different channel diameters were developed. It was based on an equal-channel step matrix with a channel diameter of 15 mm, which was built in the "Kompas" software package. Also, equal-channel step matrices of the following designs were built:

- matrix with a narrowing of the intermediate (inclined) channel of the matrix from 15 mm to 13 mm;
- gradual narrowing of the intermediate channel of the matrix from 15 mm to 14 mm, and then the output channel of the matrix from 14 mm to 13 mm;
- narrowing the output channel of the matrix from 15 mm to 13 mm.

For all four matrices the following identical conditions were set: the length and diameter of the workpiece were 100 mm and 15 mm respectively, workpiece material was AISI 1016 steel, the mesh of the billet contained 25000 elements, the temperature of the billet was 1100°C, the temperature of the die was 200°C, the heat transfer coefficient was 0.7; the coefficient of friction was 0.25; pressing speed was 5 mm/s.

3. Simulation results

To identify the effect of the construction of an equal-channel step matrix on the stress-strain state of the metal at different values of the diameters of the matrix channels, the following indicators of the stress-strain state were determined: equivalent strain and equivalent stress.

Thus, as a result of modeling the pressing of blanks in an equal-channel step matrix of different designs, equivalent strain distributions were determined (Figure 1). It is found that in all matrices, the largest equivalent strain is realized in the second intermediate channel and in the third output channel, but to different degrees. Thus, in a matrix with the same channels, intensive equivalent strains are realized only in a small area on the inner side of the inclined and output channels, in the same area this parameter reaches the maximum value (Figure 1, a). In an equal-channel step matrix with narrowing only the output channel of the matrix, this area increases and also appears from the outside (Figure 1, d). In the matrix with narrowing sloping channel matrix from 15 mm to 13 mm (Figure 1, b) and in the matrix, with a gradual narrowing of the first intermediate channel matrix from 15 mm to 14 mm, and then the output channel of the matrix from 14 mm to 13 mm (Figure 1) in the output channel due to the additional backpressure the zone of intensive equivalent strain occupies almost the entire volume of the workpiece.

If we consider the value of the equivalent strain in volume of the deformable metal, during the pressing of the workpiece in the matrix channels with the same average value on a sloping plot is 0.5–0.7; and in the output channel, by passing the metal through a second junction of the channel matrix the mean value of the equivalent strain grows to 1.2 to 1.3. In the matrix with the only narrowing of the output channel value in the second channel is 0.7–0.9; and in the output channel is 1.3–1.5, and in some places reaches 2. In a matrix with a narrowing of the inclined channel of the matrix from 15 mm to 13 mm and in a matrix with a gradual narrowing of the intermediate channel of the matrix from 15 mm to 14 mm, and then the output channel of the matrix from 14 mm to 13 mm, the values of equivalent strain are almost identical. With the expense of creating additional backpressure by narrowing the channel matrix according to the schemes in the second channel strain increases significantly, and is run from 0.7 to 2.0; and the third channel value of the runway is reduced and the strain has a value from 1.6 to 2.0.

The nature of the equivalent stress distribution shows that the highest value in all matrices is realized in the joints of the matrix channels, while the most uniform distribution of equivalent stresses is observed when using an equal-channel step matrix with a narrowing inclined channel (Figure 2b) and in a matrix with a gradual narrowing of the inclined channel first, and then the output (Figure 2c).

IOP Conf. Series: Materials Science and Engineering **969** (2020) 012086 doi:10.1088/1757-899X/969/1/012086

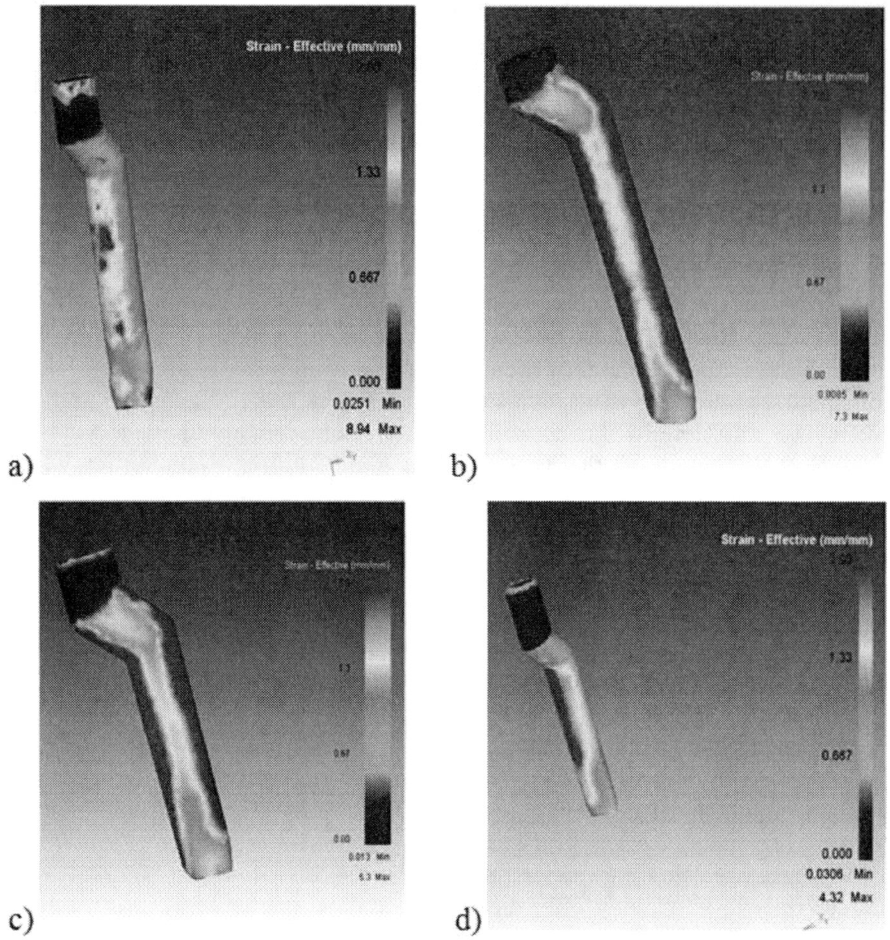

Figure 1. Distribution of equivalent strain during pressing in an equal-channel step matrix of various designs: a – matrix with 15 mm channels; b – narrowing of the intermediate (inclined) channel from 15 mm to 13 mm; c – gradual narrowing of the intermediate channel from 15 mm to 14 mm, and then the output channel from 14 mm to 13 mm; d – narrowing of the output channel from 15 mm to 13 mm.

When using an equal-channel step matrix with a narrowing inclined channel, the highest equivalent stress is 250 MPa, and the average equivalent stress on the inclined section is 200–220 MPa (Figure 2b). In the output channel, the equivalent stress is reduced to a minimum. When using an equal-channel step matrix with a gradual narrowing of the inclined channel first, and then the output channel, the maximum value also reaches 250 MPa, on the inclined section it is equal to 200–215 MPa, in the output channel due to a smooth narrowing of the channel, the equivalent stress is gradually reduced from 190 MPa to 150 MPa (Figure 2c). When using an equal-channel step matrix without narrowing, the maximum stress reaches 200 MPa, on the inclined section, the equivalent stress is in the range of 130-180 MPa, in the output channel, the equivalent stresses are reduced to a minimum as well as in the equal-channel step matrix with a narrowing inclined channel (Figure 2a). In an equal-channel step matrix only with the narrowing of the output channel, the equivalent stress reaches 230 MPa, on the inclined section it has a large run–up and is 140–215 MPa, and in the output channel, due to the narrowing of this channel, a more uniform distribution is observed, and the run–up is reduced to 20 MPa, i.e. in this section, the equivalent stress is already 130–150 MPa (Figure 2d).

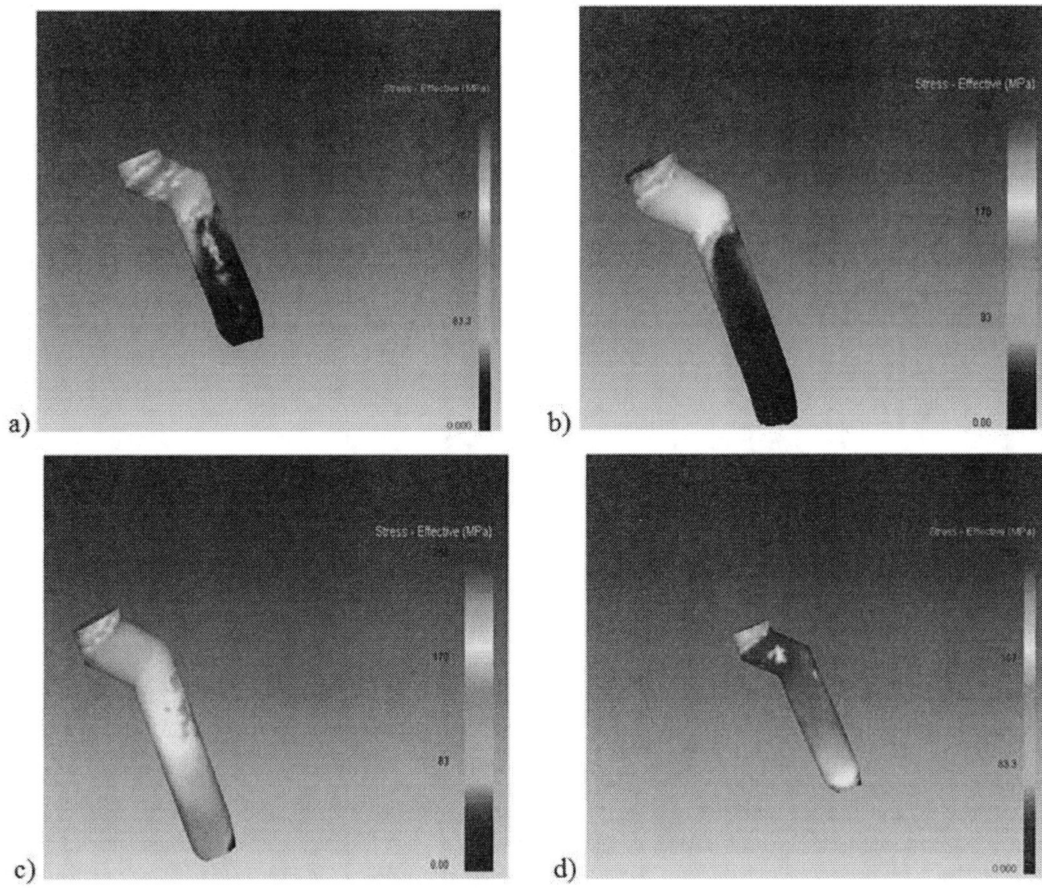

Figure 2. Distribution of equivalent stresses during pressing in an equal-channel step matrix of various designs: a – matrix with 15 mm channels; b – narrowing of the intermediate (inclined) channel from 15 mm to 13 mm; c – gradual narrowing of the intermediate channel from 15 mm to 14 mm, and then the output channel from 14 mm to 13 mm; d – narrowing of the output channel from 15 mm to 13 mm.

Analysis of the effect of the design of an equal-channel step matrix on the deformation force showed that any narrowing of the matrix channels leads to an increase in the effort spent on pressing compared to pressing in the matrix without narrowing (Figure 3). the greatest effort is observed when using an equal-channel step matrix with a gradual narrowing of the first intermediate channel of the matrix from 15 mm to 14 mm, and then the output channel of the matrix from 14 mm to 13 mm. The maximum pressing force in this equal-channel step matrix reaches 151 kN (Figure 3c). The deformation process in an equal-channel step matrix with a narrowing of the output channel of the matrix from 15 mm to 13 mm also requires significant energy consumption (Figure 3d). When using this matrix, the pressing force reached 120 kN. When using a matrix with a narrowing of the inclined channel of an equal-channel step matrix, the pressing force is not much higher than the pressing force in an equal-channel step matrix without a narrowing (Figure 3a, b). So if you are using angular equally channel step matrix with narrowing sloping channel matrix from 15 mm to 13 mm the stress is equal to 91 kN, and the use of a matrix with a channel 15 mm pressing force is equal to 81 kN.

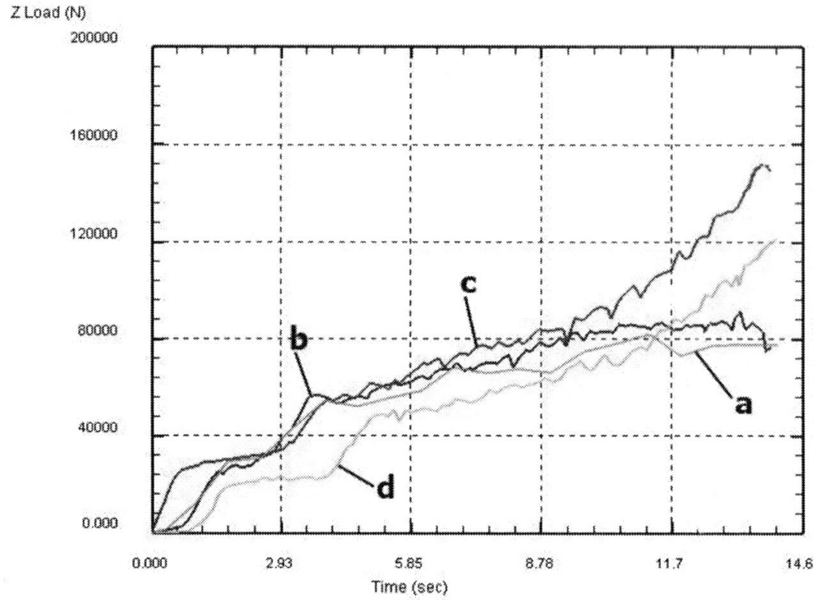

Figure 3. Graphs of the pressing force: a – matrix with 15 mm channels; b – narrowing of the intermediate (inclined) channel from 15 mm to 13 mm; c - gradual narrowing of the intermediate channel from 15 mm to 14 mm, and then the output channel from 14 mm to 13 mm; d – narrowing of the output channel from 15 mm to 13 mm.

4. Conclusions

In the course of the theoretical study of the pressing process of round cross-section billets in equal-channel step matrices of various designs by modeling in the "DEFORM" software package, it was found that the backpressure created by the narrowing of the matrix channels has a favorable effect on the stress-strain state of the metal, which in turn allows for better elaboration of the structure of the deformed metal. But the analysis also showed that when using an equal-channel step matrix with narrowing channels, the deformation force increases, which cannot be attributed to the positive sides of these matrix designs. Therefore, only the choice of rational parameters of the matrix design will allow you to obtain a metal with an ultra-fine-grained structure due to the uniform distribution of deformation and stress over the entire volume of the deformed metal and its intensive processing at low energy costs. In our case, the best results were obtained when using an equal-channel step matrix with a narrowing of the intermediate (inclined) channel of the matrix.

References

[1] Segal V M 1995 Materials processing by simple shear *Materials Science and Engineering A* **197** 157
[2] Valiev R Z 2011 Development of equal-channel angular pressing for obtaining ultrafine-grained metals and alloys *Metally* **1** 15
[3] Segal V M 2010 Equal-channel angular pressing of metal materials: achievements and directions of development *Metally* **1** 3
[4] Vishnu P, Raj Mohan R, Krishna Sangeethaa E, Raghuraman S and Venkatraman R 2020 A review on processing of aluminium and its alloys through Equal Channel Angular Pressing die *Materials Today: Proceedings* **21** 212
[5] Li B, Teng B and Chen G 2019 Microstructure evolution and mechanical properties of Mg-Gd-Y-Zn-Zr alloy during equal channel angular pressing *Materials Science and Engineering A*

744 396

[6] Luo P 2018 Development of High Ductility and Adequate Strength in Pure Titanium Recycled from Chips by Multi-Pass Equal Channel Angular Pressing *Materials Transactions* **59(1)** 53

[7] Raab G I, Kulyasov G V, Polozovsky V A and Valiev R Z 2002 RU Patent **2181314**

[8] Naizabekov A B, Ashkeev Zh A, Lezhnev S N and Toleuova A R 2005 Billet deformation in uniform-channel stepped die *Steel in Translation* **35(2)** 37

[9] Lezhnev S, Volokitina I and Koinov T 2014 Research of influence equal channel angular pressing on the microstructure of copper *J. Chem. Technol. Metall.* **49(6)** 621

[10] Raab G I 2005 Plastic flow at equal channel angular processing in parallel channels *Materials Science and Engineering A* **410–411** 230

[11] Lapovok R E 2010 Role of backpressure in angular pressing *Metally* **1** 44

[12] Raab G I, Makarychev K N and Valiev R Z 2005 Features of the stress-strain state in equidistant angular compression with backpressure *Physics and technology of high pressures* **15(1)** 72

[13] Raab G I 2009 *Development of scientific bases of technologies of intensive plastic deformation and creation of equipment according to the scheme of equal-channel angular pressing for obtaining ultrafine-grained metal semi-finished products: dissertation for the degree of doctor of engineering. sciences* (Ufa: UGATU)

[14] Ashkeev Zh A, Bukanov Zh U and Karakeeva G E 2012 Combination of casting and punching in equal-channel step matrices *Foundry production* **5** 30

[15] Stolyarov V V 2013 Deformational methods of structure grinding *Bulletin of scientific and technical development* **4** 29

[16] Stolyarov V V 2003 Ultrafine-grained Al-5 wt.% Fe alloy processed by ECAP with backpressure *Materials Science and Engineering A* **357** 159

[17] Betekhtin V I, Tabachnikova E D, Kadomtsev A G, Narykova M V and Lapovok R 2011 Influence of backpressure during equal channel angular pressing on the formation of nanoporosity in ultrafine-grained copper *Letters to the Journal of technical physics* **37(16)** 5

[18] Krasilnikov N A 2005 Strength and plasticity of copper obtained by equal-channel angular compression with backpressure *Metally* **3** 35

[19] Lapovok R, Tomus D, Estrin Y, Mang J, Lowe T C 2009 Evolution of nanoscale porosity during equal-channel angular pressing of titanium *Acta. Mater.* **57** 2909

[20] Botkin A V, Valiev R Z, Dubinina S V and Varenik E V 2011 Deformation and force parameters of angular pressing of a cylindrical metal billet *Forging and stamping production* **8** 3

[21] Naizabekov A B and Andreyashchenko V A 2012 Investigation of the influence of external factors in angular pressing *Engineering technology* **5** 18

IOP Conf. Series: Materials Science and Engineering **969** (2020) 012087 doi:10.1088/1757-899X/969/1/012087

The Structure of Annealed Hafnium Bronze Subjected to Severe Plastic Deformation by High Pressure Torsion

A Stolbovsky[1,2], V Popov[2] and R Falahutdinov[2]

[1]Ural Federal University named after the first President of Russia B. N. Yeltsin, 19, Mira St., Ekaterinburg, Russia
[2]M. N. Miheev Institute of Metal Physics, UB RAS, 18, S. Kovalevskoy, Ekaterinburg, 620108, Russia

E-mail: stolbovsky@imp.uran.ru

Abstract. The structure and mechanical properties of low-alloyed hafnium bronze (Cu-0.78 wt. % Hf) under severe plastic deformation by high-pressure torsion (HPT) by 1, 3 and 5 revolutions of anvils at room temperature have been studied by electron microscopy and microhardness measurements. In the initial annealed state Hf is practically completely bonded in intermetallic compounds. The structure of bronze specimens deformed by HPT is stable and does not undergo any changes after unloading and long room temperature ageing. Under the deformation a dispersed submicrocrystalline structure gradient along specimen radius is formed. Average crystallite size in the radius middle is 200 nm after 1 revolution and 120 nm after 5 revolutions of anvils. It is demonstrated that in all the specimens studied including those deformed by 5 revolutions the structure is not uniform which is confirmed by microhardness measurements. Such behavior of hafnium bronze with Hf bonded in precipitates has much in common with the behavior of pure copper. However, the room temperature stability of the structures obtained by HPT demonstrates similar behavior with that of low-alloyed tin bronze.

1. Introduction

Pure copper has always been one of the most widely used materials for electrical engineering as it combines high electro- and heat-conductivity with good plasticity. Particularly, copper is practically indispensable for details of contact groups in electro-technical articles or heat exchangers operating under enhanced loads. However, its application as a constructional material is limited by its not high enough strength and relatively low melting temperature.

Constant search of possibilities to enhance mechanical characteristics of copper for its wider application as a constructional material with unique properties has been carried on for several last decades [1]. And only after an appearance of a new approach to creation of materials with high density of internal defects [1–15], copper was suggested as a valuable constructional material. However, copper articles have still certain principal drawbacks such as low thermal stability of the structure [10, 12–15].

Impurities and doping elements are known to increase the recrystallization temperature of copper even in case of their very low concentration. Doping of copper with elements forming solid solutions with high concentration, such as, for example, tin, is not effective for an enhancement of thermal stability of the structure [16]. It is believed [17–26], that doping of copper with elements demonstrating the lowest solubility in it (Zr, Hf, etc.) enables to achieve the highest temperature of the beginning of recrystallization. At the same time, their concentration increasing upper 0.05 % not always results in noticeable increase of recrystallization temperature [17, 18].

The goal of the present work was to study hafnium bronze with minimal content of the doping element in solid solution subjected to severe plastic deformation by HPT in order to determine the structure formed and the level of strength characteristics.

2. Materials and methods

The composition of hafnium bronze used in the present study is given in Table 1.

Table 1. Chemical composition of hafnium bronze, wt. %

Element	**Hf**	**O**	**Fe**	**Ni**	**Zn**	**Sn**
Content, wt. %	0.78~	0.28~	0.001~	0.005~	0.005~	0.005

The hafnium bronze after melting was subjected to hot forging at 600°C. After that it was mechanically treated and annealed for homogenization at 800°C, 1 h with following cooling in a tubular vacuum furnace enabling to carry out annealing at the pressure of 10^{-2}–10^{-3} Pa.

Deformation was carried out in open Bridgeman anvils the diameter of 10 mm at room temperature, by 1, 3 and 5 revolutions of anvils, at the pressure of 6 GPa and angle speed of 0.3 revs/min.

The structure of as-deformed and annealed specimens was studied by transmission electron microscopy in Philips-CM30 SuperTwin and JEM-200CX electron microscopes.

Microhardness after the HPT was measured in a special unit to Neophot-21 optical microscope, at a load of 40 g, by the techniques described in [27].

3. Results and discussion

In the initial annealed state the bronze had polycrystalline structure with crystallite sizes of 20-40 μm, that is, much smaller than in case of pure copper and tin bronze [16]. Besides, there were second phase particles in the structure, the sizes of 1-4 μm, which is in agreement with the results obtained in [18], where hafnium bronze with 0.9 wt. % Hf was studied and in the initial annealed state it contained Cu_5Hf particles the sizes of ~3 μm. Thus, according to the phase diagram one could expect that after such treatment the main part of hafnium would be bonded in Cu_5Hf intermetallic compound. Microhardness in the initial annealed state was 760 ± 20 MPa, which is very close to its values for polycrystalline copper of commercial purity [16].

It should be noted that the calculated strain under the HPT depends on the distance from the rotation axis, and deformation is distributed non-uniformly along specimen radius increasing from center to periphery. According to the data of TEM studies, after the HPT the structure of hafnium bronze is gradient along disc radius; it is dispersed submicrocrystalline, with mainly high-angle boundaries between crystallites. After the deformation by 1 revolution of anvils, in the central part of the disc, where the strain is minimal and the structure is less fragmented, the average size of crystallites is about 200 nm (Figure 1 a, b). Specific contrast inside crystallites indicating high level of internal stresses is clearly seen in TEM images. Some crystallite boundaries are low-angle, resembling dislocation pile-ups, though thin and straight high-angle boundaries are also present. In some electron diffraction patterns there are closely located reflections forming radially extended aggregates in the Debye rings, but the number of such reflections is small. Such electron diffraction patterns can be considered as transitional from point-wise characteristic of coarse-grained structure to ring-wise indicating the submicrocrystalline structure formation.

With increasing distance from specimen center the strain increases, and the structure gets more dispersed, with smaller average crystallite sizes. The gradient character of microstructure is confirmed by microhardness measurements along specimen radius. Thus, after 1 revolution of anvils the microhardness is 1740 ± 60 MPa in specimen center, 1820 ± 20 MPa in the radius middle and 2050 ± 40 at the edge (Figure 2a). With strain increasing (increased number of revolutions) the microhardness is increasing throughout the specimen, and its values are equalizing along the radius. This behavior is quite expectable, but in the case under consideration the values of microhardness are still somewhat different even after the HPT by 5 revolutions. Thus, in hafnium bronze the structure is gradient and depends on the distance from specimen center even after the highest deformation by HPT.

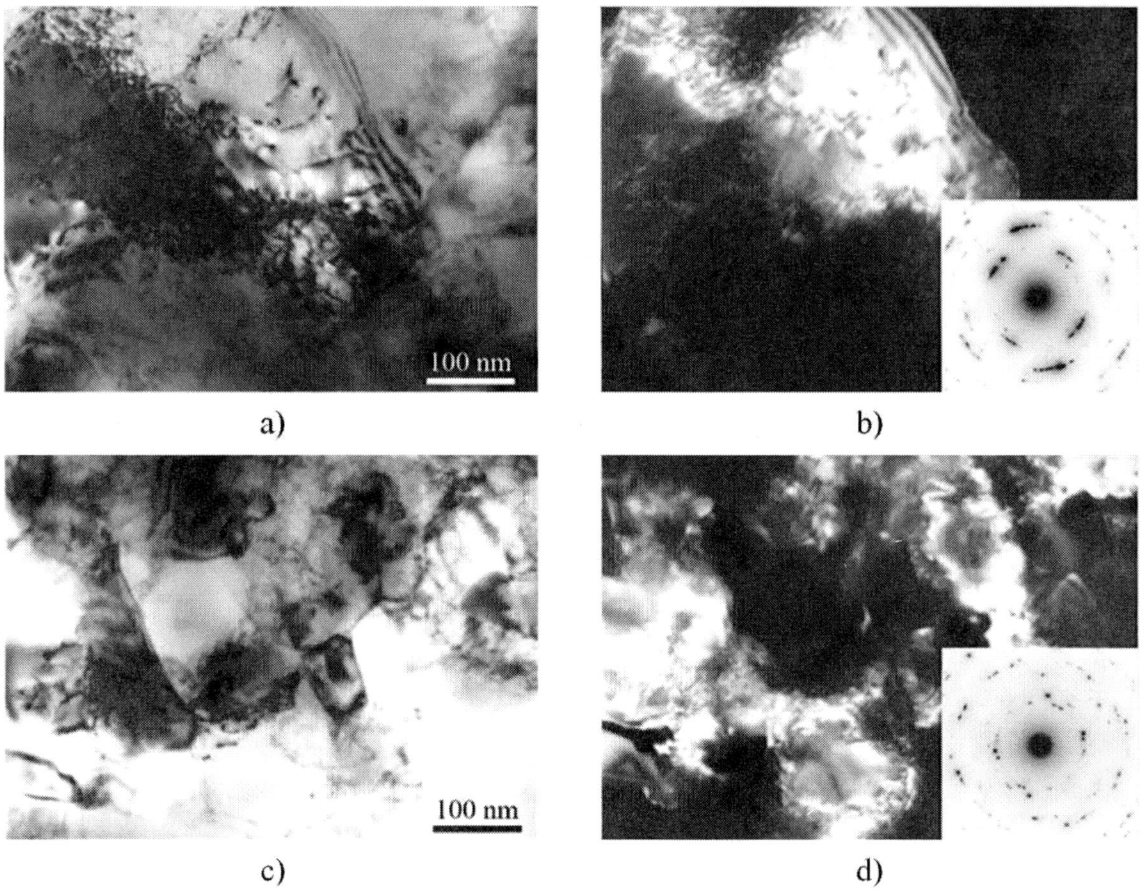

Figure 1. Electron micrographs of the structure and electron diffraction patterns in central parts of hafnium bronze specimens after 1 revolution (a, b) and 3 revolutions (c, d) of anvils: (a, c) – bright-field images, (b, d) – dark-field images with electron diffraction patterns in inserts.

Since after the HPT by 3 and 5 revolutions the microhardness is more uniform along the specimen radius than after 1 revolution, one could expect more uniform structure as well. According to TEM data, the structure is really more dispersed and uniform (Figure 1 c, d). After 3 revolutions the average size of crystallites in the radius middle is 150 nm. Crystallite boundaries are curved, and inside the crystallites there is a specific contrast characteristic of the highly stresses state, which indirectly indicates high level of internal stresses and peculiar non-equilibrium state of boundaries. However, practically in all the electron diffraction patterns the Debye rings are still not populated with the high enough number of reflections (Figure 2d).

As in the periphery of specimens after 1, 3 and 5 revolutions the microhardness increases practically up to one and the same level of 2050 ± 40 MPa (Figure 2a), one could expect similar microstructure in these areas.

The average size of crystallites in the periphery of specimens deformed by 5 revolutions is about 120 nm. However, even in this case the structure is not uniform and is very similar to that obtained after the HPT with lower deformation degree. In the electron diffraction patterns the Debye rings are not densely populated with uniformly distributed reflections. However, in some periphery areas of discs after this treatment there are elements of structure indicating the relaxation processes. There are some grains with thin straight boundaries, crystallites of elongated shape and areas with non-uniform distribution of contrast inside crystallites. The relaxation processes directly under the deformation can limit both grain refinement and microhardness increasing in this material.

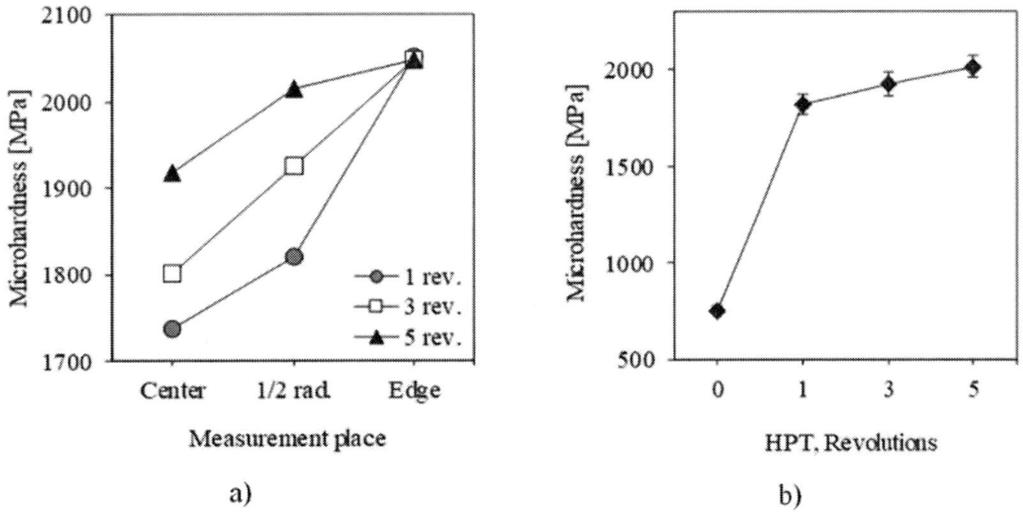

Figure 2. Microhardness along the radii of specimens deformed by HPT with different number of revolutions (a) and its dependence on the number of revolutions (in the radius middle) (b).

The strengthening of hafnium bronze with strain increasing is demonstrated by Figure 2b, in which the microhardness is shown dependently on the number of revolutions. The microhardness increases gradually, but its growth is not as pronounced as in, for example, tin bronze [28], and the values are intermediate between those obtained for the tin bronze and the commercial purity copper of M1 standard. In hafnium bronze pronounced strengthening is achieved after only 1 revolution of anvils, as in the case of tin bronze. However, with further increasing of strain some elements indicating relaxation processes appear in the structure, as it was observed in copper specimens, and the microhardness increasing is slowed down. At the same time, the structures obtained in hafnium bronze are stable at room temperature, as in tin bronze, whereas in copper specimens after unloading the structure changes at room temperature ageing.

It can be concluded that HPT even by 1 revolution enables to achieve more pronounced refinement of the structure of hafnium bronze compared to its deformation by ECAP [19,23], due to higher strain and in spite of the gradient changing of the structure along specimens radii. Besides, the structure after HPT compared to that after ECAP is more uniform in sizes, with dominating equiaxed crystallites and without pronounced deformation bands and extended twins characteristic of the ECAP-ed specimens.

4. Summary

Formation of structure and properties of hafnium bronze homogenized by annealing at 800°C and subjected to severe plastic deformation by high-pressure torsion have been studied.

It is demonstrated that HPT results in much more pronounced refinement of the structure of the hafnium bronze compared to ECAP. The structure of specimens after HPT is not only more dispersed, but also more uniform, with dominating equiaxed crystallites.

It has been established that the structure of the bronze after HPT is stable at room temperature and does not undergo any changes after unloading and prolonged ageing.

According to the data obtained, after the HPT by 1 revolution of anvils the structure of hafnium bronze is gradient along disk radius, it is dispersed and submicrocrystalline, with average crystallite size of 200 nm (in the radius middle), with dominating high-angle misorientation between crystallites. With the strain increasing (when the distance from the discs center is increased) the structure gets more dispersed. The average size of crystallites after the HPT by 5 revolutions is about 120 nm in the periphery parts of disks. However, even after 5 revolutions the structure is still not uniform along the radius, which is confirmed by microhardness measurements, and considerable scattering in crystallite sizes is observed. These features indicate the proceeding of relaxation directly under the deformation, which can result in limitation of grain refinement and microhardness increasing under such treatment of this material.

It can be concluded that the behavior of hafnium bronze with the doping element bonded in precipitates has much in common with that of pure copper, namely, the non-uniform structure is observed even after the HPT by 5 revolutions. On the other hand, from the viewpoint of the stability of the obtained structures at room temperature, the behavior of the hafnium bronze is similar to that of low-alloyed tin bronze.

References

[1] Valiev R, Zhilyaev A and Langdon T 2014 *Bulk Nanostructured Materials: Fundamentals and Applications* (Hoboken, New Jersey, USA: TMS, Wiley)
[2] Gleiter H 2000 *Acta Materialia* **48** 1–29
[3] Wang Y, Chen M, Sheng H and Ma E 2002 *Journal of Materials Research* **17**(12) 3004–3007
[4] Gertsman V, Birringer R, Valiev R and Gleiter H 1994 *Scripta Materialia* **30** 229
[5] Erbel S 1979 *Metals Technology* **6**(1) 482-486
[6] Liao X, Zhao Y, Zhu Y, Valiev R and Gunderov D 2004 *Journal of Applied Physics* **96**(1) 636–640
[7] Dalla Torre F, Lapovok R, Sandlin J, Thomson P, Davies C and Pereloma E 2004 *Acta Materialia* **52(16)** 4819–4832
[8] Valiev R, Kozlov E, Ivanov Yu , Lian J, Nazarov A and Baudelet B 1994 *Acta Metallurgica et Materialia* **42(7)** 2467-2475
[9] Alexandrov I, Dubravina A and Kim H 2002 *Defect and Diffusion Forum* **208–209** 229–232
[10] Brandstetter S, Zhang Kai, Escuadro A, .Weertman J and Van Swygenhoven H 2008 *Scripta Materialia* **58(1)** 61–64
[11] Horita Z, Langdon T 2005 *Materials Science and Engineering: A* **410–411** 422–425
[12] Islamgaliev R, Chmelik F and Kuzel R. 1997 *Materials Science and Engineering: A* **237(1)** 43–51
[13] Akhmadeev N, Kobelev N, Mulyukov R, Soifer Ya and Valiev R 1993 *Acta Metallurgica et Materialia* **41(4)** 1041–1046
[14] Popov V, Stolbovsky A, Popova E and Pilyugin V 2010 *Defect and Diffusion Forum* **297–301** 1312–1321
[15] Stolbovsky A, Popov V, Popova E and Pilyugin V 2014 *Bulletin of the Russian Academy of Sciences: Physics* **78** 908–916
[16] Stolbovsky A, Popov V, Popova E and Falakhutdinov R 2017 *Diagnostics, Resource and Mechanics of materials and structures* **(1)** 13–22
[17] Shangina D, Bochvar N and Dobatkin S 2011 *Materials Science Forum* **667–669** 301–306
[18] Shangina D, Bochvar N and Dobatkin S. 2012 *Journal of Materials Science* **47** 7764–7769
[19] Shangina D, Maksimenkova Y, Bochvar N, Serebryany V, Raab G, Vinogradov A, Skrotzki W and Dobatkin S 2014 *Advanced Materials Research* **922** 651–656
[20] Dobatkin S, Shangina D, Bochvar N and Janeček M 2014 *Materials Science and Engineering: A* **598** 288-292
[21] Shangina D, Gubicza J, Dodony E, Bochvar N, Straumal P, Tabachkova N and Dobatkin S 2014 *Journal of Materials Science* **49** 6674–6681
[22] Dobatkin S, Gubicza J, Shangina D, Bochvar N and abachkova N 2015 *Materials Letters* **153** 5–9
[23] Dobatkin S, Bochvar N and Shangina D 2015 *Advanced Engineering Materials* **17** 1862–1868
[24] Shangina D, Bochvar N, Gorshenkov M, Yanar H, Purcek D and Dobatkin S 2016 *Materials Science and Engineering: A* **650** 63–66

[25] Purcek G, Yanar H, Demirtas M, Alemdag Y, Shangina D and Dobatkin S 2016 *Materials Science and Engineering: A* **649** 114–122
[26] Shangina D, Maksimenkova Y, Bochvar N, Serebryany V, Raab G, Vinogradov A, Skrotzki W and Dobatkin S 2016 *Journal of Materials Science* **51** 5493–5501
[27] Stolbovsky A, Popov V and Popova E 2015 *Diagnostics, Resource and Mechanics of materials and structures* **(5)** 118–132
[28] Popov V, Popova E, Stolbovsky A and Falahutdinov R 2018 *Physics of Metals and Metallography* **119** 358–367

Acknowledgments
The electron microscopy studies were carried out on the equipment of the Collective Use Center in the Testing Center of Nanotechnologies and Advanced Materials of IMP UB RAS.

The research was carried out within the State Assignment (theme “Function”, State registration number AAAA-A19-119012990095-0).

IOP Conf. Series: Materials Science and Engineering **969** (2020) 012088 doi:10.1088/1757-899X/969/1/012088

Research of Radial Flattening of Rolls in the Deformation Zone during Sheet Rolling

A P Pellenen, R A Lisovsky and M A Zhludov

South Ural State University, 76, Lenin prospect, Chelyabinsk, 454080, Russia

E-mail: APPellenen@list.ru

Abstract. The radial flattening of the rolls in the deformation zone is one of the reasons leading to the appearance of a strip minimum thickness during sheet rolling. Working rolls come into contact at the barrels edges as a result of pressing the strip into their core. At the same time, the rolling force increases significantly during rolling, reaching limit values for the rolling mill. A further reduction in thickness is not possible for this reason. Therefore, an accurate determination of the rolls radial flattening of is necessary for developing modes of belt deformation. The rolls radial deformation during the rolling of the belts in various modes was investigated, the existing methods for determining it in the work for the symmetric process and asymmetric rolling processes with different peripheral speeds of the work rolls were tested at the DUO-200 KhPL SUSU (NRU) experimental mill.

1. Introduction

In theory and practice of rolling production, it is known that one of the reasons leading to the appearance of a strip minimum thickness during rolling is the rolls radial flattening in the deformation zone. The work rolls cores come into contact at the bodies edges as a result of indentation of the strip in them. In this case, for example, during ordinary (symmetrical, SR) rolling, the rolling force increases significantly, and in asymmetric rolling with different peripheral speeds of the work rolls, particular during the rolling-drawing process (RD) [1, 2], the rolls contact at the bodies edges heat rolls, thermal stresses and chip rolls. There is such a specific defect as ribbing, which makes it impossible to further rolling due to deterioration in product quality.

2. Methods

The work aim is to study the rolls radial deformation in the deformation zone during sheet rolling using symmetric and asymmetric rolling, assessment of existing methods for its determination.

Two methods are used in the practical experimental study of rolls radial deformation at present. The essence of the first method is that during the deformation of the strip in the rolls, small lead “pilot” strips are simultaneously rolled [3]. The difference in the thickness of the strip and the "pilot" determine the total radial rolls deformation in the deformation zone. The second research method is based on a change in the electrical resistance of a wire isolated from the roll core located in a small diametrical hole in the work roll [4]. The electrical resistance of the wire changes when its length changes.

IOP Conf. Series: Materials Science and Engineering **969** (2020) 012088 doi:10.1088/1757-899X/969/1/012088

The advantage of this method is the ability to obtain changes in radial deformation along the length of the deformation zone. However, to certain disadvantages of this method can be attributed the complexity of the work roll design, the violation of its continuity.

Rolling "pilots" near the strip does not cause difficulties. In order to minimize the effect on the experimental study results of the rolls elastic deformation during the rolling of "pilots", it is necessary to select the initial "pilots" thickness such that it does not exceed the strip thickness at the exit from the deformation zone in 1.1–1.2 times. In this case, the elastic deformation of the rolls in contact with the "pilots" is close to zero, as the experimental rolling shows. The use of micrometers for measuring the strip and "pilots" thickness with an accuracy of 0.5 μm allows for high measurement results accuracy.

The research technique and processing of experimental data was as follows.

Annealed strips 40 mm wide of CuZn36 brass, 040A10 and 321S31 steels were rolled with front and rear tension at the Duo-200 mill of SPSU SUSU (NRU) with front and rear tension. The initial strips thickness: brass CuZn36 and steel 321S31 – 0.5 mm, steel 040A10 – 0.33 and 0.5 mm. Hardness of rolls is 59–61 HRz.

Strip with different draws in the aisle were rolled in the SR and RD modes. During rolling, peripheral roll rotation speeds ratio (K_v) was changing from 1 to $K_v=\lambda$ (λ is the ratio of the strip thicknesses to (h_0) and after rolling (h_1), $K_v=V_{r1}/V_{r0}$; V_{r1}, V_{r0} are the peripheral speeds of the rolls, $V_{r1} > V_{r0}$).

The following parameters were recorded during rolling: the rolling force (P), the forward creep of the strips, their thicknesses before and after rolling. The rolling force was measured by dynamometer mounted under the mill screws. Measurement of the strips forward creep (i_1, i_0) relative to the rolls was carried out according to the punch method [5, 6] (here i_1, i_0 is the lead on the leading V_{r1} and driven V_{r0} rolls, respectively).

The strips thickness in the middle and the "pilots" was measured using a measuring bracket with a type "LIZ" GOST 10197 indicator with a 1 μm scale division.

Strips rolling was carried out simultaneously with the rolling of four "pilots". Two "pilots" rolled at a distance of 5 mm from the strip edges, and two "pilots" at the roll body edges. "Pilots" feeded in the rolls by the wires. Rolling four "pilots" allows you to control the roll gap shape, i.e. possible crossings and deflection of the rolls.

3. Results

The results were processed as follows. The rolls radial deformation was determined as the arithmetic mean of the left and right edges of the strip by the equation:

$$\Delta R_r = 0,5(\Delta R_{right} + \Delta R_{left}) \tag{1}$$

where ΔR_{right}, ΔR_{left} - roll radial deformation at the right and left edges, respectively.

The radial deformation at each edge of the strip was determined by the equation:

$$\Delta R_{right} = 0,5(h_1 - h_{pr}); \Delta R_{right} = 0,5(h_1 - h_{pl}) \tag{2}$$

where h_{pr}, h_{pl} are the "pilots" thicknesses at the right and left edges.

As is known, in indirect measurements, the measured quantity A is related to the measured parameters a_i by the dependence:

$$A = f(a_i ... a_m) \tag{3}$$

The parameters values $a_i ... a_m$ are usually obtained by direct measurements.

In this case, the radial deformation target value of the rolls is found from expressions (1) and (2).

The measured parameters (arguments) are changed randomly, with indirect measurements. In this regard, a special method is used to process the results – the reduction method, in which the results of indirect measurements lead to a number of direct consistent measurements. Statistical processing of experimental data was performed according to the method described in [7].

The result of indirect measurement and calculated by the equation:

$$\overline{A} = \sum_{k=1}^{n} \frac{A_k}{n} \qquad (4)$$

where n is the number of individual values of the measured, A_k is the individual value of the measurand obtained by substituting the kth combination of consistent measurement results.

The root-mean-square deviation of the indirect measurement result is found by the equation:

$$S(\overline{A}) = \sum_{k}^{n}\left(\frac{(A_k - \overline{A})}{n(n-1)}\right)^{1/2} \qquad (5)$$

Confidence limits of measurement result random error are calculated by the equation:

$$\varepsilon(p) = tS(\overline{A}) \qquad (6)$$

where t is the Student's distribution coefficient.

Empirical regression lines are plotted in Figure 1–3 based on the statistical processing results. Subsequently, the experimental data on the study of rolls radial flattening were processed using the least squares method, the parameters of the theoretical regression line were determined, and its correlation analysis was performed.

Analysis of the results and conclusions:

1. The table shows the influence of rolling process technological parameters on the rolling force and the rolls radial deformation (Figures 1–3). The presented results confirm the theoretical and experimental laws on the influence of the mismatch of the rolls rotation speeds, the strip drawing in the passage, its strength characteristics on the rolling force.

The rolling force, for example, for brass CuZn36 decreases by more than 2 times (Figure 1) with a change in the parameter K_v=1 to K_v=λ (condition RD, i_1= 0, i_0= 33%). The same trend is observed for other materials (table 1).

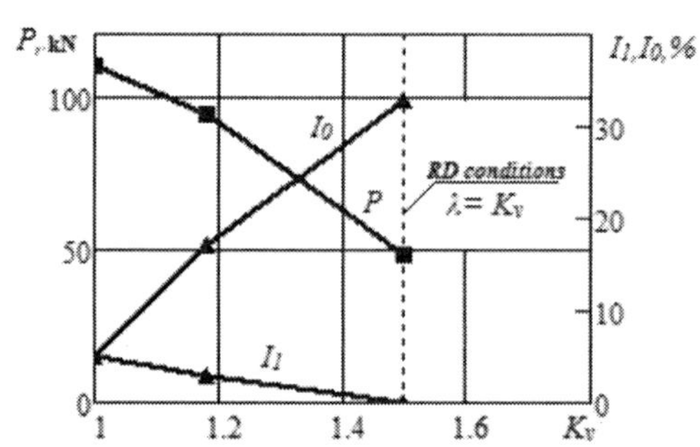

Figure 1. Effect of K_v on rolling force and forward creep (CuZn36 brass; λ = 1.5; h_0 = 0.33 mm).

The observed results (Figure 2) shows the effect of drawing on the rolling force during SR and RD. The efficiency of the RD process to reduce the rolling force increases with increasing λ. For example at λ = 1.07 R_{sp}/R_{pv} = 1.3, at λ > 1, R_{sp}/R_{pv} = 3 for steel 321S31. A similar dependence to reduce the rolling force is also characteristic of other materials.

2. The nature of the change in the radial deformation of the rolls almost corresponds to the nature of the change in the rolling force. A close relationship between the rolls radial deformation is characteristic of both SR and any asymmetric rolling process.

With an increase in the parameter K_v (Figure 3), the radial flattening of the rolls decreases almost by the same amount as the rolling force decreases during RD compared with SR.

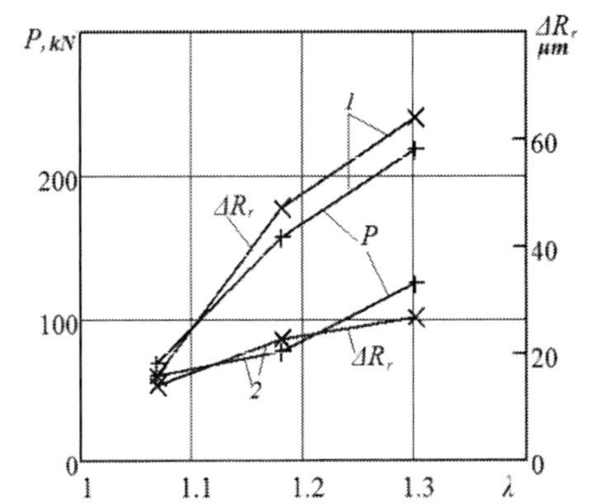

Figure 2. The effect of λ on P and ΔR_r (steel 321S31); h_0 x b = 0.5 x 40 mm; 1 – SR; 2 – DR).

Similarly, the rolls radial deformation increases in the case of both RD and SR with an increase in drawing. It can be seen from the information shown in Figure 2 that for small drawing (up to $\lambda = 1.1$), the radial deformation of the rolls differs slightly (R_{sr}/R_{dr} = 1.17 ... 1.3, R_{sr}/R_{dr} = 1.16 - 1.35). The radial deformations in case of a RD are, usually, 2.5 to 3 times less than in case of a SR with drawing $\lambda > 1.3$, which corresponds to a decrease in rolling force during RD as compared to SR.

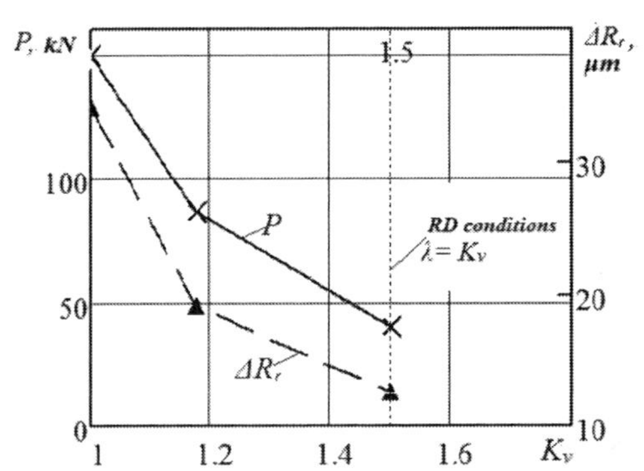

Figure 3. Effect of K_v on ΔR_r and P (040A10 steel; $\lambda = 1.5$; h_0 x b = 0.3 x 40 mm).

3. The experimental results presented in Figures 1–3 and in the table are described by the linear regression equation using the least squares method. For the conditions of the experimental mill Duo-200, the equation is:

$$\Delta R_r = 2.32 + 0.008P / B \quad (7)$$

where P is the rolling force (kN), B is the strip width (m).

The correlation analysis [8] of expression (7) confirmed the close relationship between ΔR_r and P (r $\Delta R_r P$ = 1), which indicates the possibility of using this expression to calculate the rolls radial deformation for the Duo–200 mill when designing deformation modes [9–11].

4. The experimental data obtained in this work make evaluate using known methods for engineering calculations possible for determining the radial flattening of rolls. This is especially important for designing rolling thin strips and foils deformation modes because, the rolling process should be carried out without the rolls contact at the bodies' edges in asymmetric rolling.

Mostly, the calculation methods for determining the rolls radial flattening are based on the Hertz problem solution of compressing two cylinders whose axes are parallel, and normal contact stresses on the roll contact area are distributed according to an elliptic dependence.

Kovalsky B. S. obtained the equation, using the Hertz method [12, 13]:

$$\Delta R_r = 2\theta_r \frac{P}{B}(ln\frac{2R}{a} + 0.407) \tag{8}$$

where $\theta_r = \frac{1-v_r^2}{\pi E_r}$ (v_r, E_r – Poisson's ratio and the roll material modulus of elasticity), a – half width of the cylinder contact area.

It is fundamental to determine the half-width of the rolls contact area or the deformation zone half-length (a = 0.5 lc) in the equation (8). The Hertz problem of two cylinders did not imply the presence of a plastically deformed strip between the rolls. In addition, there are zones of elastic compression and strip restoration when sheet cold rolling at the entrance and exit of the deformation zone. This case, as is known, affects the width of the contact area, i.e. the length of the deformation zone.

Nevertheless, methods for calculating the length of the deformation zone in symmetric sheet rolling have been developed, approximating real diagrams of normal contact stresses by elliptic distribution. However, it can be assumed that for any method of sheet rolling (SR, AR, DR), the distribution of normal contact stresses along the length of the deformation zone can also be approximated by an elliptic distribution.

In this case, we use known equations to determine the deformation zone length.

So, according to Tselikov-Hitchcock [14, 15], without the strip elastic deformation we have:

$$l_c = (R\Delta h(1 + 16q\frac{\theta_r}{\Delta h}))^{1/2} \tag{9}$$

where Δh is the draught of the strip, $q=P/B$.

The equation [16] is proposed by Dinnik A. A.:

$$l_c = (R(1 + 16q\theta_v(1 + \frac{\theta_P}{\theta_v})))^{1/2} \tag{10}$$

where $\theta_P = \frac{1-v_P^2}{\pi E_P}$ (here v_P, E_P - Poisson's ratio and elastic modulus of the strip material, respectively).

To determine the length of the deformation zone, we use the expression [12, 13] obtained by Polukhin V.P.:

$$l_c = (R\Delta h + C_1(\theta_r + \theta_P)Rq)^{1/2} \tag{11}$$

where $C_1 = 8 + 4(\frac{\Delta h}{\theta_r q(1+n)} + 4)^{1/2}, n = \frac{\theta_P}{\theta_r}$

An equation (12) for calculating the rolls radial flattening is obtained using expressions (8, 9) in [17, 18]:

$$\Delta R_r = \theta_r q(ln\frac{R}{16\theta_r q + \Delta h} + 32\frac{\theta_r q}{16\theta_r q + \Delta h} + 1{,}587) \tag{12}$$

The working roll radial deformation was calculated for various kinematic variants and deformation modes (see table). An analysis of the calculated and experimental data shows that the closest calculated results compared with the experimental ones were obtained using the Grigorjan - Zheleznov method equation (12). The largest deviation of various materials, processes and deformation modes is 25%. Other methods, in particular according to expression (11), give a deviation from experimental data up to 67%.

Thus, it is advisable to calculate the rolls radial deformation in the deformation zone using equation (12) for symmetric and asymmetric rolling processes, including DR.

Table 1. Experimental and calculated radial data rolls flattening during rolling at the mill Duo – 200.

Material	h_0 mm	λ	Process	K_v	q kN mm	ΔR_{re} µm	$\frac{\Delta R_r^1}{\Delta R_{re}}$	$\frac{\Delta R_r^2}{\Delta R_{re}}$	$\frac{\Delta R_r^3}{\Delta R_{re}}$	$\frac{\Delta R_r^4}{\Delta R_{re}}$
CuZn36	0.5	1.07	DR	1.07	60	6.0	1.18	1.12	1.39	1.14
CuZn36	0.5	1.18	DR	1.18	100	12.8	0.84	0.8	1.0	0.8
CuZn36	0.5	1.5	DR	1.5	150	16.2	0.93	0.89	1.12	0.82
040A10	0.5	1.07	DR	1.07	70	7.0	0.86	1.12	1.41	1.14
040A10	0.5	1.18	DR	1.18	115	9.0	0.75	1.33	1.67	1.25
040A10	0.5	1.5	DR	1.5	175	14.0	0.83	1.2	1.3	1.12
321S31	0.5	1.07	DR	1.07	135	17.5	0.86	0.83	1.05	0.88
321S31	0.5	1.18	DR	1.18	200	20.5	1.01	0.98	1.23	0.91
321S31	0.5	1.5	DR	1.5	400	33.0	0.86	1.21	1.43	1.1
321S31	0.5	1.07	SR	1.0	130	16.5	0.89	0.87	1.08	0.9
321S31	0.5	1.18	SR	1.0	445	42.0	1.06	1.06	1.28	1.07
321S31	0.5	1.3	SR	1.0	600	58.0	1.22	1.19	1.2	1.0
CuZn36	0.3	1.5	SR	1.0	275	23.0	1.22	1.13	1.43	1.1
CuZn36	0.3	1.5	AR	1.18	235	20.0	1.2	1.13	1.4	1.14
CuZn36	0.3	1.5	DR	1.5	120	12.0	1.1	1.0	1.23	1.05

[1]ΔR_r – l_c according to Tselikov–Hitchcock, ΔR_r – according to Kovalsky B. S.
[2]ΔR_r – l_c according to Dinnik A. A., ΔR_r – according to Kovalsky B. S.
[3]ΔR_r – l_c according to Polukhin V. P., ΔR_r – according to Kovalsky B. S.
[4]ΔR_r – according to Grigoryan–Zheleznov

4. Conclusion

The rolls radial flattening during symmetric and asymmetric rolling, including PV, was experimentally studied for various materials and deformation modes using the "pilots" method.

The lowest rolls radial deformation in compare with symmetric process appears during DR (in 2-3 times).

The nature of the change in the rolls radial deformation is almost proportional to the change in the rolling force.

A comparison of different calculation methods and experimental data on the rolls radial deformation during asymmetric and symmetric rolling indicates the possibility of using the Grigorjan - Zheleznov method in designing the deformation modes for sheet rolling.

References

[1] Vydrin V N and Ageev L M 1971 *Theory and Technology of Rolling: Collection of Research Papers* (Chelyabinsk: Publsher ChPI) p 3
[2] Radionov A A, Radionova L V, Kharitonov V A and Malakhov O S 2005 *Electrical systems and complexes* **10**
[3] Gallay Ya S and Meriin I M 1961 *Ferrous Metallurgy* **10**
[4] Polukhin P I, Nikolaev V A, Polukhin V P, Zinoviev A V and Kosarimov E N 1974 *Contact interaction of metal and tool during rolling* (Moscow: Metallurgy)
[5] Tselikov A I, Nikitin G S, Rokotyan S E 1980 *Theory of lengthwise rolling* (Moscow: Metallurgy)
[6] Gasiyarov V R., Usaty D Yu, Radionov A A, Radionova L V 2011 *Fourth All-Russian Multi-Conference on Management Problems "MKPU-2011"*

IOP Conf. Series: Materials Science and Engineering **969** (2020) 012088 doi:10.1088/1757-899X/969/1/012088

[7] Gmurman V E 1977 *Theory of Probability and Mathematical Statistics* (Moscow: Higher School)
[8] Radionova L V, Lisovskiy R A, Lisovskaya T A 2016 *Russian Internet Journal of Industrial Engineering* 4
[9] Radionova L V, Faizov S R, Lisovskiy R A, Lisovskaya T A, Zhludov M A, Makarov A V 2017 *Russian Internet Journal of Industrial Engineering* **5**
[10] Radionova L V, Lisovskaya T A, Lisovskiy R A 2018 *Solid State Phenomena* **284**
[11] Selivanov M N 1987 *Measurement Quality: Methodological Reference Book* (Leningrad: Lenizdat)
[12] Polukhin P I, Zheleznov Yu D and Polukhin V P 1967 *Sheet Rolling and Roll Service* (Moscow: Metallurgy)
[13] Polukhin V P, Khloponin V N and Sigitov E V 1975 *Algorithms for calculating the basic parameters of rolling mills* (Moscow: Metallurgy)
[14] Tselikov A I 1962 *Calculation of efforts in rolling mills theory* (Moscow: Metallurgizdat)
[15] Tselikov A I and Grishkov A I 1970 *Theory of Rolling* (Moscow: Metallurgy)
[16] Dinnik A A 1967 *Metal Forming* (Moscow: Metallurgy) p 221
[17] Grigoryan G G, Zheleznov Yu D, Cherniy V A, Kuznetsov L A and Zhuravsky A G 1975 *Algorithms for calculating the basic parameters of rolling mills* (Moscow: Metallurgy)
[18] Kovalsky B S 1960 *Moscow: BMSTU Journal of Mechanical Engineering* **6**

Deformation Cone of Pilger Rolling for Forward and Reverse Motion

G I Koval and T G Karimova

South Ural State University, Lenin prospect 76, Russia, Chelyabinsk 454080

E-mail: kovalgi@susu.ru

Abstract. Two mutually perpendicular pairs of rolls show the distinctive features of the existing pilger rolling processes from the pilger rolling process developed at SUSU. The peripheral velocity vectors of them are placed in different directions from the transition zone formation position from the initial billet to the finished profile (deformation cone). The change in the deformation cone shape with the billet free ends during the pilger rolling process by two mutually perpendicular pairs of rolls, circumferential velocity vectors placed in different directions, are considered. It is accepted that during forward rolling, the billet feed is equal to m, and during reverse rolling it is equal to $m\lambda$ to ensure equality of displaceable feed volumes when describing the changes sequence in the deformation cone parameters. The gradual supply volumes flattening with a forward and reverse stroke to design values is shown. This was obtained by considering several cycles of pilger rolling by sequentially increasing the feed volumes during reverse rolling and reducing the feed volumes during forward rolling.

1. Introduction

Currently, the pilger rolling process, also called pilgering and periodic rolling, is carried out on pilger mills, mills cold-rolling tube (HPT), roller type cold rolling tube mill (HPTR), planetary, pendulum and rolling-forging mills [1–14]. The pilger rolling process is carried out simultaneously by two, three and four rolls on these mills. In this case, the pilger rolling process at these mills is carried out in direct motion, with the exception of pilger rolling mills with rotating rolls designed at the South Ural State University (SUSU) [5–7] and HPT [8–13] mills that can perform pilger rolling in reverse.

A new pilger rolling process developed at SUSU [15, 16] provides pilger rolling of one pair of rolls in direct motion and the other with a pair of rolls in reverse motion. The vectors of circumferential velocities of mutually perpendicular pairs of rolls are directed in different directions to achieve it.

It is necessary to know the deformation cone geometry (transitional section from the initial billet to the finished profile), which is the initial profile specified in the rolls at each deformation pilger. This is necessary for developing rolls calibration and determining the technological parameters of the instantaneous deformation zone during rolling by two mutually perpendicular roll pairs, one of which it deforms in direct motion, and the other pair of rolls in reverse motion.

2. Methods

Consider the change in the deformation cone shape during the rolling process with the free ends of the billet. We begin the consideration on the condition that the same deformation cones 1, 2 are rolled out on the billet with the same height of the original billet and the resulting profile from both pairs of rolls (Figure 1).

At the same time, we take into account that during direct rolling, the billet feed is equal to m, and during reverse rolling mλ to ensure equality of displaced feed volumes with direct rolling. The total displaced volume in the forward and reverse stroke is $mH_0B_0 = m\lambda H_1B_1$.

We begin rolling with pair of rolls that deform by a direct course on the side B_0, B_1 with feeding m. The deformation cone Z is rolled out on the billet after rolling with the indicated pair of rolls. Suppose that the front side of the billet from the finished profile receives a linear displacement equal to half the displaced volume minus the feed itself, that is $\frac{m(\lambda-1)}{2}$. The deformation cone on the side H_0, H_1 will take a position 4.

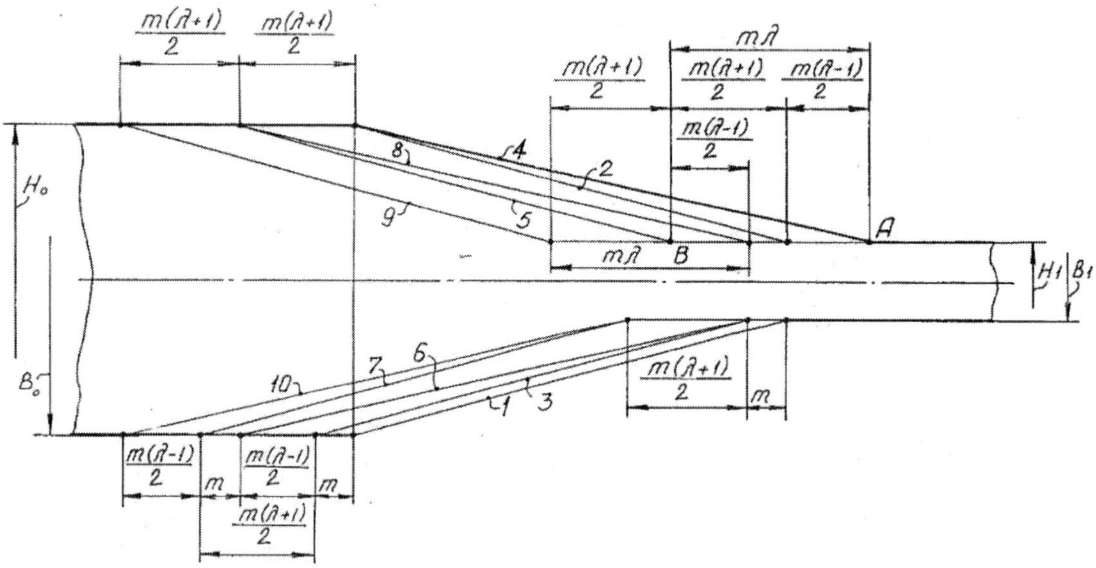

Figure 1. Changing in the shape of the deformation cone.

A pair of rolls, deforming backward along the H_0, H_1 side, will begin to deform with a rolling section from point A to point B. Distance between these points are mλ. The pair of rolls on the side H_0, H_1 of the billet roll out the deformation cone 5 after deformation. Let us assume that the back end of the billet will also receive an offset equal to half the displaced volume minus the feed itself $\frac{m(\lambda-1)}{2}$. The deformation cone on the side B_0, B_1 will take position 6.

A pair of rolls, deforming in direct motion, after feeding the billet by a value of m and rolls out the deformation cone 7 on the side B_0, B_1. The front end of the billet will receive linear displacement $\frac{m(\lambda-1)}{2}$ and the deformation cone on the side H_0, H_1 will take position 8.

Then a pair of rolls, deforming in reverse, rolls out the deformation cone 9, the back end of the billet will offset by an amount $\frac{m(\lambda-1)}{2}$ and the deformation cone on the side B_0, B_1 will take position 10.

Further rolling is similar to that described above.

Thus, after rolling in reverse (before rolling in direct motion), the deformation cone from the side of the roll deforming inverse will have the form 2 (Figure 2), and the deformation cone from the side of the roll that deforms in direct motion will have the form 4.

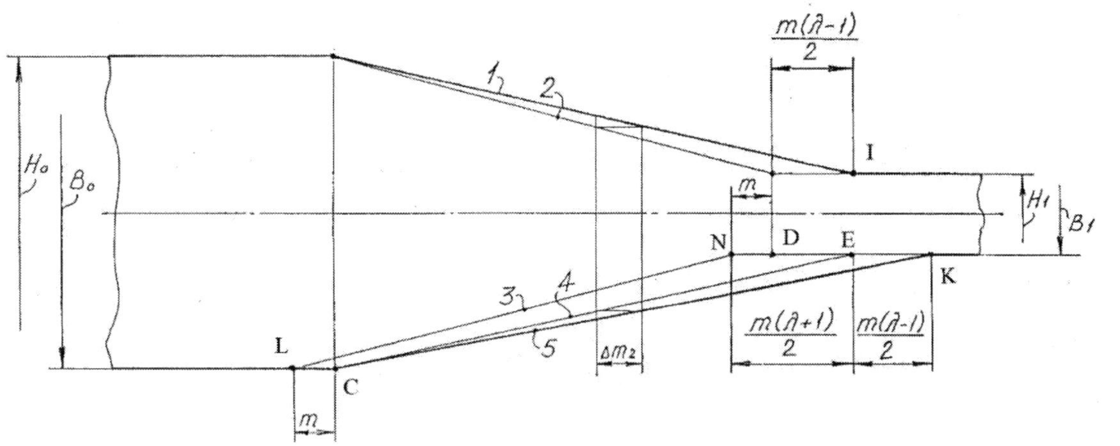

Figure 2. The deformation cone of straight pilger rolling.

During deformation by a direct stroke, the metal of the deformation cones will be displaced towards the final product. Instantaneous linear displacement values are Δm_1. As a result of this, the metal entry lines into the deforming in direct motion rolls will take position 5, and the deformation cone on the side of the roll deforming in reverse motion will take position 1. The position of line 1 determines the width of the instantaneous zone of deformation at the metal entry in the rolls, deforming in direct motion. The deformation cone from the side of the roll, deforming in direct motion, after deformation in direct motion will take position 3.

The deformation cone after rolling in direct motion (before reverse rolling in) from the side of the roll deforming in direct motion will be

have the form 3 (Figure 3), and the deformation cone from the roll side deforming backward will have the form 5.

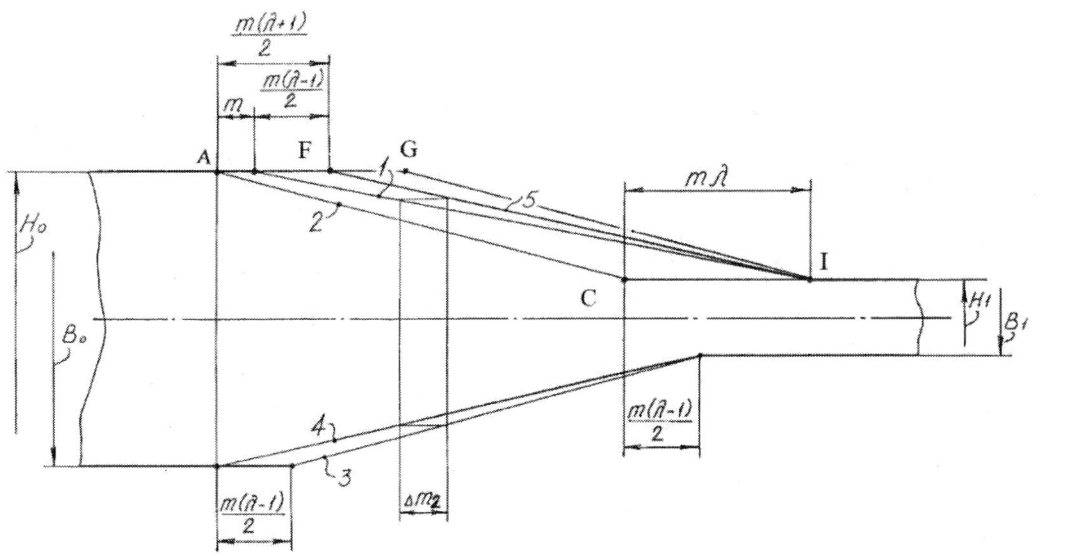

Figure 3. The deformation cone during reverse pilger rolling.

During the reverse rolling deformation, the deformation cones metal 5 and 3 will be displaced towards the initial billet. The instantaneous inline offset value is denoted by Δm_2. The metal entry into the rolls lines, deforming backward, will occupy position 1. The deformation cone from the roll side, deforming in direct motion, will occupy position 4. The position of line 4 determines the instantaneous deformation zone width at the metal entrance in the rolls, deforming backward.

3. Results

The metal forming analysis during the considered pilger rolling process shows that during direct rolling (Figure 2), the LCEN supply volume exceeds the design supply volume $V = mH_0B_0$ per volume limited by the CDE line. AFIC supply volume is less than the design supply volume $V=m\lambda H_1B_1$ per volume limited by the FGI line during reverse rolling (Figure 3).

Then, when rolling in direct motion, the linear displacement of the metal will be greater than the design $\frac{m(\lambda-1)}{2}$ and point I will move to the side of the finished profile by an amount exceeding $\frac{m(\lambda-1)}{2}$, increasing the feed volume, which is deformed when rolling in reverse.

The linear offset will be less than the design and point B will move to the side of the original billet by an amount less $\frac{m(\lambda-1)}{2}$ when rolling backwards. Then the feed volume, deformable during direct rolling, will decrease.

Direct and reverse feed volumes are leveled and equal to the designed feed volumes in subsequent rolling cycles by successively increasing feed volumes during reverse rolling and decreasing feed volumes during direct rolling.

4. Conclusion

1. The distinctive features of the developed method of pilger rolling are shown by two mutually perpendicular pairs of rolls, the peripheral velocity vectors of which are directed in different directions.
2. The patterns of the change in the deformation cones shape during pilger rolling, both direct and reverse, have been established. The use of them is necessary in calculating the rolls calibration and the instantaneous deformation zone parameters.
3. It is shown that the pilger rolling process, carried out sequentially by two pairs of mutually perpendicular rolls, the peripheral velocity vectors of which are directed in different directions, has the ability to equalize the metal supply volumes deformed by each pair of rolls to their design values.

References

[1] Teterin P K 1978 *Theory of Deformed Bar Rolling* (Moscow: Metallurgizdat)
[2] Emelyanenko P T 1949 *Theory of pilgering and skew rolling* (Moscow: Metallurgizdat)
[3] Tselikov A I 1962 *Theory of pressure calculation in rolling mills* (Moscow: Metallurgizdat)
[4] Shevakin Yu F 1963 *Calibration and forces during cold rolling of pipes* (Moscow: Metallurgizdat)
[5] Koval G I, Dremin V G, Fomenko V V, Bubnov P S and Plyusnina T G 2003 *Bulletin of South Ural State University. Metallurgy Series* **2** 62–66
[6] Koval G I 1997 *Izvestiya. Ferrous Metallurgy* **3** 29–31
[7] Koval G I 1999 *Production of rolled products* **7** 43–46
[8] Orlov G A, Orlova E V 2013 *Izvestiya. Ferrous Metallurgy* **11** 65–66
[9] Krivoruchko B I, Mironenko V A, Savotikov A S 2011 *Rolling Production* **6** 24–29
[10] Krivoruchko B I, Mironenko V A 2008 *Nonferrous metallurgy* **11** 34–36
[11] Tereshenko A A, Belikov Yu M, Golovchenko A P 2009 *Ferrous metallurgy* **4** 66–69
[12] Orlov G A, Smelchakov I V 2013 *Ferrous Metallurgy* **7** 69–70
[13] Orlov G A 2017 *Metallurg* **1** 75–78

[14] Bakanov V M 2001 *Sheet rolling on rolling mills of a new design (theory, calculation and optimization)* (Moscow: Intermet Engineering)
[15] Koval G I, Karimova T G 2005 *Method of Pilger Rolling* RU Patent **2252880**
[16] Koval G I 2005 *Bulletin of South Ural State University. Metallurgy Series* **50** 72–76

Investigation of the Influence of Dynamic Parameters on the Formation, Structure and Properties of Bimetallic Compounds during Explosion Welding with Simultaneous Stamping

V I Kuz'min, V I Lysak and E V Kuz'min

Volgograd state technical university, 28, Lenin Avenue, Volgograd, 400005, Russia

E-mail: e.v.kuzmin@yandex.ru

Abstract. This article reports on the study of the influence of dynamic parameters on the formation, structure and properties of bimetallic compounds during explosion welding with simultaneous stamping. The results of the study of the impact of the collision speed and contact speed on the strength, wave parameters and the amount of molten metal obtained by the explosion of a bimetallic compound of the iron + steel 3 alloy are presented. It is shown that the formation of a bimetal during explosion welding is significantly influenced by impact velocity, with an increase in which there is an increase in the values of the maximum deflection and radial deformation.

1. Introduction

One of the ways to improve the efficiency of explosion welding is to develop new promising energy-saving technologies that allow not only to expand the range of manufacturing flat bimetallic blanks, but also to obtain finished bimetallic products of complex configuration, such as cylindrical or spherical shape [1-5]. In explosion welding as a result of a powerful charge, the negative energy of motion and deformation of the welded bimetal can be used rationally for the simultaneous stamping of a bimetallic billet. This will eliminate the use of expensive press equipment and increase the efficiency of explosion welding [6-10]. In the process of explosion welding, the main dynamic parameters responsible for the formation of a durable bimetallic welded joint are the velocity of the impact V_i and the velocity of the collision point V_c [11-16]. Therefore, to obtain high-quality bimetallic compounds with a given profile, it is important to study the combined effect of dynamic parameters on the quality of welding and the formation of bimetals.

The aim of this work was to study the effect of velocity of the impact and velocity of the collision point on the formation, structure and properties of the bimetallic compounds at explosion welding with simultaneous stamping.

2. Materials and methods

During the research, the main material used was a pair of armco-iron + St3 5 + 10 mm thick, which has good weldability and does not cause difficulties in the manufacture of samples for mechanical testing and metallography. Explosion welding with simultaneous stamping was carried out according to a scheme with a parallel arrangement of plates mounted on a metal matrix with a cylindrical profile (Figure 1).

The detonation rate of the explosive was controlled by an electrocontact method [17] with time recording using electronic counting frequency meters. After explosion welding with simultaneous stamping, measurements were made of the cylindrical profile of the resulting bimetallic billet to determine the radial deformation ε_r and maximum deflection f_{max}. The radial deformation of the stamped bimetal was determined using the method of applying "coordinate grids" [18]. Metallographic studies and determination of the tensile strength of the layers σ_{pull} of welded joints were carried out according to standard test procedures for bimetallic and composite materials [19-22].

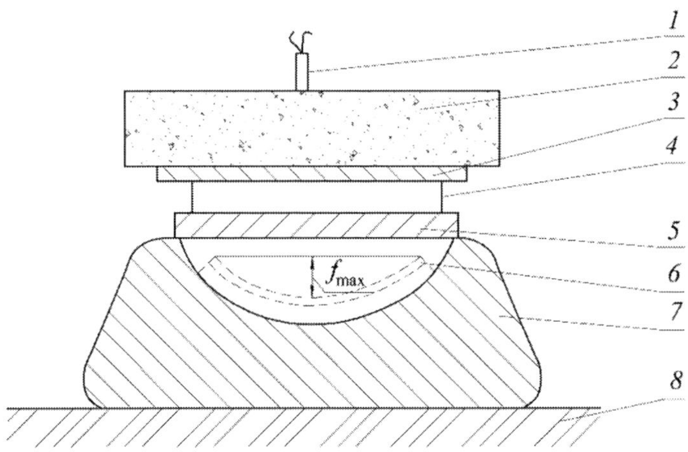

Figure 1. Explosion welding pattern with simultaneous stamping: 1 – electric detonator 2 – explosive charge; 3 – throwing plate; 4 – stop; 5 – fixed plate; 6 – bimetal after stamping; 7 – metal matrix; 8 – sand.

3. Results and discussion

The obtained results of research of influence of dynamic parameters showed that with increasing velocity of the impact in the range V_i = 290 ... 570 m/s at constant velocity of the collision point V_c = const the tensile strength of the connection layers σ_{pull} armco-iron + St3 first increases to strength balance and higher, and then remains constant at V_i > 500 m/s is reduced (Figure 2, a).

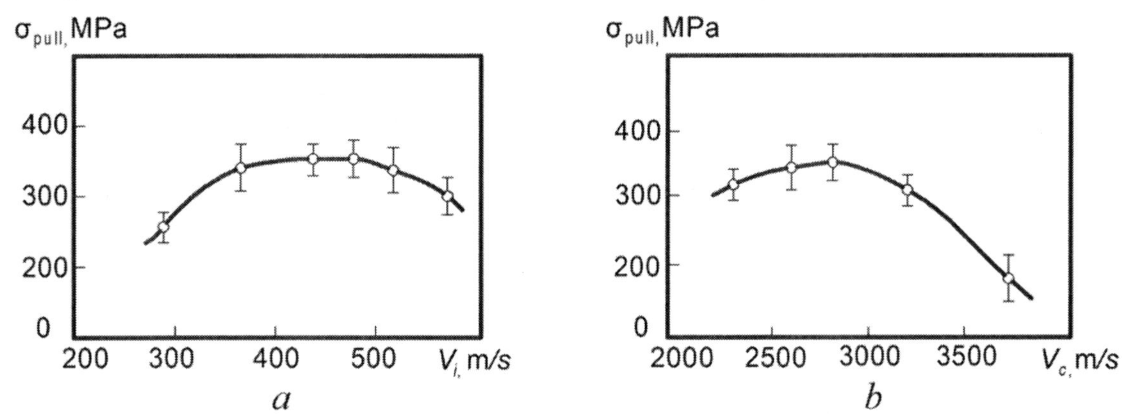

Figure 2. Influence of the velocity of the impact V_i (a) and the velocity of the collision point V_c (b) on the strength σ_{pull} of the armco-iron+St3 compound.

Metallographic studies of the boundary of the armco-iron + St3 compound have shown that with an increase in velocity of the impact V_i, there is an increase in the amount of fused metal K_{fus} and wave parameters: height $2a$, length λ (Figure 3, 4). At the same time, the presence of low joint strength at low values of V_i is explained by a small amount of plastic deformation of the contact metal layers, which is not enough to form a strong welded joint (Figure 4, a). The decrease in joint strength at high impact velocities is associated with excessive plastic deformation of the metal in the heat-affected zone, which leads to the formation of microcracks and a large amount of fused metal K_{fus} (Figure 4, b).

A similar tendency in the separation strength of the layers σ_{pull} is also observed with an increase in velocity of the collision point in the range V_c = 2240 ... 3750 m/s with a constant value of velocity of the impact V_i = const (Figure 2, b).

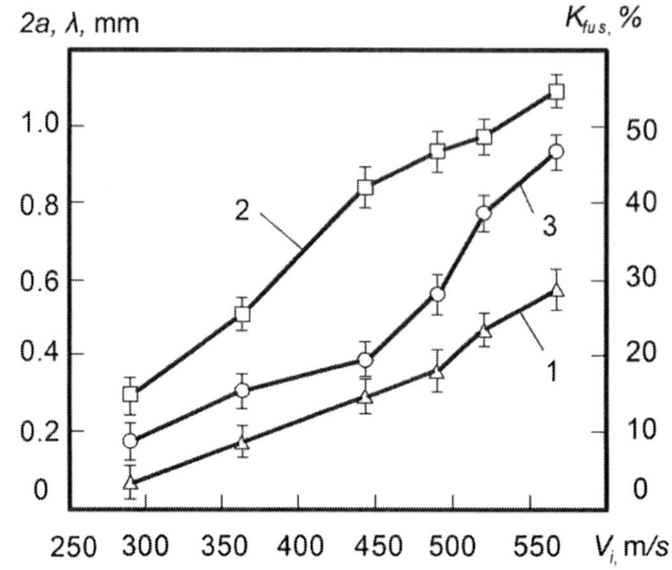

Figure 3. Effect of impact velocity V_i for changing of the height $2a$ (1), λ wavelength (2) and the amount of fused metal K_{fus} (3) a compound of armco-iron + St3.

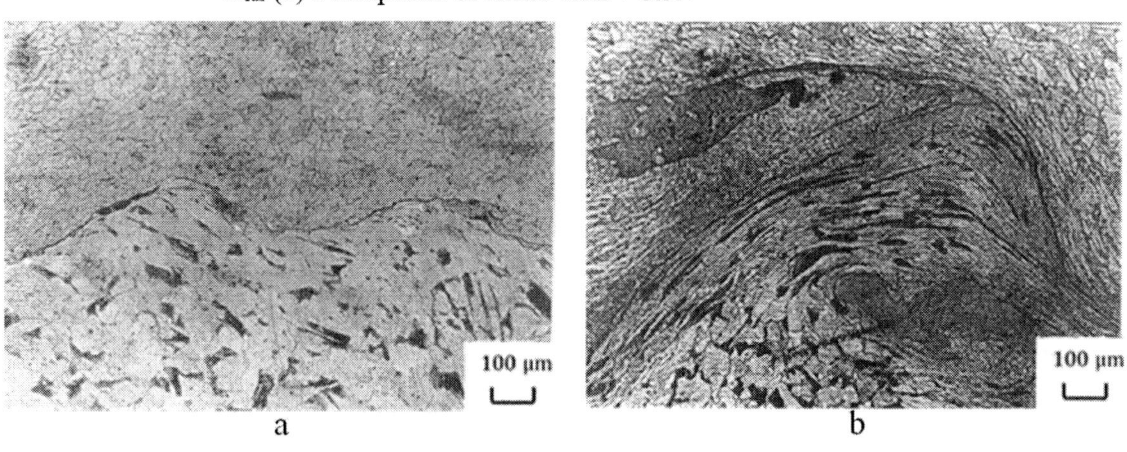

Figure 4. Microstructure of a bond region armco iron + St3 for different values of velocity of the impact: a - V_i = 290 m/s; b - V_i = 570 m/s.

Metallographic studies have shown that a sharp decrease in the separation strength of the layers σ_{pull} of the armco-iron+St3 compound at high value of velocity of the collision point is due to a significant increase in the amount of brittle fused metal K_{fus} (Figure 5, 6).

At the same time, with increasing velocity of the collision point, the wave sizes $2a$, λ decrease and at $V_c > 3200$ m/s they are practically absent, and a continuous layer of brittle melts is observed (Figure 6, b).

The results of experiments to study the influence of dynamic parameters on the formation (stamping) of bimetallic compounds showed that an increase in the velocity of the impact V_i practically does not significantly affect the value of the maximum deflection f_{max} and radial deformation ε_r of the bimetal. With an increase in velocity of the collision point V_c, on the contrary, there is a significant increase in the maximum deflection f_{max} and the radial deformation ε_r of the bimetal (Figure 7).

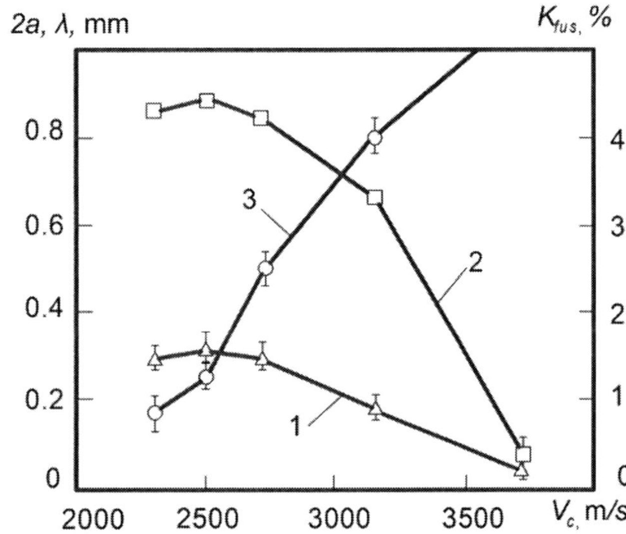

Figure 5. Effect of velocity of the collision point V_c for changing of the height $2a$ (1), λ wavelength (2) and the amount of fused metal K_{fus} (3) a compound of armco-iron + St3.

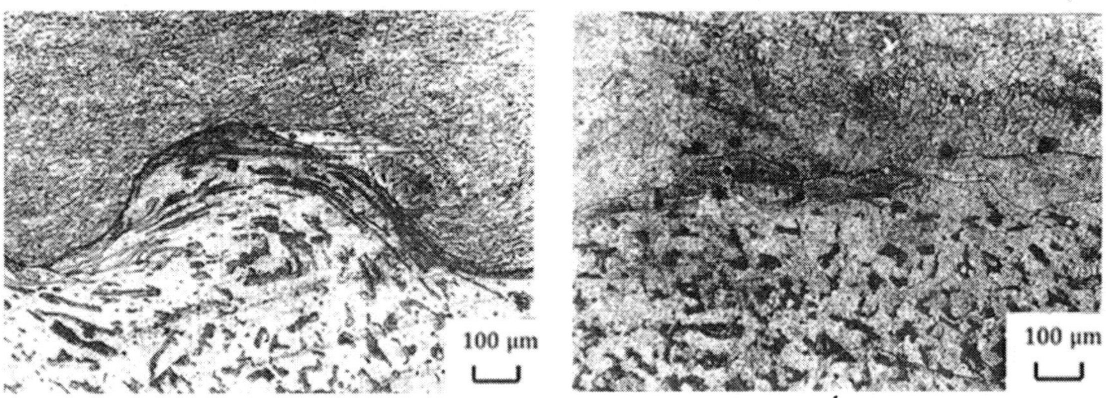

a b

Figure 6. Microstructure of a bond region armco iron + St3 for different values of velocity of the collision point: a - V_c = 2520 m/s; b - V_c = 3750 m/s.

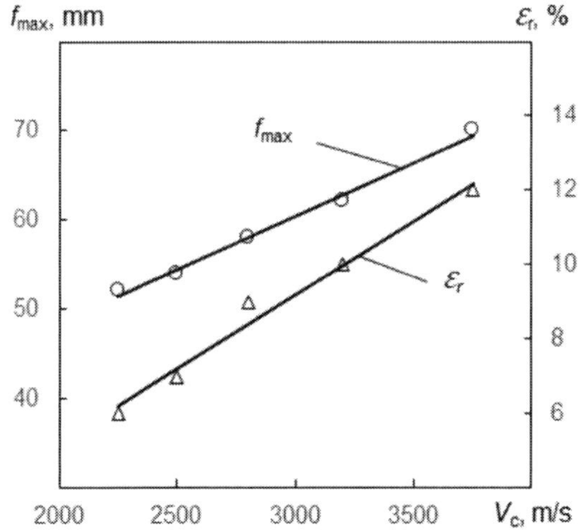

Figure 7. Influence of velocity of the collision point Vc on the maximum deflection fmax and radial deformation εr of armco iron + St3 bimetal.

The different influence of dynamic parameters is probably due to the redistribution of part of the charge energy in the direction of increasing the energy of the package movement spent on stamping the bimetal, which should be clarified during further research.

The obtained results of the research served as the basis for the development of new manufacturing processes using combined explosion welding with simultaneous stamping of bimetallic bronze-steel inserts of sliding bearings and pump linings of drilling equipment.

4. Conclusions

1. It is experimentally established that with an increase in velocity of the impact V_i and the velocity of the collision point V_c, there is an increase in the separation strength of the layers σ_{pull} of the armco-iron+St3 compound and an increase in the amount of molten metal, while the wave sizes 2a, λ increase with an increase in V_i, and with an increase in V_c, on the contrary, decrease.
2. Research of influence of dynamic parameters on the formation showed that the change in velocity of the impact V_i also has a significant impact on the stamping, while increasing the velocity of the collision point V_c there is a significant increase of the maximum deflection f_{max} and the radial deformation ε_r bimetal.

References

[1] Kuz'min V I, Lysak V I and Kuz'min E V 2020 Regularities in Bimetal Joint Formation at Explosion Welding with Simultaneous Stamping *Materials Science Forum* **989** 733

[2] Findik F 2011 Recent developments in explosive welding *Materials and Design* **32(3)** 1081

[3] Zorik V Y 2002 Theoretical grounds of designing technological system for manufacture of complex sheet parts by explosive stamping **11** 29

[4] Lepikhin P P, Romashchenko V A, Babich Y N, Beiner O S and Demenko V F 2005 Evaluation of the dynamic strength of cylindrical and conic matrices of finite length for stamping by high explosives *Strength of Materials* **37** 146

[5] Zlobin B S, Silvestrov V V, Shtertser A A, Plastinin A V and Kiselev V V 2014 Enhancement of explosive welding possibilities by the use of emulsion explosive *Archives of Metallurgy and Materials* **59** 1587

[6] Dibbo A 1997 Explosion bonding and forming provides solutions to increasingly severe material requirements *Metallurgy* **64** 357

[7] Mynors D and Zhang B. 2002 Applications and capabilities of explosive forming *Journal of Materials Processing Technology* **125** 1

[8] Fiedler T, Borovinšek M, Hokamoto K and Vesenjak M 2015 High-performance thermal capacitors made by explosion forming *International Journal of Heat and Mass Transfer* **83** 366

[9] Lysak V I and Kuzmin S V 2015 Energy balance during explosive welding *Journal of Materials Processing Technology* **222** 356

[10] Akbari Mousavi S A A, Riahi M and Hagh Parast A 2007 Experimental and numerical analyses of explosive free forming *Journal of Materials Processing Technology* **187-188**

[11] Petushkov V G, Simonov V A, Sedykh V S and Fadeenko Yu I 1995 *Explosion Welding Criteria* (USA: Harwood Academic Publishers)

[12] Crossland B 1982 *Explosive Welding of Metals and Its Application* (Oxford: Clarendon press)

[13] Blazinski T Z 1983 *Explosive welding, forming and compaction* (New York: Appl. Sci.)

[14] Fedotova N L, Chudakov I B, Korms I A, Saikov I V and Arutyunyan N A 2017 Preparation of Bimetal with Good Damping Properties *Metallurgist* **61** 63

[15] Kriventsov A N, Baluev A V and Kuz'min V I 1999 A method of calculating explosion welding parameters *Welding International* **11** 897

[16] Rybin V V, Greenberg B A, Antonova O V, Elkina O A and Ivanov M A 2009 Formation of Vortices during Explosion Welding *Physics of Metals and Metallography* **108** 353

[17] Kuz'min V I, Kriventsov A N and Baluev A V 2000 Kinetics of flight and determination of the velocity of movement and collision of a packet in explosion welding *Welding International* **14** 661

[18] Lysak V I and Kuzmin S V 2003 *Explosive welding of metal layered composite materials* (Kiev: E.O. Paton Electric Welding Institute of the NASU)
[19] Trykov Y P, Trudov A F and Stepanishchev I B 2002 Structure and Properties of Explosion-Welded Joints of Carbon Steels *Metal Science and Heat Treatment* **44** 524
[20] Kuz'min V I, Lysak V I, Kriventsov A N and Iakovlev M A 2004 Critical conditions of the formation and failure of welded joints in explosive welding *Welding International* **18** 223
[21] Greenberg B A, Ivanov M A, Inozemtsev A V, Patselov A M and Pushkin M S 2015 Microheterogeneous Structure of Local Melted Zones in the Process of Explosive Welding, *Metallurgical and Materials Transactions* **46** 3569
[22] Bataev I A, Bataev A A, Mali V I, Bataev V A and Balaganskii I A 2014 Structural changes of surface layers of steel plates in the process of explosive welding *Metal Science and Heat Treatment* **55** 509

Research of Plasma Parameters Formed in the Laser Welding Area in Vacuum

I Yu Letyagin, K V Felikan and V Ya Belenkiy

Perm National Research Polytechnic University, 29, Komsomolsky avenue, Perm, 614990, Russia

E-mail: letyagin@pstu.ru

Abstract. The use of laser technology in the manufacturing processes of mechanical engineering is gradually increasing. However, laser welding cannot compete with electron beam welding because the processes of the interaction of the laser beam with the plasma arising above the zone of the laser beam affecting the metal are not sufficiently studied. Of particular interest is the use of laser welding in vacuum. The study of the secondary emission signal in the zone of the laser beam affecting the metal allows one to obtain information about the plasma cloud in the zone of laser welding in vacuum. To study the secondary emission current in the laser welding zone, a computer system with an analog-to-digital interface can be used. The registered secondary emission current made it possible to estimate such parameters of the plasma arising in the laser welding zone in vacuum as the plasma concentration, electron temperature, and plasma potential.

1. Introduction

In recent years, laser welding has been increasingly used in industry in the manufacture of critical parts. A high concentration of heat exposure, high growth and temperature reduction rates in the treatment zone, as well as the possibility of rapid formation of a weld pool in a given volume, make it possible to efficiently use laser radiation to implement the welding process. Laser welding in shielding gases using powerful technological lasers is increasingly used in industrial production [1–3]. However, with a high power of the laser beam, the absorption of its energy by a plasma torch forming near the metal surface in the welding zone significantly reduces the efficiency of the thermal effect of the laser beam on the metal [4–6].

One of the promising directions of the development of laser technologies in the manufacture of critical large thickness products is welding with a concentrated laser beam in vacuum. This process has several advantages over its closest competitor by electron beam welding. The laser beam is not sensitive to magnetic fields and, consequently, the presence of residual magnetization in the object to be welded do not affect the quality of the process. The use of a laser beam for deep penetration also opens up fundamentally new possibilities for controlling the hydrodynamics of a weld pool using constant and variable magnetic fields at the same time. Electron beam welding requires a residual vapor pressure in the chamber 10^{-1} – 10^{-2} Pa. Laser welding in an environment with reduced pressure exerts its advantages already at pressures of about 1 kPa and pressures of about 100 Pa gives results almost indistinguishable from the results at 10^{-1} Pa [7–10]. This advantage allows you to abandon the use of molecular pumps and significantly increases the manufacturability of the process. In addition, it greatly simplifies the creation of mobile local forevacuum systems that can be used for welding large-sized products. Laser welding in vacuum allows you to get defect-free welds with a high ratio of the depth of the weld to its width. It has been established that when conducting the welding process with a laser beam in a vacuum, the penetration depth can be increased by a factor of 2–3 compared with laser welding in the atmosphere. An additional factor holding back the development of vacuum laser welding technology is the need to verify the physical models of related processes.

The processes of plasma formation under the influence of a powerful laser beam on a metal at atmospheric pressure are quite well studied [11–13]. When laser welding in vacuum in the zone of exposure to the laser beam as a result of ionization of metal vapor, plasma also appears, which, unlike a plasma torch during laser welding in a protective gas medium, does not absorb the energy of laser radiation and, at the same time, is a conducting medium for secondary emission current arising in the zone of laser welding in vacuum as a result of intense thermionic emission from the zone of the laser beam affecting the metal [14–16].

The formation of the secondary-emission signal in the zone of the action of a powerful laser beam on a metal in a vacuum is of considerable interest from the point of view of implementing operational control of the laser process vacuum welding [15, 16] The magnitude of this signal substantially depends on the conditions of current flow in the gap between the welded product and the charged particle collector installed above the welding zone to register the secondary emission signal. However, to date, there is no information on the parameters of the plasma cloud in the zone of laser welding in vacuum.

In accordance with this, the goal of this work was to study the plasma parameters in the zone of the laser beam affecting the metal in vacuum and the conditions for the passage of the secondary-emission current recorded in it to monitor the process of laser welding in vacuum.

2. Research Methodology

The study of plasma parameters was carried out by recording the magnitude of the secondary-emission current using a charged particle collector installed above the zone of the laser beam affecting the metal in vacuum and selecting current from the plasma when the positive collector potential changed. The experimental design is shown in Figure 1.

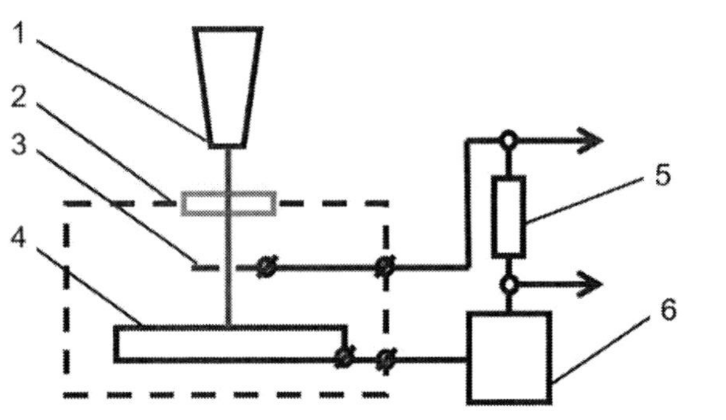

Figure 1. The registration scheme of the secondary emission current during laser welding in vacuum: 1 – installation for laser welding model ALFA-300; 2 – vacuum chamber; 3 – collector of charged particles; 4 – welded product; 5 – load resistor; 6 – bias voltage source.

The secondary emission current was recorded by creating an external electric circuit for plasma charges. This circuit contained an adjustable bias voltage source on the collector, which allowed changing the positive potential of the collector in the range 0…50 V and a resistor, the voltage drop on which was detected using a computer information-measuring system equipped with an analog-to-digital interface based on the multichannel analog-to-digital converter E20–10 from L–Card. The experiments were carried out on ALFA–300 laser welding machines equipped with a technological vacuum chamber with a quartz glass window in the upper lid of the laser beam input chamber. As the material of the processed product was used austenitic steel grade 12X18H10T.

As is known [12], one of the main factors causing plasma formation in the zone where a powerful laser beam acts on a metal is the ionization of metal vapors when interacting with powerful laser radiation due to inhibitory absorption when an electromagnetic wave interacts with free electrons. The emission of these electrons from a metal occurs both as a result of thermionic emission from a metal heated by a laser beam, and due to photoelectron emission when the metal is absorbed by laser radiation.

In this case, the charged particle collector can be considered as an electrode placed in a "foreign" plasma, and the current-voltage characteristic of the collector – work piece gap can be analyzed as a curve of the current to the Langmuir probe in the plasma. Using such a curve constructed on a logarithmic scale, the plasma concentration and electron temperature can be calculated using the expressions valid for the Maxwellian electron energy distribution [17]:

$$T_e = \frac{e}{k} \cdot \left[\frac{d}{dU} (\ln j_p) \right]^{-1} \tag{1}$$

$$n_e = \frac{j_{pm}}{e} \cdot \left(\frac{2\pi m_e}{kT_e} \right)^{1/2} \tag{2}$$

where U – is the voltage between the products in contact with the plasma electrode; $j_p = I_p \cdot S^{-1}$ is the density of the recorded current in the plasma; I_p – detectable secondary emission current; S – is the surface area of the electrode (in this case, the collector of charged particles) in contact with the plasma; j_{pm} – is the saturation current density (at the electrode potential equal to the plasma potential).

When a powerful laser beam acts on a metal in a vacuum, the current recorded by the charged particle collector is oscillatory in nature [16] In accordance with this, the average value (constant component) of this current was determined by processing the secondary-emission current registration files recorded during the experiments using the Mathcad program. During processing, the recorded signal was cleaned from associated interference by filtering at a frequency of 50 Hz. The direct and inverse Fourier transforms were performed. The received filtered signal was divided into sections. For each plot, extremes were determined with further calculation of the average amplitude.

3. Results and discussion

In Figure 2 shows the dependence of the average value of the secondary emission current recorded by the collector of charged particles in the zone of the laser beam affecting the metal in vacuum on the voltage between the collector and the product, built on a semi-logarithmic scale $\ln(I_p)$ when this voltage varies in the range of 5…30 V. The area of the charged particle collector was 180 mm^2.

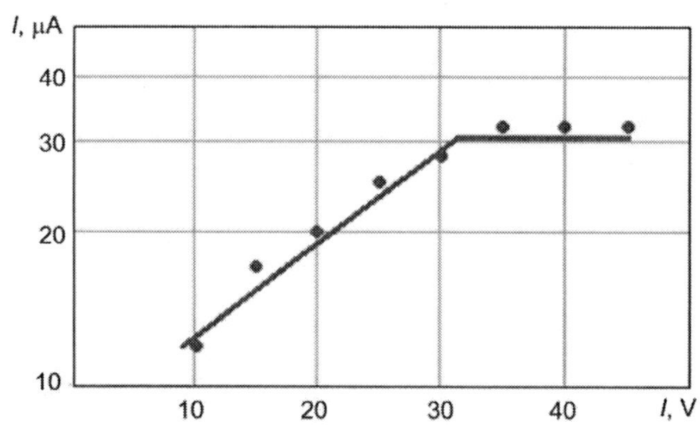

Figure 2. The semi-logarithmic dependence of the average value of the secondary emission current on the voltage between the work piece and the charged particle collector.

This dependence was analyzed as an electronic branch of the probe characteristic. The saturation of the electron current from the plasma characterizes the equality of the potentials of the collector of charged particles, taking current from the plasma, and thus, the plasma potential is 30...33 V.

The graph shows that the dependence of the density of the average value of the secondary emission current on the voltage between the charged particle collector and the electron current product being processed has a character close to linear. However, the fact that the secondary-emission signal is practically absent at a positive potential of the charged particle collector of less than 4...8 V indicates a violation of the linearity of the electron current curve and, accordingly, a deviation of the electron energy distribution function in the plasma from the Maxwellian one in the high energies. These deviations may be due to a distortion of the plasma parameters due to the selection of current from the plasma to the charged collector electrode or to the presence of a high-energy "tail" in the distribution function. Nevertheless, as shown in [18], the effective temperature of plasma electrons can also be estimated if the energy distribution function of electrons deviates from the Maxwellian one. In this case, to estimate the effective plasma temperature, a part of the electron branch of the probe characteristic corresponding to potentials close to the plasma potential should be used.

As a result of calculating the plasma parameters in the zone of influence of a powerful concentrated laser beam on a metal in vacuum using expressions (1) and (2), $T_e = 1.2 \cdot 10^{10}$ К и $n_e = 6 \cdot 10^9$ m^{-3}. It should be noted that the estimates made are of a qualitative nature, since the collector of charged particles has a significant area of contact with the plasma and selects the current, acting on the plasma and changing its parameters. High electron plasma temperatures are caused in this region by heating as a result of energy dissipation during the interaction of the electromagnetic wave of laser radiation with free plasma electrons.

The above results show that, as applied to the conditions for recording the secondary emission current generated by thermionic emission into plasma during laser welding in vacuum, the charged particle collector can be considered as an electrode operating in the probe mode. The magnitude of the potential drop in the space charge layer near this electrode is determined by the ratio of the density of the recorded secondary emission current j_p to the total current density from plasma j [19]:

$$\Delta U = \frac{kT}{e} \cdot \ln \frac{j_p}{j} \tag{3}$$

When a potential equal to the plasma potential is applied to the electrode, the detected secondary emission current is saturated, associated with the complete selection of the current from the plasma.

4. Conclusion

For research, a technique was used to register the secondary emission current between the charged particle collector and the surface of the workpiece during laser welding in vacuum. A charged particle collector can be considered as an electrode operating in probe mode. The magnitude of the secondary emission current is dependent on the voltage between the charged particle collector and the product. This site is analyzed as an electronic branch of the current-voltage characteristics of the Langmuir plasma. A change in the potential applied to the electrode made it possible to saturate the secondary-emission current, therefore, to fix the plasma potential above the processing zone of the product with a laser beam in vacuum. Thus, the study of the dependence of the secondary emission current in plasma during laser welding in vacuum on the voltage between the charged particle collector and the workpiece made it possible to determine the parameters of the plasma arising in the laser welding zone in vacuum: concentration, electron temperature, and plasma potential.

References

[1] Gapontsev V P 2003 Fiber lasers burst a laser industry *Proc. of the 12th International Laser Physics Workshop*
[2] Holzer M, Rominger V and Havrilla D 2010 Latest Results in Industrial Welding of Thick Sheets with High Power Trudisk Lasers and Optimized Peripheral Components *Physics Procedia of Lasers in Manufacturing*
[3] Schlueter H 2005 Advances in Industrial Power Lasers *Proc. SPIE* **5777** 8–15
[4] Weiler S 2009 Disk lasers for industry *Photonics* **3** 10–13
[5] Petrovskiy V N, Prokopova N M and Shcheglov P Yu 2010 Spectral diagnostics of a vapor-plasma plume produced during welding with a high-power ytterbium fiber laser *Laser Phys.*

Lett. **7** **396**
[6] Harutyunyan R V, Baranov V Yu, Bolshov V A 1985 *The effect of laser radiation on materials* (Moscow: Science)
[7] Arata Y, Abe N, Oda T and Tsujii N 1985 Fundamental phenomena during vacuum laser welding *Proc. of ICALEO '85 Materials Processing Symp. Laser Inst. of America* **44** 1–7
[8] Katayama S, Abe Y, Mizutani M and Kawahito Y 2011 Deep Penetration Welding with High-Power Laser under Vacuum *Transactions of JWRI* **40(1)** 15–19
[9] Katayama S, Yohei A, Mizutani M and Kawahito Y 2011 Development of deep penetration welding technology with high brightness laser under vacuum *Physics Procedia* **12** 75–80
[10] Reisgen U, Olschok S and Jakobs S 2013 A comparison of electron beam welding with laser beam welding in vacuum *9th International Conference Beam Technology*
[11] Scheglov P Yu and Uspensky S A 2011 Investigation of the vapor-plasma torch during welding with a high-power fiber laser *Scientific session of MEPhI* **2**
[12] Petrovsky V N and Uspensky S A 2011 Investigation of a vapor-plasma welding torch when welding with a powerful ytterbium fiber laser *Nuclear Physics Engineering* **2** 159–165
[13] Ishide T, Shono S, Ohmae T, Yoshida H and Shinmi A 1987 Fundamental Study of Laser Plasma Reduction Method in High Power CO_2 Laser Welding *Proc. of the Int. Conf. on Laser Advanced Materials Processing* 187-191
[14] Letyagin I Yu, Belenkiy V Y and Trushnikov D N 2018 On the connection between the energy parameters of secondary emission signals from the laser beam welding zone in vacuum with the parameters of metal penetration *Journal of Physics: Conference Series* **1109(1) 012013** 1–9
[15] Letyagin I Yu, Trushnikov D N, Belenkiy V Ya 2018 Investigation of secondary emission signals from the laser beam exposure zone during laser welding in vacuum *Bulletin of Perm National Research Polytechnic University. Engineering, Mechanical engineering, materials science* **20** **3** 106–114
[16] Letyagin I Yu, Belenkiy V Ya, Lyamin Y V, Erikov A P 2019 The study of processes in the area of the laser beam on the metal during laser welding in vacuum *Bulletin of the Perm Federal Research Center* **3** 20–29
[17] Kozlov O V 1969 *An electric probe in a plasma* (Moscow: Atomizdat)
[18] Ershov A P, Dovzhenko V A, Kuzovnikov A A 1981 On processing the current-voltage characteristics of a Langmuir probe in a non-Maxwellian plasma *Physics of plasma* **7(3)** 609–613
[19] Granovsky V L *Electric current in the gas* (Moscow: Science)

Acknowledgments
The reported study was partially supported by the Government of Perm Krai, research project No. C-26/794.

Research of the Influence of the Rarefaction Degree on the Metal Protection Quality and Geometric Welding Parameters During Arc Welding in the Controlled Atmosphere

I V Sitnikov, D N Trushnikov and G L Permyakov

Department of Welding Production, Metrology and Materials Technology, Perm National Research Polytechnic University, 29, Komsomolsky ave., Perm, 614990, Russia

E-mail: stu459@mail.ru

Abstract. This article is devoted to the urgent issue of welding metals and alloys that are widely used in modern aerospace industry. The paper describes the effect of the degree of rarefaction of the controlled atmosphere on the quality of protection of the weld metal and the weld zone, as well as on the change in the geometric parameters of the weld during arc welding in a controlled atmosphere. The main advantages of arc welding in a controlled atmosphere of refractory, heat-resistant, heat-resistant, corrosion-resistant metals and their alloys are indicated. The welding modes used and the welds obtained with these modes are presented. The influence of the degree of rarefaction on the change in the quality of protection of the weld metal and the heat-affected zone is described. The degree of rarefaction of the controlled atmosphere ranged from 200 to 1000 mbar. Dependencies are obtained that describe the effect of the degree of rarefaction on the geometric parameters of welds.

1. Introduction

Today, in the aerospace industry, refractory, heat-resistant, heat-resistant, and corrosion-resistant metals and their alloys are increasingly used. Heating and melting of these materials in air is practically impossible, mainly due to their high chemical activity and propensity for gas saturation, due to their high affinity for atmospheric gases. The natural instability of most metals in their elemental form causes them to react with oxygen in the air when heated during welding. In addition to the reaction with the metal, the components of the air, oxygen and nitrogen can dissolve in the molten weld pool. Hydrogen gas, although not a normal component of air, can be present in the atmosphere around the arc, since high temperatures can destroy moisture, fat, and other hydrogen-containing materials. Various methods of protection are used to eliminate the harmful effects of the air atmosphere during heating and melting [1–9].

The most common welding method for these materials is arc welding in vacuum or inert gas. When welding these materials, the most affordable means of protection against the negative effects of the atmosphere on molten metal is the use of local methods of protection. For this, argon-arc burners, gas lenses and additional blowing of the root of the seam with inert gas are used. The main disadvantage of such methods of protection is poor protection of the metal of the heat-affected zone or even its absence, which does not favorably affect the mechanical properties of the welded joint as a whole [10–18].

IOP Conf. Series: Materials Science and Engineering **969** (2020) 012092 doi:10.1088/1757-899X/969/1/012092

A more advanced method of protection is welding using protective microcams. When using this method, not only molten metal baths are protected, but also most of the metal of the heat-affected zone heated to high temperatures. The main advantage of such chambers is the creation of a protective atmosphere with a minimum content of oxygen and nitrogen, as a result of which providing more stable mechanical characteristics. In chambers with a controlled atmosphere, it is possible to weld with a consumable and non-consumable electrode, manually or automatically [19, 20].

2. Methods and materials

In this paper, we assess the effect of the degree of rarefaction of the controlled atmosphere on the quality of metal protection and the parameters of the weld during arc welding in a controlled atmosphere.

For the experiments, we used an experimental setup for the vacuum-arc layer-by-layer synthesis of metal products, developed on the basis of the electron-beam installation (EBI-M). The installation includes: a vacuum chamber with a volume of ~ 1 m3, equipped with a four coordinate CNC moving mechanism, a control panel for a vacuum system with vacuum gauges; plasma gas supply system; universal welding inverter with a water-cooled torch; liquid cooling unit.

A universal welding inverter FLAMA TIG 200E AC / DC PULSE was used as a power source. As a torch, a standard welding torch for welding with a non-consumable electrode was used. To cool the burner, the Miller Coolmate 3 liquid cooling unit was used.

The chamber was evacuated, followed by filling with premium grade argon (99.993%) until a residual pressure of 1000 mbar was reached in increments of 100 mbar. After that, welding passes were carried out (without the use of filler material) on 3 mm thick plates of steel 12Cr18Ni10Ti (analogs of AISI 304, AISI 316, AISI 321) in a protective atmosphere at various values of residual pressure. The chemical composition of steel 12Cr18Ni10Ti is presented in table 1. The parameters of the welding mode are presented in table 2.

Table 1. Chemical composition of steel 12Cr18Ni10Ti.

Element	C	Fe	Cr	Ni	Ti	Mn	Si	Cu	P	S
Content, %	≤0.12	67	17-19	9-11	0.4-1	≤2	≤0.80	≤0.3	≤0.035	≤0.02

Table 2. Welding modes.

Arc current, A	Welding speed, mm/s	Arc gap, mm	Used electrode	Diameter of electrode, mm
80	7.5	3	WL-20	3.2

3. Results

The obtained welding results are presented in Figure 1. At a residual pressure of 100 mbar, it was not possible to initiate the welding process. Additionally, a study was conducted on a RHEN-602 hydrogen analyzer to determine the hydrogen content in welds. The measurement results are presented in Figure 1.

Next, macro sections were made from the welds obtained. The appearance of macro sections is shown in Figures 2–10.

On the obtained microsections, the main parameters of the welds were measured and the coefficients of their shape φ_w were calculated. The main parameters were taken the width and depth of penetration of the weld. The shape factor φ_w of the weld is the ratio of depth to penetration width. The obtained measurement and calculation results are presented in table 3.

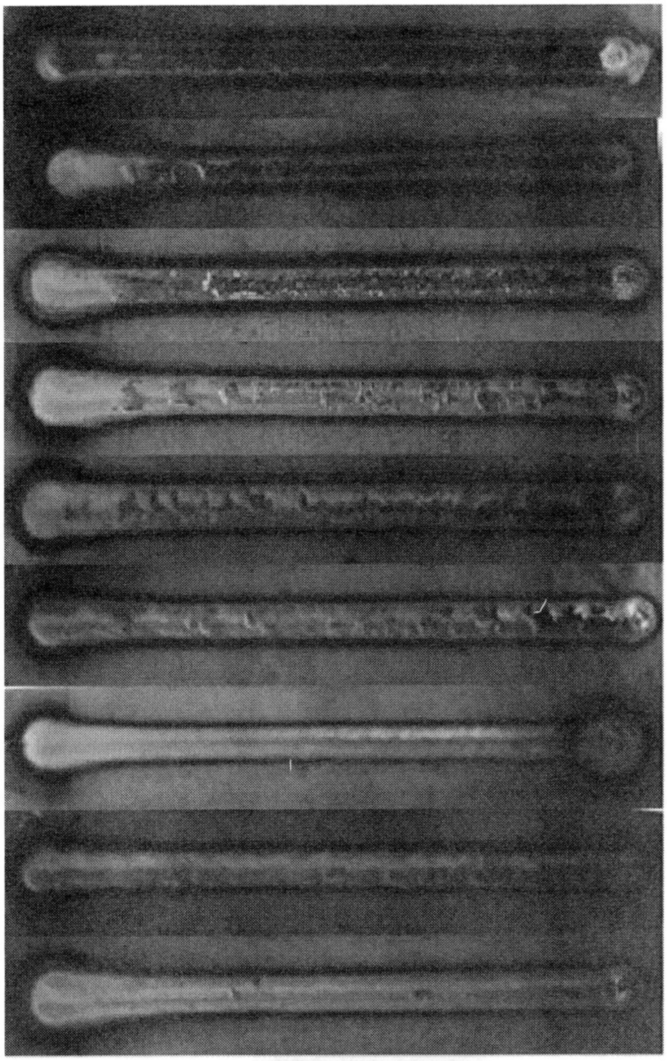

Figure 1. Results of the effect of the degree of rarefaction on the quality of protection of the weld: 1 – chamber pressure 200 mbar, mass fraction of H_2 0.0004 %; 2 – chamber pressure 300 mbar, mass fraction of H_2 0.0004 %; 3 – chamber pressure 400 mbar, mass fraction of H_2 0.0003 %; 4 – chamber pressure 500 mbar, mass fraction of H_2 0.0002 %; 5 – chamber pressure 600 mbar, mass fraction of H_2 0.0003 %; 6 – chamber pressure 700 mbar, mass fraction of H_2 0.0002 %; 7 – chamber pressure 800 mbar, mass fraction of H_2 0.0002 %; 8 – chamber pressure 900 mbar, mass fraction of H_2 0.0001 %; 9 – chamber pressure 1000 mbar, mass fraction of H_2 0.0001 %.

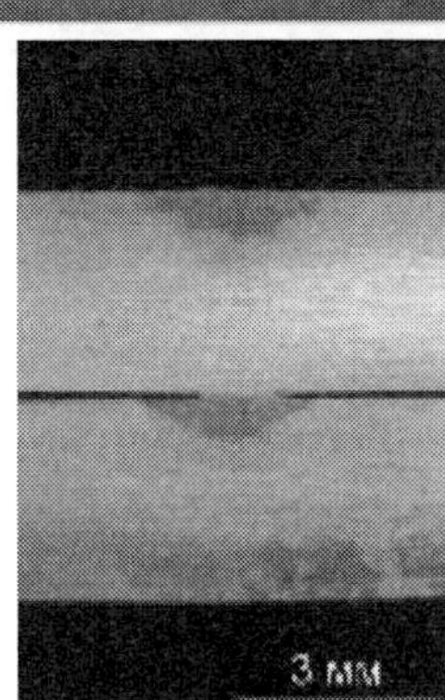

Figure 2. Macro section of a weld obtained at a pressure in a chamber with a controlled atmosphere of 200 mbar.

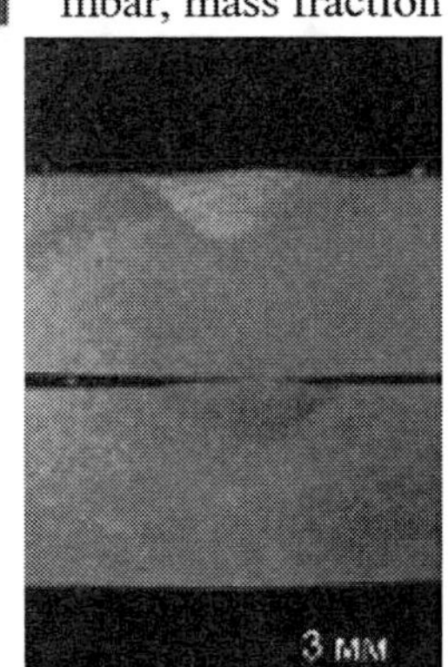

Figure 3. Macro section of a weld obtained at a pressure in a chamber with a controlled atmosphere of 300 mbar.

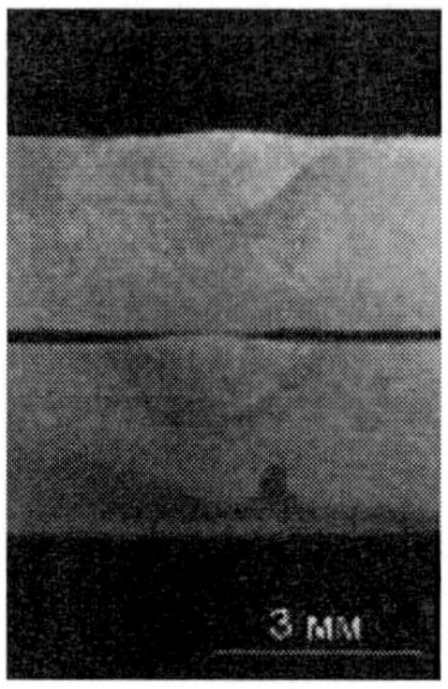

Figure 4. Macro section of a weld obtained at a pressure in a chamber with a controlled atmosphere of 400 mbar.

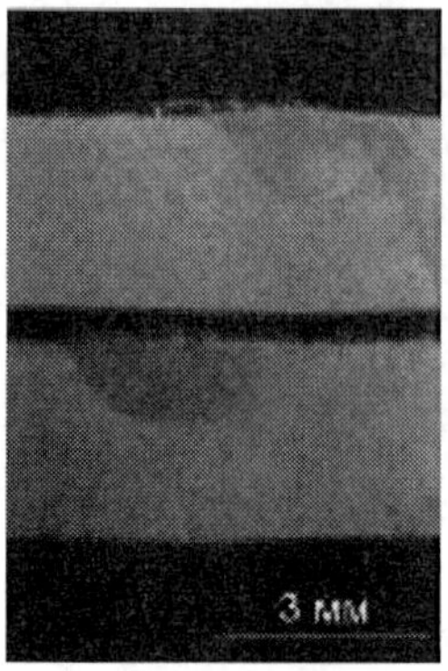

Figure 5. Macro section of a weld obtained at a pressure in a chamber with a controlled atmosphere of 500 mbar.

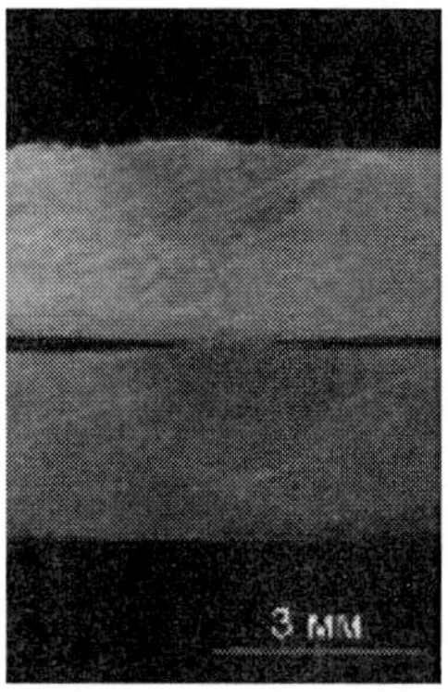

Figure 6. Macro section of a weld obtained at a pressure in a chamber with a controlled atmosphere of 600 mbar.

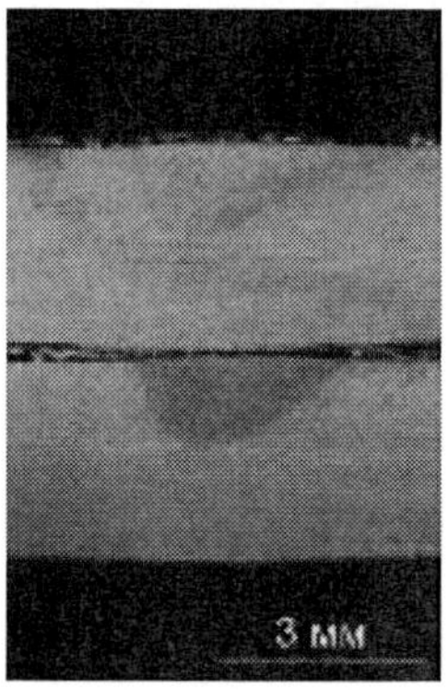

Figure 7. Macro section of a weld obtained at a pressure in a chamber with a controlled atmosphere of 700 mbar.

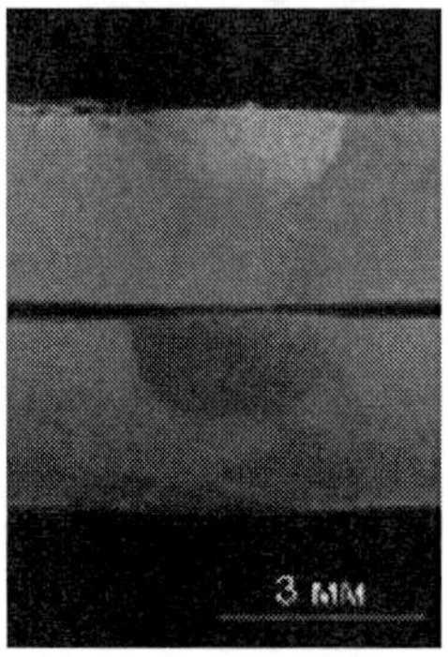

Figure 8. Macro section of a weld obtained at a pressure in a chamber with a controlled atmosphere of 800 mbar.

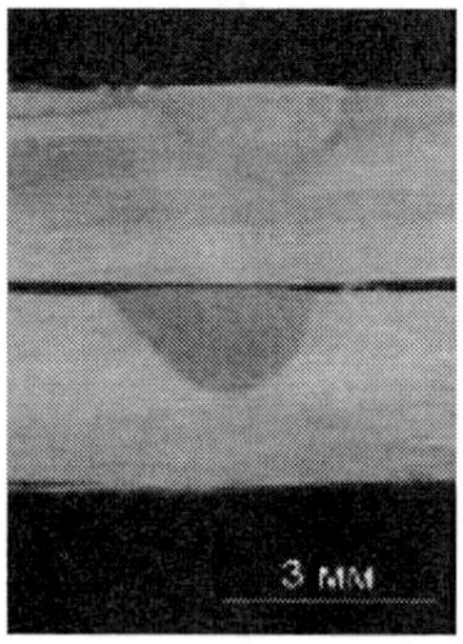

Figure 9. Macro section of a weld obtained at a pressure in a chamber with a controlled atmosphere of 900 mbar.

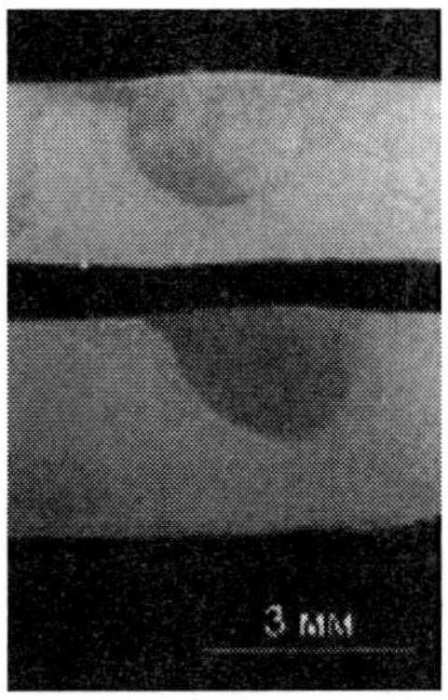

Figure 10. Macro section of a weld obtained at a pressure in a chamber with a controlled atmosphere of 1000 mbar.

Table 3. Values of parameters and shape factors of welded passages.

No.	Pressure, mbar	Width of penetration, mm	Depth of penetration, mm	The shape factor seam φ_w
1	200	2.61	0.67	0.257
2	300	3.01	1.00	0.332
3	400	3.08	1.34	0.435
4	500	3.15	1.37	0.435
5	600	3.01	1.41	0.468
6	700	3.15	1.41	0.448
7	800	3.15	1.61	0.511
8	900	2.88	1.64	0.569
9	1000	3.15	1.94	0.616

Based on the obtained measurement and calculation results, graphs of the dependences of the width and depth of penetration, as well as the shape coefficient of the weld φ_w on the pressure in the chamber, are plotted, the dependency graphs are shown in Figures 11, 12, 13.

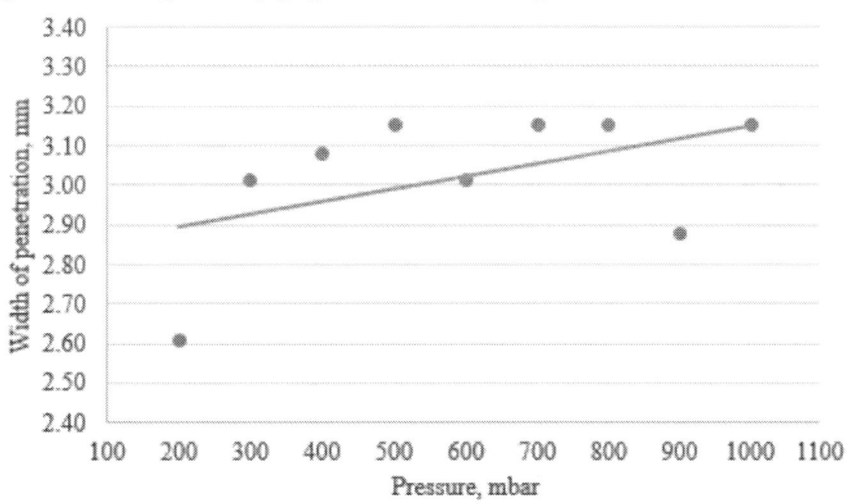

Figure 11. Graph of penetration width versus chamber pressure.

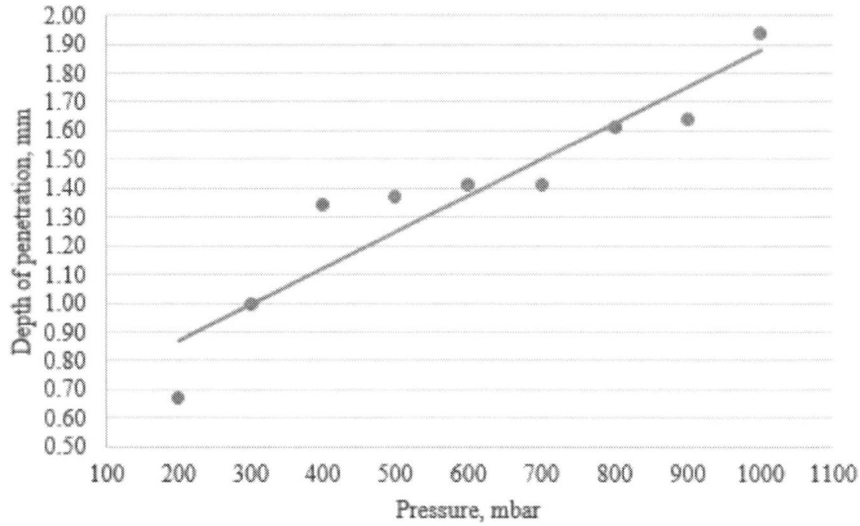

Figure 12. Graph of penetration depth versus chamber pressure.

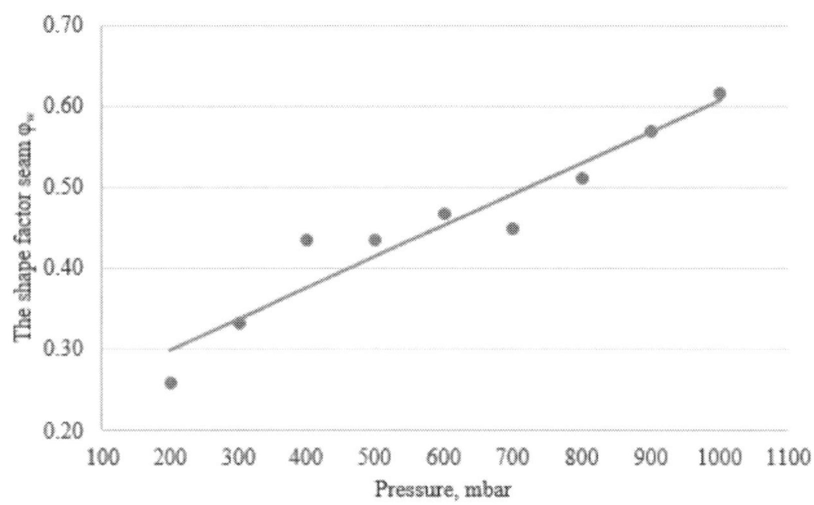

Figure 13. The dependence of the joint shape factor φ_w on the pressure in the chamber.

4. Discussion

From the obtained experimental results, it is seen that with an increase in the pressure of the protective gas in the chamber and its equalization with atmospheric, an improvement in the quality of metal protection is observed. At a pressure of 200 mbar, a large number of oxide films and deposits are observed in the area of the heat-affected zone on the surface of the welding arc.

After increasing the pressure to 400 mbar, the quality of protection improves, this can be judged by the reduction of soot in the heat-affected zone, it is practically absent. The surface of the seam is less contaminated, but also has a yellow tint, indicating poor metal protection.

At a pressure of 800 mbar, the degree of protection improves, oxide films are completely absent on the surface of the weld, but this protection is not enough for the metal of the heat-affected zone, a thin strip of soot along the weld indicates this.

After equalization of the chamber pressure with atmospheric (1000 mbar), the metal protection becomes satisfactory, however, a small deposit is present in the zone of metal heating to high temperatures. This is due to the fact that, in the presence of even a small admixture of atmospheric gas in argon, the degree of protection is significantly reduced. To improve protection, it is advisable to use additional purification of argon poured into the chamber.

As the pressure in the chamber increases, the penetrating ability of the arc increases by more than 1 mm. The penetration width also tends to increase as the pressure in the chamber approaches atmospheric pressure, this increase is not significant and is within 0.5 mm. Accordingly, the joint shape factor φ_w will increase, since it is directly related to the width and depth of penetration.

5. Conclusion

The influence of the degree of rarefaction of the controlled atmosphere on the quality of protection of the weld metal and the heat-affected zone is assessed. With increasing pressure of a controlled atmosphere, the quality of protection improves. When the residual pressure of the controlled atmosphere is close to atmospheric, the quality of protection of the weld metal and the heat-affected zone is satisfactory.

Dependencies are obtained that describe the effect of the degree of rarefaction on the parameters and shape factors of the welded passages. With increasing pressure increases the depth and width of penetration.

References

[1] Zubkov L B 1987 *Space metal: (All about titanium)* (Moscow: Nauka) p 128
[2] Savitsky E M and Klyachko V S 1978 *Metals of the space age* (Moscow: Metallurgy) p 120
[3] Nerovny V M and Yampolsky V M 2002 *Welding arc processes in a vacuum* (Moscow: Mechanical Engineering) p 274
[4] Kou S 2002 *Welding metallurgy* (New York: Willey Inter Science Publisher)
[5] Kim H J, Frost H R and Olson D L 1998 Electrochemical oxygen transfer during direct current arc welding *Weld J.* 488–493
[6] 1994 *Abington Welding Training Module* (Welding Institute of Canada, Abington Publishing)
[7] Kacar R and Kokemli K 2005 Effect of controlled atmosphere on the mig-mag arc weldment properties *Materials & Design* **26 6** 508–516
[8] Sato Y, Hayakawa N and Kuwana T 1994 Effect of oxygen on microstructure and mechanical properties of steel weld metal *Department of Materials Processing, Faculty of Engineering Tohoku University* 1–2
[9] Sato Y and Kuwana T 1995 Oxygen Absorption in Iron and Steel Weld Metal *ISIJ International* **35 10** 1162–1169
[10] Yampolsky V M, Antonov A A and Bogdanov S F 1969 Some features of the process of arc welding in vacuum by a non-consumable electrode *Proceedings of the University N. E. Bauman* **132** 38–44
[11] Yampolsky V M 1980 Arc welding and surfacing in a vacuum *Proceedings of the University N. E. Bauman* **337** 131–137
[12] Ishtykov Yu V, Kazakov Yu V and Aksyonov N A 1972 Welding with a tungsten electrode at a reduced pressure in the chamber *Automatic Welding* **2** 42–43
[13] Moskalev B I 1969 *Hollow cathode discharge* (Moscow: Energy) p 184
[14] Permyakov G L, Trushnikov D N, Mindibaev M K and Koleva E G 2019 Numerical model of a hollow cathode arc discharge formation in vacuum *IOP Conference Series: Materials Science and Engineering* **759**
[15] Raizer Yu P and Allen J E 1991 *Gas Discharge Physics* (Berlin: Springer-Verlag) p 450
[16] Lafferty J M and Cobine J D 1980 *Vacuum Arcs: Theory and Applications* (Wiley-Interscience publication) p 372
[17] Granovsky V L 1971 *Electric discharge in gases* (Moscow: Nauka) p 593

[18] Kesasv I G 1968 *Cathode processes of an electric arc* (Moscow: Nauka) p 244
[19] Leskov G I 1970 *Electric welding arc* (Moscow: Mechanical Engineering) p 335
[20] Nikolaev G A and Olshansky N A 1975 *Special welding methods* (Moscow: Engineering) p 232

Acknowledgments
The studies were financially supported by the Ministry of Science and Higher Education of the Russian Federation on state assignment FSNM-2020-0028, the Government of the Perm Territory as part of agreement S-26/787 of 12/21/2017 (studying the current state of the issue) and the Russian Foundation for Basic Research under project No. 18-08-01016 A (experimental studies).

IOP Conf. Series: Materials Science and Engineering **969** (2020) 012093 doi:10.1088/1757-899X/969/1/012093

Laser Powder Cladding Complex: Principles of Advanced Automated Control

R V Chkalov and D G Chkalova

Vladimir State University, 87, Gorky Street, Vladimir, 600000, Russia

E-mail: j.larenax@gmail.com

Abstract. Currently, the laser powder cladding process is one of the most promising methods for producing wear and corrosion resistant sealing surfaces of high hardness. To solve the problem of high-quality automated production of coatings, that meet specified requirements, on the surfaces of critical parts, it is necessary to identify key factors, as well as determine the contribution of each of them in the process of laser powder cladding. In view of the foregoing, the urgent task is to develop an integrated system that allows monitoring and recording parameters of ongoing physical processes, as well as ensuring reproducibility of experimental results. Since such an installation includes many disparate technical units and transition modules of various integration levels, each of which has its own functionality, a specialized software package has been developed, the purpose of which is to combine individual components into a single organized and controlled system. As an approbation of the experimental complex work, the results of laser cladding process are shown.

1. Introduction

The use of focused laser radiation for cladding metals can significantly expand the technological capabilities of the process, increase its productivity and improve the quality of resulting coatings. In comparison with traditional methods of powder cladding [1–3], the following key advantages of such approach can be distinguished:

- General increase in operation productivity due to decrease in mixing zone of substrate material and clad layer [4], and as a result, a decrease in thickness of applied intermediate layer, in complexity of cladding process and subsequent machining;
- Improving the processability, efficiency, increasing the adhesion resistance of applied material to substrate, reducing the porosity and the number of non-metallic inclusions characteristic of processing methods using fluxes [5–7];
- The ability to identify defective workpieces at cladding stage by developing an effective real-time process control system [8–10].

Since the laser cladding process is caused by the appearance of a number of concomitant physical phenomena that directly affect both geometry of clad bead and characteristics of coating being formed, it is of interest to develop a control system that enables automated correction of processing mode in real time, based on data recorded by sensors in areas of laser exposure.

2. Automated laser powder cladding complex

In general case, the process of laser powder cladding includes the following mechanism of action. The powder mixture enters the laser radiation propagation region, where, under the influence of radiation, material is melted, followed by deposition on the surface of the workpiece (figure 1) [11]. The process can take place both using shielding gas in processing area and in a chamber isolated from atmospheric air. Moving the cladding head is carried out by positioners of various types, in most cases, robotic manipulators are used to deliver the deposited mixture. The combination of technological equipment elements with a robotic manipulator into a single industrial complex is called a laser robotic cladding complex. The main variable parameters of laser cladding are powder feed rate, laser radiation power, as well as the speed of cladding head movement. Changing these values leads to a change in geometry of clad layer, in mixing coefficient of substrate material and deposited powder, the area of propagation of temperature front in a volume of material.

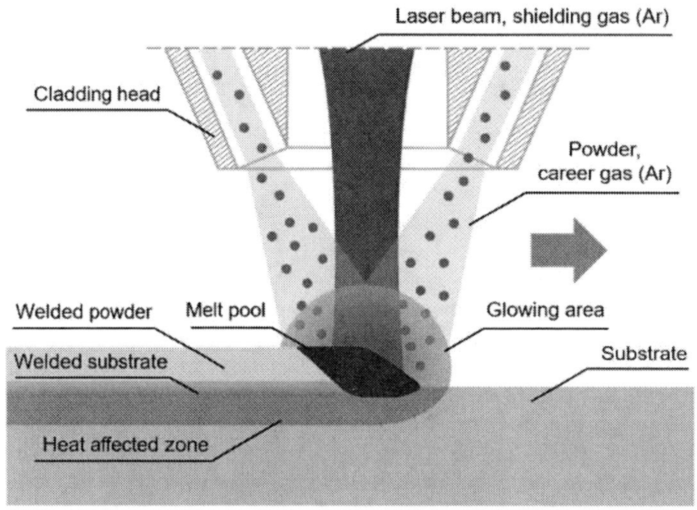

Figure 1. Laser powder cladding process.

To automate the coating process on surface of critical parts, it is necessary to identify key factors that influence the process of laser cladding, as well as determine the contribution of each of them. So, the work [12] describes the inheritance of powder material structure in cladding process of 03X17H14M3 steel. The work [13] examined the problems of crack formation during direct laser cladding SS316. A negative effect of sulfur, phosphorus and silicon on the resistance to cracking during direct laser cladding was revealed. In addition to material properties, processing mode also directly affects laser cladding result. The location of the sample being processed relative to the focusing optical system focal plane and the nozzle part is of great importance [13,14]. Both the height of the deposited structure and the uniformity of the clad bead depend on the focal plane orientation relative to the sample and the region of powder flow convergence. In [15], a study was made of the influence of cladding on the formation of defects and properties of deposited materials. It was found that rapid heat dissipation contributes to obtaining a fine-grained structure with better mechanical properties compared to samples subject to prolonged heat exposure [12–14].

In view of foregoing, an urgent task is to create an integrated system for monitoring technological processing parameters for evaluating the effectiveness of powder cladding process in real time. To monitor the progress of different types of laser technological process, various hardware and software solutions are used [16–17]. The complex under development allows monitoring and fixing the parameters of ongoing physical processes, as well as ensuring reproducibility of results. Among other things, this approach allows to collect an array of experimental data, which could allow a comparison of the results with the modeling data presented in modern scientific papers [18–22].

The complex of automated laser powder cladding includes the following main technical components: a cladding head equipped with a set of temperature sensors (pyrometers), a video recording system and a triangulation sensor; a system for spatial positioning of the head, made in the form of an industrial robotic manipulator, a system for the part pre-heating, which is a tubular heater located in a heat-insulating casing, as well as an operator terminal that controls the setup (figure 2).

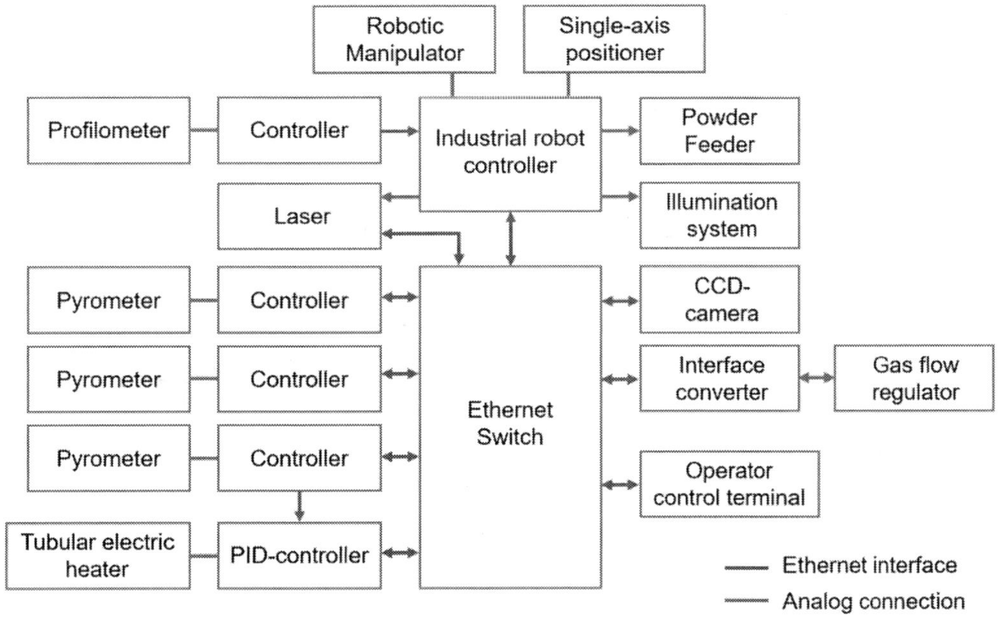

Figure 2. Scheme of the robotic laser cladding complex.

One of the most structurally complex elements of the technological complex is a cladding head. The configuration of cladding head is the most important factor, since it directly determines the physics of the process. The presented complex is equipped with a multi-jet powder supply system with the possibility of programmed control of transport gas flow, powder feed rate, and laser radiation power. The head is additionally equipped with a set of sensors for monitoring the cladding process, for which a system of brackets has been developed that allows recording devices to be located at a distance of 150–200 mm from the processing zone to prevent damage caused by cladding products. The combination of sensors used allows to obtain comprehensive information about the nature of physical processes occurring in processing area, on the basis of which it is possible to adjust the parameters of laser cladding in real time.

3. Experimental results

At the current stage, a series of experiments is being conducted on automated correction of the laser radiation power based on data obtained from a CCD camera (figure 3). Laser cladding of separately located tracks and cladding of tracks with overlap was performed. The obtained samples were examined for the dependence of the formed tracks geometric characteristics on the exposure parameters. The data obtained were used to adjust the laser complex control algorithm.

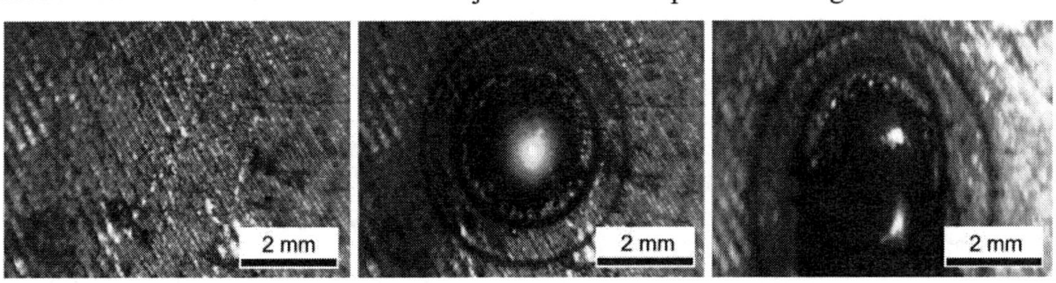

Figure 3. Images of laser cladding area obtained by CCD camera.

As a result of optimization of the control program, it was found that the use of an CCD camera to correct the processing mode based on data on deposited tracks geometric characteristics (track width) allows constant control and maintenance of a given width of clad layer by varying the parameters of laser power and cladding speed. Maintaining the geometric characteristics of the weld beads must be performed constantly, since during the multilayer cladding the thermophysical characteristics of the processed substrate change, which leads to a change in geometry of weld bead.

Figure 4 shows the end section of deposited tracks made of AISI 316 steel onto a substrate surface of the same material with a 4 mm overlap. The basic parameters were: laser radiation power – 2800 W, powder consumption – 15 g / min, cladding head speed – 6 mm / s. In this experiment, the geometric characteristics of the deposited coating were corrected by changing the laser radiation power.

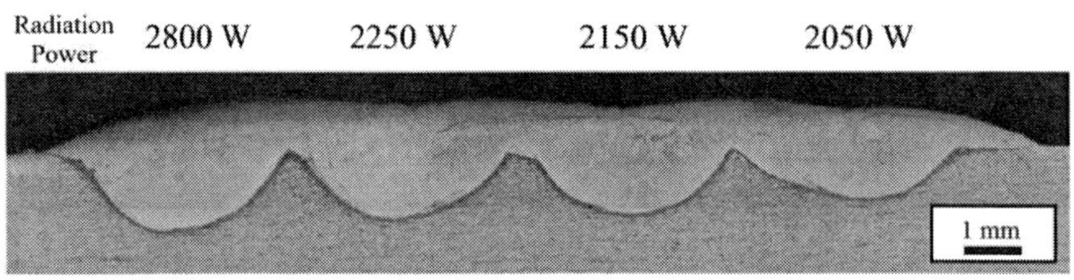

Figure 4. Cross section of the sample.

4. Conclusion

The use of focused laser radiation for welding metals can significantly expand the technological capabilities of welding processes, increase their productivity and improve the quality of the welded joints, which makes this method of powder cladding the most promising. The developed laser cladding complex allows solving a wide range of scientific and applied tasks of selecting optimal coating conditions by structuring and analyzing the data of physical processes recorded in the field of laser exposure.

References

[1] Zhao C, Tian F, Peng H R and Hou J Y 2002 Non-transferred arc plasma cladding of Stellite Ni60 alloy on steel *Surf. Coatings Technol.* 80–84

[2] Turichin G, Kuznetsov M, Sokolov M and Salminen A 2015 Hybrid Laser Arc Welding of X80 Steel: Influence of Welding Speed and Preheating on the Microstructure and Mechanical Properties *Physics Procedia* 35–44

[3] Turichin G, Kuznetsov M, Tsibulskiy I and Firsova A 2017 Hybrid Laser-Arc Welding of the High-Strength Shipbuilding Steels: Equipment and Technology *Physics Procedia* 156–163

[4] Pajukoski H, Näkki J, Thieme S, Tuominen J, Nowotny S and Vuoristo P 2016 High performance corrosion resistant coatings by novel coaxial cold- and hot-wire laser cladding methods *J. Laser Appl.*

[5] Lei J, Shi C, Zhou S, Gu Z and Zhang L C 2018 Enhanced corrosion and wear resistance properties of carbon fiber reinforced Ni-based composite coating by laser cladding *Surf. Coatings Technol.* 274–285

[6] Turichin G A, Zemlyakov E V., Pozdeeva E Y, Tuominen J and Vuoristo P 2012 Technological possibilities of laser cladding with the help of powerful fiber lasers *Met. Sci. Heat Treat.* 139–144

[7] Baidridge T, Poling G, Foroozmehr E, Kovacevic R, Metz T, Kadekar V and Gupta M C 2013 Laser cladding of Inconel 690 on Inconel 600 superalloy for corrosion protection in nuclear applications *Opt. Lasers Eng.* 65–71

[8] Purtonen T, Kalliosaari A and Salminen A 2014 Monitoring and adaptive control of laser processes *Physics Procedia* 1218–1231

[9] Doubenskaia M, Bertrand P and Smurov I 2004 Optical monitoring of Nd:YAG laser cladding *Thin Solid Films* 477–485
[10] Smurov I, Doubenskaia M, Grigoriev S and Nazarov A 2012 Optical monitoring in laser cladding of Ti6Al4V *Journal of Thermal Spray Technology* 1357–1362
[11] Vilar R 2014 Laser Powder Deposition *Comprehensive Materials Processing*
[12] Sklyar M O, Turichin G A, Klimova O G, Zotov O G and Topalov I K 2016 Microstructure of 316L stainless steel components produced by direct laser deposition *Steel Transl.*
[13] Yu J, Rombouts M and Maes G 2013 Cracking behavior and mechanical properties of austenitic stainless steel parts produced by laser metal deposition *Mater. Des.*
[14] Zhang K, Wang S, Liu W and Shang X 2014 Characterization of stainless steel parts by Laser Metal Deposition Shaping *Mater. Des.*
[15] Yadollahi A, Shamsaei N, Thompson S M and Seely D W 2015 Effects of process time interval and heat treatment on the mechanical and microstructural properties of direct laser deposited 316L stainless steel *Mater. Sci. Eng. A*
[16] Chkalov R, Khorkov K, Kochuev D, Davydov N, Davydov N, Prokoshev V and Kostrov V 2018 Computerized laser complex for monitoring and controlling of the precision micromachining processes *Proceedings of the International Conferences on WWW/Internet 2018 and Applied Computing 2018* 395–399
[17] Chkalov R, Khorkov K and Prokoshev V 2019 Development and application possibilities of multifunctional femtosecond laser complex for precision processing *International Conference on Industrial Engineering, Applications and Manufacturing* 1–5
[18] Salonitis K, D'Alvise L, Schoinochoritis B and Chantzis D 2016 Additive manufacturing and post-processing simulation: laser cladding followed by high speed machining *Int. J. Adv. Manuf. Technol.* 2401–2411
[19] Nie P, Ojo O A and Li Z 2014 Modeling analysis of laser cladding of a nickel-based superalloy *Surf. Coatings Technol.*
[20] Farahmand P and Kovacevic R 2014 An experimental-numerical investigation of heat distribution and stress field in single- and multi-track laser cladding by a high-power direct diode laser *Opt. Laser Technol.*
[21] Nenadl O, Ocelík V, Palavra A and De Hosson J T M 2014 The prediction of coating geometry from main processing parameters in laser cladding *Physics Procedia*
[22] Erfanmanesh M, Abdollah-Pour H, Mohammadian-Semnani H and Shoja-Razavi R 2017 An empirical-statistical model for laser cladding of WC-12Co powder on AISI 321 stainless steel *Opt. Laser Technol.*

IOP Conf. Series: Materials Science and Engineering **969** (2020) 012094 doi:10.1088/1757-899X/969/1/012094

An Investigation of the Laser Welding Process for Dual-Phase Steel via Regression Analysis

Dawei Zhao[1,2], Mikhail Ivanov[1] and Yuanxun Wang[3]

[1]Department of Welding Engineering, Institution of Engineering and Technology, South Ural State University, 76, Lenin Prospect, Chelyabinsk, 454080, Russia
[2]State Key Laboratory for Strength and Vibration of Mechanical Structures, Xi'an Jiaotong University, 28, Xianning Road, Xi'an, 710054, China
[3]Department of Mechanics, Huazhong University of Science and Technology, 1037, Luoyu Road, Wuhan, 430074, China

E-mail: ivanovma@susu.ru

Abstract In this work, a systematic investigation was undertaken to explore the effects of welding process parameters on the mechanical performances of the welding joints in the laser welding process for DP600. Welding experiments were arranged by a uniform experimental design method with four control factors (laser power, welding speed, focal point position, and side-blowing shield gas flow). The tensile strength of the welding joints was used to quantify the welding quality. A mathematical model based on stepwise regression analysis was employed to correlate the welding process parameters and the tensile strength. The effects of the welding process parameters on the welding quality were discussed. The genetic algorithm was then employed to select the optimum welding parameters. The verification test results proved that the method proposed in this paper could effectively evaluate and optimize the welding quality within the range of process parameters, which could enhance the welding performance in the laser welding process as feasibly and effectively as possible.

1. Introduction

Fuel consumption and gas emission are the main subjects of global interests in these years, one of the promising strategies is using high strength steel to reduce the weight of vehicles. Dual-phase steel, as a new generation of steel, has the advantages of high strength, good ductility and satisfying welding performances compared with traditional kinds of steel. Thus, it has been increasingly employed in the manufacturing industry. Laser welding is widely used in aviation, automobile, shipbuilding industry and considered to be a promising welding technique, as it has the advantages of deep penetration, high productivity and little distortion. As is known to all, laser welding is a complicated thermal-electrical process that implicates mechanical, electrical, thermal and metallurgical factors and the weld quality is highly affected by various process conditions, noise, and errors [1]. During the practical production, it is not an easy task for skilled operators to select suitable process parameter combinations to obtain sound welding quality. Exploring an effective approach to enhance and optimize the quality of laser welding joints for dual-phase steels has become a key issue [2].

There are some literature reports concerning modeling the welding process and establishing the mathematical models correlating welding quality and technological parameters via the Taguchi method, response surface methodology, gray relational analysis and soft computing. Moradi et al. [3] employed the central composite design to investigate the effects of laser power, scanning speed and focal plane position on the tensile strength for the AISI 410 sheet and the optimum levels of parameters were finally determined. Yuce et al. [4] studied influences of laser power, welding speed and focal position on the mechanical and microstructural properties of the welding joints using a Taguchi approach, the optimum parameters for the maximum tensile strength with the minimum heat input were also predicted. Yaakob et al. [5] implemented the response surface methodology to establish a mathematical model correlating the welding parameters and tensile strength. The optimized welding parameters were obtained and improvement in welding quality was achieved due to complete welding penetration using the optimized welding parameters. Prabakaran et al. [6] employed the grey relational analysis to characterize the multiple quality characteristics of welded joints and figured out that the optimal welded joint had good mechanical properties after post-weld heat treatment. Ai et al. [7] developed the relationships between the weld bead integrity and process parameters by the genetic algorithm (GA) optimized backpropagation neural network. The results indicated that the laser power was the most significant factor in the weld bead integrity and weld area. Kannan et al. [8] modeled the relationship between welding speeds, shielding gas and fusion zone, bead width, depth of penetration, pore area via an artificial neural network. The results indicated that the artificial neural network model could reveal the relationship between the process parameters and the weld geometry. The proposed approach provided a deeper understanding of the welding process.

The side-blowing shield gas is a significant laser welding process parameter. Whereas, there is little academic reference available concerning the effect of the side-blowing shield gas on the welding quality, not to mention the interaction effects of the side-blowing shield gas and other welding parameters. To overcome these shortcomings, first, the welding power, welding speed, focal position, and side-blowing shield gas were selected as the welding parameters, the tensile strength of the welding joints was treated as the welding quality index. The welding experiments were carried out based on uniform experimental design. The empirical mathematical model quantifying the relationship between the process parameters and the tensile strength was obtained by using stepwise regression analysis. Variance analysis was also conducted to reveal whether the regression model was robust and satisfactory. The effects of the welding parameters on the welding quality were investigated based on the developed model. Finally, the GA was proposed to calculate the optimal process parameters and the verification test was also performed to check the validity of the model.

2. Experimental detail

This paper was planned to optimize the welding parameters of DP600 dual-phase sheets with thicknesses of 1.7 mm. The unique properties of DP600 include high strength, good weldability, and corrosion resistance [9]; its major chemical compositions (percent by weight) are Si 1%, Mn 1.52%, P 0.015%, C 0.079%, S 0.0049%, Al 0.023%, N 0.0037%. The DP600 sheets were cut into 120 mm × 40 mm and installed as butt weld joints.

A fiber laser welding machine composed of the YLR-4000 fiber laser system was employed to conduct the welding experiments. The laser power, welding speed, focal position and quantity of shielding gas flow were considered to be the key factors affecting the welding quality. The welding experiments were arranged based on a uniform experimental design [10]. The universal tensile testing machine was used to perform tensile-shear tests and the loading speed was set at 10 mm/min. The tensile strength was employed to reveal the welding quality and it was obtained from the load-displacement curve. Table 1 presents the results of the welding experiments.

Table 1. Experimental results based on uniform experimental design.

No.	Laser power P (kW)	Welding speed V (mm/s)	Shielding gas flow Q (m^3/h)	Focal position F (mm)	Tensile strength R (kN)
1	0.9	12	2.3	1	26.02
2	1	18	2	0.5	37.90
3	1.1	24	1.7	0	39.00
4	1.2	8	2.5	–0.5	34.96
5	1.3	14	2.2	–1	37.39
6	1.4	20	1.9	1	38.98
7	1.5	26	1.6	0.5	39.63
8	1.5	10	2.4	0	36.33
9	1.7	16	2.1	–0.5	38.04
10	1.8	22	1.8	–1	41.06

3. Results and discussion

3.1. *Failure modes of the welding joints*

Figure 1 indicates three kinds of failure modes of the welding samples in the tensile tests. They are respectively sample 1 in Figure 1 (a), sample 7 in Figure 1 (b) and sample 10 in Figure 1 (c). The sample in Figure 1 (a) indicates it fails in the welding zone, while the welding joints fail in the heat-affected zone in Figure 1 (b). As for the sample in Figure 1 (a), the surface of the observed joint is not smooth and appearance defects could be found in the welding seam, which indicates the welding joints present poor permeability. The welding heat supplied to the welding zone is relatively low under the welding process parameters and the metal in the welding zone fails to melt completely, the tensile strength is far less than that of the base metal. In this case, the crack appears in the welding seam and its tensile strength is only about 382.71 MPa, which is far less than of the DP600. The welding sample in Figure 1 (b) presents good mechanical performances and the maximum tensile force is about 550 MPa, which is still less than the value of the base metal DP600. It also indicated that crack gradually develops from the heat-affected zone until the whole welding joints fail. The welding joints of Figure 1 (c) presents the best quality. The tensile strength of the welding zone is much larger than that of the base material, which indicates the failure occurs in the base material rather than the welding seam; the tensile strength is about 600 MPa [11].

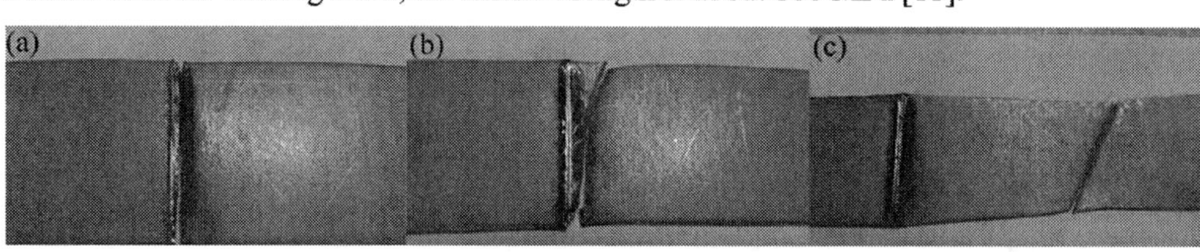

Figure 1. Three types of failure fractures in the tensile tests.

Figure 2 presents the load-displacement curves for the three kinds of welding joints in the tensile test. It seems that the mechanical performances of the welded joints are quite different from each other. The maximum displacement of sample 10 is the largest and its failure energy is also much larger than others, which indicates its failure mode is a plastic failure. Sample 1 reveals the worst mechanical performances with small tensile strength and failure energy, in this case, it is a brittle fracture, since the welding heat input is not sufficient enough, the welding seam size is not large enough so that it can not tolerate large tensile strength. It can be concluded that the mechanical properties of sample 10 are the best.

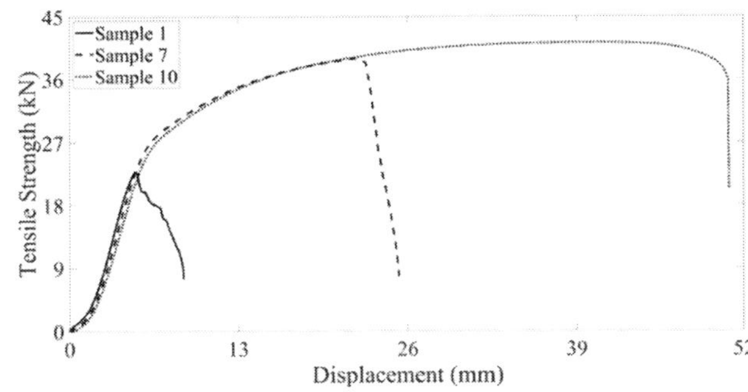

Figure 2. Mechanical performances of the welding joints with different failure fractures.

3.2. Polynomial regression analysis

Nonlinear polynomial regression is a statistical and mathematical method quantifying the relationship between two variables. The mapping relationship between the variable y and independent variables $x_1, x_2, \ldots, x_n$ can be expressed as follows [12]:

$$y=f(x_1, x_2, \ldots, x_n) \tag{1}$$

It is necessary to estimate the true relationship between dependent variables $f(x_1, x_2, \ldots, x_n)$ and independent variables $x_1, x_2, \ldots, x_n$. In general, a group of the general polynomial model is usually employed as follows:

$$f(x_1,x_2,\cdots,x_n)=a_0+\sum_i^n a_i x_i+\sum_i^n a_{ii}x_i^2+\sum_i^n a_{ij}x_i x_j+\cdots+\varepsilon \tag{2}$$

where ε is the model error, the least square regression method is usually employed to obtain the regression coefficients based on experimental results.

In this study, the mathematical model characterizing the relationship between welding parameters (laser power, welding speed, focal position, and shielding gas flow) and the tensile strength of the welding joints for DP600 was attempted to figure out utilizing regression method. Table 2 lists the analysis of variance results of the linear mathematical model via software Matlab2017b. The linear regression model was rejected as its P-value is larger than 0.05 [13].

Table 2. Variance analysis results of the linear regression model.

Term	Regression coefficient	Upper confidence limit	Lower confidence interval
Constant	0	0	0
P	4.5373	–5.4195	14.4940
V	0.8174	0.3745	1.2603
Q	8.2722	3.8827	12.6617
F	–1.8130	–5.6742	2.0482
RMSE	0.6862	F-value	4.3745
P-value	0.0590	R-Squared	8.2827

Figure 3 is a residual graph of the linear regression model. The vertical line indicates the residual and its confidence interval of the regression model. If the confidence interval of the residual does not contain 0 points, the corresponding experimental result is abnormal [14]. According to Figure 3, it can be concluded that the first set of the data is abnormal, in this case, it should be rejected.

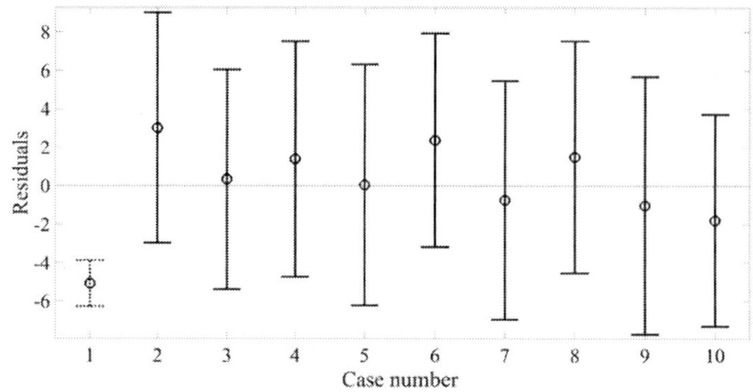

Figure 3. The residual graph of the linear regression model.

The stepwise regression method is a statistical method that can select several important variables to construct the regression equation. In this study, the stepwise quadratic regression analysis is carried out using the software MATLAB 2017b, and the P values of terms Q, F, PQ, QF, P^2, V^2, Q^2, F^2 are larger than 0.1, and they should be eliminated [15]. The final results of the analysis of variance for the second-order regression model are presented in Table 3.

Table 3. Variance analysis results of the quadratic regression model.

Term	Coefficient	t-stat	P-value	Confidence interval	
Constant	39.5825	47.2694	0.0004	35.9796	43.1855
P	–6.2659	–10.6040	0.0088	–8.8083	–3.7235
V	–0.6806	–12.1810	0.0067	–0.9210	–0.4402
PV	0.4317	13.5830	0.0054	0.2949	0.5684
PF	4.5510	12.3130	0.0065	2.9607	6.1414
VQ	0.2767	16.5000	0.0037	0.2045	0.3488
VF	–0.3574	–12.8533	0.0060	–0.4770	–0.2378

Notes: R^2 =0.9996; Adj R^2=0.9985; RMSE=0.07112; F value =861.88; P value=0.001159.

The determination R^2 and the adjusted R^2 characterize the fitting degree of the regression model [16]. A relative lager R^2 indicates that the model is sound and satisfying. The maximum error of estimated tensile strength is about 0.164 kN testifying that the model developed by the stepwise analysis is valid and flexible. Based on the above analysis, the regression coefficients of the fitted regression model for the laser welding process of DP600 were achieved and described as follows:

$$R=39.5825-6.2659P-0.6806V+0.4317PV+4.5510PF+0.2767VQ-0.3574VF \quad (3)$$

Subjected to

$$0.9\ \text{kW} \leq P \leq 1.8\ \text{kW},\ 8\ \text{mm/s} \leq V \leq 26\ \text{mm/s},\ -1\ \text{mm} \leq F \leq 1\ \text{mm},\ 1.6\ \text{m}^3\text{/h} \leq Q \leq 2.5\ \text{m}^3\text{/h} \quad (4)$$

3.3. *Welding process parameters analysis and optimization*

Figure 4 is a contour plot of the tensile strength with different welding speeds and side-blown gas flows. While the welding power is kept at 1.35 kW and the focal position is 0 mm. The tensile strength of the welding joints increases with the increase of the side-blown gas flow and welding speed. The higher the welding speed is, the less the heat input supplied to the welding zone will be absorbed, in this case, the welding penetration and the weld size will reduce, and the mechanical properties are poor [17]. However, too slow welding speed not only leads to welding defects such as weld beads but also reduces production efficiency. The side blowing gas flow prevents the molten metal from chemically reacting with the air during the welding process, it also affects the welding quality by changing the area and height of the plasma. The side blow gas flow blows the plasma away, the larger the gas flow value is, the smaller the plasma amount will be, and the less loss of the welding heat input will be dissipated [18]. So larger side blow gas flow can effectively improve the welding penetration and the welding seam size, thus the welding quality will be improved.

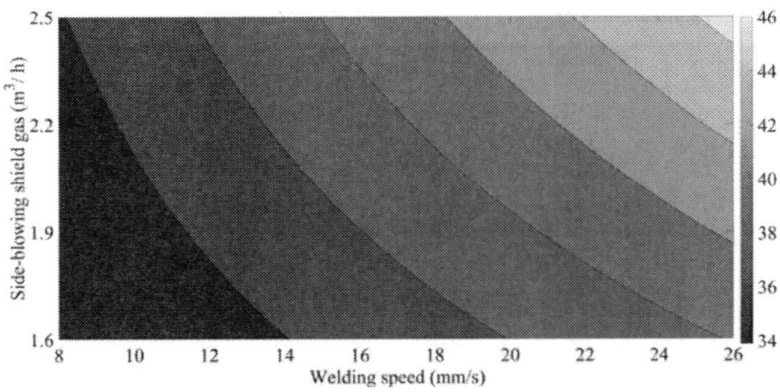

Figure 4. The interaction effect of the welding speed and side-blowing shield gas flow on tensile strength.

The GA was employed to optimize welding process parameters in this study. Unfortunately, there is not a theoretical basis available for determining GA parameters, and it is generally determined by trial and error [19]. The population size was chosen to be 40, the maxima hereditary algebra was 50, each code length was 20, the generation gap was 0.95, the crossover probability was 0.7, and the mutation probability was 0.01. The tensile strength of the welding joints changes continuously in the GA calculation progresses and eventually reaches a fixed value. The result obtained by the GA is the maximum tensile strength of the welding joints within the welding process parameters investigated in this study, which is 49.86 kN. The corresponding combination of welding process parameters is laser welding power of 1.716 kW, welding speed of 24.979 mm/s, side blowing gas flow of 2.409 m^3/h, and focal position of –0.998 mm.

The confirmation experiment aims to certificate the feasibility and reproducibility of the results determined by the regression model and the optimization method [20]. In this paper, the test was conducted using the optimum levels of the welding parameters obtained by the GA to justify whether the welding quality is improved, as tabulated in Table 4. From Table 4, we can see that the mechanical behavior of welded joints is higher than the maximum shear strength in Table 1. It proved the parameter combination obtained by the proposed approach performed well for the welding quality optimization problem.

Table 4. Results of the verify experiment.

Optimization method	P (kW)	V (mm/s)	Q (m^3/h)	F (mm)	R (kN)	R in Table 1 (kN)	R_{pred} (kN)
GA	1.7	25	2.4	–1	49.56	41.06	49.86

4. Conclusions

1. The developed quadratic polynomial regression model for quantifying the relationship between the input variables (welding power, welding speed, side blowing gas flow, and focal position) and the tensile strength is robust and flexible.
2. Larger welding speed and side blowing gas flow together with appropriate welding power and focal position are desirable. The optimum parameters for the maximum achievable tensile strength in this investigation were found to be as follows: laser welding power of 1.7 kW, welding speed of 25 mm/s, side blowing gas flow of 2.4 m^3/h and focal position of –1 mm.

3. Both the welding speed and the side blowing gas flow have significant effects on the welding quality by affecting the welding heat input supplied to the welding zone. Their effects on the welding quality are quite different.

References

[1] Farrokhi F, Endelt B and Kristiansen M 2019 *Optics & Laser Technology* **111** 671
[2] Prabakaran M P and Kannan G R 2017 *Ferroelectrics* **519** 223
[3] Moradi M, Arabi H and Shamsborhan M 2020 *Optik* **202** 163619
[4] Yuce C, Tutar M, Karpat F, Yavuz N and Tekin G 2017 *Strojniski Vestnik* **63** 510
[5] Yaakob K I, Ishak M, Quazi M M and Salleh M N M 2019 *Measurement* **135** 452
[6] Prabakaran M P and Kannan G R 2019 *Optics & Laser Technology* **112** 314
[7] Ai Y, Shao X, Jiang P, Li P, Liu Y and Liu W 2016 *Optics and Lasers in Engineering* **86** 62
[8] Casalino G, Facchini F, Mortello M and Mummolo G 2016 *IFAC-PapersOnLine* **49** 378
[9] Marwan K O H and Kaçar R 2018 *Materials Testing* **60** 40
[10] Fang K T, Lin D K, Winker P and Zhang Y 2000 *Technometrics* **42** 237
[11] Zdravecká E and Slota J 2019 *Metals* **9** 91
[12] Vedrtnam A, Singh G and Kumar A 2018 *Defence Technology* **14** 204
[13] Vyas M, Jain M, Pareek K and Garg A 2019 *Measurement* **148** 106904
[14] Leardi R 2009 *Analytica Chimica Acta* **652** 161
[15] Shihab S K 2018 *Arabian Journal for Science and Engineering* **43** 5017
[16] Kumar S S, Murugan N and Ramachandran K K 2019 *Measurement* **137** 257
[17] Xu K and Zhang SQ 2018 *Key Engineering Materials* **765** 204
[18] Gu X, Liu Y, Li W, Han Y and Zheng K 2019 *Materials* **12** 4207
[19] Sada S O 2020 *Cogent Engineering* **7** 1741310
[20] Kumar S and Singh R 2019 *Measurement* **148** 106924

Acknowledgments

The authors are grateful for the financial support provided by the Natural Science Foundation of Shandong Province (ZR2016EEM47/ZR2018PEE004) open projects of State Key Laboratory for Strength and Vibration of Mechanical Structures (SV2019-KF-39).

IOP Conf. Series: Materials Science and Engineering **969** (2020) 012095 doi:10.1088/1757-899X/969/1/012095

Effect of Electron-Beam Welding Speed on Weld Metal Mechanical Properties of 5V Titanium Alloy

E V Terentyev, D A Zhgut and K T Borodavkina

National Research University "Moscow Power Engineering Institute", 14, Krasnokazarmennaya str., Moscow, 111250, Russia

E-mail: TerenrtyevYV@mpei.ru

Abstract. The results of metallographic studies of 5V titanium alloy welds obtained with the range of electron-beam welding speeds from 20 to 120 m/h are presented. It was established that at all welding speeds some identical microstructure of a weld metal is formed, which can differ only in sizes and shapes of the primary crystals. The results of Vickers hardness, ultimate tensile stress and yield stress estimation by sample-free indentation methods are shown. The weld metal hardness distribution in the transverse direction is homogeneous while from the top to the root an increase in hardness values at all welding speed modes is occurred. It was also revealed that both the hardness and strength of the weld metal grow with the rising welding speed, herewith the increase in these properties in the top of the weld is more significant than in the root. According to the analysis of the interrelation between structure and mechanical properties it was concluded that the main reason of mechanical properties fluctuations is a change in the weld metal chemical composition as a result of the volatile components evaporation during welding.

1. Introduction

A 5V titanium alloy is applied in welded units of steam-turbine plants due to a successful combination of operational and heat resistance properties [1, 2]. The number of researches [3–8] has been devoted to solve technological challenges as well as improving performance and quality of titanium alloy weld joints. Based on these studies, it is necessary to use high-quality blanks and develop welding technologies taking into account design features to achieve homogeneous defect-free weld joints.

According to the tensile tests results of titanium alloys weld joints [8], a part of them obtained by electron beam welding (EBW) always break down in the parent metal while specimens made by argon arc welding (AAW) are broken along weld joints or heat treatment zones (HAZ). The values of ultimate tensile stress (UTS) in such cases are not lower than the minimum ones for the parent metal. In addition, the use of EBW for making weld joints of large thickness causes minimum welding deformations because of a small weld width. Vacuum as a protective atmosphere also allows to attain high quality of weld metals. Under these conditions, the EBW can be considered as the most preferable method for 5V titanium alloy welding.

Typically, the welding modes parameters selection is carried out primarily on the basis of technological requirements and minimizing the probability of defect occurrence [4–6, 9–11]. For example, the welding speed has a significant effect on hot cracks formation [12, 13], on weld shape formation (for the full penetration welding) [6, 10, 11, 14], on the probability of specific structures formation in welds [15] and pores occurrence [16]. On the other hand, the welding speed has an impact on the welding thermal cycle, so it determines the structure and chemical composition of the weld. In particular, a welding speed growth leads to a change in the crystallization conditions and a decrease in both the crystals size and the loss of alloys volatile components [14, 17, 18].

At the same time, for some alloys it is possible to get structures without any initial defects using a wide range of welding speeds. Choosing the welding speed it is more rational to be based on providing the best mechanical properties for the metal under study. In [19–21], the results of researches concerning welding conditions parameters effect on welded joints mechanical properties are presented. However, in the scientific and technical literature there is a lack of information regarding the effect of EBW speed on the mechanical properties of the 5V titanium alloy weld metal. For instance, in [22] the data relating to the hardness distribution in weld joints of 5V alloy are given. Also in [14] the results of investigation of 5V titanium alloy welds obtained at welding speeds of 20 m/h, 60 m/h and 120 m/h are performed. It was established that the mechanical properties of the weld metal increase with the welding speed rising. However, a number of experiments were insufficient to determine this conclusion reliably. In addition, it is important to understand the reasons for such phenomenon and to establish the interrelation between mechanical properties and structure parameters.

Thus, the main aim of this study is to establish a quantitative dependence of the weld metal mechanical properties on the EBW speed taking into account the microstructure parameters of a 5V titanium alloy.

2. Research methods

The studies were performed on 5V titanium alloy welded joints obtained by EBW without full penetration. Welding modes are presented in Table 1. The thickness of the welded plates was 32 mm. The chemical composition and mechanical properties according to the quality certificates are shown in tables 2 and 3.

Table 1. The EBW modes.

Accelerating voltage U, kV	Welding speed v, m/h	Beam current I, mA	Penetration depth H, mm	Mean width B, mm
60	20	100	29.4	2.68
	40	135	26.8	2.16
	60	170	28.1	1.79
	90	215	26.4	1.63
	120	250	27.7	1.43

Table 2. The chemical composition of 5V titanium alloy.

Elements, wt. %												
Ti	Al	V	Mo	Zr	Cr	Si	Fe	Mn	O	N	C	Others
Base	4.8	1.45	1.05	0.08	0.05	0.04	0.1	0.05	0.06	0.01	0.07	0.14

Table 3. The mechanical properties of 5V titanium alloy.

Ultimate tensile stress R_U, MPa	Yield stress $R_{0.2}$, MPa	Total elongation δ, %	Total cross section reduction ψ, %	Impact strength KCU, kJ/m^2
880	860	9	22	540

Four microsections were made from each weld joint. The first microsection was used to study the microstructure, the second one to define the microhardness and the other two to determine the UTS and yield stress (YS) using the indentation method.

These microsections were prepared according to the generally accepted methodology: stepwise grinding on abrasive silicon carbide paper with a grain size from P180 to P1000, then polishing using a diamond suspension with particle sizes of 9 μm and 1 μm and final polishing using a colloidal suspension with the size of oxides silicon of 0.06 microns on synthetic velvet. The first microsection was etched during 5–10 seconds in a reagent of the composition: hydrofluoric acid - 5 ml, nitric acid - 5 ml, water - 90 ml to reveal the microstructure.

The microstructure was investigated by an Observer Z1m Carl Zeiss optical microscope. Grain sizes were determined at a magnification of x100 in accordance with Russian State Standard GOST 5639 using the method of grains counting.

The weld metal hardness was defined by the Vickers method on an Instron Tukon 2500 hardness tester under the load of 5 kg (49.05 N). For each weld joint microsection the indents were made with a constant step, in 3 lines over the entire height of the weld. The scheme of measuring hardness using the second microsection is represented in figure 1.

The UTS (R_u) and YS ($R_{0.2}$) were determined by the equations:

$$R_{0,2} = k_{0,2} \cdot HB_{0,2} \quad (1)$$

$$R_U = k_U \cdot HB_{\max} \quad (2)$$

where k_U, $k_{0,2}$ - correlation coefficients, HB_{max}, $HB_{0,2}$ - ultimate hardness and hardness at yield stress, MPa.

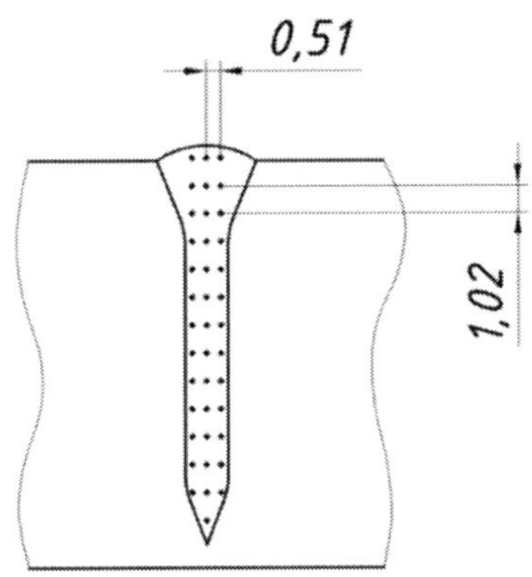

Figure 1. The weld joint hardness measurements scheme.

The procedure of determining the specific characteristics of hardness (HB_{max}, $HB_{0,2}$) is regulated by Russian State Standards GOST 22761 and GOST 22762. However, the correlation coefficients presented in these Standards are given only for carbon and low alloy steels. Therefore, in this study the predefined experimental coefficients for 5V alloy $k_U = 0.323$ and $k_{0.2} = 0.427$ were applied.

3. Results

The weld metal microstructure at all welding modes is martensite (α'-phase). The boundaries of the primary crystals are thin and clear (figure 2). The weld metal has a specific columnar structure with crystals growing from fused grains of the base metal to the weld joint centre in the direction opposite to the heat sink. Moreover, the grains in the central part are almost equiaxed since the heat removed through the weld pool rear wall and these crystals grew in the direction of welding. With an increase in the welding speed, such crystals became longer while a number of equiaxed crystals decreased due to the formation of elongated weld pool and prevailing heat dissipation through the side walls.

It is worth noting that the size of the primary crystals in the weld top is significantly larger than in the rest part of the weld. Such phenomenon can be explained by a lower cooling rate and a greater weld width which is quite typical for weld joints made by EBW. According to figure 2, the primary grains size corresponds to No. 3 according to GOST 5639 scale in all cases. The main difference consists in the form of crystals. At low speeds the grains are equiaxed (figure 2a), while at high speeds the crystals are elongated (figure 2c). In the weld top the size of the primary grains differs considerably.

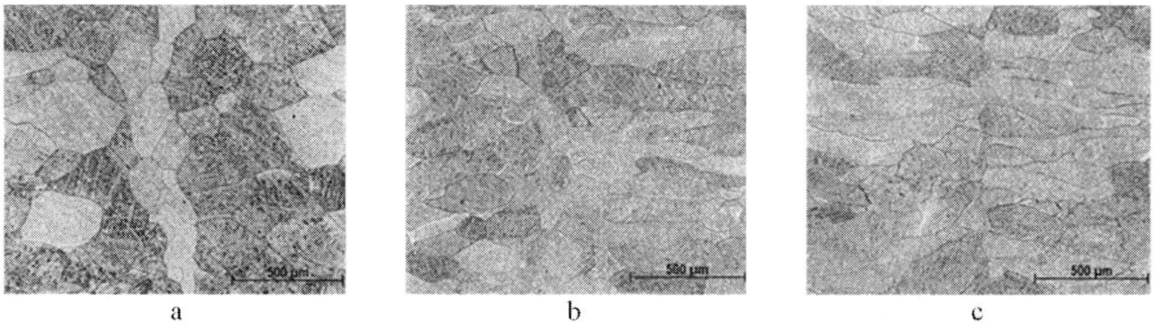

Figure 2. The microstructure of the central part of weld metal obtained at the welding speed of 20 m/h (a), 60 m/h (b) and 120 m/h (c), x200.

The grain number gradually decreases from 3 to 0 with a welding speed increasing that can be explained by considerably different crystallization conditions. Figure 3 shows the dependence of the mean grain size in the central part of the weld and in the weld top on the welding speed.

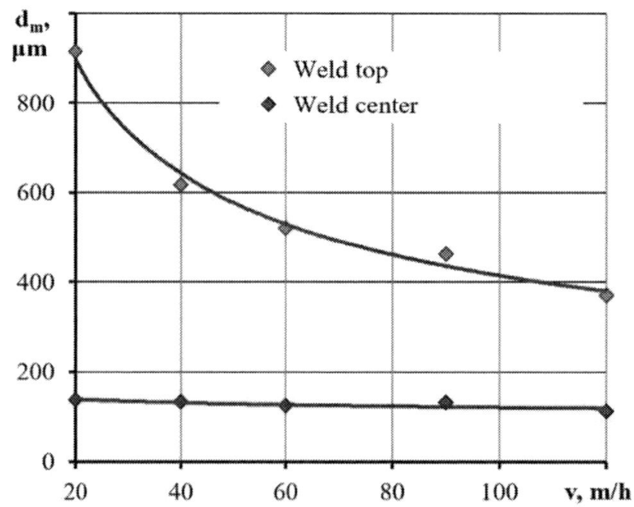

Figure 3. The dependence of the mean grain size d_m on the EBW speed.

Figure 4 shows the relation between the mechanical properties of the weld metal (hardness, UTS and YS) and welding speed. The UTS was determined in the top and central part of the weld, YS was defined only in the central part of the weld. The mean hardness was established by making 70-75 indents in each weld joint, while for the UTS and YS it was required to conduct 3…5 indents.

UTS in the weld top rises by 10% with an increase of the EBW speed from 20 m/h to 120 m/h whereas in the central part of the weld a growth does not exceed 5%. It should also be noted that with an increase in the EBW speed from 20 m/h to 120 m/h the gap between the UTS values in the weld top and weld central part decreases gradually from 7% to 2%. The difference between the weld metal YS and UTS values is quite small which is typical for the 5V titanium alloy and remains constant for all welding speeds.

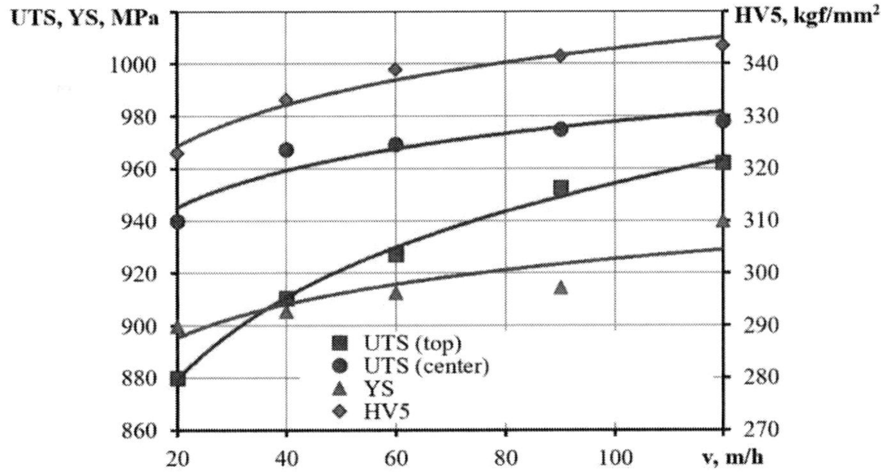

Figure 4. The impact of EBW speed on the weld metal mechanical properties.

Hardness test results show that the hardness of the weld metal in the transverse direction is constant. Namely, the hardness values in adjacent columns (figure 1) differ by less than 1% in all specimens. Moreover, there is a clear decrease in hardness in the weld top compared with the rest part of the weld. In figure 5 the distribution of hardness along the weld depth is presented.

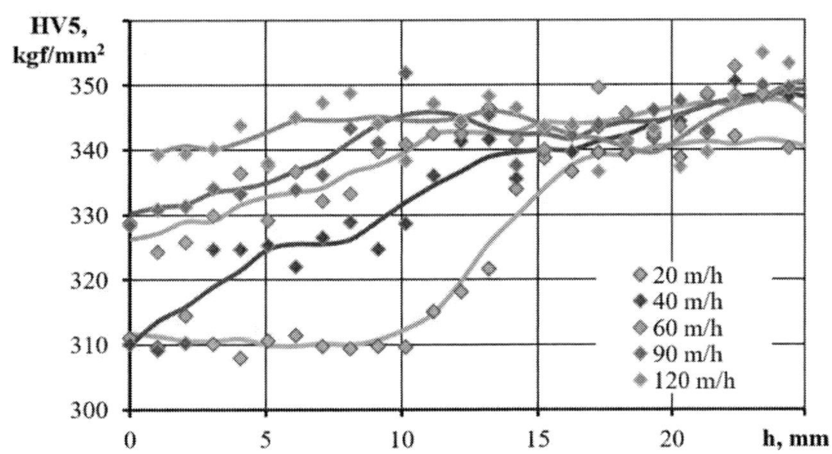

Figure 5. The distribution of hardness along the weld depth obtained for different welding speeds.

The distribution of hardness along the weld height shows that hardness in the root is almost identical at all modes and fluctuates from 340 to 350 HV5 the hardness begins to decline smoothly closer to the weld top at welding speeds of 40–120 m/h. At the EBW speed of 20 m/h a rapid hardness fall to 310 HV5 at a depth of 11–14 mm occurs. At the same time, it should be taken into account that the primary boundaries of such large crystals are unlikely to make a great contribution to the martensitic structure.

4. Discussion of the results

The analysis of mechanical properties and weld structure results does not allow to establish a constant patterns. Two main factors – structure and chemical composition – influence on weld metal mechanical properties. The structure of the weld metal is equal for all welding modes, only a slight difference in the sizes and shapes of the primary crystals is observed. The largest difference in the primary crystals sizes is noticed in the weld top. Therefore, the mechanical properties in the weld top differ from the rest part of the weld significantly.

In order to set the effect of grain-boundary hardening on the weld metal hardness, the relation between hardness and the number of primary grains under the indent was made. Even in the central and root parts of the weld joints the size of the primary grains is large and commensurate with the indents which were carried out by Vickers hardness tests under the load of 5 kg (49.05 N). The study of the area around the indents showed that the indenter interacted with 1–9 primary crystals or on average with 2–4 ones. In this case, the processing of the results represented that the hardness does not depend on the number of grains under the indent. Thus, the difference in the size of the primary grains cannot be the main reason of weld metal mechanical properties variation.

It should be viewed that during EBW the chemical composition of the weld metal could change due to the evaporation of volatile components.

The aluminium contained in the 5V titanium alloy has the highest saturation vapor pressure among the other components at the melting temperature of titanium. Thus, substantial evaporation of aluminium can affect the weld metal mechanical properties adversely. Moreover, the longer the metal is in a liquid state, the more significant the chemical composition changes. It is obvious that the metal stays in the liquid state longer at lower welding speeds. According to this, aluminium content becomes less and hence the strength properties decrease too. The experimental results fully confirm this assumption.

5. Conclusion

The weld metal microstructure is identical at all welding speeds, some differences were observed only in the sizes and shapes of the primary crystals. The largest difference in the primary crystals sizes was noticed in the top of the weld, while the mean grain size was almost the same in the weld central part and in the root.

With an increase in the EBW speed from 20 m/h to 120 m/h, the UTS rose from 880 MPa to 960 MPa in the weld top and from 940 MPa to 980 MPa in the central part. The YS grew from 900 MPa to 940 MPa.

The hardness of weld joints varied in height significantly. The lower the welding speed, the higher the difference in mechanical properties. Thus, at a speed of 20 m/h the hardness from the root to the top of the weld declined from 340 to 310 HV5 and at 120 m/h from 350 to 340 HV5.

Apparently, the main reason for weld metal poor mechanical properties at low welding speeds was the intense evaporation of aluminium and a decrease in its concentration in the weld metal. Finally, to obtain high and uniform mechanical properties over the entire height of the weld, it is preferable to use rather high welding speeds.

References

[1] Ivanova L A, Kozlova I R, Kudryavtsev A S and Chudakov E V 2009 *J. Mat. Sci.* **59(3)** 152–161
[2] Gorynin I V, Leonov V P, Kudryavtsev A S, Ivanova L A, Travin V V and Lysenko L V 2014 *J. Mat. Sci.* **78(2)** 48–62
[3] Kudryavtsev A S and Panotsky D A 2009 *J. Mat. Sci.* **59(3)** 344–350
[4] Leonov V P, Mikhailov V I, Sakharov I Yu and Kuznetsov S V 2015 *J. Mat. Sci.* **81(1)** 219–228
[5] Khanzhin A V, Fedorov A M and Sakharov I Yu 2011 *J. Mat. Sci.* **66(2)** 117–122
[6] Yolkin V N, Makarov S V, Panfilov A P, Dragunov V K, Sliva A P and Terentyev E V 2015 *Int. Conf. Electron Beam Weld. and Related Techn.* 450–458
[7] Ivanova L A, Ilyin A V, Leonov V P, Mizetskiy A V, Sakharov I Yu and Khatuntsev A N 2008 *J. Mat. Sci.* **56(4)** 37–53
[8] Ivanova L A, Kuznetsov S V, Mikhailov V I, Pas' O A, Sakharov I Yu and Khatuntsev A N 2009 *J. Mat. Sci.* **59(3)** 387–396
[9] Isaev S L 2014 *Bul. of Samara State Techn. Univ. Techn. Sci.* **43(3)** 81–84
[10] Terentyev E V, Dragunov V K, Sliva A P and Scherbakov A V 2015 *Weld. Int.* **29(2)** 150–154

[11] Dragunov V K, Terentev E V, Sliva A P, Goncharov A L and Marchenkov A Yu 2017 *Weld. Int.* **31(4)** 307–311
[12] Trukhanov K Yu and Tsar'kov A V 2014 *Weld. and diagnostics* **1** 27–31
[13] Krivonosova E A, Shchitsyn Yu D, Akulova S N, Myshkina A V, Neulybin S D and Belinin D S 2018 *Bull. of Perm National Research Polytechnic Univ. Mech. Eng., Mat. Sci.* **20(2)** 12–19
[14] Terentyev E V 2014 *Improving the quality of weld joints during electron beam welding with full penetration* (Moscow: Moscow Bauman State University Press) p 160
[15] Amara E H 2011 *Lasers in Eng.* **22(5–6)** 299–308
[16] Ovchinnikov V V and Egorov R V 2006 *Bull. of Ivanovo State Energy Univ.* **4** 1–5
[17] Dragunov V K, Terentyev E V, Goncharov A L, Marchenkov A Yu and Goryachkina M V 2015 *Int. Conf. Electron Beam Weld. and Related Techn.* 347–360
[18] Dragunov V K, Terentev E V, Goncharov A L and Marchenkov A Yu 2016 *Weld. Int.* **30(12)** 948–952
[19] Nesterenkov V M, Kravchuk L A and Arkhangelsky Y A 2013 *Automatic Weld.* **725(9)** 39–42
[20] Shiganov I N, Shakhov S V, Tarasenko L V and Plokhikh A I 2005 *Mech. Eng. Techn.* **10** 23–8
[21] Gureeva M A, Grushko O E, Ovchinnikov V V and Shiganov I N 2008 *Met. Techn.* **3** 11–19
[22] Demidov A N, Karimbekov M A and Marchenkov A Yu 2015 *Int. Conf. Electron Beam Weld. and Related Techn.* 181–186

Acknowledgements
This study conducted by Moscow Power Engineering Institute was financially supported by the Ministry of Science and Higher Education of the Russian Federation (project No. FSWF-2020-0023).

Effect of Welding Speed on Electron Beam Welding Thermal Efficiency

E V Terentyev, A P Sliva and Yu V Santalova

National Research University "Moscow Power Engineering Institute", 14, Krasnokazarmennaya str., Moscow, 111250, Russia

E-mail: TerentyevYV@mpei.ru

Abstract. The results of thermal efficiency evaluation during electron beam welding of alloys with different thermophysical properties in the range of welding speeds from 20 m/h to 120 m/h are presented. It was shown that the thermal efficiency increased in accordance with the rising welding speed. Also, the thermal efficiency greatly depends on the thermal conductivity: the lower the thermal conductivity, the higher the thermal efficiency. Thus, with an increase in welding speed from 20 m/h to 120 m/h, the thermal efficiency grew steadily for all three considered types of alloys: values for a titanium alloy 5V went up from 54% to 67%, while for the steel 40Kh13 values rose from 46% to 63% and the most significant increment related to duralumin D16, the thermal efficiency for this alloy increased from 18% to 40%. The equation which established the ratio between the thermal efficiency, welding speed and thermal conductivity for electron beam welding with full penetration was derived.

1. Introduction

One of the most important parameters of welding modes is the thermal efficiency η_T. This parameter is used for preliminary calculations of welding modes [1–3] and evaluation of thermal energy efficiency during welding. It is known [1–3] that the thermal efficiency depends substantially on thermophysical properties of the material, the welding speed and parameters of the heat source.

Both empirical and theoretical methods for determining thermal efficiency were described in literature. The paper [3] often provides the thermal efficiency value for a fast-moving linear heat source, which is equal to 48.4% as the maximum possible at the infinite welding speed, while the experimental thermal efficiency values in some cases exceed 50% [4, 5]. Analysis of cylindrical and flat heat sources models interprets the realization of the thermal efficiency more than 90% [6, 7].

The following equation [3] is generally used for the experimental determination of η_T:

$$\eta_T = v \cdot F_w \cdot \frac{S_m}{q_{eff}} \quad (1)$$

where v – welding speed; F_{ws} – cross-sectional area of the weld joint; S_m – enthalpy of the metal at the melting temperature; q_{eff} – effective beam power.

In [8] it is proposed to approximate the dependence of the penetration depth on the speed and on the beam power for electron-beam welding (EBW) by the following equation:

IOP Conf. Series: Materials Science and Engineering **969** (2020) 012096 doi:10.1088/1757-899X/969/1/012096

$$\eta_T = \frac{s_m \cdot \pi \cdot d}{4} \cdot \left(\frac{A_1}{(q_{eff} / v)^2} + \frac{A_2}{q_{eff} / v} + A_3 \right), \quad (2)$$

where A_1, A_2 and A_3 – empirical coefficients that are determined in the paper on a large number of experimental data in the beam power range from 1 to 8 kW and welding speed from 3.6 to 28.8 m/h.

However, this formula characterizes only the dependence on the linear energy q_{eff}/v and does not directly connected with the welding speed. It can lead to dramatic errors, since the same treatment q_{eff}/v can be obtained within wide limits of welding speeds.

According to [9], where experimental effect of thermal efficiency on welding speed is presented, the thermal efficiency values reach a maximum at welding speed near 15-20 m/h. The increase in thermal efficiency is explained by the fact that with a growth in the welding speed, the proportion of heat factor spent on heating the base metal decreases; and the decline in efficiency after reaching the peak is due to the fact that the weld joint width becomes thinner than the diameter of the heat source and the peripheral zone of the beam spends energy only on heating the heat affected zone (HAZ). Such dependence of thermal efficiency on welding speed can be formed by welding with a low concentration of power density and large diameters of the heat source. In such conditions of electron beam welding with full penetration, when the beam diameter is less than 1 mm, such parameter distribution is unlikely.

Therefore, currently in the scientific and technical literature there is insufficient experimental data on the effect of welding speed on the thermal efficiency during EBW with full penetration. That is why, the purpose of this study is to obtain experimental dependencies of thermal efficiency on the EBW speed for materials with various thermophysical properties.

2. Methodology

Welded joints specimens of titanium alloy 5V, steel 40Kh13 and duralumin D16, which were obtained by EBW without full penetration were used to determine thermal efficiency. The technology of their production is described in [10]. Welding modes are shown in Table 1. The thickness of the plates to be welded was 32 mm, 27 mm and 20 mm respectively.

Estimation of thermal efficiency according to experimental data was done by the formula (1). Three microsections were made from each weld joint. The cross-sectional areas of weld joints F_w on the etched microsections were determined by an Observer Z1m microscope from Carl Zeiss. Panoramic pictures were made using standard AxioVision software and then the cross-sectional area of the weld joints was obtained by contouring the weld joint profile.

Table 1. Welding modes.

Material	Accelerating voltage U, kV	Welding speed v, m/h	Beam current I, mA	Penetration depth H, mm	Weld width B_a, mm
5V titanium alloy	60	20	100	29.4	2.68
		40	135	26.8	2.16
		60	170	28.1	1.79
		90	215	26.4	1.63
		120	250	27.7	1.43
40Kh13 steel		20	115	25.4	2.10
		60	200	24.8	1.52
		120	350	26.6	1.37
D16 aluminum alloy		20	65	17.1	2.34
		60	100	19.4	1.77
		120	140	16.0	1.90

The heat factor of the melt metal at the melting temperature S_m was defined as the sum of the integral of the dependence of the specific heat capacity on the temperature and the specific heat of melting. For 5V and D16 alloys no dependencies of specific heat capacity on temperature is represented in open-source literature. Therefore, taking into account the identity of heat capacity values for pure metals and their alloys at a room temperature, data for pure titanium and aluminium were used in calculations [11, 12]. The dependence of specific heat capacity on temperature is given only up to 1073 K for steel 40Kh13 [13], thus, in such cases the data for pure iron were used [14]. The effect of specific heat capacity on temperature for different types of metals which was used in calculations are shown in figure 1.

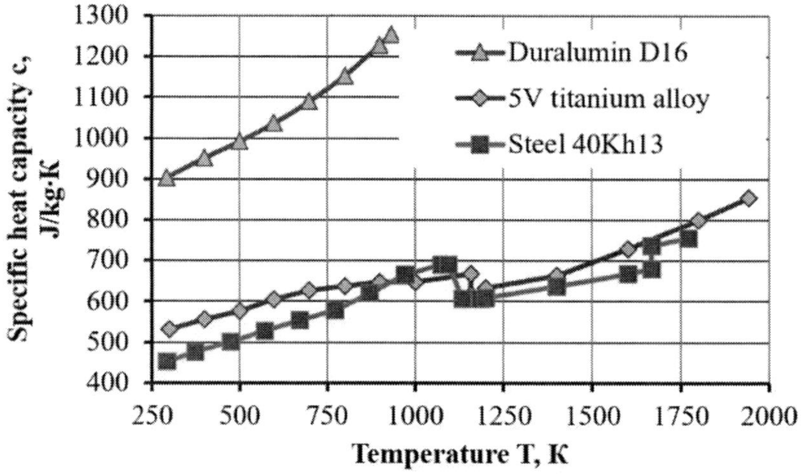

Figure 1. Dependency of specific heat capacity of materials on temperature.

There are no values of the specific heat of fusion were found in the scientific and technical literature for the alloys under consideration, so values for pure metals were used in the calculations. The values of the physical properties of the metal used in the calculations are presented in Table 2.

Table 2. Physical properties of metals.

Properties	*5V*	*40Kh13*	*D16*
Density ρ, kg/m³ at 300 K	4500 [12] for Ti	7870 [12] for Fe	2697 [12] for Al
Specific thermal capacity *c*, J/kg·K at 300 K	530,8 [11] for Ti	452 [13]	883 [15] 903,7 [12] for Al
Specific heat of melting L_m, J/kg	470 000 [16] for Ti	270 000 [16] for Fe	393 000 [16] for Al
Melting temperature T_m, K	1 944 [17] for Ti	1 773 [17]	775 (solidus) [17]

The effective beam power was determined by the formula:

$$q_{eff} = U \cdot I \cdot \eta_{eff} \quad (3)$$

where U – accelerating voltage; I – beam current; η_{eff} – efficiency of the electron beam, which is equaled to 95%.

Thus, thermal efficiency was estimated for each microsection based on the weld area sizes.

3. Results of the research

Results of thermal efficiency determination by experimental method are shown in Figure 2 by charts. As can be seen from the charts, the thermal efficiency is highly dependent on both material properties and welding speed. Thus, the maximum value η_T = 67% was obtained for the titanium alloy 5V at v = 120 m/h and the minimum value equaled to η_T = 18% for the aluminum alloy at v=20 m/h. It should be noted that the thermal efficiency increases gradually with rising welding speed and the peak registered in paper [3] was not observed.

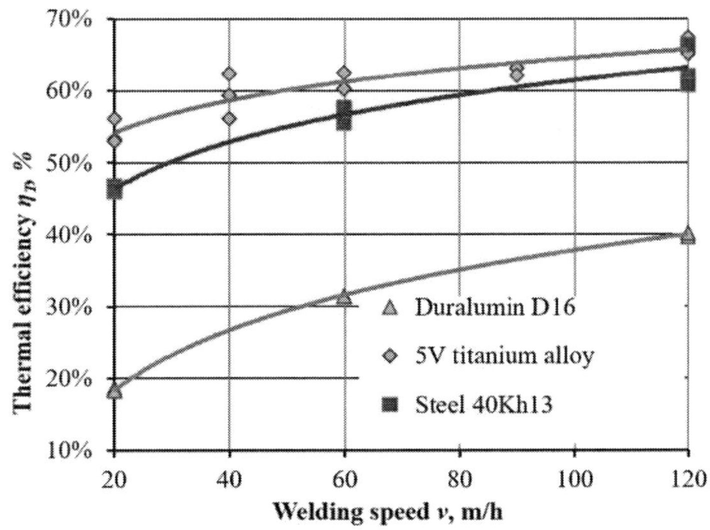

Figure 2. Dependence of thermal efficiency on welding speed for different materials.

4. Discussion of results

The presence of a maximum on the dependence of thermal efficiency on the welding speed depends on the distribution of the power density over the beam section. Thereby, during welding with a well-focused narrow beam almost all the power is spent on heating the welding bath, while when welding with a distributed heat source with an increase in the welding speed, a part of the beam power at the periphery will be spent on heating the base metal, thereby reducing the efficiency [9]. Thus, for the EBW with full penetration the thermal efficiency increases continuously with rising welding speed, provided that the diameter of the beam remains smaller than the weld width.

Apparently, at the most commonly used EBW speeds it is difficult to achieve the thermal efficiency to a value of 90%, whereas the theoretical value of which is published in papers [6, 7]. At the same time, it is obvious that the model of the linear heat source [18] also does not accurately reflect the energy balance character at EBW, especially at high welding speeds. It is associated with a multiple overestimation of temperature near the linear source area [19]. Therefore, existing analytical models of thermal processes during welding do not allow to make a precise assessment of the EBW thermal efficiency.

The analysis of the charts in Figure 2 shows that the higher the thermal conductivity of the material, the lower the thermal efficiency. The coefficient of thermal conductivity compared to the welding speed affects the thermal efficiency in the opposite way. Plotting the dependence of thermal efficiency on the ratio of the welding speed to the coefficient of thermal conductivity v/a (Figure 3) shows that the obtained results of thermal conductivity for all materials can be described by a single logarithmic ratio:

$$q_{eff} = A \cdot \ln\left(\frac{v}{a}\right) - B, \tag{4}$$

where A and B – empirical coefficients, a – thermal conductivity.

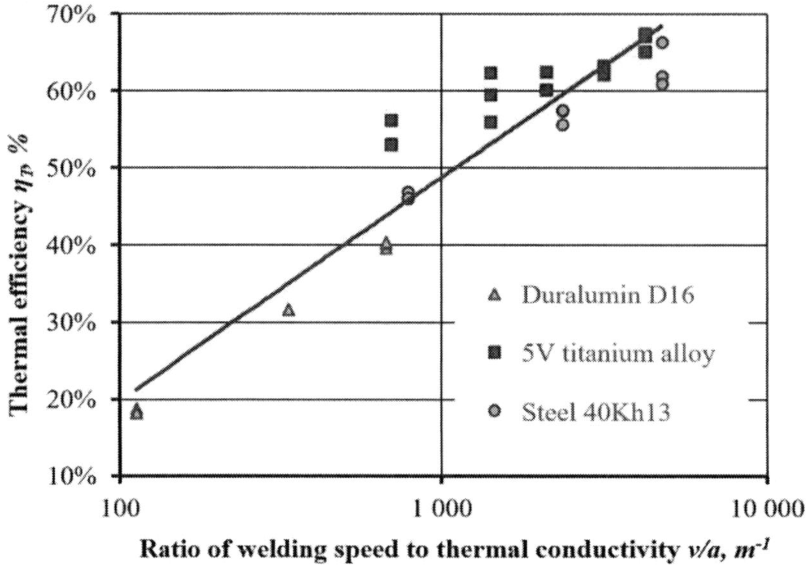

Figure 3. Dependence of thermal efficiency on ratio of welding speed to thermal conductivity coefficient.

Provided that the welding speed has a dimension of m/s and a thermal conductivity coefficient of m^2/s, the approximation of the results obtained in this work allows to estimate the following coefficients:

$$A = 0.1263$$

$$B = 0.3852$$

In this case, the correlation ratio was R^2=0.9015. Certainly, to get more accurate coefficient, more experimental thermal efficiency data for EBW materials with different thermophysical properties should be performed. However, the obtained equation allows to make a satisfactory estimation of the thermal efficiency of the EBW process with full penetration based on welding speed and thermal conductivity coefficient.

The main factors that reduce the accuracy of determining thermal efficiency by an experimental method should also be mentioned. Firstly, there is a lack of data concerning thermophysical properties of industrial alloys, that is why we have to use approximate dependencies of specific heat capacity on temperature, to use data on latent melting heat for pure metals, not to take into account latent heat of phase transformations in a solid state for polymorphic metals. Secondly, the effective efficiency in the calculation was 95%, although its value may differ significantly depending on the EBW modes and external conditions. When elaborating the coefficients in equation (4), the main problem is the dependence of the thermal conductivity coefficient on the temperature and structure of the alloy.

5. Conclusions

It has been found that during EBW with full penetration the thermal efficiency increases gradually for all alloys under study. Thus, with an increase in welding speed from 20 m/h to 120 m/h, the thermal efficiency grew steadily for all three considered types of alloys: values for a titanium alloy 5V went up from 54% to 67%, while for the steel 40Kh13 values rose from 46% to 63% and the most significant increment related to duralumin D16, the thermal efficiency for this alloy increased from 18% to 40%.

Increasing the thermal conductivity of alloys leads to a decrease in thermal efficiency.

The equation which establishes for EBW with full penetration the relationship between thermal efficiency, welding speed and thermal conductivity has been proposed.

References

[1] Rykalin N N, Uglov A A, Zuev I V and Kokora A N 1985 *Laser and Electron Beam Processing of Mate Rials: Handbook* (Moscow: Mechanical engineering) p 496
[2] Zuev I V, Uglov A A and Yanchuk L M 1974 *Welding Production* **2** 7–9
[3] Zuev I V 1998 *Treatment of materials with concentrated energy flows* (Moscow: Moscow Power Engineering Institute) p 162
[4] Zuev I V, Buruis D and Rodyakina R V 1991 *Proc. of Conf. "Concentrated energy flows in material processing and bonding* (Penza: Penza State Polytechnical University)
[5] Rodyakina R V 1996 *Experimental and calculated determination of the optimal area of parameters of the electron beam welding process based on the minimum heat deposition Dissertation of candidate of technical sciences* (Moscow: Moscow Power Engineering Institute) p 192
[6] Arutyunyan R V, Bolshov L A and Vasiliev A D 1989 *Reports of the USSR Academy of Sciences* **307(5)** 1115–1118
[7] Arutyunyan R V, Bolshov L A, Vasiliev A D, Malyuta D D and Sebrant A Yu 1989 *Effects of laser radiation on materials* (Moscow: Nauka) p 367
[8] Zuev I V, Rodyakina R V and Pigolkin G M 1997 *Welding Production* **7** 12
[9] Berezovsky B M 1979 *Automatic welding* **10** 18–21
[10] Dragunov V K, Terentyev E V, Sliva A P, Goncharov A L and Zhgut D A 2019 *IOP Conference Series: Materials Science and Engineering* **681(1)** 012011
[11] Tsvikker U 1979 *Titanium and its alloys* (Moscow: Metallurgy) p 512
[12] Zinovyev V E 1989 *Thermophysical properties of metals at high temperatures. Reference book* (Moscow: Metallurgy) p 384
[13] Zubchenko A S, Koloskov M M, Kashirsky Yu V 2003 *Steels and alloys. Reference book* (Moscow: Mechanical engineering) p 784
[14] Hultgren R et al. 1973 *Selected values of the thermodynamic properties of the elements* (USA, Metals Park, Ohio: American Society for Metals) p 636
[15] Kazantsev E I 1975 *Industrial furnaces. Calculation and Design Reference Manual* (Moscow: Metallurgy) p 368
[16] Chirkin V S 1968 *Thermophysical properties of nuclear equipment materials: Handbook* (Moscow: Atomizdat) p 464
[17] Babichev A P, Babushkina N A, Bratkovsky A M 1991 *Physical Quantities: Handbook* (Moscow: Energoatomizdat) p 1232
[18] Nerovny V M, Konovalov A V, Yakushin B F, Makarov E L and Kurkin A S 2016 *Welding process theory: textbook for universities* (Moscow: Bauman Moscow State Technical University) p 702
[19] Terentyev E V 2014 2014 *Improving the quality of weld joints during electron beam welding with full penetration* (Moscow: Moscow Bauman State University Press) p 160
[20] Lastovirya V N, Krylov K A and Leonov A V 2020 *Int. Conf. Electron Beam Weld. and Related Techn.* 74–82

Acknowledgements

This study conducted by Moscow Power Engineering Institute was financially supported by the Ministry of Science and Higher Education of the Russian Federation (project No. FSWF-2020-0023).

IOP Conf. Series: Materials Science and Engineering **969** (2020) 012097 doi:10.1088/1757-899X/969/1/012097

Dependence of Weld Penetration Shape on Energy Efficiency in Electron Beam Welding Process

A L Goncharov, A P Sliva and I A Kharitonov

National Research University "Moscow Power Engineering Institute", 14, Krasnokazarmennaya str., Moscow, 111250, Russia

E-mail: GoncharovAL@mpei.ru

Abstract. Investigation of relation between parameters of weld penetration shape and energy efficiency of electron beam welding process was performed in the paper. Analysis of welds cross sections of 316 L(N) steel weld joints obtained by the electron beam welding in the wide range of welding regimes is carried out. It is shown that the dependences of the thermal efficiency on the penetration area at different welding speeds are linear. According to the results, the thermal efficiency coefficient may exceed the ultimate theoretical value for linear heat source. Beam sweep does not affect essentially on the rate of thermal efficiency coefficient, since sweep application leads to an insignificant change in the area of the boundary of the weld pool with the base metal, and therefore, the amount of energy removed through this boundary also changes insignificantly and the penetration area remains almost constant.

1. Introduction

Since the appearance of electron beam welding (EBW), a large number of experimental and theoretical studies have been devoted to establishing the relationships between welding parameters and geometric characteristics of the weld [1-18]. Majority of researchers considered various thermal models in which penetration efficiency was determined mainly by the law of conservation of energy.

All parameters of the EBW process affect the shape and geometry of the weld, the main characteristics of which are the penetration depth and weld width. The heat input of electron beam welding is not a determining criterion for the quantitative assessment of the penetration process in contrast to arc welding methods [19], because at the same welding heat input value it is possible to obtain a different penetration depth. This phenomenon is caused by the peculiarity of dagger-shape penetration during EBW. The formation of the weld depends not only on the thermal power of the source, but also on the power density [4]. With the same source power, its specific power can be distributed on the heated surface in different ways, because the power density depends on the focusing current, the distance between the welding gun and the object as well as the type of the beam sweep. Oscillation of the beam and its sweep along various trajectories with a high frequency is a flexible technique for controlling the power density in the processing zone [20].

The relationship of the EBW parameters with the characteristics of the penetration zone for the momentary volume of molten metal according to the model [5] has the form:

$$V_w \cdot F_m - \frac{\eta_t \cdot q_{eff}}{S_m}, \tag{1}$$

where V_w - welding speed, m/s; F_m - cross-sectional melt area, m^2; η_t - thermal efficiency; q_{eff} - effective power of the source, which reaches the heated surface, W; S_m - heat capacity corresponding to the melting temperature, J/m^3.

Effective power is estimated by the formula:

$$q_{eff} = I_b \cdot U_{acc} \cdot \eta_{eff}, \tag{2}$$

where I_b - beam current, mA; U_{acc} - accelerating voltage, kV; η_{eff} - efficiency, taking into account the loss of kinetic energy of electrons during its transition to heat energy.

The volumetric heat capacity when the metal is heated from room temperature to the melting point, is determined by the expression:

$$S_m = \rho \cdot (c \cdot \Delta T_m + L_m), \tag{3}$$

where ρ - metal density, kg/m^3; c - specific heat capacity of metal, J/(kg·°C); $\Delta T_m = T_m - T_0$ - increase from the initial temperature T_0 to the melting point, °C; L_m - specific heat of fusion, J/kg.

The cross-sectional area of the weld is determined by equation:

$$F_m = B \cdot H, \tag{4}$$

where B - average weld width, m; H - penetration depth, m.

The average weld width during welding in continuous mode is found from the ratio:

$$B = \sqrt{\frac{4 \cdot d \cdot q_{eff} \cdot \eta_t}{\pi \cdot S_m \cdot H \cdot V_w}}, \tag{5}$$

where d - beam diameter, m.

The average width is determined at the level of H/e. The beam diameter is defined by the equation:

$$d = \frac{4 \cdot q_{eff}}{\pi \cdot H \cdot V_w \cdot S_{ev}}, \tag{6}$$

where S_{ev} - heat capacity corresponding to the evaporation temperature, J/cm^3.

The volumetric heat capacity of evaporation is expressed:

$$S_{ev} = \rho \cdot (c \cdot \Delta T_{ev} + L_{ev}), \tag{7}$$

Where $\Delta T_{ev} = T_{ev} - T_0$ - increase from the initial temperature to the temperature of evaporation (boiling), °C; L_{ev} - specific temperature of evaporation, J/kg.

The thermal efficiency of penetration is mainly determined by the energy conversion efficiency, which, in turn, depends on variety of factors: the source power, welding speed, type of welded structure and thermal properties of the materials being welded. The thermal efficiency of base metal penetration is determined by the ratio between heat capacity of the base metal melted per unit time and the effective power of the electron beam, i.e. thermal efficiency expresses the share of power that directed to the metal melting. The rest part of the energy aimed at heat removing by the mechanism of thermal conductivity without its melting [4]. For a linear heat source in a plate [4], the value of thermal efficiency increases with a growth in the dimensionless criterion, which can be found by the formula:

$$\varepsilon_2 = \frac{q}{\delta \cdot a \cdot \rho \cdot h_m}, \tag{8}$$

where δ - thickness of the welded plate, m; a - coefficient of thermal diffusivity, m^2/s; h_m - heat capacity, J/kg.

According to this model, the value of the theoretical efficiency cannot exceed 0.484 in the case of extremely powerful linear heat sources.

Currently, there is no single method for determining the welding mode due to the complexity and instability of the processes taking place in the vapor-gas channel, as well as the difference in the characteristics of the welding guns and power sources. All formulas characterize only the relationship between the EBW parameters and the penetration geometry.

The main aim of the study is to establish a ratio between characteristics of the penetration shape and energy efficiency of the EBW and determine the values of thermal efficiency under various parameters and regimes of deep penetration welding.

2. Equipment and research technique

The research was based on the data obtained during EBW on the ELA 60/60 electron-beam machine, developed by the Paton Electric Welding Institute (Ukraine) and ELA-40I developed by JSC "NITI Progress" (Russia). Accelerating voltage during all experiments was 60 kV; beam current varied from 100 to 780 mA; welding speed changed from 15 to 200 m/h. The material under study was AISI 316L(N) steel.

The calculation of the welding mode parameters for circular welds and plates was done in MATHCAD and Microsoft Excel programs with accordance to formulas 1-7. The values of thermal constants for stainless steel were used during estimation of the heat capacity corresponding to the melting temperature S_m = 8.915 J/mm^3; the heat capacity corresponding to the evaporation temperature S_{ev} = 60.88 J/mm^3.

Thermal efficiency was computed from formula (1), taking as a basis the parameters of the welding mode:

$$\eta_t = \frac{V_w \cdot F_m \cdot S_m}{\eta_{eff} \cdot I_b \cdot U_{acc}}, \tag{9}$$

Area F and penetration depth H were found using WeldShape software. The beam diameter d was determined by formula (6).

The initial data for the calculation were: the accelerating voltage U_{acc} = 60 kV; efficiency η_{eff} = 0.95. All calculations and measurements were performed on macrosections. The data of 48 macrosections used in analysis. Those specimens were obtained with no full penetration during welding in the lower position. Other specimens obtained by the full penetration mode were excluded from 70 processed macrosections, because for them the effective power q_{eff} could not be determined precisely.

3. Results and discussion

Several welded joints macrosections are shown in figure 1. The results of measurements and calculations are presented in the table 1.

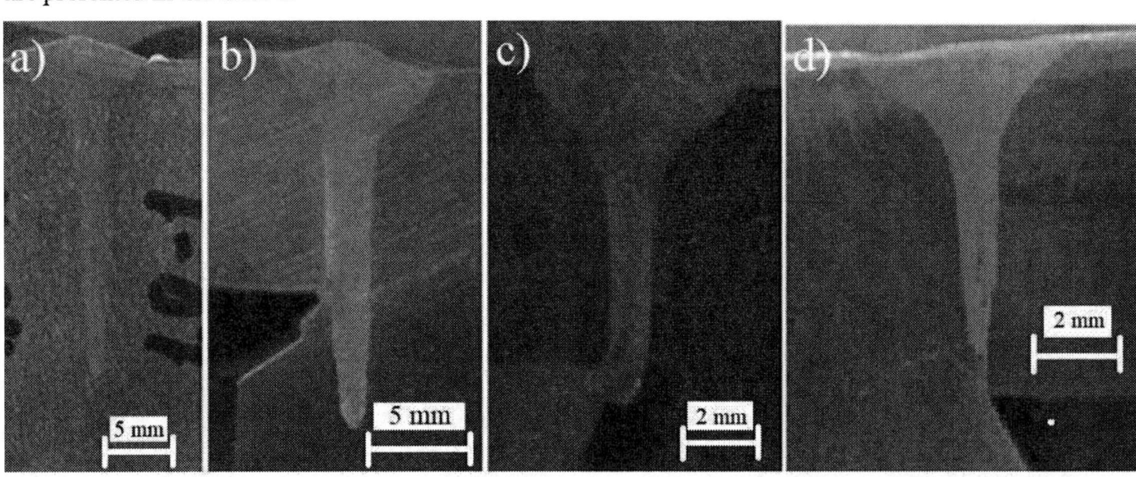

Figure 1. Macrosections: No. 12 (a); No. 20 (b); No. 39 (c); No. 49 (d).

Table 1. Results of calculating the penetration area for various EBW modes.

No.	Welding speed V_w, m/h	Beam current I_b, mA	Focusing current I_f, mA	Beam diameter d, mm	Penetration area F_m, mm^2	Thermal efficiency η_t	Weld depth H, mm	Weld width B, mm	Weld shape factor, H/B
1	30	169	620	1.084	54.649	0.421	22.312	1.838	12.142
2	30	172	610	1.034	61.116	0.463	23.802	1.838	12.948
12	30	183	620	1.236	63.778	0.454	21.177	2.176	9.731
18	30	202	620	1.417	77.943	0.503	20.387	2.626	7.764
20	100	120	966	0.706	9.686	0.351	7.294	1.093	6.673
22	100	170	966	0.754	15.339	0.392	9.677	1.233	7.845
24	100	190	970	0.676	18.010	0.412	12.062	1.134	10.638
39	100	105	958	0,592	16.429	0.680	7.617	1.275	5.973
41	100	100	957	0.446	13.058	0.567	9.630	0.877	10.978
43	100	100	940	0.641	12.864	0.559	6.692	1.253	5.343
49	100	100	965	0.769	8.752	0.380	5.579	1.239	4.503
51	100	100	958	0.404	11.652	0.506	10.618	0.751	14.137
53	15	780	740	2.592	417.454	0.349	86.104	4.001	21.521
55	15	780	750	2.667	389.082	0.325	83.685	3.973	21.066
58	15	780	760	2.690	382.413	0.320	82,949	3.976	20.861
60	15	780	760	2.582	449.210	0.375	86.433	4.131	20.921
66	15	780	780	2.570	515.489	0.431	86.838	4.408	19.698
71	35	100	942	1.408	18.037	0.274	8.709	1.926	4.522
74	35	125	932	1.501	28.312	0.344	10.214	2.300	4.441

Experimental results show that the maximum efficiency factor obtained during welding is 0.68. For seven macrosections the thermal efficiency exceeded 0.484 which is 14.6% of the total number of specimens. When plotting the graphs, the values of the mean thermal efficiency for each mode were used. This feature is explained by fluctuations in the penetration depth.

Figure 2 shows the dependence of the thermal efficiency on the weld shape factor *H/B*. The approximation of the experimental points by a parabolic function shows that the maximum of the approximated value of the thermal efficiency is close to the maximum theoretical and corresponds to a weld shape factor of about 15. For various investigated modes, the value of thermal efficiency ranges from 0.274 to 0.680 (Table 2).

The efficiency versus the amount of energy per unit area is shown in Figure 3. The approximation of the experimental results by a linear function contradicts the data presented in [6]. It means that for certain welding conditions, the use of a linear heat source model to describe the EBW process gives low values of thermal efficiency.

The dependences of the thermal efficiency on the penetration area at various welding speeds are shown in figure 4. These dependences are linear, thus for a constant welding speed in a wide range of electron beam power, the thermal efficiency is inversely proportional to the welding heat input and expression (9) is valid.

The beam sweep does not affect noticeably on the value of the thermal efficiency (figure 5); in both cases a parabolic dependence is observed. The use of a beam sweep along a circular path leads to a change in the shape of the weld and does not significantly affect the value of the thermal efficiency. Apparently, it can be justified by the fact that the beam sweep causes a slight change in the area of the border between weld pool and base metal. As a result, the value of the energy removed through this border also changes negligibly, while the penetration area remains practically constant.

Table 2. The results of determining the thermal efficiency in the investigated modes of EBW.

No.	Thermal efficiency η_t	Weld depth *H*, mm	Weld width *B*, mm	Weld shape factor, *H/B*
1	0.421	22.312	1.838	12.142
2	0.463	23.802	1.838	12.948
12	0.454	21.177	2.176	9.731
18	0.503	20.387	2.626	7.764
20	0.351	7.294	1.093	6.673
22	0.392	9.677	1.233	7.845
24	0.412	12.062	1.134	10.638
39	0.680	7.617	1.275	5.973
41	0.567	9.630	0.877	10.978
43	0.559	6.692	1.253	5.343
49	0.380	5.579	1.239	4.503
51	0.506	10.618	0.751	14.137
53	0.349	86.104	4.001	21.521
55	0.325	83.685	3.973	21.066
58	0.320	82.949	3.976	20.861
60	0.375	86.433	4.131	20.921
66	0.431	86.838	4.408	19.698
71	0.274	8.709	1.926	4.522
74	0.344	10.214	2.300	4.441

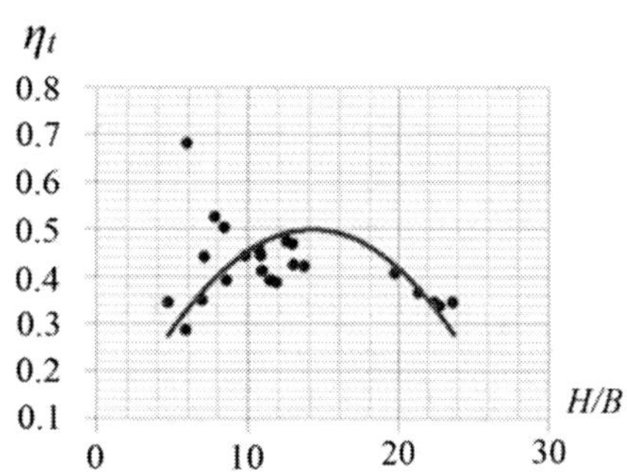

Figure 2. Dependence of thermal efficiency η_t on the weld shape factor *H/B*.

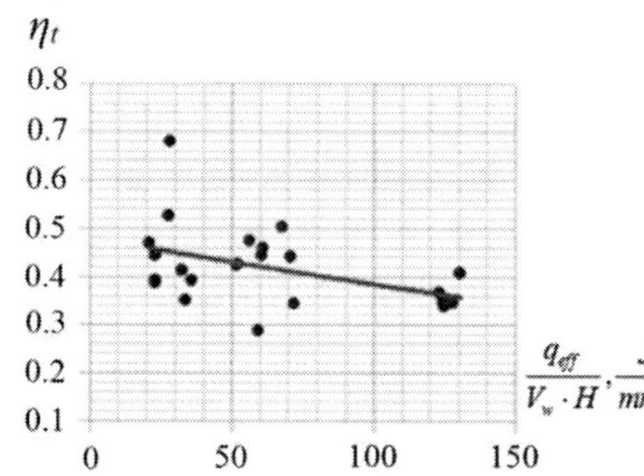

Figure 3. Dependence of thermal efficiency η_t on the amount of energy per unit area.

4. Conclusions

1. The results of the experiments make possible to confirm that the values of the thermal efficiency for EBW can significantly exceed the theoretical values of the efficiency factor for a linear heat source. The maximum value of thermal efficiency was obtained by an EBW of 316L(N) steel with a thickness of 8 mm at a speed of 100 m/h and accounted for 0.68.
2. Changing the shape of penetration during welding with the deep penetration mode using a beam sweep along a circular trajectory does not significantly affect the efficiency of the EBW process.

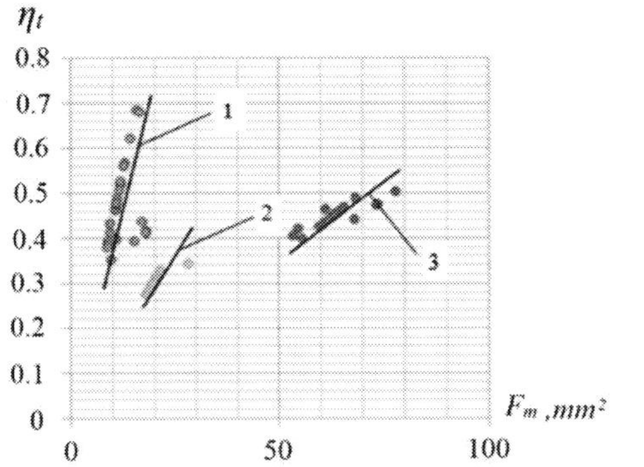

Figure 4. Dependences of thermal efficiency η_t on the penetration area: 1 – welding speed of 100 m/h; 2 – welding speed of 35 m/h; 3 - welding speed of 30 m/h.

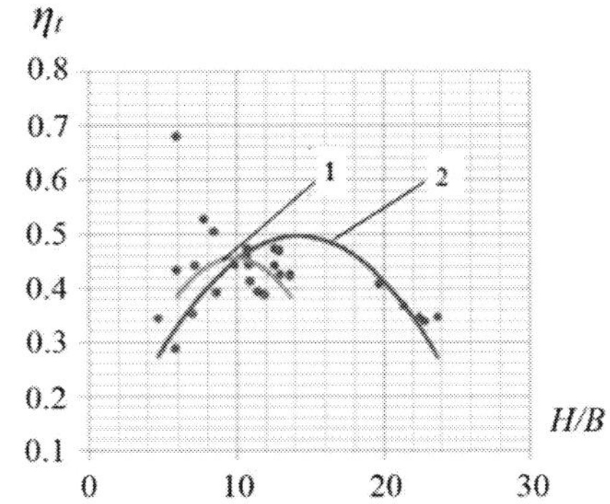

Figure 5. Dependence of thermal efficiency η_t on the weld shape factor *H/B*: 1 – with a beam sweep; 2 – no beam sweep.

References

[1] Hasimoto G and Matsude F 1965 *Trans. of Nat. Res. Inst. Met. J.* **7(3)** 96–109
[2] Gong H and Giedt W 1970 *Welding J.* **6** 259–266
[3] Olshansky N A and Guradzha G D 1972 *Welding Production* **8** 11–13
[4] Rykalin N N, Zuev I V and Uglov A A 1978 *Mechanical engineering* (Moscow: Mashinostroenie) p 239
[5] Zuev I V 1998 *Materials Processing by Concentrated Energy Flows* (Moscow: MPEI) p 162
[6] Zuev I V, Uglov A A and Yanchuk L M 1974 *Welding Production* **2** 7–9
[7] Martynov V N, Khokhlovskii A S and Sliva A P 2009 *Welding Production* **12** 22–25
[8] Martynov V N, Gribkov M S, Kozhechenko A S and Shcherbakov A V 2015 *Russian Metallurgy (Metally)* **9** 30–35
[9] Kharitonov I A 2019 *IOP Conf. Ser.: Mater. Sci. Eng.* **681** 012024
[10] Dragunov V K 2019 *IOP Conf. Ser.: Mater. Sci. Eng.* **681** 012011
[11] Leskov G I, Zhivago L I and Nesterenko V M 1978 *Electron-beam welding* **1978** 178–180
[12] Dragunov V K, Terentyev E V, Sliva A P, Goncharov A L and Marchenkov A Yu 2017 *Welding Int.* **31(4)** 307–311
[13] Lopatko A P, Kartashov L G, Tkachev L G 1977 *5thAll-Union Conf. on Electron Beam Welding* (Kiev: Naukova Dumka) 16–19
[14] Terentyev E V, Dragunov V K, Sliva A P and Scherbakov A V 2015 *Welding Int.* **29(2)** 150–154
[15] Shcherbakov A V, Rodyakina R V, Kozhechenko A S, Gaponova D A, Goncharov A L and Dragunov V K 2017 *Techn. Phys. Let.* **43(11)** 958–960
[16] Nazarenko O K 1982 *Automatic welding* **1** 33–39
[17] Dragunov V K, Muravyova T P and Rodionov Yu A 1990 *Welding production* **4** 2–4
[18] Kharitonov I A, Martynov V N, Shcherbakov A V, Dragunov V K and Gaponova D A 2017 *Proc. Int. Conf. Electron Beam Welding and Related Techn.* (Moscow: Moscow Power Engineering Institute) 257–265
[19] Volchenko V N, Yampolsky V M and Vinokurov V A 1988 *Welding processes theory* (Moscow: Higher School) p 559
[20] Kaidalov A A and Nazarenko O K 1973 *Electronic processing of materials* **3** 9–13

Acknowledgments

The research was carried out in National Research University "Moscow Power Engineering Institute" with the financial support of the Russian Science Foundation (project No. 18-19-00652).

IOP Conf. Series: Materials Science and Engineering **969** (2020) 012098 doi:10.1088/1757-899X/969/1/012098

Process Design for the Wire Arc Additive Manufacturing of a Compressor Impeller

A A Kulikov, A V Sidorova and A E Balanovskiy

Irkutsk National Research Technical University, 83, Lermontov street, Irkutsk, 664074, Russia

Email: the.tosha2013@gmail.com

Abstract. Additive manufacturing or 3D printing slowly and steadily replaces traditional subtractive manufacturing, where parts are manufactured by sequentially removing pieces of the workpiece material. In 3D printing, the material is added in layers, which enables producing parts of complex geometric shapes made from expensive metals with higher productivity and economic efficiency. In order to perform 3D printing on an industrial scale, it is necessary to combine robotics with 3D printing software. This article presents the results of the process design for 3D printing of a compressor impeller prototype. A detailed methodology of customization of the KUKA industrial robot for the 3D printing process is given. Once the impeller model was created in the CAD environment, it was cut into layers in the slicer program. Special software for offline programming of industrial robots was used to translate the program into the robot language and perform 3D printing using an experimental robotic complex. As a result, a prototype of a low-carbon steel compressor impeller was printed using a welding machine and steel wire. The developed technological process was successful and proved that an industrial robot combined with a welding machine can be used as an industrial 3D printer for printing real machine parts.

1. Introduction

The compressor impeller is a mechanism for creating torque rod (see figure 1). The principle of its operation is used in aircraft engines. Piston-type mechanisms have long been become obsolete, as jet engines are lighter, more efficient, and run on cheaper fuel. The fixed blading can be adjusted to different mechanisms.

The impeller provides maximum thrust with a minimum fan diameter, so the demand for such mechanisms is very high. Currently, high-speed bladed electric motors are widely used in aircraft modeling of the F4 category (the model copies a jet aircraft).

Impeller motors are also used for engines of jet skis, boats, and other water transport. The system gives higher thrust and extraordinary efficiency with maximum acceleration and performance.

With the special design of the impeller, it is possible to achieve minimal noise and get rid of the loss of power of the working turbine. The body of the impeller engine is smaller than that of the propeller, while the useful power remains the same. The impeller (mounted on the rotor) is a multi-blade screw in an annular channel. Air drawn into the impeller under high pressure has some weight, so there is a jet thrust as a result of the movement of air masses. The force moves the machine or moves the working medium [1, 2].

Impellers are used in a variety of ways. These are not only large industrial turbines' engines, cooling towers or compressors but also small mechanisms, such as aquarium filters, pumps, engines of dishwashers, water jets.

Figure 1. Compressor impeller.

High-quality alloys of heat-resistant steel, stainless steel of AISI 316 and AISI 304 grades, as well as various aluminum alloys serve as materials for the manufacture of impellers.

The workpieces undergo a complex of machining work for removing burrs, notches, and other defects. Turning and milling machines are used for this purpose. After that, the blades undergo heat treatment such as tempering and normalization. A set of tests for strength and hardness then follows.

The traditional technological process of manufacturing impellers consists mainly of machining operations (see figure 2). As a result, during the manufacturing process, about 50% of the material of the initial workpiece is wasted on metal shavings. Therefore, this process is economically unprofitable, especially when impellers are supposed to be made of expensive alloys.

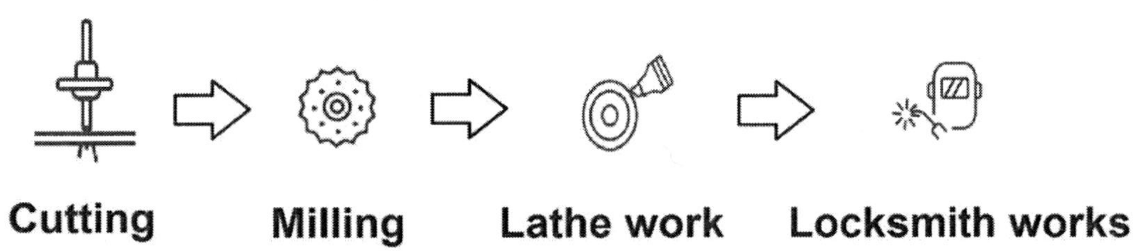

Figure 2. The traditional manufacturing process of an impeller.

3D printing, on the contrary, enables using up to 100% of the feedstock material, while only about 10% of the mass of the final workpiece is lost during finishing machining. Moreover, the use of 3D printing helps to significantly reduce the number of manufacturing operations of the impeller (see figure 3), which makes this process both cost-effective and high-performance [3].

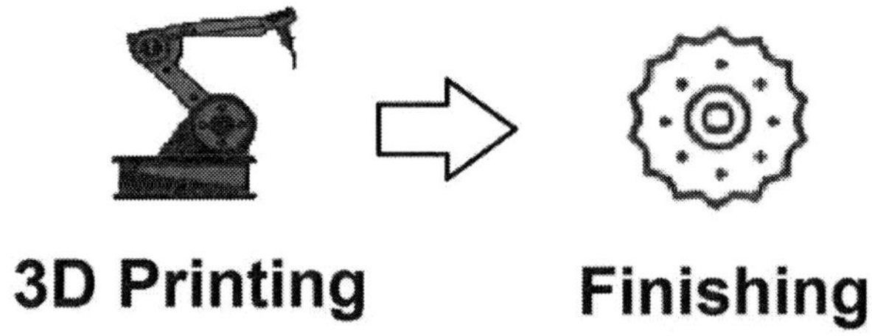

Figure 3. The alternative manufacturing process of an impeller.

An industrial robot manipulator in conjunction with a welding machine is used as a 3D printer. 3D printing is performed using the wire arc additive manufacturing (WAAM) technology, where welding wire is used as a feedstock material that is melted using an electric arc. Once the printing process is complete, the part is also subjected to heat treatment to improve its mechanical properties.

The purpose of this project is to design a technological process for 3D printing of the compressor impeller. This article describes the process itself, starting from the CAD designing and ending with the final part. The result of the study is a printed prototype of the low-carbon low-alloy steel impeller.

2. Materials and methods

The welding wire Sv-08G2S with a diameter of 1.2 mm is used as a feedstock material. The wire was deposited on a Fe37-3FN structural carbon steel plate of 16 mm thickness.

The robotic complex installed at the Irkutsk National Research Technical University is used as an industrial 3D printer (see figure 4).

The robotic complex consists of: an industrial robot KUKA KR 210 R2700 prime (KUKA robotics, Germany), a Lorch SpeedPulse S3 mobil welding machine and a CO2 protective gas cylinder.

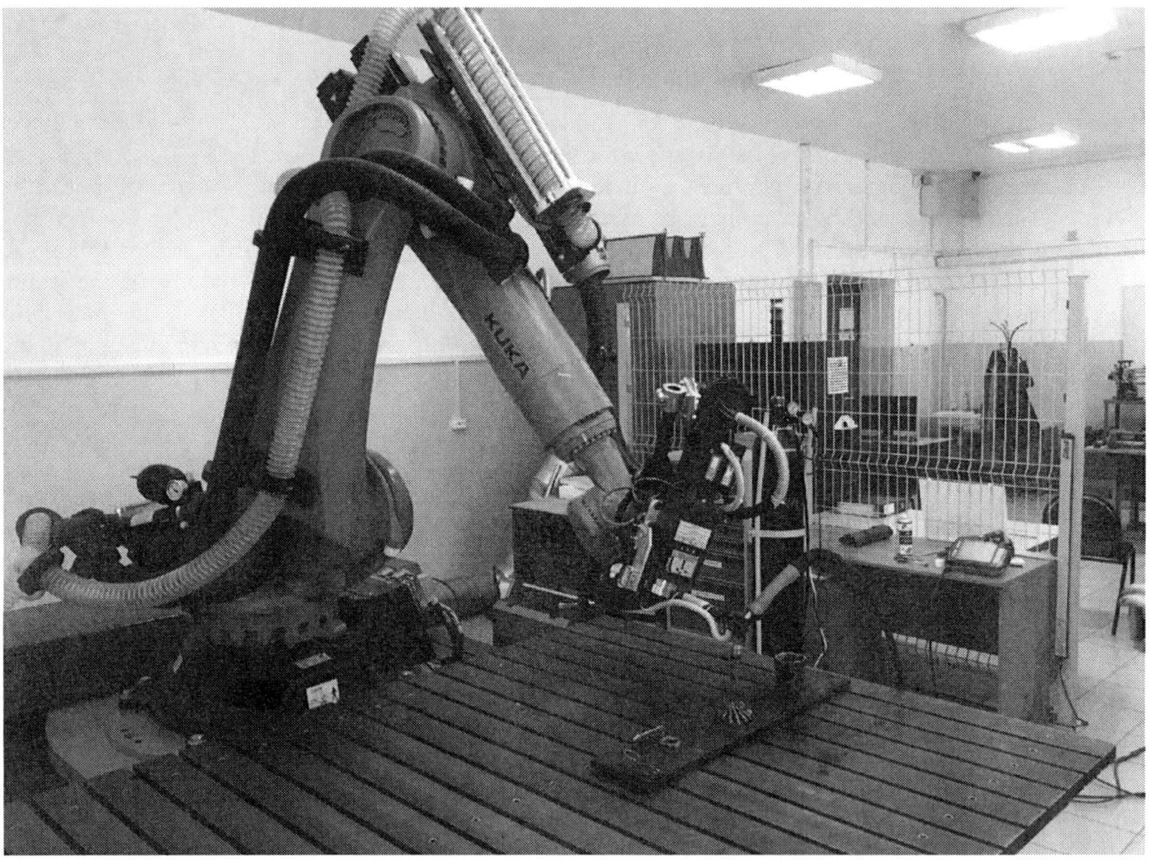

Figure 4. WAAM robotic complex installed at the INRTU.

The wire is deposited using a gas metal arc welding method (GMAW). This method is also known as a metal inert gas (MIG) welding or metal active gas (MIG) welding process. In this method of welding, an electric arc generates between the wire and the substrate metal, resulting in the wire to be melted [4]. To protect the molten metal from air contaminants, the welding grade CO2 shielding gas is used with a gas flow rate of 10 l/min.

To create a 3D model of the impeller, the Autodesk Fusion 360 CAD software was used. The Ultimaker Cura slicer was used for slicing the model into layers. Then, the RoboDK software was used to simulate the printing process and program the KUKA industrial robot.

3. Results

In order to print the impeller, it is necessary to create a 3D model of it at first (see figure 5). The model was created on the basis of the compressor impeller presented in the reference [5]. The size and shape of the impeller have been slightly changed to facilitate the printing process.

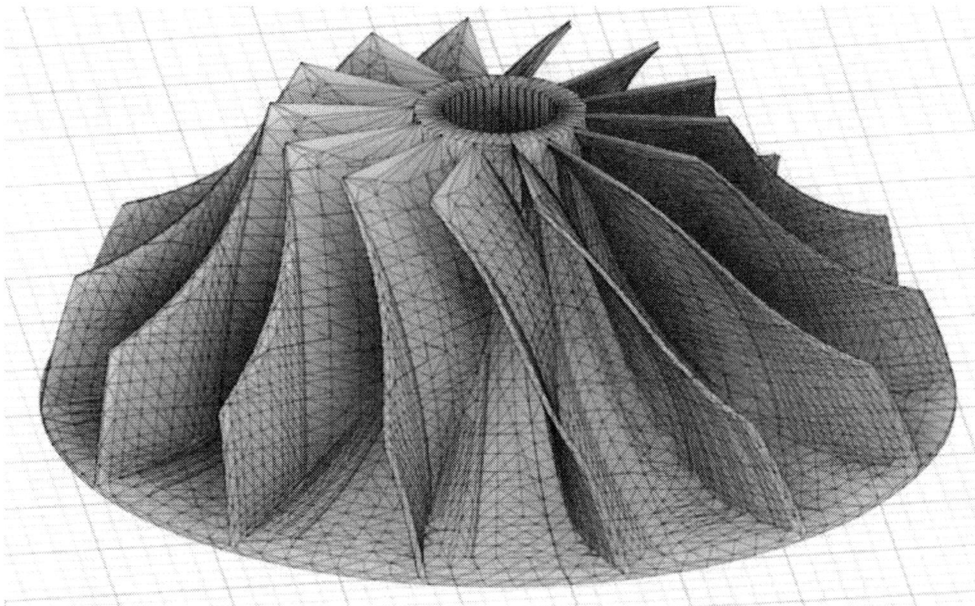

Figure 5. 3D model of the compressor impeller.

Once the model is created, it needs to be saved in STL format, which is recognized by the slicer program. Then the model is loaded into the slicer for further slicing into layers. The impeller size has been scaled down by 60% to reduce the printing time. The resulting model has a diameter of 80mm and a height of 30mm. With these geometric dimensions, printing takes about 43 minutes. After changing the geometric dimensions, configuring the print parameters follows.

First of all, it is necessary to specify the diameter of the wire used. For printing the impeller, the steel wire SV-08G2S with a diameter of 1.2 mm is used. The further parameters to be specified are the thickness (height) and width of the layer. In preliminary studies not included in this article, the following values were obtained: the average layer thickness is 2 mm, and the width of the deposited bead is 10 mm. These values were obtained during the deposition of wall structures at a travel speed of 3 mm/sec. This speed proved to be the most optimal for 3D printing using steel wire.

The next step is to select an algorithm for the movement of the welding torch during the printing process. This parameter is the most important because it determines the quality of the future part. We can also choose where to start printing each new layer. This is necessary in order to avoid the formation of such a defect as a seam. Seam forms if printing each new layer starts at the same point. Also, the formation of this defect could be caused by a slight slowdown of torch speed before the new layer, which is caused by software restrictions. Therefore, it is necessary to choose the correct starting point for printing each layer, so that seam will be invisible, or can be easily removed by finishing machining.

Since each layer of the impeller has a closed contour, and the diameter of the circles decreases gradually with each layer, the spiral algorithm is most preferable for printing this part. In a spiral movement, only the outer contour of the wall will be printed. The height of the torch will gradually increase as the layer thickness increases. In this way, the spiral is created along the contour of the model. Movements of the torch from one layer to another are excluded since the torch will gradually increase the height along the way.

After selecting the torch movement algorithm, adjustments of the print speed follows. As mentioned earlier, it was found experimentally that the most optimal torch travel speed for steel wire deposition is 3 mm/sec.

Once all the necessary printing parameters are set, slicing follows. After slicing, the layered model can be seen (see figure 6). The print time is also calculated automatically. In this case, the prototype of the impeller will be printed for 43 minutes.

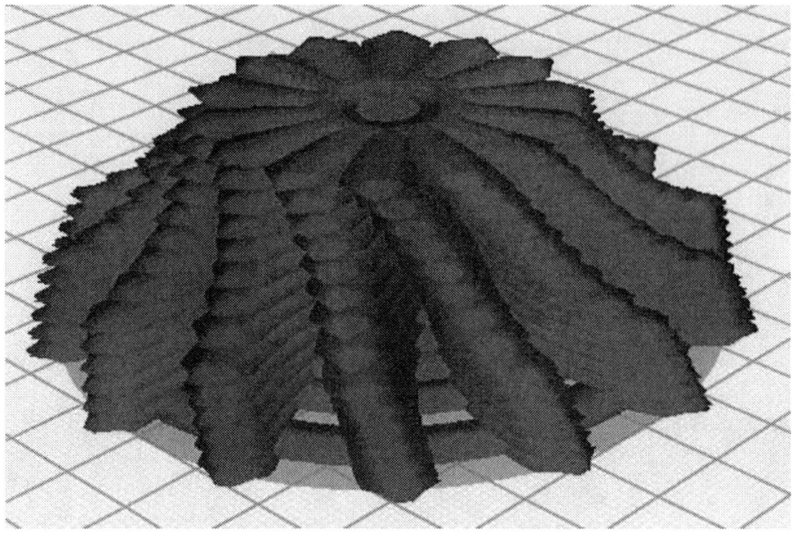

Figure 6. 3D model of the impeller after slicing.

The next step is to save the slicing model of the impeller in the g-code format. G-code is the formal name of the programming language for computer numerical control (CNC) machines. This language is also supported by most modern 3D printers. The program in the g-code format is necessary for creating the trajectory of the torch movement during the deposition of metal wire. The G-code file is then loaded into the software environment for simulating and programming industrial robots. At first, it is necessary to create a robotic cell in an environment that includes a workplace, an industrial robot, and a work tool.

Once the g-code file is generated, the simulation of the printing process continues in the software for the offline programming of industrial robots. To use industrial robots as 3 or 5-axis 3D printers, offline programming tools are required to translate 3D printer programs into robot programs. This software enables achieving the same results with industrial robots as with 3D printers. Based on the slicing model of the impeller, the robot's trajectory is created and the final printing program is generated (see figure 7). Then the printing program is converted to a program written in the robot language.

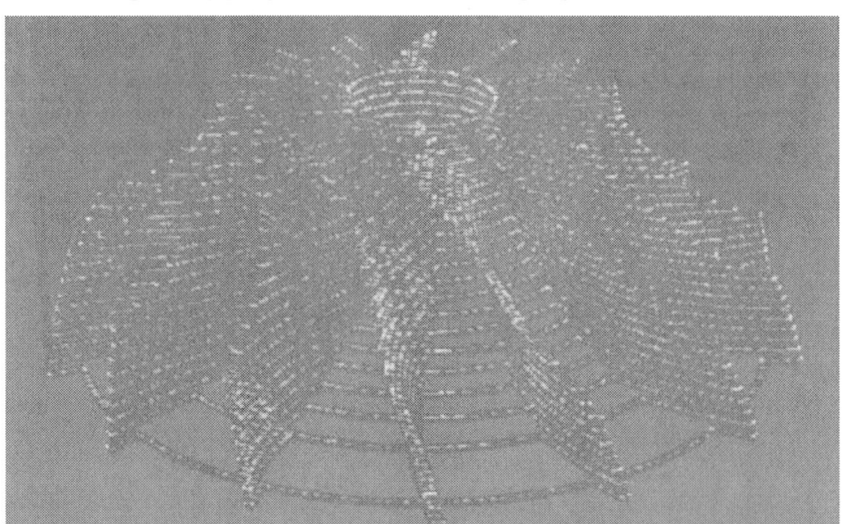

Figure 7. Generated torch path.

Then the simulation of the printing process can be started (see figure 8). All points that the robot can reach will be colored green, and unreachable points will be colored red. In order to make the points accessible to the robot, they must be rotated relative to the desired tool path. According to the loaded model, the program automatically generates the trajectory of the torch movement.

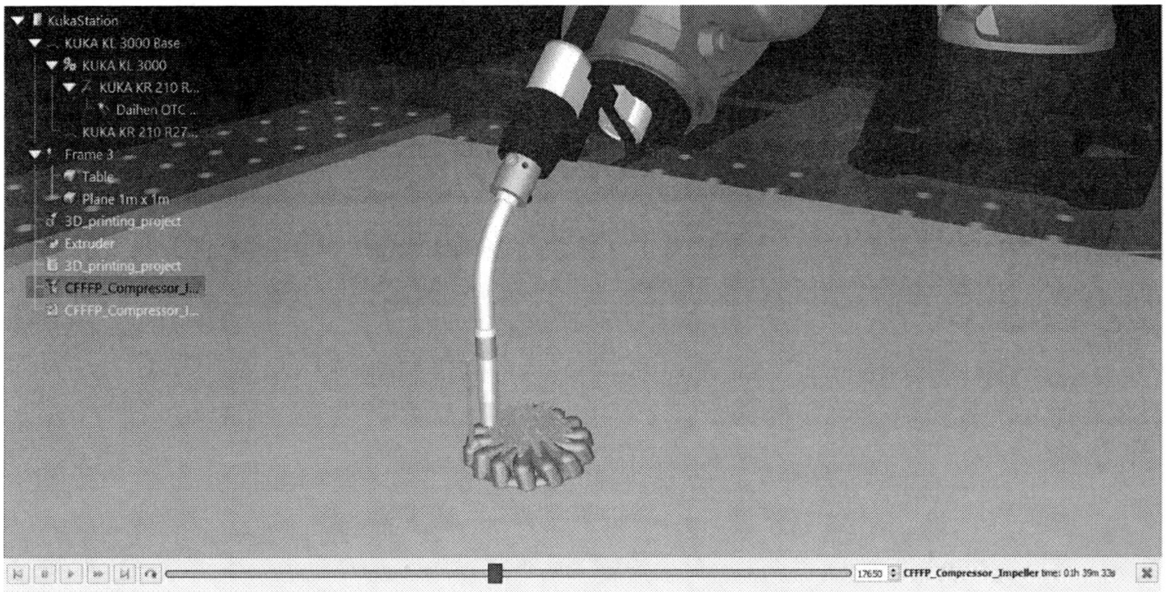

Figure 8. Simulation of the printing process.

To perform a 3D printing project, it is required to define the coordinate system and the working tool. The coordinate system of the impeller must be combined with the coordinate system of the print workspace in order for the coordinate system of the actual working environment to be correctly defined. In addition, approach and retract commands must be created for the robot's starting and ending movements. To minimize the movement of the robot joints and keep the orientation of the working tool as constant as possible, it is necessary to use a tool orientation algorithm that ensures minimal changes in the orientation of the tool during printing. Positioning the robot requires determining the center point of the robot tool using target points located in Cartesian space.

The software automatically optimizes the robot's trajectory, avoiding singularities, axis constraints, and collisions. Next, it is required to enter the coordinates of the working area of the robot and the working tool from the real working environment, after which the software will once again check whether this program can be executed. After completing all calculations and checks, simulation of the printing process can be followed and it can be seen how the 3D printing process will proceed.

Once the simulation is complete, it can be seen what the impeller will look like at the end of the printing process (see figure 9).

After the successful simulation of the virtual program, the final program written in the robot language is generated. The post-processor converts the simulation to the corresponding robotic program. The features of each particular robot are taken into account by the post-processor. By default, each robot has a standard post-processor defined by the programming software. All KUKA industrial robots use the KRL language. The post-processors have the complete versatility to create highly specialized robotic manufacturing programs. Post-processors can be easily modified for each specific task. Since the robot controller has restrictions on the file size or the number of lines of code, long programs cannot be transferred to the robot. In particular, 3D printing programs can contain hundreds of thousands of lines of code. To overcome this obstacle, long programs need to be split into subprograms that will run sequentially after the main program is started.

IOP Conf. Series: Materials Science and Engineering **969** (2020) 012098 doi:10.1088/1757-899X/969/1/012098

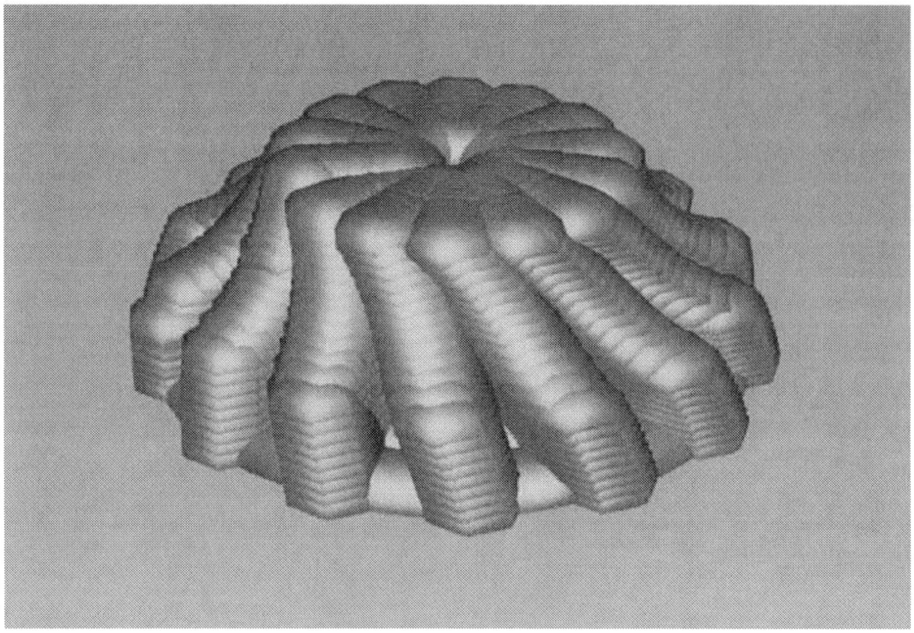

Figure 9. Impeller after the printing simulation.

Once the program has been generated, it can either be sent to the robot directly from the simulator interface or transferred to a flash drive and loaded into the robot control cabinet. The program can be transferred directly from the computer to the robot using the file transfer protocol or other special protocols. To start transferring, the robot's IP address needs to be determined and the file transfer protocol settings need to be configured in the robot connection menu.

After loading the program on the robot, it must be checked on the teach pendant to detect programming errors. Commands can also be added or deleted on the teach pendant. If the program has been checked successfully and the torch trajectory is correct, the welding machine can be switching on to set welding parameters. Then the printing process itself can be started. The printed prototype of a steel impeller is shown at figure 10.

Figure 10. Printed impeller prototype.

The impeller has the following dimensions: diameter is 80 mm, height is 30 mm, the number of layers is 15, and the thickness of one layer is 2 mm.

4. Discussion

3D printing software in conjunction with software for offline programming of industrial robots helped to create and successfully print a steel prototype of a compressor impeller. However, in order to 3D print such parts at the industrial level, it is necessary to improve the robotic complex by equipping it with additional sensors and software tools of narrow specialization. For example, it is preferable to implement a technical vision system to remotely monitor the printing process. This is primarily required in order to control the printing process in real-time to prevent the formation of possible defects and deviations from the geometric dimensions of the part specified during the CAD modelling. In addition, the welding current, voltage, and wire feed rate must be monitored and measured during the printing process. To do this, welding machines must have feedback to be capable of in situ adjusting the welding parameters.

It also requires developing highly specialized software specifically designed for 3D printing using WAAM technology. Standard 3D printing software does not take into account all the peculiarities of complex metallurgical and welding processes, which are of great importance for 3D printing of metal products. The thermal welding cycle impact and changes in the microstructure directly affect the quality and mechanical properties of the deposited parts.

During this project, a scaled-down steel prototype of the compressor impeller was printed. This was done solely in order to reduce the printing time and prove the possibility of using a robotic complex in conjunction with 3D printing software to perform 3D printing utilizing the new WAAM technology. This technological process has been successful in terms of the optimization and implementation of the printing process. However, the quality of the deposited part can only be judged visually, since no tests of mechanical properties have yet been conducted. The results of the study of the steel microstructure of the deposited impeller, as well as the results of tests, will be presented in future publications.

The objectives of future research are to develop a technological process for printing parts made of expensive metals, such as titanium and nickel alloys. Steel can still be used for printing prototypes of parts, but the manufacturing of real functional parts requires a transition to new materials [6–20]. It is also necessary to develop a process for printing objects of more complex geometric shapes with larger dimensions.

5. Conclusion

Additive manufacturing is a promising area of research, as it enables producing parts of complex geometric shapes with significant savings in the amount of additive material used. Robotics, in turn, opens new opportunities for 3D printing at the industrial level. Robots can be used as industrial 3D printers, and the flexibility and adaptability of their programming capabilities correspond to the trends and demands of small, medium and large-scale manufacturing.

This article describes the process design for 3D printing of a compressor impeller steel prototype. The method of programming an industrial robot from creating a CAD model to printing a real part is described in detail. Since there is currently no specialized software for the wire arc additive manufacturing, the only solution is to optimize the existing software for 3D printers. The slicer enables cutting the CAD model into layers and setting printing parameters. Software for offline programming of industrial robots gives an opportunity to use industrial robots as 3D printers. The g-code generated in the slicer is easily recognized in the software and then translated into the robot's language.

The available software settings help to create programs for 3D printing both simple hollow shapes, such as a cylinder and square and quite complex objects, such as a compressor impeller. The technological process developed for printing the impeller was successful and proved that the robotic complex in conjunction with 3D printing software can be used for 3D printing of metal parts using WAAM technology.

References

[1] Frigne P and Braembussche R V D 1984 Distinction Between Different Types of Impeller and Diffuser Rotating Stall in a Centrifugal Compressor With Vaneless Diffuser *Journal of Engineering for Gas Turbines and Power* **106(2)** 468–474 DOI: 10.1115/1.3239589

[2] Krain H 1981 A Study on Centrifugal Impeller and Diffuser Flow *Journal of Engineering for Power* **103(4)** 688–697 DOI: 10.1115/1.3230791

[3] Wu B, Pan Z, Ding D, Cuiuri D, Li H 2018 A review of the wire arc additive manufacturing of metals: properties, defects and quality improvement *Journal of Manufacturing Processes* **35** 127–139 DOI: 10.1016/j.jmapro.2018.08.001

[4] Ding D, Pan Z, Cuiuri D and Li H 2015 Wire-feed additive manufacturing of metal components: technologies, developments and future interests *The International Journal of Advanced Manufacturing Technology* **81(1–4)** 465–481 DOI: 10.1007/s00170-015-7077-3

[5] Gravele H E 1978 *Compressor impeller and method of manufacture* U.S. Patent **4,093,401**

[6] Wu B, Ding D, Pan Z, Cuiuri D, Li H, Han J and Fei Z 2017 Effects of heat accumulation on the arc characteristics and metal transfer behavior in Wire Arc Additive Manufacturing of Ti6Al4V *Journal of Materials Processing Technology* **250** 304–312 DOI: 10.1016/j.jmatprotec.2017.07.037

[7] Wang F, Williams S and Rush M 2011 Morphology investigation on direct current pulsed gas tungsten arc welded additive layer manufactured Ti6Al4V alloy *The International Journal of Advanced Manufacturing Technology* **57(5–8)** 597–603 DOI: 10.1007/s00170-011-3299-1

[8] Cong B, Ding J and Williams S 2014 Effect of arc mode in cold metal transfer process on porosity of additively manufactured Al-6.3%Cu alloy *The International Journal of Advanced Manufacturing Technology* **76(9–12)** 1593–1606 DOI: 10.1007/s00170-014-6346-x

[9] Xu F, Lv Y, Xu B, Liu Y, Shu F and He P 2013 Effect of deposition strategy on the microstructure and mechanical properties of Inconel 625 superalloy fabricated by pulsed plasma arc deposition *Materials & Design* **45** 446–455 DOI: 10.1016/j.matdes.2012.07.013

[10] Baufeld B 2011 Mechanical Properties of INCONEL 718 Parts Manufactured by Shaped Metal Deposition (SMD) *Journal of Materials Engineering and Performance* **21(7)** 1416–1421 DOI: 10.1007/s11665-011-0009-y

[11] Haden C, Zeng G, Carter F, Ruhl C, Krick B and Harlow D 2017 Wire and arc additive manufactured steel: Tensile and wear properties *Additive Manufacturing* **16** 115–123 DOI: 10.1016/j.addma.2017.05.010

[12] Ding D, Pan Z, Duin S V, Li H and Shen C 2016 Fabricating Superior NiAl Bronze Components through Wire Arc Additive Manufacturing *Materials* **9(8)** 652 DOI: 10.3390/ma9080652

[13] Shi X, Ma S, Liu C, Wu Q, Lu J, Liu Y and Shi W 2017 Selective laser melting-wire arc additive manufacturing hybrid fabrication of Ti-6Al-4V alloy: Microstructure and mechanical properties *Materials Science and Engineering: A* **684** 196–204 Online. Available: https://doi.org/10.1016/j.msea.2016.12.065

[14] Horgar A, Fostervoll H, Nyhus B, Ren X, Eriksson M and Akselsen O M 2018 Additive manufacturing using WAAM with AA5183 wire *Journal of Materials Processing Technology* **259** 68–74 Online. Available: https://doi.org/10.1016/j.jmatprotec.2018.04.014

[15] Zhang C, Li Y, Gao M and Zeng X 2018 Wire arc additive manufacturing of Al-6Mg alloy using variable polarity cold metal transfer arc as power source *Materials Science and Engineering: A* **711** 415–423 Online. Available: https://doi.org/10.1016/j.msea.2017.11.084

[16] Wang J, Pan Z, Yang G, Han J, Chen X and Li H 2019 Location dependence of microstructure, phase transformation temperature and mechanical properties on Ni-rich NiTi alloy fabricated by wire arc additive manufacturing *Materials Science and Engineering: A* **749** 218–222 Online. Available: https://doi.org/10.1016/j.msea.2019.02.029

[17] Ding D, Pan Z, van Duin S, Li H and Shen C 2016 Fabricating Superior NiAl Bronze Components through Wire Arc Additive Manufacturing *Materials* **9(8)** 652 Online. Available: https://doi.org/10.3390/ma9080652

[18] Wu B, Qiu Z, Pan Z, Carpenter K, Wang T 2020 Enhanced interface strength in steel-nickel bimetallic component fabricated using wire arc additive manufacturing with interweaving deposition strategy *Journal of Materials Science & Technology* **52** 226–234 Online. Available: https://doi.org/10.1016/j.jmst.2020.04.019

[19] Ahsan M R U, Tanvir A N M, Seo G-J, Bates B, 2020 Heat-treatment effects on a bimetallic additively-manufactured structure (BAMS) of the low-carbon steel and austenitic-stainless

steel *Additive Manufacturing* **32** 101036 Online. Available: https://doi.org/10.1016/j.addma.2020.101036
[20] Zeng Z, Cong B, Oliveira J P, Ke W 2020 Wire and arc additive manufacturing of a Ni-rich NiTi shape memory alloy: Microstructure and mechanical properties *Additive Manufacturing* **32** 101051 Online. Available: https://doi.org/10.1016/j.addma.2020.101051

Acknowledgments
A. A. Kulikov created a 3D model of the impeller, developed a technological process for 3D printing, and wrote the paper. A. V. Sidorova helped with the operating and programming of the KUKA robot. A.E. Balanovskiy carried out the overall project management and participated in the discussion of the results.

IOP Conf. Series: Materials Science and Engineering **969** (2020) 012099 doi:10.1088/1757-899X/969/1/012099

Investigation of the Effect of the Lighting Step on the Strength of a Ceramic Product in 3D Printing

D Yerezhep[1,2], P Piterskov[2,3] and G S Pobirokhin[3]

[1]Scientific Research Institute of Experimental and Theoretical Physics, Al-Farabi Kazakh National University, al-Farabi Ave., 71, Almaty, 050040, Kazakhstan
[2]ITMO University, 49, Kronversky Ave., St. Petersburg, 197101, Russia
[3]Additive Fabrication LLC, 12 lit A, Sveaborgskaya street, St. Petersburg, 196105, Russia

E-mail: piterskov@mail.ru

Abstract. The subject of this study is to assess the effect of such a 3D printing parameter by the SLA method as the lighting step (the distance between the lines of the beam motion when lighting the layer) on the mechanical properties of a ceramic object. For the study, the lighting step of 50 μm, 75 μm and 100 μm was used. The dependencies such as the Mises stresses, the displacements of the layers of the printed object relative to each other, and also the pressure arising in the layer were compared. To exclude edge effects, dependences were evaluated in the range from 500 μm to 1500 μm of the surface of the object layer, the layer thickness was 100 μm. These studies are necessary to optimize the 3D printing process, to minimize internal stresses and pressures in the printed object. The use of optimal printing parameters will improve the results of heat treatment and obtain objects with the required surface quality.

1. Introduction

Modern companies in leading industries are constantly looking for solutions that allow them to gain an advantage in the global market. The main ways to improve their work may be the optimal planning of production processes based on modeling [1–2] and scheduling, automation and robotics, constant monitoring of production tasks, as well as other technologies included in the Industry 4.0 strategy. One of the increasingly popular technologies is additive manufacturing (AM), especially 3D printing. 3D printing is an example of rapid prototyping technologies (or additive manufacturing), which are becoming increasingly popular because of their practicality and the constant reduction in the price of printing equipment and materials [3]. The fastest prototyping technologies are still under development and are characterized by the high cost of equipment and materials. However, they can produce complex, high-quality, high-strength elements of complex shape, which can be made of various materials with certain properties (for example, various types of plastic, biomaterials, metal, composites, and even concrete [3–4]).

One of the most interesting areas of additive technology is 3D printing with ceramics. This area is interesting for its high strength and temperature properties of the material, as well as a wide range of applications (aerospace, radio electronics, medicine, nuclear industry, general engineering and jewelry industry) [4–5].

Of all the methods for 3D printing with ceramics, the most interesting and popular is laser stereolithography (SLA). The quality of printed parts can be assessed by mechanical or thermal properties, as well as tactile and visual perception [3]. Considering other 3D ceramic printing technologies (FDM, DLP), parts printed using the SLA method have the highest surface quality due to the relatively small thickness of the layers (20–100 μm), high printing accuracy (10-50 μm) and features of the technology of applying the layer (moving the laser beam along the surface of the layer of the future product along a predetermined path) [6].

Problems with the mechanical properties (strength parameters) of printed products are usually associated not only with the mechanical properties of the material itself, but also with the weakness of the interconnection of the layers, uncontrolled shrinkage during cooling, the inaccuracy of the mechanical elements of the 3D printer (moving the platform, laser), as well as directly with print parameters. These parameters include: lighting strategy (strategy of laser beam movement) and interrelated parameters (beam diameter, distance between shading lines, exposure duration, etc.) [3,7,8].

2. Simulation model

Most of the articles devoted to the study of the influence of printing parameters (lighting strategy and related parameters) on the strength characteristics of a product are associated with 3D printing with photopolymers according to the SLA or FDM method, where the product does not require subsequent heat treatment [7–12].

In this paper, using modeling methods, we study the influence of printing parameters on the strength characteristics of the product at the green stage (before heat treatment). The main parameter of 3D printing is the lighting step. In this study we used the lighting step equal to 50 μm, 75 μm and 100 μm. The beam diameter, the distance between the lighting lines, and the exposure duration were unchanged in a numerical experiment. For modeling, we used the following properties of a material based on corundum grade F1000 (fraction 7 μm) and a photosensitive polymer (70% – alumina powder, 30% – liquid photopolymer): density ρ – 3600 kg / m3, Young's modulus E – 2500 MPa, coefficient Poisson η – 0.26 [13]. 3D models with different printing parameters are shown in Figure 1.

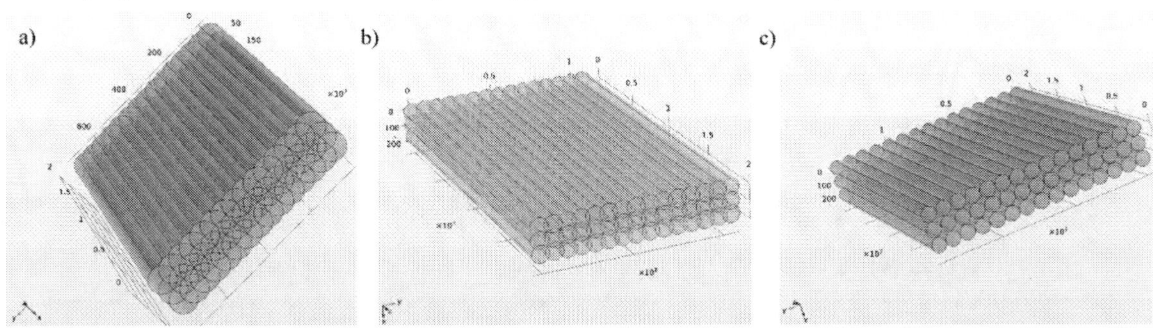

Figure 1. 3D-models with different lighting step; step: a) – 50 μm, b) – 75 μm, c) – 100 μm.

Hooke's Law of Linear Elasticity is written as:

$$\sigma_{ij} = \sum_{kl} C_{ijkl} \cdot \varepsilon_{kl} \tag{1}$$

The stress tensor σ and the strain tensor ε are second-order tensors, while the defining tensor C is a fourth-order tensor [14–17]. Due to the symmetry of the stress and strain tensors, Hooke's law can be represented in matrix form. Where the strain tensor ε is written as:

$$\varepsilon_x = \frac{\sigma_x}{E} - \frac{\mu}{E}\sigma_y - \frac{\mu}{E}\sigma_z, \tag{2}$$

$$\varepsilon_y = \frac{\sigma_y}{E} - \frac{\mu}{E}\sigma_x - \frac{\mu}{E}\sigma_z, \tag{3}$$

$$\varepsilon_z = \frac{\sigma_z}{E} - \frac{\mu}{E}\sigma_x - \frac{\mu}{E}\sigma_y, \tag{4}$$

where E is Young's modulus, μ is Poisson's ratio.

3. Analysis of the results

The results of 3D modeling are shown in Figures 2–4. The dependences of the Von Mises stresses, displacement, and pressure between the first and second laser beam paths along the surface of the third layer were considered.

As can be seen in Figures 2, the Von Mises stresses dependence varies from 1.25 to 5.49 GN/m^2. You can see that there is a significant difference in the Von Mises stresses between a step of 100 μm and 50–75 μm. If at a step of 100 μm, we observe Von Mises stresses ranging from 5.43 to 5.48 GN/m^2, then in a step of 50–75 μm we see stresses in the range from 1.27 to 1.55 GN/m^2.

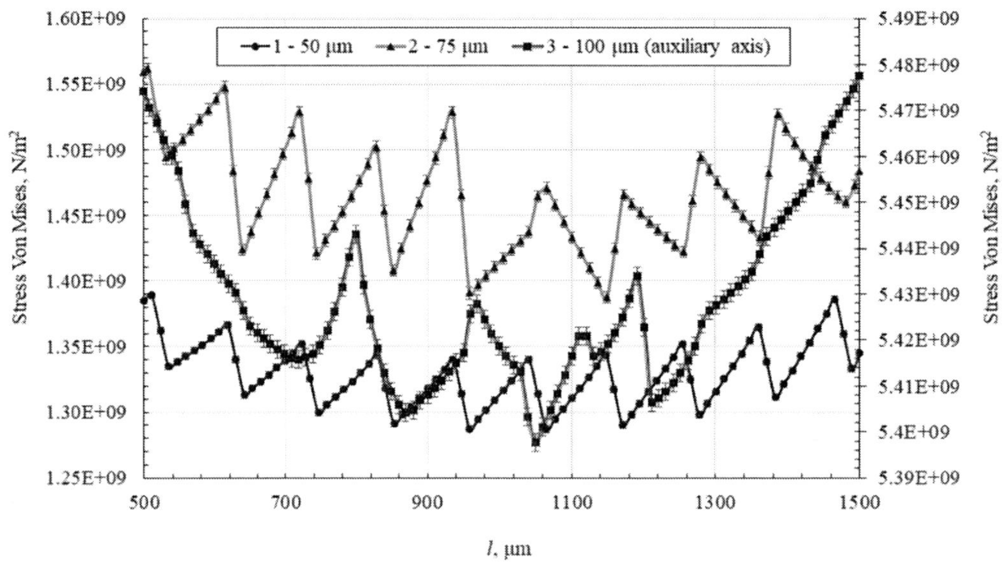

Figure 2. Von Mises stress versus sample length plot; step: 1 – 50 μm, 2 – 75 μm, 100 μm.

Consider the displacements that occur, which are shown in Figure 3, with a different lighting step. We see a similar situation, as with von Mises stresses. At a step of 100 μm, we observe a significant displacement. At a lighting step of 100 μm, the plane displacement varies in the range from 65.8 μm to 67 μm. The displacement at a lighting step of 75 μm varies from 20 μm to 25 μm, and at a step of 50 μm, on average, is 18 μm. We can draw a preliminary conclusion that the lighting step of 100 μm is not permissible, however, for an accurate conclusion, it is necessary to analyze the surface pressure arising between the layers.

The pressure between the first and second line of the laser beam along the surface of the third layer is shown in Figure 4. On the one hand, at a filling step of 100 μm, the surface pressure is not stable and varies over a wide range from -2.97 to -2.88 GPa.

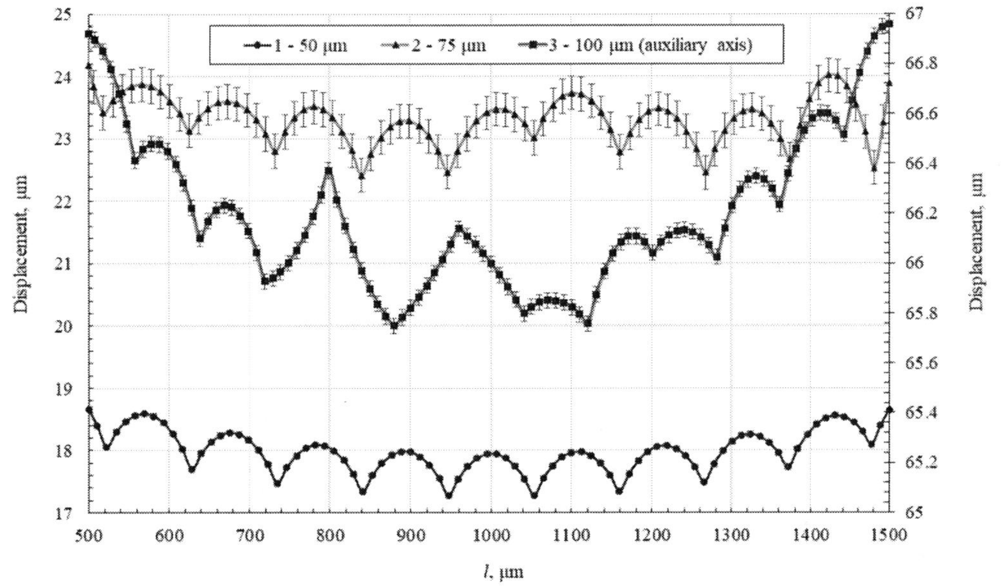

Figure 3. Dependence of displacement on sample length; step: 1 – 50 μm, 2 – 75 μm, 100 μm.

On the other hand, at a lighting step of 50-75 μm, we observe a different degree of pressure values. At a lighting step of 75 μm, the pressure values range from -0.9 to - 0.8 GPa, and at a step of 50 μm, the pressure is on average -0.7 GPa.

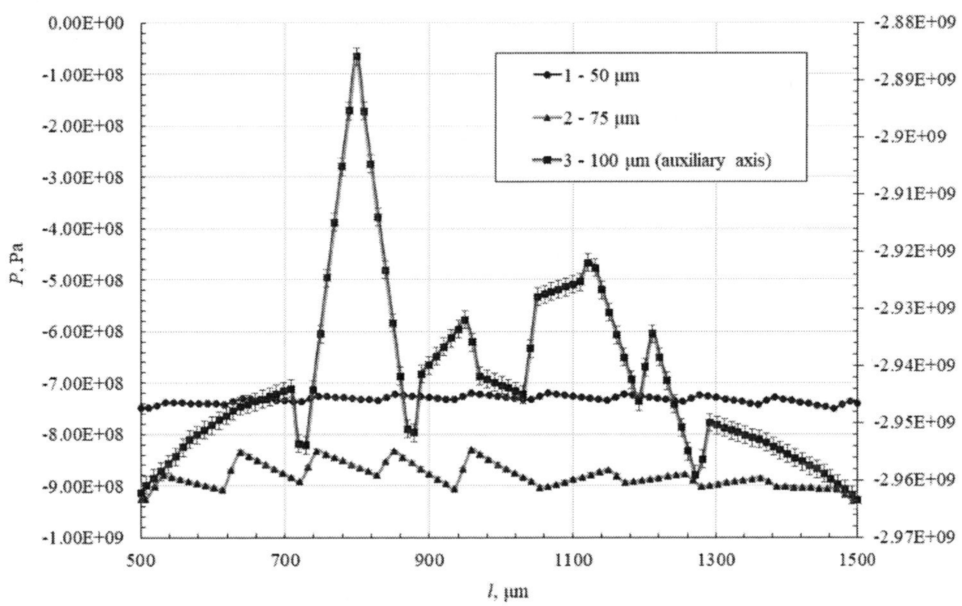

Figure 4. The dependence of pressure on the sample length; step: 1 – 50 μm, 2 – 75 μm, 100 μm.

4. Discussion

To assess the effect of the lighting step on the mechanical properties of the 3D printed product, the following dependences on the sample length were examined: von Mises stress (Figure 2), displacement

IOP Conf. Series: Materials Science and Engineering **969** (2020) 012099 doi:10.1088/1757-899X/969/1/012099

(interlayer) (Figure 3), and pressure (Figure 4). The total length of the sample in the study is 2000 µm, however, to exclude edge effects in the experiment, the range from 500 µm to 1500 µm was taken into account [18–22].

Based on the obtained dependencies, the following facts can be distinguished:

- When analyzing the stress plot of Von Mises, it can be seen that the sample with the lighting step of 50 µm has the lowest stresses, which is associated with better polymerization of the layer due to the superposition of the beam (a new beam passage affects the previous half the beam diameter).
- The smallest absolute value of the displacement, as well as the amplitude of its change, indicates the most optimal 3D printing results when using this lighting step (50 µm).
- At a lighting step of 50 µm, the pressure is higher than at 75 µm, but the pressure is almost the same over the entire length of the sample.
- The cardinal difference on the plot for the 100 µm lighting step is associated with the smallest number of intermolecular bonds arising between the beam paths, and a sharp pressure jump at several points on the plot indicates that by the middle of the sample a small number of these bonds affect the pressure increase.

5. Conclusion

Based on the results obtained during the simulation, it was concluded that the lighting step of 100 µm is not suitable for 3D printing of products, since the studied dependencies for this step are much higher and not stable depending on the area of the 3D printed object.

On the other hand, the dependences obtained for the lighting step of 50 µm are smallest in size (the smallest magnitude of the stresses and displacement, the minimum amplitude of the pressure change). Using a 75 µm lighting step can increase the speed of building an object. However, the difference in printing speed is insignificant (20-30%), while the stresses that occur in the object with a lighting step of 75 µm can provoke cracks, leashes and subsequent destruction of the printed object at the stage of heat treatment. Thus, for the material composition used (ceramic paste based on alumina powder and liquid photosensitive polymer) on an AF 200 Universal 3D printer (software used on a 3D printer - Triangulatica), a 50 µm lighting step is the most optimal and will improve the strength characteristics of the object at the stage of "rown stage" (finished product after heat treatment).

References

[1] Burduk A and Jagodziński M 2015 Assessment of Production System Stability with the Use of the FMEA Analysis and Simulation Models *Lecture Notes in Computer Science* **9375** 216–223 DOI: 10.1007/978-3-319-24834-9_26

[2] Kampa A, Gołda G and Paprocka I 2017 Discrete Event Simulation Method as a Tool for Improvement of Manufacturing Systems *Computers* **6(1)** 10 DOI: 10.3390/computers6010010

[3] Ćwikła G, Grabowik C, Kalinowski K, Paprocka I and Ociepka P 2017 The influence of printing parameters on selected mechanical properties of FDM/FFF 3D-printed parts *IOP Conf. Ser.: Mater. Sci. Eng.* **227** 012033 DOI: 10.1088/1757-899X/227/1/012033

[4] Chen Z, Li Z, Li J, Liu C, Liu C, Li Y, Wang P 2019 3D printing of ceramics: A review *Journal of the European Ceramic Society* **39(4)** 661–687 DOI: 10.1016/j.jeurceramsoc.2018.11.013

[5] Griffith M L and Halloran J W 1996 Freeform Fabrication of Ceramics via Stereolithography *Journal of the American Ceramic Society* **79(10)** 2601–2608

[6] Piterskov P, Egorov S, Boyko E and Grigoriev M 2020 Investigation of the influence of 3D printing modes with ceramics and sintering on the shrinkage process of thin-walled models *IOP Conf. Ser.: Mater. Sci. Eng.* **826** 012006 DOI: 10.1088/1757-899X/826/1/012006

[7] Moradi M, Moghadam K M, Shamsborhan M and Bodaghi M 2020 The Synergic Effects of FDM 3D Printing Parameters on Mechanical Behaviors of Bronze Poly Lactic Acid Composites *J. Compos. Sci.* **4** 17 DOI: 10.3390/jcs4010017

[8] Kumara N, Jain P K, Tandon P and Pandey P M 2018 The effect of process parameters on tensile behavior of 3D printed flexible parts of ethylene vinyl acetate (EVA) *Journal of Manufacturing Processes* **35** 317–326 DOI: 10.1016/j.jmapro.2018.08.013

[9] Godec D, Cano S, Holzer C and Gonzalez-Gutierrez J 2020 Optimization of the 3D Printing Parameters for Tensile Properties of Specimens Produced by Fused Filament Fabrication of 17-4PH Stainless Steel *Materials* **13** 774 DOI: 10.3390/ma13030774

[10] Tay W Y D, Li M Y and Tan M J 2019 Effect of printing parameters in 3D concrete printing: Printing region and support structures *Journal of Materials Processing Tech.* **271** 261–270

[11] Yerezhep D, Umutbaeva N R, Piterskov P, Baranov A Y and Yerezhep A 2019 The study of temperature fields of the WBC object in three-dimensional formulation using numerical methods *J. Phys.: Conf. Ser.* **1333** 032091 DOI: 10.1088/1742-6596/1333/3/032091

[12] Yerezhep D, Yemelyanov A, Kuznetsov Y, Perminov A, Pronin V A, Minikaev A F and Yerezhep A 2019 Computer analysis of the thermal state of the object after cryotherapy *J. Phys.: Conf. Ser.* **1333** 032092 DOI: 10.1088/1742-6596/1333/3/032092

[13] Freemantle M 1988 Chemistry in action *Journal of Chemical Education* **65(9)** 243 DOI: 10.1021/ed065pA243.1

[14] Itaya Y, Hanai H, Kobayashi N, Nakagawa T 2020 Drying-Induced Strain-Stress and Deformation of Thin Ceramic Plate *ChemEngineering* **4(9)** DOI: 10.3390/chemengineering4010009

[15] Kowalski S J, Rajewska K, Rybicki A, 2004 Mechanical Effects in Saturated Capillary-Porous Materials during Convective and Microwave Drying *Dry. Technol.* **22** 2291–2308

[16] Holmes D M, Tegeler F, Clegg W J 2008 Stresses and strains in colloidal films during lateral drying *J. Eur. Ceram. Soc.* **28** 1381–1387

[17] Lur'e S A and Solyaev Y O 2013 Simulation of the Mechanical Properties of Nanostructured Porous Ceramics *Russ. Metall.* **2013(4)** 272–281 DOI:10.1134/S0036029513040083

[18] Sagar B M, Swami M C, Permeshwar P 2015 Design and Analysis of Mono Composite Leaf Spring by Varying Thickness using FEA *IOSR J. Mechan. Civil Eng* **12** 99–103

[19] Ozmen B, Altiok B, Guzel A, Kocyigit I and Atamer S 2015 A novel methodology with testing and simulation for the durability of leaf springs based on measured load collectives *Procedia Engineering* **101(C)** 363–371 DOI: 10.1016/j.proeng.2015.02.044

[20] Kessentini A, Mohammed Sayeed Ahmed G, Madiouli J 2019 Design Optimization and FE Analysis of 3D Printed Carbon PEEK Based Mono Leaf Spring *Micromachines* **10(5)** 279

[21] Ji L B and Zhou T R 2010 Finite Element Simulation of Temperature field in Fused Deposition

[22] Modeling *Advanced Materials Research* **97(101)** 2585–2588 DOI: 10.4028/www.scientific.net/amr.97-101.2585

[23] Xiaa H, Lub J and Tryggvason G 2019 A numerical study of the effect of viscoelastic stresses in fused filament fabrication *Computer Methods in Applied Mechanics and Engineering* **346** 242–259 DOI: 10.1016/j.cma.2018.11.031

Acknowledgments
This work was financially supported by Additive Fabrication LLC.

A Method of Managing Technological Heredity in the Additive Growing of Objects from Cold-Resistant Materials on a Machine with Technology for Electric Arc Welding

D Shatagin, Y Kabaldin and P Kolchin

Department of technology and equipment of mechanical engineering, Nizhny Novgorod State Technical University n. a. R. E. Alekseev, 24, Minin St., Nizhny Novgorod, 603950, Russia

E-mail: uru.40@mail.ru

Abstract. The article considers the issue of managing print parameters to minimize the impact of technological heredity on the layers of the formed object. An analysis of the stability of the 3D printing process was carried out using acoustic emission data. The fractal dimensions of the acoustic emission attractor for a stable and unstable 3D printing process have been obtained. A method for controlling process stability by tracking 3D printing modes and assessing their stability using an artificial neural network is proposed. The mechanical properties of printed samples were evaluated under normal conditions and at low temperatures in the Arctic and the far north.

1. Introduction

Today, the leadership of our country is particularly interested in the development of the northern regions. The tasks set in the framework of federal programs for the development of the Arctic and the Far North of the Russian Federation dictate urgent tasks to the scientific community. In particular, the development of technologies capable of producing products capable of withstanding the low temperatures that prevail in these regions.

A direction is actively developing for the study of cold-resistant materials, which can provide the specified strength characteristics of structures and withstand low temperatures [1, 2]. At the same time, the direction of technologies for the production of structures from these materials is becoming relevant, one of the promising areas here is additive growing, in other words, 3D printing of objects using digital models.

At present, the most widely used for 3D printing with metals are alloyed metal powders (SLM), laser powder surfacing (DMD) and electric arc welding (WAAM), the latter two being considered the most flexible ones that can be integrated into CNC machines [3–9]. At the same time, the processes are still quite uncontrollable and depend on many factors, including technological heredity, which the next layer perceives from the previous one. In this case, technological heredity is understood as how each subsequent forming layer, a single roller (or track) takes on macro and micro defects from the previous layer already laid.

2. Aim and objectives

The aim of the work is the development of technology for the additive growing of products from cold-resistant materials on CNC machines. As part of the goal, tasks were formulated, which include: researching the manifestation of technological heredity and its influence on the quality of the formed object, the reflection of defects in printing modes, the search for ways to control heredity in 3D printing and assessing the mechanical properties of formed objects at low temperatures.

3. Equipment

To achieve the goals and objectives, an experimental bench was assembled, shown in Figure 1 [6], which is a hybrid complex consisting of a portal-type vertical milling machine upgraded in such a way that, in addition to the function of mechanical processing, it could carry out additive growing of workpieces by arc welding technology. This function was made possible thanks to the installation of a print head, which is installed in the machine spindle and the MIG 200 semiautomatic welding machine, whose control system was integrated into the machine control system. In order to fix the print modes and acoustic emission signals, the following were installed on the stand: SC145-600 sensor for measuring the arc current, SV025 sensor for measuring the arc voltage, GT200 acoustic emission sensor, which were connected to the PC via NI 6356 analog-to-digital converters and NI 9205 and NI cDAQ-9188 data acquisition bus. A pyrometer was also used to control temperature and prevent excessive metal overheating in the print zone.

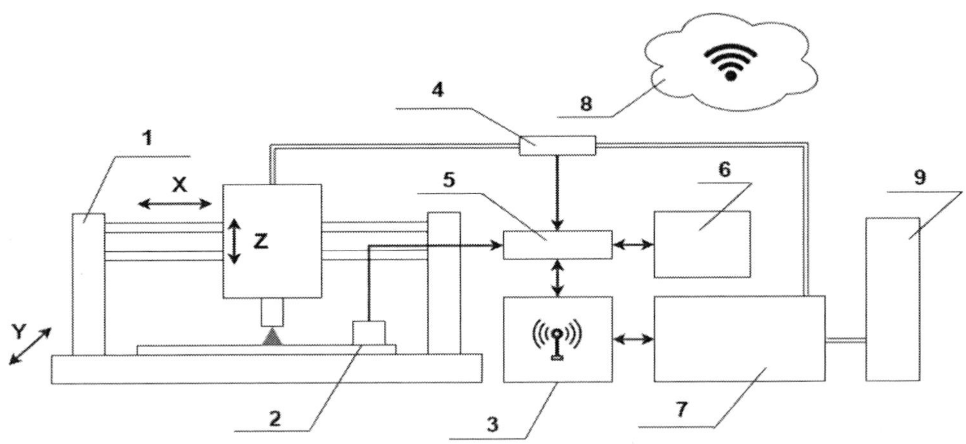

Figure 1. An experimental bench for 3D surfacing on the basis of a CNC machine (1–3 coordinate machine with portal type CNC), 2 – a broadband acoustic emission sensor GT350, 3 – a PC with a wireless access system to the Microsoft Azure cloud service, 4 – power sensors current and voltage, 5 – ADC / DAC, 6 – CNC system, 7 – welding semiautomatic device, 8 – Microsoft Azure cloud service, 9 – shielding gas cylinders).

4. Experimental studies

In the process of experimental work, 09G2S steel was printed, which is widely used in the northern regions of the country. From the selected material, 3D printing of walls with a thickness of one roller was carried out, the length of the walls was 250–300 mm, height 150–180 mm, depending on the samples obtained in the future, as well as 3D printing of the "body" part was performed, the workpiece and the final part are presented in the figure 2. Printing of each of the samples was carried out on a different combination of modes, for this, the values of the welding current strength, voltage and printing speed were changed.

IOP Conf. Series: Materials Science and Engineering **969** (2020) 012100 doi:10.1088/1757-899X/969/1/012100

(a)

(b)

Figure 2. 3D printing results: a) blanks and b) final part after machining.

To fix technological heredity during the formation of one of the samples, a situation was simulated in which a single roller was printed on a previously formed part of the wall with a temperature of more than 500 ° C. Due to excess heat in the surfacing zone, the stability of the system was violated, which led to excessive melting of the previous layer and leakage of metal from the weld pool, resulting in a violation of the geometry of the layer with the formation of a depression, as shown in Figure 3. Following the layer the next roller was laid under similar conditions, which led to an increase in the defect in geometry and the growth of the depression along the length and height. After this, the conditions were changed and the next roller was printed at a lower wall temperature, of the order of 300 ° C, while the spreading of the weld pool in the place of the formed underestimation ceased, and when printing subsequent layers, the technological defect was leveled into a flat layer. Thus, due to timely local changes in technological parameters during additive growth by electric arc welding, minimization or complete elimination of the influence of technological heredity on the geometry of the formed object is possible. It should also be separately noted that the obtained sample showed mechanical properties identical to those obtained at stable conditions, which indicates that technological heredity can be minimized at the same level of microstructure and mechanical properties.

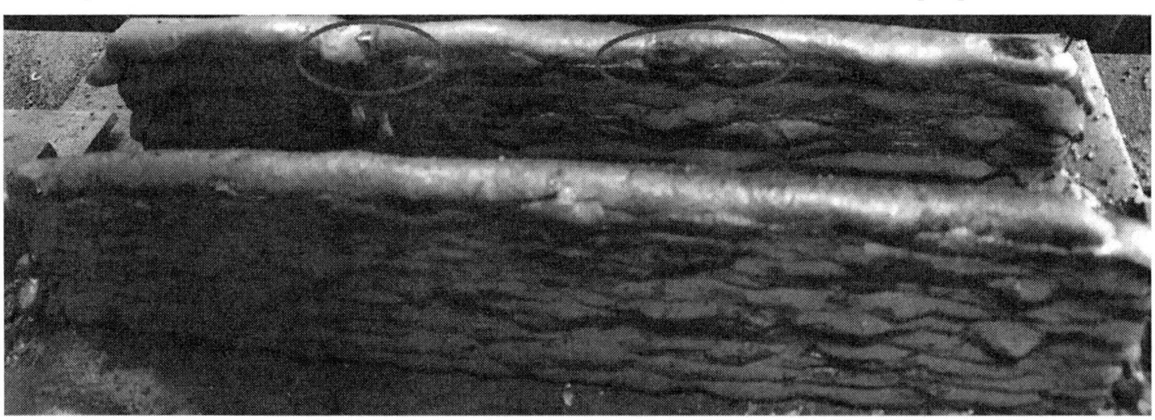

Figure 3. Sample with a defect in geometry (trough).

The next step was an analysis of the printing modes during the formation of the defective roller. So it was revealed that in the place where the metal leakage occurred, the welding current decreased, the graph of the printing modes depending on whether a defect forms or a stable process is shown in Figure 4. Such a change in the welding current was observed on all layers where the defect occurs of this kind, while there was a tendency to decrease it when smoothing the defect on the wall profile. Then, the acoustic emission signal was analyzed in the area with stable mode and in the area with the defect. To do this, the signal was pre-processed using digital bandpass and Wavelet filters. After that, acoustic emission attractors, shown in Figure 5, were reconstructed and their fractal dimensions were calculated. From previous studies it is known that the smaller the fractal dimension, the more stable the process is [9], in this case the results also confirm this, for the stable mode the attractor dimension was Df = 3.1, and for the unstable mode the fractal dimension increased to Df = 3.8.

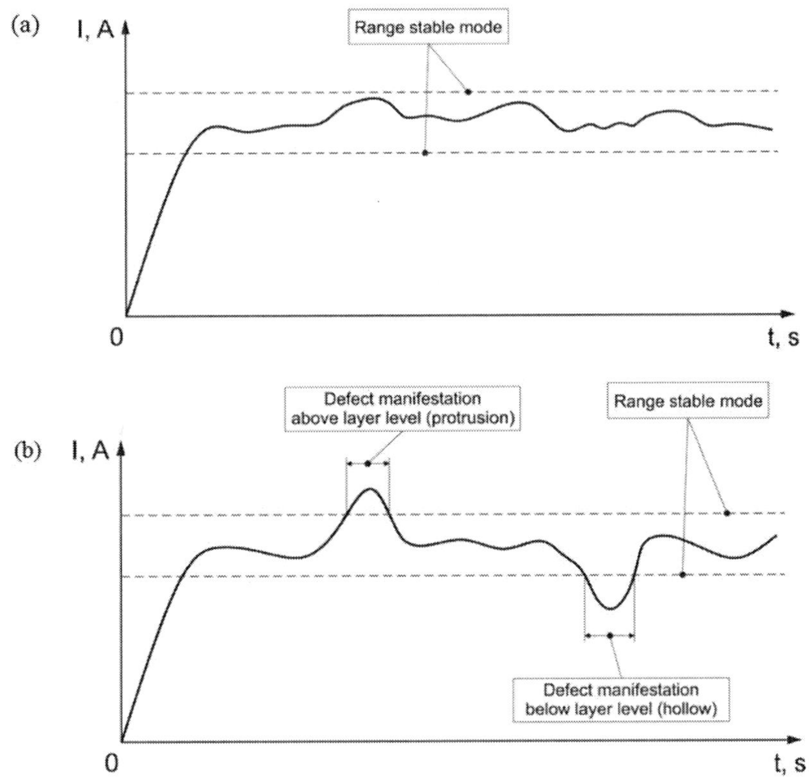

Figure 4. Graphs of the regime of electric arc growing: a) without a defect, b) with a defect.

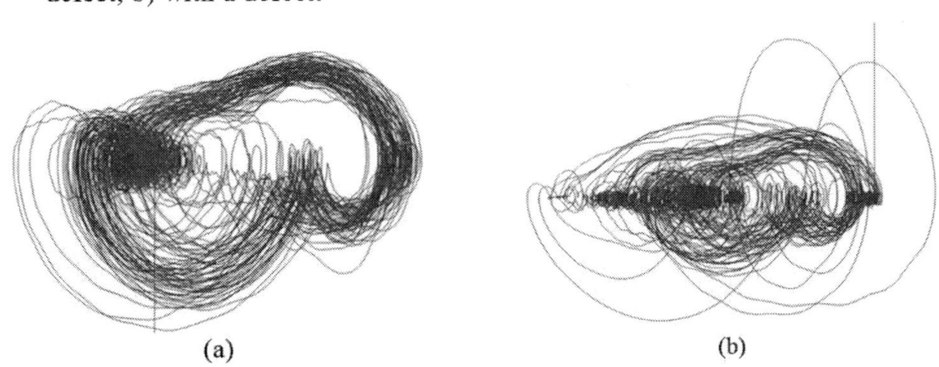

Figure 5. 3D printing system attractors: a) attractor of a stable arc b) attractor in case of loss of stability.

The experiment demonstrates the need to develop adaptive systems for controlling the process of 3D printing of objects, which would allow, depending on changing conditions, to automatically adjust the modes, maintain stability and eliminate technological heredity.

During the additive cultivation of the object, a large amount of statistical data is generated that can be used to diagnose 3D printing processes with cold-resistant materials. For the purpose of their application, the apparatus of artificial neural networks [6, 7] was taken and the structure of ANN is formed. Figure 6. The created ANN is aimed at classifying the state of the additive growth process to further maintain process stability. The data source for the ANN was data from current, voltage, and acoustic emission sensors whose waveforms are shown in Figure 7. A data vector characterizing a time series representing pulses with a time shift is fed to the input of the developed neural network. The described approach made it possible to assess the stability of the system taking into account the history of dynamic events, as well as to identify scenarios of loss of stability and to classify defects. At the output of the neural network, there are 2 neurons that show the class of the defect during its manifestation and the degree of process stability. In this case, three states are assigned to the stability of the system: stable, unstable, and a transition state from stable to unstable. Neural network training took place on the Microsoft Azure cloud service. In the formation of training samples of neural network models, [10–16] methods of data mining were used.

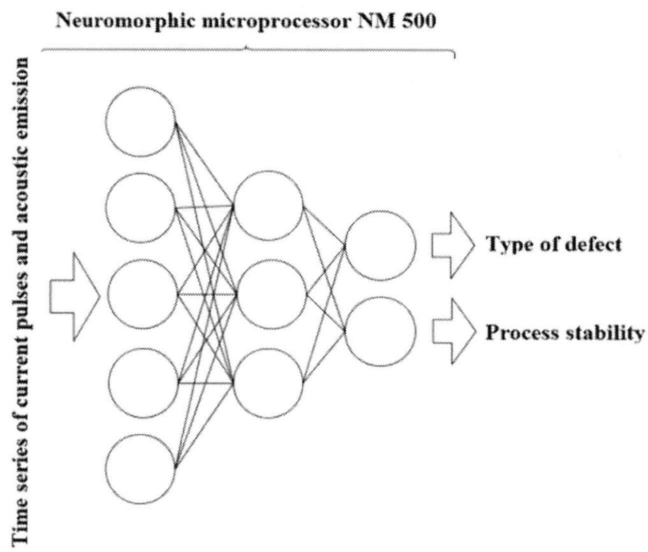

Figure 6. Artificial neural network for the classification of the dynamic state of 3D printing processes.

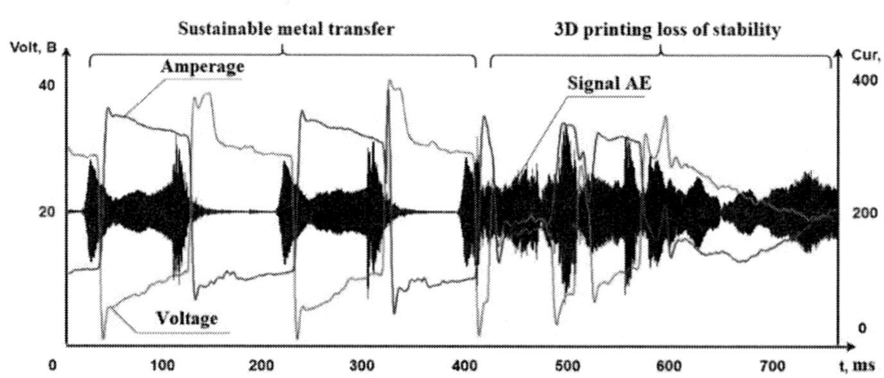

Figure 7. Oscillograms of current, voltage and acoustic emission signal during 3D printing by electric arc welding on a CNC machine.

This approach was chosen to search for patterns and trends in large volumes of data that continuously flow throughout the 3D printing process. As a rule, such regularities are difficult, and sometimes impossible to detect in the usual statistical analysis, since the relationship between the data can be quite complicated. In particular, the following Data Mining technologies were applied: neural network clustering and classification of data (self-organizing maps of Kohonnen and k-nearest neighbors), exploratory data analysis, to study the statistical properties of available samples (distribution of variables, outliers, the need for transformation, etc.) and identifying the nature of the relationship between response and predictors, estimating model parameters and diagnosing them using numerical resampling methods (cross-validation (CV), bootstrap), analysis of the contribution of individual predictors and optimal selection combinations of them, ranking of several alternative models and fine-tuning their most important parameters. After training, the neural network for classifying status and defects (Figure 6) was downloaded from a cloud service to a PC. Important here is the response time of the neural network in order to maintain the stability of the system, since the processes that occur during additive growth proceed quickly and continuously change. Especially for this, the neural network was hardware implemented, based on the Neuro Mem NM 500 neuro processor, made using 110 nm technology, and combining 576 neurons. There is no central processor in this neuroprocessor, so an additional control processor will be needed to supply the neurons with process data. Communication is carried out on a parallel bi-directional 26-bit bus. NM500 is a chain of neurons that are connected to a common parallel bus and are connected to each other. The developed nero network on the NM500 is capable of working in two main modes: training and recognition.

The next step was the integration of the neural processor NM 500 into the data acquisition and processing system. [17, 18, 19]. For this, the Neuro Shield debug board was used. This board made it possible to ensure the interaction of the neuroprocessor with the integrated FPGA via a 28-pin data bus. In addition, the selected board has protocols for exchanging data according to I2C and SPI standards, which allows it to interact with external devices (IIoT) [20]. As part of the ongoing work, the Raspberry Pi platform was used as an IIoT device. Using this approach to the diagnosis of the dynamic state of 3D printing processes, it was possible to implement high-precision classification of multidimensional time series of a sensor system in real time.

The next stage was the testing of the developed system. Which was carried out on the basis of the choice of optimal modes and control of the surfacing process in the manufacture of experimental samples. To select the optimal modes, the obtained ANN was used (Figure 6). At the same time, this ANN worked in deconvolution mode using the back propagation method of error. The essence of the method is to select the necessary state of the output neurons, fix the weight coefficients of the intermediate layers and automatically adjust the weights of the input layer. As part of the study, the output values of the neurons were selected as "stable" and "0", which corresponds to the absence of defects. Then the initialization of the neurons of the input layer occurred, until the learning error became minimal. At the end of the balance adjustment, time series of current, voltage, and acoustic emission were obtained based on the given stability conditions. After, the obtained time series were passed through an existing database on a cloud service using the correlation function. By the maximum of the correlation function, the closest in composition time series in the database and the corresponding modes were determined. Thus, optimal surfacing modes were selected that correspond to the most stable process.

Then experimental blanks were obtained during additive growth of which the modes obtained by ANN were applied and allowing to maintain a stable process. After this, the blanks without being removed from the table were machined on the machine on which 3D printing was performed.

5. Analysis of experimental studies

The next step was the evaluation of the microstructure of the samples and their comparison with the blanks obtained from sheet metal. The analyzed structures are presented in Figure 8. Based on the obtained images, one can observe the differences between the 09G2S structures obtained by different methods. A distinctive feature of the structure obtained by electric arc growing is its fine grain size, this is due to the fact that the metal of the previous layers undergoes cyclic heat treatment when applying the next layer. At the same time, we note that a hundred fine-grained structure is more preferable from the point of view of cold resistance. It is also seen from the pictures that the material has discontinuities at the grain boundaries, which can lead to a change in the mechanical properties of the material not for the better.

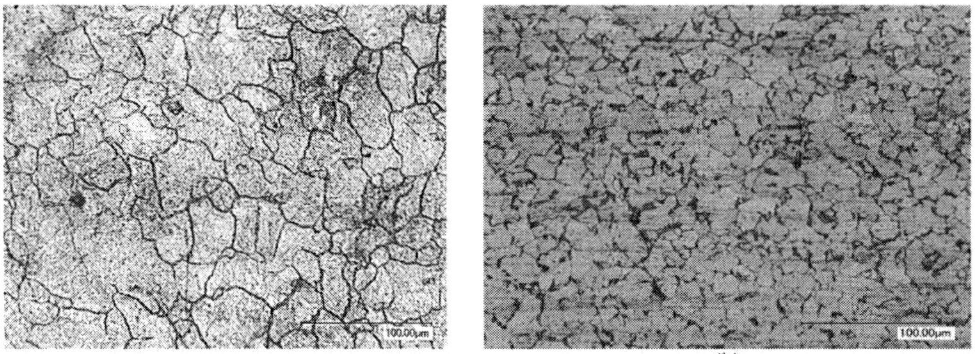

(a) (b)

Figure 8. Microstructures of steel 09G2S: a) obtained by rolling and past annealing, b) obtained by electric arc welding

Next, an assessment was made of the mechanical properties of printed blanks from 09G2S, followed by cutting to standard samples according to GOST 1497-84. In particular, a tensile test under normal conditions. The test results are presented in table 1.

Table 1. Testing the quality of recognition of a viscous component using ANN.

Mechanical indicators, MPa	δ,%	$\sigma_{0,2}$	$\sigma_{в}$
Samples cut along the direction of surfacing	27	361	506
Samples cut across the weld direction	22	352	502
Samples obtained by machining from sheet metal	23	325	490

The bottom line in the table shows the values for samples obtained from sheet metal. In this case, comparing the results, we can see that the mechanical properties according to all three criteria are higher for the images obtained by the method of electric arc growing. It can also be seen that the samples obtained by electric arc growing have anisotropy of properties depending on the direction of tension, so the relative elongation of the samples obtained along the deposition direction is 18% higher than that of the samples obtained across the layers, and differences in yield strength are also observed.

Next, shock bending tests were carried out in the temperature range from –80° C to + 20° C, the results are shown in Figure 9. The tests were also carried out in comparison with sheet metal samples. From the results obtained, it can be seen that the nature of the change in toughness values is the same for samples obtained from rolled products and 3D printing. In this case, the impact strength values themselves for samples from rolled metal sheets after annealing are 15 ... 20% higher than for samples obtained by electric arc growing in the entire temperature range.

6. Conclusion

Thus, we can conclude that the process of additive growth by electric arc welding can be applied to the production of workpieces, including for parts operating at low temperatures. At the same time, the 3D printing process requires constant monitoring and maintaining stability in order to eliminate technological heredity, for which systems incorporating artificial neural networks can be applied.

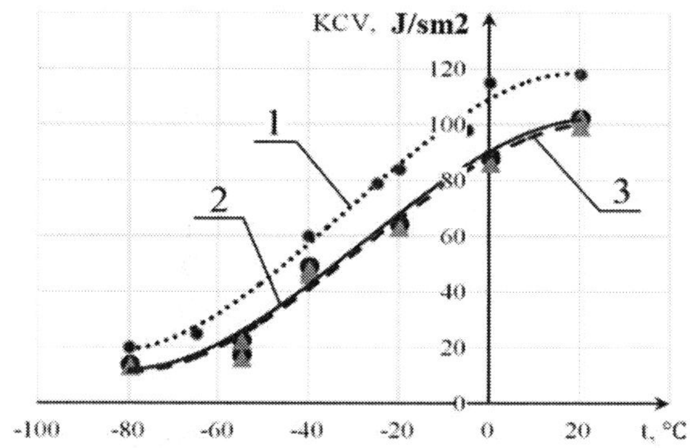

Figure 9. The results of tests for impact bending of steel samples 09G2S: 1 - from sheet metal with subsequent annealing; 2 - cut along the direction of cladding; 3 - cut across the direction of cladding.

References

[1] Solntsev Yu P 2005 *Cold-resistant steels and alloys: Textbook for high schools* (St. Petersburg: Chemistry) p 480

[2] Solntsev Yu P 2008 *Materials for low and cryogenic temperatures: Encyclopedic reference book* (St. Petersburg: Himizdat) p 768

[3] Elistratova A A and Korshakevich I S 2015 3D printing technologies: advantages and disadvantages *Actual problems of aviation and astronautics* **1** 557–559

[4] Kabaldin Y G, Kolchin P V, Shatagin D A, Anosov M S and Chursin A A 2019 Digital twin for 3D printing on CNC machines *Russian Engineering Research* **39** **10** 848–851

[5] Kabaldin Yu G and Kolchin P V 2018 Development of hybrid technology for 3D printing on CNC machines with an intelligent system for optimizing the process of surfacing a workpiece for machining *Science Week SPbPU materials of a scientific conference with international participation* 344–347

[6] Flynn Joseph M, Shokrani Chaharsooghi Alborz, Newman Stephen T, and Dhokia Vimal 2016 Hybrid additive and subtractive machine tools - research and industrial developments *International Journal of Manufacturing Technology and Management* **101** 79–101 Online. Available: https://doi.org/10.1016/j.ijmachtools.2015.11.00.007

[7] Li J, Alkahari M, Rosli N, Hasan R, Sudin M and Ramli F 2019 Review of Wire Arc Additive Manufacturing for 3D Metal Printing *Int. J. Automation Technol.* **13** **3** 346–353

[8] Kabaldin Yu G, Kolchin P V, Shatagin D A and Kiselev A V 2018 *The method of 3D printing on CNC equipment with intelligent optimization of modes* RU Patent **2018125837**

[9] Kabaldin Yu G, Shatagin D A, Anosov M S, Kolchin P V and Kuzmishina A M 2018 *Artificial intelligence and cyber-physical machining systems in digital production* (Nizhny Novgorod: Nizhny Novgorod State Technical University) p 271

[10] Kabaldin Yu G, Bilenko S V and Sery S V 2003 *Management of dynamic processes in technological systems of machining based on artificial intelligence* (Komsomolsk-on-Amur: Komsomolsk-on-Amur State Technical University) p 201

[11] Shitikov V K and Mastitsky S 2017 *Classification, regression and other Data Mining algorithms using R* p 351

[12] White T 2013 *U13 Hadoop: A Comprehensive Guide* (St. Petersburg: Peter) p 672

[13] Online. Available: https://en.wikipedia.org/wiki/Decision_tree_learning

[14] Online. Available: https://en.wikipedia.org/wiki/Cross-validation_(statistics)

[15] Online. Available: https://en.wikipedia.org/wiki/Bootstrap_aggregating

[16] Online. Available: https://en.wikipedia.org/wiki/Boosting_(machine_learning)

[17] Online. Available: https://en.wikipedia.org/wiki/AI_accelerator

[18] https://en.wikipedia.org/wiki/Machine_learning (accessed 12.15.18)

[19] https://en.wikipedia.org/wiki/Neuromorphic_engineering (accessed 12.15.18)

[20] Online. Available: https://en.wikipedia.org/wiki/Industrial_internet_of_things

Acknowledgments
This work has been supported by the grants from the President of the Russian Federation to young Russian scientists-candidates of sciences No. 075-15-2020-098.

Changes in the Structure and Properties of Nickel-Zinc Spinel Nanoferrites Series for 3D-Printing

I I Shamray[1], V Yu Buz'ko[1,2] and A I Goryachko[1]

[1]Laboratory of Electromagnetic Nanomaterials, Kuban State University, Krasnodar, 350040, Russia

[2]Center of Perspective Technologies and Nanomaterials, Kuban State Technological University, Krasnodar, 350040, Russia

E-mail: Buzkonmr@mail.ru

Abstract. The family of nickel-zinc nanoferrites with general formula NixZn1–xFe2O4 for 3D printing were synthesized via pyrochemical urea-nitrate method and annealed at temperatures of up to 770 K. The ferrite powder microstructure was studied by SEM and XRD methods. Average nanocrystallite sizes of 21 to 32 nm and average nanoparticles sizes of 22 to 50 nm were determined. The DC electrical conductivity of NixZn1–xFe2O4 clearly shows the change in conduction behavior of samples with increase in nickel content. The observed low electrical conductivity with values at 0.1–0.3 and 0.8-0.9 parts of nickel ion for NixZn1–xFe2O4 is due to change in cation distribution in spinel structure. The microwave absorbing properties of ferrites at 9.15 GHz were analyzed by FMR and maximum microwave absorption is that of compositions NixZn1–xFe2O4 with x = 0.15–0.30. Additive manufacturing FDM 3D-printing process with PLA-nanoferrite filament for fabricating magnetic components also was demonstrated.

1. Introduction

The nickel-zinc ferrites with general formula $Ni_xZn_{1-x}Fe_2O_4$ are magnetic oxide semiconductors and valuable ceramic magnetic material. The nanosized nickel-zinc ferrites attract attention due to the potential of their many applications as magnetic nanomaterials, microwave absorbing materials, for protection against high frequency electromagnetic radiation and high-sensitive nanomaterials for different sensors.

Nowadays, magnetic cores and electrical conductor windings that make up magnetic components are fabricated separately in multiple, complex steps. The nanopowders of $Ni_xZn_{1-x}Fe_2O_4$ are perspective ceramic fillers for advanced composite magnetic materials for 3D printing of machine components for mechanical engineering, machinery construction, electromechanical instruments, electromagnetic control system.

Numerous works [1–9] have presented data on the effect of substitution of nickel ions by zinc ions revealed on the electronic and electromagnetic properties of nickel-zinc nanosized ferrites. These studies show that the direct band gaps are decreased from $ZnFe_2O_4$ to $NiFe_2O_4$ [1]. Also upon an increase in the nickel content in $Ni_xZn_{1-x}Fe_2O_4$ ferrite its magnetic properties grow and reach their maxima at nickel molar fractions of 0.5 to 0.7 [10, 11]. A further increase in nickel content reduces the magnetic properties of the nickel-zinc ferrite. Compositions of $Ni_xZn_{1-x}Fe_2O_4$ with moderate nickel content are characterized by weak magnetic properties [3], and zinc nanoferrite at room temperature is weakly ferromagnetic [12] or paramagnetic [10, 12, 13, 14]. The FMR resonance field decrease at the increase nickel content in $Ni_xZn_{1-x}Fe_2O_4$ ferrite [10]. These properties can be attributed to the existence of a crystalline lattice of normal spinel in zinc nanoferrite. This type of spinel is characterized by weak interaction between its tetrahedral and octahedral sublattices and therefore has no ferrimagnetic properties. In real zinc nanoferrite samples the both types normal and inverse spinel are coexist.

The goal of this work was to investigate the effect composition of nanosized nickel-zinc ferrites with general formula $Ni_xZn_{1-x}Fe_2O_4$ on the electronic and electromagnetic properties.

2. Experiment

A series of nickel-zinc spinel-type ferrites described by general formula of $Ni_xZn_{1-x}Fe_2O_4$ (where x=0–1, with increments of 0.05) was synthesized via low temperature pyrochemical urea-nitrate processing. All reagents were weighted in stoichiometrical ratios, ground to a homogeneous paste in a porcelain mortar, and heated in porcelain crucibles from an initial temperature of 370 K with a gradual increase at a rate of about 15 K/min. After reaching 570–580 K, the reaction paste self-ignited and nitrate salts decomposed to nanosized ferrite powders. The produced raw nanoferrites powders were calcined for 30 min at 770 K for the final decomposition of nitrate salts residue and urea, and then cooled in air to ambient temperature and ground in a mortar to homogeneous powders comprised of microgranules with sizes of 2 to 5 μm. An advantage of used urea-nitrate method is the synthesis of nanosized ferrites with good magnetic properties at relatively low reaction temperatures, compared to the solid phase ceramic route sintering of powdered oxides or metal carbonates.

The crystalline microstructure of prepared $Ni_xZn_{1-x}Fe_2O_4$ powders was studied via scanning electron microscopy (SEM) using a JEOL JSM 7500F microscope. The DC resistivity of nanoferrite powder samples was measured at room temperature using the two-probe method with UT-601 tester in plastic polyethylene tubes with internal diameter 2 mm and compacted ferrite powder length 5 cm under pressure contact at 1 ton/cm^2. The DC resistivity values were averaged from 5 measurements.

The samples of $Ni_xZn_{1-x}Fe_2O_4$ specimens (where x = 0, 0.15, 0.25, 0.5, 0.75, 0.85, 1.0) were studied via X-ray phase analysis using a Shimadzu XRD-7000 X-ray diffractometer by using CuK α radiation source with a wavelength λ=1.5406Å at 2θ angle range from 10 to 70 with a step size of 0.02° and a count time of 1 s per step. The diffraction patterns do suggest for the all samples the presence of a unique spinel cubic phase. All the detectable peaks can be indexed to the phase of cubic spinel structure that are found as in the standard reference data (JCPDS card PDF#10-0325 for $NiFe_2O_4$ and PDF#89-1012 or #22-1012 for $ZnFe_2O_4$). The average nanocrystallite size in $Ni_xZn_{1-x}Fe_2O_4$ samples were calculated using the Scherrer equation.

The UV–vis & NIR diffuse reflectance spectra of nanosized powders $Ni_xZn_{1-x}Fe_2O_4$ were recorded using Hitachi U-3900 spectrophotometer with 2-chanel integrating sphere at room temperature. The optical band gap can be calculated by following relation [15]:

$$(\alpha h\nu)^k = A(h\nu - E_g) \quad (1)$$

where α is optical absorption coefficient, A is constant and exponent k depends on the nature of transition (for indirect band gap k = 1/2 and for k = 2 direct band gap). The optical absorption coefficient was calculated by following relation [16]:

$$\alpha = (R_{max} - R_{min}) / (R - R_{min}) \quad (2)$$

where R is optical diffusion reflectance.

Straight lines are obtained for both k = 1/2 and 2 by plotting the above equation, indicating the indirect and direct nature of band gap. The value of band gap is obtained by extrapolating the linear fitted curves of $(ahv)^2$ and $(ahv)^{1/2}$ to energy axis at α = 0.

A magnetic and microwave absorption properties of the full series of powdered $Ni_xZn_{1-x}Fe_2O_4$ nanoferrites was studied via FMR method using a JEOL JES FA-300 EPR spectrometer in the X-range at 9.15 GHz.

3. Results and discussion

According to our results all studied ferrite samples have the crystalline lattice of cubic spinel. The calculated lattice parameter systematically varied 8.351 to 8.458 Å for $Ni_xZn_{1-x}Fe_2O_4$ specimens with zinc ion fraction from 0 to 1. The Zn^{2+} ions have larger ionic radius (0.82 Å) than Fe^{3+} (0.65 Å) and Ni^{2+} (0.74 Å). For spinel ferrites the Zn^{2+} ions successively replace Fe^{3+} ions on tetrahedral A-site. This results in an increase of lattice parameter with the content of zinc in $Ni_xZn_{1-x}Fe_2O_4$. Average crystallite sizes as a function of nickel content were $D(x = 0) = 28.8$ nm, $D(x = 0.15) = 30.2$ nm, $D(x = 0.25) = 29.4$ nm, $D(x = 0.5) = 31.8$ nm, $D(x = 0.75) = 31.4$ nm, $D(x = 0.85) = 22.6$ nm, $D(x = 1.0) = 21.0$ nm.

Micrographs of several prepared $Ni_xZn_{1-x}Fe_2O_4$ nanoferrites are presented in figure 1.

Figure 1. Typical SEM micrographs of $Ni_xZn_{1-x}Fe_2O_4$ samples.

Based on an analysis of the microimages, the estimated average ferrite nanoparticle sizes were found to range from 22 to 50 nm, depending on the ferrite composition. This result clearly reflects association of nanocrystallites in nanoparticles of prepared nickel-zinc ferrites. The minimum size of the nanoparticles was those of the $Ni_{0.75}Zn_{0.25}Fe_2O_4$ specimen and the maximum size of the nanoparticles was observed for $Ni_{0.25}Zn_{0.75}Fe_2O_4$ sample.

The UV-Vis diffuse reflectance spectra of all studied $Ni_xZn_{1-x}Fe_2O_4$ samples clearly show that the $Ni_xZn_{1-x}Fe_2O_4$ nanoparticles can absorb energy of photons in the wavelength range of 240 nm to 450 nm. Optical band gaps was calculated from the Tauc plot confirms the semiconducting nature of the studied nickel-zinc ferrites samples. From the figure 2 it can be seen that the calculated band gap values of the $Ni_xZn_{1-x}Fe_2O_4$ samples are in the range of 1.6 eV to 2.0 eV which is comparable to the reported values [10-20]. For example the direct and indirect band gaps for our $NiFe_2O_4$ sample are obtained to be 1.69 eV and 1.49 eV respectively and are comparable with experimental investigations direct gap E_g from 1.55 to 2.85 eV [14–22] and theoretical values of direct and indirect band gaps 2.4 and 1.6 eV [23]. The direct and indirect band gaps for our $ZnFe_2O_4$ sample are obtained to be 1.99 eV and 1.77 eV respectively and are comparable with experimental investigations direct gap E_g from 1.66 to 2.43 eV [6, 24–27] and theoretical value 1.93 eV [28].

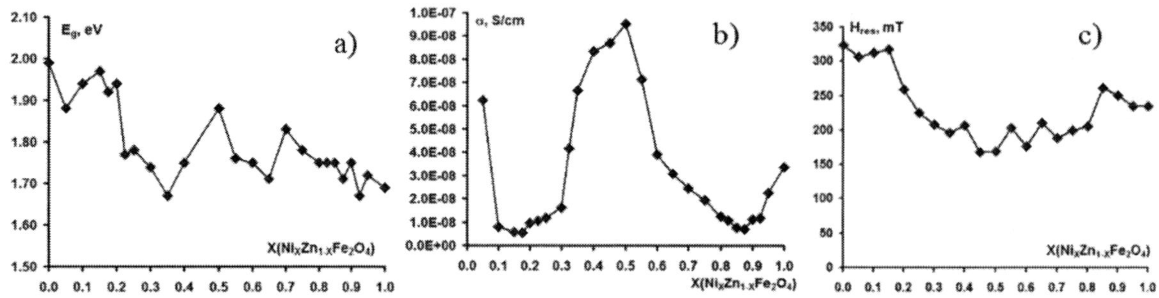

Figure 2. Calculated band gap, DC conductivity and FMR resonance field of the $Ni_xZn_{1-x}Fe_2O_4$ samples.

DC resistivity studies are one of the useful characterization techniques to understand the conductivity mechanism in oxide semiconductors. The nature of electrical conductivity with increase in nickel content in $Ni_xZn_{1-x}Fe_2O_4$ clearly shows the change in conduction behavior of the sample with different composition. The anomalous behavior of electrical conductivity with values at 0.1–0.3 and 0.8–0.9 parts of nickel ion for $Ni_xZn_{1-x}Fe_2O_4$ (figure 2b) is due to change in cation distribution in cubic spinel structure.

The FMR resonance field of $Ni_xZn_{1-x}Fe_2O_4$ ferrite versus composition (Figure 2c) decrease with the nickel content grows from 0 to 0.5, then increase at nickel fraction >0.5 up to 0.85 and also decreased at nickel fraction >0.85 in $Ni_xZn_{1-x}Fe_2O_4$. The results from FMR spectroscopy demonstrate the strong dependence of magnetic and microwave absorption properties on the composition of the considered nickel-zinc ferrites. Upon an increase in the nickel content in ferrites, the values of the magnetic resonance field diminishes, indicating a resonance shift to lower frequencies. The absorption line width also varies at different fractions of nickel in $Ni_xZn_{1-x}Fe_2O_4$. The magnetic field absorption line width sharply increased e with the nickel content grows from 0 to 0.2, then approximately constant in 0.2-0.5 nickel part range, also increase at nickel fraction >0.5 up to 0.6 and also decreased at nickel fraction >0.6 in $Ni_xZn_{1-x}Fe_2O_4$. Thus, the maximum absorption line width of synthesized nickel-zinc nanoferrites range is belong to compositions with molar fractions of nickel $x = 0.20 - 0.65$. So, we can see that the composition of nanosized nickel-zinc ferrite greatly affects its magnetic and microwave absorption properties. Compositions of $Ni_xZn_{1-x}Fe_2O_4$ ferrite with approximately equal contents of nickel and zinc ions are characterized by the lowest FMR resonance field (Figure 2c). Also was found that the maximum microwave absorption properties of synthesized nickel-zinc nanoferrites that calculated via integral of FMR absorption line normalize to sample mass is achieved at unexpectedly minor content of nickel ions in the ferrite at the $x = 0.15$–0.30 range. The obtained in this study microwave electromagnetic properties of nickel-zinc ferrites are in good agreement with the previous FMR studies of nickel-zinc nanoferrites [29].

Additive manufacturing FDM 3D-printing process with for fabricating magnetic components also was studied in this work. The ferrite fraction in plastic for 3D printing is desired to be as high as possible to provide the highest magnetic permeability for across a printed component. However, the ferrite fraction in plastic must be low enough so as to enable the material to exit the heated extrusion nozzle of the 3D printer. Polylactide (PLA) polymer filament (FDplast, Russia) which is a readily available modeling plastic was chosen as the matrix polymer in our investigation due to its well-known ease of printing, low cost and wide availability. We carried out optimization between composite filament extrusion production, printing parameters, and reliable operation, the maximum fraction is 35% weight ferrite. We make note that at higher ferrite nanopowder fractions the filament tends towards brittle behavior in feeding and becomes difficult to print. The printability of the composite filament was assessed by printing simple blocks using the FDM process. The composite filament is driven by a stepper motor into a temperature controlled nozzle and molten material is extruded and deposited onto the build platform in a layer according to the pre-programmed pattern. Figure 3 shows printed blocks of pure PLAABS and PLA + 35 weight ferrite of $Ni_{0.5}Zn_{0.5}Fe_2O_4$ and results of complex dielectric permittivity and magnetic permeability measurement.

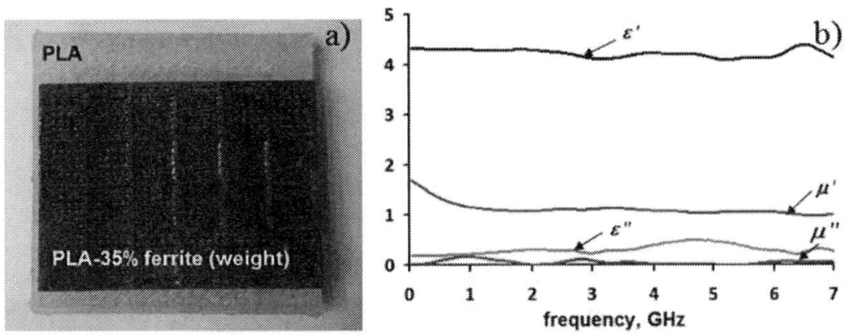

Figure 3. Photographs of printed PLA (top) and PLA + 35 weight of $Ni_{0.5}Zn_{0.5}Fe_2O_4$ (bottom) blocks and frequency dispersions of complex permittivity and permeability for printed PLA + 35 weight of $Ni_{0.5}Zn_{0.5}Fe_2O_4$

4. Conclusions

The full series of nickel-zinc ferrites with general formula $Ni_xZn_{1-x}Fe_2O_4$ (where $x = 0–1$) were synthesized and annealed at temperatures of up to 770 K. The prepared ferrite powder crystal structure and microstructure was studied using SEM and XRD. Average nanocrystallite sizes of 21 to 32 nm were determined according to XRD, while average nanoparticle sizes of 22 to 50 nm were determined according to SEM micrograph analysis. The optical diffuse reflectance measurements were carried out at room temperature gives the values of calculated band gap for the $Ni_xZn_{1-x}Fe_2O_4$ samples are in the range of 1.6 eV to 2.0 eV. The DC electrical conductivity of $Ni_xZn_{1-x}Fe_2O_4$ shows the change of cation distribution in cubic spinel structure with increase in nickel ion content.

The FMR resonance field of $Ni_xZn_{1-x}Fe_2O_4$ ferrite samples versus composition decrease in the nickel content grows from 0 to 0.5 and increase at nickel fraction >0.5 up to 0.85 in $Ni_xZn_{1-x}Fe_2O_4$. The maximum absorption line width is that of compositions with molar fractions of nickel $x = 0.20–0.65$. It was found that the maximum microwave absorption of $Ni_xZn_{1-x}Fe_2O_4$ nanoferrites in high-frequency electromagnetic wave range is achieved at minor content of nickel ions in the ferrite in the $x = 0.15–0.30$ range.

Additive manufacturing FDM 3D-printing process with PLA-nanoferrite filament for fabricating magnetic components also was demonstrated.

References

[1] Joshi G P, Saxena N S, Mangal R, Mishra A, Sharma T P 2003 *Bull. Mater. Sci.* **4** 387
[2] Zhang H E, Zhang B F, Wang G F, Dong X H, Gao Y 2007 *J. Magn. Magn. Mater.* **1** 126
[3] Kambale R C, Adhate N R, Chougule B K, Kolekar Y D 2010 *Journal of Alloys and Compounds* **491** 372
[4] Rao P B C, Setty S P 2010 *International Journal of Engineering Science and Technology* **8** 3351
[5] Thomas J J, Krishnan S, Sridharan K, Philip R, Kalarikkal N 2012 *Materials Research Bulletin* **8** 1855
[6] Tehrani F S, Daadmehr V, Rezakhani A T, Akbarnejad R H, Gholipour S 2012 *J. Supercond. Nov. Magn.* **25** 2443
[7] Chakrabarty S, Pal M, Dutta A 2015 *Materials Chemistry and Physics* **25** 1
[8] Mandal S K, Singh S, Dey P, Roy J N, Mandal P R, Nath T K 2016 *Journal of Alloys and Compounds* **656** 887
[9] Vijaya J J, Bououdina M 2016 *Journal of Nanoscience and Nanotechnology* **16** 689
[10] Priyadharsini P, Pradeep A, Sambasiva P, Chandrasekaran G 2009 *Mater. Chem. Phys.* **116** 207
[11] Costa A, Silva V J, Cornejo D R, Morelli M R, Kiminami R H, Gama G A 2008 *J. Magn. Magn. Mater.* **14** 370
[12] Daruka Prasad B, Nagabhushana H, Thyagarajan K, Nagabhushana B M, Jnaneshwara D M, Sharma S C, Chakradhar R P S 2014 *Journal of Magnetism and Magnetic Materials* **358** 132

[13] Gul I H, Ahmed W, Maqsood A J 2008 *Magn. Magn. Mater.* **320** 270
[14] Typek J, Wardal K, Guskos N, Sibera D, Narkiewicz U 2014 *IEEE Transactions on Magnetics* **50** 1
[15] Tauc J 1968 *Materials Research Bulletin* **3** 37
[16] Davis E A, Mott N F 1979 *Philosophical Magazine* A **179** 903
[17] Bazhan Z, Ghodsi F E, Mazloom J 2016 *Appl. Phys.* A. **122** 1
[18] Joshi S, Kumar M, Chhoker S 2014 *Journal of Molecular Structure* **1076** 55
[19] Dolia S N, Sharma R, Sharma M P 2006 *Indian Journal of Pure & Applied Physics* **44** 774
[20] Domınguez-Arvizu J L, Jimenez-Miramontes J A, Salinas-Gutierrez J M 2017 *International journal of hydrogen energy* **42** 30242
[21] Naidu K C B, Madhuri W 2017 *Bull. Mater. Sci.* **40** 417
[22] Rafique M Y, Ellahi M, Iqbal M Z 2016 *Materials Letters* **162** 269
[23] Sun Q C, Sims H, Mazumdar D, Ma J X, Holinsworth B S, O'Neal K R, Kim G, Butler W H, Gupta A, Musfeldt J L 2012 *Physical Review* B **86** 1
[24] Tholkappiyan R and Vishista K 2015 *Int. J. Mater. Res.* **2** 127
[25] Chavan S, Babrekar M K, More S S, Jadhav K M 2010 *Journal of Alloys and Compounds* **507** 21
[26] Gao D, Shi Z, Xu Y, Zhang J, Yang G, Zhang J, Wang X, Xue D 2010 *Nanoscale Res Lett.* **5** 1289
[27] Saravanan M and Sabari Girisun T C 2017 *Applied Surface Science* **392** 904
[28] Ziaeia V and Bredow T 2017 *The European Physical Journal* B **2** 1
[29] Vyzulin S A, Buz'ko V Yu, D A Kalikintseva D A, E L Miroshnichenko E L 2018 *Bulletin of the Russian Academy of Sciences: Physics* **82** 105

Acknowledgments
This work was supported by the State task of Ministry of Science and Higher Education of the Russian Federation of the Russian Federation (project no. FZEN-2020-0022 04.06.2020) and carried out using the equipment of the Common Use Centre 'Diagnostics of Nanomaterials Structure and Properties' of Kuban State University.

IOP Conf. Series: Materials Science and Engineering **969** (2020) 012102 doi:10.1088/1757-899X/969/1/012102

Development of a Mathematical Model of an Arc Plasma Discharge in Vacuum, Describing the Joint Solution of the Navier-Stokes Equations, Heat Transfer, with Equations of Plasma Theory

I V Sitnikov, D N Trushnikov and G L Permyakov

Department of Welding Production, Metrology and Materials Technology, Perm National Research Polytechnic University, 29, Komsomolsky ave., Perm, 614990, Russia

E-mail: stu459@mail.ru

Abstract. This article is devoted to the description of arc processes in a vacuum, relevant today for the melting of wire materials in the implementation of hybrid technologies of layer-by-layer synthesis. The directions of using an electric arc as a source of heat in creating arc additive technologies are indicated. One of the most promising methods for creating an arc discharge in vacuum using a non-consumable thermoelectronic hollow cathode is presented. A two-dimensional model has been developed for studying a hollow-cathode arc discharge in vacuum, where argon was used as the plasma-forming gas. All calculations were carried out in an axisymmetric formulation. To obtain a numerical solution, the COMSOL Multiphysics application software package was used. The obtained simulation results showed a satisfactory agreement with the literature and experimental data.

1. Introduction

The use of arc and plasma sources for melting wire material in the implementation of hybrid technologies of layer-by-layer synthesis has been actively developing in the world in recent years.

A promising area today is combining the capabilities of technologies from companies such as Sciaky [1, 2] and NorskTitanium, which will combine the effectiveness of vacuum protection and the properties of an electric arc as a heat source to create arc additive technologies in vacuum on their basis. Moreover, the degree of rarefaction in a vacuum chamber can be significantly less than when using an electron beam as a heat source, and the quality of protection will be higher than when using shielding gases or chambers with a controlled atmosphere, as with traditional arc methods.

The process of generating an arc discharge in a rarefied atmosphere for welding or single-layer surfacing is known [3, 4]. In work [5], welding with a consumable electrode in a low pressure argon medium is considered. In [6], the issues of the formation of metal compounds by arc welding methods in high vacuum are considered. Known results are primarily preliminary and represent mainly experimental results.

Speaking of layer-by-layer synthesis of materials, the process of arc melting of wire material in vacuum has not yet been considered. One of the factors that complicate the development of this technology is the lack of mathematical models. With the development of the computational capabilities of modern technical means in modeling processes associated with arc welding, the results presented in [7–9] were obtained, however, such studies for processes in a rarefied medium are practically absent. At the same time, the formation of metal in this case is fundamentally different in many ways, starting with the features of the formation of a plasma discharge [10, 11], ending with the pressure of the plasma arc on the melt [12] and increased evaporation under conditions of uncertainty in the boiling point [13]. In describing the arc discharge in a rarefied atmosphere, the model developed in [14] for a non-self-sustained plasma discharge in electron beam welding will be used.

2. Mathematical model

One of the most promising methods for creating an arc discharge in vacuum is a circuit with a non-consumable thermoelectronic hollow cathode. The industrial application of a hollow cathode arc (HCA) discharge requires an understanding of the processes of plasma interaction with the cathode and anode, as well as related processes. (HCA) discharge can stably exist even at very low pressures, which is especially important when working with chemically active alloys [15–18]. Earlier, in [19], a mathematical model was developed for the formation of an arc discharge at low pressures using a hollow cathode. The mathematical formulation of the problem of forming an arc discharge in vacuum at low pressures (100 Pa and below) is based on solving a system of transport equations for the concentration and average energy of electrons, taking into account the description of mass transfer of heavy plasma particles (ions, neutral unexcited and excited atoms). At the same time, the distribution of neutral particles, which is used as input in the calculation, was taken from general considerations. Plasma properties are important in modeling the formation of an arc discharge based on the solution of the heat equations and the Navier-Stokes equations together with the equations of electrodynamics. In the proposed work, the models are combined to determine the transfer directly in the course of a joint solution.

Figure 1 shows the design scheme (the design scheme is rotated 180° relative to that used in practice). The calculations were performed in cylindrical coordinates in an axisymmetric formulation. Parameters of the hollow cathode: inner radius rc = 1,5 mm, cathode thickness δc = 1 mm, length lc = 10 mm.

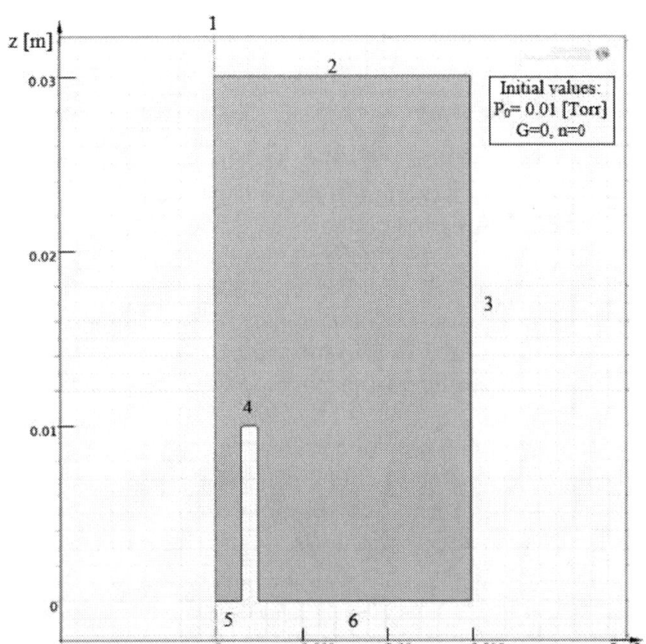

Figure 1. Schematic of the HCA discharge model: 1 – axis of symmetry, 2 – anode (U=0), 3 and 6 – gas outlet (S=100 [l/s]), 4 – cathode (U=U0), 5 – gas inlet (Qm).

IOP Conf. Series: Materials Science and Engineering **969** (2020) 012102 doi:10.1088/1757-899X/969/1/012102

2.1. Governing equations

The electron density and mean electron energy are calculated by solving the drift-diffusion equations:

$$\frac{\partial n_e}{\partial t} + \nabla \cdot \mathbf{\Gamma}_e + (\mathbf{u}_g \cdot \nabla) \cdot n_e = \mathbf{R}_e, \quad (1)$$

$$\frac{\partial n_e}{\partial t} + \nabla \cdot \mathbf{\Gamma}_\varepsilon + \mathbf{\Gamma}_e \cdot \mathbf{E} + (\mathbf{u}_g \cdot \nabla) \cdot n_\varepsilon = \mathbf{R}_\varepsilon, \quad (2)$$

where n_e and n_ε are the electron density and electron energy density, respectively, $\boldsymbol{\Gamma}_e$ and $\boldsymbol{\Gamma}_\varepsilon$ are the electron flux and the electron flux energy.

The R_e component determines the emission or loss of electrons and the electron impact reactions presented in Table 1 [20, 21]. In these equations, R_ε represents the change in electron energy in these reactions, $\boldsymbol{u}_g$ is the flow rate of the neutral gas, $\boldsymbol{E}$ is the electric field strength.

Table 1. Plasma chemical reactions.

No.	Formula	Data od reaction	Type of interaction
1	$e + Ar \rightarrow Ar + e$	cross-section	elastic
2	$e + Ar \rightarrow Ar^* + e$	cross-section	excitation
3	$e + Ar^* \rightarrow Ar + e$	cross-section	excitation
4	$e + Ar^* \rightarrow Ar^+ + 2e$	cross-section	ionization
5	$e + Ar \rightarrow Ar^+ + 2e$	cross-section	ionization

The mass transfer equation for a multicomponent mixture is used to describe the mass transfer of heavy plasma particles [22, 23]:

$$\rho \frac{\partial \omega_k}{\partial t} + \rho \cdot (\mathbf{u}_g \cdot \nabla) \cdot \omega_k + \nabla \cdot \mathbf{\Gamma}_k = \mathbf{R}_k, \quad (3)$$

where ρ is the mass density of the working gas, ω_k is the mass fraction of the kth component, $\boldsymbol{R}_k$ s the collision component of heavy particles obtained from plasma-chemical reactions in Table 1, $\boldsymbol{\Gamma}_k$ is the flux of the kth component.

To calculate the electric field strength, the Poisson's equation is used:

$$\varepsilon_0 \nabla \cdot (\nabla U) = \varepsilon_0 \nabla \cdot \mathbf{E} = \rho \quad (4)$$

where U is the electrostatic potential of the plasma, ρ is the volume charge density, ε_0 is the dielectric constant of the vacuum.

A plasma discharge is accompanied by the emission of secondary electrons from the surface of the cathode. The boundary conditions take into account the loss of charge on the walls of the chamber and on the surface of the product as a result of chaotic motion resulting from thermionic effects [23]:

$$-\mathbf{n} \cdot \mathbf{\Gamma}_e = \left(\frac{1}{2 v_{e,t} n_e} \right) + n_e \mu_e \mathbf{E} \cdot \mathbf{n} - \mathbf{n} \cdot \mathbf{\Gamma}_t \quad (5)$$

and for the electron energy flux:

$$-\mathbf{n} \cdot \mathbf{\Gamma}_\varepsilon = \left(\frac{1}{2 v_{e,t} n_\varepsilon} \right) + n_\varepsilon \mu_\varepsilon \mathbf{E} \cdot \mathbf{n} \quad (6)$$

where $v_{e,t}$ is the thermal velocity of electrons, $\mathbf{\Gamma}_t = \boldsymbol{J}_t/e$ is the flux of thermal emission of electrons and $\boldsymbol{n}$ is normal, μ_e and μ_ε are the mobility and energy coefficients of electrons. The boundary condition for the electron flux, electron energy flux, and heavy particle flux at the boundaries 2, 3, 4, 5, 6 is described in detail in [24].

A neutral gas flow, considered as a laminar flow, is described by the system of Navier-Stokes equations. The mass continuity equation and the momentum balance equation have the following form:

$$\frac{\partial \rho}{\partial t} + \nabla \cdot (\rho \mathbf{u}_g) = 0 \tag{7}$$

$$\rho \frac{\partial \mathbf{u}_g}{\partial t} + \rho(\mathbf{u}_g \cdot \nabla) = \nabla \cdot \left[-p\mathbf{I} + \mu_d(\nabla \mathbf{u}_g + (\nabla \mathbf{u}_g)^T - \frac{2}{3}\mu_d(\nabla \mathbf{u}_g)\mathbf{I} \right] \tag{8}$$

where p is the pressure of the mixture of the gaseous medium, μ_d is the dynamic viscosity, $\boldsymbol{I}$ is the tensor unit. The right term of the equation is the force of the pressure gradient and the tensor of viscous stresses, respectively. The superscript T represents the displacement of the matrix.

The heat transfer equation is a classic convection-diffusion equation, it looks like this:

$$\rho c_{p,g} \frac{\partial T_g}{\partial t} + \rho c_{p,g} u_g \cdot \nabla T_g + \nabla(-k_g \cdot \nabla T_g) = Q_{eh} + Q_{J,ion} \tag{9}$$

where T_g is the temperature of the neutral gas, $c_{p,g}$, and k_g are the specific heat at constant pressure and the thermal conductivity of the working gas, respectively. The terms of the heat source include the expression of the exchange of energy between electrons and heavy particles:

$$Q_{eh} = \frac{3}{2} k_B \left(\frac{2m_e}{m_i} \right) \cdot (T_e - T_g) v_{eh} \tag{10}$$

and the expression of the Joule heating of ions:

$$Q_{J,ion} = J_{ion} \cdot E \tag{11}$$

In equation (10), the coefficient k_B is the Boltzmann constant, m_e and m_i are the mass of one electron and one heavy particle, v_{eh} is the total frequency of collisions between electrons and heavy particles, which can be obtained by plasma chemical reactions. In equation (11), J_{ion} is the ion current density at which only (Ar +) ions are considered in this study [24].

The gas flow in the hollow cathode and directly at the exit from it is described by an intermediate flow regime called the «slip mode» (Knudsen number 0.01 <Kn <0.1). In the intermediate mode, the Navier-Stokes equations are still applicable, but special modified boundary conditions are applied. When moving away from the end of the hollow cathode towards the surface of the product, the jet spreads in cross section, the flow is characterized by an intermediate Knudsen number (Kn ~ 1). When moving away from the weld zone, the flow passes the molecular flow regime (Kn >> 1) and can be described using the Monte Carlo method. In the deposition zone (the cathode cavity and the zone below it), an intermediate mode (Kn ~ 1) is realized. In the studies known in the open literature, in this case both the description by the "sliding mode" using the Navier-Stokes equations and the molecular flow description are used, sometimes correcting the obtained solutions by introducing some calibration coefficients. A numerical experiment showed a satisfactory convergence between the results obtained using both the first and second approaches.

3. Results

The calculations were carried out in an axisymmetric cylindrical formulation. To obtain a numerical solution, we used the COMSOL 4.4 application software package, the Plasma Module, Molecular Flow, and Rarefied Flow modules. The simulation results show a satisfactory agreement with the literature and experimental data.

The pressure inside the hollow cathode depends on the flow rate of the supplied gas and decreases almost linearly as it approaches the outlet (Figure 2). In the zone of interaction between the plasma flow and the anode, the distribution is close to Gaussian (Figure 3).

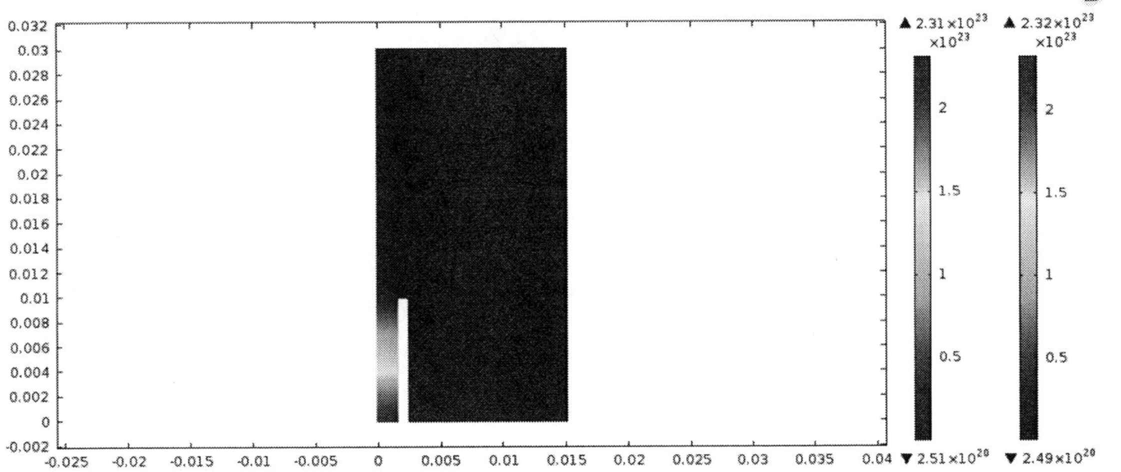

Figure 2. Pressure in the cavity of the hollow cathode.

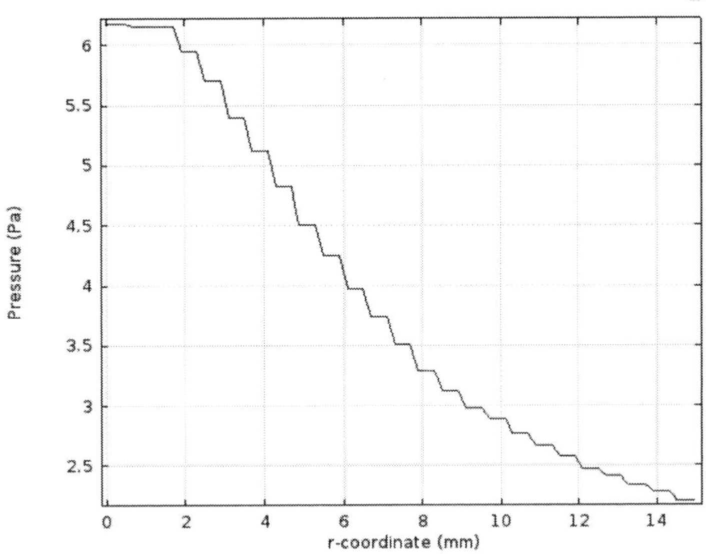

Figure 3. Distribution of plasma pressure on the surface of the anode, calculated as part of the description of the flow by free molecular flow.

The distribution of the concentration of neutral particles obtained from the description of the gas flow is an input parameter for the description of the plasma using equations (1) - (11). The values of the energy released as a result of solving equations (1) - (11) are used in solving the problem of describing a gas flow. Solving the system of equations (1) - (11) also allows you to calculate the transfer coefficients, such as plasma conductivity, diffusion coefficients, etc.

Figure 4 shows the calculated distribution of electron density in the case of HCA discharge in vacuum with a microflow of a plasma-forming gas (argon) through the cathode cavity. The maximum electron density is near the end of the hollow cathode [19].

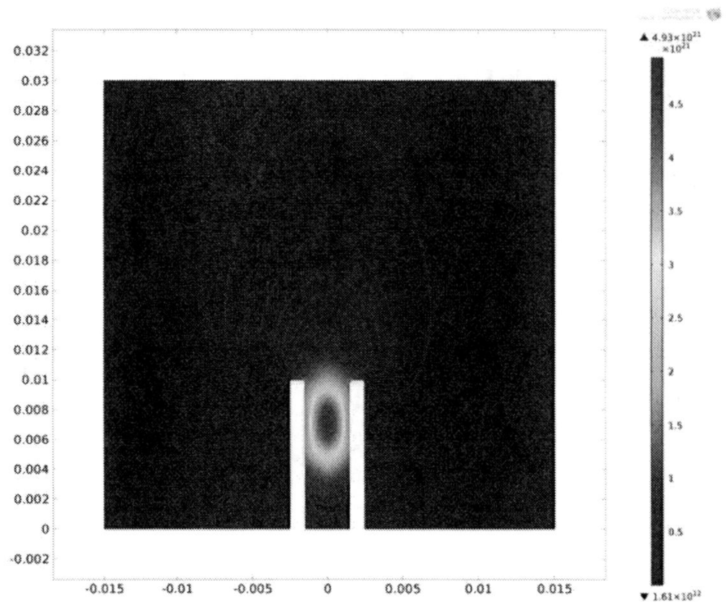

Figure 4. Distribution of electron density at HCA discharge (discharge current I_d = 96 A, electric potential U_0 = 30 V, gas flow rate Q_m = 3 mg / s).

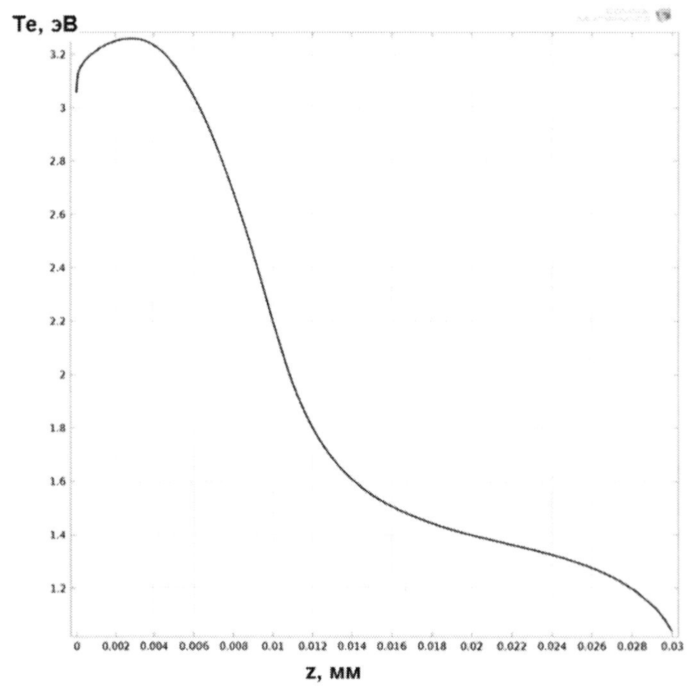

Figure 5. Change in electron temperature (eV) in the space under the hollow cathode.

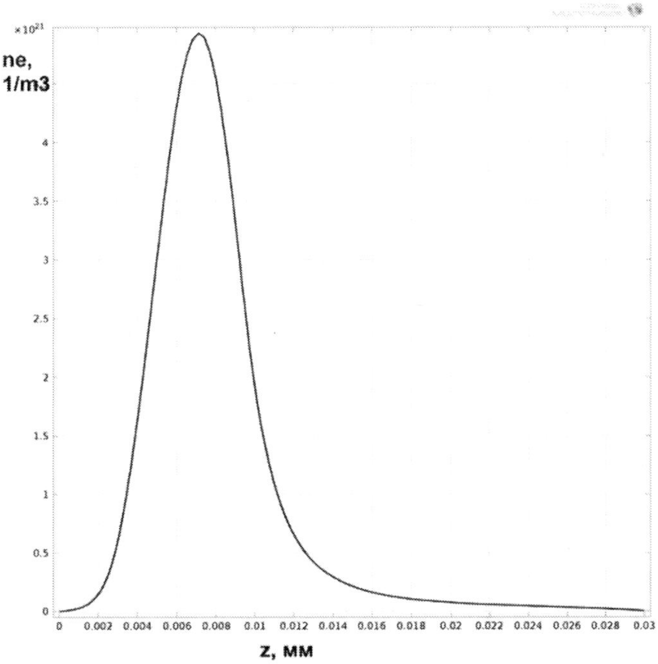

Figure 6. Change in plasma concentration in the cavity of the hollow cathode and under it.

Figures 5, 6 show the distribution of electron temperature and plasma concentration in the cavity of the hollow cathode and below it along the axis of symmetry.

4. Conclusion

A mathematical model has been developed that describes a method for creating an arc discharge in vacuum using a non-consumable thermoelectronic hollow cathode.

As a result of the simulation, the following results were obtained:

- - data on the pressure in the cavity of the hollow cathode;
- - distribution of plasma pressure on the surface of the anode;
- - distribution of electron density in HCA discharge;
- - data on changes in electron temperature in the space under the hollow cathode;
- - data on changes in plasma concentration in the cavity of the hollow cathode and under it.

The solution of the associated mathematical problem for an arc plasma discharge in vacuum, which describes the joint solution of the Navier-Stokes equations, heat transfer, and equations of plasma theory, was obtained for the first time.

References

[1] Sciaky Inc. Electron Beam Additive Manufacturing (EBAM) - Advantages of Wire AM vs. Powder AM Online. Available: http://additivemanufacturing.com/2015/10/14/electron-beam-additive-manufacturing-ebamadvantages- of-wire-am-vs-powder-am/

[2] SciakyInc, Electron Beam Additive Manufacturing (EBAM) Online. Available: http://www.sciaky.com/images/ pdfs/product-sheets/Sciaky-EBAM-Technology.pdf

[3] Toya H and Hiedaand K 2006 Preliminary Study on Arc Welding in Vacuum *Preliminary Study on Arc Welding in Vacuum International Symposium on Discharges and Electrical Insulation in Vacuum* **2** 762–765

[4] Geavit Musa, Nicolae Betiu, Ion Mustata, Baltog Alexandra 1983 Low Voltage Arc Weldingin Vacuum *Rev. Roum. Phys.* **28(10)** 907–908
[5] Suitaetal Y 1994 *Welding International* **8(4)** 269–273
[6] Toyaetal H 2002 *20th Int. Symp. On Discharges and Electrical Insulation in Vacuum* 327–330
[7] Fan H G and Kovacevic R 2004 A unified model of transport phenomena in gas metal arc welding including electrode, arc plasma and molten pool *Journal of Physics D: Applied Physics* **37 18**
[8] Wu C S, Hu Q X, Gao J Q 2009 An adaptive heat source model for finite – element analysis of keyhole plasma arc welding *Computational Materials Science* **46 1** 167–172
[9] Wang H X, Wei Y H, Yang C L 2007 Numerical simulation of variable polarity vertical-up plasma arc welding process *Computational materials science* **38 4** 571–587
[10] Greenwood Allan 1994 *Vacuum Switchgear* (IEE Power Engineering Series) (Pergamon Press)
[11] Musa Geavit, Baltog Alexandra 1986 Electrical and Spectral Characteristics of a Heated Cathode Discharge in metal Vapors *Contrib. Plasma Phys.* **26(3)** 171–177
[12] Sass-Tisovskaya M 2009 *Plasma arc welding simulation with Open FOAM* (Chalmers University of Technology)
[13] Trushnikov D N 2016 Determination of temperature in the penetration channel in electron beam welding *Welding International* **30 2** 133–136
[14] Trushnikov D N and Mladenov G M 2015 Numerical model of the plasma formation at electron beam welding *Journal of Applied Physics* **117 1**
[15] Raizer Yu P and Allen J E 1991 *Gas Discharge Physics* (Berlin Springer-Verlag) p 450
[16] Lafferty J M and Cobine J D 1980 *Vacuum Arcs: Theory and Applications* (Wiley-Interscience publication) p 372
[17] Granovsky V L 1971 *Electric discharge in gases* (Moscow: Nauka) p 593
[18] Kesasv I G 1968 *Cathode processes of an electric arc* (Moscow: Nauka) p 244
[19] Permyakov G L, Trushnikov D N, Mindibaev M K, Koleva E G 2019 Numerical model of a hollow cathode arc discharge formation in vacuum *IOP Conference Series: Materials Science and Engineering* **759**
[20] Raju G G 2006 *Gaseous Electronics: Theory and Practice* (Boca Raton, FL: CRC PR INC)
[21] Lee C and Lieberman M A 1995 Global model of Ar, O2, Cl2, and Ar/O2 high-density plasma discharges *J. Vac. Sci. Technol.* **A 13** 368–380
[22] Bird R B, Stewart W E, Lightfoot E N 2002 *Transport Phenomena* (John Wiley and Sons)
[23] Hagelaar G J M 2000 *Modeling of microdischarges for display technology* (Technische Universiteit Eindhoven) p 109
[24] Shiyuan Cong, Runhui Wu, Lei Mu, Jizhong Sun, Dezhen Wang 2019 A 2D model for low-pressure hollow cathode arc discharges in argon *J. Phys. D: Appl. Phys.* **52**

Acknowledgments
The studies were financially supported by the Ministry of Science and Higher Education of the Russian Federation on state assignment FSNM-2020-0028, the Government of the Perm Territory as part of agreement S-26/787 of 12/21/2017 (studying the current state of the issue) and the Russian Foundation for Basic Research under project No. 18-08-01016 A (experimental studies).

IOP Conf. Series: Materials Science and Engineering **969** (2020) 012103 doi:10.1088/1757-899X/969/1/012103

Effect of Temperature Field on Mechanical Properties of Direct Laser Deposited Ti-6Al-4V Alloy

M O Gushchina[1,2], S Yu Ivanov[1,2] and A M Vildanov[2]

[1]Institute of Laser and Welding Technologies, St. Petersburg State Marine Technical University, 3, Lotsmanskaja, St. Petersburg, 190121, Russia
[2]Department of Laser Technologies, Peter the Great St. Petersburg Polytechnic University, 29, Polytechnicheskaya, St. Petersburg, 195251, Russia

E-mail: skmar.spb@gmail.com

Abstract. The mechanical and service properties of Ti-6Al-4V alloy parts produced by direct laser deposition (DLD) depend on the thermal cycle parameters. The temperature field during deposition is significantly affected not only by the parameters of the process, but also by the interpass dwell time and length of the deposited layers. The aim of the article is to establish the relationship between mechanical properties of deposited Ti-6Al-4V samples and DLD thermal cycle parameters. Numerical simulation was used in order to establish relationship between temperature field parameter and the process parameters. The nonlinear three-dimensional heat conduction problem was solved by the finite element method. It is shown that an increase in the dwell time between passes from 5 to 10 seconds leads to a significant decrease in the inter-pass temperature and an increase in the cooling rate. This leads to the metastable structure formation of the deposited layers of Ti-6Al-4V that consists mainly of a nonequilibrium α'-phase which hardness is higher than 390 HV. Without dwell time an equilibrium α + β structure with hardness of 360 HV and higher elongation is formed.

1. Introduction

Additive manufacturing (AM) of titanium alloys is one of the most interesting areas of 3D metal printing. Titanium alloys have unique properties that are combined with hard workability by mechanical methods and a tendency to grain growth during casting [1]. Additive manufacturing methods solve these problems, which leads to decreasing costs and efficiency increased [2]. Despite, the novelty of 3D printing technology requires solving the problems that arise during the transition to the AM [3]. For example, a large number of research works of laser AM of titanium α + β alloys confirm the metastable structures formation due to high heating and cooling rates [4, 5]. The formation of a nonequilibrium martensitic α'structure reduces ductility and can significantly complicate the deposition of large-sized parts. Since during the deposition process there is a significant accumulation of residual stresses and failure of the part [6, 7].

The formation of the metastable phase in titanium alloys with β-stabilizers is associated with the diffusionless transformation of β-α' from β-phase region when a cooling rate is higher than 20 °C/s [8, 9]. It has been established that the cooling rate in laser additive manufacturing can reaches 10^6 K/s. However, subsequent heating from the higher and next layers also has a significant effect on the structure formation. For example, in work [10] was found that the initial completely martensitic structure, which forms at a cooling rate more than 200 °C/s during β transformation after thermal influence from the subsequent and next layers partially goes into equilibrium and the final structure consists of a mixture of α'+ α+β, and the content of the equilibrium component increases with a decrease in the cooling rate from the β-region. The formation of the thermal cycle during process AM depend on many different parameters

It is shown that the shape and dimensions of Ti-6Al-4V products in the DLD process influence on the mechanical properties and structure [11]. The influence of beam oscillations on the structure and properties formation was demonstrated in [12, 13], as a result established that the use of a Z-shaped oscillation led to a change in the structure from nonequilibrium martensitic to equilibrium two-phase, which accordingly led to an increase in the mechanical characteristics of the deposited samples from Ti-6Al-4V.

The authors Jingjing Yang et al. [14] presented that in the SLM Ti-6Al-4V process, the content and morphology of the α'-phase can be controlled by changing the distance between the bead as well as scanning speeds to control the time and type of thermal cycles. Additive DLD technology are most promising for the manufacturing of large-sized parts. Because of this, studies of thermal cycles are required that would correspond to different cooling times. As the dimensions of the part increase, the time between the application of adjacent and subsequent layers will increase, which will significantly change the thermal cycle [15]. Thermal cycle change will affect the structure, which will lead to a change in the properties of deposited samples from Ti-Al-4V.

The aim of the article is to establish the relationship between mechanical properties of deposited Ti-6Al-4V samples and DLD thermal cycle parameters.

2. Materials and methods

2.1. Experimental procedure

The material used in this study is ASTM B 265 grade 5 Ti6Al4V titanium alloy powder with a fraction of 45-90 microns produced by PREP (Plasma Rotating Electrode Process). The experimental trials were carried out using a robotic complex for direct metal deposition based on IPG fiber laser. The laser system has the maximum power of 3000 W, a Gaussian beam profile. The complex also include a six-axis robot and two-axial positioner [16]. Working tool is a laser-weld head with coaxial four-jet nozzle for powder feeding. The process was carried out in chamber with argon atmosphere, the oxygen level in the chamber was 2000 ppm

A 50 layers wall from Ti-6Al-4V powder was deposited on the 12 mm thick Ti-6Al-4V rolled plate (Figire 1). Seven passes formed each deposited layer of 80 mm length. Each layer in the plate build consisted of seven beads with a horizontal overlap distance of 2 mm. A vertical step size of 0.80 mm per layer was used. The powder, carried by the inert gas through four radially symmetrical nozzles, is injected into the molten pool. The process parameters were the following: a beam power of 2200 W, a beam diameter of 3 mm and a forward speed of 20 mm s^{-1}, a coaxial gas flow of 25 l min^{-1}, transport gas flow of 4 l min^{-1}, powder flow rate of 14.84 g min^{-1}. The layer height resulted to 0.73 mm leading to a total build height of 28 mm. Three different dwell times of 0, 5 and 10 s was analysed.

Mechanical properties were measured using a universal testing machine Zwick/Roell Z250 Allround series. Cylindrical samples of standard dimension were used for uniaxial tensile test. For detailed study of the defects location in structure, the samples were cut in a cross section, after there were etched in a reagent: 93 ml H2O + 2mlHF + 5ml HNO3. Metallographic studies of the samples were carried out using an optical microscope DMI 5000 (Leica) with the "Tixomet" software. Micro hardness tester FM-310 was used to study microhardness of the samples.

2.2. Simulation procedure

Deposition of material during DLD was simulated using the so-called element birth technique [17]. In this method, elements to be deposited are deactivated at the beginning and then gradually activated or "born" into the solution domain. The pass shape was as a cuboid. An 8-node axisymmetric elements were employed in 3D model (Figure 1). A laser beam thermal efficiency was assigned a value of 0.45 based on [18]. An internal volumetric heat source with uniform density was applied to model heat input for each pass. To account for heat losses, convective heat transfer on the all surfaces exposed to air was modeled. The convective heat transfer coefficient is taken as 18 W m-2 K-1 [19]. The temperature-dependent thermophysical properties for Ti-6Al-4V including the enthalpy, thermal conductivity, density were taken from [20]. Solid-to-liquid phase transformation releases the latent heat of fusion, resulting in an increase in the enthalpy.

3. Results and discussion

3.1. Analysis of temperature field

Interpass dwell time has a decisive influence on the temperature field if the length of the compares parts is the same. As can be seen in Figure 1, the increase in the dwell time leads to a significant decrease in the interpass temperature. The average interpass temperature decreases by 300 °C from 540 °C when the dwell time increases to 10 s. The deposited pass undergoes multiple reheating from subsequent passes. Magnitude of the reheating temperature depends of distance between the point of interest and axis of newly deposited pass. As can be seen, the high interpass temperature provides a significantly higher peak reheating temperature. Thus, in the absence of dwell time, the peak values of reheating correspond to 1550-990-790 °C (Figure 1a) while if the dwell time equals 10 seconds: 1390-760-525 °C respectively (Figure 1b).

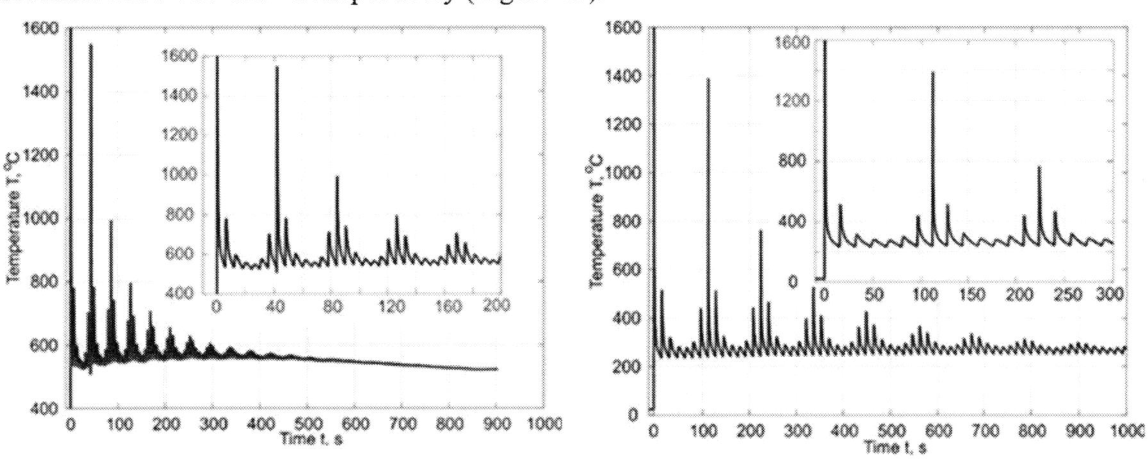

Figure 1. Thermal cycles of the 40th layers deposited with 0 (a) and 10 s dwell time.

Microstructure of the just deposited pass depends on the cooling rate. The analysis of thermal cycles showed a significant change in cooling rate with increasing distance from the substrate (Figure 2). The cooling rate of the first layer passes practically does not depend on the interpass dwell time. It is explained by the significant heat dissipation into the cold substrate, which prevents the heat accumulation. In this case an initial microstructure of the deposited layer must contain martensite, according to [21]. As the distance from the substrate increases, the dwell time effect becomes more noticeable. Cooling rate of the 40th layer deposited without dwell time is 1.7 time lower than that of the first layer.

The final microstructure is significantly affected by reheating of the subsequent passes. Reheating to high temperatures leads to activation of diffusion processes associated with phase transformation and tempering of nonequilibrium phases. The effect of diffusion processes can be indirectly estimated by the value of metal exposure time (Figure 3). As in the case of cooling rates, the holding time of the first layer higher than the different cut-off temperatures practically does not depend on dwell time (Figure 3a). A significant difference is observed at a distance from the substrate. Without dwell time the total exposure time above 600 °C is 54 s. If the dwell time equals to 2.5 s, it is only 7.5 s.

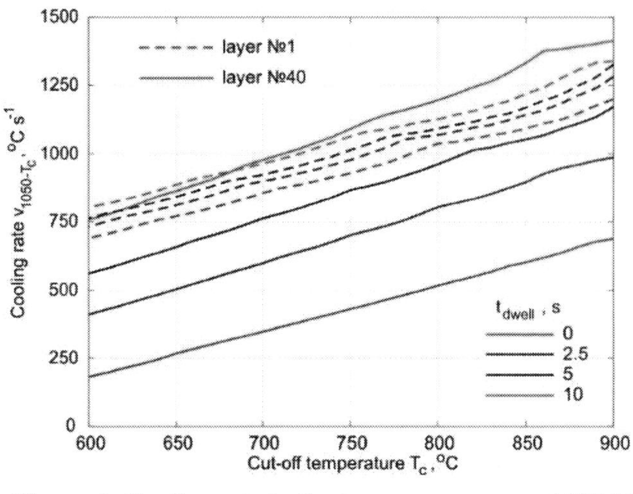

Figure 2. Cooling rate in the temperature range 1050-T_c for the 1st and 40th layer.

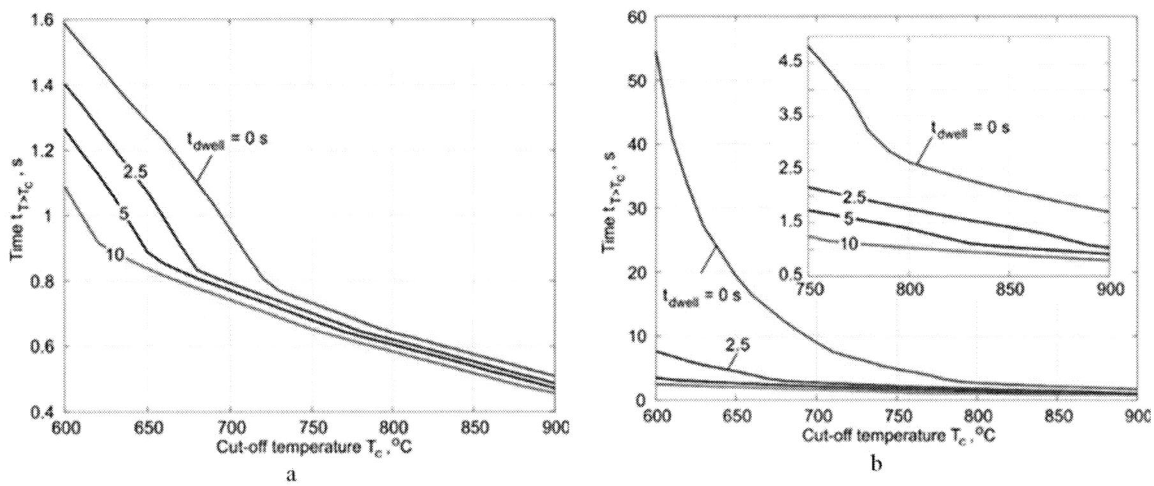

Figure 3. Total exposure time higher than cut-off temperature T_c of the 1st (a) and 40th (b) layer for the different interpass dwell time.

3.2. Microstructure of deposited Ti-6Al-4V samples

Reducing of the interpass dwell time leads to a significant change in the structure. In the sample without dwell time, an equilibrium structure is formed (Figure 4a). It is consisting of β grains with fine Widmanstatten α-phase structure, with plates widths on the order of one micron. This observation present that the cooling rate of the Ti-6Al-4V samples from temperature above the β transus, while rapid, was not sufficiently fast to allow the formation of martensite α' phase. In samples with dwell time of 5 and 10 seconds, a nonequilibrium martensitic structure are observed unlike the microstructures of samples with dwell time of 0 seconds (Figure 4b,c). Thin acicular martensitic needles located orthogonal to each other. Moreover, reduce of a dwell time the ratio of the length and width of the martensitic needles decreases.

Based on diagram shown on figure 4 and prediction thermal cycles it is follow, that cooling rate low than 410 ^{0}s-1 corresponding to the model sample deposited with dwell time of 0 seconds.

Comparison of the elongation values obtained in this study (Table1) for samples deposited with using different dwell time presented that with 0 second dwell time have value in two time higher respect to values for 10 second dwell time. Microhardness is reduce with dwell time decreasing and significantly reduced when dwell time is 0 sec. This difference is likely due thermal cycles as well as cooling rate change.

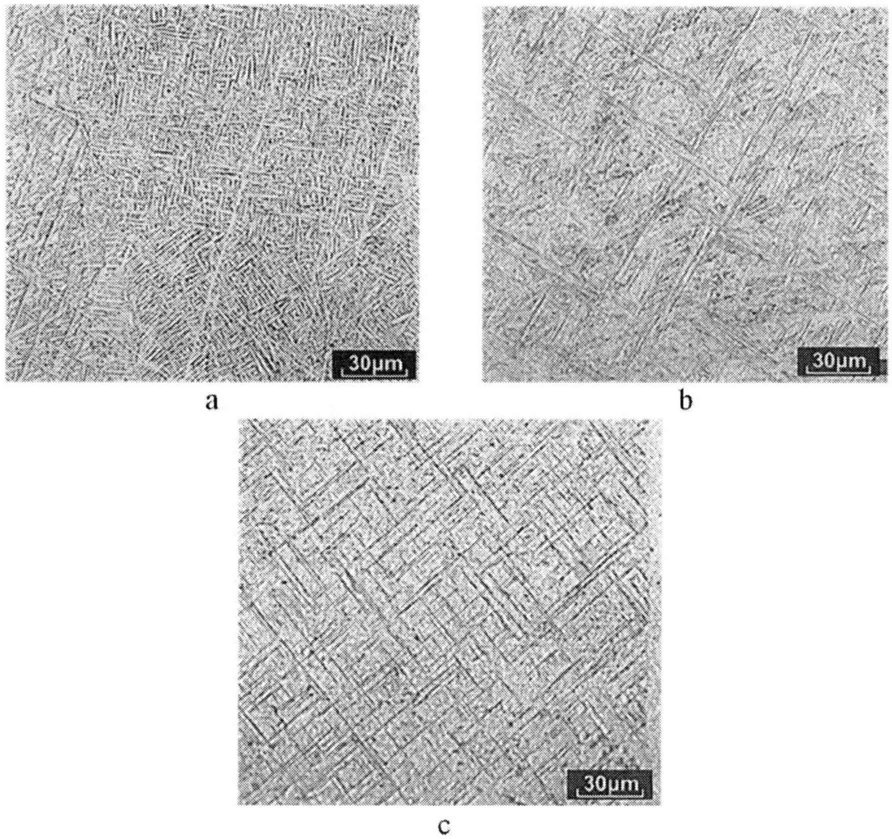

Figure 4. Optical micrographs of deposited Ti-6Al-4V samples (a) dwell time=0 s, (b) dwell time=5 s, (c) dwell time=10 s

Table 1. Mechanical tests result.

Sample	Tensile strength, MPa	Yield strength, MPa	Elongonation, %	Microhardness, $HV_{0,03}$	Dwell time, sec
1	1114	1097	12,5	379	0
2	1140	1070	6,2	397	5
3	1199	1140	4,8	398	10

4. Results

A Ti–6Al–4V model samples was fabricated using a directed laser deposition AM process with using different value of interpass dwell time. The primary conclusions from this study are as follows:

The mechanical properties in the deposited samples improved when dwell time value is 0 second due to lower cooling rate, which allow to equilibrium $\alpha+\beta$ formation. Increasing dwell time value lead to higher cooling rate, that effect on metastable α' phase formation with different relationship in $\alpha+\beta$ phase composition. Therefore, ductility values is increase in two times, microhardness is decrease at least on 40 HV.

IOP Conf. Series: Materials Science and Engineering **969** (2020) 012103 doi:10.1088/1757-899X/969/1/012103

References

[1] Banerjee D and Williams J C 2013*Acta Mater.* **61** 844–879

[2] Shamsaei Nima, Yadollahi Aref, Bian Linkan, Thompson Scott M 2015 *Add. Manuf.* **8** 12–35

[3] Shunyu Liu, Yung C, Shin 2019 Mater. and Design **164** 107552

[4] Kazantseva N, Krakhmalev P, Thuvander M, Yadroitsev I, Vinogradova N, Ezhov I 2018 *Mater. Character.* **146** 101–112

[5] Kyung-Min Hong and Yung C. Shin 2016 *J. of Mater. Process. Tech.* **237** 420–429

[6] Parry L, Ashcroft I A, Wildman 2016 R D *Add. Manuf.* **12** 1–15

[7] Ivanov S, Zemlyakov E, Babkin K, Turichin G, Karpov I Em V, Rylov S. 2019 *Proced. Manufact.* **36** 240–248

[8] Gil Mur F X, Rodriguez D, Planell J A 1996 *J. of Alloys and Comp.* **234** 287–89

[9] Ahmed T and Rack H J 1998 *Mater. Sci. and Engineer. A* **243** 206–211

[10] Lia Frederick, Park Joshua Z, Keist Jayme S, Sanjay Joshi, Martukanitz Richard P 2018 *Mater.Sci. and Engineer. A* **717** 1–10

[11] Keist Jayme S and Palmer Todd A 2016 *Mater. & Design* **106** 482–494

[12] Shalnova S A, Klimova-Korsmik O G, Turichin G A, Gushchina M *2020 Solid St. Phen.* **299** 716–722

[13] Vildanov A M, Babkin K D, Zemlyakov E V, Gushchina M O 2018 *J. of Phys: Conf. Ser.* **1109** 012059

[14] Jingjing Yang, Hanchen Yu, Jie Yin, Ming Gao, Zemin Wang, Xiaoyan Zeng 2016 *Mater. & Design* **108** 308–318

[15] Stankevich S, Gumenyuk A, Strasse A, Rethmeier M 2019 *Key Engineer. Mater.* **822**

[16] Turichin G A, Somonov V V, Babkin K D, Zemlyakov E V, Klimova O G 2016 *Equip.and Mater.* **125** 012009

[17] Em V T, Ivanov S Y, Karpov I D, Rylov S A, Zemlyakov E V, Babkin K D 2018 *J. of Phys. Conf. Ser.* **1109** 012049

[18] Kwon H, Baek W K, Kim MS, ShinW S, Yoh J 2012 *J Opt. Lasers Eng.* **50** 114–121

[19] Gouge M and Michaleris P 2017 *Thermo-Mechanical Modeling of Additive Manufacturing* (Butterworth-Heinemann) p 294

[20] Mills K C 2002 *Recommended values of thermophysical properties for selected commercial alloys* (Cambridge: Woodhead Publishing)

[21] M. Majdic, G. Zeigler 1973 *Zeit. Metall.* **64** 751–758

Acknowledgments

The reported study was funded by RFBR according to the research project №.19-38-50083\19.

Direct Energy Deposition of Cu-Fe System Functionally Graded Structures

K I Makarenko and I V Shishkovsky

Center for Design, Manufacturing and Materials, Skolkovo Institute of Science and Technology, 30, build. 1, Bolshoy boulevard, Moscow, 121205, Russia

E-mail: konstantin.makarenko@skoltech.ru

Abstract. The paper demonstrates the results of microstructure, microhardness and elasticity analysis of the functionally graded (FG) specimens with multilayer structure created of stainless steel and aluminium bronze powder materials via direct energy deposition (DED) laser technology. Increase of microhardness (up to 266 HV) and Young's modulus of elasticity (up to 43.2 GPa) along with growth of the dendritic crystals in the gradient structures are observed. The results of numerical simulation demonstrate stress distribution in FG Cu-Fe system structure with a sharp interface. The results of the research can be used for 3D-printing of the aerospace industry details created from two kinds of material with rather different thermomechanical properties.

1. Introduction. Theoretical part, scholarly importance and problem statement

A problem of functionally graded materials (FGM) synthesis (via inter alia methods of rapid prototyping) was first time investigated near the end of the XX – beginning of the XXI centuries [1–3]. Classification of FGM by factor of volume (thin FGM and bulk FGM) [4] is the most common and obvious among others. The first group – thin FGM – are combinations of several layers, segments or multimaterial coatings deposited on a flat or curved surface; FGM of the second group are complete tridimensional gradient parts, which are more complex and resource-consuming in manufacturing than FGM of the first group [4]. A direct energy deposition (DED) technology [5–7] is convenient, processable and efficacious way to produce bulk FGM with required gradient of physical and chemical characteristics [8–11]. One of the first mentions of usage of DED for FGM synthesis is 23 years old [12]. The common techniques of FGM synthesis are a straight joining, a method of a gradient transition and a method of an intermediate segment [8]. A straight joining is a simplest conjunction method with a sharp transition between two pure initial metals. A method of a gradient transition supposes producing of interjacent layers with incrementally variable concentration of original materials between a pure metal A and a pure metal B. A method of an intermediate section is similar to the previous one, but is notable for using of a segment of the third pure material deposited amongst first and second ones. Advantage of this technique is exception of forming of unnecessary phases like intermetallics, which can easily appear in the method of a gradient transition [8].

There are several important groups of conjoining metallic materials in FG structures, which have the highest significance in mechanical engineering (including aerospace, milling and nuclear industries), medicine, food industry and so on.

- The **first** group is Ni-based FG superalloys (such as Diamalloy 1005 [13]) which are combinations of Ni, Cr and Al. These superalloys are commonly used for tools working in elevated temperatures like airfoils of turbines [14]. Such intermetallic phases as NiAl and Ni_3Al, which have good oxidation resistance in intense heat conditions, even under influence of hot flows of various gases, find their application as protective coverings in aerospace and power mechanical engineering, including components of gas turbine units and engines for missiles of diverse purposes. The advantages of these materials are their moderate density, unique mechanical characteristics under high temperatures, chemical and erosion resistance within a wide temperature range [13].
- The **second** group is Ti-based FG superalloys (Co-V-Ta-Ti, Co-Ni-Al-Ti, Nb-Ti, Ti-B-Cu, Ti-Al-V-C, Ti-W-C and so on), Ti & TiO_2 [15], and aluminides of titanium. They are also widely used in aerospace and power industries, above all in high temperature conditions [10, 16]. The area of application of these FGM is larger in comparison with pure Ti alloys. The advantages of Ti-based superalloys are higher rigidity, hardness, thermal stability, and heat resistance [17, 18]. The results of investigation tests performed with TiB_2-Cu FGM in conditions similar to missile operation with thermal shock influence demonstrated absence of brittle failure or cracking [19]. Ti alloys-based FG structures, as well as technically pure titanium, due to their biocompatibility also find their application in endoprosthetics [20].
- The **third** group is aluminium-based FGM such as AlSi40 and Al [21] where particles of Si serve as hard reinforcements of an inhomogeneous Al-based structure. The resulted material has enhanced mechanical and chemical resistance properties in comparison with initial components [21].
- The **fourth** important group is Fe-Al system (such as Fe_3Al and SS 316L) FG alloys. These compounds have many special characteristics [22]: corrosion, creep and heat resistance, significant mechanical strength in high temperature conditions [23–26]. Manufacturing of these FGM is not so expensive partly owing to reasonable cost of substrates [27–29].
- The **fifth** group is CoCrMo multimaterial alloys, which play an important role in biomedicine. They are used, in particular, in producing of implants for TKR and THR (total knee and total hip replacement) [30].
- The **sixth** group is FG system of Ni-Cr-B-Si and steel such as SS 316L [31]. Ni-Cr-B-Si alloys are widely used in corrosive and high temperature conditions. These FGM have many different solid inclusions and are applicable for hardfacing [32].
- The **seventh** group is Ti-Fe system FGM [33, 34]. They combine low density, high mechanical strength and heat resistance of Ti with specific properties of steel (such as processability and lower price) [33].
- The **eighth** group we should mention is Ni-Cu (including Inconel-Cu) and Fe-Cu (including steel-bronze) FG systems [35, 36, 4]. Combining of two aerospace alloys such as, for example, GRCop-84 and Inconel 718 allow to enhance thermophysical properties of a resulted bimetallic structure [35]. Copper allow combining good oxidation resistance in aggressive atmospheres (such as alkaline and salt) with high values of heat and temperature conductivity coefficients [36]. Nickel is a common component of high temperature alloys. Perfect notch toughness, mechanical strength, and corrosion resistance in elevated temperature conditions characterize them. High temperature strength of Ni and heat conductivity of Cu provide usage of these FGM in extreme temperature conditions [36]. FGM sintered using tool steel H13 and copper can be used for purposes of casting industry as materials for injection moulding tools due to high mechanical strength, wear resistance and thermal conductivity [4]. The stainless steel and Cu combination find its application for producing of food processing, steam turbine and power nuclear plants, electronic components and so on because of mutually supportive characteristics, such as electrical and heat conductivity of copper along with good corrosion resistance and manufacturability of stainless steel [37].

IOP Conf. Series: Materials Science and Engineering **969** (2020) 012104 doi:10.1088/1757-899X/969/1/012104

FGM of the eighth group are widely used in manufacturing of space industry parts due to the mentioned physical and exploitation properties of copper and advantages of nickel alloys and stainless steel. The first steps in research of FGM usage for space industry were performed 18 years ago [19]. There is a description of FGM application for producing of thermal barrier materials for space shuttles, creating of air and gas vanes, thrust chambers, piston tops, nosetips, and so on. A radial laser deposition additive technology [38] allows producing axially symmetric gradients, such as metallic parts for spaceship elements created of carbonaceous filaments. Such gradient parts can be fabricated radially from the center of a sample to the outside.

Our research is devoted to synthesis of FG structures associated with the eighth group. The key problem is analysis of the reasons of cracking occurring in FGM created from aluminium bronze and stainless steel via DED, investigation of microstructure, microhardness and elasticity of the specimens and performing of numerical simulation of the stress-strain state forming in nonequilibrium conditions.

2. Materials and equipment

Deposition process was performed with stainless steel AISI 316L and aluminium bronze with 10% content of aluminium via the InssTek MX-1000 machine in the DMT (direct metal tooling) mode. Laser source was an IPG Photonics 1 kW ytterbium doped fiber laser (1064 nm). Focus of a beam was on 1 mm level under the surface of the specimens. Average laser radiation power was equal to 308 W (450 W maximal) for stainless steel and to 500 W (750 W maximal) for aluminium bronze. Power of laser radiation was chosen in accordance with the bronze infrared radiation absorption coefficient which was lower than the same one of stainless steel. Velocity of laser source was equal to 0.85 m/min, powder rate – 3.5 g/min, cooling time between layers – 5 s. We used Ar as a shielding and feeding gas. There was no surface preheat in the experiments. Configuration of single layer tracks formation is shown in the figure 1. Width of a single track was equal to 800 μm. Covering of the tracks was equal to 300 μm.

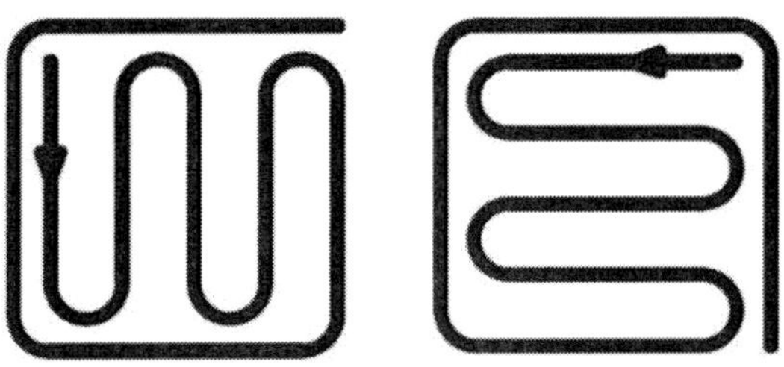

Figure 1. Configuration of the tracks of a single layer (left picture – odd layers, right picture – even).

Microstructure research was performed via an optical microscope Carl Zeiss and a scanning electronic microscope Quattro SEM. All samples were previously treated with the etch (the etchant - H_2SO_4 with etanol and $CuCl_2$, etch time – 5…7 s). Microhardness mechanical tests and stiffness measurements were performed in automatic mode via Nanovea mechanical hardness tester PB1000 with a triangular Bercovich diamond pyramid as an indentor. Number of test points was equal to 9 in each experiment, indention force was equal to 3 N, load-relaxation cycle time was approximately 200 s per each measurement. Resulted values of microhardness were recalculated in a Vickers's scale automatically by means of Nanovea equipment and software.

3. Results and discussion

Two different series of experiments were performed in our research.

3.1. Experiment №1.

The FG experimental specimen type №1 was created of 25 layers of steel, 1 layer of bronze and 25 layers of steel again (figure 2). The results of the experiment demonstrated cracking in the interface between steel and bronze. It was proved via optical microscopy (figure 2).

The results of microhardness and stiffness measurements (figure 3, table 1) of the specimen type №1 demonstrated local increase of microhardness on a border between steel and bronze. Supposedly the reason of this embrittlement was intermetallic growth (presence of intermetallics was proved by X-ray

IOP Conf. Series: Materials Science and Engineering **969** (2020) 012104 doi:10.1088/1757-899X/969/1/012104

diffraction analysis). Internal stresses and deformations in the transitional zone caused by difference between temperature expansion coefficients of materials (up to $19.6 \cdot 10^{-6}$ °C^{-1} for AISI 316L and at the most 17.5 °C^{-1} for aluminium bronze) were the reason of crack nucleus occurring on the grains of intermetallics.

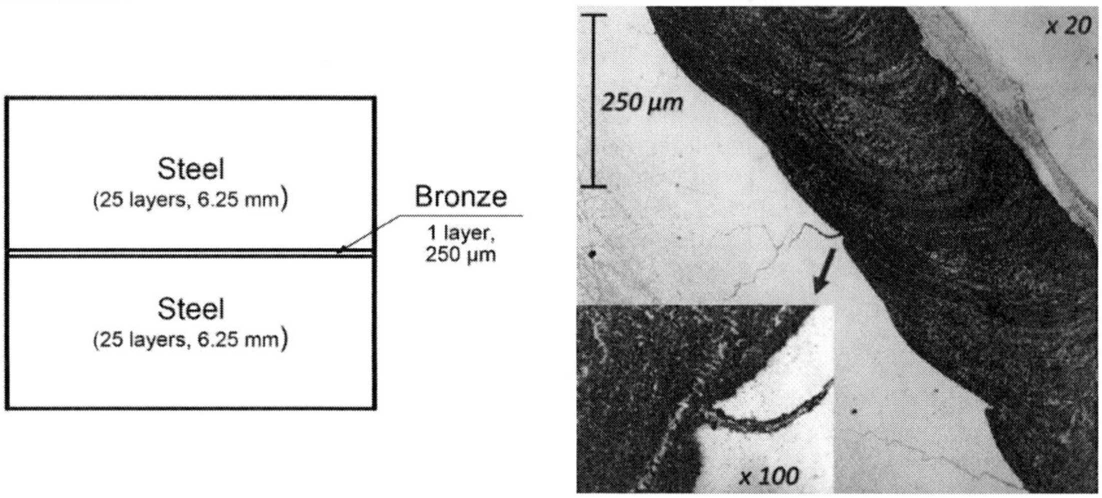

Figure 2. The experimental specimen type №1. In the right picture: dark area is bronze, bottom part of the specimen is leftward.

The results of microhardness and stiffness measurements (figure 3, table 1) of the specimen type №1 demonstrated local increase of microhardness on a border between steel and bronze. Supposedly the reason of this embrittlement was intermetallic growth (presence of intermetallics was proved by X-ray diffraction analysis). Internal stresses and deformations in the transitional zone caused by difference between temperature expansion coefficients of materials (up to $19.6 \cdot 10^{-6}$ °C^{-1} for AISI 316L and at the most 17.5 °C^{-1} for aluminium bronze) were the reason of crack nucleus occurring on the grains of intermetallics.

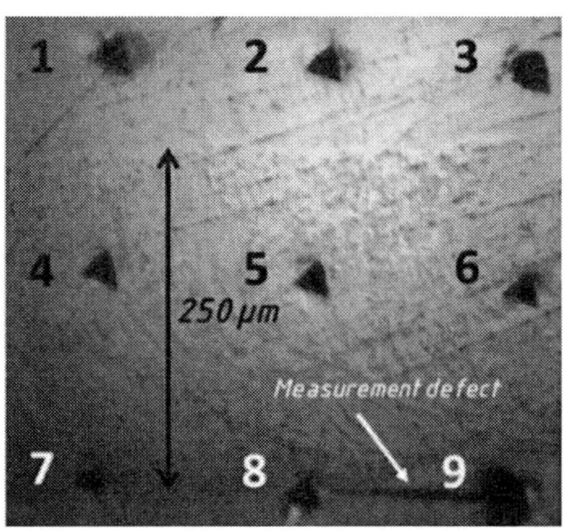

Figure 3. The microhardness measurement pattern of the experimental specimen type №1.

Table 1. The results of microhardness and stiffness measurements of the experimental specimen №1.

	1	2	3	4	5	6	7	8	9
Microhardness, HV	133	171	143	179	196	195	223	177	186
Young's modulus of elasticity, GPa	32.6	34.9	29.8	34.2	33.8	33.7	36.7	33.5	34.4

3.2. Experiment №2.

The specific multilayer transition technique (figure 4) was used to cease cracking on the same treatment modes as in the experiment №1. Optical microscopy analysis proved absence of cracks (figure 4).

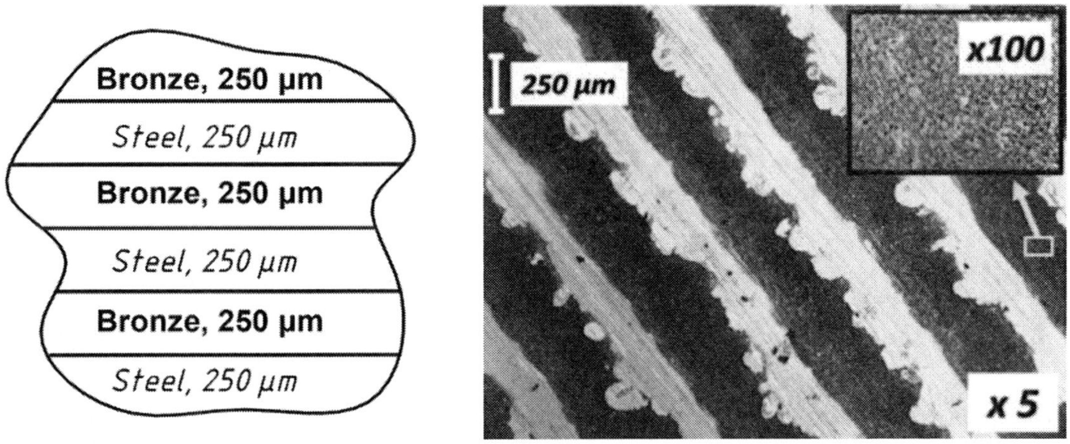

Figure 4. Experimental specimen type №2: multilayer bronze (dark layers) - steel (light layers) structure. Bottom part of the specimen is leftward in the right picture.

The microhardness research demonstrated even higher values of microhardness and lower elasticity (figure 5, table 2) in comparison with experiment №1. However, these phenomena didn't cause cracking.

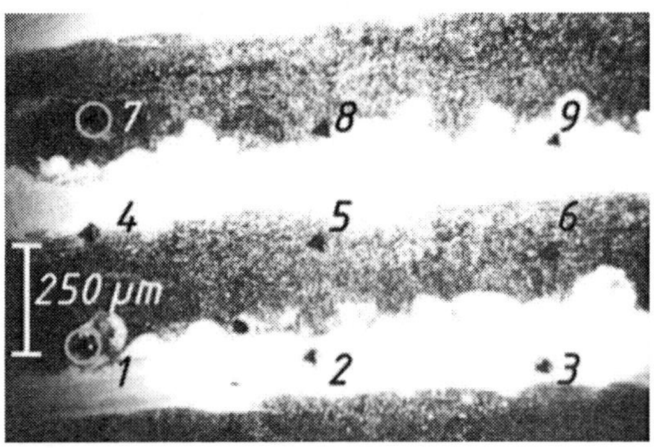

Figure 5. A microhardness measurement pattern of the experimental specimen type №2.

Table 2. The results of microhardness and stiffness measurements of the experimental specimen №2.

	1	2	3	4	5	6	7	8	9
Microhardness, HV	170	266	257	242	191	198	203	188	238
Young's modulus of elasticity, GPa	37.8	43.2	42.1	40.3	35.6	36.2	37.1	36.6	40.6

Scanning electronic microscopy of the specimen type №2 allowed to observe dendritic structure (figure 6, left picture - dendritic crystals appear as herringbone dark patterns, two of which are marked with orange ellipses) and indicated nanometer porosity (figure 6, right picture - porosity is seen as a small black disks, some of which are marked with the red circles) which had no influence on the strength properties.

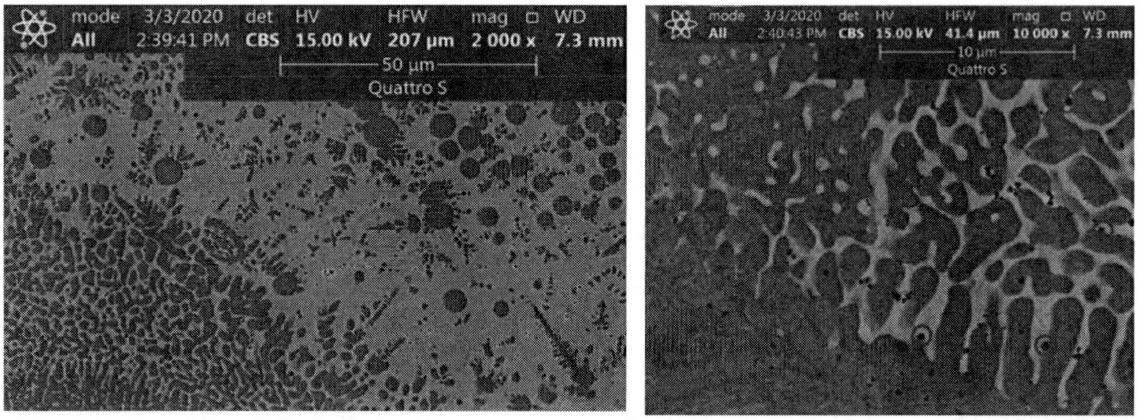

Figure 6. Dendrites (left picture) and nanometer porosity (right picture) of the specimen type №2.

3.3. Numerical simulation

A numerical simulation in our research was performed via Dassault Systèmes Abaqus software. A purpose of analysis was confirmation that stress concentrators (which were the factor of cracking) were located in the interface area between steel and bronze in the specimen type №1. We researched a stress-strain state of the 1 mm diameter elementary analytical cylinder (imaginatively cut from the central area of the specimen to exclude edge effects from consideration). Influence of the temperature gradient was substituted by equivalent mechanical stresses. One tail of the cylinder was encastred (boundary conditions in finite elements analysis were: $F_{x0} = F_{y0} = F_{z0} = M_{x0} = M_{y0} = M_{z0} = 0$); plane distributed tensile stresses, caused by inequal heating, acted at another tail; compressive stresses from metal around the cylinder acted at its envelope (figure 7). The value of all stresses (tensile and compressive) for a first approximation was set equal to 500 MPa [39, 40]. Exact value of resultant stresses did not play a significant role in the simulation results. Mechanical characteristics of materials were: of steel - mass density 7850 kg/m^3, Young's modulus $2.1 \cdot 10^5$ MPa, Poissons ratio 0.3; of bronze - mass density 8960 kg/m^3, Young's modulus $1.1 \cdot 10^5$ MPa, Poissons ratio 0.35. Parameters of mesh: $9 \cdot 10^{-5}$ m approximate global size of seeds, 0.1 maximum deviation factor, linear geometric order C3D8R element type (an 8-node linear brick). The results of numerical simulation (figure 8) proved that mechanical stresses in the specimen type №1 had their ultimate values in the interface between steel and bronze (the red circles in the figure 8). The model in figure 8 is demonstrated with local angular cross section which shows distribution of stresses inside the cylinder. Blue area in the base of cylinder is a zone with the lowest value of mechanical stresses (what was certainly predictable because of boundary conditions). A light-green area in the figure 8 demonstrates a low-stressed zone.

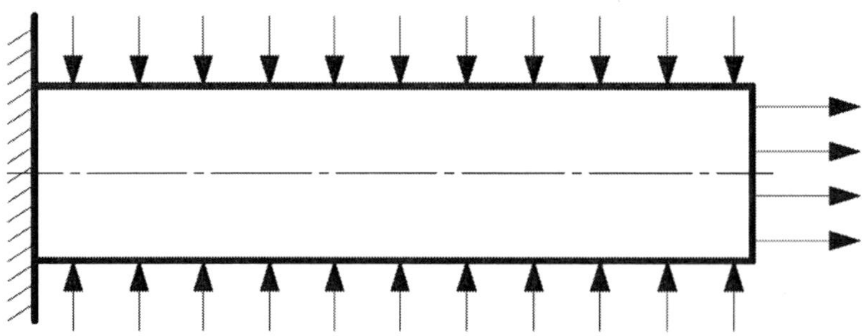

Figure 7. The loading pattern of the elementary cylinder (not to scale).

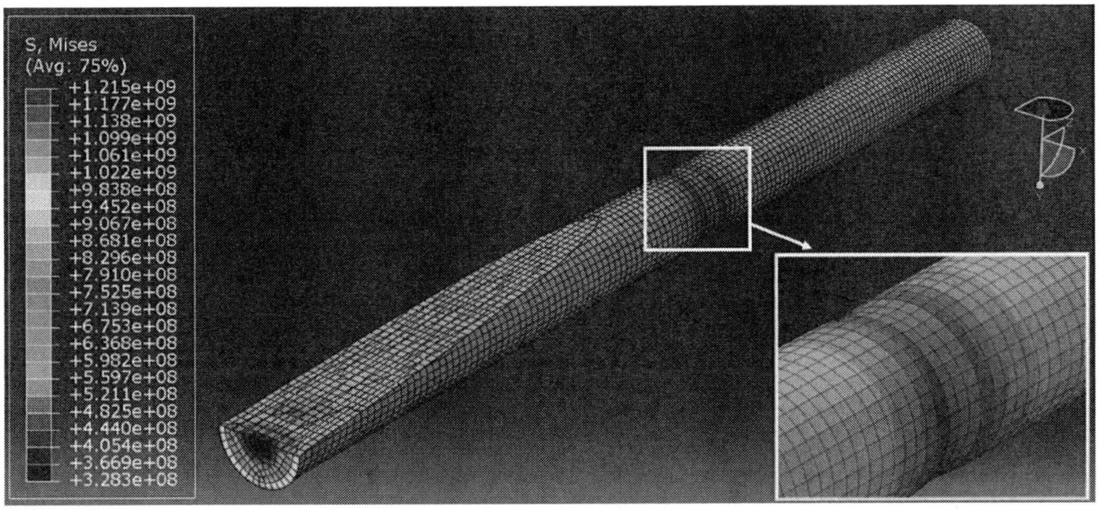

Figure 8. The 3D-stress analysis results.

4. Conclusion. Key results and practical significance

A difference between thermal expansion coefficients of the materials along with rapid intermetallic growth, increase of microhardness (up to 223 HV), increase of elasticity modulus (up to 36.7 GPa), stress concentration and local embrittlement of the FG laser deposited specimens with two sharp transitions created of stainless steel AISI 316L and aluminium bronze in the DMT mode via the straight joining scheme lead to cracking on a border between stainless steel and aluminium bronze. These effects can be suppressed by producing (on the same regimes) a multilayer structure with alternating of steel and bronze layers, which will get comparatively more favourable thermal history during the process of laser treatment. Such multilayer structures can have even a higher level of microhardness (up to 266 HV) and Young's modulus of elasticity (up to 43.2 GPa), but nevertheless, cracking of them can be entirely excluded because of reassignment of internal stresses and deformations in a volume of a detail. Porosity of such multilayer structures has nanometer level size and doesn't effect on their mechanical strength properties. These phenomena and manufacturing method have a possibility to be used for purposes of space industry FG parts manufacturing of stainless steel and aluminium bronze via DED.

References

[1] Shishkovsky I V 2001 Synthesis of functional gradient parts via RP methods *Rapid prototyping J.* **7** **4** 207–211
[2] Koizumi M 1997 FGM activities in Japan *Compos.* **B 28** 1–4
[3] Kieback B, Neubrand A and Riedel H 2003 Processing techniques for functionally graded materials *Mater. Sci. and Eng.* **A 362** 81–105
[4] Articek U, Milfelner M and Anzel I 2013 Synthesis of functionally graded material H13/Cu by LENS technology *Adv. Prod. Eng. Manag.* **8** **3** 169–176
[5] Smurov I, Doubenskaia M, Grigoriev S and Kotoban D 2014 Selective laser melting and direct metal deposition: from process fundamentals towards advanced products *2014 Int. Conf. Laser Optics*
[6] Smurov I Y 2008 *Surf. Coat. Technol.* **202** 4496–4502
[7] Ribeiro S B K, Mariani F E and Coehlo R T 2020 A study of different deposition strategies in direct energy deposition (DED) processes *Proc. Manuf.* **48** 663–670
[8] Yan L, Chen Y and Liou F 2020 *Addit. Manuf.* **31** 100901
[9] Meng W, Zhang Wenhao, Zhang Wang, Yin X and Cui B 2020 *Opt. Las. Technol.* **131** 106451
[10] Shishkovsky I, Missemer F and Smurov I 2012 *Phys. Proc.* **39** 382–391

[11] Oyelola O, Grawforth P, M'Saoubi R and Clare T A 2018 *Addit. Manuf.* **24** 20–29
[12] Griffith M L, Harwell L D, Romero T J, Schlienger E, Atwood C L and Smugeresky J E 1997 Multi-material processing by LENS *Int. Solid Freeform Fabrication Symp.* 387–393
[13] Shishkovsky I, Kakovkina N and Missemer F 2016 *Laser Eng.* **33** 1–15
[14] Vilar R et al 2009 *Acta Mater.* **57(18)** pp 5292–302
[15] Balla V K, DeVasConCellos P D, Xue W, Bose S and Bandyopadhyay A 2009 *Acta Biomater.* **5** 1831–37
[16] Srivastava D, Chang I T H and Loretto M H 2001 *Intermetallics* **9** 1003–1013
[17] Liu W and DuPont J N 2004 *Metallurg. Mater. Transact.* **A 35 13** 1133–1140
[18] Pimenova N V and Starr T L 2006 *Electromech. Acta* **51** 2042–2049
[19] He X, Han J and Zhang X 2002 Preparation and ablating behaviour of FGM used in a heat flux rocket engine *34th COSPAR Scientif. Assembly of the 2nd World Space Congress*
[20] Moussa A A and Yadav R 2017 *J. Medic. Eng.* **2017** 3069351
[21] Pei Y T and De Hosson J TH M 2000 *Acta Mater.* **48** 2617–2624
[22] Pęska M, Karczewski K, Rzeszotarska M and Polański M 2020 *Mater.* **13** 531
[23] Łyszkowski R 2015 *Mater.* **8** 1499–1512
[24] Novák P, Zelinková M, Šerák J, Michalcová A, Novák M and Vojtěch D 2011 *Intermetallics* **19** 1306–1312
[25] Morris D G, Gutierrez-Urrutia I and Muñoz-Morris M A 2008 *Int. J. Plast* **24** 1205–1223
[26] Schmitt A, Kumar K S, Kauffmann A, Li X, Stein F and Heilmaier M 2017 *Intermetallics* **90** 180–187
[27] Liu Y, Chong X, Jiang Y, Zhou R and Feng J 2017 *Phys.* **B 506** 1–11
[28] Shen C, Liss K, Pan Z, Wang Z, Li X and Li H 2018 *Intermetallics* **92** 101–107
[29] Ruan Y, Zan N, Zhu H Z, Zhou K and Wei B 2017 *J. Alloys Compd.* **701** 676–681
[30] Sahasrabudhe H, Bose S and Bandyopadhyay A 2018 *Acta Biomater.* **66** 118–128
[31] Banait S M, Paul C P, Jinoop A N, Kumar H, Pawade R S and Bindra K S 2020 *Opt. Las. Technol.* **121** 105787
[32] Amado J M, Montero J, Tobar M J and Yáñez A 2012 *Phys. Proc.* **39** 362–367
[33] Li W, Karnati S, Kriewall C, Liou F, Newkirk J, Taminger K M B and Seufzer W J 2017 *Addit. Manuf.* **14** 95–104
[34] Reichardt A, Dillon R P, Borgonia J P, Shapiro A A, McEnerney B W, Momose T and Hosemann P 2016 *Mater. Design* **104** 404–413
[35] Onuike B, Heer B and Bandyopadhyay A 2018 *Addit. Manuf.* **21** 133–140
[36] Karnati S, Sparks T E, Liou F, Newkirk J W, Taminger K M B and Seufzer W J 2015 Laser metal deposition of functionally gradient materials from elemental copper and nickel powders *Proc. of the 26th Solid Freeform Fabrication Symp.* 789–802
[37] Moharana B R, Sahu S K, Maiti A, Sahoo S K and Moharana T K 2020 An experimental study on joining of AISI 304 SS to Cu by Nd-YAG laser welding process *Mater. Today: Proc.*
[38] Hofmann D C, Roberts S, Otis R, Kolodziejska J, Dillon R P, Suh J, Shapiro A A, Liu Z and Borgonia J 2014 *Sci. Rep.* **4** 5357
[39] Chen C, Xiao Z, Zhu H and Zeng X 2020 *J. of Mater. Processing Technol.* **284** 116726
[40] Saboori A, Piscopo G, Lai M, Salmi A, Biamino S 2020 *Mater. Sci. Eng.* **A 780** 139179

Acknowledgments
Authors are grateful to their colleagues: Oleg Dubinin, Petr Shornikov, Stanislav Evlashin and Yulia Kuzminova, who made a great contribution to the research.

IOP Conf. Series: Materials Science and Engineering **969** (2020) 012105 doi:10.1088/1757-899X/969/1/012105

Enhancement of Deposition Process Controlling in Electron Beam Metal Wire Deposition Method

A V Shcherbakov, R V Rodyakina and R R Klyushin

National Research University "Moscow Power Engineering Institute", 14, Krasnokazarmennaya str., Moscow, 111250, Russia

E-mail: RodiakinaRV@mpei.ru

Abstract. The urgency of improving control systems for electron beam metal wire deposition method is proved in terms of creating methods for controlling the geometric parameters of deposited layer. It is shown that to solve this problem in electron beam metal wire deposition method, it is advisable to use reflected electron signal detection when scanning the deposited layer. Using a mathematical model based on the application of continuous loss approximation and Monte-Carlo method, the probe characteristics recorded during cylindrical layer scanning are investigated. Experimental verification of the results obtained was carried out using an electron-beam technological installation. The effect of the distance between the layer and electron collector on the recorded signals is investigated. The possibility of using this method for simultaneous control of deposited layer height and width is shown.

1. Introduction

The electron beam metal wire deposition technology has been actively developed in the last decade. Technologies have been developed based on the use of a laser beam [1], an electric arc [2] and other types of electric heating, implemented in a protective gas environment. When implementing these technologies, it is extremely important to continuously monitor the deposited layer height. This is necessary for the correct selection of deposition tool lifting speed during the transition from the previous layer to the next. In addition, it is important to know the geometrical dimensions of the layer being formed – its height and width. To solve the set tasks in the implementation of the above technologies, machine vision systems [3] and 2D laser scanners [4] are used. Continuous monitoring of height and width of the layer being deposited allows adjusting the main process parameters – speed of deposition tool movement and wire feed speed, as well as stabilizing distance between the deposition tool and the layer. In the case of electron beam using as a heating source, such difficulties for optical sensors using arise as intense metal evaporation in a vacuum chamber closed volume and overheating of electronic components. An alternative solution that reduces both the cost and the technical complexity of such control system is reflected electron signals using, generated during layer surface scanning. Such solutions have been developed and applied for a long time to detect the joint of parts in electron beam welding [5, 6], and even for automatic joint tracking [7]. The joint is a narrow deep slit, when combined with which most of the reflected electrons are absorbed. Therefore, when a signal is detected, a "contrast" is formed between the joint and the surrounding surfaces of parts. When electron beam metal wire deposition method using, such a sharp contrast is not observed, and the surface on which the layer is deposited can have a very different shape. Therefore, confirmation of the efficiency of reflected electron detectors application for determining of deposited layer height and width requires additional research.

2. Proposed control method

Figure 1a shows a typical form of a single deposited layer, obtained by electron beam metal wire deposition method [8]. It can be seen that the roll in cross section has a shape close to a semicircle, and the ratio of the roll height to its width is approximately equal to one. Figure 1b shows a diagram of deposited layer scanning process. Electron beam 1 is periodically deflected onto the already formed layer 2 and crosses it at a constant speed in a direction perpendicular to the direction of metal's deposition. A magnetic deflection coil is used for beam's deflection. During scanning periods, the metal heating process is interrupted, however, this should not affect heat and mass transfer, since time period spent on scanning is small (no more than a few milliseconds). It is important that plasma processes do not affect measured signal during scanning. When interacting with layer surface, flows of backscattered electrons 3 will be created, some of which will fall on the collector 4. The flux intensity and angular distribution of backscattered electrons will depend on the inclination angle of metal surface to the beam [9]. The re-reflection of electrons from layer surface will also play a role. The reflected electrons current flowing through the measuring resistor 5 will be small (usually 0.01-1% of the beam current), therefore, amplifier 6 is needed to amplify the current signal.

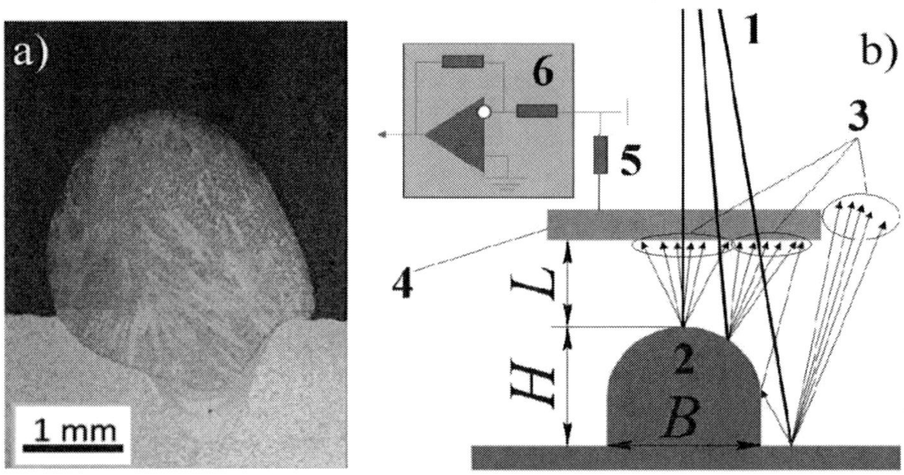

Figure 1. Layers shape deposited in the Electron Beam Free Form Fabrication process (EBF3 process) (a) and diagram of layer surface scanning process with detection of backscattered electron signal (b): 1 – scanning beam, 2 – deposited layer, 3 – backscattered electrons, 4 – collector (or several collectors), 5 – measuring shunt, 6 – signal amplifier.

Figure 1 also shows such geometrical parameters of deposited layer as its width B and height H. The reflected electron detector is installed at a distance L from the deposited layer and is rigidly attached to an electron gun with a wire feeder.

3. Simulation of scanning process and verification

To study the influence of layer geometric parameters on the recorded signal, a mathematical model based on the continuous loss approximation was used [9]. The stopping power, or electron energy loss per unit path length, was determined according to the expression

$$-\frac{dE}{dS}=-2\pi e^4 N\frac{Z\cdot\rho}{A\cdot E}\ln\left(\frac{1.166E}{J}\right), \tag{1}$$

where E is electron energy current value, keV, S is electron path, cm, e is electron charge, C; other parameters refer to the material: Z is atomic number, ρ is density, g/cm^3, A is atomic mass, g/mol, J is ionization potential, keV, which is determined by the expression:

$$J=\left(9.76Z+58.5Z^{0.19}\right)\cdot 10^{-3}. \tag{2}$$

In accordance with the indicated approximation, it is assumed that only elastic scattering acts lead to electrons deflection at significant angles. The elastic scattering cross section was calculated by the expression

$$\sigma_E=5.21\cdot 10^{-21}\frac{Z^2}{E^2}\cdot\frac{4\pi}{\alpha(1+\alpha)}\cdot\left(\frac{E+511}{E+1024}\right)^2, \tag{3}$$

where α is screening parameter determined by the Bishop formula

$$\alpha=3.4\cdot 10^{-3}\frac{Z^{0.67}}{E}. \tag{4}$$

The most probable mean free path is

$$\lambda_E=\frac{A}{N\cdot\rho\cdot\sigma_E}. \tag{5}$$

The electron beam in the model is represented in the form of a finite number of electrons having equal energies and distributed over a circle in accordance with normal distribution law [10]. It is assumed that the electron experiences elastic scattering at some point P1, arriving in it from the point P0. The main task of modeling is to calculate the coordinate of the point P2, where the electron arrives. To calculate this coordinate, it is necessary to determine electron path s, as well as solid scattering angle ϕ and azimuthal scattering angle ψ [8, 9]. In accordance with Monte Carlo method, electron path between elastic scattering events is determined by the expression

$$s_E=-\lambda_E\cdot\ln\gamma, \tag{6}$$

where γ is a random value selected from the range from 0 to 1. The next step is to calculate the scattering angles. Based on the Rutherford model, we use the Reimer-Krefting expression to determine cosine of elastic scattering angle:

$$\cos\phi=1-\frac{2\alpha\gamma}{1+\alpha-\gamma} \tag{7}$$

The azimuth angle is determined randomly:

$$\psi=2\pi\gamma. \tag{8}$$

Expression (1) shows that electron energy loss per unit path length will depend on the current value of particle energy, as well as on the material properties. Electron trajectory range between acts of elastic scattering leading to deflection at significant angles (and the angle itself) will also depend on the material properties. Using expressions (1)–(8), a computer program was developed to simulate electron beam (with an energy of 60 keV) scattering process when interacting with a steel layer (calculations were carried out for pure iron) [11, 13, 16, 17]. To simulate scanning, beam angle of incidence on the layer θ was changing. Layer height and distance between layer surface and collector were taken as variable parameters. The number of calculated particle trajectories is 10 000. The simulation results are shown in figure 2.

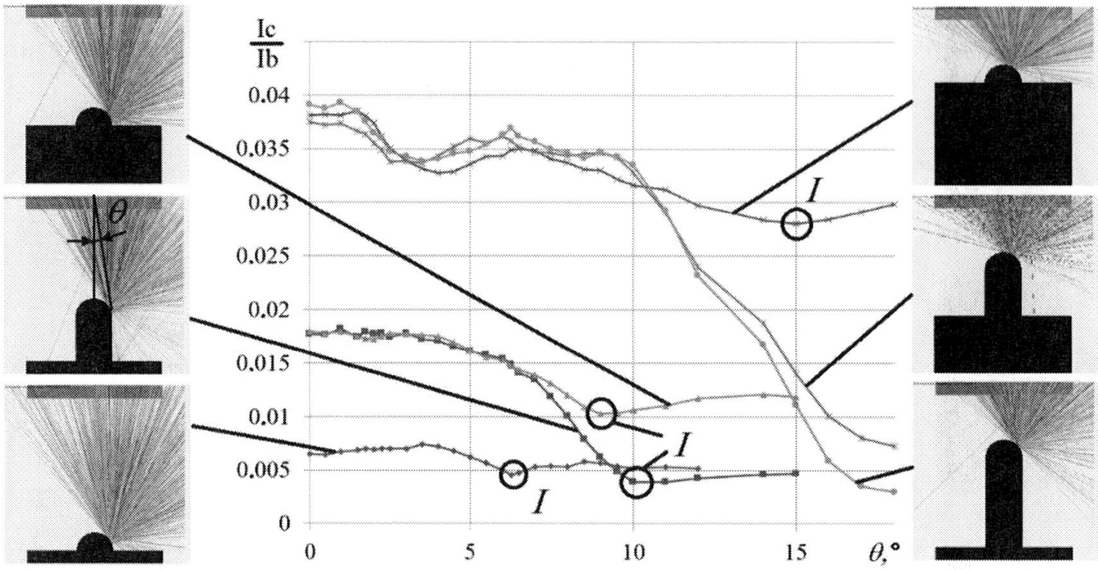

Figure 2. Dependences of the ratio between collector and beam currents (I_c/I_b) on beam deflection angle θ obtained using the model.

The experimental results presented in figure 3 also show that the probe characteristics can be used to determine the distance between detector and layer surface, as well as layer width. The minimum values of recorded signal can be interpreted as corresponding to beam position near the vertical edges of the layer being formed. The waveforms shown in figure 3 correspond to the case when beam deflection angle is small, and distance between collector and layer is not less than 100 mm. However, in figures 2 and 3, the regions where detector current passes through the minimum are visible, they marked as I. These regions correspond to such time moments at which electron beam crosses layer border, and the reflection of electrons from vertical surface leads to a decrease in the current recorded by the detector [12, 14, 15].

The beam deflection angle in the real case does not exceed 5 degrees, therefore, displacement of the minimum signal regions I with a change in the distance between detector and layer surface is insignificant. However, the signal amplitude changes, and when the layer is positioned close to the detector, it will be maximum. It is advisable to use two detectors to determine the distance between layer surface and electron gun and layer width. The first detector should be narrow (narrower than the layer). It is used to determine the distance between electron gun and layer surface. The second detector can be wider and is used to measure layer width. The recorded current decreased almost to zero due to the fact that the beam was deflected beyond the substrate – to the tooling located below. The described control method is well combined with other proposed methods based on the measurement of probe signals [18, 19, 20].

Conclusions

1. A method based on reflected electron signal measurement is proposed to control the distance between electron gun and layer surface. This method is also suitable for measuring layer width.
2. It is shown that by minimizing scanning time, the method is suitable for continuous measurement of deposited layer height and width. This is extremely important for building a control system for the additive forming process.

3. This control method can be implemented by means of a neural network control system and introduced into the numerical program control system of an electron-beam technological installation.

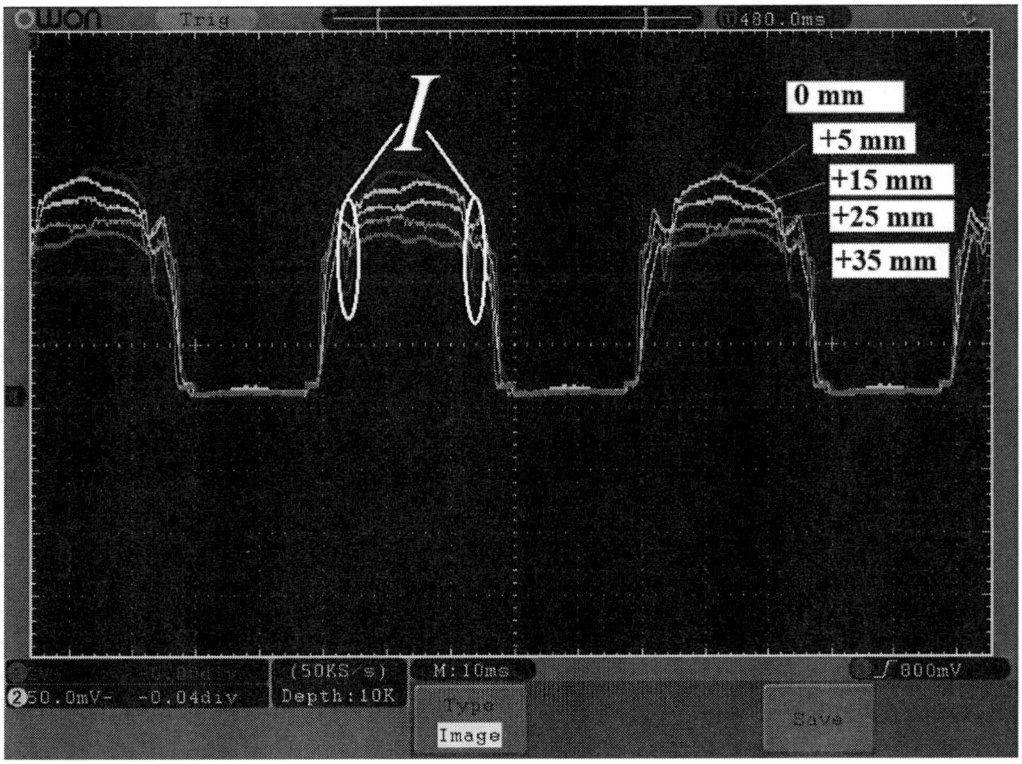

Figure 3. Waveforms obtained experimentally when scanning the deposited layer at various values of relative displacement between layer surface and detector (position 0 mm corresponds to a distance of about 100 mm).

References

[1] Meiners W, Wissenbach K and Gasser K 1996 *Shaped body especially prototype or replacement part production* (Germany: Patent DE19649865C1)
[2] Busachi A, Erkoyuncu J, Colegrove P, Martina F and Ding J 2015 *Procedia CIRP* **37** 48–53
[3] Taminger K M, Domack C S, Zalameda J N, Taminger B L, Hafley R A and Burke E R 2016 *Proc. SPIE* **9861** 986102
[4] Zalameda J N, Burke ER, Hafley R A, Taminger K M, Domack C S, Brewer A and Martin R E 2013 *Proc. SPIE* **8705** 87050M
[5] Gegel M, Nisbett A, Bristow D and Landers R 2016 *Proc. of the 27th Annual Int. Solid Freeform Fabrication Symposium*
[6] Oltean S 2018 *Proc. Manufacturing* **22** 605–612
[7] Druzhinina A A, Laptenok V D, Murygin A V and Laptenok P V 2016 *IOP Conf. Ser.: Mater. Sci. Eng.* **155** 012009
[8] Węglowski M St, Błach S and Phillips A 2016 *Vacuum* **130** 72–92
[9] Adam V, Claub U, Dobeneck D, Krüssel T and Löwer T 2011 *Electron Beam Welding. The fundamentals of a fascinating technology* (Germany: Pro-Beam AG & Co) p 98
[10] Davé V 1995 *Electron Beam (EB)-Assisted Materials Fabrication/ Ph.D. Thessis in Materials Engineering* (USA: Massachusetts Institute of Technology) p 280
[11] Fuchs J, Schneider C and Enzinger N 2018 *Welding in the World* **62** 267–275
[12] Dragunov V K 2019 *IOP Conf. Ser: Mater. Sci. Eng.* **681** 012008
[13] Węglowski M, Błacha S, Jachym R, Dutkiewicz J and Rogal Ł 2018 *E+E* **53(3–4)** 74–78
[14] Gudenko A and Sliva A 2018 *J. Phys.: Conf. Ser.* **1109** 012037

[15] Shcherbakov A V, Gaponova D A and Rodyakina R V 2020 *Proc. of the 5th Int. Conf. on Ind. Eng.* **2** 1127–1135
[16] Goldstein J, Newbury D E, Joy D C, Lyman C E, Echlin P, Lifshin E, Sawyer L and Michael J R 2003 *Scanning Electron Microscopy and X-Ray Microanalysis. Third Edition.* (USA: Springer) p 689
[17] Joy D C 1995 *Monte Carlo Modeling for Electron Microscopy and Microanalysis* (New York: Oxford University Press) p 216
[18] Shcherbakov A V, Martynov V N, Kharitonov I A, Gaponova D A, Rodyakina R V and Dragunov V K 2018 *Russian Electrical Eng.* **89** 249–254
[19] Shcherbakov A V, Rodyakina R V, Kozhechenko A S, Gaponova D A, Goncharov A L and Dragunov V K 2017 *Tech. Phys. Letters* **43** 958–960
[20] Shcherbakov A V, Goncharov A L, Kozhechenko A S, Pogrebisskii M Ya, Rubtsov V P and Dragunov V K 2016 *Russian Electrical Eng.* **87** 487–492

Acknowledgments
This study conducted by Moscow Power Engineering Institute was financially supported by the Ministry of Science and Higher Education of the Russian Federation (project No. FSWF-2020-0023).

IOP Conf. Series: Materials Science and Engineering **969** (2020) 012106 doi:10.1088/1757-899X/969/1/012106

Cold Spray Deposition of Copper/Tungsten Composite Coatings

M Doubenskaia[1], Yu S Latfulina [2] and M N Samodurova [2]

[1]Lyon University, ENISE, LTDS Laboratory, UMR CNRS 5513, 58 rue Jean Parot, Saint-Etienne Cedex 2, 42023 France
[2]Processes and Machines of Metal Forming by Pressure Department, South Ural State University, 76, Lenin prospect, Chelyabinsk, 454080, Russia

E-mail: samodurovamn@susu.ru

Abstract. This work is devoted to the study of the microstructure and the properties of the Cu-W composite coatings deposited using cold spray technology. This process allows fabrication of high-quality coatings on large surfaces of different types of materials. The technology is widely used and investigated due to technological simplicity, wide range of sprayed materials and operating temperatures. However, there are some technological problems which appear when materials with low melting temperatures are sprayed. During softening, the particles adhere to the nozzle walls and block spraying. By the way, it is possible to solve the problem adding some ceramics particles to the spraying mixture of powders. As a result, the formed coating will have better mechanical properties and the technological problem will be solved. The analysis of the general results of the microstructure studies and experiments, and comparison with other coating methods are given. Moreover, the experimental procedure of samples preparation and used equipment are given. It is shown the microstructure before and after heat treatment of the samples sprayed, microhardness measurements, thermal conductivity measurements.

1. Introduction

Tungsten copper (W-Cu) composites are widespread refractory material [1–4]. They are prospective materials for such areas of manufacturing like arcing resistance electrodes, heavy duty electronic contacts, electrical contacts and electrodes, electro-forging and welding dies, heat sinks, packaging material and thermal management devices owing to their outstanding properties [7].

For use in these areas of production [4], the materials should have the highest possible thermal conductivity and low coefficient of thermal expansion (CTE) with the possibility of its regulation, the same as for an electronic substrate such as gallium arsenide (CTE=5.8×10^{-6}/K). Tungsten/copper composites have priority over other materials as the best candidates for heat sinking because they have high thermal conductivity (~200 W/m K) and low CTE (~6.5×10^{-6}/K).

The production of tungsten/copper coatings by infiltration is very expensive due to the need to use high temperatures for tungsten copper impregnation [5].

The cold spraying process is more suitable in terms of production costs due to the lack of high temperatures and does not change the properties of the sprayed surface. Therefore, H.-K. Kang and S.B. Kang [5, 6] studied the cold spraying of tungsten/copper composition (Figure 1) and compared it with the plasma spraying method. As a result, copper oxidized during the plasma spraying process and deteriorated the coating properties, while copper did not get oxidized after the cold spraying process (Figure 2).

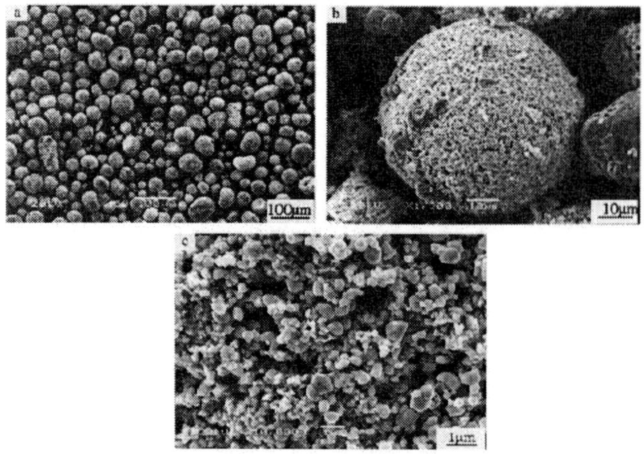

Figure 1. SEM images of agglomerated W/Cu composite feedstock (a) ×150, (b) ×1300, and (c) ×10,000.

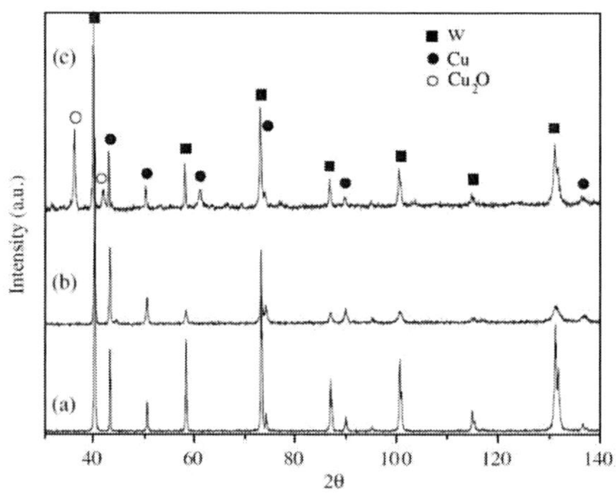

Figure 2. XRD patterns of (a) feedstock powder, (b) cold-sprayed and (c) plasma-sprayed deposits.

Microstructural analysis also showed the dependence of the amount of porosity on the tungsten content in the composition. The more tungsten is in a certain area of coating, the greater the number and size of pores. (Figure 3).

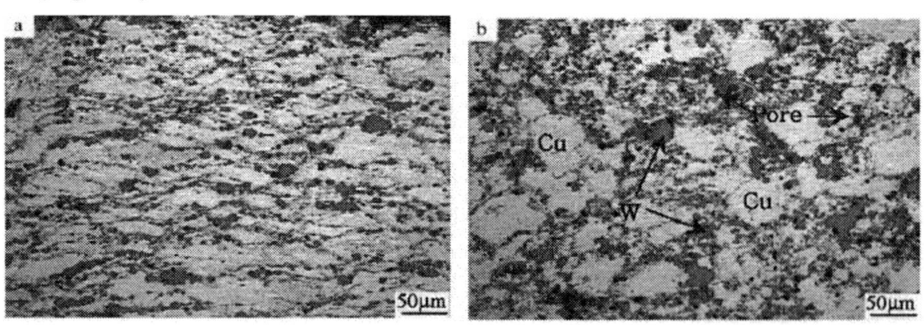

Figure 3. Optical images of cold-sprayed deposit: (a) cross-section and (b) in-plane section.

IOP Conf. Series: Materials Science and Engineering **969** (2020) 012106 doi:10.1088/1757-899X/969/1/012106

The content of tungsten in the coating thickness of about 100-200 μm was the maximum (Figure 4), then showed its decline after increasing the thickness of the coating, which showed the ineffectiveness of applying coatings thicker than 200 μm.

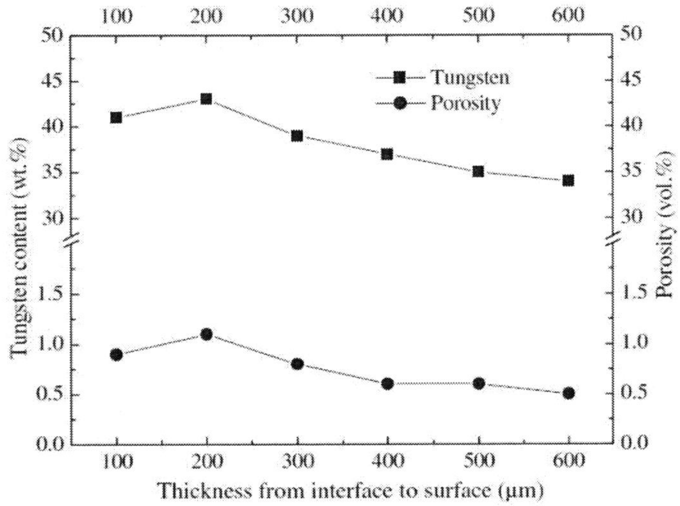

Figure 4. Tungsten content and porosity versus location in the deposited layer.

It can be seen that cold spray technology is promising for the production of coatings on different materials. In contrast to other additive manufacturing methods, cold spraying technology does not use high temperatures and metal melting as in selective laser melting. This means that metals are not subject to thermal distortion, and for the manufacture of parts does not require the use of protective inert gas medium or vacuum seal, which allows you to create much larger structures [8–11]. In the manufacture of products using cold spraying technology, such advantages become apparent as the ability to create any shape and size of products, and capable of fusing dissimilar metals to create hybrid metal parts – materials such as titanium alloys, copper, zinc, stainless steel, aluminium, nickel, even hastelloy and inconel can be sprayed together.

2. Materials and methods

5 mixtures of different composition were taken. Compositions of mixtures are presented in Table 1.

Table 1. Compositions of mixtures.

Copper, wt.%	Tungsten, wt.%
95	5
75	25
50	50
25	75
10	90

Commercial cold spray equipment CGT KINETIKS 4000 with Type 40 spray nozzle was applied for coating deposition. Deposition was performed on aluminum plates under parameters listed in Table 2.

Thick coatings (more than 1 mm) were applied in order to minimize an influence of substrate on the kinetics of coating formation. Coating composition was defined using image analysis. The image analysis was performed using cross-section images obtained by scanning electron microscopy.

Table 2. Spraying parameters.

Gas type	Gas pressure, bar	Gas temperature °C	Powder feeder speed mm/s	Manipulation speed, mm/s	Track overlapping, mm	Number of layers	Spraying distance, mm
Nitrogen	30	400 °C	8	20	3	2-10	30

The obtained samples were prepared for further metallographic studies by using the BRILLANT 230 and Buehler SimpliMet 1000. The sample preparation system "Buehler Phoenix 4000" was used for surface grinding and polishing as the standard procedure.

The specimens were etched in order to observe the microstructure of the composite coatings. The specimens were studied by a Zeiss AxioCam ICc 3 Microscope. For more detailed studies, the scanning electron microscope (SEM) Tescan was used. SEM analyses were performed using a 15 kV beam voltage. The specimens were analyzed under magnifications in a range of ×50 to ×2000 at two modes: back-scattered electron (BSE) to observe a phase contrast and secondary electron (SE) to get high-resolution images of a sample surface, revealing details less than 1 nm in size [12].

3. Results and discussion

3.1. Deposition efficiency

Images of obtained cross-sections are presented in Figure 5. Hard particles (light) of tungsten are distributed in soft matrix of copper.

During the chemical analysis the percentage of tungsten in the coating was found (Table 3).

Table 3. The amount of tungsten in coatings after cold spray.

Composition	Amount of tungsten, wt.%
Cu + 5% W	5
Cu + 25% W	9
Cu + 50% W	15
Cu + 75% W	30
Cu + 90% W	49

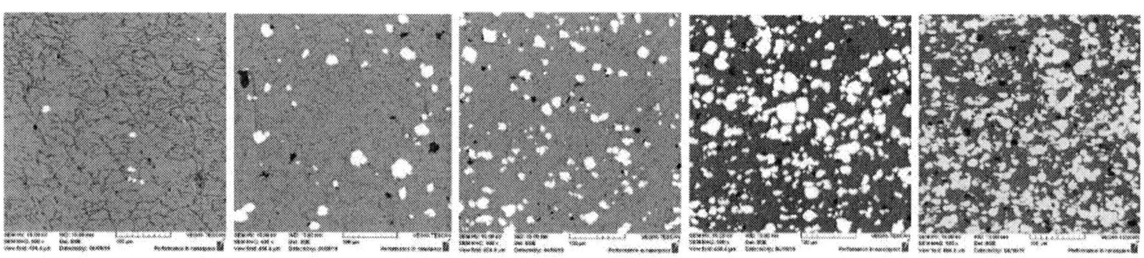

(a) (b) (c) (d) (e)

Figure 5. SEM images of coating cross-section deposited using different blends; (a) – Cu + 5% W, (b) – Cu + 25% W, (c) – Cu + 50% W, (d) – Cu + 75% W, (e) – Cu + 90% W.

From the experimental results it can be seen that the maximum concentration of tungsten does not exceed 48–50%. After the chemical analysis, it becomes obvious that the maximum content of tungsten in the mixture will not exceed 50%. The sharp drop in the effectiveness of the coating deposition is noticeable after an increase in the tungsten content of more than 10–15% of the total weight of the composition. This phenomenon is due to the high hardness of tungsten particles. If the hard particle meets the surface covered by previously bonded hard particle, the impact energy is not high enough for bonding and hard particle rebounds.

3.2. *Microhardness results*

The Vickers hardness test [13] was performed on the well-polished cross-sections of the specimens. To get an average hardness level the measurements were carried out in 5 equal lines consisted of 50 indentations with the 350-μm-distance between two of them in the direction from the top of the composite coating to the steel surface. The microhardness level of the two samples was examined by the Buehler MicroMet 5100. The load was 25 g. The indenter is a diamond in the form of a square-based pyramid.

Microhardness measurement was carried out for samples before the application of heat treatment and after the application of heat treatment. In Table 4 results of the measurement are presented. Average value of microhardness is mentioned.

Table 4. Average value of microhardness.

Composition	Heat treatment	Microhardness, HV
Cu + 5%W		109
Cu + 25%W	-	100.75
Cu + 50%W		126.95
Cu + 5%W		85.7
Cu + 25%W	400°C, 2 h	83.45
Cu + 50%W		93
Cu + 5%W		63.8
Cu + 25%W	600°C, 2 h	58.75
Cu + 50%W		74.9

It can be seen that microhardness decreases during heat treatment. The higher the heat treatment temperature, the more intense the drop in the microhardness values. The relative error of the experiment in microhardness is 15%.

Apparently, the hydrogen content in the initial copper powder, which is the reason for the increase in porosity after heat treatment of the samples, decreases the microhardness with increasing heat treatment temperature.

As the coating thickness increases, the porosity decreases due to the overlapping of the pores by subsequent metal layers, therefore the coating thickness must be such that it is generally non-porous or with a minimum number of pores. The coating must have a strong adhesion with the base metal and not peel off under mechanical action. Therefore, depending on the requirements of the coating, it is necessary to choose the most optimal coating thickness. But this drop in microhardness can be useful in further machining the surface, if necessary.

3.3. *Microstructure after heat treatment*

The purpose of heat treatment in this case is to improve the mechanical properties of the coating material and eliminate boundaries between copper grains for obtaining of homogeneous structure and chemical composition.

During heat treatment, copper and its alloys can interact with oxygen, hydrogen, water vapor and other compounds (CH_4, etc.). When ingots and semi-finished products are heated in a reducing environment of H_2, H_2O, CH_4, and other compounds containing hydrogen, they dissociate, the resulting atomic hydrogen quickly diffuses into the metal, since the diffusion coefficient of hydrogen in copper is quite large. Hydrogen interacts with copper oxide in copper by the reaction $CuO + H_2 = 2Cu + 2H_2O$. Water vapor and other gases released are insoluble in copper and create high pressure, leading to discontinuities along the grain boundaries and voids in the grain body. The discontinuities then develop into cracks, which is especially pronounced during hot pressure treatment [15].

The evolved gas may include steam and sulphur dioxide as well as hydrogen. The composition and quantity of the gas evolved is determined mainly by the following factors:

- the alloy composition in respect of components which form very stable oxides, thereby reducing the activity of dissolved oxygen;
- the opportunity for hydrogen absorption from the environment;
- whether the metal is melted under oxidizing or reducing conditions;
- deoxidation and degassing practices;
- the degree of sulphur contamination, if any [16].

Copper and alloys based on it actively interact with oxygen and water vapor at elevated temperatures. In this regard, during the heat treatment of copper products and its alloys, protective atmospheres are often used.

In Figures 6, 7 and 8 it can be seen 3 composition without heat treatment and with heat treatment at 400° and 600°C. In all three compositions the grain boundaries became less visible after heat treatment, and porosity appeared at the grain boundaries. The air between the copper particles when exposed to high temperatures appear to be subjected to high pressure intergranular space, which led to the formation of pores. The occurrence of this phenomenon in the longer term may lead to embrittlement of the coating and the formation of defects in it, leading to the destruction of the material.

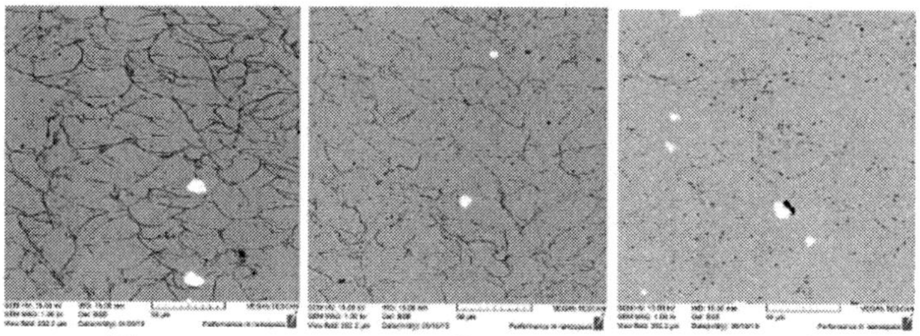

(a) (b) (c)

Figure 6. Composition Cu+5%W: (a) – without heat treatment; (b) – after heat treatment at 400°C; (c) – after heat treatment at 600°C.

To avoid such a defect in especially important products it is necessary to control the content of oxygen and other gases in the original composition.

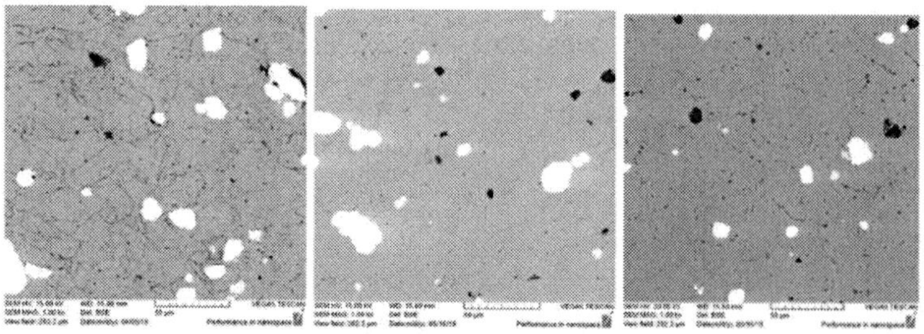

(a) (b) (c)

Figure 7. Composition Cu+25%W: (a) – without heat treatment; (b) – after heat treatment at 400°C; (c) – after heat treatment at 600°C.

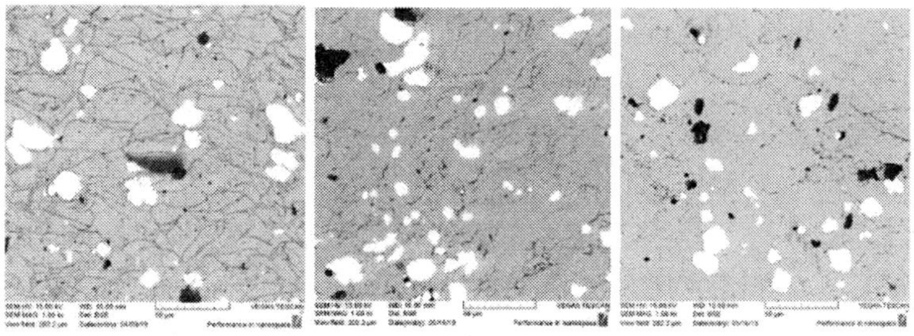

Figure 8. Composition Cu+50%W: (a) – without heat treatment; (b) – after heat treatment at 400°C; (c) – after heat treatment at 600°C.

3.4. Thermal conductivity measurements

The effective use of materials used for the manufacture of coatings in various areas is largely determined by the accuracy of information on their thermophysical properties, and in particular, on thermal conductivity in a wide range of concentrations and temperatures.

The study of the physicochemical properties of these materials in a wide range of changes in temperature and concentration is necessary for a deeper understanding of temperature-concentration changes in their structure when solving technological problems.

The measurements were carried out by an indirect method. It includes the measurement of conductivity by measuring the specific heat, thermal diffusivity and density.

Mass heat capacity was measured by C80 Setaram calorimeter [14]. The C80 calorimeter is one of the world's most versatile calorimeters. It uses the principle of Calvet, which makes it possible to work in scanning and isothermal temperature conditions in a wide range of temperatures from room temperature to 300°C and is capable of detecting low-energy thermal phenomena. Designed in particular for measuring the heat dissipation of reactions and mixtures.

In order to compare the theoretically and experimentally obtained values of the thermal conductivity of the objects of study, a numerical model has been used.

Thermal conductivity was measured for two samples: Cu+25%W and Cu+50%W.

The first step of measurements is thermal diffusivity measurement. In Table 5 the results of the measurements for two samples at room temperature and 150°C are presented.

Table 5. Diffusivity measurements.

Composition	Temperature of the experiment, °C	Apparent thermal diffusivity, $10^{-6}\cdot m^2/s$	Uncertainty, $10^{-6}\cdot m^2/s$
Cu+25%W	24	89.5	6.3
	150	85.3	6.0
Cu+50%W	24	85.1	6.0
	150	81.2	5.7

Apparent thermal diffusivity of a pure copper is $117\cdot10^{-6}\cdot m^2/s$ [18].

The next step is measurement of mass thermal capacity of the samples. In Table 6 the results of the measurements are presented.

Apparent mass thermal capacity of pure copper is 385 J/kg/K at room temperature [18].

With using of values of density from Table 7 it could be possible to calculate thermal conductivity of the samples (Table 8).

Apparent volumetric mass of pure copper is 8933 kg/m^3 at room temperature [18].

Table 6. Mass thermal capacity measurements.

Composition	Maximum temperature of the enthalpic peak, °C	Apparent mass thermal capacity, J/kg/K	Uncertainty, J/kg/K
Cu+25%W	28.3	357	18
	149	375	19
Cu+50%W	28.3	325	16
	149	344	17

Table 7. Density measurements.

Composition	Measurement conditions		Apparent volumetric mass, kg/m^3
	T, °C	Relative humidity, %	
Cu+25%W	21.4	38.5	9245
Cu+50%W	21.4	38.5	9956

Table 8. Thermal conductivity results.

Composition	T, °C	Average apparent thermal conductivity, W/m·K
Cu+25%W	26	295.2
	150	296.2
Cu+50%W	26	275.1
	150	278.4

Thermal conductivity of pure copper:

- λ(300K) = 401 W/m.K;
- λ(400K) = 393 W/m.K;
- λ(600K) = 379 W/m.K [18].

Thermal conductivity can be calculated in different ways. Several mathematical models have been created for this purpose. Hasselman and Johnson presented the most recent model of them. [17]. This model shows that the TC of a composite with continuous matrix and dilute volume fractions of spherical reinforcement is expressed as

$$K_C = \frac{K_m\left[2\left(\frac{K_p}{K_m} - \frac{K_p}{ah_c} - 1\right)V_p + \frac{K_p}{K_m} + 2\frac{K_p}{ah_c} + 2\right]}{\left(1 - \frac{K_p}{K_m} + \frac{K_p}{ah_c}\right)V_p + \frac{K_p}{K_m} + 2\frac{K_p}{ah_c} + 2} \tag{1}$$

where K is the thermal conductivity, V the volume fraction of reinforcement, and the subscripts c, m and p refer to the composite, matrix and reinforcement, respectively; a is the radius of the spherical reinforcement and hc is the thermal boundary conductance. In the absence of an interfacial thermal resistance, i.e., $hc = \infty$, an ideal thermal conductivity can be obtained by equation and the equation becomes agreeable to the expression of Maxwell [18]. In this work, the thermal conductivity of two composites sprayed with Cu+25%W and Cu+50%W powder with 12 μm tugsten was estimated by equation under the condition of $hc = \infty$. The concentration of tungsten taken for calculation was 9% and 15% respectively (the concentration of tungsten after the cold spraying process). At room temperature, the thermal conductivity of tugsten is chosen as 162.8 W/mK. The matrix of the composite sprayed is considered as copper, and its thermal conductivity is 401 W/mK (equable to the thermal conductivity of pure copper). At 150°C copper is also considered as the matrix material in the composite sprayed with Cu+50%W powder, and has 393 W/mK thermal conductivity, and for tungsten it is 155.9 W/mK thermal conductivity. Substitution of these data into equation yields 375.55 and 368.06 W/mK for the composites from Cu+25%W powder at room temperature and at 150°C respectively and for Cu+50%W powder thermal conductivities are 356.08 and 348.98 W/mK at room temperature and at 150°C respectively.

The comparison between calculated and experimentally measured data is presented in Table 9.

Table 9. Comparison between experimental and numerical model data.

Composition	Temperature, °C	Experimental data, W/mK	Numerical model data, W/mK
Cu+25%W	Room temperature	295.2	375.55
	150	296.2	368.06
Cu+50%W	Room temperature	275.1	356.08
	150	278.4	348.98

These values are higher than the measured values. Hence, the influence of the intense porosity and oxidation at grain boundaries on the thermal conductivity is very obvious for the composites.

4. Conclusions

From the experimental results it can be seen that the maximum concentration of tungsten does not exceed 48–50%.

The sharp drop in the effectiveness of the coating deposition is noticeable after an increase in the tungsten content of more than 10–15% of the total weight of the composition. This phenomenon is due to the high hardness of tungsten particles. If the hard particle meets the surface covered by previously bonded hard particle, the impact energy is not high enough for bonding and hard particle rebounds.

The hydrogen content in the initial copper powder, which is the reason for the increase in porosity after heat treatment of the samples, decreases the microhardness with increasing heat treatment temperature.

As the coating thickness increases, the porosity decreases due to the overlapping of the pores by subsequent metal layers, therefore the coating thickness must be such that it is generally non-porous or with a minimum number of pores. The coating must have a strong adhesion with the base metal and not peel off under mechanical action. Therefore, depending on the requirements of the coating, it is necessary to choose the most optimal coating thickness.

Copper and alloys based on it actively interact with oxygen and water vapor at elevated temperatures. In this regard, during the heat treatment of copper products and its alloys, protective atmospheres are often used.

In all compositions the grain boundaries became less visible after heat treatment, and porosity appeared at the grain boundaries. The air between the copper particles when exposed to high temperatures appear to be subjected to high pressure intergranular space, which led to the formation of pores. The occurrence of this phenomenon in the longer term may lead to embrittlement of the coating and the formation of defects in it, leading to the destruction of the material. To avoid such a defect in especially important products it is necessary to control the content of oxygen and other gases in the original composition.

The values obtained from numerical model for calculation of thermal conductivity are higher than the measured values. Hence, the influence of the intense porosity and oxidation at grain boundaries on the thermal conductivity is very obvious for the composites.

References

[1] Alkhimov A P, Klinkov S V and Kosarev V F 2010 *Cold gazdynamic spraying. Theory and practice* (Moscow: Fizmatlit) p 535
[2] Robitaille F, Yandouzi M, Hind S and Jodoin B 2009 *Surf. Coat. Technol.* **203** 2954–2960
[3] Lupoi R and O'Neill W 2010 *Surf. Coat. Technol.* **205** 2167–2173
[4] Zweben C 1992 *The Journal of The Minerals, Metals & Materials Society* **44–7** 15–23
[5] Kang H-K and Kang S B 2003 *Scripta Materialia* **49** 1169–1174
[6] Kang S B, Kang H-K and Euh K 2005 *Materials Science Forum* **475–479** 945–948
[7] Dong L L, Ahangarkani M, Chen W G and Zhang Y S 2018 *Int. J. of Refractory Metals and Hard Materials* **75** 30–42
[8] 2012 *Cold spray. A guide to best practice* (Arbegast Materials Processing and Joining Lab: South Dakota School of Mines and Technology, United States)
[9] Hussain T, McCartney D G, Shipway P H and Zhang D 2009 *J. Therm. Spray Technol.* **18(3)** 364–379
[10] Huang R, Ma W and Fukanuma H 2014 *Surface and Coatings Technology* **258** 832–841
[11] Sova A, Maestracci R, Jeandin M, Bertrand Ph and Smurov I 2016 *Surface and Coatings Technology*
[12] Goldstein J and Newbury D 2003 *Scanning electron microscopy and X-ray microanalysis* (Kluwer) p 689
[13] ISO 6507-1:2005 *Metallic materials. Vickers hardness test. Part 1: Test method*
[14] Calvet Calorimeter C80 From ambient to 300 °C by Setaram 2019 Online. Available: SETARAM Instrumentation: www.setaram.com
[15] Akulov A I 2003 *Technology and equipment for fusion welding and thermal cutting* (Moscow: Bauman Moscow State Technical University)
[16] Talbot D E J 1975 *International Metallurgical Reviews* **20(1)** 166–184
[17] Manchang G, Suk B K and Kwangjun E 2004 *Scripta Materialia* **52** 51–56
[18] Incropera F P, DeWitt D P, Bergman T L and Lavine A S 2011 *Fundamentals of Heat and Mass Transfer* (John Wiley & Sons) p 1024

Acknowledgments

The work was supported by Russian Science Foundation (project No. 20-19-00304).

IOP Conf. Series: Materials Science and Engineering **969** (2020) 012107 doi:10.1088/1757-899X/969/1/012107

Mathematical Modelling of Low Temperature Solder Direct Extrusion

L V Radionova, S R Faizov and A E Sarafanov

Processes and Machines of Metal Forming Department, South Ural State University, 76, Lenin prospect, Chelyabinsk, 454080, Russia

E-mail: radionovalv@susu.ru

Abstract. Defining of process technological parameters is very important for low temperature solder extrusion. A mathematical model of low temperature solder wire and rods direct extrusion, which allows determining the energy-power process parameters, has been developed. The validation of the mathematical model for adequacy was checked during a laboratory experiment on a hydraulic press. In laboratory the force and speed of pressing were measured and recorded using the data collection system installed on it. The validation of the model for adequacy was carried out by pressing Ø8.00 and Ø15 mm rods and Ø2.00 mm wire and showed fine precision of calculations with experimental measurements. The error did not exceed 10%. The resulting mathematical model was used for analytical studies of Sn-In alloy bars and wires direct pressing technological modes. Calculations have shown that a decrease in the finished rod (wire) diameter from 16.00 to 2.00 mm while maintaining the workpiece size Ø30.00 mm leads to an increase in the pressing force from 86 kN to 131 kN at the initial moment of pressing. It increases the elongation ratio from 4 to 256. The pressing force of Ø8.00 mm rod increases from 25 to 171 kN after an increase in the diameter of the workpiece from 12.0 to 40.0 mm. Analytical studies on the model showed that it is possible not only to study the pressing process in order to understand the mechanisms of forming the mechanical and operational properties of the finished rod (wire), but also to design resource-saving pressing modes for various product gauges, to carry out substantiated selection of the required equipment and accessories.

1. Introduction

The works devoted to the methods and technologies of soldering using lead-free solders have been actively developed recently [1–6]. The reason for this is that the European Union's RoHS (Restriction of use of Certain Hazardous Substances) directive limited the use of lead in new electrical and electronic equipment to no more than 0.01% in 2006. These limits are established not on the weight of a product, assembly or component, but on each individual homogeneous material [7].

A number of lead-free solders of the Sn–Cu, Sn–Ag, Sn–Ag–Cu systems have been developed in recent years to solve this problem. Lead-free technologies are distinguished by a significant increase in soldering temperatures from 230 °C for Sn63Pb37 (melting point 183 °C) to 265 ... 270 °C for the most low-melting, generally recognized lead-free solder Sn3Ag0.7Cu (melting temperature 217 °C). Such a temperature load requires not only a change of solders, but also a change of fluxes, base materials of printed-circuit boards, renewal of technological equipment, and revision of specific energy consumption. This leads to the need for investments in the renovation of production, the use of more expensive heat-resistant materials, and the restructuring of technologies. However, low melting point solders have also been developed and applied in recent years. These are alloys of the Sn–In, Sn–Bi systems.

Low-melting Sn–In solder consist of 51–53% indium and the rest of tin. This solder is one of the most commonly used in 2016 (after the introduction of restrictions on the use of cadmium and lead) in semiconductor technology.

The Processes and Machines of Metal Forming Department at SUSU has been developing and mastering the technology for the production of solder Sn-In alloy wire and rod under the conditions of low-rate production since 2019 [8, 9].

The purpose of this study is to develop a mathematical model for direct pressing of solder from low-melting materials, which makes it possible to determine the energy-power modes of pressing wires and bars of various diameters.

2. Mathematical Description of the Extrusion Power Mode

Figure 1 shows a diagram of the deformation zone during pressing. The energy for pressing is supplied to the extrusion ram by force F, while the total work of process A is the sum of the plastic deformation work A_{def} and the friction forces work A_{fr}.

$$A = A_{def} + A_{fr}. \tag{1}$$

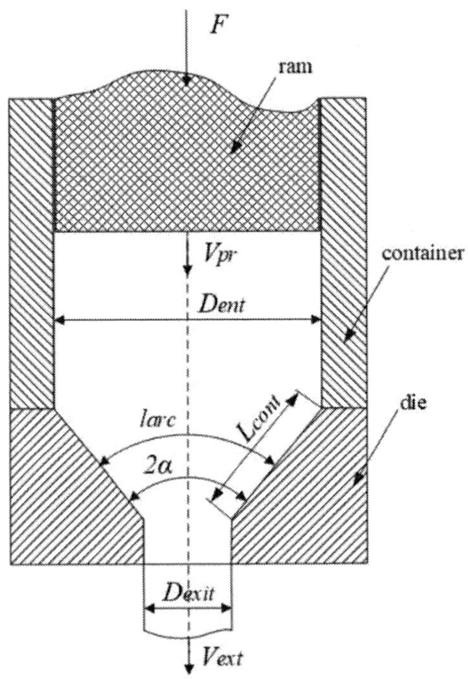

Figure 1. Scheme of the deformation zone during extrusion.

The extrusion energy consumption (while neglecting the sliding friction forces on the contact surface of the ram and the container) is proposed to be estimated as a sum of forces in [10, 11]

$$F = F_{def} + F_{fr}, \tag{2}$$

where F_{def} – forming (plastic deformation) force; F_{fr} – efforts to overcome sliding friction forces on the processed metal with the container and with the die contact surface. The derivation of the equation for determining the plastic deformation required force can be based on the Fink dependence. This dependence determines the work of metal shaping, spent on the implementation of plastic deformations, determined by the initial and final dimensions of the processed metal (without taking into account the peculiarities of the plastic deformation conditions)

$$A_F = \theta \cdot \sigma_{av} \cdot \ln \mu \tag{3}$$

where θ – volume of processed metal; μ – extrusion ratio

$$\mu = \frac{D_{ent}^2}{D_{exit}^2}, \tag{4}$$

D_{ent}=D_{con} – billet diameter at the entrance to the die equal to the container diameter; D_{con} – diameter of the wire (rod) at the exit from the die; σ_{av} –the average value of the processed metal deformation resistance, which can be determined as (5), under the assumption that the specific resistivity of the billet metal deformation is equal to the average value of the yield stress

$$\sigma_{av} = \sqrt{\sigma \cdot \sigma_{exit}}, \tag{5}$$

σ – billet material yield strength; σ_{exit} – metal yield stress at the exit from the die.

A distinctive feature of the considered extrusion process is the value of the deformation zone shape parameter Δ_B=l_{arc}/L_{cont}= <<*1*, where l_{arc} is the length of the arc perpendicular to the tool contact line in the middle of the deformation zone; L_{cont} - the length of the tool contact line (Figure 1).

The value Δ_B is calculated by the equation [12]:

$$\Delta_B = \frac{\alpha}{\frac{\mu-1}{\mu} \cdot \left(1 + \sqrt{1 - \frac{\mu-1}{\mu}}\right)^2}, \tag{6}$$

where α – half-angle of the die.

The shape parameter Δ_B<<1 leads to an increase in power conditions similarly to how it is accepted in sheet rolling [13, 14]. It should be taken into account when calculating:

$$A_{def} = \frac{1}{\Delta_B} \cdot A_F = \frac{1}{\Delta_B} \cdot \theta \cdot \sigma_{av} \cdot \ln \mu. \tag{7}$$

Taking into account that the deformation power on one side is

$$N_{def} = F_{def} \cdot V_{pr}, \tag{8}$$

where V_{pr} – pressing speed (movement speed of the ram), on the other hand, can be found as

$$N_{def} = \frac{dA_{def}}{dt} = \frac{1}{\Delta_B} \cdot \sigma_{av} \cdot \ln \mu \cdot \frac{d\theta}{dt} = \frac{1}{\Delta_B} \cdot \sigma_{av} \cdot \ln \mu \cdot S_{ent} \cdot V_{pr}, \tag{9}$$

then provided that the area of the round billet

$$S_{ent} = \frac{\pi \cdot D_{ent}^2}{4} \tag{10}$$

the equation for calculating the deformation force is:

$$F_{def} = \frac{\pi \cdot D_{ent}^2}{4} \cdot \frac{1}{\Delta_B} \cdot \sigma_{av} \cdot \ln \mu. \tag{11}$$

The friction force F_{frI} is determined by the geometry of the deformation zone (Figure 1) and consists of three components – the effort aimed at overcoming the processed metal sliding friction against the container F_{frI}, the surface of the die F_{frII} and the parallel land of the die F_{frIII} [15]

$$F_{fr} = F_{frI} + F_{frII} + F_{frIII} \tag{12}$$

The formation of friction forces according to Sybel's law is

$$\tau = f \cdot \sigma \tag{13}$$

Components of friction forces can be found:

$$F_{frI} = S_{con} \cdot \tau = \pi \cdot D_{con} \cdot L_{bil} \cdot f \cdot \sigma; \tag{14}$$

$$F_{frII} = S_{die} \cdot \tau = \frac{\pi}{4 \cdot \sin\alpha} \cdot (D_{con}^{\ 2} - D_{exit}^{\ 2}) \cdot f \cdot \sigma_{\tau}; \tag{15}$$

$$F_{frIII} = S_{cal} \cdot \tau = \pi \cdot D_{exit} \cdot L_{cal} \cdot f \cdot \sigma_{exit}, \tag{16}$$

where f – friction coefficient; S_{con}, S_{die}, S_{cal} – lateral surface area of the container inner liner, die and its parallel land; D_{con} – container diameter; σ_{τ} – plastic shear strength

$$\sigma_{\tau} = \frac{\sigma_{av}}{2} = \frac{\sqrt{\sigma \cdot \sigma_{exit}}}{2}, \tag{17}$$

L_{bil} – length of pressed billet; L_{cal} – length of die parallel land.

The mathematical model of direct extrusion is implemented in the Excel mathematical table. An example of calculating the power conditions for pressing rods ∅ 2.00; 8.00 and 15.00 mm from Sn-In alloy are given in table. 1.

Table 1. Results of mathematical model calculation.

Parameter	Unit	Value		
		Exper.1	Exper.2	Exper.3
Initial data				
Diameter of billet, D_{ent}	mm	30	30	20
Diameter of the wire (rod), D_{exit}	mm	15	8	2
Billet length, L	mm	120	120	120
Container diameter, D_{con}	mm	31	32	21.8
Pressing speed, V_{pr}	mm/s	12	3.5	0.5
Billet material yield strength, σ	MPa	10.5	10.5	10.5
Half-angle of the die, α	°	33.5	40	18.4
Length of die parallel land, L_{cal}	mm	5	5	5
Friction coefficient, f	-	0.5	0.5	0.5
Calculation results				
Extrusion ratio, μ	-	4.3	16.0	118.8
Extrusion speed, V_{ext}	mm/s	51	56	59
Force at the start of pressing, F_{start}	kN	92.8	106	105
Force at the end of pressing, F_{end}	kN	41.0	55.6	71.9

3. Experimental Studies

Validation of a model for adequacy was carried out in the laboratory of the Processes and Machines of Metal Forming Department experiment by Pressure on a hydraulic press model D2428 equipped with a data collection and storage system (Figure 2). The press allows us to determine the speed and pressing force after modernization, which included replacing the oil pump (8) and equipping it with pressure sensors and the stroke (2) with displacement sensors. The press can work both with the included data collection system (10), displaying the pressing diagram (Figure 3) on the monitor (9) and recording the measurement results into the database, or without it. The press is controlled from the control panel (5) with the data collection system turned off, and the change in pressure can be observed on the pressure gauge (6).

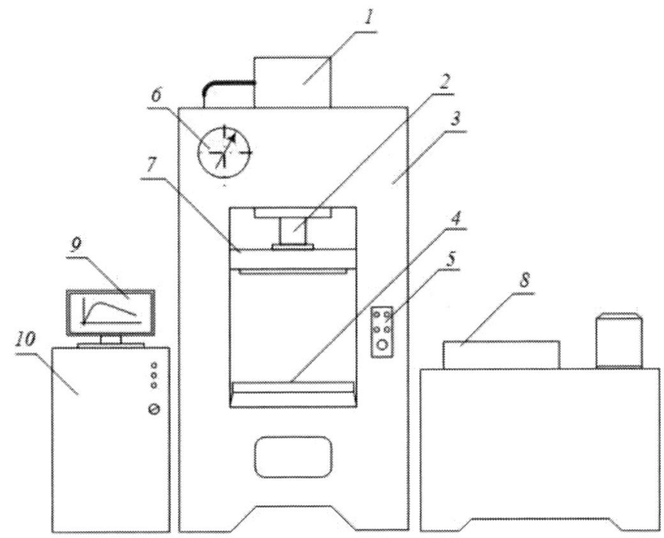

Figure 2. Scheme of laboratory hydraulic press: 1 – main cylinder; 2 – stroke; 3 – housing; 4 – bench; 5 – control desk; 6 – manometer; 7 – slide; 8 – oil pump; 9 – monitor; 10 – block of control and data collecting.

Wire (rods) were obtained from cast sections during experiment in accordance with the conditions presented in table 2.

Table 2. Experimental results.

Parameter	Unit	Value		
		Exper.1	Exper.2	Exper.3
Force at the start of pressing, F_{start}	kN	85	116	100
Force at the end of pressing, F_{end}	kN	42	61.5	70

Figure 3 shows the results of measuring the pressing force for a billet Ø 30.00 mm from alloy into Sn-In a bar Ø 8.00 mm. The pressing force decreases with the billet length decreasing in the container. Reduction of losses for overcoming friction forces is a consequence of a decrease in the surface between the billet and the container.

The fine precision of the calculated energy-power parameters (Table 1) and experimental results (Table 2) is quite high, the error does not exceed 10%. This allows us to conclude about the adequacy of the compiled mathematical model and the possibility of its further use for the rods and wires direct extrusion process analytical studies.

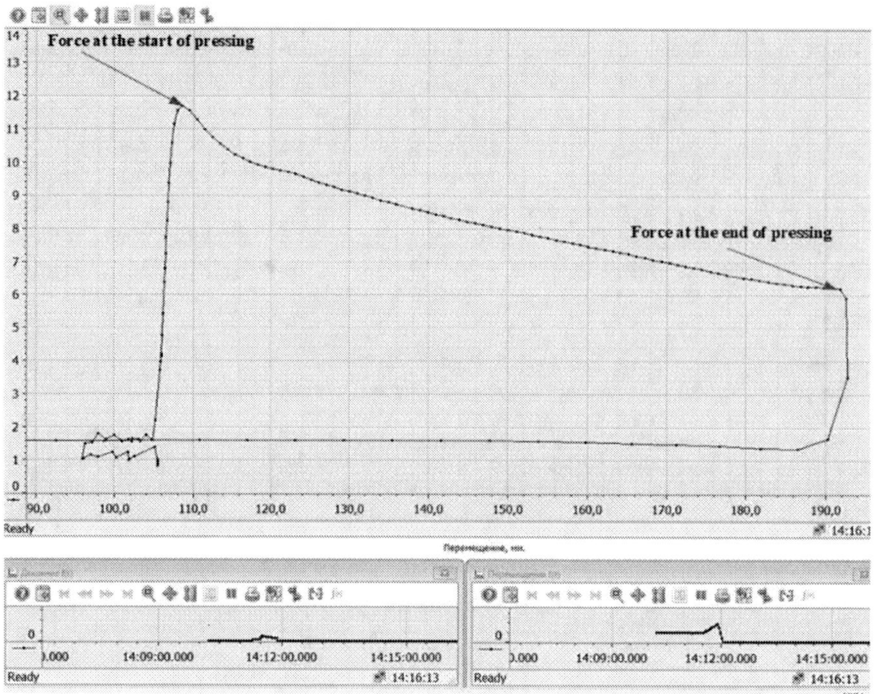

Figure 3. Screenshot with the results of measuring the rod pressing force Ø8.00 mm made of Sn-In alloy.

4. Analytical Study

The developed mathematical model can be used for operational calculation and extrusion modes selection for wires and rods of various assortments.

As an example, let us consider how the energy-power indexes change when extruding a various diameter rods from a billet Ø30.00 mm at a pressing speed V_{pr} = 3.5 mm/s. The simulation results are shown in Table 3.

Table 3. Results of power extruding parameters mathematical modeling of various diameters rod (wire) from billet Ø30.00 mm.

Diameter of the wire (rod) D_{exit}, mm	Extrusion ratio, μ	Extrusion speed, V_{ext}, mm/s	Force at the start of pressing F_{start}, kN	Force at the end of pressing F_{end}, kN
16.00	4.0	0.14	86	36
14.00	5.22	0.18	91	41
12.00	7.11	0.25	96	46
10.00	10.24	0.36	101	51
8.00	16	0.56	106	56
6.00	28.44	1.00	112	61
4.00	64	2.24	119	68
2.00	256	8.96	131	80

A decrease in the finished rod (wire) diameter from 16.00 to 2.00 mm while maintaining the billet size leads to an increase in the pressing force from 86 kN to 131 kN at the initial moment of pressing, as follows from the table. The reason for this is that the stretch ratio increases significantly (from 4 to 256).

The calculations of the Ø8.00 bar pressing force with a change in the billet diameter (V_{pr} = 3.5 mm/s) are presented in Table 4.

Table 4. Results of energy-power parameters mathematical modeling of rod (wire) a Ø8.00 mm extrusion from a different diameters billet.

Diameter of billet D_{ent}, mm	Extrusion ratio, μ	Extrusion speed, V_{ext}, mm/s	Force at the start of pressing F_{start}, kN	Force at the end of pressing F_{end}, kN
40.00	27.6	0.97	171	103
36.00	22.6	0.79	143	82
32.00	18.1	0.63	118	64
28.00	14.1	0.49	95	48
24.00	10.6	0.37	74	35
20.00	7.6	0.27	56	23
16.00	5.1	0.18	39	14
12.00	3.1	0.11	25	7

The pressing force with an increase in the diameter of the billet from 12.0 to 40.0 mm increases as follows from the calculation results. It is quite predictable even without the use of a model, but a mathematical model is needed in order to estimate the quantitative value. The presence of an adequate mathematical model allows not only to reasonably select equipment for production, depending on the planned assortment, but also to optimize the tooling design for the extrusion process. So, we see that even a 10-tonnes press can be used to produce a bar of Ø8.00 mm. However, this is only possible with billet with a less than 28.00 mm diameter.

5. Conclusions

A mathematical model of low temperature solder direct extrusion has been developed. It allows determining the energy-power parameters of the process. Testing the mathematical model adequacy in the laboratory of the Processes and Machines of Metal Forming Department of SUSU on a hydraulic press equipped with a data collection and storage system showed fine precision. The error in both the pressing force and the temperature of the metal at the exit from the die did not exceed 10%, which is quite acceptable for analytical studies of the process and the extrusion technological modes choice.

Analytical studies on the model showed that it is possible not only to study the extrusion process in order to understand the mechanisms of finished rod (wire) mechanical and operational properties formation, but also to select resource-saving pressing modes for various product mixes, to make a reasonable choice of the required equipment and tooling.

References

[1] Medvedev A M 2004 *Electronic components* (Moscow: ID Electronics)
[2] Medvedev A M 2006 *Components and Technologies* (Moscow: Finestreet)
[3] Grigoriev V 2001 *Electronic components* (Moscow: ID Electronics)
[4] Zenin V V, Belyaev V N, Segal Yu E, Kolbenkov A A 2003 *Microelectronics* (Moscow: Nauka Publishes)
[5] Bleifreiloten 2000 *Silber und Kupfer statt Blei. Krempelsauer. Elector* (BRD 5)
[6] Shapiro L 2007 *Bulletin of Electronics* **2**
[7] Shapiro L 2006 *Electronic components* (Moscow: ID Electronics)

[8] Glebov L A, Radionova L V, Faizov S S 2019 *Magnitogorsk Rolling Practice: materials of the IV youth scientific-practical conference* 54–56
[9] Radionova L V 2019 *Proceedings of the XII Congress of distributors* **I** 76–79
[10] Kargin V R 2015 *Theory and technology of extrusion of aluminum alloys: textbook. Allowance*
[11] V R Kargin, A F Grechnikova, Y A Erisov, Yu S Starostin (Samara: Publishing House of SSAU)
[12] Loginov Y N 2016 *Pressing as a method of intense deformation of metals and alloys: a training manual* (Yekaterinburg: UrFU) p 156
[13] Backofen W 1977 *Deformation processes* (Moscow: Metallurgy Publ) p 288
[14] Tselikov A I 1982 *Rolling theory* (Moscow: Metallurgy Publ) p 335
[15] Zheleznov Y D, Polukhin P I 1967 *Sheet rolling and roll service* (Moscow: Metallurgy Publ) p 388
[16] Radionova L V, Kharitonov V A, Zyuzin V I 2001 *Steel in Translation* **12** 49–50

Acknowledgments
The work was carried out with the financial support of the Ministry of Science and Higher Education of the Russian Federation within the framework of a subsidy for financial support for the fulfillment of a state task (fundamental scientific research), contract № FENU-2020-0020 (2020071GZ).

IOP Conf. Series: Materials Science and Engineering **969** (2020) 012108 doi:10.1088/1757-899X/969/1/012108

Improving the Performance Characteristics of the Drawing Machine Intermediate Block Working Surface by Additive Technologies

L V Radionova, M N Samodurova and B A Chaplygin

Department of Processes and Machines for Metal Forming, South Ural State University, 76, Lenin prospekt, Chelyabinsk, 454080, Russia

E-mail: radionovalv@susu.ru

Abstract. This paper presents the study results of the drawing machine intermediate block surface hardening treatment by the methods of additive technologies. Laser surfacing of powder material Fe-4,5Cr-4,5Mo-5,5W-4V on the intermediate block working surface a wear-resistant layer with a hardness of 52-54 HRC with a thickness of more than 1.5 mm. The laser complex LaserClad 4 C R120 based on the industrial robotic manipulator KUKA KR-120 together with the positioner is able to melt powder material only on the working surface of the intermediate block, thereby significantly reducing its cost and labor intensity during its manufacture. The applying of detonation spraying with subsequent penetration for this type of product proved to be ineffective. The detonation coating itself with a powder material based on tungsten carbide turned out to be uniform and had a hardness in the range of 61-65 HRC. But rather large pores and cracks appeared in it after penetration, which is associated with a significant difference in the coefficients of thermal linear expansion of steel 41Cr4 (14·10−6 1/K) and an alloy based on tungsten carbide (3,9·10−6 1/K).

1. Introduction

Drawing is the main method for producing wire, both commercial and in the form of a blank for the production of hardware, for example: nails, mesh, fasteners and ropes [1–7]. Drawing mills of various types are used as the main equipment for the wire manufacture [8]. All of them are united by the presence of such structural elements as an intermediate block, a "soap box" with a drawing holder, a die, a motor, a gearbox and a stand.

A single and multiple drawing mill was designed and manufactured at the department "Processes and Machines for Metal Forming" for research purposes, as well as for the training of specialists in the hardware production field (Figure 1) [9, 10].

The designers are faced with the question of how and from what material intermediate block should be make when designing a drawing machine. The drawing machine intermediate block is one of the most loaded units. The surface of which is subject to wear during operation, in addition, its design features require machining, which significantly increases its cost. The block surface must be smooth and hard so that the wire can slide along the block to carry out the drawing process, and the lower turn can push the accumulated turns up along the block surface. In this regard, a thick-walled tube made of 41Cr4 steel was chosen as a workpiece for production, it was supposed to use a hardening treatment such as surfacing to create a functional working layer on its surface.

Figure 1. General view of the SUSU straight-flow mill.

2. Materials and methods

The capabilities of the SUSU Laboratory for Mechanics, Laser Processes and Digital Production Technologies namely the KLS complex based on LaserClad 4 C R120, equipped with an LS-4 ytterbium fiber laser, an industrial robot-manipulator KUKA KR-120 and detonation complex CCDS2000 [11] were used to carry out cladding and obtain a wear-resistant surface layer of the intermediate block.

The purpose of this work is to develop a technology for obtaining a functional surface layer of a drawing machine intermediate block with increased operational properties.

The technology for obtaining a functional layer was tested on modeling samples (Figure 3) since the intermediate block (Figure 2) is a rather massive product, its diameter is 400 mm, and its weight is more than 50 kg.

Figure 2. Intermediate block.

Two methods of surface hardening were considered to obtain a functional surface layer: detonation spraying followed by laser treatment of the applied layer and laser cladding of a powder material.

Castoline powder 55586C (WC-10Co-4Cr) was chosen for detonation spraying, and EuTroLoy® 16606 powder (Fe-4.5Cr-4.5Mo-5.5W-4V) for direct laser fusion. Tungsten carbide powder (WC-10Co-4Cr) is used to protect against wear and corrosion at temperatures below 480 °C. Fe-4,5Cr-4,5Mo-5,5W-4V powder has excellent wear resistance when working up to 500 ° C.

The surface of the samples (substrate) made of steel grade 41Cr4 (Figure 3), before the application of the hardening layer, was subjected to cleaning in an abrasive-blasting cab.

Thirty five shots were made with a powder based on tungsten carbide WC-10Co-4Cr during detonation hardening with subsequent fusion penetration onto the substrate. The resulting coating was uniform. Metallographic sections were made to study the microstructure and determine the microhardness. The samples were etched with a 3% solution of nitric acid in ethyl alcohol.

Figure 3. Samples (substrate) from steel grade 41Cr4.

3. Results and discussion

Metallographic examination showed that the thickness of the coating is within 0.25 mm (Figure 4) and that the coating is very dense and uniform. The adhesion zone of the coating to the base metal is shown in Figure 5. The performance characteristics of the surface layer were evaluated through hardness. Determination of microhardness was carried out using a universal stationary hardness tester HV-1000 (Table 1). The measurement results and the conversion of the obtained values into HRC units are shown in Table 1.

The resulting layer was melted on a laser facility after detonation spraying. It was revealed that cracks and pores are present in the layer in the course of metallographic studies (Figure 6).

Figure 4. Detonation spraying layer (x100).

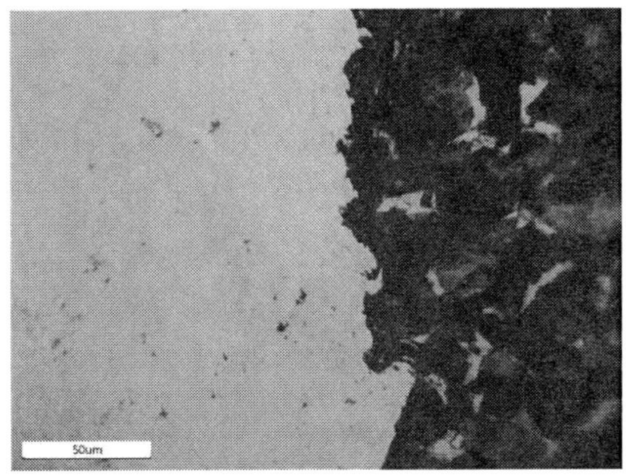

Figure 5. Area of the base metal and the applied layer adhesion (x500).

Table 1. The results of the surface layer obtained by detonation spraying hardness measuring.

Measurement number	1	2	3	4	5
Hardness, HV	840	733	763	763	727
Hardness, HRC	65,3	61,8	62,5	62,5	61,5

The cracks formation reason is a significant difference in the coefficients of thermal linear expansion of steel 41Cr4 (14 × 10−6 1/K) and an alloy based on tungsten carbide (3.9 × 10−6 1/K) in our opinion [12–14]. The fused layer thickness over the entire surface varies from 0.41 mm to 0.67 mm (Figure 6). The microhardness measurements were carried out according to the scheme shown in Figure 7, and the results of these measurements are shown in Table 2.

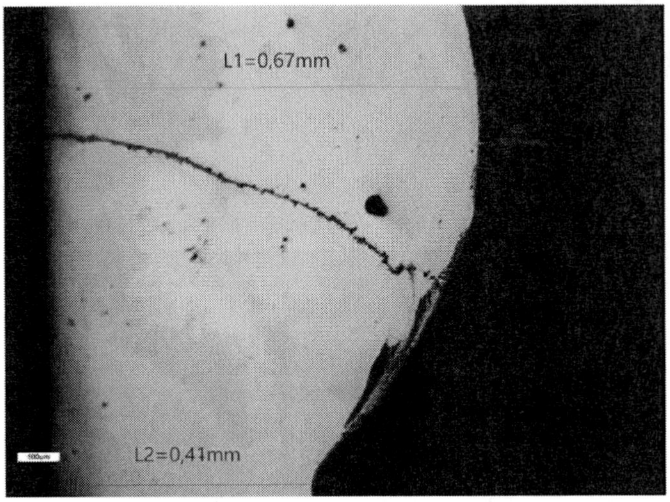

Figure 6. Remelted layer (x100).

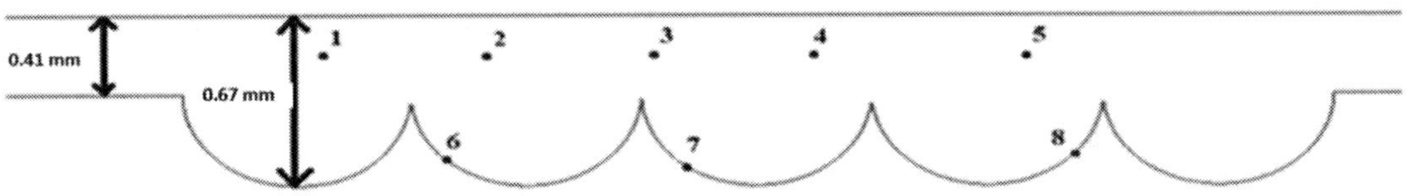

Figure 7. Scheme of microhardness measurement after melting of the layer applied by detonation spraying.

Table 2. The results of detonation sputtering surface layer with followed by fusion hardness measuring the of the.

	Applied layer					Mixing zone		
Measurement number	1	2	3	4	5	6	7	8
Hardness, HV	857	876	903	822	723	527	541	551
Hardness, HRC	65.9	66.4	67	64.7	61.4	51.1	51.7	52.3

Direct laser fusion of the powder material was used for the third hardening method. The metal powder Fe-4.5Cr-4.5Mo-5.5W-4V was deposited on the substrate using the previously developed mode [15] (Table 3).

Table 3. Powder alloy Fe-4.5Cr-4.5Mo-5.5W-4V welding mode.

Collimator position	Power (Watt)	RPM%	V (mm/s)	Spot d, mm	Displacement
0	1200	12	12	3	1.4

The results of the obtained layer metallographic studies showed that there are no cracks and microcracks in the structure, as well as the deposited layer porosity does not exceed 3% (Figure 8). The thickness of the deposited layer is 0.73… 0.76 mm (Figure 8). The thickness of the directional layer can be increased, since there are no technological obstacles for this. The microstructure of the deposited layer is shown in Figure 9.

Figure 8. Deposited layer before etching (x100).

A base material layer of steel grade 41Cr4 subjected to thermal action during the cladding is formed under the deposited layer [16]. The microstructure of this area is shown in Fig. 10. The thickness of the heat-affected zone is about 0.6 mm. A martensite zone with a thickness of about 0.3 mm is formed under the deposited layer in this case. Therefore, the hardened layer is about 1.5 mm in total. The results of the deposited layer hardness measuring are given in table. 4.

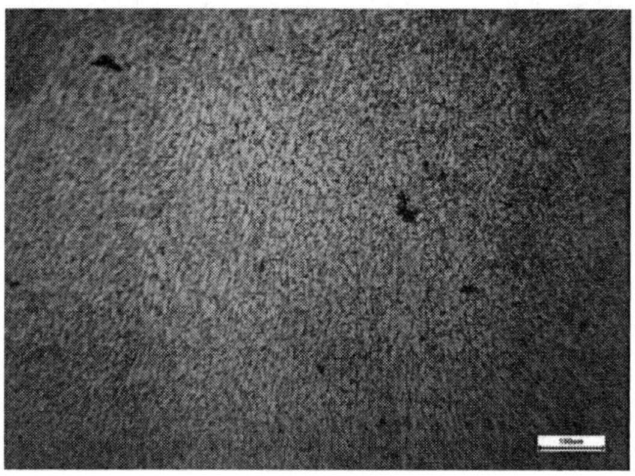

Figure 9. Microstructure of the deposited layer (x100).

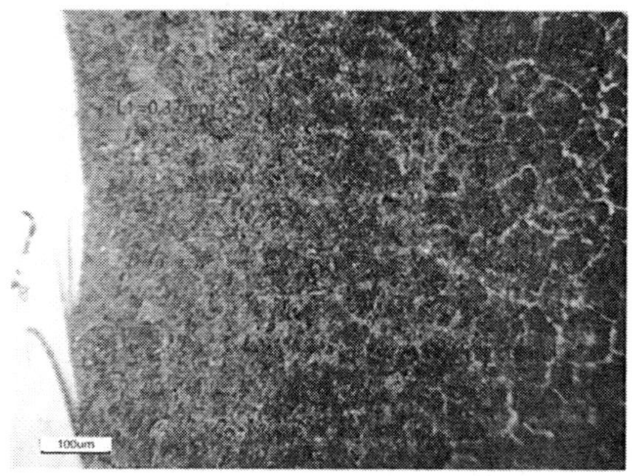

Figure 10. Microstructure of the heat-affected zone (x100).

Table 4. The results of measuring the hardness of the surface layer, obtained by laser cladding.

Measurement number	1	2	3	4	5
Hardness, HV	570	555	555	551	551
Hardness, HRC	53.6	52.4	52.4	52.3	52.3

4. Conclusions

It was found that it is advisable to use laser surfacing of a powder material of the composition Fe-4.5Cr-4.5Mo-5.5W-4V for increasing the wear resistance of the working surface of the drawing machine intermediate block. It provides a hardened layer with a hardness of 52–54 HRC to at least 1.5 mm depth. The laser complex LaserClad 4 C R120 based on the industrial robotic manipulator KUKA KR-120 together with the positioner is able to melt powder material only on the working surface of the intermediate block, thereby significantly reducing its cost and labor intensity during its manufacture.

The applying of detonation spraying with subsequent penetration for this type of product proved to be ineffective. The detonation coating itself with a powder material based on tungsten carbide turned out to be uniform and had a hardness in the range of 61–65 HRC. But rather large pores and cracks appeared in it after penetration, which is associated with a significant difference in the coefficients of thermal linear expansion of steel 41Cr4 (14·10–6 1/K) and an alloy based on tungsten carbide (3.9·10–6 1/K).

References

[1] Gun G S 2015 *Bulletin of the South Ural State University Ser Metallurgy* **15 2** 34–48
[2] Delarbre D and Montmitonnet P 1999 *Journal of Materials Processing Technology* **91** 95–104
[3] Hassan M A, Takakura N and Yamaguchi K 2003 *Journal of Materials Processing Technology* **139** 408–413
[4] Petrescu D, Savage S C and Hodgson P D 2002 *Journal of Materials Processing Technology* **125/126** 361–368
[5] Cho H, Jo H H, Lee S G, Kim B M and Kim Y J 2002 *Journal of Materials Processing Technology* **130/131** 416–420
[6] Li L, Zhou J and Duszczyk P 2003 *Modelling and Simulation in Materials Science and Engineering* **11** 401–416
[7] Lo S W and Lu Y H 2002 *Journal of Materials Processing Technology* **123(2)** 212–218
[8] Radionova L V and Radionov A A 2013 *Russian Internet Journal of Industrial Engineering* **1** 3–11
[9] Radionova L V, Faizov S R, Lisovskiy R A and Lisovskaya T A 2017 *Russian Internet Journal of Industrial Engineering* **5** 68-73
[10] Samodurova M, Radionova L, Logachev I, Shaburova N, Samoilova O, Trofimov E, Zakirov R, Pashkeev K and Myasoedov V 2019 *Materials* **12**
[11] Samodurova M, Radionova L, Shaburova N, Samoilova O, Trofimov E, Zakirov R, Pashkeev K, Myasoedov V and Erdakov I 2020 *Materials* **13**
[12] Thomas R L S, Scully J R and Gangloff R P 2003 *Metallurgical and Materials Transactions* **34** 327–344
[13] Berns Hans and Werner Theisen 2008 *Ferrous Materials* (Berlin: Springer) p 418
[14] Yi J 2016 *Materials Science Forum* **850**
[15] Baitimerov R M, Lykov P A, Radionova L V and Safonov E V 2017 *Bulletin of the South Ural State University Ser Mechanical Engineering Industry* **17** 36–40
[16] Gusev A I 2016 IOP *Conference Series: Materials Science and Engineering* **150**

Acknowledgments

The work was carried out with the financial support of the Ministry of Science and Higher Education of the Russian Federation within the framework of a subsidy for financial support for the fulfillment of a state task (fundamental scientific research), contract № FENU-2020-0020 (2020071GZ).

AUTHOR INDEX

Institute of Physics
Dirac House, Temple Back
Bristol BS1 6BE UK

ISSN: 1757-8981
ISBN 978-1-7138-4789-2
9 781713 847892

CURRAN ASSOCIATES INC.
proceedings.com